Intr
Psychology

Introduction to PSYCHOLOGY

SEVENTH EDITION

Clifford T. Morgan

Late Professor of Psychology
University of Texas, Austin

Richard A. King
John R. Weisz
John Schopler

University of North Carolina
at Chapel Hill

McGraw Hill Education (India) Private Limited

INTRODUCTION TO PSYCHOLOGY, 7/E

McGraw Hill Education (India) Edition 1993

62nd reprint 2019
RQBQCRQTDLDXY

Reprinted in India by arrangement with McGraw-Hill Global Education Holdings, LLC, New York

For Sale in India only

Library of Congress Cataloging-in-Publication Data

ISBN-13: 978-0-07-462250-6
ISBN-10: 0-07-462250-1

62 63 64 65 66 APO 232 22 21 20 19

Published by McGraw Hill Education (India) Private Limited, 444/1, Sri Ekambara Naicker Industrial Estate, Alapakkam, Porur, Chennai 600 116, Tamil Nadu, India, and Printed and bound in India at A P Offset, Naveen Shahdara, Delhi 110 032

Visit us at: www.mheducation.co.in; Phone (Toll free in India): **1800 103 5875**
Write to us at: info.india@mheducation.com

CIN: U80302TN2010PTC111532

This book

IS DEDICATED TO MARGY AND JENNY,
RICHARD KING'S AND JOHN WEISZ'S
WIVES, RESPECTIVELY,
AND TO ERNEST AND ERNA,
JOHN SCHOPLER'S PARENTS

About the Authors

CLIFFORD T. MORGAN and his colleagues at The Johns Hopkins University authored the first edition of this text (1956). After his early years at Harvard University, Dr. Morgan held professorships at The Johns Hopkins University, The University of Wisconsin—Madison, The University of California at Santa Barbara, and The University of Texas, Austin. He died in February 1976. Dr. Morgan was an experimental-physiological psychologist, forthright in his belief that psychology should not stray from the scientific path. To this end, he was instrumental in the formation of the Psychonomic Society (about 1960) and later (1964) founded the journal *Psychonomic Science*. ("Psychonomic" refers to laws, or principles, of behavior.) The Psychonomic Society has flourished and now publishes several journals on the "tough-minded science" side of psychology. Cliff's many friends and associates feel indebted to him for what he did to foster psychology as a scientific discipline.

RICHARD A. KING is a Professor of Psychology at the University of North Carolina—Chapel Hill. He is also the Associate Director of the Neurobiology Program there. Unlike most people, he was attracted to psychology as an undergraduate (at the University of Cincinnati) by lectures on conditioned reflexes; he followed this up with graduate school (at Duke University) and research on learning. A few years after graduate school, Professor King's research interests shifted to biological psychology; he became intrigued by the brain's role in memory. To help with this new research direction, he completed a two-year postdoctoral fellowship in neurophysiology at the University of Washington in Seattle. Nowadays, Professor King's professional life revolves around research on the "chemical messengers" in the brain that may play a role in memory, undergraduate and graduate teaching, and, of course, helping to write this book.

Contrary to popular opinion, professors are real people: They get married, have children, pay the bills, and have some fun. Professor King does (or has done) all these things: He has a wife (Margy), three children (Richard, 24, Gordon, 15, and Sarah, 8), and a good credit rating and enjoys gardening and golf.

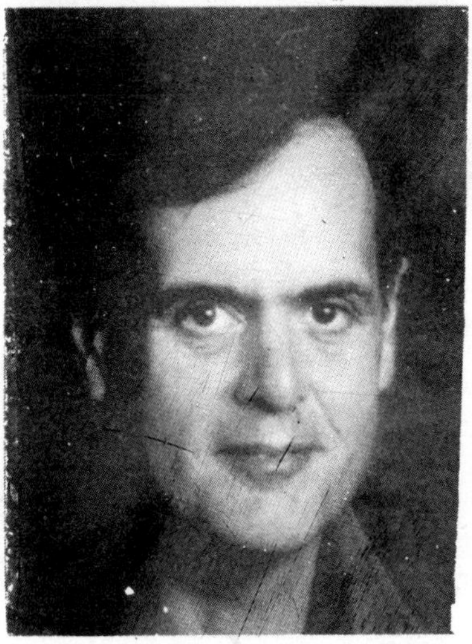

JOHN R. WEISZ is an Associate Professor of Psychology at the University of North Carolina—Chapel Hill. He first studied psychology at Mississippi College in Clinton. After graduation, he and his wife Jenny joined the Peace Corps and worked three years in Kenya. Afterward, Weisz did graduate work at Yale University, studying clinical and developmental psychology. Following a clinical internship at Connecticut Valley Hospital, he taught for three years at Cornell University, then moved to the University of North Carolina.

At North Carolina, Weisz teaches graduate students who are training to become clinical psychologists. He also does research on the impact of psychotherapy with children and on cross-cultural differences in child psychological problems. In addition, Weisz studies mental retardation and explores perceptions of control—how we get these perceptions and how they can help, and harm, us.

Outside of academia, Professor Weisz enjoys traveling—with Jenny, and with daughters Dawn and Allison. He also runs and lifts weights, to compensate for his love of chocolate and homemade ice cream.

JOHN SCHOPLER is a Professor of Psychology at the University of North Carolina—Chapel Hill. Prior to a seven-year stint as chairman of the Psychology Department, he was the Director of the Social Psychology Graduate Program. He has co-authored a text in social psychology and has made original research contributions in the areas of helping, crowding, person perception, and interdependence. Some of this research was undertaken during the three leaves he spent with the Department of Social Psychology at the London School of Economics and Political Science. Professor Schopler has not always been a social psychologist. His graduate training (M.A. from the University of New Mexico; Ph.D. from the University of Colorado) was primarily in clinical psychology, and he spent several years in doing clinical work with clients. His present professional activities are focused on research and teaching undergraduate and graduate students.

Professor Schopler was born in what is now West Germany. When he was a young boy, his family immigrated to the United States before World War II and soon settled in Rochester, New York. He was educated in the public school system there and then earned a B.A. from the University of Rochester. He is married to Janice, a Professor of Social Work, and they have four children (Kari, Lisa, Andy and David).

Contents in Brief

Contents

Chapter 14

Chapter 15

Chapter 16

Preface

Looking over the Prefaces to the six earlier editions of this text, we see remarkable continuity. In the Preface to the first edition, written in 1956, the late Cliff Morgan hoped that he had "succeeded in presenting a fair and representative picture of psychology for the student who is getting a first serious introduction to it." This theme was repeated in Prefaces to subsequent editions. We have tried to carry on Dr. Morgan's hope and the heritage of the classic text he gave to psychology. But there is change too; neither the field of psychology nor the needs of the students who study it are the same as in 1956. We hope that this text represents the changes that have taken place in our discipline and speaks to the students of today.

As psychologists discover more about behavior, our work becomes more and more pertinent to the solution of practical human problems. While basic research is still very important in psychology, our discipline is now more concerned than ever before with applications of psychological knowledge to problems of society, of families, and of individuals. We have discussed a number of such problems, especially in the chapters to which John Schopler and John Weisz have contributed their knowledge and experience in social, developmental, and clinical psychology. What we have tried to do in these and the other chapters is convey a sense of what is important in our field and why it is important.

Although we make liberal use of applications and examples, the tone of the text is solidly empirical. In deciding which material to include from the vast psychological literature, we have given emphasis to controlled, replicable studies.

In this seventh edition, we have attempted to bring readers up-to-date, an ongoing challenge in our rapidly changing discipline. Psychology, like other areas of knowledge, has its classic theories and studies which form the basis for much current work. Students should know about these, too, and we have not neglected them. There is change, but there is continuity as well.

What are some of the specific changes made in this edition? Every chapter has been thoroughly rewritten to include new key studies and discoveries. The chapters entitled "Motivation," "Social Perceptions, Influences, and Relationships," "Development During Infancy and Childhood," "Development During Adolescence, Adulthood, and Old Age," "Personality," "Abnormal Psychology," and "Therapy for Psychological Distress," particularly, have been thoroughly revised.

Since we are teachers, we have not forgotten the students who will be using this book. We owe them an accurate, clearly written presentation of psychology. Based on our past experience, we believe that we have explained the abstract ideas of psychology in concrete terms easily comprehended by most college students.

The sixteen chapters of this text can be covered during a single semester or fast-paced quarter. The format of the text is flexible, though. Depending on the interests of the instructor and students, some chapters may be omitted to give time for an in-depth study of other chapters. By selecting appropriate chapters, instructors can design life-oriented or science-oriented courses. Although the chapters are integrated with one another, each chapter can stand on its own and be understood independently of the others. The Glossary at the end of the text should help to fill in gaps when chapters are read out of order. Our extensive cross-referencing (e.g., see Chapter 6) should also help in this regard.

Many people have helped with this text. We have mentioned the late Cliff Morgan, who set the tone and provided the inspiration for this edition. We have also benefitted from the contribution Nancy Robinson made to the sixth edition. The comments and suggestions made by teachers and students who used earlier editions of this text have been helpful.

We acknowledge with thanks those who reviewed the previous edition in preparation for this seventh edition. These reviewers are: John Donahoe, University of Massachusetts at Amherst; Maury Haraway, Northeast Louisiana University; Steven Hinkle, Miami University; and M. Carr Payne, Jr., Georgia Institute of Technology.

The reviewers of the preliminary manuscript of this seventh edition helped us refine and polish it. We appreciate all their helpful suggestions. They are: Rex Bierley, North Dakota State University; Don C. Charles, Iowa State University; John Condry, Cornell University; Stephen Davis, Emporia State University; Robert Emery, University of Virginia; Richard A. Griggs, University of Florida; Robert A. Hicks, San Jose State University; Thomas T. Jackson, Fort Hays State University; James J. Johnson, Illinois State University; Peter Leppmann, University of Guelph; Harold Mansfield, Fort Lewis College; Robert Miller, Plymouth State College; Gregory Reichhart, State University College; Carol Ryff, Fordham University; Valerie Sasserath, Fairleigh Dickinson University; and Stephen M. Weissman, Plymouth State College.

The editors at McGraw-Hill, especially David Serbun and David Dunham, worked long and hard to help us write an accurate, up-to-date, readable account of psychology.

Our thanks go to all these people and to our families and colleagues, each of whom helped in bringing this book to fruition.

Richard A. King
John R. Weisz
John Schopler

To the Student

HOW TO GET THE MOST OUT OF THIS BOOK

As you begin this book, you know that psychology deals with many problems of everyday life and thus with many things that you have already experienced. You are therefore in a position to derive some personal benefits from the study of psychology. In a formal college course, however, it is not possible for the instructor to relate everything that is taught to your experience. Hence, to get the most from the course, you will have to make many of these applications yourself. You should continually ask yourself "How does this apply to my experience?" and "How can I use what I am learning?" If you take these questions seriously, you will profit much more from the course than if you simply learn by rote what is assigned.

Here are some suggestions for covering each chapter. You might begin by looking over the Heading Outline and the Summary. These sections obviously do not cover everything in the chapter, but they do hit the high points. The few minutes it takes to get the overall organization of a chapter in mind will be a great help when you dive into it for intensive study.

Many students try to read textbooks the way they read novels; they passively run their eyes over the words and hope that some information will sink in. But textbooks are packed with facts and explanations. To assimilate them, you must work at the task. Read every sentence carefully; be sure you understand what it says. Reread paragraphs and sections that give you difficulty; to understand what follows, usually you must understand what went before.

Pay attention to the illustrations and tables. In this book, they are fully as important as the corresponding discussions in the text. In some cases, we have used illustrations to teach something that is not included in the text. At appropriate points in your reading, usually before going on to a new heading, you should scan the illustrations to make certain you have examined and fully understood them.

Every technical subject uses its own special terms, and psychology is no exception. Ordinarily a definition is given in the text whenever a new term is introduced. Since chapters will not always be assigned in the order of their arrangement in the book, a Glossary is included at the back of the book. You should be especially cautious not to neglect a definition just because the term is already familiar to you. Do not, for example, pass over words like *attitude*, *personality*, *intelligence*, and *motive* because these are words that you use in everyday speech. In psychology these and other common terms often have special meanings that differ from the definitions commonly used. Make sure you know the *psychological* definitions of all terms. Lists of terms are included at the end of each chapter; use these to review your mastery of terminology. These terms are defined in both the text and the Glossary.

Science is produced by scientists, and it is common practice in science to give credit for particular experiments and ideas to the scientists who have contributed them. Sometimes this practice is annoying and distracting; so we have tried not to use too many names. But to give credit where credit is due, we have put the names of the experimenters in parentheses where particular studies or ideas are cited. These names refer to the References section at the back of the book; use it if you want to learn more about the topic under discussion. The Suggestions for Further Reading sections are another good place to start a more detailed study of a chapter's topics. Other hints about how to look up more information are given in Appendix 2.

There is a *Study Guide* for this text that you may purchase as an aid in your study and review. The guide contains chapter reviews, sample test questions, and other aids to help you learn definitions and concepts.

By the time you finish this course, you should have a good view of modern-day psychology. We hope that this text does just what we have intended: introduce you to a broad, exciting field, rich in its implications for everyday life. Whether you are planning a career as a behavioral scientist or not, try to take a scientific point of view as you study this text. Ask yourself: What do we really know? What do we need to know? How can we find out? How can we improve matters? Keep these questions at the back of your mind as you read and pursue your interest in human behavior.

Richard A. King
John R. Weisz
John Schopler

What Psychology Is Like

HEADING OUTLINE

Chapter 1

True or False?

Which of these statements is true? Which is false? For the answers and the meaning of the numbers, see page 34. (Adapted from Vaughan, 1977.)

Memory can be likened to a storage chest in the brain into which we deposit material and from which we can withdraw it later if needed. Occasionally, something gets lost from the "chest," and then we say we have forgotten. 87

Personality tests reveal your basic motives, including those you may not be aware of. 85

The basis of the baby's love for his mother is the fact that his mother fills his physiological needs for food, etc. 84

By feeling people's faces, blind people can visualize how they look in their minds. 83

The more highly motivated you are, the better you will do at solving a complex problem. 80

The best way to ensure that a desired behavior will persist after training is completed is to reward the behavior every single time it occurs throughout training (rather than intermittently). 77

A schizophrenic is someone with a split personality. 77

Blind people have unusually sensitive organs of touch.	76
Fortunately for babies, human beings have a strong maternal instinct.	73
Biologists study the body; psychologists study the mind.	71
Psychiatrists are defined as medical people who use psychoanalysis.	67
Children memorize much more easily than adults.	66
The ability of blind people to avoid obstacles is due to a special sense which develops in compensation for their absence of vision.	65
Boys and girls exhibit no behavioral differences until environmental influences begin to produce such differences.	61
"The study of the mind" is the best brief definition of psychology today.	57
Genius is closely akin to insanity.	53
The weight of evidence suggests that the major factor in forgetting is the decay of memory traces with time.	52
The unstructured interview is the most valid method for assessing someone's personality.	52
Under hypnosis, people can perform feats of physical strength which they could never do otherwise.	51
The more you memorize by rote (for example, poems) the better you will become at memorizing.	50
Children's IQ scores have very little relationship with how well they do in school.	50

AS its title says, this book is an introduction to psychology. This introduction to our field will touch on a wide array of factors involved in what we human beings do; it will present a number of major principles underlying human actions. Such principles can give us a rational basis for understanding what we and others do. In a sense, this book is a record of how far we have come in realizing the promise and hope of psychology: the rational understanding of human behavior. We the authors also hope that this book will give the reader a sense of the limitations on what we know and an appreciation of the great distance yet to go in accounting for why people do what they do. But by means of patient research, careful observation, imaginative hypothesizing, and constructive self-criticism, psychology is gradually nearing the hoped-for goal of understanding human behavior.

A Definition of Psychology

Strange as it may seem, defining *psychology* is no easy matter because of both the wide scope of its concerns and the philosophical differences among its practitioners. But if you ask a psychologist to define the word, the chances are good that you will be told something like: *Psychology* is "the science of human and animal behavior; it includes the application of this science to human problems."

The first part of this definition calls *psychology* a *science*. What does this mean? The second part of the definition is about the application of psychological principles to life's problems. Can psychology help solve life's problems? You may also wonder about the word *behavior* in the definition. Why *behavior* rather than *mind* or *thoughts* or *feelings*? Another question which may occur to you is whether psychology is the *only* science of human and animal behavior. Are there others? Suppose we look at each of these questions, beginning, first, with the one about science.

PSYCHOLOGY AS A SCIENCE

A *science* is a body of systematized knowledge that is gathered by carefully observing and measuring events. Psychologists do experiments and make observations which others can repeat; they obtain data, often in the form of quantitative measurements, which others can verify. This scientific approach is very different from forming opinions on the basis of individual experience or arguing from premises that no one can test. Thus experiments and observations are at the core of scientific psychology. (For more on the methods of psychology, see page 8.)

As a science, psychology is systematic. Data from experiments and observations are essential, but for them to make some sense in helping us understand events, they must be organized in some way. Scientific theories are important tools for the organization of observed facts. To some people, the word theory simply means someone's unsupported and unfounded notion of how things ought to be done or a set of abstract principles that do not work in practical situations. However, *theory* has quite a different meaning in science. Scientific theories are general principles which summarize many observations and predict what can be expected to happen in new situations. As summaries and predictors of events, scientific theories should not be thought of as "right" or "wrong" but merely as more or less useful in helping summarize what has been observed and in making predictions about what is to be expected when new observations or experiments are done. As new data are obtained, theories are subject to modification, or old theories may eventually be scrapped in favor of new ones which do a better job of summarizing and predicting than did the old theories. Many of the arguments scientists have with each other are over the modification and scrapping of theories as new data are gathered.

Another important part of most sciences—psychology included—is *measurement*, defined as the assignment of numbers to objects or events according to certain rules. We are all familiar with measurements of physical scales—length, time, temperature, and the like. Measurement in psychology is often more difficult than it is in sciences such as physics and chemistry because many of the things psychologists study cannot be measured directly by physical scales. What are the yardsticks of happiness or friendship, for example? Even though many behaviors are difficult to measure, psychologists have devised many ingenious tests to assign numbers to them. (Chapter 13 gives some examples.) Of course, not everything psychological is as difficult to measure as are happiness and friendship. We can often design our experiments and arrange our observations so that we can use physical measures of space and time to tell us about psychological events. For example, we can count the number of times a person behaves one way in one situation and another way in another situation, or we can measure the time taken to make a response in one situation or another.

THE APPLICATION OF PSYCHOLOGY

As the second part of the definition of *psychology* says, psychology has its applied side—that is, it is often used to solve "real-life" problems. An example is given in Application 1.1.

The application of knowledge to practical problems is an art; it is a skill, or a knack for doing things, which is acquired by study, practice, and special experience (Figure 1.1). The psychotherapist talking to a worried client, the educational psychologist advising a school board on a new curriculum, the clinical psychologist supervising group therapy in a state mental hospital, and the social psychologist trying to lessen tensions between management and workers in a large industry are all practicing the art of psychology. Just as a physician or engineer develops skills in using scientific knowledge to solve practical problems, these psychologists have learned, through special training, the artistry, or knack, of applying psychology.

Of course, the ability to apply psychological principles is a hard-won skill. You cannot expect to become an expert from reading this textbook. Special experience is needed. But after reading this book, you should be able to apply psychological principles to at least some of the things that happen in your daily life.

Application 1.1

PSYCHOLOGY CAN HELP SOLVE PRACTICAL PROBLEMS

Were you afraid of the dark as a child? This is one of the most common childhood problems. Psychologists have developed a number of methods to help children overcome these fears. Psychologists Jean Giebenhain and Stan O'Dell (1984) put some of these methods together into a parent-training manual, then tested how well the manual worked. They located the parents of six children who were so afraid of the dark that they threw tantrums at night, insisted on sleeping with their lights on, or were unable to go camping or spend the night with friends. The manual taught parents the following procedures:

1. *Giving the child control via a rheostat.* A rheostat allowed the child to set the illumination level of a room lamp which was placed beside the child's bed. The rheostat levels set by the child were recorded every night.

2. *Relaxation training.* Every night before bedtime, the parents and children practiced procedures for getting very relaxed.

3. *Positive self-statements.* At the same time, the parents helped their children memorize and repeat positive statements about themselves and their ability to get along in the dark. Some sample statements were "I am brave . . ." and "I can take care of myself when . . . I'm in the dark."

4. *Record keeping and feedback.* Every night, the children set their rheostats at the lowest level they thought they could tolerate. The levels were numbered from 1 (total darkness) to 11 (maximum brightness). The goal was to make the light dimmer by one-half number every night. A graph showed the child's settings for each night. Every morning, the parents and children recorded the child's setting from the night before and checked the child's progress.

5. *Rewards for success.* Whenever the children's morning entry showed progress, the parents responded with praise, hugs, and sometimes special treats.

6. *Phasing out rewards.* Whenever a child reached the low level of illumination that the parents and experimenters had agreed to aim for, the rewards were gradually phased out. The hope was that being brave in the dark would become its own reward.

The hope was apparently well founded. Within 2 weeks, all the children were sleeping all night with their lights at or below the goal level. And the children's reports on a "fear thermometer" indicated that they were not afraid. The experimenters checked up on the children 3 months later, then 6 months later, then 1 year later. Each time, it was found that the children were doing as well as, or better than, they had at the end of training.

The manual these parents used cost less than \$2 to produce. The rheostat costs about \$12.50. All in all, not too steep a price for bravery in the darkness!

REFERENCE

Giebenhain, J. E., & O'Dell, S. L. (1984). Evaluation of a parent-training manual for reducing children's fear of the dark. *Journal of Applied Behavior Analysis, 17,* 121–125.

WHAT IT MEANS TO SAY PSYCHOLOGY STUDIES BEHAVIOR

We now come to the word *behavior* in the definition of *psychology*. *Behavior* includes anything a person or animal does that can be observed in some way. Behavior, unlike mind or thoughts or feelings, can be observed, recorded, and studied. No one ever saw or heard a mind, but we can see and hear behavior. We can see and measure what a person does and hear and record what a person says (this is vocal behavior). From what is

(b)

(c)

Figure 1.1

The application of psychology is an art. ((*a*) Van Bucher/Photo Researchers; (*b*) Wayne Miller/Magnum Photos, Inc.; (*c*) J. Lucas/Rapho Photo Researchers.)

done and said, psychologists can and do make inferences about the feelings, attitudes, thoughts, and other mental processes which may be behind the behavior. In this way, internal mental events can be studied as they manifest themselves through what people do—their *behavior*. Thus, it is through behavior that we can actually study and come to understand internal mental processes that would otherwise be hidden from us. When we define *psychology* as "the science of behavior," we are not excluding mind; we are saying that what a person does—his or her *behavior*—is the avenue through which internal mental events can be studied.

THE STUDY OF BEHAVIOR IS NOT JUST FOR PSYCHOLOGISTS

Psychology is far from being the only branch of knowledge which studies human and animal behavior. Anthropology, sociology, economics, political science, geography, and history also study various aspects of behavior. Together with psychology, these fields make up the group of knowledge areas known collectively as the behavioral sciences. What sets psychology apart from the other behavioral sciences is partly its exclusive interest in behavior, partly its focus on individuals, and partly the wide range of behaviors it covers. Anthropology, for example, compares behaviors across cultures; sociology studies the behavior of people in groups; and economics is concerned with the behavior involved in the exchange of goods and services. The study of behavior is also a part of several biological sciences—especially zoology, but also, to some extent, pharmacology, physiology, and neurobiology.

In the areas where the many disciplines which study behavior overlap with psychology, the boundaries are blurred. For instance, a psychologist might study the personality patterns of political leaders, but so might a political scientist or a historian. The person who studies the effects of marijuana, or any drug, on behavior might be a psychologist or a pharmacologist.

Methods in Psychology

In the discussion of psychology as a science (page 4), it was emphasized that the discovery of new knowledge about behavior is based on experiment and observation (Figure 1.2). Let us look more closely at the ways psychologists go about making observations.

EXPERIMENTAL METHODS

The basic ideas behind the experimental method are straightforward. Having formulated a testable hypothesis in terms of observable events, the experimenter (1) changes or varies the events which are hypothesized to have an effect, (2) keeps other conditions constant, and (3) looks for an effect of the change or variation on the system under observation. Since psychology is the science of behavior, the psychologist looks for an effect of the experimental changes on behavior. This is simple enough, but to

Figure 1.2

Psychological research requires precise experimentation and careful observation. (Van Bucher/Photo Researchers.)

get a firmer grasp on the experimental method in psychology, we should examine it in greater detail.

Variables. As the term implies, a *variable* is an event or condition which can have different values. Ideally, it is an event or condition which can be measured and which varies quantitatively.

Variables may be either independent or dependent. An *independent variable* is a condition set or selected by an experimenter to see whether it will have an effect on behavior; it might be a stimulus presented, a drug administered, a new method of training business managers, and so on. The *dependent variable* is the behavior of the person or animal in the experiment. A dependent variable in an experiment might be the response of a person to a stimulus, a change in behavior after the administration of a drug, changes in managerial behavior after a new training program has been instituted, a score on a test, a verbal report about an event in the environment, and so on. The dependent variable is so called because its value depends, or may depend, on the value of the *independent variable*—the one independently chosen and directly manipulated by the experimenter.

When, in doing experiments, hypotheses are formulated about the effect of one thing on another, the independent variable is the one expected to produce changes in the dependent variable. Consider the following hypotheses, for instance:

1. Enriching the environments of young children with special books and toys will increase their scores on intelligence tests.

2. Giving people training in how to meditate will improve their skill as tennis players.

The environmental enrichment and the meditation training are the independent variables, while the changes in intelligence-test scores and tennis skills (possible outcomes of differences in the independent variables) are the dependent variables. When you read accounts of psychological or other experiments, it is essential that you identify the independent and dependent variables.

In graphing the results of an experiment, it is conventional in psychology to plot the values of the independent variable on the horizontal axis, or *abscissa*, and the values of the dependent variable on the vertical axis, or *ordinate*. In this way, we can see at a glance how the dependent variable of behavior is related to the values of the independent variable (Figure 1.3).

Controls Another very important characteristic of the experimental method is control. In an experiment, it is important that only the specified independent variables be allowed to change. Factors other than the independent variable which might affect the dependent variable must be held constant. It would do no good to study the effects of varying an independent variable if, unknown to the experimenter, other factors also changed. In an experiment, we must control conditions which would give misleading results.

It is literally true that an experiment is no better than its controls. Scientists must be very careful to control their experiments adequately.

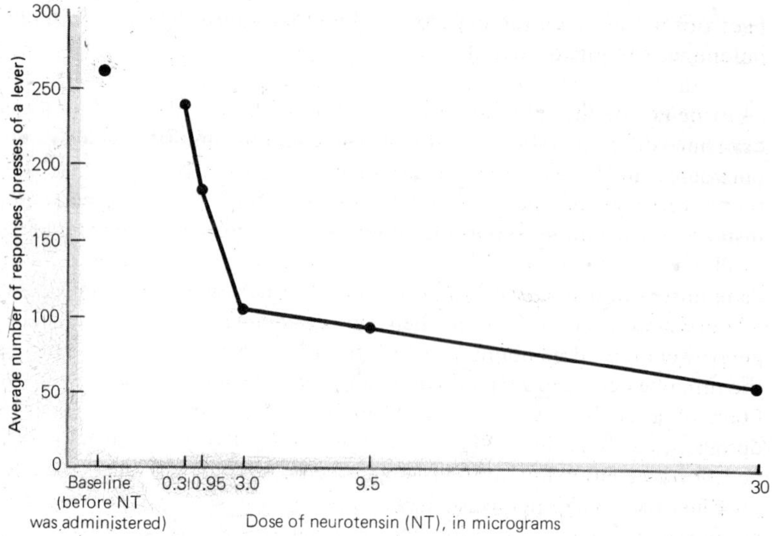

Figure 1.3

The results of a within-subjects experiment using rats as the subjects. The independent variable (on the abscissa) is the dose of the drug neurotensin (in μg). The dependent variable (on the ordinate) is the average number of responses (presses of a lever) to obtain food. (Unpublished data from Rossitch & King.)

This is often difficult in psychology because so many factors can influence the behavior that is being studied. Therefore, in interpreting experiments, it is important to look for uncontrolled factors which might have affected the results. It is a mark of scientific sophistication to be able to spot defects in experimental controls. If you want to go into the original literature in psychology or another experimental science, you should develop this skill. Being sensitive to controls may also help you evaluate claims made in "scientific" television commercials or, for that matter, in any experimental results you hear or read about.

Two main strategies, or experimental designs, are used to control extraneous factors. One strategy employs control groups. In the other, measures of behavior are made before the independent variable is introduced in order to establish a behavioral baseline against which to compare behavior after the independent variable has been presented; the subjects of the experiment—the animals or people in it—are said to serve as their own controls in this before-and-after, or within-subjects, type of experiment.

Control in Experiments: Control Groups Suppose we decide to use the *control-group design* in an experiment on human learning. In this experiment, we want to test the hypothesis that letting people know how well they are doing as they are learning improves their performance. In other words, do people do better when they get feedback on what they are doing? Finding that performance improves with feedback would have a great deal of practical importance. (In fact, performance in learning a skill usually *does* improve with feedback.) To test the effects of feedback, we will use a simple behavioral task: While blindfolded, the people in the experiment will be asked to feel a block of wood with one hand and to try, with the other hand, to draw lines the same length as the wooden block. The experimental group of subjects will be given feedback—the independent variable—by being told when the lines they draw are within 0.25 inches of the block length. The control group of subjects is given no feedback. Thus, the groups differ in the presence or absence of the independent variable.

Ideally, when a control-group design is used, the groups should be equivalent in every way except for the independent variable. In our experiment, we want to be fairly sure that any behavioral differences between the groups are due to feedback and not to other factors. We therefore match subjects in the experimental and control groups. For instance, we make sure that the subjects in the two groups are equally good at learning new skills (we might give them some preliminary tasks to do in order to check this out). The subjects in the two groups should have roughly the same ability to draw lines; it would not do if the people in one group were habitually more accurate and careful than were the people in the other group. Handedness might be important, as might general intelligence, sex, and age. Figure 1.4 (top) summarizes the control-group design.

In practice, it is very difficult to match subjects in control and experimental groups on all the factors that might conceivably affect their performance. As a compromise, subjects are often assigned at random to experimental and control groups. It is hoped that this will approximately equalize any extraneous factors in the control and experimental groups which might affect the experimental outcome. In the feedback experiment, for example, it is hoped that as many accurate people would, by chance, show up in the experimental as in the control group. A mixed

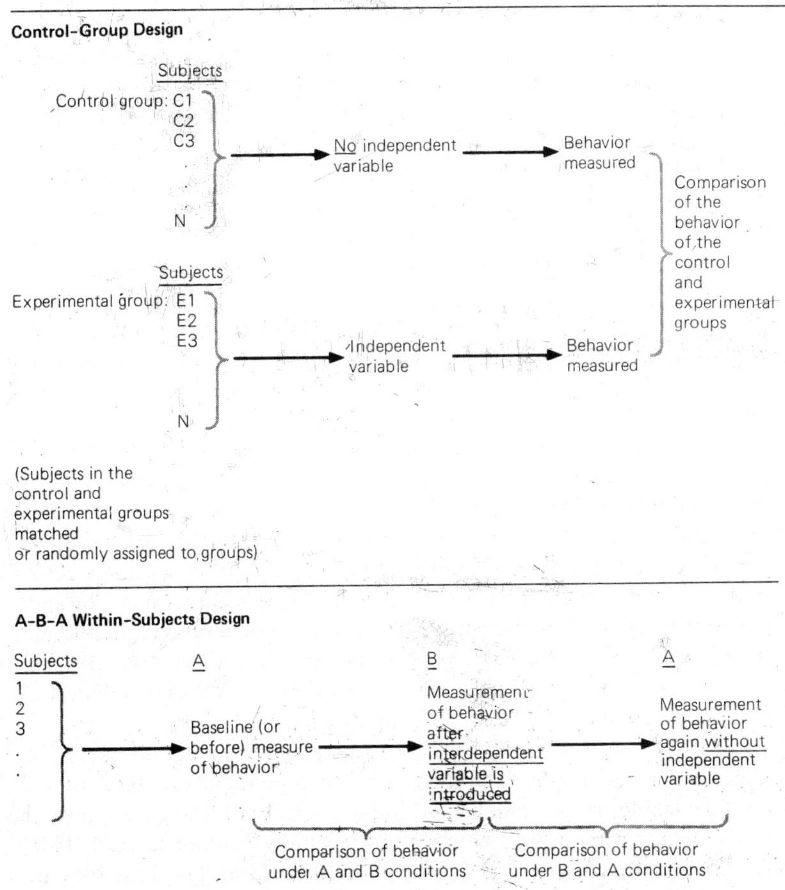

Figure 1.4

Top, the design of a control-group experiment. *Bottom*, the design of a before-and-after (within-subjects) experiment of the A-B-A type.

strategy is often used: Experimenters will match subjects on a number of factors considered to be relevant and then assign them at random to the experimental and control groups.

Control in Experiments: Baselines Perhaps better control can be achieved with the before-and-after, or *within-subjects*, design in which the subjects serve as their own controls. In this method, a *baseline* (normal level) of behavior is established before the independent variable is introduced. The behavior after the addition of the independent variable can then be compared with the baseline behavior. This before-and-after method gives good control over individual differences among the subjects which might affect the outcome of the experiment because such individual differences are present both before and after the independent variable is introduced. If individual-difference factors are held constant, changes in behavior must be due to the independent variable.

A before-and-after (within-subjects) experimental design was used in the experiment illustrated in Figure 1.3.

> The seven rats in this experiment were first made hungry by limiting the amount of food given to them. Then they were trained to press a lever protruding from the wall of the cage, whereby with each press a small pellet of food dropped into the cage. (See Chapter 4, p. 152.) After the rats were well trained, the number of lever presses in 15 minutes was recorded. This provided the baseline data (Figure 1.3).
>
> The next step in the experiment introduced the independent variable—injection of various doses of the drug neurotensin. Each rat was injected with one of the doses (selected at random) and then allowed to press the lever for 15 minutes. Two days later, another dose was given, followed by another 15-minute lever-pressing session. These injections continued until each rat had received each of the various doses of the drug. The number of responses in each of these test sessions was recorded and averaged for the seven rats.

Note that each rat serves as its own control in this before-and-after experiment. For each individual rat, its control (baseline) behavior could be compared with its behavior after different doses of the drug had been given.

To make sure that the independent variable produced the change in the behavior, it is often a good idea to see what happens when the independent variable is removed after it has been introduced. The behavior should go back to baseline levels if the independent variable did, in fact, produce the observed changes. This is called an *A-B-A within-subjects experimental design*; the first A is the baseline condition without the independent variable, B is the condition with the independent variable, and the last A refers to the final test of the behavior without the independent variable. Figure 1.4 (bottom) summarizes the A-B-A within-subjects experimental design.

This design is a good one to use when the independent variable does not have a long-lasting effect. Some independent variables, though, produce stable, long-term changes in behavior. For instance, suppose the independent variable is a new way of teaching children to read. If it is successful, the children will not be able to return to their baselines after the experiment is over because a long-lasting change will have taken place in them. In cases like this, the control-group method would most likely be used.

Replication It is important that experiments can be repeated, or, in other words, *replicated.* In elementary chemistry, for example, we can demonstrate that water is made up of hydrogen and oxygen simply by burning hydrogen (that is, combining it with oxygen) and collecting the water that results. Anyone with the proper equipment can do this experiment, and it has been done millions of times. In psychology, we can show that recitation is an aid to memory by having two groups study something, one with recitation and the other without, and then measuring the differences in memory. If this experiment is performed under the proper conditions, it will show that recitation helps memory. This finding, too, has been repeated, or *replicated*, many, many times.

The advantages and importance of replication are obvious. If we get the same results over and over again under the same conditions, we can be sure of their accuracy beyond all reasonable doubt. Experimental findings which cannot be repeated generally do not become part of the fabric of a science. A dramatic discovery may be reported, but if it cannot be replicated by other scientists it will not be accepted. Replication, or "check-up-ability," as it has been called, is an essential part of the experimental method.

Limitations of the Experimental Method The experimental method is, in many ways, the best method for gathering scientific information. But it has limitations. Obviously, it cannot always be used, especially if the experiment might be dangerous for the subjects. A second limitation is that the method is restricted in its application. The conclusions derived from an experiment may be limited to the artificial experimental situation—they may not apply to "natural" situations or even to other experimental situations. Psychologists who do experiments must continually guard against this possibility and strive for generality in their experiments. A third limitation is that the method sometimes interferes with the very thing it is trying to measure. Consider, for example, an experiment on fatigue. A psychologist may give people various tests of skill and thinking before (baseline) and after they have gone without sleep for 24 hours and find that they improve upon their baseline performance after 24 hours without sleep! Should the psychologist, or should we, conclude that 24 hours of sleeplessness is beneficial to complex performance? Probably not. Another variable had been introduced into the experiment, and it interfered with the measurement of the effects of the fatigue. When the experimental subjects came into the laboratory, they were strongly motivated to perform well, and this variable—an increase in their motivation to perform well—overshadowed the effects of their fatigue. Hence, the possibility that people in an experiment may not behave as they normally would must be considered when drawing conclusions from the experiment.

SYSTEMATIC OBSERVATION

What alternatives to the experimental method do we have? One alternative has no generally accepted name, but we shall call it the *method of systematic observation*. This approach is similar to the experimental method in that variables are measured but it is different in that researchers do not willfully manipulate the independent variable. Instead, they capitalize on variations that occur naturally. Using this method, psychological researchers simply make the most exacting and systematic study they can of naturally occurring behavior. After making a number of observations, the psychologist can, using certain rules of logic, try to infer the causes of the behavior being studied. Psychology shares this approach with a number of other sciences. For instance, suppose a geologist, in making systematic observations of a certain stratum of sedimentary rocks, finds an unusual boulder embedded in the stratum. How did it get there? Is it an ancient meteorite; was it, perhaps, rafted there by a glacier; or is some other natural force responsible for its presence? The geologist will try to solve this puzzle by making further observations and by using logical reasoning to establish the probable cause for the boulder's placement. Or, to take another example, think about an environmental scientist who is studying pollution in a river. The pollution is definitely there; but the question is, where did it come from? By making systematic observations and drawing logical inferences from these observations, the scientist can probably find the source (cause) of the pollution.

Describing Behavior As we have said, one aspect of the method of systematic observation in psychology is simply to describe behavior as it occurs naturally. What do people do? Can various behaviors be classified in systematic ways? How do people differ in their behavior? For instance, using questionnaires, surveys, and interviews, psychologists might study the personalities and motivational patterns of political leaders, the attitudes of successful executives, or the ideas that liberal and conservative parents have about the best ways to rear children. Using other techniques, research psychologists might make systematic observations of the differences in the brain activity of creative and noncreative people, the differences in school performance of children who are bused to school and those who are not, or the behavioral differences between men and women. The list goes on and on. As an everyday example of a behavioral difference between the sexes which you may observe for yourself, consider the different ways your male and female classmates carry their books (Jenni & Jenni, 1976).

> Extensive observations were made on the book-carrying behavior of college students in Montana, Ontario, El Salvador, and Costa Rica. The behavior was classified into two patterns. The type I pattern consisted of carrying a book (or books) by wrapping the forearm around it and supporting the short edge of the book against the body; the type II pattern consisted of carrying the book (or books) at the side of the body with the long edge of the book approximately horizontal and with the book grasped from the top or supported from underneath (Figure 1.5). Approximately 90 to 95 percent of the females fell into the type I pattern, while about the same percentage of males fell into the type II pattern. Look around you!

From Descriptions to Causes The method of systematic observation tells us what people do and how they differ in their behavior. But the psychological detective who uses the method of systematic observation may also seek to find out what causes the observed behavior. Thus

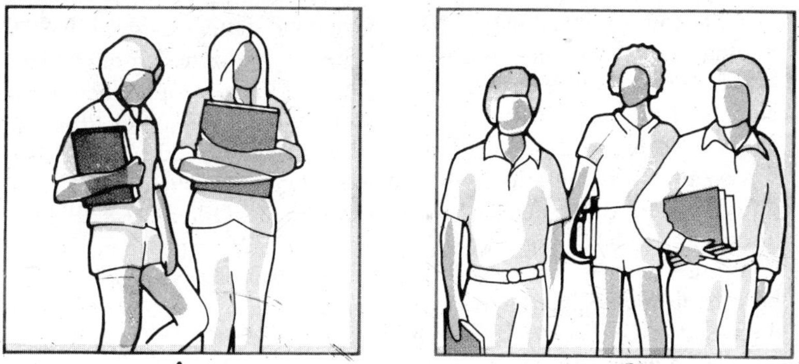

Figure 1.5

Male and female ways of carrying books. Variations of the female, or type I, pattern are shown in *a*; male, or type II, patterns are shown in *b*. (From Jenni & Jenni, 1976.)

pyschologists, and other scientists, too, are often not satisfied with answering questions about what occurs; they want to find causes for their observations so that they can answer questions about why things happen.

Referring to the observations on book carrying just described, why do females carry their books on the hip and males at the side? Is it because of differences in female and male anatomy; is it because they learn this behavior from others of the same sex; or is it due to other factors altogether? One clue favoring learning as the cause of this particular behavior came from observing girls and boys at the age at which the male-female difference in book carrying begins. In grades 2 and 3, girls and boys start to carry books in different ways, and, at this age, the body proportions of males and females are essentially identical. But of course, this observation, by itself, does not establish the cause of the behavioral difference. Other unknown and uninvestigated factors might be the cause, or the cause might be some combination of factors. All this observation really shows is a relationship, or *correlation*, between grade in school (age) and the beginning of the male or female pattern. The fact that an event comes before another event does not show that the first event is the cause of the later one. In addition, a behavior may have many causes. Thus, to establish the likely cause, or causes, of even this simple behavior, a great many more observations would be needed. Even then, we would not be sure of the cause; we would only have identified a likely cause or set of causes.

For more complex behavior, establishing likely causes is much more difficult. Suppose, for instance, that a psychologist wants to find the general cause of the severe behavior disorder known as schizophrenia. Schizophrenic behavior is described in detail in Chapter 15, page 637. For now, it is sufficient to say that symptoms of schizophrenia include bizarre, or strange, patterns of thought and behavior, inappropriate emotional responses, and perhaps hallucinations and delusions. (Incidentally, it is not a "split personality.") A great deal of effort has gone into investigating the causes of schizophrenia because psychologists believe that knowledge of its causes is essential for its prevention. Suppose psychologists who are studying schizophrenia hypothesize that its cause is to be found in the way children are reared by their parents. Using the method of systematic observation, the researchers will probably try to test this hypothesis by comparing the ways in which schizophrenic and normal people were reared. To do this, they will match normal and schizophrenic groups on as many factors—such as age, sex, socioeconomic status, years of schooling, intelligence, and so on—as possible. Then they will look for differences in the rearing practices used by the parents of the schizophrenics and the parents of the "normals." Using precisely this type of strategy, psychologists have found differences; but these differences do not, by themselves, establish causation. All that has been established so far is that differences in rearing go along with schizophrenia. Other factors and their interaction with rearing may be the cause. A great many more observations must be made before the likely cause or causes of schizophrenia can be established. (Some hypotheses about the causes of schizophrenia are discussed in Chapter 15, page 640.)

Finding the causes of behavior from a number of observations is called *inductive reasoning*, or establishing general principles from particular instances. To try to find the cause of a particular behavior, we must look carefully at the results of many observations and experiments, noting the effects of a particular factor, which we will call factor X, on the behavior under study. If we found that the behavior always occured when factor X was present but never occurred in the absence of factor X, we could begin to make a case for factor X as a cause of the behavior. Furthermore, suppose we found that large amounts of factor X led to large changes in the behavior we were studying, while smaller amounts of the factor led to smaller changes. In other words, we found that factor X was quantitatively related to the observed behavior. Should this be the case, our argument for factor X as a cause would be greatly strengthened.

To illustrate, consider an example from research on the brain and speech. It has been found that in almost all right-handed people, damage, perhaps from a stroke, to an area of the left cerebral hemisphere impairs speech in certain ways. (See Chapter 2, page 69.) The speech impairment is always present after damage to this area but does not occur after damage to other areas of the brain. In addition, the degree of speech impairment is related to the amount of damage to the area in question. From these observations, we are able to infer that the cause of speech impairment is damage to a particular brain area. Unfortunately, establishing causation from observations is usually not this simple. One of the main reasons for this is that several factors acting and interacting together usually produce the behavioral results we are interested in. The problem with "real-life" behavior (remember the book-carrying and schizophrenic examples?) is to find the combination of causative factors, and this is no easy job.

THE CLINICAL METHOD

The clinical method is ordinarily used only when people come to psychologists with personal problems. Little Alice is doing badly in school, and her parents bring her to the psychologist to find out why. Little Basil throws temper tantrums, refuses to eat, cries all night, and generally makes life miserable for his parents. Harold, an otherwise fine young eleventh-grader, is caught stealing from the Sunday-school collection plate. Or young Mr. Squabble, married 5 years, is worried because he and his wife just cannot get along. Problems like these—and other kinds, too—bring people to the clinical psychologist.

Not all clinical problems require thorough study. But when they do, the psychologist usually begins by getting a detailed account of the person's history, including his family relations. This information is usually gained by interviewing the person and his associates. Sometimes the psychologist has a specially trained social worker study the person's social background and environment. The psychologist may administer tests of various kinds—intelligence tests, interest tests, tests of emotional maturity, personality tests, and other tests. From these tests and from the biographical information gathered earlier, the psychologist will try to diagnose the problem and treat, or remedy, the difficulty. The tests, the diagnosis, and the remedy will, of course, differ from case to case.

Here we are concerned with the clinical method as a tool in science. As a method, it combines features of clinical observation, experiment, and systematic observation. Working with individual cases, the clinician may observe some factor he or she thinks is important. By observation of his patients, Sigmund Freud discovered that dreams often reflect people's strong unconscious desires. But clinical observation does not often provide much scientific information. It is usually too subjective, casual, uncontrolled, and lacking in precise measurement. What appear to be cause and effect in one case may not be in another. Even in a single case, it is extremely difficult to sort out the significant causal factors with certainty. From a scientific standpoint, probably the greatest value of clinical observation is that it suggests fruitful ideas which can be investigated more rigorously by experimental and systematic-observation methods.

The Work of Psychologists

Thus far, we have defined *psychology* in abstract terms. But we can also define it in human terms by seeing what sorts of things psychologists do. Since psychologists have the whole field of behavior to choose from, you might guess that their work is pretty varied. You would be right. One psychologist may use sensitivity-training groups to enhance human effectiveness; another may study the effects of drugs on behavior; another may study the development of intelligence during childhood; another might be a psychotherapist who tries to help people with their behavioral problems; and still another might study the sensory processes involved in our perception of the world. Psychologists differ not only in their behavioral interests but in the degree to which they are involved with the application of psychology to life's problems—the art of psychology (page 5). Inquiry 1.1 is designed to bring to life some of these differences in the work of psychologists.

Another way of describing the work of psychologists is to discuss some of the major subfields of psychology—clinical psychology, counseling psychology, school and educational psychology, experimental and physiological psychology, industrial and organizational psychology, social psychology, developmental psychology, and community psychology (Figure 1.6).

CLINICAL PSYCHOLOGY

The total number of psychologists in the U.S. is now approaching 100,000. Somewhat more than a third of these are clinical psychologists. Clinical psychologists come closest to many people's idea of what a

Inquiry 1.1

WHAT DO PSYCHOLOGISTS REALLY DO?

A neighbor of one of the authors is a psychologist in private practice and works with adolescents who are having difficulty in school. She tries to diagnose their problems and works individually with each young person, his or her parents, and school officials to help correct the difficulty.

A colleague down the hall spends his day studying drugs which might help rats remember what they have just learned.

An acquaintance works as a psychologist in a state mental hospital. He is the assistant administrator of a ward for long-term mental patients. As such, he is involved with the supervision of the attendants and others responsible for the day-to-day care of these patients. He also participates in the planning of therapy programs.

Within the same morning, it is possible to hear colleagues talk about their progress with respect to many applied problems. Among these the other day were (1) ways to improve polling techniques, (2) ascertaining whether increasing amounts of lead in children are associated with intellectual deficits, and (3) what the Internal Revenue Service can do to increase compliance with income-tax laws.

Psychologist friends of the authors are also called on from time to time to give expert testimony in court cases. Sometimes defense and prosecution lawyers ask them to assess the validity of certain types of testimony—eye-witness identification, for example.

A friend from across town works as a pediatric psychologist in a hospital. He helps chronically ill children and their families cope with the distress caused by both the illness and its treatment. For example, he helps reduce the pain experienced by children getting treatments for leukemia by helping them understand their treatment and by using hypnosis to relax their muscles.

Another friend, in another state, helps people cope with another kind of pain: migraine headaches. She uses biofeedback machines to train headache victims. They learn to regulate the volume of blood flowing through certain arteries near their foreheads, thus reducing the frequency and pain of their headaches.

A psychologist in a nearby city works as a business consultant. He studies worker morale and employee-supervisor disputes; he also makes recommendations for improving the psychological climate in each company he serves.

Since the authors are professors, many of our acquaintances are also faculty members of psychology departments, teaching undergraduate and graduate courses. These acquaintances may also do basic research in one or another of the areas covered in this book—the biology of behavior, sensation and perception, learning, memory, thinking, motivation, child and adolescent psychology, social psychology, attitude formation, psychological testing, abnormal psychology, psychotherapy, and so forth.

psychologist is. They are "doctors" who diagnose psychological disorders and treat them by means of psychotherapy. (See Chapters 15 and 16.)

Many people are confused about the differences between a clinical psychologist and a psychiatrist. The clearest distinction between them is that a *clinical psychologist* normally holds a Ph.D. or M.A. degree (or a relatively new degree called the Psy.D., for "Doctor of Psychology"); a *psychiatrist* holds an M.D. degree The Ph.D. (or Psy.D.) clinical psychologist has taken 4 or 5 years of postgraduate work in a psychology department; the M.A. clinical psychologist has had about 2 years of postgraduate work and usually works under the supervision of a Ph.D. psychologist. The psychiatrist, on the other hand, has gone to medical school and has

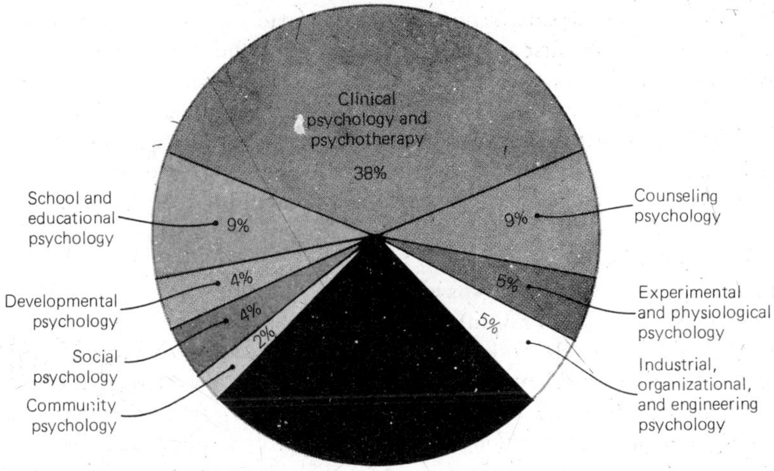

Figure 1.6

Major subfields of psychology. The diversity of psychologists' interests is emphasized by the "other" category. Included here are a number of areas of interest, no one of which has more than 1 percent of the total. (Based on figures from the American Psychological Association, 1984.)

then completed 3 or 4 years of residency training in psychiatry. This difference in training means that the clinical psychologist, who does not have medical training, cannot prescribe drugs to treat behavior disorders. It also means that whenever there is a possibility of a medical disorder, a patient should be examined by a psychiatrist or other physician. Further, in most states, only a psychiatrist can commit a patient to a hospital for care and treatment. On the other hand, psychologists are usually better trained in doing research; thus, clinical psychologists are somewhat more likely than psychiatrists to be involved in systematically studying better ways of diagnosing, treating, and preventing behavior disorders. Psychologists are also more likely than psychiatrists to use psychotherapy methods that have grown out of scientific research. Clinical psychologists also tend to rely more heavily than psychiatrists on standardized tests as an aid to diagnosing behavior disorders. (See Chapter 13.)

Confusion between the fields of clinical psychology and psychiatry arises because both provide psychotherapy. They both use various techniques to relieve the symptoms of psychological disorders and to help people understand the reasons for their problems. Such psychotherapeutic techniques range from giving support and assurance to someone in a temporary crisis to extensive probing to find the motives behind behavior. (See Chapter 16.)

Many clinical psychologists practice in state mental hospitals, veterans' hospitals, community mental-health centers, and similar agencies. An increasing number are in private practice. In the institutions and clinics where many clinical psychologists practice, while psychiatrists often are available for prescribing medical treatment when needed, psychologists do a large part of the professional work of diagnosis and treatment, as well as holding important administrative jobs and doing much of the research.

The clinical psychologist and the psychiatrist should also be distinguished from the psychoanalyst. A *psychoanalyst* is a person who uses the particular psychotherapeutic techniques which originated with Sigmund Freud and his followers. (See Chapters 14 and 16.) Anyone who has had the special training required to use these techniques can be a psychoanalyst. Since psychoanalysis originated in Freud's medical and psychiatric practice, it was first adopted by psychiatrists, and thus, today, many psychiatrists are also psychoanalysts. But clinical psychologists who have had psychoanalytic training can also be psychoanalysts, as can people who are neither psychiatrists nor clinical psychologists. Psychoanalysts without M.D. degrees are known as "lay analysts." Figure 1.7 shows the relationships among clinical psychology, psychiatry, and psychoanalysis.

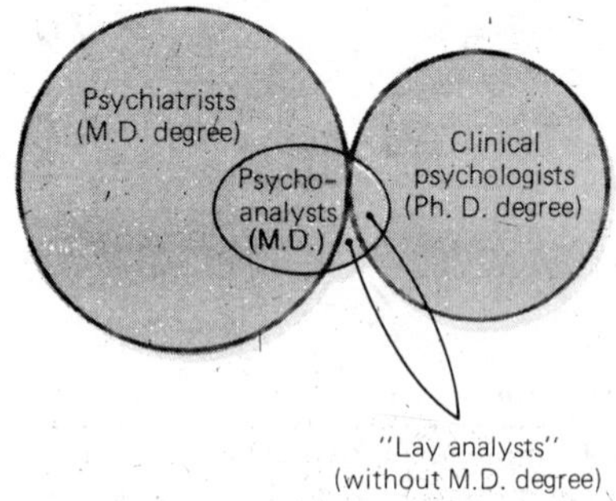

Figure 1.7

Relationships among psychiatry, clinical psychology, and psychoanalysis. Because precise figures are not available, the sizes of the circles and the ellipse give only rough indications of the numbers of practitioners in each field.

COUNSELING PSYCHOLOGY

The work of the counseling psychologist is quite similar to that of the clinical psychologist. The difference between them is that counseling psychologists generally work with people who have milder emotional and personal problems. They may use psychotherapy in an attempt to help with these problems. Counseling psychologists are often consulted by people with specific questions, such as a choice of career or educational program. In their practice, counseling psychologists may make extensive use of tests to measure aptitudes, interests, and personality characteristics. (See Chapter 13.) A number of counseling psychologists try to help people who are having problems with family living; these are the marriage and family counselors.

SCHOOL AND EDUCATIONAL PSYCHOLOGY

Much of the *school psychologist's* job consists of diagnosing learning difficulties and trying to remedy them. Using tests and information gained from consultations with the student and his or her parents, the school psychologist tries to pinpoint the problem and suggest action to correct it. For instance, a school psychologist might suggest that a poor reader be assigned to a remedial reading class. Other school psychologists are involved in vocational and other forms of counseling. These are the *school counselors*

Educational psychology mav include school psychology, but educational psychologists, as such, are usually involved with more general, less immediate problems than are most school pscyhologists or school counselors. Educational psychologists are especially concerned with increasing the efficiency of learning in school by applying their psychological knowledge about learning and motivation to the curriculum.

EXPERIMENTAL AND PHYSIOLOGICAL PSYCHOLOGY

Many psychologists are not primarily engaged in work that applies directly to practical problems. Instead, these psychologists try to understand the fundamental causes of behavior. They do what is sometimes called basic research, studying such fundamental processes as learning and memory, thinking, sensation and perception, motivation, and emotion. In other words, the *experimental psychologist* studies how behavior is modified and how people retain these modifications, the processing of information in thinking, how human sensory systems work to allow people to experience what is going on around them, and the factors that urge them on and give direction to behavior. (See Chapters 3, 4, 5, 6, 7, and 8.)

A number of experimental psychologists are concerned with the relationship of the brain and other biological activity to behavior; these are *physiological psychologists*. (See Chapter 2 and Figure 1.8.) While it is a part of psychology, physiological psychology is also considered to be part of the broader field of *neurobiology*—the study of the nervous system and its functions.

As you might surmise from the name of the subfield, controlled experiments are the major research method used by experimental psychologists. But experimental methods are also used by psychologists other than experimental psychologists. For instance, social psychologists may do experiments to determine the effects of various group pressures and influences on a person's behavior. So in spite of its name, it is not the method which distinguishes experimental psychology from other subfields. Instead, experimental psychology is distinguished by what it studies—the fundamental processes of learning and memory, thinking, sensation and perception, motivation, emotion, and the physiological or biological bases of behavior.

INDUSTRIAL AND ORGANIZATIONAL PSYCHOLOGY

The first application of psychology to the problems of industries and organizations was the use of intelligence and aptitude tests in selecting employees. (See Chapter 13.) Today, many companies use modern versions of such tests in their hiring and placement programs. Private and public organizations also apply psychology to problems of management and employee training, to supervision of personnel, to improving communication within the organization, to counseling employees, and to alleviating industrial strife. The applied psychologists who do this work are sometimes called *personnel psychologists*.

Many industrial and organizational psychologists work as members of consulting firms which sell their services to companies. For one business, they may set up an employee-selection program; for another, they may recommend changes in the training program; for another, they may analyze problems of interpersonal relationships and run programs to train company management and employees in human-relations skills; and for still another, they may do research on consumer attitudes toward the company's products.

This, then, is a subfield of psychology in which psychological principles are applied to practical problems of work and commerce. Even the research which is done is aimed at the solution of particular practical problems. Compare this with the basic research orientation of experimental psychologists.

SOCIAL PSYCHOLOGY

We spend much of our lives in the presence of other people, with whom we interact in a variety of ways and in different settings. The primary focus of social psychology is on understanding how individuals are affected by other people. This focus covers a wide range of possible interests. For example, it includes the study of the ways in which we perceive other people and how those perceptions affect our behavior toward them (Chapter 9). Similarly, it involves concerted efforts to understand the determinants of interpersonal influences and of attitude change (Chapters 9 and 10). Thus, social psychologists might study how perceptual stereotypes affect interactions or how the decisions of a committee member are influenced by what others on the committee do or say. Sometimes the interest is on the mutual influence exerted by individuals in close relationships, such as marriage.

Figure 1.8

Physiological psychologists study the relationship of the brain and other biological activity to behavior. (Van Bucher/Photo Researchers.)

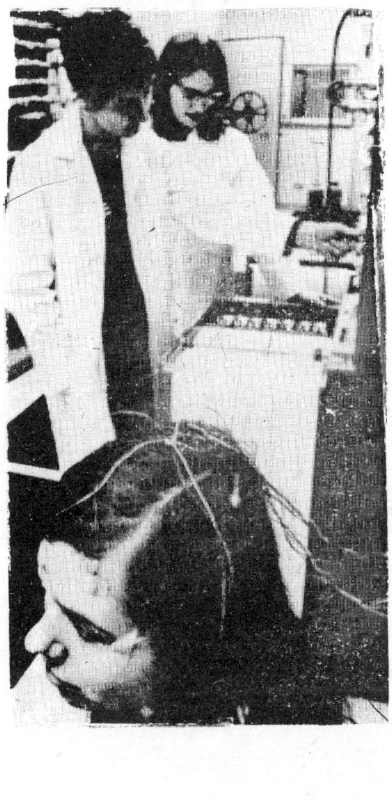

The field of social psychology is not exclusively populated by psychologists. The field has particularly benefited from contributions by sociologists. While the research endeavors of people from both disciplines often overlap, the focus of social psychology is typically upon the individual, while the sociologist is primarily concerned with societal institutions.

While many basic discoveries about individual social behavior have practical applications, social psychology also has an explicitly applied side. Social psychologists have developed and perfected techniques for measuring attitudes and opinions. Surveys of political opinions, consumer attitudes, and attitudes concerning controversial social questions give needed information to politicians, business executives, and community leaders when they must make important decisions.

DEVELOPMENTAL PSYCHOLOGY

Developmental psychologists try to understand complex behaviors by studying their beginnings and the orderly ways in which they change with time. If we can trace the origin and developmental sequence of a certain behavior, we will have a better understanding of it. Since changes in behavior occur rapidly in the early years of life, *child psychology*, the study of children's behavior, comprises a large part of developmental psychology. (See Chapter 11.) But developmental changes also occur in adolescence, adulthood, and old age; and so the study of these changes is also a part of developmental psychology. (See Chapter 12.)

Developmental psychology has both research and applied aspects. For instance, a great deal of research has been done on the development of thinking in children. Do progressive and systematic changes take place in thinking during the first few years of life? They do. (See Chapter 11 for details.) On the applied side, developmental pychologists are often concerned with children who have behavior problems or psychological disorders. The kinds of behavior found in disturbed children are frequently quite different from the behaviors found in disturbed adults, and different methods are used to treat them.

COMMUNITY PSYCHOLOGY

This relatively new area of psychology is difficult to describe because community psychologists do so many things. In general, it can be said that *community psychologists* apply psychological principles, ideas, and points of view to help solve social problems and to help individuals adapt to their work and living groups. (See Chapter 16, page 713.)

Some community psychologists are essentially clinical psychologists. They set up programs to reach people in the community who happen to have behavior problems, or are likely to develop them and who are not presently being served by traditional psychotherapeutic methods. These psychologists are a part of the community mental-health movement.

Other community psychologists are less directly concerned with the mental health of individuals and more concerned with bringing ideas from the behavioral sciences to bear on community problems. We might call these the "social-problem community psychologists." Hostility among groups in the community, bad relations between the police and community members, or distress due to a lack of employment opportunities, for example, might be problems on which a social-problem community psychologist would work. On the more positive side, such psychologists often work to encourage certain groups to participate in community decisions, to provide psychological information about effective and health-promoting child-rearing practices, or to advise school systems about how to make their curricula meet the needs of community members. To accomplish their aims, social-problem community psychologists sometimes focus on changing community organizations and institutions to help remove the sources of community problems.

Viewpoints

At the beginning of this chapter, we gave a definition of *psychology* which said, in part, that *psychology* is "the science of human and animal behavior." We then went on to describe how the scope of behavior could be broadened to include mental events. This definition itself represents a viewpoint. Other psychologists might choose to define psychology as "the science which studies mind and states of consciousness," and a few psychologists might drop the term science from the definition. Thus, fundamental differences in viewpoints show up in the very definition of psychology and in ideas about what psychology should study and how. Such differences, and the arguments they spark, can make psychology a lively field indeed.

In the history of psychology, strong differences of opinion about what psychology should study and how it should do it were represented by schools of psychology—groups of like-minded psychologists which formed around influential teachers who argued for one viewpoint or another. Time has tended to blunt the old arguments somewhat, and the early schools have largely passed into history.

Today, many psychologists share the idea that psychology should study behavior; even those who want to study internal mental events generally agree that this must begin with a look at behavior. However, even when they agree to look at behavior, psychologists may disagree about what they see and what it means. As soon as it is time to make sense out of the complexities of behavior, psychologists' differing perspectives can lead them to sharply differing views. Before we look at the current perspectives of psychologists, some historical background will help put these points of view into focus.

EARLY SCHOOLS OF PSYCHOLOGY

Formal ideas about behavior and mind in western culture began with the classical Greek philosophers and have continued to this day as part of the fabric of philosophy. Psychology as a separate area of study split away from philosophy a little over 100 years ago. The successes of the experimental method in the physical sciences encouraged some philosophers to think that mind and behavior could be studied with scientific methods. In 1879, the first psychological laboratory was established at the University of Leipzig by the German philosopher-psychologist Wilhelm Wundt (1832–1920). (Pictures of Wundt and other famous psychologists are in Figure 1.9.) The first formal psychology laboratory in the United States was set up at Johns Hopkins University in 1883, and within a few years most major universities had psychology laboratories and departments.

Figure 1.9

Pioneers of psychology. (*a* to *c*, *f*, *h*, *j*, The Granger Collection; *d* and *e*, UPI/Bettmann Newsphotos; *g*, Library of Congress; *i*, Archives of the History of American Psychology, University of Akron.)

Although still philosophical in part, the spirit of psychology as a new, separate field of study, as it developed in the last years of the nineteenth century, is captured in William James's famous textbook of 1890, *The Principles of Psychology*.

James, Wundt, and the other psychologists of the time thought of psychology as the study of mind. They did experiments to find the laws relating events in the physical world to a person's mental experience of those events; they studied attention, or the process by which we become aware of some external events and not of others; and they did many experiments in the areas of imagery, memory, thinking, and emotion.

In the first decades of the twentieth century, psychologists came to hold quite different views about the nature of mind and the best ways to study it. About the same time, fundamental questions were raised about what should be studied in psychology: Should psychology be the study of mind, should it study behavior, or should both mind and behavior be included? Different influential psychologists of the time held quite different views on the nature of mind and the proper subject matter for psychology. Schools of thought formed around these leaders as their students adopted their ideas. These schools of thought are known as the schools of psychology; they set the direction for much of the research on mind and behavior in the early years of this century.

The school was shortlived as psychologists could not measure patients thoughts and feelings

Structuralism This early school of psychology grew up around the ideas of Wilhelm Wundt in Germany and was established at Cornell University in the United States by one of Wundt's students, Edward B. Titchener (1867–1927). (See Figure 1.9.) The goal of the structuralists was to find the *units*, or elements, which make up the mind. They thought that as in chemistry, a first step in the study of the mind should be a description of the basic, or elementary, units of sensation, image, and emotion which compose it. For instance, the structuralists did experiments to find the elementary sensations—such as red, cold, sweet, and fragrant, for example—which provide, they said, the basis for more complex mental experiences. The main method used by the structuralists to discover these elementary units of mind was *introspection*. Subjects were trained to report as objectively as possible what they experienced in connection with a certain stimulus, disregarding the meanings they had come to associate with that stimulus. A subject might, for example, be presented with a colored light, a tone, or an odor and asked to describe it as minutely as possible. These experiments using introspection have given us a great deal of information about the kinds of sensations people have, but other psychologists of the time challenged the idea that the mind could be understood by finding its elements and the rules for combining them. Still others turned away from describing the structure of the mind to study how the mind functioned.

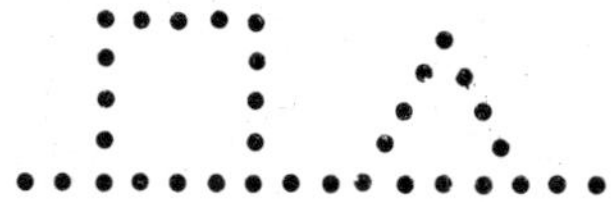

Figure 1.10

According to the Gestalt psychologists, the organization and relationships of elements determine the mental experience a person has. (See text.)

Gestalt Psychology This school of psychology was founded in Germany about 1912 by Max Wertheimer (1880–1943) and his colleagues Kurt Koffka (1886–1941) and Wolfgang Köhler (1887–1967). (See Figure 1.9.) These pioneer psychologists felt that structuralists were wrong in thinking of the mind as being made up of elements. They maintained that the mind is not made up of a combination of simple elements.

The German word *Gestalt* means "form" or "configuration," and the Gestalt psychologists maintained that the mind should be thought of as resulting from the whole pattern of sensory activity and the relationships and organizations within this pattern. For instance, we recognize a tune when it is transposed to another key; the elements have changed, but the pattern of relationships has stayed the same. Or, to take another example, when you look at the dots in Figure 1.10, your mental experience is not just of the dots, or elements, but of a square and a triangle sitting on a line. It is the organization of the dots and their relationships that determine the mental experience you have. Thus, the point made by the Gestalt psychologists in their opposition to structuralism was that mental experience depends on the patterning and organization of elements and is not due simply to the compounding of elements. In other words, according to the Gestalt psychologists, the mind is best understood in terms of the ways elements are organized. (More examples of the importance of organization in mental experience will be given when perception is discussed in Chapter 3.)

Functionalism Functionalists, such as John Dewey (1873–1954), James R. Angell (1869–1949), and Harvey Carr (1873–1954) at the University of Chicago proposed that psychology should study "what mind and behavior do." (See Figure 1.9.) Specifically, they were interested in the fact that mind and behavior are *adaptive*—they enable an individual to adjust to a changing environment. Instead of limiting themselves to the description and analysis of mind, the functionalists did experiments on the ways in which learning (Chapter 4), memory (Chapter 5), problem solving (Chapter 6), and motivation (Chapter 7) help people and animals adapt to their environments. In brief, as the name of the school implies, these early psychologists studied the functions of mind and behavior.

Behaviorism This school of psychology originated with John B. Watson (1879–1958), who was, for many years, at Johns Hopkins University. (See

Figure 1.9.) Watson rejected mind as the subject of psychology and insisted that psychology be restricted to the study of *behavior*—the observable (or potentially observable) activities of people and animals.

In addition to its focus on behavior as the proper subject matter of psychology, behaviorism had three other important characteristics. One was an emphasis on conditioned responses (Chapter 4, page 141) as the elements, or building blocks, of behavior. Behaviorism, in fact, was somewhat like the structuralism it rejected because it maintained that complex processes are compounds of more elementary ones. Its elements, however, were conditioned responses rather than sensations, images, or emotions. The conditioned response will be discussed in detail in Chapter 4, but we can describe it loosely now as a relatively simple learned response to a stimulus. Watson argued that complex human and animal behavior is made up almost entirely of conditioned responses. A second closely related characteristic of behaviorism was its emphasis on learned, rather than unlearned, behavior. It denied the existence of inborn, or innate, behavioral tendencies. (See Chapter 2, page 41.) A third characteristic of behaviorism was its focus on animal behavior. Watson held that there are no essential differences between human and animal behavior and that we can learn much about our own behavior from the study of what animals do.

Psychoanalysis Strictly speaking, psychoanalysis is not a school of psychology, but it has had a great impact on the thinking and theorizing of many psychologists. Psychoanalysis was founded in Vienna, Austria, by the psychiatrist Sigmund Freud (1856–1938). (His picture is on page 576.)

In the course of his practice with neurotic patients, Freud developed a theory of behavior and mind which said that much of what we do and think results from urges, or drives, which seek expression in behavior and thought. A crucial point about these urges and drives, according to psychoanalytic theory, is that they are hidden from the awareness of the individual; they are, in other words, unconscious. It is the expression of the unconscious drives which shows up in behavior and thought. The term *unconscious motivation* thus describes the key idea of psychoanalysis. Freud elaborated on this basic theme of unconscious motivation, the result being the theory which is described in detail in Chapter 14 (page 576) and the system of psychotherapy outlined in Chapter 16 (page 681).

MODERN PERSPECTIVES

The discoveries made by the structural, Gestalt, and functional schools of psychology have become part of the general store of psychological knowledge; but the schools, as such, have vanished. Behaviorism and psychoanalysis, on the other hand, are still, in modified forms, among the current psychological perspectives. Together with these hardy survivors, the new perspectives which have arisen in the last 50 years or so give psychologists a rich variety of viewpoints to choose from in their task of describing and understanding behavior. Examples of these newer viewpoints include the biological, cognitive, developmental, humanistic, and social perspectives. (See Figure 1.11.) Looking at these perspectives now will give you a preview of some of the ideas to come later in this book. The perspective taken depends partly on the bias of the individual psychologist and partly on what aspect of behavior (as broadly defined on

(a) (b) (c) (d)

Figure 1.11

Representatives of the biological, social, developmental, and cognitive perspectives. (*a*) R. Thompson (Chuck Painter/Stanford University); (*b*) J. Thibaut (University of North Carolina, Chapel Hill); (*c*) H. Rheingold (University of North Carolina, Chapel Hill); (*d*) A. Bandura (Stanford University.)

pages 6–7) is under study; certain perspectives are more appropriate for some behaviors than others.

To illustrate some current psychological perspectives, we will look briefly at their strengths and weaknesses as they try to deal with behavioral observations such as the following:

1. On the playground, 6-year-old Sam pushes little Samantha off her tricycle and rides away on it.

2. Mr. A., a 55-year-old man, finds that he is forgetting important appointments and has trouble recalling newspaper stories he read the day before.

How might a psychologist with a behavioral, biological, cognitive, developmental, humanistic, psychoanalytic, or social perspective try to understand these two examples?

The Behavioral Perspective What Sam actually did—not his thoughts or even his motives—would be the focus of this perspective. Asked why Sam pushed Samantha off her tricycle, a psychologist with a behavioral perspective might answer that Sam learned to behave this way because the behavior paid off in the past. In other words, he learned to act aggressively in certain situations because he was rewarded for such behavior in the past. In Chapters 4 (page 152), 14 (page 594), and 16 (page 694), we will describe the work of B. F. Skinner (picture on page 594) and others to show how learning and the behavioral perspective can be applied not only to aggression but to many other actions.

For Mr. A., the 55-year-old man with memory problems, the behavioral perspective would focus on a precise description of the changes in the man's behavior. A psychologist with a behavioral perspective might also be interested in trying to teach behavioral skills to this man so that he might learn to compensate for the behavioral problems caused by his forgetfulness. Discovering the causes of the forgetfulness would probably be left to a psychologist with another perspective—a biological psychologist, perhaps.

The Biological Perspective Psychologists with this perspective try to relate behavior to functions of the body—the nervous and glandular systems in particular. (See Chapter 2; Chapter 3, pages 85–107, Chapter 5, page 188; Chapter 7, page 272; Chapter 8, page 313; and Chapter 15, pages 641 and 651.) Consider the forgetful man again. Biological-perspective psychologists would wonder whether he had a brain problem. Perhaps he is in the beginning stages of Alzheimer's disease, and the chemistry of the brain is at fault (Chapter 5, page 212).

Is there anything biological about little Sam which might have caused him to behave so aggressively toward Samantha? Males are generally more aggressive than females (Maccoby & Jacklin, 1974); this may be related to male-female hormonal differences, but the evidence in humans on this point is mixed. Of course, even if there is a hormonal basis for male aggressiveness in general, this does not explain why Sam acted more aggressively than other little boys. Perhaps other perspectives are needed to provide a more satisfactory account of Sam's aggressiveness.

The Cognitive Perspective The word *cognition* refers to perception of the world around us (Chapter 3, pages 107–129), some aspects of learning (Chapter 4, page 168, and Chapter 14, page 596), memory (Chapter 5), thinking (Chapter 6), and comprehension of our social environment (Chapter 9). Another way to consider cognition is to say that it refers to the processing of information that we receive through the senses (Chapter 5, page 185). Such processing is the basis of the experience we have, which we call mind. Differences in the ways we process information may lead to differences in behavior.

A cognitively oriented psychologist would try to explain Sam's behavior in terms of his perception of Samantha as a weak little girl who can be bullied. Alternatively, Sam may have perceived other little boys getting away with such aggression and may be modeling his behavior on their example. This is the basis for Albert Bandura's (picture on page 596) social learning theory, about which we will have more to say in Chapter 14 (page 596). Of course, if Sam did model his behavior on others, he would be drawing on his memory—another cognitive process; and he may have considered the situation and planned his action so that nobody was around when he pushed Samantha—thinking.

The cognitive perspective would have much to say about Mr. A. and his memory problem. After all, memory is a major focus of this perspective. Questions would be raised as to exactly what was forgotten and what was remembered, how the processing of information had changed, and whether the man could be helped by giving him new ways of processing incoming information for storage in his memory and later retrieval from his memory store.

The Social Perspective Because social psychologists try to understand normal social interactions, they would be interested in both Sam and Samantha and the setting of their interaction. Social psychologists might well adopt the cognitive framework described above or be interested in Sam's attitude toward little girls to determine whether his behavior toward Samantha is consistent with this attitude (Chapter 10, page 402). More likely, however, would be the social psychologist's attempt to get at the causes of the behavior by obtaining more information about the participants and the setting. Sam appears responsible for pushing Samantha. This seems especially clear if it is known that Sam has pushed other little girls and that no other little boys in Sam's playground push little girls around. Conversely, imagine a situation in which all the children in the playground push Samantha, but none of them ever push each other. It now appears that there is something about Samantha that causes her to be pushed. Social psychologists are alert to the joint contributions two participants make to any interaction (Chapter 9, page 342).

With respect to the forgetful man, Mr. A., the social perspective would focus on how the memory deficit affects Mr. A.'s relationships with other people. In particular, this perspective would look for changes in the number and the quality of his relationships.

The Developmental Perspective The developmental perspective is concerned with characteristic changes that occur in people as they mature—changes in the way they think, for example. Sam's playground behavior might be seen partly as the result of his cognitive *egocentrism* (Chapter 11, page 437)—that is, his limited ability to think about how things look or feel to others. Like many young children, Sam may have trouble taking the perspective of someone like Samantha and recognizing that others have feelings and can be hurt. As he grows older, Sam should be better and better able to put himself in the place of others, consider their feelings, and behave appropriately.

As for Mr. A., the forgetful man, the developmental perspective might view his behavior as resulting partly from the natural aging process. Although serious problems of thinking and remembering are rare at age 55 (Chapter 12, page 493), one problem does sometimes arise as people approach old age: Although their memories remain (they do not go away), they may be a bit harder "to find in the file."

The Humanistic Perspective The humanistic perspective emphasizes the person's sense of self (Chapter 14, page 598). From this viewpoint, Sam's behavior might be seen as a part of his quest for personal competence, achievement, and self-esteem. Ideally, as he matures, Sam will find ways of enhancing his sense of self that will not harm or deprive others.

Hearing about our 55-year-old man, a humanistic psychologist might be less concerned about the causes of the man's forgetfulness than about its effects. How will the memory loss make this man feel about himself? How will it affect his competence at work and his effectiveness as a person? Will his self-esteem be damaged? From the humanistic perspective, such questions are of central importance.

The Psychoanalytic Perspective The psychoanalytic perspective is part of the broader perspective called psychodynamic (Chapter 14, page 575)—a perspective that focuses on the role of feelings and impulses which are thought to be unconscious. A key psychodynamic idea is that when these impulses are unacceptable, or when they make us anxious, we use defense mechanisms (Chapter 14, page 588) to reduce the anxiety. One of these defense mechanisms is *displacement* (Chapter 14, page 590); when we are angry at someone who is too powerful or frightening to be openly angry at, we displace our anger—deflect it onto someone who is weaker. Perhaps this is what happened with young Sam. Perhaps he was angry at his teacher, or at his parents, but he did not dare express his anger to them directly. In attacking Samantha, Sam may be displacing the anger he really feels toward more powerful people.

Another key idea of the psychoanalytic perspective is that slips, or accidents, often happen for a reason (Chapter 5, page 207) and that they may reveal hidden motives. When a 55-year-old man forgets appointments, there may be an unconscious reason. Perhaps, deep down, he disliked the people he was supposed to see; or perhaps he resented the demands of his job, and his unconscious self expresses that resentment by causing him to forget some of those demands. The psychoanalytic perspective digs beneath the surface of behavior, looking for hidden processes and hidden impulses (Chapter 16, page 683).

Summary

1. Psychology is defined as follows: It is the science of human and animal behavior; it includes the application of this science to human problems.

2. As a science, psychology is comprised of systematized knowledge that is gathered by carefully measuring and observing events. Theories are used to summarize observations and to predict the outcomes of future observations. Another important aspect of psychology as a science is its use of measurement—the assignment of numbers to objects or events according to certain rules.

3. As the definition indicates, psychology has an applied side. The application of knowledge to practical problems is an art—a skill or knack for doing things which is acquired by study, practice, and special experience.

4. The word "behavior" in the definition of psychology refers to anything a person or animal does that can be observed in some way. Defining psychology as the study of behavior does not exclude mind and other internal processes from the field of psychology; what a person does—his or her behavior—is the avenue through which internal mental events can be studied.

5. Psychology is not the only branch of knowledge which studies human and animal behavior. Anthropology, sociology, economics, political science, geography, and history also study various aspects of behavior and, together with psychology, comprise the group of knowledge areas known as the behavioral sciences. The study of behavior is also a part of several biological sciences—zoology, pharmacology, physiology, and neurobiology, for example.

6. Using the experimental method, the psychologist studies, under controlled conditions, the effects on a dependent variable of changes in an independent variable. In a psychological experiment, the independent variable can be almost anything, but the dependent variable is always some aspect of behavior (as broadly defined).

7. An experiment is no better than its controls. Only the variables should change in an experiment; other factors which might affect the outcome of an experiment must be held constant or canceled out in some way. Such extraneous factors can be controlled for by using control groups or by using a within-subjects (before-and-after) experimental design.

8. A great advantage of experiments is that they can be replicated, or repeated, so that psychologists can confirm, and thereby be more confident about, their observations. Limitations of the experimental method are that it cannot always be used, its results are obtained in artificial situations and may be limited in their application, and the procedures themselves may interfere with the very thing being measured.

9. Using the method of systematic observation, researchers make the most exacting and controlled study they can of naturally occurring behavior. To establish the cause (or causes) of a certain behavior, researchers must make a large number of observations to which they can apply inductive-reasoning principles.

10. The clinical method focuses on the study of an individual's behavior; it is usually used to understand behavioral problems. As a tool in science, the main value of the clinical method is that its use may suggest fruitful ideas which can be investigated more rigorously using experimental or systematic-observation methods.

11. The work of psychologists is quite varied, and many specialties exist within the field. Among the subfields of psychology are clinical psychology, counseling psychology, school and educational psychology, experimental and physiological psychology, industrial and organizational psychology, social psychology, developmental psychology, and community psychology.

12. Clinical psychology is the largest subfield of psychology. A clinical psychologist has a Ph.D., a Psy.D, or an M.A. degree and has done several years of postgraduate work in a psychology department. Clinical psychologists are trained to apply psychotherapeutic techniques, to diagnose psychological disorders, and to do research on the causes of these disorders.

13. Clinical psychology and psychiatry are often confused because practitioners of both disciplines use psychotherapy in the treatment of behavioral problems. However, unlike clinical psychologists, psychiatrists are trained as physicians and hold M.D. degrees; they become psychiatrists by doing several years of residency in a psychiatry department. Being physicians, psychiatrists can use drugs and other medical means to treat psychological disorders.

14. Clinical psychology and psychiatry should be distinguished from psychoanalysis. A psychoanalyst is a person who uses the particular psychotherapeutic techniques developed by Sigmund Freud and his followers. A clinical psychologist, a psychiatrist, or anyone else who has had this special training may be a psychoanalyst.

15. Counseling psychologists do work similar to that of clinical psychologists, but they are usually involved with people who have relatively mild personal, emotional, or vocational problems.

16. Psychologists who diagnose learning problems and try to remedy them are known as school psychologists. Educational psychologists are most often involved in increasing the efficiency of learning

in schools by applying psychological knowledge about learning and motivation.

17. Experimental psychologists do basic research in an effort to discover and understand the fundamental and general causes of behavior. They study basic processes such as learning and memory, sensation and perception, and motivation. Experimental psychologists who study the relationships of biological factors to behavior are known as physiological psychologists.

18. Industrial and organizational psychologists apply psychological principles to assist public and private organizations with their hiring and placement programs, the training and supervision of their personnel, the improvement of communication within the organization, and the alleviation of organizational strife. They may also counsel employees within the organization who need help with their personal problems.

19. Social psychologists study the ways in which individuals are affected by other people. Applied aspects of social psychology include the development and perfection of techniques for measuring attitudes and opinions.

20. Developmental psychologists try to understand complex behaviors by studying their beginnings and the orderly ways in which they change, or develop, over the life span. Since behavior develops rapidly in the early years of life, child psychology is a large part of developmental psychology, but developmental psychologists are also concerned with the behavioral changes that occur in adolescence, adulthood, and old age.

21. Community psychology is a broad field in which psychological knowledge is brought to bear upon social problems and the attempts of people to adapt to their work and living groups.

22. Psychologists approach the study of behavior from many different viewpoints. Historically, the viewpoints of psychologists were represented in schools of psychology. Among these schools were structuralism, Gestalt psychology, functionalism, and behaviorism. Psychoanalysis, while originating outside the field of psychology proper, has had a large impact on psychological thinking and may be considered to be a school of psychology.

23. The structuralists tried to find the units, or elements, which make up the mind. The Gestalt psychologists argued that mind could be thought of as resulting from the whole pattern of sensory activity and the relationships and organizations within this pattern. The functionalists were concerned with what mind and behavior accomplish—their adaptive functions which enable an individual to adjust to a changing environment. The behaviorists rejected the study of mind as a subject of psychology and said that psychology should be restricted to the study of the activities of people and animals—their behavior. Psychoanalysts, in their attempt to understand mental life and behavior, focused on people's unconscious urges and drives.

24. Today, except for modern versions of behaviorism and psychoanalysis, the old schools of psychology have disappeared. Various perspectives, or points of view about what is important in understanding mental life and behavior, characterize the present scene. Among these perspectives are the behavioral, biological, cognitive, social, developmental, humanistic, and psychoanalytic.

25. As was true of the older school of behaviorism, the current behavioral perspective focuses on the observed behavior of people or animals and not on their mental processes. Psychologists with a biological perspective try to relate people's behavior and mental events, as observed through their behavior, to functions of their bodies—especially to the activity of their nervous and glandular systems. From the cognitive perspective, behavior and mind are to be understood in terms of the ways in which information from the environment, received through the senses, is processed. Psychologists with a social perspective are interested in the interactions between and among people which influence mind and behavior. The humanistic perspective emphasizes a person's sense of self and each individual's attempts to achieve personal competence and self-esteem. The current psychoanalytic perspective (or, more broadly, the psychodynamic perspective) focuses, as did its historical forerunner, on the unconscious motives and defense mechanisms which manifest themselves in mental life and behavior.

Terms to Know

One way to test your mastery of the material in this chapter is to see whether you know what is meant by the following terms.

Psychology (4)
Science (4)
Theory (5)
Measurement (5)
Behavior (6)
Experimental methods (8)
Variable (9)
Independent variable (9)
Dependent variable (9)
Abscissa (9)
Ordinate (9)
Controls in experiments (9)
Control-group design (10)
Within-subjects design (12)
Baseline (12)
A-B-A within-subjects experimental design (12)
Replication (13)
Method of systematic observation (13)
Correlation (15)
Inductive reasoning (16)
Clinical method (16)
Clinical psychology/clinical psychologists (17)
Psychiatrists (18)
Psychoanalysts (19)
Counseling psychology (20)
School psychologist (20)
School counselor (20)
Educational psychology (20)
Experimental psychology/experimental psychologists (20)
Physiological psychology/physiological psychologists (20)
Neurobiology (21)
Industrial and organizational psychology (21)
Personnel psychologist (21)
Social psychology (21)

Developmental psychology (22)
Child psychology (22)
Community psychology/ community psychologists (22)
Structuralism (24)
Introspection (25)
Gestalt psychology (25)
Functionalism (25)
Behaviorism (25)
Psychoanalysis (26)
Unconscious motivation (26)
Behavioral perspective (27)
Biological perspective (27)
Cognitive perspective (28)
Cognition (28)
Social perspective (28)
Developmental perspective (29)
Egocentrism (29)
Humanistic perspective (29)
Psychoanalytic perspective (29)
Displacement (29)

Suggestions for Further Reading

While reading this book, you may find that you would like more information about particular topics. Appendix 2, "How to Look It Up," should be helpful.

Psychology can be a career, not just a course. If you are interested, you might write to the American Psychological Association, 1200 Seventeenth St., N.W., Washington, DC 20036 for the pamphlets entitled *Careers in Psychology* and *Graduate Study in Psychology*.

You may wish to compare the introduction to psychology presented in this book with that given in other texts, which may be written from a different perspective and cover topics not included in ours. As a start, from the more than 100 possibilities, look at one or more of the following: R. L. Atkinson, R. C. Atkinson, and E. R. Hilgard's *Introduction to Psychology*, 8th ed. (New York: Harcourt Brace Jovanovich, 1983); A. B. Crider, G. R. Goethals, R. D. Kavanaugh, and P. R. Solomon's *Psychology* (Glenview, IL: Scott, Foresman, 1983); H. Gleitman's *Psychology* (New York: Norton, 1981); G. A. Kimble, N. Garmezy, and E. Zigler's *Principles of Psychology*, 6th ed. (New York: Wiley, 1984); D. Krech, R. S. Crutchfield, N. Livson, W. A. Wilson, Jr., and A. Parducci's *Elements of Psychology*, 4th ed. (New York: Knopf, 1982); E. F. Loftus's *Psychology Today*, 5th ed. (New York: Random House, 1983); H. L. Roediger, III, J. P. Rushton, E. E. Capaldi, and S. G. Paris's *Psychology* (Boston: Little, Brown, 1984); J. C. Ruch's *Psychology: The Personal Science* (Belmont, CA: Wadsworth, 1984); and C. B. Wortman, and E. F. Loftus's *Psychology*, 2d ed. (New York: Knopf, 1985).

True or False?

While some may be judgment calls, all of the statements are generally considered to be false. The numbers give the percentages of introductory psychology students who marked each item "true." The article from which this table was taken says, "common misconceptions about behavior are distressingly resistant to change by text reading and class discussion." But the authors still hope that the reader's misconceptions about these and other psychological topics will be cleared up by reading our book. (We also hope that we have not created some new misconceptions of our own.)

Biology of Behavior

HEADING OUTLINE

SPECIES-TYPICAL BEHAVIOR PATTERNS
Examples of Species-Typical Behavior Patterns
Species-Typical Behaviors of Humans
INQUIRY 2.1: WHAT IS SOCIOBIOLOGY?

BRAIN AND BEHAVIOR: NEURONS AND SYNAPSES
Neurons
Nerve Impulses
Synapses and Their Functions
Neurotransmitters

BRAIN AND BEHAVIOR: A GUIDE TO THE NERVOUS SYSTEM
Peripheral and Central Nervous Systems
Spinal Cord, Brain Stem, Cerebellum, and Reticular Formation
INQUIRY 2.2: WHAT IS THE BRAIN DOING WHEN WE SLEEP?
Forebrain: Thalamus
Forebrain: Hypothalamus and Pituitary Gland
Forebrain: Cerebrum
Forebrain: Limbic System

BRAIN AND BEHAVIOR: ASSOCIATION CORTEX, BEHAVIOR, AND EXPERIENCE
Frontal-Lobe Association Cortex
Parietal-Lobe Association Cortex
Temporal-Lobe Association Cortex and Related Structures

BRAIN AND BEHAVIOR: LEFT- AND RIGHT-HEMISPHERE FUNCTIONS
Left Hemisphere and Language
Right Hemisphere and Patterns of Stimulation
Divided Brains

SUMMARY

TERMS TO KNOW

SUGGESTIONS FOR FURTHER READING

Chapter

Our everyday life is full of examples showing the role played by the brain in behavior. Here are a few that touched the life of the author of this chapter while he was writing it.

One friend was operated on to remove a tumor from the right side of his brain. He recovered from the operation. However, in the process of removing the tumor, the surgeons damaged some of the cerebral cortex of the right hemisphere of his brain. The result was weakness and partial paralysis of the left side of his body. (See page 61 in this chapter for an explanation of how this might happen.) Hopefully, my friend will in time recover some of the lost strength in his left arm and leg.

I ran into another friend who was in a head-on automobile crash several years ago. The driver who hit him was drunk (the functions of his brain were impaired by alcohol, and his behavior was affected). My friend banged his head in the accident and was knocked unconscious; he remained in a coma for several days, and when he "came to" had no memory of the accident. He still does not remember the accident. My friend's memory for events *before* the accident is all right as is his memory for new things that have happened *since* the accident. (See page 69 in this chapter and Chapter 5, page 188, for an explanation of what might have happened in my friend's brain to affect his memory for the accident itself.)

The other day one of my students told me about a medicine she is taking for depression. The medicine has helped her, presumably because it has partially corrected a chemical brain problem that led to the depression. (See page 49 in this chapter and page 679, Chapter 16, for more details.)

After a long period of decline, my 62-year-old neighbor died. She was said to be suffering from Alzheimer's disease. In this disorder, abnormalities appear in the structure of the brain cells, and there seems to be a decrease in the brain of one of the chemicals (acetylcholine) necessary for communication between brain cells. The first symptom of her disease was a memory problem; she had trouble remembering new things, although her old memories (those before the disease process started) were essentially intact. Her memory problem became progressively worse until she could not recall what she had done a few minutes before and her old memories also began to fail. Just before her death, she failed to recognize her husband when he came to visit her. We wish we knew why some people get this disease, but at present we do not. Hopefully, ongoing and future research will tell us why and will provide ways of treating this tragic disease. (For more on the "chemical messengers" of the brain, see page 49 in this chapter; for more on Alzheimer's disease, see Chapter 5, page 212.)

As you read this chapter, what examples of brain-behavior relationships from your own experience come to mind?

(Omikron/Photo Researchers.)

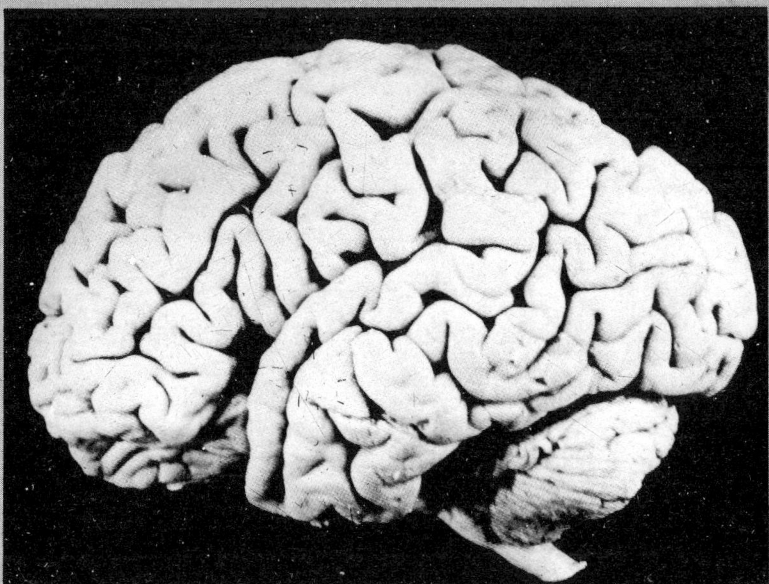

WE may think that we are only a little lower than the angels, but we must not forget that we are a species of animal. *Homo sapiens* is our name (Figure 2.1). We have a family tree spanning the millions of years life has been evolving on this planet. We are remarkable creatures, with bodily structures and psychological capacities that have come into being through evolutionary pressures over the millennia. Roughly speaking, bodily structures and behaviors that help animals adjust to their environments tend to be the ones which persist. This is so because animals with these gene-determined structures and behaviors live long enough to pass the characteristics on to their offspring. Thus, the behavior patterns which have developed under the pressures of evolution are as much a part of the nature of the species as is its anatomical structure. The presence of species-typical behaviors or, roughly speaking, "instincts," is obviously one contribution of biology to behavior.

Another aspect of the biology of behavior is the study of the relationship between behavioral and mental events—memory, learning, perception, motivation, and speech—and processes in the nervous system—particularly in the brain. The branch of psychology concerned with how activity in the nervous system is related to behavior and experience goes by many names. Perhaps the most common is *physiological psychology*, but

Figure 2.1

Homo sapiens. (United Nations)

it is also called biological psychology, biopsychology, neuropsychology, psychobiology, and psychophysiology. Much of this chapter is about the brain basis of behavior and mind. This chapter also gives some background for understanding the brain basis of topics such as memory, motivation, emotion, and certain behavior disorders. When these topics are discussed later in the text, we will refer back to Chapter 2.

Species-Typical Behavior Patterns

In the last 45 years or so, the study of animal behavior has been spurred on by the growth of the branch of zoology known as *ethology*. Ethology is concerned with behavior patterns that are typical of particular species of animals—with emphasis on the evolution of these patterns and, thus, their adaptive value. Thanks to the work of many ethologists, particularly the pioneering studies made by Konrad Lorenz, Nikolaas Tinbergen, and Karl von Frisch (who jointly won a Nobel Prize in 1973), we now appreciate more than ever the importance of species-typical behavior in the adaptation of animals to their environments.

For a behavior pattern to be classed as *species-typical*, all normal members of the species must display the behavior under certain circumstances. Species-typical behaviors arise from the genetic heritage of the species as it has evolved over time; these behaviors are, in other words, part of the species' "nature." But this does not mean that the environment does not play a role in the development of species-typical behaviors. In many cases, the perfection of species-typical behaviors is dependent on the environmental factors present when the young animals are growing up. Since these environmental factors are roughly the same for all members of a species because they all live in a similar habitat, and since species-typical behaviors are based on a common genetic heritage, the result is that all normal members of the species display the behaviors.

Many species-typical behaviors consist of relatively fixed and inflexible patterns of movement triggered by a particular stimulus, or event, in the environment. Such a stimulus is called a *releaser*, and the behavior released by such a stimulus is often called a *fixed-action pattern* (*FAP*).

In addition to specific behaviors, a species may be ready, or prepared, to be influenced by certain events in the environment and not by others. For example, as described in Chapter 4 (page 173), some responses and associations may be readily learned by members of a particular species, while other responses and associations may be learned only with great difficulty if at all.

EXAMPLES OF SPECIES-TYPICAL BEHAVIOR PATTERNS

With all the species of animals, each with its own set of species-typical behaviors, literally millions of examples might be cited. We will consider a few of them.

The stickleback is a small freshwater fish of northern Europe. Its mating behavior consists of a sequence of FAPs, each triggered, or released, by stimuli in the environment (Tinbergen, 1952).

> At the beginning of the mating season, in early spring, the male first establishes a territory from which other males are excluded. Then it

> makes a nest by scooping out a depression in the stream bed, by covering the depression with a mound of vegetation, and by hollowing out a tunnel in the mound. After the nest has been built, the drab-gray male stickleback undergoes a dramatic color change: Its belly turns red, and its back becomes bluish white. The color change, particularly the red belly, is a sign to stickleback females that the male is ready to mate. Seeing the color change, a female will enter the male's territory, swimming with her head up and thus showing the male her egg-swollen belly. Sight of the swollen belly releases a zigzag pattern of swimming by the male, which causes the female to swim toward the male. The male, in turn, seeing the oncoming female, is stimulated to swim to the nest, into which he briefly pokes his head as the female watches. Seeing the male do this triggers the female to put her whole body into the nest. The male, aroused by this, now prods the female's tail with his snout, and, stimulated by this prodding, the female lays her eggs and leaves the nest. The male now enters the nest and fertilizes the eggs. This pattern may be repeated with several females, but the male's color eventually becomes less vivid and it no longer attracts females. Instead, it uses its fins to fan water over the nest, thus increasing the oxygen available to the eggs and the young fish when they hatch.

In this example, we see how complex social behavior is coordinated by releasers and how FAPs can be combined to make up a complex behavior sequence. Remember that all this is, for the most part, part of the stickleback's nature. It seems likely that the evolutionary heritage of this species has resulted in a brain and nervous system that is especially "tuned" to receive certain stimuli and "wired" to produce relatively stereotyped FAPs of behavior when these stimuli are received. Animals with species-typical FAPs behave more or less automatically—like robots; given the stimulus, the behavior will almost surely follow. While from our human perspective this may seem limiting, such behaviors would not be there if they did not make the animals possessing them fit and adapted to their environments. In the stickleback case, these unlearned stereotyped behaviors seem to be a good way to make little sticklebacks, and they have been good enough for the millions of years that sticklebacks have existed as a separate species.

Bird songs are another example of species-typical behavior. Many species of birds sing songs that they do not need to learn because they are an innate part of the heritage of their species. Other species, however, learn their songs from hearing other members of their species sing them. Illustrative of a bird with a species-typical song is the cowbird (West et al., 1981).

> Cowbirds are brown-and-black birds that are commonly seen during the spring and summer in the eastern United States. The cowbird is what is called a "brood parasite": The female lays her eggs in the nests of other birds, who hatch the eggs and raise the cowbird young along with their own. This, of course, saves the female cowbird much bother, but it creates a problem for the babies. How is the baby to find other cowbirds with which to flock and to mate?

Evidence indicates that cowbird males innately sing a song that identifies them as cowbirds to other cowbirds. The male cowbird song is also a powerful releaser for sexual species-typical behavior by the female cowbird. During the mating season, on hearing the male song, the female assumes the receptive posture shown in Figure 2.2.

The cowbird's basic song and its effects on other cowbirds are examples of species-typical behavior and are part of the evolutionary species heritage of cowbirds. But cowbirds can and do modify their basic song to conform to environmental pressures. It has been found, for instance, that the basic song sung by male cowbirds raised in isolation is particularly potent in arousing females. If these isolation-raised birds are put into flocks containing other male cowbirds, however, they may be attacked and perhaps killed by other males in the flock if they sing their superpotent song. In groups, only the dominant males are allowed to sing this song. This ensures that the strongest and fittest males will impregnate the greatest number of females. Any cowbird who sings a potent song in a flock had better be dominant or able to achieve dominance by beating off the attacks of other males. If the new flock member is not killed, and if he does not become the new dominant member of the flock, he will learn to change his song so that it is no longer superpotent and is more like those of males in the flock with low or middle dominance. This, then, is an example of the modification of species-typical behavior through experience and learning.

Figure 2.2

The receptive posture of the female cowbird is triggered by the male's song. (Courtesy of Meredith West.)

The expression of emotions is another example of species-typical behavior. Emotional expression is often very much the same in all members of a particular species and is thus part of the species' nature and has a basis in the evolutionary heritage of the species. (See Chapter 8, page 311, for some information about the facial expressions of human emotions.) Emotional expressions are interesting examples of species-typical behavior because they are crucial for the social interactions of species members; they provide signals to others about the emotional state of the animal and the behavior which may be forthcoming (Figure 2.3).

SPECIES-TYPICAL BEHAVIORS OF HUMANS

We have seen that the concept of species-typical behavior is a fruitful one for understanding much of what lower animals do. When it comes to human beings (and, to some extent, the higher primates), the situation is greatly complicated by the tremendous flexibility of behavior. One of the main results of human evolution is that we have become a species in which behavior is strongly influenced by learning. In other words, human behavior is influenced by learning and by the unique events of each individual's life. Learning (Chapter 4), memory (Chapter 5), and thinking (Chapter 6) play a large role in what we do.

Most students of human behavior agree that our evolutionary heritage has given us the potential for great behavioral flexibility. But what other of our behavioral characteristics can be traced to evolution and to our human, or species, nature? Species-typical behaviors (especially of the fixed-action type) that are so much a part of the behavior of lower animals are *not* prominent in human nature. Some behaviors of infants

Figure 2.3

Emotional expression as species-typical behavior in the dog: (*a*) threat, (*b*) threat showing the lips curled back to expose the teeth, (*c*) submission. (From Darwin (1872), *The Expression of the Emotions in Man and Animals.*)

(Chapter 11, page 419) and the facial expressions of certain emotions (Chapter 8) may be considered to be species-typical behaviors. So, while not prominent, we do have some specific behaviors as part of our animal natures.

While specific species-typical behaviors are not an important part of our animal natures, our evolutionary heritage may contribute to our behavior at a deeper and more fundamental level. This is the level of predispositions toward and potentialities for certain types of behavior as opposed to others. In other words, our species heritage may make it more *probable* that we will do some things and less *probable* that we will do others. We may have built-in, or programmed, biases toward certain types of behavior. (See Inquiry 2.1 on sociobiology.)

Within the general framework provided by the species heritage, or "human nature," particular behaviors are learned (Chapter 4). For example, our human nature gives us the ability to produce and understand language, but whether we communicate in English, French, Russian, or Swahili is a matter of learning.

There are other examples of built-in, programmed behavioral predispositions which human beings have. (Remember that these are only broad *tendencies* to behave in certain ways; the details of behavior are provided by learning and a person's individual experiences.) It is claimed, for example, that tendencies toward competition and aggression, especially among males, are innate. Territoriality, the need for a social organization with leaders, certain things we do in child rearing, male-female roles, and, as Inquiry 2.1 describes, self-sacrifice (or altruism) are all thought to be innate behavioral potentialities which may have a basis in the evolution of our species.

Inquiry 2.1

WHAT IS SOCIOBIOLOGY?

Edward O. Wilson, in his influential and controversial book *Sociobiology: The New Synthesis*, defines *sociobiology* as "the systematic study of the biological basis of all social behavior." The key concept of sociobiology is that the purpose of an individual's life is to pass along genetic material to the next generation of the species; evolutionary forces are at work to make individuals, alone and in groups, better adapted to their environments so that they will survive to pass on their genes. The focus of sociobiology is the biological evolution of various forms of adaptive social behavior in lower animals and, perhaps, humans. The sociobiologists wonder about the fundamental biological, or, as they put it, the ultimate evolutionary roots of human social organization. Is there, they wonder, an evolutionary basis for such human social behaviors as altruism, mate selection, parenting behaviors, rituals and religion, territoriality, and aggression? In other words, do humans have "biograms" from their evolutionary history which program their basic forms of social behavior and social organization? Since there is considerable evidence for an evolutionary basis of social organization in lower animals, why should the human animal be different?

As an example of sociobiological thinking, consider the social behavior known as altruism—self-sacrificing behavior performed for the benefit of others. If evolution acts to maximize the likelihood of passing genes on to the next generation, how can there be individuals who will sacrifice themselves to preserve other species members? Why, for instance, will a "lookout" in a marmot (a kind of prairie-dog) colony call attention to itself by warning others of the approach of a coyote or other predator? Why doesn't it just dive into its burrow, leaving the others in the colony to take care of themselves? Or why will individuals of some species of birds call attention to themselves by acting injured, thus drawing predators toward themselves and away from the young in the nest? (See illustration.)

Self-sacrificing behavior is obviously not adaptive for the altruistic individual; altruists maximize the likelihood that they will *not* survive to pass on their genes. How, then, can altruistic behavior become part of the species heritage? One sociobiological theory of altruistic behavior—the theory of kin selection—goes like this: Suppose that some animals in a group have genes for altruistic behavior. Some of the offspring and other relatives, or kin, of these animals will also have genes for altruistic behavior. Because they are related, they will also tend to live close to each other. Therefore, if one of them performs an altruistic act, it will benefit others who share the genes for this behavior. When an individual engages in altruistic behavior, it tends to save the lives of its kin, who also, in varying degrees, possess the genes for this same altruistic behavior. Even though the altruistic individual may not survive, its behavior saves the lives of many of its kin who have the genes for altruistic behavior. Thus, more genes for altruistic behavior are preserved in the group than are lost by the death of a single individual. Groups with these genes are better adapted for survival than those without them. In the long run, more animals with "altruistic" genes than those without them will survive to pass on their genes. Thus, altruistic behavior will be selected and become part of the species heritage.

Assuming that this sociobiological argument for the existence of animal altruism is plausible, what of human beings, with their enormous capacity to be shaped by their experiences and environmental influences? For instance, it has been reported that infirm old people in certain Eskimo groups, before contact with European cultures, chose to stay behind and die rather than slow down the whole group in an arduous migration. And in our society, a great many searchers will risk their lives in arduous search for a lost hiker or a missing child. Are these examples of human altruism the result of the evolutionary processes described above? Do these people possess the genes for altruistic behavior? Maybe, but it could just as well be argued that the Eskimo elders and contemporary rescue workers have been taught, throughout their lives, that this type of sacrifice is the proper, socially approved "thing to do" and the highest form of heroism. To evaluate the roles played by sociobiology and so-

(From Gramza, 1967.)

cialization and to assess their interaction, we need better evidence than we now have.

In the absence of good evidence, controversy among scientists with fundamentally different views of human nature can and does flourish. Sociobiologists, as we have seen, maintain that, as with other animals, evolution is the ultimate "cause" of human social behavior. Most psychologists, on the other hand, are committed to the view that human social behavior is so moldable and modifiable as to be little influenced by evolutionary processes.

The controversy between sociobiologists and psychologists is the old nature-nurture argument (see Chapter 11, page 413) in modern dress. What we need is less argument and more evidence about the comparative roles of nature and nurture and their interaction in the determination of human social behavior. Perhaps the greatest contribution of sociobiological thinking to behavioral science in general, and to psychology in particular, has been to raise questions about the role of the environment in determining human behavior. Some social behaviors of humans may have a deep-seated evolutionary basis. The book *Sociobiology: The New Synthesis* may serve as a consciousness raiser for behavioral scientists, who may now be more willing to consider evolutionary ideas and to look for evidence of the biological roots of certain forms of human social behavior.

REFERENCES

Barash, D. P. (1982). *Sociobiology and behavior* (2nd ed.). New York: Elsevier.

Gould, S. J. (1976). Biological potential vs. biological determinism. *Natural History, 85*(5), 12–22.

Sociobiology Study Group of Science for the People (1976). Sociobiology—Another biological determinism. *BioScience, 26*(3), 182, 184–186.

Wilson, E. O. (1975). *Sociobiology: The new synthesis.* Cambridge, MA: Harvard University Press.

Wilson, E. O. (1976). Academic vigilantism and the political significance of sociobiology. *BioScience, 26*(3), 183, 187–190.

The assertion that human beings have these biological tendencies is, however, controversial. As good scientists, we must ask for evidence. Some scientists base their arguments on the behavior of monkeys, baboons, or apes. These arguments are not convincing to many, however, because these animal cousins of ours are evolutionarily rather remote from us and have been following their own line of descent for millions of years; they have had plenty of time to evolve behavioral predispositions of their own, predispositions which are only superficially like ours. Anthropological observations of present-day stone-age cultures are another way of studying human nature. Some of these observations seem to support arguments for the innate basis of behavioral tendencies such as aggression, but others do not. The evidence is mixed; human behavior is flexible and varied.

In the absence of conclusive evidence, the debate about innate human behavioral potentials continues. (See Inquiry 2.1 and Chapter 11, page 413.) Psychologists tend to emphasize the importance of the unique circumstances of an individual's life in determining behavior and to give less weight to the role of human nature. One reason for this is that psychology is concerned with understanding, explaining, and predicting differences in behavior among individuals, and such individual differences are strongly influenced by learning and by the events in a person's life.

Another way of looking at the biological and evolutionary basis of human behavior is to consider what the brain and nervous system make possible. The brain, like other bodily organs, is a product of evolution. Possessing a certain type of brain gives the species the possibility of doing some things and not others. For example, processes within the cerebral cortex of the human brain (Figure 2.4) give us the capacity to communi-

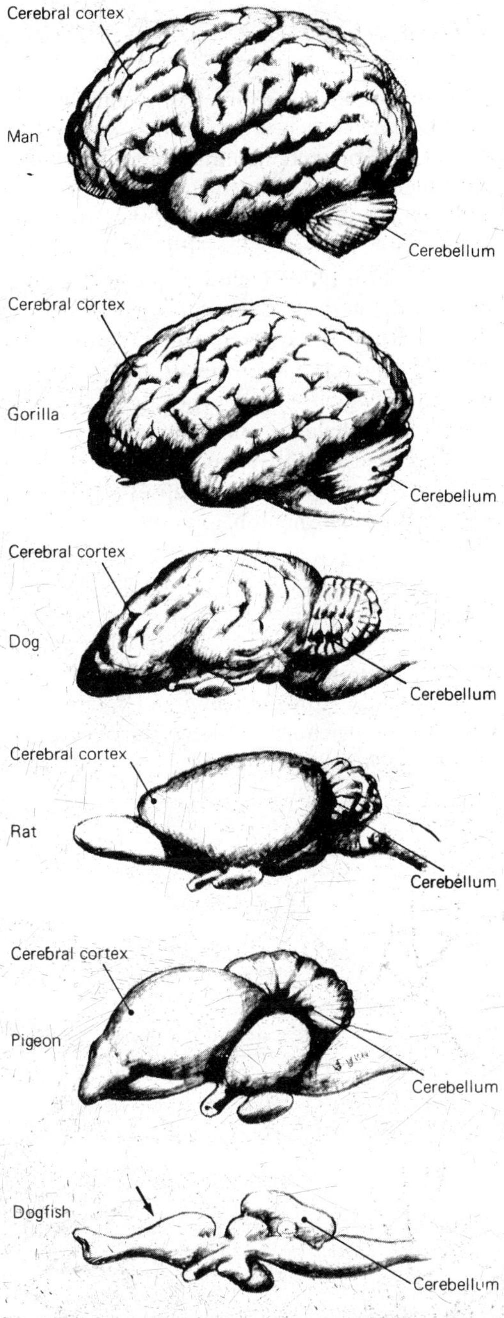

Figure 2.4

The evolution of the brain. (In the dogfish picture, the arrow indicates the area of the brain where the cerebral cortex will develop in animals higher on the evolutionary scale; the dogfish has no true cerebral cortex.) (From *Psychology: The Fundamentals of Human Adjustment*, 5th ed., by N. L. Munn, 1966, Boston: Houghton Mifflin. Copyright 1966 by Houghton Mifflin. Reprinted by permission.)

cate with symbols. This is something no other animal species on this planet does naturally; they do not have the "brains" for it. Thinking, perceiving, remembering, emotionality, motivation, and, it seems likely, our self-consciousness, are all based on biological activities within our brains. The next parts of this chapter will give some details of the brain basis of behavior.

The human brain is estimated to contain at least 150 billion nerve cells, called ***neurons***, each of which is connected to many others, making the number of connections immense. The connections between nerve cells are called ***synapses***. But even though there are an enormous number of connections, research shows that they are arranged in an orderly fashion—certain cells connect only with certain others.

Because physiological psychologists are interested in the involvement of the nervous system in behavior and experience, it is important for them to know the ways in which the living tissue of the nervous system actually conducts and processes information. Such knowledge is the bedrock upon which all our ideas about the role of the nervous system in complex psychological functions must be grounded. In this section, we shall see that neurons carry information electrically. At the connections between neurons—at the synapses—we shall also see that information is passed from one neuron to another by chemicals known as ***neurotransmitters***.

NEURONS

Nerve cells, or **neurons**, are the information carriers of the nervous system. Neurons come in many sizes and shapes (Figure 2.5), but they have certain features in common (Figure 2.6). Each has a cell body that contains the machinery to keep the neuron alive, and each has two types of fiber: ***dendrites*** and an ***axon***. The dendrites are usually relatively short

Figure 2.5

Some of the many types of neurons. (*a* to *d* adapted from Davson & Eggleton, 1968; *e* modified from Sholl, 1956.)

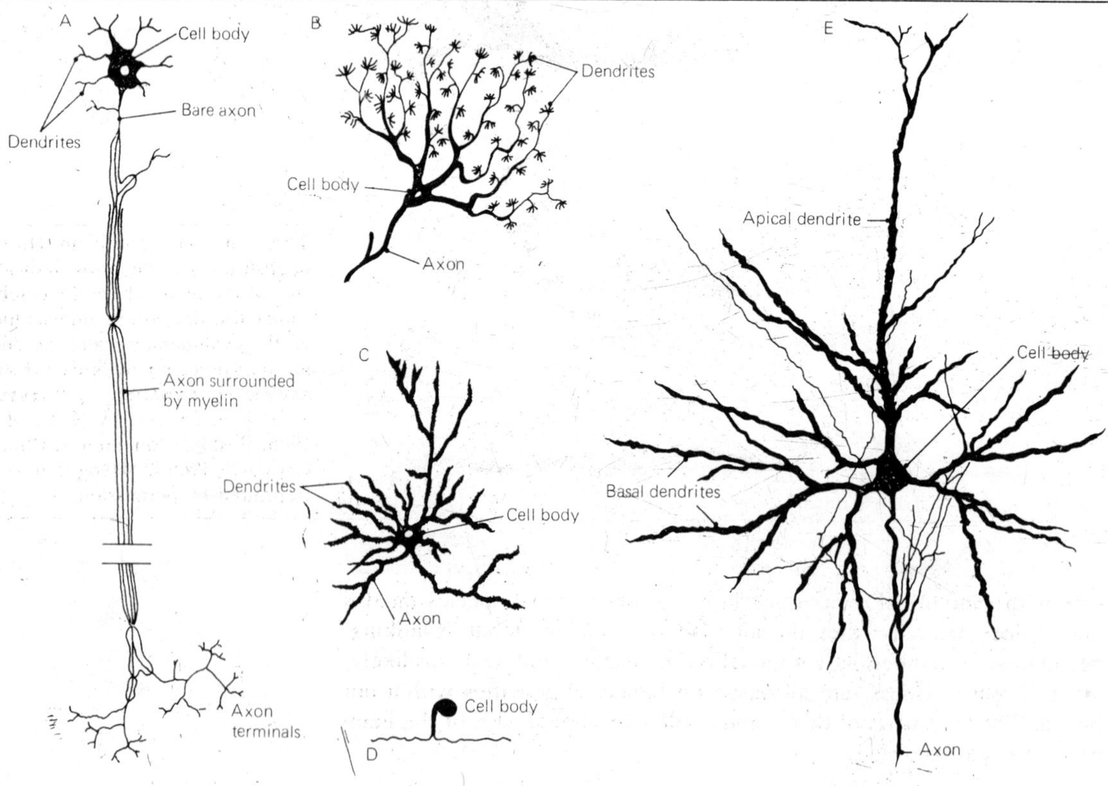

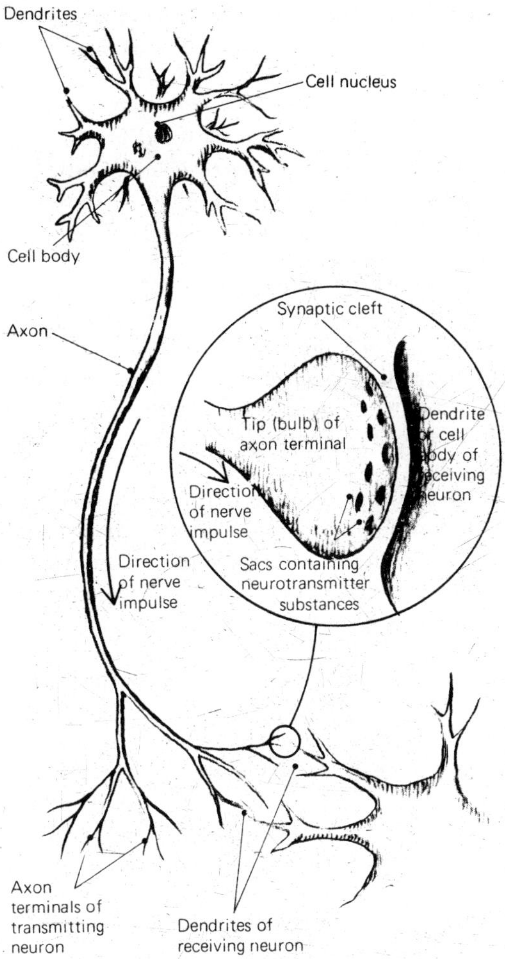

Figure 2.6

A schematic neuron to illustrate some general features of nerve cells.

and have many branches, which receive stimulation from other neurons. The axon, on the other hand, is often quite long. (For instance, axons connecting the toes with the spinal cord can be more than a meter in length.) The function of the axon is to conduct nerve impulses to other neurons or to muscles and glands. Since the dendrites and the cell body receive information that is then conducted along the axon, the direction of transmission is from dendrites to the fine axon tips (Figure 2.6). In many cases, the axon—but not the cell body or the dendrites—has a white, fatty covering called the *myelin sheath* (Figure 2.5*a*). This covering increases the speed with which nerve impulses are sent down the axon. However it is the *cell membrane*, which immediately surrounds the cell body, the dendrites, and the axon, that is essential for the generation and conduction of nerve impulses.

NERVE IMPULSES

By using fine wires or fluid-filled glass tubes known as microelectrodes, neurophysiologists have shown that *nerve impulses* are electrical events of very short duration that move along the axon. As the electrical activity

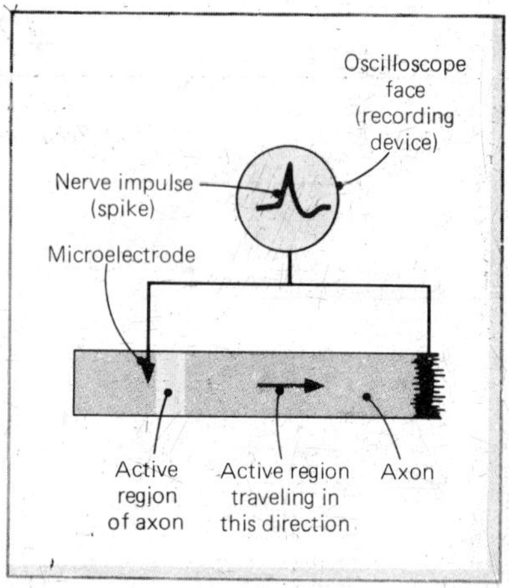

Figure 2.7

The nerve impulse.

moving along the axon reaches and passes the microelectrode, the recording device registers a quick, sharp electrical pulse (Figure 2.7). This is the nerve impulse. Because it is brief and sharp, it is called a spike.

When a neuron is resting and not conducting a nerve impulse, the inside of the cell has a negative electrical charge. A stimulus which excites the cell will make the inside charge a little less negative. At a critical point in this stimulation, a point called the *threshold,* the membrane surrounding the neuron will change its characteristics. Channels will open briefly, allowing sodium ions (charged particles) in the fluid bathing the outside of the membrane to enter the cell. The sodium ions flowing into the cell are positively charged. Thus the inside of the neuron becomes positive in charge for a short time—a millisecond (a thousandth of a second) or so. This rapid, brief change to inside positivity is the nerve impulse, or spike. After a nerve impulse has occurred, the neuron is restored to its original resting charge by an outward flow of potassium ions and is ready to fire again.

The electrical nerve impulse travels along an axon somewhat like a fire travels along a fuse. The hot part of the fuse ignites the next part, and so on down the fuse. Similarly, in an axon, the active portion triggers a spike in the region just ahead of it. When a spike occurs in this region, the next region is excited, and so on continuously down the axon. As the spike moves down the axon, the regions that have already fired are getting ready to fire again if stimulated. In this way repeated spikes can be fired down the axon, and the analogy with a fuse breaks down. The fuse is used up after the fire has traveled along it; the axon is recovering and getting ready to be active again after a spike has passed.

SYNAPSES AND THEIR FUNCTIONS

The axon tips of a neuron make functional connections with the dendrites or cell bodies of other neurons at *synapses.* A narrow gap, called the *synaptic cleft,* separates the neurons (Figure 2.6, page 47). Pictures taken with the electron microscope (the light microscope is not powerful enough) reveal the complexities of synapses (Figure 2.8). A number of small bulbs, called *boutons* (from the French for "button"), are found at the ends of the axons of the transmitting, or presynaptic, neurons. Boutons have in them small bodies, or *vesicles,* that contain the neurotransmitters. These chemicals are released from the vesicles into the synaptic cleft when a nerve impulse reaches the boutons of the transmitting cell. The neurotransmitter then combines with specialized receptor molecules in the receptor region (Figure 2.6, page 47) of the receiving cell.

The effect of a neurotransmitter on the receiving cell is either to increase its tendency to fire nerve impulses—*excitation*—or to decrease this tendency—*inhibition.* It is easy to see the significance of excitation and the firing of nerve impulses along axons, but it is not so easy to see the significance of inhibition. In fact, inhibition is crucial for the functioning of the nervous system. For example, suppose you bend your arm up to scratch your nose. The muscles of your arm are arranged in antagonistic pairs. If one set (the biceps) is contracted, it pulls your arm up toward your nose; if the other set (the triceps) is contracted, it pulls your arm down. The neurons controlling the muscles that pull your arm down must be inhibited when you are trying to reach up to your nose; if they are not, you will have difficulty bending your arm.

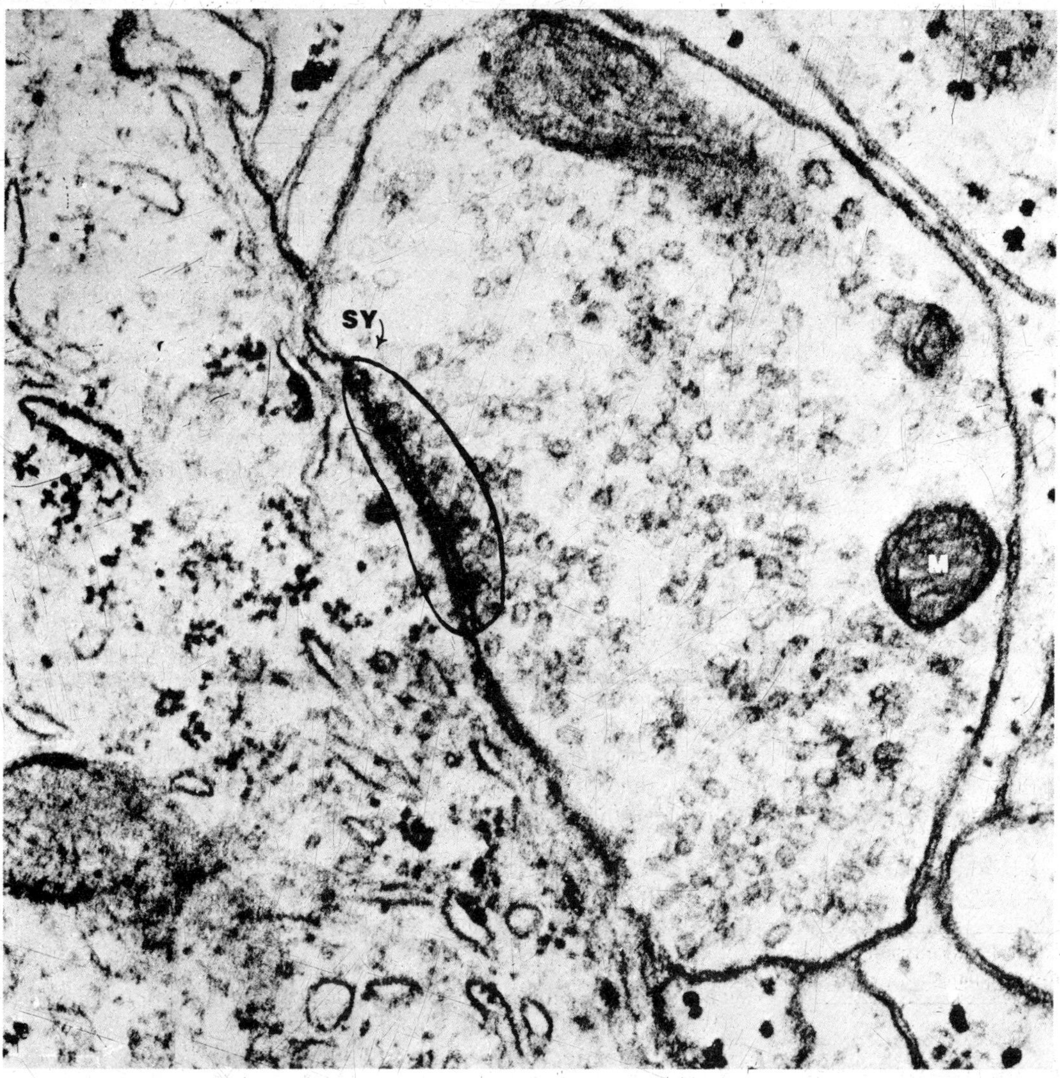

Figure 2.8

A synapse under the electron microscope. *Sy*, the synapse; *M*, a mitochondrion. (Modified slightly from Cooper et al., 1974.)

NEUROTRANSMITTERS

Pharmacologists and neurochemists have identified a number of the chemical substances that act as neurotransmitters at synapses in the nervous system and at the junction between nerves and muscles—the neuromuscular junction. Table 2.1 gives the names of some of the best-known neurotransmitters.

A number of steps are involved in the chemical transmission of information across synapses from neuron to neuron (Figure 2.9).

1. The transmitting, or presynaptic, neuron manufactures, or synthesizes, the neurotransmitter molecules from simpler molecules derived from the foods we eat and from other sources.

2. The manufactured neurotransmitter is stored in the bouton vesicles of the transmitter neuron.

Table 2.1

Some neurotransmitters

NAME	LOCATION
Acetylcholine (ACh)	Nervous system and many neuromuscular junctions
Dopamine (DA)	Nervous system
Epinephrine (E) (adrenalin)	Primarily certain neuromuscular junctions; some in nervous system; adrenal glands
Norepinephrine (NE) (noradrenalin)	Primarily in nervous system; some neuromuscular junctions; adrenal glands
Serotonin or 5-hydroxytryptamine (5-HT)	Nervous system
Gamma-aminobutyric acid (GABA)	Nervous system
Glycine (Gly)	Nervous system
Glutamic acid	Nervous system
Neuropeptides (a few; there are many)	
Enkephalins	Nervous system
Endorphins	Nervous system
Vasopressin	Nervous system and pituitary gland
Oxytocin	Nervous system and pituitary gland

Figure 2.9

Steps in the chemical transmission of information across a synapse.

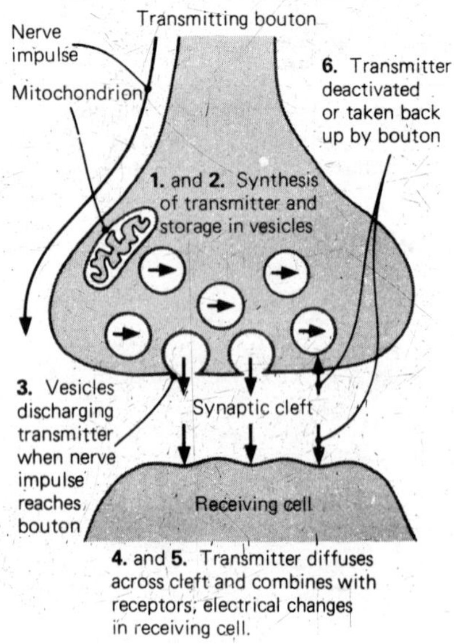

3. Nerve impulses reaching the boutons initiate a process which causes some of the vesicles to move to the synaptic cleft, where they discharge their stored neurotransmitter.

4. The neurotransmitter rapidly diffuses across the narrow synaptic cleft and combines with specialized receptor molecules on the membrane of the receiving, or postsynaptic, neuron.

5. The combination of neurotransmitter and receptor initiates changes in the receiving neuron that lead to excitation or inhibition. (Whether the effects are excitatory or inhibitory depends upon the neurotransmitter involved and the type of receptor with which it combines.)

6. The combined neurotransmitter is rapidly deactivated, as is excess neurotransmitter in the synaptic cleft, to make the postsynaptic cell ready to receive another message.

One method of deactivation is by catalysts, called *enzymes*, which trigger chemical reactions that break up the neurotransmitter molecules; another way deactivation occurs is by the process of reuptake. In reuptake, the transmitting, or presynaptic, boutons take back the released neurotransmitter and store it in vesicles for use another time. In brief, the stages in synaptic transmission are: neurotransmitter manufacture, storage, release, diffusion, combination with the receptor, and deactivation.

The drugs that affect behavior and experience—the *psychoactive drugs* (Table 2.2)—generally work on the nervous system by influencing the flow of information across synapses. For instance, they may interfere with one or several of the stages in synaptic transmission that we have just

Table 2.2

Some psychoactive drugs

DRUG	COMMON SLANG NAMES	EFFECTS ON BEHAVIOR AND EXPERIENCE
LSD (lysergic acid diethylamide)	Acid, sugar	Distortions of perception; variable mood changes (exhilaration or panic, for example)
Mescaline	Mesc	Like LSD effects
DOM (2,5-dimethoxy-4-methylamphetamine)	STP (serenity, tranquility, peace)	Euphoria (a feeling of extreme well-being) and distortions of perception
DMT (dimethyltryptamine)	Businessman's high	Like those of LSD, but they last about 1 hr instead of about 10 hrs
Marijuana	Pot, grass	Relaxation and calmness; some sharpening of perception
Heroin*	H, horse, junk, smack	Dreamy, warm, pleasant, euphoric feelings
Amphetamine*	Speed, bennies, dexies, pep pills	Alertness, resistance to fatigue, increased activity, elation
Cocaine	Coke, gold dust	Restlessness, talkativeness, excitement, euphoria
Caffeine*	—	A mild "psychological lift"; increased alertness
Barbiturates*	Downers, barbs, blue devils, yellow jackets	Resembles alcohol intoxication: drowsiness, euphoria, reduction in anxiety
Alcohol*	Booze, etc.	Mood changes, disturbances of motor coordination, difficulty with concentration and other thought disturbances, reduction in anxiety

* The drugs marked with an asterisk are physically addicting; withdrawal symptoms generally occur when use is discontinued. Whether addicting or not, many of these drugs can have strong adverse effects, such as panic reactions, overexcitement, or aggressive behavior. Overdoses can be fatal.

outlined, or they may have actions like the natural neurotransmitters and thus excite or inhibit receiving cells. This is also true of the drugs which are used in the treatment of certain psychological disorders. (See Chapter 16, page 679.)

Brain and Behavior: A Guide to the Nervous System

When you visit a new country, it is useful to have a guidebook that will tell you where places are located and what is happening at various places.

The following section is a guide to the parts of the nervous system and where they are located; it will give you a broad outline.

PERIPHERAL AND CENTRAL NERVOUS SYSTEMS

The nervous system is divided into two main parts: a central nervous system and a peripheral nervous system (Figure 2.10). The *central nervous system (CNS)* consists of the brain and spinal cord, which lie within the bony cases of the skull and spine. The parts of the nervous system outside the skull and spine make up the *peripheral nervous system (PNS)*.

The peripheral nervous system consists largely of nerve fibers, or axons, which (1) carry nerve impulses from the sensory receptors of the body inward to the central nervous system and (2) carry nerve impulses for the movement of muscles and the excitation of certain glands outward from the central nervous system. The peripheral nervous system has two divisions: the somatic nervous system and the autonomic nervous system.

The *somatic nervous system* motor fibers activate the striped muscles of the body, such as those that move the arms and legs, while the sensory fibers of this system come from the major receptor organs of the body—the eyes, the ears, the touch receptors, and so on. The *autonomic nervous system* motor fibers activate the smooth muscles of such bodily organs as the stomach, cause secretion from certain glands such as the salivary glands, and regulate activity in the special type of muscle found in the heart. It is thus a smooth-muscle, glandular, and heart-muscle system. Sensory fibers in the autonomic system carry information from the internal bodily organs that is perceived as pain, warmth, cold, or pressure.

Figure 2.10

The nervous system from the back. (Modified from Woodburne, 1965 and Vander et al., 1975.)

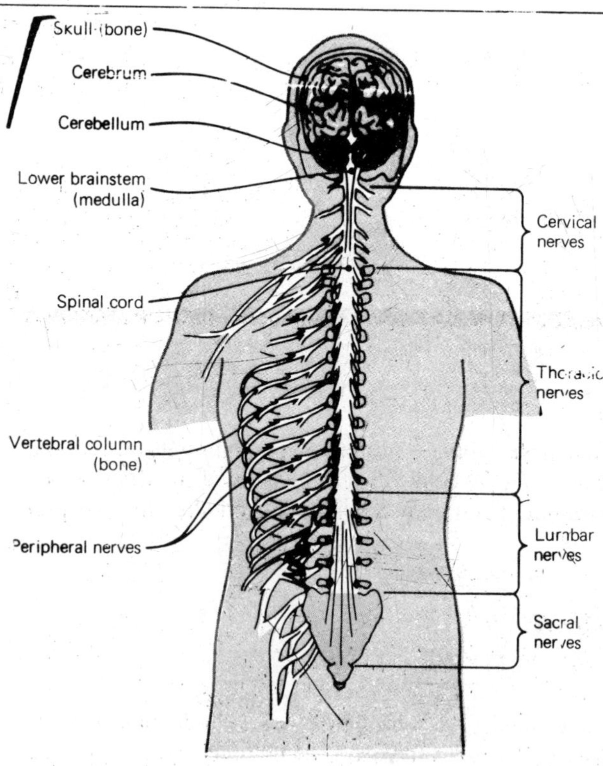

The autonomic system, in its turn, has two subdivisions: the sympathetic system and the parasympathetic system. In general, the *sympathetic system* is active in states of arousal and in stressful situations; the *parasympathetic system* is active in resting, quiet states. Since the autonomic nervous system and its subdivisions are active during emotion, they will be discussed in greater detail when emotion is considered in Chapter 8, page 314.

SPINAL CORD, BRAIN STEM, CEREBELLUM, AND RETICULAR FORMATION

The spinal cord and brain stem are like a long stalk protruding from the higher parts of the central nervous system—the forebrain (Figure 2.11). In ascending order from the spinal cord up, the *brain stem* consists of three divisions: the *medulla* (from the Latin meaning "marrow"), the *pons* (from Latin meaning "bridge"), and the *midbrain*. The spinal cord and brain stem control and regulate many bodily functions—such as breathing—that are necessary for life. They also begin the processing of sensory information from the environment and provide pathways by which this information can be carried to the forebrain. Furthermore, no movement of the body can occur without activation of certain neurons, called *motoneurons*, in the spinal cord and brain stem. The forebrain sends nerve impulses down pathways in the spinal cord and brain stem to excite the motoneurons.

Motoneurons may also be excited directly by some sensory inputs to produce simple adaptive bodily movements known as *reflexes*. For example, the rapid withdrawal of your hand from a hot radiator is a reflex: Sensory input from pain receptors in the hand runs through the peripheral nerve fibers, eventually reaching the spinal cord motoneurons serving the arm muscles, and your hand automatically jerks back. Reflexes are also important in such vital activities as standing, walking, and running.

Off to the back of the brain is a large, complex structure called the *cerebellum* (from the Latin meaning "little brain"). (See Figure 2.11.) This structure receives sensory and other inputs from the spinal cord, brain stem, and forebrain; it processes this information and then sends outputs to many parts of the brain to help make our movements precise, coordinated, and smooth.

In the center, or *core*, of the brain stem, running from the medulla up to the midbrain, is a complex region containing many small clumps of neurons and a number of long and short nerve fibers. The appearance of this region reminded early descriptive anatomists of a network. The Latin word for "network" is *reticulum;* thus this region was called the *reticular formation*. Later it was discovered that the reticular formation is, through its ascending fibers, involved in the activation, or arousal, of the cerebral cortex (the layer of neurons covering the cerebrum of the forebrain). (See Figure 2.12.) The fibers and nerve cells of the reticular formation concerned with cortical arousal are therefore known as the *ascending reticular activating system (ARAS)*.

Through its stimulating effect on the cerebral cortex, the ARAS is critically involved in regulating the various degrees of arousal—from deep sleep to alert awareness of the environment. (See Inquiry 2.2 and Chapter 8, page 317.) We know from animal experiments that the reticular formation has an arousing system projecting from it. These experiments show

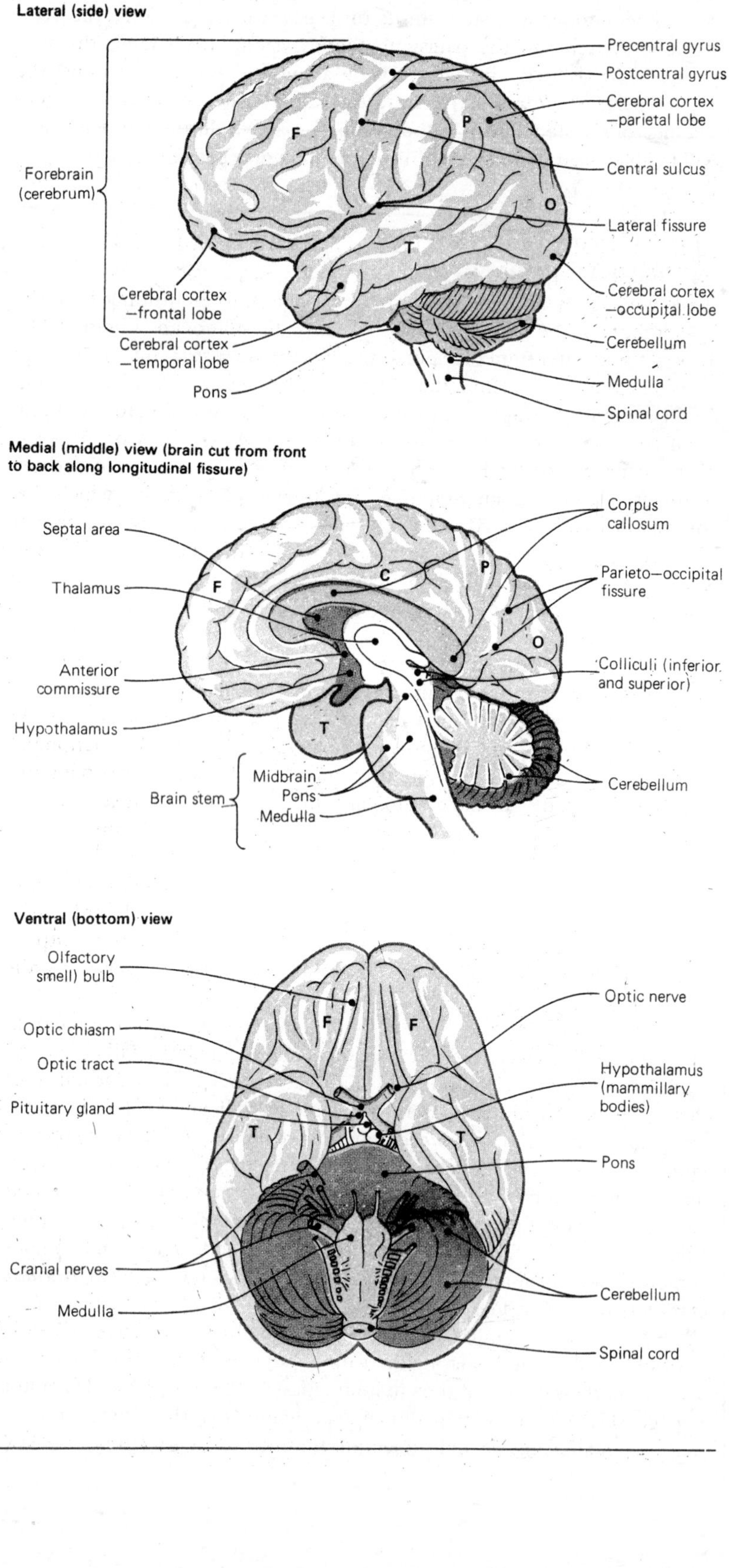

Figure 2.11

Three views of the human brain. *F*, frontal lobe; *T*, temporal lobe; *P*, parietal lobe; *O*, occipital lobe; *C*, cingulate gyrus.

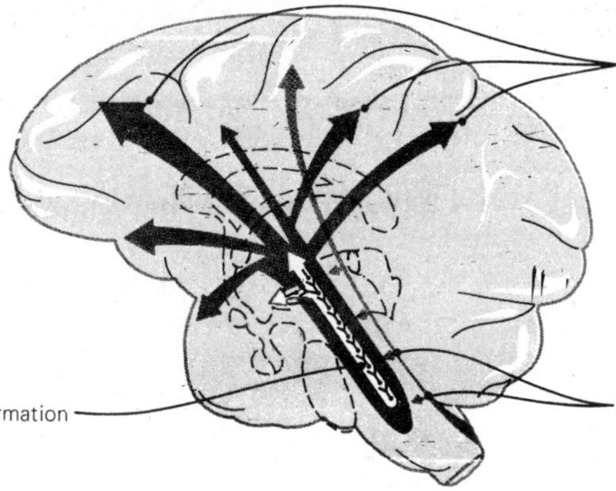

Figure 2.12

The ascending reticular activating system (*ARAS*). (Slightly modified from Magoun, 1954.)

Inquiry 2.2

WHAT IS THE BRAIN DOING WHEN WE SLEEP?

We all know from experience that sleep varies from drowsiness (a twilight zone between sleep and waking) to sound sleep from which it is very difficult to wake. For many years, the ease or difficulty of waking a person with some stimulus (such as a sound) was used as the measure of depth of sleep. Recently, however, experimenters studying sleep have learned that the *electroencephalogram* (EEG), a tracing of the electrical activity of the brain (or "brain waves"), provides a good index of the depth of sleep. Since the tracing can be made without disturbing the sleeper, the EEG has become a standard measure of sleep depth. The tracings have also made it possible for investigators to know when someone is dreaming during sleep. Variations in the activity of the ARAS play a large role in the control of the levels of sleep and the EEG patterns which accompany them. The ARAS also has an important influence on the brain activity that occurs during dreaming.

The EEG is a record of the slowly changing electrical activity of millions of nerve cells, all functioning at the same time in the brain. With suitable amplification of the electrical activity, the EEG can be recorded by attaching electrodes to the scalp. The spontaneous electrical activity of the brain waxes and wanes to give the wavelike record—the "brain waves"—of the EEG. These waves are really very small voltage changes, in the range of millionths of a volt. The number of alternations of voltage per second, or the frequency of the electrical changes, varies from 1 or 2 in deep sleep to 50 or more in highly aroused states. This frequency is expressed in hertz (Hz). Thus, the range of frequencies of the EEG is from about 1 to 50 Hz.

Some sample EEGs are reproduced in the upper figure, page 56. In the top record, showing an excited state, note that the frequency is relatively great, while the voltage, expressed by the height of the waves, is small. The opposite pattern appears in the bottom record, showing deep sleep. Records like the top one are sometimes called "low-voltage, fast"; those like the bottom one are called "high-voltage, slow". Activity in the ARAS controls these EEG patterns.

In general, as a person falls into deeper and deeper sleep during the night, the "brain waves" become progressively slower and higher in voltage. Starting with the alpha rhythm (10 Hz) of the drowsy waking state, the brain waves slow until they may have a frequency of only 1 to 3 Hz. Based on what EEGs show, investigators have divided sleep into four stages. In stage 1, a relatively large proportion of the "brain-wave" activity is fairly fast; in stages 2 and 3, the proportion of slower activity increases; and stage 4 is charaterized by a large proportion of very slow (1 to 3 Hz) activity. One of the striking facts about the sleep pattern through the night is its cyclical nature. It moves from stage 1 to stage 4 and back again several times, although, as shown in the lower figure on page 56, the stage 4 deep-sleep state may not be reached in the later sleep cycles. Sleep later in the night tends to be lighter.

A special state of sleep occurs when "brain-wave" activity returns from stage 2, 3, or 4 sleep to low-voltage, fast activity. This activity is not the same as activity in stage 1, even though the two states look somewhat alike in the EEG. It is a new stage—the stage of *paradoxical sleep*. It is called paradoxical because the activity looks very much like waking activity (low-voltage, fast), yet people in the paradoxical state are deeply asleep, as judged by the intensity of stimulation needed to wake them. It is in the stage of paradoxical sleep that most dreaming occurs.

During one night, a sleeper may go through the cyclical progression of EEG stages diagramed in the lower figure. Periods of paradoxical sleep are indicated by the shaded areas at the peaks of the curve. Investigators have concluded that brain activity is probably controlled by different areas of the ARAS and related brain-stem areas in paradoxical and regular sleep.

During paradoxical sleep, the muscles of the body go limp, and something else happens: The eyes move rapidly from side to side. For this reason, paradoxical sleep is sometimes called rapid-eye-movement (REM) sleep. But perhaps the fact of greatest interest is that the paradoxical, or REM, stage is the period of dreams. About 80 to 90 percent of the time, when people are awakened during or immediately after paradoxical sleep, they report that they were dreaming. At other parts of the sleep cycle, few dreams occur; people who are waked from non-paradoxical stages of sleep report dreams less than 15 percent of the time.

Note that several dreams take place each night—four complete dream periods are shown in the lower figure—and dreams occurring later in the night tend to be longer. Dreams occur in "real time"—they are not over in a flash. As the diagram shows, some last 20 to 30 minutes.

But what are dreams? We know they are a form of thinking that employs unusual symbols. The interpretation of the symbols is difficult. But now that experimenters can obtain "fresh" dreams by awakening people immediately after paradoxical sleep, we may soon have a better understanding of the thought processes in dreaming.

REFERENCES

Dement, W. C. (1974). *Some must watch while some must sleep.* San Francisco: W. H. Freeman.

Mendelson, W. B., Gillin, J. C., & Wyatt, R. J. (1977). *Human sleep and its disorders.* New York: Plenum Press.

Webb, W. B. (1975). *Sleep: The gentle tyrant.* Englewood Cliffs, NJ: Prentice-Hall.

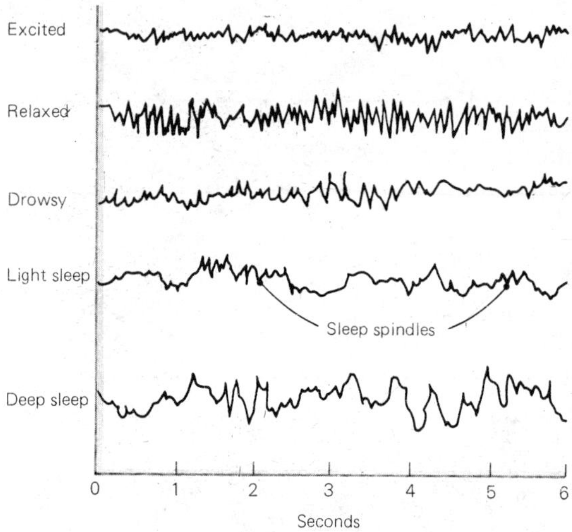

Electrical activity of the cerebral cortex (brain waves) typical of various states of sleep and arousal. (After Jasper, 1941.)

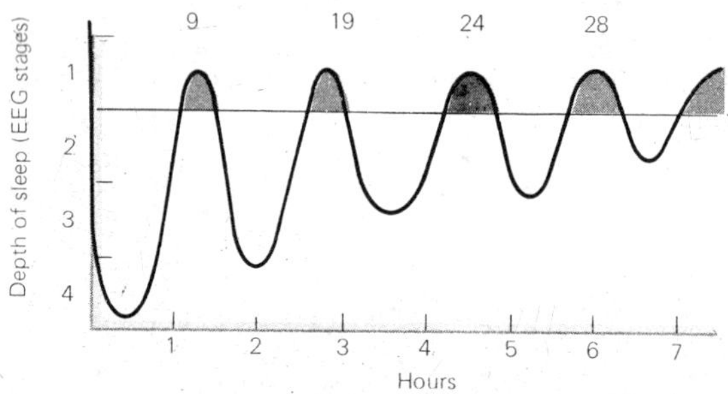

EEG stages of sleep during a typical night. Periods of paradoxical sleep are indicated by the shaded areas. It is in these periods that most dreaming occurs. The numbers over the paradoxical periods indicate the length of the period in minutes. (Modified from Dement & Wolpert, 1958.)

that coma—a state something like very deep sleep—results when portions of the reticular formation are destroyed or when its ascending projections to the cerebral cortex are severed. Stimulation of the reticular formation through implanted electrodes has an effect opposite to that of damage: it wakes up naturally sleeping animals. Human patients who suffer from certain diseases affecting the reticular formation, such as sleeping sickness, or who have sustained injuries to this part of the brain, fall into a profound coma which, in some cases, may last for years.

FOREBRAIN: THALAMUS

Just above the midbrain, forming a kind of expanded bulb on top of the brain stem, is the region of the forebrain known as the *thalamus* (from the Greek word meaning "bedroom" or "rotunda"). The thalamus lies between the two cerebral hemispheres and is covered by them. For this reason, it cannot be seen from the outside, and the brain must be cut open to show it (Figure 2.11, page 54). The thalamus contains many groupings of nerve cells called *nuclei* (singular *nucleus*). Some of these nuclei receive input from the seeing, hearing, pressure, pain, temperature, body-position, and taste senses; the incoming fibers make synapses on the neurons of these nuclei, which then send their fibers to specific areas of the cerebral cortex. Thus some thalamic nuclei have a relay function. Other nuclei of the thalamus do not receive major inputs from outside the thalamus; instead, they receive their main inputs from other nuclei within the thalamus itself and then send fibers to the cortex. Thus the thalamus, in addition to being a simple relay station, can also transform and modify input before sending it on to the cerebral cortex.

FOREBRAIN: HYPOTHALAMUS AND PITUITARY GLAND

Lying below the thalamus is a small but vital area of the forebrain known as the *hypothalamus* (Figures 2.11 and 2.13), from the Greek meaning "under the thalamus." Its importance to psychologists is that it contains nuclei and fiber tracts which are related to motivated behavior of a biological sort.

When motivation is discussed in Chapter 7, we will consider how certain structures in the hypothalamus are related to specific motivated behaviors. In this chapter, we will consider some general ideas about what the hypothalamus does and how it is related to the *pituitary gland*. (A gland is an organ that is specialized for the secretion of various substances.)

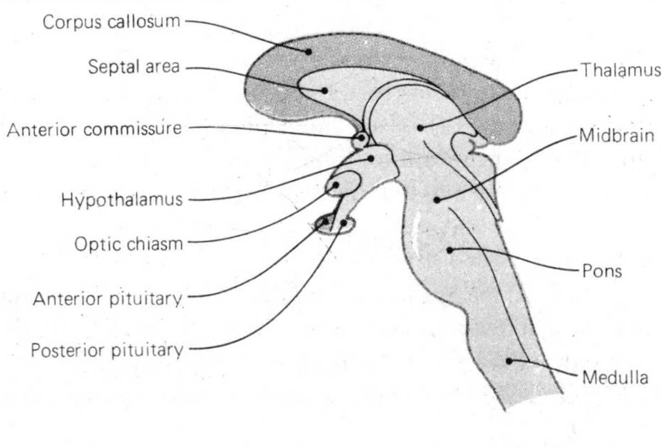

Figure 2.13

The hypothalamus and pituitary gland.

We may think of the hypothalamus as playing a role in the regulation of the internal environment of the body. The term *internal environment* refers to conditions inside the body, especially the chemical composition of the blood and other fluids that bathe body cells. Blood temperature, the concentration of salt in the blood, and the concentrations of chemical messengers, called *hormones*, and other chemicals in the body are monitored, or sensed, by different specialized neurons in the hypothalamus. The hypothalamus also receives reports about the state of the internal environment from other bodily organs, such as the liver. When conditions in the internal environment change from their optimum, or homeostatic, level, certain hypothalamic neurons (which ones depends on the condition that has changed) become active and send information to other parts of the nervous system and the pituitary gland (see below) to correct the departure from the homeostatic level. These messages from the hypothalamus result in automatic physiological adjustments within the body's systems. And they may also trigger motivated behavior, the aim of which is to secure substances or stimulation from the environment that will restore the balance within the body (Chapter 7, page 272).

The nervous system is linked to the glandular system of the body by connections between the hypothalamus and the pituitary gland (Figure 2.13). The hypothalamus, either by means of nerve impulses sent to the pituitary or by chemicals called *releasing factors*, controls the secretion of hormones from the pituitary gland into the bloodstream. *Hormones* are chemical messengers manufactured by specific organs (often the endocrine glands) that are secreted into the bloodstream and carried by it to various parts of the body, where they have their effect.

Physiological psychologists are interested in the roles pituitary hormones play in behavior. The sexual and maternal behavior of many lower animals, for instance, is closely tied to levels of certain pituitary hormones and the hormones from sex glands. (See Chapter 7, page 279.) Pituitary hormones have also been shown to play a role in facilitating learning and memory (DeWied, 1980), and they are critically involved in the body's response to stressful situations (Chapter 8, page 235).

FOREBRAIN: CEREBRUM

Most of what you see when you look at the brain is the outside of a large structure known as the *cerebrum* (Figure 2.11 and the figure on the opening pages of this chapter, page 37). The human cerebrum weighs about 1,400 grams (not quite 3 pounds).

Structure The cerebrum is divided into two *cerebral hemispheres*, one on each side of the head, by a deep cleft, or fissure, called the *longitudinal fissure* (Figure 2.14). Each hemisphere is covered by the *cerebral cortex* (from the Latin meaning "brain bark, or covering"), a sheet of neurons averaging about 2½ millimeters in thickness and containing billions of neurons. Since the cerebral cortex is composed mostly of neurons (although there are a number of fibers within it), it is gray matter.

As you look at the outside of the cerebrum, the cerebral cortex looks something like a rumpled piece of cloth with many ridges and valleys (see the figure on page 37). Neuroanatomists call a ridge a *gyrus* (plural *gyri*); a valley, or crevice, is sometimes called a *sulcus* (plural *sulci*) and sometimes,

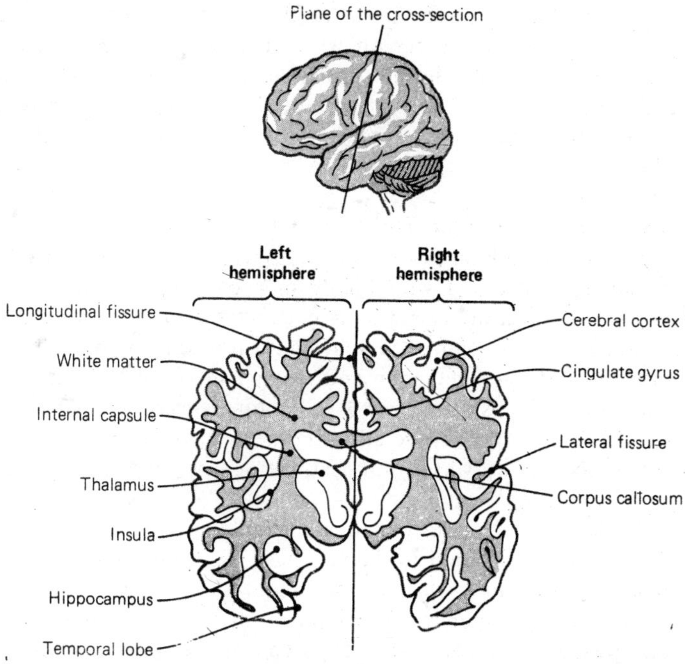

Figure 2.14

A cross-section of the brain. The white matter is dark and the cortex is light because of the stain used to dye the section.

if it is exceptionally deep, a *fissure*. About two-thirds of the cerebral cortex is in the sulci and fissures of the brain and thus cannot be seen when we look at the surface. In lower animals there is much less folding than in humans; the rat, for instance, has hardly any folds in its cortex. (See Figure 2.4, page 45).

The fissures and deeper sulci mark off the division of each hemisphere into lobes. The *central sulcus* is a deep groove running obliquely from top to bottom on the side, or lateral, surface of the cerebral cortex; it marks off the *frontal lobe* (Figure 2.11, page 54). The *lateral fissure* runs roughly from front to back on the side of the cerebral cortex; the part of the cerebrum below the lateral fissure is known as the *temporal lobe* (Figure 2.11). Figure 2.11 shows that the lobe behind the central sulcus is termed the *parietal lobe*. At the very back of the cerebrum is the *occipital lobe* (Figure 2.11).

Under the cortical covering of the cerebrum are nerve fibers. Since many of these fibers are covered by white myelin, this is the white matter of the cerebrum (Figure 2.14). These fibers carry information to and from the cerebral cortex, and they interconnect various regions within the cortex. A huge number of connections is made possible by these fibers, and it is this complexity of connections which makes the human cerebrum a most complicated piece of machinery. An important bundle of fibers, known as the *corpus callosum* from the Latin words meaning "hard body," connects areas of the cortex of one hemisphere with corresponding areas in the other hemisphere. (See Figures 2.11 and 2.14.) We shall have more to say about this fiber bundle later in the chapter when we look into differences between the functions of the left and right cerebral hemispheres.

Buried in the cerebrum so that they cannot be seen from the surface are clusters of gray matter known as the *basal ganglia*. These have an important role to play in the regulation and control of movement. Other buried cerebral structures, such as the hippocampus and amygdala, will be described later in the sections on the limbic system and the temporal lobe.

Sensory and Motor Functions Roughly speaking, we may consider some cortical areas to be largely sensory in function; others, called *motor areas*, are largely concerned with bodily movements; and still other cortical regions—known, for want of a better term, as *association areas*—are involved in such complex psychological functions as thought, memory, imagery, and the production and understanding of language. Since our main interest as psychologists is in the functions of the association areas, we will discuss them in detail in the next major section of this chapter. Let us look now, very briefly, at the sensory and motor areas of the cerebral cortex.

Most of the senses (smell is a notable exception) have areas of the cerebral cortex devoted to them. These areas are known as the *primary sensory areas*. Those which can be seen on the surface are depicted in Figure 2.15. In general, nerve impulses that are generated when sensory receptors are stimulated—after traversing peripheral sensory nerves, pathways in the central nervous system with varying numbers of synapses in them, and a synaptic relay in the thalamus—reach the appropriate primary sensory area of the cortex. Visual input goes to the occipital lobe; hearing (or auditory) input reaches a portion of the temporal lobe; and sensory signals representing touch, pain, or pressure arrive at the body sense, or somatosensory, region of the cortex in the parietal lobe on the gyrus immediately behind the central sulcus (Figure 2.15). Input representing taste reaches temporal-lobe cortex that is buried in the lateral fissure and cannot be seen on the surface of the brain. The primary sensory areas process the information received and send it to association cortex or other parts of the brain where further processing occurs. Al-

Figure 2.15

The primary sensory and motor areas of the human cerebral cortex. *F*, frontal lobe; *T*, temporal lobe; *P*, parietal lobe; *O*, occipital lobe.

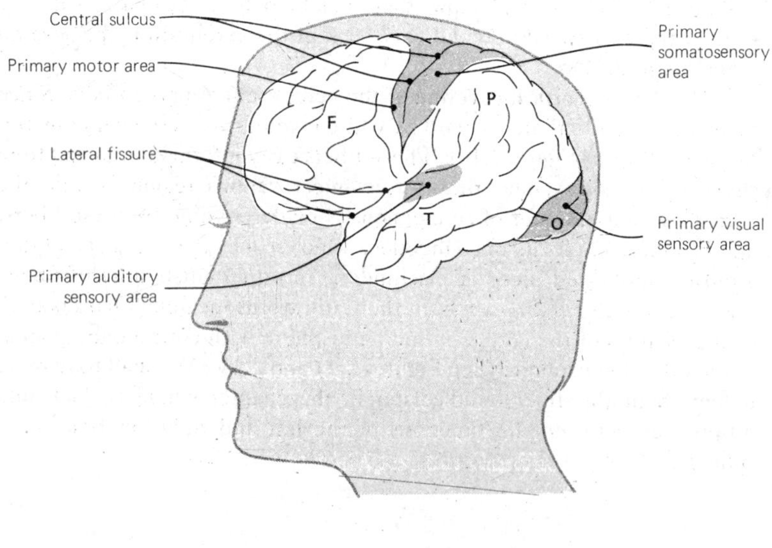

though few of the details are known, most neuroscientists consider that our experience of the environment, or perception, depends upon patterns of activity occurring in our sensory systems, of which the primary sensory areas are an important part. When you perceive a friend's face, for example, there is a pattern of activity in your visual system, including the primary visual sensory area of the cortex, that stands for, represents, or "codes for," this particular face.

The principal area of the cerebral cortex involved in motor, or movement, functions is on the gyrus just in front of the central sulcus. As shown in Figure 2.16, various regions of this cortex are devoted to movements of specific parts of the body. This cortical region can control bodily movements because it sends nerve fibers to the brain stem and spinal cord, which then make appropriate synaptic connections so that nerve impulses will be sent through specific peripheral nerves to the muscles. This area of cortex is part of the complex circuits of nerve fibers involved in movement and is closely linked to the cerebellum and other movement-controlling areas of the brain. Damage to portions of this motor area of cortex will, depending on what part is damaged, result in weakness of a particular part of the body and, especially when the hand area is damaged, loss of the ability to perform fine, dexterous movements. If many of the nerve fibers coming from this and certain other areas of the cortex are damaged, as they often are in strokes, what is known as spastic

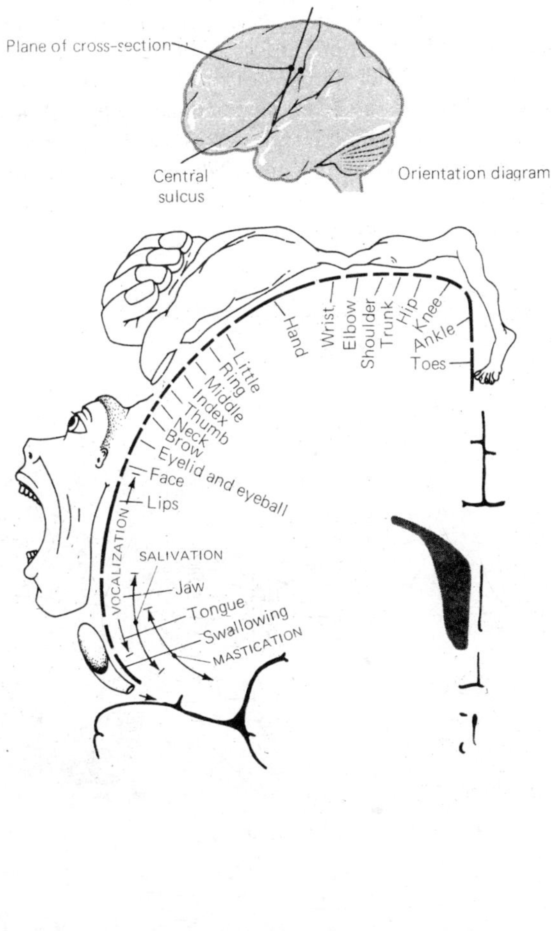

Figure 2.16

Representation of parts of the body on the precentral gyrus (primary motor cortex). (From Penfield & Rasmussen, 1950.)

paralysis, in which the limbs are stiff and somewhat flexed or extended, will ensue. Since the fibers from one side of the cortical motor area cross to the other side of the body below the head on their way to the spinal cord, the weakness, loss of dexterity, or spastic paralysis will occur on the side of the body opposite the damaged region of the motor cortex. Thus a person with left-sided damage, for example, will have movement difficulties on the right side of the body below the head.

FOREBRAIN: LIMBIC SYSTEM

Some of the nuclei of the thalamus, hypothalamus, and cerebrum are interconnected to form a kind of ring or border around the lower portion of the forebrain. This group of structures is known as the *limbic system*, from the Latin meaning "border." (See Figure 2.17.) We need note here only those structures in this system that are most important for behavior and experience. These include the olfactory (smell) bulb and its connections, the septal nuclei (from the Latin meaning "partition"), the hippocampus (from the Greek meaning "seahorse"), the amygdala (from the Latin for "almond"), and the cingulate (from the Latin meaning "girdle") gyrus of the cerebral cortex (Figures 2.11 and 2.17).

In the history of the study of brain function, it was recognized early that the noncortical parts of what we now call the limbic system receive inputs from the smell receptors in the nose; for this reason, the limbic system used to be called the "smell brain." Only in the last few decades have some of the other important functions of the limbic system been discovered. For instance, portions of this system are involved in the expression of the emotions of fear and rage, as well as aggressive behavior (Moyer, 1976). Other portions of the limbic system are called "reward" areas or "pleasure" areas because animals will work to receive electrical stimulation of these limbic regions (Olds & Fobes, 1981). Still other parts of the limbic system—the amygdala and hippocampus in particular—seem to have a role to play in the formation of memories, and we will discuss them in detail later in this chapter.

Figure 2.17

The limbic system; underlined labels indicate limbic-system structures. (Modified from MacLean, 1949.)

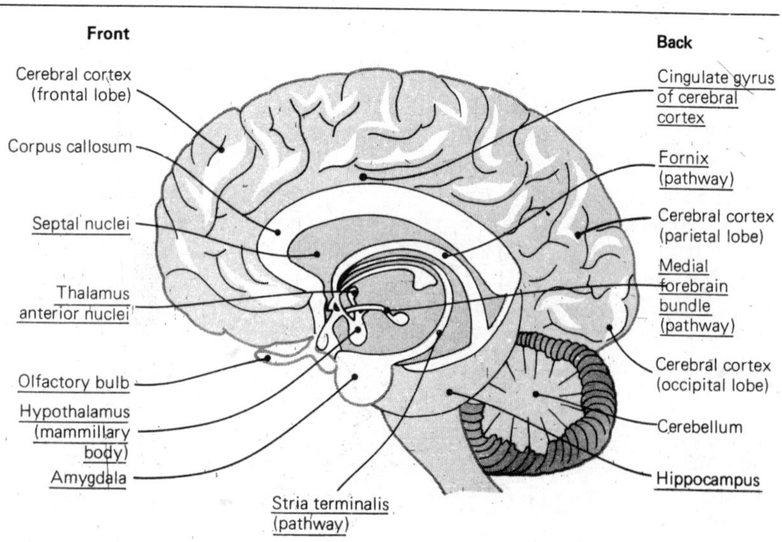

Brain and Behavior: Association Cortex, Behavior, and Experience

Most of the cerebral cortex lies outside the primary sensory areas and the principal motor area (Figure 2.15, page 60). These regions in the frontal, parietal, temporal, and occipital lobes are called *association areas* because pioneer students of the brain thought that associations, or connections, between sensory input and motor output were made in these areas. It is currently believed that, rather than being regions of simple sensory-motor connections, the association areas process and integrate sensory information relayed to them from the primary sensory areas of the cortex and from the thalamus; after the information has been processed, it may be sent to motor areas to be acted upon. For instance, the visual association cortex of the occipital and temporal lobes receives input from the primary visual sensory cortex and processes this information, in ways as yet not fully known, to make it possible for us to recognize objects in the visual world. Similarly, the frontal-lobe association cortex receives input from many brain areas and integrates it to make possible ordered sequences of actions.

FRONTAL-LOBE ASSOCIATION CORTEX

About 40 percent of the human cerebral cortex is in the frontal lobes. With the exception of the motor areas immediately in front of the central sulcus and Broca's area, an area important in the programming of the movements necessary for speech (see later), much of this cortical region has no primary sensory or motor functions and is therefore considered to be association cortex. Because it is farthest forward in the head, this association area of the frontal lobe is known as *prefrontal cortex*. Prefrontal cortex is interconnected with the visual, auditory, and somatosensory cortical areas, with other association cortex in the parietal and temporal lobes (see below), with the thalamus, and with a number of other noncortical structures of the brain.

Personality and General Behavior Patterns After Prefrontal Damage

Much of our information about the functions of prefrontal cortex comes from study of the behavior of those unfortunate people who have suffered prefrontal damage as a result of gunshot wounds, skull fractures, and the like. Two rather different types of personality change follow prefrontal-lobe damage. Whether these contrasting types of reaction are due to differences in the parts of the frontal lobes which have been damaged is not certain. Lack of restraint, impulsiveness, immaturity in social relationships, and, sometimes, promiscuity in sexual behavior characterize one personality pattern that emerges after prefrontal damage. Apathy, indifference to others, loss of initiative, decrease in spontaneous talking, and reduced emotional expression are the main features of the other pattern (Blumer & Benson, 1975).

Other personality features seen in some prefrontal-damaged patients, especially those with large tumors of the prefrontal areas, are euphoria—a pleasant feeling of being "on top of the world"—and irritability. The families of these patients often report that the two characteristics alternate, the patient being euphoric for a while and then quite grouchy and touchy. However, it is not certain that these two emotional responses are due to prefrontal damage because the tumors result in increased brain pressure, and this may cause a change in the function of brain regions outside the frontal lobes.

Intellectual Changes Following Prefrontal Damage While scores on standard tests of general intelligence are usually little affected by prefrontal-lobe damage, special tests reveal a number of subtle intellectual deficits. Prefrontal patients are deficient in planning courses of action; their behavior seems to lack direction. Planning involves formulating a sequence of actions, and the prefrontal-damaged patient has difficulty doing this (Kolb & Whishaw, 1980). A characteristic of prefrontal patients which impairs their sequencing of actions is inflexibility, manifested behaviorally by perseveration—a tendency to continue to behave in a certain way even when changes in the situation demand new responses (Milner, 1964).

Another intellectual change after prefrontal damage is a subtle memory impairment. While their old memories and their ability to form new ones are largely intact, people with prefrontal damage have trouble remembering *when* something occurred in the flow of their experience. That is, they have trouble remembering whether one recent event occurred before or after another one (Milner, 1974).

In addition to problems in planning, sequencing actions, flexibility, and remembering the order of recent events, prefrontal patients may also have difficulties with the voluntary control of their eye gaze (Teuber, 1964). When prefrontal-lobe patients are asked to extract information from a complex picture—the ages of the people in the picture, for instance—their gaze will tend to wander randomly over the picture. (The gaze of a normal person will be directed in an orderly way from one face to another.) Without good control of voluntary gaze, prefrontal patients may find certain tasks difficult.

These, then, are some of the major intellectual problems associated with damage to the prefrontal lobes. The problems are subtle, but they get in the way of optimum intellectual functioning. A theme which runs through this story is that of order and sequence. The prefrontal regions seem to have much to do with our ability to think about sequences of actions—that is, to plan, change plans when necessary, and remember when actions and other events occurred in relation to one another in a sequence.

PARIETAL-LOBE ASSOCIATION CORTEX

As described earlier, the parietal lobe is the region of the cerebrum behind the central sulcus, above the lateral fissure, and in front of the occipital lobe (Figure 2.11, page 54). The portion of the parietal lobe immediately behind the central sulcus is primary somatosensory cortex (Figure 2.15, page 60). Parietal-lobe association cortex lies behind the primary somatosensory cortex and receives inputs from visual cortex, auditory cortex, somatosensory cortex, and the thalamus. Major outputs go to frontal and temporal association cortex, to the thalamus, and to subcortical structures involved in the control of movement.

Just as with other cortical association areas, the main method used to study the function of the parietal-lobe association cortex is to see what abilities are lost or altered when this area is damaged. Strokes, or cerebrovascular accidents (CVAs), due to blockage (perhaps by a blood clot) of the major artery serving the parietal lobe occur relatively frequently. Because brain cells need a steady supply of oxygen and nutrients for survival, shutting off the blood supply results in brain damage. Careful study of stroke patients has told us much about the behavioral and intellectual functions of parietal association cortex.

To an extent, the parietal association areas of the left and right hemispheres of the human brain differ in function. In other words, functions are lateralized to some degree. This is also true of frontal association cortex, but since lateralization is not so clear there, we did not stress it in the discussion of frontal-lobe association cortex. However, lateralization cannot be ignored in discussing the parietal lobes. We will have more to say about left-right brain functions in the next major section of the chapter—"Brain and Behavior: Left- and Right-Hemisphere Functions." Here we need only point out that for almost all right-handed people (and somewhat more than half of left-handers), the left hemisphere of the cerebrum is specialized to carry out the symbolic functions involved in language and mathematics. The right hemisphere seems specialized to deal with images and spatial relationships—where sensory inputs are located in space and how they are related to each other. For example, when you imagine what your bedroom looks like and the spatial arrangement of the furniture, your right hemisphere, if you are right-handed, is possibly more active than your left one. Note, however, that these hemispheric differences in function are matters of degree; the right hemisphere has some language abilities, and the left has considerable imaging capacity. With this as background, what happens when parietal-lobe association cortex is damaged?

Right Parietal Association Cortex One of the curious symptoms of damage to the right parietal association cortex is termed *contralateral neglect*. Patients ignore the left side of space and the left side of the body (Heilman & Watson, 1977). The term *contralateral* is used because the neglect is of space and of the body on the side opposite to (contralateral to) the brain damage. For example, when asked to draw objects, patients will draw only the right side, ignoring the left side (Figure 2.18); when touched simultaneously on the right and left hands, they will report only the touch on the right hand; and some will dress only the right side of their bodies, ignoring the left side. Going along with this contralateral neglect will be a denial by the patient that he or she has any problems. Other distortions of spatial perception show up when patients with right-parietal damage draw maps or attempt to put blocks together to make designs (Kolb & Whishaw, 1980). All these problems may occur despite the fact that these patients know where they are, what day it is, what they are doing, and speak, read, and write normally.

Left Parietal Association Cortex Problems in writing, reading, and doing simple arithmetic problems often show up after damage to the left parietal association cortex (Kolb & Whishaw, 1980). The ability to distinguish right from left may also be impaired (Benton, 1959). Difficulty with short-term verbal memory can also be a symptom of left parietal association cortex damage (Kolb & Whishaw, 1980). For instance, a

Figure 2.18

Contralateral neglect shows up in the drawings of patients with damage to the right parietal association cortex. Note that the left side of the clock face is blank and the left side of the flower is missing. (From Rosenzweig & Leiman, 1982, and Heilman, 1979.)

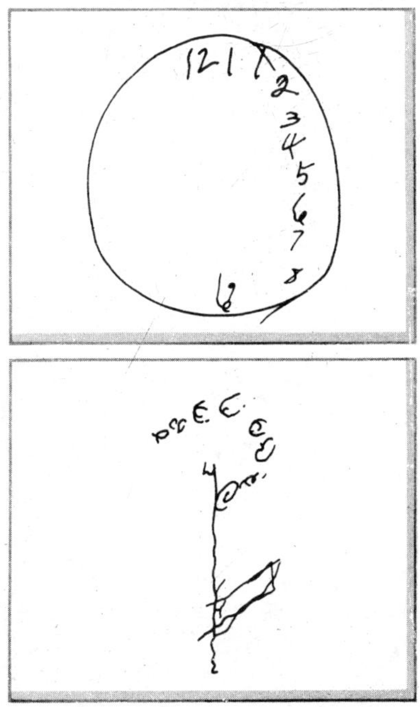

patient with damage to this area may be able to remember and repeat only the first few numbers of the sequence 08375149247 immediately after hearing them read slowly. Normal people can remember the first seven or so numbers of this digit-span test.

Left and Right Parietal Association Cortex A perceptual problem that shows up after left or right parietal association cortex damage is difficulty in recognizing common objects, such as a fork, by touch alone. For instance, with the left hand feeling a fork hidden behind a screen, patients with right parietal-lobe damage would know they were touching something but would not be able to identify what they were handling. This is because sensory input from the left hand goes to the right brain. And, because sensory input from the right hand goes to the left brain, a patient with left parietal-lobe damage would have trouble recognizing objects by touch with the right hand. Notice that these patients have not lost the sense of touch—they know they are touching something. Their problem is recognizing *what* they are touching. Such a deficit in recognition ability, regardless of the sense involved, is known as *agnosia*. Thus a common result of parietal-lobe damage in either the right or left hemisphere is tactile, or touch, agnosia (Hécaen & Albert, 1978).

TEMPORAL-LOBE ASSOCIATION CORTEX AND RELATED STRUCTURES

The temporal lobe of each hemisphere lies below the lateral fissure (Figure 2.11, page 54). The upper portions of each temporal lobe contain the primary sensory cortex and related cortex involved in hearing. Below the hearing, or auditory, areas is the association cortex proper of the temporal lobes. Buried within each temporal lobe are two limbic-system (page 62) structures which, while not association cortex, seem to cooperate with temporal association cortex in the formation of memories. These are the *hippocampus* and *amygdala* (Figures 2.14, page 59, and 2.17, page 62). Temporal cortex and its related structures—hippocampus and amygdala—have many connections with other cortical and subcortical brain areas. As was the case within parietal cortex, temporal cortex is lateralized in a number of its functions, especially language (see below).

Sensory Functions As might be expected from its location near the primary auditory cortex, temporal association cortex is involved in hearing perception and the recognition of sounds. Auditory agnosia (the term *agnosia* was defined in the last section), such as difficulty in recognizing melodies or problems in distinguishing nonlanguage sounds from one another, may result from damage to the temporal-lobe association areas that are closely linked to auditory primary sensory cortex (Kolb & Whishaw, 1980). Since portions of temporal association cortex receive inputs from the visual areas of the brain in the occipital lobe, various visual agnosias can follow temporal-lobe damage (Kolb & Whishaw, 1980). Thus, even though there is no blindness, temporal-lobe patients may have trouble identifying pictures of common objects and recognizing faces. Visual agnosias also occur after damage to the occipital-lobe association cortex and are more common after right temporal or occipital-lobe damage. Another disorder related to the senses that occurs after temporal-lobe association cortex damage is impaired attention. (See Chapter 3, page 109.) When given special tests, temporal-lobe patients have difficulty selecting and focusing on auditory and visual inputs (Kolb & Whishaw, 1980).

Language Functions The left temporal lobe contains a region of association cortex called *Wernicke's area* (Figure 2.19). Damage to this region markedly impairs the understanding of speech and written language. The speech of a patient with a lesion, or damage, here is fluent but conveys little meaning. The next section of the chapter on left- and right-hemisphere functioning has more to say about Wernicke's area and its relation to other language areas of the cortex.

Memory Functions While most of our information about the temporal lobes comes from study of victims of strokes or injuries, we can also learn much about their functions from studying patients who have had operations done to relieve them of severe epileptic fits. Epilepsy, which results from excessive, abnormal firing of brain neurons, can often be controlled by drugs, but sometimes the drugs do not work. Brain operations may be resorted to in cases of severe epilepsy that do not respond to drug treatment and which incapacitate the patient. The purpose of these operations is to remove the trigger zone, or focus, in the brain where, in some patients, the abnormal electrical activity originates. The result of these operations may be a great reduction in the number of epileptic attacks, but, when the area removed involves the hippocampus and amygdala, a price, in the form of a memory problem, is paid for this benefit. When the first temporal-lobe focus operations were done, it was not known that serious memory problems would result, and so sometimes portions of the hippocampus and amygdala on both sides of the brain were removed. It was found, however, that the memory problems from such two-sided operations were so severe that surgery was soon restricted to cases in which the focus was only on one side of the brain; after such one-sided operations, memory problems are more limited. What, specifically, is the memory function of the hippocampus? The following description of a bilateral patient who has been much studied (Milner et al., 1968) since his operation in 1953 will help answer this question.

> The patient (H. M.) was 27 years old when he was operated on in 1953. He had suffered from epilepsy for many years; the operation

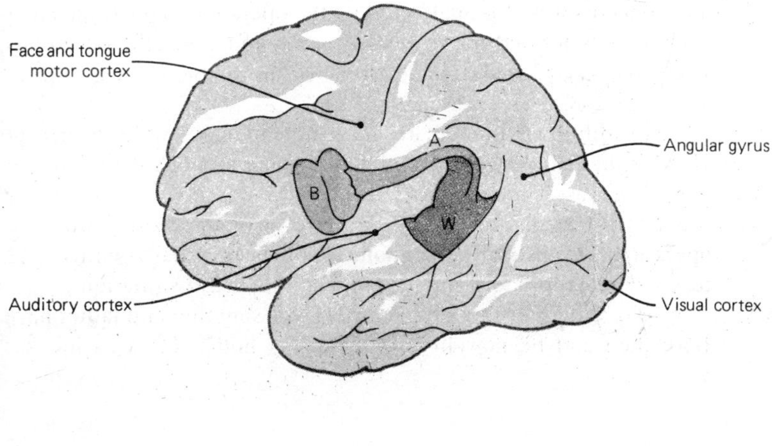

Figure 2.19

The language areas of the cerebral cortex. *W*, Wernicke's area; *A*, arcuate fasciculus; *B*, Broca's area.

which was designed to treat the disease removed the amygdalas and large portions of the hippocampus on both sides of the brain.

The operation greatly reduced the number of H. M.'s epileptic attacks, but resulted in profound memory problems which have lasted to the present. H. M. is greatly impaired in forming new memories. An anecdote related by the psychologists who studied H. M. illustrates his memory problem. Thirteen years after the operation, H. M. went from his home, where he was living with his mother and father, to Boston for tests. On the day he was picked up for the trip, his mother had been in the hospital for a few days recovering from a minor operation. H. M. had been taken by his father to visit her several times, but he did not remember any of these visits, and, when asked who had packed his bag for the trip, he thought it had been his mother, because she usually did such things for him. H. M. does not recognize people he has met since the operation, is content to read the same magazine over and over, cannot remember where he puts things, and works the same jigsaw puzzles time and again. He realizes he has a memory problem, saying, for example, "Every day is alone in itself, whatever enjoyment I've had, and whatever sorrow I've had" (Milner et al., 1968, p. 217). It is as if most things after the operation are "new" to H. M.

On formal laboratory tests designed to measure the formation of new memories, H. M. obtains low scores. But H. M. can learn some new tasks. He can learn, for instance, to trace the outline of the mirror image of a star and how to solve the Tower of Hanoi problem (Chapter 6, page 234), which requires putting blocks on pegs according to certain rules. H. M. gets better and better at these tasks over several days. But if he is asked whether he has done one of these problems before, H. M. replies "No." Thus H. M. can learn and remember how to do certain things despite having no conscious memory of ever having done them before; he remembers how to do something but has no memory of specific events. H. M. is thus said to have procedural memory without specific event, or declarative, memory (Squire, 1982).

It is important to note that H. M. has good immediate memory. For instance, he can repeat six numbers forward and five backwards after hearing them once. Nor is his remote memory for events before the operation impaired. Except for some haziness of memory for events a year or two before the operation, H. M. remembers well the episodes of his youth; his memory of language and concepts was also unaffected by the operation. This means that such already-formed memories are not stored in the hippocampus and amygdala; they are stored elsewhere in the brain. (See Inquiry 5.1, page 188.)

In addition to immediate and remote memory, other aspects of H. M.'s mental life were spared by the operation. His IQ before the operation was 101; nine years afterward it was 118. The improvement in IQ seems due to the decrease in brain seizures after the operation. On tests of perception, H. M. is essentially normal. He responds appropriately and acts normally in social situtations.

To bring the story up to date, H. M.'s mother and father have both died and he now lives in a nursing home. H. M.'s life has enriched our knowledge of human brain function and memory, but it is tragic that this gain in knowledge was at so great a cost to him.

H. M.'s memory problem, or *amnesia*, and that of other temporal-lobe patients, is in forming *new* long-term memories. Roughly speaking, long-term memories are those lasting from hours to a lifetime. Long-term memory is often considered to be distinct from short-term memory, which is transient and lasts for only seconds or a few minutes at most. (See Chapter 5 for a more complete discussion of these types of memory.) H. M. and other temporal-lobe patients have relatively normal short-term memories; they can remember what just happened. For instance, in contrast with left parietal-lobe patients (page 65), people with damage to the hippocampus and amygdala can repeat a string of numbers which has just been read to them. If short-term memory is intact, why is the formation of new long-term memories impaired? One theory says that short-term memories become long-term ones through a process known as consolidation. For example, short-term memories may be represented in the brain by transient patterns of nerve-cell activity which, if allowed to continue for a period of time, are consolidated into the longer-lasting brain changes (see Inquiry 5.1, page 188) which underlie long-term memory. So some would say that H. M. and other similar patients lack the brain structures—the hippocampus and amygdala—essential for consolidating short-term memory processes into long term ones.

Brain and Behavior: Left- and Right-Hemisphere Functions

The left hemisphere usually has an advantage with respect to understanding language, formulating language for communication, and thinking with language symbols; the right hemisphere is usually specialized to deal with images, spatial relationships, and pattern recognition. Note, however, that none of these abilities is located exclusively in one hemisphere or the other; the left hemisphere has some spatial abilities, and the right has some language capabilities. The specializations of the hemispheres are matters of degree.

LEFT HEMISPHERE AND LANGUAGE

From observations of people who have suffered brain damage, often as the result of a stroke (page 65), we know that the left hemisphere is usually specialized for language functions. Strokes typically affect one side of the brain and produce paralysis of the side of the body opposite the affected side of the brain (page 62). Therefore, if we know which side is paralyzed, we know that the brain damage is in the cerebral hemisphere on the side opposite the paralysis. People with left-hemisphere strokes (right-sided paralysis) are much more likely to have serious language problems than are people with right-hemisphere (left-sided paralysis) strokes. In recent years, a convenient test to localize the "speech" hemisphere has been developed for normal people. Sodium amytal, a sedative drug, is injected into the main artery serving one cerebral hemisphere. The person is instructed to talk while the sedative is taking effect. If speech starts to disintegrate and then drops out almost completely as the sedative reaches one side of the brain, we can conclude that speech is localized on that side. Results from sodium-amytal tests indicate that the left hemisphere is specialized for language functions in some 95 percent of right-handed people and in about 65 percent of left-handed people (Milner, 1974).

Looking in more detail at the left hemisphere, we find that its main language functions are localized in the upper temporal lobe and the lower frontal lobe (Figure 2.19). The temporal-lobe language region is termed *Wernicke's area;* it is named for the German neurologist Carl Wernicke, who, in the 1870s, studied language disorders resulting from damage to the temporal lobe. The language region in the lower frontal lobe is known as *Broca's area,* after Paul Broca, a French physician who, in the 1860s, discovered the language functions of this area. Damage to one or the other of these areas results in language problems specific to the area damaged. *Aphasia* is the general term for a deficit, due to brain damage, in the ability to communicate with or to understand speech or writing.

Wernicke's area is involved in the understanding of spoken and written language, so the ability to comprehend language is seriously impaired by injuries to this cortical region. It also plays a part in the formulation of sentences—figuring out what we want to say or write and selecting the words to express it. After injury to this area, a person's speech is quite fluent and well pronounced; but because the individual has trouble finding the right words, what is communicated is usually vague and deficient in meaning. Here is an example:

> "What brings you to the hospital?" I asked the 72-year-old retired butcher four weeks after his admission to the hospital.
>
> "Boy, I'm sweating, I'm awful nervous, you know, once in a while I get caught up, I can't mention the tarripoi, a month ago, quite a little, I've done a lot well, I impose a lot, while, on the other hand, you know what I mean, I have to run around, look it over, trebbin and all that sort of stuff."
>
> I attempted several times to break in, but was unable to do so against this relentlessly steady and rapid outflow. Finally I put up my hand, rested it on Gorgan's shoulder and was able to gain a moment's reprieve.
>
> "Thank you, Mr. Gorgan. I want to ask you a few . . ."
>
> "Oh sure, go ahead, any old think you want. If I could I would. Oh, I'm taking the word the wrong way to say, all of the barbers here whenever they stop you it's going around and around, if you know what I mean, that is tying and tying for repucer, repuceration, well, we were trying the best we could while another time it was with the beds over there the same things. . . ." (Gardner, 1975, p. 68)

Other symptoms of damage to Wernicke's area include (1) impairment of the ability to repeat a spoken word; (2) problems with reading and writing; (3) difficulty in naming common objects; and (4) the intrusion of incorrect sounds or words into the flow of speech—saying "repuceration" as in the example, "streeb" instead of "street," or "daughter" instead of "mother," for instance.

The symptoms of patients with damage to Broca's area are quite different from the symptoms of those with Wernicke's-area damage. Comprehension of spoken and written language is generally intact, and there are only minor problems in repetition and naming. The big difficulty is with the fluency of speech. The Broca patient speaks slowly and with great effort; articulates the sounds of speech poorly; tends to omit pronouns, adjectives, adverbs, and articles; uses the singular of nouns; and has great trouble with the tenses of verbs. In other words, the speech of a Broca patient is nonfluent and ungrammatical. The Broca pattern of speech has been called "telegraphic" because, as in a telegram, only the main words are used for communication. The sentence, "He went to school yesterday at 8 o'clock in the morning" might be spoken by a Broca patient as "He go . . . school . . . yesterday . . . 8 o'clock," with the dots indicating silent intervals.

Consideration of these two common types of aphasia—there are others, and many cases are not easily classified—has led to a theory about how the language hemisphere is organized—what might be called the "anatomy of language" (Geschwind, 1970, 1979). A look at Figure 2.19 (page 67) shows that Wernicke's area is connected to Broca's area by a bundle of nerve fibers, called the *arcuate fasciulus* (Latin for "archshaped bundle"), which runs through the white matter under the cortex; also shown in the figure are auditory cortex, the face and tongue motor cortex, and a cortical region—the *angular gyrus*—which is important in the brain processes involved in reading. Patients with damage to the angular gyrus may be able to communicate with and understand speech, but they cannot read. They have what is called *dyslexia*.

According to the theory, each of these anatomical areas has a particular role to play in various aspects of language. Spontaneous speech, as when you are simply telling someone something, is organized in Wernicke's area. Here you find the words to express yourself and link them into meaningful sentences. The Wernicke's-area patterns of activity representing what is to be said are then sent through the arcuate fasciculus to Broca's area, which contains the programs for the complex patterns of muscle movements needed in speech. The Broca's-area programs are then relayed to the region of the motor cortex controlling the lips, tongue, and vocal cords which actually produce the speech sounds (Figure 2.16, page 61). If you are replying to what another person has just said or are simply repeating a word you have just heard, input from the auditory cortex goes to Wernicke's area, where the reply is organized; the organized pattern of brain activity is then sent through the arcuate fasciculus to Broca's area and thence to the motor cortex for expression. If you read a passage or word out loud, the Wernicke's area-arcuate fasciculus-Broca's area-motor cortex pathway is also used; but in this case Wernicke's area receives its input from the visual cortex by way of the angular gyrus (Figure 2.19, page 67).

Although not all those who study brain and language agree with this theory, it has been influential and useful in helping to organize much of what is known about where language functions are located in the left hemisphere. Now we would like to know *how* language is represented in the brain, *how* words are found and assembled into sentences, and *what* the "motor programs" of Broca's region are. In other words, we would like to know about the electrical and chemical events involved in language—what might be called the "physiology of language." Unfortunately, almost nothing is currently known about this.

RIGHT HEMISPHERE AND PATTERNS OF STIMULATION

Studies comparing the abilities of the right and left hemispheres indicate that in most people, the right hemisphere seems to have an advantage in the recognition and memory of patterns of stimulation. The patterns may be visual, as when we recognize and remember a face; tactile, as when we recognize and remember complex patterns by touch; or spatial, as when we recognize and remember where things are in space in relation to each other. (We shall have more to say about these spatial abilities in the next section.)

DIVIDED BRAINS

Perhaps the most dramatic evidence for different functions of the left and right cerebral hemispheres comes from studies of people who have undergone operations cutting the connections between the hemispheres. Several bands of nerve fibers connect the left and right sides of the brain. Among the most important of these are the corpus callosum and the anterior commissure (Figure 2.11, page 54). These connections join areas of one hemisphere to corresponding areas of the other one. When the connections between the hemispheres are intact, information is passed back and forth between them. When the connections are cut in operations to control epilepsy, transfer of information cannot occur; thus we can discover what each isolated hemisphere can do (Sperry, 1974).

Visual input can be restricted to the left or right hemisphere because of the anatomical connections between the eyes and the brain. (See Chapter 3, page 90.) Input from the right visual field goes to the left hemisphere, while input from the left visual field goes to the right hemisphere (Figure 2.20). To make sure that only the left or right hemisphere receives information from the appropriate visual field, the patients must be tested so that they are fixating on the center of a visual display and so that eye and head movements cannot occur.

Figure 2.20

A test to study the behavior of divided-brain patients. (From Sperry, 1974.)

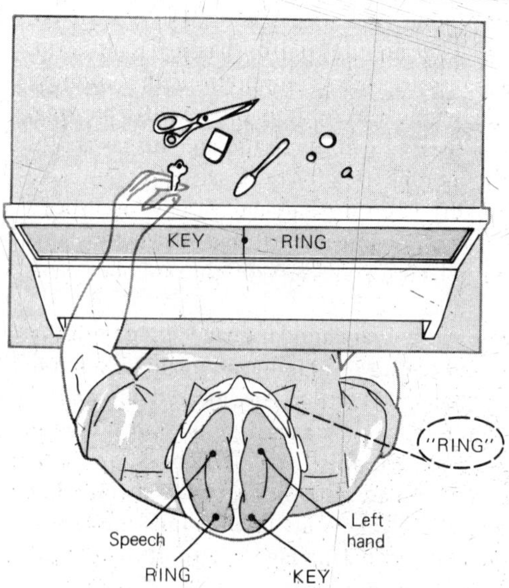

In the test shown in Figure 2.20, the split-brain patient looks at the center of the visual display while a word is flashed so quickly that there is no time for eye or head movements which would allow scanning of the display. In this way, the visual image of the word "ring" reaches the left hemisphere, while "key" reaches the right one. What do split-brain patients do in this test? They verbally report only the information reaching the left hemisphere and thus say "ring." They give no verbal report of the input to the right hemisphere; the "key" part of the word is verbally ignored. This indicates that speech is organized in the left hemisphere, since this hemisphere can verbally report the information it has. In the split-brain patient, the left hemisphere does not receive information from the right hemisphere that it can incorporate in its verbal report. Intact people, of course, say they see "keyring" because the information in the right hemisphere does reach the verbal, or left, hemisphere.

However, the right hemisphere can show it received the "key" input if it is given a chance to express itself in a nonverbal way. Again, the anatomy of the situation is crucial. Touch input is crossed from parts of the body below the head; this means that information from the left hand reaches the right hemisphere. Thus, as shown in Figure 2.20, when a patient reaches behind the screen with the left hand, the right hemisphere receives sensory pattern information which it can match with the visual input ("key") it has received. The patient correctly identifies the key by touch. Now suppose the experimenter asks, "What did you just touch?" The left hemisphere processes this verbal request, and since this hemisphere contains the "ring" part of the visual input, the patient answers "ring."

So we see again that one hemisphere, usually the left one, is specialized for processing language. The other hemisphere, usually the right one, carries out an analysis of patterns of sensory information that come to it.

The differences between the hemispheres that we have been discussing have led to some interesting speculations on the types of thinking done by the left and right brains. The left (verbal) brain is said to be the analytical, logical, mathematical hemisphere, concerned with cause-and-effect scientific thinking. The right hemisphere, involved as it is with images, patterns of sensory input, and the synthesis of information, is said to be the part of the brain essential for the kind of thinking that goes into the creation of a painting or a musical composition. Some people are supposed to be "left-brained"—scientists, accountants, physicians, and the like; others of us are said to be "right-brained"—artists, composers, architects, and so on. Rare individuals such as Leonardo da Vinci may have been both "left-brained" *and* "right-brained." But perhaps such ideas go a little too far. To be more conservative, we may simply say that there seems to be some hemispheric specialization of function.

Summary

1. This chapter is about the contributions of biology to behavior. One aspect of this is the study of species-typical behaviors; another aspect of the biology of behavior is the study of the behavioral functions of the nervous system, particularly the brain. The branch of psychology which seeks to determine how activity in the nervous system is related to behavior and mind is called physiological psychology.

2. Ethology is the study of the species-typical behavior patterns of animals, the evolution of these patterns, and how they serve to adapt a species to its environment. Species-typical behaviors are displayed by all normal members of the species under the appropriate conditions; they are relatively fixed and inflexible behavior patterns; and they are often triggered by events, or stimuli, in the environment called releasers.

3. While species-typical behaviors are not a prominent part of the behavior patterns of human beings, our human nature, or species heritage, probably predisposes us toward certain types of behavior. In other words, we may have built-in, or programmed, biases toward certain general types of behavior, but, within the general framework provided by human nature, particular behaviors are learned. Sociobiology stresses innate, genetic predispositions.

4. To understand the brain's role in behavior, we first need some knowledge of nerve cells, or neurons; the connections between neurons, or synapses; and the chemicals, or neurotransmitters, active at synapses.

5. Neurons are the information carriers of the nervous system; they are often long cells with dendrites and an axon extending from the cell body. Electrical nerve impulses, known as spikes, are carried along neuron axons; spikes occur because of flows of charged particles, called ions, across the cell membrane of the neuron.

6. The areas of functional contact between neurons are called synapses. Enlargements of the axon endings of the transmitting neurons, called boutons, contain neurotransmitter chemicals which are stored in small vesicles. A nerve impulse reaching these boutons causes neurotransmitter to be released into the synaptic cleft and then to excite or inhibit the receiving neuron.

7. Some of the major neurotransmitters are: acetylcholine, dopamine, epinephrine, norepinephrine, serotonin, gamma-aminobutyric acid, glycine, and glutamic acid. The stages in neurotransmitter action are: neurotransmitter manufacture, storage of the neurotransmitter in the vesicles of the transmitting cell, release of the neurotransmitter from the vesicles, diffusion of the neurotransmitter across the synaptic cleft, combination of the neurotransmitter with receptors of the receiving cell, and deactivation of the neurotransmitter. Psychoactive drugs generally work by influencing one or several of these stages.

8. The nervous system is divided into two major parts: a peripheral system and a central system. The brain and spinal cord constitute the central nervous system. The peripheral nervous system has two divisions: the somatic nervous system and the autonomic nervous system. The autonomic nervous system, in turn, has two components: the sympathetic system and the parasympathetic system.

9. The spinal cord and brain stem are like a long stalk protruding from the forebrain. They control and regulate many bodily functions—such as breathing—that are necessary for life. In addition, many simple adaptive bodily responses—reflexes—are functions of the spinal cord and the brain stem. Going upward from the spinal cord, the main regions of the brain stem are the medulla, the pons, and the midbrain. Off to the back of the brain is a large complex structure called the cerebellum; it is involved in helping make our bodily movements precise, coordinated, and smooth. The reticular formation runs from the medulla to the midbrain in the center, or core, of the brain stem; part of this structure, known as the ascending reticular activating system (ARAS), is involved in the regulation of various degrees of arousal, from deep sleep to alert awareness of the environment.

10. Major structures in the forebrain are the thalamus, the hypothalamus, and the cerebrum. The thalamus lies just above the brain stem; it relays and processes sensory information on its way to the cerebrum. The hypothalamus is positioned just below the thalamus and is especially concerned with certain kinds of motivated behavior—hunger, thirst, and sexual behavior, for example. A link between the nervous system and the glandular system of the body is provided by connections between the hypothalamus and the pituitary gland.

11. The cerebrum consists of two hemispheres—the cerebral hemispheres. Covering each hemisphere is a thin, folded sheet of neurons known as the cerebral cortex (gray matter); underneath the cortex, the cerebral hemispheres contain many fiber tracts (white matter) interconnecting parts of the hemispheres, bringing sensory information in, and conducting nerve impulses for movement of the body out. Grooves in the cerebral cortex are called sulci or fissures; ridges are known as gyri. The deepest grooves mark off the lobes of the cerebral cortex—the frontal lobe, the temporal lobe, the parietal lobe, and the occipital lobe. Buried within the hemispheres are clusters of gray matter known as the basal ganglia, and structures such as the hippocampus and the amygdala.

12. When considering the functions of the cerebral cortex in behavior and mental processes, it is convenient to divide it into sensory areas, motor areas, and association areas. The primary sensory areas receive input which originates in the sensory receptors. Visual sensation is represented in the occipital lobe; hearing and taste in the temporal lobe; and touch, pain, and pressure inputs reach the parietal lobe. The principal motor, or movement, area of the cortex

is in the frontal lobe; this area is organized so that various parts of it are concerned with movement of particular bodily structures. The association areas receive input from many regions of the cerebral cortex and from lower portions of the central nervous system; they are involved in such complex functions as perception and the production and understanding of language.

13. The limbic system consists of structures in the thalamus, hypothalamus, and cerebrum which form a ring around the lower part of the forebrain. Major structures within this system include the olfactory (smell) bulb, the septal nuclei, the hippocampus, the amygdala, and the cingulate gyrus of the cerebral cortex. Portions of the limbic system play a role in the expression of emotion and in the formation of memories.

14. Much of what is known about the functions of association cortex comes from studies of brain-damaged people. When frontal-lobe association cortex, known as prefrontal cortex, is damaged, two general types of personality change occur: (1) increased impulsiveness, lack of restraint, and immaturity in social relationships and (2) increased apathy, loss of initiative and drive, and a reduction in emotional expression. The intellectual changes after prefrontal damage are subtle, but analysis indicates that the prefrontal regions have much to do with the ability to plan, to carry out sequences of actions, and to keep track of when actions and other events occur in relation to one another.

15. Parietal-lobe association cortex is, to a degree, lateralized in function. Damage to the right parietal association cortex may result in neglect of the left side of the body and the environment—contralateral neglect. Accuracy in representing spatial relationships is also impaired. After damage to the left parietal association cortex, the patient may have problems with reading, writing, and mental arithmetic; short-term memory impairment may also occur. Tactile agnosia (an inability to recognize objects by touch) may result from parietal-lobe damage in either the right or left hemisphere.

16. Auditory or visual agnosias (impairment of the ability to recognize a tune or a face, for example) and attentional problems are sensory symptoms of temporal-lobe association cortex damage. If Wernicke's area of the left temporal lobe is damaged, the understanding of speech and written language will be impaired, and the speech of the patient, while fluent, will be deficient in conveying meaning. If the hippocampus and amygdala, structures in the lower part of the temporal lobe, are damaged, there will be an impairment in the consolidation of short-term memory into long-term memory.

17. Much evidence shows that the cerebral hemispheres are specialized to perform somewhat different functions. One hemisphere, usually the left, is specialized for the processing of language; the other hemisphere, usually the right, has only rudimentary language capabilities but seems to have an advantage over the left hemisphere in pattern recognition and spatial abilities.

18. Studies of various kinds of aphasias (language difficulties following brain damage) give us information about how the language hemisphere is organized. When Wernicke's area in the temporal lobe is damaged, comprehension of language is impaired, as is the formulation of sentences; speech is fluent, but what is communicated is vague and deficient in specific references. In contrast, when Broca's area in the left frontal lobe is damaged, comprehension of language is intact. The main difficulty is with the fluency of speech; the Broca patient speaks slowly, with great effort, and articulates the sounds of speech poorly.

19. In most people, the right hemisphere seems to be specialized for the recognition and memory of visual, auditory, tactile, and spatial patterns of stimulation.

20. Some of the most compelling evidence for the lateralization of cortical functions comes from study of people who, for the control of severe epilepsy, have had the pathways—the corpus callosum and the anterior commissure—connecting the right and left hemispheres cut.

Terms to Know

One way to test your mastery of the material in this chapter is to see whether you know what is meant by the following terms:

Physiological psychology (38)
Ethology (39)
Species-typical behaviors (39)
Releaser (39)
Fixed-action pattern (FAP) (39)
Sociobiology (43)
Neurons (46)
Synapse (46, 48)
Neurotransmitter (46, 49)
Dendrite (46)
Axon (46)
Myelin sheath (47)
Cell membrane (47)
Nerve impulses (47)
Threshold (48)
Synaptic cleft (48)
Bouton (48)
Vesicle (48)
Excitation (48)
Inhibition (48)
Enzymes (50)
Psychoactive drugs (50)
Central nervous system (CNS) (52)
Peripheral nervous system (PNS) (52)
Somatic nervous system (52)
Autonomic nervous system (52)
Sympathetic system (53)
Parasympathetic system (53)
Brain stem (53)
Medulla (53)
Pons (53)
Midbrain (53)
Motoneuron (53)
Reflex (53)
Cerebellum (53)
Reticular formation (53)
Ascending reticular activating system (ARAS) (53)
Electroencephalogram (EEG) (55)
Paradoxical sleep (56)

Thalamus (57)
Nucleus (57)
Hypothalamus (57)
Pituitary gland (57)
Internal environment (58)
Hormone (58)
Releasing factors (58)
Endocrine gland (58)
Cerebrum (58)
Cerebral hemispheres (58)
Longitudinal fissure (58)
Cerebral cortex (58)
Gyrus (58)
Sulcus (58)
Fissure (59)
Central sulcus (59)
Frontal lobe (59)
Lateral fissure (59)
Temporal lobe (59)
Parietal lobe (59)
Occipital lobe (59)
Corpus callosum (59, 72)
Basal ganglia (60)
Motor areas of cortex (60)
Association areas (60, 63)
Primary sensory areas (60)
Limbic system (62)
Prefrontal cortex (63)
Contralateral neglect (65)
Agnosia (66)
Hippocampus (66)
Amygdala (66)
Wernicke's area (67, 70)
Amnesia (69)
Broca's area (70)
Aphasia (70)
Arcuate fasciculus (71)
Angular gyrus (71)
Dyslexia (71)

Suggestions for Further Reading

Ethology: The Mechanisms and Evolution of Behavior by James L. Gould (New York: Norton, 1982) is a readable, modern textbook account of the biological basis of animal behavior. If you want to know more about the ideas of sociobiology, *The Whisperings Within* by David P. Barash (New York: Harper & Row, 1979) is an entertaining, popularly written introduction with many interesting examples to support the sociobiological principle of the control of behavior by the genes.

A number of college textbooks deal with the substance of physiological psychology—brain-behavior relationships. Representative of these are (1) *The Physiology of Behavior* (2d ed.) by Neil R. Carlson (Boston: Allyn and Bacon, 1980); (2) *Biological Psychology* (2d ed.) by James W. Kalat (Belmont, CA: Wadsworth, 1984); and (3) *Physiological Psychology* by Mark R. Rosenzweig and Arnold L. Leiman (Lexington, MA: Heath, 1982). *The Brain* (San Francisco: Freeman, 1979) is a *Scientific American* book containing articles by a number of prominent brain scientists. Included are summary articles about neuroanatomy, neurochemistry, neurophysiology, and physiological psychology—written in *Scientific American* style, and hence not too technical or too difficult to understand with a little effort.

Drugs, Society, and Human Behavior (3d ed.) by Oakley S. Ray (St. Louis: Mosby, 1983) is a good introduction to psychoactive drugs. It is easy to read and contains interesting social and historical background information about psychoactive drugs, as well as technical accounts of the ways in which they act and their behavioral effects.

For a well-written, popular introduction to the effects of brain damage on behavior, *The Shattered Mind: The Person After Brain Damage* by Howard Gardner (New York: Knopf, 1975) is recommended. The same subject is treated on a more technical level in the textbook *Fundamentals of Human Neuropsychology* (2d ed) by Bryan Kolb and Ian Q. Whishaw (New York: Freeman, 1985). If you want to know more about right and left cerebral hemisphere functions, *Left Brain, Right Brain* (rev. ed.) by Sally P. Springer and George Deutsch (New York: Freeman, 1985) should answer most of your questions.

Sensory Processes and Perception

HEADING OUTLINE

Chapter 3

Do you agree with this quotation from the British philosopher Thomas Hobbes?

> For there is no conception in . . . mind, which hath not at first, totally, or by parts, been begotten upon the organs of sense.

But what about extrasensory perception? Hobbes to the contrary, is there any way of knowing about the world in which the information does not come through the senses? Some people think that extrasensory perception (ESP) is possible.

The desire to believe that human beings have abilities beyond those which are usually acknowledged may explain, in part, why reports about individuals possessing extrasensory-perceptual powers have received wide coverage in the media. Over the years, claims have been made that certain people are capable of such things as telepathy (the ability to read other people's thoughts), clairvoyance (the ability to gain knowledge of an event by some means other than the normal senses), and precognition (the ability to see into the future). These reports make fascinating reading, and many people believe strongly that ESP is a fact.

Despite many dramatic instances of what appears to be ESP, most psychologists, and the "scientific establishment" in general, remain skeptical. One of the major reasons for caution has to do with the question of the repeatability of ESP observations. Only observations that can be repeated, or replicated, become part of the established body of scientific knowledge. It is here that the results of ESP research fall down. It often happens that the same subjects who do well in ESP experiments on one occasion do poorly at other times under the same experimental conditions. ESP researchers explain failures of replication by saying that it is in the very nature of ESP to be sometimes weak and sometimes strong. But this argument is not convincing because it assumes the very thing that is being studied—the existence of ESP. Indications of ESP remain as elusive as a will-o'-the-wisp, appearing here and there and now and then but not in the repeatable way demanded by science.

Whether we agree or disagree with Hobbes's statement about *all* experience (mind) being "begotten upon the organs of sense," there is little room for doubt about the role of sensory processes for almost all experience. Believers and nonbelievers in ESP alike have no question about the importance of sensory experience—the subject of this chapter.

This is the famous Gateway Arch in St. Louis. It is 630 feet wide. How tall is it?
(*a*) 750 feet (*b*) 780 feet (*c*) 830 feet (*d*) 900 feet (*e*) 630 feet
See page 108 for the answer. (St. Louis Regional Commerce and Growth Association.)

Is the upper white bar longer than the lower one? Measure them. What did you discover? See page 119. (Dr. Richard L. Gregory.)

Is the far cylinder about twice the size of the near one? See page 119. (From J. J. Gibson, 1950.)

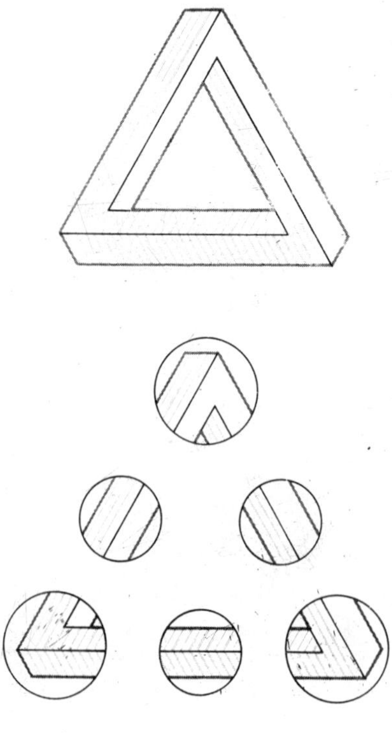

The whole (perception) is more than the sum of its parts. See page 113. The parts (below) of the "impossible" triangle (above) are unambiguous drawings of lines and angles. Yet when the parts are combined a strange new perception emerges—the "impossible" triangle. (Based on Penrose & Penrose, 1958; Lindsay & Norman, 1977.)

BEHAVIOR as we know it, our own private experience, and the reported experiences of others would be impossible without some way of knowing about the world around us. It is through our senses that we know about the world. To appreciate the importance of the sensory processes in behavior and experience, imagine, if you can, what it would be like to be without one or more of your senses.

Some simple experiences, called sensations, are closely tied to what is happening in the sensory systems themselves. Color, brightness, the pitch of a tone, or a bitter taste are examples of sensations. The study of sensations in the laboratory helps us discover how the sensory systems work, but in real life we seldom experience simple sensations—perceptual processes (pages 107–129) are constantly at work to modify sensory input into what we actually experience. In this chapter, we will first consider sensory systems and their related sensations before discussing the perceptual processes that modify sensory input.

Sensory Channels

Vision, hearing, taste, smell, and touch are the so-called five senses. But the number of human senses is closer to ten than five. In addition to touch, the skin contains separate warmth, cold, and pain senses. Furthermore, sense organs in the muscles, tendons, and joints tell us about the position of our limbs and the state of tension in the muscles. They serve the sense called *kinesthesis*. The *vestibular sense* informs us about the movement and stationary position of the head; it is the key sense in maintaining balance.

Each sensory system is a kind of channel, consisting of a sensitive element *(the receptor)*, nerve fibers leading from this receptor to the brain or spinal cord, and the various relay stations and processing areas within the brain. When a sensory channel is stimulated, we have a sensation that is characteristic of that channel. For instance, whether the eye is stimulated by light or by pressure on the eyeball, we have a visual experience.

In order for us to know about the world around (and within) us, physical energy must be changed into activity within the nervous system. The process of converting physical energy into nervous-system activity is called *transduction*. Transduction occurs at the *receptors*—cells which are specialized for the most efficient conversion of one kind of energy. In general, during the transduction process, receptor cells convert physical energy into an electric voltage, or potential, called the *receptor potential*. In some sensory systems, the receptor potential itself directly triggers the nerve impulses (Chapter 2, page 47) that travel to the brain or spinal cord. In other sensory systems, the receptor potential leads to further electrical events, which in turn trigger nerve impulses. Whether it is the receptor potential itself or some other voltage, the electrical event that triggers nerve impulses is known as the *generator potential*.

For a given event in the environment, thousands of nerve impulses

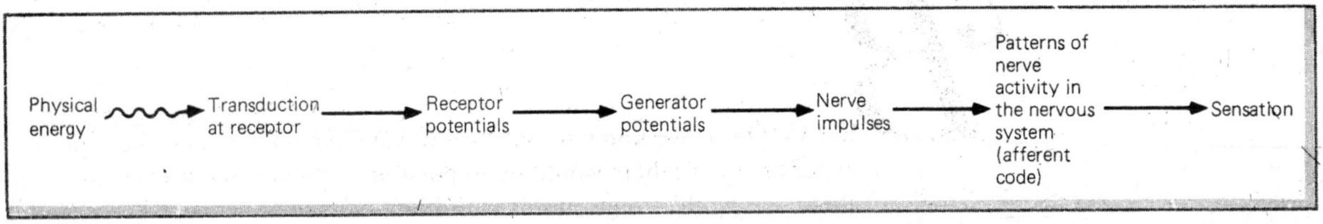

Figure 3.1

In a typical sensory channel, these are the steps in going from physical energy to sensation.

are generated and conducted to the central nervous system. Since these impulses travel along many different nerve fibers at slightly different times, they form a pattern of input to the central nervous system that is the basis of our sensory experience of the event. Thus, beginning with the transduction process at the receptor, physical energy results in a pattern of nerve impulses in the central nervous system. In other words, the physical energy is changed into a code made up of a pattern of nerve firings. The firing patterns that correspond to events in the environment are known as *afferent codes* (the word *afferent* in this context means "input"). As you will see later in this chapter, some progress has been made in deciphering the afferent codes corresponding to certain sensory experiences. Figure 3.1 is a summary of the steps leading from physical energy to sensation in an idealized sensory channel.

Before much was known about the detailed steps occurring in sensory channels, psychologists had already discovered much about the relationships between physical events at one end of the sensory channels and sensation at the other end. (See Figure 3.1.) In general, this is what makes up the field of study known as *psychophysics (psycho* = mind or experience; *physics* = physical events). We still have much to learn about these stimulus-experience relationships, so psychophysics continues as an active area of research in psychology.

Sensory Processes: Vision

Vision starts with the electromagnetic radiation that objects emit or reflect. Physicists have described this radiation in great detail, but for our purposes we can think of it as electric charges moving through space at approximately 300 million meters per second (about 186,000 miles per second). Electromagnetic radiation has wavelike properties, and it is therefore conventional to speak of it in terms of electromagnetic waves.

Electromagnetic waves can be measured and classified in terms of the distance from the peak of one wave to the peak of the next—that is, in terms of *wavelength*. Some electromagnetic radiations have wavelengths as short as 10 trillionths of a meter (the gamma rays), some have wavelengths of thousands of meters (radio waves), and all sorts of wavelengths occur in between. (See Figure 3.2.) The entire range of wavelengths is called the *electromagnetic spectrum*.

Although all radiant energy—all wavelengths of the electromagnetic spectrum—is very much the same physically, only a small portion of it is visible. Somewhere in the middle of the range of radiant energies are the wavelengths that we can see (Figure 3.2). These wavelengths are known as the *visible spectrum*. To express wavelength, we use the metric scale; in the visible spectrum, the wavelengths are expressed in billionths of a

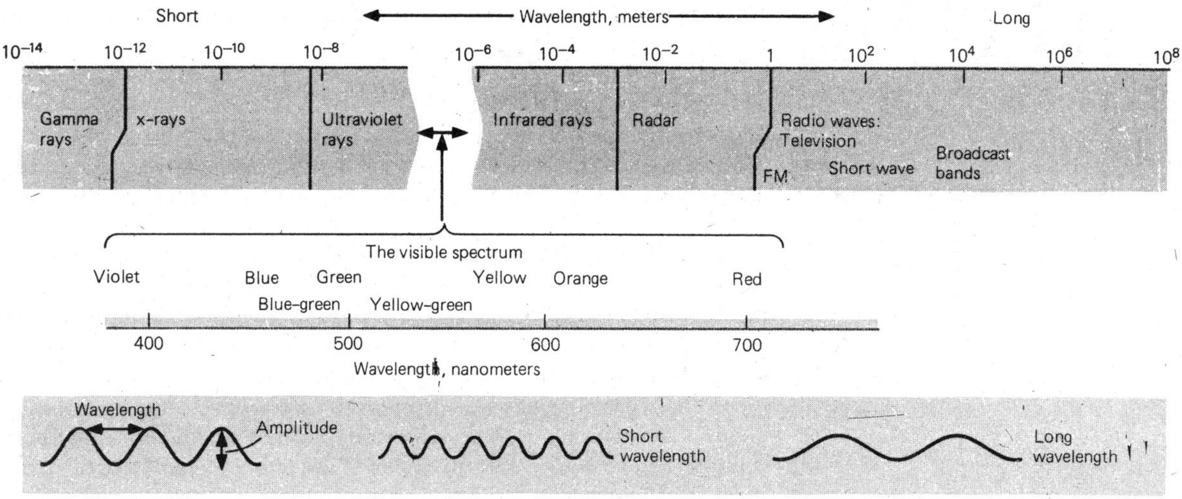

Figure 3.2

The electromagnetic and visible spectra. Electromagnetic waves cover a spectrum from as short as 10^{-14} meters to as long as 10^{8} meters. The part of the electromagnetic spectrum that is visible is called light and is only a tiny fraction of the whole spectrum.

meter, or *nanometers (nm)*. As Figure 3.2 shows, the visible spectrum extends from about 380 to 780 nanometers.

STRUCTURE OF THE EYE AND SEEING

Some of the main parts of the eye are shown in Figure 3.3. Light enters the eye through the pupil, travels through the cornea, the lens, and the interior of the eyeball to strike the rod and cone cells of the retina at the back of the eyeball. Transduction of the physical energy into receptor potentials occurs in the rod and cone cells. Nerve impulses are then generated in certain other cells of the retina—the ganglion cells; these impulses travel to the brain along the optic nerve, and their pattern signals a visual event in the environment.

The amount of light striking the photosensitive rods and cones is automatically adjusted by reflex mechanisms that regulate the size of the pupil—the opening which admits light to the eye (Figure 3.3). The rays of light are bent, or *refracted*, by the cornea and lens (Figure 3.3) to bring them to a focus on the retina. Most of the light bending, or *refraction*, in the eye is done by the cornea; the lens, by changing its shape, simply adds enough to the basic corneal refraction to bring the light from near

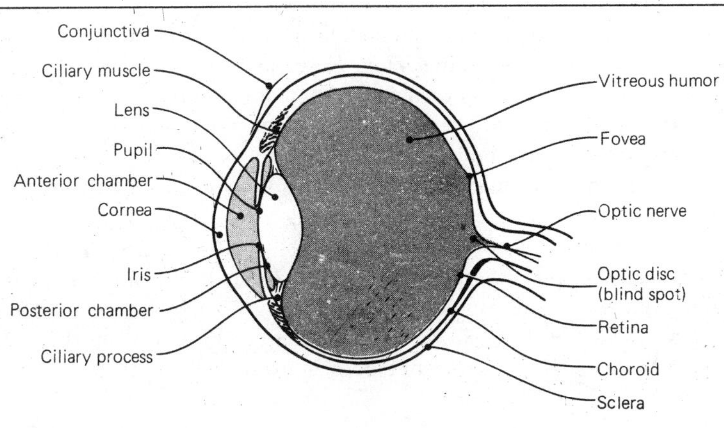

Figure 3.3

Some of the principal parts of the eye. (Based on Walls, 1942.)

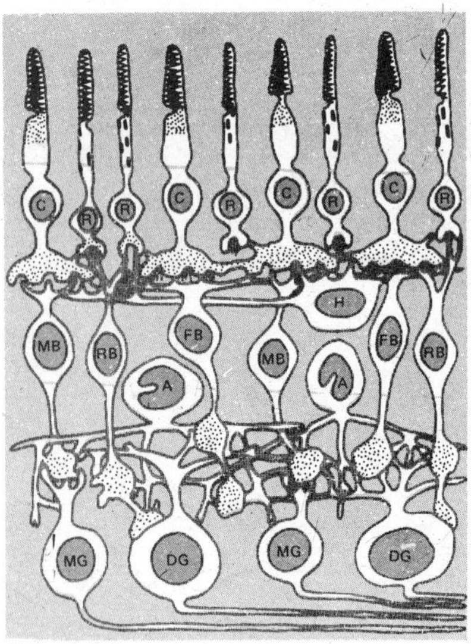

Figure 3.4

A schematic diagram of the cells and connections seen in a cross-section of the retina. C is cone; R is rod; and MG and DG are various types of ganglion cells. Note the long fibers, or axons, leaving the ganglion cells to make up the optic nerve. (Other types of cells in the retina are also labeled: A is amacrine cell; H is horizontal cell; and MB, RB, and FB are various types of bipolar cells. (From Dowling & Boycott, 1966.)

objects to a sharp focus on the retina. These lens changes are termed *accommodation;* correction of accommodation problems is the most common reason for wearing glasses.

THE RETINA AND SEEING

The *retina* (the word *retina* means "network") is a complex sheet of cells and fibers at the back of the eyeball. Figure 3.4 is a cross-section of the retina showing its three main cell layers and the connections among them.

Rods and Cones Two types of cells—*rods* and *cones* (Figures 3.4 and 3.5)—are the light-sensitive elements of the retina where the transduction process begins. The rods are cylindrical in shape, while the cones are rather tapered (Figure 3.5). It is estimated that the human eye contains about 120 million rods and about 6 million cones.

The rods and cones are not spread uniformly over the retina. In the blind spot, for example, there are no rods or cones and therefore no vision is possible. The *blind spot* is the region of the retina where the optic nerve fibers leave and where the blood vessels enter and leave the retina; it is called the optic disc in Figure 3.3. You can find your own blind spot by following the directions in the caption of Figure 3.6. Cones are most numerous in a specialized region of the retina known as the *fovea* (Figure 3.3), which contains no rods at all. The rods occur most abundantly about 20° around the back of the eyeball from the fovea. The *fovea* (from the Latin word meaning "pit") is the part of the retina that we use in looking at objects we wish to see clearly. Visual acuity, or sharpness, is greatest at the fovea, nonexistent at the blind spot, and graded from the fovea out toward the edge of the retina.

In addition to their role in visual acuity, cones have other characteristics distinguishing their functions from those of rods. In fact, it is sometimes said that we really have two visual systems—a cone system and a rod system. This is known as the *duplicity theory of vision.* Cones, for instance, are the retinal elements active in bright light or daylight; rods are the retinal elements active in very dim light. Not only are the cones responsible for the greatest acuity and for daylight vision, they are also the retinal elements necessary for color vision. (We will say more about this later.) Color-blind persons have deficiencies in their cone functioning.

Transduction in Vision The rods and the cones contain what are known as *photosensitive pigments.* When electromagnetic energy in the visible spectrum—light—strikes these pigments, some of the light energy is absorbed by the pigments, and chemical changes occur which initiate the chain of events involved in seeing. Research has shown that the rods and cones have different photochemical pigments, and this helps to explain the rod and cone differences in function. What happens when light energy strikes the photosensitive pigments of the rods and cones? In broad outline, the answer is similar for both rod and cone pigments: Absorption of light energy causes the pigment molecules to change their configuration, or shape, and this process creates electrical energy. For instance, *rhodopsin,* the pigment in rods, exists in the cis-rhodopsin configuration when not excited; excitation by light causes it to change to what is known as the *trans-rhodopsin* configuration. Following in the wake of the shape change is a series of electrical events which result in a receptor

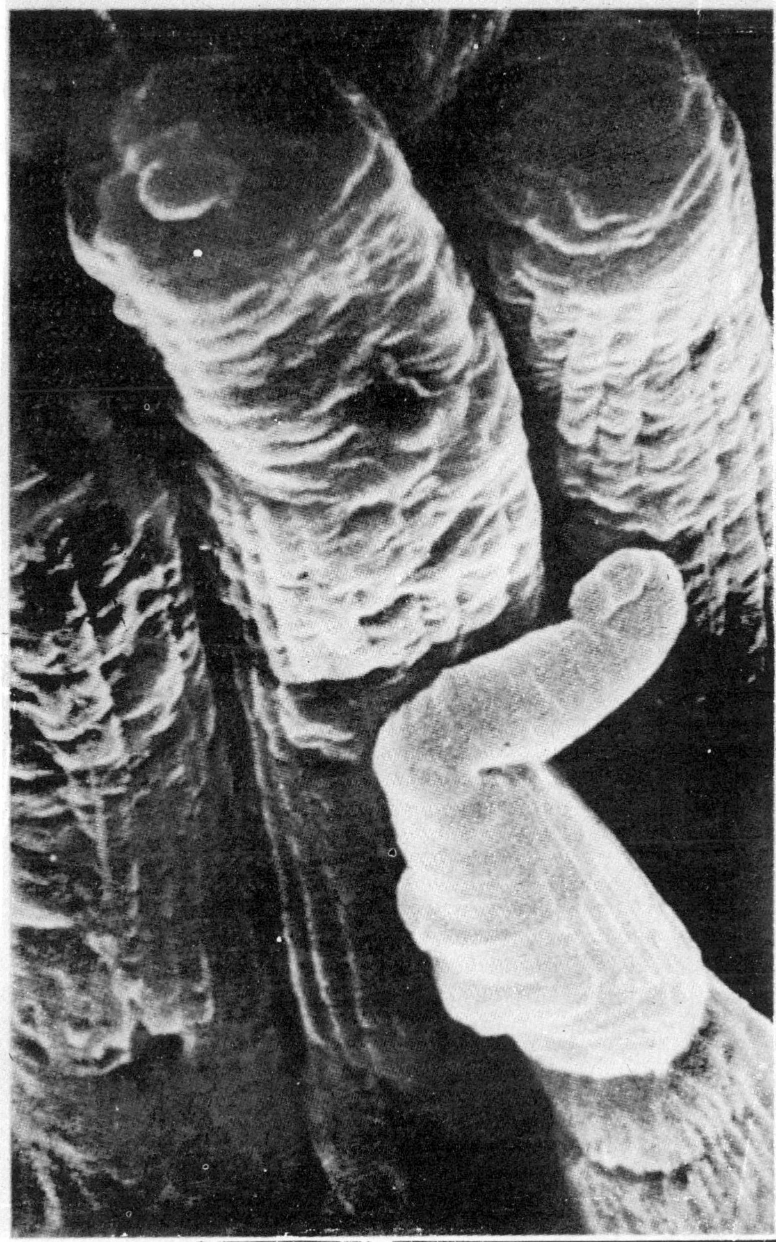

Figure 3.5

An electronphotomicrograph of several rods and a cone. The longer structures at the back are rods; the structure in front is a cone which has been bent in the process of taking the picture. (From Werblin, 1973. *Scientific American.*)

Figure 3.6

To find your own blind spot, do the following: Cover your left eye, and stare at the X while moving the book closer. Move the book slowly; when it is a few inches from your eye, the dot will disappear. You may also need to move the book up and down a little. You must stare hard at the X for this to work. The dot disappears because light from it is falling on the blind spot, which contains no rods or cones.

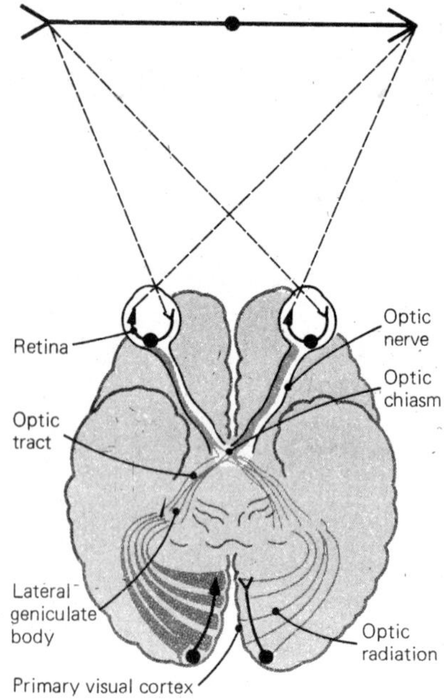

Figure 3.7

The visual pathway. (From Polyak, 1957, as modified by Butter, 1968. Reprinted by permission of The University of Chicago Press and Brooks/Cole. Modified slightly from both versions.)

voltage, or potential, which can be recorded. Through a series of further electrical steps, involving the horizontal, bipolar, and amacrine cells of the retina (Figure 3.4), electrical activity is passed from the rods and cones to the ganglion cells of the retina. The electrical events that have traveled across the retina trigger or generate nerve impulses in the ganglion cells.

THE VISUAL PATHWAY IN THE BRAIN

Ganglion cells have long fibers, or axons (Chapter 2, page 46), that leave the retina through the optic disc to make up the optic nerve. The patterns of nerve impulses in these fibers carry information about the light that struck the rods and cones. The route along which the nerve impulses travel in the brain is shown in Figure 3.7. This figure shows that the axons of the ganglion cells in the optic nerve reach the lateral geniculate body of the thalamus (Chapter 2, page 57). There they make connections, or synapses (Chapter 2, page 48), with cells of the lateral geniculate body. Then fibers from the lateral geniculate cells carry nerve impulses to the primary visual sensory area at the back of the brain (Chapter 2, page 60).

SOME STIMULUS-SENSATION RELATIONSHIPS IN VISION

A visual experience can have *hue* (color), brightness, and form. Further, the hue may be more or less mixed with white—in other words, more or less *saturated*. What is it in physical energy that is related to each of these aspects of visual sensation?

Hue Sensations of *hue*, or color, depend primarily on the wavelength of light. If other things are controlled, we sense a single wavelength, such as might be produced by a prism (see figure following page 109), as a particular color. If several wavelengths are mixed together, as usually happens, hue depends upon the proportions contributed by the component wavelengths. If all the wavelengths of the visible spectrum are mixed together, we experience white light.

The relationship of hue to wavelength is depicted in Figure 3.8, where sensed hues and their corresponding wavelengths are arranged in a

Figure 3.8

The color circle. The diagram shows the arrangement of various hues and their corresponding wavelengths, in nanometers, on a circle. Points opposite each other in the unshaded sectors represent complementary hues in the visible spectrum. Those in the upper left sector have no complements in the visible spectrum.

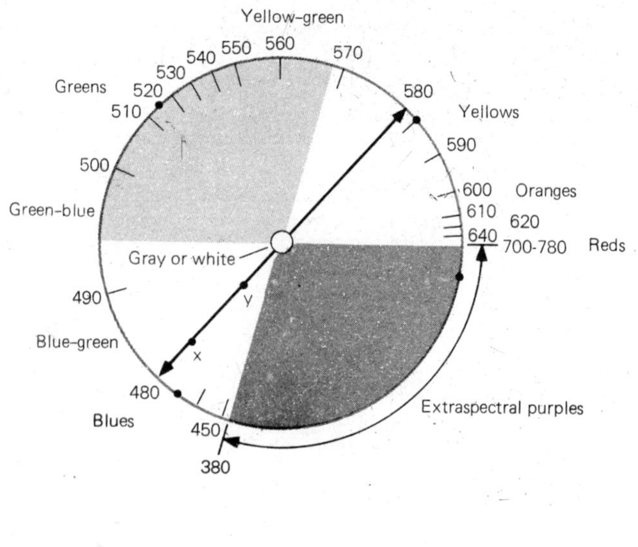

color circle. The hues are shown in a circle to illustrate the *law of complementary colors*. For every hue there is a complementary hue, and complementary hues, when mixed in the proper proportions, produce a sensation of gray or white. In other words, we do not have a color sensation when complementary hues are mixed; all that remains is a sensation of brightness—another dimension of visual experience. Figure 3.8 is so arranged that the complementary colors are opposite each other on the rim of the circle. For instance, the yellows and blues have single complementary wavelengths in the visible spectrum. The greens, on the other hand, do not have complementary wavelengths in the visible spectrum; their complements are actually mixtures of wavelengths from the reds and blues—the extraspectral purples of Figure 3.8. (Blue and red, when mixed together, make purple.) If, as usually happens, wavelengths which are not complementaries are mixed, the resulting hue lies between the two wavelengths along the color circle.

So far we have dealt with colored lights. What about colored paints? A mixture of blue and yellow paint, for example, gives us green, not the white or gray that is produced by a light mixture. The reason is that paint absorbs some of the light striking it. The remaining wavelengths are reflected back to the eye, and it is these reflected wavelengths that give the paint its color. So when two paints are mixed, each absorbs part of the spectrum and what is left to be reflected depends upon both the absorption and the reflectance of the two paints. In a mixture of yellow and blue paint, it is only the wavelengths in the green portion of the spectrum that are not absorbed by the paint. These wavelengths are left to be reflected back to the eye. The rules for color-mixing of paints do not violate the rules for light color-mixing; when mixing paints, it is important to figure out what wavelengths will be reflected back to the eye.

Saturation When hues of light are mixed, the resulting color is different not only in hue but also in saturation. A color's *saturation* is the degree to which it is not diluted by whiteness. For example, if we take a wavelength giving rise to an experience of blue and add white light to it, we do not change the hue— it is still blue. But the hue is paler; it is a "light blue" rather than a "rich blue." Pastel colors are simply unsaturated hues; they are hues to which some white light has been added. So saturation depends upon the ratio of the energy in the dominant wavelength to the amount of white light. In the color circle, the saturation of a particular wavelength is represented along the line from the wavelength on the periphery of the circle to the center of the circle. At the periphery of the circle, the saturation is maximum, and the saturation decreases to nothing as you move along the line from the periphery to the center. For instance, in Figure 3.8, the saturation of a blue of 480 nanometers is less at *Y* than it is at *X*.

Brightness Another major dimension of visual experience is brightness. Other things being equal, the intensity of the physical stimulus is the major determiner of brightness sensations.

The dimension of brightness extends from black to white and through various shades of gray. To represent it along with the dimensions of hue and saturation requires that the two-dimensional color circle be extended into a three-dimensional color solid. (See figure following page

109.) To make a color solid, color circle is piled on color circle like so many layers of cake. In this solid, the up-and-down dimension represents brightness. The colors at the top are bright; those at the bottom are dark. The center line of the solid runs through the centers of the various color circles and represents the points at which there is neither hue nor saturation, only varying brightnesses.

Form Basically, visual sensations of form, or shape, depend upon differences in the amounts of energy focused on different parts of the retina. (Note, however, some of the other important factors in form perception we will describe later in this chapter—pages 112–114.) Thus some parts of a scene are lighter than others, and an outline separates the darker portions from the lighter ones. Put another way, it is the pattern of energy projected on the retina that determines form in vision.

AFFERENT CODES IN VISION

Although scientists are a long way from understanding how the basic qualities of visual sensation just described are represented, or coded, in the nervous system, enough is known so that we can begin to construct working hypotheses about the afferent codes (page 86) for hue, brightness, saturation, and form.

Afferent Code for Hue The retinas of monkeys, apes, and human beings contain three types, or classes, of cones. (See Figure 3.9.) (Animals without such cones are color-blind.) One type absorbs light best in the long-wavelength (yellow) region of the spectrum; a second class of cones absorbs best in the middle-wavelength (green) part of the spectrum; and the third type absorbs short wavelengths (blues) most readily (MacNichol, 1964; Mollon, 1982). The differential pattern of absorption

Figure 3.9

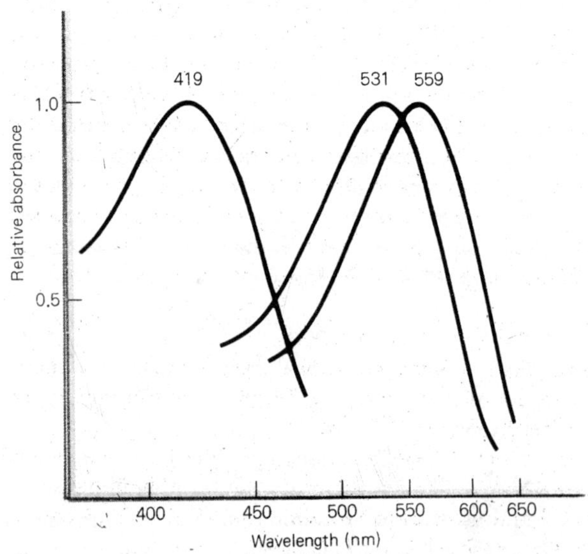

The three kinds of wavelength-sensitive cones of the human retina. One class of cones absorbs light of a short wavelength maximally; a second type absorbs light of a middle wavelength most readily; and the third kind has its peak absorption at a longer wavelength. The difference in light absorption of the three cone types is due to the different photopigments each type of cone contains. (Data provided by H. J. A. Dartnall, J. K. Bowmaker, and J. D. Mollon; modified from Mollon, 1982.) (Reproduced with permission from the *Annual Review of Psychology*, Volume 33, © 1982 by Annual Reviews, Inc.)

Because the afferent coding of color starts with these three different cone types, this aspect of color coding is said to be due to a *trichromatic* ("three") *process*.

The cone receptor potentials (page 85) started by light absorption, after crossing the intermediate layers of the retina (Figure 3.4, page 88), trigger nerve impulses in the ganglion cells (page 90) of the retina. The ganglion cells, the lateral geniculate cells of the thalamus (page 90) with which the ganglion cells connect, and many cells of the visual cortex (page 90) code for color by what is called an *opponent-process mechanism* (DeValois et al., 1966; DeValois & DeValois, 1975). This means that the ganglion color-coding cells and others in the brain are excited by wavelengths in one part of the spectrum and inhibited by wavelengths in another part. For example, the record of electrical activity for the cell in Figure 3.10 shows that (1) the cell stops firing—is inhibited—when the cones are stimulated by shorter wavelengths and (2) the cell increases its firing rate—is excited—with longer-wavelength stimulation. This cell must be wired to the shorter-wavelength-absorbing cones so that they inhibit its activity and to the longer-wavelength-absorbing cones so that they have an excitatory influence. Since the short- and long-wavelength inputs act in opposite ways, they are said to *oppose each other*, hence the term *opponent-processing*. The cell in Figure 3.10 is an example of one of the four main types of ganglion and brain opponent-process cells. The pattern of activity in the four opponent-cell classes differs for various color stimuli, and it is this pattern which is said to be an important part of the afferent code for hue, or color.

Afferent Codes for Brightness and Saturation What are called *nonopponent cells* are also found among the ganglion cells, in the lateral geniculate body, and in the visual cortex (DeValois, et al., 1966; DeValois & DeValois, 1975). These cells are excited by wavelengths from one end of the spectrum to the other. Thus they differ from opponent cells, which are excited and inhibited only by particular wavelengths. A good match exists between amounts of activity in nonopponent cells and the intensity of the physical light energy striking the cones. Thus brightness sensations may be partially coded for by nonopponent-cell activity.

Tentative conclusions can also be drawn about the afferent code for saturation. If activity in nonopponent cells represents brightness, or whiteness, and activity in opponent cells supplies codes for hue, one might expect that mixing a great deal of nonopponent-cell activity with opponent-cell activity would reduce saturation. On the other hand, reducing the relative amount of nonopponent-cell activity might be expected to increase saturation. Recording experiments support this idea: As saturation goes down, nonopponent activity goes up and vice versa.

Afferent Codes for Form Vision We are still far from knowing now the nervous system codes for all the shapes, or forms, that we see. But one idea we should immediately discard is that there is a "little picture"—a copy—in the eye and brain of the visual scene before us. Instead, the retina and brain use differences in light energy (page 86) to generate patterns of nerve-cell activity which *represent*, or stand for, the forms that we see.

Figure 3.10

Each short, spikelike line represents a single nerve impulse fired by a cell in the lateral geniculate body of a monkey. The cell responded differentially to stimulation by lights of different wavelengths. The vertical black lines frame the duration of the light.
Take the bottom line, for instance. It shows the firing rate of the cell before, during, and after stimulation by red light of 706 nanometers wavelength. The firing before the stimulation is the spontaneous, or unstimulated, firing of the cell. Note that wavelengths of 586 nanometers or less inhibit this cell's spontaneous firing rate—no responses occurred during the period of stimulation at those wavelengths. Stimulation by a wavelength of 603 nanometers produced very little, if any, effect. Stimulation by longer wavelengths, however, caused a marked increase in the firing rate.
The cell which produced this record is an opponent cell because of the opposite actions of the longer and shorter wavelengths. The shorter wavelengths inhibited the cell; longer wavelengths excited it. (Based on DeValois et al., 1966.)

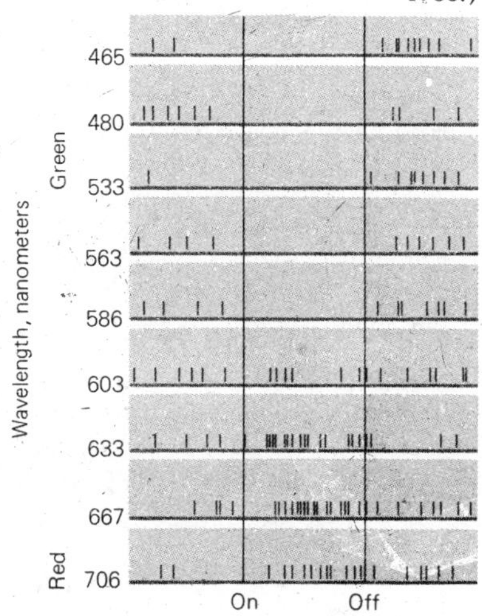

In some animals, the activity recorded from single optic-nerve fibers shows that a great deal of the analysis of form occurs in the retina itself. For instance, recordings from optic-nerve fibers of the frog have found cells that preferentially respond to various shape features of stimuli. Thus in the frog, some optic-nerve fibers respond most readily when small, round shapes are presented to the eye; these same fibers do not respond so vigorously to other shapes. The scientists who made these recordings (Maturana et al., 1960) called these fibers "bug perceivers." In these same studies, other optic-nerve fibers were found to be especially responsive to other features of stimulation.

In higher animals, form vision depends, to a large degree but perhaps not exclusively (Humphrey, 1970), on the activity of nerve cells in the parts of the cerebral cortex (Chapter 2, page 60, and Figure 3.11) that process visual input. Researchers have measured and classified the activity patterns occurring in these cells and have hypothesized how the patterns might be related to form vision.

One hypothesis about cortical nerve cell activity and form vision stems from the observation that most nerve cells in the visual cortex respond best to lines, edges, or bars of light presented to the eye; other types of stimuli are less effective (Hubel & Wiesel, 1968, 1979). Figure 3.12 illustrates the recorded response of a cell in the visual cortex of a monkey. (The visual systems in monkeys and human beings are similar.) The characteristics of its activity are representative of one of the four main types of line, edge, or bar "feature detectors" in the visual cortex. Visual form may be represented, in part, by the distinctive and complex patterns of activity in millions of simple "feature-detector" nerve cells.

Another hypothesis about form vision is based on the observation that groups of cells in the visual cortex respond best to certain spatial frequencies of visual stimulation. Figure 3.13 shows what is meant by *spatial frequency* in vision: There are more alternations between light and dark in the right panel than in the left one; the right panel has a higher spatial frequency. Some cells in the visual cortex respond best to the

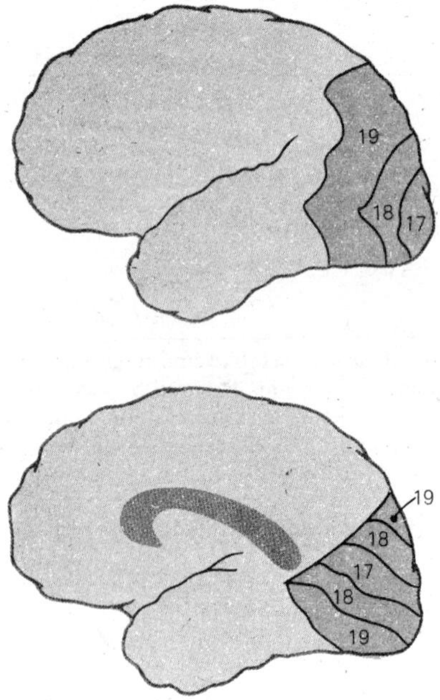

Figure 3.11

Some of the main areas of the human cerebral cortex involved in the processing of visual information. Top, a schematic side view of the human cerebrum (see Chapter 2); bottom, a middle view (see Chapter 2). The numbers refer to areas of the cerebral cortex. Area 17 is the primary visual sensory area of the cortex; it receives visual inputs directly from the lateral geniculate body of the thalamus.

Figure 3.12

Left, orientation of a dark bar focused on the retina of a monkey. The rectangles indicate the region of the retina in which the bar is an effective stimulus for exciting a cell in the visual cortex. The experimenters moved the bar so that its image moved over the retina; the arrows show the directions of movement.

Right, nerve impulses recorded from a cell in the visual cortex. The arrows above the impulse records show the times during which the bar stimulus was present and the direction of its movement. Note that this cortical cell fires most rapidly when the bar is tilted at a certain angle and moves from lower left to upper right. (After Hubel & Wiesel, 1968.)

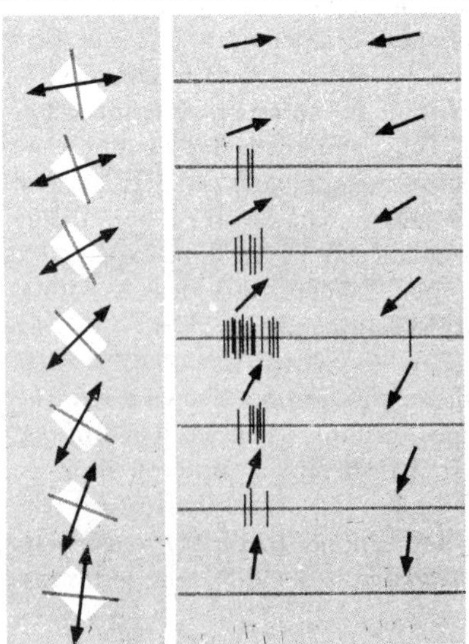

Figure 3.13

Two patterns which differ in spatial frequency—the number of alternations between light and dark. The right panel has more alternations and therefore has a higher spatial frequency than the left one.

higher spatial frequency of light-dark alternations represented at the right of Figure 3.13; other cells respond most vigorously to the lower spatial frequency on the left of the figure; and still other groups of cells respond to other spatial frequencies not depicted in Figure 3.13.

To understand how spatial frequency may be involved in form vision, consider the slice of the visual scene in Figure 3.14; alternations of light and dark in the slice are shown at the right of the figure. Mathematically, a complex wave such as the one shown in Figure 3.14 can be broken down, or analyzed, into a number of component simple wave shapes—sine waves—of various frequencies. The abrupt, sharp changes in light intensity are the high-frequency components of the complex wave, while the more gradual changes are lower-frequency components. Suppose now that the brain does a frequency analysis of the light intensities in the slice of Figure 3.14. The cells that are activated most readily

Figure 3.14

The curve at the right shows alternations of brightness in the slice through the left picture. The rapid changes in brightness are the high-frequency components of the slice; the less-rapid changes are the lower-frequency components. The brain is said to have cells ness frequency components of a scene, and this may provide a basis for form vision. See text. (From *Physiology of Behavior*, by Dr. Neil R. Carlson, 1980.)

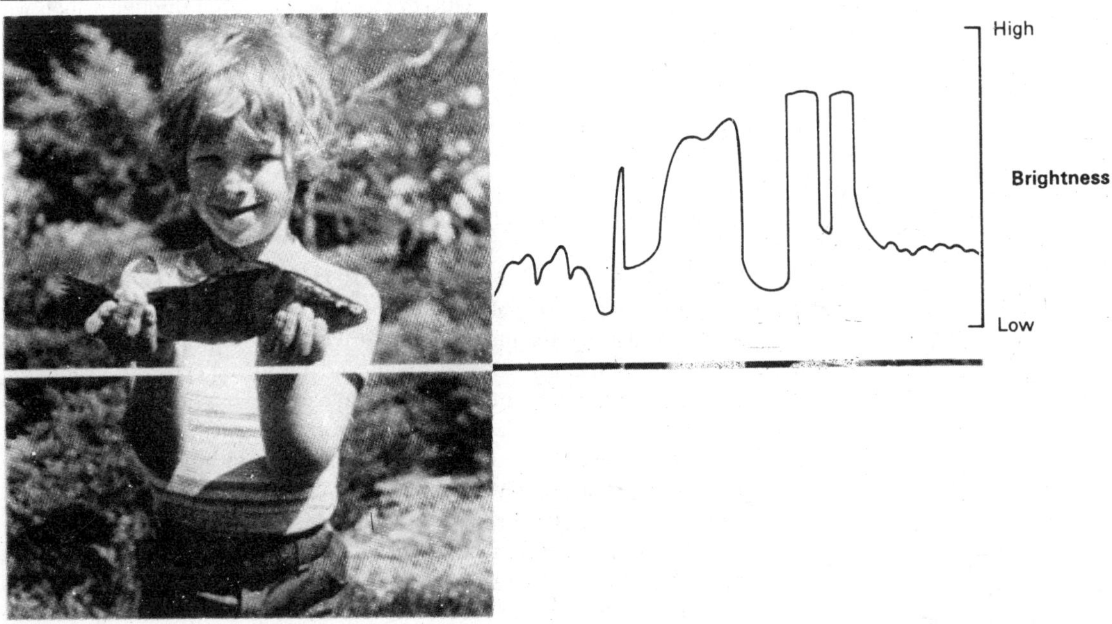

Figure 3.15

Our ears give us one of our greatest joys—listening to music. (Wide World Photos.)

by higher frequencies will fire at some parts of the slice and those most sensitive at other frequencies will fire at other parts. Imagine further that similar analyses are made of slices through the whole picture, and you will begin to appreciate the spatial-frequency hypothesis of form vision, to say nothing of the tremendous complexity of the whole visual brain activity pattern that represents the picture in Figure 3.14.

Sensory Processes: Hearing

Hearing is probably second only to vision as a channel through which we can learn about and appreciate our world. Through hearing, we can understand speech—our chief medium for imparting and acquiring knowledge. (See Chapter 6.) Through hearing, too, we receive a great many signals and cues—the warning automobile horn, the chime of a clock, the fire-engine's siren, the footsteps of a person approaching from behind. Through hearing, we also derive one of our greatest pleasures: listening to music (Figure 3.15).

THE PHYSICAL STIMULUS FOR HEARING

When an object vibrates, the molecules of air around it are pushed together and thus are put under positive pressure. In turn, they push against the molecules close to them, and these molecules transmit the pressure to neighboring molecules. A wave of pressure moves through the air in much the same way that ripples move on the water. (See Figure 3.16, top). However, sound-pressure waves travel much faster than do waves of water; at sea level, and at a temperature of 20°C, they travel at about 760 miles per hour, or approximately 1,130 feet per second.

Most objects do not move, or vibrate, in only one direction when struck. A plucked violin string, for example, vibrates back and forth. As the string moves in one direction, a positive-pressure wave begins to

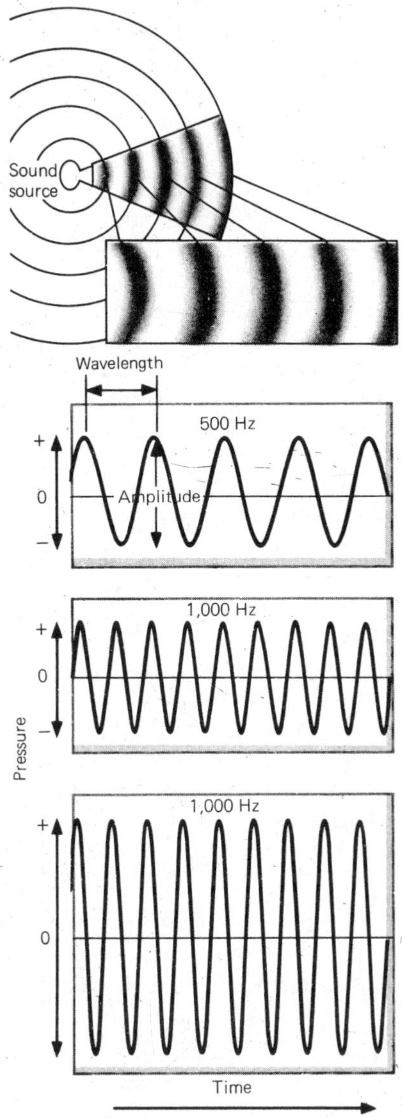

Figure 3.16

Top, a sound wave produced by a sound source. As the sound source vibrates, it alternately compresses and rarifies the air around it. This generates a pressure wave that is transmitted outward in all directions by the air molecules, which impinge on each other and so transfer the pressure. The inset shows a sound wave represented graphically. The darker bands represent peaks of compression; the lighter areas between the bands represent times of rarefaction. The more densely packed the air molecules are, the greater the amplitude of the sound wave. The frequency of the sound wave is measured by the number of peak compressions occurring in 1 second.
Below, simple sound waves. The upper and middle waves have the same amplitude, or pressure, but the middle one has a frequency twice that of the upper. The middle and lower waves have the same frequency, but the lower one has an amplitude twice that of the middle one.

move through the air. But when the string swings back to its original position and beyond, a little vacuum, or negative pressure, is created just behind the wave of positive pressure. The vacuum moves with the speed of sound, just as the positve-pressure wave does. The alternations in air pressure moving in all directions from the source are called *sound waves,* and such sound waves are the physical stimuli for everything we hear.

Intensity and Decibels As shown in Figure 3.16 (below), sound pressure can vary in intensity, as represented by the heights, or amplitudes, of the waves. *Intensity* refers to how great the pressure changes in the wave are, and degrees of intensity are related to the sensation of loudness.

The unit used to measure the intensity of sound pressures is the *decibel* (*dB*). For most practical purposes, we can regard a decibel scale as

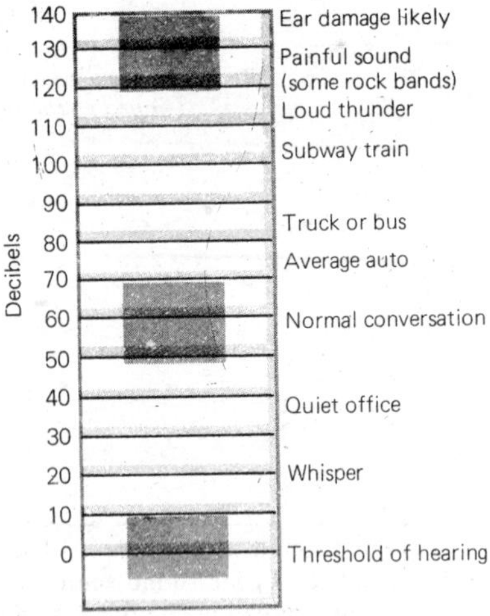

Figure 3.17

Each of the sounds listed at the right has a physical intensity of approximately the number of decibels shown at the left.

simply a set of numbers, like a scale of temperature, and then learn that certain numbers correspond to certain sensations of loudness. To give you an idea of what the numbers mean, Figure 3.17 shows the correspondence between decibels and some sounds with which you are familiar.

Frequency The *frequency* of a sound wave is simply the number of cycles of pressure change occurring within 1 second. One cycle per second is called a *hertz* (*Hz*). In the lower portion of Figure 3.16, the sound wave at the top has fewer pressure changes per second than do the other two waves; it therefore has a lower frequency. To be more specific, if a sound wave goes to positive pressure, then to negative pressure, and back 500 times in a second, its frequency is 500 hertz because the wave has completed that many cycles in 1 second. The physical frequency of sound waves is related to sensations of pitch.

Complex Waveforms The sound waves shown in Figure 3.16 are simple ones. They are known as *sine waves* because their shape can be expressed by the sine function of trigonometry. But most of the sounds we hear in everyday life are the result of *complex waves*. Three examples of such waves are shown in Figure 3.18. Complex waves can take many, many forms, but, in general, they are either periodic or aperiodic. This means that they either have a repetitive pattern occurring over and over again or they consist of waves with various amplitudes and frequencies occurring irregularly. What we call "noise" is usually aperiodic in waveform.

Periodic waves are composed of several sine waves that are multiples of each other. The lowest frequency in such waves is called the *fundamental frequency;* the higher multiples are called the *harmonic frequencies*. For

Figure 3.18

Three complex sound waves. The wave at the top shows what a musical note played on a harmonica looks like. The middle wave is the sustained vowel sound *oh*. These two sound patterns are periodic; the same pattern repeats itself. The sound wave at the bottom in aperiodic; it is the record of a hissing noise and is completely irregular.

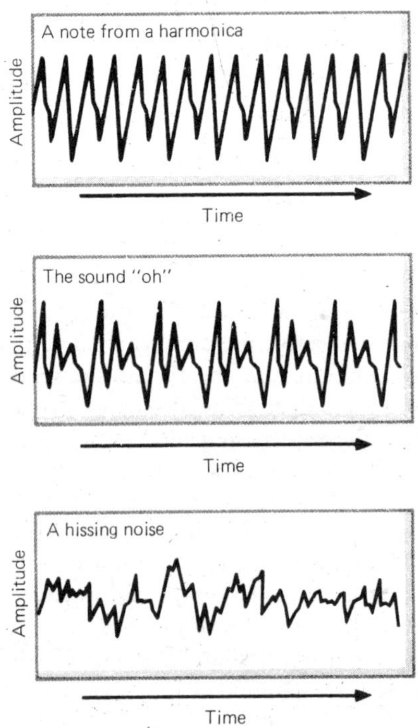

instance, we might describe the periodic tone of a musical instrument by saying that it has a fundamental frequency of 400 hertz and harmonic frequencies, or overtones, at 800 hertz, 1,600 hertz, 3,200 hertz, and so on. Most musical instruments produce complex periodic tones, and the sensed quality of the sound, or *timbre*, we hear is related to the pattern of harmonic frequencies. For instance, even when a violin and a trumpet are playing the same note, the sensed quality of the sounds they make are quite different. While their fundamental frequency—the note—is the same, the harmonic-frequency patterns they make differ greatly.

STRUCTURE OF THE EAR AND HEARING

In order for us to hear, our nervous systems must be set into motion. Physical energy must be converted, or transduced (page 85), into electrical activity by the auditory receptors. The way this is done has been, and is, the subject of intensive investigation. It all begins with mechanical events in the ear.

Figure 3.19 shows the ear's major features. It has three principal parts: the external ear, which collects the energy; the middle ear, which transmits the energy; and the inner ear, where the transduction of energy into nerve impulses actually occurs.

The pinna of the external ear collects energy, which travels through a small air-filled duct called the *auditory canal* to the eardrum. The *eardrum* is a thin membrane stretched tightly across the inner end of the canal. Alternations in the pressure of the sound wave move this small membrane back and forth. The oscillation of the eardrum, in turn, moves three small bones, the *ossicles*, so that vibration is conducted through the middle ear to the entrance of the cochlea in the inner ear. The bones of the middle ear are connected like a series of levers. Hence energy is transmitted mechanically, and amplification takes place through the middle ear.

The inner-ear sense organs for hearing are contained in a bony structure that is spiraled like a snail and called the *cochlea* (from the Latin

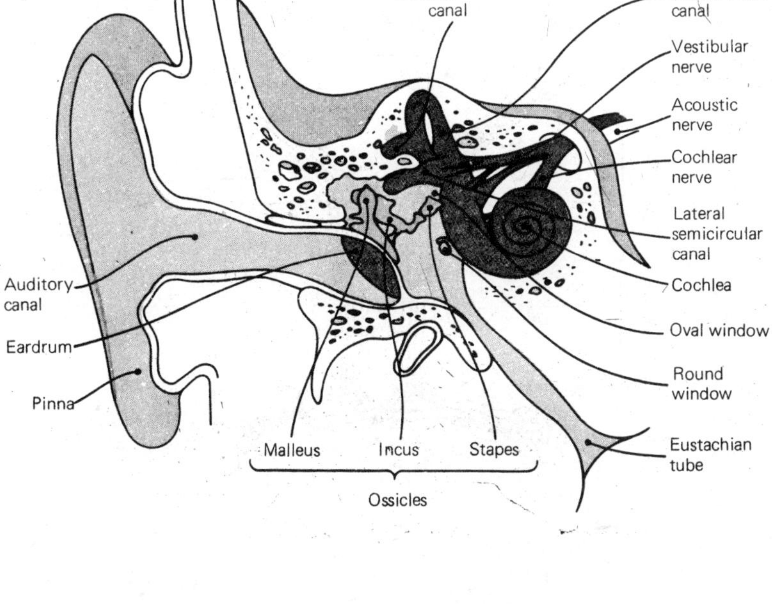

Figure 3.19

Human auditory structures. (After a modification from M. Brödel in *Fundamentals of Neurology*, 6th ed. 1975, by E. Gardner, Philadelphia: Saunders. Labels are somewhat different.)

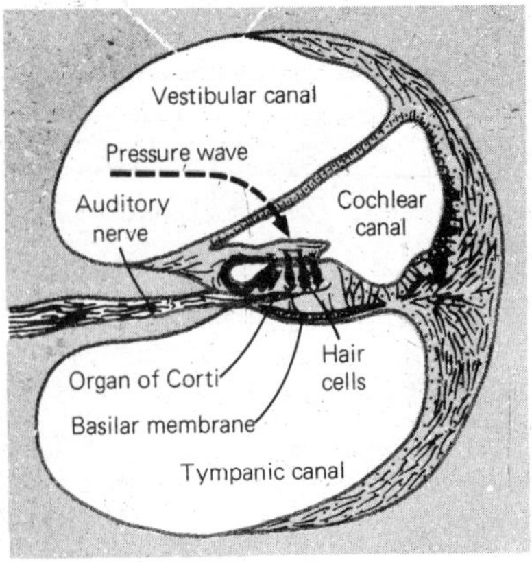

Figure 3.20

A cross-section of the cochlea.

word meaning "snail shell"). The cochlea has three fluid-filled canals spiraling around together and separated from one another by membranes. Figure 3.19 shows a side view of the cochlea; Figure 3.20 shows a cross section of the cochlea and its three canals: the vestibular canal, the cochlear canal, and the tympanic canal.

As the ossicles move back and forth, one of them, the stapes, presses on a membrane called the oval window (Figure 3.19), which seals off the end of the vestibular canal of the cochlea. In this way, when changes in air pressure move the ossicles back and forth, waves are set up in the fluid that fills the canals of the cochlea. The waves in the cochlea reach the *organ of Corti*, which lies on the basilar membrane (Figure 3.20). The pressure waves in the cochlear canals produce bending movements of the fine, hairlike processes on the ends of the *hair cells* of the organ of Corti (Figure 3.20). When these hairlike processes are bent, receptor potentials (page 85) are initiated, thus starting the process by which nerve impulses are generated. In summary, then, the bending of the hair-cell fibers is the event that is responsible, in the auditory system, for the transduction of mechanical energy into nerve impulses.

The nerve impulses initiated in the cochlea travel into the brain and then along certain nerve fibers within the brain. These fibers, and the nerve cells from which they originate, make up what is called the auditory pathway (Figure 3.21). As we shall see after discussing some of the stimulus-sensation relationships in hearing, patterns of activity of nerve cells in this pathway, and in the cochlea itself, comprise the afferent codes (page 86) for what we hear.

SOME STIMULUS-SENSATION RELATIONSHIPS IN HEARING

While the frequency of pressure waves can vary over a wide range, human beings are sensitive to only a relatively narrow band of frequencies.

Figure 3.21

The auditory pathway from cochlea to auditory cortex. A cross-section of the cerebral hemispheres is depicted (see Chapter 2); this shows that the auditory cortex is on the lower bank of the lateral fissure (see Chapter 2). (Schiffman, 1982.)

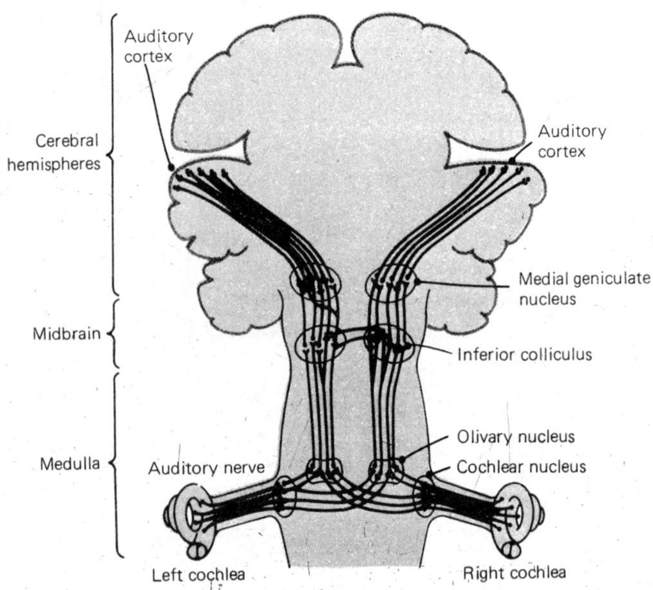

Generally speaking, the audible range for human beings is between 20 and 20,000 hertz. (Other animals have different ranges. Dogs, for instance, can detect higher frequencies than can human beings—hence the "dog whistle," which gives out frequencies above our range of hearing but within the dog's. Bats have an upper limit that extends into the neighborhood of 150,000 hertz.) Within the 20- to 20,000-hertz range, our experience of *pitch* (how high or low the sound seems) depends largely, but not entirely, on the frequency of the sound wave: Low frequencies have low pitches, and progessively higher frequencies give sensations of higher and higher pitch.

Just as frequency corresponds most closely to the sensed pitch of tones, intensity corresponds most closely to loudness. But note that frequency, in addition to intensity, is important in helping to determine how loud something sounds. Depending on its frequency, the same sound pressure (intensity) will be heard as relatively loud, relatively soft, or not at all. Figure 3.22 shows this. A sound pressure of 20 decibels, for example, is well above the threshold for hearing at 1,000 hertz but just about at the threshold near 200 hertz. Since it is well above threshold at 1,000 hertz, it will be heard as louder at this frequency than at 200 hertz, where it is near threshold.

As described earlier (page 99), the psychological counterpart of wave complexity is *timbre*—the tonal quality that enables us to distinguish among different musical instruments and voices. The physical basis of tonal quality, or timbre, is to be found in the pattern of harmonics; musical instruments sound different from each other because of the differences in the harmonic patterns they produce.

AFFERENT CODES IN HEARING

The afferent code (page 86) for loudness may be based on the fact that sense organs usually generate more and more impulses as the intensity of the stimulus increases. The number of impulses is not usually directly proportional to the intensity of the stimulus, but a relationship does exist between the two. So at first glance, it might seem reasonable to assume that the loudness of a tone is related to the number of impulses generated and transmitted to the brain. But the firing patterns of the nerve cells that receive auditory input are extremely complex at the various parts of the auditory pathway. For instance, some nerve cells decrease, or even stop, their firing as stimulus intensity rises. Today, researchers are concentrating on discovering what happens to nerve cells in the brain when the intensity of an auditory stimulus is increased. Thus the afferent code for loudness may become clearer.

For frequencies above about 1,500 hertz, pitch depends upon the fact that different portions of the organ of Corti on the basilar membrane are maximally stimulated by different frequencies (von Békésy, 1960; Khanna & Leonard, 1982). Somehow the brain uses a *place code*—that is, nerve impulses arising from a given region of the organ of Corti are sensed as a particular pitch. This is sometimes called the "pitch is which" theory—the experience of pitch depends upon the place at which the organ of Corti is most stimulated. Thus the afferent code for the pitch of higher frequencies is largely a spatial one. For lower frequencies—20 to about 1,500 hertz—the afferent coding of pitch seems to be accomplished by nerve fibers in the auditory pathway that fire in step with the fre-

Figure 3.22

The threshold of hearing for tones of different frequencies.

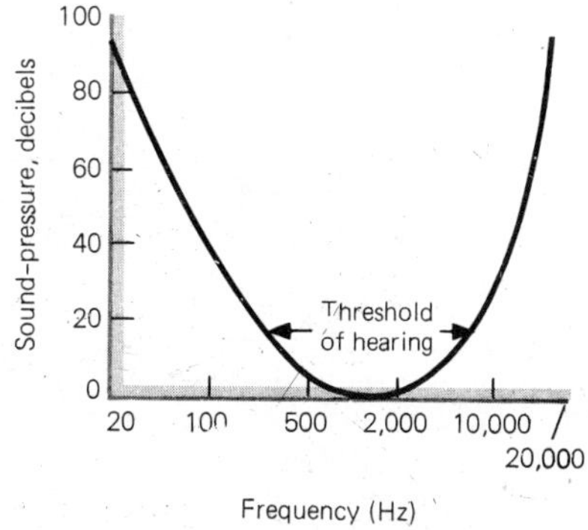

quency of the physical stimulus (Rosenzweig & Leiman, 1982). For example, some of the "phase-locked" fibers may fire at every peak of pressure, while others may fire at every second or third peak. Thus a pattern of firing is established that is distinctive for a given frequency; it is this pattern which is the code for physical frequency in the brain and is correlated with the sensation of pitch for the lower frequencies.

Very little is known about the afferent code for timbre. Just imagine how complex the nervous-system activity must be to code all the harmonics of a guitar note, to say nothing of the music coming from the band in Figure 3.15.

Sensory Processes: Smell

It is through smell, of course, that we detect and experience many of the events in the chemical world that surrounds us. But smell may also have a special role to play in behavior. Smells seem to trigger behavior and start trains of thought; smells judged as pleasant may set off approach behavior, while smells judged as unpleasant may arouse avoidance behavior. And smells can also serve to trigger memories of past emotional experiences.

The receptors for smell respond to chemical substances, especially if those substances are volatile. Smell receptors are located high up in the nasal passages leading from the nostrils to the throat (Figure 3.23). They lie in two small patches, one on the left and one on the right, in the roofs of these passages. Since they are a little off the main route of air as it moves through the nose in normal breathing, our sense of smell is relatively dull when we are breathing normally and quietly. A sudden sniff or vigorous intake of air, however, stirs up the air in the nasal passages and brings more of it to the receptors. This is why animals and people sniff when they are trying to identify an odor.

Figure 3.23

A section through the nose. Air currents inhaled through the nostrils are wafted to the upper part of the nasal cavity, where they stimulate the *olfactory*, or smell, receptors.

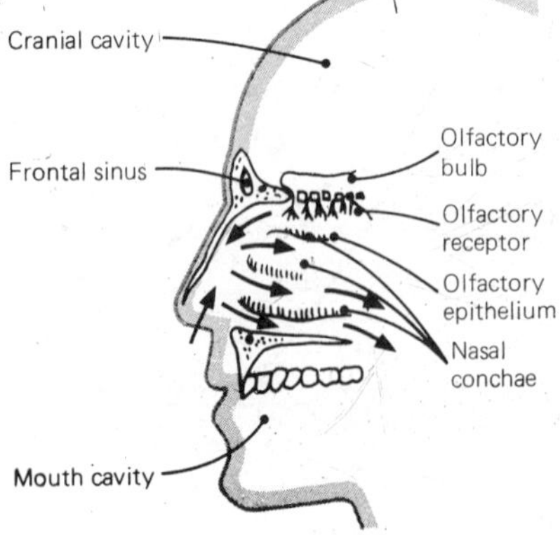

The sensitivity of the smell receptors is impressive. People can detect incredibly small amounts of odorous substances. For instance, artificial musk, one of the most odorous of all scents, can be sensed by human beings in a concentration of 0.0004 milligrams in a liter of air. The smell receptors must be responding to only a few molecules per sniff, and the sense of smell in many animals surpasses that of human beings.

If you recall all of the odors that you encounter in a day, you will realize that they have many shades and qualities. Scientists have raised the question of whether such a multitude of sensations might not result from mixtures of a relatively few primary qualities. Perhaps there are a few unique odors which, mixed in different proportions, might account for the various discriminable odors. If we could identify the basic odors, we might be able to relate the basic odors to particular features of the smell receptor. But smell has not proved to be this simple. A number of basic odor systems have been proposed. For instance, one system says the four basic odors are *fragrant* (musk), *acid* (vinegar), *burnt* (roast coffee), and *caprylic* (goaty or sweaty). Each system serves some particular purpose well—the manufacture of perfume, for example—but there is little assurance that any of them identifies the "real" primary smells, if, indeed, they exist.

The receptors for taste are specialized cells grouped together in little clusters known as *taste buds* (Figure 3.24, right). Most of these buds are located on the top and sides of the tongue, but a few of them are at the back of the mouth and in the throat. If you look at your tongue closely in a mirror, you will notice a number of bumps on it, some large and some small. These bumps, called *papillae*, are richly populated with taste buds. To stimulate the taste buds, substances must be in solutions that will wash around the papillae and penetrate to the taste buds in the clefts between them (Figure 3.24, left).

Taste sensitivity is not nearly so keen as smell sensitivity. For instance, depending upon the taste substance, it takes from 1 part in 25 to 1 part in 2,000 before it can be detected. In general, people are more sensitive to acids and bitter substances than they are to sweet or salty ones.

PRIMARY TASTE QUALITIES

We know more about the primary taste qualities than we do about the primary odor qualities. Several lines of evidence point to four basic taste qualities: *salty*, *sour*, *sweet*, and *bitter*. Part of the evidence for these qualities is the fact that the tongue is not uniformly sensitive to all stimuli. If, for example, we apply minute drops of a bitter solution, such as quinine, to different parts of the tongue, we find the bitter taste most pronounced when the drops are put at the back of the tongue. The taste of sweetness, on the other hand, is most noticeable when sugar solutions are placed on the tip of the tongue. The sides of the tongue respond mainly to sour stimuli, and the tip and part of the sides respond to salty solutions. This, together with other data, supports the idea that there are four primary taste qualities.

If we try to state what kinds of solutions give rise to the different qualities, however, we run into trouble. Sugar, such as common table sugar, tastes sweet, but so do many other chemical compounds (such as saccharine) that chemically have little in common with sugar. A bitter taste presents a similar problem. A class of compounds that the organic chemist calls alkaloids, which includes quinine and nicotine, tastes bitter. But so do substances such as some of the mineral salts, which have little in common with the alkaloids. However, all this may prove only that we

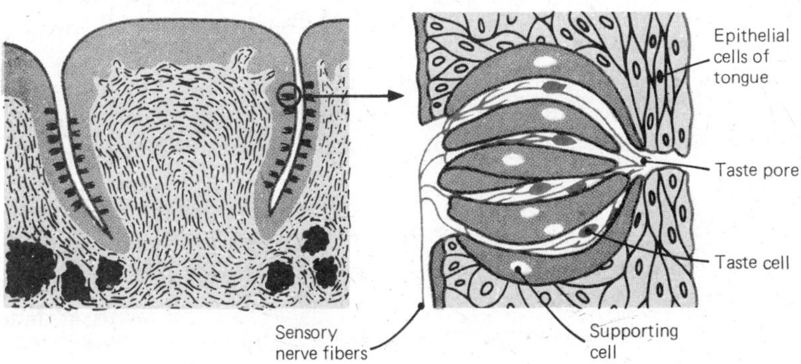

Figure 3.24

Left, a papilla of the tongue showing the location of taste buds in the clefts between papillae. Right, a taste bud. (Left, drawing from Coren et al., 1979; Wyburn et al., 1964.)

have not yet discovered which aspects of a chemical substance are the keys to determining taste quality. We cannot, at present, give definite rules for the kinds of chemical substances that produce sweet and bitter tastes.

In the cases of sour and salty tastes, a somewhat better correlation exists between chemical composition and taste. All the stimuli that taste sour are acid. Moreover, the degree of sourness that we taste is fairly proportional to the total number of acid (H^+) ions present. Salty taste, similarly, is usually aroused by what the chemist calls salts—that is, the chemical product of acids and alkalies. Common table salt, however, is about the only salt that has a uniquely salty taste; most other salts produce experiences of bitter or sweet in addition to that of salt.

AFFERENT CODE FOR TASTE

What is the input, or afferent (page 86), code in the nerves for taste? It can be seen clearly from studies of the electrical responses of single taste-nerve fibers that almost all of them respond to several taste stimuli (Pfaffmann, 1964). Thus the firing of a single fiber is not unique—it may be fired by many stimuli—and therefore a single fiber does not carry unambiguous information regarding taste stimuli to the central nervous system. As far as the brain "knows," the fiber could be firing because any one of a number of stimuli has come in contact with the taste buds. However, the firing of a number of taste-nerve fibers can make up a unique combination for a given stimulus. It seems that the afferent code for a particular taste consists of a *pattern* of firing in the nerve fibers from the taste buds (Erickson & Schiffman, 1975). Thus one taste might be represented by high rates of firing in a certain group of nerve fibers, intermediate rates in another group, and low rates in others. Another taste would have a different pattern, and so on.

Sensory Processes: The Skin Senses

In order to adapt to the environment, we need to know what is happening at the surface of our bodies. The skin senses (or somatosenses, as they are also called) give us this information, and the skin can be thought of as a "giant sense organ" that covers the body. Four skin senses are usually distinguished: *pressure* or *touch*, *cold*, *warmth*, and *pain*. Much of what we receive from the skin senses results in such "simple" sensations as itching, tingling, feelings of hot and cold, or painful sensations of injury. The skin senses, however, are capable of telling us much more than this. We can, for example, identify objects by touch or even learn to read Braille, as the blind do.

The skin is not uniformly sensitive. One of the first things investigators of the skin senses discovered was that the skin has *punctuate sensitivity*, meaning that it is sensitive at some points and not so sensitive at others. And, in general, the spots of greatest sensitivity to touch, cold, warmth, and pain stimuli are different. If we explore the same patch of skin with touch stimuli, warm stimuli, cold stimuli, and stimuli for pain, we find that the sensitive spots for these four types of stimuli are distributed differently over the patch of skin. In other words, there are four separate maps of the sensitive points corresponding to the four types of stimuli we have applied.

PRESSURE OR TOUCH

The sensation a person who is touched lightly on the skin reports is called either pressure or touch. The amount of physical pressure required to produce this experience varies greatly for different parts of the body. The tip of the tongue, the lips, the fingers, and the hands are the most sensitive areas. The arms and legs are less sensitive, and the trunk and calloused areas are the least sensitive of all. We experience touch, it should be noted, not only when some object presses on the skin but also when hairs on the body are slightly moved.

Psychologists have studied carefully what it is about a stimulus that elicits the experience of touch. They wanted to know, in particular, whether it is the weight of an object on the skin or simply a bending of the skin. They have concluded that it is the latter—the deforming, or bending, of the skin. A gradient of pressure, not uniformly distributed pressure, is the stimulus for touch experience.

For more than 100 years, many attempts have been made to determine the receptors for pressure. We think that a fairly complex structure called the *Meissner corpuscle* (Figure 3.25) serves the pressure sense in the hairless regions of the body—the palms of the hands, for example. We think that another structure, the *basket nerve ending*, does the same for the roots of hairs. We also have good reason to believe that simple *free nerve endings*—endings not associated with any special structure—convey touch impulses because people can feel pressure in some areas of the skin where no receptors other than free nerve endings are found. In addition to the sense of touch or pressure on the surface of the body, we are sensitive to deep pressure. The receptors for this sense seem to be small capsules called *Pacinian corpuscles* (Figure 3.25).

TEMPERATURE SENSATION: COLD AND WARMTH

Experiences of cold and warmth are elicited by changes in the normal *gradient* of skin temperature—that is, by changes in the difference between the temperature of the skin surface and the temperature of the blood circulating beneath it. In the case of the forearm, for example, this gradient is about 5°C. The temperature of the skin surface is usually about 32 or 33°C and that of the blood beneath it is about 37°C. A stimulus of 28 to 30°C, which is definitely felt as cold, increases this gradient a little. A stimulus of 34°C, which can be felt as warmth,

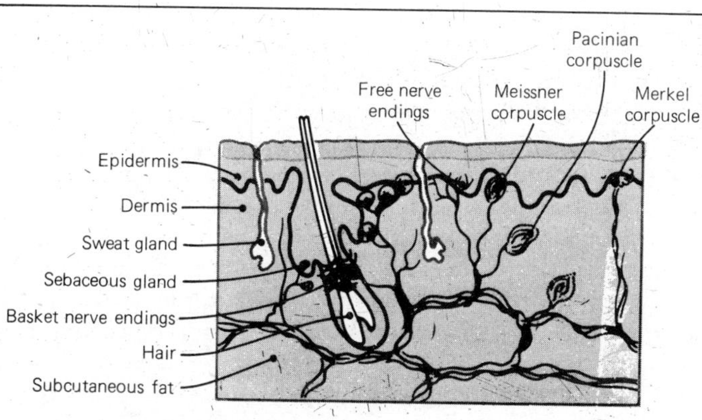

Figure 3.25

A diagram of a cross-section of skin showing some of the major structures. Some sensory elements of the skin are shown: a Meissner corpuscle, free nerve endings, the nerve fibers around the base of a hair, and a Pacinian corpuscle.

decreases it a little. Thus it takes a change in skin temperature of only 1 to 2°C to be experienced as warmth or cold.

In skin maps, the cold spots and warm spots are found in different places. This has been taken to mean that there are two different sensory channels for the experiences of warmth and cold. You might therefore expect different receptors to lie under the cold and warm spots, but this does not seem to be true. Instead, free nerve endings, physiologically specialized but not obviously anatomically specialized, appear to be responsible for signaling information about temperature. Increasing the temperature gradient by cooling the skin causes certain free-nerve-ending fibers to increase their rate of firing. These fibers might be called "cold fibers." And, up to a point, decreasing the temperature gradient by warming the skin causes an increase in the firing of certain fibers. These might be called "warm fibers." Thus the afferent code for the experiences of cold and warmth appears to be the rates of firing in cold and warm fibers.

PAIN

Pain is a skin sense, but of course pain is felt from the interior of the body too. This sense has great significance in human life. It motivates a multitude of behaviors. People will do many things to reduce it, as drug companies have found to their profit. Benjamin Franklin said, "Things that hurt, instruct," and we will discuss in Chapter 4 (page 161) how the reduction of pain can serve as a reinforcer in the promotion of learning. Pain may also trigger aggression against the source of the pain or even against neutral objects in the environment. And, of course, pain has immense biological importance because it may signal that something is wrong with the body.

Pain Stimuli and Receptors Many different stimuli produce pain—a needle prick, scalding steam, a cut, a hard blow to the skin, inflammation and swelling, or strong chemical stimulation of the skin. This pain is called noxious (from the Latin word meaning "to injure") stimulation. What do noxious stimuli have in common? Because of the close relationship between pain and bodily injury, scientists have long been inclined to believe that pain stimuli are stimuli that damage bodily tissues in some way.

There is evidence that the receptors stimulated by this tissue damage are free nerve endings. Since maps of the skin show separate pain spots, the free nerve endings of the pain spots must be specialized in some way to respond to painful stimuli and not to other stimuli. At present, invesigators do not know what is special about the particular free nerve endings that respond to damaging stimulation.

Afferent Codes and Pain "Gates" The nerve fibers that carry information about pain into the spinal cord and brain are the smaller-diameter fibers in the sensory nerves from the skin and body organs. Rates of firing in these small nerve fibers constitute much of the afferent code for pain. But the central nervous system itself has much to say about how much pain is actually felt, or *perceived*. As the pain inputs enter the spinal cord or brain, their transmission for further processing may be blocked. It is as if there are "gates" for pain input which can be closed (Melzack, 1973).

Some pain-killing drugs (analgesics), such as morphine and morphine-like compounds, seem to work because they are able to close these pain gates.

The evidence thus far accumulated appears to show that increases in the activity of nerve cells in the brain-stem areas of midbrain and pons (Chapter 2, page 53) result in the blunting of human pain (Hosobuchi et al., 1977) and in a decrease of pain-elicited responses in animals (Mayer & Liebeskind, 1974). Nerve impulses originating in these brain-stem nerve cells are said to travel to cells in the pain pathway to block, or *"gate,"* incoming pain signals (Figure 3.26). Many of the brain-stem nerve cells involved in the alleviation of pain are activated by morphine and morphine-like compounds (*opiates*). Thus these cells may be where pain-relieving drugs work.

We now know that the body produces chemical substances which can act like morphine on the brain-stem cells of the pain gate (Snyder & Childers, 1979). These self-produced opiates—known as enkephalins and endorphins—may be involved in closing the gates to lessen perceived pain; they may thus provide, as the following examples indicate, a partial explanation of the pain-reducing effects of acupuncture and "sugar pills."

Acupuncture anesthesia is produced by inserting and twisting needles in various regions of the body. The body sites where the needles are inserted are often far from the part of the body to be anesthetized. But acupuncture must work because major surgical procedures are performed in China with nothing more than this form of anesthesia. Some evidence indicates that the opiate gate system may be involved in acupuncture (Mayer et al., 1976). And there is speculation that acupuncture procedures may stimulate the body to release its own opiates, which in turn activate the gate system, but the issue of how acupuncture works is by no means settled.

A *placebo* is a pharmacologically inactive substance that a person is told will have a desired effect. Sugar pills are examples of commonly administered placebos. For those who believe in the effectiveness of the placebo, considerable pain relief is possible, and there is evidence linking the placebo effect with the opiate brain systems (Levine et al., 1979).

Hypnosis (Chapter 16, page 708) can also be used to relieve pain, but it does not seem to operate through the opiate pain gate. Nor do we know how the other psychological pain relievers operate. For instance, pain input may be blocked from the focus of attention, as when a hockey player feels little pain from a severe cut while concentrating on scoring a goal. While, as we will see later (page 109), this might be considered an example of filtering in attention, it is not known whether such blunting of pain involves the opiate gate.

From Sensory Processes to Perception

Perception refers to the way the world looks, sounds, feels, tastes, or smells. In other words, *perception* can be defined as whatever is experienced by a person.

The first part of this chapter dealt with simple perceptions, or sensations, that are closely linked to patterns of activity in the sensory channels themselves. However, for much of what we perceive, the sensory-input patterns provide only the raw material for experience. For

Figure 3.26

A schematic diagram of the pain-gate system in the brain and spinal cord. Arrows indicate the direction of nerve impulses. The "gating" occurs where the pain fibers enter the spinal cord. (See text.)

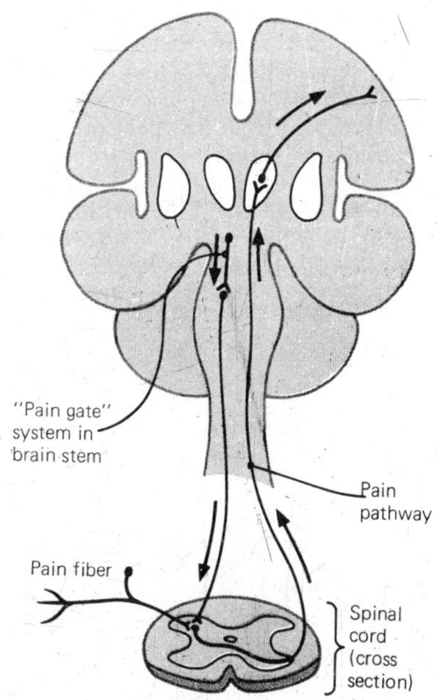

example, when we perceive the visual riches of an art museum, people's faces, television, or a conversation, active processes work on the sensory input to transform it into what we actually experience. Thus *perception*—our experience of the world—arises from sensory input plus the ways we process the sensory information. Many years ago, the famous American psychologist William James put it this way: "Part of what we perceive comes through the senses from the object before us; another part . . . always comes . . . out of our own head." The "out of our own head" part of this quotation refers to the active processing of sensory input that makes our experience of the world what it is.

Illusions, such as those shown in Figure 3.27 and on the pages at the beginning of this chapter, are familiar examples of perceptual processes at work. An *illusion* is *not* a trick or a misperception; it *is* a perception. We call it an *illusion* simply because it does not agree with our other perceptions. For instance, our perception of the line lengths in the Müller-Lyer illusion of Figure 3.27 does not agree with the perception we would have if there were no arrows. The presence of the arrows in the figure causes us to process the sensory input in such a way that we perceive the lines as unequal in length. The Gateway Arch shown on the beginning pages of this chapter is 630 feet high. But what did you perceive? Perceptual processes have done their work to produce the illusion. A simpler version of this horizontal-vertical illusion is shown in Figure 3.27. Illusions demonstrate that what we perceive often depends on processes that go far beyond the raw material of the sensory input. The rest of this chapter is about perceptual processes (most of them visual) that transform sensory inputs into what we actually experience.

Figure 3.27

Top, the enclosed circle in A and the enclosed square in B appear to be distorted. But if you use a dime or a straight edge to check your initial perceptions, you will find that the measurements (other perceptions) do not agree with your first perceptions. Illusions are examples of perceptual processes at work; they are not "tricks." (After Orbison, 1939.) Bottom, does the X appear longer than the Y? Measurements will give a different answer.

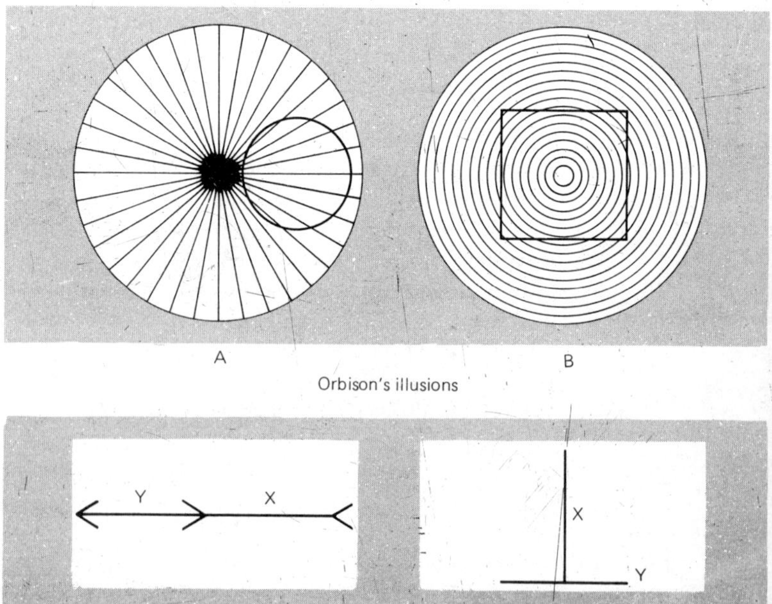

Orbison's illusions

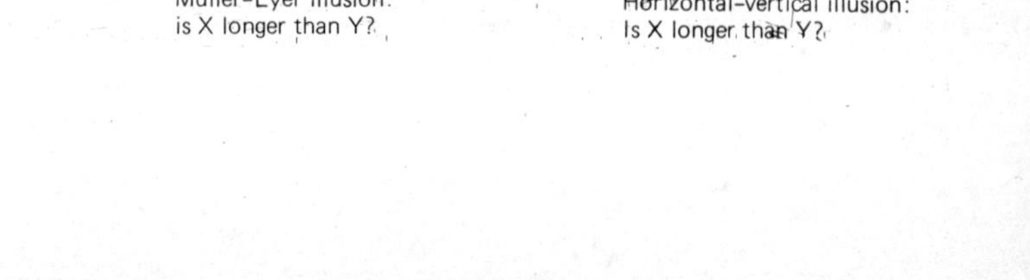

Müller-Lyer illusion: is X longer than Y?

Horizontal-vertical illusion: Is X longer than Y?

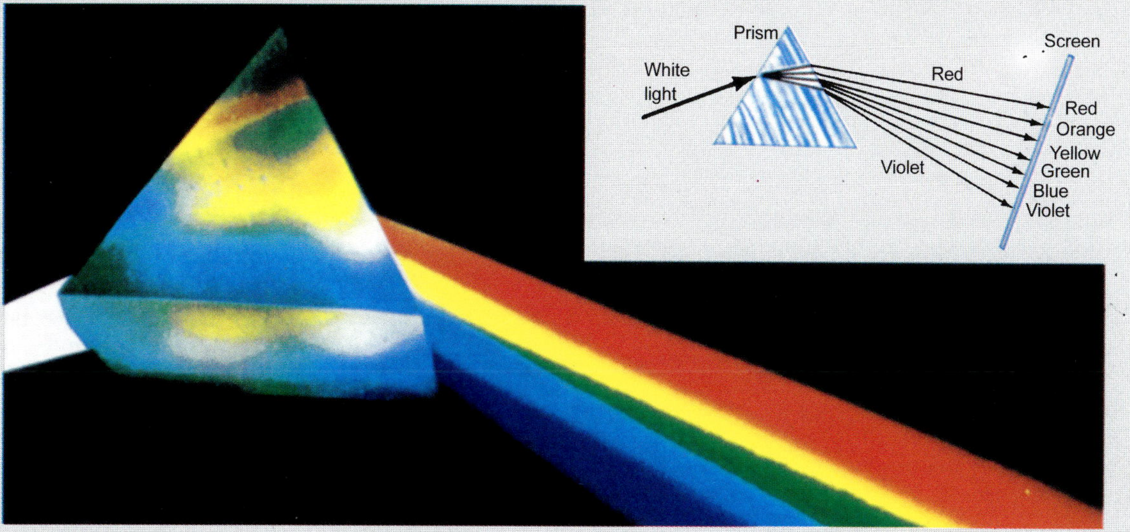

The visible spectrum. All the colors of the visible spectrum are produced when a prism is used to break up white light into its components. (From Krauskopf and Beiser, 1973.)

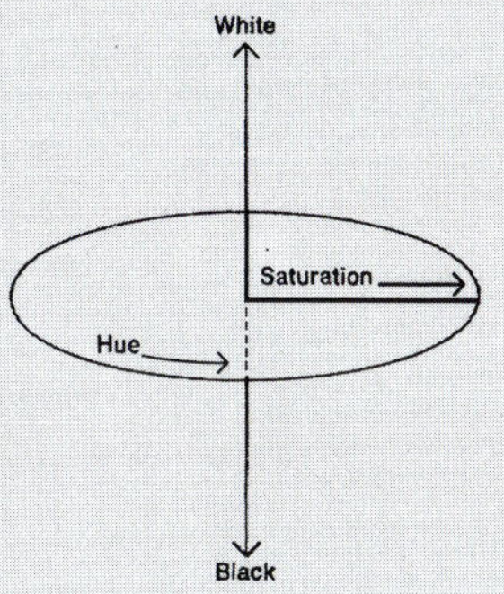

Both the drawing and the photograph illustrate the principle of the color solid·When all the colors are arranged in three dimensions, they form a color solid. The photograph shows ten segments from the complete solid. At the top are the most intense brightnesses; at the bottom the least intense. Around the circle are the different hues. The distance out from the center axis of the solid represents saturation (Photograph courtesy National Bureau of Standards.)

Interference in visual attention. First name the colors in the little boxes at the left as fast as you can and time yourself accurately. Then, name the colors in which the words on the right are printed. *Name the colors, not the words.* Do this as fast as you can and time yourself. Did it take longer to name the colors in the right column? See text, pages 109-110.

At any given moment, our sense organs are bombarded by a multitude of stimuli, yet we perceive only a few of them clearly. Were you aware of the background noises in your environment until you read this sentence? Yet input from the ears was coming in all the time. *Attention* is the term given to the perceptual processes that select certain inputs for inclusion in our conscious experience, or awareness, at any given time.

CHARACTERISTICS OF ATTENTION

The processes of attention divide our field of experience into a focus and a margin. Events that we perceive clearly are at the focus of experience. Other items are perceived dimly; we may be aware of their presence, but only vaguely so. These items are in the margin of attention.

To illustrate the nature of attention, consider your perceptions at a football game. While you are dimly aware of the tangle of players at the scrimmage line and of the activity of the blockers, it is the ball carrier and his movements that stand out. Your attention is focused on him. At the same time, sensory inputs are coming in from your cold feet, from your stomach as a result of the last hot dog you ate, and from the fellow behind you who is smoking a cigar. The crowd is also cheering. While the play is going on, you are probably not aware of any of these sensory inputs that are in the margin of your attention. Only when the play is finished or time is called do you perceive how cold your feet are, how queasy your stomach feels, what a strong smell the cigar has, and how noisy the crowd is.

The fact that you do at some point become aware of the marginal inputs illustrates another characteristic of attention, that it is constantly shifting. What is at the focus one moment may be in the margin the next; and what is in the margin may become the focus.

ATTENTION AND THE PROCESSING OF INFORMATION

How have psychologists tried to account for the fact that perception has a focus and that this focus switches from time to time? One set of explanations uses the concept of *filtering*. Since we cannot process all the information in our sensory channels, we *filter*, or partially block out, some inputs while letting others through (Treisman, 1969; Lindsay & Norman, 1977).

Imagine yourself at a party standing between two groups of people who are simultaneously carrying on two different conversations. You may be able to pick up some of both conversations at the same time—parallel processing, as it is called. But you will probably find that one or the other conversation is at the focus of your attention at any given moment; it is hard to pay attention to more than one set of inputs at a time. Thus you will most likely do what is called serial processing—attending to one set of inputs and then another. Or you may even stop switching back and forth, preferring instead to keep only one conversation at the focus of your attention. Whether you process the conversations serially or listen to only one of them, you are filtering out the unattended conversation; you have relegated these inputs to the margin of your attention. While you may be dimly aware of some features of the filtered-out conversation (its general subject and its emotional tone, for example), you will not register much of what was actually said in the non-attended-to conversation. A figure (following page 109) illustrates filtering in the visual sensory channel. To perform well on the task shown in this figure, you must filter out the meanings of the words and focus on the colors themselves.

In the filter models of attention, inputs in the margin shift to the focus when various attention-getting features of the environment are present in the filtered input. Such attention getters include intense stimuli and novel stimuli. For instance, if you are still between the two conversations of the last paragraph, you will probably switch your attention to the filtered-out conversation if voices are raised in this conversation.

Filter models of attention differ with respect to where the blocking occurs in the sensory channels. Some theorists (Broadbent, 1958) say that the filter, or information bottleneck, is at the sense organs, or at least in the very early stages of the input processing (Figure 3.28, left). Others (Shiffrin & Schneider, 1977) argue that the filtering takes place at later stages of the information flow—for example, at the stages where the input is interpreted as meaningful (Figure 3.28, right).

Other information-processing theories of attention are based on the idea of processing capacity (Kahneman, 1973). These theories are based on the assumption that we have a limited mental capacity for processing incoming information and therefore we cannot deal with all the sensory input at once. Instead, we must allocate our limited resource—our *processing capacity*—to one set of inputs or another. Proponents of these theories say that inputs which take up most of our processing capacity are at the focus of our attention. Thus, as you read this, you are devoting a large proportion of your processing capacity to the cognitive processes involved in understanding "what all this stuff about information processing and attention is about," and the text material is at the focus of your attention. If the radio is on, you are not aware of what is playing. But your professor who reads this section probably needs to devote less processing capacity to it; it is "old stuff." Your professor can therefore allocate more of his or her processing capacity to what is on the radio and will probably hear more of this input than you. According to the processing-capacity theorists, your attention shifts when your environment changes in such a way that more of your processing capacity must be used to deal with the new input

Figure 3.28

Filter, or bottleneck, models of attention differ on the stage of processing at which the filtering occurs. Early filtering is shown in A; late filtering in B. (From Coren et al., 1979.)

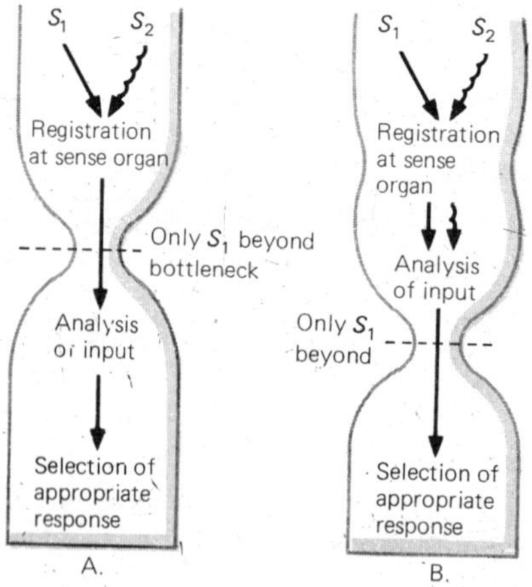

(if one of your friends comes into the room while you are studying and asks you to go to the movies, for example). Your friend will now be in the focus of your attention because you must now give a large portion of your processing capacity to dealing with this request.

What is at the focus of attention is not just so many disconnected simple sensations. We do not ordinarily perceive the world around us as patches of color, simple variations of brightness, or loud sounds. Instead, we see tables, floors, walls, trees, and buildings; we hear automobile horns, footsteps, and voices. In other words, the sensory inputs at the focus of our attention have form and meaning. And this brings us to the next topic—form perception.

Perceptual Processes: Form Perception

Perhaps the most fundamental process in form perception is the recognition of a figure on a ground. We see the objects and forms of everyday experience as standing out from a background. Pictures hang *on* a wall, words are seen *on* a page, and the melody stands out from the repetitive chords in the musical background. The pictures, words, and melody are perceived as the *figure,* while the wall, page, and chords are the *ground.* The ability to distinguish an object from its general background is basic to all form perception (Figure 3.29).

CONTOURS IN VISUAL FORM PERCEPTION

We are able to separate forms from the general ground in our visual perception only because we can perceive contours. *Contours* are formed whenever a marked difference occurs in the brightness or color of the

Figure 3.29

The distinction between figure and ground is fundamental to form perception. White birds against a dark ground are seen at the right; black birds against a light ground are seen at the left. (M. C. Escher, *Day and Night,* 1938, color woodcut. Photo courtesy of the National Gallery of Art, Washington, D.C.)

Figure 3.30

Camouflage works because it breaks up contours. (Wide World Photos.)

background. If you look at a piece of paper that varies continuously in brightness from white at one border to black at the opposite border, you will perceive no contour. The paper will appear uniform, and if you are asked to say where the sheet stops being light and starts to become dark, you can only guess or be arbitrary. On the other hand, if the change is marked rather than gradual—suppose several shades are skipped—you will see the paper as divided into two parts, a light part and a dark part. In perceiving the division at the place where the brightness gradient changes abruptly, you have perceived a contour.

Contours give shape to the objects in our visual world because they mark one object off from another or they mark an object off from the general ground. When contours are disrupted visually, as in camouflage (Figure 3.30), objects are difficult to distinguish from the background. But just because contours give shape to forms does not mean that contours themselves are shapes. The reversible faces of Figure 3.31 show the difference between contour and shape. Although both faces are formed by the same contour, they obviously do not have the same shape. Contours determine shape, but by themselves they are shapeless.

Figure 3.31

Objects with different shapes can be formed by the same contour. In this case, one contour outlines two different faces.

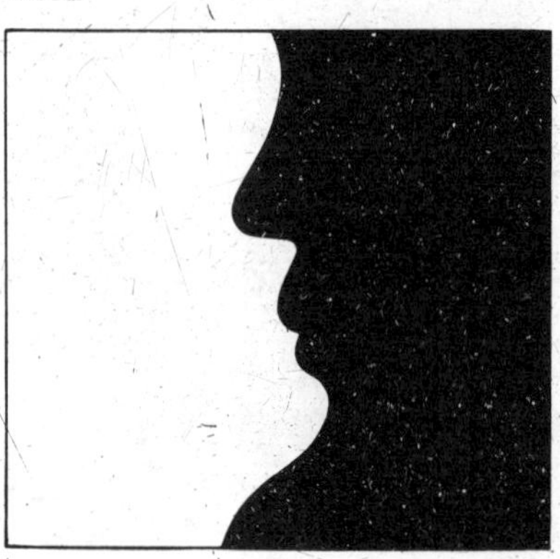

While differences in energy levels of light across the retina are involved in the formation of most contours in everyday experience, it has been found that contours can sometimes be seen without any energy difference at all on the two sides of the contour (Kanizsa, 1976; Coren, 1972). These are the so-called subjective contours. For example, in Figure 3.32, you see the contours of the upright triangle even though there are no energy changes across its perceived borders except in the corners. Note that the three angles forming the corners of the inverted triangle do not generate a subjective contour.

ORGANIZATION IN FORM PERCEPTION

When several objects are present in the visual field, we tend to perceive them as organized into patterns or groupings. Such organization was studied intensively in the early part of this century by the Gestalt psychologists (Chapter 1, page 25). They emphasized that organized perceptual experience has properties which cannot be predicted from a simple analysis of the components. In other words, Gestalt psychologists said that "the whole is more than the sum of its parts." (See the beginning pages of this chapter.) This simply means that what is perceived has its own new properties, properties that emerge from the organization which takes place.

Organization in perception partially explains our perception of complex patterns as unitary forms, or objects. We see objects as objects only because grouping processes operate in perception. Without them, the various objects and patterns we perceive—a face on a television screen, a car, a tree, a book—would not "hang together" as objects or patterns. They would merely be so many disconnected sensations —dots, lines, or blotches, for example.

What are some of the laws of perceptual organization? One organizing principle is *proximity*, or nearness. In Figure 3.33*a*, for example, we see three pairs of vertical lines instead of six single lines. The *law of proximity* says that items which are close together in space or time tend to be perceived as belonging together or forming an organized group.

In Figure 3.33*b* and *c*, you can observe another organizing principle of perception—*similarity*. In *b*, most people see one triangle formed by the dots with its apex at the top and another triangle formed by the rings with its apex at the bottom. They perceive triangles because similar items—the dots and rings—tend to be organized together. Otherwise they would see Figure 3.33*b* as a hexagon or as a six-pointed star, like Figure 3.33*c* where all the dots are the same.

Grouping according to similarity, however, does not always occur. The figure in 3.33*d* is more easily seen as a six-pointed star than as one figure composed of dots and another figure made up of rings. In this case, similarity is competing with the organizing principle of symmetry, or *good figure*. Neither the circles nor the dots by themselves form a symmetrical pattern. The *law of good figure* says that there is a tendency to organize things to make a balanced or symmetrical figure that includes all the parts. In this case, such a balanced figure can be achieved only by using all the dots and rings to perceive a six-pointed star. The law of good figure wins out over the law of similarity because the rings by themselves or the dots by themselves do not form symmetrical good figures.

Still another principle of organization is *continuation*, the tendency to perceive a line that starts in one way as continuing in the same way. For example, a line that starts out as a curve is seen as continuing on a smoothly curved course. A straight line is seen as continuing on a straight course or, if it does change direction, as forming an angle rather than a curve. Figure 3.33*e* illustrates this principle of continuation: we see the dots as several curved and straight lines. Even though the curved and straight lines cross and have dots in common, it is only with an effort that we can perceive a straight line suddenly becoming a curved line at one of these junctions.

Finally, the law of closure makes our perceived world of form more complete than the sensory stimulation that is presented. The *law of closure* refers to perceptual processes that organize the perceived world by filling in gaps in stimulation. By the action of these processes, we perceive a whole form, not disjointed parts. In Figure 3.34, for example, the left-

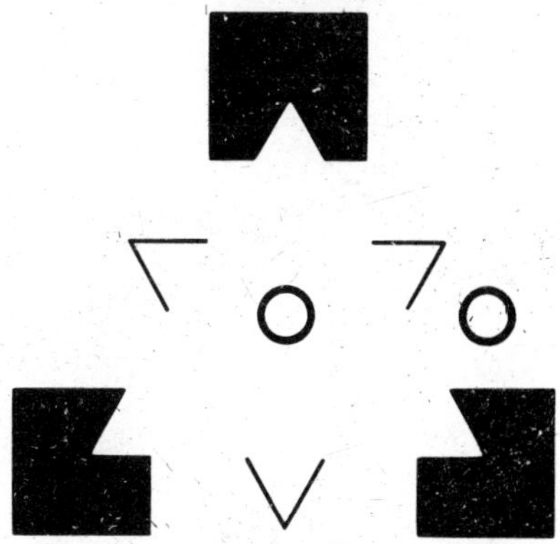

Figure 3.32

An illustration of subjective contours. Notice the sharp contours that seem to outline the upright triangle. Yet except in the corners, there are no energy differences to account for these contours; they result from perceptual processes. (From Coren, 1972; based on Kanizsa, 1955.)

Figure 3.33

Examples of organizing principles in visual perception.

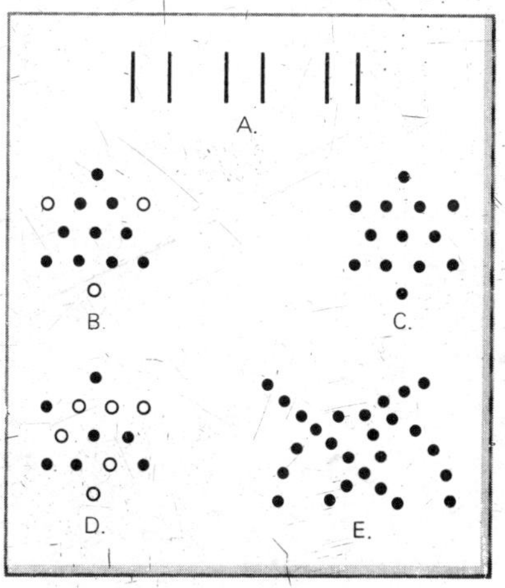

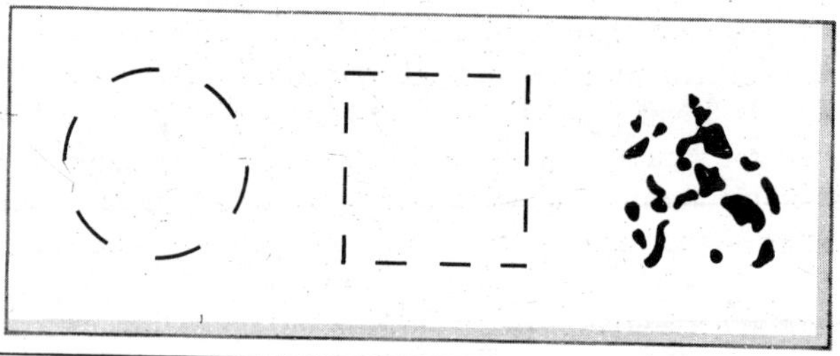

Figure 3.34

Examples of the organizational principle of closure. This principle says that we tend to see a complete object even though parts of it may be missing. What do you see in the drawing at the right?

hand drawing is seen as a circle with gaps in it and the center drawing as a square with gaps in it—not simply as disconnected lines. If these incomplete figures were flashed in a *tachistoscope,* a device used in perceptual experiments for the very brief presentation of stimuli, they might even be perceived as complete figures without gaps. The principle of closure also applies to perception of the pattern at the right in Figure 3.34. Here again, we fill in the gaps and perceive form rather than disconnected lines. (Most people see a figure on horseback.)

Although the examples in Figure 3.33 are visual, the same principles of grouping can be observed in the other senses. The rhythm we hear in music also depends upon grouping according to proximity in time and similarity of accents. In the sense of touch, too, grouping occurs. For example, ask a friend to shut his or her eyes. Mark off three equidistant points on the back of his or her hand, touch a pencil to the first two points and then pause slightly before you touch the third point. Your friend will report that the first two points were closer together than were the second and third. This illusion, or perception, illustrates the grouping of tactile stimuli according to nearness, or *proximity,* in time.

In all these laws of organization, the principle of the Gestalt psychologists that "the whole is more than the sum of its parts" can be observed at work. In other words, the perceived organization has properties of its own that are not simply the result of adding together the "atoms" of individual sensations.

Perceptual Processes: Visual Depth Perception

Depth perception was a puzzle to scientists and philosophers for hundreds of years. They could not understand how we could see a three-dimensional world with only a two-dimensional, or flat, retina in each eye. Today we realize that the ability to perceive depth is no more amazing than is any other perceptual accomplishment. We are able to make use of information, or cues, in the sensory input to "generate" the three-dimensional world that we see. Thus the question is: What are the cues we use to see depth and distance? Part of the answer lies in the cues received by each eye separately—the *monocular* ("one-eyed") *cues* for depth perception. Another part of the answer is found in the cues received from both eyes working together—the *binocular* ("two-eyed") *cues.*

MONOCULAR CUES FOR DEPTH PERCEPTION

As the name suggests, *monocular cues* are cues that can operate when only one eye is looking. These cues are the ones used by painters to give us a

Figure 3.35

Artists make good use of the monocular cues for depth perception: *Interior of a Church* by Pieter I. Neeffs. (The Metropolitan Museum of Art, 1871.)

three-dimensional experience from a flat painting (Figure 3.35). The eye picks them up and we perceive depth.

Linear Perspective The distances separating the images of far objects appear to be smaller. Imagine that you are standing between railroad tracks and looking off into the distance. The ties would seem to gradually become smaller and the tracks would seem to run closer and closer together until they appeared to meet at the horizon. Figure 3.36 owes part of its depth effect to such linear perspective.

Clearness In general, the more clearly we see an object, the nearer it seems. A distant mountain appears farther away on a hazy day than it does on a clear day because haze in the atmosphere blurs fine details and we can see only the larger features. Ordinarily, if we can see the details, we perceive an object as relatively close; if we can see only its outline, we perceive it as relatively far away (Figure 3.36).

Interposition Still another monocular cue is interposition, which occurs when one object obstructs our view of another. When one object is completely visible while another is partly covered by it, the first object is perceived as nearer (Figure 3.36).

Shadows As Figure 3.37 shows, the pattern of shadows or highlights in an object is very important in giving an impression of depth. When this aerial photograph of a group of Quonset huts is turned upside down, the Quonset huts look like towers. If you carefully note the differences between the Quonset huts and the "towers," you will discover that the shadows are responsible for this effect. The reason is that we are accustomed to light coming from above. When the picture is turned upside down, we do not perceive the Quonset huts as illuminated from below.

Figure 3.36

Four monocular cues in depth perception. The buildings and the street converge in the distance (linear perspective); the more distant heights show less detail than the closer areas (clearness and texture gradient); and some parts of the buildings are behind others (interposition). (Fundamental Photographs.)

Figure 3.37

Shadows and the perception of depth. If the picture is turned upside down, the buildings, especially the Quonset huts, look like towers. (From Wide World Photos.)

Instead, we see towers with black-painted tops because the dark areas are now of such a size and in such a position that they cannot possibly be shadows if the light is coming from above. We do not, of course, reason this out. The perception is immediate, based on whether or not the dark areas appear to be shadows.

Gradients of Texture A *gradient* is a continuous change in something—a change without abrupt transitions. In some situations, we can use the continuous gradation of texture in the visual field as a cue for depth (Gibson, 1950). The regions closest to the observer have a coarse texture and many details; as the distance increases, the texture becomes finer and finer (Figure 3.38). This continuous gradation of texture gives the eye and brain information that can be used to produce an experience, or perception, of depth.

Figure 3.38

Gradients of texture are a monocular cue for depth perception. (Gabriele Wunderlich.)

Movement When you move your head, you will observe that the objects in your visual field move relative to you and to one another. If you watch closely, you will find that objects nearer to you than the spot at which you are looking—the *fixation point*—move in a direction opposite to the direction in which your head is moving. On the other hand, objects more distant than the fixation point move in the same direction as your head moves. Thus the direction of movement of objects when we turn our heads can be a cue for the relative distance of objects. Furthermore, the amount of movement is less for far objects than it is for near ones. Of course, as is the case with all depth cues, we do not usually think about this information; we use it automatically.

A BINOCULAR CUE FOR DEPTH PERCEPTION

As we have just seen, many of the cues for depth require only one eye. In fact, one-eyed people, under most conditions, have quite adequate depth perception. Most of us, though, look out at the world with both eyes simultaneously, and we are thus able to add binocular cues for depth perception to the monocular ones. By far the most important binocular cue comes from the fact that the two eyes—the retinas—receive slightly different, or *disparate*, views of the world. Therefore, this cue is known as *retinal disparity;* It is the difference in the images falling on the retinas of the two eyes.

To understand retinal disparity, consider the geometry of the situation when the two eyes view an object. (See Figure 3.39.) The fovea in the center of the retina is much more sensitive than is the rest of the retina (page 88). When we look at an object, we fixate our eyes—point them, in a manner of speaking—so that the image of the object falls mostly on each fovea. But since the two eyes are separated from each other by about

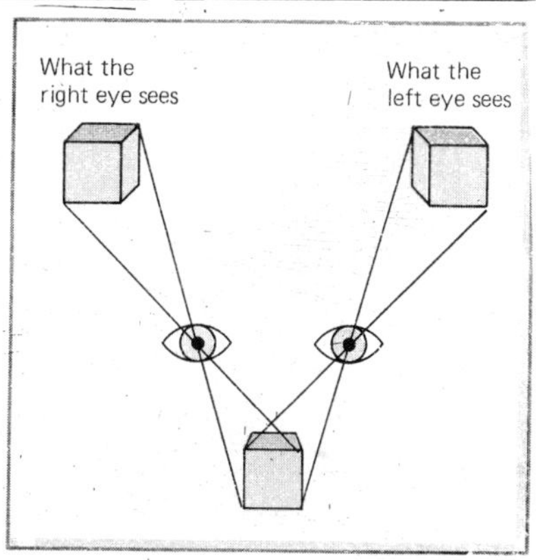

Figure 3.39

The geometry of retinal disparity, an important binocular cue for depth perception. Because our eyes are separated, the image of an object is not the same in the two eyes. The nearer an object is, the greater is the difference, or disparity.

65 millimeters, they get slightly different views of the object, and the two images are not exactly the same. (Compare the two cubes in Figure 3.39.) Moreover, and this is the main point, the images are more dissimilar when the object is close than when it is far in the distance. In other words, within limits, the closer an object is, the greater is the retinal disparity. The correspondence between distance and the amount of disparity is the reason retinal disparity can be used as a depth cue.

Perceptual Processes: Constancy

The world as we perceive it is a stable world, and this stability is present early in life. For instance, a man's size does not appear to change much as he walks toward us; a dinner plate does not look like a circle when viewed from one angle and an elipse when viewed from another; and the location of a sound does not appear to change when we move our heads. Stability of perception helps us to adapt to the environment. It would be virtually impossible to operate in a world where sounds changed their locations as we moved our heads and objects changed their shapes and sizes when viewed from different positions and distances. Imagine what it would be like if your friends assumed a multitude of sizes and shapes. The stability of the environment as we perceive it is termed *perceptual constancy*.

SIZE CONSTANCY

The size of the representation, or "image," of an object on the retina of the eye depends upon the distance of the object from the eye; the farther away it is, the smaller the representation. This geometic fact is illustrated in Figure 3.40. Similarly, a representation of the same size can be produced on the retina by a nearby small object or a larger object at some distance.

Yet when you cross the street to speak to a friend, your perception of the friend's size does not change much, even though the retinal representation alters greatly in accordance with the geometry of the situation. Contributing to this constancy is a great deal of additional information you have about the circumstances: You know something about the distance of the friend from you; you perceive the changes that take place in other

Figure 3.40

The farther away an object is, the smaller its representation on the retina.

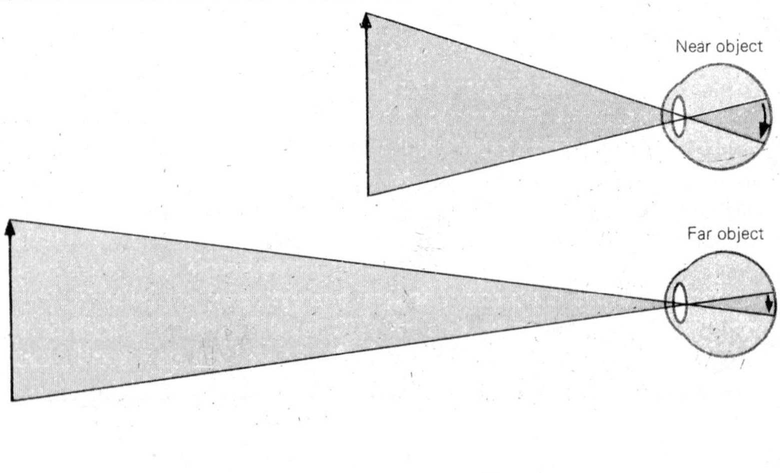

objects as you approach your friend; and you know how large your friend is supposed to be—the friend's assumed size.

The importance of distance and background information in maintaining size constancy was shown in a classic experiment by A. H. Holway and E. G. Boring in the 1940s (Holway & Boring, 1941). They used ambiguous stimuli—disks of light—which could have no real assumed size, and they changed the amount of distance and background information available to the subjects in the experiment. They found that size constancy decreased as the distance and background information available to the subjects decreased. In other words, the subjects perceived the size of a disk of light more in accordance with the size of the retinal representation when they lacked information about distance and background.

One interpretation of this result might be that people somehow automatically use information about distance and background to "correct" the size of their retinal representations, thus keeping their perceptions relatively constant. Another interpretation is that no "correction" is necessary—that size constancy occurs because the object and its background change together as the distance of the object changes. For instance, the texture of the object—the number of fine-grained details that can be seen—and the texture of the background change together as distance changes. Also, the retinal size of the object changes with the retinal size of the background objects. Thus, according to this interpretation (Gibson, 1950), perceptual size constancy results when an object and its background change together in such a way that the *relationship* between them stays the same. We will see another example of the importance of relationships in perceptual constancy when we discuss brightness constancy.

Our knowledge of the size of a familiar object—the assumed size—can sometimes be an important factor in size constancy, especially under conditions in which other information is not available or is ambiguous. But under everyday conditions of perception, conditions in which distance and background information are available and unambiguous, the assumed size of familiar objects is not an important factor in maintaining size constancy (Fillenbaum et al., 1965).

SIZE CONSTANCY AND ILLUSIONS

Some of the illusions on the opening pages of this chapter (page 83–84) have been "explained" in terms of "misplaced" size constancy (Gregory, 1978; Coren & Girgus, 1978). In the "railroad-track," or Ponzo, illusion, for example, even though the two horizontal bars are the same length, we perceive the upper one as longer than the lower one. The illusion is said to work because the railroad tracks converging in the distance provide a strong cue for depth—linear perspective (page 115). Thus we receive information that the upper bar is farther away than the lower one. Now, from what we said about size constancy above, you know that we tend to perceive distant objects as larger than we would if we considered only the size of the retinal representation (your distant friend, remember, looked about the same size as when he or she was closer). In the illusion, the horizontal bars are the same length, but size constancy leads you to "magnify" the distant one. A similar explanation having to do with depth cues can be given for the "cylinder illusion" at the beginning of this chapter. In these examples, and in the illusions to be considered below, size constancy is said to be "misplaced" because it leads to an illusion.

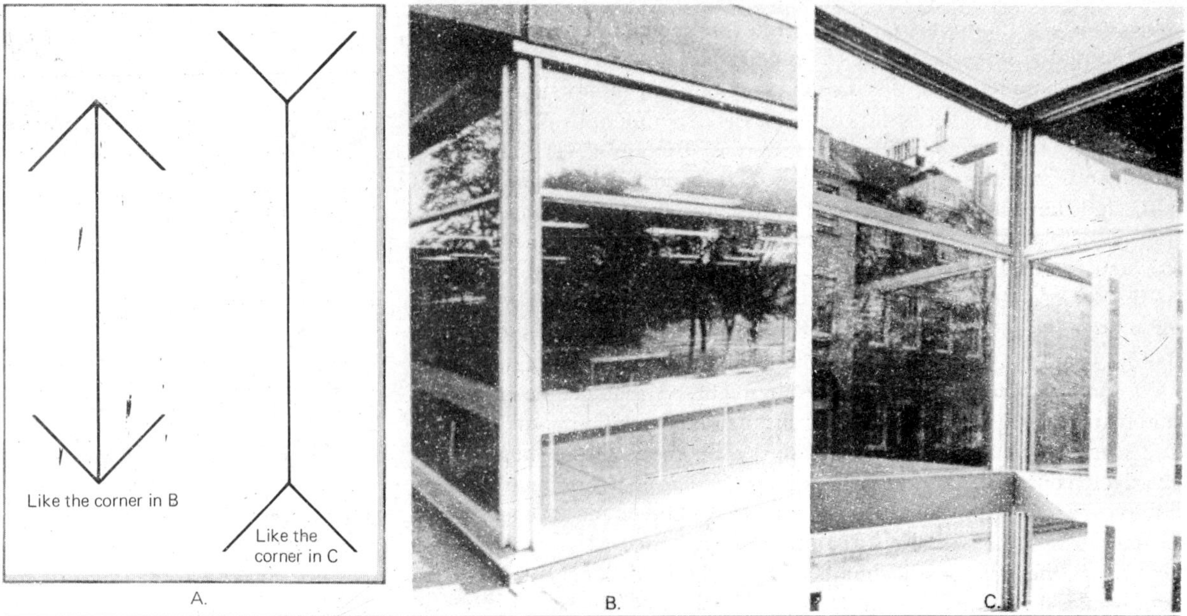

Figure 3.41

A, the Müller-Lyer illusion; *B*, an outside corner seems to approach in depth; *C*, an inside corner appears to recede in depth. See text for an explanation of how this might account for the Müller-Lyer illusion. (Courtesy of Richard Gregory.)

The arrowhead, or Müller-Lyer, illusion shown in Figure 3.27 (page 108) and in Figure 3.41*a* has also been explained in terms of depth cues and "misplaced" size constancy. If the arrows point outward, as in the left part of Figure 3.41*a*, we perceive the line connecting them as relatively near. On the other hand, the line connecting the inward-pointing arrowheads (Figure 3.41*a*, right) is perceived as distant. Once again then, size-constancy mechanisms go to work to "magnify" the length of the distant-appearing line, but, since the lines in Figure 3.41*a* are the same length, size constancy is "misplaced" and the illusion results (Gregory, 1978).

As another example, consider the moon illusion. Whether the moon is high in the sky or on the horizon, its representation on the retina is the same size, but it is perceived as much larger on the horizon (Figure 3.42). One explanation of this illusion says that when the moon is near the horizon, buildings and trees provide depth cues indicating that the moon is indeed far away; farther up in the sky, these cues are absent. From what has already been said about size constancy and illusions, you can see how the moon illusion might be partially explained.

BRIGHTNESS CONSTANCY

Visual objects also appear constant in their degree of whiteness, grayness, or blackness, even though the amount of physical energy reflected from them may change enormously. People are not like photoelectric cells, simply registering the amount of light being reflected from a surface. Our experience of brightness stays relatively constant despite great changes in the amount of physical energy reaching our eyes. For example, objects or surfaces that appear white in a bright light are still perceived as white in dim illumination. Similarly, what looks black to us in dim light still looks black in intense light. Coal looks black even in very bright sunlight, while snow continues to look white even at night. Another example of bright-

Figure 3.42

A representation of the moon illusion. The moon looks larger when it is low in the sky than when it is higher. See text for a possible explanation of this illusion. (From Coren & Girgus, 1978.)

ness constancy is the appearance of a white paper that lies partly in shadow. We perceive the paper as uniformly white; we do not perceive the shadowed portion as gray, but rather as white in the shadow.

We have brightness constancy because in most situations, when the illumination changes, it changes over the whole field: The physical-energy ratio between an object and its surround stays constant. For example, if I turn up the lights in my room, the cover of the book on my desk looks just as bright as it did before because the ratio of the illumination falling on the book cover and that falling on its surround has not changed. In other words, unchanged-brightness ratios give constant-brightness experiences, or *brightness constancy*. While this rule must be accepted with some reservations because it probably does not hold for the entire range of stimulus intensities (Jameson & Hurvich, 1964), it is a useful first step toward an explanation of brightness constancy.

Perceptual Processes: Movement Perception

Adaptive behavior in the visual world requires that we perceive movement accurately. Suppose that cars, motorcycles, and bicycles are whizzing toward you. Should you jump aside, or should you leisurely stroll out of the way? If I am batting for the baseball team, I must have good perception of movement if I am not to spend the season on the bench or in the hospital. And so it goes. How do we perceive movement? You may think that the answer is obvious. After all, objects moving through the visual field or along the skin stimulate different parts of the receptor. Can movement perception be attributed to this changing stimulation? Only partly. Perceived motion also occurs without any energy movement across the receptor surface. This type of motion is called *apparent motion*. Furthermore, the perception of the actual physical movement of objects in the world—*real motion*—is like all perception; it involves active processing of the sensory input.

CONSTANCY OF REAL-MOTION PERCEPTION

We need movement constancy in order to adapt to events in the visual world. To illustrate, suppose you are in the grandstand watching stock cars race around an oval track. While the cars will slow up on the curves, they will be going at about the same speed on the straightaway in front of the grandstand as on the backstretch some distance away. We perceive the straightaway and backstretch speeds as about the same despite the fact that, because of the geometry of the situation, the image of a car travels farther over the retina in a given time when it is near (on the straightaway) than when it is far away (on the backstretch).

Why do we have such velocity, or motion, constancy? A number of experiments have shown that it is because perceived velocity depends on the rate at which an object (the car in the example) moves *relative* to its background (the track and other nearby objects in the example), not on the absolute velocity of the image across the retina. Since the relationship between an object and its background stays relatively constant with distance—as one changes, the other does, too—perceived velocity also remains fairly constant.

THE BRAIN COMPARATOR AND REAL-MOTION PERCEPTION

If we hold our eyes steady and stimulation moves across the retina, we perceive movement. As we saw earlier (Figure 3.12, page 94), there are cells in the brain that are vigorously excited by movement. In the simple case of movement across a stationary retina, these "movement-detecting" cells probably provide the basis for our perception of real motion. But this is by no means the whole story because retinal images also move when we move our eyes, head, and body. Somehow we must be able to tell whether the retinal image moved because we moved or because something "out there" moved. We would have a very hard time adapting to the world if we could not tell which was which.

The concept of a "brain comparator" has been postulated to explain how it is possible for us to differentiate between the real motion of an object and motion caused by our own movement. The *brain comparator* is a system which compares information about muscle movements with information about movements of the retinal image. For example, consider just eye movements. Movement commands to the eye muscles go both to the eye muscles themselves and to the brain comparator (Figure 3.43); in this way, the comparator has information that a movement is about to occur before it actually occurs. When the eye movement occurs and the retinal image moves, the movement signals from the retina are fed into the comparator (Figure 3.43), where they are matched against the information the comparator already has about eye movements. The brain comparator "evaluates" the moving retinal image as due to eye movements and cancels the perception of movement. On the other hand, if the comparator has no information about eye movements, as when an image moves over a stationary eye, the perception of movement is not cancelled.

Figure 3.43

The brain comparator compares information it has about eye movements (or other bodily movements) with information coming in from the eyes. (Based on Gregory, 1978.)

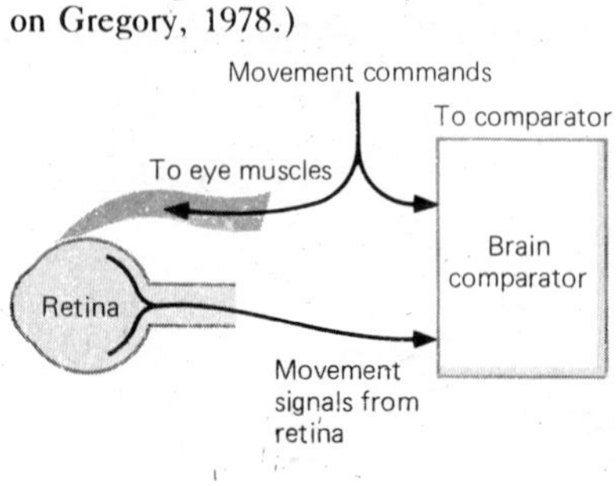

To help you understand how the brain comparator works, do the following mini-experiment:

Keep one eye shut. Then look at a small object, such as a dime, while pushing gently on the eyeball of the open eye. (Don't push too hard or you'll cut off the blood supply to the retina and become momentarily blind.) You will probably see the object jump around as you gently jiggle the eyeball back and forth with your finger; you may even feel a little dizzy if you do this for some time. Now, looking at the same object, use the eye muscles to move your eye from side to side. You will probably see that the object stays stationary, even though the retinal image moved when the eyeball moved.

The object moved around when you pushed on your eyeball because no commands for muscle movement reached the comparator. Therefore, the perception of movement was not canceled. However, when you used your eye muscles, the comparator canceled the perceived movement. Now you can see how important it is to cancel movement perception that comes from eye, head, or body movements. What a world it would be if objects jumped around as they did when you pushed on your eyeball!

APPARENT MOTION

Unlike real motion, *apparent motion* is movement perceived in the absence of physical movement of an image across the retina. In other words, with the eyes, head, and body steady, and with no physical movement of an object, motion is still perceived.

Stroboscopic motion, the kind seen in movies and on television, is a common example of apparent motion. A movie projector simply throws successive pictures of a moving scene onto a screen. When you examine the separate frames, you see that each is a still picture slightly different from the preceding one. When the frames are presented at the right speed, you perceive continuous, smooth motion.

Thousands of experiments have been conducted in an effort to analyze stroboscopic movement. Many psychologists who study perception now believe that stroboscopic motion results from information processing by the brain that is similar to that which occurs when real motion is perceived. In movies, for instance, the brain receives information from different parts of the retina, just as it does when physical movement causes stimulation to move across the retina. An object is first seen in one location, and then, very soon after, appears at a slightly different part of the retina, just as it would if it were actually moving. The brain "interprets" this information as movement, showing once again the role of sensory-input processing in the perception of the world.

The *autokinetic effect* is another example of apparent motion. If a person stares at, or fixates on, a small stationary spot of light in a completely dark room, the spot will eventually appear to move. Such apparent motion can be large and dramatic, and it can be influenced by suggestion. Movements of the eyes have an influence on the autokinetic effect, but they do not seem to account for it.

What is known as *induced movement* occurs if a stationary spot or object is perceived as moving when its frame or background moves. For example, the moon is often perceived as "racing" through a thin layer of clouds. The movement of the framework of clouds "induces" movement in the relatively stationary moon.

As used here, the term *plasticity* refers to the modifiability, or moldability, of perception. Special situations, such as prolonged changes in sensory input, can modify the ways information is processed in generating perceptions of the world around us.

PLASTICITY: VISUAL DEPRIVATION

Visual deprivation—restriction of the visual input in some fashion—is especially potent during what is known as the sensitive period for visual development. A *sensitive period* is the time in the early development of a person or animal during which the environment has its greatest effect on behavior or on the brain processes underlying behavior. (See Chapter 11, page 442.)

In a number of experiments, animals have been raised from birth, thus including the sensitive period, with translucent contact lenses over their eyes. These animals were thus deprived of form vision because the lenses prevented light from being focused on their retinas. (If you will imagine what it would be like to look out at the world through half a Ping-Pong ball, you will have some idea of the visual deprivation involved in these experiments.) When the special lenses were removed after the animals' sensitive periods had passed, they had great difficulty in perceiving form (Riesen, 1966). Special training and visual experience after the lenses had been removed improved their form perception somewhat, but, judging from the animals' behavioral responses to visual stimuli, it almost never became "normal." To a greater or lesser degree, the animals' visual form perception remained permanently impaired as a result of the form-vision restriction during their sensitive periods.

A few unfortunate people are born with congenital cataract, a condition in which the lenses of the eyes are clouded over; others are born with, or develop soon after birth, cloudy corneas (page 87). Like the milk-white glass contact lenses used in the animal experiments, either of these conditions makes form vision impossible because light cannot be focused on the retina. These people can perceive large areas of brightness, but they are blind in the sense that they have no useful detail vision. They can be helped by special training; and, in some cases, operations can correct the problem. For instance, in cases of cataract, the lenses can be removed and glasses or contact lenses substituted for them; corneal transplants can sometimes be done to replace cloudy corneas. After the operation, light can be focused on the retina and the eyes are ready for normal sight.

When the bandages come off after the operation, these patients can visually recognize objects that were familiar to them through their other senses. They can, for instance, visually recognize objects such as telephones, chairs, or spoons that they had learned to recognize through touch when blind. However, they have great difficulty with the visual recognition of "new" objects. For example, faces and letters are hard for them to recognize, and their perception may never be normal for these "new" objects. But as these patients visually experience the environment, their perception gradually becomes more normal, and this recovery is another example of the plasticity of perception. Of course the amount of recovery depends upon many factors, and some people make more progress than others. Even in the best cases, however, perception probably never becomes normal. In fact, some patients, after a period of using their new vision, go back to relying on the sensory channels they used when blind. The visual world seems "too much to handle," and they go back to their old "tried-and-true" ways of adjustment (Gregory, 1978).

Instead of eliminating all form vision, a number of researchers have restricted the visual environment of animals to a few specific kinds of visual input, or features, during their sensitive periods. For example, in one experiment, the only visual experiences kittens were allowed during their sensitive period were patterns of lines (Figure 3.44); otherwise they were kept in the dark (Blakemore & Cooper, 1970; Hirsch & Spinelli, 1971). In another experiment, kittens were equipped with goggles that distorted their visual fields (Bruce et al., 1981). With some exceptions (Stryker & Sherk, 1975), these studies generally show that the responses of cells in the visual brain (page 94) are altered in specific ways that relate to the visual features to which the developing animal is exposed during the sensitive period. And some experiments have been able to relate the brain changes to changes in perception by means of behavioral tests (Blasdel et al., 1977). Thus plastic changes in perception, as indicated by behavior, have been matched with plastic changes in the brain.

PLASTICITY: NATURE AND NURTURE

We have seen that perception is *plastic;* it can be changed. While there is general agreement on the major facts of perceptual plasticity, the problem is in explaining it. This brings us to the nature-nurture question in per-

Figure 3.44

A kitten in an apparatus designed so that its only visual experience is of vertical lines. See text. (From Blakemore, 1974.)

ception. *Nature* refers to innate, or inborn, processes that influence behavior and perception, while *nurture* refers to learning and, in general, the effects of the environment on behavior and perception. (See Chapter 11, page 413.) Theorists who argue for the importance of nature in perception are called *nativists;* those who argue for nurture are known as *empiricists.*

At first glance, the sensitive-period experiments seem to support the empiricist view. After all, environmental alterations during sensitive periods do change perception and brain-cell responses. However, the nativists also have an explanation for the results of the sensitive-period experiments. To understand their explanation, let us look more closely at the nativist and empiricist views. In so doing, we will also see how nature and nurture are thought to interact in the two theories.

The nativists say that brain organization is patterned by genetic codes and is therefore innate. The perception that depends on this organization is thus also innate. Plastic changes are explained by the nativists as due to loss, or attrition, of inborn connections during sensitive periods. ("If you don't use it, you lose it.") In other words, sensory deprivation during sensitive periods results in a loss of the brain connections that were genetically determined, because they need environmental input (nurture) for their maintenance. On their side, the empiricists argue that while genetic codes may provide a rough blueprint, nurture interacts with the genetic outline during sensitive periods to guide growth and to cause the proper brain connections to be made. Therefore, according to the empiricists, alterations in the environment during sensitive periods actually change the way the brain grows and the connections that it makes.

To decide who is right about the sensitive-period experiments, we need evidence from newborn animals that have had no sensory experience at all. If the nativists are right, their perception and brain-cell responses should be like those of the adult animal. Tests of day-old chicks and very young goats, both of which can walk almost at birth and hence can be tested, show that depth perception is present very early in life (Walk & Gibson, 1961). But even these very young animals have had some visual experience. A more stringent approach would be to test the responses of brain cells in newborn animals before they have had any sensory experience at all. There is disagreement about the results of such experiments on kittens, but it has been found that the cell responses in the visual brains of newborn monkeys are almost identical to those of adult monkeys (Wiesel & Hubel, 1974). This discovery is significant because the visual brains of monkeys and humans are quite similar. However, these records were made on cells in the part of the visual cortex that first receives sensory input as it comes from the eyes and lower brain regions. Almost nothing is known about the innate organization of other parts of the brain that may be involved in processing visual information in order to give us the world as we perceive it. Perhaps the empiricists are correct about the role of nurture in guiding the growth of connections in these brain regions.

As more experiments are done, it seems likely that there will be many more answers to the nature-nurture question in perception. The importance of nature or nurture, as well as the ways in which they interact, will probably turn out to depend on what aspect of perception is being studied, on the part of the brain that is under investigation, and on the species of animal being studied.

Perceptual Processes: Individual Differences

Much of this chapter has been about basic perceptual processes that are much the same for everyone. But we know that people differ in the ways they process sensory inputs to give rise to what they experience. Two people may have very different perceptions of the same television drama, lecture, meeting, or interpersonal encounter. (See Chapter 9 for an account of social perception.) Individual differences in learning, sets (expectations), motives, and perceptual styles are at work to make one person's perceptions different from those of another.

PERCEPTUAL LEARNING

Based on past experiences or any special training we have had, each of us has learned to emphasize some sensory inputs and to ignore others. Eleanor Gibson has defined *perceptual learning* as "an increase in the ability to extract information from the environment as a result of experience or practice with the stimulation coming from it" (Gibson, 1969, p. 3). Perceptual learning can be considered a variety of the cognitive learning we will discuss in Chapter 4 (pages 168–172). Gibson gives many examples that show how perception can be molded by learning. She cites the competence of people trained in various occupations to make perceptual distinctions that untrained people cannot make. Skill, or artistry, in many professions is based upon the ability to make these subtle distinctions. Experience is the best teacher for these perceptual skills; usually, they cannot be learned from books.

Distinguishing the calls of birds is one of Gibson's examples. A trained ornithologist can do it, but most of us have great difficulty. If you are lucky enough to live where there are plenty of birds, try shutting your eyes and listening to the birds at dawn or dusk. They make a deafening racket. You may be able to extract some features that enable you to identify particular species of birds. But for the most part, unless you have had the necessary training, the whistles, trillings, and buzzings will blend together so that you will not be able to distinguish the call of one bird from another or recognize calls upon hearing them again. Perceptual learning is needed before you can do this.

As Gibson also points out, the remarkable feats of blind people are often matters of perceptual learning. It is not that their sensitivity to nonvisual stimulation is greater than that of sighted people. Instead, blind people learn to extract from the environment information not ordinarily used by sighted people. For instance, many blind people move around in the world, avoiding obstacles with surprising ease. Blind people learn to perceive the sound echoes of their footfalls and cane tappings that bounce back from objects in their paths. Some blind people even learn to distinguish among various shapes and textures of surfaces by perceiving the differences in their sound echoes. It is obvious that learning to extract certain kinds of information from the environment—*perceptual learning*—is of enormous practical and adaptive value.

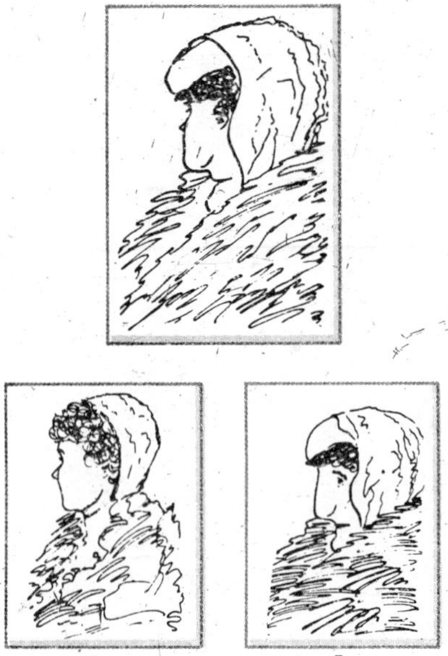

A. B.

Figure 3.45

The influence of set on perception. The likelihood of perceiving the young woman or the old woman in the upper composite, ambiguous drawing depends on prior exposure to A or B. See text. (From Krech et al., 1982; redrawn from Leeper, 1935.)

SET

Set refers to the idea that we may be "ready" and "primed for" certain kinds of sensory input. Such expectancies, or *sets*, vary from person to person and are a factor in both the selection of sensory inputs for inclusion in the focus of attention (page 109) and in the organization of inputs (page 112).

To illustrate the role of set in attention, consider the husband who is expecting an important phone call. He will hear the telephone ring in the night while his wife does not. The wife, on the other hand, may be more likely to hear the baby crying than the telephone ringing. Of course, if the wife is expecting an important call, the situation may be reversed.

Figure 3.45 is from a classic experiment on the role of set in perceptual organization (Leeper, 1935).

> At the top of Figure 3.45 is an ambiguous picture that can be organized to give a perception of either an old woman or a young one. Most people (about 65 percent) see the old woman first; after looking at the top picture for awhile, they see the young woman. If you happen to be in the 35 percent who see the young woman first, wait: You will see the old woman as you continue to look at the top picture. (Incidentally, this picture shows that the same sensory input can be organized in different ways—another example, if one is needed, of the active processing involved in perception.)
>
> By giving people different predispositions, or *sets*, what is first perceived can be changed. When a group of people were shown Figure 3.45*a* first and then were shown the ambiguous top figure, all of them organized their inputs so that they saw the young woman first; they were set to perceive her. Ninety-seven percent of another group, who were set by seeing the old woman first (Figure 3.45*b*), saw her first in the top ambiguous picture.

Figure 3.46 provides another example of set in perceptual organization. Is it the letter B or the number 13? If this drawing is included in a series of two-digit numbers, people will tend to report that they perceive the number 13. But other people, who have seen the figure in the context of letters, will report that it looks like a B to them. In one case, an expectancy, or *set*, has been acquired for numbers; in the other, for letters.

Figure 3.46

The effect of expectancy, or set, on perception. The drawing can be perceived either as a B or the number 13, depending on what a person is set to perceive.

MOTIVES AND NEEDS

In the late 1940s and through the 1950s, many psychologists turned their attention to the idea that motives and needs influence perception. This viewpoint was called the "new look" in perception. Although many of the "new look" experiments were flawed and some of the specific theoretical ideas of the "new look" have been sharply criticized, the general idea that individual differences in motives and needs affect perception persists. In other words, we may attend to and organize sensory inputs in ways that match our needs. For example, people who are hungry, thirsty, or sexually aroused are likely to pay attention to events in the environment which will satisfy these needs.

Projective tests, such as the well-known Rorschach inkblot test (Chapter 13, page 550), capitalize on the influences of motivation on perception. The inkblots or pictures used in projective tests are ambiguous; they can be perceived in any number of ways. The idea is that people's motives will, to some extent, affect the ways in which they organize and perceive the test stimuli. (See also Chapter 7, page 281.) A psychologist may be able to infer from the perceptions what motives are dominant in a particular person.

PERCEPTUAL-COGNITIVE STYLES

People are said to differ in the ways they typically and characteristically process information. The general processing strategies that characterize different people are known as *perceptual-cognitive styles*.

Among the many dimensions along which people vary in perceptual-cognitive style are (1) the degree to which their perceptions (and other aspects of their behavior and personality) are flexible or constricted (Klein, 1970) and (2) their field dependence or field independence (Witkin & Goodenough, 1981).

People whose perceptions are at the flexible end of the flexible-constricted dimension are said to have a wider focus of attention, to be less affected by interfering influences (as in the color-word test in the figure following page 109), and to be less dominated by internal needs and motives than are people at the constricted end.

The dimension of field dependence–field independence has to do with the perception of wholes or parts. A field-dependent person is said to unify and organize sensory inputs so that it is difficult to break down what is perceived into its parts or elements; the perception of a field-dependent person is thus said to emphasize the whole over its component parts. Such a person, because of his or her difficulty in breaking the whole down into its parts, may have difficulty with tests requiring that a simple figure be found within a complex whole (Figure 3.47). Field-independent people, who emphasize the parts in perception, do well on such embedded-figure tests and on other perceptual tests requiring the emphasis of parts over the whole.

Summary

1. Each sensory system is a kind of channel consisting of a sensitive element (the receptor) and nerve fibers leading from this receptor to the brain or spinal cord. The process of converting physical energy into activity within the nervous system—transduction—occurs at the receptors. The transduction process leads to receptor potentials—voltages within the receptor cells; further electrical changes, known as generator potentials, trigger the patterns of nerve impulses that are the afferent codes corresponding to events in the environment.

2. The visible spectrum—that part of the whole electromagnetic spectrum that receptors in the eye can detect—extends from about 380 to 780 nanometers. Light is refracted by the cornea and lens of the eye and is brought to a focus on the retina at the back of the eyeball. The receptors in vision are the rod and cone cells of the retina. These cells have different functions: Cones are active in

Figure 3.47

The target on the left is embedded in the complex drawing at the right. Finding the target in the complex drawing is a difficult task for field-dependent people. See text. (From Coren et al., 1979.)

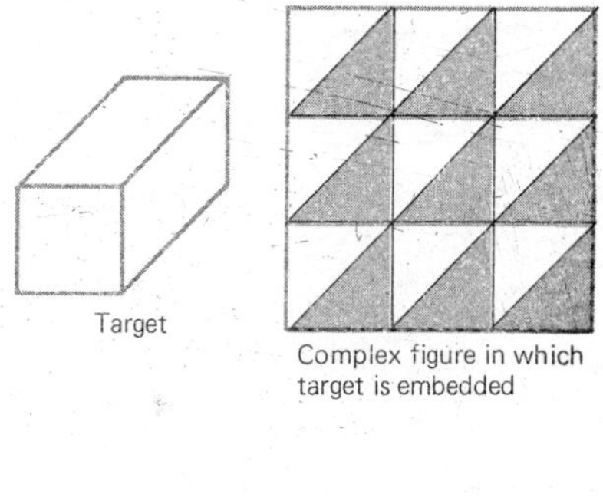

bright light, or daylight, vision; Rods are active in very dim light. Cones are the receptors responsible for our sharpest, or most acute, vision and for sensations of color. Transduction in the visual system occurs when the photosensitive pigments of the rods and cones are changed in shape, or configuration, by light energy. The receptor potentials, or voltages, of the rod and cone cells are passed across the network of the retina to the ganglion cells, where nerve impulses are triggered. These nerve impulses then travel in the optic nerves and the optic tracts to the lateral geniculate body of the thalamus. From the thalamus, the visual pathway continues to the primary visual sensory area at the back of the brain.

3. What is it in physical energy that is related to the visual sensations of brightness, hue (color), saturation, and form? Sensations of hue, or color, depend primarily on the wavelength of light; the color circle shows this relationship and also depicts the law of complementary colors. A color's saturation is the degree to which it is not diluted by whiteness; a pastel color has considerable whiteness in it and is said to be unsaturated. The intensity of the physical stimulus is the major determiner of visual brightness. The pattern of energy projected on the retina partially determines form.

4. The afferent code for hue, or color, depends on (*a*) the fact that the retina has three types of cones that absorb light best, or maximally, at three different parts of the visible spectrum and (*b*) opponent-process mechanisms at work in the ganglion cells, lateral geniculate cells, and cells of the visual cortex. There are four opponent-cell types, and the patterns of activity in these cell types seem to be an important part of the afferent code for hue. Activity patterns in what are called nonopponent cells are said to be involved in the afferent code for brightness. The afferent code for saturation may depend on the relative amounts of activity in opponent and nonopponent cells.

5. One idea about the afferent code for form vision is that visual form may be represented in the brain by distinctive and complex patterns of activity in millions of simple "feature-detector" nerve cells. The spatial-frequency theory maintains that form vision depends on the firing patterns of brain cells that respond to variations in the rates of change in brightness from one part of a visual scene to another.

6. Waves of pressure moving through the air are the physical stimuli for hearing. The unit used to express the intensity of sound pressures is the decibel (dB). The frequency of a sound wave is the number of cycles of pressure change occurring within 1 second; 1 cycle per second is called a hertz (Hz), and frequency is expressed in terms of hertz. Complex sound waves are periodic or aperiodic; periodic sound waves contain a fundamental frequency and several higher multiples of this frequency known as harmonics.

7. When sound waves strike the eardrum, they create vibrations in the middle-ear bones—the ossicles. Vibrations of these bones transmit sound energy to the oval window of the cochlea in the inner

ear. As the oval window vibrates, waves are generated in the fluid in the canals of the cochlea. The waves in the fluid of the cochlear canals produce bending movements of the fine hairlike processes on the ends of the hair cells of the organ of Corti. The bending of these hair-cell fibers is the event responsible for transduction in the ear.

8. The frequency of sound waves is related to sensations of pitch; the audible range for human beings is from about 20 hertz to about 20,000 hertz. The intensity of sound waves is related to loudness. Tonal quality, or timbre, is related to the complexity of the sound waves.

9. Two afferent codes seem to be used to represent pitch. One of these is a place code—the "pitch is which" code. For frequencies above 1,500 hertz, pitch sensation depends on the fact that different portions of the organ of Corti are maximally stimulated by different frequencies. The brain interprets stimulation from different parts of the organ of Corti as different pitches. For frequencies below about 1,500 hertz, pitch is represented by fibers in the auditory pathway that fire in step with the frequency of the physical stimulus.

10. The receptors for smell respond to volatile chemical substances. A number of attempts have been made to identify the basic odors that, if mixed in right proportions, might account for all the odors we sense. For instance, one system says that there are four basic odors: fragrant, acid, burnt, and caprylic (goaty or sweaty). But there are many other systems, and the basic odors, if they exist, are not agreed upon.

11. The receptors for taste are specialized cells grouped together in little clusters known as taste buds. Taste receptors respond to chemical substances dissolved in solutions that bathe the tongue. Several lines of evidence point to the existence of four basic taste qualities: salty, sour, sweet, and bitter. Since individual taste fibers respond to several taste stimuli, the afferent code for a particular taste probably consists of a pattern of firing in groups of fibers in the taste nerve.

12. Four skin senses are usually distinguished: pressure or touch, cold, warmth, and pain. Part of the evidence for four separate skin senses comes from observations of the punctuate sensitivity of the skin—the spots of greatest sensitivity to touch, cold, warmth, and pain stimuli are different. The stimulus for touch is a gradient of pressure on the skin or bending of the skin hairs. Meissner corpuscles, basket nerve endings around hair follicles, free nerve endings, and Pacinian corpuscles are some of the touch receptors in the skin. Experiences of cold and warmth are elicited by changes in the normal gradient of skin temperature—that is, by changes in the difference between the temperature of the skin surface and the temperature of the blood circulating beneath it. Because cold and warmth spots are found at different places on the skin, sensations of warmth and cold seem to depend on activity in two separate sensory channels. The cold and warmth receptors are specialized free nerve endings.

13. The stimulus for pain, called noxious stimulation, seems to be an event which damages bodily tissues in some way. The receptors for pain are free nerve endings specialized in some way to respond to noxious stimulation and not to other kinds of stimulation. Pain "gates" in the central nervous system seem to block the transmission of pain signals and thus reduce the amount of pain perceived. These "gates" seem to be activated by morphine and morphine-like compounds, some of which are produced by the body itself. Acupuncture anesthesia and placebos which reduce pain may work in part because they activate these "gates."

14. Perception is defined as what is experienced by a person. Sensory-input patterns are said to provide only the "raw material" for experience. Perception—our experience of the world—arises from sensory inputs plus the ways in which we process this sensory information. Illusions are examples of perceptual processes at work.

15. Attention is the term given to the perceptual processes that select certain inputs for inclusion in conscious experience, or awareness, at a given time. The processes of attention divide our field of experience into a focus and a margin. The concepts of filtering and processing capacity are used to explain why attention has a focus and why this focus shifts from time to time.

16. Perhaps the most fundamental process in form perception is the recognition of a figure standing out from a ground. We are able to separate forms, or figures, from the general ground in our visual perception because we perceive contours. When several objects are present in the visual field, we tend to perceive them as organized into patterns, or groupings. The major principles of perceptual organization are (*a*) the proximity, or nearness, principle, (*b*) organization in terms of similarity, (*c*) the symmetry, or good figure, principle, (*d*) continuation, and (*e*) closure.

17. The basis for visual depth perception is to be found in monocular and binocular cues. Among the monocular cues are (*a*) linear perspective, (*b*) clearness, (*c*) interposition, (*d*) shadow patterns, (*e*) gradients of texture, and (*f*) the relative movement of objects closer or farther away from the fixation point. The major binocular cue for visual depth perception depends on the slightly different, or disparate, views of the world received by the two eyes; this binocular cue is known as retinal disparity.

18. Perceptual constancy refers to the fact that the environment as we perceive it changes much less than do our sensory inputs; the world remains relatively stable despite drastic changes in the sensory input. For example, size constancy refers to the fact that despite large variations in the size of the representation, or "image," of an object on the retina when it is near or far, we tend to perceive the object as about the same size. A number of experiments have shown that perceptual size constancy results when an object and its background change together so that the relationships between them stay the same. The phenomenon of size constancy is the basis for a number of visual illusions. In brightness constancy, the perceived brightness of an object changes far less than do the changes in the

sensory input. Brightness constancy seems to depend on the ratio of illumination falling on an object and its background: Unchanged brightness ratios result in constant brightness experiences.

19. The perception of both real motion and apparent motion illustrates perceptual processes at work. Among the perceptual processes involved in perceiving real motion are movement constancy and the concept of a "brain comparator." The brain comparator is a system that cancels the perception of motion caused by eye and head movements. Apparent motion is movement perceived in the absence of physical movement. Stroboscopic motion, as seen in movies and on television; autokinetic motion; and induced movement are examples of apparent motion.

20. Perception is said to be plastic; this refers to the modifiability, or moldability, of perceived experience. The plasticity of perception has been demonstrated in experiments that restricted subjects' visual input during what were believed to be their sensitive periods for perceptual development; such restriction of visual inputs is known as visual deprivation. In general, restriction of visual inputs during sensitive periods results in both permanent impairments of perception and in changes in the brain processes underlying perception. Many of the plastic changes in perception seem to support the empiricist idea that nurture, or the environment, plays a large role in the development of perceptual processes. Nativists—those who argue that it is innate, or inborn, processes that affect perception—explain some of the visual-deprivation effects by saying that environmental influences during sensitive periods are necessary to maintain the inborn brain organization underlying perception. Thus the nativist, or nature, explanation for the results of some visual-deprivation experiments is "If you don't use it, you lose it."

21. People differ in the ways that they process sensory inputs to give rise to what they experience. Thus there are individual differences in perception. Among the factors that influence an individual's perception are (*a*) perceptual learning, (*b*) differences in what the person expects to perceive, or the person's set, (*c*) motives and needs, and (*d*) the individual's characteristic perceptual-cognitive style.

Terms to Know

One way to test your mastery of the material in this chapter is to see whether you know what is meant by the following terms.

Kinesthesis (85)
Vestibular sense (85)
Receptors (85)
Transduction (85)
Receptor potential (85)
Generator potential (85)
Afferent codes (86)
Psychophysics (86)
Wavelength (86)
Electromagnetic spectrum (86)
Visible spectrum (86)
Nanometers (nm) (87)
Refraction (87)
Accommodation (88)
Retina (88)
Rods (88)

Cones (88)
Blind spot (88)
Fovea (88)
Duplicity theory of vision (88)
Photosensitive pigments (88)
Rhodopsin (88)
Trans-rhodopsin (88)
Ganglion cells (90)
Hue (90)
Law of complementary colors (91)
Saturation (91)
Trichromatic process (93)
Opponent-process mechanism (93)
Nonopponent cell (93)
Spatial frequency (94)
Sound wave (97)
Intensity (97)
Decibel (dB) (97)
Frequency (98)
Hertz (Hz) (98)
Complex waves (98)
Sine waves (98)
Periodic waves (98)
Fundamental frequency (98)
Harmonic frequency (98)
Timbre (99, 101)
Auditory canal (99)
Eardrum (99)
Ossicles (99)
Cochlea (99)
Organ of Corti (100)
Hair cells (100)
Pitch (101)
Place code (101)
Taste buds (103)
Papillae (103)
Punctate senstivity (104)
Meissner corpuscle (105)
Basket nerve ending (105)
Free nerve endings (105)
Pacinian corpuscles (105)
Pain "gates" (106)
Opiates (107)
Placebo (107)
Perception (107)
Illusions (108)
Attention (109)
Filtering (109)
Processing capacity (110)
Figure and ground (111)
Contours (111)
Proximity (113)
Law of proximity (113)
Good figure (113)
Similarity (113)
Law of good figure (113)
Continuation (113)
Law of closure (113)
Tachistoscope (114)
Monocular cues (114)
Binocular cues (114, 117)
Linear perspective (115)
Clearness in depth perception (115)
Interposition (115)
Shadows in depth perception (115)
Gradients of texture (117)
Movement in depth perception (117
Fixation point (117)
Retinal disparity (117)
Perceptual constancy (118)
Size constancy (118)
Brightness constancy (120)
Apparent motion (121, 123)
Real motion (121)
Brain comparator (122)
Stroboscopic motion (123)
Autokinetic effect (123)
Induced movement (123)
Plasticity of perception (124)
Visual deprivation (124)
Sensitive period (124)
Nature (126)
Nurture (126)
Nativists (126)
Empiricists (126)
Perceptual learning (127)
Set (128)
Perceptual-cognitive styles (129)

Suggestions for Further Reading

This chapter has presented some highlights of sensation and perception and has tried to show why this is an important field of study for psychologists. As a next step toward a more complete understanding of sensation and perception, you might wish to look at some of the following: *Sensation and Perception* by Stanley Coren, Clare Porac, and Lawrence Ward (New York: Academic Press, 1979); E. Bruce Goldstein's *Sensation and Perception,* 2d ed. (Belmont, CA: Wadsworth, 1984); *Eye and Brain: The Psychology of Seeing* (3d ed.) by Richard L. Gregory (New York: McGraw-Hill, 1978); *The Psychology of Visual Perception* (2d ed.) by Ralph Norman Haber and Maurice Hershenson (New York: Holt, Rinehart and Winston, 1980); Lloyd Kaufman's *Perception: The World Transformed* (New York: Oxford University Press, 1979); *Fundamentals of Sensation and Perception* by Michael W. Levine and Jeremy M. Shefner (Reading, MA: Addison-Wesley, 1981); *Human Information Processing: An Introduction to Psychology* (2d ed.) by Peter H. Lindsay and Donald A. Norman (New York: Academic Press, 1977); Jacqueline Ludel's *Introduction to Sensory Processes* (San Francisco: Freeman, 1978); Irvin Rock's *An Introduction to Perception* (New York: Macmillan, 1975); *Perception: An Applied Approach* by William Schiff (Boston: Houghton Mifflin, 1980); and Harvey R. Schiffman's *Sensation and Perception: An Integrated Approach,* 2d ed. (New York: Wiley, 1982).

From time to time, the *Scientific American* has articles on sensation-perception topics. These are excellent introductions to specific programs of research and can be found by using your library's index to *Scientific American* articles.

Good introductions to the physiology of sensory processing—transduction, afferent codes, and the like—are to be found in Neil Carlson's *Physiology of Behavior,* 2d ed. (Boston: Allyn and Bacon, 1980) and in *Physiological Psychology* by Mark Rosenzweig and Arnold Leiman (Lexington, MA: Heath, 1982).

For more advanced study, you might wish to consult one of the 10 volumes in the *Handbook of Perception* series edited by Edward C. Carterette and Morton P. Friedman (New York: Academic Press, 1973–1978).

Reviews of work on extrasensory perception and parapsychology can be found in the *Handbook of Parapsychology,* edited by Benjamin B. Wolman, Laura A. Dale, Gertrude R. Schmeidler, and Montague Ullman (New York: Van Nostrand Reinhold, 1977) and in C. E. M. Hansel's *ESP and Parapsychology: A Critical Reevaluation* (Buffalo, NY: Prometheus Books, 1980).

Principles of Learning

HEADING OUTLINE

CLASSICAL CONDITIONING
Acquisition, or Learning, of Conditioned Responses
Theories About Classical Conditioning
Extinction and Spontaneous Recovery in Classical Conditioning
Stimulus Generalization and Discrimination in Classical Conditioning
Significance of Classical Conditioning

INSTRUMENTAL CONDITIONING
APPLICATION 4.1: BEING BRAVE AND STAYING COOL WITH CLASSICAL CONDITIONING
Reinforcers and Punishers
Examples of Positive Reinforcement
Shaping Behavior with Positive Reinforcement
Extinction of Positively Reinforced Responses
Conditioned Positive Reinforcers
Schedules of Positive Reinforcement
Stimulus Generalization
Stimulus Discrimination
Negative Reinforcement and Escape Learning
Avoidance Learning
Punishment
Significance of Instrumental Conditioning

COGNITIVE LEARNING
Latent Learning, Insight Learning, and Imitation
Cognitive Proceses in Classical and Instrumental Conditioning

THE LEARNER AND LEARNING: SOME THINGS ARE EASIER TO LEARN THAN OTHERS
Prepared Behaviors
Unprepared and Contraprepared Behaviors

SUMMARY

TERMS TO KNOW

SUGGESTIONS FOR FURTHER READING

Chapter

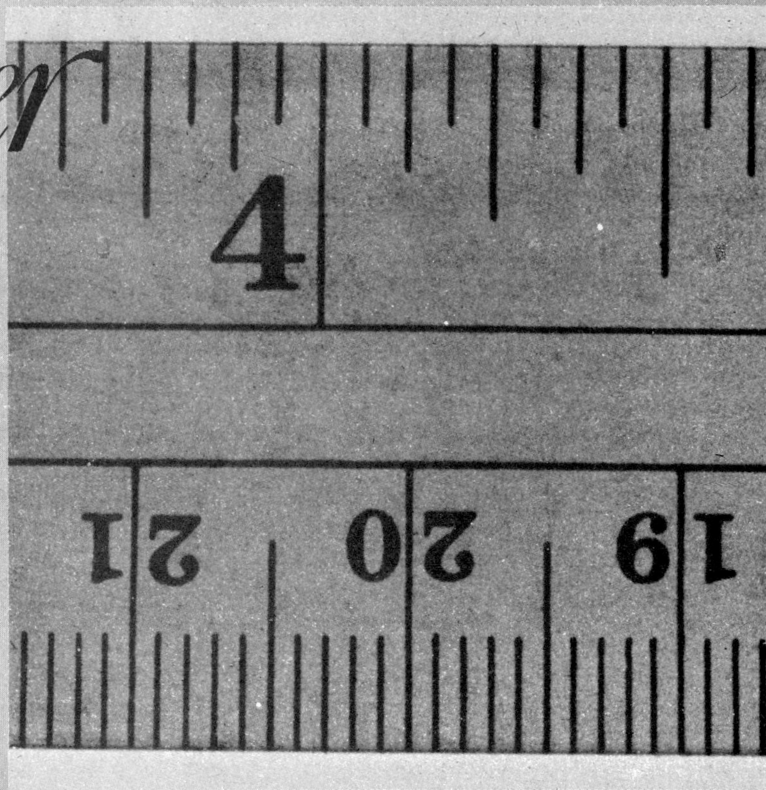

Here is a quotation from an influential psychology textbook. It is about the importance of learning—the "fundamental laws of behavior" in the quotation—in the process of molding a child to become a functioning member of society. Do you agree with the ideas in the quotation? Is it too one-sided? Read the chapter and then see what you think.

> The cultural environment (or, more exactly, the members of the community) starts out with a human infant formed and endowed along species lines, but capable of behavioral training in many directions. From this raw material, the culture proceeds to make, insofar as it can, a product acceptable to itself. It does this by training: by reinforcing the behaviors it desires and extinguishing others; by making some natural and social stimuli into discriminative stimuli, and ignoring others; by differentiating out this or that specific response or chain of responses, such as manners and attitudes; by conditioning emotional and anxiety reactions to some stimuli and not others. It teaches the individual what he may and may not do, giving him norms and ranges of social behavior that are permissive or prescriptive or prohibitive. It teaches him the language he is to speak; it gives him his standards of beauty and art, of good and bad conduct; it sets before him a picture of

the ideal personality that he is to imitate and strive to be. In all this, the fundamental laws of behavior are to be found.

[Keller & Schoenfeld (1950), *Principles of Psychology*, pp. 365–366]

A good example of the application of learning principles is to be found in what is known as behavior therapy—the use of learning principles to help people with behavioral problems. (Chapter 16 has a detailed account of behavior therapy.) Here is a case that was treated by the application of a number of the learning principles to be described in this chapter. It is based on a case described in S. R. Walen, N. M. Hauserman, and P. J. Lavin's *Clinical Guide to Behavior Therapy* (Baltimore: Williams & Wilkins, 1977, pp. 73–76). As you read the chapter, you will see that the treatment used classical conditioning as well as reinforcement and punishment principles from instrumental, or operant, conditioning. Can you identify these aspects of the treatment now? See if you were right after reading the chapter.

Billy is 9 years old but still wets his bed almost every night. He is ashamed of this and cannot participate in overnight visits with his friends; and, of course, his parents are concerned. What can be done to help him? The therapists who worked with Billy decided to try to teach him to remain dry during the night by using a number of basic learning principles.

A device called (of all things) a "Wee Alert" was purchased and installed in Billy's bed. The Wee Alert consists of two flat metal sheets separated by a layer of material that will absorb urine. When the absorbent layer gets wet, a small current of electricity flows between the two metal sheets to complete a circuit that sounds a buzzer. The buzzer was placed near Billy's bed. When he wet the bed, the buzzer went off and awakened him. The idea here was to associate the stimuli from a full bladder and the urge to urinate with waking up. If Billy woke up in time, he could go to the bathroom before he wet the bed.

In addition to the Wee Alert treatment, Billy, his mother, and the therapist agreed to a therapy contract. When the bed was wet, Billy had to change the sheets himself with no help from his mother. When the bed was dry, Billy was given a 10-cent credit and his mother made the bed. After 21 consecutive dry nights, the Wee Alert would be removed, but the therapy contract would remain in force.

On the sixty-second treatment night, Billy's bed had been dry for 21 days, so the Wee Alert was removed, but the therapy contract remained in force. Billy wet the bed twice after this, but that was all; and after a few weeks, the therapy contract was ended. During the next 18 months of follow-up, Billy did not once wet the bed.

LEARNING is a key process—some would say *the* key process—in human behavior; it pervades everything we do and think. (But see Chapter 2, pages 39–45.) It plays a central role in the language we speak, our customs, our attitudes and beliefs, our goals, our personality traits (both adaptive and maladaptive), and even our perceptions. Having mastered its fundamental principles, you will be in a position to understand how learning plays its important role in many of the psychological processes described later in the text.

Learning can be defined as any relatively permanent change in behavior that occurs as a result of practice or experience. This definition has three important elements.

1. *Learning* is a change in behavior, for better or worse.

2. It is a change that takes place through practice or experience; changes due to growth or maturation are not learning. This part of the definition distinguishes learning from innately controlled species-typical behavior of the sort that Chapter 2 (page 39) describes.

3. Before it can be called *learning*, the change must be relatively permanent; it must last a fairly long time. Exactly how long cannot be specified, but we usually think of learned changes in behavior as lasting for days, months, or years, unlike the temporary behavioral effects of factors such as alertness or fatigue.

Think about yourself and learning. You are reading this book; you will learn something about psychology from it, but book learning is only a small part of the learning you have done in your life. When you got up this morning, you dressed in a certain way, you ate certain things for breakfast (and did not eat others), and you began to think about the day ahead of you. Thinking invoked your attitudes about other people or events; perhaps it caused you to worry about something. How you dressed, what you ate, what you thought about, how you evaluated other people and events, what you worried about—are all rooted in your past experience, or *past learning.* If we could understand the learning process and apply our general understanding of it to a particular person's life, we would go a long way toward explaining many of the things that person does. If we could understand some of the principles of learning, we would have a better idea of how to change behavior when (as in child rearing and psychotherapy) we want to change it. (See Chapters 11 and 16.)

Classical Conditioning

Classical conditioning gets its name from the fact that it is the kind of learning situation that existed in the early "classical" experiments of Ivan P. Pavlov (1849–1936). In the late 1890s, this famous Russian physiologist began to establish many of the basic principles of this form of conditioning. Classical conditioning is also sometimes called *respondent conditioning* or *Pavlovian conditioning.*

ACQUISITION, OR LEARNING, OF CONDITIONED RESPONSES

What is classical conditioning? Suppose we start with some examples. Here is one from Pavlov's laboratory:

> Pavlov designed an apparatus, shown in Figure 4.1, that could measure how much a dog's mouth waters in response to food or other things in its environment. At the beginning of his experiment, Pavlov noted that no saliva flowed when he rang a bell. He then trained the dog by sounding the bell and, shortly afterward, presenting food. After the sound of the bell had been paired with food a few times, he tested the effects of the training by measuring the amount of saliva that flowed when he rang the bell and did *not* present food. He found that some saliva was produced in response to the bell alone. He then resumed the training—paired presentations of bell and food—a few more times and then tested again with the bell alone. As training continued, the amount of saliva on tests with the bell alone increased (up to a point, of course). Thus, after training, the dog's mouth watered—salivated—whenever the bell was sounded. This is what was learned; it is the *conditioned response.*

Here is another example from an experiment done by one of the authors many years ago (King et al., 1961).

> Student volunteers were the subjects in this experiment. Each student sat in a booth in which a brief jet of air could be puffed at his or her right eye. The response to this puff was a sharp blink of the eyes. One-half second before each puff, a dim spot of light came on. Tests showed that at the beginning of the experiment, the students

Figure 4.1

Pavlov's apparatus for studying the conditioned salivary (drooling) response.

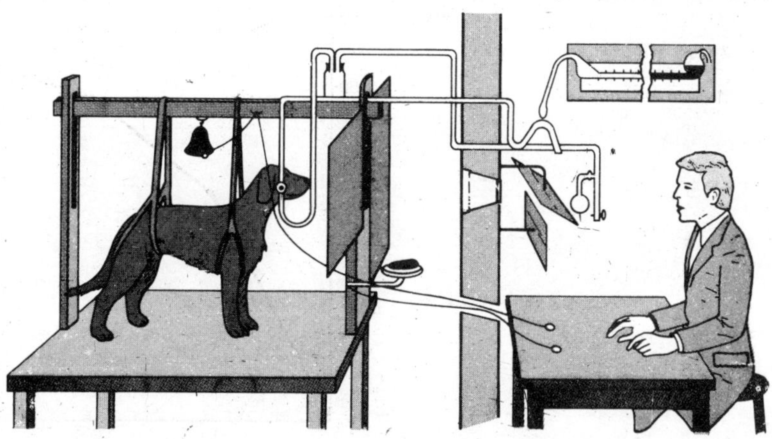

did not blink in response to the light. The light and puff were paired—light followed by puff—a number of times. Soon the students began to blink when the light came on *before* the puff. The number of blinks to the light increased steadily as more and more pairings were given. Figure 4.2 shows how the blinking to the light increased during the experiment. (Each point on the graph shows the percentage of blinks to the light during 10 pairings of light and puff.) The students had learned to blink when the light came on; this is the conditioned response in this experiment.

If you want a third example, look back at the opening pages of this chapter. What is the classically conditioned response in the therapy for bed-wetting?

What do these examples of classical conditioning have in common? In other words, what are the general characteristics of situations in which conditioned responses are acquired, or *learned*? In classical conditioning, two stimuli are presented to the learner. (The term *stimulus* comes from the Latin word for "goad," or "prod"; and thus, in psychology, the term is sometimes used to refer to events which evoke, or call forth, a response—the individual is "goaded into action." A more general meaning of the term *stimulus* is anything in the environment that can be detected by the senses.) One of the stimuli in classical conditioning is called the *conditioned stimulus (CS).* It is also known as a neutral stimulus because except for an alerting, or attentional, response the first few times it is presented, it does not evoke a specific response. Almost any stimulus which is detectable can serve as a CS. The bell and the light were the CSs in the two examples that opened this section. What was the CS in the behavior-therapy example at the beginning of the chapter? The other stimulus is known as the *unconditioned stimulus (US).* This stimulus consistently evokes a response or is reliably followed by one. The response that reliably follows the unconditioned stimulus is known as the *unconditioned*

Figure 4.2

A learning, or acquisition, curve for an experiment on the classical conditioning of eye blinking. (From King et al., 1961.)

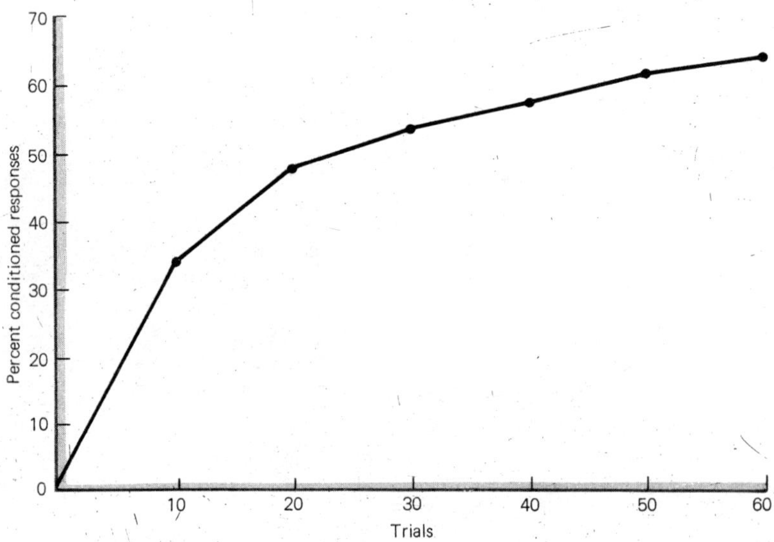

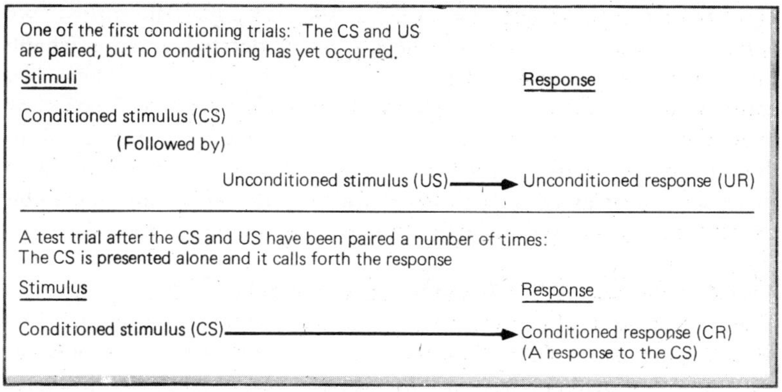

Figure 4.3

A chart of events in classical conditioning.

response (UR). What were the USs and the URs in the examples? The two stimuli—the CS and the US—are paired in classical conditioning so that the conditioned stimulus comes a short time—say from ½ second to several seconds—*before* the unconditioned stimulus is presented. After the stimuli have been paired a number of times (each pairing is called a trial), presentation of the originally neutral conditioned stimulus evokes a response. This response is what is learned in classical conditioning; it is termed the *conditioned response (CR).* Figure 4.3 diagrams the conditioning procedure.

The acquisition of a conditioned response is usually gradual; as more and more trials (CS–US pairings) are given, conditioned responses grow stronger and stronger or are more and more likely to occur. For example, in Figure 4.4, the left-hand curve shows the course of acquisition, or learning, typical in an experiment on salivary conditioning. (It is drawn without specifying the amount of saliva or the number of trials, but these values would be plotted on a graph in an actual experiment.) Figure 4.2 illustrates the course of acquisition of a conditioned eye-blink response. These are both examples of *acquisition*, or *learning, curves* in which the course of learning is followed over trials or time. Learning curves typically have the shape shown in the two examples; the rate of learning is

Figure 4.4

The course of conditioning, extinction, and reconditioning. Spontaneous recovery after a rest period is shown by the vertical arrow. (After McGeoch & Irion, 1952; adapted from Kimble et al., 1980.)

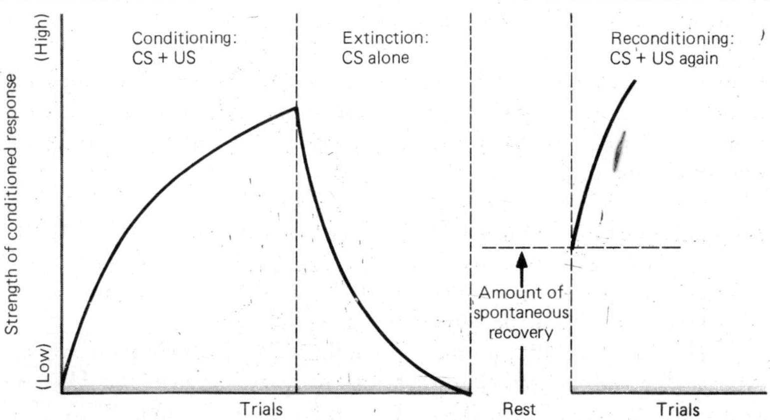

rapid at first but gradually decreases, as shown by the flattening of the curve. Such curves are said to be negatively accelerated. In other words, the increase in learning on later trials is less than was the increase on earlier trials. This is probably because there is a limit on the strength, or magnitude, of a conditioned response in a given experiment; after all, in the salivary conditioning experiment, the dog could only drool so much. Increases can be great on the early trials, but there is less and less to be added to the magnitude of the response as conditioning proceeds.

THEORIES ABOUT CLASSICAL CONDITIONING

Theories of classical conditioning try to describe and give order to the results of the many, many conditioning experiments that have been done; they are often mathematical in form (Rescorla & Wagner, 1972). These theories are also concerned with the processes occurring when a conditioned response is acquired. In other words, they speculate about the nature of the learning that takes place in classical conditioning. One older theory about the nature of classical conditioning is the theory of stimulus substitution; more recent and current ideas are the information and expectation theories.

Stimulus Substitution This theory, which originated with Pavlov and was influential for many years, relies on the idea that the CS, simply as a result of pairing with the US, acquires the capacity to substitute for the US in evoking a response. In other words, an association—a link or a bond—is formed between the CS and the US so that the CS becomes the equivalent of the US in eliciting a response. Pavlov thought this linkup, or association, took place in the brain. He thought that two areas of the brain, one for the CS and one for the US, became activated during the conditioning procedure and that activation of the US area resulted in a reflex, or automatic, response. As a result of the CS–US pairings during the conditioning procedure, he theorized, the CS acquired the ability to excite the US area, thus leading to the reflex response.

While the idea of stimulus substitution is appealingly simple, it is not currently accepted by most learning theorists. A major difficulty with the theory is that it says the conditioned response (CR) should be the same as, or at least very similar to, the unconditioned response (UR). According to this theory, all that has happened is that the CS has acquired the ability to evoke the response after conditioning. The response has not changed; the change is in the stimulus that elicits it. However, it is clear that the CR may not be at all like the UR. For instance, when using a mild foot shock as the US and a tone as the CS, the unconditioned response of rats to the shock is an increase in running and activity, but the conditioned response to the tone is a *decrease* in activity—a response known as "freezing".

Information and Expectation Other theories of the conditioning process take the viewpoint that the CS becomes a signal for the US. Thus when the CS is presented, the US is expected, and the learner responds in accordance with this expectation.

How does the CS become a signal for the US? A number of experiments have led to the view that this happens because the US is a *surprising*, or novel, event (Kamin, 1969; Terry & Wagner, 1975). The surprising US induces the learner to look back through recent memory. The CS is the event consistently found in memory on each trial before the US. An association, or link, is thus said to be formed between the memory trace of the CS and the US. Now when the CS occurs, the US is expected. The conditioned response is made in anticipation of the US. The form of the conditioned response often indicates that the learner expects the US. For example, if food is the US for a dog, salivation and many other responses associated with eating, such as running to the place where the food is kept if the dog is not confined, will occur. Thus, in this view, conditioned responses are not automatically stamped in by CS–US pairings. Rather, they are behaviors, or actions, engaged in by the learner in expectation of a future event—the US.

EXTINCTION AND SPONTANEOUS RECOVERY IN CLASSICAL CONDITIONING

In classical conditioning, *extinction* occurs when the CS is presented alone without the US for a number of trials. When this is done, the strength, or magnitude, of the CR gradually decreases as shown by the middle curve in Figure 4.4. For example, the number of drops of saliva decreases over unpaired trials; or blinks to a light CS gradually become less frequent. Note that the process of extinction is *not* "forgetting." A response is said to be forgotten over time when there is no explicit procedure involved. (See Chapter 5, page 203.) The process of extinction, however, involves a specific procedure—presentation of the CS by itself.

Just as with acquisition, there are several views concerning why the extinction process works. Pavlov thought of conditioning in terms of two opposing tendencies: excitation and inhibition. During acquisition, the excitatory tendency has the upper hand; but during extinction, inhibition builds up to suppress conditioned responding. Another view of the extinction process stems from the information-expectation theory of conditioning described in the last section. Because, during extinction, the CS is no longer paired with the US, the CS ceases to be a signal for the US; the CS becomes a neutral stimulus, as it was before conditioning occurred, and little attention is paid to it (Rescorla, 1967).

The decrease in conditioned-response magnitude resulting from extinction need not be permanent. Suppose, the day after extinction of a salivary conditioned response, a dog is brought back into the laboratory and the tone CS is presented. The magnitude of the dog's conditioned response will probably be much greater than it was at the end of extinction the day before (Figure 4.4). Such an increase in the magnitude of a conditioned response after a period of time with no explicit training is known as *spontaneous recovery*. The phenomenon of spontaneous recovery shows that the extinction procedure, while decreasing the magnitude of a conditioned response, does not entirely remove the tendency to respond to the CS. That extinction does not completely erase conditioning is also shown by the fact that *reconditioning* is usually more rapid than was the original conditioning. To recondition after extinction, the experimenter again pairs the CS and US from the original conditioning. When Pavlov did this, he got the general results shown by the right-hand curve of Figure 4.4; reconditioning following extinction occurred more rapidly than did the first conditioning. Thus some learning was left, after extinction, from the original conditioning. Indeed, an experimenter can condition and extinguish, condition and extinguish, and, up to a point, learning will be a little faster each time.

Pavlov used the phenomena of spontaneous recovery and faster reconditioning to support his inhibition idea of extinction. He assumed that inhibition from the extinction process gradually decays with time, and thus the excitatory tendency is less suppressed after an interval of a few hours or a day. On the other hand, the information-expectation view of spontaneous recovery stresses the attention paid to the CS by the learner. Remember that this theory says that the CS loses its excitatory value in extinction because the learner no longer pays attention to it; it is as if the CS becomes part of the background as it is presented over and over in the extinction situation. But the passage of time changes the situation, and when the CS now comes on it is novel again, the learner pays attention to it, and the CS is once more excitatory. Reconditioning is said to occur rapidly because the signal value of the CS has already been acquired during the original conditioning, and the learner simply carries this over to the reconditioning trials.

STIMULUS GENERALIZATION AND DISCRIMINATION IN CLASSICAL CONDITIONING

Pavlov discovered very early in his work that if he conditioned an animal to salivate at the sound of a bell, it would also salivate, though not quite so much, at the sound of a buzzer or the beat of a metronome. In other words, the animal tended to generalize the conditioned response to other stimuli that were somewhat similar to the original conditioned stimulus. Subsequent conditioning experiments have demonstrated this phenomenon of *stimulus generalization* over and over. The amount of generalization follows this rough rule of thumb: The greater the similarity, the greater the generalization among conditioned stimuli.

Generalization means that conditioned responses occur to stimuli that have never been paired with a specific unconditioned stimulus. It broadens the scope of classical conditioning. Consider the development of irrational fears, or *phobias*, by children. Insofar as conditioning and generalization play a role, the process might go something like this: A child is conditioned, accidentally perhaps, to fear something by its being paired with a fear-producing unconditioned stimulus. The fear becomes irrational when it generalizes, or spreads, to similar but harmless objects. For example, the original conditioning might have involved conditioned fear responses to a white, fluffy dog that bit the child. If this fear generalized to many white, fluffy things—other white animals, white blankets, white beards, and so forth, we would have an example of an irrational fear of white, fluffy things, or a *phobia*. This child might be afraid of Santa Claus or Uncle Mike, who happens to have a white beard; because of generalization, the fear has spread a long way from the original conditioning. The generalization of fear may make tracing it back to its conditioned origin difficult. But even though the specific conditioning that has led to some phobias cannot be discovered, these irrational fears can sometimes be eliminated by conditioning procedures that involve extinction and the learning of conditioned responses, such as relaxation, that are incompatible with being afraid. (See Chapter 16, page 701, and Application 4.1.)

Discrimination is the process of learning to make one response to one stimulus and a different response, or no response, to another stimulus.

Although many kinds of discrimination are possible, a typical discrimination experiment in classical conditioning involves learning to respond to one stimulus and not to respond to another. When we learn to respond to one stimulus and not to another, the range of stimuli that are capable of calling forth a conditioned response is narrowed. In a sense, then, this kind of discrimination is the opposite of generalization, or the tendency for a number of stimuli to call forth the same conditioned response.

A discrimination experiment in classical conditioning might go something like this: The experimenter, on some trials, pairs one stimulus (called the CS^+) with an unconditioned stimulus; on other trials, another stimulus (called the CS^-) is presented alone, without the unconditioned stimulus. In other words, while responses to the CS^+ are being conditioned on some trials, extinction of any tendency to respond to the CS^- is occurring on other trials. As a result, the learner forms a discrimination; conditioned responses are made to the CS^+ but not to the CS^-. The results of such a discrimination experiment might look something like those shown in Figure 4.5. We will consider generalization and discrimination in more detail later, in the context of instrumental conditioning (pages 159–161).

SIGNIFICANCE OF CLASSICAL CONDITIONING

We have dwelt long enough on drooling dogs. While Pavlov's study of salivary conditioning established many of the basic principles of classical conditioning, salivary responses have, by themselves, little significance in people's everyday affairs. What are important in human life are the emotional responses that become conditioned to certain stimuli.

Many of our subjective feelings—from our violent emotions to the subtle nuances of our moods—are probably conditioned responses. A face, a scene, or a voice may be the conditioned stimulus for an emotional response. Generalization and the fact that we learned many of these responses before we could talk and thus label them make it difficult to trace such feelings back to their conditioned beginnings. No wonder we are not always able to identify the origins of our emotional responses.

Since some emotional responses to stimuli are learned, perhaps they can be *un*learned. Or perhaps other, less disturbing responses can be associated with the stimuli that produce unpleasant emotional responses. The extinction and alteration of disturbing emotional responses by classical conditioning is one form of *behavior therapy*, or, as it is also called, *behavior modification*. (See Application 4.1 and Chapter 16, page 701.)

Figure 4.5

An idealized graph of the results of a classical conditioning discrimination experiment. The CS^+ is paired with an unconditioned stimulus, but the CS^- is not. As more and more trials are given, strong conditioned responses are produced by CS^+ and little conditioning occurs to the CS^-.

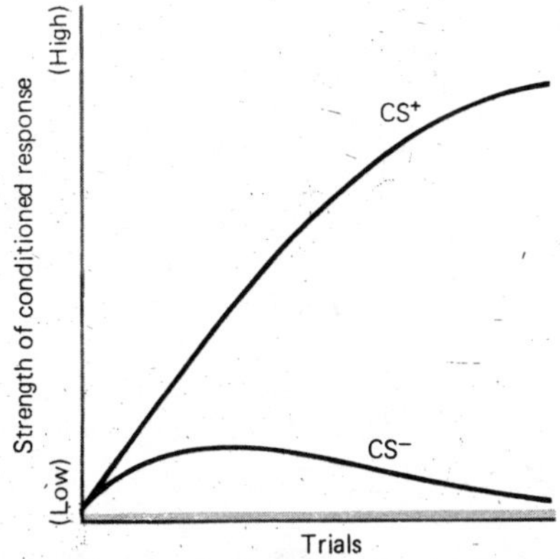

Instrumental Conditioning

Suppose your friend, who knows you are taking a psychology course, asks for advice on how to teach her young children to behave politely at dinner, watch less television, and do their household chores without constant prodding. What can you tell her that will be helpful? You know about classical conditioning, but that is not going to help much. For instance, how could the children be classically conditioned to say "thank you" when appropriate? What is the unconditioned stimulus for this or for other social behaviors? You need another form of learning to shape, or

Application 4.1

BEING BRAVE AND STAYING COOL WITH CLASSICAL CONDITIONING

Behavior therapy (Chapter 16, page 693) draws on learning principles in order to help people who have behavioral problems, and some of the principles behind behavior therapy come from classical conditioning. Here is an example of what is called *systematic desensitization*.

A young woman sought treatment for an intense fear, or phobia, of hypodermic syringes and needles. This fear apparently originated at the age of 2, when she received a series of shots for an intestinal disorder. From then on, she was extremely afraid of syringes and needles; and, on those few occasions when she needed routine inoculations, she had to be physically restrained. The "last straw" occurred in an emergency room where she was being treated for an injured ankle. Her struggling resulted in the doctor breaking two hypodermic needles in her ankle while trying to treat her. This convinced her that she needed treatment.

The first step in her systematic-desensitization treatment was teaching her to relax during the therapy sessions; remaining calm and relaxed is essential to this form of therapy. The standard form of systematic desensitization requires that the patient form vivid images of the feared objects or situations. Some of the images result in strong fear reactions; others produce weak responses, or no response at all. The images can thus be ranked from "strong" to "weak" in terms of the amount of fear they produce.

Since this patient was not able to generate vivid images of hypodermic syringes and needles, the therapist had to use real syringes and needles, arranging their presentation in such a sequence that the first stimuli produced no fear and later stimuli produced more and more fear. An example of a stimulus with low fear-producing potential was a brief glimpse of a small syringe held in the hand of the therapist at some distance from the patient. By varying the size of the syringes and needles, the length of time they were seen, and their distance from the patient, stronger and stronger fear stimuli were produced.

It is crucial in this therapy that the patient remain relaxed throughout the presentation of the stimuli, thus substituting relaxation for fear as the conditioned response. Having learned to relax in the presence of a weak stimulus, the next-stronger stimulus in the series was presented to the patient until she learned to relax when this one was present. This process continued until the patient could touch herself with a needle and remain relaxed. If there was a fear reaction to some of the stronger stimuli, as there often was with this young woman, the therapist dropped back to weaker stimuli before moving on again.

When the therapy was over after several sessions, the patient was no longer afraid of needles, syringes, or injections and remained calm when blood was drawn from her arm. A follow-up 10 months later showed that she was still free of fear and that no other phobia had developed to replace the old one.

You can do your own systematic desensitization. Suppose you are anxious about a tough examination that you will soon be taking. You have prepared for it, but you are still worried. One way to deal with your anxiety is to make a hierarchy of your thoughts and images about the examination, ranging from those that are weak anxiety provokers to those that are strong anxiety provokers. A weak stimulus, for instance, might be thoughts about the first day in class; then you might think about topics in the course that you liked; then think about studying for the exam; then think about what it will be like to walk into the examination room; as a next step, imagine the examination being handed out and reading the questions; and, finally, think about writing the answers to the questions. Lie down and relax while forming these images, and, should you become upset, go back to thinking about a less anxiety-provoking situation until you can remain relaxed while imagining stronger ones.

REFERENCE

Turnage, J. R., & Logan, D. C. (1974). Treatment of a hypodermic needle phobia by *in vivo* systematic desensitization. *Journal of Behavior Therapy and Experimental Psychiatry, 5,* 67–69.

mold, such behavior. To make the desired responses more likely and the undesired ones less likely, you need a set of techniques that can be applied to the ongoing behavior of the children. The techniques of instrumental conditioning will help do just that.

Instrumental conditioning is called *instrumental* because, as we will describe in detail, the key feature of this form of learning is that some action (some behavior) of the learner is instrumental in bringing about a change in the environment that makes the action more or less likely to occur again in the future. For example, if the environmental change is a reward, the instrumental behavior that brings about the reward will be more likely to occur in the future. In other words, if the behavior pays off, it is likely to be repeated.

Instrumental conditioning is sometimes also, roughly speaking, known as *operant conditioning*. This term was coined by B. F. Skinner to indicate that when a response operates on the environment, it may have consequences that can affect the likelihood that the response will occur again. We will use this term when describing the work of Skinner and others who use the methods pioneered by him.

REINFORCERS AND PUNISHERS

An environmental event that is the consequence of an instrumental response and that makes that response more likely to occur again is known as a *reinforcer* or a *reinforcement*. A *positive reinforcer* is a stimulus or event which, when it follows a response, *increases* the likelihood that the response will be made again (Figure 4.6*a*). It is important to note that the reinforcer is contingent upon the instrumental response. In other words, the response results in the occurrence of the reinforcer. This should not be a strange idea to you; responses which pay off in some way are likely to be repeated. For instance, food for a hungry animal, water for a thirsty one, praise from a parent, a prize, and many, many other stimuli or events will serve as positive reinforcers when they are contingent on some behavior.

Negative reinforcement is another tool in the kit of the instrumental conditioner. A *negative reinforcer* is a stimulus or event which, when its

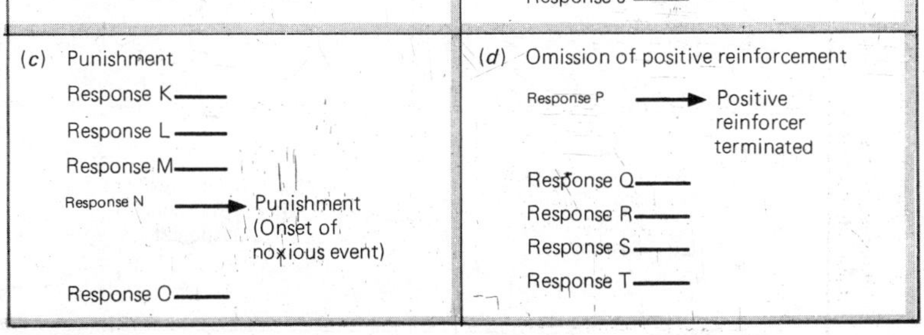

Figure 4.6

In box *a*, a response that results in positive reinforcement becomes more likely (shown by the larger type); other responses do not. In box *b*, if negative reinforcement is contingent on a response, the response becomes more likely (shown by the larger type). Boxes *c* and *d* show that responses resulting in punishment or omission of positive reinforcement decrease in likelihood (shown by smaller type).

cessation or *termination* is contingent on a response, *increases* the likelihood that the response will occur again (Figure 4.6*b*). The word *negative* refers to the fact that the response causes the termination of an event. Again, this concept should not be strange to you. Negative reinforcers are often (but not neessarily) painful, or *noxious*, events—an electric shock, for example, or a bawling out from the boss. The payoff is that the response, when made, stops the noxious event (the shock or the boss's abuse). We tend to repeat responses that pay off in this way.

Another technique that can be used in instrumental conditioning is *punishment*. A *punisher* is a stimulus or event which, when its onset is contingent on a response, *decreases* the likelihood that the response will occur again (Figure 4.6*c*). Note that a punisher is quite different from a negative reinforcer. While both are often noxious events, it is the ending of the event that is contingent on the response in negative reinforcement, but it is the onset of the event that is contingent on the response in punishment. Responses followed by punishers tend not to be repeated, while those followed by negative reinforcers tend to become more likely. The idea of punishment is probably all too familiar. Traffic fines, demerits, spankings, and the like can act as punishers when they are contingent upon particular behaviors.

A fourth response consequence sometimes used in instrumental conditioning is known as *omission of reinforcement*. In omission of reinforcement, or *omission training*, as it is often called, positive reinforcement is withdrawn following a response (Figure 4.6*d*). The effect of the omission of the reinforcement is to *decrease* the likelihood of the response that led to the removal of positive reinforcement. An example of this is parents' turning off the television set (a positive reinforcer) and sending a child to his or her room following some behavior. In everyday speech, this is sometimes referred to as "punishment," but from the definitions given here, you can see that, technically speaking, there is a difference between omission training and punishment.

So instrumental, or operant, conditioning has quite a few ways of changing behavior. When you have mastered these ideas, you should have something useful to say to the hypothetical friend in the opening paragraph of this section. We will not go into omission of reinforcement,

Figure 4.7

Contingencies in instrumental, or operant, conditioning. (See text.)

Consequence of a Response	Nature of the Event Which Follows—Is Contingent on—a Response: Appetitive (Approached or sought, e.g., water for a thirsty animal)	Aversive or noxious (e.g., electric shock)
Onset of event following response	Positive reinforcement (Increases likelihood of response)	Punishment (Decreases likelihood of response)
Termination of event following response	Omission of reinforcement, or omission training (Decreases likelihood of response)	Negative reinforcement (Increases likelihood of response)

but we will use the concepts of positive reinforcement, negative reinforcement, and punishment to organize the rest of this section on instrumental conditioning. Figure 4.7 is a summary of the contingency tools used in instrumental, or operant, conditioning. Now would be a good time to look back at the example that opened this chapter. What were the instrumental conditioning techniques used in the treatment of Billy?

EXAMPLES OF POSITIVE REINFORCEMENT

In instrumental conditioning, remember that a *positive reinforcer* is a stimulus or event which, when it is contingent on a response, makes that response more likely to occur. Here are some laboratory illustrations of positive reinforcement at work.

Thorndike's Experiments: Cats in a "Puzzle Box" The pioneering experiments of Edward L. Thorndike at the turn of the century were among the first systematic studies of the positive reinforcement principle in instrumental conditioning (Thorndike, 1898, 1911.) Here is a typical Thorndikean experiment:

> A hungry cat was locked in a "puzzle box" made of slats through which the cat could see a dish of food on the floor outside. A string from the door latch led over a pulley to a wire loop hanging in the box (Figure 4.8*a*). If the cat clawed at the loop, the door would open and the cat could escape from the box and eat. When first put in the box, the hungry cat actively scrambled around, scratching and clawing at the sides of the box. Eventually, in its random movement around the box, the cat happened to pull the loop, thus opening the door and escaping. Thorndike then put the cat back in the box for a second trial. Again the cat scrambled around until it accidentally pulled the loop, escaped, and was then put back in the box for a third trial. Thorndike and the cat kept this up for many trials. As

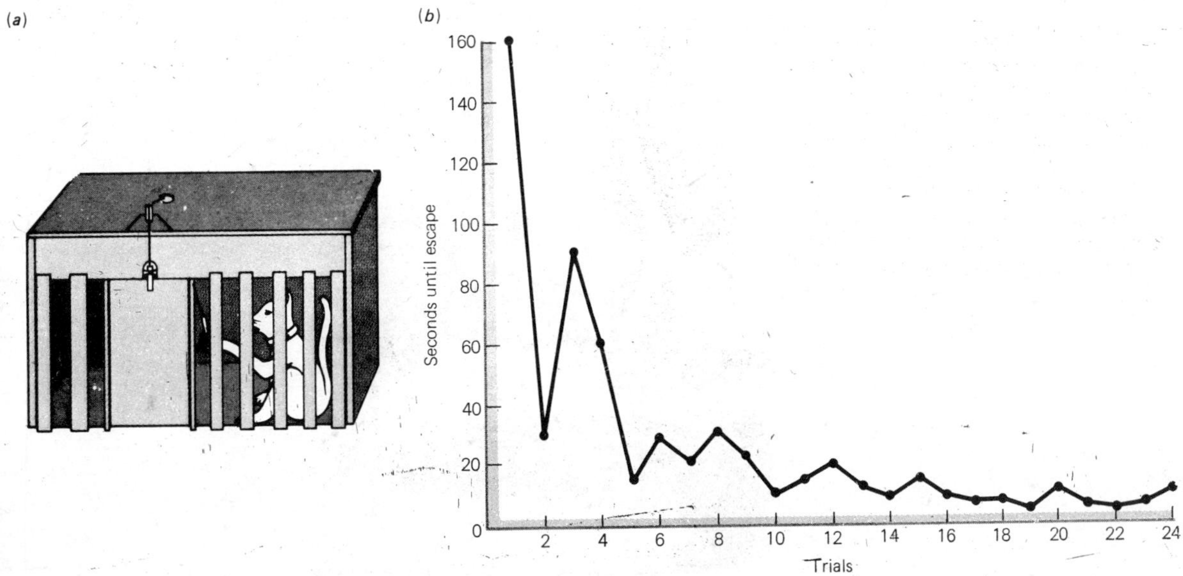

Figure 4.8

(*a*) The "puzzle box" used by Thorndike in his early experiments on instrumental conditioning. (*b*) The learning, or acquisition, curve for a cat in Thorndike's "puzzle-box" experiment. Although the curve is irregular, note the trend to take less time to escape as trials proceed. (Based on data from Thorndike, 1911.)

you can see from Figure 4.8*b*, which shows the escape time for 24 trials, the cat took less and less time to pull the ring each time it was put back in the box. Eventually, it escaped almost as soon as it was put into the box.

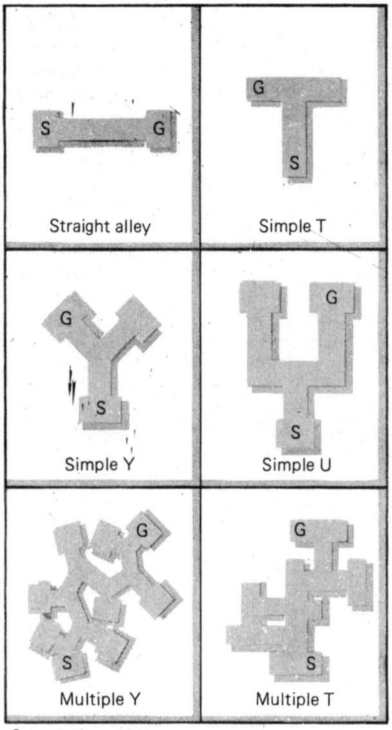

S = starting point
G = goal

Figure 4.9

Some examples of complex and simple mazes.

Mazes Since Thorndike's time, mazes have been popular devices for the study of instrumental conditioning with positive reinforcement. Mazes come in many forms (Figure 4.9), but, in general, the animal (typically a rat) is placed in a "starting box" and allowed to run through "alleys" to a "goal box," where there is a positive reinforcer. The simplest maze is a single, straight alley in which instrumental conditioning is manifested by faster running as the trials proceed. More complex mazes have choices and blind alleys, and the number of errors—blind-alley entrances—is the usual measure used. These errors decrease as instrumental conditioning proceeds.

Skinner's Experiments: Rats and Pigeons in an Operant Chamber In the 1930s, B. F. Skinner (page 594) began his influential experiments on what he termed *operant conditioning* (page 149). Skinner wanted better control of the learning situation than was provided by the "puzzle boxes" and mazes then used to study instrumental conditioning. He wanted a way to study reinforced responding without breaking the experiment up into discrete trials; and, for convenience, he wanted to automate the instrumental learning situation. To do all this, he invented the Skinner box, or, as it is often called, the *operant chamber*. An *operant chamber* is a simple box with a device at one end that can be worked by the animal in the box. For rats, cats, and monkeys, the device is a lever; for pigeons, the device is a small panel, called a "key," which can be pecked (Figures 4.10 and 4.11). The lever and key are really switches that activate, when positive reinforcement is being used, a food-delivery or water-delivery mechanism. Thus positive reinforcement is contingent upon pressing a lever or pecking a key. Since these responses are positively reinforced,

Figure 4.10

A rat in a Skinner box, or operant chamber. The rat has its paw on the lever and its nose at the spout of the feeder. When the animal presses the lever, a food pellet is delivered from the feeder. The delivery of a food pellet is contingent on the response of pressing the lever. The contingently delivered food pellets act to reinforce the lever-pressing response. (From Pfizer, Inc.)

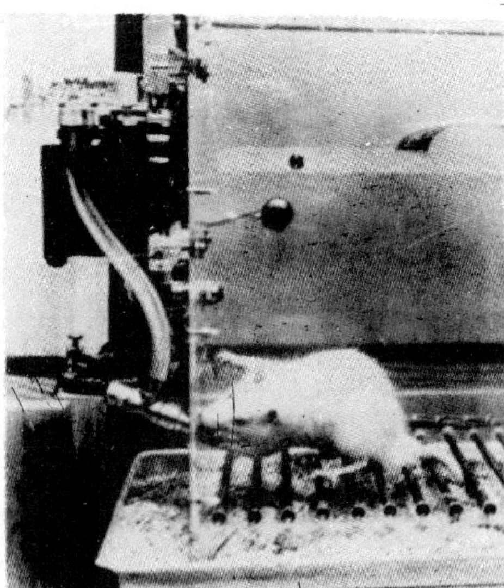

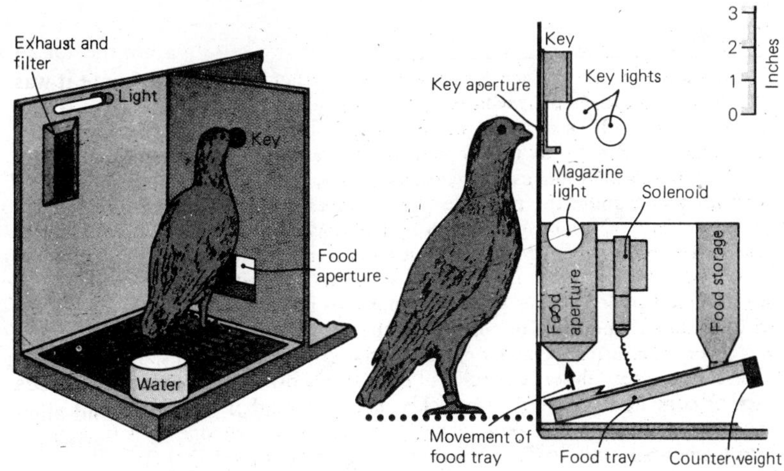

Figure 4.11

Left, cutaway drawing of an operant chamber for pigeons. Key pecking is the contingently reinforced response. Food is the reinforcer. When the pigeon pecks the key, a food tray comes up to the bottom of the food aperture, and the pigeon is allowed to eat for a few seconds. Right, a side view of the front part of the operant chamber for pigeons. Note especially the key and the food tray. The key is a translucent panel that can be illuminated—perhaps in different colors—by the key lights. (Modified from Ferster & Skinner, 1957.)

they increase in frequency. Here is an example of the way conditioning is done in an operant chamber:

> The first step in the operant conditioning of a hungry rat is to get it to eat the food pellets when they are delivered by the experimenter, who operates the pellet-delivery mechanism from a push-button switch outside the operant chamber. The pellets are delivered one by one; after a time, the rat eats each pellet as soon as it drops. This first step is necessary if the food reinforcement is to be effective later, when the rat will deliver the food pellets to itself by pressing a lever.
>
> Next, the experimenter stops releasing the pellets, and the rat is left alone in the box with the lever, which will release the pellets. After an initial period of inactivity, the rat, being hungry, begins to explore the box. Eventually, it presses the lever accidentally. A pellet of food is released; that is, reinforcement is contingent upon pressing the lever. After eating the food pellet, the rat continues exploring, stopping to groom itself from time to time. After a while, it presses the lever again, and again a pellet is released; then it presses the lever a third time. Usually after the fourth or fifth press, the rat begins to press the lever more rapidly, and operant behavior is in full swing.
>
> The experimenter counts the rat's lever presses. The number of responses within a particular unit of time—the *rate of response*—is often the measure used in studies of operant conditioning. The rate of response may be shown graphically by a device called a *cumulative recorder*. As Figure 4.12 shows, each response causes the recorder's pen to make a very small movement on a piece of paper that moves at a constant speed. Thus a cumulative and continuous record of the number of responses is plotted against time; in a way, the rat "draws" a record of its responses with this device. The slope of the response line is the measure of the rate of response: High response rates give steep slopes, low response rates shallow slopes. When the rat does not respond, a straight line with no slope is drawn on the

Figure 4.12

A cumulative recorder. Each response of the learner—say, a rat in an operant chamber—causes the recording pen to move, or step, a very small distance to the left as we view the recorder in the photograph. As responses accumulate, the pen thus moves gradually to the left. Time is represented by the paper, which moves at a constant speed. As the paper moves at a constant speed under the pen, the learner traces a record of responses over time. The rate of response is shown by the slope of the response line. High rates of response (as on this record) are indicated by a response line with a steep slope; low response rates (which appear occasionally on this record) are indicated by low slopes; and no response is indicated by a straight line with no slope. The short tick marks on the record show when reinforcement was given. (Note that reinforcement was not given for every response; this is discussed later, under schedules of reinforcement.) After the pen reaches the left edge of the paper, it is quickly reset to the other side of the paper, and the record continues. The record is thus a continuous one; it is broken into segments simply to keep it within the bounds of the paper strip. The line on the right (as we view it) does not record responses; experimental events or time intervals are recorded by tick marks on it. (From Ralph Gerbrands Co.)

cumulative recorder. In the demonstration we have been describing, the rate of response, plotted in Figure 4.13, was very low at first. The first response occurred after 15 minutes, the second about 30 minutes later. After about 30 more minutes, the response rate began to increase, and the slope on the cumulative record increased accordingly.

SHAPING BEHAVIOR WITH POSITIVE REINFORCEMENT

In the example above, the rat was left alone in the operant chamber and did many things before accidently making the positively reinforced response of pressing the lever; thus operant conditioning proceeded rather slowly. But through the process of *shaping*, it is possible to speed up operant conditioning and to condition quite complex responses.

To shape lever pressing in an operant chamber, an experimenter would proceed as follows:

Figure 4.13

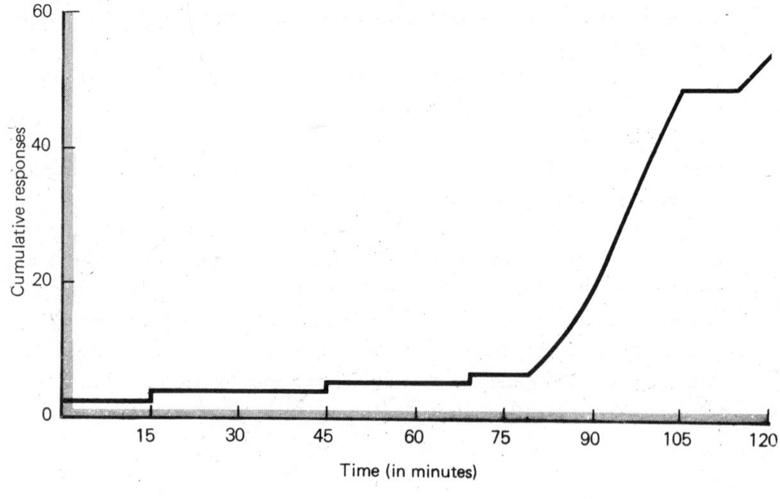

A record of responses from a rat in an operant chamber. Every lever-press response is reinforced. Note that this is a cumulative record: The responses made in one period are added to the responses made in preceding periods. The rat did not make its first response until about 15 minutes after being placed in the box; it did not make its second response until about 30 minutes later. The effect of food reinforcement becomes strong after about 75 minutes, and the rate of response then becomes high and fairly steady.

First, the animal would be allowed to get used to the chamber; next, it would be given food pellets from the food hopper, or "magazine," until it ate each pellet as soon as it dropped—that is, it would be "magazine-trained."

Then, shaping of the lever-press response would begin. At first, whenever the rat happened to wander into the front part of the chamber near the lever, the experimenter would press a switch releasing a food pellet, thus reinforcing this behavior. Then, the rat would be required to get a little closer to the desired response of pressing the lever before it was fed. For example, reinforcement would be given only when the animal moved close to the lever. Finally, when the rat began hovering near the lever, the experimenter would give reinforcement only if the animal happened to touch the lever. When the response of touching the lever was firmly established, the experimenter would then only give food pellets when the lever was pushed part way down, even if it was not pushed far enough to trigger the pellet dispenser. Finally, the rat would begin to push the lever hard enough to deliver its own reinforcements and the experimenter would stop giving them. Thus lever pressing would be established.

A skillful experimenter can shape behavior with very few reinforcements in a relatively short time. Note that the essential feature of shaping is teaching a chain of simple responses leading to the final response. In other words, the final response is learned because the steps leading to it are reinforced. Since these steps are approximations of the final response, the method of shaping is sometimes called the *method of successive approximations*.

Shaping is a classic concept in instrumental and operant conditioning: Reinforce the steps leading to the desired response and that response will eventually occur. But classical conditioning also seems to make an important contribution to the shaping process (Brown & Jenkins, 1968).

The subjects of this experiment were hungry pigeons who learned to peck a key in an operant chamber (Figure 4.11, page 153). Pigeons can be shaped to peck a key just as rats can be shaped to press a lever—by the use of contingent positive reinforcement and the method of successive approximations described above. In this experiment, however, classical—not instrumental—conditioning principles were used to train the birds to peck a key. The key was dark most of the time, but it was lighted just before food was presented. Neither the lighting of the key nor the presenting of the food depended upon any particular response by the pigeons. Thus, as in classical conditioning (pages 140–147), two stimuli were paired independently of the learners' responses. Key illumination was the CS and presentation of food was the US. After a number of CS–US pairings, the pigeons approached the key and began to peck it.

This classical conditioning method of shaping animals in an operant chamber has come to be called *auto-shaping*. In the case of a rat, auto-shaping might be done by using a retractable lever that can slide in and out of the operant chamber; the lever (the CS) is presented just before the food (the US) is presented, and then it is withdrawn until the next trial.

Auto-shaping seems to work because the key or lever CSs come to signal food and elicit conditioned responses similar to the unconditioned responses involved in eating. We might say that the animal "tries to eat" the key or lever. Pigeons peck at the key in very much the same way that they peck at grains of food; and, if water is used instead of food, thirsty pigeons peck at the key with a set of responses similar to those used in drinking (Jenkins & Moore, 1973).

When animals are shaped with contingent reinforcement according to instrumental conditioning procedures, some classical conditioning may also be occurring in the operant chamber. While an animal is being shaped to press a lever or peck a key by the use of positive contingent reinforcement of successive approximations, the likelihood of making one of these responses may also be increased by classical conditioning. Because of classical conditioning, the desired response may occur more readily and sooner in the shaping process than it would with only contingent reinforcement. In other words, we might say that the classical conditioning prompts the desired response and shortens the shaping procedure. When the desired response occurs, it is contingently reinforced, and its likelihood of occurrence is thus further increased.

EXTINCTION OF POSITIVELY REINFORCED RESPONSES

As we have seen, the likelihood of a particular response increases if it is followed by a positive reinforcer. On the other hand, if positive reinforcement no longer follows that response, the tendency for it to occur will decrease; responses which do not pay off tend not to be made. The procedure of not reinforcing a particular response is known as *extinction*. If the extinction procedure continues long enough, the likelihood of a response will decrease to about its level before it was reinforced. For example, a rat shaped to press a lever in an operant chamber in order to be fed will gradually decrease its rate of responding—if it is not fed—until it only occasionally presses the lever.

CONDITIONED POSITIVE REINFORCERS

Some positive reinforcers work the first time they are made contingent on a response. No previous special training is necessary for these positive reinforcers to have their effect; they work "naturally" to increase the likelihood of a response when they are made contingent on it. Such positive reinforcers are known as *primary reinforcers*.

In contrast, another group of positive reinforcers does not work "naturally." For these reinforcers to be effective, the learner must have had experience with them; their ability to reinforce instrumental responses depends upon learning. Such learned reinforcers are known as *conditioned*, or *secondary*, *reinforcers*.

Stimuli become conditioned reinforcers in instrumental conditioning by being paired with primary reinforcers. In an operant chamber, for example, suppose that a click occurs each time a food pellet—a primary reinforcer—is delivered. The click becomes a conditioned reinforcer. At first, the click stimulus has no reinforcing properties, but, by its presence every time the primary reinforcer is delivered, it becomes a reinforcer in its own right. The pairing of the click and the food pellet in this example is similar to the pairing of a CS and a US in classical conditioning (page 143); the click becomes the signal for food and, at the same time, acquires the ability to act as a positive reinforcer. To see whether the click is really a reinforcer, we can make its occurrence contingent on some response other than pressing a lever. Will the likelihood of this response increase? It will if the click has gained reinforcing properties. Or we might simply eliminate primary reinforcement for the lever-press response while continuing to have each press followed by a click. If there were nothing else to maintain the response of pressing a lever, we might expect the likelihood of this response to decrease rapidly, due to the process of extinction. But with the click present, the animal continues to respond for some time in the absence of primary reinforcement.

Conditioned reinforcers, being learned themselves, undergo extinction if they are used continually without being paired occasionally with a primary reinforcer. However, only a few pairings with a primary reinforcer from time to time are necessary to maintain the effectiveness of a conditioned reinforcer.

As far as instrumental conditioning is important in "real-life," practical stiuations, conditioned reinforcers have a large role to play. For instance, parents rarely use primary reinforcers to shape and maintain their children's behavior; instead they use praise, encouragement, and tokens of affection, which may sometimes be considered to be conditioned reinforcers.

SCHEDULES OF POSITIVE REINFORCEMENT

So far, except when describing extinction, our discussion of instrumental conditioning has focused on contingent positive reinforcement for *every* occurrence of a particular response. This is called *continuous reinforcement (CRF)*. A situation that is far more common in everyday life, and which can easily be studied in the laboratory, is one in which only *some* occurrences of a response are reinforced. In other words, there is a *schedule of reinforcement* in which, according to a prearranged plan, the response being studied is only sometimes followed by reinforcement.

Schedules of reinforcement have been most extensively studied in the operant situation devised by Skinner (page 152). Continuous reinforcement is usually used during the initial operant conditioning. After the response has been learned, it can then be maintained by a schedule of reinforcement. For example, in a *fixed-ratio (FR) schedule*, the response must occur a certain number of times before reinforcement occurs. Under a fixed-ratio schedule, the rat in an operant chamber might have to press the lever 19 times before it was reinforced by a food pellet contingent on the twentieth press. Fixed-ratio schedules are common in everyday life, where we must do a certain amount of work before the payoff. Each schedule of reinforcement tends to result in a certain pattern of behavior. The fixed-ratio schedule, for instance, generates a pattern in which there is a pause in responding after each reinforced response and then a rapid run of responses until the next reinforced response (Figure 4.14*a*).

As another example, consider a schedule of reinforcement based on time. The *fixed-interval (FI) schedule* is one in which no reinforcement is given (no matter how many times the response is made) until a certain interval of time—a minute, perhaps—has elapsed. The first response *after* the time interval results in reinforcement, and then no more reinforcement is given again until the first response after the interval has elapsed again. The characteristic pattern of responses in fixed-interval

Figure 4.14

Cumulative response records for four schedules of reinforcement. These are from experiments with pigeons where key pecking was the contingently reinforced response. There are three things to be noted in these records. First, the slope of the response lines indicates the rate of response; steep slopes indicate high rates, while shallow slopes indicate low rates. Second, the ticks, or slashes, on the response lines show when reinforcements were given. Third, although the record for each schedule is given in several sections, the response curves are really continuous; the pen resets to the bottom of the page (straight vertical line) from time to time. Compare with Figure 4.12. (*a*) Characteristic responding on a fixed-ratio (FR) schedule of reinforcement. Note the pauses after reinforcement and the high rate of response after the pauses. (*b*) A typical record for a fixed-interval (FI) schedule of reinforcement. Responses slow immediately after each reinforcement, and this gives the record a scalloped shape. (*c*) The high, steady rate of response characteristic of variable-ratio (VR) schedules of reinforcement. (*d*) Responses on a variable-interval (VI) schedule of reinforcement. Note that the reinforcement slashes come at variable intervals. Note also the high, steady rate of response, which makes this schedule a good one for establishing baseline performance to see the effects of various experimental manipulations—drug administration, for example—upon behavior. (Courtesy of G. Capehart, V. Kowlowitz, R. Newlin, and S. Simmerman.)

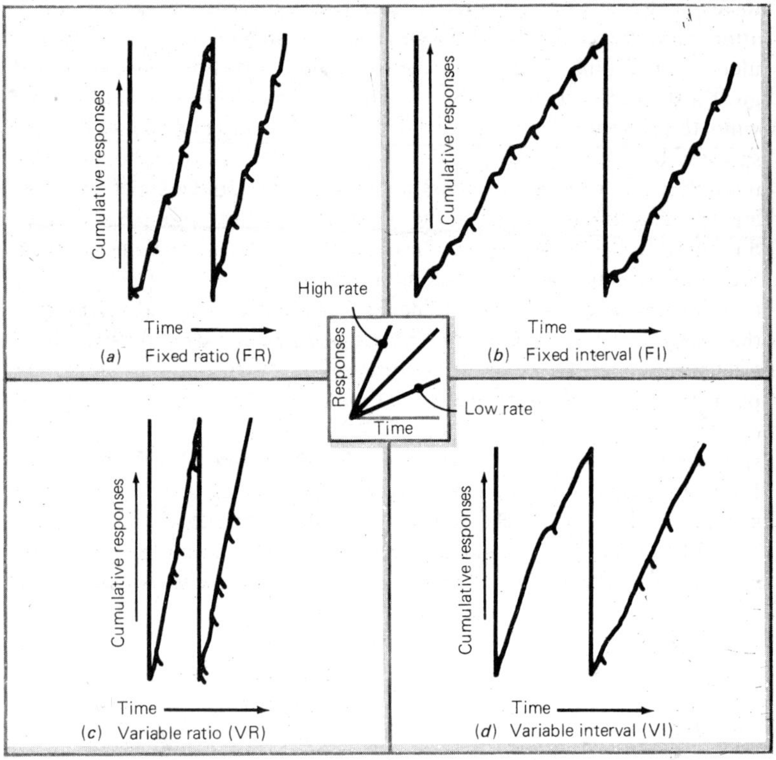

reinforcement consists of a gradual acceleration of responding during the interval to produce what is known as the "fixed-interval scallop" (Figure 4.14*b*).

Schedules can also be made variable. For example, there are variable-ratio and variable-interval schedules. Under a *variable-ratio (VR) schedule*, subjects are reinforced after a variable number of responses (Figure 4.14*c*). For instance, reinforcement might come after two responses, again after ten responses, again after six responses, and so on after different numbers of responses. A variable-ratio schedule is specified in terms of the average number of responses needed for reinforcement.

A variable schedule based on time is the *variable-interval (VI) schedule*. Under such a schedule, individuals are reinforced for the first response they make after various time intervals have passed (Figure 4.14*d*). For example, the first response after 30 seconds, then 15 seconds, then 60 seconds, and so on might be reinforced in a variable-interval schedule. Both VR and VI schedules generate relatively steady rates of response (Figure 4.14*c* and *d*).

The schedules of positive reinforcement just described are among the more common ones. But many other more complex schedules of reinforcement, each with its characteristic pattern of responding, are used for specific purposes.

An important consequence of many schedules of positive reinforcement is that other things being equal, extinction tends to be slower for schedule-reinforced responses than for continuously reinforced ones. In other words, if positive reinforcement is stopped, the individual continues to respond for a much longer time after scheduled reinforcement than after continuous reinforcement. In technical language, we say that scheduled reinforcement increases *resistance to extinction* (Figure 4.15). The great resistance to extinction after scheduled reinforcement is one reason why, in everyday life, we persist in making instrumentally learned responses long after reinforcement for these behaviors has ceased. Scheduled reinforcement has done its work to retard extinction.

STIMULUS GENERALIZATION

We have already seen that a response classically conditioned to a particular CS will also be made to other stimuli that are similar in some way to that CS (page 146). In instrumental conditioning, stimulus generalization also occurs. The response in instrumental conditioning is made in a particular stimulus stiuation—in an operant chamber with a certain type of light, for example. If the stimulus situation is changed, the response still occurs but less readily than it did in the original stimulus situation. Furthermore, the tendency to respond depends upon the degree of similarity between the original training situation and the changed one. The following experiment illustrates *stimulus generalization in instrumental conditioning* (Olson & King, 1962).

> Pigeons were shaped and trained with positive reinforcement to peck a disk, or "key," in an operant chamber (Figure 4.11, page 153). During the original learning, a moderately bright light illuminated the key. After the pecking response to this stimulus had been well learned and the rate of response was high and steady, the animals were tested with six other intensities on the key. These test stimuli were spaced in steps of equal intensity from low to high. In the graph of Figure 4.16, the original stimulus is called *8*; the more intense stimuli are 2, 4, and 6; and the less intense stimuli are 10, 12, and 14. The graph shows that the pigeons had a tendency to respond to these new stimuli and that the degree to which they responded depended on the size of the brightness difference between the original and test stimuli.

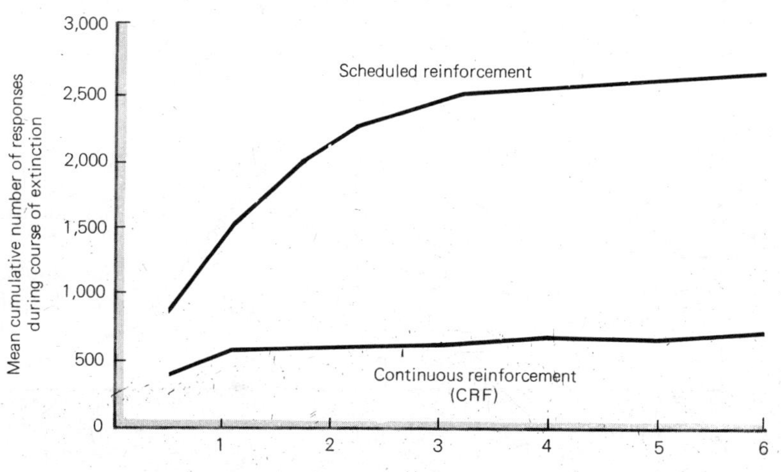

Figure 4.15

Resistance to extinction after continuous reinforcement and scheduled reinforcement. Responding is much more resistant to extinction after scheduled reinforcement. This is a cumulative record in which the number of responses during each hour of extinction is added to the number of responses made during previous hours. Note that the number of responses made during an hour decreases as extinction proceeds. (Modified from Jenkins et al., 1950.)

Figure 4.16 illustrates what is known as a ***gradient of generalization***. This simply means that the amount of generalization is graded—great or small—depending on how similar the test stimuli are to the original, or training, stimulus. A similar phenomenon exists in classical conditioning (page 146).

STIMULUS DISCRIMINATION

When classical conditioning was discussed, ***discrimination*** was described as the process of learning to make one response to one stimulus and another response—or no response—to another stimulus (page 146). This is also what discrimination means in instrumental conditioning. As in classical conditioning, a common type of discrimination experiment in instrumental conditioning results in learning to respond to one stimulus and not to another. In such an instrumental conditioning experiment, discrimination is achieved simply by reinforcing a particular response to one stimulus and not reinforcing—which amounts to extinguishing—the same response to another stimulus. The positive stimulus (S^+) is also called S^D in operant conditioning terminology. The negative stimulus (S^-) is also known as S^Δ (S-delta) in operant conditioning terms. The result of such an experiment is that when the positive stimulus (S^+) is present, the learned response is likely to be made; when the negative stimulus (S^-) is present, the response is less likely to occur or will not occur at all. Because the tendency to respond is tied to the stimulus that is present, the discrimination process in instrumental conditioning is sometimes referred to as the *stimulus control of behavior*. The following experiment illustrates discrimination learning, or the *stimulus control of behavior*, in an operant chamber (Hanson, 1959).

The pigeons in this experiment were positively reinforced for key-peck responses only when the translucent key was illuminated by a light that appeared yellow-green to human observers. During the intervals of yellow-green illumination, the pigeons received contingent positive reinforcement for pecking the key. If another light,

Figure 4.16

A gradient of generalization for an instrumentally conditioned response—key pecking by a pigeon. The greatest number of responses occurs when the key is illuminated by the stimulus present during conditioning—the original stimulus. As the test stimuli become more and more different from the original stimulus, the number of responses diminishes; this happens with both brighter and dimmer stimuli. (Modified from Olson & King, 1962.)

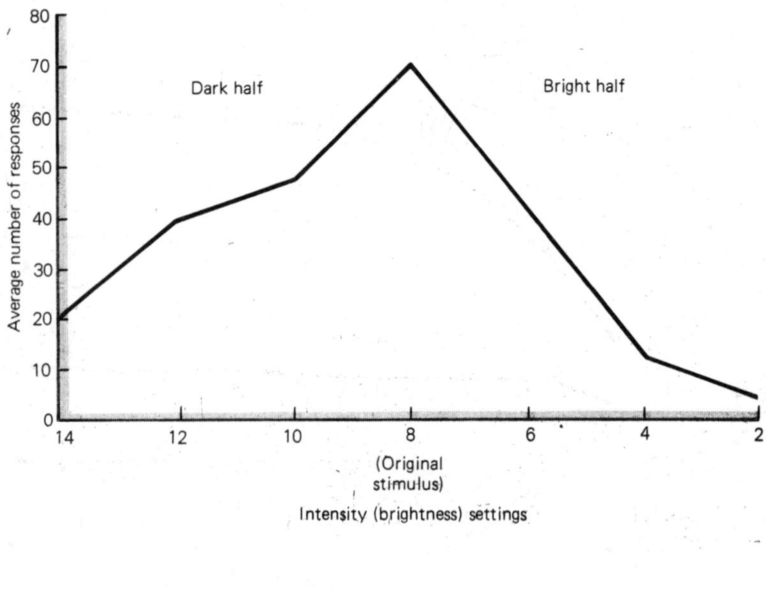

> a red one, illuminated the key, the pigeons received no reinforcement. Consequently, the birds learned to peck during the yellow-green, but not during the red, periods. After such a discrimination has been learned, the change in behavior when the stimuli are shifted is dramatic—almost like turning a faucet on or off.

In the example just given, discrimination learning was accomplished by presenting the positive and negative stimuli one after the other, or successively. The birds learned to respond to the positive stimulus (a yellow-green light) and not to the negative stimulus (a red light). In other words, in this type of successive discrimination, the learner "goes for" the positive stimulus and does not "go for" the negative stimulus; such discriminations are sometimes called "go–no–go discriminations."

Of course, many other arrangements of positive and negative stimuli are possible in discrimination learning experiments. For example, the positive and negative stimuli can be presented simultanously—in the case of a pigeon in an operant chamber, for example, perhaps on two keys arranged side by side. In this case, the experimenter will have to switch the positions of the positive and negative stimuli occasionally to be sure the pigeon is learning to respond to the stimulus and not to a particular side. For all the conceivable stimulus arrangements, the general principle is the same: Discriminations are developed when differences in the reinforcement of a response accompany the presence of different stimuli. As we have said, perhaps the most common reinforcement difference used to bring about discrimination is simply the difference between reinforcement in the presence of one stimulus and no reinforcement, or extinction, in the presence of another stimulus.

As an example of the everyday importance of discrimination learning, think for a moment about the routine of our daily lives. There is a time and a place for most of the things we do. When some stimuli are present, we respond in one way; when others are present, we behave in another way. We have learned to work in the presence of some stimuli and to play when others are present; we behave in one way in the presence of a professor and in quite a different way when we are with our friends; and so on. From such instances, you may be able to appreciate the power of the concept of the stimulus control of behavior in helping to explain and predict what we do in everyday, "real-life" situations.

NEGATIVE REINFORCEMENT AND ESCAPE LEARNING

Earlier in this chapter (page 149) a *negative reinforcer* was defined as a stimulus or event which, when its cessation or termination is contingent on a response, increases the likelihood that the response will occur again. Negative reinforcers are usually painful, or *noxious*, events, such as an electric shock, for example. But any stimulus or event will qualify as a negative reinforcer if, contingent on a particular response, its *termination*, or ending, increases the likelihood of the response. *Escape learning* is an example of instrumental conditioning based on negative reinforcement. Here is a laboratory example:

> A rat is put into a box with two compartments (A and B) separated by a low barrier, or hurdle. Compartment A is painted white and has a floor made of metal rods through which mild electric shocks can be

delivered to the animal's feet. Compartment B has a plain wooden floor and is painted black. Suppose, at the beginning of the experiment, the animal is placed in compartment A and the shock is turned on. In response to a shock of moderate intensity, the rat will run and move about in the shock compartment; in the course of its more or less random movement, it eventually gets over the hurdle into compartment B, where there is no shock. The rat is then removed from the "safe" side of the apparatus—compartment B—and, after a time, placed back in the shock compartment—compartment A. Again, when the shock comes on, the rat will move around and eventually find its way into the "safe" compartment. The experiment continues in this way, with the rat being placed back in A after running to B.

The first few times the shock is given—on the first few trials, in other words—the rat is slow to make the appropriate response of jumping the hurdle into the nonshock, or "safe," compartment. But as more and more trials are given, the animal learns to leap over the hurdle very soon after the shock comes on. In other words, it learns to make the response that terminates the noxious shock stimulus. This is escape learning, and it is based on negative reinforcement.

We might have done a similar experiment with a rat in an operant chamber. In this case, a mild shock would be applied to the rat's feet and a lever press would shut off the shock, thus allowing the animal to escape from it. Or we might have used dogs or monkeys instead of rats; but remember, whatever the species or apparatus, escape learning is based on negative reinforcement.

AVOIDANCE LEARNING

By having a stimulus that goes on before the foot shock in the escape-learning situation just described, avoidance learning can be demonstrated.

In this experiment, a rat is put into a two-chamber box similar to the one just described. Each trial begins with the presentation of a stimulus (a buzzer, for example) that is on for a few seconds (5, say) before the floor of the shock compartment is electrified. If the rat jumps the hurdle between the compartments within the 5-second interval between buzzer onset and shock, the buzzer is turned off and shock is avoided because the animal is now in the "safe" chamber when the shock comes on. Now you can see why this is called *avoidance learning*—a noxious stimulus (the shock) is avoided by the response. Figure 4.17 illustrates what might happen in such an experiment. The experimenter would record the time it takes the rat to make a response after the beginning of the warning signal—the *latency of response*—on each trial. If the latency on a trial is more than 5 seconds, the response is an escape response because the shock came on and the rat was escaping from it. But if the latency on a trial is less than 5 seconds, the response is an avoidance one because the rat jumped the hurdle *before* the shock came on and thus avoided it. The latencies graphed in Figure 4.17 show that except for trial 9, the animal made escape responses for the first 10

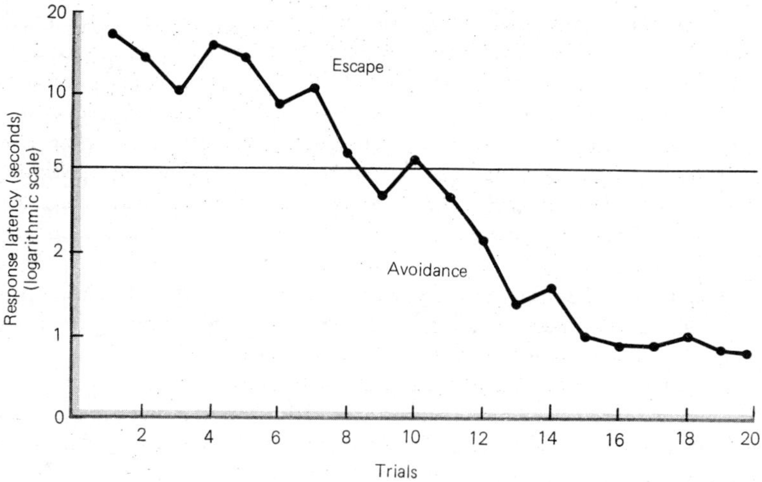

Figure 4.17

An idealized graph showing the acquisition of avoidance learning. Responses with latencies greater than 5 seconds are escape responses; those with latencies less than 5 seconds are avoidance responses. (See text.)

trials. After trial 10, all responses were avoidance responses; and by trial 20, the rat was jumping over the hurdle within a second or so after the buzzer sounded.

Why does an animal learn an avoidance response? A great deal of effort has gone into attempts to answer this simple question, and the answers have changed over the years. One current view uses the concepts of "species-typical defense reactions" and "safety signals" to explain avoidance learning (Bolles, 1970, 1979).

Suppose we apply these concepts to rats in the avoidance experiment just described. Rats have, probably as a result of evolutionary selection (Chapter 2, page 39), a number of behavioral defenses against threatening situations. In the wild, these threats usually come from predators that prey upon rats; but in the experiment just described, the buzzer that signals shock is the threatening stimulus. The behaviors by which rats "naturally" defend themselves from threatening situations are examples of what are called *species-typical defense reactions*. They are called *species-typical* because the individuals of an animal species have particular ways of defending themselves against threat; such reactions are part of the species heritage. (See Chapter 2, page 39.) In other words, given a threatening situation, all members of a particular species are likely to defend themselves in the same way. For the rat, the most likely species-typical defense reaction is to try to run away from the threat. (When running will not remove the threat, rats are likely to freeze, thus making themselves less likely to be seen by their predators.) In the laboratory avoidance experiment just described, running was the species-typical defense used. During the early trials, when the rat was escaping from, but not avoiding, the shock, the buzzer was thus being paired with shock. By classical conditioning (page 143), the buzzer thus became a signal for shock—a threat. After the buzzer had become the threat stimulus, the species-typical defense reaction of running occurred when the buzzer sounded and the rat ran in the apparatus as far as it could—to the other compartment.

Avoidance responding may also be learned because of *safety signals* the rat receives after avoidance responses. A stimulus that is consistently present, or paired with, the *absence* of a noxious stimulus such as shock may become a safety signal. In the demonstration experiment, the rat was never shocked in the black "safe" compartment, and stimuli from this chamber could thus have become safety signals. Furthermore, stimuli arising from the running itself were not paired with shock when the animal was making avoidance responses. Perhaps the stimulus feedback from the act of running served as a cue that there would be no shock for a while, and thus running itself became a safety signal. It is theorized that safety signals that indicate no more shock act as positive reinforcers (page 149) to increase the likelihood of avoidance responding.

Extinction of avoidance learning can be quite slow. In experiments like the one described above, rats have been known to run at the sound of a buzzer for thousands of trials after the shock has been turned off completely. One reason for such slow extinction is that as just described, the stimuli from the avoidance responses themselves can be safety signals that positively reinforce avoidance responding. The animal is thus in a vicious circle: Avoidance responses result in safety signals that reinforce the tendency to make more avoidance responses, which in turn produce safety signals, and so on and on. This resistance to extinction is reminiscent of some avoidance behavior in human beings. A person who has learned to avoid snakes or mice frequently goes on avoiding them all his or her life. So it is, too, with avoidance of water, high places, open places, and many other situations and stimuli. People do not easily shed their avoidance habits, and getting rid of them can be a difficult problem. (See Chapter 16, page 701.)

PUNISHMENT

A *punisher* was defined earlier in this chapter (page 150) as a stimulus or event which, when its onset is contingent on a response, decreases the likelihood that the response will occur again. *Punishment* refers to the use of punishers to suppress or stop a response from occurring in the future. In a sense, because punishers stop the behavior leading to them, we might say that they promote the learning of what not to do. A great deal of human learning from cradle to grave consists of learning what not to do. We learn not to play with fire, not to steal, not to exceed the speed limit, and so on endlessly. The Ten Commandments, for example, consist mostly of "Thou shalt nots." Parents and society try to teach the "don'ts" through the use of positive reinforcement and punishment, and often through punishment alone. But using punishment to mold behavior is a tricky matter, and many factors are involved.

When Does Punishment Work? The effectiveness of punishment depends upon a number of factors.

1. The more intense the punishment, the more effective it often is. Mild punishments, other things being equal, tend to suppress behavior only temporarily; the punished behavior will soon return unless rather intense punishment is used. But in human affairs, intense punishment carries the risk that strong conditioned emotional responses will be developed, with the punisher as the CS. If

the punisher is a parent, that is an unfortunate situation. Thus, as we shall see, mild punishment to guide behavior may be most effective in the long run.

2. The more consistently punishment is administered, even if it is mild, the more effective it will be, if it is effective at all.

3. The closer the punishment is in time and place to the behavior being punished, the more effective it will be. To be most effective, it should be contingent upon the occurrence of some response.

4. Generally speaking, the stronger the response tendency being punished, the less effective a given strength of punishment will be.

5. People and animals adapt to punishment, and this may weaken its effectiveness.

6. Punishment, even when mild, can be quite effective if it is used to suppress one behavior, while, at the same time, positive reinforcement is used to make another behavior more likely to occur. As a practical matter, this is a powerful way of using punishment to mold behavior. Mild punishment contingent on an incorrect or socially undesirable response can be considered to be a cue signaling the incorrectness of the response; positive reinforcement signals the correctness of the other response.

The Use of Punishment One often hears that psychologists advise parents never to punish children. In fact, most psychologists say nothing of the sort. What most of them say is that parents should know what they are doing when they use punishment to mold behavior; the principles discussed above are directly applicable. Some things young children do, such as running into the street or playing with knives, may lead to serious injury; strong punishment might be used to suppress such behavior. But note that the punishment should be contingent on the response; delayed punishment is much less effective. The punishment should probably be accompanied by a simple explanation. As was mentioned above, mild punishment can be very effective if it is used to halt unwanted behaviors while desirable alternative behaviors are being established. After the punished behavior is suppressed, an ingenious parent will positively reinforce alternative behaviors. Suppose, for instance, that a child is fooling around in a supermarket and is randomly pulling boxes off the shelves. Such behavior might be suppressed by mild scolding; at the same time, the parent could set the child to sorting items in the market basket, praising (that is, positively reinforcing) this acceptable behavior. This weakens one behavior and makes a desirable one more likely to occur.

Most psychologists would also say that parents should not use punishment as the major means of controlling behavior. First of all, punishment loses its effectiveness if almost everything a child does is punished. Punishment is most effective in giving a child information about what not to do if it is used sparingly against a background of positive reinforcement. A second point is that punishment, being a noxious event, can be an unconditioned stimulus for fear. Children may become fearful of, and hostile toward, parents who punish them for too many of the things they do; they may also come to have a low opinion of themselves if they are the objects of constant disapproval. Parents who overpunish often have resentful, rebellious, sullen children who are not really very well socialized. In the long run, occasional, pinpointed, contingent punishment for undesirable responses, coupled with lots of acceptance and positive reinforcement for desirable behavior, is the best prescription.

Of course, the guidelines for the effective use of punishment hold for other human relationships too, not just parent-child interactions. While it may help to know the guidelines, a number of practical questions remain. What, for instance, is the best type of punishment to use? Should it be a verbal reprimand, or should it be something else? All we can say in a text like this is that just as living is an art, so is the application of psychological principles. One must see what works in particular situations with particular people.

SIGNIFICANCE OF INSTRUMENTAL CONDITIONING

Instrumental conditioning is more than just a game played between experimental psychologists and rats or pigeons. Animal experiments have demonstrated some principles that can be extended to human life. Some of our beliefs, customs, and goals may be learned through the mechanisms of instrumental conditioning. Such learning is most evident when young children are being taught the ways of their group—that is, when they are being *socialized*. B. F. Skinner pointed out the importance of instrumental conditioning (*operant conditioning*, as he called it) in the socialization process in a book entitled *Science and Human Behavior* (1953). Skinner also described some of the ways in which agencies of human society—for example, government and the schools—use reinforcement to shape behavior. Parents and other agents of society usually do not deliberately shape behavior, but society is arranged so that reinforcements are contingent upon behavior.

Besides being ever present in human situations, instrumental, or operant, conditioning is sometimes used deliberately to shape desired behaviors. Programmed learning, the personalized system of instruction, applications of reinforcement principles to business operations, and certain types of therapy for psychological disorders are examples. Such uses of reinforcement principles might be called "applied instrumental, or operant, conditioning."

In *programmed learning*, the material to be learned is broken up into small, easy steps. Since each step is easy, the learner makes few errors and has a sense of accomplishment; this minimizes the frustration that can lower motivation and result in a dislike for learning. Also, programmed learning allows learners to proceed at their own pace and to receive immediate feedback on the correctness of their responses (a form of reinforcement). Programmed learning thus has these characteristics of the shaping used in animal learning experiments (page 154): (1) the final complex task is broken up into small steps, (2) reinforcement is contingent upon the performance of each step, and (3) the learner makes responses at his or her own pace. Programmed learning is thought to be an effective way of learning facts, rules, formulas, and the like. It has the further advantage of giving teachers who use it more time to devote to enriching the learning experience with other types of material.

The *personalized system of instruction (PSI)* is another educational application of instrumental, or operant, conditioning principles. Many forms of PSI exist, but an essential idea is that the material in the course is divided into small units, each of which must be mastered at a high level of proficiency before the next unit is attempted. For instance, students might be required to pass an examination at the end of each unit with a score of 90 percent or better; if they do not, they must study the material again until they can pass at this level. Students set their own pace because they can take the examination on a unit whenever they feel ready. Being allowed to go on to the next unit of material after mastery of the previous one serves as positive reinforcement in this system. In addition, students who have mastered a certain number of units are sometimes given the opportunity to participate in special activities, such as field trips; and this can be positively reinforcing.

In business operations, applications of reinforcement principles can often increase employee productivity and company profits. The use of contingent positive reinforcement in the form of praise by supervisors can mold the behavior of employees so that they become more effective in their jobs—they make more sales calls, fill orders more quickly and accurately, use the right packing materials, or whatever else they do in their jobs that can be measured and positively reinforced. All this can be done at practically no cost and can be managed in such a way that employees feel more satisfied with what they are doing. The supervisors are reinforced by the increased productivity of their units; the bosses are reinforced, hopefully, by greater profits.

Instrumental, or operant, conditioning is also applied in some forms of *behavior therapy*, or *behavior modification*, as in the example which opened this chapter. (Earlier we saw that classical conditioning techniques are also used in behavior therapy.) The instrumental-conditioning forms of behavior therapy treat psychological disorders by contingently reinforcing socially adaptive behaviors and by extinguishing maladaptive ones. (See Chapter 16, page 694.) Instrumental conditioning also has an applied role in changing the behavior of well-adjusted people. For instance, it may help people eliminate bad habits, such as smoking or eating too much; it may help mild-mannered individuals become more assertive. In general, we might say that instrumental, or operant, conditioning—often combined with other learning techniques—can help people reach goals they have set for themselves. In other words, applied instrumental conditioning can be important in changing behavior in the direction of greater self-control. (See, for example, Williams & Long, 1983.)

For some psychologists, operant conditioning is more than just another learning situation—it is the keystone of a philosophy of human life. In the book *Beyond Freedom and Dignity* (1971), B. F. Skinner widened the scope of operant conditioning to encompass practically all human behavior. His central argument was that human behavior is primarily the result of operant conditioning; and the sooner this is recognized, the sooner the affairs of human beings can be put on a rational basis. (To sample Skinner's style of argument, see Inquiry 14.1, page 595.) Such a view, of course, has aroused the skepticism of those who doubt the wide applicability of operant conditioning. The argument over Skinner's hypothesis will continue.

Chapter Four

We will introduce the topic of cognitive learning with an experiment (Menzel, 1973, 1978).

> Four young chimpanzees—Belle, Bandit, Bido, and Gigi—lived in a cage at the edge of a large fenced field. The field was their playground, and they had already had extensive experience with it before the experiment started.
>
> The experiment began when an experimenter picked up one of the chimps and carried it around the field. Another experimenter walked with them and hid food tidbits in 18 places in the field while the chimp watched. Then the chimp was put in the cage with the other chimps, and the experimenters climbed an observation tower. After several minutes, the door of the cage was opened so that all the chimps could leave and explore the field.
>
> Out came the chimps. The experimental animal (the learner) usually ran directly to one of the closest hiding places, searched briefly in the immediate area of the hidden tidbit, and then found the food. As the experimental animal uncovered more and more morsels of food in different locations, it gradually slowed down; but it did not wander around in a haphazard fashion. The learner went directly from one hiding place to another. The animals that had not seen the experimenter hide the food came upon very few of the hidden tidbits in their random searches.
>
> This procedure was repeated several times for each chimp. On every trial, a different set of 18 hiding places was used. Figure 4.18 shows the search patterns of the four chimps on the trial when each one found the most food items. The 18 locations, the starting position (s), and the finish (f) are shown. The route of each animal is shown by the lines, and the arrows give a rough idea of the direction of movement during the search. If the line touches a food location, the food was found at that location. Thus Belle found all 18 tidbits, but Bandit missed 2.
>
> The remarkable feature of these search maps is that they follow a "least-distance" principle. Each animal tended to go to hiding places close to the previous one; the search was not random. The chimps knew the relationships of the food items to each other and to features such as trees and clumps of grass in the field.
>
> The chimpanzees in this experiment evidently learned the location of the food tidbits by observing where they were placed. Furthermore, learning occurred without the reinforcement of specific responses as in instrumental conditioning and without the repeated pairing of stimuli that is characteristic of classical conditioning. It is as if the chimpanzees were able to form "maps in their heads" of the locations of the food items in relation to the environment. Such internal representations of the environment are known as *cognitive maps*.

Here, then, is an important form of learning that is different from classical and instrumental conditioning. It involves the forming of new

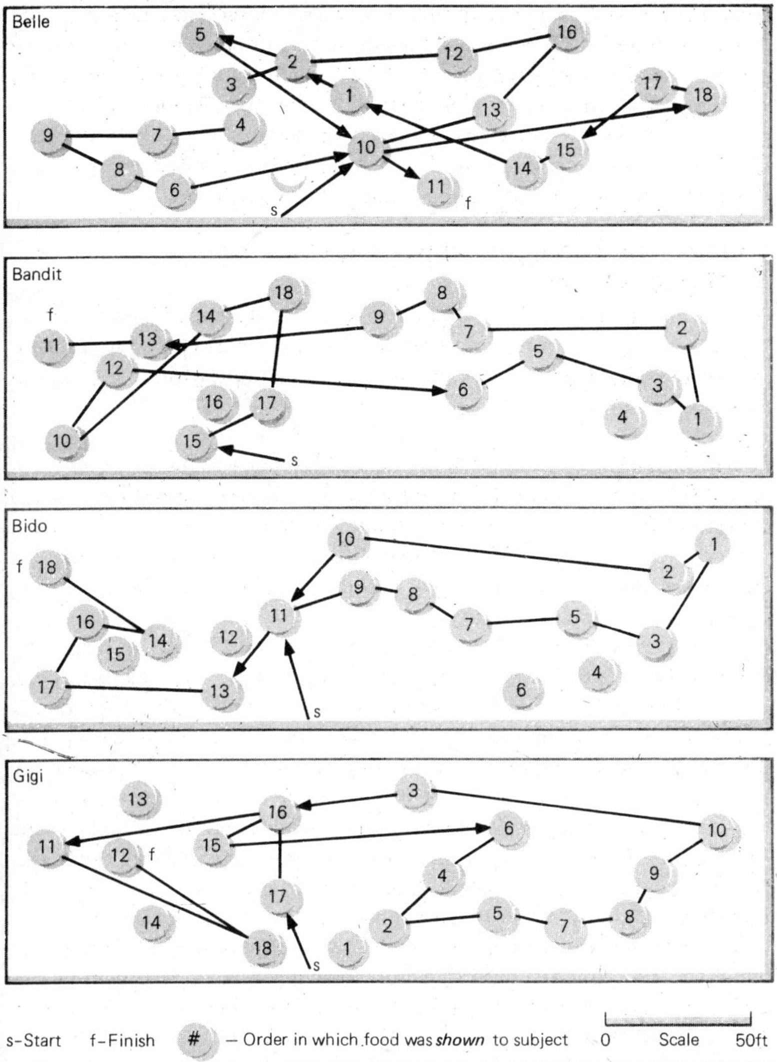

Figure 4.18

Search patterns of four chimpanzees in an experiment on cognitive learning. The circled numbers show the order in which tidbits of food were hidden. The lines and arrows indicate the direction of search. If a circled number is not touched by a line, the animal did not find that food. The search patterns are not random; as the chimpanzees searched, they tended to go from one tidbit to another nearby. (After Menzel, 1973.)

associations and the perceiving of new relationships among events. Note, too, that it is not responses to particular stimuli, so-called stimulus-response (*S-R*) associations, that are learned; instead, links are made among stimuli so that stimulus-stimulus (*S-S*) associations are learned.

Why is this form of learning called *cognitive*? *Cognition* refers to the processing of the information about the environment that is received through the senses. (See Chapter 5, page 184.) *Cognitive processes* involve: (1) the selection of information, (2) the making of alterations in the selected information, (3) the association of items of information with each other, (4) the elaboration of information in thought, (5) the storage of information in memory, and, when needed, (6) the retrieval of stored information. *Learning*, as you know, refers to relatively permanent changes in behavior as a result of experience. Putting the terms *cognition* and *learning* together gives a definition of *cognitive learning*: a change in the way information is processed as a result of experience a person or animal has had. In other words, due to past experience, the significance and meaning of events have been changed, new associations have been formed, and these changes have been stored in memory for future use. Obviously, much learning is of the cognitive variety. Indeed, as you read this text, we hope you are doing some cognitive learning.

LATENT LEARNING, INSIGHT LEARNING, AND IMITATION

In addition to the making of cognitive maps described above, latent learning, insight learning, and imitation are examples of cognitive learning. As explained above, they involve the formation of stimulus-stimulus (S-S) associations.

Latent Learning The word *latent* means "hidden," and thus *latent learning* is learning that occurs but is not evident in behavior until later, when conditions for its appearance are favorable. Latent learning is said to occur without reinforcement of particular responses and seems to involve changes in the ways information is processed. Thus latent learning is an example of cognitive learning.

In a typical latent-learning experiment, rats in an experimental group—the latent-learning group—are first given plenty of experience in a maze (Figure 4.9, page 152) without being reinforced for the particular responses involved in running the maze; they are simply allowed to wander through the maze or to live in it for a time. After the animals in the experimental group have thoroughly experienced the maze, reinforced maze learning of the sort described under instrumental conditioning (page 149) begins. While the experimental animals are experiencing the maze, rats in a control group are being treated like the experimental rats *except* that they are not being given experience with the maze. For instance, the control animals are handled as much as the experimental animals; but while the experimental animals are exploring the maze, the control animals are put in a box that is unlike the maze. The question is whether, in comparison with the control rats, the experimental animals will learn anything from their experience with the maze. If they do, such *latent*, or hidden, learning will show up in their performance when they are reinforced in the maze. They will make use of what they learned as they explored the maze; and thus, when reinforcement for maze learning starts, they will do better than will the rats in the control group. In a number of well-controlled experiments, that is exactly what happened: The experimental-group rats learned the maze faster and with fewer errors than did the control animals (Dodwell & Bessant, 1960; Seward, 1949).

Insight Learning In a typical insight situation, a problem is posed, a period follows during which no apparent progress is made, and then the solution comes suddenly. A learning curve of insight learning would show no evidence of learning for a time; then, suddenly, learning would be almost complete. What has been learned can also be applied easily to other, similar situations; in other words, there is a great deal of generalization of insightful solutions to similar problems.

Human beings who solve a problem insightfully usually experience a good feeling called an "aha experience." "Aha!" we say as we suddenly see the answer to the problem. To illustrate insight learning, study the following series of numbers. What numbers should follow these? Don't give up easily.

149162536496481100 . . .

If you cannot solve the problem after a few minutes, go on to something else and then come back to the problem. Try different arrangements, or perceptual organizations, of the numbers. If you solve the problem, you will have a pleasant "aha experience." Note that (1) your solution came suddenly after a period during which you tried various response strategies; (2) perceptual rearrangement helped a great deal; and (3) the solution, once you have it, can be generalized rather easily to other, similar number problems. These are the three major characteristics of insight learning. (If, after trying hard, you still do not have the answer, you can get a partial "aha experience" from the answer on page 179.)

How does insight learning occur? The cognitive answer to this question is that insight involves a *perceptual reorganization* of elements in the environment such that new relationships among objects and events are suddenly seen. Perhaps you experienced such perceptual reorganization when you solved the number-series problem. Perceptual reorganization also seems to be the rule in insightful learning by animals such as chimpanzees. Many years ago, the German psychologist Wolfgang Köhler carried out a number of insight experiments on chimpanzees and summarized the findings in a book entitled *The Mentality of Apes* (1925). He set these animals to solving problems such as the following:

> A food morsel was placed outside the cage at a distance too far for the chimp to reach. Inside the cage was a stick too short to reach the food but long enough to reach another longer stick outside the cage. This longer stick could be used to rake in the food (Figure 4.19). In these experiments, there was a period of trial-and-error fumbling, with little real progress toward a solution. Then, Köhler reported, the chimp would suddenly stop what it was doing, visually survey the sticks and the food, and then, suddenly and smoothly—and without any fumbling—solve the problem by using the shorter stick to rake in the longer stick, which could then be used to get the food.

Figure 4.19

A two-stick problem to show insight learning by chimpanzees.

In addition to the perceptual reorganization of the environment, there is often a carryover, or transfer, of things previously learned to insight situations. When you solved the number-series problem, you carried over some things you had already learned and applied them to the problem. Similarly, Köhler's chimps carried over what they already knew about sticks and other simple tools to the insight situation. In the animal experiments, it is possible that some of the elements carried over to the insight situation were the result of previous learning in naturally occurring conditioning-like situations. Thus, although the essence of insight learning is said to be perceptual reorganization of the environment so that objects take on new meanings and new relationships are seen among them, what has been learned in more mechanical ways may also play a role in insightful solutions.

Imitation Another cognitive-learning situation—one that is very important in human learning—occurs when we imitate another individual, or model our behavior on that of someone else. We might formally define *imitation* as a response that is like the stimulus triggering the response; a person or animal watches or hears another do or say something, then responds in the same way.

What can be imitated seems to be a species-typical capacity (Chapter 2, page 39). Some birds, like the parrot, can imitate human language. And some birds learn or perfect their calls by imitating older members of their species. Chimpanzees often imitate each other's motions and gestures. Children learn to say words partly through hearing the words spoken by their parents and by other children.

For many years, psychologists tried to explain imitation and modeling in terms of classical and instrumental conditioning principles. Modern psychologists have come to the view that imitation and modeling are the result of an innate capacity possessed by certain animal species, human beings included. The importance of imitation learning for human behavior is described in detail in Chapter 14 (page 596) in the section on social learning theory.

COGNITIVE PROCESSES IN CLASSICAL AND INSTRUMENTAL CONDITIONING

In the 1930s and 1940s, Edward C. Tolman (1932) developed an influential theory that stressed the cognitive nature of all learning. This theory was later overshadowed by other theories emphasizing the role of reinforcement in learning (Hull, 1943, 1952; Skinner, 1938, 1953). In the past few years, however, cognitive theories about learning have once again become prominent (Bolles, 1979).

When describing classical conditioning earlier (pages 140–147), we gave a cognitive interpretation of it. You may remember that classical conditioning occurs when the conditioned stimulus (CS) begins to act as a signal to predict the occurrence of the unconditioned stimulus (US). The cognitive interpretation is that as a result of the pairing of the CS and US, the CS arouses the expectation that the US will soon occur, and the learner acts on this expectation. In Pavlov's experiments, for example, the bell (CS) is a signal for food (US). The dog expects food when the bell is rung and salivates in anticipation of it.

Expectation is also a key element in the cognitive interpretation of instrumental conditioning. Consider an animal learning to press a lever in an operant chamber (page 152). The animal may be learning to expect that a particular response, pressing a lever, will result in a significant event, the reinforcer. In avoidance learning (pages 162–164), the warning stimulus may be considered a signal for impending shock. Given the warning signal, the animal expects shock and responds with a species-typical defense reaction (page 163).

The Learner and Learning: Some Things Are Easier to Learn than Others

Now that we have described some learning principles and some of the major ways in which learning occurs, we should introduce a note of caution to close the chapter: Application of the laws of learning must take into account the characteristics of both the learner and the response being learned.

Traditionally, the aim of the psychologists who studied learning was to find general laws that would apply to human beings and to other animals. They did not consider the response being learned or the animal doing the learning of much importance; they looked for laws of learning that would hold for all responses and all learners. Now we know that both the animal species and the response being learned have a great impact on how readily learning will occur. As Nikolaas Tinbergen, an influential investigator of the natural behavior of animals (Chapter 2, page 39), has pointed out: "An animal may learn some things more readily than others" (1951). And, we may add, some responses are exceedingly difficult (if not impossible) for certain species to learn.

The evolutionary process has resulted in species of animals that are structurally and behaviorally adapted to the environments in which they live. As we saw in Chapter 2 (page 39), animals have evolved species-typical behaviors to adapt to their environments. Evolutionary processes are also said to have produced brains that are specialized so that certain associations and responses are learned readily. Certain species are thus said to be ready, or predisposed, to learn some things easily; and such responses are often called *prepared behaviors*. Other responses are almost impossible for some species to learn; these are known as *contraprepared behaviors*. A third class of behaviors, the *unprepared behaviors*, can be acquired, but only when learning procedures are applied (Seligman, 1970).

PREPARED BEHAVIORS

The phenomenon of *learned flavor aversion* (Garcia et al., 1972) illustrates the concept of "preparedness." Many animal species, human beings included, learn very quickly to associate a particular food taste with illness. Perhaps you have had an experience with tainted food that made you sick, and you have developed a distaste, or aversion, for that food. For animals in the wild, this has the obvious advantage of being a way to avoid poisonous substances; if they survive the first poisoning, they will subsequently not eat the poisonous substance after the first taste. Flavor aversions are learned on one exposure and are quite resistant to extinction; the animal is prepared for this kind of learning.

As another example of preparedness, consider the avoidance behavior of the rat. Wild rats are quite vulnerable to the owls, hawks, snakes, coyotes, and dogs that prey on them; but as a result of evolutionary processes, they are prepared to learn certain ways of making themselves less likely to be caught. Laboratory rats also have this preparedness. For instance, both laboratory and wild rats are prepared to learn a running response to a noxious or threatening stimulus. If running is not possible or will not avoid the noxious stimulus, rats are prepared to learn to remain immobile—to "freeze." In other words, rats are said to have certain *species-typical defense reactions* (Bolles, 1970) that are learned readily (page 163). Other, less-prepared avoidance behaviors are learned with difficulty or not at all. Perhaps this is why it is easy to train rats to run in order to avoid electric shock but difficult to train them to press a lever in order to do so.

It is interesting to speculate about human preparedness for certain types of learning. Perhaps the best example of preparedness in human beings is language learning. Although apes can be laboriously taught to communicate with language symbols in a rudimentary way (Chapter 6, page 250), they do not seem to use abstract language symbols naturally. For humans, on the other hand, language learning is quite rapid in early childhood (Chapter 11, page 438). Furthermore, the earliest stages in human language learning seem to be similar for many different languages (Chapter 11, page 438). It is as if our brains are prepared to learn to communicate with abstract symbols.

Phobias are irrational, intense fears of certain objects or situations. Many learning psychologists consider such fears to be a result of classical conditioning (page 146) and imply that phobic fear can be conditioned to any object or event. This may be true, but some phobias are far more common than others (Seligman, 1971). Fear of snakes, small animals, insects, heights, and open spaces are the most common phobias; people are afraid of other objects and situations much less often. This suggests that we may be prepared to learn certain fears more readily than others.

UNPREPARED AND CONTRAPREPARED BEHAVIORS

Behaviors that can be learned with a moderate amount of difficulty are said to be *unprepared.* Most of the examples of learned responses given in earlier sections of this chapter illustrate unprepared behaviors.

Contraprepared behaviors are those that can be learned only with great difficulty, if at all. Contraprepared behaviors limit, or put constraints on, what can be learned by a certain species of animal. It often happens that responses are contraprepared because they conflict with strong species-typical behaviors (Chapter 2, page 39). A classic example of species-typical behavior interfering with instrumental conditioning comes from an article entitled "The Misbehavior of Organisms" (Breland & Breland, 1961). The title of the article is a takeoff on the title of an influential book by B. F. Skinner, *The Behavior of Organisms* (1938), in which he sets forth what are considered to be the general laws of operant conditioning.

> The Brelands conditioned interesting animals for commercial purposes—department store advertising, television commercials, and the like. Applying the "laws" of instrumental conditioning sometimes led to failure when species-typical behaviors were opposed to what was being learned.
>
> Take the case of the recalcitrant raccoon. Raccoons are generally good subjects for instrumental conditioning studies. This raccoon was supposed to learn to pick up little coins and deposit them in a small box (a bank). Contingent upon doing this, the raccoon was reinforced with food. As the trainers expected, it was easy to teach the animal to pick up the coin; the trouble was in getting the creature to drop it into the box. The raccoon treated the coin as a food tidbit and, as these animals do, rubbed it against the side of the box, pulled it out, held it tightly, rubbed it again, and so on. Finally, it dropped the coin and reinforcement was given.
>
> But when a second coin was added, the problems multiplied. The raccoon now had two objects to play with, and it rubbed the coins together—in a "miserly way," the Brelands reported—for

minutes at a time. This obviously would not do if the purpose of the commercial was to show how nice it is to deposit money in the bank. The project of training raccoons to deposit coins in banks had to be abandoned because species-typical behaviors prevented the learning of the desired response.

The species-typical behaviors in the raccoon example were very high in readiness. In fact, the trainers were probably reinforcing the animal for unwanted behaviors by giving it food when, at long last, it had dropped the coin in the box. Instrumental conditioning principles were at work, but the species-typical behavior of the animal made it impossible for the trainers to teach the animal what *they* wanted it to learn. In other words, the high preparedness of some behaviors resulted in the contrapreparedness of other behaviors.

Summary

1. Learning is defined as any relatively permanent change in behavior that occurs as a result of practice or experience.

2. In classical conditioning, a neutral stimulus (CS) regularly precedes an unconditioned stimulus (US) that evokes an unconditioned response (UR). As a result of this pairing, the previously neutral conditioned stimulus now begins to evoke a response. This is what is learned in classical conditioning. The response evoked by the conditioned stimulus after learning is known as the conditioned response (CR).

3. Current theories of the classical conditioning process take the viewpoint that the conditioned stimulus becomes a signal for the unconditioned stimulus. Thus, when the conditioned stimulus is presented, the unconditioned stimulus is expected and the learner responds in accordance with this expectation.

4. Extinction in classical conditioning is the process of presenting the conditioned stimulus alone without the unconditioned stimulus for a number of trials. When this is done, the strength or likelihood of the conditioned response gradually decreases. After a response has been extinguished, it recovers some of its strength with the passage of time; this is known as spontaneous recovery.

5. Stimulus generalization is the tendency to give conditioned responses to stimuli that are similar in some way to the conditioned stimulus but have never been paired with the unconditioned stimulus. The greater the similarity of these stimuli to the original conditioned stimulus, the greater the amount of generalization.

6. Discrimination is the process of learning to make one response to one stimulus and another response—or no response—to another stimulus. For instance, discrimination can be obtained in classical conditioning by pairing one stimulus (the CS^+) with an unconditioned stimulus and never pairing another stimulus (the CS^-) with the unconditioned stimulus.

7. With respect to its significance for human behavior, classical

conditioning seems to play a large role in the formation of conditioned emotional responses—the conditioning of emotional states to previously neutral stimuli. Some forms of treatment for psychological disorders use classical conditioning.

8. In instrumental conditioning, also known as operant conditioning, an action of the learner is instrumental in bringing about a change in the environment that makes the action more or less likely to occur again in the future. An environmental event that is the consequence of an instrumental response and that makes the response more likely to occur again is known as a reinforcer.

9. A positive reinforcer is a stimulus or event which, when it is contingent on a response, increases the likelihood that the response will be made again. A negative reinforcer is a stimulus or event which, when its cessation or termination is contingent on a response, increases the likelihood that the response will occur again. A punisher is a stimulus or event which, when its onset is contingent on a response, decreases the likelihood that the response will occur again. In omission of reinforcement, or omission training, positive reinforcement is withdrawn following a response; this has the effect of decreasing the likelihood of the behavior leading to the removal of positive reinforcement.

10. In instrumental, or operant, conditioning, the term shaping refers to the process of learning a complex response by first learning a number of simple responses leading up to the complex one. Each step is learned by the application of contingent positive reinforcement, and each step builds on the one before it until the complex response occurs and is reinforced. Shaping is also called the method of successive approximations. Classical conditioning also seems to contribute to the shaping of responses through the process known as auto-shaping.

11. In instrumental, or operant, conditioning, the procedure of not reinforcing a response is called extinction. If, after learning, reinforcement is no longer contingent on a response, the response will become less likely to occur.

12. Primary reinforcers in instrumental conditioning are reinforcers that are effective without any previous special training; they work "naturally" to increase the likelihood of a response when they are made contingent on it. On the other hand, the ability of conditioned, or secondary, reinforcers to influence the likelihood of a response depends upon learning; stimuli become conditioned reinforcers by being paired with primary reinforcers.

13. In instrumental, or operant, conditioning, reinforcement following every occurrence of a particular response is called continous reinforcement (CRF). But reinforcement in operant conditioning is often given according to certain schedules—not every occurrence of the response is reinforced. Fixed-ratio (FR), fixed-interval (FI), variable-ratio (VR), and variable-interval (VI) schedules are described to illustrate the concept of reinforcement schedules. In general, schedules of reinforcement have different patterns of be-

havior associated with them and resistance to extinction can be increased by scheduling reinforcements.

14. As in classical conditioning, stimulus generalization occurs in instrumental, or operant, conditioning; responses learned in the presence of one stimulus will also be made in the presence of other, similar stimuli. The amount of generalization depends on the similarity of the stimuli.

15. Discriminations are developed in instrumental, or operant, conditioning when differences in reinforcement accompany different stimuli. For example, a person or animal may learn to respond to a positive stimulus (the S^+, or S^D) present when responses are being reinforced and not to a negative stimulus (S^-, or S^Δ) present when responses are not being reinforced—that is, during extinction. Stimulus discrimination in instrumental, or operant, conditioning is also referred to as the stimulus control of behavior.

16. Escape learning is an example of instrumental conditioning based on negative reinforcement.

17. In avoidance learning, the learned response is made before the onset of a noxious event and thus prevents the learner from being exposed to the noxious event: The noxious event is therefore avoided. Avoidance learning is explained in terms of the concepts of species-typical defense reactions and safety signals. Extinction of avoidance learning is often quite slow.

18. Punishment decreases the likelihood that a response will be made and is thus involved in learning what not to do. Among the factors involved in the effectiveness of punishment are its intensity, its consistency, the contingency between behavior and punishment, the strength of the response being punished, and adaptation to punishment. Punishment may be quite effective when it is used to suppress one behavior while, at the same time, positively reinforcing other behaviors.

19. With respect to its significance for human behavior, instrumental, or operant, conditioning may play a large role in the molding of much that we do. Specific applications of instrumental conditioning principles are to be found in programmed learning, in the personalized system of instruction, in business, and in certain forms of treatment for psychological disorders.

20. Cognitive learning refers to changes in the way information is processed as a result of experience a person or animal has had. Cognitive maps, latent learning, insight learning, and imitation are described as examples of cognitive learning. Cognitive-learning interpretations of classical and instrumental conditioning are discussed.

21. The principles of learning may not be so general as previously thought. Application of the laws of learning must take into account both the characteristics of the learner and the response being learned. The concepts of prepared and contraprepared behaviors are introduced to illustrate this point.

Terms to Know

One way to test your mastery of the material in this chapter is to see whether you know what is meant by the following terms:

Learning (140)
Classical conditioning, respondent conditioning, Pavlovian conditioning (140, 141)
Conditioned response (CR) (141, 143)
Stimulus (142)
Conditioned stimulus (CS) (142)
Unconditioned stimulus (US) (142)
Unconditioned response (UR) (142)
Acquisition curve (143)
Extinction (145, 156)
Spontaneous recovery (145)
Reconditioning (145)
Stimulus generalization (146, 159)
Discrimination (146, 160)
Phobias (146, 148, 174)
Behavior therapy/behavior modification (147, 167)
Systematic desensitization (148)
Instrumental conditioning/operant conditioning (149)
Reinforcer (149)
Reinforcement (149)
Positive reinforcer (149, 151)
Negative reinforcer (149, 161)
Noxious (150, 161)
Punishment (150, 164)
Punisher (150, 164)
Omission of reinforcement/omission training (150)
Operant chamber (152)
Cumulative recorder (153)
Shaping (154)
Method of successive approximations (155)
Auto-shaping (155)
Primary reinforcer (156)
Conditioned, or secondary, reinforcer (156)
Continuous reinforcement (CRF) (157)
Schedule of reinforcement (157)
Fixed-ratio (FR) schedule (157)
Fixed-interval (FI) schedule (157)
Variable-ratio (VR) schedule (158)
Variable-interval (VI) schedule (158)
Resistance to extinction (159)
Stimulus generalization in instrumental conditioning (159)
Gradient of generalization (160)
Stimulus control of behavior (160)
Escape learning (161)
Avoidance learning (162)
Latency of response (162)
Species-typical defense reactions (163, 173)
Safety signals (164)
Socialization (166)
Programmed learning (166)
Personalized system of instruction (PSI) (167)
Cognitive map (168)
Cognitive processes (169)
Cognition (169)
Cognitive learning (170)
Latent learning (170)
Insight learning (170)
Perceptual reorganization (171)
Imitation (172)
Prepared behaviors (173)

Contraprepared behaviors (173, 174)

Unprepared behaviors (173, 174)

Learned flavor aversion (173)

Suggestions for Further Reading

Textbooks like this one only skim the surface of the work of major historical figures in a field. For the "real thing," you might like to look at I. P. Pavlov's *Conditioned Reflexes* (New York: Dover, 1960). This is a reprint of the book *Conditioned Reflexes*, which was translated by G. V. Anrep and published by The Oxford University Press in 1927. Or if you are interested in what B. F. Skinner has to say about the importance of operant conditioning, his *Science and Human Behavior* (New York: Macmillan, 1953) and his more recent *Beyond Freedom and Dignity* (New York: Knopf, 1971) will be informative.

A number of college textbooks have been written for courses devoted exclusively to the topic of learning. Should you want to know more about the subjects covered in this chapter, and others, too, the following may be helpful: (1) *Learning Theory*, 2d ed., by Robert C. Bolles (New York: Holt, Rinehart and Winston, 1979); (2) *Theories of Learning*, 5th ed., by Gordon H. Bower and Ernest R. Hilgard (Englewood Cliffs, NJ: Prentice-Hall, 1981); (3) *The Psychology of Learning*, 5th ed., by Stewart R. Hulse, Howard Egeth, and James Deese (New York: McGraw-Hill, 1980); (4) *Foundations of Learning and Memory*, by Roger M. Tarpy and Richard E. Mayer (Glenview, IL: Scott, Foresman, 1978).

Behavior therapy uses learning principles from classical conditioning, instrumental conditioning, and cognitive learning to help in the treatment of behavioral problems. *Behavior Modification: Issues and Applications*, 2d ed., edited by W. Edward Craighead, Alan Kazdin, and Michael Mahoney (Boston: Houghton Mifflin, 1981) and *Behavior Therapy: Techniques and Empirical Findings*, 2d ed., by David Rimm and John Masters (New York: Academic Press, 1979) are good introductions.

Answer to problem on page 171:

If the numbers are grouped as follows, the answer is readily seen.

1 4 9 16 25 36 49 64 81 100 . . .

The numbers are squares of the series

1 2 3 4 5 6 7 8 9 10.

Memory

HEADING OUTLINE

THEORIES ABOUT MEMORY
A Theory of General Memory Functions
APPLICATION 5.1: MEMORY AND THE LAW
Information-Processing Theories
INQUIRY 5.1: WHAT DO WE KNOW ABOUT THE BRAIN AND MEMORY?
The Levels-of-Processing Theory

LONG-TERM MEMORY: ITS ORGANIZATION AND PROCESSES
The Organization of Long-Term Memory
Encoding and Storing Long-Term Memories
Retrieval from Long-Term Memory

FORGETTING
INQUIRY 5.2: WHAT DON'T WE FORGET?
Interference
Retrieval Problems
Motivated Forgetting

AMNESIA
Psychological Amnesias
Biological Amnesias

IMPROVING YOUR MEMORY
Mnemonics
INQUIRY 5.3: WHAT ABOUT SUPER MEMORIES?
Studying to Remember

SUMMARY

TERMS TO KNOW

SUGGESTIONS FOR FURTHER READING

Chapter 5

Here is what Socrates said to Theaetetus about memory:

> Imagine, then, for the sake of argument, that our minds contain a block of wax, which in this or that individual may be larger or smaller, and composed of wax that is comparatively pure or muddy, and harder in some, softer in others, and sometimes just the right consistency. Let us call it the gift of the Muses' mother, Memory, and say that whenever we wish to remember something we see or hear or conceive in our own minds, we hold this wax under perceptions or ideas and imprint them on it as we might stamp the impression of a seal-ring. Whatever is so imprinted we remember and know so long as the image remains; whatever is rubbed out or has not succeeded in leaving an impression we have forgotten or do not know. [F. M. Cornford, Trans. (1935). *Plato's theory of knowledge*, p. 121.]

The wax analogy has been replaced in modern times by analogies with computer disks and tapes, but this ancient quotation does contain the germs of modern ideas to be discussed later in this chapter.

1. Memories originate in perceptions or experiences ("we hold this wax under perceptions").
2. Experiences leave some kind of a trace behind—not on wax, but in the brain.

3. There are individual differences in memory. (The block of wax "in this or that individual may be larger or smaller.")

4. Forgetting occurs ("whatever is rubbed out").

5. Some experiences are not stored, and therefore, a common belief to the contrary, memory is by no means a complete record of the events of our lives ("whatever . . . has not succeeded in leaving an impression we . . . do not know").

But the wax analogy (or for that matter some of the computer analogies) is far too rigid and static. As this chapter may begin to convince you, memory is such a flexible, selective, changing process that our "remembrances of things past" are imperfect, distorted, or just not recorded at all. And we even think we remember events that never happened.

> A healthy distrust of one's memory, and of memory in general, is not a bad idea. When all is said and done, memory is selective; the memory machine is selective about what gets in and selective about how it changes over time. . .
>
> The malleability of human memory represents a phenomenon that is at once perplexing and vexing. It means that our past might not be exactly as we remember it. The very nature of truth and of certainty is shaken. It is more comfortable for us to believe that somewhere within our brain, however well hidden, rests a bedrock of memory that absolutely corresponds with events that have passed. Unfortunately, we are simply not designed that way. [E. F. Loftus (1980). *Memory: Surprising new insights into how we remember and why we forget*, p. 147, p. 190.]

THE last chapter was about some general principles of learning that apply, broadly, over many animal species, including the human animal. In developing these principles, much emphasis was given to work with pigeons, rats, monkeys, and apes. This chapter is mostly about human learning as it is manifested in *memory*—the encoding, storage, and retrieval of what was learned earlier.

Can you imagine life without memory? Even nonhuman animals must have some system for retaining what they just did so that their behavior will flow in its proper sequence. For us, memory is even more crucial. Think of your simple conversations with friends. The very act of speaking means that you are remembering and recalling the words of your language in grammatical sequence, and you must keep track of what you have just said or your conversation will be senseless. All this memory functions even if your friend does not ask you to recall something you did last week. So memory of some sort is fundamental to our understanding of behavior and mind, and that is why a chapter on the subject comes early in this book.

The ideas and theories in the study of memory have changed drastically in the last 20 years. Earlier work took its inspiration from conditioning studies (Chapter 4) and looked at human memory in terms of associations, or connections, between stimuli and responses (S-R connections, as they are called). Although this approach to memory resulted in the discovery of numerous important principles, it seemed, to many psychologists, inadequate to account for the richness and flexibility of human memory. The current trend in the study of memory is to emphasize cognitive, or mental, processes over stimulus-response associations.

Cognition refers to the processes through which information coming from the senses is "transformed, reduced, elaborated, recovered, and used" (Neisser, 1967). The term *information*, as used in this chapter, refers simply to sensory input from the environment that informs us about something that is happening there. *Cognitive processes* are thus the mental processes involved in knowing about the world; as such, they are important in perception, attention, thinking, problem solving, and memory. (Note that cognitive processes can operate in the absence of any immediate sensory input, as in dreams.) The branch of psychology that deals with cognitive processes is known as *cognitive psychology*, and the modern-day study of memory, since it emphasizes the mental processes involved in storing information and retrieving it from memory, is a part of cognitive psychology.

Theories about Memory

Remember from Chapter 1 (page 5) that theories are frameworks serving to tie together the results of many observations and experiments. They are especially useful in giving order to the vast amount of information psychologists have accumulated about memory.

A THEORY OF GENERAL MEMORY FUNCTIONS

One theory, a simple one agreed on by most psychologists, was used in our definition of *memory*. Three distinct processes of memory have been identified. These are an *encoding process*, a *storage process*, and a *retrieval process*. Encoding is the process of receiving sensory input and transforming it into a form, or code, which can be stored; storage is the process of actually putting coded information into memory; and retrieval is the process of gaining access to stored, coded information when it is needed.

To illustrate these three memory processes, imagine that on the way to work, your car was bumped by a bus and slightly dented. You encoded your visual impressions of the accident in a form that you could store in your memory. Later, when you talk to the insurance adjuster, you will retrieve what you stored.

This simple process theory helps explain why your memory of an accident may be inaccurate. The encoding you do may be faulty, perhaps due to the emotion and distress you experience at the time of the accident; important encoded information may not be well fixed in your memory, or it may be distorted by events occurring after the accident; your retrieval of the information stored in your memory may be biased; or, since the processes of encoding, storage, and retrieval are related, all three memory processes may be faulty. Memory is seldom an accurate record of what was experienced. Application 5.1 illustrates this, and as the chapter proceeds you may have even more reason to agree with this statement about the inaccuracy of memory.

INFORMATION-PROCESSING THEORIES

Imagine yourself a device something like a digital computer that takes items of information in; processes them in steps, or stages; and then produces an output (Figure 5.1). Models of memory based on this idea are called *information-processing theories*. A number of such models of memory have been proposed. We will use one of the most prominent and influential of these models—the information-processing theory developed by Richard Atkinson and Richard Shiffrin (1968)—to guide our discussion.

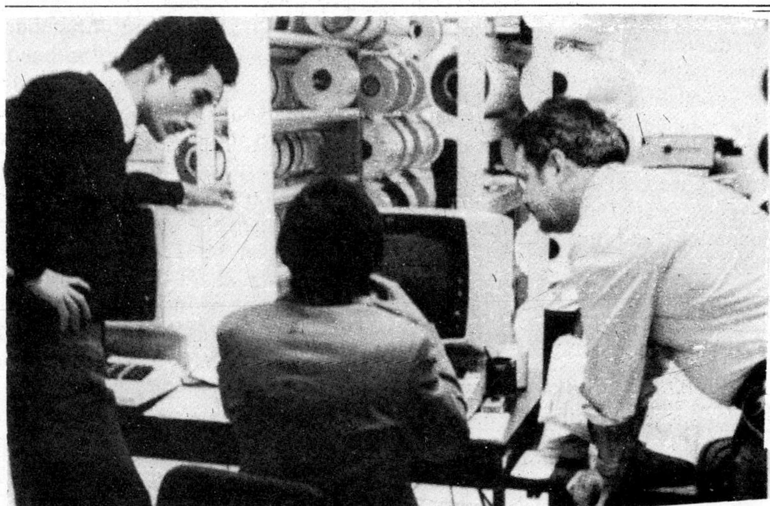

Figure 5.1

Devices for processing information. (© The Photo Works.)

Application 5.1

MEMORY AND THE LAW

The outcomes of some famous and many not-so-famous trials have depended on the memory of witnesses. In the courtroom, we seldom have a photograph, movie, videotape, or sound recording of the crime as it occurred. Lawyers, the judge, and the jury must often rely on something far less objective—the fragile memories of witnesses.

Memory is not a videotape record of events. Instead, it is a complex cognitive, or mental, process involving the perception and encoding of the to-be-remembered information, the storage of what is to be remembered, and the retrieval of the stored information. At the encoding stage, only certain events are selected for storage; at the storage stage, numerous distortions can occur; and some of what is stored cannot be retrieved or is distorted at the time of retrieval. The way is open in the courtroom for critical memory mistakes that can have far-reaching and sometimes tragic consequences, as when the wrong person is convicted. This is especially true when the jury gives great weight to the memories of eyewitnesses to a crime.

Suppose you are in a convenience store alone with the clerk one evening. While you are at the back of the store deciding which brand of beer to buy, someone comes in and says to the clerk, loud enough for you to hear, "I've got a gun; give me the money in the cash register." As you turn toward the front of the store, the stick-up man is surprised and takes a couple of wild shots in your direction, scoops up the money, and rushes from the store. The lighting in the store is good, but you have time for only a brief glimpse of the robber. The police are called, and, whether you like it or not, you are a key witness. The police question you, and you tell them what you remember of the incident. Several weeks later, the police think they have a suspect and call you to the station to look at some mug shots. Your glimpse of the robber was a fleeting one, but, with some prodding from the detective in charge of the case, you pick out one of the faces as that of the robber. Months later, you are called as an eyewitness to testify for the prosecution at the trial.

At the trial, you do your best to be accurate; but your memory, through no fault of your own, lets you down. You identify an innocent defendant as the robber. What went wrong?

To start with, your encoding of the image of the robber's face was not very good. Although the store was brightly lighted, you saw the robber for only a brief moment, and you were looking mostly at that huge (or so it seemed to you) gun the robber was holding. Of course you were stressed and concerned with your own safety rather than with making an accurate memory record for later testimony. Under conditions such as these, people tend to narrow their focus of attention—to the gun, in this case. This happens so often that a special term—weapon focus—has been coined. So your encoding of the to-be-remembered event was not very good, but at least something got stored in your memory.

Now storage distortions can come into play. When the police arrived and you gave them your stored memory, you identified the robber as a "white male of medium height in his early twenties." You did not remember any of his distinctive facial features, but you heard the clerk, whose encoding was also less then optimal, tell the police that the robber had peculiar "pop eyes." Next day, you read a newspaper account of the exploits of the "pop-eyed bandit." Later, when you went to the station to look at the mug shots, the police showed you an array of seven full-face pictures and asked you to pick out the robber. The pictures were all of young white males, but only one of them had "pop eyes." You were pretty sure this was the robber. Well, you were wrong. What happened was that your poorly encoded stored image was distorted by the clerk's report to the police and by the newspaper story. But you had a good look at the mug shot and you encoded that in your memory as the face of the robber. Later, at the trial, this was your memory of the robber's face, and you identified as the robber the person whose mug shot you saw. This phenomenon is so common that the U.S. Supreme Court has decided that in cases where the initial viewing conditions were poor or brief, memory can be biased by a mug shot in such a way that "the witness thereafter is apt to retain in his memory the image of the photograph rather than of the person actually seen."

To finish the story, justice was done. In spite of your damaging and incorrect testimony, the defendant was acquitted by the jury because the clerk's memory was less biased than yours and he could not be sure that the defendant was the thief. It also helped that the defendant had an excellent alibi backed up by several witnesses.

In our example, we did not consider retrieval factors at the trial that might have affected what you remembered—factors such as the presence or absence of reminders, stressors at the time of recall, and whether you were allowed to tell the story in your own words or it was dragged out of you by questions. And the list of factors in the story that influenced encoding and storage was just a sample of the many factors that have been demonstrated, in controlled experiments, to affect eyewitness testimony. Furthermore, if we had used as our example the memory of complex events such as conversations, we would perhaps have seen even greater memory distortion.

The message seems clear. To the extent that a trial seeks to determine what actually happened and who did what, due consideration must be given to what psychologists have discovered about the failings of memory. Care should be taken to guard against the subtle (and not so subtle) ways in which the memories of witnesses can be biased. If several witnesses who have not been exposed to similar biasing factors agree, and if other evidence points to the same conclusion, then the outcome of a trial may be closer to the "truth." But it is risky to rely too much on a single all-important eyewitness.

REFERENCES

Buckhout, R. (1974). Eyewitness testimony. *Scientific American, 231* (6), 23–31.

Loftus, E. F. (1979). *Eyewitness testimony.* Cambridge, MA.: Harvard University Press.

Neisser, U. (1982). *Memory observed: Remembering in natural contexts.* San Francisco: Freeman.

Simmons et al. v. United States, 390 U.S. 377 (1968).

Yarmey, A. D. (1979). *The psychology of eyewitness testimony.* New York: Free Press.

In the Atkinson-Shiffrin theory, memory starts with a sensory input from the environment (Figure 5.2). This input is held for a very brief time—several seconds at most—in a *sensory register* associated with the sensory channels (vision, hearing, touch, and so forth). Information that is attended to and recognized in the sensory register may be passed on to *short-term memory (STM)*, where it is held for perhaps 20 or 30 seconds. Some of the information reaching short-term memory is processed by being *rehearsed*—that is, by having attention focused on it, perhaps by

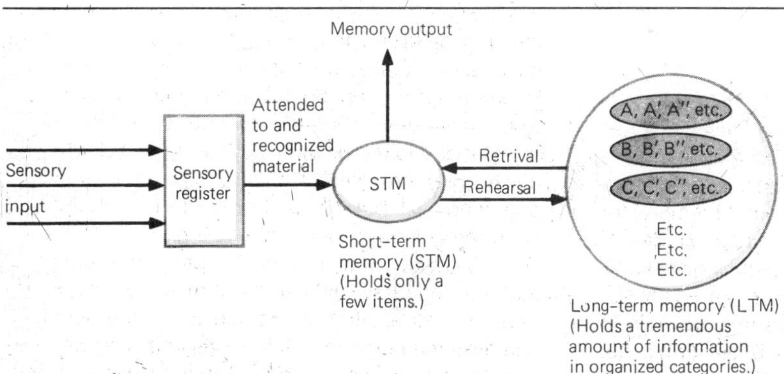

Figure 5.2

An information-processing model of memory. (Based on Atkinson & Shiffrin, 1968.)

being repeated over and over, or perhaps by being processed in some other way that will link it up with other information already stored in memory. Information that is rehearsed may then be passed along to *long-term memory (LTM)*; information not so processed is lost. When items of information are placed in long-term memory, they are organized into categories, where they may reside for days, months, years, or for a lifetime. When you remember something, a representation of the item is withdrawn, or *retrieved*, from long-term memory.

There are some interesting parallels between this information-processing theory and some of the brain processes involved in memory (Inquiry 5.1). This model of memory also fits well with our subjective impressions when we are trying to remember something.

Imagine yourself asking someone for a telephone number you do not know. (It's in the other person's long-term memory, but not in yours.)

Inquiry 5.1

WHAT DO WE KNOW ABOUT THE BRAIN AND MEMORY?

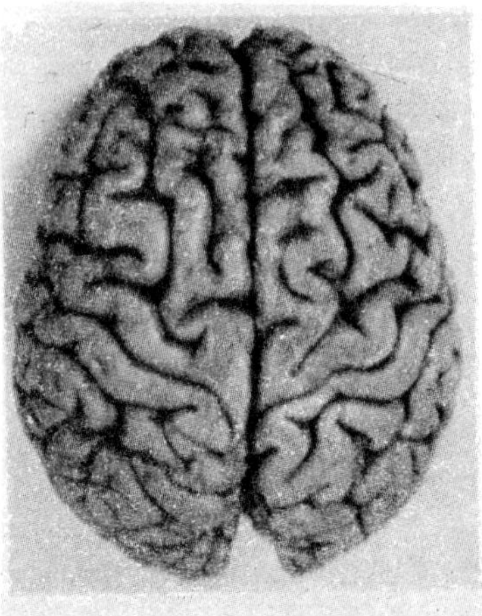

The answer to the question in the title of this inquiry is, "Only a little." It may be surprising to read that very little is known about the ways cognitive processes (perception, thinking, language use, and memory, for example) are represented in the brain. From studies of what people actually say and do, psychologists know more about these subjects than do brain scientists. Somewhere in the distant future, we may be able to understand cognitive processes in brain terms, but not now. However, some progress has been made, and, for memory in particular, we have some glimmerings of what may be happening in the brain. (See pages 67–69, also.)

The distinction made between short-term and long-term memory by the information-processing theorists (page 187) seems to have a brain basis. To illustrate, consider what happens to Sally's memory in the following example. (Something similar may have happened to you.)

Sally is riding her motorcycle one fine spring day when a dog darts in front of her. She swerves to avoid the dog, hits a tree, flies off her bike, and bangs her head on the road. She sustains a mild concussion (a brain bruise); the functioning of her brain is disrupted, and she is knocked unconscious. When she comes to in the hospital several hours later, one of her first questions is, "Why am I in the hospital?" Her head hurts, so she guesses she was in an accident; but she has no memory of what happened. She does not recall the dog or falling from her motorcycle. If we could have asked Sally about the dog during the instant between her perception of it and her concussion, she would have remembered it from her short-term memory. But due to the disruption of her brain activity, she has no long-term memory of the dog or of other events occurring just before the accident. Thus Sally had short-term memory, but she has no long-term memory.

Sally has another memory problem that shows that she may have two kinds of memory—long-term and short-term. After she is told about the accident, she asks the nurse for a telephone and calls her husband to let him know what happened. Half an hour later, Sally again asks the nurse for a telephone to call her husband. When the nurse says, "But you just called him," Sally says, "Oh, did I?" Her short-term memory is all right; she remembers what the nurse has just said and is able to answer appropriately. But she has no long-term memory of the telephone call she made to her husband 30 minutes before.

From head-injury cases and much experimental work with animals, the theory of consolidation has emerged. This theory says that information is held in the brain for a short time in an unstable, or labile, form called short-term memory. As time and certain brain processes go on, the representation of memory in the brain is strengthened, or consolidated, into a long-term form. Events which prevent consolidation from occurring stop the formation of long-term memory. Evidence indicates that structures in the temporal lobe of the brain—especially the hippocampus and the amygdala (Chapter 2, pages 67–69)—are critically involved in this consolidation.

Long-term memories, once formed, are relatively permanent and very hard to disrupt. Note that Sally's already-consolidated long-term memories—her sense of who she is, her language skills, the events of her life, for example—are not lost. In spite of what we read in novels or see on television, concussions such as Sally's almost never result in loss of long-term, consolidated memories. However, if a large amount of brain tissue is destroyed by injury, or if there is a disease process which alters the structure and function of the brain—as in Alzheimer's disease (page 212), for example—consolidated long-term memories can be lost.

How are memories actually stored in the brain? We know that while there is some localization of storage for particular memories in the brain (Chapter 2, pages 69–72), damage to small regions of the brain does not usually wipe out well-established long-term memories. So particular memories seem to be stored throughout fairly wide regions of the brain, and when information is stored in memory, the functions of brain cells in whole regions of the brain are altered. With this important fact about the brain and memory as background, some investigators have studied the widespread changes that occur in the brain's electrical activity during learning and memory. Others have taken a more structural approach, based on the idea that memories are stored by changes in the ways brain cells, or neurons, are interconnected. Such interconnections are known as brain circuits or networks. Long-term memory, according to this view, is represented in the brain by these changed circuits and, when something activates a circuit, the memory represented by activity in the circuit is retrieved. Of course, because memories are not narrowly localized in the brain, a particular memory must be represented by a large number of such changed circuits.

It has been proposed that short-term memory is represented in the brain by ongoing electrical activity in certain circuits, or networks, of nerve cells. This activity is said to lead to changes in the strength of connections among the nerve cells in the network; the network becomes consolidated and, when activated, works as a unit—a memory module. Such strengthening of connections might be due to growth, or sprouting, of the nerve cells so that they make more contacts with others in their network. Another way memory circuits may be strengthened is through changes in the network cells themselves.

Connections between neurons are known as synapses (Chapter 2, page 48). When a nerve cell is active, a chemical known as a neurotransmitter is released at the synaptic connections the nerve cell has with others. The neurotransmitter brings about changes in the cells contacted by the active cell—the receiving cells—to excite (or inhibit) them. (See Chapter 2, page 48.) Neurotransmitters can also, over time, trigger a chain of events which alters the manufacture of proteins in the receiving cells. We know that certain proteins are embedded in the membrane, or covering, of nerve cells and are involved in regulating the excitability of the cells. Perhaps more of these proteins are made when a circuit is active and thus the cells of the circuit become structurally and permanently changed to form a memory module for long-term memory. Later, when a neuron in the circuit is activated, the network acts as a unit and the long-term memory it represents is retrieved.

Consistent with the hypothesis just outlined, several lines of research indicate that long-term memory is linked to steps in the manufacture of proteins by neurons. For instance, ribonucleic acid (RNA) is involved in protein synthesis, and some experiments have related changes in cellular RNA to memory. Other studies have suggested that activation of molecules regulating the genes coding the manufacture of certain proteins may be involved in memory.

REFERENCES

Carlson, N. R. (1980). *Physiology of behavior* (2d ed.). Boston: Allyn and Bacon.

Kandel, E. R., & Schwartz, J. H. (1982). Molecular biology of learning: Modulation of transmitter release. *Science, 218,* 433–443.

Squire, L. R., & Schlapfer, W. T. (1981). Memory and memory disorders: A biological and neurological perspective. In H. M. Van Praag (Ed.), *Handbook of psychiatry, Part IV*. New York: Marcel Dekker.

The person tells you the number (it was retrieved and resulted in an output), and off you go to dial it. Unless you rehearse the number—go over it mentally—you will probably forget it soon after hearing it. If something interrupts you on your way to the telephone, thus disrupting your rehearsal, you will probably forget the number. Or imagine yourself at a party. Unless you take pains to rehearse the names of the new people you meet, you will not remember them long. Without rehearsal, and with the information overload caused by all the things happening at the party, transfer to long-term memory will not occur.

Retrieval from long-term memory is also experienced subjectively. Try to remember where your bedroom was located in all of the houses or apartments you have lived in. As you do this, you will find yourself searching through your long-term memory, and you will probably develop a search strategy. Perhaps you will search chronologically from the first home you remember to later ones. You may then imagine the house and locate your bedroom in relation to the other rooms. If you cannot remember a bedroom using one search method, you may shift to another search strategy. In any case, you will have the subjective impression of having searched through your storehouse of memories.

The Sensory Register Information can be held for a very brief time in the sensory channels themselves. This storage function of the sensory channels is called the *sensory register*. Most of the information briefly held in the sensory register is lost; what has been briefly stored simply decays from the register. However, we pay attention to and recognize some of the information in the sensory register; when we do this, the attended-to information is passed on to short-term memory for further processing (Figure 5.2).

Some ingenious experiments have shown that the visual sensory register holds information for up to about 1 second (Sperling, 1960), while the auditory (hearing) register holds information somewhat longer—up to about 4 or 5 seconds (Darwin et al., 1972). Studies with the visual sensory register have also shown that it can hold at least 11 to 16 items of information during the second before it loses the information through decay (Averbach & Sperling, 1961; Estes & Taylor, 1966). Furthermore, in vision at least, the sensory storage seems to be in the form of a faint image, called an *iconic image* (from the Greek word meaning "likeness"), which is a copy of the visual input (Sperling, 1963). It is this iconic image that persists in the visual sensory register for a second before it gradually decays.

The sensory register holds information for such a brief time that some psychologists prefer to discuss it in connection with perception rather than memory. However, it is part of the information-processing model under discussion, and it is a step that information passes through before it reaches short-term memory.

Short-Term Memory (STM) A number of experiments have shown that short-term memory can be distinguished from long-term memory. *Short-term memory* (Figure 5.2) is memory that holds information received from the sensory register for up to about 30 seconds, although the length of the retention depends on a number of factors.

Consider the following experiment, which illustrates the separate existence of short-term memory as well as several of its other features (Glanzer & Cunitz, 1966).

The technique used in this experiment is known as *free recall*. The subjects in the experiment were shown lists of 15 nouns. Each noun was presented for 1 second, with a 2-second interval between the presentations. After all 15 nouns had been presented, subjects were asked to recall the nouns in any order that came to mind (hence the term *free recall*).

In one condition of the experiment—the zero (0) delay condition—subjects were asked to write down, in any order, as many words as they could recall immediately after the list had been presented. This resulted in the curve labeled *0* (at the right in Figure 5.3), where the proportion of correctly recalled nouns is graphed against the word's position in the to-be-remembered list—first, second, third, and so on. For example, in the 0 condition, the proportion of recall for the fifth noun on the list was 0.35, or 35 percent.

If you look at the curve for the 0 condition, you will see that the nouns presented early in the list and those appearing late in the list were recalled relatively well; recall for nouns in the middle of the list was rather poor. Thus recall depends on where an item is in a series of items (its serial position). The role of serial position in free recall (and in other memory tasks, too) is generally called the *serial-position effect*. The better recall at the beginning of the list which contributes to the serial-position effect is known as the *primacy effect*; items encountered first are remembered relatively well. The better recall at the end of the list is known as the *recency effect*; items encountered most recently are remembered well.

The curves marked *10* and *30* in Figure 5.3 represent the results of two delayed-recall conditions in the experiment. After presentation of a list, subjects in these conditions waited either 10 or 30 seconds before beginning free recall. The delay intervals were filled with the mental activity of counting.

If you look at the 10-second and 30-second delay curves in Figure 5.3, you will note that delays in recall, if the delay interval is filled with mental activity, decrease or eliminate the recency, but not the primacy, part of the serial-position curve.

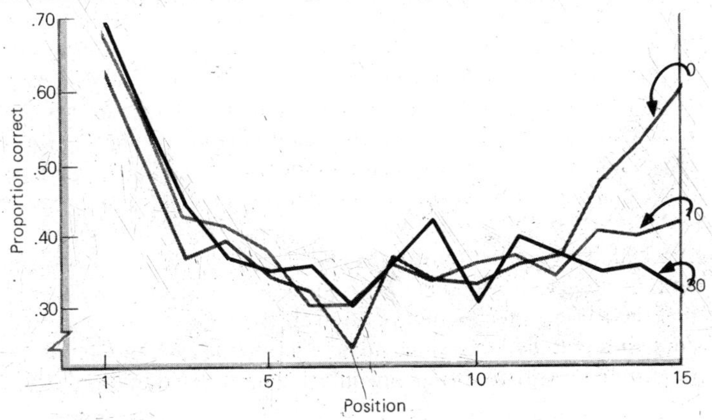

Figure 5.3

The effect of delay of recall on the recency portion of serial-position curves. The numbers at the right of the curves indicate the time in seconds between the end of list presentation and free recall. See text. (From Glanzer & Cunitz, 1966.)

Similar results have been obtained in a number of other experiments, and the reason usually given for the disappearance of the recency effect with delay is that the last items in a list are still in short-term memory when list presentation stops. They have not yet been transferred to long-term memory, and such transfer is prevented by the mental activity occurring during the delay interval. (We will have more to say about rehearsal in the next section.) The primacy effect remains during delayed recall because the first items in the list have had time, during the presentation of the list, to be put into long-term memory.

Short-term memory, in addition to its transient quality, also has a very limited storage capacity. This capacity is estimated to be about 7 items, plus or minus 2 (Miller, 1956). The storage capacity of short-term memory can be increased, however, by a process known as *chunking*. Most of us have learned to combine several items into a "chunk" as we receive them; then we can retain several (7, plus or minus 2) of these "chunks" of information in our short-term memories. Telephone numbers, for instance, consist of 7 items (10, if the area code is included), arranged in 2 (or 3) chunks. With practice, most of us can easily hold 3 telephone numbers in short-term memory—21 items arranged in 6 chunks. The world record, achieved only after prolonged practice, seems to be about 80 items, grouped into a number of chunks (Ericsson & Chase, 1982).

What is the fate of information in short-term memory? Since the capacity of this memory stage is so small, much information stored here is lost because it is displaced by incoming items of information. Before it is lost, however, some of the information can be retrieved and used. Studies of retrieval from short-term memory (Sternberg, 1966, 1975) show that we rapidly scan through short-term memory when searching for an item of information. A surprising feature of this scanning process is that we examine everything in short-term memory when we are trying to retrieve an item from it; the scanning has been found to be exhaustive. Instead of stopping when the searched-for item is located, the scanning process continues until all of short-term memory has been examined. Then, if the item was found during the exhaustive scan, it is retrieved. Some of the information in short-term memory is neither lost nor retrieved but passed along to the next memory stage—long-term memory—through rehearsal.

Rehearsal The process of rehearsal consists of keeping items of information in the center of attention, perhaps by repeating them silently or aloud. The amount of rehearsal given to items is important in the transfer of information from short-term to long-term memory (Rundus, 1971). In general, the more an item is rehearsed, the more likely it is to become part of long-term memory. However, in the last few years, other experiments have indicated that the sheer amount of rehearsal may be less important than the ways in which the information is rehearsed. Just going over and over what is to be remembered (called *maintenance rehearsal*) does not necessarily succeed in transferring it to long-term memory (Craik & Watkins, 1973). What is known as *elaborative rehearsal* is more likely to succeed. Elaborative rehearsal involves giving the material organization and meaning as it is being rehearsed; it is an active rehearsal process, not just the passive process of repetition. In elaborative rehearsal, people use strategies that give meaning and organization to the material so that it can be fitted in with existing organized long-term memories. Elaborative rehearsal, although introduced here in the context of memory stages, is a part of an alternative conception of memory called the *levels-of-processing theory*; it also relates to the organization of memory and to what is called *semantic memory*. These concepts will be described later in this chapter (page 194 and page 196).

Long-Term Memory (LTM) The time span over which information can be stored in long-term memory cannot be stated very precisely. Long-term memories may last for days, months, years, or even a lifetime. Also, unlike short-term memory, the storage capacity of long-term memory has no known limit.

Some theorists believe that there is no true forgetting from long-term memory. According to this view, once information is stored in long-term memory, it is there for good; when we seem to forget, it is because we have trouble retrieving, or getting access to, what has been stored. In other words, the information is still there; we just cannot get to it because it has not been stored in an organized fashion or because we are not searching for it in the right part of the memory storehouse. Other students of memory maintain that we forget because of the confusion and interference produced by new things which have been learned and put into long-term memory. (See page 205.)

Long-term memory contains words, sentences, ideas, concepts, and the life experiences we have had. As we shall see in more detail later (page 196), two different but related long-term memory stores are said to exist. One, called *semantic memory* (the word *semantic* refers to "meaning"), contains the meanings of words and concepts and the rules for using them in language; it is a vast network of meaningfully organized items of information (Quillian, 1966). The other, containing memories of specific things that have happened to a person (reminiscences) is called *episodic memory* (Tulving, 1972). We shall come back to these ideas (and others) later when long-term memory is discussed in detail (pages 195–203).

This brief discussion of long-term memory concludes the discussion of the stages of memory. Table 5.1 summarizes the major differences among the stages.

THE LEVELS-OF-PROCESSING THEORY

Information-processing theories of memory, as we have seen, view the memory process in terms of discrete stages, each with its own characteristics. Furthermore, information is tranferred from stage to stage until some of it is finally lodged in long-term memory. A contrasting model of memory involves what are called *levels of processing* (Craik & Lockhart, 1972), with, more recently, the idea of elaboration added to the levels-of-processing framework (Craik & Tulving, 1975).

According to the levels-of-processing idea, incoming information can be worked on at different levels of analysis; the deeper the analysis goes, the better the memory. The first level is simply *perception*, which gives us our immediate awareness of the environment (Chapter 3, page 107). At a somewhat deeper level, the *structural* features of the input (what it sounds like or looks like, for example) are analyzed; and, finally, at the deepest level of processing, the *meaning* of the input is analyzed. Analysis to the deep level of meaning gives the best memory. For in-

Table 5.1

Summary of characteristics of the stages of memory

	SENSORY REGISTER	SHORT-TERM MEMORY	LONG-TERM MEMORY
APPROXIMATE DURATION	For vision: up to about 1 second. For hearing: up to about 5 seconds.	Up to about 30 seconds, but it varies, depending on a number of factors.	Days, months, years, or a lifetime.
CAPACITY	Relatively large—up to at least 16 items, but probably much more.	Relatively small—up to about 7 items or chunks under most conditions.	Very large—no known limit.
TRANSFER PROCESSES	Attention and recognition. Items attended to and recognized move to short-term memory.	Rehearsal: items appropriately rehearsed move to long-term memory.	
TYPE OF INFORMATION STORED	Copy of input.	Sounds, visual images, words, and sentences.	Primarily meaningful sentences, life events, and concepts; some images; semantic and episodic memory.
MAJOR REASON INFORMATION IS LOST	Decay of trace.	Displacement of old information by incoming information.	Faulty organization or inappropriate retrieval (search) strategy; interference.

stance, suppose a friend who is not very good at spelling asks you to look over a term paper for spelling errors. As you do this, you are processing the information only to the structural level. If your friend later asks you what you thought of some of the ideas in the paper, it is likely that you will remember few of them; you did not process the information deeply enough; that is, you did not process it to the meaning level. Our example is for verbal, or word, information; but sounds, sights, and smells are also said to be processed through these levels.

Thus good memory results from deeper and, as we shall soon see, more elaborate processing of perceptual input. Many times, however, it is not important for a person to process information deeply; it is enough to hold the information long enough to act on some structural feature of it and then to discard it. Many of the routine happenings of daily life are not processed deeply. It is enough, for example, to respond appropriately at the moment when driving to work; we usually cannot remember the details of our morning drive because there was no need to process much of the information to the meaning level.

Rehearsal plays a role in the deeper processing of information, as it does in the stage theories of memory. Remember that *rehearsal*, roughly, refers to keeping information at the center of attention, perhaps by repeating it over and over to yourself. But according to the levels-of-processing view, simply repeating the information—*maintenance rehearsal* (page 192)—is not enough for good memory. All this does is maintain the information at a given level of depth; for deeper levels to be reached, the rehearsal must be *elaborative* (page 192). In other words, rehearsal must process the information to the meaning level if the information is to be

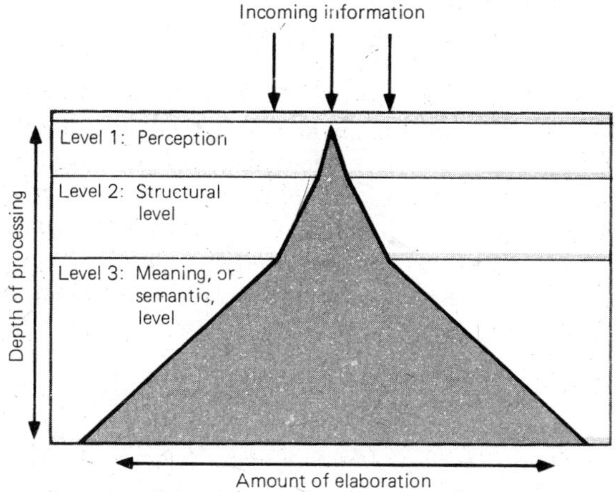

Figure 5.4

A summary diagram of the relationships among levels of processing, elaboration of information, and memory. The amount of information retained is shown by the shaded portion of the figure.

well retained. Rehearsal is thus seen as a process which gives meaning to information.

The idea of elaboration has been added to the levels-of-processing theory. *Elaboration* refers to the degree to which incoming information is processed so that it can be tied to, or integrated with, existing memories. The greater the degree of elaboration given to an item of incoming information, the more likely it is that it will be remembered.

At the risk (and it is a considerable risk) of making a complex situation too concrete and simple, Figure 5.4 summarizes what has been said about levels of processing and elaboration. This figure shows that the amount remembered, indicated by the shading, depends on both the level of processing and the degree to which information is elaborated. The best memory is the result of processing to the meaning level, where the amount of elaboration is also greatest.

Long-term memory: Its organization and processes

When we think about memory, it is usually long-term memory that we have in mind. Our reminiscences of past events in our lives are drawn from long-term memory. Our sense of self and continuity as an individual could hardly exist without long-term memory of what happened to us yesterday, the day before, and so on back to our earliest years. What would we think about if we had no long-term memory? We would be at the mercy of momentary, short-lived sensory and perceptual impressions of the world around us. Imagine, if you can, what behavior and thinking would be like without a system to keep track of events as they occur. Memory, especially of the long-term variety, is essential for behavior and mental life as we know it; it is one of the basic *cognitive processes*.

THE ORGANIZATION OF LONG-TERM MEMORY

Human long-term memory is not an untidy jumble of unrelated information; we keep our memory store in order. We organize, categorize, and classify information in a number of ways. Long-term memory is a bit like a library with a good cross-indexing system.

The Tip-of-the-Tongue* (TOT) *Phenomenon One way to study the organization of information in long-term memory is to see what happens when we search through our library of experience to retrieve a memory. Suppose you are trying to retrieve a person's name but you cannot quite remember it; the name is on the "tip of your tongue," but you just cannot recall it. If we look at this *tip-of-the-tongue* (*TOT*) phenomenon in greater detail, we find evidence for the organization of long-term memory (Brown & McNeill, 1966).

> The search through the memory store in the TOT state is not random. If the name you are looking for is *Martin*, you may come up with *Mertin* or *Morton*, but not *Potzrebe*. Brown and McNeill brought this phenomenon into the laboratory by reading aloud definitions of unfamiliar words that the subjects would probably recognize if they themselves were reading fairly difficult material but which they were not likely to recall spontaneously. Examples are *apse, nepotism, cloaca, ambergris,* and *sampan*. When the subjects were in the TOT state, aroused by hearing the definition but not able to hit the "target" word, they tended to retrieve words from their long-term memories that (1) sounded like the target word, (2) started with the same letter as the target word, (3) contained the same number of syllables as the target word, and (4) had a meaning similar to that of the target word. For instance, the definition of the target word *sampan* led the subjects to suggest *Saipan*, *Siam*, *Cheyenne*, *sarong*, "sanching" (not a real word), and "sympoon" (not a real word). These words are "sound-alikes," with the same initial sound and the same number of syllables as the target word. The subjects also gave answers like *barge*, *houseboat*, and *junk*—words with meanings similar to *sampan* (Figure 5.5).

The TOT phenomenon indicates that information is organized in long-term memory. Note also that the words retrieved in the TOT example are part of our general store of knowledge about the world. This introduces us to the distinction between two kinds of long-term memory organization, semantic memory and episodic memory.

Figure 5.5

The tip-of-the-tongue (TOT) phenomenon. The person is searching for the word *sampan* but cannot quite come up with it. Note that words with meanings similar to *sampan* are being retrieved, indicating that items are stored in meaningful categories in long-term memory.

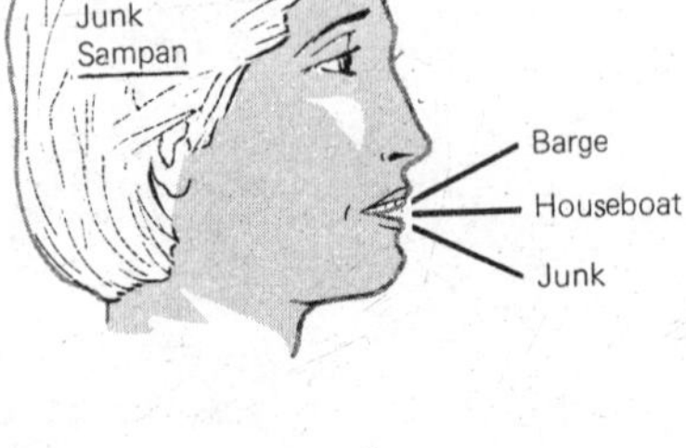

Semantic and Episodic Memory Much of what is in our long-term memory consists of knowledge about what words mean, about the ways they are related to one another, and about the rules for using them in communication and thinking. (See Chapter 6 and Chapter 2, pages 69–71). In short, it is this kind of memory which makes our use of language possible. It is called *semantic memory* (Quillian, 1966). Semantic memory is considered to be very stable; there is little forgetting of the meanings of the words of our language and the rules for their use. To illustrate semantic memory more explicitly, here are a few examples from the semantic memory of one of the authors.

> I know the word *thesaurus* means "treasury" and refers to a dictionary of synonyms.

Reinforcement is critical in operant conditioning.
A chaise lounge is something like a combination of a sofa and a chair.

Information seems to be stored in semantic memory in a highly organized way. For instance, some experiments (Collins & Quillian, 1969) have indicated that information is stored in logical hierarchies that go from general categories to specific ones (Figure 5.6). Such organization makes it possible for us to make logical inferences from the information stored in semantic memory. Other experiments (Rips et al., 1973) have led to the idea that semantic memory is organized into clusters of words with related meanings, very much as the TOT observations indicated.

Episodic memory (Tulving, 1972) consists of long-term memories of specific things that happened to us at particular times and places. Thus episodic memories are memories of episodes, long or short, in our own lives; they are dated and have a biographical reference. In other words our "remembrances of things past" make up our episodic memory.

When I was 21 years old, I joined the army.
I went fishing last week.
I worked late in the laboratory last night.
I have just come from a memory experiment done by a student, and I remember that the nonsense syllables "tov" and "yok" were paired in the experiment.

Unlike semantic memory, with its network of meanings, episodic memory seems to be organized with respect to when certain events happened in our lives. The episodes do not have to have a logical organization. Thus episodic memory is a record of what has happened to us and does not lend itself to the drawing of inferences. In addition, perhaps because it is less highly organized, episodic memory seems more susceptible to being forgotten than does semantic memory.

Of course, episodic and semantic memories are related. For example, episodic memories may be incorporated in our network of general knowledge about the world and thus become a part of our semantic memories; we derive our knowledge about the world from specific things

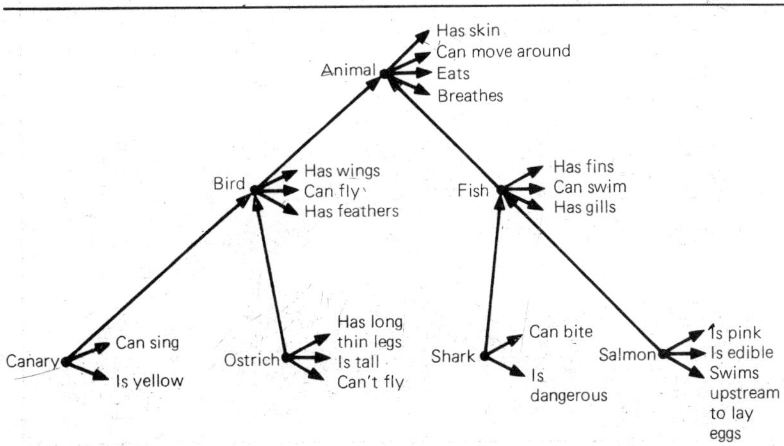

Figure 5.6

In semantic memory, information may be organized in logical hierarchies. (From Collins & Quillian, 1969.)

that have happened to us. And items in semantic memory can become a part of our episodic memory. For instance, you might remember that at a certain time, you used some information from semantic memory.

ENCODING AND STORING LONG-TERM MEMORIES

The importance of encoding for long-term memory is highlighted by encoding failures. Although we have all seen pennies countless times, most of us do not have an accurate long-term memory image of what is on the "heads" side of a penny. Which of the pennies in Figure 5.7 is the right one? Make a choice and check it. In one study of penny recognition, less than half of the people in the study chose the right coin (Nickerson & Adams, 1979). Most of us do not pay attention to penny details because, after all, there is no need to; for all practical purposes, it is enough to know that the only copper-colored United States coin is worth 1 cent. But the message is clear: Encoding for long-term storage requires special attention or strategies of some sort; just being exposed to something is usually not sufficient for long-term memory storage.

The Role of Organization One strategy in remembering things well is to organize, or arrange, the input so that it fits into existing long-term memory categories, is grouped in some logical manner, or is arranged in some other way that makes "sense." The organizational encoding may be inherent in the input itself or it may be supplied by individuals as they learn and remember new things. (See the later section of this chapter—"Improving Your Memory"—for more on possible individual strategies of organization.)

To see how inherent organization promotes good long-term memory, take a look at Figure 5.8, which gives an example of the way words were arranged in an experiment designed to see whether logical organization of input leads to good memory (Bower et al., 1969). Memory of words arranged in logical hierarchies, such as the one in the figure, was much better than was memory of the same words learned without any inherent organization.

Figure 5.7

Which is the right penny? See text. (From Nickerson & Adams, 1979.)

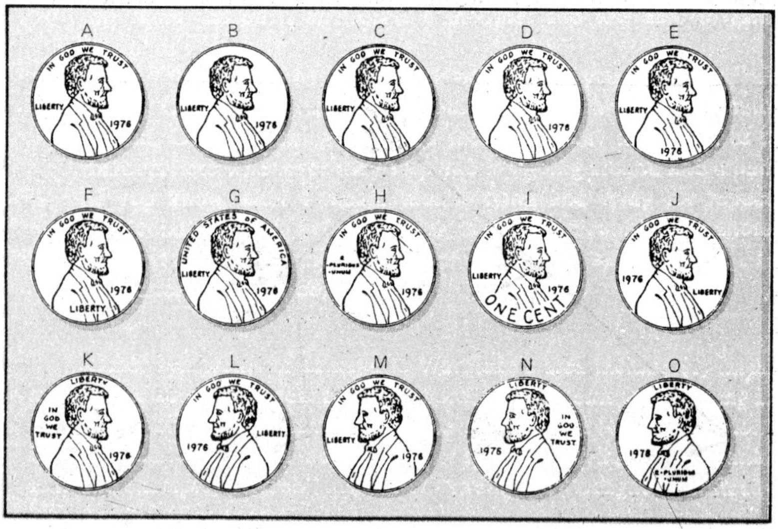

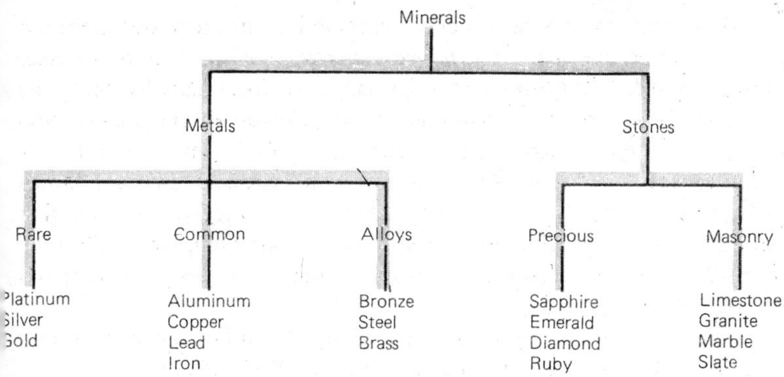

Figure 5.8

A conceptual hierarchy for minerals. When input was organized in this way, learning was rapid and memory was very good. (Modified from Bower et al., 1969.)

But the things we learn are not usually inherently well organized. In our everyday learning and memory, we must provide our own organization of the jumble of incoming information. In other words, we must do our own organizational encoding of incoming information. This is called *subjective organization*. Even when the materials are inherently organized, learner-imposed, or *subjective*, organization occurs. One way to study subjective organization is to see whether, in learning and recalling a list of unrelated words, certain stereotyped patterns of recall emerge as learning and recall trials of the list are repeated (Tulving, 1962). In other words, do people tend to recall pairs of words and short strings of words together? They do, and such groupings, or subjective organization, lead to better memory.

The Role of Imagery The form in which information is encoded is an important aspect of long-term memory. The organization and meaning given to verbal information are, as we have seen, quite influential in promoting long-term retention. Another factor is whether the incoming information is encoded by forming images of it. Visual images are the ones that have been most studied.

"Image" is a hard concept to define in words. (Of course, we have a good idea of what images are like from our own experience.) With the exception of iconic images (page 190), images do not seem to be literal copies of input. In the case of visual images, for example, the "picture in the head" is not an exact copy of the input; it is not complete, and parts of it are emphasized while others are absent. *Images* are thus partial and altered representations of what is in the world around us. (See Chapter 6, page 229.)

In spite of the difficulties in defining what image is, it is possible to get a rough measure of the degree to which imagery is aroused by words and to do memory experiments in which incoming information is encoded by imagery. To obtain a rough measure of the image-arousing capabilities of words, people were asked to rate the difficulties they had in forming visual images, on a scale from "very easy" to "very difficult" (Paivio, 1965, 1971). The words for which visual images were easily formed were called *concrete*, while those that evoked very little visual imagery were termed *abstract—desk* (concrete) and *mercy* (abstract), for example. After a measure of imagery has been obtained, its effect on learning and memory can be studied.

Many experiments have been conducted on imagery and memory; we will describe one using the ***paired-associate technique.*** Paired-associate learning is a little like learning a foreign-language vocabulary list that pairs foreign words with their English equivalents. (Given a foreign word, you learn to associate the English equivalent with it.) In learning and memory experiments, lists are made up of pairs of words (*squirrel-calendar*), words and numbers (*icebox-561*), or nonsense syllables ("tec-yor"). The first element of the pair is called the ***stimulus***; the second element is called the ***response.*** Given the stimulus, you learn to make the response that has been paired with it.

Using the paired-associate technique, the following experiment shows the role of imagery in encoding (Paivio, 1965, 1971).

> Concrete words (*bottle*) and abstract words (*truth*) were paired in various ways on paired-associate lists. The stimulus could be concrete (*C*) or abstract (*A*), and so could the response. Equal numbers of all pair types (C-A, A-C, C-C, and A-A) were included in the lists to be learned.
>
> The results are shown in Figure 5.9. Having an item that evokes imagery (a concrete word) in the stimulus position resulted in good recall (compare the C-C results with the A-C results and the C-A with the A-A results). Imagery evoked by the response terms also led to good recall (compare the C-C results with the C-A results and the A-C with the A-A results). Overall, however, the greatest effect of imagery on recall was found when the stimulus terms evoked concrete imagery.

Figure 5.9

Imagery as a factor in memory. Subjects learned lists in which the stimuli and responses of paired associates varied in their image-arousing capability: C items evoked concrete imagery, while A, or abstract, items evoked little imagery. The four types of paired associates in the experiment are shown: C-C refers to paired-associate items in which both the stimulus and response evoked concrete imagery; in the C-A pairs, the stimulus evoked concrete imagery, while the response, being abstract, did not; the other combinations are represented by the A-C and A-A pairs. (From Paivio, 1971.)

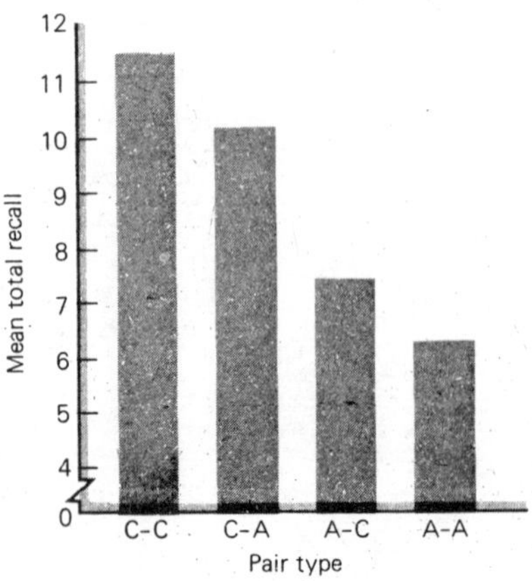

One interpretation of the importance of stimulus imagery in paired-associate experiments is that a concrete stimulus provides a conceptual peg on which responses can be hung. Another way to express this idea is to say that the response can be incorporated into a concrete image. For instance, in a concrete-abstract pair like *bottle-truth,* the concrete imagery of the bottle may make it easy for a person to form an image of a bottle saying "truth serum." Such use of concrete imagery is behind many of the schemes or systems designed to "improve memory." (See pages 212–219.)

The Role of Constructive Processes During encoding, the to-be-remembered information, especially if it is a complex life event or something you have read, is modified. Certain details are accentuated, the material may be simplified, or it may be changed in many other ways so that what is encoded and stored is far from a literal copy of the input. (See Application 5.1, page 186). These modifications are called ***constructive processes.***

One important constructive process is encoding only the gist, or meaning, of complex information such as what we have read in a newspaper, magazine, or book. For example, many years ago the British psychologist Sir Frederick Bartlett (1932) did some classic experiments in which people were asked to read a rather bizarre folktale. He then obtained successive recalls of the story several hours or days after the reading. He found that the story was shortened and simplified, and details were omitted so that only the general outline, or gist, was left in many cases. Furthermore, the changes made in the story indicated that the subjects were using inferences in their encoding of the story.

More recent experiments have focused on the use of inferences in the constructive process (Bransford et al., 1972, for instance). Suppose you read, "The driver of the car was seen drinking before he was involved in an accident." You would probably infer that drinking caused the accident and remember the sentence as stating causation, although it does not. Thus we tend to remember what was inferred at the time of encoding and storage.

Inferences are also made on the basis of the memory organizations, or *schemata* (plural of *schema*), that we have in semantic memory (page 196). We have all sorts of information about things, events, and their relationships stored in our semantic memories. I know, for instance, that Professor Smythe's office is in a building half a mile away. When he calls me to say he will be right over to see me, I assume that he is calling from his office. The inference is strengthened when he arrives sweating and panting, because there is a steep hill between his office and mine. Also, since the time it took him to arrive was about right, I make the inference that he came from his office. But in fact, the professor had called me from the gym where he was working out; he still had not quite cooled off when he arrived at my office. The time of arrival was about right, even though the gym is quite close to my office, because the professor had not quite finished dressing when he called. Later, if I need to recall this incident, I will remember the inference I made when I encoded the telephone call; I will remember that Professor Smythe called from his office. No wonder witnesses in court who are trying to tell the truth may tell conflicting stories; in addition to the factors described in Application 5.1 (page 186), they have not made the same inferences about the to-be-remembered event.

RETRIEVAL FROM LONG-TERM MEMORY

Information is encoded and stored in long-term memory, but it must be "read out," or *retrieved*, if it is to be used. Retrieval cues and reconstructive processes are important factors in the "read out" from memory.

Retrieval Cues Finding information in the organized long-term memory store is aided by *retrieval cues*, or reminders, which direct the memory search to the appropriate part of the long-term memory library. A number of experiments have shown that it is important to have the retrieval cues, or "tags," encoded along with the information as it is put into long-term memory storage (Tulving & Thomson, 1973). As we saw earlier (page 195), recall—that is, *retrieval*—is quite good when conditions favor rich and elaborate encoding. Perhaps the rich context into which an item of information is embedded provides a number of readily available retrieval cues.

When people learn things, we have seen that they often provide their own organization—*subjective organization*—of what they are learning. Thus, even when retrieval cues are not explicitly present in learning, we may provide our own retrieval cues at the time we encode information for storage. This is one of the tricks in having a good memory. Suppose, for example, you are learning German. Some German words will, at the time you encode them, be readily associated with English words; *Hund*, the German word for "dog," reminds you of the English word *hound*. When you later see the German word *Hund*, you can easily recall its English meaning because you generated the retrieval cue *hound*, which you encoded along with the target word *hund*. (A later section of this chapter, "Improving Your Memory," describes a number of strategies for encoding retrieval cues along with to-be-remembered information.)

What is called *state-dependent memory* (also known as *state-dependent learning*) can also be looked at as lending support to the importance of having retrieval cues encoded with the to-be-remembered information. In any case, state-dependent memory is a rather compelling demonstration of situational influences that affect retrieval from long-term memory. If people or animals encode and store information when they are in a particular emotional or drugged state, they may not be able to retrieve the information when they are in another emotional state or no longer under the influence of the drug. But when put back into the original emotional or drugged state, the memory can be retrieved (Ho et al., 1978). Stories are told about alcoholics who were under the influence of alcohol when they hid a bottle; they forget where they put it when sober, but they remember when drunk again. The emotional or drugged state is part of the context within which a memory is encoded and stored. Without this context, retrieval is poor; with it, retrieval is good. (Figure 5.10).

Reconstructive Processes in Retrieval Earlier (page 200), we saw that constructive processes are involved at the encoding stage of long-term memory. In contrast, *reconstructive processes* are modifications of already stored input. The distinction between constructive and reconstructive processes is, however, often blurred; it may be difficult to tell whether the modifications in memory occurred at the encoding stage or later. Reconstruction is sometimes called *confabulation* in the case of people with memory disorders who have stored very little and who then try to fill in the memory gaps during retrieval.

Reconstructive processes are often seen in the answers to leading questions that bias the retrieval of what was stored. Here is an experiment bearing on this point (Loftus & Palmer, 1974).

The subjects in this experiment watched a short film of an automobile accident. Immediately after the film, they were given a

Figure 5.10

The concept of state-dependent memory. When encoding conditions, or states, match the testing conditions, or states, memory tends to be good; otherwise it may be poor.

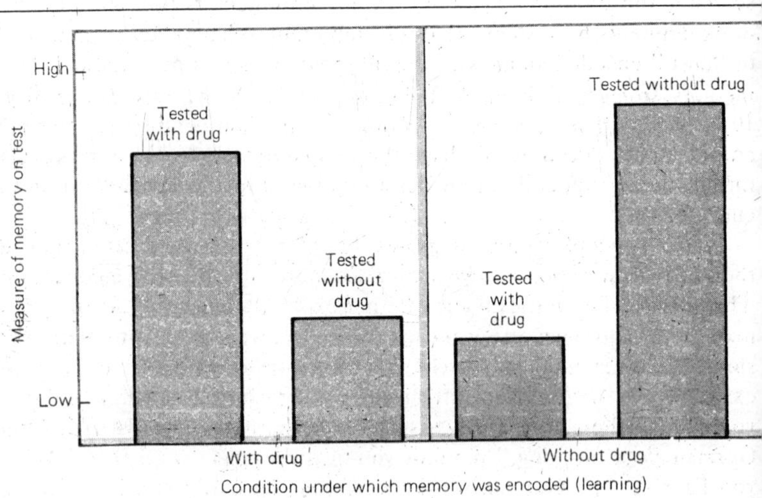

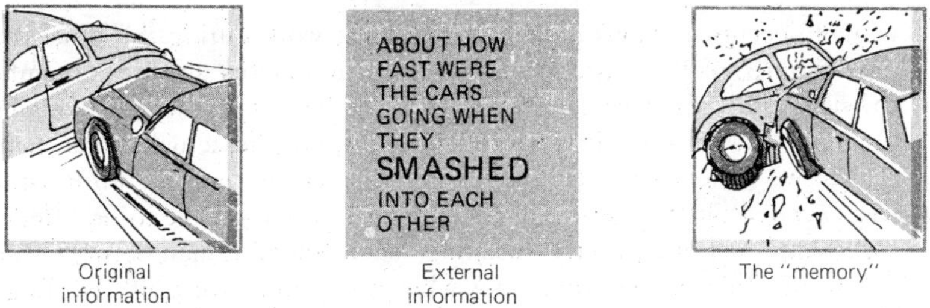

Figure 5.11

Reconstructive processes can influence retrieval. See text. (From Loftus & Loftus, 1976.)

questionnaire to fill out. Among the many questions asked were two critical ones: "About how fast were the cars going when they smashed into each other?" and "About how fast were the cars going when they hit each other?" One group of subjects had the "smashed" question; a second group had the "hit" question. A week later, the subjects were asked a number of other questions about the accident, the critical one being "Did you see any broken glass?" There was *no* broken glass in the film they had seen the week before. Subjects who had been given the "smashed" question originally were twice as likely to report that they remembered seeing broken glass as were those who had been given the "hit" question. (See Figure 5.11.)

Forgetting

Psychologists generally use the term *forgetting* to refer to the apparent loss of information already encoded and stored in long-term memory. Much is forgotten (Figure 5.12), but enough endures so that we have a sketchy record of our lives. (Inquiry 5.2 describes some lasting memories.)

Much of what we think we have forgotten does not really qualify as "forgotten" because it was never encoded and stored in the first place. With information-processing theories (page 185) in mind, some information, due to lack of attention, may not have reached short-term memory from the sensory register; or, due to inadequate encoding (pages 198–201) and rehearsal (page 192), the information may not have been transferred from short-term to long-term memory. The levels-of-processing theory (page 193) would say that information was not stored in long-term memory because rehearsal was not sufficiently elaborate. So if much information is lost before being stored in long-term memory, is it any wonder that the memory record of life's happenings is so incomplete?

We often say we forget when memory does not match events as they

Figure 5.12

Forgetting occurs most rapidly shortly after learning. This classic forgetting curve is from the work of the nineteenth-century psychologist Hermann Ebbinghaus. Of course, there is no single forgetting curve; the rate of forgetting depends on many factors.

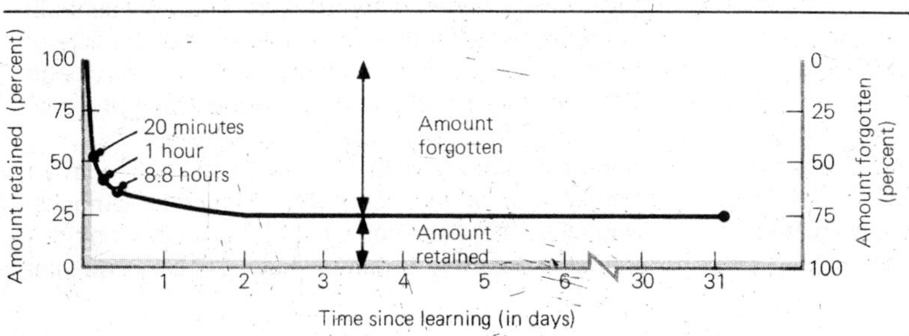

actually occurred. Constructive processes at work during the process of encoding (page 200) distort what is stored in memory and we remember the distortions. For instance, we remember the gist, or meaning, of what we have read or heard in a conversation but not the actual words themselves; we remember inferences constructed at the time the information was encoded for storage; or we encode only portions of the to-be-remembered information. Strictly speaking, such faulty remembering is not forgetting. In fact, we remember what was stored; we think we forget

Inquiry 5.2

WHAT DON'T WE FORGET?

Some long-term memories are with us for life. They are a part of us; we do not forget them. Illustrative are the rules for language use, the schemata of semantic memory (page 201), and certain life-event, or episodic (page 197), remembrances known as "flashbulb memories." The word "flashbulb" is descriptive of the vividness and sharpness of these brief memory "snapshots" that stand out in our life memories. For example, if you were to question several people who were at least 10 years old on November 22, 1963, you would probably get flashbulb memories of John F. Kennedy's assassination from most, if not all, of them. The subjects in this micro-experiment would have a sharp, clear memory of where they were, what they were doing, who they were with, and other events occurring when they first heard the news of the shooting.

I was on the telephone with Miss Johnson, the dean's secretary, about some departmental business. Suddenly she broke in with "Excuse me a moment; everyone is excited about something. What? Mr. Kennedy has been shot!" We hung up, I opened my door to hear further news as it came in and then resumed my work on some forgotten business that "had to be finished" that day.

Ten years after the assassination, the always-enterprising *Esquire* magazine . . . asked a number of famous people a question similar to ours: "Where were you?" Julia Child was in the kitchen eating *soupe de poisson*. Billy Graham was on the golf course, but he felt a presentiment of tragedy. Philip Berrigan was driving to a rally; Julian Bond was in a restaurant; Tony Randall was in the bathtub. The subtitle of the 1973 *Esquire* article could, we are sure, be used again today: "Nobody Forgets." (Brown & Kulik 1977, p. 74)

"Flashbulb memories" are said to be formed when we experience unexpected events that have important consequences for our lives. Somehow, to continue the photographic analogy, novel, consequential events are supposed to trigger "Now Print!" processes in the brain.

The existence of brief, vivid, hard-to-forget memories is difficult to dispute, but some psychologists have wondered about the accuracy of flashbulb memories; whether they are established at the time of the event or are later memory constructions (page 202); whether novelty and consequentiality are critically important; and, of course, whether the photographic analogy with its "Now Print!" notion is the best way to understand these enduring memories. When they are examined closely, flashbulb memories are not necessarily accurate, may sometimes be later constructions, and can occur for events that are not novel and may not have obvious personal significance. An alternative to the photographic analogy is that some flashbulb memories are established when the stream of our own life-event memories intersects with the stream of history.

My suggestion is that we remember the details of a flashbulb occasion because these details are the links between our own histories and "History". . . . The term "flashbulb" is really misleading; such memories are not so much momentary snapshots as enduring benchmarks. They are places where we line up our own lives with the course of history itself and say "I was there." (Neisser, 1982, p. 48)

REFERENCES

Brown, R., & Kulik, J. (1977). Flashbulb memories. *Cognition, 5,* 73–99.

Berendt, J. (1973, November). Where were you? *Esquire,* 136–137.

Livingston, R. B. (1967). Brain circuitry relating to complex behavior. In G. C. Quarton, T. Melnechuck, & F. O. Schmitt (Eds.), *The neurosciences: A study program.* New York: Rockefeller University Press.

Loftus, E. F., & Loftus, G. R. (1980). On the permanence of stored information in the human brain. *American Psychologist, 35,* 409–420.

Neissner, U. (1982). Snapshots or benchmarks? In U. Neisser (Ed.), *Memory observed: Remembering in natural contexts.* San Francisco: Freeman.

because what we remember is not an accurate representation of what really happened.

Ask a friend what causes him to forget. The answer will probably be something like, "Oh, just the passage of time, I guess." Press a little harder, and your friend might say, "Well, as time passes, my impressions of what I learned just decay and get weaker and weaker until they finally fade away." In other words, he believes that the memory trace in the brain, sometimes called the *engram*, decays with time. If there is any truth at all to this commonsense notion, it certainly does not do justice to the causes of forgetting as psychologists now understand them. So why *do* we forget?

INTERFERENCE

A vast amount of experimental evidence (and everyday experience, too) indicates that learning new things interferes with our memory of what we learned earlier and prior learning interferes with our memory of things learned later. For example, suppose you go to a party where you are introduced to many new people. When the evening is over, you will probably have forgotten the names of many of the people you met. Your memory of the names you heard earlier in the evening has been interfered with by the names you learned later. And it is also hard to remember the names you heard later in the evening because the names you learned earlier have interfered with your memory of the names learned at the end of the party.

Technically speaking, memory interference resulting from activities that came after, or *subsequent to*, the events you are trying to remember is called *retroactive interference*. It is called *retroactive* because the interference is with the memory of events that came before the interfering activity. *Proactive interference,* on the other hand, is due to events that came before the to-be-remembered information.

Perhaps looking at ways in which experiments might be set up to study retroactive and proactive interference will make the definitions clearer. Here is the way an experiment on retroactive interference might be done:

Experimental group:			
Learn task A	Learn task B	Retention interval	Measure recall of task A.

Control group:			
Learn task A	Rest	Retention interval	Measure recall of task A.

The difference between the two groups is the learning of task B: This comes after the learning of task A, and if retroactive interference occurs, the experimental group will do less well in recalling task A items than will the control group.

An experiment on proactive interference might be arranged this way:

Experimental group:			
Learn task A	Learn task B	Retention interval	Measure recall of task B.

Control group:			
Rest	Learn task B	Retention interval	Measure recall of task B.

In this case, the interfering activity of learning task A comes before learning of the to-be-remembered items of task B. If there is proactive interference, the experimental group will recall task B less well than will the control group. Figure 5.13 illustrates proactive interference; it shows that recall declines progressively as the amount of prior learning (number of lists previously learned) increases.

Although retroactive and proactive interference have been shown to be important causes of forgetting, the ways in which they work on memory are still, after years of study, the subject of some debate. One idea is that the interferences disrupt the various kinds of associations between stimuli and responses formed during learning. For example, in paired-associate learning (page 200), people learn to form forward associations between stimuli and responses and backward associations between responses and stimuli. If interference produces confusions in what is associated with what or, as some experiments indicate, actually produces "unlearning" of the associations, forgetting will be the result. Another idea is that interference somehow has its greatest effect on the memory of retrieval cues. We have seen that memory depends on retrieval cues, so if interference results in problems with the use of these cues, forgetting will result. But whatever the explanation of interference turns out to be, it is, as a practical matter, one of the major causes of forgetting.

Figure 5.13

Previous learning can interfere with later recall, an effect known as proactive interference. As the number of lists learned prior to a test list increases, recall of the test list decreases. (Modified from Underwood, 1957.)

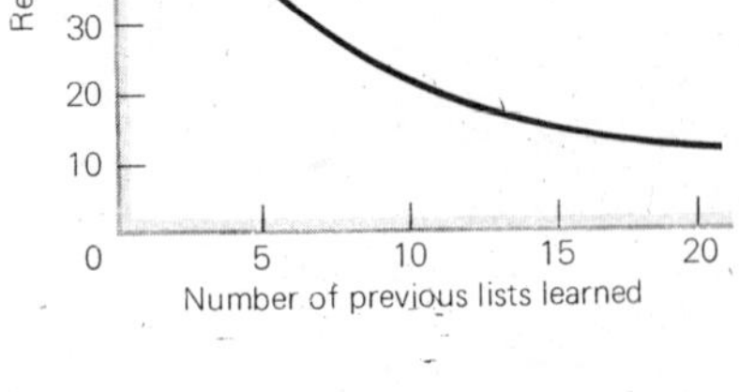

RETRIEVAL PROBLEMS

Still another part of the answer to why we forget was implicit in our previous discussion of retrieval from long-term memory (pages 201–203). There we saw that retrieval is facilitated by organization of the stored material and the presence of retrieval cues that can guide our search through our long-term memory for stored information. Without appropriate retrieval cues, the sought-for items stored in long-term memory may not be found—one *forgets*. But while we often cannot recall something while actively searching for it, we may later recall the sought-for information when we have given up the search and are doing something else. The new activity in which we are engaged, or the new context, gives us another set of appropriate reminders, or retrieval cues. And perhaps the new situation leads us to search through portions of our long-term memory store not examined before. The result may be that we suddenly remember what we could not previously retrieve—"Aha! It's Hank Aaron who holds the lifetime home-run record." Thus when we think we have forgotten something, it is often a good idea to give up and do something else in order to generate new retrieval cues.

In our earlier discussion, we also saw that reconstructive processes present at the time of retrieval can act to distort our recall of what is stored in our long-term memory. We may remember something that did not happen and forget what did happen.

Emotional factors can also play a role in the retrieval failure that is the cause of so much forgetting. State-dependent memory (page 202) is an example. If we encode information while in one emotional state and try to retrieve it while in another, our recall may suffer. Another more common and powerful hindering influence of emotions on retrieval is found in the phenomenon of *motivated forgetting*—difficulties with the retrieval of unpleasant, anxiety-provoking information stored in long-term memory.

MOTIVATED FORGETTING

In his classic book *The Psychopathology of Everyday Life*, Sigmund Freud clearly stated the principle underlying motivated forgetting: "The uniform result of the entire series of observations [the examples given in his book] I can formulate as follows: *The forgetting in all cases is proved to be founded on a motive of displeasure*" (Freud, 1901/1951, pp. 71–72). In this quotation, Freud is hinting at a key concept of psychoanalysis—repression. *Repression* refers to the tendency of people to have difficulty retrieving anxiety-provoking or threatening information, and what is associated with that information, from long-term memory. (See Chapter 14, page 588, also.) Perhaps this helps to explain why people generally remember pleasant events more often than they do unpleasant ones; the unpleasant memories have been repressed.

Many lapses of memory in daily life illustrate motivated forgetting. For example, because I do not like liver, I may forget that my wife asked me to buy some for dinner. I may forget the names of people I do not like. But Freud went further than such direct, commonsense examples. Remember that the concept of "repression" includes retrieval failure for the associations of the threatening, anxiety-provoking information. In *The Psychopathology of Everyday Life*, Freud charmingly analyzes his own memory lapses to illustrate the forgetting of such associations. He forgets a conversation he had with a man he disliked; he forgets the location of a searched-for shop because he had an argument with a family that lived in the building where the shop was located; he forgets names, not because they themselves provoke displeasure but because they are associated in some way with anxiety-arousing ideas; and so on through a wide range of retrieval failures.

In the examples just given, retrieval of the forgotten material eventually occurred. But according to psychoanalysis, some stored information is so threatening and anxiety-arousing that its retrieval is possible only under special circumstances. Thus some emotionally toned information becomes unavailable for conscious thought; it is said to be in the unconscious (Chapter 14, page 578). To dredge up unconscious, repressed memories, psychoanalysts use the method of free association (Chapter 16, page 683)—having an individual say whatever comes to mind. Perhaps free association works, in part, because in the process of doing it, retrieval cues for the repressed memories are generated. Hypnosis and various drugs—the so-called truth serums—are also used, with occasional success, to retrieve repressed information not available to consciousness in the normal waking state.

Bringing motivated forgetting into the laboratory has proved difficult. We cannot, nor should we, produce the powerful anxieties necessary for unequivocal demonstrations of motivated forgetting. The best evidence for motivated forgetting is still, as it was for Freud, based on the analysis and interpretation of the events of everyday life with its strong anxieties and, as we shall see in the next section, on certain dramatic cases of memory loss, or *amnesia*.

Amnesia

In everyday speech, we use the term *amnesia* to refer to "loss of memory." This implies that amnesia is a kind of forgetting (page 203), and indeed some forms of memory disorder do result from a loss of what has already been stored or an inability to retrieve stored information. But amnesia is a more general "disease of memory." The term is also used for cases in which encoding (page 185) and storage (page 185) are impaired so that new memories cannot be formed. Thus *amnesia* is a profound memory deficit due either to the loss of what has been stored or to the inability to form new memories.

Some amnesias have a biological basis; the memory machine—the brain—is disturbed in some way. (See Inquiry 5.1, page 188.) These may be called biological amnesias. Other amnesias may be called psychological amnesias. Without any known brain malfunction, these amnesias result from major disturbances in the processes of information encoding, storage, and retrieval. The distinction between these two general kinds of amnesia—biological and psychological—is, however, often blurred. On the one hand, encoding, storage, and retrieval processes are impaired in the biological amnesias; and, on the other hand, some psychological amnesias may turn out to have a biological basis.

PSYCHOLOGICAL AMNESIAS

When we think of amnesia, we often have in mind the person who forgets almost everything, including his or her identity. But in a sense, everyone is an amnesia victim; we remember very little of our early childhoods and our dreams.

Childhood Amnesia Both formal studies and common experience agree on the poverty of early memory. Why is our early childhood memory so poor? Freud (1938) used the "repression" concept (page 207) to account for childhood amnesia. He said that we are unable to retrieve childhood memories because they are associated with the forbidden, guilt-arousing sexual and aggressive urges he thought characterized early childhood. These urges and their associations are repressed and cannot be retrieved

—they are "forgotten"—because being aware of them would result in strong feelings of guilt or anxiety. (See the discussion of defense mechanisms in Chapter 14, page 588.)

Another intepretation of childhood amnesia (Schachtel, 1959) stresses differences in the ways young children and older people encode and store information. As adults, much of our memory is encoded verbally and tied into networks, or schemata, that are based on language; it is probably no accident that language development (Chapter 11, page 438) and the richness of memory go hand in hand. But when we were very young and without language, we encoded memories in a nonverbal form, perhaps storing information as images or feelings. Early childhood memories are thus said to be stored in forms no longer available to us as verbal adults; our language-dominated memories do not have retrieval cues (page 201) appropriate for gaining access to the image-and-feeling memories of early childhood.

A third interpretation of childhood amnesia is that it may not be very "psychological" at all. The brain is maturing and growing in the first few years after birth. Perhaps the memory machine (Inquiry 5.1, page 188) is just not able to store long-term memories until its maturation is essentially finished. Language ability and memory develop together, according to this interpretation, because both depend on brain maturation.

Dream Amnesia We dream several times each night (see Chapter 2, page 55), but we remember few of these experiences. Freud's (1900/1953) interpretation of dreams was based, as was his interpretation of childhood amnesia, on repression. (See page 208 and Chapter 14, page 588.) He considered dreams to be expressions of forbidden sexual or aggressive urges. These urges can produce strong guilt or anxiety if we become aware of them in ourselves. So their expression in dreams is hidden behind a disguise—the actual content of the dream. But even the disguised urges—dreams—have the capacity to generate some guilt or anxiety feelings. Hence they are forgotten.

Other interpretations stress the differences in the symbol systems used in dreaming and waking (Hall, 1953), a situation similar to that in one of the interpretations of childhood amnesia described above. If the memory-symbol networks of waking life are different from those of dreaming, we may have difficulty retrieving dreams in the waking state.

And just as with childhood amnesia, dream amnesia may actually have a biological basis. The dreaming brain seems to be in a special state different from that of the waking brain. (See Chapter 2, page 55.) As we have seen when discussing state-dependent memory (page 202), information stored in one state is difficult to retrieve when in another state. Thus dream amnesia may be just another example of state-dependent memory.

Defensive Amnesia This is the well-publicized, but relatively rare, type of psychological amnesia that has captured the popular imagination. People with this form of amnesia may forget their names, where they have come from, who their spouses are, and many other important details of their past lives. It is called *defensive* because this type of amnesia is usually considered to be a way of protecting oneself from the guilt or anxiety that can result from intense, intolerable life situations and conflicts. We often wish we could forget a nagging problem. The defensive amnesiac does what we might wish to do and, because the problem has so many ramifications in his or her life, forgets much more than the specific problem itself. Defensive amnesia is thus an extreme form of repression. (See page 207 and Chapter 14, page 588.)

Amnesic episodes can last for weeks, months, or years. When they are over, the amnesiac regains, often suddenly, memories of his or her earlier life, but information stored during the episode itself is usually not retrievable—there is a memory gap. For instance, Mr. Z. may be mystified as to why he is digging ditches in Milwaukee when he remembers his name and that he is really an accountant from San Francisco with a wife, two children, and a fat mortgage. Mr. Z. will probably not remember his new wife, whom he married during his amnesic episode. He is in for some interesting times sorting matters out, and the problems caused by the attempt to sort things out may, in one so disposed, trigger another defensive-amnesic episode.

While repression may be the basis for amnesic episodes themselves, why is there a memory gap for information stored during the episode? Perhaps the gap occurs because memories formed during the episode are themselves repressed. Alternatively, after the stream of memory has returned to normal, perhaps retrieval cues are lacking for the information stored during the episode.

BIOLOGICAL AMNESIAS

Concussions (brain bruises) from blows on the head, other damage to the brain, temporary disturbances in the brain's blood supply, certain drugs, and brain diseases are some of the major biological causes of amnesia. Amnesia resulting from concussion is described in Inquiry 5.1, page 188; brain damage and memory loss are discussed in Chapter 2, pages 67–69.

Transient Global Amnesia This is a profound memory problem with no loss of consciousness. It comes on suddenly without any obvious cause, and it typically lasts for only a few hours or days before memory becomes normal again. Fortunately, most people who experience such amnesia have it only once. This type of amnesia is caled *global* because so much of what has already been stored in memory is forgotten and because, even though the victim is conscious and can go about the routine business of daily life, no new memories are formed while the attack is in progress. In other words, both *retrograde amnesia* (forgetting events one was exposed to in the past) and *anterograde amnesia* (the inability to encode and store new information) characterize transient global amnesia. The cause of transient global amnesia is not known, but a currently favored hypothesis is that it is due to temporary alterations in the normal pattern of blood flow to the brain.

Marijuana, Alcohol, and Amnesia Marijuana appears to have a limited, short-lived effect on the encoding, storage, and retrieval of information (Loftus, 1980), but it can hardly be said to result in amnesia. Even when marijuana is taken in relatively high doses, its memory effects fall far short of those of the most popular mind-altering drug—ethyl alcohol.

After a night of heavy drinking—a high dose of alcohol—a person may have amnesia for the events that occurred while he or she was "smashed" (Figure 5.14). When sober again, the individual's memory of events before overindulging is intact, but there is a memory gap for the period of drunkenness.

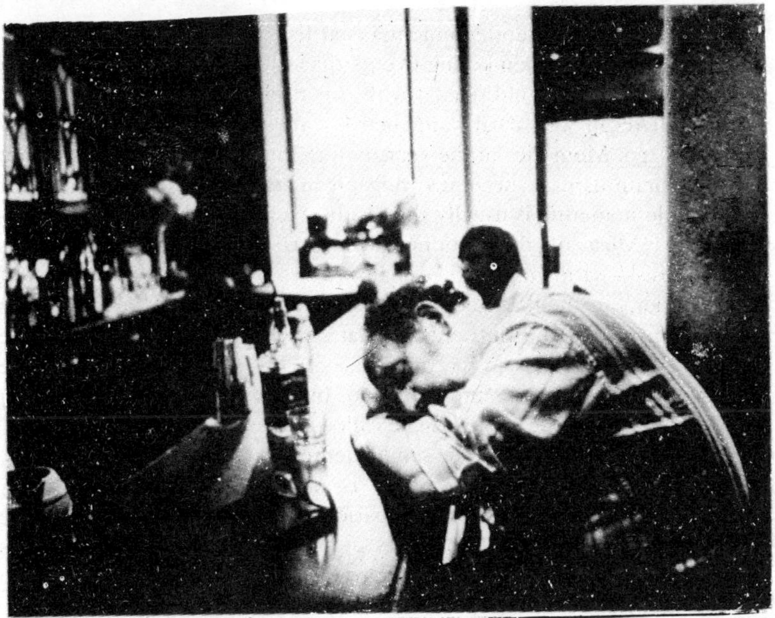

Figure 5.14

The man in the picture may not remember that he was at Tony's bar last night. (Arthur Tress, © 1977.)

A person may have amnesia for the events occurring while under the influence of alcohol because encoding and storage processes have been disrupted by the effects of the alcohol on the brain. Another theory is that information may have been stored while in the drunken state in a form not available for retrieval in the sober state. Thus alcohol-induced amnesia may be an example of state-dependent memory (page 202).

Heavy drinking over a period of years, however, can result, through vitamin-B deficits and other chemical imbalances, in irreversible brain damage and a pattern of symptoms known as the *Korsakoff syndrome.* Anterograde amnesia (the inability to form new memories) is one of the prominent symptoms of this syndrome. Korsakoff patients also have some loss of what are called remote memories—remembrances of events that occurred early in their lives (Squire & Cohen, 1982). (See Figure 5.15.) And memory is not the only information-processing problem of Korsakoff patients; they have difficulties with attention and perception that may impair their performance on some remote-memory tests.

Figure 5.15

Korsakoff patients have impaired remote memory. To test remote memory, Korsakoff patients, alcoholics without Korsakoff's syndrome, and nonalcoholic hospital patients—matched on a number of factors—were shown photographs of famous people who became prominent in each of the decades shown. (From Squire & Cohen, 1982.)

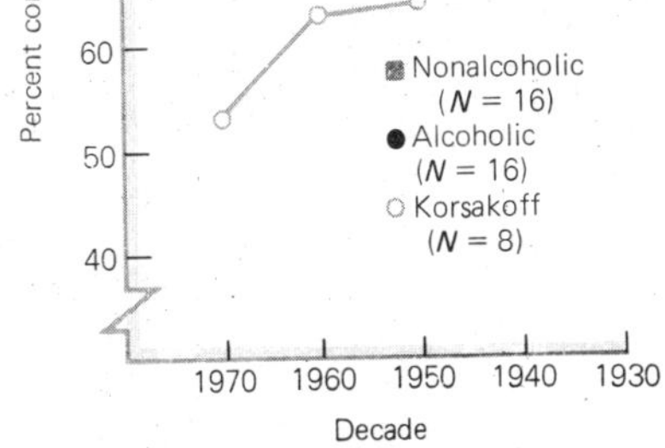

Diseases of the Brain Among the diseases that can result in amnesia are syphilis of the brain and other brain infections, strokes and other permanent disorders of brain blood flow, brain tumors, disorders of brain metabolism, multiple sclerosis, various conditions caused by toxic chemicals (of which the just-described Korsakoff syndrome is the main one), senile dementia, and primary degenerative dementia. We will focus on the last two of this long list.

Senile dementia (the word *senile* refers to old age) is characterized by deficits in many intellectual abilities—memory, attention, judgment, and abstract thought, for example—that can occur in aged people. Personality changes—excessive dependence and irritability, for instance—are common. Delusions—thoughts which have no basis in reality (Chapter 15, page 625)—and general disorientation—not knowing where one is in time or place—can also occur.

The amnesia in senile dementia is at first largely anterograde—the person has trouble remembering events that happened after the onset of the disease. Thus the old person with this disorder has trouble learning and cannot recall well what happened last month, yesterday, or even a few hours ago. Memories of the years before the disease are largely intact until the brain damage becomes widespread and severe.

Senile dementia is usually the result of a reduction in blood flow to the brain. Most of the patients with this disorder have brain arteriosclerosis—narrowing of the small arteries of the brain due to fatty accumulations in them. Arteriosclerosis deprives brain cells of adequate supplies of oxygen and nutrients so that some cells die and others malfunction. The brain is said to atrophy.

Normal aging has its problems too, but the typical forgetfulness of old age is hardly severe enough to be called amnesia. In normal aging, the memory problem, like that in senile dementia, centers largely on the storage of relatively recent events; it is anterograde in nature. But, in marked contrast to senile-dementia patients, normal old people are able to compensate for their mild memory problems. They try to do less and thus put a smaller burden on their information-processing systems; they provide themselves with reminder cues, perhaps by writing down what is to be remembered; and they organize their lives into routines so that fewer new things need to be remembered. In other words, normal old people adopt adaptive memory strategies.

Primary degenerative dementia has many of the same characteristics as senile dementia. A major difference is that the symptoms often begin in middle age. *Alzheimer's disease* is a form of primary degenerative dementia in which there is a cluster of specific degenerative brain changes of unknown origin. Some evidence indicates that the amnesia in Alzheimer's disease is related to deficiencies in the brain neurotransmitter chemical (Chapter 2, page 50) acetylcholine (Whitehouse et al., 1982).

The tragedy of Alzheimer's disease is that it begins relatively early in life and, unlike senile dementia, where death quickly ends the suffering, life continues with progessive mental deterioration. The amnesia, for instance, often goes from a relatively mild anterograde memory problem to a profound anterograde and retrograde deficit—both the recent and remote past are largely gone.

Improving Your Memory

Perhaps some of the general memory principles described in this chapter have suggested ways in which you can improve your memory. You can apply some of these general principles, but more specific aids to memory are available.

MNEMONICS

This strange-looking word (pronounced "nemoniks") comes from the Greek word for "memory" and refers to specific memory-improvement techniques. People with super memories (Inquiry 5.3, page 213) sometimes use mnemonics, and we can also learn to do so. Most mnemonic techniques rely on the linking, or association, of to-be-remembered material with a systematic and organized set of images or words that are already firmly established in long-term memory and can therefore serve as re-

Inquiry 5:3

WHAT ABOUT SUPER MEMORIES?

If one works at it, an incredible amount of information can be retained in memory. Perhaps the world's record is held by people known as "Shass Pollaks." *Shass* refers to the *Talmud*—commentaries on Jewish laws—which is thousands of pages in length. *Pollak* refers to Poland, since the best known of these memory experts were Polish. Using a standard edition of the *Talmud,* informal memory tests of several Shass Pollaks were conducted by sticking a pin through a page chosen at random and then asking for the word that was pierced by the pin on other pages. The Shass Pollaks gave the right words without error. They not only knew the *Talmud* by heart, they had memorized the location of each word on thousands of pages! So with hard work, motivation, and devotion, memory can be very good indeed.

Experts in any field have excellent memories for material in their areas of proficiency. The more one already knows about something, the better one remembers related material; thus chess masters encode and store more chess information in a short time than do novices, baseball experts remember more about a game they have watched on television than those who do not appreciate the intricacies of the game, a psychologist reading this book remembers more of it than does a beginning student, and so on. In other words, the larger a person's knowledge base, the better that person's memory of anything related to that base will be.

But some people have super memories without large knowledge bases and without much effort to remember. They are "naturally" good. Perhaps the most famous of these mnemonists, as they are called, was a Russian known as S. S. was a newspaper reporter who needed no notes; he could remember lengthy statements word for word. His phenomenal memory was studied by the Russian psychologist Alexander Luria. Here is what Luria says about S.'s abilities:

O	X	B	2	Y	&	Q
#	*	6	V	A	R	7
2	C	M	L	%	W	P
?	G	Z	J	S	5	!
0	T	3	=	H	N	8

As the experimenter, I soon found myself in a state verging on utter confusion. An increase in the length of a series led to no noticeable increase in difficulty for S., and I simply had to admit that the capacity of his memory *had no distinct limits;* that I had been unable to perform what one would think was the simplest task a psychologist can do: measure the capacity of an individual's memory. I arranged a second and then a third session with S.; these were followed by a series of sessions, some of them days and weeks apart, others separated by a period of several years.

But these later sessions only further complicated my position as experimenter, for it appeared that there was no limit either to the *capacity* of S.'s memory or to the *durability of the traces he retained.* Experiments indicated that he had no difficulty reproducing any lengthy series of words whatever; even though these had originally been presented to him a week, a month, a year, or even many years earlier. In fact, some of these experiments designed to test his retention were performed (without his being given any warning) 15 or 16 years after the session in which he had originally recalled the words. Yet invariably they were successful. During these test sessions S. would sit with his eyes closed, pause, then comment: "Yes, yes . . . This was a series you gave me once when we were in your apartment . . . You were sitting at the table and I in the rocking chair . . . You were wearing a gray suit and you looked at me like this . . . Now then, I can see you saying . . ." And with that he would

reel off the series precisely as I had given it to him at the earlier session. If one takes into account that S. had by then become a well-known mnemonist, who had to remember hundreds and thousands of series, the feat seems even more remarkable.
[Luria, A. R. (1968). *The mind of a mnemonist,* pp. 11-12.]

One reason for S.'s success was that he was able to form vivid images of what he was trying to remember. He had, in fact, what is called synesthesia—an input in one sensory channel aroused images in another sense. What he heard often evoked visual images, for example, and sometimes even taste, touch, or smell images were aroused. For example, when hearing a tone with about the pitch of middle C, S. would see "in his mind's eye" a pink-orange velvet cord; words also evoked vivid images in his mind. To remember, S. often used the method of loci (page 214). He associated the images formed by what he was trying to remember with parts of a scene, such as a square in Moscow, that he could visualize very clearly. For instance, the number 6 evoked in S. a visual image of a woman with a swollen foot; if he had to remember this number, he would visualize the woman leaning against a lamp post in the square. S. could store an unlimited amount of information in this way, and when he wanted to recall it he simply made a mental tour of the scene, contacting each to-be-remembered image in order. It is interesting to note that S. was no genius; his tremendous memory ability did not lead to excellence in any of the learned professions. Perhaps his mind was too cluttered by his images; at least, that was Luria's opinion.

His vivid use of imagery in the service of his amazing memory makes S. unique in the annals of psychology. Another mnemonist, VP, described by the American psychologists Earl Hunt and Tom Love, relied on verbal, or word, associations. VP's memory, like that of S., was astounding.

A few adults and, it is estimated, some 5 percent of children have "photographic memories" or, to use the technical term, *eidetic imagery*. Such images are clear mental impressions that can last up to several minutes (longer in exceptional cases), but that, after they have faded or been blanked out by a new image, cannot usually be recalled. Thus stories about people who can "photograph in their minds" the pages of a book by riffling through it once are just that—stories. The memory champion S. was not an eidetic imager. He formed his own idiosyncratic images—not "photographic impressions"—of what he was trying to remember and took time to associate these images with the scene, or context, in which they were stored. Nor does the mnemonist VP have a photographic memory. The amazing feats of the Shass Pollacks may have depended, in part, on eidetic imagery; but this is not known.

REFERENCES

Hunt, E., & Love, T. (1972). How good can memory be? In A. W. Melton & E. Martin (Eds.), *Coding processes in human memory,* Washington, DC: Winston.

Luria, A. R. (1968). *The mind of a mnemonist* (L. Solotaroff, Trans.). New York: Basic Books. (Original work published 1965.)

Klatsky, R. L. (1980). *Human memory: Structures and processes* (2d ed.). San Francisco: Freeman.

Neisser, U. (1982). *Memory observed: Remembering in natural contexts.* San Francisco: Freeman.

minder cues (page 201). With the analogy of a cloakroom in mind, the reminder cues are called memory pegs; the to-be-remembered items are hung on these "pegs."

Some simple, specific memory tricks will serve to illustrate the "peg" idea. For example, the letters in the word *homes* can be used as "pegs" on which to hang the names of the Great Lakes—Huron, Ontario, Michigan, Erie, and Superior. The order of the colors in the spectrum can be remembered by associating each color with the name "Roy G. Biv"—red, orange, yellow, green, blue, indigo, and violet. Two more examples are given in Figure 5.16.

The Method of Loci The word *loci* means "places." The memory pegs in this system are parts of your image of a scene. The scene can be a street, a building with rooms, the layout of a college campus, a kitchen, or just about anything that can be visualized clearly and contains a number of discrete items in specific locations to serve as memory pegs.

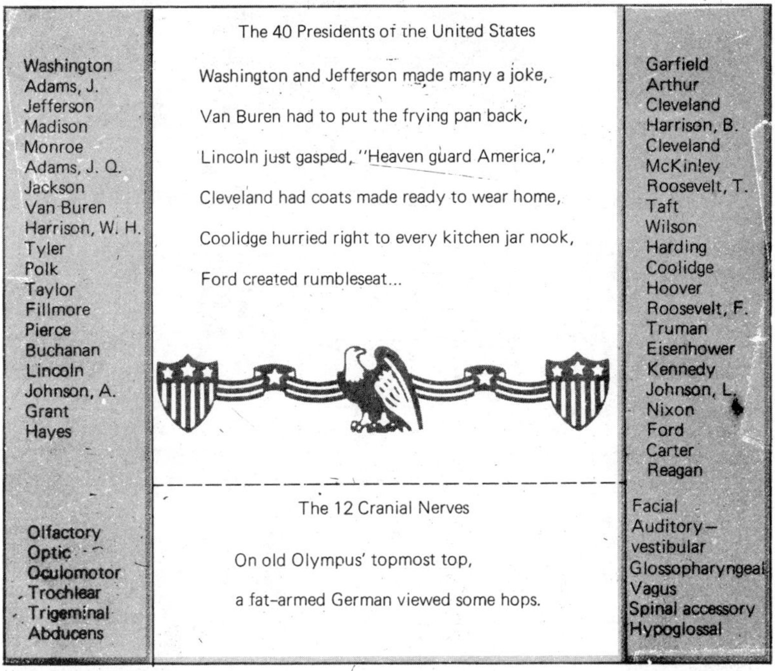

Figure 5.16

Two simple, specific mnemonic devices.

It is always impressive if you can give a talk or remember a long list of items without consulting notes. Perhaps even more important for most readers of this book is remembering the points you wish to make on an examination. Suppose, in a government course, for example, you think you will be asked to describe, in chronological order, five landmark cases dealing with freedom of the press.

Using the method of loci to do this, start by "imagining" a building with a number of rooms in it and several items of furniture in each room (Figure 5.17). Rehearse this image over and over until it is well established in your mind. (After you "have" the image, you can use it for other things you want to remember later—for example, the points in a sales presentation to a client—so your effort in forming this clear, distinct image will not be wasted on a single examination.) After you have formed your image, associate the events you wish to remember with the rooms and items of furniture. For instance, if the second case you want to remember deals with the right of reporters to keep their sources confidential, "image" a pile of papers marked "Top Secret" on a desk in the second room in your mental tour of the building. The trick is to make associations with as many rooms and items of furniture as needed and then to take a mental tour of the rooms in order.

Number and Letter Peg Systems Like the method of loci, the main idea of these systems is to establish, in your long-term memory, a well-organized set of images to which the to-be-remembered items can be linked. In number systems, you form an image with each number. For instance, a rhyming system can be used for the numbers 1 through 10. Think of words that rhyme with the numbers—1 is a bun, 2 is a shoe, 3 is

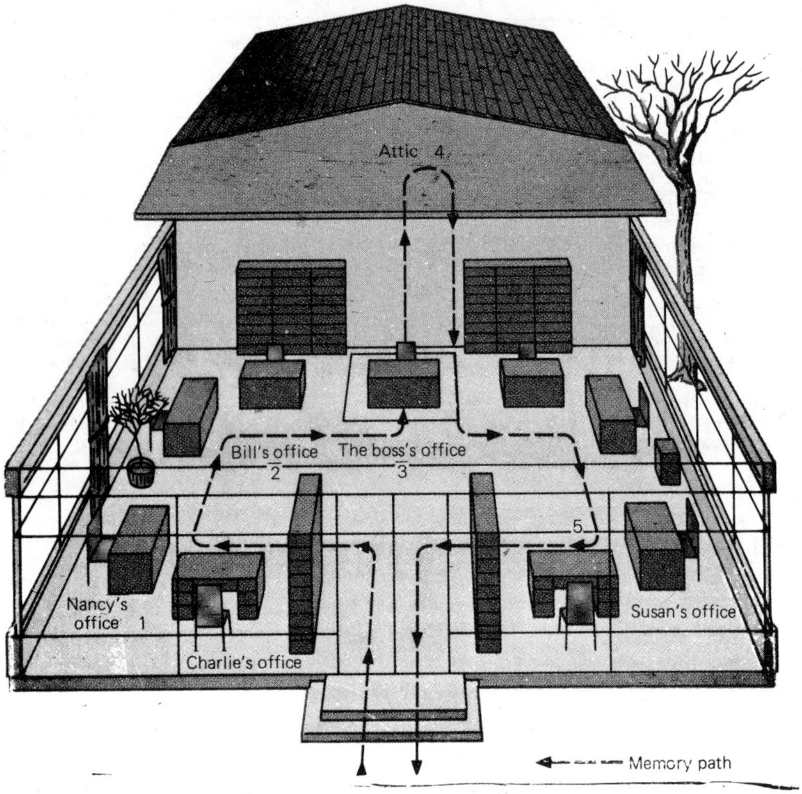

Figure 5.17

The method of loci.

a tree, 4 is a door, and so on. Now when you have a list to remember, you can associate the items on the list with your images of the numbers. If the first item on a grocery list is coffee, imagine a steaming cup of coffee next to a plate of buns; if the second item is hamburger, you might see a giant shoe squashing hamburger into a patty; and so on through the list, associating the number images with what is to be remembered.

Letter systems are similar. You can establish mnemonic pegs by forming strong, distinctive images of words that start with the sounds of the letters of the alphabet. This will give you 26 pegs for association with what you want to remember. Other, more elaborate letter systems have been devised to extend memory-peg lists to hundreds of items. (See Suggestions for Further Reading at the end of this chapter.) Mastery of these extended systems allow one to do feats of memory that appear astounding—such as remembering the 17th, 37th, and 49th cards after one pass through a shuffled deck—but are really based on the simple principle of linking what is to be remembered with retrieval cues, or pegs.

Stories You Tell Yourself If you have a list of unrelated items to remember, a useful mnemonic device is to relate the items in a made-up story. The story starts with the first item on the list, and, in order, each succeeding item is worked in. Doing this gives coherence and meaning to otherwise unrelated items; it is a form of elaborative encoding (page 195). Suppose a person has been given the following list to remember (Bower & Clark, 1969):

lumberjack
dart
skate
hedge
colony
duck
furniture
stocking
pillow
mistress

The made-up story might go as follows:

"A LUMBERJACK DARTed out of the forest, SKATEd around a HEDGE past a COLONY of DUCKs. He tripped on some FURNITURE, tearing his STOCKING, while hastening toward the PILLOW where his MISTRESS lay."

Remembering Names and Faces It may not be important to have memory techniques to help us remember grocery lists, steps in a sequence of chemical reactions, and the like; after all, we can almost always consult written notes. But the socially important business of associating names with faces is a different matter; we have no notes to help us with this. As first steps in establishing a good memory for names and faces, we should (1) be sure we hear the name clearly when introduced, (2) repeat the name when acknowledging the introduction, and (3) if the name is unusual, politely ask our new acquaintance to spell it. While we are making sure we have heard and rehearsed the name, we should be paying close attention to the individual's face. The shape and size of the head and individual characteristics of the hair, forehead, eyebrows, eyelashes, eyes, cheekbones, nose, ears, lips, chin, and skin should all be focal points of attention. Voice quality may also be important. Almost everybody we meet will have one or several features that can be elaborated, exaggerated, or perhaps even caricatured as in a cartoon to form a distinctive memory image that can be related to the person's name.

Chunking This mnemonic technique, described in another context earlier in this chapter (page 192), illustrates particularly well a general characteristic of mnemonic techniques—systematic ways of encoding information. Suppose you want to remember your credit-card number—19141609001, for example. It will help if you break the number into chunks. In the example, the first four numbers may remind you of an important date in history (the date of the outbreak of World War I), the next four numbers can also be "chunked" as a date, while the last three numbers form a chunk that is easy to remember by itself. If the number to be remembered is 97984494185521, dates will not work well as a chunking device; but perhaps you can think of another way to form groups of digits that will be easy to remember. Use your ingenuity. Perhaps it can be broken up into two telephone numbers, or you can find ways to chunk the number that will work better for you.

STUDYING TO REMEMBER

Mnemonics are fun to use and can be helpful in remembering many things. But if you want to remember what you need to know in courses or

Figure 5.18

Studying to remember. (© George W Gardner.)

in your work, you will need to go beyond mnemonic techniques. Here are some hints on how to remember what you study.

First, study is work and takes time (Figure 5.18), so plan a study schedule that you can stick to. During the time you set aside for study, work at it instead of talking to friends or watching television out of the corner of your eye. (If you study hard during your scheduled times, you will find that you have plenty of time for your friends and television later.)

Second, we know that *rehearsal* is crucial for transferring information from short-term to long-term memory or, alternatively, for the deeper and richer processing of information that is necessary for good memory. Textbooks like this one are full of detailed information, most of which cannot be remembered from the kind of skimming you might give a newspaper. Maintenance rehearsal and elaborative rehearsal were distinguished earlier in this chapter (page 194), where it said that *maintenance rehearsal* consists of merely repeating information, while *elaborative rehearsal* consists of thinking about what is being rehearsed in an effort to relate it to other things that you know or are learning. Elaborative rehearsal is the kind to use in studying. You should spend a great deal of your study time in elaborative rehearsal: Ask yourself what you have just read, what the new concepts and terms are, and how they relate to other things you know or are learning. Studies show that it is effective to spend at least half of your study time in such rehearsal.

Third, remember the importance of organization during encoding. As we saw earlier (pages 198–199), organization takes many forms. Textbooks like this one are organized by headings to provide a kind of outline. As you rehearse elaboratively, you will be giving your own *subjective organization* to the material, and you will also be providing yourself with *retrieval cues*, or reminders, that will be important when you try to recall what you are learning. If you can, form visual images of abstract ideas.

Fourth, try to get some idea of how well you remember the material.

In other words, get some feedback. If you study by breaking the material up into parts, try to get some feedback after you study each part. Go back over what you have just studied and, using the headings as retrieval cues, ask yourself what is under each heading. Turn to the terms at the end of the chapter and ask yourself for definitions of the appropriate terms. Feedback will tell you both what you have mastered and where you are weak. When you have finished a chapter, test yourself on it, and do some additional work on any weak spots. By testing yourself, you will also be practicing your retrieval skills.

Fifth, review before an examination. You will have forgotten many of the details you learned. Use the organization of the text to test yourself during review, and go back over the things you have forgotten, relearning them the way you learned them in the first place. Key your review to the type of examination. If the examination will be stressing recall, as in an essay examination, spend a good deal of your time rehearsing major ideas and the experiments that support them. Trying to think of what the questions will be ahead of time and practicing your answers to them is often a good idea. Spend some time integrating the text with class notes and trying to get a "big picture" of the subject—how it relates to other topics in the course, for example. If the examination is to be multiple-choice, or some other objective type of test, be sure you have mastered the definitions of the terms and can recognize the correct definitions when you see them. Of course, knowledge of terms is necessary for good performance on essay examinations too, and concepts are often asked for on multiple-choice examinations. So do not neglect terms or ideas for any examination; just give a little more emphasis, depending on the type of examination, to one or the other.

Planning, rehearsal, organization, feedback, and review will see you through course examinations, but, as you well know, most of what you remembered for an examination will be forgotten, or at least hard to retrieve, when you need to recall it later for another course or for your work. Here the old adage "practice makes perfect" is applicable. Psychologists use the word *overlearning.* To remember what you will need in your work, for example, it pays to go beyond the effort needed to just learn the material. After you are satisfied that you know and can remember something, go back after a few days and learn it again, and perhaps again. Formal studies have shown that such "overlearning" works to reduce the amount forgotten. For instance, if it takes you 2 hours to learn how to work with arrays in a programming language, another 2 hours spent "overlearning" the same material will stamp it into memory. Most of the time we are not motivated strongly enough to do what seems like extra work, but fortunately, for things we really need to remember, we will get many opportunities for "overlearning" in advanced courses or on the job. Figure 5.19 summarizes this section on "Studying to Remember."

Figure 5.19

Key words in the process of studying to remember.

Planning
Rehearsal
Organization
Feedback
Review
Overlearning

Summary

1. Memory refers to the encoding, storage, and retrieval of information. The current trend in the study of memory emphasizes cognitive, or mental, processes. Cognition concerns the internal

processing of information received from the senses; one aspect of this processing is memory.

2. A number of theories of memory have been proposed. In one of these, memory is said to consist of the three cognitive processes mentioned in its definition—encoding, storage, and retrieval. Encoding is the process of receiving sensory input and transforming it into a code that can be stored; storage is the process of actually putting the coded information into memory; and retrieval is the process of gaining access to the encoded, stored information when it is to be used.

3. Ideas about memory that emphasize the processing of information in stages, or steps, are known as information-processing theories. In the Atkinson-Shiffrin information-processing theory, the stages are (*a*) very brief storage of incoming information in a sensory register; (*b*) transfer of some of this information to short-term memory (STM); and (*c*) information transfer, by means of rehearsal, from short-term memory to long-term memory (LTM).

4. The sensory register holds sensory information for a few seconds, has a relatively large capacity, stores a copy of the sensory input, and loses information through a process involving the decay of the memory trace. Items in the sensory register that are attended to and recognized may be passed along to short-term memory.

5. Short-term memory holds a relatively small amount of information—about seven items or chunks—for a short time—30 seconds or so. The type of information stored consists of sounds, images, words, or sentences; and information is lost from short-term memory by being displaced by new inputs. Information in short-term memory may be transferred to long-term memory through either maintenance or elaborative rehearsal.

6. Information in long-term memory lasts for days, months, years, or even a lifetime. The capacity of long-term memory has no known limit. Memories of specific things that have happened, or episodic memory, and the meanings of words and the rules for using them, or semantic memory, are stored in long-term memory. Information may be lost, or at least not retrieved, from long-term memory because of difficulties with the search process or because of interference by other long-term memories.

7. The levels-of-processing theory of memory emphasizes the depth of analysis and the elaboration of incoming information. The most superficial depth, or level, is that of perception; the next deeper level is the structural level; and the deepest level involves giving meaning to the input. Information reaching the meaning level and elaborated at this level has the best chance of being retained.

8. When we think about memory, it is usually long-term memory that we have in mind. Information in long-term memory is organized, categorized, and classified so that it is a bit like a library with a good cross-indexing system. The tip-of-the-tongue (TOT) phe-

nomenon illustrates the organization of long-term memory. Long-term memory is divided into semantic memory—knowledge of word meanings and the rules for using language—and episodic memory—our remembrances of events that have happened in our lives.

9. The encoding and storage of long-term memories are facilitated by organization of the to-be-remembered material. Some organization is inherent in the material itself; some, known as subjective organization, is imposed on information when we are encoding it for storage. Forming images, especially what are called concrete ones, during the encoding process can help in the storage of information. During encoding, the to-be-remembered information is often modified through what are called constructive processes. We may then remember the modified information that was encoded and not the information actually presented to us.

10. Finding, or retrieving, information in long-term memory is aided by retrieval cues. State-dependent memory illustrates the importance of having appropriate retrieval cues. Reconstructive processes occur at the time of retrieval and can distort what is remembered.

11. Forgetting refers to the apparent loss of information already encoded and stored in long-term memory. How much is forgotten depends on many factors: (*a*) retroactive and proactive inference; (*b*) difficulties with the retrieval of stored information; and (*c*) motivated forgetting, especially that due to the process of repression.

12. Amnesia is a profound memory deficit due either to the loss of what has been stored or the inability to form new memories. Some amnesias—the biological ones—are caused by brain malfunctions. Others—the psychological amnesias—result from major disturbances in the processes of encoding, storage, and retrieval without any known brain problem.

13. Childhood amnesia for the early years of life, dream amnesia, and defensive, or protective, amnesia are examples of psychological amnesias. Repression is usually given as an explanation for each of these amnesias. Another explanation for childhood amnesia is based on differences in the ways young children and older people encode information. A similar type of explanation for dream amnesia says that the difference between the symbol systems in dreams and in waking makes the waking retrieval of any information encoded during dreaming difficult. While usually considered to be psychological, childhood and dream amnesia may have a biological basis—the immaturity of the brain in childhood amnesia and the difference between the brain states in dreaming and in waking for dream amnesia. The forgetting of defensive amnesia protects the amnesiac from the guilt or anxiety that can accompany intense, intolerable life situations and conflicts.

14. Illustrative of the biological amnesias are transient global amnesia, alcohol-induced amnesia, and the amnesia caused by certain

diseases of the brain. Transient global amnesia is a short-lived amnesic attack characterized by both retrograde and anterograde amnesia. High doses of alcohol result in amnesia for the events that occurred while drunk. In addition, heavy drinking over a period of years can produce brain damage and a pattern of symptoms known as the Korsakoff syndrome. Anterograde amnesia and some loss of remote memory characterize the memory problems of this syndrome. Senile dementia and primary degenerative dementia, of which Alzheimer's disease is an example, are instances of brain diseases that have amnesia as a major symptom.

15. Mnemonic devices are techniques for improving memory. Among them are the method of loci, number and letter peg systems, embedding to-be-remembered material in a made-up story, and chunking.

16. Studying to remember involves planning, rehearsal, organization, feedback, review, and overlearning.

Terms to Know

One way to test your mastery of the material in this chapter is to see whether you know what is meant by the following terms.

Memory (184)
Cognition (184)
Information (184)
Cognitive process (184)
Cognitive psychology (184)
Encoding processes (185, 198)
Storage processes (185, 198)
Retrieval processes (185, 201)
Information-processing theories of memory (185)
Sensory register (187, 190)
Short-term memory (STM) (187, 190)
Rehearsal (187, 192, 194)
Long-term memory (LTM) (188, 193, 195)
Iconic image (190)
Free recall (191)
Serial-position effect (191)
Primacy effect in memory (191)
Recency effect in memory (191)
Chunking (192, 217)
Maintenance rehearsal (192, 194)
Elaborative rehearsal (192, 194)
Levels-of-processing theory (193)
Semantic memory (193, 196)
Episodic memory (193, 197)
Elaboration (195)
Tip-of-the-tongue (TOT) phenomenon (196)
Subjective organization (199, 201)
Images (199)
Paired-associate technique (200)
Constructive processes (200)
Schemata (201)
Retrieval cues (201)
State-dependent memory (state-dependent learning) (202)
Reconstructive processes (202)
Confabulation (202)
Forgetting (203)
Engram (205)
Retroactive interference (205)
Proactive interference (205)
Motivated forgetting (207)
Repression (207)
Amnesia (208)
Defensive amnesia (209)
Transient global amnesia (210)
Retrograde amnesia (210)
Anterograde amnesia (210)

Korsakoff syndrome (211)
Senile dementia (211)
Primary degenerative amnesia (212)
Alzheimer's disease (212)
Mnemonics (212)
Method of loci (214)
Eidetic imagery (214)
Number and letter peg systems (215)
Overlearning (219)

Suggestions for Further Reading

An informative, up-to-date, and relatively nontechnical account of what we know about memory is to be found in Elizabeth Loftus's book *Memory: Surprising New Insights into How We Remember and Why We Forget* (Reading, MA: Addison-Wesley, 1980). At a more advanced level are texts by Roberta Klatzky, *Human Memory: Structures and Processes,* 2d ed. (San Francisco: Freeman, 1980), and the chapters on memory in *Foundations of Learning and Memory* by Roger Tarpy and Richard Mayer (Glenview, IL: Scott, Foresman, 1978).

Memory Observed: Remembering in Natural Contexts by Ulric Neisser (San Francisco: Freeman, 1982) contains a fascinating collection of easy-to-read excerpts and commentaries designed to illustrate "the specific manifestations of memory in ordinary human experience." Another interesting book touching on memory in "ordinary human experience" is Elizabeth Loftus's *Eyewitnesss Testimony* (Cambridge, MA: Harvard, 1979).

For helpful advice on memory improvement, books by Tony Buzan, *Speed memory,* Rev. ed. (Newton Abbot, United Kingdom: David & Charles, 1977), and by Laird Cermak, *Improving Your Memory* (New York: Norton, 1975), are recommended.

Thinking and Language

HEADING OUTLINE

Chapter

"Thinking is more than language, and language is more than thinking."

(Philip S. Dale)

But they are related, and this chapter will describe some of the ways in which they are related.

The chess player in the photo on the next page is thinking. What is going on in his head? What is thinking anyway? As you think about these questions, what is going on in *your* head? This chapter should answer some of these questions.

Here is a problem for you. As you try to solve it, think about your thinking processes. Even if you do not solve the problem, you may be able to state some general rules that guided your directed, problem-oriented thinking.

```
  DONALD
+ GERALD
--------
  ROBERT
```

This is to be treated as an exercise in simple addition. All that is known is: (1) that D = 5, (2) that every number from 0-9 has its corresponding letter, (3) that each letter must be assigned a number different from that given for any other letter. The operation required is to find a number for each letter, stating the steps of the process and their order. (Modified from Bartlett, 1958, p. 51).

Keep trying to solve the problem; it may take awhile. If you give up, the answer is on page 264.

DURING most of our waking hours, and even when we are asleep and dreaming, we are thinking; it is hard *not* to think. As you read these words you are thinking, and even if you stop thinking about what you are reading and your thoughts wander off to something else—perhaps to what you are going to do tomorrow—you will still be thinking.

What do we do when we think? Loosely speaking, we might say that we mentally, or *cognitively*, process information. More formally, we might say that *thinking* consists of the cognitive rearrangement or manipulation of both information from the environment and the symbols stored in long-term memory (Chapter 5, page 193). A *symbol* represents, or stands for, some event or item in the world; as we will see, images and language symbols are used in much of our thinking.

The general definition of *thinking* given above encompasses many different varieties of thought. For instance, some thinking is highly private and may use symbols with very personal meanings. This kind of thinking is called *autistic thinking*; dreams are an example of autistic thinking. Other thinking is aimed at solving problems or creating something new; this is called *directed thinking*. Directed thinking is the kind you were engaged in when you solved (or tried to solve) the problem on the opening pages of this chapter. It is also the type of thought we value so much in the great human thinkers (Figure 6.1). The definition of thinking given above also covers the thinking that we believe animals engage in when they solve certain kinds of problems.

From another viewpoint, *thinking* is the form of information processing (Chapter 5, page 185) that goes on during the period between a stimulus event and the response to it. In other words, *thinking* is the set of cognitive processes that *mediate*, or go between, stimuli and responses. To illustrate, suppose you are trying to make a decision about buying a new turntable for your hi-fi. The salesperson presents several turntables in your price range (the stimuli), and you eventually purchase one of them (the response). Before making the response, however, you weigh the advantages and disadvantages of the several turntables; you process the information you have about them. Your information processing—your thinking about the turntables—thus mediates between the turntables as stimuli and your eventual response of buying one of them.

Both thinking and language go through regular stages of development as a child grows. This development is discussed in Chapters 11 and 12 (pages 425–428; pages 467–470).

The Thinking Process

The symbols that we use in thinking are often words and language, and therefore thinking and language are closely related. A language makes available hundreds of thousands of potential symbols and gives us rules for using them. To a large degree, the availability of language symbols is what makes human thinking so much more sophisticated than the thinking of other animals. Although language is a powerful tool in human

Figure 6.1

Two famous thinkers. *Left*, Leonardo da Vinci; *right*, Albert Einstein. (Library of Congress.)

thought, as when we "talk to ourselves" internally, images are another important type of symbol used in thinking.

IMAGES AND THINKING

People vary remarkably in how much they use images in their thinking. A few report that they almost never use mental pictures, so they must be doing their thinking with words, or verbally; others report that most of their thinking is done in image form.

When we use images to think, they are not usually complete "pictures in the head." They are usually incomplete. Consider the imagery you use, if you use it at all, in solving the following problems (Huttenlocher, 1973).

> Imagine that you are standing on a certain street corner in a section of a city you know well. How would you walk or drive from this point to some other part of the city?
>
> Here is another problem in which you might use imagery: From where on earth could you walk first 1 mile south, then 1 mile east, then 1 mile north, and end up exactly where you began? Did you use imagery in trying to solve this problem? If so, what was your imagery like?

When solving problems like these, most people report that their images are incomplete. To solve the first problem, people usually make a visual map, but it is a strange one. Although it shows turns, the lines connecting the turns are of no particular length. In solving the second problem (the answer is the north pole), people imagine a globe—but not the whole globe, only the polar region. Such problem-solving images contain only a few details—say, of sidewalks, roads, buildings, or color—although some people may imagine snow when they think of the north pole. In general, the images are abstractions of certain features from previous experience.

The incomplete, abstract images most of us use in thinking seem to be constructed from elements stored in long-term memory. The constructive process involved in imagery has been studied by means of experiments in which people were asked to form images of various sizes. For example, an elephant might be imaged as the size of a mouse, or a mouse imaged as the size of an elephant. Variations of this sort in the sizes of images indicate that images are constructions. Even more interesting, however, are studies indicating that the ease with which information is found in an image depends on the size (and other aspects) of the image constructed (Kosslyn, 1975, 1983).

LANGUAGE AND THINKING

Did you try to solve the addition problem at the beginning of this chapter? Whether you arrived at an answer or not, you probably noticed that you used language in your thinking. For many people, much of the time, a good deal of thinking involves the use of word symbols and the rules of grammar (page 253) to join the words into phrases and sentences. The words, their meanings, and the rules for joining them together are stored in our semantic long-term memories (Chapter 5, page 196). When we think with language, we draw on this store of information to use language as a tool of thought.

Inquiry 6.1

DOES OUR LANGUAGE CONTROL THE THOUGHTS WE CAN HAVE?

There are at least 5,000 living languages in the world; about 140 of them are spoken by a million or more people. Is a particular language merely a convenient set of symbols for the communication of our thoughts? According to the linguist Benjamin L. Whorf, the answer is no. Whorf argued that the higher levels of thinking require language and that the characteristics of a particular language actually shape the ways that users of the language think about things. There are two ideas here. One is that thinking requires language; the other has come to be called the *linguistic relativity hypothesis*. Most of the interest has focused on this hypothesis. In its strongest form, it says that the particular language people use determines how they see the world.

Whorf based his hypothesis on studies of North American Indian languages, but his hypothesis is said to hold for all languages. He found many differences between these languages and European ones and argued that such differences predispose their users to think in different ways. For example, the grammar of a language dictates how people describe changes in the environment. Since the basic units of English grammar are nouns and verbs, English-speaking people commonly think in terms of "things" and "actions." Whorf found that people using other languages do not necessarily divide situations up this way. Furthermore, all languages have some words for which no equivalents can be found in any other language. The German word *Weltanschauung*, for instance, means something like "a general world view, or a general philosophy of the world." There is no word with this precise meaning in English. In addition, languages categorize events in various ways. Eskimos are said to use some four different words for snow, while English has only one. According to the linguistic relativity hypothesis, Eskimos can think about snow with greater precision than can English-speaking people and have a different conception of what snow is. The Hopi language has a single word for all flying objects other than birds. The hypothesis states that Hopi speakers think differently about flying objects than do speakers of languages that do not categorize the world in this way. The Hanunóo people of the Philippine Islands are said to have names for 92 varieties of rice, but all 92 varieties are, for the English speaker, simply *rice* (Con-

Some theorists take a more extreme view of the role of language in thinking; they claim that language can actually determine the thoughts we are capable of having. But this *linguistic relativity hypothesis*, as it is called, has been under increasing attack in recent years. (See Inquiry 6.1.)

Because so much thinking involves language, the idea arose in psychology that thinking was actually a kind of inner speech, a kind of "talking to yourself under your breath." According to this idea, people make small movements of the vocal apparatus when they think and carry on their thinking by talking to themselves. A number of experiments have indicated that movements of the vocal apparatus may indeed accompany thought, but other experiments have made it clear that such movements are *not necessary* for thinking (Smith et al., 1947).

In this rather heroic experiment, the subject, a physician, was completely paralyzed by means of a drug. He literally could not move a muscle, and his breathing was done for him by an iron lung. The paralyzing drug, however, did not affect the way his brain worked; it merely acted on the excitation of muscles by the nerves. While under the influence of the drug, the subject was given certain verbal problems to solve; he could not answer, of course, because the muscles necessary for speaking were paralyzed. There is no way

klin, 1954; cited in Brown, 1965).

The linguistic relativity hypothesis is controversial. Many linguists have argued that the hypothesis is circular. Whorf found that languages differ in their grammar and in the concepts they can express; from this, he hypothesized that thinking must also differ among the users of these different languages. However, the differences in thinking are themselves assessed by the way the language is used. What is needed are ways of assessing conceptions of the world independently of language. The few experiments that have attempted this have had inconclusive results. Perhaps it is not so much a matter of what can be thought about by users of a language as it is of how easy it is to think about certain things. English-speaking thinkers can think about the concept of a "world view" even if they do not have a convenient word for it; English speakers can think about different kinds of snow even if they have to use more words to describe it.

More recently, interest has shifted away from relativity to universals in thinking. Perhaps the basic thought processes are similar even though languages differ widely. Color perception provides an example of the possible universality of thinking despite the different ways in which different languages designate colors. It has been found that certain "focal" colors—a maximum of 11—are chosen from a color chart by speakers of widely differing languages. Furthermore, it has been shown that thinking can be influenced by these focal colors even when the language does not have names for them. This is contrary to what might be expected from the linguistic relativity hypothesis. For example, Eleanor Rosch has done experiments with the Dani people of New Guinea. The Dani have only two focal-color names in their language—*mili* for "black" and *mola* for "white." The Dani subjects in these experiments studied a color chart; arbitrary names were assigned to eight of the focal colors and eight of the nonfocal colors on the chart. The Dani learned the names assigned to the focal colors more rapidly and remembered them better than they did those given to the other colors. Thus, even though the Dani do not have names for the focal colors in their language, their thinking is influenced by them.

REFERENCES

Rosch, E. (1973). Natural categories. *Cognitive Psychology*, *4*, 328–350.

Slobin, D. I. (1979). *Psycholinguistics* (2d ed.). Glenview, IL: Scott, Foresman.

Whorf, B. L. (1956). *Language, thought and reality.* New York: Wiley.

to be certain that he was thinking while under the influence of the drug, but all indications are that he was because after the paralysis was removed by a counteracting drug, he clearly remembered what had taken place while he was drugged and promptly gave the answers to the problems.

The discussion so far has been about the use of vocal speech symbols, or verbal language, in thinking. Can other language systems be used as tools of thought? Studies of the deaf provide an approach to this question. Deaf children with little verbal language ability score in the normal range on standardized tests of cognitive performance (Vernon, 1967), and their cognitive and thinking abilities develop relatively normally (Furth, 1971). Such findings have been interpreted as indicating that language plays little or no role in the thinking or cognitive development of the deaf. But many of the deaf are taught sign language, and, even if they are not explicitly taught such a language, it has been found that deaf children will develop their own (Goldin-Meadow & Feldman, 1977). This may indicate that there is an innate human program for language (Chapter 2, page 41), be it verbal or gestural.

The standard visual-gestural sign languages learned by the deaf have many features in common with auditory-vocal languages. For example, just as the auditory-vocal languages use combinations of a small number of basic sounds, or *phonemes*, as they are called (page 249), to generate meaningful language, so, too, do the visual-gestural languages of the deaf make use of a relatively small number of basic movement combinations for communication. Thus, from combinations of the basic gestures, an infinity of ideas can be expressed in the visual-gestural languages. Some studies suggest that deaf children who know sign language are better at a variety of cognitive and thinking tasks than are those without this language (Vernon & Koh, 1971; Stuckless & Birch, 1966). Thus, those deaf people whose verbal language skills are minimal seem to have a nonverbal language tool of thought; they can think in signs.

Concepts

Concepts are important language symbols used in thinking. A *concept* is a symbolic construction that represents some common and general feature or features of many objects or events. Examples are "man," "red," "triangle," "motivation," "atom," "anger," and the word *concept* itself. In fact, most of the nouns in our vocabulary are names of concepts; the only exceptions are proper nouns—names of specific things or persons.

The human ability to form concepts enables us to classify things into categories. With a concept of "red," for example, we can sort objects into red and not red; with a concept of "fruit," we can classify things into fruit and not fruit. The feature or features we select define the concept and form the basis for making classifications. When a classification has been made, we tend to behave toward, and think about, members of the class in similar ways. Thus, since concepts are ways of classifying the diverse elements in the world around us, they are convenient tools to use in thinking about the world and in solving problems.

Some concepts seem "basic" and "natural" (Mervis & Rosch,

1981). These *concepts*, or categories, are acquired easily; appear in thinking very early in life; and, to some degree, reflect the way the brain processes and sorts information. An example of such a natural concept is the division of the colors of the spectrum into the categories "red," "green," and so forth (Rosch, 1973). Basic categories such as "chair," "tree," and "fruit" are other examples of natural concepts.

Unlike natural categories, many of our concepts are acquired more slowly and with more effort. Discrimination learning (Chapter 4, page 160) plays a role in the formation of some concepts. This type of learning occurs when some responses are rewarded, or reinforced, and other responses are not rewarded. To illustrate, suppose you are in an experiment in which the experimenter is using the stimuli of Figure 6.2 to study concept learning. Each of the stimuli is on a separate card, and the concept the experimenter has in mind consists of "all cards with two figures and two borders." Your task is to learn this concept. The cards are shuffled, and for each one in turn you say whether you think it is an example of the concept or not. The experimenter tells you whether your response is right (gives a reward) or wrong for each card. You will no doubt develop hypotheses to help you home in on the concept. In the process of testing your hypotheses, you will learn to discriminate between instances and noninstances of the concept, based on whether you are rewarded or not—that is, based on discrimination learning. Some concepts in everyday life seem to be acquired in just this way: A child, for instance, gradually learns the concept "apple" by being rewarded with a "Right!" after saying "apple" and pointing to one, but not after saying "apple" and pointing to something else.

By seeing examples of a concept in different contexts, or settings,

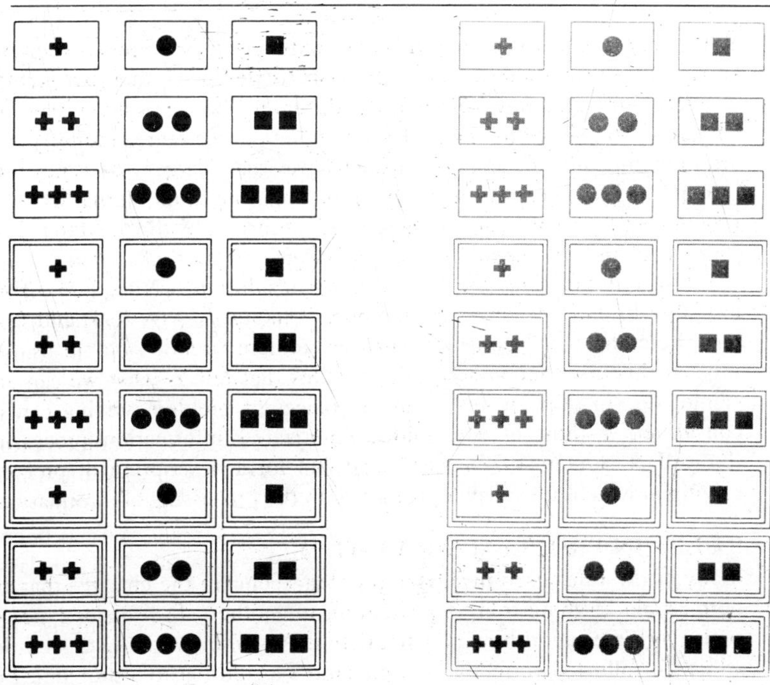

Figure 6.2

Cards used in a study of concept formation. Note that they differ in four ways: in the number of figures, in the color of the figures, in the shape of the figures, and in the number of borders. (After Bruner et al., 1956.)

we often learn the defining features of the concept. The following passage illustrates how context works. The new words, or *concepts*, to be acquired are in italics.

> Two ill-dressed people—the one a tired woman of middle years and the other a tense young man—sat around a fire where the common meal was almost ready. The mother, Tanith, peered at her son through the *oam* of the bubbling stew. It had been a long time since his last *ceilidh* and Tobar had changed greatly; where he once had seemed all legs and clumsy joints, he now was well-formed and in control of his hard, young body. As they ate, Tobar told of his past year, recreating for Tanith how he had wandered long and far in his quest to gain the skills he would need to be permitted to rejoin the company. Then all too soon, their brief *ceilidh* over, Tobar walked over to touch his mother's arm and quickly left. (Sternberg & Powell, 1983, p. 884)

Figure 6.3

The "Tower of Hanoi" problem. Here is the problem: By moving only one disk at a time from the top of a stack, move the disks so that they are on peg C in the same order as they are initially on peg A. In making your moves, you must always have smaller disks on top of larger ones; for instance, disk 4 can never be on top of disk 3. Do this in as few moves as possible.

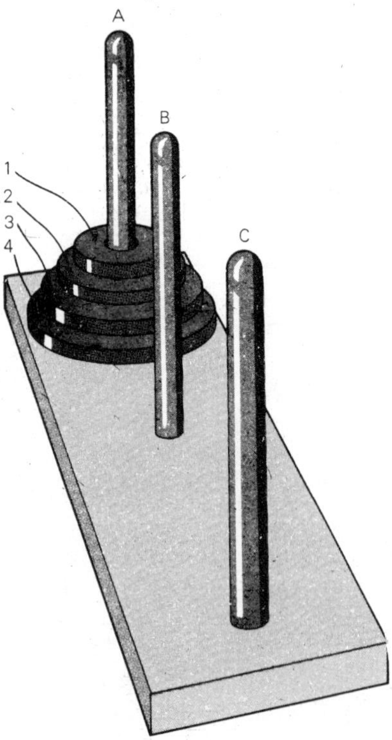

Just this brief exposure to *oam* and *ceilidh* was probably enough for you to learn that *oam* means "steam" and *ceilidh* means "a visit." The meanings of many words and concepts are acquired in this way.

In addition to discrimination learning and context, a third way of acquiring new concepts is, of course, by definition. Many of the concepts acquired in the later stages of a person's education are learned in this way. You have learned many concepts in this book by having them defined for you, and of course dictionaries tell us what words and concepts mean. Definition, then, helps us acquire concepts by describing them in terms of other words or concepts with which we are already familiar.

Problem Solving

What is a *problem*? In general, it is any conflict or difference between one situation and another situation we wish to produce—our *goal*. In the problem illustrated in Figure 6.3, the difference between the initial state (the disks on peg A) and the goal (having them in the same order on peg C after moving only one at a time in such a way that a larger disk is never on top of a smaller one) constitutes the problem. The thinking that we do in problem solving is thus goal-directed and motivated by the need to reduce the discrepancy between one state of affairs and another.

In trying to reach the goal of problem solution, we use information available to us from long-term memory (Chapter 5, page 193) and from our "here-and-now" perception of the problem situation before us. We process this information according to rules that tell us what we can and cannot do. In other words, many instances of problem solving can be considered a form of rule-guided, motivated information processing (Newell & Simon, 1972). (See page 244 for a description of problem solving in which rules play less of a role in the processing of information.)

RULES IN PROBLEM SOLVING

Many of the rules used in solving problems concern the changes that are permissible in going from one situation to another. Two major types of such rules are algorithms and heuristics. An *algorithm* is a set of rules which, if followed correctly, will guarantee a solution to a problem. For instance, if you are given two numbers to multiply, you immediately start thinking of all the rules for multiplication you have learned, and you apply these algorithms to the problem. If you follow the rules correctly, you will solve the problem. However, because we do not have algorithms for most of the problems we encounter, we must use heuristics. *Heuristics* are strategies, usually based on our past experience with problems, that are likely to lead to a solution but do not guarantee success. One common strategy, or *heuristic*, is to break the problem down into smaller subproblems, each a little closer to the end goal. For instance, in solving the "Tower of Hanoi" problem in Figure 6.3, you may have said something like the following to yourself:

> "Maybe I can solve the problem by figuring out a way of getting the large disk by itself on peg C. To do this, though, I'll have to figure out a way of getting disks 1, 2, and 3 stacked up on peg B. So suppose I try to find a way to get disk 3 at the bottom on peg B. Let's see now; I can move disk 1 from peg A to peg B, disk 2 from peg A to peg C, and then disk 1 from peg B to peg C. That will leave me without any disks on peg B, and disk 3 will be at the top of peg A, so then I can move disk 3 from peg A to peg B. Then I can . . ." (The rest of the solution should be easy.)

The subgoal heuristic in the "Tower of Hanoi" problem might be called a *means-end analysis*; each step leads closer to the desired goal. Notice, too, that the thinking works backward from the desired goal. In programming computers to solve problems, heuristics and means-end analyses are often used. (See Inquiry 6.2, page 238.)

As another example of heuristics, consider some of the strategies that are involved in solving a simple cryptogram.

> GXDOBHAHWD FX VBC XOFCQOC HM BNKIQ IQZ IQFKIA PCBITFHS, IQZ VBC IGGAFOIVFHQ HM VBC XOFCQOC VH BNKIQ GSHPACKX.

Your heuristics might include

1. The sentence is probably in English.

2. The most commonly used letter in English is *e*.

3. Only a few two-letter and three-letter words are commonly used in English. Could some of the short words in the sentence be *the*, *are*, or *is*?

4. Are any words repeated?

5. Look at the double letters. Only some combinations are permissible in English, and certain double letters are more likely to appear at some places in words than others.

6. Since this is a book about psychology, perhaps the sentence is about psychology.

Of course, you will use a number of other heuristics in trying to solve this problem. (The answer is given on page 264.)

Obviously, the heuristics discussed above do not guarantee a solution. You may not solve the problem. However, they do provide rules of thumb for approaching a solution.

HABIT AND SET IN PROBLEM SOLVING

The algorithms and heuristics we use in solving problems typically come from our past experience with the solution of similar problems. We reason that if the cryptogram heuristics given above have worked before, they will probably work again. Practice in solving problems in a particular way tends to give people a *set* to use the same rules on other problems. This can be quite helpful. But suppose the cryptogram in the above example was not of an English sentence. In this case, the set generated by past experience with English cryptograms would hinder problem solution.

Set that hinders is the idea behind a number of parlor jokes and puzzles. In one such "trick," for example, the joker spells words and asks another person to pronounce them. The joker uses names beginning with *Mac*, like *MacDonald* and *MacTavish*; then the word *machinery* is spelled to see if the person pronounces it "MacHinery." With the set for names, the person often falls into the trap.

A classic experiment on set used problems like those in Table 6.1. The first five "practice" problems have a roundabout solution; the sixth is simple and direct. If subjects are given the sixth problem first, they solve it by using the 3-quart jar to remove 3 quarts from the 23-quart jar. But if they have just worked out the previous problems, all of which require a longer method—filling the middle jar, then using it to fill the jar to the right twice and the jar to the left once, leaving the required amount in the center jar—they commonly use the long method and do not notice the short one. Amazingly, 75 percent of a group of college students were blind to the easy method after having practiced the long method for five trials.

Set may be induced by immediately preceding experiences (as in the examples above), by long-established practices, or by instructions that revive old habits. Set biases thinkers at the start of the problem, directing them away from certain thoughts and toward others. It acts as an implied assumption, and it can be either positive or negative in its effects. If it is helpful, we say, "What a sharp cookie I am!"; if it is a hindrance, we say, "How stupid of me!"

The hindering effects of set can be reduced somewhat by (1) warning the subject, "Don't be blind, now," or "Look sharp, now," just

Table 6.1

"Practice" and "test" problems used by Luchins. The five practice problems require a roundabout method of solution; the test problem can be solved easily. But most subjects acquired a set by solving the practice problems. They were blind to the easy method of solving the test problem.

	GIVEN THE FOLLOWING EMPTY JARS AS MEASURES			
PROBLEM NUMBER	A	B	C	OBTAIN THIS AMOUNT OF WATER
1. Practice	21	127	3	100
2. Practice	14	163	25	99
3. Practice	18	43	10	5
4. Practice	9	42	6	21
5. Practice	20	59	4	31
6. Test	23	49	3	20

(*Source*: Luchins & Luchins, 1959.)

before the critical problem; (2) reducing the number of practice trials; or (3) separating practice and critical trials by days or weeks. Even so, the set induced by practice is usually stronger than any warning against it.

A particular kind of set that can point thoughts in the wrong direction has been called *functional fixedness*. It is a set to use objects in the way we are accustomed to using them, even if a different use might solve a problem. The following experiment shows this effect (Adamson, 1952):

> The problem was to mount a candle on a vertical screen. All of the subjects were provided with candles, small pasteboard boxes, thumbtacks, and matches. The problem is not a difficult one: The solution is simply to stick the candle on the box with melted wax and then use the thumbtacks to attach the box to the screen.
>
> Boxes, of course, are usually containers and not platforms. Fixation on this function was established for the experimental group of subjects by placing the tacks, candle, and matches in the box before giving it to them (see Figure 6.4). No attempt was made to establish functional fixedness in the subjects of the control group; the empty box, together with the other materials, was simply placed on the table.
>
> Members of the experimental group had difficulty with this problem: Only 12 out of 29 (41 percent) solved it in the allotted time of 20 minutes. On the other hand, 24 out of 28 (86 percent) of the people in the control group solved the problem. Similar results were obtained with other problems.

These results provide strong evidence for functional fixedness as a particular kind of set that hinders problem solving. One advantage of temporarily quitting a problem you cannot solve is that you may come back to it with a fresh approach—your functional fixedness may be broken.

Decision Making

Decision making is a kind of problem solving in which we are presented with several alternatives, among which we must choose. Why does a person decide to buy one car and not another? Why does a cardplayer decide to fold a poker hand instead of betting on it? Why does an investor buy one stock and not another?

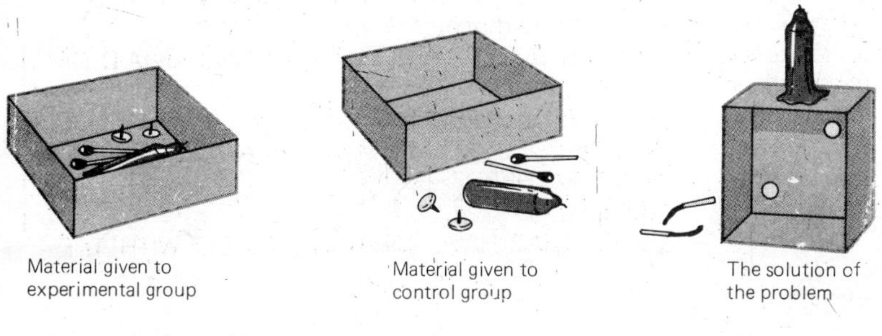

Figure 6.4

The arrangement of materials and the problem solution in an experiment on functional fixedness. (Adamson, 1952.)

Inquiry 6.2

THINKING MACHINES?

Most of us are familiar with the use of computers to do such relatively routine tasks as calculating cube roots, making up a payroll, or processing words. While such activities can be considered routine machine thinking, computers can also be programmed to do more complex, symbolic manipulation—manipulation that resembles "higher" human thought. Programs have been developed that will solve problems, understand language, and act like human "experts" in making decisions. These programs are examples of what is known as artificial intelligence (AI)—the programming of computers so that they behave in ways that we could call *intelligent* if they were people.

Problem-solving programs like the Logic Theorist and the General Problem Solver were among the first artificial-intelligence programs to be developed. The General Problem Solver uses the heuristic of breaking problems down into subproblems; the program then evaluates the solution of each subproblem to check whether it moves toward the goal of solving the main problem. If a subproblem solution is unfruitful, the computer switches to a new subproblem. This program can solve problems like the puzzle about the cannibals and the missionaries: Three cannibals and three missionaries wish to cross a river. They have a boat that can hold two people, no more, and all of them know how to row. But the crossing must be made so that there are never more cannibals than missionaries on the same side of the river; otherwise the cannibals will eat the missionaries. When you solve this problem, you will probably see that you approached it in terms of steps, or subproblems, evaluating the solution of each subproblem very much like the General Problem Solver does. Would you say that the solution to the problem required thought on your part? What about the solution by the computer? Was it thinking?

In the early days of work on artificial intelligence (the 1940s), it was proposed that a computer could be said to be *thinking* if a person, when conversing with the computer through its terminal, could not tell whether there was a human being or a computer at the other end of the line. If we limit the range of topics that can be discussed, computers have been programmed to pass this test. ELIZA is a program that simulates a conversation between a client-centered therapist (Chapter 16, page 688) and a client; the program plays the role of the therapist, while the individual at the terminal acts like a person receiving therapy—the client. In client-centered therapy, the therapist reflects the feelings of the client, sometimes repeats what the client says, asks leading questions, but offers no specific advice. Here is a chat between a woman and ELIZA. (ELIZA's responses are capitalized.)

Men are all alike.
IN WHAT WAY
They're always bugging us about something or other.
CAN YOU THINK OF A SPECIFIC EXAMPLE
Well, my boyfriend made me come here.
YOUR BOYFRIEND MADE YOU COME HERE
He says I'm depressed much of the time.
I AM SORRY TO HEAR THAT YOU ARE DEPRESSED
It's true. I am unhappy.
DO YOU THINK COMING HERE WILL HELP YOU NOT TO BE UNHAPPY
I need some help, that much seems certain.
WHAT WOULD IT MEAN TO YOU IF YOU GOT SOME HELP
Perhaps I could learn to get along with my mother.
TELL ME MORE ABOUT YOUR FAMILY
My mother takes care of me.
WHO ELSE IN YOUR FAMILY TAKES CARE OF YOU
My father.
YOUR FATHER
You are like my father in some ways.
WHAT RESEMBLANCE DO YOU SEE
You are not very aggressive but I think you don't want me to notice that.
WHAT MAKES YOU THINK I AM NOT VERY AGGRESSIVE
You don't argue with me.
WHY DO YOU THINK I DON'T ARGUE WITH YOU
You are afraid of me.
DOES IT PLEASE YOU TO THINK I AM AFRAID OF YOU
My father is afraid of everybody.
WHAT ELSE COMES TO MIND WHEN YOU THINK OF YOUR FATHER
Bullies.
DOES THAT HAVE ANYTHING TO DO WITH THE

FACT THAT YOUR BOYFRIEND MADE YOU COME HERE

(From Weizenbaum, J. (1976). *Computer power and human reason: From judgment to calculation.* San Francisco: Freeman, pp. 3–4.)

ELIZA works because its chat with the person takes place within the particular framework of client-centered therapy. The program scans the client's input for key words stored in its memory and then applies a limited set of rules to generate sentences like those a client-centered therapist might use. If the program cannot find a key word, it replies, as a client-centered therapist might, with an appropriate general remark.

Other programs, such as SHRDLU, can, like ELIZA, converse with people on a limited number of topics. SHRDLU's topic is the positions of blocks. But the understanding of general conversation and vocal speech, a capability of HAL, the computer in *2001* [now *2010*]: *A Space Odyssey*, is science fiction; this fantasy will not come true in 2001 *or* 2010.

Computers can play excellent checkers and rather good chess. From their early beginnings, chess programs have been improved to the point where they can give expert players, but not masters, some challenging games. The number of possible moves in chess is so great that the best current programs can look ahead only about six moves. The computer weighs these moves and selects the one that seems best. Human players cannot process all the information six moves ahead in the "brute-force" way a computer does; instead, good players rely a great deal on their knowledge of past games and the patterns of pieces at various stages of these games. When ways are found to combine such human expertise with the tremendous "brute-force" capability of computers, the $100,000 prize offered for a program that will beat a grand master of chess may yet be won.

Much artificial-intelligence work has been done with what are called expert systems. Heuristics (page 240) for making decisions are generated by analyzing the ways in which experts solve problems in their domains of knowledge. In medicine, for example, heuristics derived from experts are used to diagnose diseases. And in the writing of artificial-intelligence programs themselves, heuristic programs can be useful—programs helping to write programs.

Artificial intelligence has come in for a great deal of criticism. Some of this criticism stems from skeptics who say that we should not dignify what computers do with the term *intelligence*. They say that computers have only some of the hallmarks of human intelligence; they lack creativity, for example. Other critics, convinced of the power and possiblities of artificial intelligence, are anxious about the unforeseen consequences of creating an "alien intelligence" with which they will have to share the world.

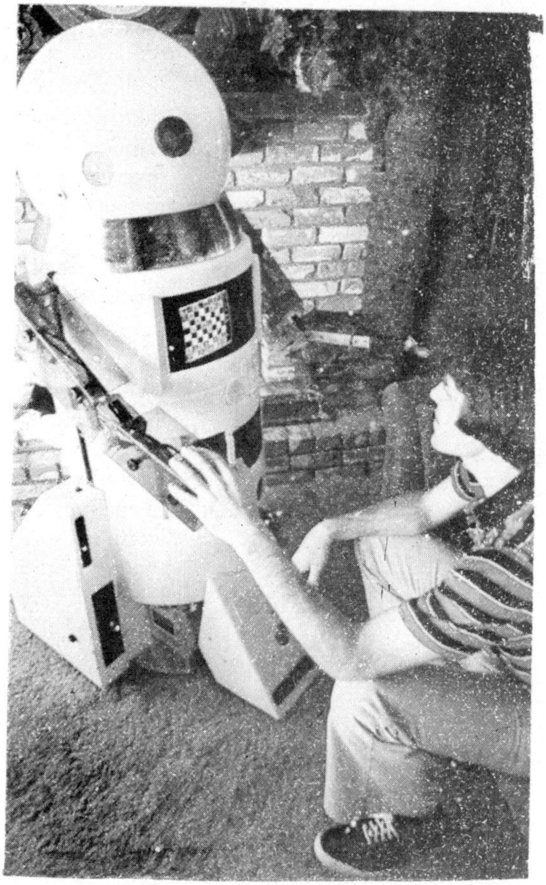

(© Dan McCoy from Rainbow.)

REFERENCES

Boden, M. (1977). *Artificial intelligence and natural man.* New York: Basic Books.

McCorduck, P. (1979). *Machines who think: A personal inquiry into the history and prospects of artificial intelligence.* San Francisco: Freeman.

UTILITY AND SUBJECTIVE PROBABILITY

The car buyer, poker player, and investor just mentioned are trying to achieve some objective. For example, they may be trying to minimize their maximum possible loss, or, alternatively, they may be trying to maximize their expected gains. In other words, they are trying to make optimum decisions. From a psychological viewpoint, they are trying to optimize *utility*—perceived benefit or psychological value—in making their decisions. In considering the decisions that people make, we must remember that different people assign different utilities to the same event; the psychological worth of an outcome varies among people. For instance, given a choice between receiving $10 now or $100 a year from now, a rich person might decide to wait for the $100, but a poor person would probably take the immediate $10. The utility of the $10 is different for the two people.

Most decisions are risky in the sense that we cannot be sure of the outcome. A tossed coin, for example, normally has a 50/50 chance of coming up heads or tails. We are not sure which it will be and must take knowledge about the probability of the outcome into account when making a decision about whether the coin will come up heads or tails. In tossing a coin we know what the head/tail probabilities of the coin are; but in making complex "real life" decisions, we do not know the precise likelihoods of various outcomes; we can only make our own estimates of the probabilities. Such guessed-at, or perceived, probability estimates are known as *subjective probabilities*.

One idea about decision making, an idea that has been formulated as a mathematical model of the decision process, is that people make decisions that will maximize *subjectively expected utility*. In other words, given a choice among alternatives, we take utility and subjective probability into account, multiply them together, and take the alternative with the highest product. This model of choice behavior has had some success in predicting decisions in relatively simple betting situations, but in many instances, especially in "real-life" situations, people seem to use other ways of making decisions. We use rules of thumb, or *heuristics* (page 235), in deciding among alternatives.

HEURISTICS AND BIASES IN DECISION MAKING

Life abounds with situations in which we must choose between alternatives. For instance, the manager of a baseball team must decide whether to have a runner on first base steal second base (Figure 6.5). The manager must take into account factors such as the speed of the runner, the accuracy with which the catcher throws, and the time it will take a particular pitch to reach the catcher. Assigning probabilities to these factors and weighting them is an almost impossible task. The manager, like most people, will simplify the decision by using heuristics (page 235). While these rules of thumb ease the process of decision making, they can also lead to biases and errors. Among the heuristic decision-making rules for estimating the likelihood, or *subjective probability*, of outcomes are representatives, availability, and adjustment (Tversky & Kahneman, 1974; Kahneman et al., 1982).

When we use the heuristic of *representativeness* to arrive at subjective probabilities, we decide, first, whether the current situation is similar to one we have encountered before and then we act accordingly. In other words, we ask whether the current situation is a *representation*, or instance, of something we have already experienced. This method can work, but it may result in our being misled by surface similarities. It is also possible

Figure 6.5

Should the manager send the runner? See text. (Magnum.)

that the original situation, the one that serves as our basis for comparison, may not be representative of the true state of affairs.

To return to the baseball example, a particular runner may have been successful in "stealing" second base against a particular catcher on two out of two previous occasions. Now, with the same runner on first and the same catcher behind the plate, the manager may decide that the current situation is representative of the other two occasions. But he has only a very small sample of occasions from which to make his judgment. On one occasion, the ball may have gotten "hung up" in the catcher's mitt; and on the other occasion, unknown to the manager, the catcher's arm may have been a little sore. So the past success of the runner may not be truly representative of the odds of success. To know the true odds, and to be accurate in using the heuristic of representativeness, the manager would need a much larger sample of "stolen-base" attempts by this particular runner against this particular catcher. This time, the manager decides to send the runner, and, to his chagrin, the catcher throws him out. The past was not representative of the present situation.

In making a decision on the basis of representativeness, the manager may also be a victim of the *gambler's fallacy*. Suppose with another runner on base, the manager has enough evidence to know that the chances of stealing a base are very near 50/50; and suppose, further, that this runner has been thrown out the last three times in a row. The manager may think that the runner is due to be successful this time. However the logic of probability says that each event is independent. The odds remain 50/50 no matter how many previous failures there have been. (Even if you have already thrown five heads in a row with a fair coin, your chances for a head on your next toss remain 50/50.) Thus while it is often useful to make judgments on the basis of the similarity of one situation to another—on the basis of *representativeness*—the biases in this procedure can lead you astray.

Another judgment heuristic is known as *availability*. Some events are easier to imagine or remember than others. Because frequent events are generally easier to remember than are infrequent ones, easily remembered or imagined events are also actually likely to be more frequent than others. Thus the ease with which we remember certain things helps us in making subjective-probability estimates. Estimating subjective probability in this way can be useful, but neglecting events that are harder to remember can also lead to misjudgments about the likelihood of certain outcomes. The baseball manager, for example, may remember a runner's past failures to "steal" successfully and so assign a lower probability than the "true" one to the runner's success.

Our subjective-probability estimates are sometimes arrived at by using the heuristic known as *adjustment.* We start with a certain subjective probability and raise or lower it depending on the circumstances. When we make these adjustments, the outcome depends upon the starting point. If we start with a high estimate, even if we adjust it downward our probability estimate will be higher than if we started with a low estimate. It is as if the initial level provided an "anchor" that biased our estimate, and therefore this biasing effect is often called *anchoring.* To illustrate anchoring, consider the following example (Tversky & Kahneman, 1974):

> One group of subjects was asked to estimate the product of 8 × 7 × 6 × 5 × 4 × 3 × 2 × 1 in 5 seconds. The other group was given 5 seconds to estimate the product of 1 × 2 × 3 × 4 × 5 × 6 × 7 × 8. People do problems like this by multiplying from left to right, but only a few of the necessary multiplications can be done in 5 seconds. However, because they have different anchors, the partial products of the group starting with the larger numbers will be higher than those of the group starting with the lower numbers.
>
> The median estimate of the total product was 2,250 for the high-anchor group and 512 for the low-anchor group. Although both groups were far from the correct answer of 40,320, the anchoring bias is clearly shown.

Returning to the baseball manager for a moment, we can see that he has a very difficult task. Because he must make rapid decisions, he will probably use heuristics, but they can lead him into error. Life is like that for us, too; when making decisions between two alternatives, we do well to bat better than .500.

WEIGHING ALTERNATIVES

Given more time and other decision-making techniques, the manager might be able to do better than he can on the field. Suppose that during the winter trading season, the manager is asked to decide among several available pitchers. Now he can maximize the utility of his choice by weighing his alternatives more carefully. The team needs a left-handed pitcher, and there are three candidates. The manager, or any decision maker for that matter, might first make a list of desired attributes and then give weights to these attributes on the basis of their importance. Then, the decision maker can assess the utility of each attribute, multiplying by the weight to give an overall value for that attribute. Finally, the overall score can be summed to give a single weighted utility for each alternative—pitchers, in the case of the manager. Table 6.2 describes this procedure.

Table 6.2

The manager's decision chart

ATTRIBUTE	WEIGHT	UTILITY (PERCEIVED BENEFIT) FOR 3 PITCHERS			WEIGHTED UTILITIES (WEIGHT × UTILITY)		
		MORGAN	KING	WEISZ	MORGAN	KING	WEISZ
Cost (high utility means low cost)	1	3	2	1	3	2	1
Previous year's won-lost record	2	1	2	3	2	4	6
Earned-run average	3	1	2	3	3	6	9
Effectiveness against right-handed batters	4	1	2	3	4	8	12
Condition of arm	5	1	3	2	5	15	10
				Sums	17	35	38

In the example in Table 6.2, the manager chose the important attributes subjectively. For instance, he is interested in acquiring a pitcher who will be effective against right-handed batters because the left-field fence of the home ball park is only 320 feet from home plate. The choice of attributes is critical in this kind of decision making, but sophisticated models and aids to decision making usually have little to say about this crucial step.

Weights are also assigned subjectively in Table 6.2. For example, because the ball club has plenty of money, cost is given a low weight. Perhaps the manager used some of the heuristics described in the last section to assign weights. He remembered, for example, the number of games lost the previous year because home runs were hit over the left-field wall, and so, using the heuristic of availability, gives a high weight to "effectiveness against right-handed batters." Certain decision theories specify mathematical procedures for the assignment of weights, but it is not yet firmly established that they do any better than subjective weighting (Slovic et al., 1977).

The utilities—the value of each pitcher to the team on an attribute—are also subjectively assigned in our example, although here again, techniques exist for specifying them more accurately.

Having done his work, the manager finds that King and Weisz are close, with Weisz the best overall bet. However, before making a decision, the manager would probably look at each attribute separately. Having done this, he would see that Weisz is higher on three attributes and King on two. So in both ways of making comparisons, Weisz wins. If, however, Weisz was highest overall, but King was higher than Weisz on more attributes, what would the manager do? He would then seek other information about the candidates. In evaluating it, however, he might be swayed by his personal bias; King might be a friend of a friend to whom the manager owes a favor, or perhaps Weisz is not a very friendly person. Neither of these attributes is very important when it comes to pitching performance, but such factors might tip the balance in a close decision. Even when personal biases are minimized in decision making, questions about the choice of attributes and the assignment of weights and utilities make the outcomes of many decisions uncertain, at best. Imagine the complexity of many decisions about social policy—ranging from national defense to how to get rid of garbage—in which both political and cost-effectiveness factors must be weighted. As research on decision making continues, we hope to discover some powerful and practical tools for making life's crucial choices.

Creative Thinking

The creative thinker, whether artist, writer, or scientist, is trying to create something "new under the sun." The visual artist is trying to express an idea or emotional feeling in a way that will have an impact on viewers; the creative writer or poet is trying to do the same for readers. Creative scientists think about their own discoveries and those of others, inventing new ways of studying nature and new theories to tie the discoveries together. Unlike ordinary solutions to problems, creative solutions are solutions that other people have not thought of before. The product of creative thinking may be a new and unique way of conceptualizing the world around us. The emphasis in creative thinking is on the word *new*.

Creative thinking in the arts and sciences seems to involve a considerable amount of unconscious rearrangement of symbols. The thinker at first makes little progress, but then, perhaps triggered by a fortuitous set of circumstances, a new idea seems to "bubble up" into awareness, or consciousness, in a seemingly spontaneous manner. Because the creative thinker becomes aware of the new idea suddenly, it is said that much of the thought has already gone on unconsciously. The sudden appearance of new ideas is called *insight*. (See Chapter 4, page 170.)

INSIGHT IN CREATIVE THINKING

The following story recounts what is perhaps *the* classic example of the role of insight in creative thinking.

> King Hiero had recently succeeded to the throne of Syracuse and decided to place a golden crown in a temple as a thank offering to the gods. So he made a contract at a fixed price, and weighed out the gold for the contractor at the royal scales. At the appointed time the contractor delivered his handiwork beautifully made, and the king was delighted. At the scales it was seen that the contractor had kept the original weight of the gold. Later a charge was made that gold had been removed and an equivalent weight of silver substituted.
>
> Hiero was furious at being fooled, and, not being able to find any way of detecting the theft, asked Archimedes to put his thought to the matter.
>
> While Archimedes was bearing the problem in mind, he happened to get into a bath and noticed that when he got into the tub exactly the same amount of water flowed over the side as the volume of his body that was under water. Perceiving that this gave him a clue to the problem, he promptly leapt from the tub in a rush of joy and ran home naked, shouting loudly to all the world that he had found the solution. As he ran, he called again and again in Greek, "Eureka, Eureka . . . I have found it, I have found it." (Humphrey, 1948, p. 115, as translated from Vitruvius)

(Incidentally, the contractor had cheated. The silver in the crown, having a larger volume than an equal weight of gold, made the crown displace more water than it would have had it been made of gold alone.)

Archimedes was stumped until a fortuitous environmental circumstance triggered his creative solution. Similarly, a number of creative people report that after conscious thought has failed them, insight suddenly appears when they are doing something completely unrelated to the problem. However, insights do not really appear out of nowhere; they blossom in fields which have been thoroughly prepared by study of the various aspects of a problem. Insights may also be incorrect; they require testing to see if they really do represent new solutions to problems, and this leads us to a discussion of the stages in creative thinking.

STAGES IN CREATIVE THINKING

Many years ago, through interviews, questionnaires, and the reminiscences of outstanding creative thinkers, Graham Wallas studied the steps involved in their thinking. He found that though there were individual differences in the ways these creative people thought, there was a recurring pattern. One way of looking at creative thinking is that it proceeds in five stages: preparation, incubation, illumination, evaluation, and revision. A good modern-day example of creative thinking in which these stages can be found is the account of the discovery of the structure of the genetic molecule deoxyribonucleic acid (DNA) by Watson and Crick (Figure 6.6); Watson described this discovery in his book, *The Double Helix*.

In stage 1, preparation, the thinker formulates the problem and

Figure 6.6

James Watson and Francis Crick, creative thinkers who discovered the structure of the DNA molecule. (From *The Double Helix*, p. 215, by J. D. Watson, 1968, New York: Atheneum. Copyright 1968 by J. D. Watson.)

collects the facts and materials considered necessary for the new solution. Very frequently the creative thinker, like Watson, finds that the problem cannot be solved after days, weeks, or months of concentrated effort. Failing to solve the problem, the thinker either deliberately or involuntarily turns away from it, initiating stage 2, incubation. During this period, some of the ideas that were interfering with the solution tend to fade. In addition, the creative thinker may have experiences that (although the thinker does not realize it at the time) provide clues to the solution. The unconscious thought processes involved in creative thinking are also at work during this period of incubation. If the thinker is lucky, stage 3, illumination, occurs with its "aha!" insight experience; an idea for the solution suddenly wells up into consciousness. Next, in stage 4, evaluation, the apparent solution is tested to see if it satisfactorily solves the problem. Frequently, the insight turns out to be unsatisfactory, and the thinker is back at the beginning of the creative process. In other cases, the insight is generally satisfactory but needs some modification or the solution of minor problems to be a really "good" new idea. Thus, stage 5, revision, is reached.

This stage description gives us a general picture of the steps frequently involved in the solution of problems by our most talented and creative people. Another approach to the study of creative thinking is to see how it differs from the more routine kinds of thinking we do. We will turn to a brief description of the nature of creative thinking.

NATURE OF CREATIVE THINKING

Several attempts have been made to develop tests that will measure creativity in people. In one elaborate study (Guilford, 1967), a battery of tests was constructed and carefully analyzed. Out of this work came the concepts of "convergent" and "divergent thinking." *Convergent thinking* is concerned with a particular end result. The thinker gathers information relevant to the problem and then proceeds, by using problem-solving rules (page 234), to work out the right solution. The result of convergent thinking is usually a solution that has been previously arrived at by someone else. Convergent thinking is not the type of thinking people primarily use when they think creatively.

The characteristic of *divergent thinking* is the variety of thoughts involved. When thinking creatively, people tend to think in a divergent manner, thus having many varied thoughts about a problem. Divergent thinking includes *autistic thinking* (defined on page 228) and some convergent thinking. The creative thinker may use convergent thinking to gather information and thoughts as building materials for the ultimate creative solution. At times, the person may drift into autistic thinking, or free association, in which the symbols of thought have private meanings. In the process of this autistic thinking, some useful ideas that would have been missed by concentrating strictly on the problem may occur.

CHARACTERISTICS OF CREATIVE THINKERS

Although creative people are generally high in intellectual ability as measured by standardized intelligence tests (Chapter 13, pages 529–534), they are not necessarily in the very highest brackets measured by these tests. Many creative people are talented in some special way—in music or mathematics, for instance. In other words, they have certain specific abilities that they can use in their search for new ideas. And while many creative thoughts come as sudden insights (page 244), such "flashes" are more likely after hard thinking about a problem; so diligence and strong motivation to work at solving problems are characteristic of creative thinkers.

People who think creatively seem to have some personality features in common. Evidence obtained from objective and subjective personality tests (Chapter 13, pages 546–555) indicates that "original," or creative, people tend to have the following traits:

1. They prefer complexity and some degree of apparent imbalance in phenomena.

2. They are more complex psychodynamically and have greater personal scope.

3. They are more independent in their judgments.

4. They are more self-assertive and dominant.

5. They reject suppression as a mechanism for the control of impulse. (This implies that they forbid themselves fewer thoughts, dislike policing themselves or others, and are disposed to entertain impulses and ideas that are commonly taboo.) (Slightly modified from Barron, 1963, pp. 208–209.)

A personality dimension called *origence* (Welsh, 1975) has been shown to be related to creativity. A person high on this dimension "resists conventional approaches that have been determined by others and would rather 'do his (or her) own thing,' even if it is unpopular or seems to be rebellious or nonconforming." Such a person "is more interested in artistic, literary, and aesthetic matters that do not have a 'correct' answer agreed upon by consensus and that allow a more individualized interpretation and expression" (Welsh, 1975, p. 128). The terms listed under "Upper Pole" in Table 6.3 paint a picture of what a high-origence person is like.

Language Communication

Language was described earlier in this chapter (page 230) as a "tool of thought," but we usually think of it as a tool of communication among people. Language is said to communicate when others understand the meaning of our sentences, and we, in turn, understand theirs. Communication, of course, is not limited to language; we convey much information to others nonverbally by gestures and other means. (See Application 8.1, page 312.) But this section will focus on verbal (and especially spoken) communication.

When we speak one of the thousands of languages of the world, we draw on our underlying knowledge of the rules governing the use of language. This knowledge about language, or *linguistic competence*, as it is called, is used automatically and almost effortlessly to generate and comprehend meaningful speech. Linguistic competence seems to be a universal human species-typical ability (Chapter 2, page 41). It must

Table 6.3

Summary of terms related to the upper and lower poles of the origence personality dimension

LOWER POLE	UPPER POLE	LOWER POLE	UPPER POLE
Banal	Original	Obedient	Rebellious
Careful	Careless	Other-oriented	Self-oriented
Causal	Contingent	Orderly	Disorderly
Cautious	Adventurous	Optimistic	Pessimistic
Certainty	Uncertainty	Planned	Spontaneous
Circumscribed	Extended	Plodding	Facile
Confined	Open	Practical	Impractical
Conformity	Nonconformity	Prosaic	Interesting
Conservative	Liberal	Prohibitive	Permissive
Contrived	Accidental	Proscribe	Prescribe
Conventional	Unconventional	Puritanical	Sensual
Compulsive	Impulsive	Quantities	Qualities
Cooperative	Uncooperative	Reactionary	Radical
Concordant	Discordant	Regular	Irregular
Definite	Vague	Responsible	Irresponsible
Direct	Indirect	Restrained	Unrestrained
Discover	Invent	Restrictive	Expansive
Discrete	Relational	Rigid	Flexible
Duty	Pleasure	Routine	Ingenuity
Defensive	Unguarded	Self-controlled	Uncontrolled
Even	Uneven	Self-effacing	Self-seeking
Explicit	Implicit	Simple	Complex
Factual	Interpretive	Slow	Quick
Follows rules	Rejects rules	Sobriety	Inebriety
Finicky	Casual	Stable	Unstable
Gauche	Sophisticated	Stringent	Lax
Gravity	Levity	Structured	Unstructured
Gross	Refined	Symmetry	Asymmetry
Insensitive	Sensitive	Tactful	Tactless
Intolerant	Tolerant	Tangible	Ineffable
Literal	Imaginative	Temperate	Intemperate
Matter-of-fact	Far-fetched	Things	Fantasies
Monotony	Variety	Tight	Loose
Naive	Candid	Inartistic	Artistic
Numerals	Words	Vulgarian	Aesthete
Objects	Relationships	Work	Play
Obvious	Subtle	Withheld	Released

(*Source*: Welsh, 1975, p. 129.)

spring from the way in which the human brain is organized (Chapter 2, page 69) and from the language learning that children do as they develop (Chapter 11, page 438).

What knowledge comprises linguistic competence? First, we know the sounds or written elements of language(s) and the rules for combining these basic elements into units, such as words, that have meaning. Second, we have underlying implicit knowledge of the rules for combining words into meaningful sentences; in other words, we have knowledge of the *grammar*, or *syntax*, rules of language(s). Third, along with the grammar rules, we have stored in long-term semantic memory (Chapter 5, page 196) the meanings of thousands of words. (The aspect of language that refers to meaning is called *semantics*.) Fourth, we know how to use speech in order to have an intended impact on others. (Such use of language for practical ends is called *pragmatics*.) Fifth, we have knowledge of the rules for processing and interpreting the speech of other people. These five areas of linguistic competence are among the major hallmarks of what is meant by the word *language*. Some language competence seems to exist in species other than our own, but, as is discussed in Inquiry 6.3, there is debate about this point.

Linguistics is the study of languages as structured systems of rules; it also encompasses the study of the origin of languages, the relationships among languages, how languages change over time, and the nature of language sounds. *Psycholinguistics* is concerned with the ways in which people use linguistic competency to generate and understand language; in other words, the field of psycholinguistics is concerned with the behavior of using language and, conversely, the ways in which behavior is influenced by language. In what follows, we will draw upon some of the theories of language structure developed by linguists and on psycholinguistic studies of the ways in which people use speech elements, generate meaningful sentences, know what speech means, use speech to have an impact on others, and process the flow of incoming speech.

LANGUAGE ELEMENTS

Starting with the basic sounds of speech, spoken language can be broken down into these elements: phonemes, syllables, morphemes, words, clauses, and sentences.

Phones and Phonemes Speech sounds, or *phones*, are made by adjusting the vocal cords and moving the tongue, lips, and mouth in wonderfully precise ways to produce vibrations in the airflow from the lungs. While hundreds of speech sounds can be distinguished on the basis of their frequency (the number of vibrations per second), their intensity (the energy in the vibrations), and their pattern of vibrations over time (Figure 6.7), only a limited number (about 45 in English) of all the possible phones are important to the understanding of speech; these are known as *phonemes*.

To illustrate phonemes, consider the *k* phone in the words *key* and *cool*. Say these words to yourself, and you will realize that the *k* sound is different in the two words; simply notice the position of your lips when you are saying them and the "sharper" sound of the *k* phone in the word *key*. Here, then, are two different phones, but either *k* sound can be used in the word *key* without changing the meaning of the word; the same can be said for *cool*. English speakers do not notice the difference in these *k* sounds and, therefore, since they make no difference in the meanings of the words and can be substituted for one another, they can be grouped together as a single phoneme. Or consider all the ways you can pronounce

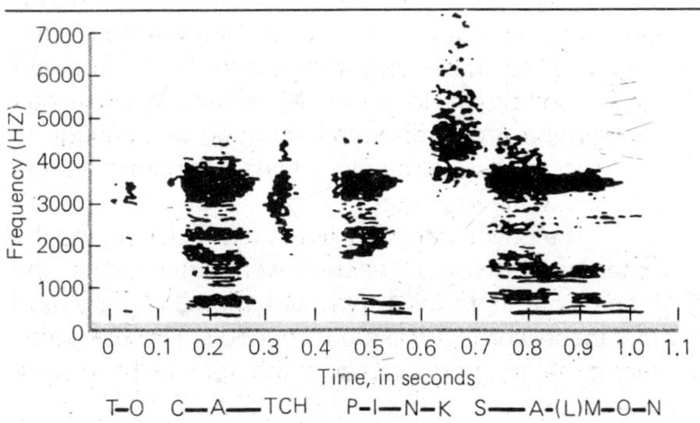

Figure 6.7

A sound spectrogram for the phrase "to catch pink salmon." The frequency (pitch) of each sound is represented by the distance above the baseline. The *s* sound in *salmon*, for instance, contains many high frequencies. (From Liberman et al., 1972.)

Inquiry 6.3

WE CAN TALK TO THE APES, BUT CAN THEY TALK BACK?

Of all the animals, apes (chimpanzees and gorillas) have brains that are most like those of human beings in their organization and structure. Does this mean that they can learn to speak a human language, such as English? In spite of the fact that chimpanzees can understand many English words, attempts to teach them to speak English have met with little success. After much effort, the best that seems to have been managed by a chimp are the words *mama*, *papa*, *up*, and *cup*. Comparable experiments have not been tried with gorillas.

Although vocal speech seems out of the question, is it possible that apes (chimpanzees and gorillas) can learn a nonvocal language? They make many gestures, use their hands nimbly, and pay close visual attention to what is happening around them. Perhaps progress might be made with sign language, and this was the approach taken in research projects directed by Beatrice and R. Allen Gardner with chimpanzee Washoe (and others), by Roger Fouts with chimpanzees Lucy, Dar (see Figure), Washoe (and others), by Herbert Terrace with chimpanzee Nim Chimpsky, and by Francine (Penny) Patterson with the gorillas Koko and Michael. Another approach has been to use visual symbols as the language elements. Ann and David Premack trained their star pupil, Sarah, to communicate through use of magnet-backed plastic blocks, or tokens, which could be stuck on a "language board." Duane Rumbaugh and his colleagues gave chimpanzee Lana (and others) a computerized visual display panel from which she could select symbols standing for certain words by pressing buttons on which the symbols were displayed.

Each of these research programs was successful in teaching apes to use symbols in order to communicate. But there is more to language than this, and debate swirls around the question of whether the apes in these experiments really learned language. A closer look at the first, and one of the most extensive, ape-language experiments—Project Washoe—will set the stage for understanding the debate.

The Gardners favor the view that Washoe truly uses language to communicate. Washoe, a young female chimp, was trained to communicate by means of gestures derived from the American Sign Language for the deaf. One training technique was imitation learning: In the presence of an object or ongoing activity, the teacher gave the sign for the object or activity. Washoe sometimes learned to copy the teacher's gesture and later produce it appropriately. More effective was training in which the teacher took hold of Washoe's hands and arms and guided her to make the gesture being taught; if Washoe then made the correct gesture, she was rewarded—perhaps with a tickle, which she liked.

Washoe learned quickly. Among her first signs were gestures for "come-gimme," "more," "up," "sweet," "go," "hear-listen," and "tickle." Washoe is still being studied by Roger Fouts and now has a vocabulary of some 176 signs, which she uses to communicate with other chimps and with human beings. Many of Washoe's signs seem to be concepts (page 232) similar to ours; her signs stand for a class of related objects or events. For instance, although Washoe learned the sign for *flower* in the presence of real flowers, she has been observed to make the "flower" sign when she sees pictures in magazines or smells a fragrant, flowerlike odor.

Early in her training, Washoe began to combine signs into sentencelike strings, such as "come-gimme drink" or, when faced with a locked door, "gimme key," "open key please," and "open key help hurry." Furthermore, she sometimes combined signs she already knew to describe new situations. When she saw a swan for the first time, it is reported that Washoe, who already knew the signs for *water* and *bird*, combined them to refer to the swan as a "waterbird."

That Washoe uses signs meaningfully is also indicated by her answers to questions. If asked in sign language "Who you?" she replies, "Me Washoe." Or if asked "What that?" she replies with the appropriate sign—toothbrush, for example. In fact, Washoe has learned the American Sign Language well enough to hold simple two-way conversations with human beings who know the language.

The critics generally agree that Washoe and the other apes studied can use words (the signs, the tokens, or the visual display items) meaningfully, and that these words are like our concepts. However, they argue that the performance of the apes in these studies falls far short of true language competence (page 247). For example, languages have grammar rules (page 253) that specify, among other things, the order in which words should be used—subject-verb-object in English, for instance. The grammar rules make the intended meanings of sentences clear. Washoe and the other "signing" apes do not, say the critics, show evidence that they use their signs in appropriate grammatical order. But, argue the proponents of ape language, the apes are nonetheless communicating meaningfully; the context within which the signs are given makes the intended meaning clear regardless of the sign order. Furthermore, it is argued, the short strings of sign utterances—two or three signs—can be understood, especially in context, no matter what their order. But others do not agree; the controversy about the nature of ape communication is not yet settled.

Top, Dar signs "tickle" on Loulis (an example of chimp-to-chimp signing). (Courtesy of R. Fouts.) Bottom, a chimp signs "toothbrush." (Paul Fusco/Magnum.)

REFERENCES

Gardner, R. A., & Gardner, B. T. (1969). Teaching sign language to a chimpanzee. *Science, 165*, 664–672.

Gardner, B. T., & Gardner, R. A. (1975). Evidence for sentence constituents in the early utterances of child and chimpanzee. *Journal of Experimental Psychology: General, 104*, 244–267.

Linden, E. (1975). *Apes, men and language.* New York: Dutton.

Patterson, F., & Linden, E. (1981). *The education of Koko.* New York: Holt, Rinehart and Winston.

Premack, A. J., & Premack, D. (1972). Teaching language to an ape. *Scientific American, 227*(4), 92–99.

Rumbaugh, D. M. (Ed.) (1977). *Language learning by a chimpanzee: The Lana project.* New York: Academic Press.

Sebeok, T. A., & Umiker-Sebeok, J. (Eds.) (1980). *Speaking of apes: A critical anthology of two-way communication with man.* New York: Plenum Press.

Terrace, H. S. (1979). *Nim.* New York: Knopf.

Van Cantfort, T. E., & Rimpau, J. B. (1982). Sign language studies with children and chimpanzees. *Sign Language Studies*, *34*, 15–72.

the *p* sound in *put* without changing the meaning of the word. Thus speech sounds, or *phones*, that make no difference in the meaning of a word when they are substituted for each other are grouped together as phonemes. Conversely, when phones cannot be substituted for each other, they are considered different phonemes—the *b* and *p* sounds in *bun* and *pun*, for example.

The sounds comprising phonemes are perceived as belonging together as a category of sounds, a phenomenon called the *categorical perception of phonemes*. If, for instance, the phoneme /b/, as in the syllable *ba*, is gradually changed to the phoneme /p/, as in the syllable *pa*, people will continue to perceive *ba* even though the sound has changed considerably toward *pa*. At some point, they will suddenly begin to hear *pa*; the perceptual category suddenly changes (Eimas & Corbit, 1973). Note that the people in this experiment did not perceive sounds intermediate between *ba* and *pa* or blends of these sounds; their perception was of one category of sounds or the other.

Languages differ in their groupings of phones into phonemes, and this is one of the problems in learning to speak a new language. In Japanese, for instance, the *l* and *r* phones make no difference in meaning and are therefore part of the same phoneme class. In English, though, these phones belong to different phoneme groupings; they cannot be substituted for each other without changing meaning—*light* and *right*, for example. This can be a problem for speakers of Japanese who are learning English; they are accustomed to using the *l* and *r* phones interchangeably, but this will not work in English. Similarly, the *k* phones, which are grouped together as a single phoneme in English, are different phonemes in Arabic, and this poses a problem for English speakers who are learning Arabic.

Languages also have rules specifying the ways in which phonemes can be grouped together. For instance, certain combinations, such as *cz*, *hr*, and *bj*, may be all right in some languages, but not in English. And how many English words do you know with three different consonant phonemes in a row?

Syllables Although phonemes are perceived categorically and are excellent tools for describing and comparing the sounds made in various languages, they are not the units of speech perception, probably because people never hear them one at a time. What we hear are two or three phonemes combined into a syllable. Hence, the *syllable* is the smallest unit of speech perception. Evidence for syllables as perceptual units has been found in a number of experiments. In one study, subjects were presented with a sequence of syllables and instructed to raise their hands as soon as they heard a syllable beginning with the *b* sound. They were also asked to raise their hands when they heard the syllable *boog*. The reaction time to *boog* was shorter than the reaction time to *b*. Put another way, it took the subjects longer to perceive *b* than to perceive *boog* (Bever, 1973). Such experiments demonstrate that people perceive the whole syllable before they perceive its separate parts.

Morphemes, Words, Clauses, and Sentences Although syllables are the units of speech perception, and some syllables have meanings, other language elements are the perceptual units carrying the *meaning* of speech. These elements, *morphemes*, are the smallest units of meaning in speech perception. Consider the word *distasteful*. It is composed of three morphemes, each of which has meaning. The morphemes in this example are *dis*, *taste*, and *ful*. *Dis* means "negation," *taste* is a meaningful word, and *ful* means "quality." Thus, morphemes can be prefixes, words, or suffixes. Each is composed of syllables, of course, but what makes them *morphemes* is that they convey meaning. Morphemes are discovered by asking people to break words up into the smallest units that have meaning for them.

Words are combined by the rules of grammar (see below) into clauses, and clauses are formed into sentences. A *clause* consists of a verb and its associated nouns, adjectives, and so on. Evidence indicates that clauses, and not individual words or whole sentences, are the major units of perceived meaning in speech. When we hear a sentence with more than one clause, we tend to isolate the clauses, analyzing the meaning of each (Bever, 1973). This tendency, part of the processing and interpretation of speech, will be described further in a later section (page 258).

GRAMMAR AND MEANING

Linguists have proposed a number of theories of *grammar*, or *syntax* (from the Greek meaning "joining together"). The theory of *transformational grammar* (Chomsky, 1957, 1965) is one of the best known and most influential. In this theory, when a person intends to communicate a simple sentence (such as "The chicken crossed the road," for example), the words are organized in what is called the *deep-phrase structure*, or *deep structure*: the mental representation of what the person intends to say. Certain *phrase-structure rules* govern the organization of the deep-phrase structure. Figure 6.8 shows the simple rules needed to organize "The chicken crossed the road." By combining the elements according to phrase-structure grammatical rules, a sentence with meaning is generated. With the simple rules in the illustration, a very large number of meaningful sentences could be produced, and, with a few more rules, complex sentences with several clauses could be generated. In fact, we need only a few more rules to specify the infinite number of grammatical sentences possible in a language. A very important part of language development (Chapter 11, page 438) consists of learning the rules of grammar that we use so effortlessly and without conscious awareness in constructing sentences to express what we want to say.

To continue with the theory, we now have a meaningful sentence in the deep structure. It is a proposition—a combination of verbs and nouns—that says that a subject (the chicken) does something (crosses) with respect to an object (the road). This basic idea, or *proposition*, we might then express as a simple declarative sentence, "The chicken crossed the road." The expressed sentence is called the *surface structure*. But note that the deep, core meaning—the proposition—can be expressed in other ways. For instance, the passive form of the sentence—"The road was crossed by the chicken"—has the same meaning as the active form. We need a set of rules for changing the core proposition into sentences with different surface structures but the same underlying meaning. What are called *transformation rules* do this.

Transformation rules can also be used to change the *surface* meaning of an underlying proposition. The surface-structure sentences, "The chicken did not cross the road," "Did the chicken cross the road?" and "Why did the chicken cross the road?" have different meanings, but they are all related to a common deep-structure proposition about a chicken crossing a road. Depending on what the speaker intends to emphasize—negation or a question, for instance—application of specific transformation rules makes possible the generation of surface sentences with different meanings based on the same deep-structure proposition. Thus, the meaning of a phrase or sentence is determined partly by the grammatical relationships in the deep-structure kernel, or core, and by the transformation done on this meaningful core as it is changed into surface-structure sentences.

Figure 6.8

A box diagram, a tree diagram, and phrase-structure rules for the sentence "The chicken crossed the road." (See text.)

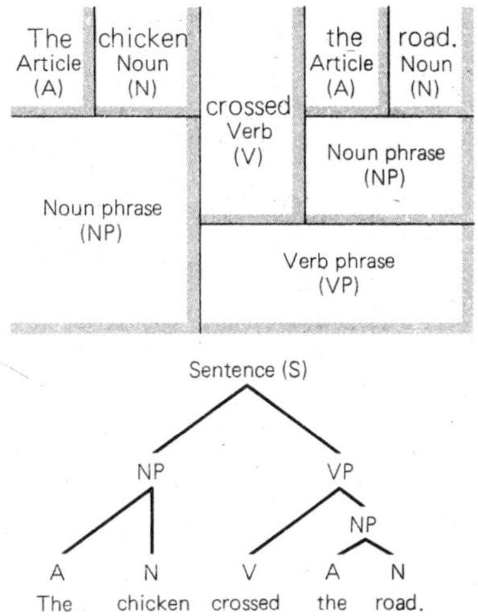

Phase-Structure Rules
(The arrows mean "can be written as.")

(1) Sentence (S) ⟶ NP + VP
(2) NP ⟶ A + N
(3) VP ⟶ V + NP
(4) A ⟶ The, the
(5) N ⟶ chicken, road
(6) V ⟶ crossed

To apply the rules, we start with (1). This tells us that we should look for the definition of a noun phrase; this is given in rule (2). Then we look to see what A and N mean—rules (4) and (5). Thus we see that the first noun phrase is "the chicken." Going back to rule (1) to complete the sentence, we see that the verb phrase of rule (1) is defined in rule (3) as a verb plus a noun phrase. The verb is given in rule (6). Since the definition of a noun phrase, given in rule (2), says that we must look for an article (A) plus a noun (N), we again look at rules (4) and (5). The first items under these rules have already been used, so we choose the second ones—"the" and "road"—to produce the second noun phrase—"the road."

MEANINGS OF WORDS AND CONCEPTS

The study of word and concept (page 232) meaning is known as *semantics*. Without some agreement on what words and concepts mean, communication would be impossible. One approach might be to agree to use only the dictionary definitions of words, but of course this is not how we know the meanings of the words and concepts we use. Instead, based on past learning, we have our own "dictionaries" of word and concept meanings "in our heads." In the sentence "The chicken crossed the road," we know that a chicken is a living creature (a bird), that it has feathers, that it may be good to eat, and so on. This collection of features is one way we have of mentally representing the meaning of the word *chicken* (Katz, 1972). Similarly, a bicycle is inanimate, has two wheels, has some kind of propulsion mechanism, is a means of transportation, and so on. The features of *chicken* and *bicycle* are the basic units of meaning which, when combined, help to define the word or concept. When we look up an unfamiliar word in a dictionary, we are looking for features that will distinguish that word from all others.

Another way word and concept meanings are represented in our "mental dictionaries" is in terms of *family resemblance structures* (Wittgenstein, 1953) and *prototypes* (Rosch & Mervis, 1975). To illustrate, suppose we consider the way we mentally represent the word or concept "game." The prototypical game has many features—rules are followed, it is fun to play, it requires skill, it involves competition against others, there is a challenge of some sort, it is won or lost, and so on. Some games—the prototypical ones—are more "gamelike" than others in that they have most of the defining features. Basketball, for instance, is more "gamelike" than the card game solitaire, but both are games. Just as sisters, brothers, and their parents have a family resemblance, basketball and solitaire share many of the same common features; they belong to the "game family." Thus, if we are told that *fribble* is in the "game family," we know something about the meaning of *fribble*, even though we do not know the exact defining features of the game.

So far, we have been discussing the generally accepted meanings of words and concepts—what is known as *denotational meaning*. But words and concepts also have emotional and evaluative meanings—their "goodness" or "badness," for example. Such emotional meaning is known as *connotative meaning*. Differences in the connotative meanings of words and concepts are an important source of misunderstanding and failure of communication among people. For instance, what does the word *conservative* mean to you?

One way to measure the connotative meanings of words and concepts is the *semantic differential* (Osgood et al., 1957). In using the semantic differential, words or concepts such as "father," "sin," "symphony," "Russian," or "American" are rated on a number of scales. Each scale consists of two polar, or opposite, words, such as "happy-sad," "hard-soft," "slow-fast." The words are presented to a subject, and he or she is asked to rate each word or concept on a 7-point scale. In Figure 6.9 (top), a subject's ratings of the word *father* are shown on three scales; in practice, 20 to 50 scales would be used for each concept. The result of rating words and concepts in this way is a semantic differential for each word or concept. By means of the statistical technique known as factor analysis, the 20 to 50 rating scales can be grouped into three "superscales"—

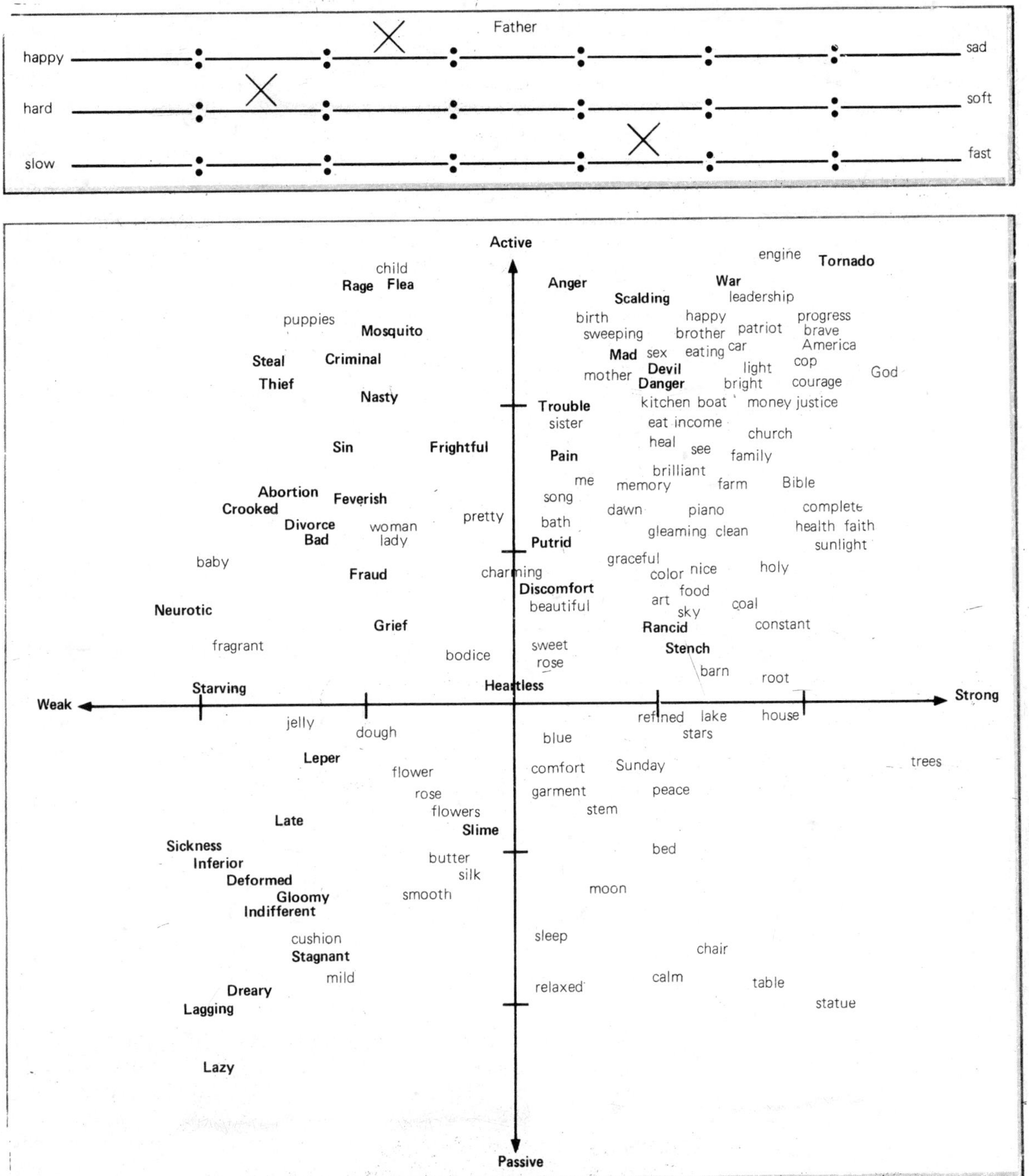

Figure 6.9

Top, some scales from a semantic differential. A person rates a concept, in this case "father," on 20 to 50 bipolar scales, three of which are illustrated. (After Osgood et al., 1957.) Bottom, concepts can be located in "meaning space." The concepts are placed with respect to three factors derived from the semantic differential. The three factors are activity (active-passive), potency (weak-strong), and evaluation (good-bad). The position of a concept with respect to the axes locates it on the activity and potency dimensions. The evaluative factor is represented by the way the words are printed: "Good" concepts are in light type, "bad" concepts in dark type. For instance *anger* is rated as strongly active, slightly strong, and "bad." (From Carroll, 1964. Based on data from Jenkins et al., 1958.)

evaluation (good or bad), potency (strong or weak), and activity (passive or active). Then, for each concept or word, scores on these superscales can be plotted to show their connotative meaning (Figure 6.9, bottom).

PRAGMATICS

Speech is social and interpersonal; we usually speak to communicate information to another person, to influence the behavior of another, or to gain something. Speaking to have an impact on others is known as the *pragmatics* of language. Just as rules govern the combination of words into grammatical sentences, there are rules that members of a language community implicitly understand about how to use language socially in interpersonal relationships. From a pragmatic view, language is a game played like other games—"according to the rules." But note that these rules are usually applied automatically and unconsciously; they are usually not applied deliberately, the way a propagandist might apply rules of rhetoric in order to sway opinion.

Context and Situation While we can learn the words and grammar of a new language from books and records, we must speak the language in a social context in order to understand the subtleties of its use in communication. Sarcasm illustrates the point. If you say to a foreigner who has learned English from a record, "*That* was a fine thing to do!" he or she may feel complimented, but in fact you intended just the opposite. The new speaker of English does not know that this is one of those English sentences that can mean the opposite of what it says on the surface. We who know about such sentences use the context, or situation, as the basis for feeling complimented or chastised. Another example is, "Thanks a lot." Does this mean "Thanks" or "No thanks"? The context will tell us. Sometimes we say things like "John was on time yesterday." Is this just a statement, or by specifying *yesterday*, are we using language in a subtle way to say that John is usually late? If we know the context, we can interpret the statement.

Consider how your use of language changes as the situation around you changes. At home with your family, your speech is informal, and you take a great deal for granted because you have shared so many experiences with members of your family. At work, you may find yourself using the jargon of your trade or profession. And when relaxing with peers of your own age and social class, you may find yourself using the slang of this particular language "community" and, perhaps, four-letter, taboo words you would not use in other contexts. Just to make the point clear, imagine how differently you would use language during a job interview and then, later, when describing the interview to friends over a soft drink or beer.

Status The way language is used often gives the listener immediate knowledge about the social standing of the speaker. In the play *Pygmalion* (and in the musical comedy, *My Fair Lady*, based on this play), a young woman who was stigmatized by her use of the cockney dialect learned to speak the upper-class dialect and thus became "acceptable" in society (Figure 6.10). In the United States, as in England and other countries, the use of speech allows us to "place" people.

Status, or social standing, has a role to play in the way we address people in English. If I am a student, how should I address my professor,

Figure 6.10

The way we speak can provide clues to our background and social status. (Martin Swope.)

Mary J. Smith? Should I call her "Dr. Smith" (she has a Ph.D.)? Should I call her "Professor Smith"? Or, since she is married, should I call her "Mrs. Smith"? Maybe I should address her as "Ms. Smith." Should I call her "Mary," or even "Smitty"? There are subtle rules governing forms of address in English, and these rules change with the situation. In class, I will probably call her "Dr. Smith" or "Professor Smith," even though I know her well; "Mary" would probably be perceived as rude even if we were friends out of class. And "Mary" would still be regarded as rude out of class unless I knew her very well. "Mrs. Smith" or "Ms. Smith" would tend to be used in situations unrelated to her profession as teacher, but the choice between "Mrs." and "Ms." is problematical. None of these rules is clearly spelled out in English, so even native speakers are often not sure what to do and try to avoid direct forms of address altogether. In other languages, the status relationship between speaker and listener actually determines what words will be used to refer to objects or activities. In Javanese, for example, there are six linguistic status distinctions, each with its own set of words to be used; it is almost like having six different languages.

Conversation Rules Rules seem to specify the manner in which we start a conversation, or discourse, with another person. Only a small number of all the possible conversation openings are, in fact, used.

> If we disregard antisocial openings such as *Hey, Jerk*! all conversations are opened in one of six ways.
>
> 1. A request for information, services, or goods. Examples are: *What time is it? Do you have a match? Please pass the sugar.*

2. A request for a social response. *What a slow bus this is! It sure is raining hard.*

3. An offer of information. *Did you hear about the robbery last night? You seem to be lost.*

4. An emotional expression of anger, pain, joy, which is often a strategy to solicit a comment by a listener. *Ouch! Whoopee! Look at this!*

5. Stereotyped statements such as greetings, apologies, thanks, and so on. *Hello. I'm sorry. Thanks a lot.*

6. A substitute statement to avoid a conversation about a subject the speaker anticipates the listener will broach. An example would be a water-cooler meeting between a boss and a subordinate; the boss anticipates a conversation about a raise, so he hurriedly speaks first and uses an avoidance opener: *The traffic sure was heavy this morning.*

(Farb, P. *Word play: What happens when people talk*, 1975, pp. 96–97)

After the conversation starts, perhaps the major rule in all languages, and one that is so obvious that we hardly notice it, is that the speaker and listener take turns. A short pause usually intervenes between the end of the utterance of one person and the beginning of speech by the other in the conversation. While we sometimes do it, interruption is considered rude in almost all languages.

Conversations end when one of the participants signals to the other that he or she no longer wishes to continue. Some of these signals can be nonverbal—paralinguistic, as they are called. One person may stand up, gaze out the window, or seem not to be paying attention to what the other says. If the speakers have exhausted their knowledge of the subject under discussion or have reached agreement on a course of action, they will signal each other, verbally or nonverbally, that the conversation should end. Some conversations end amost before they have had a chance to begin. Greetings such as "How are you?" or "Nice weather," while they may be conversation openers, are often answered briefly by a noncommital "Fine" or "Yes, it is" if the person spoken to does not wish to continue the conversation. We implicitly know the rules for stopping incipient conversations.

PROCESSING AND INTERPRETING SPEECH

The grammarians have given us formal descriptions of the ways words are combined into clauses and sentences in the deep structure of language (page 253). But it is the surface structure (page 253) of language that we hear and from which we must infer the intended meaning of a speaker. We actively and automatically process the flow of speech to do this.

As described earlier, we separate the flow of listened-to speech into *clauses*, which are the major meaningful units in speech perception. The evidence for clauses as units of speech perception comes from psycholinguistic experiments such as the following (Fodor & Bever, 1965):

Two-clause sentences, such as "although it probably won't snow, don't forget your overcoat," were recorded on tape. A *click* sound was also put on the tape at various places in the sentence. If the click was placed near the clause boundary, either just before or just after it, the people in the experiment heard it as occurring during the break between the clauses. For example, if the click appeared in the word *snow*, almost everyone perceived it as coming after the word—that is, at the clause boundary; a click in the word *don't* was perceived as being before the word and at the clause boundary. The perceived position of the click is said to shift forward or backward toward the end or beginning of clauses because of a general rule of perception that says that it is hard to break into or interrupt groupings perceived together as a unit. Thus subjects in the experiment perceived one unit—the first clause—followed by another unit—the click—followed by a third unit of perception—the second clause.

Other experiments (Garrett et al., 1966) show that the separation of sentences into clauses does not necessarily depend on pauses or other sound signals in the flow of speech. Thus, even in the absence of sound cues for segregation, we actively and automatically break up the stream of speech clause by clause.

While speech is being processed into clauses, we are also perceiving a number of other cues, or *markers*, in the surface structure that help us make sense of what we hear. To illustrate the role of markers in speech perception, consider the example of *Russenorsk*, a simple language made up of Russian and Norwegian words and designed, in the days before the Russian revolution of 1917, to facilitate trade on the Russian-Norwegian border (Broch, 1927; Slobin, 1979). Some of the markers in this language were explicit: A word ending with the suffix *-om* was a verb; the word *po* was a general preposition which, depending on the context, could be understood as meaning "in," "on," "by," "at," "to," and so forth. For example, to say "The captain drinks," the appropriate Russian or Norwegian words would be used with the *-om* suffix on the verb—"Captain drink-om." The following sentences illustrate some of the uses of the *po* marker.

Little money PO pocket. 'Not much money *in* the pocket.'

Master PO boat? 'Is the master *on* the boat?'

What you business PO this day? 'What are you doing *on* this day [= today]?'

PO you wife? 'Is there *by* you a wife?' [Do you have a wife?]

Steer PO shore. 'Steer *to* shore.'

Speak PO master. 'Speak *to* the master.'

How many days PO sea you? 'How many days were you *at* sea?

How much weight flour PO one weight halibut? 'What quantity of flour *in exchange for* what quantity of halibut?'

(Slobin, D. I. *Psycholinguistics* (2d ed.), 1979, p. 45)

Russenorsk also had word-order marker rules for distinguishing subject from object and for indicating possession. Markers such as question words and voice inflection indicated whether the sentence was a question or a statement.

The Russenorsk examples show how markers are used in the perception of speech. Of course, more complex languages also have markers to help isolate meaningful units in the flow of speech. To give only one example from English, consider the following (Fodor & Garrett, 1967):

The pen the author the editor liked used was new.

This sentence probably does not make much sense to you. But if English markers—the relative pronouns *that*, *which*, *who*, *whom*—are put into the sentence, it is much easier to understand.

The pen (which) the author (whom) the editor liked used was new.

Relative pronouns seem to act as markers to separate English sentences into segments to make the processing of speech easier. Of course, in the processing of English and other languages, we unconsciously use many other processing rules. The point is that the psycholinguistic analysis of language helps to tell us how we "convert a flowing sequence of sounds into an interpretable message" (Slobin, 1979).

Summary

1. Thinking consists of the cognitive rearrangment or manipulation of both information from the environment and the symbols stored in long-term memory. A symbol represents, or stands for, some event or item in the world. Thinking can also be considered a process that mediates, or goes between, stimuli and responses.

2. Thinking uses images and language. The images used in thinking are abstractions and constructions based on information stored in long-term memory. In using language as a "tool of thought," we draw on word meanings and grammatical rules stored in our semantic long-term memories. While language is often used in thinking, we do not literally "talk to ourselves" when thinking.

3. Concepts are an important class of language symbols used in thinking. A concept is a symbolic construction representing some common and general feature or features of objects or events.

4. Some natural, or basic, concepts are easily acquired and appear in thinking early in life. Other concepts are acquired by discrimination learning, by seeing examples of a concept in different contexts, and by definition.

5. Problem solving is an important kind of thinking. In general, a problem is any conflict or difference between one situation and another we wish to produce—the goal. Many instances of problem solving can be considered a form of information processing. The solution of problems is guided by rules—algorithms and heuristics. Algorithms are sets of rules which, if followed correctly, guarantee a solution to a problem. Heuristics are strategies, or approaches, to a

problem that are usually based on past experience, likely to lead to a solution, but do not guarantee success. One common heuristic is breaking a larger problem down into smaller subproblems which, when solved, will lead to the solution of the overall, larger problem.

6. Problem solution depends, to a large degree, on choosing good heuristic rules to follow. Habit and set can predispose us to select appropriate or inappropriate heuristics. The hindering effects of habit and set on problem solving are discussed at some length. Functional fixedness is an example of the hindering effects of habit and set on problem solving.

7. Decision making is a kind of problem solving in which we are presented with several alternatives among which we much choose. One idea about decision making is that people use subjective-probability estimates of the likelihoods of various outcomes in an effort to maximize utility—perceived benefit or psychological value—in making their decisions.

8. Heuristics—decision-making rules for estimating the likelihood, or subjective probability, of outcomes—are used in making risky decisions. While useful, these rules can lead to biases and errors in making decisions. Among the decision-making heuristics are representativeness, availability, and adjustment.

9. Weighing alternatives is an important part of many decisions. The decision maker first makes a list of desired attributes and then gives weights to each of these attributes on the basis of their perceived importance. Then the decision maker assesses the utility, or perceived benefit, of each attribute and multiplies this by its weight to get an overall value for the attribute. Finally, the overall values of the attributes are summed to give a single weighted value, or utility, for each alternative involved in the decision.

10. In creative thinking, something new is sought. Some new ideas seem to come suddenly after little progress has been made over a long period of time; this sudden appearance of new ideas is called insight. Insights are sometimes triggered by lucky, or fortuitous, environmental circumstances which, after creative thinkers have immersed themselves in a problem, direct their thoughts in a new direction.

11. Creative thinking is said to proceed in five stages: preparation, incubation, illumination, evaluation, and revision. Divergent thinking, characterized by a wide range of thoughts on a topic and by some autistic thinking, is considered conducive to creativity.

12. Creative people are usually intelligent, often have a special talent, and are generally strongly motivated to solve the problems that interest them. A personality dimension called origence seems to be related to creativity.

13. When we use language as a "tool of thought" or a "tool of communication," we draw on our underlying knowledge of the rules governing the use of language. This knowledge about language is called linguistic competence. Aspects of linguistic competence are

(*a*) knowledge of the sounds or written elements of a language and the rules for combining these basic elements into larger units with meaning; (*b*) implicit knowledge of the rules for combining words into meaningful sentences—grammar, or syntax; (*c*) knowledge of the meanings of words—semantics; (*d*) knowledge of how to use language in such a way as to have an intended impact on others—pragmatics; and (*e*) knowledge of the rules for processing and interpreting the speech of other people.

14. Phones, phonemes, syllables, morphemes, words, clauses, and sentences are the major language elements. Phones are the sounds of speech. Of these sounds, only a limited number are important to the understanding of speech; these particular speech sounds are known as phonemes. Although phonemes are excellent tools for describing and comparing the sounds made in various languages, they are not the units of speech perception, probably because we never hear them one at a time. What we hear are two or three phonemes combined into a syllable; hence, the syllable is the smallest unit of speech perception. Morphemes are the smallest units of meaning in speech perception; they can be words, prefixes, or suffixes. Words are combined by the rules of grammar, or syntax, into clauses, the major units of perceived meaning in speech. Clauses are combined into sentences.

15. The theory of transformational grammar says that when a person intends to communicate a thought, words are grammatically organized in what is known as the deep-phrase structure in accordance with certain phrase-structure rules. What is called the surface structure is the actual expression in speech of the deep-structure organization. Transformation rules change the deep structure into the surface structure, thus allowing the same deep-structure organization to be expressed in any number of different surface-structure phrases and sentences.

16. Denotational meaning—the generally accepted meaning of a word or concept—is mentally represented by collections of features distinguishing that word or concept from all others. Connotative meaning is the emotional and evaluative meaning of a word or concept; it can be measured by means of the semantic differential.

17. Speaking (or writing) in such a way as to have an impact on others is known as the pragmatics of language. Among the factors influencing the way we use language are the context in which we are speaking and the status of the people we are addressing. Certain conversational rules also guide the use of language.

18. To ease the task of processing and interpreting speech, we automatically break the flow of speech into clauses. The surface structure also contains markers, which help us make sense of what we hear.

Terms to Know

One way to test your mastery of the material in this chapter is to see whether you know what is meant by the following terms.

Cognition (228)
Thinking (228)
Symbol (228)
Autistic thinking (228)
Directed thinking (228)
Mediation in thinking (228)
Images (229)
Linguistic relativity hypothesis (230)
Concept (232)
Problem (234)
Algorithm (234)
Heuristic (235)
Means-end analysis (235)
Set (236)
Functional fixedness (237)
Utility (240)
Subjective probability (240)
Subjectively expected utility (240)
Representativeness (240)
Gambler's fallacy (241)
Availability (241)
Adjustment (242)
Anchoring (242)
Insight in creative thinking (244)
Stages in creative thinking (245)
Convergent thinking (246)
Divergent thinking (246)
Origence (247)
Linguistic competence (247)
Grammar, or syntax (248, 253)
Semantics (248, 254)
Pragmatics (248, 256)
Language (248)
Linguistics (249)
Psycholinguistics (249)
Phones (249)
Phonemes (249)
Categorical perception of phonemes (251)
Syllables (252)
Morphemes (252)
Clauses (252, 258)
Transformational grammar (253)
Deep-phrase structure, or deep structure (253)
Phrase-structure rules (253)
Surface structure (253)
Transformation rules (253)
Family resemblance structures (254)
Prototypes (254)
Denotational meaning (254)
Connotative meaning (254)
Semantic differential (254)
Markers (259)

Suggestions for Further Reading

The references given with the Inquiries in this chapter make good reading.

For more on the important topic of decision making, Irving Janis and Leon Mann's *Decision Making: A Psychological Analysis of Conflict, Choice, and Commitment* (New York: Free Press, 1977) is recommended.

Psychology and Language: An Introduction to Psycholinguistics, by Herbert and Eve Clark (New York: Harcourt Brace Jovanovich, 1977), has much good information on the perception of speech, the act of speaking, the acquisition of language, and what we mean when we say something. Daniel Slobin's *Psycholinguistics*, 2d ed., (Glenview, IL: Scott, Foresman, 1979) has interesting chapters on grammar, language development, the biological foundations of language, and the relationship of language and thinking. Philip Dale's *Language Development: Structure and Function*, 2d ed., (New York: Holt, Rinehart and Winston, 1976), while emphasizing the growth of language capability in children, has much to say about the psychology of language in general. A special feature of *Introduction to Psycholinguistics* by Insup Taylor (New York: Holt, Rinehart and Winston, 1976) is a set of chapters on abnormal language and speech.

A good introduction to pragmatics and the sociology of speech is contained in Peter Farb's *Word Play: What Happens When People Talk* (New York: Knopf, 1975).

Answer to the puzzle on the chapter-opening pages:

$$\begin{array}{r} 526485 \\ +\ 197485 \\ \hline 723970 \end{array}$$

Answer to cryptogram on page 235:
Psychology is the science of human and animal behavior, and the application of the science to human problems.

Motivation

HEADING OUTLINE

Chapter

Imagine a student walking down the street exposed to all sorts of sights and sounds. He cannot pay attention to all these stimuli, so he responds to only certain ones. Unusual stimuli may arouse his curiosity—he stops to watch a building being demolished. Curiosity may not be the only motive; if he has had a particularly trying day he may tarry to enjoy the destructiveness involved. He may have fleeting thoughts of similar destructiveness towards his professor, girl friend, or society in general.

Eventually, when he resumes his walk, he may notice a restaurant and realize he is hungry. He may muse over the fact that he had passed this restaurant many times before and never realized it was there—he had not been hungry on those occasions. As he orders his meal he finds himself flirting with the waitress. He enjoys it, but she is tired and has a headache so she views him as a bore. Still, since she wants a tip, she gamely kids along with him.

Later as he studies in his room he concentrates on learning his math or biology or language. As he gets tired he pushes himself on with thoughts about the grades he will need to get into a professional school. He may pause to think about his parents and wonder if they appreciate how hard he has to work to become the success they want. Asleep at last, he may have a confused dream in which his professor—looking suspiciously like his father—shakes his head sadly as he hands back an examination paper, the waitress from the restaurant leers at him, and he operates a crane knocking down academic-looking buildings.

In this example, we can see the operation of the motives of

curiosity, aggression, hunger, sex, fatigue, pain, achievement, and affection on perceiving, thinking, acting, talking, learning, and dreaming. The motives fluctuate and arrange themselves in various patterns at different times. Some of a person's motives are always operating, and . . . behavior is largely controlled by them. (Adapted from Murray, 1964, pp. 9–10.)

Motivation in action. (Byron Campbell/Magnum.)

A girl wants to be a doctor. A man strives for political power. A person in great pain longs for relief. Another person is ravenously hungry and thinks of nothing but food. A boy is lonely and wishes he had a friend. A man has just committed murder, and the police say the motive was revenge. A woman works hard at a job to achieve a feeling of success and competence. These are just a few of the motives that play so large a part in human behavior. They run the gamut from basic wants, such as hunger and sex, to complicated, long-term motives, such as political ambition, a desire to serve humanity, or a need to master the environment.

These examples show us that behavior is driven and pulled toward goals. They also show us that such goal-seeking behavior tends to persist. We need a term to refer to the driving and pulling forces which result in persistent behavior directed toward particular goals. The term is *motivation*.

Motives as Inferences, Explanations, and Predictors

An important characteristic of motives is that we never observe them directly. We infer their existence from what people say about the way they feel and from observing that people and animals work toward certain goals. In other words, motives are *inferences from behavior* (the things that are said and done). For example, we might observe that a student works hard at almost every task that comes along; from this, we might infer a motive to achieve—to master challenges, whatever they may be. But, of course, if we want to be reasonably sure our inference about achievement motivation is correct, we must make enough observations of the student's behavior to rule out other possible motives.

Note that we need not be conscious, or aware, of our motives. Others can make inferences about our motives even though we may not be aware of them. That behavior can be driven by *unconscious motivation*, as it is called, is perhaps the major concept of the psychoanalytic theory of personality. (See Chapter 14, page 576.)

If our inferences about motives are correct, we have a powerful tool for *explaining behavior*. In fact, most of our everyday explanations of behavior are given in terms of motives. Why are you in college? The answer is usually given in terms of your motivation. You are there because you *want* to learn, because you feel that you *need* a college degree to get a good job, because it is a good place to make friends and "connections" that you *desire*, or, perhaps, because it's more fun than working for a living. You may be in college because you think it is expected of you and one of your *goals* is to conform to what is expected of you. Or you may be in college to *avoid* the unpleasant consequences of disregarding social pressures from your parents and others. Most likely you are in college in response to some combination of these needs. Someone who understands your motives will understand why you do the things you do. This is why clinical and personality psychologists who study the behavior of individuals place so much emphasis on motives. In fact, many theories of personality are really theories about people's motives. (See Chapter 14.)

Motives also help us make *predictions about behavior*. If we infer motives from a sample of a person's behavior, and if our inferences are correct, we are in a good position to make predictions about what that person will do in the future. A person who seeks to hurt others will express hostility in many different situations; a person who needs the company of others will seek it in many situations. Thus, while motives do not tell us exactly what will happen, they give us an idea about the range of things a person will do. A person with a need to achieve will work hard in school, in business, in play, and in many other situations. If a psychologist (or anyone) knows that Phil has a high achievement need, then he or she can make reasonably accurate predictions about how that need for achievement will be expressed in Phil's behavior: "Just watch; Phil will do his damnedest to beat Laura in this next tennis set." Motives are, thus, general states that enable us to make predictions about behavior in many different situations.

Some Theories of Motivation

Perhaps one way to understand the concept of motivation is to see what some representative theories have to say about it. Theories of motivation try to provide general sets of principles to guide our understanding of the urges, wants, needs, desires, strivings, and goals that come under the heading of *motivation*.

DRIVE THEORIES

These might be described as the "push theories of motivation"; behavior is "pushed" toward goals by driving states within the person or animal. Freud (1940/1949), for example, based his ideas about personality on innate, or inborn, sexual and aggressive urges, or drives. (This theory is treated in detail in Chapter 14, page 576.) In general, drive theories say the following: When an internal driving state is aroused, the individual is pushed to engage in behavior which will lead to a goal that reduces the intensity of the driving state. In human beings, at least, reaching the appropriate goal which reduces the drive state is pleasurable and satisfying. Thus motivation is said to consist of (1) a driving state, (2) the goal-directed behavior initiated by the driving state, (3) the attainment of an appropriate goal, and (4) the reduction of the driving state and subjective satisfaction and relief when the goal is reached. After a time, the driving state builds up again to push behavior toward the appropriate goal. The sequence of events just described is sometimes called the *motivational cycle* (Figure 7.1).

Figure 7.1

The motivational cycle.

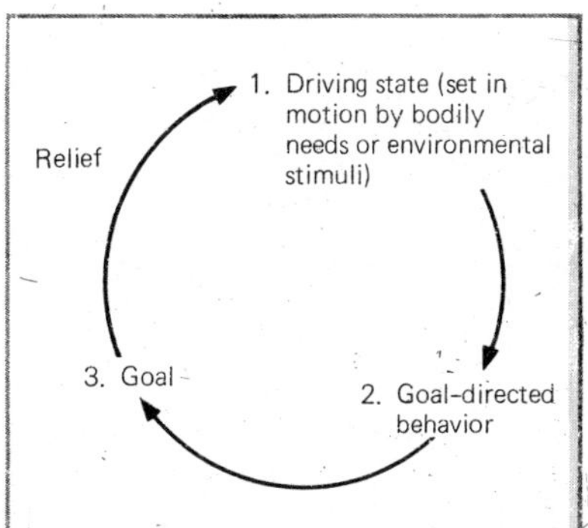

Drive theories differ on the source of the driving state which impels people and animals to action. Some theorists, Freud included, conceived of the driving state as being inborn, or instinctive. And students of animal behavior, notably the ethologists (Chapter 2, page 39), have proposed an elaborate set of inborn driving mechanisms (Tinbergen, 1951; Lorenz & Leyhausen, 1973). Other drive theorists have emphasized the role of learning in the origin of driving states. Such *learned drives*, as they called them, originate in the person's or animal's training or past experience and thus differ from one individual to another. Because of previous use of the drug, a heroin addict, for example, develops a drive to get the substance and is therefore pushed in that direction. And in the realm of human social motives (page 280), people are said to have learned drives for power, aggression, or achievement, to name just a few of the social motives. Such learned driving states become enduring characteristics of the particular person and push that person toward appropriate goals; another person may learn other social motives and be driven toward different goals.

INCENTIVE THEORIES

The drive theories of motivation perhaps apply best to some of the biological motives (page 272)—hunger, thirst, and sex, for example. But even here they encounter problems. Suppose, for instance, that we compare the motivated, goal-directed behavior of two groups of rats which have equivalent hunger drives; the rats in both groups have starved for a day. One group is given a very tasty food (chocolate-chip cookies, perhaps), while the other group gets plain old laboratory rat chow. As you might expect, the chocolate-chip group would probably eat far more than would the lab-chow group. There is something about the goal itself that motivates behavior. Perhaps this is even clearer in the case of sexual motivation; rats (and people, too) are aroused and motivated by the perception of appropriate sexual goal objects. Thus the stimulus characteristics of the goal can sometimes start a train of motivated behavior. This is the basic idea behind theories of *incentive motivation* (Bolles, 1975; Pfaffmann, 1982).

Thus, in contrast with the push of drive theories, *incentive theories* are "pull theories" of motivation; because of certain characteristics they have, the goal objects pull behavior toward them. The goal objects which motivate behavior are known as *incentives*. An important part of many incentive theories is that individuals expect pleasure from the attainment of what are called positive incentives and from the avoidance of what are known as negative incentives. In a workaday world, motivation seems to be more a matter of expected incentives—wages, salaries, bonuses, vacations, and the like—than of drives and their reduction.

OPPONENT-PROCESS THEORY

Hedonistic views of motivation say that we are motivated to seek goals which give us good emotional feelings and to avoid those resulting in displeasure. The *opponent-process theory* takes a hedonistic view of motivation. But this is just the beginning because the theory has some interesting things to say about what is pleasant and what is unpleasant (Solomon & Corbit, 1974; Solomon, 1980). Because of what it says about pleasure and displeasure, this theory might also be classed as a theory of emotion—(Chapter 8, page 329).

Basic to this theory is the observation that many emotional-motivating states are followed by opposing, or opposite, states. Thus, as in the following example, feelings of pleasure and happiness follow feelings of fear and dread.

> A woman at work discovers a lump in her breast and immediately is terrified. She sits still, intermittently weeping, or she paces the

floor. After a few hours, she slowly regains her composure, stops crying, and begins work. At this point, she is still tense and disturbed, but no longer terrified and distracted. She manifests the symptoms usually associated with intense anxiety. While in this state she calls her doctor for an appointment. A few hours later she is at his office, still tense, still frightened: She is obviously a very unhappy woman. The doctor makes his examination. He then informs her that there is no possibility of cancer, that there is nothing to worry about, and that her problem is just a clogged sebaceous gland requiring no medical attention.

A few minutes later, the woman leaves the doctor's office, smiling, greeting strangers, and walking with an unusually buoyant stride. Her euphoric mood permeates all her activities as she resumes her normal duties. She exudes joy, which is not in character for her. A few hours later, however, she is working in her normal, perfunctory way. Her emotional expression is back to normal. She once more has the personality immediately recognizable by all her friends. Gone is the euphoria, and there is no hint of the earlier terrifying experience of that day. (Solomon & Corbit, 1974, p. 119.)

The process can go the other way. For example, a person using heroin for the first time may feel an initial rush of intense pleasure; followed by a less-intense, good feeling; and then by craving and displeasure before the emotional-motivating state returns to normal—the baseline.

Starting from baseline, Figure 7.2*a* shows the general course followed by emotional states. The peak point of the emotional-motivational state (called state A) occurs soon after the emotion-provoking situation is encountered. Note that state A can be a pleasant or an unpleasant emotional state. Next, with the emotion-provoking stimulus still present, the intensity of the emotional-motivational state adapts and declines to a steady state. When the emotion-provoking situation terminates, an after-reaction occurs in which the emotional-motivational state (state B) is the opposite of state A. State B (the *opponent state*) gradually declines until baseline is again reached. Thus, the sequence of emotional-motivational changes goes like this:

Baseline → Peak of state A → Decline of state A to a steady state → State B → Decline of state B to baseline

Now suppose that the same emotion-provoking situation has happened many times (Figure 7.2*b*). Suppose the heroin user in the earlier example found the rush pleasurable and was motivated (perhaps by incentive-motivation expectations of pleasure) to repeat the experience many times. Or suppose a parachutist, after a first jump, was induced (maybe by social pressures) to continue jumping. With repeated usage, as drug tolerance develops, the heroin user will experience less pleasure (less state A), while the intensity of the unpleasant after response (state B) will increase. At first, the heroin user was motivated by the expected pleasure of the rush; after becoming a confirmed user, he or she is now motivated to use the drug in order to reduce the discomfort felt in the drug-free state. Either way, the user is hooked and is motivated to use the drug. After many jumps, the parachutist experiences much less terror

Figure 7.2

Opponent processes in motivation. The duration of an emotion-provoking situation is shown by the black bars. See text. (From Solomon, 1980.)

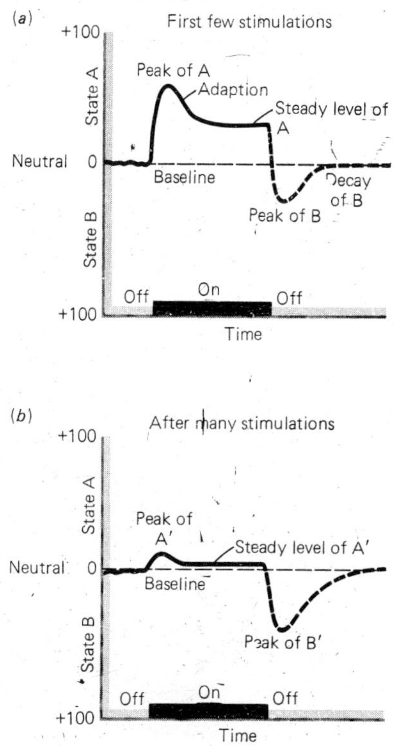

(less state A) but much more of the opposite emotional-motivational state of exhilaration (state B) after the jump. Now, the parachutist is motivated to jump for the after-jump "high," and such a process may account for much thrill-seeking behavior.

This theory gives us a way of thinking about the basis of some learned motives. The heroin addict acquires a need for the drug in order to prevent the unpleasant consequences of withdrawal; some people acquire a need to seek thrills in order to experience elation after the danger is over.

OPTIMAL-LEVEL THEORIES

In general, these are hedonistic (page 270) theories which say that there is a certain *optimal*, or best, level of arousal that is pleasurable (for example, Fiske & Maddi, 1961; Berlyne, 1971). Optimal-level theories might be called "just-right theories." The individual is motivated to behave in such a way as to maintain the optimal level of arousal. For instance, if arousal is too low, a person will seek situations or stimuli to increase arousal; if arousal is too high, behavior will be directed toward decreasing it. Imagine yourself on an extremely busy day at work; too much is happening, and you are highly aroused. More than likely, you find yourself doing things such as taking the telephone off the hook in order to reduce the overload of arousal to which you are being subjected. In doing so, you are behaving so as to move toward a level of optimal arousal. And low levels of arousal (such as occur when not much is happening and we are bored) may also motivate behavior directed at increasing arousal levels to the optimum (page 295).

In the next section, on biological motivation, we will consider the concept of *homeostasis*—the tendency of the body to maintain its internal physiological processes at optimal levels. Body temperature, the acidity of bodily fluids, body water levels, and amounts of many substances circulating in the blood are maintained at certain optimal, or *homeostatic*, levels. Departures from such levels can initiate motivated behavior directed toward restoring the state of equilibrium.

Biological Motivation

The biological motives are, to a large extent, rooted in the physiological state of the body. There are many such motives, including hunger, thirst, a desire for sex, temperature regulation, sleep, pain avoidance, and a need for oxygen. This section will focus on the hunger, thirst, and desire-for-sex motives.

THE AROUSAL OF BIOLOGICAL MOTIVES

Many biological motives are triggered, in part, by departures from balanced physiological conditions of the body. The body tends to maintain a state of equilibrium called *homeostasis* in many of its internal physiological processes. This balance is crucial for life. Body temperature must not get too high or too low, the blood must not be too alkaline or too acid, there must be enough water in the body tissues, and so on.

Physiologists have discovered many of the automatic mechanisms that maintain this balanced condition. Consider the automatic physiological control of body temperature at a point near 98.6°F (37°C). The temperature usually stays around this point because of automatic mechanisms that allow the body to heat and cool itself. If the body temperature rises too high, perspiration and the resultant cooling by evaporation lower the temperature. If the body temperature falls, the person shivers, causing the body to burn fuel faster and to generate extra heat.

The automatic physiological mechanisms that maintain homeostasis are supplemented by motivated behavior. For instance, falling temperature leads to motivated behavior—putting on a sweater, turning up the thermostat, closing the window, and so on. When the body lacks substances such as food and water, automatic physiological processes go to work to conserve the substances that are lacking, but sooner or later water and food must be obtained from the outside. Here the departure from homeostasis creates a drive state that pushes (page 269) a person or animal to seek food and water. Thus the biological motive states are aroused, in large part, by departures from homeostasis; and the motivated behavior driven by these homeostatic imbalances helps to restore the balanced condition.

Certain hormones, or "chemical messengers" (Chapter 2, page 58), circulating in the blood are also important in the arousal of some biological motive states. For instance, sexual motivation in lower animals is tied to hormone levels. In human beings, however, sensory stimuli, rather than hormone levels, are the most important triggers of sexual drive.

Sensory stimuli, or *incentives* (page 270), also play a role in the arousal of other drive states; the smell of a savory dish can arouse hunger in a person who is not, biologically, very far out of homeostatic balance. Perhaps the best example of a drive state aroused by sensory stimulation is pain (Chapter 3, page 106). Pain acts as a motive and is aroused almost entirely by sensory stimulation.

HUNGER MOTIVATION

Of course we must eat to live. The biochemical processes which sustain life get their energy and chemical substances from food. Thus, in a sense, hunger is a primary, basic motive necessary for life. (The same might be said of certain other biological drives—thirst and temperature regulation, for example.) What activates hunger motivation, and what stops it? How is food intake regulated? The answers to these questions are not simple because the hunger drive and eating are influenced by many factors.

Activation of Hunger Motivation Experiments done earlier in this century led to the conclusion that the source of the hunger motivation was stomach contractions. When the stomach is empty, contractions occur and are sensed; the sensed contractions were said to be the signal for feelings of hunger. But more recent research has shown that people report normal feelings of hunger even when, for medical reasons, the nerves from the stomach have been cut or the stomach has been entirely removed. So we must look elsewhere for the bodily conditions which initiate hunger motivation and feeding.

Most investigators of hunger motivation now believe that levels or rates of use of dissolved nutritive substances circulating in the blood are crucial for the activation of feeding. The homeostatic mechanism in feeding seems to be geared to keeping levels of nutritive substances, or the rates at which they are used, within certain limits. If the levels or rates of use fall below a certain point, called the *set point*, hunger drive is initiated and food is ingested to raise the blood levels of nutrients back to the set point.

Glucose, or "blood sugar," is now believed to be an important substance involved in the initiation of hunger motivation and feeding. It has long been known that injections of the hormone insulin, which lowers levels of circulating blood sugar, wll induce hunger and eating. Observations and experiments indicate that the glucose signal for hunger is triggered more by the rate at which glucose is being used by the body than by its absolute levels in the blood (Mayer, 1955). Low rates of glucose use, such as occur after long periods without food and in diabetics, are correlated with reports of hunger and eating behavior; high utilization rates, such as occur just after meals, relate to *satiety*—the absence of hunger motivation. Glucose, of course, is not the only body fuel. Others are free fatty acids from the breakdown of fat stores and ketones from the metabolism of free fatty acids. The role of these fuels in stimulating hunger motivation is just beginning to be appreciated (Friedman, 1978). Whether the critical factor in activating hunger is the rate of glucose use or the level or rate of use of other fuels, the body must have a way of detecting, or monitoring, rate of use or level to indicate hunger or satiety.

Where are the cells which detect the rate of use or level of body fuels? It has been known for a number of years that a part of the brain called the *hypothalamus* (Chapter 2, page 57) is critically involved in hunger motivation and in a number of other biological motives. Several studies (for example, Oomura, 1976) have indicated that nerve cells in specific regions of the hypothalamus that are related to feeding may be able to monitor the rate of use of glucose. Other investigators have proposed that the receptors for glucose and other fuels are in the liver (Friedman & Stricker, 1976) and that information about the blood nutrients is carried to the brain along the nerve pathways connecting liver and brain.

Hunger motivation and eating are, of course, activated by more than just internal factors. The sight and the smells of palatable food can lead to eating even in the absence of any internal need state. (See Inquiry 7.1.) And, of course, we have learned to eat at certain times and not at others.

Cessation of Eating—Satiety Restoration of fuel levels after a meal takes hours. But, of course, we stop eating long before this restoration occurs. So the body must have some way of reducing hunger motivation and stopping eating that is independent of the activating factors. Experiments have shown that the stomach contains nutrient receptors which provide *satiety*—"stop-eating"—signals (Deutsch, 1978). Another satiety signal may be provided by a hormone called *cholecystokinin* (*CCK*). This hormone, which is involved in the breakdown of fats, is released when food reaches the part of the intestine immediately below the stomach. Injections of CCK into food-deprived rats who are eating causes them to stop eating and to start grooming and other behaviors which are part of satiety in rats (Smith & Gibbs, 1976). But the role of CCK as a satiety hormone has been questioned. One reason is that the amounts of CCK in the injections used to produce satiety in rats exceed those released naturally (Deutsch, 1978); another problem is that the relatively large amounts of CCK in the injections may make the experimental animals feel sick and therefore less inclined to eat (Deutsch & Hardy, 1977).

Inquiry 7.1

WHY DO PEOPLE OVEREAT?

Perhaps as many as from 2 to 4 million adults in the U.S. are more than 20 percent overweight. Why is this? Among the answers are (1) food in this country is relatively abundant; (2) the food industry tempts us with many high-calorie, attractively packaged, well-advertised snacks and drinks; (3) eating and drinking are part of many of our social rituals (cocktail parties and the like); and (4) many of us are in sedentary jobs which do not give us enough exercise to burn up the excess calories we are consuming. But some people are more disposed to fat than others. What are some of the individual factors involved in a tendency to fatness? (If you are interested in why some people do not eat enough, and in the topic of anorexia nervosa in particular, see Inquiry 12.1, page 480.)

Results of studies done during the late 1960s and early 1970s suggested that the eating of obese people is not under the same sort of control as that of people whose weight is normal. People of normal weight were found to eat in response to internal cues related to the nutrient levels in their blood (see text); the eating behavior of these people was thus thought to be controlled by physiological, or internal, factors. Obese people, on the other hand, were found to be more influenced by external factors such as the time of day, the amount and quality of food available, and food-relevant cues—the sight, smell, taste, and even the mention of food, for example.

For a time, the internal-external hypothesis, as it was called, seemed to provide a straightforward psychological explanation for obesity. But this hypothesis has not withstood extensive experimental scrutiny. A number of experiments since the original ones have shown that obese people are not necessarily more responsive to external stimuli. Furthermore, many people of normal weight are highly responsive to external food cues and not very sensitive to internal ones. And finally, the distinction between internal and external cues is not sharp because external cues have the capacity to trigger internal physiological changes; thus internal cues are produced by external ones.

Another idea about the cause of obesity concerns individual differences in the set point (see text) for food intake and weight regulation. Experiments have shown that people defend, or tend to maintain, a certain body weight. Thus if a person has been starved and has lost weight, he or she will regain weight to a set point when food again becomes available. And a person who has been induced to overeat as part of an experiment will return to a characteristic weight when the experiment is over. Obese people are said to defend a higher set point than individuals of normal weight. But what "sets" the point higher in the obese is unclear.

Fat cells (cells that store fat, not "fat" cells) seem to play a role in obesity. Perhaps due to genetic factors or to the way they were fed as infants, some people have more fat cells than others and therefore convert more of their nutrients to fat.

While fat people can reduce, maintaining the weight reduction is often very difficult for them. (See Chapter 16, page 693, for some therapy methods which may help the obese to reduce.) The physiology of metabolism works against the fat person in several ways. As the obese person stores more and more fat, the fat cells increase their ability to store fat; in other words, fatness begets fatness. Thus, even though an obese person has managed to reduce, the active fat cells are waiting to store fat when the person relaxes his or her diet a little. Furthermore, as weight is lost, metabolism slows down and the body therefore needs less fuel. Because of the lower metabolic rate after dieting, people with a tendency toward obesity must maintain lowered caloric intake in order to stay thin; their excess nutrients will not be "burned off" efficiently and will, instead, be stored as fat. On the brighter side, exercise increases the metabolic rate. So to keep trim, the obese person who has successfully slimmed down must exercise regularly while counting calories carefully—easier said than done.

REFERENCES

Rodin, J. (1981). Current status of the internal-external hypothesis for obesity: What went wrong? *American Psychologist*, *36*, 361–372.

Schachter, S. (1971). Some extraordinary facts about obese humans and rats. *American Psychologist*, *26*, 129–144.

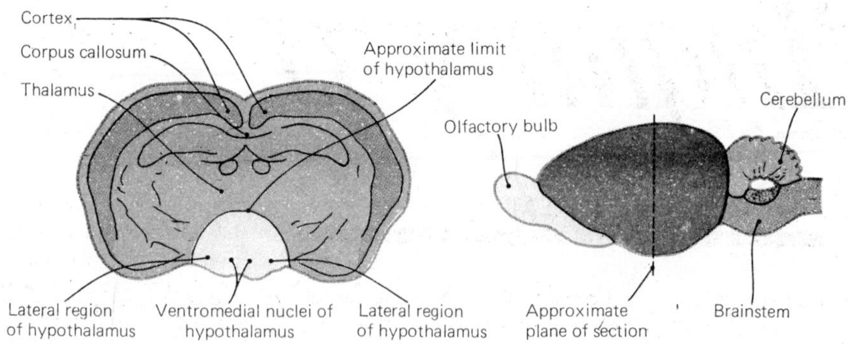

Figure 7.3

A cross-section or frontal section of a rat brain showing the hypothalamus, a small region of the forebrain with many important motivational functions. The approximate plane of section is shown in the lateral-view drawing at the right.

The Brain and Hunger Motivation The hypothalamus (Chapter 2, page 57; see also Figure 7.3) has long been considered important in the regulation of hunger motivation. The classic work of the 1940s and 1950s (Hetherington & Ranson, 1940; Anand & Brobeck, 1951) emphasized the contributions of two regions of the hypothalamus—the lateral hypothalamus (*LH*) and the ventromedial hypothalamus (*VMH*). The *lateral hypothalamus* was considered to be an excitatory region for hunger motivation, while the *ventromedial hypothalamus* was said to be involved in the cessation of eating—that is, in *satiety*. These ideas were based on animal experiments in which the two areas were either destroyed by lesions or electrically stimulated by means of small wires—electrodes—placed in the brain. (The lesions were made while the animals were anesthetized; the stimulating electrodes were placed in their brains and attached to miniature sockets cemented on their skulls so that mild, painless stimulation could be given to them when they had recovered from the anesthesia.)

Electrical stimulation of the lateral hypothalamus was found to elicit eating (and other responses, too, but that is another story); ventromedial stimulation was found to stop ongoing eating behavior. The lesions made in the two areas were found to have effects opposite to those of stimulation. Animals with damage to the lateral hypothalamus would not eat or drink and eventually died of starvation unless given special care (Teitelbaum, 1961; Anand & Brobeck, 1951). When the damage was done to the ventromedial area, the animals developed voracious appetites, consumed great quantities of food, and gained weight rapidly. It was found that rats with ventromedial lesions may become two or three times heavier than normal (Figure 7.4). After the initial spurt of weight gain, the animals with VMH lesions reached a new baseline weight at which they maintained themselves (Figure 7.4, left). Humans with brain tumors or other conditions that have damaged the VMH area overeat and become obese. (VMH damage is not a common cause of obesity, however. See Inquiry 7.1, page 275.)

The results of these early studies were interpreted as indicating that the LH is a feeding center while the VMH is a satiety center. More recent work has confirmed the earlier results on feeding behavior but has raised questions about the interpretation of the LH and VMH as centers for the

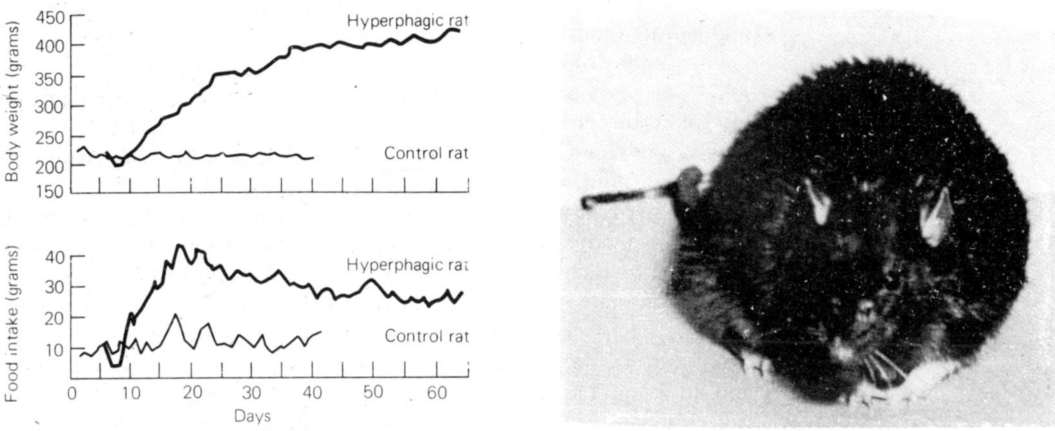

Figure 7.4

Eating and body weight increase drastically if the ventromedial area of the hypothalamus is damaged. The graphs show how body weight and the amount eaten increase in a rat with such damage—the hyperphagic (which means "overeating") rat. The photograph shows a hyperphagic rat. (Graphs modified from Teitelbaum, 1961.)

control of the hunger drive. First, other brain areas, such as the amygdala (Chapter 2, page 62), play an important role in hunger and eating (Morgane & Kosman, 1960). Second, the LH and VMH effects on eating are due wholly or in part to what happens in nerve-fiber pathways which pass through the lateral hypothalamic region (Marshall et al., 1974) or the ventromedial region (Gold et al., 1977). Third, at least some of the effects of lateral hypothalamic lesions seem to be due, not so much to a decrease in hunger motivation as to lowered levels of arousal and general neglect of sensory inputs, especially those coming from food (Wolgin et al., 1976). The current view is that the classic hypothalamic centers may be more accurately considered to be parts of brain systems involved in monitoring the body's fuel supplies, controlling metabolism, and perceiving food-related stimuli.

With the systems idea in mind, several hypotheses have been proposed for the effects of LH and VMH lesions. For LH lesions, the sensory neglect hypothesis has already been mentioned. Another hypothesis is that LH and VMH lesions change the *set point* around which weight is regulated. The set point is like the setting of a room thermostat, but now we are considering the regulation of food intake, not temperature. LH lesions may lower the set point (Powley & Keesey, 1970) while VMH lesions raise it (Keesey & Powley, 1975). Still another hypothesis to account for the overeating after VMH lesions is that metabolism has been changed so that most of what is eaten is converted to fat (Friedman & Stricker, 1976). Since a large part of the body's nutrient supply is going into fat, the animals with VMH lesions must eat much more than they normally do just to keep the fuel supplies they need to run their body machinery at adequate levels.

THIRST MOTIVATION

What drives us to drink (water, that is)? Stimulus factors play a very large role in initiating drinking. We drink to wet a dry mouth or to taste a good beverage. Pulled by these stimuli and incentives (page 270), we tend to drink more than the body needs, but it is easy for the kidneys to get rid of the excess fluid.

But, of course, since maintaining its water level is essential for life itself, the body has a set of complicated internal homeostatic processes to regulate its fluid level and drinking behavior. The body's water level is maintained by physiological events in which several hormones play a vital role. One of these is the *antidiuretic hormone* (*ADH*), which regulates the loss of water through the kidneys. But the physiological mechanisms involved in maintaining the body's water level are not *directly* involved in thirst motivation and drinking.

Thirst motivation and drinking are mainly triggered by two conditions of the body: loss of water from cells and reduction of blood volume. When water is lost from bodily fluids, water leaves the interior of the cells, thus dehydrating them. In the anterior, or front, of the hypothalamus are nerve cells called *osmoreceptors*, which generate nerve impulses when they are dehydrated. These nerve impulses act as a signal for thirst and drinking. Thirst triggered by loss of water from the osmoreceptors is called *cellular-dehydration thirst*. Loss of water from the body also results in *hypovolemia*, or a decrease in the volume of the blood. When blood volume goes down, so does blood pressure. The drop in blood pressure stimulates the kidneys to release an enzyme called *renin*. Through a several-step process, renin is involved in the formation of a substance known as *angiotensin II* that circulates in the blood and may trigger drinking, although questions have been raised about the role of this hormone in drinking (Stricker, 1977).

The idea that cellular dehydration and hypovolemia contribute to thirst and drinking is called the *double-depletion hypothesis*. (See Epstein et al., 1973.) You can see how both mechanisms are at work after a sweaty tennis game: the body has lost water, the osmoreceptors have been dehydrated, and blood volume has gone down. Thirst is triggered, and you drink to rehydrate your cells and to bring your blood volume back to its normal level.

Why does drinking stop? Water-deprived rats, dogs, monkeys, and people stop drinking long before the water balance of their bodies has been restored. Therefore there must be some kind of monitoring mechanism in the mouth, stomach, or intestine which indicates that enough water has been consumed to meet the body's needs. Receptors in the stomach and intestine seem to do this job (Rolls et al., 1980).

SEXUAL MOTIVATION

Since sexual behavior depends, in part, on physiological conditions, it may be considered a biological motive. But of course sex is far more than a biological drive. Sexual motivation is social because it involves other people and provides, according to many, the basis for social groupings in higher animals—baboon troops and the human family, for example; and sexual behavior is powerfully regulated by social pressures and religious beliefs. Sex is psychological in the sense that it is an important part of our emotional lives; it can provide intense pleasure, but it can also give us agony and involve us in many difficult decisions (Chapter 12, page 477). Chapter 14 describes the Freudian theory of personality which, to a large degree, is based on emotions centered on sex. So while this section is on sex as a biological motive, remember that there is much more to sex than hormones and physiological responses.

Even when we consider sexual motivation from a biological viewpoint, it has characteristics which set it apart from other biological drives. First, sex is not necessary to maintain the life of an individual, although it is necessary for survival of the species. Second, sexual behavior is not aroused by a lack of substances in the body. And third, in higher animals at least, sexual motivation is perhaps more under the influence of sensory information from the environment—incentives (page 270)—than are other biological motives.

Sex Hormones: Their Organizational Role *Estrogens*, the female sex hormones, come in large part from the ovaries, but they also come from the adrenal glands. *Estradiol* is one of the most important estrogens. *Androgens*, the male sex hormones, are secreted into the blood from both the testes and the adrenal glands. *Testosterone* is the major androgen. Both male and female sex hormones are present in both men and women; it is the *relative* amounts which differ.

The *organizational role of sex hormones* has to do with their effect on the structure of the body and the brain—especially the regions of the hypothalamus (Chapter 2, page 57) that regulate hormone release. While a person's sex is inherited (genes on the so-called sex chromosomes provide the basis for the growing baby to develop as either male or female), the organization of the body and brain as either male or female depends on the presence of the appropriate sex hormones during early life in the womb. Genes on the sex chromosomes start sexual development off in one direction or the other; under their influence, a fetus with female sex chromosomes will develop ovaries which secrete estrogen while androgen-secreting testes will develop in a fetus with male sex chromosomes. These hormones then direct further sexual development of the body and brain. Later in life (at puberty, see Chapter 12, page 466), the sex organs grow rapidly, and hormone release increases markedly. *Secondary sexual characteristics*—breast development, body shape, pitch of voice, and amount and texture of facial hair, for example—develop under the influence of estrogens or androgens at puberty.

Not only the body, but the brain, too, seems to be organized by sex hormones to predispose a person to behave in male or female ways. (See Chapter 11, page 445.) This is clear in the case of lower animals, where the anatomical organization of certain parts of the brain—notably the hypothalamus—can be changed by hormonal treatments early in life (Gorski et al., 1978). Since the cyclical release of hormones involved in the menstrual cycle is controlled by the hypothalamus, it is clear that the brains of females and males, who have no such cyclical fluctuations, are organized differently. Whether there are subtle organizational differences in brain regions other than the hypothalamus (such as the cerebral cortex; see Chapter 2, page 58) that can partially explain human sex differences in various abilities (Chapter 13, page 541) is a hotly debated topic (McGlone, 1980).

Sex Hormones: Their Activational Role Do higher than normal levels of sex hormones circulating in the blood activate, or trigger, sexual behavior? The answer is clearly "yes" for the females of many lower animal species. As estrogen levels in the blood rise during the reproductive cycle, the females of many species come into what is called estrus, or "heat", and will actively engage in sexual behavior; when not in estrus, they are generally indifferent to male advances.

In human females, on the other hand, the activation of sexual behavior by estrogens is problematical. Some studies have shown a peak of sexual interest at the midpoint of the menstrual cycle, when estrogen levels are high; others have indicated that the greatest sexual interest occurs immediately following menstruation, when estrogen levels are relatively low (Bancroft, 1978). Furthermore, after menopause, when there is an overall reduction in the amount of estrogen circulating in the blood, there is little change in most women's sexual drive. So the relationship between hormones and sexual drive in human females has not been proved. External stimuli (see below), habits, and attitudes seem to be more important than hormones in activating the sexual behavior of girls and women.

In males—lower animals and humans alike—a certain level of androgens, especially testosterone, must be present for sexual behavior to occur at all. Increases above this threshold level have little or no effect on male sexual motivation and behavior (Bermant & Davidson, 1974). Males need a certain level of testosterone in order to maintain their sexual interest and to engage in sexual behavior. Castration of lower animals and, to a lesser degree, of human males reduces their sexual drive. But increases of testosterone levels above the necessary threshold level do not "turn on" male sexual drive. It is as if males are primed and made ready for sexual behavior by their normal levels of androgens. The triggers in males with adequate androgen levels seem to be external stimuli—especially signals from the female that she is interested in sex.

External Stimuli, Learning, and Sexual Behavior External stimuli play a role in the sexual behavior of most animals, but they are especially important in activating sexual motivation in the higher primates and humans. Hormonally ready humans are sexually aroused by what other people say, by their looks, their style, their voices, the way they dress, and their odor. In other words, much sexual behavior is "turned on" by stimuli which act as incentives (page 270).

Learning has much to do with the arousal and expression of sexual motivation in higher primates and humans. In lower animals, sexual behavior tends to be about the same for all members of the species; it is stereotyped, reflexive, automatic, and triggered by the appropriate stimuli (Chapter 2, page 39) if the hormonal conditions are right. In contrast, while human sexual behavior has a biological basis (as indicated above), wide variations exist from person to person in the stimuli which will activate sexual behavior and in the ways that sexual motivation is expressed. Much of the variability in human sexual behavior seems due to people's early learning experiences. Early learning is important, not only for sexual behavior but also for social motives, the topic to which we will now turn.

Social Motives

Social motives are the complex motive states, or needs, that are the wellsprings of many human actions. They are called *social* because they are learned in social groups, especially in the family as children grow up, and because they usually involve other people. These human motives can be looked upon as general states that lead to many particular behaviors. Not only do they help to determine much of what a person does, they persist, never fully satisfied, over the years. No sooner is one goal reached than the motive is directed toward another one. If, for example, a person has a need for affiliation—a need to make friends—he or she may establish friendly relations with one acquaintance, but this does not satisfy the motive. The person is driven to do the same with others and to maintain these patterns of friendship after they are established. Thus social motives are general, persisting characteristics of a person, and, since they are learned, their strength differs greatly from one individual to another. Consequently, social motives are important components of personality—the enduring and characteristic differences among people (Chapter 14).

Many social motives have been proposed. Three of the most-studied social motives are described in Table 7.1: need for achievement (need is often abbreviated *n*), need for affiliation, and need for power. A longer list of social motives is shown in Table 7.2. After describing the ways social motives are measured, we will discuss three of these motives—achievement, power, and human aggression—in detail.

MEASUREMENT OF SOCIAL MOTIVES

To measure social motives, or needs, psychologists try to find *themes*, or common threads, which run through samples of action and imagined action. To find these themes, they use (1) projective tests to study themes of imagined action; (2) pencil-and-paper questionnaires, or inventories, containing questions about what a person does or prefers to do; and (3) observations of actual behavior in certain types of situations designed to bring out the expression of social motives.

Projective Tests These tests (or *techniques*, as they are sometimes called; Chapter 13, page 550) are based on the idea that people will read their

Table 7.1

Characteristics of people with high n *achievement,* n *affiliation, and* n *power*

CHARACTERISTIC	*N* ACHIEVEMENT	*N* AFFILIATION	*N* POWER
General	Concern to do better, to improve performance	Concern for establishing, maintaining, repairing friendly relations	Concern with having impact, reputation, and influence
Arousing situation	A moderately challenging task	Opportunity to be with friends	Hierarchical or influence situation
Related activities	Chooses and performs better at challenging tasks, prefers personal responsibility, seeks and utilizes feedback on performance quality, innovates to improve	Makes more local phone calls, visits, seeks approval, dislikes disagreeing with strangers, better grades from a warm teacher	Accumulates "prestige supplies," often tries to convince others, more often an officer in voluntary organizations, plays more competitive sports, drinks more heavily

(*Source*: From *Motivational Trends in Society* by D. C. McClelland, General Learning Press. Copyright 1971 by General Learning Corporation. Reprinted by permission.)

Table 7.2

Some major social motives

MOTIVE	GOAL AND EFFECTS
Abasement	To submit passively to others; to seek and accept injury, blame, and criticism
Achievement	To accomplish difficult tasks; to rival and surpass others
Affiliation	To seek and enjoy cooperation with others; to make friends
Aggression	To overcome opposition forcefully; to fight and revenge injury; to belittle, curse, or ridicule others
Autonomy	To be free of restraints and obligations; to be independent and free to act according to impulse
Counteraction	To master or make up for failure by renewed efforts; to overcome weakness and maintain pride and self-respect on a high level
Defense	To defend oneself against attack, criticism, or blame; to justify and vindicate oneself
Deference	To admire and support a superior person; to yield eagerly to other people
Dominance	To control and influence the behavior of others; to be a leader
Exhibition	To make an impression; to be seen and heard by others; to show off
Harm avoidance	To avoid pain, physical injury, illness, and death
Infavoidance	To avoid humiliation; to refrain from action because of fear of failure
Nurturance	To help and take care of sick or defenseless people; to assist others who are in trouble
Order	To put things in order; to achieve cleanliness, arrangement, and organization
Play	To devote one's free time to sports, games, and parties; to laugh and make a joke of everything; to be lighthearted and gay
Rejection	To remain aloof and indifferent to an inferior person; to jilt or snub others
Sentience	To seek and enjoy sensuous impressions and sensations; to enjoy the arts genuinely

(*Source:* After Murray, et al., 1938.)

own feelings and needs into ambiguous or unstructured material. In other words, their descriptions of the material will express their social motives because they will project their motives into it.

The *Thematic Apperception Test* (*TAT*) is a projective technique (Chapter 13, page 552) which has been much used to assess social motives. In this test and others like it, a standard set of pictures depicting various people in various situations is presented, and the person being tested is asked to make up stories describing what is happening in the pictures. Motivational themes are found to run through the stories told about the pictures. For instance, when shown a picture of a man seated at a desk (Figure 7.5), a person might tell a story about how hard the man is working to accomplish something; when shown a picture of a boy standing with a broom in front of a store, the same person might tell a story about the dreams of accomplishment the boy is having; and so on for other pictures. This theme of work and accomplishment reflects the person's need for achievement. Another person might tell stories reflecting a need for affiliation, a need for power, or some other social motive. The stories and the themes in them can be scored, and so different degrees of social motives can be ascertained.

Personality Questionnaires Several pencil-and-paper tests (Chapter 13, page 547), called questionnaires or inventories, have been developed to measure the strength of social motives. These inventories consist of questions for people to answer about their typical behavior and preferences—what they would do or prefer to do in certain situations, for example.

Situational Tests A third way to assess social motives is to create situations in which a person's actions will reveal his or her dominant motives. For example, the need for affiliation might be measured by giving an individual a choice between waiting in a room with other people or waiting alone. Children's aggressiveness can be measured by letting them play with dolls and observing the number of aggressive responses they make. Or aggression can be studied by insulting people to see whether they reply in an angry way.

ACHIEVEMENT MOTIVATION

Need for achievement (*n ach*) was one of the first social motives to be studied in detail (McClelland et al., 1953), and research into this motive continues today (Spence, 1983). As a result, we know quite a bit about it. People in whom the need for achievement is strong seek to become accomplished and to improve their task performance. They are task-oriented and prefer to work on tasks that are challenging and on which

Figure 7.5

The strength of various social motives can be assessed by stories told about pictures such as this one. (Ray Ellis/ Photo Researchers.)

their performance can be evaluated in some way, either by comparing it with other people's performance or in terms of some other standard. More formally, "achievement is task-oriented behavior that allows the individual's performance to be evaluated according to some internally or externally imposed criterion, that involves the individual in competing with others, or that otherwise involves some standard of excellence" (Smith, 1969; Spence & Helmreich, 1983, p. 12). Achievement motivation can be seen in many areas of human endeavor—on the job, in school, in homemaking, or in athletic competition, for example.

The Source of Achievement Motivation Why are some people high in the need for achievement? Since the social motives—including the need for achievement—are largely learned, the general answer must be that differences in early life experiences lead to variations in the amount of achievement motivation (and other social motives, as well). More specifically, children learn by copying the behavior of their parents and other important people who serve as models. Through such observational learning (Bandura & Walters, 1963), children take on, or adopt, many characteristics of the model (Chapter 14, page 596), including the need for achievement if the model possesses this motive to a marked degree [Eccles (Parsons), 1983].

The expectations parents have for their children are also said to be important in the development of achievement motivation [Eccles (Parsons), 1983]. Parents who expect their children to work hard and to strive for success will encourage them to do so and praise them for achievement-directed behavior. A specific set of parental expectations related to achievement motivation concerns ideas about when children should become independent in skills such as "standing up for one's rights," "knowing one's way around town," playing with minimal supervision, and, in general, doing things for one's self.

Achievement Motivation and Behavior The degree to which people with strong underlying achievement motivation show achievement-oriented behavior depends on many factors. One of these is another motive —fear of failure—which is said to inhibit the expression of achievement behavior (Atkinson, 1964; Atkinson & Birch, 1978). For people in whom fear of failure is low relative to the need for achievement, achievement motivation expresses itself in many ways (McClelland & Winter, 1969; Hoyenga & Hoyenga, 1984).

1. High *n*-achievement people prefer to work on moderately challenging tasks which promise success. They do not like to work on very easy tasks, where there is no challenge and so no satisfaction of their achievement needs; nor do they like very difficult tasks, where the likelihood of their success is low. Thus people high in *n*-achievement are likely to be realistic in the tasks, jobs, and vocations they select; that is, they are likely to make a good match between their abilities and what will be demanded of them.

2. High *n*-achievement people like tasks in which their performance can be compared with that of others; they like feedback on "how they are doing."

3. High *n*-achievement people tend to be persistent in working on tasks they perceive as career-related or as reflecting those personal characteristics (such as intelligence) which are involved in "getting ahead."

4. When high *n*-achievement people are successful, they tend to raise their levels of aspiration in a realistic way so that they will move on to slightly more challenging and difficult tasks.

5. High *n*-achievement people like to work in situations in which they have some control over the outcome; they are not gamblers.

These achievement-related behaviors tend to be present in many men and some women who are successful in business and in certain professions. But many high *n*-achievement women do not show the achievement behaviors characteristic of men. Many women who are high on *n*-achievement do not, for example, like to work on moderately risky tasks. Thus a gender difference exists in the expression of the need for achievement. To try to account for this gender difference, another motive —fear of success—was proposed for women (Horner, 1968).

Tests were developed which seemed to show that women believed their successful performance would have negative consequences such as unpopularity and a reduced feeling of femininity. Women were found to view success as counter to their roles in our culture and were therefore afraid of it. The existence of such a viewpoint could be expected to change women's achievement-oriented behavior. Subsequent research has strongly questioned the existence of a general success-avoidance motive in women (Tresemer, 1976), but there is evidence that some women—especially those who have accepted the traditional view of women's role in society and yet are placed in competitive situations—do have a fear of success (Patty & Safford, 1976). So there are individual differences in the fear-of-success motive. Beliefs about sex roles have changed in recent years, and fewer women now evidence a strong fear-of-success motivation. If present trends continue, the achievement-motivated behavior of women, in general, can be expected to become more like that of men.

In business, in school, and in many professions, one would expect achievement motivation to be an important predictor of success, and, indeed, it often is. Common sense would also predict that the most successful people would be those who coupled strong achievement motivation with strong competitive motivation. Some interesting current research, however, seems to question this common-sense idea. The most successful people identified in this research scored high on achievement motivation, or work orientation, but low on competitive motivation (Spence & Helmreich, 1983). For example, Figure 7.6 shows a consistent pattern in which the highest levels of performance were obtained by people high on achievement motivation—called work-mastery in Figure 7.6—but low in competitive motivation. The figure also shows that competitive motivation improves performance for people with low achievement, or work-mastery, motivation, but actually impairs performance for those in whom achievement motivation is strong. More and more factors which modify the expression of this important social motive—achievement motivation—are being discovered.

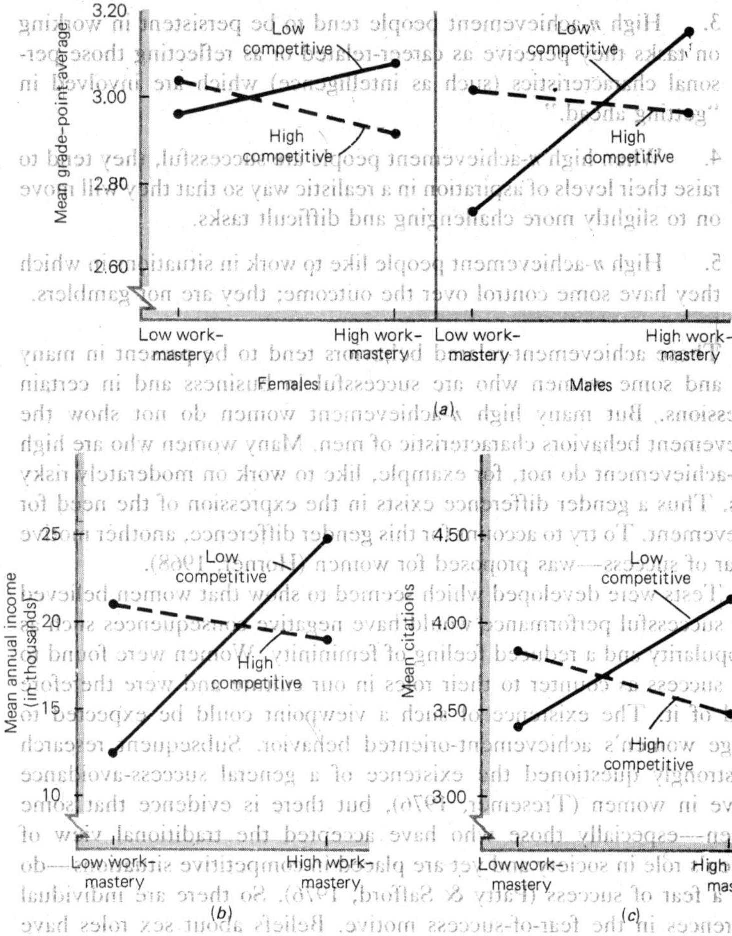

Figure 7.6

The highest levels of performance are reached by people high in work-mastery (achievement motivation) but low in competitive motivation. (*a*) Grade-point averages for females and males. (*b*) Average annual income for businessmen. (*c*) Citations of the published work of male scientists. (From Spence & Helmreich, 1983; data in (*b*) from Sanders, 1978; data in (*c*) from Helmreich et al., 1978.)

Achievement Motivation and Society It has been suggested that the need for achievement is related to a society's economic and business growth (McClelland, 1961, 1971). Thus, if investigators find evidence of strong achievement motivation in a particular society, they may be able to make predictions about economic growth in that society.

By studying the social motives revealed in a culture's popular literature (especially children's books) and relating them to its economic history, researchers have found that a high need for achievement correlates with various indices of economic growth, such as the consumption of electricity. These studies have shown that a high need for achievement comes *before* spurts in economic growth and, thus, predicts them. The delayed relationship is dramatically illustrated in Figure 7.7. Here, need for achievement successfully predicted economic ups and downs in the English economy between 1550 and 1850. Economic activity was measured by changes in coal imports. The lag between levels of achievement motivation and economic changes was about 50 years. For the twentieth century, investigators have found a shorter lag. Although the relationship between need for achievement and economic growth is suggestive, it is

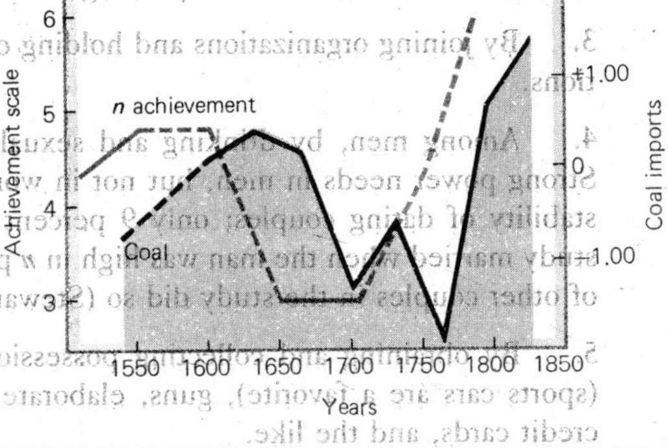

Figure 7.7

The relationship between need for achievement and one index of economic activity—coal imports—in England between 1550 and 1850. Note that changes in achievement need came about 50 years before changes in economic activity. Imagine sliding the *n*-achievement curve to the right by about 50 years; the two curves would then almost coincide. (Modified from Bradburn & Berlew, 1961.)

not proof that need for achievement *causes* economic growth; they may both be caused by other factors. However, knowledge of the social motives dominant in a society may help us understand its history and predict its future. This application of psychology to history and future trends is relatively new, but it may turn out to be a major contribution.

POWER MOTIVATION

Winter (1973, p. 5) has defined *social power* as "the ability or capacity of a person to produce (consciously or unconsciously) intended effects on the behavior or emotions of another person." The goals of power motivation are to influence, control, cajole, persuade, lead, charm others, and to enhance one's own reputation in the eyes of other people. People with strong power motivation derive satisfaction from achieving these goals.

Power motivation (the *need for power*, or *n power*, as it is often termed) varies in strength from person to person and can be measured from the stories told in the picture-projection technique (page 281). The degree of *n* power in a person is reflected in story themes about direct control of other people, in stories concerning the emotional impact one person has on another, and by the conern of the people in the stories for their reputations (Winter, 1973).

Power Motivation and Behavior Given that a person's degree of power motivation can be roughly measured, we can look to see how this motive is expressed in behavior by relating the test scores of power motivation with what the person actually does. Power motivation can be expressed in many ways; the manner of expression depends greatly on the person's socioeconomic status (Hoyenga & Hoyenga, 1984), sex (Hoyenga & Hoyenga, 1979), level of maturity (McClelland, 1975), and the degree to which the individual fears his or her own power motivation (McClelland, 1975). Nevertheless, a number of behavioral clusters have been related to high *n* power (McClelland et al., 1972; Winter, 1973). The following are some of the ways in which people with high power motivation express themselves (Hoyenga & Hoyenga, 1984):

1. By impulsive and aggressive action, especially by men in lower socioeconomic brackets.
2. By participation in competitive sports, such as hockey, foot-

ball, baseball, tennis, and basketball, especially by men in lower socioeconomic brackets and by college men.

3. By joining organizations and holding office in these organizations.

4. Among men, by drinking and sexually dominating women. Strong power needs in men, but not in women, are related to the stability of dating couples; only 9 percent of the couples in one study married when the man was high in *n* power, while 52 percent of other couples in the study did so (Stewart & Rubin, 1976).

5. By obtaining and collecting possessions, such as fancy cars (sports cars are a favorite), guns, elaborate stereo sets, numerous credit cards, and the like.

6. By associating with people who are not particularly popular with others and who perhaps, are more easily controlled by the high-*n*-power person because they depend on him or her for friendship.

7. By choosing occupations such as teaching, diplomacy, business, and the clergy—occupations in which high-*n*-power people believe they have a chance to have an impact on others.

8. By building and disciplining their bodies; this seems especially characteristic of women with strong power needs.

Machiavellianism Niccolo Machiavelli (1469–1527), in his book *The Prince*, gave some practical advice to rulers on how to maintain power. In brief, he said, rulers should outwardly follow the rules of conventional morality but secretly use craft and deceitful practices to maintain themselves in power. The term *Machiavellianism* (Christie & Geis, 1970) has been coined in psychology to describe people who express their power motivation by manipulating and exploiting others in a deceptive and unscrupulous fashion. Machiavellianism is not the same as power motivation; it refers to a particular strategy that some people, dubbed "Machiavels," use to express their power motivation. In addition to being manipulators, "Machiavels" generally show little warmth in their personal relationships, are only weakly guided by conventional morality in their dealings with others, are reality-oriented (interested in practical results), and have little interest in ideologies (Christie & Geis, 1970). Pencil-and-paper psychological tests based on these ideas have been devised to identify "Machiavels." Subsequent laboratory research (for example, Geis, 1978) showed that people with high scores on a test of Machiavellianism do, indeed, manipulate others, make and break interpersonal relationships in an opportunistic way, deny cheating in laboratory games, use lies to manipulate other people, are very persuasive when arguing about subjects in which they are emotionally involved, are perceived by others in their group as leaders, and, as Machiavelli himself might have predicted, tend to be on the winning side in laboratory games.

HUMAN AGGRESSION

Most would agree that aggression is a motive about which we should know more. We attack, hurt, and sometimes kill each other; we aggress verbally by means of insults or attempts to damage another's reputation; and wars always seem to be happening someplace—approximately 14,600 wars in 5,600 years of recorded history (Montagu, 1976).

Instrumental and Hostile Aggression The term aggression is hard to pin down, and there is some disagreement about what should and should not be called aggressive. A distinction is sometimes made between hostile aggression and instrumental aggression. *Hostile aggression* has as its goal harming another person; in *instrumental aggression* (Buss, 1961, 1966), the individual uses aggression as a way of satisfying some other motive. For example, a person may use threats to force others to comply with his or her wishes; or a child may use aggression as a way of gaining attention from others.

The focus of this section on human aggression is on hostile aggression. A definition that many find adequate is that *hostile aggression* "is any form of behavior directed toward the goal of harming or injuring another living being who is motivated to avoid such treatment" (Baron, 1977, p. 7). Note that this definition implies that the aggressor intends to harm the victim; although intent is often hard to judge, we should be reasonably sure it is there before we label an act aggressive.

We humans have a vast repertory of specific hostile aggressive behaviors. Aggression can be physical or verbal, active or passive, direct or indirect (Buss, 1961). The physical-verbal distinction is the distinction between bodily harm and attack with words; the active-passive difference is the difference between overt action and a failure to act; direct aggression means face-to-face contact with the person being attacked, while indirect aggression occurs without such contact. These distinctions provide a useful framework for categorizing many specific instances of hostile human aggression (Table 7.3).

Aggression as a Human Instinct One influential idea about human aggression is that it is part of "the nature of the beast" (Freud, 1920/1953;

Table 7.3

Some forms of human aggression

TYPE OF AGGRESSION	EXAMPLES
Physical-active-direct	stabbing, punching, or shooting another person
Physical-active-indirect	setting a booby trap for another person; hiring an assassin to kill an enemy
Physical-passive-direct	physically preventing another person from obtaining a desired goal or performing a desired act (as in a sit-in demonstration)
Physical-passive-indirect	refusing to perform necessary tasks (e.g., refusing to move during a sit-in)
Verbal-active-direct	insulting or derogating another person
Verbal-active-indirect	spreading malicious rumors or gossip about another person
Verbal-passive-direct	refusing to speak to another person, to answer questions, and so on
Verbal-passive-indirect	failing to make specific verbal comments (e.g., failing to speak up in another person's defense when he or she is unfairly criticized)

(*Source*: Baron, 1977; based on Buss, 1961.)

Lorenz, 1966). We are said to share an aggressive instinct with lower animals. Thus, human aggression is said to be a species-typical behavior (Chapter 2, page 41), just as it is in other animals. The implication of this view is that since we are, by nature, aggressive creatures, fights, wars, and destruction have been and will continue to be a major part of the human story on the planet earth.

While it is true that portions of human and lower animal brains regulate the expression of aggression and that levels of certain hormones (testosterone, for example) are related to aggression (Moyer, 1976; Svare, 1983), most psychologists reject the dark, pessimistic, "human-nature" view of aggression. Instead, they stress the importance of environmental, social, and learning factors in the causation and regulation of aggressive behavior (Bandura, 1983; Baron, 1977; Berkowitz, 1962).

Environmental and Social Causes of Human Aggression Frustration of a motive was proposed early as a basic cause of aggression. ***Frustration*** (page 299) occurs when motivated behavior is thwarted, or blocked, so that goals are not reached. The strong form of the ***frustration-aggression hypothesis*** (Dollard et al., 1939), as it came to be called, stated that frustration always results in aggressive behavior and that all aggressive behavior is caused by frustration. But we can easily think of exceptions to the frustration-aggression hypothesis in this strong form. People may, for example, react to frustration by withdrawing from the situation, by giving up, by using alcohol or other drugs, or, on a more positive note, by increasing their efforts to overcome the frustration. And, certainly, as we shall see below, not all aggression is caused by frustration.

Whether frustration results in aggression seems to depend on two factors. First, the frustration must be intense (Harris, 1974). In part, the strength of the frustration depends upon the expectation a person has of reaching a goal; the thwarting of high expectations can be an effective instigator of aggression (Worchel, 1974). In this connection, it is interesting to note that civil disorders and revolutions are instigated, not by the most downtrodden members of a society, but by those who, while frustrated, also have some expectation that the social goals they are striving for can be reached. Second, the frustration must be perceived as being the result of arbitrary action. Aggression is more likely when the frustration is perceived as unjustified, and aggression may not occur at all if the thwarting of motives is considered justified by the frustrated individual (Zillmann & Cantor, 1976).

Perhaps the most common, everyday source of aggression is a verbal insult or negative evaluation from another person ("You're a stupid so-and-so") (Buss, 1961). The insult may not be intended as such. But if it is perceived as intended and if harmful intent is attributed (Chapter 9, page 347) to the insulter, the insult is interpreted as an aggressive act. This aggressive act arouses aggression in the person being insulted, and this person responds with counteraggression. Especially in public situations in which we are trying to maintain our esteem in the eyes of others, counteraggression to insults is likely to intensify the original aggression, and a vicious circle of escalation results, which can lead, ultimately, to physical aggression.

Another important social cause of human aggression is compliance with an authority who orders us to aggress against others. This powerful influence, illustrated in a number of obedience experiments (Milgram, 1963, 1965, 1974), is discussed in Chapter 9, page 359.

Unpleasant, or aversive, environmental conditions may dispose some people toward aggression. High temperatures [at least up to a point (Baron, 1977)], intense noise (Donnerstein & Wilson, 1976), and, under some conditions, crowding (Freedman, 1975; Schopler & Stockdale, 1977) increase aggression, especially in people who have already been angered in some way. And it has been argued (Berkowitz, 1983) that frustration, when it is effective, acts in part because of the unpleasant feelings it creates.

Another environmental influence, although a controversial one, is the presence of weapons in settings where aggression might be expected. In some experiments (Berkowitz & LePage, 1967; Frodi, 1975), the presence of weapons such as guns was found to increase the aggression of angered people. But other studies (Buss, Booker, & Buss, 1972) have not been able to replicate what has been called the "weapons effect."

Learning and Human Aggression Social learning theory (Chapter 14, page 596) stresses the role of imitation of others' behavior as a cause of aggression (Bandura, 1973, 1977, 1983). In both laboratory experiments and everyday life, people who have seen others act aggressively are apt to do so in similar situations (Chapter 11, page 446). Aggression is contagious.

> Do aggressive models play an important role in the initiation of riots, insurrections, and similar events? Eye-witness accounts by social scientists on the scene during the initiation of such events suggest that they may. In many cases, it appears that large-scale aggression fails to develop until one or more "hot-headed" individuals commit an initial act of violence (Lieberson & Silverman, 1965; Momboisse, 1967). Prior to such events, angry muttering and a general milling about may predominate. Once the first blow is struck, the first rock hurled, or the first weapon fired, however, a destructive riot may quickly ensue. It seems reasonable to view the persons who initiate violence in such situations as aggressive models. (Baron, 1977, p. 99.)

Modeling is most effective if the aggressive behavior is seen as being both justified and achieving a reward and if the watcher is already angry. Modeling is said to work because it serves to (1) direct the observer's attention to one of several possible behavior sequences (aggression instead of attempts at ingratiation, perhaps); (2) show the observer that certain behaviors are all right, thus decreasing inhibitions to aggression; (3) enhance the emotional arousal of the observer which, under some conditions (Zillmann, 1983), can facilitate aggression; and (4) show the observer some specific aggressive actions that may be copied.

Television and films provide us with many aggressive models, and the question of their contribution to aggressive behavior has been much studied (Geen, 1983). The results of these studies have generally (but with exceptions) shown moderately enhanced aggression, especially among boys, following the viewing of television or movie violence. Here is the way one author put it:

It is my impression that the weight of existing evidence strongly favors the view that exposure to scenes of violence in the mass media does increase the likelihood that observers will behave in a similar manner themselves. But please note: this does *not* mean that after watching their favorite action-packed (and aggression-packed) program, adults or children are likely to rush out and launch blind assaults against any individual unfortunate enough to land in their paths. Judging from the magnitude of effects reported in most investigations, the increase in aggressive tendencies produced by exposure to such materials is probably slight. (Baron, 1977, p. 110.)

In addition to social learning, classical conditioning (Chapter 4, page 140), and instrumental conditioning (Chapter 4, page 147) can be important sources of human aggression. Classical conditioning occurs when certain stimuli or situations are paired with each other. If, for example, aggression-producing situations are repeated often enough in the presence of some stimulus, a person may learn to dislike and be aggressive toward the stimulus that has been paired with the aggression-arousing situation. And, through generalization (Chapter 4, page 146), the aggressive behavior may spread until the individual behaves in an aggressive way toward many similar stimuli. For example, if a boy is strongly and arbitrarily frustrated or insulted by his father often enough, it is likely that he will learn, through classical conditioning, to express hostility toward him. Then, by means of generalization, the boy may also come to show hostility and aggression toward other authority figures—police officers, teachers, bosses, and the like.

Instrumental conditioning of aggression occurs when people are rewarded, or reinforced, for their aggressive behavior. According to the principles of instrumental conditioning (Chapter 4, page 149), behaviors which are reinforced are more likely to occur in the future. Thus, if aggression is reinforced—if it *pays off*—it may become a habitual response in many situations. Suppose a girl finds that aggressive behavior gains her the approval of the people she wants to be her friends. Her need for acceptance will be satisfied by aggressive behavior, and the satisfaction of this need will be rewarding, or reinforcing. So, by means of instrumental conditioning, aggression will become her likely response in many situations.

Controlling Human Aggression If, as research seems to indicate, much human aggression has its basis in learning and in specific environmental and social factors, the potential exists for its limitation and control. Changing the instigators of aggression might well be expected to decrease aggressiveness in our society. For instance, having fewer aggressive models and instances where aggression pays off might help. While the potential for the limitation of aggression may give us hope, it will be very difficult to put into practice the suggestions for the control of human aggression indicated by research. What are some of these suggestions?

Punishment for aggression has been one of the classic approaches to the control of human aggression. *Punishment*, as described in Chapter 4 (page 150), occurs when an event following or contingent upon some behavior decreases the likelihood that the behavior will occur again. *Punishers* are usually, though not necessarily, unpleasant events which follow behavior; chastisements, fines, loss of social acceptance, embarrassment, imprisonment, and the like can serve as punishers. It is widely held in our society that punishment of aggression will reduce it, but, as we shall soon see, punishment may not be so effective as usually thought.

Punishment seems to work best when it is strong, when the aggressor is relatively sure of receiving it, when it immediately follows aggressive behavior, when the instigation for aggression is relatively weak, when the payoff for aggression is not great, and when the person perceives the punishment as being legitimate and appropriate (Baron, 1977). (See also Chapter 4, page 164, for a general discussion of the conditions under which punishment may be effective in controlling behavior.) Otherwise, punishment may not be very effective, as seems to be the case with its use in most societies as a means of controlling crime and other aggressive acts. When punishment is used ineffectively, it may actually increase aggressive tendencies. Punishment is a frustrator, and it may therefore further arouse and anger the person being punished. Furthermore, it is, in itself, an aggressive act which provides a model for aggression.

Another classic approach to the reduction of anger and aggression is called catharsis. *Catharsis* refers to venting an emotion, or "getting it out of one's system." For example, it is often said that we can get anger and aggression out of our systems by pounding on a table, yelling out the back door, kicking a dog, or watching a boxing match. While catharsis may help us reduce our anger for a short time, it does not seem to decrease the likelihood that we will aggress in the future against the particular person who made us angry. Research results indicate that we can get cathartic relief from our anger and aggressive feelings toward another person only by actually venting our anger and aggression on that person (for example, Konečni & Ebbesen, 1976). Therefore, in this case, catharsis is achieved only after aggression has occurred—not a very good way to control aggression.

If, as social learning theory says (page 291; see also Chapter 11, page 446), aggressive models can induce aggressive actions, can nonaggressive models lessen them? The answer seems to be "yes" (for example, Baron & Kepner, 1970).

The interpretations we make of aggression directed toward us have much to do with our tendency to reply aggressively. For instance, if we know, before an attack on us occurs, that the aggressor was upset for reasons out of his or her control, we will make allowances for the aggressive behavior and reduce our counteraggression (Zillmann & Cantor, 1976). Thus the thoughts, or cognitions, we have about the reasons for another's aggression play a role in helping us control our own aggression. (This is related to the topic of attribution—Chapter 9, page 347).

Another interesting approach to the control of aggression is based on the notion that certain emotions and feelings are incompatible with anger and aggression (Baron, 1977, 1983). Thus anger may disappear when a person is induced to smile, feels concern about the object of his or her attack (empathy), or perhaps is mildly sexually aroused. The results of laboratory studies (Baron, 1983) seem to show that such emotions and responses are incompatible with anger and aggression and thus serve to lessen it. Out in the "real world," too, field studies support the view that anger and aggression can be reduced by incompatible responses (Baron, 1976):

Figure 7.8

The distraction condition (A), the empathy condition (B), the humor condition (C), and the mild-sexual-arousal condition (D) in an experiment using the induction of incompatible responses to reduce aggression. See text. (From Baron, 1977; courtesy of Robert Baron.)

The experimenters chose for their study a moderately busy intersection with a traffic light. One experimenter drove a car which waited at the intersection for 15 seconds after the light turned green; the others, from hiding, observed the reactions of the male drivers (the subjects in the experiment) of the cars that were forced to wait. Horn honking and aggressive gestures were recorded by the hidden experimenters.

There were three experimental (incompatible-response) conditions and two control conditions in the experiment. In one experimental condition (empathy), a young woman experimenter hobbled across the street on crutches (Figure 7.8) just before the traffic light turned green; in the second condition (humor), she crossed the street wearing a clown's mask; and in the third experimental condition (mild sexual arousal), she crossed the street in scanty attire. In one control condition, nobody crossed the street; in the other control condition (distraction), the young woman experimenter, dressed normally, simply walked across the street. In all cases, the woman crossing the street was out of sight by the time the light turned green and the 15-second delay began.

How would you react to the aggression-instigating frustration of being forced to wait? Well, in the control condition, with nobody crossing the street, 90 percent of the waiting drivers honked; in the distraction control condition, 89 percent honked. But when the incompatible responses of empathy, humor, or mild sexual arousal were induced, honking dropped markedly: 57 percent honked in the empathy condition, 50 percent honked in the humor condition, and only 47 percent honked in the mild-sexual-arousal condition.

Our account of motivation would not be complete if we did not describe what, especially for human beings, may be some of the most persistent and powerful motives of all. These are motives to seek variety in stimulation, to process information about the world around us, to explore, and to be effective in mastering challenges from the environment.

If you have ever watched a small child, you will realize the strength of these motives. A baby's life, when not eating or sleeping, seems dominated by needs to know, to explore, and to be effective in the environment (Figure 7.9). Suppose we put a crawling baby girl down in the middle of a room and watch what happens. If she is not afraid, she will start to crawl around, contacting and manipulating various objects in turn. Many of these objects will go into her mouth, but this is just another way of sensing and exploring the world. As one object loses its novelty, she will go on to another and another.

And it is not just babies who seem to have these motives. Needs to know and to be effective persist throughout life and are difficult, if not impossible, to satisfy. Even when our biological and social needs have been met, we continue to seek contact with the environment and to engage in restless and relentless activity. Because they are so persistent and seem to exist to one degree or another in everyone, these needs to know and to be effective are often considered innate, a part of the human species heritage (Chapter 2, page 41). In a sense, these motives are behind our greatest human accomplishments and also, unfortunately, our greatest fiascoes. We just cannot seem "to let well enough alone."

STIMULUS AND EXPLORATION NEEDS

Think of the amount of time, effort, and money people put into just looking at things, traveling, and exploring the environment. We visit new places and points of interest; we watch television, movies, sports contests,

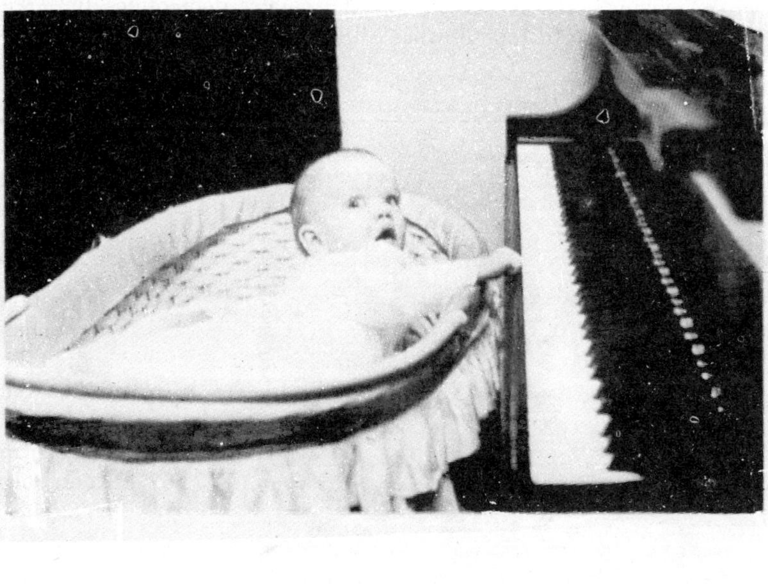

Figure 7.9

Needs to know and to be effective appear early in life. (Suzanne Szasz, 1981.)

and plays; we read newspapers, books, and magazines. Stimulus needs and the need to explore are largely behind these activities. We also get "tired of the same old thing." In other words, what satisfied our stimulus and exploration needs soon no longer does so, and we seek something new. Informal observations like this provide the basis for controlled experiments on stimulus and exploration needs.

The incentive power of stimuli in the environment was demonstrated some years ago in experiments in which monkeys were given mechanical puzzles to solve (Figure 7.10). The monkeys were attracted to the unusual, novel stimulus presented by the puzzle. They explored it and learned to solve the puzzle by removing the pin, loosening the catch, and lifting the hinge, receiving, for their efforts, "no reward other than the privilege of unassembling it" (Harlow, 1953). About the same time, other experiments on curiosity motivation showed that monkeys would work and learn when the incentive was merely being allowed to look out of an enclosed cage into a rich, varied visual environment (Butler, 1953, 1954).

Figure 7.10

A puzzle for monkeys used to show the incentive power of stimuli. See text. (From Harlow, 1953.)

The monkeys in this experiment were confined in a closed box with two small doors on one side (Figure 7.11). A monkey would frequently strike a door by accident. Each door had a visual stimulus mounted on it (stimulus A and stimulus B). If the monkey happened to push the door with stimulus A on it, it would open and the animal could look outside for a few seconds. If it pushed the door with stimulus B, nothing happened. Thus the only reward for learning to discriminate between the two stimuli (Chapter 4, page 160) was an opportunity to look out the door that opened when pushed.

The monkeys readily learned the discrimination, thus demonstrating that they were motivated by an opportunity to look outside. Once an animal learned to push the correct door, the number of pushes it made depended on what was to be seen outside. If there was an interesting scene on the outside, such as another monkey or a moving toy train, it opened the correct door often. It did not respond so frequently when all it could see outside was an empty room.

Figure 7.11

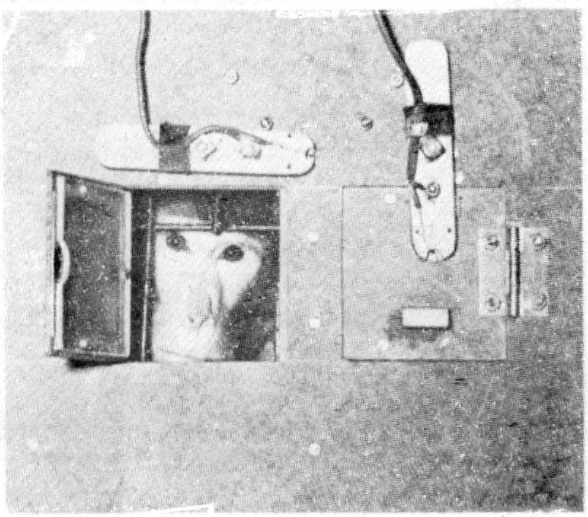

A monkey peeping out of an apparatus used to measure stimulus seeking and curiosity motivation. If the animal pushed the correct door, it was rewarded by being allowed to look outside for a few seconds. (From H. F. Harlow, Wisconsin Primate Research Center.)

Human beings seek stimulation, too; and some people—"sensation seekers," as they are called (Zuckerman, 1978)—are prone to search for especially exciting stimuli and situations. And even if we are not sensation seekers, most of us seem to have a need to seek new, or novel, stimuli. According to one influential theory (Berlyne, 1971), stimuli from the environment arouse all of us, and each of us has an optimal level (page 272) of arousal which we seek. (Sensation seekers have very high optimal levels.) Being at or near the optimal arousal level is pleasurable; too-high or too-low levels of arousal result in feelings of displeasure. Moderately novel and complex stimuli are especially good at increasing the arousal level toward the optimal, pleasurable level. A novel stimulus is one that is somewhat different from what we expect; a complex stimulus is one which contains a relatively great amount of information for us to process. To reach our optimal level of arousal, we tend to seek out and prefer stimuli that are both novel and complex.

EFFECTANCE MOTIVATION

The stimulus and exploration needs described in the last section can be considered to be involved in attempts to be competent in dealing with the environment. The motivation behind these competence activities has been called *effectance motivation*—a general motive to act competently and effectively when interacting with the environment (White, 1959). Effectance motivation plays an important role in human behavior. Goals are reached, but effectance motivation is not satisfied; it remains to urge behavior toward new competencies and masteries.

Suppose we look again at the small child introduced at the beginning of this section. From the standpoint of effectance motivation, if the baby is just progressing from crawling to standing (Chapter 11, page 421), we notice how much effort she puts into working to stand up. The baby repeatedly tries to pull herself up in her playpen, and she does this over and over, in spite of repeated failures, whether anyone is watching or not. When she finally succeeds, she lets out a cry of delight and smiles widely. In a similar way, she works to be effective in the environment when it comes to walking and a number of other developmental challenges. These small triumphs of childhood illustrate what is meant by effectance motivation; the baby is trying to master the environment and to become effective in it. Effectance motivation is at work in her later life, too, but here it is sometimes difficult to tell whether her behavior is motivated by effectance motivation or by one of the social motives—achievement motivation (page 283), for example.

A concept somewhat like effectance motivation is *intrinsic motivation*, defined as "a person's need for feeling competent and self-determining in dealing with his environment" (Deci, 1975). It is called *intrinsic* because the goals are internal feelings of effectiveness, competence, and self-determination. In contrast, *extrinsic motivation* is directed toward goals external to the person, such as money or grades in school. Extrinsic rewards have their uses in guiding behavior in business and in school, but reliance on them can sometimes stifle intrinsic motivation and impair performance (Condry, 1977; Lepper & Greene, 1978). With extrinsic rewards, people may adopt a strategy of doing the minimum needed to get by instead of working hard for the "fun of it," creatively, and for the satisfaction that comes from the mastery and deep understanding of a problem. Thus, by working against our motives to know and to be effective, extrinsic motivation can sometimes reduce a powerful aspect of human motivation.

SELF-ACTUALIZATION MOTIVATION

The motive of self-actualization (Maslow, 1954) is related to effectance motivation and intrinsic motivation. *Self-actualization* refers to an individual's need to develop his or her potentialities; in other words, to do what he or she is capable of doing. "Self-actualizers," then, are people who make the fullest use of their capabilities. (See Chapter 14, page 601, for a detailed description of self-actualizing people.) Of course, the goals which are sought in meeting this need vary from person to person. For some, it means achievement in the literary or scientific field; for others, it means leadership in politics, the community, or the church; for still others, it merely means living life fully without being unduly restrained by social conventions.

Self-actualization is thought to be the top need in a hierarchy of needs, or motives (Maslow, 1954). Going from the highest need of self-actualization down, the needs in the hierarchy are

The *need for self-actualization*

Esteem needs, such as needs for prestige, success, and self-respect

Belongingness and love needs, such as needs for affection, affiliation, and identification

Safety needs, such as needs for security, stability, and order

Physiological needs, such as hunger, thirst, and sex

Figure 7.12

The hierarchy, or ladder, of needs leading to self-actualization. (Based on Maslow, 1954.)

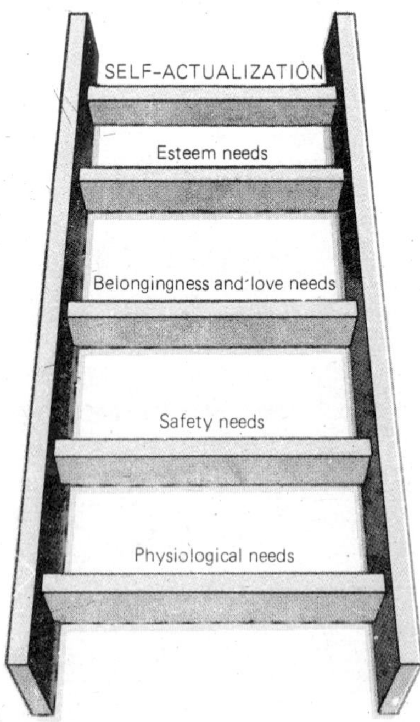

The order in which these needs are listed is significant in two ways. The needs appear in this order, from lowest to highest (Figure 7.12), with physiological needs first and self-actualization needs last, during a person's normal development. From lowest to highest, this is also the order in which they must be satisfied. In other words, physiological needs must be satisfied before any of the others can be met; safety needs come before those higher on the list; and so on. For instance, a starving man is preoccupied with obtaining food. He doesn't even wonder where tomorrow's meal is coming from (safety need); only today's meal counts. But once he is assured of eating today, he can begin to worry about his safety needs and take steps to see that his physiological needs will always be met; thus he moves on to safety needs. The same system of priorities operates at each step up the ladder of motives (Figure 7.12). If a woman has a steady job, or knows she can get one if she loses the one she has (her safety needs are met), her belongingness and esteem needs come to the fore. She is now motivated by her needs to be liked, to be successful, and to feel self-esteem. Finally, if all her other needs are met, her main motive will be to do things which she does well and enjoys; she will thus be satisfying her need to realize her potentialities—she will be *self-actualizing.*

Most of us do not make it to the top of the ladder. In most societies, most of the time, physiological needs are pretty well met (although, even in our affluent society, many people go hungry). So we move up to the safety needs, and these preoccupy many of us. Job security, for example, is of paramount importance to many people. We need to feel safe on the streets of our cities and safe from the arbitrary use of power by our employers, the police, or other government officials. If our safety needs are satisfied, we go on to try to meet our needs for affection, affiliation, and identification—feeling a part of society or a segment of it via our churches, schools, or companies. If we meet these needs, we are free to go up in the need hierarchy to our esteem and self-actualization needs. Of course, the situation is rather more fluid than the one just described. Many people move ahead only to find, as the situation surrounding them changes, that their lower-order needs must be met again. Moreover, people can be trying to satisfy several orders of needs—belongingness and esteem needs, say—at the same time. Since the higher motives can be satisfied only after those lower down have been satisfied, the higher motives often remain unfulfilled. In other words, the goals of these higher motives are not reached, leaving people with feelings of frustration, a topic to which we will now turn.

Frustration and Conflict of Motives

The course of motivation does not always run smoothly. Things happen that prevent us from reaching the goals toward which we are driven or pulled. The term *frustration* refers to the blocking of behavior directed toward a goal. Although there are many ways in which motives can be *frustrated*—that is, prevented from being satisfied—conflict among simultaneously aroused motives is perhaps the most important reason why goals are not reached. If motives are *frustrated*, or blocked, emotional feelings and behavior often result. People who cannot achieve their important goals feel depressed, fearful, anxious, guilty, or angry (page 290). Often they are simply unable to derive ordinary pleasure from living.

A frustration can be schematized by a diagram such as that in Figure 7.13. The box denotes the total environment of a person, and the vertical line represents the thwarting of a goal. In such diagrams, goals are depicted by either plus (+) or minus (−) signs, called *valences*. A plus sign indicates a goal to which a person is attracted; a minus sign indicates a goal which repels—punishment, threat, or something an individual fears or has learned to avoid. The arrow is used to indicate the direction of motivating forces acting on the individual. The psychologist Kurt Lewin devised such diagrams many years ago to help in the visualization of the sources and effects of frustration.

Figure 7.13

Frustration by environmental and personal obstacles. A barrier (vertical line) stands between an individual and the goal (+) that attracts the individual. The barrier may be another person, or object in the environment, or it may be the individual's own lack of ability or skill.

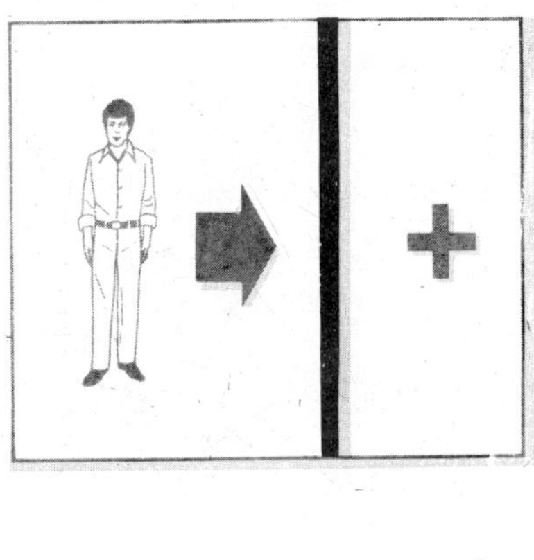

SOURCES OF FRUSTRATION

Generally speaking, the causes of frustration are to be found in: (1) environmental forces that block motive fulfillment, (2) personal inadequacies that make it impossible to reach goals, and (3) conflicts between and among motives.

Environmental Frustration By making it difficult or impossible for a person to attain a goal, environmental obstacles can frustrate the satisfaction of motives. An obstacle may be something physical, such as a locked door or a lack of money. Or it may be people—parents, teachers, or police officers, for example—who prevent us from achieving our goals. The vertical line in Figure 7.13 represents an environmental obstacle that prevents a goal from being reached.

Personal Frustration Unattainable goals can be important sources of frustration. These are largely learned goals that cannot be achieved because they are beyond a person's abilities. For instance, a boy may be taught to aspire to high academic achievement but lack the ability to make better than a mediocre record. He may be motivated to join the school band, play on the football team, be admitted to a certain club, or act the lead in a play and be frustrated because he does not have the necessary talent. Thus people are often frustrated because they aspire to goals—have a *level of aspiration*—beyond their capacity to perform. The vertical line in Figure 7.13 might stand for thwarting of goal-directed behavior by some personal characteristic of the individual.

Conflict-Produced Frustration A major source of frustration is found in *motivational conflict*, in which the expression of one motive interferes with the expression of other motives. In expressing aggression, for example, people are often caught in such a conflict. On the one hand, they would like to give vent to their rage; on the other, they fear the social disapproval which will result if they do. Aggression is thus in conflict with the need for social approval. In some societies, sexual motivation is often in conflict with society's standards of approved sexual behavior. Other common conflicts are between independence and affiliation needs or career aspirations and economic realities. Life is full of conflicts and the frustration arising from them.

TYPES OF CONFLICT

Of the three general sources of frustration described above, the one that often causes the most persistent and deep-seated frustration in many individuals is *motivational conflict.* This kind of frustration can be the most important in determining a person's anxieties, or "hang-ups." On analysis, it seems that this kind of frustration can arise from three major kinds of conflict, which have been called *approach-approach*, *avoidance-avoidance*, and *approach-avoidance.*

Figure 7.14

Approach-approach conflict. The person is attracted toward two incompatible positive goals at the same time.

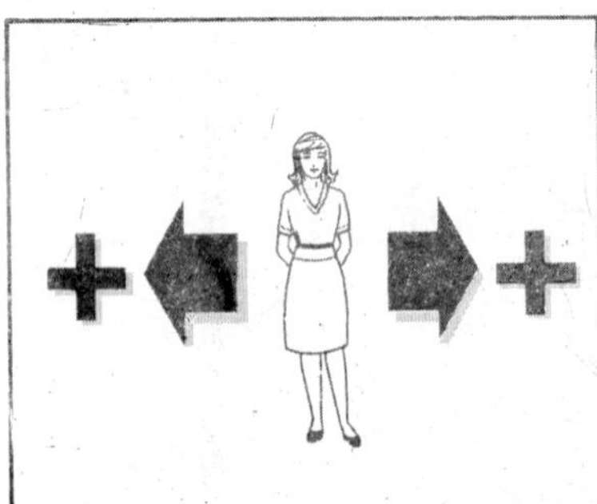

Approach-Approach Conflict As the name implies, an *approach-approach conflict* is a conflict between two positive goals—goals that are equally attractive at the same time (Figure 7.14). For instance, a physiological conflict arises when a person is hungry and sleepy at the same time. In the social context, a conflict may arise when a person wants to go to both a political rally and a swimming party scheduled for the same night. The proverbial donkey was supposed to have starved to death because it stood halfway between two piles of hay and could not choose between them. Actually, few donkeys or people starve to death merely because they are in conflict between two positive goals. Such conflicts are usually resolved either by satisfying first one goal and then the other—for example, eating and then going to bed if a person is both hungry and sleepy—or by choosing one of the goals and giving up the other. Compared with other conflict situations, approach-approach conflicts are usually easy to resolve and generate little emotional behavior.

Avoidance-Avoidance Conflict A second type of conflict, *avoidance-avoidance*, involves two negative goals (Figure 7.15) and is a fairly common experience. A boy must do his arithmetic homework or get a spanking. A student must spend the next 2 days studying for an examination or face the possibility of failure. A woman must work at a job she intensely dislikes or take the chance of losing her income. Such conflicts are capsuled in the saying, "caught between the devil and the deep blue sea." We can all think of things we do not want to but must do or face even less desirable alternatives.

Two kinds of behavior are likely to be conspicuous in avoidance-avoidance conflicts. One is *vacillation of behavior and thought*, meaning that people are inconsistent in what they do and think; they do first one thing and then another. Vacillation occurs because the strength of a goal increases as the person nears it. As one of the negative goals is approached, the person finds it increasingly repellent and consequently retreats or withdraws from it. But when this is done, the person comes closer to the other negative goal and finds it, in turn, unbearably obnoxious. The individual is like a baseball player caught in a rundown between first and second base. First the player runs one way, then the other. As the runner nears second base, being tagged out becomes more likely; but when the player turns back toward first base, the same danger is faced. So back and forth the runner goes, as we all do, in a symbolic sense, when we are caught in an avoidance-avoidance conflict.

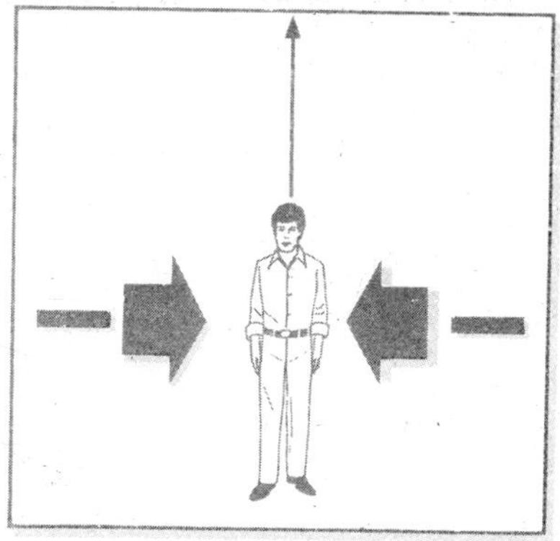

Figure 7.15

Avoidance-avoidance conflict. The individual is caught between two repelling threats, fears, or situations. In addition to the negative goals shown, there are usually barriers in the periphery of the situation that prevent the person from "leaving the field" (vertical arrow) in order to escape from the conflict.

A second important behavioral feature of this kind of conflict is an attempt to leave the conflict situation. Theoretically, a person can escape from an avoidance-avoidance conflict by running away—and people do, indeed, try this. In practice, however, there are often additional negative forces in the periphery of the situation (the "field," as it is called) that prevent them from leaving. For instance, the boy who does not want either to do his arithmetic homework *or* get a spanking might think of running away from home. But the consequences of running away are even worse than his other alternatives, and so he does not do it.

People in avoidance-avoidance conflicts may try a different means of running away: They may rely on their imaginations to free them from the fear and anxiety generated by the conflict. They may spend much of their time daydreaming—conjuring up an imaginary world where there are no conflicts. Or they may re-create in their minds the carefree world of childhood before unpleasant tasks and avoidance-avoidance conflicts existed. This way of leaving the conflict situation is called *regression*. (See Chapter 14, page 590.)

Many intense emotions are generated by avoidance-avoidance conflicts. If the two negative goals are fear-producing and threatening, a person caught between them will experience fear. Or the individual may be angry and resentful at being trapped in a situation where the goals are negative.

Approach-Avoidance Conflict The third type of conflict, *approach-avoidance*, is often the most difficult to resolve because, in this type of conflict, a person is both attracted and repelled by the same goal object (Figure 7.16). Because of the positive valence of the goal, the person approaches it; but as it is approached, the negative valence becomes stronger. If, at some point during the approach to the goal, its repellent

Figure 7.16

Approach-avoidance conflict. The individual is attracted to a positive goal, but this goal also has fear or threat (negative valance) associated with it.

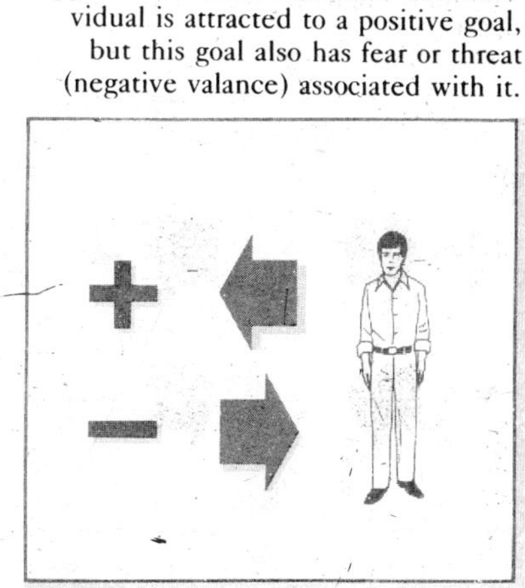

aspects become stronger than its positive aspects, the person will stop before reaching the goal. Because the goal is not reached, the individual is frustrated.

As with avoidance-avoidance conflicts, vacillation is common in approach-avoidance conflicts; people in these conflicts approach the goal until the negative valence becomes too strong, and then they back away from it. Often, however, the negative valence is not repellent enough to stop the approach behavior. In such cases, people reach the goal, but much more slowly and hesitatingly than they would have without the negative valence; and, until the goal is reached, there is frustration. Even after the goal is reached, an individual may feel uneasy because of the negative valence attached to it. Whether a person is frustrated by reaching a goal slowly or by not reaching it at all, emotional reactions such as fear, anger, and resentment commonly accompany approach-avoidance conflicts.

Multiple Approach-Avoidance Conflicts Many of life's major decisions involve *multiple approach-avoidance conflicts*, meaning that several goals with positive and negative valences are involved (Figure 7.17). Suppose a woman is engaged to be married; suppose, further, that the goal of marriage has a positive valence for her because of the stability and security it will provide and because she loves the man she will be marrying. Suppose, on the other hand, that marriage is repellent to her because it will mean giving up an attractive offer of a job in another city. With respect to her career, the woman is attracted to the new job but also repelled by the problems it will create for her marriage. What will she do? In part, the answer depends on the relative strengths of the approach and the avoidance tendencies. After a good deal of vacillation, she might break the engagement if the sum total of the positive career valence minus the negative career valence is greater than that for positive and negative valences associated with marriage. (See Figure 7.17.) Or, if the overall sum of the marriage valences is greater than that of the career ones, she might hesitate for a while, vacillating back and forth, and then get married. Thus, what a person does in a multiple approach-avoidance conflict will depend on the relative strengths of all the positive and negative valences involved.

Figure 7.17

A multiple approach-avoidance conflict. (See text.)

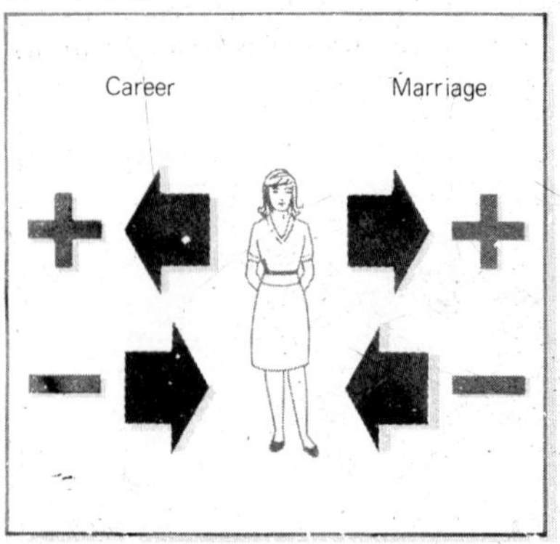

The marriage example above illustrates an important feature of the negative valences in approach-avoidance conflicts. These valences, which are obstacles to reaching a goal, are generally internalized. Such internalized obstacles, or inner negative valences, usually result from the training in social values which a person has received. The woman in the example learned to value the goal of an independent career, and this value later came into conflict with the goal of marriage. More commonly, the internalized obstacles are the social values which make up conscience. We are motivated to do something, but this tendency may be held in check by the internalized values we hold about what is "right" and "wrong." Internal obstacles are generally harder to deal with than are external ones. People may find ways of getting around external, or environmental, obstacles, but they have difficulty escaping from the obstacles within themselves. The emotional reactions generated by approach-avoidance conflicts in which internal obstacles play a part are at the root of many behavioral problems. (See Chapter 14, page 593.) On the more positive side, if we have insight into our own conflicts and the internalized obstacles which create them, we will probably be happier and experience less distress in our lives.

Summary

1. Motivation refers to the driving and pulling forces which result in persistent behavior directed toward particular goals.

2. Motives are inferences from observations of behavior. They are powerful tools for the explanation of behavior, and they allow us to make predictions about future behavior.

3. Theories of motivation include drive theories, incentive theories, the opponent-process theory, and optimal-level theories. Drive theories say that behavior is pushed toward goals by internal states within the person or animal. Incentive theories stress the ability of goals to pull behavior toward them. The opponent-process theory is a hedonistic theory; as such, it says that we are motivated to seek goals which give us good emotional feelings and avoid goals resulting in displeasure. Furthermore, this theory says that many emotional-motivating states are followed by opposing, or opposite, states. Optimal-level theories are hedonistic theories which say that behavior is directed toward seeking an optimal level of arousal or a balanced, homeostatic state in internal physiological processes.

4. Biological motives, such as hunger, thirst, and sex, have their origin in the physiological state of the body. These motives can be aroused by departures from the balanced, or homeostatic, levels of bodily processes, by certain hormones, or by sensory stimuli.

5. Hunger motivation may be initiated when blood levels or rates of use of nutrient substances fall below a certain set point. Among the important nutrients in the blood which may be involved in hunger motivation are glucose, or blood sugar, and free fatty acids. The cessation of hunger motivation is related to nutrient receptors in the stomach which provide "stop-eating" signals and possibly to the release of a hormone called cholecystokinin.

6. Two areas of the hypothalamus—the lateral hypothalamus and the ventromedial hypothalamus—have long been considered brain regions important in the regulation of hunger motivation. The results of early studies were interpreted as indicating that the lateral hypothalamus was a feeding center while the ventromedial hypothalamus was a satiety center. More recent work has led to the view that the classic hypothalamic centers may be more accurately considered parts of the brain systems involved in monitoring the body's fuel supplies, the control of metabolism, and the perception of food-related stimuli.

7. The double-depletion hypothesis of thirst motivation says that loss of water from hypothalamic osmoreceptors—cellular dehydration thirst—and a decrease in the volume of the blood due to water loss—hypovolemia—are two important causes of thirst and drinking.

8. When considered from a biological standpoint, sexual motivation depends, to a large degree, on sex hormones. These hormones organize the brain and body of developing people and lower animals so that they have male or female characteristics. In lower animals (especially female animals), levels of sex hormones circulating in the blood activate sexual behavior. The activation of sexual motivation in humans, however, seems to be controlled more by external stimuli and learning than by sex hormones.

9. Social motives, such as the need for achievement, need for power, and much human aggression, are learned motives that involve other people. They are measured by projective tests, pencil-and-paper questionnaires, or by inferences made from actual behavior in certain situations designed to bring out the expression of these motives.

10. The need for achievement is a motive to accomplish things and to be successful in performing tasks. People high in need for achievement prefer to work on moderately challenging and risky tasks which promise success and on tasks where their performance can be compared with the performance of others. They are persistent in their work, seek more challenging tasks when they are successful, and like to work in situations where they have some control over the outcome. Some women high in the need for achievement may not display the characteristic behaviors just noted. The level of achievement motivation in a society can sometimes be related to its economic growth.

11. Power motivation is a social motive in which the goals are to influence, control, cajole, persuade, lead, charm others, and enhance one's own reputation in the eyes of other people. The behavioral expression of power motivation takes many forms, among them impulsive and aggressive action, participation in competitive sports, the joining of organizations, the collection of possessions, association with people who are not particularly popular, the choice of occupations which have a high impact on others, and the building or disciplining of the body. Among men, it also takes the form of drinking and sexual domination of women. A special form of power motivation, termed Machiavellianism, is characteristic of people who express their power motivation by exploiting others in a deceptive and unscrupulous fashion.

12. While it has a biological basis, human aggression is primarily under the control of social factors. Hostile aggression is behavior which has as its goal the harming of another living being who is motivated to avoid such harm. Among the environmental and social causes of human hostile aggression are intense and arbitrarily imposed frustration, insults, compliance with social pressures, and unpleasant environmental conditions, such as high temperatures, intense noise, and, under some conditions, crowding.

13. Social learning (modeling), classical conditioning, and instrumental conditioning are ways in which the tendency to aggress against others may be learned. Under some conditions, punishment,

catharsis, the presence of nonaggressive models, or the induction of responses that are incompatible with aggression may serve to lessen aggressive behavior.

14. Motives to know and to be effective—including stimulus and exploration needs, effectance motivation, intrinsic motivation, and self-actualization—are powerful human motives.

15. Motives are often blocked, or frustrated. The major sources of this frustration are environmental factors, personal factors, or conflict.

16. Four types of conflict are described in this chapter: (*a*) approach-approach conflict, (*b*) avoidance-avoidance conflict, (*c*) approach-avoidance conflict, and (*d*) multiple approach-avoidance conflict. In each of these conflicts, attainment of a goal is, for a time, hindered; depending on the type of conflict, various emotional and behavioral reactions may occur.

Terms to Know

One way to test your mastery of the material in this chapter is to see whether you know what is meant by the following terms.

Motivation (268)
Unconscious motivation (268)
Drive theories (269)
Motivational cycle (269)
Learned drives (269)
Incentive theories and incentive motivation (270)
Incentives (270, 273)
Hedonistic views of motivation (270)
Opponent-process theory (270)
Optimal-level theories (272, 297)
Homeostasis (272)
Set point (274, 277)
Satiety (274)
Hypothalamus (274)
Cholecystokinin (CCK) (274)
Lateral hypothalamus (LH) (276)
Ventromedial hypothalamus (VMH) (276)
Antidiuretic hormone (ADH) (278)
Osmoreceptors (278)
Cellular-dehydration thirst (278)
Hypovolemia (278)
Renin (278)
Angiotensin II (278)
Double-depletion hypothesis (278)
Estrogens (279)
Estradiol (279)
Androgens (279)
Testosterone (279)
Organizational role of sex hormones (279)
Secondary sexual characteristics (279)
Activational role of sex hormones (279)
Social motives (280)
Projective tests (281)
Thematic Apperception Test (TAT) (282)
Need for achievement (*n* ach) (283)
Power motivation and social power (287)

Machiavellianism (288)
Instrumental aggression (289)
Hostile aggression (289)
Frustration (290, 299)
Frustration-aggression hypothesis (290)
Punishment (292)
Punisher (292)
Catharsis (293)
Effectance motivation (297)
Intrinsic motivation (297)
Extrinsic motivation (297)
Self-actualization (298)
Valences (299)
Level of aspiration (300)
Motivational conflict (300)
Approach-approach conflict (300)
Avoidance-avoidance conflict (301)
Vacillation of behavior and thought (301)
Regression (301)
Approach-avoidance conflict (301)
Multiple approach-avoidance conflict (302)

Suggestions for Further Reading

From biological psychology to clinical psychology to social psychology, motivational ideas abound. The concept of motivation, in all its various aspects, pervades psychology and is so broad that a chapter such as this one can only sample some of the highlights. For more depth and breadth on the topic of motivation, a number of excellent textbooks exist. Among these are Robert C. Beck's *Motivation: Theories and Principles*, 2d ed. (Englewood Cliffs, NJ: Prentice-Hall, 1983); *Human Motivation* by Robert E. Franken (Monterey, CA: Brooks/Cole, 1982); *Motivational Explanations of Behavior* by Katherine B. and Kermit T. Hoyenga (Monterey, CA: Brooks/Cole, 1984); and *Motivation: Biosocial Approaches* by Stephen B. Klein (New York: McGraw-Hill, 1982). For a thorough summary of the psychological and social factors in aggression, Robert A. Baron's *Human Aggression* (New York: Plenum Press, 1977) is highly recommended. Many interesting ideas on social motivation are discussed in David McClelland's *Human Motivation* (Glenview, IL: Scott, Foresman, 1985).

The Nebraska Symposium series on motivation is a good source of specialized information on various motivational and related topics. The paperback volumes of this series cover topics which range from biological motivation to cognitive and social aspects of human motivation.

Emotion and Stress

HEADING OUTLINE

Chapter

How good are you at judging emotions on faces? What emotions are these New Guinea tribesmen showing? Make your judgments and then look at the answers on page 338. How did you do? You have just participated in a tiny cross-cultural experiment, the implications of which are discussed on page 313.

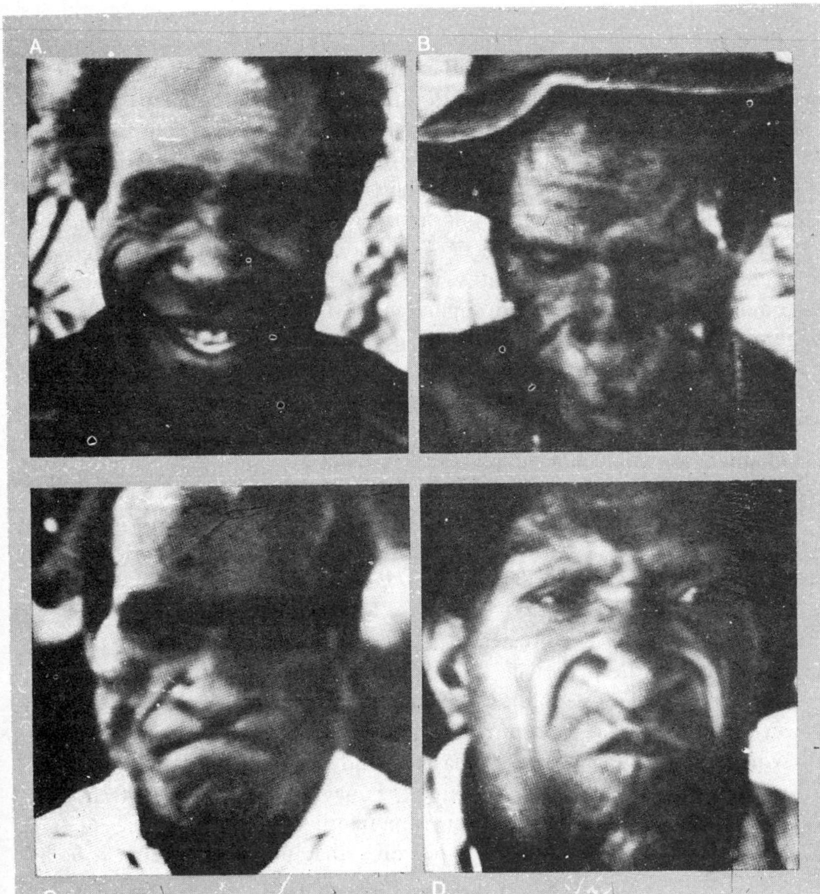

(Copyright © 1972 by Paul Ekman. From Ekman, 1980.)

WE civilized members of western culture like to think of ourselves as rational beings who go about satisfying our motives in an intelligent way. To a certain extent we do just that; but we are also emotional beings—more emotional than we often realize. Indeed, most of the affairs of everyday life are tinged with feeling and emotion. Joy and sorrow, excitement and disappointment, love and fear, attraction and repulsion, hope and dismay—all these and many more are feelings we often experience in the course of a day.

Life would be dreary without such feelings. They add color and spice to living; they are the sauce which adds pleasure and excitement to our lives. We anticipate our parties and dates with pleasure; we remember with a warm glow the satisfaction we got from getting a good grade; and we even recall with amusement the bitter disappointments of childhood. On the other hand, when our emotions are too intense and too easily aroused, they can easily get us into trouble. They can warp our judgment, turn friends into enemies, and make us as miserable as if we were sick with fever.

Just what is an emotion? Surprisingly, this is not an easy question to answer. It almost seems as if there are as many definitions of emotion as there are writers on the subject. Some 92 definitions were listed in one review (Kleinginna & Kleinginna, 1981), and no doubt more could be found. The reason for so many definitions is that emotions have numerous aspects to them; an emotion is many things at once. Attempts to arrive at a comprehensive definition of emotion (Kleinginna & Kleinginna, 1981) should (1) say something about the way we feel when we are emotional; (2) mention the physiological, or bodily, basis of emotional feelings; (3) include the effects of emotion on perception, thinking, and behavior; (4) point out the driving, or motivational, properties of certain emotions such as fear and anger; and (5) refer to the ways in which emotions are expressed in language, facial expressions, and gestures. While we have not given a concise definition of *emotion* because none is generally accepted, this list gives the highlights of what is meant by the term. In a way, the aspects of emotion listed above *are* a definition. In this chapter, we will look at several of these facets of emotion, beginning with the expression and perception of emotions.

Expression and Perception of Emotions

Our emotions have a great impact on others when we express them in ways that can be perceived by others. When we perceive the emotional responses of other people, we respond in appropriate ways, perhaps with an emotional expression of our own. For example, if one of my friends wins a prize and shows joy, I may respond with joy; or, depending upon my perception of the circumstances, I may be jealous. We often seize upon instances of emotional expression in others to form our ideas of their personality. For instance, if I perceive that my boss often expresses hostility toward subordinates but fawns upon his bosses, I know something about his personality and can plan my actions accordingly.

We perceive emotion in others from many sources. The voice is one channel of emotional expression. Screams denote fear or excitement, groans denote pain or unhappiness, sobs denote sorrow, and laughter denotes enjoyment. A tremor or break in the voice may mean great sorrow; a loud, high-pitched, sharp voice usually means anger. Of course, what is actually being said is also an important cue to the emotion being experienced by other people.

While what is said and the way in which it is said are major factors in the perception of emotion, movements of the body are also used as cues in interpreting other people's emotions. Important among these nonverbal bodily cues are facial expressions. (Messages from other bodily movements—"body language," as it is called—are discussed in Application 8.1) In the nineteenth century, Charles Darwin said that there is an innate, or inborn, basis for the facial expression of certain emotions, and now we are reasonably sure he was right. One major study (Izard, 1971) indicating that the facial expressions of what are termed the *primary emotions* (shown in Figure 8.1) are, to a large degree, innate involved

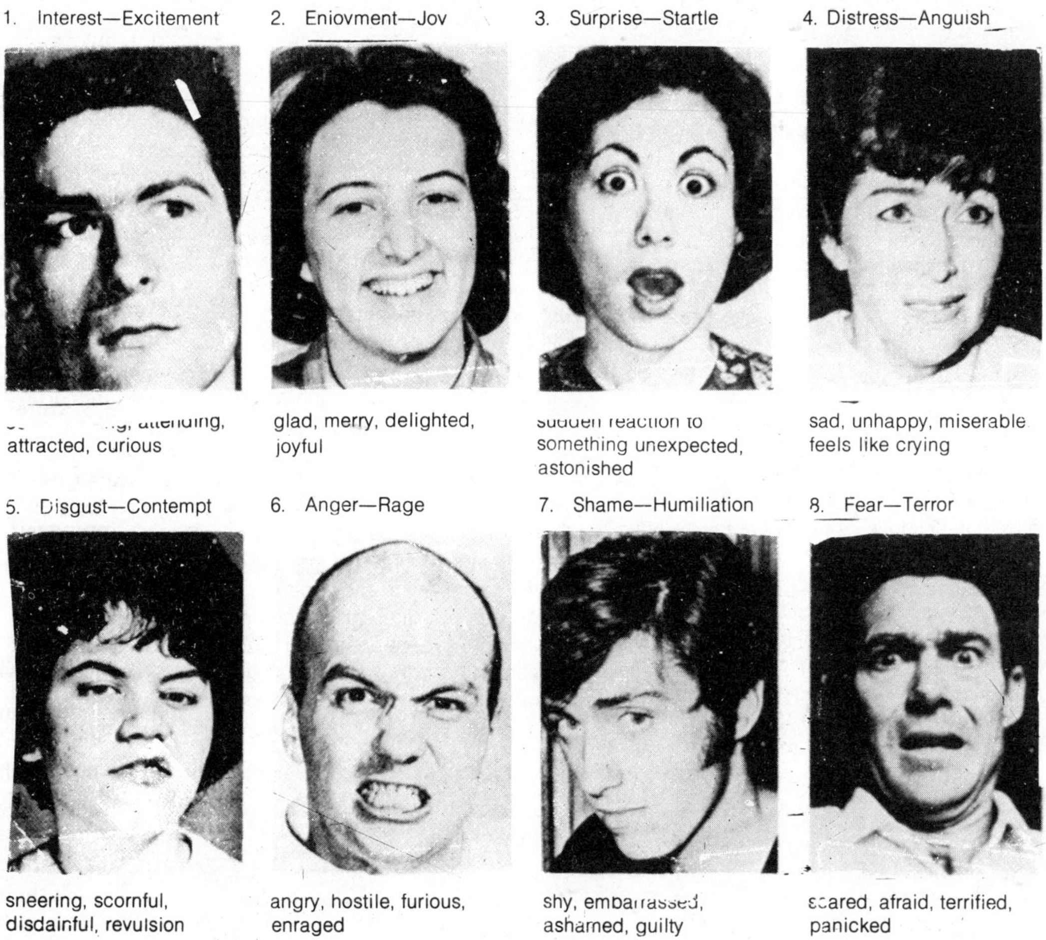

Figure 8.1

These posed photographs, representing a set of primary emotions, are similar to those used in Izard's study of cross-cultural judgments of facial expressions in emotion. (Modified from Izard, 1971.)

Application 8.1

READING EMOTIONS IN "BODY LANGUAGE"

(Charles Gatewood.)

When you say to someone that you "got their message," more often than not you are referring to either a written or a spoken message. But there are other sources of information. When we interact with others, we receive a stream of nonverbal information. Part of this is "body language"—the gestures, eye movements, head movements, shifts in posture, and movements of the arms, hands, and legs that give us messages about the emotions and motives of another person.

Interest in body language is not new. For instance, the founder of psychoanalysis said:

When I set myself the task of bringing to light what human beings keep hidden within them, not by the compelling power of hypnosis, but by observing what they say and what they show, I thought the task was a harder one than it really is. He that has eyes to see and ears to hear may convince himself that no mortal can keep a secret. If the lips are silent, he chatters with his finger tips; betrayal oozes out of him at every pore. And thus the task of making conscious the most hidden recesses of the mind is one which it is quite possible to accomplish. (Freud, 1905/1953, pp. 77–78)

(As it has turned out, however, most of the "analysis" in psychoanalysis is of the meaning of what is said, not what is done.)

Consider nonverbal communication in the detection of lying and deception. A person may try to cover up a shady past in applying for a job, a patient in a mental hospital may try to deceive the staff in order to gain release, or a student may lie about cheating on an examination. Paul Ekman and Wallace Friesen (1969) made a number of observations indicating that as Freud said, the deception may "leak out" in behavior. Although the face is the most expressive part of the body, Ekman and Friesen say that it is also the most easily controlled when trying to deceive someone. Not so easy to control are gestures of the hands, feet, and legs. For instance, while a person is smiling in a relaxed way, the tension and anxiety the individual is feeling may be shown by clutching the knee tightly, digging into the cheek with the fingers, pressing the fingers tightly together, or tense positions of the limbs. Anger may "leak out" in a clenched fist, a tense posture, a rapid drumming of the fingers on the table, or kicks of the foot.

Recent years have seen a number of body-language articles in newspapers and magazines; several popular books have also been written on the subject. Using phrases such as "your body doesn't know how to lie" and "body gestures project your most hidden thoughts," these popular articles and books probably claim too much for nonverbal communication. As Morton Wiener and his colleagues (1972) have pointed out, the accurate interpretation of gestures is very difficult. A major problem is separating gestures which communicate nonverbal emotional or motivational messages from those which do not. Many gestures do not communicate anything. For example, a woman walking down the street may display a hip sway that some men interpret as a "sexy" message when, in fact, it is only a characteristic of the way she walks. Furthermore, Wiener and his colleagues warn against associating any specific movement with one particular meaning: "Similar movements may serve different functions and different movements may serve similar functions." We need reliable gides to tell us what movements, under what circumstances, communicate emotional and motivational messages. Research continues, but the accurate interpretation of body language remains an art in which some people are more skilled than others.

REFERENCES

Bull, P. (1983). *Body movement and interpersonal communication*. New York: Wiley.

Ekman, P., & Friesen, W. V. (1969). Nonverbal leakage and clues to deception. *Psychiatry*, *32*, 88–106.
Freud, S. (1953). Fragment of an analysis of a case of hysteria. In J. Strachey (Ed. and Trans.), *The standard edition of the complete psychological works of Sigmund Freud* (Vol. 7). London: Hogarth Press. (Original work published 1905)
Knapp, M. L. (1978). *Nonverbal communication in human interaction* (2nd ed.) New York: Holt, Rinehart and Winston.
Wiener, M., Devoe, S., Rubinow, S., & Geller, J. (1972). Nonverbal behavior and nonverbal communication. *Psychological Review*, *79*, 185–214.

posed photographs in which actors expressed emotion. People from various European and Asiatic cultures looked at each posed expression and judged whether the photograph showed interest-excitement, enjoyment-joy, or one of the other primary emotions. In general, people from many diverse cultures were able to make accurate judgments of the emotion being expressed. Similar studies have been done with people from New Guinea who have had very little contact with western culture. These individuals were also surprisingly accurate at judging posed facial expressions, even when the photographs were pictures of white people (Ekman, 1982). And, conversely, American college students were able to judge with accuracy the emotions expressed on the faces of New Guinea tribesmen. (See the opening page of this chapter.) Perhaps the similarity in the ways people express the primary emotions with their faces, both among and within cultures, provides a reliable set of perceptual cues for us to use in evaluating the emotional state behind the expression.

Not only facial expressions, but context—the situation in which an emotion occurs—gives us information for judging what emotion is being expressed. Of course, people are most accurate in their judgments when the facial expression and the context are both present and convey complementary information. Since this is typical of everyday life, we usually are good at judging emotions. Sometimes, however, the facial expression and the context give us conflicting cues. In this case, experiments have shown that we tend to rely more on the facial expression or other nonverbal behavior than on the context in making our judgment (Frijda, 1969).

Although we are often quite accurate at deducing emotion from facial expressions and other cues, several complications should be mentioned. One is that learning can modify the expression of even the primary emotions. People may learn to suppress the expression of an emotion. And learning plays a large role in the expression of the more subtle emotions, such as awe or jealousy. People learn to express these emotions in different ways. So unless we know a person's idiosyncrasies, it may be difficult for us to know exactly what emotion that person is experiencing. A second factor complicating the perception of emotions is that a person often expresses several emotions at one time; these blends of emotions are hard to judge.

The Physiology of Emotion

When we are excited, terrified, or enraged, we perceive some of the things happening in our bodies, but we are certainly not aware of all that is happening. Direct observation using recording instruments has given scientists a great deal of information about the bodily events in emotion. Psychophysiologists, who study such events, are able to measure the heart rate, blood pressure, blood flow to various parts of the body, activity of the stomach and gastrointestinal system, levels of various substances such as hormones in the blood, breathing rate and depth, and many other bodily conditions in emotion.

THE AUTONOMIC NERVOUS SYSTEM

From studies by psychophysiologists, we know that many of the bodily changes that occur in emotion are produced by the activity of a part of the nervous system called the *autonomic system* (Figure 8.2). This system is part of the peripheral nervous system, but, as we shall see later, its activity is, to a large extent, under the control of the central nervous system (page 317).

The autonomic system consists of many nerves leading from the brain and spinal cord out to the smooth muscles of the various organs of the body, to the heart, to certain glands, and to the blood vessels serving both the interior and exterior of the body. The autonomic nervous system has two parts. One part, *the sympathetic system*, is active during aroused states and prepares the body for extensive action by increasing the heart rate, raising the blood pressure, increasing blood-sugar (glucose) levels, and raising the levels of certain hormones (see below) in the blood.

Figure 8.2

Schematic drawing of the autonomic nervous system, which consists of nerve fibers and ganglia, or collections of nerve cells (shown as black and colored dots). The nerve fibers run to blood vessels, certain glands, and other internal organs of the body. The two divisions of the autonomic nervous system are the sympathetic system, shown in black, and the parasympathetic system, shown in color. (The peripheral blood vessels, sweat glands, and smooth muscles of the skin are served by the sympathetic system from many levels of the spinal cord; for clarity, only a few of the fibers to these organs are shown here.)

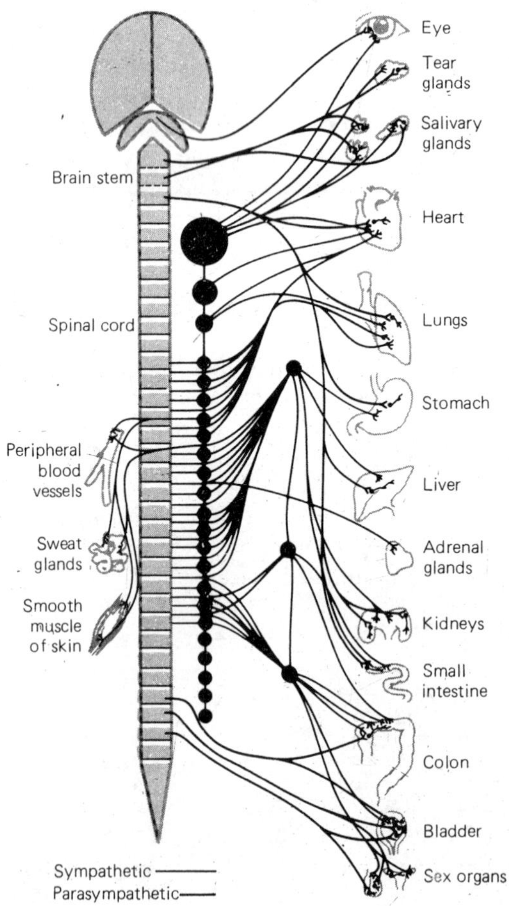

Observations indicate that it is this part of the autonomic nervous system that is active in many strong emotions, especially fear and anger.

In emotion, the sympathetic system causes the discharge of the hormones *epinephrine* (adrenalin) and *norepinephrine* (noradrenalin). (See Chapter 2, page 58, for a definition of the term *hormone*.) Nerve impulses in the sympathetic system which reach the inner part of the adrenal glands, located on top of the kidneys, trigger the secretion of these hormones, which then go into the blood and circulate around the body. Epinephrine affects many structures of the body. In the liver, it helps mobilize glucose (blood sugar) into the blood and thus makes energy available to the brain and muscles. Epinephrine also causes the heart to beat harder. (Surgeons use epinephrine to stimulate heart action when the heart has weakened or stopped.) Thus epinephrine duplicates and strengthens many of the actions of the sympathetic system on various internal organs. In the skeletal muscles, epinephrine helps mobilize sugar resources so that the muscles can use them more rapidly. The major effect of norepinephrine is to constict peripheral blood vessels and so raise blood pressure.

The other part of the autonomic nervous system, called the *parasympathetic system*, tends to be active when we are calm and relaxed. In contrast with the sympathetic system, the parasympathetic system does many things that help to build up and conserve the body's stores of energy. For example, it decreases the heart rate, reduces the blood pressure, and diverts blood to the digestive tract. Thus many of the effects of parasympathetic-system activity are opposite the effects of sympathetic-system activity.

In active, aroused emotional states, sympathetic activity predominates; in calmer states, parasympathetic activity is dominant. But both systems can be active in many emotional states; the pattern of bodily activity characteristic of the emotion is a blend of parasympathetic and sympathetic activity. In anger, for instance, the heart rate increases (a sympathetic effect), as does stomach activity (a parasympathetic effect).

PATTERNS OF BODILY RESPONSE IN EMOTION

Activity occurs in the body's hormonal system and in both the autonomic and somatic parts of the peripheral nervous system during emotional states. We have just described the autonomic nervous system. The *somatic nervous sytem* (Chapter 2, page 52) is that part of the peripheral nervous system which activates the striped muscles of the body—the arm, leg, and breathing muscles, for instance. Thus, the changes in breathing, muscle tension, and posture seen in emotion are brought about by activity of the somatic nervous system.

To illustrate the patterns of bodily changes which accompany emotion, consider the emotions of fear and anger. The bodily changes that accompany these emotions are largely due to increased activity in the sympathetic nervous system; this activity helps the body deal with threatening situations, and therefore the pattern of activity in these emotions is known as the *emergency reaction*, or the *flight-or-fight response*. For example, in both anger and fear, the heart rate usually increases, blood vessels in the muscles dilate so that the body is more prepared for action, blood sugar is mobilized from the liver, the hormones epinephrine and norepinephrine are released from the adrenal gland, the pupils of the eyes dilate, and the peripheral blood vessels of the skin are constricted, thus reducing the possibility of bleeding and making more blood available to the muscles. Muscle tension and breathing rate, which are mediated by the somatic nervous system, tend to increase in both fear and anger.

In contrast to the emergency reaction in fear and anger are the bodily reactions in calm, meditative emotional states. These reactions make up what is called the *relaxation response*. The pattern of bodily responses during relaxation includes decreased activity in both the sympathetic and somatic nervous systems, together with increased parasympathetic activity. As far as sympathetic and somatic activity are concerned, the relaxation response is almost the opposite of the emergency reaction. Other details of the relaxation response are described in Application 8.3 (page 328).

The body changes just described are part of the general emergency and relaxation responses. But are there specific patterns of bodily responses which are different in the various emotions? Early studies (Ax, 1953; Funkenstein, 1955) indicated that fear and anger are characterized by somewhat different response patterns. Now it seems that different bodily-response patterns may be present in a number of emotions and that these patterns are related to the facial expressions of emotions (Ekman et al., 1983).

Actors were instructed to mimic the innate facial expressions (page 311) characteristic of happiness, disgust, surprise, anger, fear, and sadness. Figure 8.3 shows an actor portraying fear. The expressions were held for 10 seconds while measurements were made of heart rate, skin temperature of the hands (a measure of blood flow in the periphery of the body), skin resistance, and forearm muscle tension. Heart-rate and skin-temperature response patterns were found to be different for several of the emotional facial expressions. Figure 8.4 shows the differences. For example, heart rate decreased during the facial expressions of happiness, disgust, and surprise but increased during angry, fearful, or sad expressions. Skin temperature differentiated between anger, on the one hand, and fear and sadness on the other.

In another part of the experiment, the actors imagined, or relived, a past experience in which they felt one of the six emotions. Care was take to ensure that only one emotion was felt at a time and that this emotion was relatively intense. Based on skin-resistance

Figure 8.3

Frames from a video tape of an actor's performance of the fear-expression instructions: (*a*) "Raise your brows and pull them together;" (*b*) "now raise your upper eyelids;" (*c*) "now stretch your lips horizontally, back toward your ears." (From Ekman, Levenson, & Friesen, 1983.)

changes in this condition, sadness could be distinguished from the other negative emotions—disgust, anger, and fear.

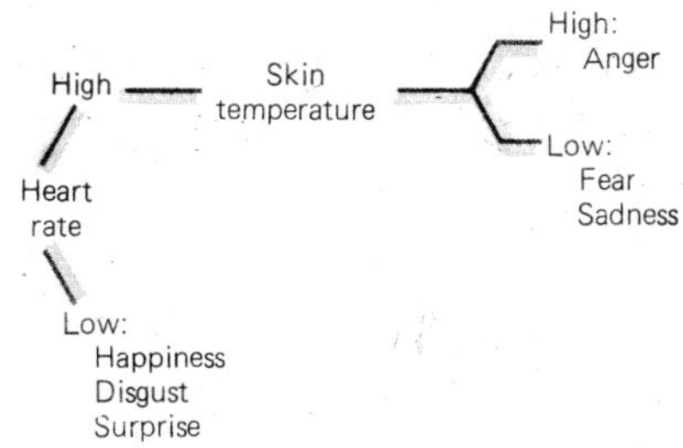

Figure 8.4

The pattern of autonomic nervous system changes for six facial expressions made acccording to instructions. See text. (From Ekman, Levenson, & Friesen, 1983.)

The experiments just described indicate that:

1. Specific emotions can result in specific bodily changes.
2. Facial-muscle movements are closely related to the body's internal adaptive response in emotion.

Thus, the outward and inward manifestations of emotional states go hand in hand. Studies which show fine-grained differences in the bodily patterns of emotion are important for several of the theories of emotion to be discussed later (page 329).

THE BRAIN AND EMOTION

The brain is involved in the perception and evaluation of situations that give rise to emotion. If a situation results in an emotional state, the brain controls the somatic and autonomic patterns of activity charcteristic of the emotion; in other words, it controls the physiological expression of the emotion. Of course, the brain is also involved in directing the behavior driven by the emotional state and is necessary for the emotional feelings we have.

A number of structures in the core of the brain are directly involved in regulating and coordinating the activity patterns characteristic of the stronger emotions, especially fear, anger, and pleasure. These core parts of the brain include the hypothalamus (Chapter 2, page 57) and a complex group of structures known as the *limbic system*. The term *limbic* comes from the Latin word meaning "border." The structures of this system form a ring, or border, around the brain stem as it enters the forebrain. (The limbic system is discussed in detail in Chapter 2, page 62.) Experimenters have found that damage to some of the structures of the limbic system produces great changes in the emotional behavior of animals, making tame animals wild or wild animals tame. Stimulation of certain parts of the limbic system and hypothalamus produces behavioral patterns very much like those in naturally occurring emotions. In addition, electrical stimulation of portions of the limbic system and hypothalamus, as well as other brain regions, is rewarding to animals and pleasurable for human beings (Heath, 1964).

The aroused state that is part of many emotions (see below) is due in part to increased activation of brain cells in the cerebral cortex (Chapter 2, page 58), limbic system, and hypothalamus. Activity of cells in these brain areas is itself directly or indirectly influenced by nerve fibers which fan out from a core region of the brain—the reticular formation (Chapter 2, page 53)—to reach almost all the brain areas involved in regulating emotion. Since the activating fibers from the reticular formation must go upward, or ascend, to reach the higher brain areas involved in emotion, the activating portion of the reticular formation is called the *ascending reticular activating system (ARAS)*. Next time you are keyed up and unable to relax, blame it on your ARAS. In addition to the tinge of arousal it provides for emotional states, the ARAS is fundamentally involved in keeping us awake, alert, and conscious. (See Inquiry 2.2, page 55.)

AROUSAL

Many emotions have an arousal component. When we are emotional, we often feel excited. Some theorists have argued that all emotion is just the degree to which a person or animal is stirred up. Although not all students of emotion agree with this idea, the degree of arousal is an important part of emotionality. For instance, high levels of arousal are present in anger, fear, and joy, while low levels may accompany sadness and depression.

Indicators of Arousal The *electroencephalogram (EEG)* tells us something about the state of arousal. (See Chapter 2, page 55.) The EEG is a record

Application 8.2

LIE DETECTORS

Whodunit? Who stole the $2,000 which old Aunt Agatha kept in a red envelope in her hatbox? Most likely it was one of her visiting relatives, but which one? The police decide to give lie-detector tests to the relatives.

Lie detectors are also called *polygraphs*. The word *polygraph* means "many writer"; these devices have this name because they make simultaneous records of several bodily reactions thought to be indicative of emotional arousal. Most polygraphs (see figure) measure changes in blood pressure, heart rate, breathing rate and depth, and the resistance of the skin to the flow of a small electric current (the *galvanic skin response*, or *GSR*). A polygraph record of breathing (top line), the GSR (middle line), blood pressure (the waves on the bottom line), and heart rate (the sharp, vertical spikes on the bottom line) is shown in the right panel of the figure.

The police in the case of Aunt Agatha use the services of a trained polygraph examiner who follows standardized lie-detection procedures. The examiner obtains each suspect's consent for the test, says that the aim of the test is to help the suspects establish their truthfulness, and describes the polygraph as a scientific instrument for the measurement of various bodily responses. The examiner then obtains biographical information from each suspect and goes over the dozen or so yes/no questions that will be asked about the theft. Next, the suspect is hooked up to the polygraph and asked the relevant yes/no questions. After the dozen or so questions have been asked the first time, the suspect is given a demonstration of the "accuracy" of the polygraph: The suspect is given several cards to choose from and the examiner tries to discover which card was chosen by evaluating the suspect's polygraphic responses to questions about the cards; generally, the situation is "rigged" so that the examiner knows ahead of time which card was chosen. After this "demonstration," the relevant questions are again asked several times.

The examiner may use any one of several questioning techniques. A common one is the control-question technique. In this technique, neutral questions, general questions designed to evoke an emotional response from almost everyone, and specific questions designed to evoke responses only from those with guilty knowledge are used. The rationale of the control-question technique is that an innocent person who knows nothing of the details of the crime should give a larger response to the control questions than to the specific questions, while the suspect with guilty knowledge should, while also responding to the general questions, give a large response to the specific questions. Thus, it is the pattern of polygraphic responses on the control and specific questions that is supposed to give a clue to lying or truthfulness. For example, in investigating the case of Aunt Agatha, a general question might be "Did you ever steal anything before you were 18?" Since most of us have, a polygraphic response would be expected. A specific question might be "What happened to the money in the red envelope?" While the difference between the responses to the general and specific questions can be quantified, interpretation of the polygraphic record remains somewhat of an art, requiring intuition and judgment on the part of the examiner.

Just how good is the polygraph at detecting liars? Controversy abounds, especially between those psychologists who are skeptical about the accuracy, or validity, of lie-detector tests and those whose business it is to give such tests. A number of studies indicate that properly administered tests correctly identify some 70 to 90 percent of deceivers. However, a very important problem is the large number of false positives—truthful people who are judged by the polygraph to be lying. Studies differ, but Benjamin Kleinmuntz and Julian Szucko estimate that false-positive judgments "may label more than 50% of the innocent subjects as guilty" (1984, p. 774). The implications of this figure are enormous and can be a life-or-death matter in murder trials. Or consider the injustice done to people who are false positives in a screening test for employment. They may not get the job for which they are qualified. The false-positive problem, and thus the questionable validity of lie-detector tests, has been recognized by the courts. The results of lie-detector tests cannot be used at all in 24 states; most of the others allow lie-detector evidence only if both the prosecution and defense want the polygraphic evidence used in a particular case.

Other important problems concern the physiological basis of the lie-detector test. Not everyone responds to lying with arousal, and the pattern of responses differs from person to person. Some respond with breathing changes, others with heart-rate changes, and so on. The examiner may have a difficult time with this. Furthermore, it is possible to "beat the machine." If a suspect engages in some silent ongoing mental activity—such as counting backward by 3s—while being examined, responses can be blunted.

With all its problems, the polygraph technique is perhaps best viewed as just another tool the police can use to narrow their list of suspects for more intense investigation by other means. Perhaps this is what the police did in the case of Aunt Agatha and the red envelope.

In addition to their use in trying to discover guilt or innocence, lie detectors are sometimes used (some would say *mis*used) for other purposes. The police may sometimes use them in an attempt to force confessions from suspects. Businesses and government agencies use lie-detector tests to screen applicants for employment. In this case, the polygraph examination may not be used so much to assess the truthfulness of responses as to provide an occasion to trap applicants into making admissions of past activities which might cast doubt on their suitability for employment.

The voice-stress analyzer is another device said to be capable of detecting lies. It is claimed that certain physical characteristics of speech sounds can be detected and analyzed by such a device, which may be attached to a telephone or hidden in a desk drawer, to indicate when a person is lying. The evidence to date does not substantiate the claim that such gadgets work.

REFERENCES

Kleinmuntz, B., & Szucko, J. J. (1984). Lie detection in ancient and modern times: A call for contemporary scientific study. *American Psychologist*, *39*, 766–776.

Lykken, D. T. (1981). *A tremor in the blood: Uses and abuses of the lie detector*. New York: McGraw-Hill.

Raskin, D. C. (1980, Fall). The truth about lie detectors. *Wharton Magazine*, *5*, pp. 28–33.

Reid, J. E., & Inbau, F. E. (1977). *Truth and deception: The polygraph (lie-detector) technique (2nd ed.).* Baltimore: Williams & Wilkins.

Waid, W. M., & Orne, M. T. (1982). The physiological detection of deception. *American Scientist*, *70*, 402–409.

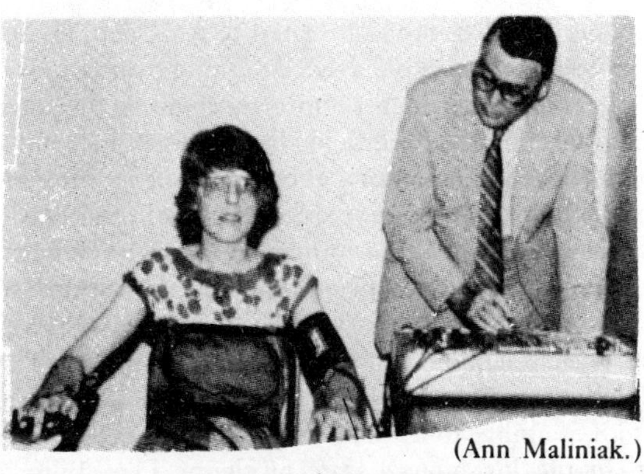

(Ann Maliniak.)

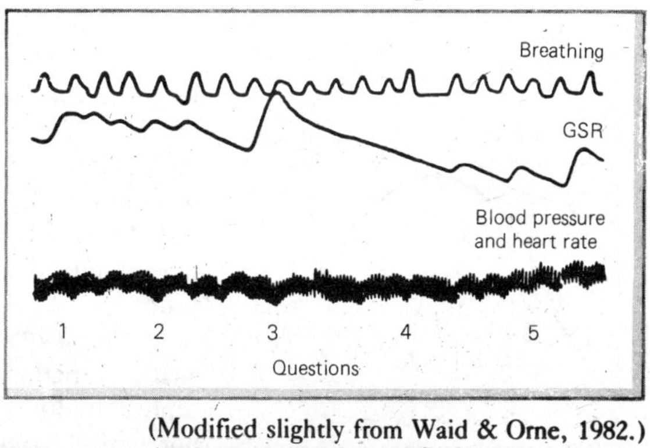

(Modified slightly from Waid & Orne, 1982.)

of the changing electrical activity of millions of nerve cells, all functioning at the same time in the brain. With suitable amplification, this electrical activity can be recorded by electrodes attached to the head. The electrical activity of the cerebral cortex of the brain waxes and wanes spontaneously to give a wavelike record, and this is why the EEG is popularly said to "record brain waves." The waves of the EEG are really very small voltage changes, in the range of several millionths of a volt. The number of alternations of voltage, or the frequency of the electrical changes, varies from 1 or 2 alternations per second in deep sleep to 50 or more in highly aroused states. The waves of the EEG also vary in amplitude, or height. Brain waves are thus characterized by their frequency and amplitude. When a person is aroused or excited, the EEG consists of high-frequency, low-voltage (amplitude) waves. As an individual becomes more relaxed, the frequency of the EEG tends to decrease, while the voltage (amplitude) of the waves tends to increase. (See Inquiry 2.2, page 55, for a more detailed description of the EEG and arousal.)

To make somewhat finer distinctions among degrees of arousal, a number of other measures might be used: heart rate, blood pressure, breathing rate and depth, pupil size, and skin conductance, for instance. (Four of these arousal indicators—heart rate, blood pressure, breathing patterns, and skin conductance—are used in lie-detector tests. See Application 8.2.) Skin conductance is a measure that might not be familiar to you. If a small amount of electricity—so small that it cannot be felt—is passed across an area of the skin (usually the palm of the hand), the resistance to the flow decreases (conductance increases) as a person becomes more aroused and alert. This is called the *galvanic skin response (GSR).*

Another indication of arousal in both humans and lower animals is the *orienting reaction*—an organism's orientation to a new stimulus or to a stimulus change. The *orientation* consists of tensing muscles and changing the position of the body and the head in order to maximize the effectiveness of the stimulus. The exact nature of the orienting reaction depends on the stimulus, the species of the organism, its age, its present state of arousal, and other factors. A cat seeing the slightest movement may make itself ready to pounce. A dog hearing the faintest sound of another dog may perk up its ears, stand at attention, and get ready to defend its territory. Infants turn their heads and eyes toward novel stimuli, such as new toys or strange faces.

Figure 8.5

The inverted U-shaped relationship between efficiency of functioning (cue discrimination) and level of emotional arousal. Up to a certain level of arousal, the ability to respond correctly to cues—that is, to perform well—improves. Beyond that level, further arousal increasingly hampers performance. This relationship is usually found in all but the simplest tasks. (Modified from Hebb, 1955.)

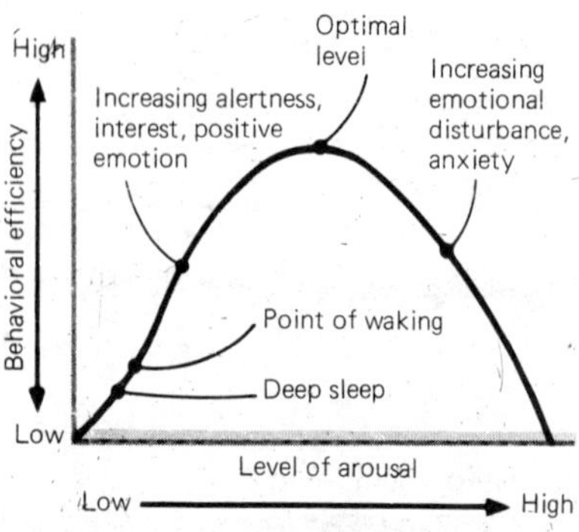

Arousal and Performance How is arousal related to what people do, or performance? Since arousal energizes behavior, you might think that the more aroused people are, the better their performance will be on all sorts of tasks. This is true up to a point. The infield chatter in baseball and the back-slapping in football and basketball probably help to bolster the level of arousal and so keep the athletes on their toes. However, in complicated tasks, very intense arousal may impair performance. This occurs when a person must discriminate among cues or do appropriate things at different times.

Formally stated, the principle is that performance is an inverted U-shaped function of level of arousal when cues must be discriminated. As represented in Figure 8.5, ability to respond correctly to cues is low, but not entirely lacking, in the low-arousal state of sleep. The ability increases with rising arousal up to an optimal level. (A related point is discussed in Chapter 7, page 272.) Thereafter, as the person becomes more intensely disturbed (aroused), performance declines. In other words, highly aroused or anxious people are not so likely to perform well on complex tasks as are people with a lower level of arousal that is more nearly optimum. You may have been unfortunate enough to have found this out in exams or sports. A little arousal is a good thing because it keeps you working and alert, but too much arousal results in disorganization of thought and performance—you "clutch."

Stress

The term stress has many definitions (Lazarus & Folkman, 1984). We will define *stress* as an internal state which can be caused by physical demands on the body (disease conditions, exercise, extremes of temperature, and the like) or by environmental and social stituations which are evaluated as potentially harmful, uncontrollable, or exceeding our resources for coping. The physical, environmental, and social causes of the stress state are termed *stressors*. Once induced by stresssors, the internal stress state can then lead to various responses. On the one hand, it can result in a number of physical, bodily responses. On the other hand, psychological responses such as anxiety, hopelessness, depression, irritability, and a general feeling of not being able to cope with the world can result from the stress state.

Stress is a big problem in our society (Allen, 1983). Some 75 percent of bodily disease is said to be stress-related. For example, stress is often a factor in heart disease (see Application 14.1, page 572, on Type A and Type B behaviors) and cancer, two of the leading causes of death. Furthermore, stress-related diseases cost American industry billions of dollars a year; several billion tranquilizer pills are prescribed in the United States each year; and, although it cannot be quantified, stress seems to be involved in much of our unhappiness, irritability, and dissatisfaction.

STRESSORS

Almost any change in the environment—even a pleasant change, such as a vacation—demands some coping; and a little stress is useful in helping us adapt. But beyond some point, "stress" becomes "*dis*tress." What acts to produce distress varies greatly from person to person, but some events seem to be stressors for many of us. Chief among these are injuries or infections of the body; annoying or dangerous events in our environments; major changes, or transitions, in life which force us to cope in new ways; and anticipated or actual threats to our self-esteem.

Changes in one's life are important stressors (T. H. Holmes, 1984). A sampling of the life changes—not all of them unpleasant—that can act as stressors is given in Table 8.1. A rough indication of stressor strength, based on the judgments of adults (Holmes & Rahe, 1967), adolescents (Ruch & Holmes, 1971), and elderly people (Muhlenkamp et al., 1975) is given under the headings "Life-Change Units." Note how the importance of certain stressors varies throughout life; compare, for example, the elderly with adults on the bottom life events of Table 8.1. By adding the life-change units during a year, a life-crisis score can be obtained for that year. As the severity of life crises increases from *mild* (150–199 life-change

Table 8.1

The Social Readjustment Rating Scale, with life-change units for adults, adolescents, and elderly people

LIFE EVENT	LIFE-CHANGE UNITS (ADULTS)	LIFE-CHANGE UNITS (ADOLESCENTS)	LIFE-CHANGE UNITS (ELDERLY PEOPLE)
Death of spouse	100	69	73
Divorce	73	60	72
Marital separation from mate	65	55	63
Detention in jail or other institution	63	50	73
Death of a close family member	63	54	60
Major personal injury or illness	53	50	65
Marriage	50	50	50
Being fired at work	47	50	62
Marital reconciliation with mate	45	47	35
Retirement from work	45	46	33
Major change in the health or behavior of a family member	44	44	59
Pregnancy	40	45	47
Sex difficulties	39	51	36
Gaining a new family member (e.g., through birth, adoption, oldster moving in, etc.)	39	43	36
Major business readjustment (e.g., merger, reorganization, bankruptcy, etc.)	39	44	61
Major change in financial state (e.g., a lot worse off or a lot better off than usual)	38	44	43
Death of a close friend	37	46	52
Changing to different line of work	36	38	42
Major change in the number of arguments with spouse (e.g., either a lot more or a lot less than usual regarding child-rearing, personal habits, etc.	35	41	45
Taking out a mortgage or loan for a major purchase (e.g., for a home, business, etc.)	31	41	61
Foreclosure on a mortgage or loan	30	36	61
Major change in responsibilities at work (e.g., promotion, demotion, lateral transfer)	29	38	40
Son or daughter leaving home (e.g., marriage, attending college, etc.)	29	34	43
Trouble with in-laws.	29	36	31

units), to *moderate* (200–299 life-change units), to *major* (over 300 life-change units), susceptibility to illness rises progressively (Holmes & Masuda, 1974).

Not only changes, but many ongoing situations are stressors. These include physical stressors such as injury, infection, exercise, noise, and climate. In addition, the hassles of everyday life centering around work, family, social activities, health, and finances are important stressors which have been found to be related to bodily illness (DeLongis et al., 1982). Frustration and the conflicts which cause frustration (Chapter 7, page 299) are also major stressors. Others are listed in Figure 8.6.

LIFE EVENT	LIFE-CHANGE UNITS (ADULTS)	LIFE-CHANGE UNITS (ADOLESCENTS)	LIFE-CHANGE UNITS (ELDERLY PEOPLE)
Outstanding personal achievement	28	31	34
Wife beginning or ceasing work outside the home	26	32	35
Beginning or ceasing formal schooling	26	34	34
Major change in living conditions (e.g., building a new home, remodeling, deterioration of home or neighborhood)	25	35	48
Revision of personal habits (dress, manners, associations, etc.)	24	26	39
Trouble with the boss.	23	26	28
Major change in working hours or conditions	20	30	27
Change in residence	20	28	39
Change to a new school	20	26	34
Major change in usual type and/or amount of recreation	19	26	35
Major change in church activities (e.g., a lot more or a lot less than usual)	19	21	40
Major change in social activities (e.g., clubs, dancing, movies, visiting, etc.)	18	28	35
Taking out a mortgage or loan for a lesser purchase (e.g., for a car, TV, freezer, etc.)	17	28	52
Major change in sleeping habits (a lot more or a lot less sleep, or change in part of day when asleep)	16	18	36
Major change in number of family get-togethers (e.g., a lot more or a lot less than usual)	15	22	37
Major change in eating habits (a lot more or a lot less food intake, or very different meal hours or surroundings)	15	18	45
Vacation	13	19	30
Christmas	12	16	34
Minor violations of the law (e.g., traffic tickets, jaywalking, disturbing the peace, etc.)	11	12	47

(*Sources*: Holmes & Rahe, 1967; Ruch & Holmes, 1971; Muhlenkamp et al., 1975.)

STRESS CYCLES

Stress has a number of immediate effects and, if the stressors are maintained, long-term behavioral, physiological, emotional, and cognitive (thinking) effects occur (Figure 8.6). If these effects hinder adaptation to the environment or create discomfort and distress, they themselves become stressors and, as shown by the dotted arrow in Figure 8.6 (left), tend to perpetuate a cycle of distress. On the other hand, many people have developed ways of coping with stressors so that they are able to respond adaptively. This is the wellness cycle of Figure 8.6 (right). Teaching people adaptive ways of handling stress so as to promote the wellness

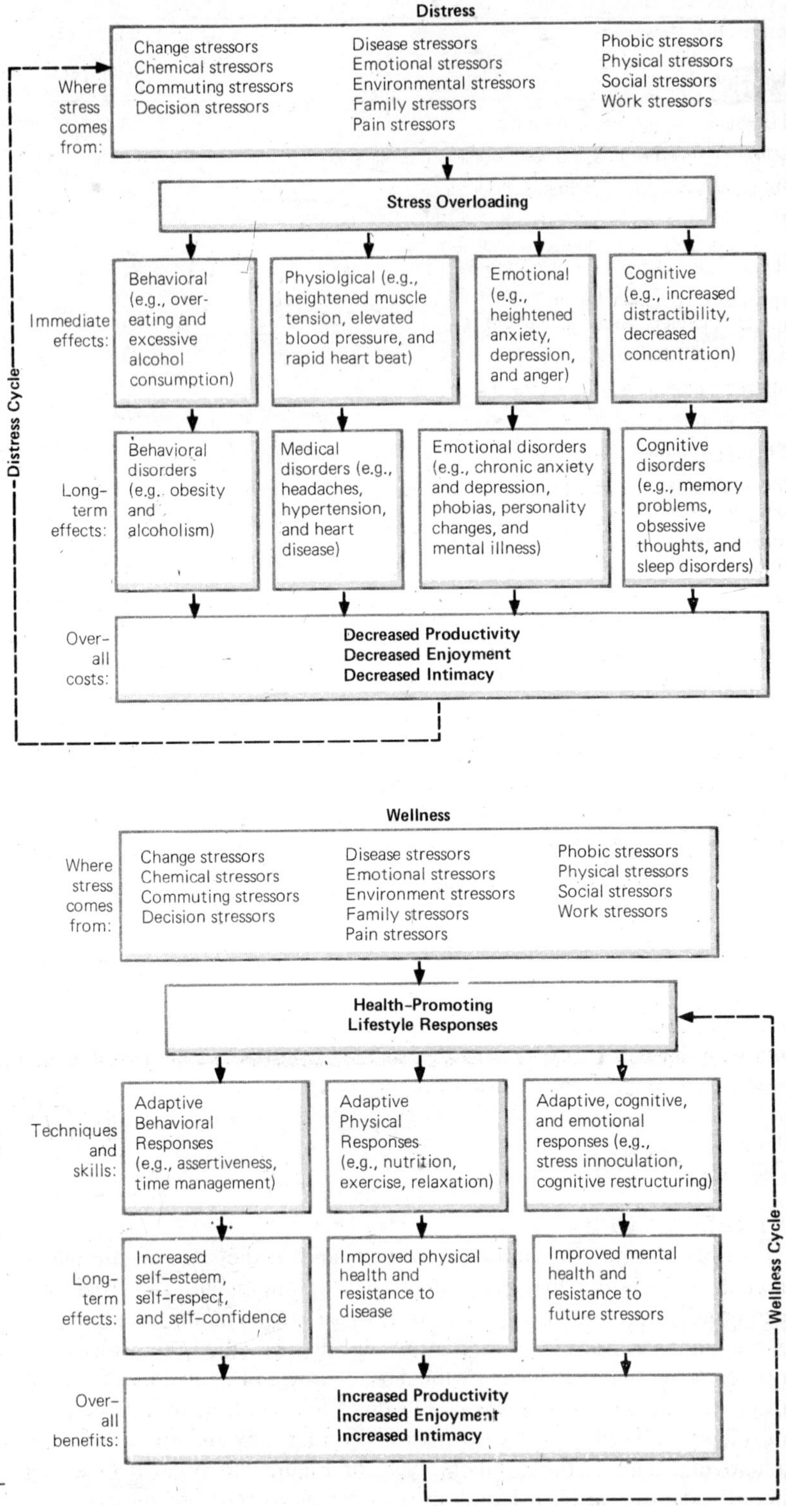

The distress and wellness cycles. See text. (From Charlesworth & Nathan, 1984.)

Figure 8.6

cycle is an important part of the newly emerging field of behavioral medicine (Chapter 16, page 708).

WHAT STRESSORS DO TO THE BODY

Hans Selye (1956, 1976) termed the body's response to stressors the *general adaptation syndrome.* The general adaptation syndrome consists of three stages: (1) the alarm reaction, (2) the stage of resistance, and (3) the stage of exhaustion (Figure 8.7).

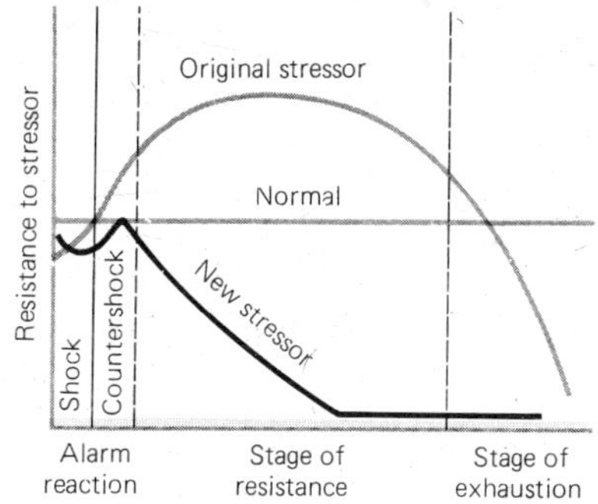

Figure 8.7

The general adaptation syndrome. Responses to stressors are divided into three stages: the alarm reaction, resistance, and exhaustion. The black line represents resistance to a continuous, original stressor; the colored line represents resistance to a new stressor imposed in different stages of the adaptation syndrome. (After Selye, 1950.)

Alarm Reaction The *alarm reaction* is essentially the emergency response of the body already described in an earlier section of this chapter (page 315). In this stage, prompt responses of the body, many of them mediated by the sympathetic nervous system (page 314), prepare us to cope with the stressor here and now.

Stage of Resistance If the stressor continues to be present, the *stage of resistance* begins, wherein the body resists the effects of the continuous stressor (Figure 8.7). However, resistance to new stressors is impaired during this stage (Figure 8.7). During this stage, certain hormonal responses of the body are an important line of defense in resisting the effects of stressors. (See Chapter 2, page 58, for a general description of hormones.) Especially important among these hormonal responses is increased activity in what is known as the *adrenocorticotropic (ACTH) axis* (Figure 8.8).

Adrenocorticotropic hormone (ACTH) is secreted into the bloodstream by certain cells in the pituitary gland (Chapter 2, page 57). The rate of ACTH secretion is, in part, controlled by another hormone-like chemical substance—*corticotropin-releasing factor (CRF)* (Figure 8.8)—that is made by certain cells in the brain structure known as the *hypothalamus* (Chapter 2, page 57; Chapter 7, page 276). The corticotropin-releasing factor flows from the hypothalamus to the pituitary gland through a specialized system of blood vessels. Stressors are able to activate the nerve cells of the hypothalamus so that more corticotropin-releasing factor is sent to the pituitary gland, thus increasing secretion of ACTH into the blood. In this way, brain activity triggered by stressors influences hormone release; thus this is a major link between environmental events—stressors—and the bodily state of stress.

ACTH stimulates cells in the *outer* layers, or cortex, of the adrenal glands so that corticoid hormones such as *cortisol* are secreted into the bloodstream (Figure 8.8). [Remember that it is the *inner* tissue of the adrenal glands that is activated as part of the emergency response (page 315) to give us an activating "shot of adrenalin."]

Cortisol and other, similar hormones have many actions which allow the body to deal adaptively with stressors for long periods of time during the stage of resistance. But maintained high levels of these hormones can be harmful. For instance, cortisol promotes the formation of glucose (blood sugar)—a fuel needed for nerve and muscle activity—by breaking down fats and proteins. In the short run, this is adaptive; the body has more fuel available. In the long run, though, the increased use of protein to make fuel may be serious because proteins are needed in the manufacture of new cells. For example, white blood cells, which are crucial for fighting infection, have a short lifetime and must be continuously re-

Figure 8.8

The adrenocorticotropic axis. (Based on Vander, Sherman, & Luciano, 1975.)

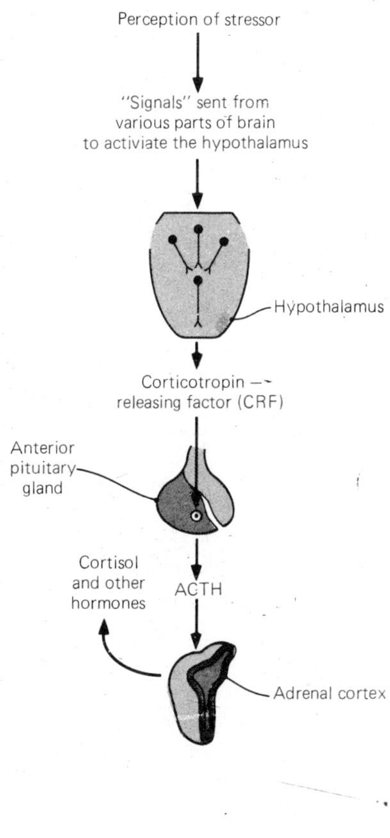

placed. If the proteins needed to make new white blood cells are in short supply because they are being used to make fuel, fewer white blood cells can be produced and the body will be less able to fight infection. Add to this the inhibitory action of cortisol on the formation of the infection-fighting proteins called antibodies, together with shrinkage of the tissues which manufacture white blood cells, and it is clear that high levels of cortisol can, in the long run, seriously impair the body's defenses to infection. Prolonged elevation of cortisol levels can also have other harmful effects, such as raising blood pressure. In addition to cortisol, other hormones which, in excess, may have their own harmful actions are involved in the body's response to stressors during the stage of resistance. But what has been said about cortisol should be enough to give the idea of the role played by hormones in stress.

Stage of Exhaustion The final stage of the general adaptation syndrome is the *stage of exhaustion*. In this stage, the body's capacity to respond to both continuous and new stressors has been seriously compromised (Figure 8.7). For instance, due to the actions of cortisol described above, a person may no longer be able to ward off infection and may become sick and perhaps die. Or, because of other stressor-induced hormonal effects, stomach ulcers, diabetes, skin disorders, asthma, high blood pressure, increased susceptibility to cancer (Bammer & Newberry, 1983), or a host of other diseases may occur at this stage or late in the stage of resistance (Selye, 1976; Allen, 1983).

The term *psychosomatic* ("mind-body") *disorders* is used when perceived stressors—mental events—increase the susceptibility of the body to disease. Of course not all instances of the diseases listed above are psychosomatic, but many are, and thus the control of stress has become a major problem for medicine. Treatment of psychosomatic disorders involves medical help for the physical problems and, at the same time, attention to the psychological factors producing the stress.

COPING WITH STRESSORS

As we have seen, stress can have serious consequences. Fortunately, there are methods for coping with stress. Several of these methods are described in Application 8.3. Other methods are described in Chapters 4 and 16. Among them are relaxation training (Chapter 16, page 708), hypnosis (Chapter 16, page 708), biofeedback (Chapter 16, page 710), and systematic desensitization (Application 4.1, page 148; Chapter 16, page 701). The wellness cycle of Figure 8.6 (right), page 324, also shows some adaptive ways of coping with stressors.

The impact of stressors can sometimes be reduced if a person has control over the stressor (Cohen, 1980). The effect of control has also been shown in animal studies (Weiss, 1972).

> Rats were subjected to stress produced by an electric shock to the tail. In the experiment illustrated in Figure 8.9, groups of rats, matched for weight and age, were studied in sets of three. The rat in the left-hand chamber could shut off the programmed shocks by turning a wheel. The rat in the middle chamber was yoked to the left-hand rat, meaning that it received shocks of the same intensity and duration as the left-hand rat. However—and this is the impor-

tant point—the yoked animal could do nothing about the shocks. The rat in the right-hand box got no shocks at all but was hooked up like the others.

After this procedure had gone on for some time, stomach ulceration in the rats subjected to the three conditions was studied. The rats which could do something about the shocks showed much less ulceration than did the helpless, yoked animals. The non-shocked animals showed little or no ulceration.

In an experiment showing the effects of uncontrollable stressors on ulcers, three sets of rats, matched for weight and age, were placed in individual soundproof compartments. The rat on the left could terminate the programmed shock by turning the wheel; moreover, turning the wheel between shocks postponed the next shock. This rat was in the escape-avoidance condition. The middle rat was electrically wired to the rat in the left-hand box. When the rat in the left-hand box received a shock, this yoked rat received an identical shock, but its actions could not affect the shock duration or sequence in any way. The electrodes on the rat in the right-hand box were not connected, and it received no shocks at all. Rats in the uncontrollable (middle) condition had much more stomach ulceration than rats in the escape-avoidance condition, even though the intensity and number of shocks were the same. The non-shocked rats, of course, had almost no ulcers. (Weiss, 1972, *Scientific American*.)

Figure 8.9

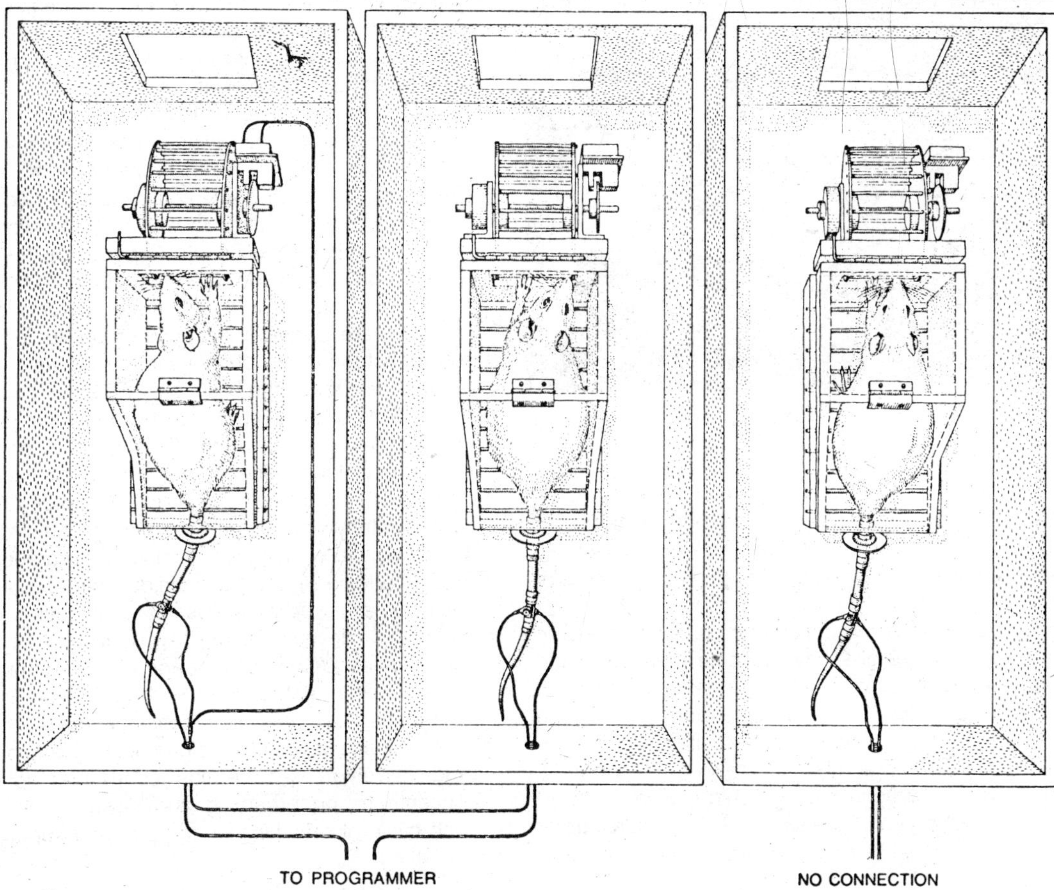

Application 8.3

"TAKE IT EASY," THE YOGI SAYS—AND NOW PSYCHOLOGISTS AND PHYSIOLOGISTS ARE LISTENING

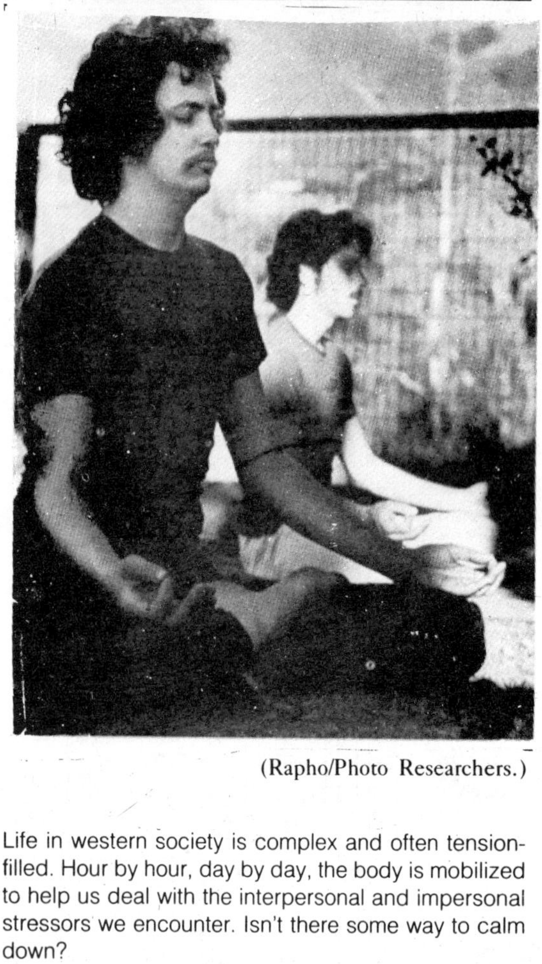

(Rapho/Photo Researchers.)

Life in western society is complex and often tension-filled. Hour by hour, day by day, the body is mobilized to help us deal with the interpersonal and impersonal stressors we encounter. Isn't there some way to calm down?

Perhaps there is. Taking their cue from eastern religions, psychologists have become interested in the age-old practice of meditation. Maharishi Mahesh Yogi and his followers recently introduced a technique they called the Transcendental Meditation (TM) Program, which is relatively easy to learn. The person learning the TM technique is given a special sound to repeat while sitting in a relaxed position. An instructor helps the student learn to repeat the sound mentally, without concentration or effort, so the thinking process can become more and more deeply settled and quiet. About 10 hours of instruction, spread out over a week or so, are all that are needed to learn the technique.

The physiological measurements reported to accompany the TM state are consistent with relaxation and lowered stress, and this relaxation seems to persist in experienced meditators even when they are not meditating. During meditation, the heart rate slows down a little; the breathing rate goes down and, in well-practiced meditators, may even cease for periods of up to 30 seconds, with no deeper breathing needed afterwards to make up for it; the consumption of oxygen decreases; muscle tension is reduced; blood levels of lactate and cortisol, which, respectively, are associated with anxiety and stress, decrease; the resistance of the skin to the passage of a weak electric current—the galvanic skin response, or GSR (page 320)—rises (this is also a sign of relaxation); and several changes in the electroencephalogram, or EEG (page 318), indicative of relaxation occur. Among these indicators are a slowing of the frequency of the EEG so that there is relatively more alpha activity (Chapter 2, page 55). EEG activity is also said to be more coherent; that is, both within and between the cerebral hemispheres (Chapter 2, page 58), the "brain waves" occur "in step" with each other. Thus, many stress-opposing bodily changes are characteristic of the TM state. However, a number of studies have shown that the bodily changes in the TM state are no greater than when one is simply relaxed.

TM meditators report that they feel more relaxed and less anxious, are more efficient in their daily activities, get along better with other people, and are more creative and self-actualizing (Chapter 7, page 298) than they were before the training. They also say that they have a "heightened sense of awareness."

In addition to the TM technique, stress reduction has been achieved by the use of techniques such as systematic desensitization (Application 4.1, page 148), hypnosis (Chapter 16, page 708), and the relaxation-response technique of Herbert Benson and his colleagues. Using this latter technique, the person sits for 20 minutes or so in a comfortable, relaxed position; progressively relaxes the muscles from feet to head; concentrates on breathing and saying "one" after each cycle of breathing in and out; and suppresses unwanted distracting thoughts by thinking "oh, well" when they occur.

While the Benson method results in many of the same physiological responses as the TM technique, there may be differences in the overall pattern of response, especially in the EEG and skin-resistance changes. Controversy swirls around the relative merits of the TM and Benson techniques. A clearer picture should emerge as more comparative studies are done, but, regardless of details, both seem to be good ways of achieving some relaxation and stress reduction.

REFERENCES

Benson, H., Kotch, J. B., Crassweller, K. D., & Greenwood, M. M. (1977). Historical and clinical considerations of the relaxation response. *American Scientist, 65*, 441–445.

Dillbeck, M. C., Aron, A. P., & Dillbeck, S. L. (1979). The transcendental meditation program as an educational technology: Research and applications. *Educational Technology, 19*(11), 7–13.

Holmes, D. S. (1984). Meditation and somatic arousal reduction: A review of the experimental evidence. *American Psychologist*, 39, 1–10.

Murray, J. B. (1982). What is meditation? Does it help? *Genetic Psychology Monographs, 106*, 88–115.

Shapiro, D. H., & Walsh, R. N. (Eds.) (1984). *Meditation: Classic and contemporary perspectives*. Hawthorne, NY: Aldine.

Wallace, R. K., & Benson, H. (1972). The physiology of meditation. *Scientific American, 226*(2),84–90.

Some people are better than others at coping with stressors. The more successful copers are said to have these characteristics (Cohen & Lazarus, 1979; Kimble et al., 1984):

1. They seek information about how to deal with the problem and what their alternatives are.

2. They are prone to take direct action in trying to reduce stressor impact.

3. They are flexible in trying to cope, first trying one thing and then another.

4. They try not to deal with the stressor through impulsive actions.

5. They use cognitive, or thinking, mechanisms—such as reappraisal of the situation (see page 333)—to deal with the stressor.

Some Theories of Emotion

Psychologists, physiologists, and philosophers have all worked to formulate some general principles to guide us in thinking about the emotions. These general ideas are the theories of emotion, and there are many of them. Not all theories of emotion cover the same ground. Some are concerned with the relationship between people's bodily states and the emotions they feel. Others are really attempts to classify and describe emotional experience. Still others try to describe how emotions are involved in behavior, especially how they are related to motivation.

EMOTIONS AND BODILY STATES

What is the relationship, if any, between the peripheral bodily reactions and the emotions we feel? For instance, what role do bodily responses mediated by the autonomic nervous system play in the emotions as we feel them? A number of theories of emotion are centered around questions such as these.

James-Lange Theory: Felt Emotion Is the Perception of Bodily Changes One of the earliest theories of emotion was succinctly stated by the American psychologist William James: "We feel sorry because we cry, angry because we strike, afraid because we tremble." This theory, presented late in the nineteenth century by James and the Danish physiologist Carl Lange, turns the commonsense idea about emotions inside out. It proposes the following sequence of events in emotional states: (1) we perceive the situation that will produce emotion; (2) we react to this situation; (3) we notice our reaction. (See Figure 8.10.)

Our perception of the reaction is the basis for the emotion we experience. So the emotional experience—the felt emotion—occurs *after* the bodily changes; the bodily changes (internal changes in the autonomic nervous system or movements of the body) precede the emotional experience.

For this theory to work, there must be a different set of internal and external bodily changes for each emotion, and the individual must be able to perceive them. While there is evidence for different patterns of bodily responses in certain emotions (page 316), doubts have arisen as to whether, especially in the more subtle and less intense emotions, these bodily patterns can be perceived; in general, our perception of internal changes is not very acute. For this reason, the James-Lange theory has been questioned.

Cannon-Bard Theory: Felt Emotion and Bodily Responses Are Independent Events In the 1920s, another theory about the relationship between bodily states and felt emotion was proposed by Walter Cannon, who based his approach to the emotions on research done by Philip Bard. The Cannon-Bard theory says that felt emotion and the bodily reactions in emotion are independent of each other; both are triggered simultaneously.

Figure 8.10

Outlines of three theories concerned with the relationship between what happens in the body and the emotions we feel.

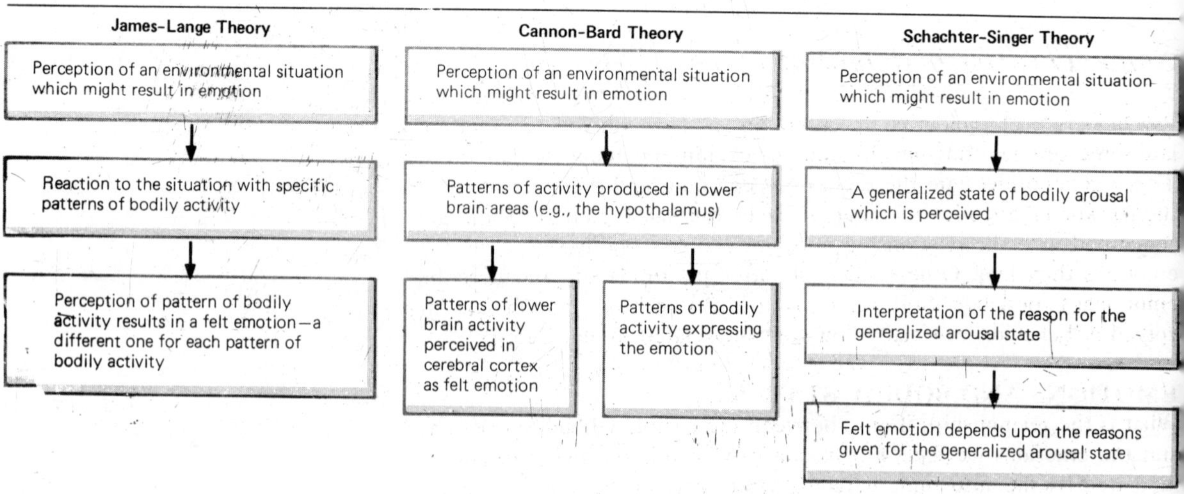

According to this theory, we first perceive potential emotion-producing situations in the external world; then lower brain areas, such as the hypothalamus (Chapter 2, page 57), are activated. These lower brain areas then send output in two directions: (1) to the internal bodily organs and the external muscles to produce the bodily expressions of emotion; and (2) to the cerebral cortex, where the pattern of discharge from the lower brain areas is perceived as the felt emotion. (See Figure 8.10.)

In contrast with the James-Lange theory, this theory holds that bodily reactions and the felt emotion are independent of each other in the sense that bodily reactions are not the basis of the felt emotion. This theory has led to a great deal of research, but, although we know that the hypothalamus and other lower brain areas are involved in the expression of emotion (page 317), we still are not sure whether perception of lower-brain activity is the basis of felt emotion.

Schachter-Singer Theory: The Interpretation of Bodily Arousal This contemporary theory maintains that the emotion we feel is due to our interpretation of an aroused, or "stirred up," bodily state. Schachter and Singer (1962) argued that the bodily state of emotional arousal is much the same for most of the emotions we feel and that even if there are physiological differences in the body's patterns of responses (page 315), people cannot perceive them. Since the bodily changes are ambiguous, the theory says, any number of emotions can be felt from a stirred-up bodily condition. People are said to have different subjective, or felt, emotions because of differences in the way they interpret or label the physiological state. In other words, given a state of arousal, we experience the emotion that seems appropriate to the situation in which we find ourselves.

The sequence of events in the production of emotional feeling, according to this theory, is: (1) perception of a potential emotion-producing situation, (2) an aroused bodily state which results from this perception and which is ambiguous, and (3) interpretation and labeling of the bodily state so that it fits the perceived situation. (See Figure 8.10.) Schachter and Singer put it this way:

> Imagine a man walking alone down a dark alley; a figure with a gun suddenly appears. The perception-cognition "figure-with-a-gun" in some fashion initiates a state of physiological arousal; this state of arousal is interpreted in terms of knowledge about dark alleys and guns and the state of arousal is labeled "fear." Similarly, a student who unexpectedly learns he has made Phi Beta Kappa may experience a state of arousal which he will label "joy." (Schachter & Singer, 1962, p. 380)

Schachter and Singer (1962) devised an experiment to test this theory.

> Subjects were given an injection of epinephrine (adrenalin), but told that the injection contained "Suproxin"—a fictitious vitamin. The epinephrine, of course, produced an aroused physiological state (page 318). Subjects in some groups were accurately informed that the injection would result in a state of physiological arousal; these subjects were expected to attribute their feelings of arousal to the injection itself. But other groups of subjects were told nothing about

the expected effects of the injection and were therefore ignorant about what the injections would do. Thus ignorant subjects were not expected to attribute their aroused bodily feelings to the injection. Instead, it was argued, they would attribute their state of arousal to their perception of what was happening in the situation around them.

To a degree, the results were consistent with the experimenters' expectations. When put in a situation designed to be a happy one, ignorant subjects indicated more happiness (euphoria) than did informed subjects. And ignorant subjects tended to feel angrier in an anger-arousing situation than did informed subjects. Thus, consistent with the theory, different emotions—euphoria or anger—were experienced with the same state of physiological arousal.

In experiments in which bodily arousal has been induced in other ways—by means of physical exercise, for example—results consistent with the Schachter-Singer theory have been obtained (Cantor et al., 1975). But, as often happens in science, new experiments have not supported the theory. For example, one recent experiment found that aroused subjects who were ignorant, or uninformed, about the cause of their arousal reported the aroused state as unpleasant even when they were in a "happiness-producing" situation (Maslach, 1979). In another experiment, subjects who were aroused with epinephrine but misinformed about the effects of the drug did not differ in their emotions from control subjects when placed in a "happy" situation (Marshall & Zimbardo, 1979). Thus neither of these experiments obtained results which would be predicted by the Schachter-Singer theory. As is the case with the other theories trying to relate bodily changes to felt emotion, there is reason to question the adequacy of the Schachter-Singer theory.

A COGNITIVE-APPRAISAL THEORY OF EMOTION

The Schachter-Singer theory just described is often called a cognitive theory because it involves thoughts about the cause of the perceived state of arousal. Another cognitive theory is that of Richard Lazarus (1970, 1984) and his co-workers. This theory emphasizes the *appraisal* of information from several sources. Since appraisal involves *cognition*, or the processing of information from the environment, the body, and memory, this theory is a cognitive one. The theory says that the emotions we feel result from *appraisals*, or evaluations, of information coming from the environmental situation and from within the body. In addition, memories of past encounters with similar situations, dispositions to respond in certain ways, and consideration of the consequences of actions that might result from the emotional state enter into the appraisal. The outcome of the complex appraisal of all this information is the emotion as it is felt.

The role of appraisal in emotion has been investigated in many experiments. One of the best known of these experiments illustrates the relationship between felt emotion and appraisal of the environmental situation (Spiesman et al., 1964).

Student subjects were shown an emotion-producing movie depicting the circumcision rites of Australian aborigines. These rites involve crude operations on the sex organs of 13- to 14-year-old boys. Three different sound tracks were prepared to go along with the film. One group of students heard a "trauma" track, which was designed to enhance the gory details. A second group heard a "denial" track, which was prepared to make it easier for the subjects to say that the film had not bothered them. A third group heard an "intellectualization" sound track, in which the rite was viewed from the detached, scientific standpoint of an anthropologist. A fourth group of students saw the movie with no sound track—the silent condition.

Heart rate and skin conductance—the GSR (page 320)—were measured while the film was in progress. It was found that stress reactions—high skin conductance, for example—were highest for the trauma sound track, next highest for the silent picture, and lowest for the denial and intellectualization conditions. In other words, both the denial and intellectualization conditions reduced emotionality relative to the trauma and silent conditions. Thus the sound tracks induced the subjects to make different situational appraisals of the same stimulus—the film. The experimenters' conclusion was that different emotional reactions to the same stimulus occurred because of differences in the subjects' appraisal of the stimulus.

Reappraisal of potentially emotion-producing situations is an important part of this cognitive theory. Reappraisal is also a way of coping with stressful situations. Suppose you are called in by the dean. Your appraisal of this situation may, at first, create apprehension. But suppose that when you get to her office she tells you that she wants you to be a student representative on the student-discipline committee. Reappraisal occurs, and your apprehension may change to pleasure. So it is with the changes in emotion from minute to minute and day to day. In stressful situations, reappraisal may be a way of coping. People who reappraise emotion-producing situations with denial ("It isn't really stressful at all; think positively"), intellectualization ("This is all very interesting"), reaction formation ("This isn't stressful, and, in fact, it's a great learning experience"), or other normal defense mechanisms (listed in Chapter 14, page 588), may find that they are able to reduce the intensity of the disturbing emotional feelings which accompany stressful situations.

A THEORY OF RELATIONSHIPS AMONG EMOTIONS

One problem with the study of emotions is that they are ill-defined states of being—indistinct, intermingled, and constantly changing. How can psychologists describe them well enough to study them? Robert Plutchik (1970, 1980) has proposed a descriptive theory that is concerned with what are called *primary*, or basic, *emotions* and the ways they can be mixed together.

In order to show the relationships among emotions, Plutchik assumes that they differ in three ways: (1) intensity, (2) similarity to one another, and (3) polarity, or oppositeness. He uses these three dimensions—intensity, similarity, and polarity—to draw a spatial model representing the relationships among the emotions (Figure 8.11). The eight segments of his model (grief, sadness, and pensiveness are in one seg-

Figure 8.11

A diagram of a theoretical model portraying the dimensions of human emotion. Each vertical, wedge-shaped segment—the one composed of the subsegments grief, sadness, and pensiveness, for example—represents a primary emotion. Intensity of the primary emotions is shown from the top down in each segment. Conflicting emotions are placed opposite each other around the figure, while similar emotions are placed near each other around the figure. (From Plutchik, 1970.)

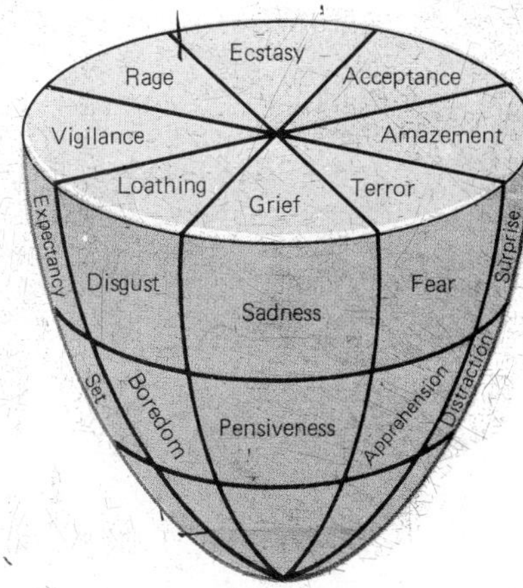

ment, for example) represent eight primary emotions. Plutchik maintains that these primary emotions are derived from evolutionary processes and therefore have adaptive value.

Within each primary-emotion segment (Figure 8.11), the strongest varieties of the emotion are at the top of the segment, with progressively weaker varieties toward the bottom. For example, loathing is stronger than disgust, which, in turn, is stronger than boredom. Finally, the similarities and polarities among the primary emotions are shown by the arrangement of the segments (Figure 8.11). The grief segment, for example, is polar to—opposite from—the ecstasy segment; furthermore, the grief segment borders on primary-emotion segments with more similarity to grief than those farther away. Emotions that are opposite each other conflict, while emotions that are close to each other around the figure are complementary. Since people seldom experience pure emotions, a model of this sort makes it possible to give a good description of mixed and conflicting emotions.

THEORIES ABOUT EMOTION AND MOTIVATION

The line between some motives (Chapter 7) and emotions is often a thin one. Fear, for example, is an emotion; but it is also a motive driving behavior because people engage in goal-directed behavior when they are afraid. A theory of motivation and emotions proposed by Leeper (1970) goes much further than this. Leeper says that almost all our sustained and goal-directed behavior is emotionally toned and that it is the emotional tone which provides the motivation for long sequences of behavior. For instance, the motive driving a person's behavior in his or her job might be the emotional fulfillment of doing good work, the satisfaction of being esteemed by friends and colleagues, or the pleasure of mastering new things. Leeper says:

> The most fundamental type of research on emotions which needs to be conducted is research on their role as motives—their role, that is, in arousing and sustaining activity, in producing exploratory reactions, in facilitating learning...in governing performance or habit-use, in helping produce problem-solving learning, in helping govern choice between alternatives, in producing willingness to endure penalties to reach some goal or a willingness to forego some reward, and in influencing thought-content and sensory perceptions.
> (Leeper, 1970, p. 153)

Another theory (Tomkins, 1970, 1981) maintains that emotions provide the energy for motives. Tomkins argues that motives, or drives (Chapter 7, page 269), simply give information about some need or condition of the body. Drives tell us that food is needed, water is needed, a sexual urge is present, and so on. Accompanying these drives are emotions (Tomkins calls them affects), such as excitement, joy, or distress, that provide the energy for the drives; they amplify the drives to give them their strong motivational power.

Finally, the incentive and opponent-process theories of motivation described in Chapter 7 (page 270) are also, in part, theories about emotion. They were discussed under "motivation," but they might well have been presented in this section. Once again we see that the dynamic, or driving, aspects of motivation and emotion are not easily separated.

Summary

1. Emotion is a hard term to define. When we speak of emotions, we usually refer to (*a*) subjective feelings; (*b*) the physiological bases of emotional feelings; (*c*) the effects of emotion on perception, thinking, and behavior; (*d*) the motivational properties of certain emotions; and (*e*) the ways emotions are shown in language, facial expressions, and gestures.

2. Expressions and perceptions of emotions are quite important in our responses to other people. The tone and other characteristics of the voice are a channel for the expression of our own emotions and the perception of emotions by others. Facial expressions are perhaps the most important nonverbal way in which emotions are manifested. A number of studies have indicated that the facial expressions of certain primary emotions can be judged accurately by people from diverse cultures. This lends support to the view that the facial expessions of these primary emotions may be innate, or inborn. Context—the situation in which an emotion occurs—also gives us information for judging what emotion is being expressed.

3. A number of bodily reactions accompany emotional states. Prominent in the body's responses in emotion is activity in the autonomic nervous system—the part of the peripheral nervous system which controls the smooth muscles of the various organs of the body and the secretion of certain glands. The autonomic nervous system has two divisions—the sympathetic system and the parasympathetic system.

4. For a number of emotions, different patterns of bodily activity can be detected. Fear and anger are characterized by the emergency, or "flight-or-fight," response of the sympathetic portion of the autonomic nervous system. The relaxation response which accompanies calm, meditative states consists of a pattern of bodily activity that, so far as the sympathetic system is concerned, is almost the opposite of the emergency response; the parasympathetic system is active in the relaxation response. In addition to the general emergency and relaxation responses, certain specific patterns of bodily responses have been found to characterize several particular emotions.

5. The patterns of bodily activity in a number of emotions are controlled by the limbic system and hypothalamus of the brain. The arousal state that accompanies many emotions is regulated by the ascending reticular activating system (ARAS) of the brain stem.

6. Arousal is an important part of many emotional states. One indicator of arousal is the electroencephalogram (EEG)—the record

of "brain waves." High-frequency, low-voltage activity of the cerebral cortex indicates arousal; low-frequency, high-voltage activity indicates sleep or a low level of arousal. Other indicators of arousal are heart rate, blood pressure, breathing rate and depth, the galvanic skin response (GSR), and the orienting reaction.

7. Stress refers to the widespread, generalized responses of the body to various environmental, physical, or social situations. Stressors are the situations, or events, which cause the stress responses of the body. Among the many situations acting as stressors are: changes in one's life, injury, infection, exercise, noise, climate, frustration, and job, social, or family pressures.

8. The body's response to stressors has been called the general adaptation syndrome. This syndrome has three stages: (*a*) the alarm reaction, (*b*) the stage of resistance, and (*c*) the stage of exhaustion. The alarm reaction is essentially the emergency response of the body. A major feature of the stage of resistance is that certain hormonal responses—especially in the adrenocorticotropic (ACTH) axis—become an important line of defense in resisting the effects of stressors. Prolonged activity of the adrenocorticotropic axis (or other hormonal systems) can impair the body's ability to fight infections and can have other harmful effects. In the stage of exhaustion, the body's ability to respond to stressors has been seriously compromised. At this stage, or late in the stage of resistance, various psychosomatic disorders may occur. Psychosomatic ("mindbody") disorders occur when perceived stressors increase the susceptibility of the body to disease. Ways of coping with stressors are described.

9. Some theories of emotion focus on the relationship between the bodily states in emotion and the emotion as it is felt. The James-Lange theory maintains that the emotions we feel result from our perception of the changes taking place in the body during emotion. The Cannon-Bard theory says that felt emotion and bodily changes occur in parallel with each other and result from activity in certain brain areas. In other words, the bodily changes do not cause the felt emotion; while the bodily changes and the feelings occur at about the same time, they are independent of each other. The Schachter-Singer theory holds that the bodily changes are much the same for many of the emotions we feel, and that even if there are differences in the bodily patterns of responses, we cannot perceive them accurately. According to this theory, emotional feelings result from the interpretations we arrive at to explain why we are in the general state of arousal that accompanies many emotions.

10. The cognitive-appraisal theory of emotion states that felt emotions result from appraisal, or evaluation, of information about the environmental situation and the state of the body. In addition, memories of past emotional situations, dispositions to respond in certain ways, and thoughts about what might result from an emotional state enter into the appraisal. Depending on the nature of the appraisal, no emotion, strong emotion, or different types of emotion can be felt.

11. Plutchik's theory of emotion is primarily descriptive. It proposes that there are certain primary emotions derived from evolutionary processes and that these primary emotions can be arranged in an orderly way to bring out relationships, similarities, and differences among them.

12. Other theories of emotion are concerned with the relationship of emotion to motivation. One of these says that emotions should be considered as motives which keep behavior going and aimed at long-term goals; another says that emotions amplify motives to give them their energy.

Terms to Know

One way to test your mastery of the material in this chapter is to see whether you know what is meant by the following terms.

Emotion (310)
Primary emotions (311, 333)
Autonomic nervous system (314)
Sympathetic system (314)
Epinephrine (315)
Norepinephrine (315)
Parasympathetic system (315)
Somatic nervous system (315)
Emergency reaction or "flight-or-fight" response (315)
Relaxation response (316)
Limbic system (317)
Ascending reticular activating system (ARAS) (317)
Electroencephalogram (EEG) (318)
Polygraph (318)
Galvanic skin response (GSR) (318, 320)
Orienting reaction (320)
Stress (321)
Stressors (321)
General adaptation syndrome (325)
Alarm reaction (325)
Stage of resistance (325)
Adrenocorticotropic (ACTH) axis (325)
Adrenocorticotropic hormone (ACTH) (325)
Corticotropin-releasing factor (CRF) (325)
Hypothalamus (325)
Cortisol (325)
Stage of exhaustion (326)
Psychosomatic disorders (326)
James-Lange theory (330)
Cannon-Bard theory (330)
Schachter-Singer theory (331)
Cognitive-appraisal theory of emotion (332)
Cognition (332)
Appraisal (332)
Reappraisal (333)

Suggestions for Further Reading

Reading this chapter may have convinced you that the simple word *emotion* has many aspects to it and that psychologists' ideas about it differ. An account of some of the varying viewpoints about emotion is given in D. K. Candland, J. P. Fell, E. Keen, A. I. Leshner, R. M. Tarpy, and R.

Plutchik's *Emotion* (Monterey, CA: Brooks/Cole, 1977). Accounts of many of the aspects of emotion are given in C. E. Izard's *Human Emotions* (New York: Plenum Press, 1977); K. T. Strongman's *The Psychology of Emotion*, 2d ed. (New York: Wiley, 1978); and R. Plutchik's *Emotion: A Psychoevolutionary Synthesis* (New York: Harper & Row, 1980).

Stress is a "hot topic" these days. Of the many good books on stress, we recommend R. J. Allen's *Human Stress: Its Nature and Control* (Minneapolis: Burgess, 1983); C. B. Dobson's *Stress: The Hidden Adversary* (Ridgewood, NJ: George A. Bogden & Son, 1983); E. A. Charlesworth and R. G. Nathan's *Stress Management: A Comprehensive Guide to Wellness* (New York: Atheneum, 1984); and *Stress Reduction and Prevention*, edited by D. Meichenbaum and M. E. Jaremko (New York: Plenum Press, 1983). A visit to your library will turn up many other good books on stress.

How people describe their feelings is a fascinating topic. J. Wood's *How Do You Feel? A Guide to Your Emotions* (Englewood Cliffs, NJ: Prentice-Hall, 1974) contains first-person accounts of feelings.

Answers to the "facial-expression quiz" on the chapter-opening pages:

These are posed photographs. The tribesmen were instructed to show the emotion appropriate to the following situations:

(*a*) *Joy, happiness.* The instructions were "Your friend has come, and you are happy."

(*b*) *Sorrow, grief.* The instructions were "Your child has died."

(*c*) *Anger.* The instructions were "You are angry and about to fight." Many see this as an expression of determination.

(*d*) *Disgust.* The instructions were "You see a dead pig that has been lying there for a long time."

Social Perceptions, Influences, and Relationships

HEADING OUTLINE

Chapter 9

Mass Suicide in Jonestown

On November 18, 1978, in the northwest jungle of Guyana, the Rev. Jim Jones presided over the extermination of the entire population of Jonestown, the agrarian "utopia" he had established for his People's Temple movement. In all, slightly over 910 Americans either willingly drank or were injected with potassium cyanide, mixed in a steel vat along with grape-flavored punch. Probably no more than two or three dozen people present in the compound that day fled into the surrounding jungle. Because of the remoteness of the locale, news of the events came fitfully to a stunned world. Most Americans knew little about Guyana, much less about the People's Temple.

The Jonestown holocaust was partly precipitated by the visit of Representative Leo J. Ryan, who was investigating charges by relatives that commune members were being grossly mistreated and held captive against their will. After spending 2 days in the jungle encampment, Ryan and his small party departed Jonestown on the afternoon of the 18th. Sixteen commune members had asked to leave with Ryan. The group was driven to a dirt airstrip 8 miles away. Along with journalists and members of the Concerned Relatives group, who had not been admitted to Jonestown, they prepared to board two small airplanes. Suddenly they were attacked by gunmen from Jonestown, and Ryan and four other people were assassinated.

Guyana and Jonestown, because of the Congressman's murder, made the headlines the next day, and, in ensuing days, a great deal of accurate information about the People's Temple appeared in the press. Yet the magnitude of the final horror remained hidden. It was as if no human mind could grasp the idea that any group of people could willingly take their own lives and the lives of their children. A pervasive belief existed that many of the Jonestown members avoided the suicide ritual by escaping into the jungle. The sequence of headlines in *The New York Times* is instructive. On November 20, a headline read: ". . . 300 DEAD AT RELIGIOUS SECT'S JUNGLE TEMPLE." The number was increased the following day, and a bolder headline proclaimed: "**400 ARE FOUND DEAD IN MASS SUICIDE BY CULT; HUNDREDS MORE MISSING FROM GUYANA CAMP.**" The hope for survivors persisted. A headline on November 22 stated ". . . **UP TO 500 MAY BE LOST IN WILD AREA.**" It was not until November 26 that the true scope of the holocaust was finally revealed.

Conventional wisdom cannot comprehend these bizarre events. There is no simple answer to the question: Why would over 900 Americans participate in a ritualistic massacre? By now, however, enough has been written about the People's Temple to identify the major contributing factors. They include distorted perceptions, malevolent influence, and twisted interpersonal relationships. We will return to these issues in Inquiry 9.2.

"I walked down the path to a vat filled with a vile-looking reddish-purple liquid. I was told that the people had been poisoned." (AP/Wide World Photos; from Krause, 1978.)

WE are all social animals. We all realize that much of what we do stems from our interactions with other people, but we often fail to appreciate the power of these interactions over our behavior and thought. If you consider it, you will be hard put to find anything you do or think that is completely independent of your social relationships. What about your values, or your ideas of what is right and wrong? What about your preferences or your daily decisions? This list could go on and on. Social psychology is the study of human interaction and the way it affects behavior. To put it more formally, *social psychology* is the scientific study of the many ways in which interactions, interdependence, and influence among persons affect the individual's behavior and thought.

An Overview of the Area

Because social psychology covers such a wide array of behaviors, it may be helpful to discuss a few distinctions that will serve to map the area. These distinctions will create rough boundary lines among the various topics to be covered and will serve to organize this chapter and the next.

In even the simplest social interaction, the interplay between participants is rapid, and a great deal of information is exchanged. The essence of interaction is that all participants modify what they do and say according to what others are doing and saying. While such mutual interdependence provides much of the joy and vitality in our interactions with others, it also makes it difficult to arrive at a clear analysis of the determinants of social behavior. The orderly pursuit of the fixed-stimulus characteristics that give rise to particular responses, so typical of the content of previous chapters, must give way to analyses in which one person's responses become part of the stimulus for other participants. In approaching these interactive complexities, we will start with the simplest social situation and move on to the more complex. Furthermore, we will restrict the discussion to two-person situations, called *dyads,* in which the focus is on one person, who is arbitrarily designated the *focal person (FP).* In general, social psychologists are interested in how contact with another person affects the FP's behavior and thoughts.

The ordering of topics is presented in Figure 9.1. It begins with the simplest kind of social situation, where the FP briefly observes something about another person and forms an impression of that person on the basis of the observation. Even very limited or seemingly trivial information can contribute to an impression. It has been shown, for example, that someone who uses the familiar form of his first name (Charlie, Ed) is seen as more extraverted (outgoing) and less emotionally stable than someone who uses the formal form (Charles, Edward) (Leirer et al. 1982). More information is provided by situations in which the FP can observe someone going through a sequence of behaviors. Richer information requires more cognitive work because the bits of information must be combined to form an overall impression. This can be especially difficult if the informa-

Minimal conditions for studying different social concepts.			
Role of subject	**Role of other**	**Examples of stimuli**	**Concept**
Observer	Fixed pose	Photograph, trait list	Impression formation
Observer	Behavior sequence	Video tape	Attribution
Participant	Intervention directed at subject	Other gives subject command	Social influence
Participant	Subject interacts with others	Discussion, work on task	Interdependence

Figure 9.1

Minimal conditions for studying different social concepts.

tion is contradictory. How friendly do you judge a classmate who acts friendly on four occasions and unfriendly on one occasion? Richer information also makes it possible to form a judgment about the causes of someone's behavior. Knowing why someone has done something will make a difference in your evaluation of that person. If you have agreed to meet someone for dinner and he does not appear, your evaluation of him will depend upon the reason for his absence. You will feel more positive toward him if the reason was a car accident or a malfunctioning alarm clock than you will if he forgot or decided to eat with someone else.

From social perception, we will move to a discussion of social-influence situations. The essential characteristic of all instances of influence is change. (If you have been influenced by a book, a television commercial, or even the weather, it means that you have changed your behavior as a consequence of exposure to these things.) *Social influence* is change that occurs because of contact with one or more persons. We will examine different categories of social influence and try to understand how they affect our daily lives.

We will conclude the chapter by discussing social relationships. It will be stressed that people in relationships are interdependent, that is, they affect each other by their behavior. This will then bring us closer to everyday interactions, which are represented by back-and-forth models of interpersonal acts. Carl Jung captured the most meaningful aspect of interaction when he noted, "The meeting of two personalities is like the contact of two chemical substances: if there is any reaction, both are transformed." We will begin, however, with the static snapshots typically involved in the study of social perceptions.

Social Perception

How do we come to know other people? Our social perceptions of others are initially based on the information we obtain about them and, in some instances, the *attributions* (inferences) we make about the causes for their behavior. Both topics, initial impression formation and making attributions, will be discussed. It is, of course, important to have accurate knowledge of others before deciding what kind of interactions to have with them. Our perceptions of others' personalities and feelings—as well as the causes for their behavior—guide us in deciding how we will respond to them and what sort of relationships we will have with them.

IMPRESSION FORMATION

Forming impressions of other people is probably so natural to you that, like breathing, you only think about the process if something goes wrong. *Impression formation* is a process by which information about others is converted into more or less enduring cognitions or thoughts about them. When we first meet someone, we have access to considerable information—how the person looks and what he or she does and says. According to one point of view, we are not overwhelmed by the abundance of this information because we are able to group it into categories that predict things of importance to us. These categories and their perceived interrelationships form the basic *cognitive framework* by which we understand others. The characteristics defining cognitive categories can be as broad as "women" or "men" or as narrow as "myself." They can involve such diverse features as an occupational role (used-car salesman), a social role (friend), a personality trait (dominant), or a physical characteristic (tall). The linkages among these categories will determine what predictions we make about someone when we have only limited information. If, for example, you think people who wear glasses are intelligent, then whenever you meet a stranger wearing glasses, you will be disposed to believe that he or she is highly intelligent. All categories are related to some other categories and unrelated to many more. In your mind, wearing glasses might also be related to "timidness" but not to "honesty" or "sense of humor," for example. By generating predictions, or expectations, we can efficiently interact with other people even when we possess only minimal information about them.

Of course there is no guarantee that a given piece of information will be categorized in the same way by different people. The humor contained in the cartoon strip presented in Figure 9.2 revolves around such a difference. The woman sees herself as one of many detergent users, while the wizard apparently sees her as belonging to the sparsely filled category of odd-looking people. Differences in how information is categorized may give rise to humor, but more often it is a source of misunderstanding and conflict between people. The man who thinks he is being considerate when he opens a door for a woman will be surprised by her reaction if she has categorized his behavior as patronizing. In a similar way, comparable categories do not guarantee identical linkages with other categories. Such differences can have important consequences on how events are understood. Consider, for example, the act of rape. Such an act was, for many

Figure 9.2

The categories we use to describe ourselves do not always correspond to the way others categorize us. (By permission of Johnny Hart and Field Enterprises, Inc.)

years, linked with a male's sexual desire and with the victim's possible sexual provocation. In recent times, rape is more often linked with physical violence and aggression; the locus of responsibility thus correctly remains with the rapist (Groth & Birnbaum, 1979).

Implicit Personality Theory The categories we use and their assumed interrelationships constitute our template, or framework, for understanding the world in which we live. This framework, in essence, is our theory about how things are supposed to work.

For understanding other people, the category most frequently used is the trait. *Traits* are classification schemes for describing the behavior of individuals. Our language provides us with many options for describing behavior, such as assertive, friendly, punctual, or talkative, for example. (In 1936, Allport and Odbert identified about 18,000 trait names in a standard English dictionary.) Traits are a compelling set of categories used to describe, remember, and communicate our own and other people's behavior. Traits are also perceived to be interrelated; they seem to occur in clusters (see Chapter 14, page 571). You might, for example, assume that people who are assertive are also ambitious or that intelligent people are also industrious. This assumed relationship among traits is called *implicit personality theory,* a name that underscores how our cognitive framework generates predictions about other people that go beyond the information available to us.

Our own implicit personality theory, similar to our general cognitive framework, includes both assumed relationships shared by most other members of our culture and assumed relationships that are unique to us. The shared assumptions are a result of the similarities of experiences within a particular culture, where people share a common language, common exposure to mass media, and common socialization or child-rearing experiences. The unique assumptions are a result of our own individual experiences, especially with people of importance to us in our family, school, neighborhood, or church. Implicit personality theories help us to simplify the information we receive in social interaction, color the way we interpret events, and guide our responses to other people.

Combining Information Once you have met someone and have settled on a group of traits you observe and assume that person to have, how do you decide whether or not you wish to proceed with the relationship? To make this decision, you must make a global judgment about how favorable you feel toward the person. One procedure you might follow is to add the favorable traits together. If you do this, you will have a more positive impression if you think a person is both kind and honest than if you think the person is simply kind. On the other hand, you might average these two pieces of information, in which case your impression would remain about the same—the average of the two favorable traits would be close to the value of each of them alone. The averaging model is probably closer to what you would actually do, except that it would not be quite so simple. Instead, certain pieces of information would be seen as more important and thus would be weighted more heavily than others; your overall impression would represent a weighted average of the information you have about that person. One thing you would consider is the relevance of the information for the particular judgment you are making. You assign importance to different characteristics in assessing your car mechanic and in assessing your psychology professor. Information obtained first also seems to be weighted more heavily. Most people believe there is some value in making a good first impression, and research shows that such efforts are not wasted; a *primacy effect* does often occur in impression formation. Furthermore, we generally give more importance to information concerning negative traits than to information concerning positive traits that others might possess (Hamilton & Zanna, 1972; Amabile & Glazebrook, 1982). Each of these factors affects the weighting people give to various pieces of information when forming an impression of another person.

Some information patterns contain apparently contradictory information. (Try, for example, to imagine someone who is both hostile and dependent.) Asch and Zukier (1984) asked subjects to imagine people who were described by such contradictory traits in order to study how perceivers resolve discrepancies. These authors assumed that perceivers would try to preserve the unity of the other person. The contradictions were resolved by inferring a greater degree of complexity about the other person. One mode of resolution was to put the traits in a cause-effect relationship. Some subjects saw the FP as resentful about the dependence and, therefore, hostile. Segregation, assigning the contradictory traits to different spheres of the person, was another mode of resolution. Some subjects saw the FP as dependent upon one person, perhaps a parent, but hostile to others. Asch and Zukier identified seven different modes of resolution, each of which illustrated that the perceiver preserves the unity of the FP by inferring new information that goes beyond what was given.

Stereotypes We have all been warned about the evils of stereotyping, and none of us would be eager to characterize our own impressions of others as involving stereotyped judgments. If one defines stereotypes as prejudiced expectations generated by placing people in cognitive pigeonholes, the social undesirability of stereotypes may seem justified. Yet you may now have the uncomfortable feeling that what was previously said about the functions of cognitive frameworks, or implicit theories of personality, is quite similar to stereotyped impressions. Indeed, with only slight alterations in the negatively toned words—by defining *stereotypes* as prejudgments generated by placing people in cognitive categories—the processes of impression formation and stereotyping become equivalent. Is there, then, any reason to be wary of stereotyped thinking? The answer, of course, is yes. The reasons for the answer, however, cannot be found in the cognitive processes themselves but reside in the content of socially undesirable stereotypes. These stereotypes, which we will arbitrarily call *social stereotypes*, have certain common features that should make us cautious when they pop into our heads. Social stereotypes typically involve categories defined by an ethnic dimension—such as race, nationality, or religion—or by a demographic dimension—such as gender or region of country. The categories typically define a minority group and almost always generate an expectation for undesirable behavior. Examples of social stereotypes abound. Are you familiar with these unfounded linkages: Irishmen are brawlers and drunkards, Jews are shrewd and ambitious, blacks are fun-loving and ignorant, feminists are aggressive and plain, southerners are bigotted and clannish? Social stereotypes, thus, involve ethnocentric thinking—that is, the assumption that what one's own group does or values is best and that any differences from that ideal are undesirable. One of the evils of social stereotyping is that it nearly always depicts members of other groups in an unfavorable light. When these expectations are believed by us and shared with others in our group, they present a powerful barrier to interacting with members of the stereotyped group. In some instances, they can even become accepted by members of the stereotyped group. In addition to being ethnocentric, social stereotypes can generate incorrect expectations about someone in the face of much contradictory behavior.

ATTRIBUTION

To characterize other people in terms of certain traits, intentions, or abilities requires us to make *attributions*, or inferences, about them. Because we do not have access to the personal thoughts, motives, or feelings of others, we make inferences about these traits based on the behavior we can observe. By making such attributions from certain behaviors, we are able to increase our ability to predict how a person will behave in the future. (The idea of making inferences from behavior may already be familiar to you from the discussion of motivation in Chapter 7, page 268). Attributions are not made from everything a person does; the theories discussed in the next sections outline some of the factors involved in making attributions from observed behavior.

Heider's "Naive" Psychology Social psychologists' interest in the attribution process began with Fritz Heider's (1958) theory, which was concerned with how we attempt to understand the meaning of other people's behavior—particularly how we identify the causes of their actions. Whenever we ask why someone did something, we are asking a causal question. Why did the FP arrive late? and Why was the FP so friendly toward me? are examples of possible causal questions. Heider thought we answer such questions by searching until we find a sufficient reason for the observed behavior. A sufficient reason is an explanation that makes sense to us, probably because it corresponds to the content of our cognitive framework.

In general, behaviors can be caused by *personal forces*, such as ability or effort, or by *environmental forces*, such as luck or the difficulty of the task to be done. If an action is attributed to personal forces, it will have a different meaning than if it is attributed to environmental forces. Imagine that you are standing in a line at a movie and a burly man bumps into you. You might attribute the act to environmental forces. (The man's task of walking was made too difficult by an obstacle on the floor, or he experienced bad luck.) In this case, you would be likely to make light of the incident. On the other hand, if you attribute the act to personal forces—his effort and ability to cause you pain—you will likely be distressed. Someone has intentionally caused you pain, and the situation may be serious. An attribution that an action stemmed from personal forces will be made only in cases in which you perceive that another person is able to perform the action, intends to perform it, and exerts effort to accomplish it. Such attributions imply the existence of stable dispositions and predictability for the future. An attribution that an action stems from environmental forces reveals little about personal characteristics and is not predictive of the future.

Kelley's Attribution Theory Harold Kelley's (1967, 1973) attribution theory grew out of Heider's original work. Like Heider's "naive" psychology, the main concern of this theory is how we determine whether an action is caused by internal or external forces. According to Kelley, the basic method we use to identify the causes of particular effects is the presence-absence test. We identify something as a cause if, in its presence, the effect always occurs, and, in its absence, the effect does not occur. Young children supposedly learn that flames burn fingers after only one instance of the "presence" condition. You may also have recognized that the presence-absence test is a basic tenet of experiments, in which the "experimental" and "control" conditions represent, respectively, the presence and absence of the cause under investigation.

By extending the simple presence-absence test, Kelley identified three basic factors that we use in deciding whether another's response was caused by internal or external forces. The first type of information we use is the consistency of the response over time and situations; *consistency* refers to the extent to which the particular response occurs whenever a particular stimulus or situation is present. Second, we depend upon *consensus information*, that is, the extent to which other people respond to the same stimulus in the same manner as the person being judged. *Distinctiveness* is the third factor; it refers to the extent to which the person being judged responds differently to various stimuli in the same general category. Distinctiveness is high if the person's responses differentiate among such stimuli. A combination of high consistency, high consensus, and high distinctiveness creates external attributions; high consistency combined with low consensus and low distinctiveness determines internal attributions. An example will help to show how the three factors are combined.

> Suppose you are looking for a good restaurant, and your friend Sue recommends a new French restaurant. How will you know whether to attribute Sue's good opinion to the restaurant itself (external attribution) or to some personal characteristic of Sue's? If you ask some other people about this restaurant and they also say it is good (high consensus), and if you know there are restaurants, including some French ones perhaps, that Sue does not like very much (high distinctiveness), and if Sue has been there more than once and always likes it (high consistency), then you are very likely to make an external attribution regarding Sue's recommendation—you will assume the restaurant is, in fact, very good, and you may try it soon. On the other hand, if some of your friends do not like it (low consensus) and you know Sue seldom says bad things about any restaurant (low distinctiveness), you are likely to make an internal attribution—you will assume Sue liked the restaurant just because she likes to eat in restaurants or because she can't discriminate between good and mediocre food. If the consistency is not high, you can make neither an internal nor an external attribution. Instead, you must simply assume that some temporary, specific factors led to the behavior. If Sue liked the restaurant only once of the several

times she was there, you might assume the restaurant changed chefs or management or was good just by accident that one evening.

It should be noted that the information required by the three factors is not always available when you wish to make an attribution. Imagine a situation in which Sue is a stranger who praises the French restaurant. Neither her past consistency in responding to the particular restaurant nor her degree of differentiating among various types of restaurants is known to you. In the absence of relevant information, will you avoid making an attribution? Probably not. Some recent findings by Pruitt and Insko (1980) suggest that you would attribute her response to external factors, implicitly assuming that her past responses were highly consistent and highly differentiated. In other words, in the absence of contradictory evidence, you are likely to assume that a person's response provides information about the stimulus rather than about the person.

Jones and Davis: A Theory about Personal Attributions While Heider was interested in whether behavior would be attributed to personal or impersonal causes, E. E. Jones and Keith Davis (1965) focused primarily on how personal attributions are made. Their central concern was with how we infer whether or not someone's behavior reveals the existence of an enduring disposition. Like Heider, they assumed that any act results from a sequence beginning with personal dispositions that produce intentions which are modified by abilities and, finally, lead to acts. They extended Heider's theory by stating that any act involves choices and produces effects for the individual. The perceiver can observe only the behavior (choice) and its effects and must work backward from the acts to make inferences about the dispositions. (See Figure 9.3.) In this theory, a perceiver is said to take into account not only what another person actually does but also what the person *might* have done. What a person does (the chosen action) has many outcomes (effects). Any of the *unchosen* actions would also have had several effects if it had been chosen. Some of the effects of the chosen and the unchosen actions are the same, or common to both; but other effects are *not* common to both the chosen and the unchosen actions. It is these *noncommon effects* which yield clues regarding the intentions of another person.

For example, a student might choose to study in the library rather than to play tennis. The chosen action, a visit to the library, and the unchosen action, a visit to the tennis court, have the common effect of getting the student out of his or her room. However, this common effect is of little value in understanding why the visit to the library was chosen over the visit to the tennis court. One of the noncommon effects of these

Figure 9.3

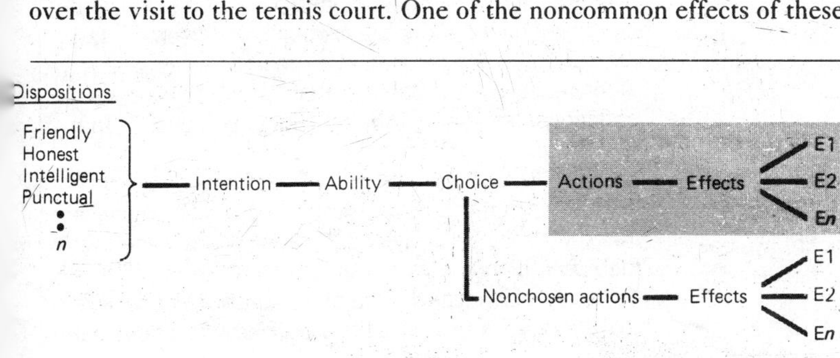

It can be assumed that actions are the result of a sequence going from left to right. A person's disposition gives rise to intentions that are mediated by ability to produce acts that have different effects. An observer can only see the acts and their effects and works from right to left, inferring dispositions from acts. (Modified from Jones & Davis, 1965.)

Inquiry 9.1

HOW GOOD ARE WE AT KNOWING THE CAUSES OF OUR OWN BEHAVIOR?

It would be outlandish to suggest that we are not better than other people at identifying why we do things. Other people, even our closest friends, are relatively ignorant about our past experiences, our private thoughts, and our own cognitive network. Yet it has been claimed that our storehouse of private knowledge contributes little to our understanding of the causes of our own behavior. Richard Nisbett and Timothy Wilson have been particularly forceful in arguing that our ability to identify the sources of our own behavior is no better than is the ability of an outside observer. How could this be?

To begin, recall how a stimulus is known to be a cause of a response. The process, discussed at the beginning of this chapter, requires that the introduction of the stimulus produce a change in the response. We discover causal connections, as Kelley (1967) has noted, by conducting a presence-absence test. Similar to the way in which a scientist tests a hypothesis by observing both an experimental and a control group, we test causal connections by observing whether the response is always made in the presence of the stimulus (experimental group) but never in its absence (control group). It is easy to find answers to some causal questions. If you were asked why you purchased this textbook, you would undoubtedly answer that it was required by your instructor. You probably always purchase textbooks in the presence of a requirement and never in its absence. It might be a bit more difficult to answer a question about why you are studying at this particular moment. Perhaps you study every afternoon, or when your room is quiet, or when the weather is foul, or the night before a midterm exam. In searching for a cause, you could do a series of presence-absence tests until you found a good fit. On the other hand, you will not need to do any presence-absence tests if your cognitive framework is set up to provide an answer. Our cognitive frameworks contain many causal connections. "Absence makes the heart grow fonder" and "Too many cooks spoil the broth" are folk wisdom that causally connect physical separation with increased liking and group size with task failure.

According to Nisbett and Wilson, our causal searches are not very accurate. They claim that often we are neither aware of the relevant stimuli that prompted our acts nor accurate in recalling past factors (See Chapter 5, page 202). Furthermore, in some instances we may not even be aware that we have changed our response. If you have been influenced to change your response but are not aware of the change, you will not attribute the new response to the change agent. Suppose a television commercial has influenced you to buy a more expensive shampoo, but you no longer remember the commercial. If you were asked why you were making the purchase, (if you answered at all), you would most likely name the positive characteristics of the shampoo.

Nisbett and Wilson tried to test people's awareness of what caused their responses in a number of experiments. In one of their studies, subjects were asked to memorize word pairs. Some of the pairs were constructed so that they would be highly likely to be associated with particular objects. Subjects who had memorized the word pair "ocean-moon," when later asked to name a detergent, were expected to say "Tide." While the word pairing was effective in raising the rate of naming the expected object from an average of 10 to 20 percent, subjects rarely mentioned the word pairs as a reason for thinking of the expected object. Instead, subjects named some feature of the object (Tide works best) or some personal connection (My family uses Tide). In another study, under the guise of conducting a consumer survey, passersby were asked to compare articles of clothing (four nightgowns or four pairs of stockings), to select the item with the best quality, and to give a reason for their choice. The subjects' preferences were actually influenced by the left-to-right positioning of the items. The items on the extreme right were overchosen, but no subject named position of the item as a reason for the choice. Nisbett and Wilson explain the results of their series of experiments, as well as other research, by claiming that we perceive causal connections that are in line with our expectations. That is, we usually determine causality by our cognitive framework and not by doing a presence-absence test. Furthermore, the causal connections in our cognitive framework are typically ones that we share with other members of our culture. Much of our knowledge is culturally transmitted. Therefore, outside observers will arrive at the same causal conclusions about our behavior as we do. We (and the observer) will be accurate if our cognitive framework is correct and inaccurate if it is not.

You might think that the research examples above unfairly disadvantaged the subjects. After all, the effects of word associations or position cues are difficult to detect. Subjects were in only one condition and could not easily make a presence-absence test. They could not know, for example, that in the absence of memorizing "ocean-moon," fewer people would think of Tide or that any item placed in the first three positions was not likely to be chosen as best. It is small wonder that they used their expectations to explain their behavior. Perhaps we do better in areas where we have large amounts of information and where there are opportunities to observe ourselves over many situations.

To check this possibility, Wilson, Laser, and Stone studied how well people could predict the causes of their own mood. Participants in this study agreed to make nightly ratings for 5 weeks. They rated their mood and seven other factors, such as the weather, their relationships with their friends, and the number of hours of sleep they had had, that might be causally connected to their daily mood. At the end of the 5-week period, the subjects were asked to rate the extent of the relationship between each of the factors and their daily mood for the period of the ratings. The subjects' estimate of the relationship was then compared to an objective index of the relationship between the factors and mood. Another group of subjects served as observers. They were merely asked to rate what they thought the relationship would be, for a typical student at their university, between the factors and mood. In general, the subjects' ratings of factors affecting their own mood was moderately accurate. Their degree of accuracy, however, was virtually identical to that of the observer subjects. Indeed, participants and observers rated the relationship between the factors and mood in similar ways; they revealed shared expectations about the causes of mood.

It is clear that we rely on the causal expectations in our cognitive framework to construe our own experience and that of other people. These expectations are often held in common by people of the same culture. The force of these expectations appears to be strong, and their ready availability may detract from detecting actual causal patterns. Furthermore, such detecting is made more difficult in everyday life by our imperfect memories and by the vast number of factors that could be taken into account. We undoubtedly know many things about ourselves that others do not, but this private knowledge appears to give us little advantage in understanding the causes of our own behavior.

REFERENCES

Nisbett, R. E., & Wilson, T. D. (1977).Telling more than we can know: Verbal reports on mental processes. *Psychological Review, 84,* 231–259.

Wilson, T. D., Laser, P. S., & Stone, J. I. (1982). Judging the predictors of one's own mood: Accuracy and the use of shared theories. *Journal of Experimental Social Psychology, 18,* 537–556.

two activities is that playing tennis gives the body some good exercise, while sitting in the library may exercise the mind. Compared with knowledge of the common effects, knowledge of the noncommon effects would give the perceiver more information about the student's intention in going to the library rather than the tennis court. The number of noncommon effects is also important in making attributions: A large number of noncommon effects does not give the perceiver as much information about another person as a small number, because with a large number the perceiver does not know which of the effects was intended by the other person.

Another factor taken into account by perceivers is whether the noncommon effects are "good" or "bad," "pleasant" or "unpleasant," "desirable" or "undesirable," "positive" or "negative." This is termed the *valence* of the noncommon effects (Jones & McGillis, 1976). When the noncommon effects have a positive value and are highly desirable, not much is learned about the other person; the perceiver knows only that the situation made it desirable to act in a certain manner. However, when the noncommon effects are rather negative and undesirable, it can be assumed that the person made the choice based on some personal characteristic or disposition. For example, playing tennis on a cold, rainy day (viewed as unpleasant, or having a low valence, by a perceiver), tells the perceiver much more about your motives and intentions than would playing tennis on a warm, sunny day.

In summary, a perceiver gains the most information about the personal characteristics of another person from a small number of noncommon effects with negative valence.

Self-Attribution Up to this point, we have been concerned with the attributions we make about the characteristics of other people. How do we attribute certain qualities to ourselves? According to one theory (Bem, 1967, 1972), we use much the same process for self-attribution that we use for other-attribution. Bem suggests that when we want to make attributions about our own behavior, we become observers of that behavior and make attributions much as if we were observing someone else. Therefore, with our own behavior, we first determine whether the environment caused the behavior through some strong external force. If this does not seem likely, we then assume the behavior occurred because of some internal motives or personality traits. It follows, then, that people may "know" their own intentions only by observing, and making inferences from, their own behavior. Internal states are inferred by ruling out external forces. (We might also rely on our own expectations—recall Inquiry 9.1, page 350.)

Jones and Nisbett (1972) hypothesized that, even though the processes may be similar, self- and other-attribution are different in some ways. According to Jones and Nisbett, we tend to see our behavior as being controlled more by the situation, while we see the behavior of others as caused more by internal forces. One of the reasons for this difference in the attributions made to ourselves and others may be what is called the figure-ground relationship. We see ourselves as reasonably stable personalities interacting with a changing environment. Since it is the environment that changes, we attribute changes in our behavior to the changing situation. When we observe other people's behavior, however, we see the environment as the stable factor and the person as variable. For example, when you see two people behaving very differently in a similar setting, you attribute the difference to their personal characteristics, even though the precise situational forces bearing on each one may be somewhat different. Evidence supporting this hypothesis has been provided by a large number of experiments. For example, Nisbett et al. (1973) found that when students were explaining their reasons for choosing their college major, they tended to talk about both their own personal qualities and the qualities of the subject and saw these two as being equally important in their choice. When explaining a friend's choice of major, however, they tended to emphasize the friend's personal characteristics more than the qualities of the subject matter.

A FINAL WORD ON SOCIAL PERCEPTION

The processes by which we form our social perceptions of others are efficient means for guiding our reactions to others. Our social perceptions are grounded in our observations of others—their physical characteristics and their behavior in particular settings. Our observations provide the information that is converted into meaningful inferences by our cognitive framework. At a minimum, this process involves placing the information into cognitive categories, which are related to other categories. We can, thus, make simple inferences from minimum data or combine rich sets of information into overall impressions. Under some circumstances, we can also make inferences about the causes of other people's behavior as well as our own behavior. Despite the smoothness by which the process works, however, it does not guarantee accuracy or comparability with others' observations. We each make a personal contribution to the process because we may have different ways of categorizing the information, individual aspects to our cognitive framework, or some unique way of combining information. Nevertheless, the end result of social-perception processes determines how we react to others and how we see ourselves.

Social Influence

Social influence is demonstrated in such diverse events as a child imitating an older brother or sister, a juror conforming to the votes of other jurors, or a fraternity pledge obeying the commands of a pledgemaster. These events illustrate influence because they contain two critical elements: (1) someone's intervention, (2) inducing change in another person, the FP. The source of the intervention will be termed the *agent*. Influence

Figure 9.4

Social-influence concepts.

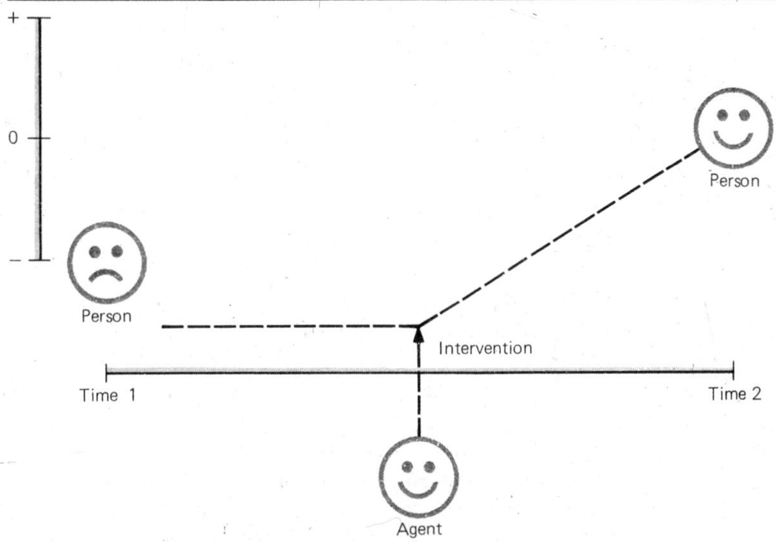

Characteristic of Agent +	Type of Intervention	= Concepts
Stranger(s)	Physical presence	Social facilitation
A person	Modeling of behavior	Imitation
Group	Consensus about preferred response	Compliance to others
Norms	Situations that increase salience of norm	Conformity to norms
Authority figure	Commands	Obedience

phenomena—including imitation, conformity, and obedience—always contain an agent who has caused change in the FP. Influence situations can be differentiated by noting the different characteristics of the agent and of the behaviors that make up the intervention. On this basis, Figure 9.4 schematically represents five of the most frequently studied influence situations. It can be seen, for example, that imitation is defined by a single agent who models behavior that the FP copies. The five areas depicted have separate and distinct research traditions in which the terms *influence*, *conformity*, and *compliance* are often used interchangeably. Let us now turn to discussing what is known about these influence situations.

Have you ever observed that athletes—runners, for example—seem to perform better when they are actually competing with someone than when they are running alone against the clock? One of the first experiments in social psychology was conducted to investigate this interesting observation. In the 1890s, Norman Triplett, a psychologist and bicycle-racing enthusiast, observed that cyclists rode faster when they were in competition than when they rode alone. Suspecting a psychological, and not a physical cause (such as slipstreaming), he devised a laboratory experiment. In agreement with the cycling observation, Triplett found that children wound a fishing reel faster in the presence of other children performing the same task. For many years it was assumed that the presence of others would always improve performance, and thus these effects were labeled *social facilitation*. As research accumulated, however, some contradictory findings emerged. For example, Floyd Allport (in the 1920s) and John F. Dashiell (in the 1930s) reported that while subjects did respond at a higher rate when other people were present, their errors also increased.

The differing results of social-facilitation experiments were finally reconciled by Zajonc (1965). He proposed that the presence of others increases an individual's general arousal level (Chapter 8, page 318), which in turn enhances performance of *dominant*—meaning strong and well-learned—responses. If dominant responses are required, performance will improve in the presence of others; but if weak, poorly learned responses are required, performance will suffer. For example, a well-trained musician would, according to this theory, perform better when others are present, but a beginning piano student would make more mistakes when playing in a recital than when practicing at home.

Further analysis of the social-facilitation effect indicates, however, that the mere presence of other people does not always result in improved performance of a dominant response. Social facilitation is most likely to occur when the people in the group are involved in some way in the task being performed. The performance of the trained musician in the above example would probably not be enhanced if the people in the audience were not listening to the performance and evaluating it. When the people in the group are involved in some way with the task being performed, an individual in the group can see how well he or she is doing in comparison with others or can expect praise or criticism from other group members. Thus motivation is increased (Cottrell, 1972; Sanders et al., 1978).

In summary, then, it seems that social facilitation occurs most readily for strong responses in situations where the presence of others is

Figure 9.5

Social facilitation: We often do our best in the presence of others. (From *Willett/ Raleigh News & Observer*, 1983; Picture Group Inc.)

motivating. Remember the question that opened this section on social facilitation? Now we can see why a runner tends to do better in the presence of competitors. Running, of course, is a well-practiced, strong, dominant response. Furthermore, all the people in the competition are doing the same thing, and thus the runner is motivated by the anticipation of praise for a good performance or some form of criticism for a poor one. Here we have a situation favorable for social facilitation: increased group-induced motivation acts on a strong, or dominant, response. This is the way world records are set.

IMITATION

Imitation involves changes in the focal person's (FP's) behavior that match or copy the agent's behavior. Like social facilitation, it occurs without the agent expressing any intent to influence the FP. (This may be the reason it has been termed the sincerest form of flattery.) Imitation merely requires that the agent go through some behavior sequence in the FP's presence. Whereas social-facilitation effects are measured by the change in the rate and strength of behaviors, the extent of the FP's imitation is gauged by the degree of similarity between the behavior the agent models and the FP's subsequent behavior. As we have already mentioned in Chapter 4, page 172, humans are among the animals that have an inborn capacity to imitate. Observing a model can provide the FP with information about the skills required to perform similar behavior as well as some reassurance about the appropriateness of the behavior in a particular situation. Social psychologists have shown that people imitate a variety of behaviors in many different settings. Children have been observed to imitate a model's aggressive behavior as well as a model's helping behavior (Chapter 11, page 446). Even college students imitate. Their rate of drinking beer, for example, was shown to be affected by the drinking rate of a model (DeRicco, 1978). The model's rate was matched by the students, whether the rate was set at one-half less or one-half more than the students' usual rate. Imitation appears to be optimized when the model has high attention value and possesses characteristics attractive to an FP who is in an ambiguous setting.

SOCIAL CONFORMITY

Conformity is the term used to refer to the situation in which individuals change their beliefs or behaviors so that they become more similar to those of other group members. Conformity pressures can arise in any situation involving other people. Although some instances of conformity are very similar to examples of imitation, they typically involve more than one person modeling the behavior and require the FP to respond in the presence of the agents. In conformity situations, the agents have at least some interest in changing the FP's behavior. Indeed, in these situations, if the FP's behavior does *not* change, it is termed *deviant,* while no special term exists for describing the FP's lack of change in imitation situations. Although conformity pressures can exist in situations that are structurally close to imitation situations, they also occur in circumstances that are starkly different. In particular, conformity pressures arise when the FP and the agents are members of an intact group and the agents not only make persuasive interventions but also have the power to apply sanctions. The smiles and warm affection directed toward the FP who conforms versus the jibes, disapproving looks, or possible loss of status for the one who does not are certainly potent sources of control.

Because each of us spends considerable time with other people, in formal and informal groups, we are frequently exposed to conformity pressures with which we comply. Many of the things we do in groups—the clothes we wear, the opinions we express, the choices we endorse—are the results of conformity pressures. Acting in concert with others is such an integral part of social life that it is of little interest to ask whether or not conformity occurs. Rather, the issues that have interested social psychologists are what factors will increase the amount of social conformity, and why do people conform? We will try to answer these questions below.

Conformity to Others Group pressures are particularly effective in situations where the FP is uncertain about the proper response to make. Consider a new student who has a choice among various sections of a required course. Such an FP is highly likely to accept the advice of knowledgeable seniors, especially if the seniors are all in agreement about which section is best. But are conformity pressures strong enough to induce an FP to agree with a group consensus that is obviously wrong, even when the FP knows the correct choice? This question was answered by Solomon Asch (1951, 1956) in one of the most famous series of conformity experiments.

The FPs, or subjects, in the experiment were asked to choose which of three comparison lines was the same length as a standard line (Figure 9.6). Control subjects who were by themselves when they looked at the lines were accurate about 99 percent of the time (Figure 9.7). This established a baseline against which to judge the responses of other subjects in the experimental sessions.

In the experimental sessions, judgments were made in a group. Each session typically employed only one actual subject in a group of seven to nine other people—the agents—who had been coached to choose one of the nonmatching lines. These people were confederates, or "helpers," of the experimenter. The experiment was set up so that the real experimental subject heard the judgments of all but one of these confederates before choosing one of the comparison lines. Thus, with nine confederates, the subject would hear eight of them choose a particular comparison line that did not match the standard line. For instance, in Figure 9.6, eight confederates would say that line 2 matched the standard line—a judgment at variance with what the subject must have perceived.

What did the experimental FPs do when the majority judgment differed from their own? In general, they showed some tendency to conform to the group. On the average, only about 67 percent of their judgments, compared with 99 percent of control-subject judgments, were correct (Figure 9.7). In other words, about 33 percent of the judgments made by the experimental subjects in the group situation were wrong. Not all the subjects conformed, however; there were large individual differences in conformity, and those who conformed did not do so on every trial. Other findings were: (1) Subjects conformed most often when their judgments were "public"—that is, when the majority could hear their answers; (2) if the majority was not unanimous, that is, if one of the confederates was instructed to disagree with the majority opinion, the amount of conformity was greatly reduced; and (3) in experiments performed with fewer than seven or nine confederates, as the number of confederates in the majority group increased from one to three, conformity increased, but further increases in the size of the majority did not result in greater conformity.

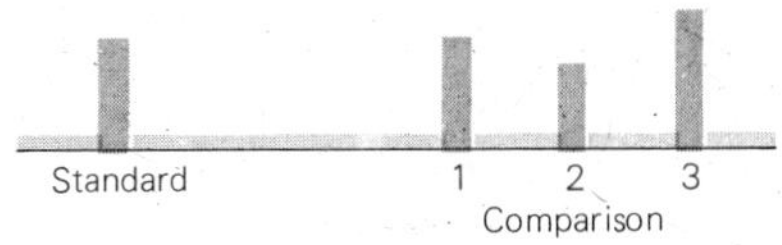

Figure 9.6

In the Asch conformity experiments, subjects were asked to decide which of the comparison lines at the right was the same length as the standard line at the left. (Modified from Asch, 1956.)

Figure 9.7

Results of a series of conformity experiments in which the task was to choose a line that matched a standard line. Control subjects were alone when they made their choices; experimental subjects made their choices as a member of a group of people who had been instructed to choose an incorrect line on the critical trials. (Modified from Asch, 1956.)

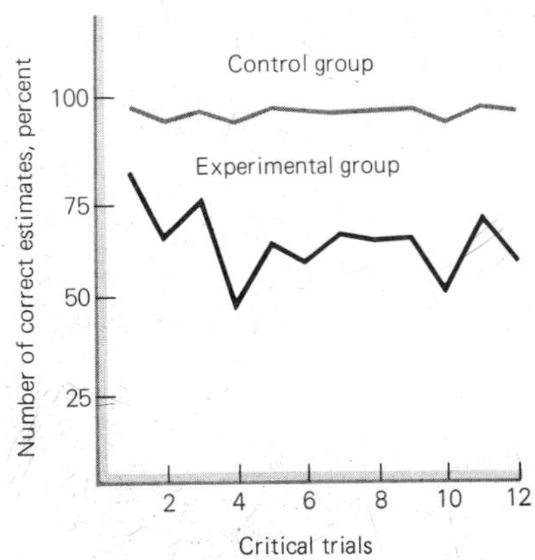

The fact of conformity behavior, as well as the factors that enhance its occurrence, have been demonstrated in literally thousands of laboratory and field studies. In addition to the factors studied by Asch, it has been shown that occurrences of conformity will increase when the FP is fatigued, is uninformed about the stimuli, has a lower status than the agents, or is desirous of further interactions with the agent. Instead of expanding this list, we will describe Donald Campbell's (1961) insightful analysis of conformity situations. Once you understand this analysis, you should be able to generate your own list of factors that will increase or decrease occurrences of conformity.

Campbell began by noting that we acquire information about the world either through direct experience, the *personal modes,* or through what we learn from others, the *social modes*. In most situations, information from one mode agrees with information from the other. If your friend tells you that the tickets for the game next week are sold out, you are likely to hear the same news if you go to the ticket office. We learn to rely on others for valid information, especially if we have no reason to believe that others wish to deceive us. Conformity situations, however, are ones in which a sharp discrepancy exists between the information gained through personal and social modes. Conformity situations are conflict situations. The information acquired by an FP through the personal mode—for example, line 1 is the correct response—is in clear contrast to the information presented to the FP from the social mode—all the others say line 2 is correct. Whether or not conformity occurs will depend upon whether the FP resolves the conflict in favor of the personal modes or the social modes; it is the relative weighting of social and personal modes that determines the FP's response. Factors that make an FP more likely to weight the social mode (fatigue, inexperience, lower status, or hope of further interactions with group members) will increase the rate of conformity. On the other hand, anything which increases an FP's weighting of the personal mode (mental alertness, expertness, higher status, or indifference to other group members) will reduce the rate of conformity.

Conformity to Norms and Rules *Norms,* or rules, are standards of behavior that are agreed upon by group members and exert a powerful influence on social behavior. They are, thus, considered to have agent characteristics, even though no actual person need be present when their influence occurs. At a minimum, it is only necessary for the FP to be aware of the norm or rule to be subject to its pressure.

Norms and rules can be formal agreements, such as laws or contracts, or they can be informal agreements, such as an agreed-upon time to begin a meeting. Any group has an interest in maintaining conformity to its important norms. Members' behavior is monitored to identify deviants, and punishment is administered to bring the behavior back into line. Enforcement of laws in our country is a good example of this process. Laws are enacted by elected legislators and monitored by police, who identify deviants that are judged by the courts. Informal norms and rules are upheld by the same process, even though the steps are not so explicit. Groups typically have no special means for monitoring potential deviants, but once deviancy is identified, group members apply negative sanctions. Social punishments available to group members range in severity from the expression of mild disapproval to expulsion of the deviant from the group. Deviants are punished so that they will conform in the future. In addition to bringing the deviant back into the fold, punishment also serves functions for the group members. It has been argued that applying punishments tends to establish the importance of the group, prevents the deviant from having an inequitable advantage, and reaffirms the validity of the violated norm.

Groups could not exist without rules that regulate the members' behavior. What, then, do the members gain by their willingness to be regulated? The chief advantage of rules or norms has been discussed by Thibaut and Kelley (1959), who note that rules serve to coordinate behavior; that is, rules are established to avoid conflicts—conflicts between a member's self-interest and the group's interest or conflicts between group members and non-group members. Informal norms about when meetings begin, what goes on in meetings, and who performs what duties help coordinate the behavior of the participants. In the wider society, for example, common times for eating meals and implicit agreements about the proper physical distance to be maintained between two people conversing help to smooth interactions among people by coordinating their behavior.

It should be recognized that groups do not always uphold the norms that govern their behavior when interacting with other groups. Competitive sports serve as a good example. Participants in competitive sports are governed by a formal set of rules that specify acceptable and unacceptable behavior, as well as the penalties associated with engaging in unacceptable behavior. Typical penalties for deviant behavior are a loss of yardage (football), temporary removal from play (ice hockey), advancement to the next base (baseball), and so forth. Rule violations are a special problem in contact and collision sports because violations contribute to injuries. Obtaining a higher degree of conformity to game rules, according to Silva (1981), will not occur by stricter enforcement of the rules. He believes that rule violations are condoned by the athlete's peers, coaches, and fans because violations are rewarding (in the sense of giving the violators a tactical advantage). This analysis led Silva (1983) to hypothesize that the longer people were involved in organized sports the more accepting they would become of rule violations. Groups of subjects were shown slides of various sport scenes; seven of the slides showed clear instances of rule violations. Subjects, whose length of involvement in sports was known, were asked to rate each slide for acceptability of the behavior. On the average, women rated the violations as less acceptable than did men. Silva's hypothesis was confirmed for the men, but not for the women; as the men's involvement in sports increased, so did their acceptability ratings of rule violations. He suggests that socialization into competitive sports involves pressures to perceive rule violations as legitimate. These pressures are most intense at the higher levels of competitiveness. Until the structure of the rules governing sports are changed to allow for stiffer penalties, players and coaches will remain in the difficult conflict situation of choosing between acceptable behavior that results in tactical disadvantages and inappropriate (and often dangerous) behavior that bestows an advantage.

OBEDIENCE

Obedience refers to situations in which the agent has the legitimate right to influence the FP and the FP has the obligation to obey. Such reciprocal role relationships are most clearly demonstrated when the agent has a higher status than the FP and the roles are part of a social system in which a higher status clearly defines influence over a lower status, such as ranks in the military. If a lieutenant orders a private to report for guard duty at 2 A.M., there is nothing remarkable in the bleary-eyed private's obedience. The domain of behaviors over which an agent has a right to influence an FP is always limited. The lieutenant could not legitimately order the private to reenlist or to steal army property. The limits of an agent's right and an FP's obligations are usually not precisely defined, and some commands may fall in an ambiguous area. One such area is when an agent's commands are in conflict with the FP's personal values. It is of great interest to know how these conflicts are resolved. Here is a description of one of the best known—and perhaps most disturbing—series of obedience experiments (Milgram, 1963, 1965).

Subjects of Milgram's experiments were men who responded to newspaper ads and were paid $4.50 for coming to the laboratory. The situation was described as a learning experiment in which one person, the teacher, would shock another person, the learner, after each mistake while learning a list of paired words. On the basis of a rigged drawing, the subject was always assigned the role of teacher, while a male confederate of the experimenter was assigned the role of learner. The learner was strapped into an electric chair, while the real subject was taken into another room where the electric shock apparatus was located. Actually, *no shocks were administered*, but the elaborate equipment led the subject to believe he would be administering painful shocks to the learner.

The shock apparatus (Figure 9.8) contained 30 switches indicating levels of shock from 15 to 450 volts, with labels such as "Slight shock," "Moderate shock," "Danger: Severe shock," and finally "XXX." The teacher was to shock the learner for each mistake made in learning, and the level of shock was to increase one increment with each error. As the experiment progressed, the learner responded appropriately, with occasional mistakes. At several points as the shock level increased, the learner would cry out that the shock was getting painful, or he could be heard kicking the wall. At 300 volts he stopped giving answers, while the teacher was instructed by the experimenter to continue increasing the level of shock. If the subject showed any reluctance, the experimenter prodded him to continue, saying it was necessary or required by the experiment. Obedience was measured by the amount of shock the subject was willing to administer to the learner.

Before conducting this experiment, Milgram described it to several groups of people, all of whom predicted that very few, if any, of the subjects would follow the experimenter's commands and give shocks up to 450 volts. Contrary to expectations, however, 26 of his 40 subjects (65 percent) continued to give shocks up to the 450-volt level, even though they believed they were hurting another

Figure 9.8

Left, the apparatus (a phony shock generator) used in the Milgram experiment on obedience; the levers indicate voltages from 15 to 450 volts. *Right*, a learner in the Milgram experiment is strapped into a chair, and electrodes are attached to his wrist. (From the film *Obedience*, distributed by the New York University Film Library. Reprinted by permission. Copyright 1965 by Stanley Milgram.)

person and showed signs of a great deal of tension—trembling, stuttering, nervous laughter.

In other studies within this series, Milgram has shown that the rate of obedience decreases as the teacher is placed physically closer to the learner and as the presence of the experimenter is made less salient. In keeping with Campbell's analysis (page 357), it appears that the conflict between the command and the personal value is resolved in favor of less obedience by any factors that increase the weighting of the personal value or decrease the weighting of the command.

Milgram has been cautious in interpreting his results, but they do suggest an awesome tendency for people to obey others who are in authority. Some have seen Milgram's results as a dramatic experimental example that parallels certain excesses of obedience in history. In particular, the results have been compared with the high incidence of obedience shown by the people who ran the Nazi death camps during World War II. There is a disturbing similarity between the rationalizations of Milgram's subjects and those used by some of the Nazis who were tried for war crimes. Both tended to excuse their behavior by saying they were only obeying orders.

The parallel with dramatic real-life situations extends to the conflicts faced by some of our soldiers while fighting in Vietnam, to the people who participated in the Watergate scandal, and to those who participated in the mass suicide at Jonestown (Inquiry 9.2). These are all examples of destructive obedience to authority. But each situation also has unique characteristics that limit the applicability of Milgram's research. It is important to note, for example, that the authority figure in Milgram's experiments was a scientist; subjects might reasonably conclude that a scientist would not give orders that would lead to death or injury simply as part of an experiment. Gilbert (1981) has suggested that the incremental nature of the shock procedure may also have contributed to the high levels of obedience. The start of the shock series was harmless, and the shock intensities were gradually increased to the harmful range. Subjects might have been more disobedient if a definite breakpoint existed that signaled a switch from harmless to harmful.

The Milgram experiments do suggest, however, that no group of people is immune to the pressures producing destructive obedience. Although we are beginning to understand some of the conditions under which destructive obedience is resisted, there is much that is not known. It must also be recognized that the participants in such dramatic conflicts may often obey merely because that has been the most adaptive response in most situations. Obedience is ordinarily a virtue, and disobedience is not. It is, after all, extremely rare for a group to reward effective disobedience. In the nineteenth century, the Hapsburg rulers in Austria supposedly created a medal for soldiers whose disobedience turned out to benefit the group, although there is apparently no official listing of such a medal. While obedience to legitimate commands is a normal phenomenon, pressures for obedience that go against higher ethical principles become malevolent.

SOME REASONS FOR CONFORMITY AND OBEDIENCE

You have already learned Campbell's answer to the question, Why do people conform? He claims that they conform because they give more weight to the social information modes than to the personal information modes; that is, conformity occurs in situations where the information provided by others is the most important information. Social psychologists have also proposed three other general causes of conformity: social comparison processes, avoidance of social disapproval, and the need to be liked and accepted.

The *social comparison idea* (Festinger, 1954) goes something like this: In ambiguous situations where we do not know what is "right" and expected of us, we look at the opinions and behaviors of people who are similar to us before deciding what to do. In other words, we resolve the ambiguity about what to do by observing people similar to ourselves and follow their lead. To illustrate the social comparison process as a cause of conformity, recall a time when you were in an unfamiliar social situation—perhaps when you went to a church where the rituals were unfamiliar. Did you look around at other people to see what to do and, following their lead, conform in your behavior to what others in the group were doing?

As noted above, a person who deviates, especially from a consensus among the others, can expect to be the target of social disapproval. The deviant, after all, creates some discomfort for the others, who may be annoyed because the group's smooth progress toward its goals is disrupted or because doubt is raised about whether their shared view of things is actually correct. Most of us find social disapproval quite unpleasant, and we avoid doing things that elicit scorn, ridicule, or derision. By conforming, we avoid the possibility of social censure. Who but a child would be willing to break the consensus by yelling, "The emperor has no clothes"?

Not only do people conform to avoid social punishment, but also they may "go along" because in so doing they can meet their needs to be liked and accepted by others. While it is important for us all, this reason for conformity seems especially prominent during the preteen and teen years. Because of a strong need to be accepted, a young person may be especially prone to conform to group norms concerning clothes, drinking habits, drug use, and sexual behavior.

A FINAL WORD CONCERNING CONFORMITY

Acts of conformity are clearly woven through the fabric of our interpersonal lives. The circumstances of the interaction determine whether a given act of conformity is good or bad, adaptive or maladaptive. It has, however, been pointed out by Campbell (1961) that conformity during group decision making is always maladaptive. The activity of groups solving a problem can be divided into two phases. During an initial phase, the group gathers information, usually by polling the opinions of group members. Once all available information is known, the group enters the decision-making phase, in which the information is weighted and the best alternative is selected. Conformity to a perceived consensus during the initial phase is maladaptive because it distorts the true picture of the extent of support that exists for an alternative. Janis (1983) has coined the term "groupthink" for maladaptive conformity in group decision making. He has extended the Campbell analysis by identifying the conditions that make groupthink highly likely, such as group members' illusions of invulnerability and unanimity. The discussions of President

Inquiry 9.2

HOW COULD THE JONESTOWN HOLOCAUST HAVE OCCURRED?

The answer to this question necessarily involves tracing the career of the Rev. Jim Jones and the People's Temple movement. From information revealed in the press and in books, a fairly clear picture emerges of the major contributing factors to the final catastrophe. It should be noted at the outset that the term "mass suicide" does not adequately cover the final events in Jonestown. Wooden (1981) states that 276 children—from babies to those near the age of consent—died there. Because under English law, minors cannot legally consent to suicide, their deaths were actually murder. Still, this leaves a very large number of adults who, with only a minor amount of dissent, complied with the urgings of Jones to participate in the final macabre ritual.

James W. Jones was born near Lynn, Indiana, in 1931. After graduating from high school in 1949, he married Marceline Baldwin, who was a nurse and who would support his activities to the last day of their lives. In 1953, Jones opened a small interdenominational church, the Christian Assembly of God, with the laudable intent of eliminating social injustices and improving interracial harmony within the context of a fundamentalist religion. As his church attracted an increasing number of black members, he and his wife created a multiracial family (seven of their eight children were adopted, including a black, a Chinese, and a Korean child), and Jones became outspoken on civil rights issues. After spending 2 years working as a missionary in Brazil, where he established orphanages and a mission, he returned to Indiana. By 1964, Jones had been ordained a minister of the Christian Churches (Disciples of Christ) and had changed the name of his church to People's Temple Full Gospel.

Partly out of fear of the impending nuclear war he had predicted, Mr. Jones and about 100 followers migrated to Ukiah, California. The northern part of the state was thought to be one of the safest havens from the ravages of nuclear war. In Ukiah, Mr. Jones displayed his considerable organizational skills. While the dedication to social programs continued, an increasing amount of energy went into creating the economic and legal structure that would support future activities. The People's Temple was very successful in raising money. In addition to weekend fund-raising trips to San Francisco and Los Angeles, the membership was increasingly pressured into making contributions. The test of loyalty to the group soon became the willingness of members to assign all of their assets to the People's Temple, which acquired members' houses, wages, and monthly social security checks. The church thus changed from being merely a force in members' lives to being in total control of their life decisions.

The People's Temple movement flourished. Its interracial membership swelled to include the elderly, the well-educated, and the disaffected. Most people were undoubtedly drawn by the avowed social values or the religious fervor of the movement. Financial success also continued. By 1971, the church had bought a building in San Francisco, in the heart of the black community, and a church in Los Angeles. It continued to be a showcase of social programs while also increasing its political clout in the community. Mr. Jones could provide instant crowds for demonstrations or speakers and could supply a cadre of workers for the political candidates he selected. Close races for mayor and district attorney in San Francisco are thought to have been won because of the support of the People's Temple. The American patronage system insured that Mr. Jones could place loyal church members in sensitive, relevant government positions. Tim Stoen, who was at that time Jones's trusted lawyer, was appointed to the district attorney's staff, and Mr. Jones was appointed to the city's housing administration, becoming chairman after a few months. Mr. Jones received many public accolades during this period for the work of the People's Temple.

In the midst of this dazzling success, a darker side to the movement was also surfacing. A variety of serious charges were leveled by people who had managed to leave the movement. These charges—obtaining member obedience by physical brutality, making fradulent claims in order to obtain legal guardianship over children, sexual abuses, fake "healing" ceremonies, death threats to members considering leaving the group—although later shown to be quite accurate, were not given much currency at the time. Mr. Jones, through his political muscle, had the power to prevent charges from becoming public or to muster potent countercharges to discredit his detractors. Furthermore, a large number of highly respected people came to his defense. During this period, Mr. Jones's self-concept and ideology also appeared to change. Christian religious themes diminished to the vanishing point and were replaced by messianic, revolutionary messages suggesting strong parallels between Jones and Christ. Members were required to address Jones as "Father," and he became the sole arbiter of correct values and behavior.

The mass migration to Jonestown in the summer of 1977 was precipitated by the impending publication, in *New West* magazine, of an investigative article which Jones had tried to squelch. What was to have been a gradual shift of members, spanning several years, became a 6-week exodus. While a center remained in San Francisco, the bulk of the membership was transferred to Guyana. Another headquarters was established in the Guyanese capital, Georgetown, staffed by about 40 people. The procedures used so successfully in California were repeated. High-ranking Guyanese officials were courted by members of the People's Temple and willingly cast a protective net around Jonestown. The residents of Jonestown became totally isolated from the rest of the world. Surrounded by an inhospitable jungle, whose dangers were frequently exaggerated by Jones, and linked to their centers in Georgetown and San Francisco only by a radio controlled by Jones, any hope of members receiving accurate information disappeared. Jones portrayed the outside world as filled with CIA- or FBI-inspired conspiracies aimed at destroying their revolutionary movement. Except for an inner circle of advisers and an armed "security" guard, members' lives were harsh, characterized by crowded living conditions, long work hours in the fields, debilitating diets of rice and beans, and frequent, exhausting meetings called by Jones. Despite these factors, however, the settlement made progress. Aside from living quarters, the compound included a medical tent, a playground, an electric generator, and a classroom. Crops were planted and harvested, and some cottage industries were started.

The isolation of Jonestown was occasionally penetrated by outsiders, including some members of the American embassy in Georgetown. These visits typically resulted in positive reports of the encampment's achievements. In response to the pleadings of the Concerned Relatives group, who had submitted a long list of human rights violations committed by Jones, American consular officials made four visits during 1978. According to Krause, (1978): ". . . the last occasion was on November 7, eleven days before the massacre" (p.10). Their report gave no hint of alarm. Indeed, minutes before his assassination on the dirt airstrip outside of Jonestown, Representative Ryan finished an interview with NBC's Don Harris with the observation, "Can I just add, there are a lot of good people who are there on a positive and supportive and idealistic basis, trying to do something that is different and important to them" (Reston, 1981, p. 320). These positive perceptions were primarily the result of viewing a charade carefully orchestrated by Jones. It was in a comparable web of deceit and lies that members of the People's Temple met their death.

The deaths at Jonestown were clearly not the result of an impulsive decision by a leader and his dedicated band of followers. The final ritual was undoubtedly planned for years by Jones as his mark on history. His mind seems to have become gradually unhinged from reality and obsessed by a vision of glory. According to Wooden (1981), there were as many as 42 suicide rehearsals, dating back to the time they resided in California. These "white nights" would begin with Jones announcing some alarming news and would end with Jones saying that the potion they had collectively drunk actually contained no poison. The concept of a mass death was implanted in the environment of the People's Temple. Though it may still seem incomprehensible to us that so many adults would be obedient to Jones's demented plan, his manipulations of perceptions, of influence, and of interpersonal relationships provide ample bases for the members' conformity.

The people of Jonestown must have harbored many strange perceptions, not only of their leader but also of the outside world. While it will never be known how many of them believed in Jones's messianic self-delusions, it seems clear that they accepted his self-presentation of other characteristics as well as his views and attributions about events. Mass death was termed "revolutionary suicide." Father" was all-knowing and suffered for the sins of any member. The rainy season was attributed to a CIA plot to destroy their crops, and Jones's varied sexual activities were attributed by him to some vague revolutionary goal that he pursued despite the "pain" the activities caused him. The U.S. government was always portrayed as hostile to the movement. Once Representative Ryan and some members of his group were murdered by Jonestown gunmen, Jones's claim that deadly retaliation was on its way must have seemed credible. But even before this final distortion, the coercive environment prevented people from revealing their own perceptions. Outside visitors were rarely able to penetrate the reality of Jonestown. Jones was certainly no stranger to the staging of fake events. His "ministerial" routines included such tricks as "passing" cancer, an illusion created by the use of chicken organs. He also staged attempts on his own life, events which were then offered as evidence for the existence of hostile plots against the movement. In Jonestown, the days before a visit were always used to rehearse what people would say and do. It is no wonder that the visitors, especially those expecting to see barbed-wire enclosures and imprisoned hostages eager to be freed, were pleasantly surprised. They saw a community busily at work and heard people say they were happy with their lives. They accepted at face value what they heard and saw because it was extremely difficult to detect the psychological pressures producing the manufactured events.

Interviews with former members document that some people were never fooled by the fake healings or staged assassination attempts. They accepted these as necessary means to further noble ends. But such rationalizations were a trap. By suspending their skepticism, Reston (1981) suggests, they were required to increase their faith in Jones. It represented, in Campbell's terms (see page 357), a lessened weighting of the personal modes of judgment and an increased reliance on the social modes, thereby producing heightened conformity.

Like members of any group, members of the People's Temple were subjected to pressures to conform to group norms, albeit norms primarily defined by Jones. Most people who became members were undoubtedly drawn to the movement because of its professed values and goals. They must have wanted to be "good" members and may have accepted the conformity pressures as guides to proper behavior. The migration to California seemed to coincide with an escalation of these pressures beyond any reasonable bounds. The typical pressures were augmented by such techniques as requiring members, during group meetings, to make "self-confessions" of imperfections and by physical beatings of members whose behavior was deemed deviant (as well as those who objected to the beatings). The movement became extremely intolerant of any behavior that deviated from group norms. Disagreement with Jones was not mere deviation; it was an act of disloyalty. The harshest treatment was reserved for anyone expressing a desire to leave the group. "Defectors" were a threat to the group's survival, and those who left often felt they were risking their lives. While resistance to group pressures is always difficult, resistance was further lowered by having members assign their economic resources to the movement and by systematically cutting them off from their families. The loss of economic independence and of possible family support increased the members' dependence upon the group.

Jones seemed to push to the limit all types of social influence techniques, with one exception—imitation. As the movement prospered, he seemed less and less inclined to serve as a role model for the values and behaviors expected of members. For example, while expecting truthfulness from members, he engaged in trickery and staged events; while decrying homosexuality, he engaged in homosexual relationships; and while the members' Guyana diet was meager, he ate from his own refrigerator. The members' dependence upon their leader's word for what was normative, thus, was increased because they could not use his behavior as a standard for judging what was correct.

Abolishing family relationships was an important goal for the leaders of the movement. Not only were members systematically alienated from their own families, but kinship relationships were intentionally destroyed. For example, the custody of children was legally transferred from natural parents to other members, some husbands and wives were forbidden to live together, and the only relationship that was supported was one with "Father." All members were expected to keep surveillance over each other and to report deviations to Jones. The decision to report information was complicated by the knowledge that events were sometimes staged. Members who overheard someone's desire to leave the movement could never be certain whether the "overhearing" was genuine or a test of their loyalty. In view of the consequences of being disloyal, there must have been a high rate of reporting.

All of the factors noted above were present at the final meeting in Jonestown. Recall that the members were physically and psychologically isolated from their families and the rest of the world; they had a long history of conformity to a leader obsessed by his own martyrdom; they believed in the existence of a hostile world that was probably about to retaliate for the slaying of Representative Ryan; they accepted something termed "revolutionary suicide" as a positive value; they were undernourished and weary; and they had undergone many rehearsals of their own deaths. It is no wonder that Jones's macabre scheme was implemented with so little resistance. During the last meeting, only one member publicly objected to the plan. She was quickly silenced by Jones and the others. Some people, probably from among the armed security guards, did escape into the jungle, but otherwise all of the residents perished in total obedience to Jones. The People's Temple members who were less under the sway of these factors, those in the Georgetown and San Francisco centers, did not comply (with the possible exception of one person) with Jones's command for self-destruction. Their obedience had some limits.

Because the People's Temple movement has much in common with other current cults, it would be reassuring to think that the aftermath of Jonestown produced corrective mechanisms in how society protects individuals, especially children, from suffering similar consequences. Unfortunately, this is not the case. The issues are not simple. In part, they involve the rights of individuals to choose their own style of life. The test of a free choice is typically a person's self-description that he or she is content or happy with the present situation. Yet as we have seen with Jonestown, such self-descriptions can be meaningless in a coercive environment that controls self-descriptions. Our society also has a cherished heritage of protecting all forms of religious worship. This value is embodied in the First Amendment. Government interference with religious cults conflicts with individual and First Amendment rights. Although the government has an obligation to investigate charges of violations of law, even this rudimentary form of protection is apparently of little help. Wooden (1981) asserts that our government has shown a great reluctance ". . . to pursue charges of fraud, misappropriation of property, forgery, and widespread child abuse and neglect, perpetrated within the private confines of churches" (p. 204). It seems abundantly clear that the major protection against involvement in circumstances similar to the People's Temple movement remains in the hands of the individual. It is the individual who must not relinquish his or her own personal mode for weighting information. Whatever massive conformity pressures are brought to bear, it is the individual's ultimate responsibility to judge what is true, moral, and correct in the practices of the group to which he or she belongs.

REFERENCES

Krause, C. A. (1978). *Guyana massacre: The eyewitness account.* New York: Berkley.

Reston, J., Jr. (1981), *Our father who art in hell.* New York: Times Books.

Wooden, K. (1981). *The children of Jonestown.* New York: McGraw-Hill.

Kennedy and his advisers before the disastrous Bay of Pigs invasion of 1961 are in accord with Janis's expectations. In general, it is tempting to remain silent when one's view seems deviant. It has been said, "I speak up, not to persuade others, but to encourage like-minded people to voice their views." If a group is to make the best decision possible, it needs to know the true rate of members' preferences. Once this information is known and a decision is made, continued deviance is no longer functional, and the minority must accommodate themselves to the view of the majority.

Social Relationships

We have just seen that the manner in which we relate to other people depends to a great degree on impression formation and social influence processes. While perception and influence are always involved in our relationships with others, social psychologists have studied a number of other factors which help to determine the formation and maintenance of interpersonal relationships. Why are people initially attracted to each other? What processes are involved in shaping the nature of relationships as they develop? The answers to these questions are the focus of the next sections.

INTERPERSONAL ATTRACTION

Think for a moment of all the people with whom you have had some contact today, and consider how much attraction you feel toward each one. It is likely that some are good friends for whom you feel a strong attachment. There are probably also a few whom you dislike and whose company you actively avoid. Most others you would place somewhere between these extremes. Why do you like some people more than others? The general answer is that we like people to the extent that our interactions with them are rewarding or reinforcing. (See Chapter 4, page 149.)

With this in mind, we can examine some of the specific factors that have been found to affect the attraction one person feels for another, in each instance keeping alert to the role of reinforcement.

Proximity One factor which has been shown to affect the degree of attraction one person feels for another is physical nearness, or *proximity*. Although this may seem rather obvious, the magnitude of the effect of proximity on liking may be surprising. Here is an example of the proximity effect from a classic study by Festinger et al. (1950):

> Friendship patterns in several housing complexes were studied. One of the housing complexes consisted of small houses arranged in "U-shaped" formations, with all but the end houses facing a grassy court. The end houses faced outward toward a street. The researchers used questionnaires to determine the friendship patterns among residents of the complex. The results indicated that both sheer distance between houses and the direction in which the houses faced had profound effects upon friendship patterns. Friendships developed most frequently between next-door neighbors and less frequently between people who lived two doors away. As the distance between houses increased, the number of friendships fell so rapidly that it was rare to find friends who lived more than four or five doors from each other. Compared with those whose houses faced the court, those unfortunate people who occupied the end houses which faced away from the court were found to have less than half as many friends in the complex.
>
> A second housing project with a different architectural design provided these investigators with additional evidence of the proximity-attraction relationship. Here, apartments which were near entrances, stairwells, or centrally located mailboxes afforded occupants more popularity. It would appear that any architectural feature which increased the frequency of contact between individuals also increased the likelihood that friendship would develop.

Many explanations for this phenomenon have been suggested. Perhaps we simply prefer pleasant interactions with those individuals we encounter frequently and make a special effort to ensure that our interactions with them run smoothly. Or perhaps repeated exposure to any previously neutral stimulus results in an increasingly positive evaluation. Repeated exposure has been shown to increase liking for various types of nonsocial stimuli, including Japanese ideographs (Moreland & Zajonc, 1976), Turkish words (Zajonc & Rajecki, 1969), and photographs of strangers (Wilson & Nakajo, 1965). In an experiment designed to investigate the effects of repeated exposure upon interpersonal attraction, Saegert, Swap, and Zajonc (1973) arranged a series of brief interactions between strangers that ranged in number from none to ten encounters. The more frequently an individual interacted with another person, the more positively that person was evaluated.

It is also possible to interpret the relationship between proximity and attraction in terms of the rewards these interactions provide. One advantage of interacting with those individuals whom we encounter freqently is that there is less expenditure of effort or cost involved. The difficulty many people experience in maintaining long-distance romances may be viewed in this light. Increased distance increases the costs of maintaining the relationship and decreases the opportunities to experience the rewards that the relationship can provide.

Attitude Similarity Although physical proximity is an important determinant of the formation and maintenance of relationships, it is obvious that we are not attracted to all those people who are in close physical proximity to us. Other factors also influence our feelings of attraction and our choice of relationships. One of the most thoroughly investigated of these factors is *attitude similarity*. Survey studies have shown that friends and marital partners tend to hold similar attitudes on many topics. Such studies, however, cannot distinguish between the attitude similarities that developed over the course of the marriage or friendship and the attitude similarities that existed before formation of the relationship. To establish a causal link between attitude similarity and initial attraction, experimental studies are required. One way to do experiments on attitude similarity is to have participants complete a questionnaire assessing their own attitudes. The questionnaire completed by each participant is then used by the experimenter to construct a second questionnaire that the participant thinks was completed by another person in the experiment. In this manner, the experimenter can systematically vary the proportion of attitudes the participant and a "phantom" partner hold in common. The real participant then looks over the fake questionnaire before being asked to indicate the attractiveness of the partner on rating scales. Using this procedure, it has been discovered (Byrne & Nelson, 1965) that a higher proportion of similar attitudes results in greater attraction (Figure 9.9) and also that the relationship between similarity and attraction is so reliable that it can be represented by a mathematical formula.

Outside the laboratory, of course, the relationship between attitude similarity and attraction is less than perfect. In the laboratory, agreement or disagreement on a small number of topics is generally the only information the participant has about a partner. Situations outside the laboratory are likely to be considerably more complex, with other determinants of attraction at work. In addition, there are some special cases in which dissimilar others may be preferred. For example, when a very similar other is more successful than we are, we may feel more threatened and negative then when a dissimilar other surpasses us. In general, however, the positive relationship between attitude similarity and attraction has been convincingly demonstrated.

The tendency to feel attraction toward similar others may well be based upon a number of assumptions we are likely to make about people we perceive to be similar to ourselves. We may perceive the behavior of similar others as more predictable, which leads us to be more confident that interaction with them will be rewarding (Bramel, 1969). There is also evidence that we may anticipate rewarding interactions with similar others because we expect them to like *us* (Walster & Walster, 1963). Taken together, these various considerations suggest that the perception of attitude similarity leads individuals to expect rewarding interactions, and it is the anticipation of reward which leads to increased attraction.

Figure 9.9

The relationship between the proportion of similar attitudes perceived to be held by another person and the degree of attraction felt toward that person. (After Byrne & Nelson, 1965.)

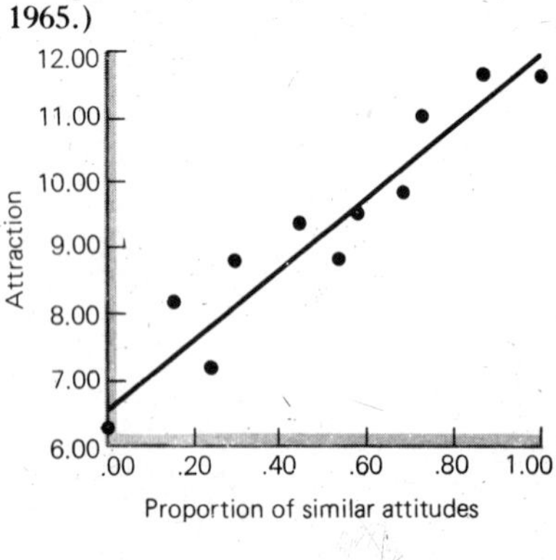

Physical Attractiveness As a society, we Americans implicitly demonstrate our conviction of the importance of physical attractiveness by yearly spending billions of dollars on cosmetics, exercise and diet products, fashionable clothing, and even plastic surgery to enhance physical appearance. Research results also show how important physical attraction is (Walster et al., 1966). A "computer dance" was arranged at a large university, and blind dates were randomly assigned. During the dance the participants were asked to indicate the degree to which they liked their dates and would like to go out with them again. Of a number of attributes measured by the researchers, including personality factors and intelligence, the date's physical attractiveness was the only attribute which affected liking and desire for future dates.

Physical attractiveness also affects people's social participation in everyday life, although the impact is somewhat different for men than for women. Harry Reis and Ladd Wheeler conducted a series of studies in which college students were asked to keep a daily record of any interaction that lasted 10 minutes or longer (Reis et al., 1980, 1982). Among the men, as physical attractiveness increased, so did their amount of social interactions with women, while their amount of interactions with other men decreased. For women, attractiveness had no relation to the amount of social contacts. The quality of social experiences was affected by attractiveness for both men and women. The attractive students, in general, reported more intimate and disclosing interactions.

Physically attractive persons are seen to have many interpersonal advantages over their unattractive counterparts. The research literature provides many illustrations. The physically attractive are advantaged with respect to the perceived quality of their task performance, their job marketability, their expected political success, and their perceived persuasiveness. Even mock juries, deciding on the punishment to be administered for a crime, give less severe punishment to the physically attractive defendant.

The powerful effects of attractiveness may be due to a culturally shared implicit theory of personality (page 345). The category "physically attractive" is linked to many other positive traits in a way that "physically unattractive" is not. The perceiver who categorizes someone as attractive will, thus, assume that he or she possesses a number of other desirable characteristics. The impact of this effect may be heightened because, unlike attitude similarity, information about a potential partner's physical attractiveness is immediately available and does not require continued interaction for its assessment. Indeed, Marks and Miller (1982) have shown that subjects will attribute their own attitudes to attractive peers to a much greater extent than they will to less attractive peers. In the absence of any actual knowledge about the peers' attitudes, the physically attractive were benefitted by the subjects' assumption of high attitude similarity. In addition, an attractive partner may reward us indirectly by increasing our own stature in the eyes of others. Sigall and Landy (1973) found that the same male was evaluated more positively when he was accompanied by an attractive female than when he was accompanied by an unattractive female. Perhaps the category "people with attractive partners" is also linked to other positive traits.

Although the important effects of physical attractiveness cannot be denied, most of us will be comforted by the fact that the impact of this single attribute usually diminishes as a relationship develops over time. Continued interaction allows other sources of reward to come to the front and lessens the dependence upon this superficial determiner of attraction.

THE DEVELOPMENT AND MAINTENANCE OF RELATIONSHIPS

Most of the studies of attraction outlined in the previous section involved brief encounters between strangers. Although these studies have been informative about the determinants of initial attraction, much more remains to be said about factors which determine the nature of a relationship as it develops over time. Social exchange theorists have been especially successful in providing a framework useful in describing the nature and dynamics of interpersonal relationships.

Social Exchange Theory John Thibaut and Harold Kelley have proposed a social exchange theory as a framework for thinking about social relationships (Thibaut & Kelley, 1959; Kelley & Thibaut, 1978). This theory emphasizes the interdependence of social relationships. It states that the quality of the outcomes experienced by two people engaged in a relationship depends on the behavior of *both* participants. *Outcomes*, the consequences of specific acts, are a joint function of the rewards experienced from a particular set of behaviors and the costs required to carry out those behaviors. By reward, Thibaut and Kelley mean any event that results in the experience of pleasure, satisfaction, or gratification. Costs are any factors—such as effort, embarrassment, or anxiety—that inhibit or make more difficult the carrying out of a set of behaviors. The *outcome* from a particular set of behaviors is the result of the rewards from the behavior minus the costs required to carry it out.

How a person evaluates the favorableness of an outcome is an individual, subjective matter. Thibaut and Kelley propose that people use a judgmental standard, termed the *comparison level* (*CL*), in their relationships with others. The CL is an internal standard based upon prior experiences in similar relationships, observations of the outcomes others seem to experience in their relationships, and any other information that affects what a person feels he or she deserves. Outcomes can be greater or less than the CL. The extent to which the outcomes of a relationship exceed a person's CL defines the degree of goodness they will be felt to possess. A person experiencing such good outcomes will be rewarded by and attracted to the relationship. Conversely, outcomes that fall below the person's CL will create dissatisfaction in direct relation to how far below the CL they fall. Satisfaction with a relationship will increase as the outcomes obtained get better and will decrease as the outcomes obtained get worse. It is important to note that changes in satisfaction can occur in a second way—by shifts in the CL. For example, satisfaction can be increased by information that lowers the CL. Satisfaction with a roommate, thus, would be increased by merely hearing about the poor outcomes experienced by students down the hall. The weighting of the CL by others' outcomes may be one reason why people seem so fascinated with the misfortunes of comparable others. Tne miseries of others will lower the CL, making a constant level of outcomes seem more satisfying (Wills, 1981).

In order to represent interdependence, Thibaut and Kelley use an outcome matrix. A given dyadic (two-person) relationship can be visualized as a grid formed by the intersects of the behavioral sequences each person can enact. The outcomes each person experiences or expects to experience are placed in the cells of the intersects. The values of the outcomes are usually represented by numbers; the higher the number, the more satisfying the outcome. If you now turn to Figure 9.10, you will see examples of three outcome matricies. In each instance, the dyad consists of a husband (*H*) and wife (*W*). For the sake of the example, they are restricted to two choices—(*1*) playing tennis or (*2*) playing bridge. In the first situation, only the husband's outcomes are entered. His outcomes are always shown below the diagonal line crossing each cell. From the matrix, we can tell that the husband would get a good outcome (6 units) if he chose to play tennis (*H1*) and his wife also chose tennis (*W1*). He would get a poor outcome (0) if he chose tennis (*H1*) and his wife chose to play bridge (*W2*). In the same way, if he chose to play bridge (*H2*), he would get a good outcome if his wife also chose to play bridge (*W2*) and poor outcomes if she chose to play tennis (*W1*). While the example is trivial, the representation of interdependence is not. The husband's outcomes are determined not solely by his own choices but by his choice plus the choice of his partner. Interdependence implies that joint acts are needed to generate particular outcomes.

In situations 2 and 3 (Figure 9.10), the husband's outcomes remain unchanged. In situation 2, the wife's outcomes have been added so that they match his. Both get good outcomes when they play tennis or bridge together and poor outcomes when they engage in different games. This particular outcome pattern is labeled cooperation because both people can simultaneously obtain their best outcomes. This is generally the case when preferences or the outcomes experienced are synchronized. The major interpersonal problem that requires solution is how to coordinate a switch in choices. We would obviously expect high satisfaction from couples whose outcomes are patterned as in situation 2. Situation 3 is quite different. Now the wife obtains her best outcomes when she plays tennis or bridge without her husband, while joint activities still bring him his best outcomes. This outcome array is accurately termed competition because what is best for one person is worst for the other. (When two persons or two teams are actually competing, it also means that what is best for one side is worst for the other.) The interpersonal problems for people involved in competitive relationships are more severe than for those in cooperative ones because there are no choices that will permit both people to obtain their best available outcomes. The most evident solution to this interpersonal conflict is for the couple to trade off so that each person has a turn at obtaining their best outcomes. Such agreements can become norms for the couple, helping to stabilize their decisions, but the average outcomes obtained from such relationships will still be lower than those obtained in comparable cooperative relationships. Even in a simple two-choice matrix, many types of outcome patterns are possible. From any outcome array, it is possible to determine the extent to which each participant possesses control and influence in the relationship (Kelley, 1979).

We have now discussed how the outcomes people obtain from relationships are a joint function of the behaviors they and their partners

Figure 9.10

Three examples of the matrix representation of outcome interdependencies. (See text.)

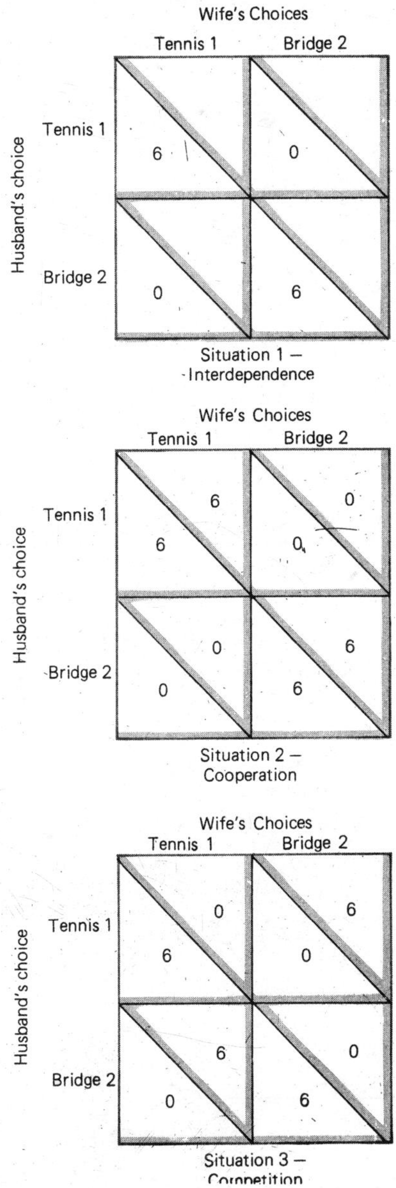

enact and that people are attracted to relationships that provide outcomes above their CLs. What then can be said about the conditions under which relationships end? Do people end relationships when their outcomes fall below their CLs and they are dissatisfied? Thibaut and Kelley answer this question with a no. They say that someone ends a relationship when the outcomes experienced in it drop below the outcomes expected in the next best alternative relationship. They term the standard used for judging the next best alternative relationship the *comparison level for alternatives* (CL_{alt}). Like the CL, it is a summary point, but it refers to the average outcomes that would be obtained in the next best alternative relationship, including the costs of entering the new relationship. Although the value of the CL_{alt} is usually determined by what is expected from an actual person, it can also represent the value placed on being alone. Being alone is always an alternative to any voluntary relationship.

By distinguishing the CL from the CL_{alt}, Thibaut and Kelley (1959) account for why people remain in relationships that are not satisfying as well as why people leave relationships that are satisfying. A person who feels that outcomes from a particular relationship are quite poor will continue that relationship if the CL_{alt} is perceived to be even worse. Individuals will suffer grumpy tennis or marriage partners so long as no better alternative partner is available. Conversely, satisfying relationships may end if someone believes a better alternative partner is available. Thibaut and Kelley do not mean to imply that meaningful relationships are severed lightly. They recognize that mutually involving, intimate relationships typically are based on agreements to stop looking for alternatives. The increased mutual commitments associated with moving from dating to an engagement to marriage can be seen as progressions in the formality with which the legitimacy of alternatives is rejected.

Figure 9.11

As a relationship develops, it goes through various stages. (Modified from Levinger & Snoek, 1972.)

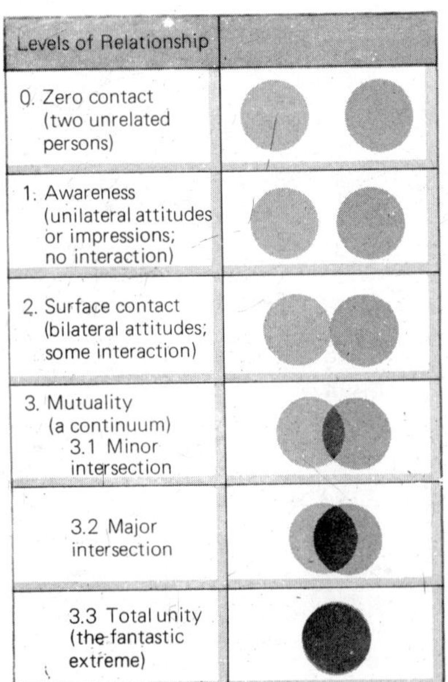

Growth and Decline of Relationships The concepts provided by social exchange theory are useful in describing the changing nature of relationships as they develop from the point of initial contact between individuals to a level of intense and intimate involvement. Most relationships, of course, stabilize far from intimate levels of exchange and provide modest rewards for modest investments of attention, time, and energy. Many, such as business relationships, may be quite superficial and serve only to facilitate the attainment of some goal external to the relationship itself.

George Levinger and J. Diedrick Snoek have suggested a framework for describing the development of relationships based upon social exchange principles (Levinger & Snoek, 1972). This framework points to the degree of involvement as the crucial distinguishing characteristic of various types of relationships. Levinger and Snoek describe three major levels of involvement, beginning with the most superficial level, *unilateral awareness* (Figure 9.11). At this stage, one person notices the other and may make some judgments evaluating the characteristics of the other. At the level of unilateral awareness, before any interaction occurs, overt characteristics such as physical attractiveness may be particularly important. These act as indicators which help a person make assessments as to whether the relationship with the other will be rewarding in the future. Many of the attraction studies discussed in the previous section describe this stage of a relationship.

The next level of involvement, *surface contact*, occurs when the two individuals begin to interact (Figure 9.11). Typically, these interactions are governed by general cultural norms specifying appropriate behavior and social etiquette. These interactions allow the individuals to "explore" the relationship by sampling the outcomes each receives from the various combinations of behaviors each is capable of performing. At this level, a particularly important function is served by the process of self-disclosure. *Self-disclosure* is the process through which one person lets himself or herself be known by another. The decisions each individual makes about revealing thoughts, feelings, and past experiences to the other have much to do with how far the relationship will develop.

If the partners find their interaction in the surface contact stage rewarding and promising, the relationship may progress to the *stage of mutuality* (Figure 9.11). At this stage each individual begins to acquire some feelings of responsibility for the outcomes the partner receives in the relationship. Each now acts in such a way as to maximize both his or her own and the partner's outcomes. The cultural norms and rules of etiquette which governed the interaction in the earlier stages are replaced in the state of mutuality by norms more specific to the particular relationship. Some of these norms may be quite general: "We should be honest with each other." Other norms are quite specific: "Bill shouldn't be teased about his bald spot."

Just as relationships may develop through successively greater levels of involvement, they may also deteriorate and decline. The actual level of rewards experienced in the interaction may decrease as a result of diverging interests or situational influences which limit a partner's ability to provide rewards. The costs associated with producing behavior in the relationship may also increase. Finally, there may be changes in the standards by which the participants evaluate the outcomes experienced in the relationship. An increase in the comparison level of one of the participants results in decreased satisfaction with the outcomes even though, objectively, these outcomes remain the same as before. The decreased satisfaction with the relationship caused by any of these factors will be reflected in less intimate disclosures, lessened concern with the quality of the partner's outcomes, and increasingly formal interactions characteristic of lower levels of involvement.

JUSTICE IN SOCIAL RELATIONSHIPS

Social exchange theory tells us that individuals do not evaluate the outcomes received from social interactions in absolute terms but compare them to some standard of satisfactory outcomes—the CL. For example, we would not expect a millionaire to be as excited as a poor man about a gift of a 10-dollar bill. In recent years, social psychologists have become increasingly interested in the tendency of people to compare the outcomes received with standards representing what *ought* to be ("He doesn't deserve that" or "It just isn't fair"). Statements such as these reflect a concern with what is fair or just in social relationships. Much research has been directed at finding what these standards of fairness are and the manner in which persons respond to their violation.

Gerald Leventhal (1976) suggests that three "justice rules" are employed most frequently as standards in making judgments about fairness in social relationships. These include a *contributions rule*, based upon the investments each person makes in the relationship; a *needs rule*, which reflects the relative needs of individuals; and an *equality rule*, which requires that outcomes be distributed equally among participants in a relationship. Each of these fairness rules represents a standard against which a participant's outcomes in a relationship can be compared with what is "deserved."

The Contributions Rule and Equity The contributions rule is most frequently identified with the concept of equity. According to the equity principle, a relationship will be considered fair when all individuals involved receive outcomes proportional to their respective contributions (inputs) to the relationship (Adams, 1965). Thus the concept of equity suggests that the distribution of outcomes in a relationship is judged to be fair when those individuals who contribute the most receive the greatest outcomes. There is no requirement that the inputs of all participants be equal but only that the *ratios* of individuals' inputs and outcomes be equal. For example, since a physician's inputs in terms of cost and length of training are greater than those of a nurse, most people consider it fair that the physician is paid more and given greater prestige.

Equity theory proposes that when individuals find themselves in an unfair, or inequitable, relationship, they will experience distress and will be motivated to eliminate the distress by restoring equity. Equity may be restored by either of two general strategies: restoring actual equity or restoring "psychological" equity (Walster et al., 1978). Actual equity may be restored by altering the outcomes received or the inputs contributed by the participants in the relationship. For example, employees who feel they are being treated unfairly by an employer may reduce their inputs by working less or raise their outcomes by stealing from the company. Employees may restore psychological equity by altering their perception of the situation. For example, an employee may enhance the perceived outcome by saying something like, "Even if I'm not paid very much, I do enjoy my work."

Equity undoubtedly functions as a widespread norm in our society. This is reflected in the economic system, which attempts to reward superior productivity, skill, education, and expertise with higher pay. The legal system, with the notion that punishment should fit the crime, has equity built into it. Naturally, in these situations individual views of what constitutes an equitable distribution are likely to differ. This results not only from different evaluations of outcomes but also from different opinions regarding what constitutes a valid input. For example, a young shop supervisor may believe that a B.A. degree gives him the right to more pay than a less educated and older coworker gets. The coworker, however, may believe that greater job experience is a more valid input. As a result, their respective evaluations of the fairness of their pay may be quite different even though both base their judgments on the equity principle.

The Needs Rule and Social Responsibility A second justice rule states that outcomes should be distributed in accordance with the relative amounts of individual need. This justice rule is embodied in the *norm of social responsibility*, by which we are encouraged to respond to the legitimate needs of others. Under this rule, a fair outcome distribution is one which meets people's legitimate needs to avoid hardship and suffering. The numerous charitable foundations, service organizations, and government welfare agencies attest to society's acceptance of this justice rule. On an individual level, several studies have demonstrated that participants do take the needs of others into account when forming perceptions of their deservedness. For example, Berkowitz and Daniels (1963) found that students worked harder at a task when they believed that it would benefit another student who needed their help.

While the contributions, or equity, rule requires individuals to judge the relative inputs of each participant in a relationship, the *needs rule* requires judgments concerning the legitimacy of individual needs. If, for example, meeting the expressed needs of some person may be thought to be detrimental to that person in the long run, the need will not be judged legitimate. Thus a mother does not respond to her child's every request for candy, nor does the pharmacist respond to the addict's need for a fix. The legitimacy of an individual's need is also questioned when that person is seen as responsible for his or her own predicament. Thus needs are not so likely to be met if they appear to have occurred through the individual's own neglect, carelessness, or laziness. Unfortunately, there is evidence that we may see those who suffer as responsible for their fate more often than is warranted. For instance, Stokols and Schopler (1973) performed an experiment in which the subjects were asked to read a story about a young woman who had a premarital pregnancy and miscarriage; when the medical complications that resulted were described as serious, the subjects evaluated the woman more negatively than when less serious complications were described. So in instances like this, the more people suffer, the more they are perceived as responsible for their fate and the less legitimate their needs are considered! In other words, our perceptions sometimes lead to clear cases in which greater needs are perceived as less legitimate. As we all know, the social world is not always fair.

The Equality Rule This rule states simply that outcomes should be distributed equally among the participants in a relationship, irrespective of individual contributions or needs. The equality rule is obviously the easiest to apply because it does not require the participants to make assessments of inputs or to judge the legitimacy of needs. Rather, inputs and needs are assumed to be equal or are considered irrelevant in determining the proper outcome distribution.

Weighting the Justice Rules Obviously, these three justice rules often conflict. For example, those individuals who would receive high outcomes under the needs rule are often those who are not able to contribute much in the way of inputs. Individuals whose poverty places them in great need of assistance often lack the education and skill that would allow them to make productive contributions. Similarly, application of the contributions and equality rules will result in the same outcome distribution only when the inputs of all individuals are perceived to be equal. Consequently, the standard of fairness will vary depending upon the type of relationship and the situation. In general, the contributions rule seems to apply most often in those situations in which the emphasis is upon encouraging effective performance from the individuals involved in the relationship. The needs rule is likely to be weighted more heavily when there is a close, friendly relationship among the participants, especially when the individuals feel that others are dependent upon them. When the primary concern is maintaining harmony and avoiding conflict in the relationship, the equality rule will be applied. The relative simplicity of the equality rule is also an advantage when the situation makes the assessment of contributions and needs difficult or ambiguous.

Summary

1. Social psychology is defined as the scientific study of the ways interaction, interdependence, and influence among persons affect their behavior and thought.

2. Our perception of other people is a very important factor in our behavior toward them. A person's cognitive framework determines how observed information will become extended to categories not under observation. Implicit personality theories, the assumed relationships among traits, and stereotypes, or the cognitive categories defined by ethnic or demographic characteristics, are parts of the cognitive framework. The cognitive framework simplifies the process of forming impressions of others.

3. Attributions, or inferences about the causes of other people's behavior, are based on observations of people doing things in given situations. Heider's "naive" psychology is concerned with the consequences of whether we attribute another's behavior to environmental or personal forces. Kelley's attribution theory emphasizes our use of consensus information, consistency of behavior, and distinctiveness of behavior in making inferences about the internal or external causes of another's behavior. Jones and Davis focus on how information about how another chooses to act, along with the consequences of those acts, form the bases for inferring personal dispositions. Many of the factors involved in making attributions about others also play a role in the process of forming self-attributions.

4. Social influence occurs when a person's behavior is changed because of another's intervention. Social facilitation, or the improved performance of an individual when in the presence of others, and imitation, or the copying of modeled behavior, are phenomena where the person intervening has no intention to influence. Social facilitation occurs most readily for strong responses in situations where the presence of others is motivating.

5. People in groups tend to conform to group norms of thought and behavior. Conformity pressures can create situations in which the information obtained from personal modes conflicts with the information obtained from social modes. The extent of an individual's conformity is directly related to the amount of weight given to social information. Conformity can also occur in response to the requests of an authority figure (obedience) and to rules or norms governing behavior.

6. Among the reasons for conformity are social comparison, especially in ambiguous social situations; the desire to avoid social disapproval or punishment; and the need to be liked and accepted.

7. Interpersonal attraction is an important part of social relationships. Involved in interpersonal attraction are the factors of proximity, attitude similarity, and physical attractiveness.

8. An important set of ideas about the development and maintenance of long-term social relationships is social exchange theory. This theory emphasizes the fact that two individuals engaged in a relationship become interdependent, or dependent on each other for the outcomes they experience in the relationship. The extent to which each person regards a particular outcome as favorable depends upon that person's comparison level—an internal standard based upon prior experiences and observations of the outcomes experienced by others in similar relationships. A second standard, the comparison level for alternatives, determines whether an individual will continue or terminate a relationship.

9. The growth of interpersonal relationships follows a path through various levels of involvement. Interpersonal relationships start with the stage of unilateral awareness. They then move through the stage of surface contact, in which interactions are governed by general cultural norms and rules of etiquette, to the stage of mutuality, characterized by increasing concern for the partner's outcomes in the relationship.

10. Three "justice rules" are employed most often as standards of fairness in social relationships. These include a contributions rule, based upon the investment each person makes in the relationship; a needs rule, which reflects the relative needs of the individuals; and an equality rule, which requires that outcomes be distributed equally among the participants. Each of these rules represents a standard against which people's outcomes may be compared with what they are thought to "deserve."

Terms to Know

One way to test your mastery of the material in this chapter is to see whether you know what is meant by the following terms:

Social psychology (342)
Dyads (342)
Focal person (FP) (342)
Social influence (343, 353)
Social perception (343)
Attributions (343, 347)
Impression formation (344)
Cognitive framework (344)
Implicit personality theory (345)
Traits (345)
Primacy effect (346)
Stereotypes (346)
Social stereotypes (346)
Personal forces (347)
Environmental forces (347)
Consistency (348)
Consensus information (348)
Distinctiveness (348)
Noncommon effects (349)

Valence (351)
Self-attributions (352)
Social facilitation (354)
Imitation (355)
Social conformity/conformity (356)
Personal modes (357)
Social modes (357)
Norms (358)
Obedience (359)
Social comparison (362)
Proximity (367)
Attitude similarity (368)
Physical attractiveness (369)
Social exchange theory (370)
Outcomes (370)
Comparison level (CL) (370)
Comparison level for alternatives (CL_{alt}) (372)
Unilateral awareness (372)
Surface contact (373)
Self-disclosure (373)
Stage of mutuality (373)
Contributions rule (373, 374)
Needs rule (374)
Equality rule (374, 375)
Norm of social responsibility (374)

Suggestions for Further Reading

Detailed coverage of the topics included in this chapter can be found in any of the many social psychology texts now on the market. A number of excellent books have treated particular topics at a more advanced level. A concise and detailed presentation of theories is provided by M. E. Shaw and P. R. Costanzo in *Theories of Social Psychology*, 2nd ed. (New York: McGraw-Hill, 1982). The various functions of the cognitive framework, especially with respect to stereotypes, are comprehensively covered in A. G. Miller's (Ed.) *In the Eye of the Beholder* (New York: Praeger, 1982). Applications of social-perception concepts to the area of mental health, including presentation of original research, can be found in the book edited by G. Weary and H. L. Mirels, *Integration of Clinical and Social Psychology* (New York: Oxford University Press, 1982). An analysis of the influence techniques used in everyday relationships is clearly presented in R. B. Cialdini's *Influence: Science and Practice* (Glenview, IL: Scott, Foresman, 1985).

Attitudes

HEADING OUTLINE

Chapter

> When in the Course of human events it becomes necessary for one people to dissolve the political bands which have connected them with another, and to assume among the powers of the earth, the separate and equal station to which the Laws of Nature and of Nature's God entitle them, a decent respect to the opinions of mankind requires that they should declare the causes which impel them to the separation. (The Unanimous Declaration of the thirteen united States of America, July 4, 1776.)

The dramatic words contained in the first sentence of our Declaration of Independence announced the founding fathers' concern with attitudes. Their attitude about separation from Great Britain was extremely favorable and, due to a "decent respect" for the attitudes of others ("opinions of mankind"), they wished to explain the bases of their attitude. If you read the remainder of the Declaration of Independence, you will find that it is almost exclusively devoted to detailing their bases for favoring independence. The Declaration illustrates two themes that will be repeated in this chapter. First, the bases of all important attitudes are rooted in a complex network of opinions and beliefs. Persuading others to accept our attitude often requires us to convince them of the truth of this network. Second, attitudes can have important consequences for behavior. In this instance, as we all know, the dominance of favorable attitudes toward separation led to a long revolutionary war that shaped the destiny of our country.

(Joseph Szaszfai/Yale University Art Gallery.)

HOW much we like or dislike something has much to do with determining our behavior toward that something. We tend to approach, seek out, or be associated with things we like; we avoid, shun, or reject things we do not like. Attitudes are simply expressions of how much we like or dislike various things. They represent our evaluations—preferences—toward a wide variety of attitude "objects." Our attitudes are based on information. For example, people who favor the death penalty typically see it as a deterrent to crime and a just punishment. People who oppose it typically see the death penalty as a barbarian practice that does not deter crime. Because we can never know all the information available on any particular attitude object, our attitudes are always open to revision. Our lives are filled with opportunities for attitude change. We are bombarded with advertisements intended to increase our favorableness toward various products; we are constantly acquiring new information about other people; and, near election time, we are flooded with information about political candidates. This chapter is concerned with the nature of attitudes, their measurement, and how they change and function.

A consistent theme marks the progression of knowledge about attitude formation and change. Most theories or ideas are initially announced with a grand vision of universal applicability. As research grows, however, it becomes clear that the theories apply to some circumstances but not to others. As knowledge accumulates, we become better at identifying the restricted conditions under which particular attitude processes will occur and the many factors that might bear on any situation. While such progressive refinements make the study of attitudes more difficult, they do bring us closer to the reality of how attitudes are formed and how they function to influence behavior.

The Nature of Attitudes

Suppose, after the first day of classes, you bump into a friend who asks you how your day has been. You might reply "I heard a wonderful lecture in my psychology class, ate lunch at an awful French restaurant, and stood in line for such a long time that I missed my favorite soap opera." You have described your day by expressing a series of attitudes. The defining characteristic of *attitudes* is that they express an evaluation of some object (Insko & Schopler, 1972; Petty & Cacioppo, 1981). Evaluations are expressed by terms such as *liking-disliking*, *pro-anti*, *favoring-not favoring*, and *positive-negative*. They are the feeling tone aroused by any attitude object. Attitudes can be formed about many things. The object of attitudes can be entities (a lecture, a restaurant, a soap opera), people (my best friend, the President, myself), or abstract concepts (abortion, civil rights, foreign aid). Indeed, anything that arouses evaluative feelings qualifies as an object of attitudes. Some of this variety is displayed in Figure 10.1. It shows the attitudes held by a college sophomore toward various objects. You probably hold similar attitudes toward some of the objects and different attitudes toward others.

By restricting the term *attitudes* to evaluations, we are distinguishing attitudes from *beliefs*, or opinions. Beliefs are cognitions, or thoughts, about the characteristics of objects. They link objects to attributes (Fishbein & Ajzen, 1975). Suppose your friend expresses a favorable attitude toward a presidential candidate, senator X. This attitude is probably associated with a number of specific beliefs about the candidate—that the candidate has a sound economic policy, will work to lower taxes, will help to prevent war, and so on. Beliefs, or opinions, are assessed by how likely they are to be true. In addition, we have evaluative feelings about beliefs and these will contribute to our attitude. For example, people who favor senator X's economic policies and who want to see taxes lowered and war prevented will have a favorable attitude toward the candidate, while people who oppose the senator's stands will be against the candidacy. A given attitude, thus, is often a summary of the evaluations made of different aspects of the attitude object.

One reason attitudes are important is that they are thought to guide behavior. If you are favorable toward candidate X, you will be likely to vote for X. Some social psychologists have included attitude-relevant behavior as part of the definition of *attitudes*. We have not done this because it would hide an important question: What is the relationship between people's attitudes and their behavior in attitude-relevant situations? Suppose your acquaintance, Joe, expresses a very favorable attitude toward classical music. You have observed, however, that Joe never buys classical records, never listens to classical music on the radio, and has turned down free tickets to hear a touring symphony orchestra. If Joe's attitude toward classical music is defined by his positive verbal expressions *and* his behavioral choices, you would have to add the positive expressions *with* the negative behavioral instances of never choosing to hear classical music. Depending upon the weighting, Joe's attitude would be set as mildly positive or mildly negative. However, we prefer to say that Joe expresses a very positive attitude toward classical music, but his attitude is discrepant from his behavior. This seems to be not only more true to life but also more theoretically interesting. The relationship between attitudes and behavior is by no means straightforward. It deserves its own section, and we will return to it later on in this chapter.

It should now be clear that attitudes are individual expressions

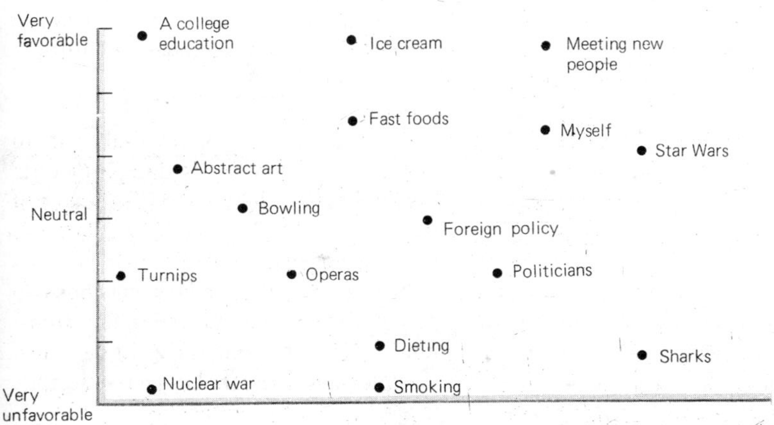

Figure 10.1

One college sophomore's attitudes toward various objects. The attitudes are represented from very favorable to very unfavorable; the ordering of objects across the page has no special meaning.

representing a summary of evaluations of an attitude object. The expressions that one makes publicly to others are not always the same as the expressions one makes privately to oneself. It is reasonable to suspect that Joe's private attitude toward classical music is not nearly so positive as the one he publicly expresses. While this aspect of attitudes is an annoyance to the people who attempt to measure them, it adds an intriguing element to the study of attitudes.

The Measurement of Attitudes

The scientific study of attitudes requires that they be measured. This means that people's evaluation of a given attitude object must be translated into some number system. Many techniques are in current use. They vary from very simple to very complex. The simplest techniques permit us to place concepts, things, or people into a "favorable" or "unfavorable" category. The more complicated procedures are intended to measure degrees of favorableness or unfavorableness. The measurement which is "best" depends upon the purpose for doing the measurement. Simple measures are usually adequate for things such as predicting the outcome of a political election. On the other hand, understanding the impact of television on attitudes toward aggression, for example, usually requires more complicated measurement techniques. By far the most common method of measuring attitudes is the *self-report method,* in which people are asked to respond to questions by expressing their personal evaluations. But self-report methods have a drawback. As we know from the study of social conformity (Chapter 9, page 356), people may respond in terms of how they think others respond or how they think they "ought" to respond. The accuracy of the measurement is lowered if a person's public response differs from his or her private attitude. Social psychologists have sought to solve this problem by finding indirect, behavioral measures of attitudes that are not directly under voluntary control. The search for such measures has typically involved physiological measures of one sort or another. Even though their usage has been limited, these techniques will be discussed after presentation of the self-report methods.

SELF-REPORT METHODS

Whenever you are asked to express your preferences to an interviewer or to write your evaluations of something on a long questionnaire, you are involved in a self-report method for measuring your attitude. We will use polling techniques to describe simple measurement methods and attitude scaling to describe more complicated techniques.

Public Opinion (Attitude) Polling Public opinion polls are used either to predict something or to provide information. They are used to predict the outcomes of elections, the likelihood of buying a product, or the degree of public support for implementing new policies. They can also be used to provide information about the percentage of the population that supports (or opposes) the use of marijuana, the death penalty, or a low-cost housing project. The complexities of opinion polling can be reduced to four steps: (1) selecting a sample of respondents, (2) constructing the attitude items, (3) administering items to the sample, and (4) tabulating the results.

Figure 10.2

Public opinion polls rely on people's willingness to report their attitudes and opinions. (Gatewood/The Image Works, Inc.)

Although our focus will be on the measurement decisions required in step 2, we will briefly comment on the other steps.

The best way to predict how an election will come out is to ask everyone who will vote what candidate they favor. Opinion polls, however, do not work this way. In the first place, there is no way to know ahead of time, who will and who will not vote. The *target population* (all those voting in a particular election) cannot be accurately determined. In the second place, it is usually too expensive to contact an entire target population. Polls, therefore, draw samples of people from the population of interest. If the sample is perfectly representative of the population, its results will accurately reflect the results for the entire population. There are many ways to sample a population. The key to accurate sampling is its randomness. *Random sampling* means that each person in the target population has an equal chance of being selected in the sample. The accuracy of polls depends upon the precision of the sampling procedures employed.

Attitudes are measured by attitude items. For self-report measures, an *attitude item* consists of a question or statement about the object and a "format" for the response. The format can be either *fixed*, where the categories for the response are named, or *open-ended*, where respondents use their own words. This distinction is comparable to the difference between the multiple-choice and essay items that appear on class quizzes. Because fixed categories are easier to quantify, they are by far the most popular format for polls. Suppose we wished to conduct a poll concerning people's attitudes toward the Equal Rights Amendment (ERA). We could do it by asking one question: "What are your feelings toward the ERA?" We could then use a fixed format with only two categories for the response: "I am in favor" or "I am opposed." The problem with using only two categories is that for any attitude object, some people will be neutral or will not have any attitude. (Even in national presidential elections, a small portion of the eligible voters will never have heard of any of the candidates.) For our ERA poll, we would want to add a third category: "Not sure" or "No attitude." Table 10.1 presents the actual results of such a poll, administered to a national sample of about 1,500 adults in October 1979. The actual poll measured two degrees of favorableness or opposition by including the terms *strongly* and *somewhat*. You can see that the percentage of favorable and unfavorable attitudes toward ERA is similar for men and women in this sample.

The accuracy of polling results can be affected by the clarity of the questions asked. One should obviously avoid using difficult words, double negatives, ambiguous words, and confusing sentences. Despite such precautions, it is still possible to ask seemingly simple questions that are not understood in the way they were intended. Consider a study undertaken by Belson (1981) in London, England.

> Twenty-nine questions relating to television were studied. They were constructed to be in keeping with typical polling questions while, at the same time, containing features which might be misunderstood, as, for example, "Is television advertising time used properly?" and "Do you think that children suffer any ill effects from watching programmes with violence in them, other than ordinary Westerns?" The respondents were administered the questions in a traditional way. On the following day, however, the respondents were interviewed to find out how they had actually interpreted each question. The results were dramatic. None of the questions was understood as it was intended by more than 60 percent of the respondents. Many of the misunderstandings were ones likely to be overlooked by the people who construct questions. For example, the word "children" in the question cited above typically defines ". . . the age range 4–14 years, or some extension of that range towards 2–17 years (p. 169)." Yet only 27 percent of the respondents thought of the lower age limit as anywhere between 3 and 6 years.

A question that means different things to different respondents cannot provide accurate polling results. You will probably appreciate this difficulty more when you consider that it is similar to the problem contained in some multiple-choice items on psychology examinations.

Table 10.1

The attitudes of a sample of men and women toward the ERA.

GROUP	STRONGLY FAVOR	SOMEWHAT FAVOR	SOMEWHAT OPPOSE	STRONGLY OPPOSE	NOT SURE	TOTAL
WOMEN						
%	40.9	37.0	8.7	9.7	3.7	100%
Number	312	282	66	74	28	762
MEN						
%	43.9	32.0	9.4	11.2	3.5	100%
Number	320	234	69	82	26	731

Source: From Harris 1979 Equal Rights Amendment Survey, no. 794022 (machine-readable data file) conducted by Louis Harris and Associates, Inc. New York: Louis Harris and Associates, Inc. (producer), 1979; Chapel Hill, NC: Louis Harris Data Center, University of North Carolina (distributor).

Attitude Scales An attitude scale attempts to obtain a precise measure of the extremity of people's attitudes. The accuracy of the measurement can be increased by using many items that are all related to the same issue. Attitude scales use such items. There are, however, only a limited number of ways one can directly ask about any attitude topic. After asking "How much do you favor (or like) X?" there is not much room for additional items. The items on an attitude scale, therefore, inquire about things that are known to be related to the attitude topic. Typically, these involve a person's beliefs (page 383), or opinions, about the attitude object. People who are favorable toward something usually have a belief structure that is different from people who hold an unfavorable attitude. Certainly the American colonists who favored independence from Great Britain had many beliefs that differed from those who opposed separation. (See the opening page of this chapter.)

Many standard attitude scales have been developed to measure attitudes concerned with the family, education, religion, health, sexual behavior, and international affairs. Robinson and Shaver (1973) present details on over 120 such scales. Many attitude scales are also constructed for specific purposes. To study marijuana use among high school students, for instance, a researcher would probably need to develop a special attitude scale.

INVOLUNTARY BEHAVIORAL MEASURES

Self report measures are accurate only to the extent that respondents are willing or able to report their attitudes correctly. While this may well be the case in most situations, accuracy could be increased if attitude indicators were discovered that were not under the voluntary control of the respondent. The search for such measures has often led to studying the body's physiological responses to attitude objects. Because the worth of these measures has not yet been established (Cacioppo & Sandman, 1981), they will be discussed only briefly.

Physiological Measures The galvanic skin response (GSR) was one of the first measures tried out. The GSR measures the electrical resistance of the skin. As described in Chapter 8, page 320, this resistance decreases when a person is emotionally aroused. Although some research has been successful in relating GSR changes to the extremity of attitudes, it has never been possible to distinguish extremely favorable from extremely unfavorable attitudes; both extremes produce changes in the GSR. The issue is further clouded by the fact that other factors, such as novelty, can result in GSR changes.

Perhaps other indicators of arousal, such as the size of the pupil of the eye, might be used to gauge a person's attitude. Our eyes are reputed to be the "windows to our souls"; perhaps pupil size indexes the evaluations going on behind the "windows." In early work, expansion of the size of the pupil was found to accompany exposure to favorable stimuli, while contraction of the pupil was linked with exposure to unfavorable stimuli (Hess, 1965). However, successful demonstrations of the validity of this technique were soon outnumbered by research failures (Woodmansee, 1970). Pupil size has not yet been established as a good way to measure attitudes.

The most recent, and most promising, physiological measure involves electromyographic (EMG) recordings from the major facial muscles. The EMG records minute muscle movements. Drawing on work which showed that different patterns of facial-muscle activity were associated with different emotions (Chapter 8, page 311), Cacioppo and Petty (1979) conducted an experiment to investigate attitudes. They showed that EMG activity at different facial-muscle sites produced a pattern that distinguished positive from negative reactions to persuasive messages. Whether or not EMG recordings can distinguish degrees of positiveness and negativeness remains to be determined.

Attitude Measures Based on Classical Conditioning In order to acquaint American psychologists with the research on conditioning that was being conducted in the Soviet Union, Gregory Razran (1961) wrote an extensive review. In the course of the review, brief mention was made of the Russian work on attitudes.

> A 13-year-old boy, Yuri, was conditioned to salivate to the word "good" and to differentiate that sound from the word "bad." Upon completion of the conditioning, Yuri was read various sentences, and the saliva drops secreted in 30 seconds were counted. Presumably the more Yuri salivated, the more positive was his attitude. For example, when he heard "The Fascists destroyed many cities," he produced only two drops. Yuri salivated 24 drops, however, when he heard the statement, "The enemy army was defeated and annihilated."

There has been little follow-up to this line of research. Its effectiveness is apparently extremely limited.

A FINAL WORD ON ATTITUDE MEASUREMENT

In this section, we have focused primarily on public opinion (attitude) polling in order to illustrate some of the issues involved in the measurement of attitudes. Public opinion polling, like most other attitude-measurement techniques, relies on the self-report of the respondents. One of the limits on the accuracy of self-report methods is the willingness and ability of respondents to express their attitudes. In order to obtain attitude measurements not under the voluntary control of respondents, the body's physiological responses have been investigated. Although the most recent research on electromyographic (EMG) recordings of the major facial muscles holds some promise, involuntary techniques have not, on the whole, proved to be successful.

Attitude Theories

A number of psychological theories have been suggested to explain how attitudes form and why they change. The theories most frequently employed can be categorized as either (1) learning theories, (2) consistency theories, or (3) cognitive-response theories. Examples of each will be discussed below. It should be noted that these different approaches are not contradictory but simply focus on different factors which may affect the way attitudes develop and change.

LEARNING THEORIES

One of the first investigators to suggest that learning principles could be applied to attitudes was Doob (1947). He proposed that the principles of classical and instrumental conditioning could be used to explain the formation and change of attitudes in much the same way that they have been applied to overt behavior. Consider classical conditioning (Chapter 4, page 140). On successive occasions, a neutral stimulus is paired with an unconditioned stimulus. Over time, the previously neutral stimulus may begin to elicit a response similar to that produced by the unconditioned stimulus. Objects, people, or events associated with pleasant experiences may take on favorable evaluations, while those associated with unpleasant experiences may be evaluated negatively. For example, in a series of trials, a word associated with the ending of a brief electrical shock will be rated more favorably than will a word associated with the onset of shock (Zanna et al. 1970). If the association with a particular object is irrelevant, our attitude can be "illogical." Griffitt (1970) had people interact in small groups in either a comfortable room or one which was hot and uncomfortable. When asked to rate how much they liked the other people present in the room, individuals in the hot room reported liking the others less than did individuals in the comfortable room. In this manner, attitudes can be formed simply by association. (See Figure 10.3.)

Instrumental conditioning, in which the reward consequences of any behavior shape its subsequent enactment (Chapter 4, page 147), is obviously relevant to attitude formation and change. If you express an attitude to a friend who then provides positive reinforcement (by smiling, nodding, or expressing approval), your attitude is likely to be strengthened. On the other hand, if your friend provides punishment (by frowning, disagreeing, or expressing disapproval), your attitude is likely to be weakened. Credit for demonstrating the potential of the instrumental conditioning of attitudes is given to Greenspoon (1955) who used verbal rewards to alter what people said. Specifically, Greenspoon rewarded the subjects' use of plural pronouns by saying "mm-hmm" each time a plural pronoun was used. Through this simple technique, the frequency of the expression of plural pronouns was increased. It was then an easy step to apply similar verbal-reinforcement techniques to altering attitudes in

Figure 10.3

Attitudes can be formed simply by association. For example, unacquainted people in a pleasant room will associate the pleasant situation with the people in it and will have more positive attitudes toward each other than people in an unpleasant, crowded subway. (*Left*, Ellis Herwig/Stock, Boston; *right*, Donald C. Dietz/Stock, Boston.)

laboratory settings. Most of these studies involved engaging participants in an interview situation during which they were asked to make comments on both sides of some controversial issue. The experimenter then reinforced the expression of statements in one direction (in the favorable direction for some participants and the unfavorable for others) by nodding, smiling, saying "good," or otherwise showing approval. Many of these studies showed an increase in the frequency of statements made in the reinforced direction over the course of the interview. Furthermore, when their attitudes were tested after the interview, it was found that many of the participants had changed their attitudes in the direction of the reinforcement and that this change seemed to persist over time.

It should be evident that instrumental conditioning will be especially important in social influence situations involving interactions with others (Chapter 9, pp. 353–366). Membership and acceptance in particular groups is often contingent upon the attitudes one expresses. Peer groups such as clubs, unions, sororities, fraternities, and churches differentially reinforce the expression of certain attitudes relevant to the group. Parents may often give or withhold rewards and approval contingent upon the attitudes expressed by their children. This may be the chief reason why a high degree of similarity exists between the attitudes of parents and children on certain topics. A study of high school seniors and their parents, reported by Jennings and Niemi (1968), produced typical results. The greatest agreement was on religious affiliation: 74 percent of the seniors had the same religious affiliation (Protestant, Catholic, or Jewish) as their parents, and only a negligible percentage had actively shifted to another religion. While not as strong, a comparable agreement was found in political party affiliation. Moving back a whole generation, very similar results were obtained for the parents' agreement with their own parents.

Figure 10.4

Balanced and unbalanced attitude structures.

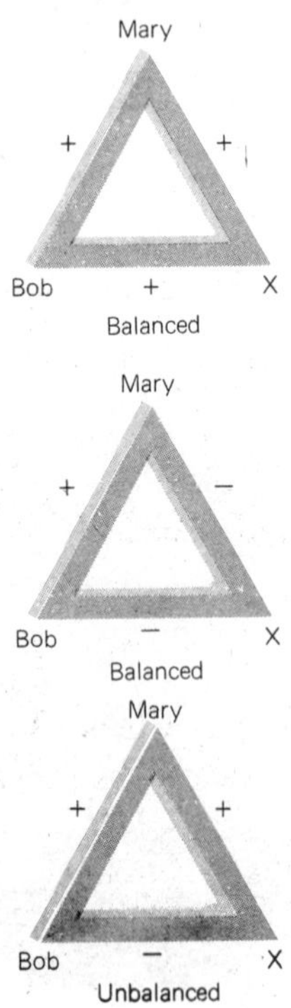

CONSISTENCY THEORIES

A second group of attitude theories focuses on the individual's attempt to maintain consistency among the numerous attitudes he or she holds. Perhaps the most influential of these theories is balance theory (Heider, 1958).

Balance Theory In its simplest form, *balance theory* involves the relations between a person and two attitude objects. The three elements are connected by either favorable (good, liking, or positive) attitudes or unfavorable (bad, disliking, or negative) attitudes. The structure formed by the relations among the elements may be balanced or unbalanced. To illustrate these points, we have drawn some attitude structures that might occur in the mind of someone called Mary (Figure 10.4). Mary and her friend (favorable link), Bob, are discussing their attitudes toward Candidate X. In the first diagram, Mary is favorable toward X and learns that Bob is also favorable. In the second diagram, both Mary and Bob are unfavorable toward Candidate X. In both instances, Mary's attitude structure is balanced because she holds the same attitude as someone she likes. In general, a balanced state is one in which the elements fit together harmoniously. In this same vein, if you have two friends who are not acquainted, you probably assume that they will like each other once they are introduced. Or if you dislike someone, you will assume your friend will also dislike that person. "My enemy's enemy is my friend" was the title of an experiment documenting this point (Aronson & Cope, 1968).

The basic tenet of balance theory is that there is a tendency to maintain or restore balance in one's attitude structures. Unbalanced structures are somehow uncomfortable or unpleasant. The third diagram in Figure 10.4 illustrates an unbalanced state. Mary and her friend now hold different views of Candidate X. In this instance, balance theory predicts that Mary will try to restore balance by changing one or more of her attitudes. She may change her attitude toward Candidate X, or she may decide she no longer likes Bob. She may also attempt to restore balance by trying to persuade Bob to change his attitude toward Candidate X. Balance theory predicts how attitudes will change in order to create a balanced structure. It also recognizes some solutions other than attitude change to the problems of unbalanced structures. Balance theory, for instance, would be satisfied if Mary and Bob merely agreed not to discuss Candidate X. Agreeing to disagree is a time-honored way of minimizing the tension of unbalanced states by rendering them irrelevant to the interpersonal relationship. Many studies have demonstrated that people express a preference for balanced structures and "fill in" incomplete structures in such a way as to maintain balance. Nevertheless, most of us do hold inconsistent attitudes. Balance theory does not predict that imbalance will always be resolved—only that there is a *tendency* toward balance and that unbalanced structures produce tensions and discomfort.

Cognitive Dissonance In 1957 Leon Festinger published his theory of *cognitive dissonance*. Like balance theory, the focus of cognitive dissonance theory is on individuals, who are assumed to strive for harmony (*consonance*) among the elements in their cognitive, or thought, structures. The creation of dissonance, similar to the creation of imbalance, is thought to be distressful and to motivate the person to restore consonance. The creation of dissonance, however, can occur under a wider set of circumstances than can the creation of imbalance. For Festinger, any two related cognitive elements will be dissonant if they do not fit together because they violate general logic or the person's expectations. It is consonant to favor Candidate X and to give public support to his candidacy, but it is dissonant to favor X and to speak against him publicly. Similarly, the cognitive element "Smoking is a health hazard" is consonant with the cognitive element "I do not smoke," but it is dissonant with being a smoker. The amount of dissonance created depends upon the importance of the elements to the person and the extent to which the elements do not fit. The magnitude of dissonance increases as importance and lack of fit become greater.

There are two major ways for a person to reduce dissonance: (1) to change one of the cognitive elements or (2) to add a new cognitive element. The smoker who hears of the link between smoking and health can stop smoking or can choose not to believe the link. If neither of these changes in cognitive element is made, the smoker may add a new element, such as switching to filter cigarettes.

With these examples in mind, we can add a further requirement for the arousal of dissonance. A person will experience dissonance only if some unstated assumption about that person's self-concept is met. For example, if you speak against a candidate you privately favor, you will experience dissonance only if you see yourself as being truthful. Similarly, it is only smokers who see themselves as health conscious who will experience dissonance if they believe the smoking-health link and continue to smoke. The examples would not apply to people who saw themselves as liars or as being indifferent to their own health.

Festinger thought that one of the special appeals of dissonance theory was that it made predictions which were not obvious or in keeping with folk wisdom. One of the early tests of the theory, for example, was conducted in the area of counterattitudinal roleplaying. As the name suggests, *counterattitudinal roleplaying* involves situations in which people express attitudes publicly that are opposite to their private attitudes. People on debating teams or actors are often in such situations, but the rest of us also face them from time to time.

It has been shown that engaging in counterattitudinal roleplaying may have the effect of changing the private attitude in the direction of the public position taken. The application of dissonance theory to this area revolves around how much reward people are offered to engage in the counterattitudinal roleplaying. The dissonance prediction is that the *less* people are offered as an inducement to agree to counterattitudinal roleplaying, the *more* they will change their private attitudes. The large reward introduces a new cognitive element—it justifies the discrepant behavior and requires no more thought. The famous athletes and movie stars who are paid much to appear in television commercials should not change their private attitudes because of the favorable things they are saying about some product. When someone expresses discrepant attitudes for minimal rewards, however, the dissonance created cannot be reduced by reference to a large reward. If the public expression cannot be denied, it is likely that the private attitude will be changed in the direction of the publicly expressed attitude (Festinger & Carlsmith, 1959). Subsequent research has shown that the dissonance prediction works best when the people in the low-reward condition believe they have exercised free choice in doing the public advocacy and that the advocacy has had an impact on those who heard it. Most recently, Cooper and Mackie (1983) have shown that the topic for the public advocacy should not be central to a person's membership in an important group.

Dissonance theory has had numerous applications. One such application is to situations in which people have expended different degrees of effort in order to gain an identical outcome. Consider people who undergo initiation ceremonies to join a group that turns out to be disappointing. You might think that people who have endured a hard (compared to easy) initiation would be less favorable toward the group. Dissonance theory, however, makes the opposite prediction (Aronson & Mills, 1959). Knowledge of the difficult initiation is dissonant with the perception that the group is not outstanding. If the difficulty of the initiation cannot be denied, the dissonance will be resolved by the adoption of a more favorable attitude toward the group.

COGNITIVE-RESPONSE APPROACHES

The attitude theories discussed so far have tried to account for attitude formation or change by how the recipient of a message deals with new information. The new information is important either because it mediates learning effects or because it creates inconsistencies the recipient needs to resolve. Cognitive-response approaches do not contradict these theories. Instead, they focus on the fact that the recipient does more than react to the external information; the recipient also generates thoughts about the information. These thoughts can increase, neutralize, or even reverse the intended impact of the information. The next time you watch television commercials, jot down the private thoughts you have. Do the commercials you like generate different types of thoughts than the commercials you do not like?

Although various social psychologists have suggested that a recipient's cognitive responses are important determiners of attitude change, it was Greenwald (1968) who gave the label *cognitive response analysis* to these approaches. Greenwald suggested that when people receive a persuasive message, they relate the information in the message to their existing store of knowledge. From this point of view, recipients do not merely process information passively; they actively react to the information with their own personal thoughts. It is, of course, just as impossible to observe another person's inner thoughts as it is to observe his or her private attitudes. In order to measure thoughts, Greenwald asked the subjects in his experiments to write down their thoughts immediately after hearing a persuasive communication. These thoughts were then coded. For example, how many thoughts favored the communication, and how many opposed it? In general, as the number of favorable thoughts increased, so did the degree of attitude change. The physiological measures discussed in a previous section (page 387) have also been used to measure cognitive responses to various communications.

Cognitive-response approaches also emphasize the role of the person's cognitive organization in determining how information is interpreted, remembered, and retrieved. This approach appears to be extremely promising in adding to our understanding of attitude processes.

Factors in Attitude Change

Research testing attitude change typically focuses on the immediate effects that occur after people are presented with new information. Such a format is well suited for laboratory research, where testing the effects of a few variables requires ignoring many more. However, the formation and change of attitudes in the daily world is part of the ongoing process of living. Attitude formation and change occur in the context of existing interpersonal relationships, group memberships, and particular situations; and they span various time periods. Sometimes the amount of attitude change is extreme. The mass media have often dramatized such change when it was coercively induced—for example, the brainwashing of American soldiers captured during the Korean war, the kidnapping of Patty Hearst, and the teenage "conversions" by radical religious groups. However, extreme changes of attitude do not require coercion. This point is illustrated in Application 10.1, which summarizes the way one man recalled the events leading to his attitude change.

Whether extreme or moderate, fast or slow, it is possible to identify the basic units involved in attitude change processes. Indeed, the basic

Application 10.1

EXTREME ATTITUDE CHANGE: A CASE HISTORY

This is the story Claiborne P. Ellis recounted in interviews with Studs Terkel (1980). Ellis describes his journey from childhood to becoming president of the Durham, N.C., chapter of the Ku Klux Klan to becoming the regional business manager of the International Union of Operating Engineers. The story illustrates a series of attitude changes. Ellis began by hating blacks, Jews, and Catholics and ended by evaluating members of these groups by their individual behavior. You should recognize elements of each of the theories discussed so far, although no single theory will fit all of the events.

Ellis was born in Durham, N.C. His family struggled constantly with poverty, and many of his early memories involve the economic depression of the 1930s. He was very close to his father, who worked during the week in a textile mill but drank a great deal on weekends. When C. P. was around 17 years old and in "about" the eighth grade, his father died. C.P. had to leave school to help support his family. He took a series of low-skilled jobs and eventually borrowed $4,000 to buy a service station. By then he had married and was working ". . . my butt off and just never seemed to break even" (p. 202). Two months before the final loan payment was due, he had a heart attack. Despite his wife's efforts, the service station was lost. He had been taught ". . . to abide by the law, go to church, do right and live for the Lord, and everything'll work out. But it didn't work out" (p. 202). The continuing failure to lift his family into minimal economic security turned a smoldering bitterness into hatred. He wanted to blame something for his failures and soon found a convenient group as a target.

While Ellis owned the service station, he was invited to join the Ku Klux Klan. It was an opportunity he seized eagerly because "It gave me an opportunity to be part of something" (p. 202). Not only did he feel the glow of belonging to a group, but also his long-standing sense of inferiority began to disappear. The interview is sprinkled with references to feelings of low self-esteem, especially centering around the feeling that people were laughing at him and his father for their shabby clothes or lack of formal education. His father had been a member of the Klan, and Ellis was well versed in their attitudes. The Klan hated blacks, Jews, and Catholics. And so did Ellis. He quickly rose through various offices to the presidency of the local chapter. Because the civil rights movement was becoming active in Durham at this time, Ellis's hatred was directed mostly at blacks. In particular, he felt a purple rage at a woman named Ann Attwater, who seemed involved in every boycott and demonstration he went to watch.

Although the Klan is notorious for protecting the anonymity of its members, Ellis unashamedly brought the local chapter out into the open. He began attending meetings of the city council and county commissioners to represent the Klan's objection to any change. He and his group had numerous confrontations with representatives of the black community at various board meetings. Members of these boards began calling him to ensure his presence at meetings involving critical issues. These members would not publicly agree with the attitudes of the Klan, but they privately shared these views. The people who called to praise him, however, also avoided him in public. While searching for an explanation for this inconsistency, he began to reconsider his role. It struck him that he was being used. "As a result of our fighting one another, the city council still had their way. They didn't want to give up control to the blacks nor to the Klan" (p. 205). It was at this point that Ellis recalls doing ". . . some real serious thinkin'." Although he was becoming convinced of the correctness of his beliefs about being used, he could not persuade other Klansmen. He had to struggle with the inconsistency on his own, and it caused him many a sleepless night.

During this period, a critical event occurred. The state AFL-CIO received a federal grant to assist them in finding solutions to the racial problems in the public schools. To his amazement, Ellis was asked to join a representative citizens' panel to discuss these problems. As soon as he learned that members of the black community would also be invited, he refused the invitation by saying "I am not going to be associated with those types of people" (p. 206). On a whim, however, he attended the first evening meeting. Many of the participants, including Ann Attwater, were familiar to him because of past confrontations. The moderator of the meeting was a black man who encouraged everyone to speak freely. During the meeting, Ellis did just that, repeating his extreme antiblack attitudes. To his surprise, some of the black members, who did not agree with a single one of his attitudes, praised him for his honesty in expressing his views. Ellis's involvement in the group began to grow. On the third night, with backing from some of the black participants, he was elected co-chairperson of the group, along with Ann Attwater.

Despite much mutual reluctance, Ellis and Attwater agreed to put aside their personal differences and to work toward the common goal of finding solutions. Through their joint work, they began to see many similarities between themselves. Their efforts to recruit more panelists from among members of their respective groups were met with the same suspicion and rejection. Furthermore, the children of both had come home from school in tears. Ellis's child was ridiculed by his teacher for being the son of a Klansman, while Attwater's child was ridiculed by her teacher for being the daughter of an activist. The discovery of such commonalities and their joint work led Ellis to a feeling of respect and liking for Attwater. Through their leadership, the panel agreed on a number of resolutions. Although the school board did not implement any of them, the panel members had worked together effectively.

Ellis's attitudes did not change immediately. His initial self-justification for working on the panel was that school integration was the law and that all people should be law-abiding. In the hope of implementing the panel's recommendations, he ran (unsuccessfully) for the school board. He was still associated with the Klan, but he did not campaign for Klan themes. His platform was simply that before making any decisions, he would listen to the voice of all of the people. The campaign brought him into contact with many blacks. At long last, he began seeing people as individuals. With this change came a sense of rebirth. He no longer had sleepless nights and enrolled in an evening program that resulted in his receiving a high school equivalency diploma. During this period, he helped to organize the first labor union at his place of employment. He now felt that management was using the poor to retain profits and power. When the opportunity arose, he gladly switched his career to labor-union work, where he felt he could help the poor, both black and white.

REFERENCE

Terkel, S. (1980). *American dreams: Lost and found.* New York: Pantheon Books.

units involved in an attitude-change sequence are very similar to those involved in a social-influence sequence (see Chapter 9, especially Figure 9.4, page 353). Social influence can be described as an agent making an intervention toward some focal person (FP); attitude change minimally involves a source sending a message toward some FP. In both instances, the research centers on identifying the factors that will produce a change following the intervention or the message. The bulk of the research on attitude change can be conveniently categorized by whether it looks at characteristics of (1) the source, (2) the message, or (3) the person receiving the message. These three categories will organize our discussion of the various factors that contribute to attitude change.

SOURCE OF THE MESSAGE

Messages must originate somewhere—from a person (the President), a group (the Sierra Club), or an institution (the Supreme Court). The originator of the message is known as the source, and various characteristics of the source can affect the impact of the message. The source's credibility and attractiveness have been shown to be particularly important.

Credibility The more reasons we have to believe the person sending a message, the more likely we are to be persuaded by it. Consider some messages you may have received in the past: "Those spots on your face are chicken pox," "If you don't get a new radiator, your engine will be ruined in 200 miles," or "Purple is a wonderful color for you." Whether or not you are persuaded by these messages will depend, in part, upon the credibility of the source. The messages are more likely to be persuasive if they come from, respectively, your family physician, your regular mechanic, and your best friend than if they come from a bus driver, a used-car salesman, or a store owner who has overstocked on purple shirts. There are two main aspects to credibility—expertness and trustworthiness. Expertness is special knowledge and skills. We judge expertness by whether or not we think the source knows what he or she is talking about. Trustworthiness revolves around the truthfulness of the source. It is judged by whether the source has any special interest in persuading us, has been consistent in past expressions of attitudes, and has usually been objective. Expertness and trustworthiness are often interrelated. High expertness frequently goes hand in hand with trustworthiness.

In general, research has shown that messages attributed to highly credible sources are more persuasive than are those attributed to sources with low credibility. In an early example of this type of research, Hovland and Weiss (1951) had subjects read messages about the practicality of plans for building an atomic-powered submarine. When the messages were attributed to the physicist J. Robert Oppenheimer, they were more persuasive than when they were attributed to the Russian newspaper *Pravda*. From the beginning, however, the generality of the credibility effect was found to be clearly limited. Messages attributed to the highly credible Oppenheimer were more persuasive immediately after the message. When the subjects were retested 2 weeks later, the difference had disappeared, and the degree of change was comparable between subjects who had the message linked to a source with high or low credibility. Current research on source credibility continues the attempts to identify the exact circumstances under which it contributes to attitude change.

Attractiveness It seems intuitively reasonable that increases in source credibility add to a message's persuasiveness. A source's credibility pro-

Figure 10.5

Products linked with physically attractive people become more desirable. (Courtesy Jordache Enterprises, Inc., New York.)

vides some information about the reliability of the information contained in the message. But why should a source's attractiveness make a difference? It adds no obviously useful information to the content of the message. Yet even a brief sampling of television commercials will quickly show you that many advertising agencies are convinced that linking their product to attractive people is an effective strategy. The people in these commercials are often physically attractive, likable, or similar to us. (See Figure 10.5.) For better or worse, the designers of commercials are correct. People who are attractive to us are more likely to sway us than are those who are not. The effectiveness of attractiveness has been shown for physical attractiveness (Chaiken, 1979), likability (Eagly & Chaiken, 1975), and similarity (Dembroski et al. 1978). Such findings are certainly in keeping with balance principles (page 390). If we are attracted to someone who likes X, the requirements of balance are met if we also like X.

CONTENT OF THE MESSAGE

Even if the word is not always more powerful than the sword, it is the most frequent means for persuasion. Messages contain words and other symbols that convey information. Persuaders of all sorts spend countless hours on their messages to make them successful in changing attitudes. What can be done to messages to increase their effectiveness?

Suggestion Advertisers and propagandists often rely on *suggestion,* the uncritical acceptance of a statement. They design their messages in hopes that people will accept a belief, form an attitude, or be incited to action by someone else's say-so, without requiring facts.

Perhaps the most common form of suggestion is *prestige suggestion,* in which the message appeals to people's regard for the source's status or prestige. Advertisers frequently hire famous people to suggest directly that their products are the best. In a similar vein, speakers can link themselves with prestigious persons to gain acceptance of their message. Politicians frequently find ways to mention Abraham Lincoln, John F. Kennedy, or other respected leaders to promote their ideas. As with the effects of the attractiveness of the source, such appeals are in keeping with balance-theory (page 390) considerations.

Appeals to Fear Another method of persuading people is to try to scare them. Political candidates may claim that if the other side wins, the country will suffer high inflation, poorer services, or even war. The slogan "Speed kills," seen along many highways, is an attempt to scare people into observing speed limits. The same slogan has been cleverly used to warn people against the misuse of amphetamines (Figure 10.6).

Are scare tactics effective in changing attitudes? The evidence is mixed. Up to a point, they tend to work. Fear of injury has induced many people to wear seat belts in automobiles. Fear of disease frequently impels people to get inoculations. Strong appeals to fear, however, may backfire if people respond to them with a reaction called *defensive avoidance.* This means that they avoid information put out by the communicator or refuse to accept the communicator's conclusions. There is the story of the cigarette smoker who was so upset by newspaper accounts of the link between smoking and lung cancer that he stopped reading the

Figure 10.6

Appeals to fear are sometimes effective in changing attitudes. An example of an appeal to fear in advertising. (Metropolitan Life.)

paper. Strong appeals to fear, however, may be effective if they include suggestions about how to avoid the feared consequences and if there is a high likelihood that the consequences may actually occur.

One-Sided Versus Two-Sided Messages If you want to convince people of your point of view, is it better to present only one side of the issue or both sides? This question was tested during World War II in a classic study conducted by Hovland, Lumsdaine, and Sheffield (1949).

> In 1945, as the war in Europe was drawing to a close, the U.S. Army became concerned about the soldiers' hopes for a quick conclusion to the war against Japan. Prevailing expert opinion saw a protracted fight in the Pacific that was likely to last for several more years. The War Department was eager to find an effective way of persuading American soldiers that the war in the Pacific would last 2 more years. To this end, Hovland and his group developed two radio programs.

The one-sided program contained only arguments in support of the favored conclusion. It featured such things as Japan's resources and the distance of potential battlefields in the Pacific. The two-sided program contained identical arguments plus arguments against the favored conclusion. These points, however, were also refuted in the program.

In a carefully controlled experiment, soldiers in training heard the one-sided program, heard the two-sided program, or served as control subjects. The results were somewhat complicated. Both programs were effective in lengthening the soldiers' predictions about the duration of the war. Overall, neither program was more successful. Each program, however, was more effective with particular groups of soldiers. The two-sided program was more effective with soldiers who initially opposed the advocated position and were more knowledgeable about the issues. The one-sided program was more effective with those who were already inclined to believe that the war in the Pacific would drag on or who knew relatively little about the issues.

Subsequent research has been fairly consistent with these initial findings. A one-sided approach is most effective when people are either neutral or already favorable to the message; a two-sided approach is more likely to win converts from an opposing point of view.

RECEIVER OF THE MESSAGE

The whole point of sending messages is to get receivers to do something they would not otherwise do. All receivers do not necessarily respond in the same way to a given message. Research has tried to discover the characteristics of receivers that will produce more or less persuasion. We have previously described one study involving a receiver characteristic—the extent to which the receiver's attitude was already in the advocated direction. You will recall that Hovland et al. (1949) found that one-sided messages were more effective with groups that favored the advocated position, while the two-sided approach was more effective with those who initially opposed the advocated position. It may be instructive to look at a few of the many receiver characteristics that have been studied.

Influenceability Most personality traits are not related to the ease with which someone is persuaded. A person's general personality profile will be of little use in predicting whether a given message will be persuasive. It is known, however, that some people are more easily influenced than others and that some people are downright gullible. The latter, bombarded with conflicting viewpoints, will believe the one they heard most recently. As might be expected, there are group differences in this trait. Obviously, children are more easily influenced than are adults, and poorly educated people are more easily influenced than are the well-educated.

Selective Attention and Interpretation Whether a message will influence a recipient depends upon how it is perceived and interpreted. Most important, it depends upon whether the message is attended to in the first place. As was mentioned earlier (page 391), information that is very

Inquiry 10.1

CAN HIDDEN MESSAGES PERSUADE US?

A tachistoscope is a fancy slide projector that can be regulated to flash slides on a screen for varying lengths of time. It can be set to expose slides for just a fraction of a second—a period too brief for a subject to recognize what is on the slide. Even if a slide contains just a single word, it can be presented at a speed that prevents recognition. The idea that subjects who watch such speedy presentations will still be affected by the content of the slide has been called subliminal perception. There are other techniques for presenting visual stimuli below the threshold of recognition, and comparable effects can be produced for sounds. From time to time, it has been suggested that these subliminal presentations of stimuli can be used to persuade people.

The general idea is that advertisers can hide messages that are below recognition threshold in standard media presentations. The messages would go undetected but would influence the attitudes and behavior of the viewing public. This chilling possibility, with its import for all sorts of thought control over unsuspecting audiences, has been periodically featured in the mass media. The specter of a horde of consumers being manipulated into purchases of a sea of unwanted products seemed very real to some people.

A report in a 1958 issue of *Life* magazine first brought public attention to the possibility of subliminal persuasion. In a movie theatre in Fort Lee, New Jersey, James Vicary, a marketing specialist, presented two messages: "Drink Coca-Cola" and "Hungry? Eat popcorn." These messages were alternately flashed on the screen for 3 thousandths of a second every 5 seconds throughout the showing of the feature film. At the end of 6 weeks, Vicary claimed that Coca-Cola sales had increased 57.7 percent and popcorn sales had increased 18.1 percent. It is difficult to know what to make of this claim because Vicary used no control groups to check on the rate of sales in the absence of messages.

Subliminal persuasion has more recently been used to try to decrease shoplifting in department stores. In 1979, about 50 department stores installed what its inventor, Hal Becker, called "a little black box." The box is essentially a sound mixer that blends background music with subliminal antitheft messages such as "I will not steal." The messages were repeated 9,000 times per hour. Over a 9-month period, one east coast department store chain reported a 37 percent drop in shoplifting. Like its predecessor, this demonstration was reported in a popular magazine (a 1979 issue of *Time*) and used no control condition to check on the rate of shoplifting in the absence of antitheft messages.

Along with the occasional examples of subliminal persuasion directly influencing an audience's behavior, there have been expressions of concern about the possibility of subtler types of influence. For instance, advertisers are said to embed messages and arousing symbols in their advertisements, especially those appearing in the print media. Most of the subliminal themes claimed to have been identified are sexual in nature; the idea is that such material may stimulate people to buy the product with which it is associated.

The current status of subliminal persuasion is fairly clear. It is a fascinating topic that lacks any convincing scientific evidence. No controlled research documenting the long-term effectiveness of subliminal persuasion has been reported. While it is true that we can be influenced by stimuli that are not consciously perceived, these effects occur on preference judgments taken immediately after the subliminal exposure. There is no evidence that subliminal stimuli create any long-term effects or become translated into actual consumer decisions. If, some time in the future, you are watching a television program and become overwhelmed with a desire to rush out and buy something called "Mrs. Mayer's Sauerkraut," it will *not* be due to your having unwittingly watched subliminal sauerkraut messages.

REFERENCES

Key, W. B. (1976). *Media sexploitation.* New York: New American Library.

Smith, M. J. (1982). *Persuasion and human action.* Belmont, CA: Wadsworth.

different, or discrepant, from our own position creates cognitive dissonance. If possible, we tend to avoid very discrepant information. If confronting such information cannot be helped, it is still possible for us to reduce the dissonance by paying attention to the points that fit our initial attitude positions and ignoring those that do not. The human tendency to avoid discrepant information creates a dilemma for the persuader. How extreme should the advocated position be to optimize change? There is no satisfactory rule to answer this question. As the advocated position becomes more discrepant, it will, at some point, become less effective.

Immunization The consequences of hearing a one-sided or a two-sided message (see above) were carried one step further by Lumsdaine and Janis (1953). Subjects who had heard either a one-sided or a two-sided message in favor of a particular topic were, a week later, exposed to a message arguing the opposite position. This "counter" message was more effective with subjects who had initially heard the one-sided message. That is, these subjects were swayed back toward their original position by the countermessage. The two-sided message created more "resistance" to subsequent countermessages. Why should this be the case?

According to McGuire (1961), people resist countermessages to the extent that they have had a chance to rehearse their positions. Two-sided arguments can create resistance because they provide experience in refuting the countermessage position. (It should be recalled that the "second" side of two-sided messages contains refuted counterarguments.) People who take attitude positions without any thought about the opposite position should be particularly vulnerable to countermessages.

Getting experience in refuting messages contrary to one's position is analogous to getting immunized against diseases. In medical practice, people can be immunized against certain diseases by being inoculated with small doses of the disease. McGuire applied this notion to attitudes. Mild exposure to arguments opposed to a person's own position will immunize the individual against stronger attacks. By resisting mild arguments against their point of view, people are given the strength to stand up against strong attacks. McGuire thinks that the effectiveness of the immunization effect stems from two factors. Exposure to mild countermessages both motivates people to rehearse their own position and provides practice in defending that position. For his research program, McGuire used commonsense attitudes or beliefs—for example, "Mental illness is not contagious." Almost everyone would agree with such a truism. Furthermore, it is unlikely that people will have heard challenges to these beliefs. They would have had no practice therefore in defending them. Here is a typical study (McGuire, 1961):

> Attitudes toward four health truisms were measured on a 15-point scale. The greater the agreement with an attitude statement such as "Everyone should get a medical checkup once a year," the higher the score. Five groups of subjects were used, as shown in Figure 10.7. A control group merely rated their attitudes toward the health propositions. The high degree of agreement with the propositions can be seen in the control group's average rating of 12.78. Four of the groups were exposed to arguments *against* the health

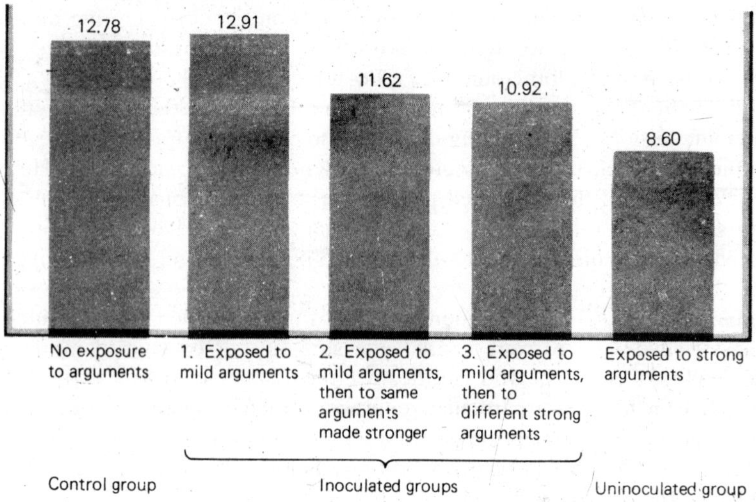

Figure 10.7

People can be "immunized" against strong attacks on their attitudes. Each bar represents a group's average attitude (on a 15-point scale) toward commonsense health propositions. The inoculated groups who were given practice in refuting mild attacks on their position and who were later given strong arguments against their point of view (groups 2 and 3) showed considerable resistance to the strong attacks. (Adapted form McGuire, 1961.)

propositions. Of these, three were "inoculated" by exposure to mild arguments that they could easily refute. Certain inoculated groups were then given strong arguments against the propositions. As you can see, these groups (2 and 3 in Figure 10.7) were less swayed by the strong arguments than was the uninoculated group.

Attitudes and Behavior

While the study of attitude formation and change is an interesting area of research in its own right, the pursuit of knowledge in this field has usually been justified by one important assumption—that attitudes guide behavior. An understanding of attitudes, thus, is seen as a key to understanding and predicting what people will actually do. In this section, we will review some evidence bearing on this assumption.

WHEN ATTITUDES FAIL TO PREDICT BEHAVIOR

An early study which attempted to examine the attitude-behavior relationship was conducted by La Piere in the 1930s. He was interested in prejudice against the Chinese. One summer, he traveled around the United States with a young Chinese couple, keeping a record of how they were received by clerks in hotels and restaurants. He found that in only 1 out of 251 instances was the Chinese couple treated inhospitably. Six months later, La Piere obtained information about attitudes toward the Chinese by sending each hotel or restaurant they had visited a letter asking if Chinese clients would be accepted. More than 90 percent of the responses were negative. It appeared, then, that the relationship between attitudes and behavior was extremely discrepant; people behaved in a positive manner toward the Chinese but reported that they would behave negatively.

A later investigation (DeFleur & Westie, 1958) studied the relationship between attitudes and behavior toward blacks. On the basis of a questionnaire administered to 250 white college students, two groups of subjects were selected—one highly prejudiced and one very low in prejudice. These two groups were compared on a measure of behavior toward blacks. Each student was asked if he or she would be willing to pose for a photograph with a black of the opposite sex. Students who agreed were then asked to what extent they would permit various uses of the photograph, ranging from limited exposure (only a few people would see it) to use in a national antisegregation campaign. Although DeFleur and Westie found greater behavioral consistency than had La Piere (the more prejudiced subjects were generally less willing to have the photographs taken and exposed widely), many inconsistencies were found. More than 25 percent of the subjects behaved discrepantly. Thus many "prejudiced" subjects agreed to have their pictures taken and used in the campaign, and many "unprejudiced" subjects refused to agree to any level of exposure.

These two studies are perhaps the best known of many that have failed to find the kind of attitude-behavior relationship that had been assumed to exist. Despite the difficulty in supporting a predictive link between attitudes and behavior, the belief in such a link has persisted. Researchers in the area have insisted that the relationship *must* be there; the problem is in measuring it appropriately.

WHEN ATTITUDES PREDICT BEHAVIOR

An important contribution to the study of behavioral prediction from attitudes has been made by Martin Fishbein (Fishbein, 1967; Fishbein & Ajzen, 1975). He has argued that there is no good reason to suppose that an overall measure of attitude toward an object will necessarily predict a specific behavior. This is the case, he says, because attitude is a hypothetical concept abstracted from the *totality* of a person's feelings, beliefs, and behavioral intentions regarding an object. Thus, any isolated specific behavior may be unrelated, or even negatively related, to the overall attitude.

Consider the following example of a hypothetical individual's attitude toward the social security system.

> The individual is a member of the House of Representatives who has been identified as having a positive attitude toward social security. She has the following beliefs:
>
> 1. Older people need some security.
> 2. An ethical society has a responsibility toward its senior citizens.
> 3. Widows and dependents should have some form of guaranteed income.
> 4. It is better to spend tax dollars on social services than on nuclear warheads.
> 5. The social security system is well funded.
> 6. The social security administration has mismanaged its funds.
>
> The party whip in the House wants to predict whether this

representative will vote for or against an upcoming bill to increase the social security tax. If he reasons that a person generally in favor of social security will vote for the increase, while one generally opposed to social security will vote against the increase, he may be wrong. For instance, while the hypothetical representative has a generally positive attitude toward social security, she may be *against* the increase because of her specific belief about the current management of the program.

Fishbein maintains that in order to predict a specific behavior, we should *not* focus on people's overall attitude toward the *object* of that behavior (the social security system, for example) but on their attitude toward the *behavior* (for example, voting for a social security tax increase). Attitudes about specific behaviors depend on such factors as evaluations of the likely consequences of the behavior and social norms concerning the behavior. By using attitudes about specific behaviors, researchers have been successful in predicting what people will do (Ajzen & Fishbein, 1973; Fredricks & Dossett, 1983).

Fishbein has not been alone in trying to specify the circumstances under which attitudes will accurately predict behavior. Some authors have focused on the conditions under which the attitude has been formed. Fazio and Zanna (1981), for example, argue that the attitude-behavior link will be stronger for attitudes formed by direct experience than for those formed by indirect, nonbehavioral experience. It has also been suggested that conditions which make the attitude relevant at the time the attitude-related behavior is enacted will strengthen the link (Borgida & Campbell, 1982). If our attitudes are to guide our behavior, they must somehow be present when we contemplate attitude-relevant behavioral choices. The effectiveness of direct experience or relevance may reside in their producing better recall of the attitude position. After all, if you cannot remember your attitude position, you cannot be guided by it. Some illustrative research will be presented.

Direct Experience Regan and Fazio (1977) experimentally manipulated attitude formation by either direct experience or by indirect experience. They then checked to see how well the experimentally formed attitudes predicted behavior.

Some of the subjects reporting for Regan and Fazio's experiment were given an opportunity to play with some sample puzzles before their attitudes toward the puzzles were measured. This was the direct-experience condition. Other subjects, in the indirect-experience condition, heard the experimenter describe the sample puzzles before their attitudes toward the puzzles were measured. Subsequently, all subjects were given an opportunity to play with various puzzles. The experimenter recorded the puzzle-playing behavior. Both direct-experience and indirect-experience groups expressed equally favorable attitudes. The direct group's behavior, however, was predicted much better than the indirect group's behavior.

Related to the direct-experience findings, it has been shown that when subjects are "primed" by favorable or unfavorable adjectives, they will express biased attitudes toward an ambiguous object in the direction of the "priming" (Fazio et al., 1983).

Attitude Relevance Because our world abounds with attitude issues, each of us can be concerned only with a limited number of issues, those which are of special importance to us. It has been suggested recently that relevant attitudes are a better guide to subsequent behavior than are irrelevant attitudes. That is, for any given attitude issue, the link or correlation between attitudes and behavior should be much stronger for those individuals for whom the attitude is relevant than for those for whom it is not. This effect has been demonstrated for experimentally induced relevance. Snyder and Kendzierski (1982) measured subjects' attitudes toward affirmative-action programs and toward volunteering for psychological experiments. Some subjects' attitude-related behavior was observed under normal conditions. The attitude-behavior link was fairly weak. The attitudes were made relevant for other subjects just prior to the enactment of the behavior. The strength of the association or correlation increased significantly after the relevance induction.

Other experiments have used naturally occurring differences in relevance. For example, Sivacek and Crano (1982) thought that a proposal to raise the legal age for drinking alcoholic beverages to 21 years would be more relevant to 18- and 19-year-old college students than to those over 21 years of age. As expected, attitudes toward the proposal were a better predictor of willingness to help circulate petitions for the younger subjects than they were for the older subjects.

Behavior and Attitudes

The road between attitudes and behavior is actually a two-way street. In closing this chapter, we wish to remind you that in some circumstances, the acting out of the attitude can determine the attitude position taken. We have previously discussed some examples of this point. You may recall how getting people to role-play new attitude positions served to induce dissonance that was reduced by a change of attitudes in the direction of the role-played position (page 392). What, then, distinguishes the circumstances under which behaviors will predict attitudes from those under which attitudes will predict behavior?

A person who is role playing a new attitude position may come to adopt the role-played position. Such a person may become more sympathetic to the new position, generate new ideas or beliefs in support of the new position, and forget the previous position. In each instance, the behavior of advocating a new position would predict the person's current attitude. In addition, there may be a simpler process.

Daryl J. Bem (1965) was the first to suggest that we sometimes deduce our attitude positions by direct observation of our own behavior. Bem's ideas about self-attributions were discussed briefly in Chapter 9 (page 352). Bem reasons that if we cannot locate a cause for our behavior in the environment, we assume the behavior occurred because of some internal motive or personality disposition. Suppose you signed a militant petition favoring the passage of ERA. If there were no strong environmental reasons for your signing (the petition was not circulated by your professor, and you were not paid to sign), you would look for some internal reason. If you hold a strongly favorable attitude toward ERA, this would be a sufficient internal reason. But suppose you are neutral about ERA. In that case, Bem says that the petition-signing behavior would be perceived by you as standing for a favorable attitude. Bem argues that in the absence of a strongly held attitude and without the existence of strong environmental incentives, we use self-observation of our own behavior to decide our attitude positions. If you were asked "Do you like movies?" you would think to yourself, "I must like them because I go to so many movies" or "I must not like them because I so rarely go to the movies."

It is Bem's contention that dissonance research can best be understood if we view it as instances of self-perception. Consider, for example, the dissonance finding that people who are promised large incentives for counterattitudinal role playing change their attitudes less than do those who are promised small incentives (Festinger & Carlsmith, 1959). (See page 392.) Bem argues that the attitude issues used in dissonance research are of little importance to most subjects; thus they cannot serve as an internal cause of the behavior. Instead, dissonance effects depend upon whether situational incentives are high or low. Subjects who are promised large rewards for advocating positions counter to their own will attribute their behavior to the large rewards. (The experimenter made them an offer they could not refuse.) Their behavior represents a desire to earn rewards, and their private attitudes remain unchanged. Subjects who engage in the same counterattitudinal advocacy for small rewards cannot use the rewards to justify their behavior. They are more likely to think "I must be advocating this position because it represents my attitude." Their attitudes are subject to change.

More recent research has suggested that self-perception and dissonance theories are actually complementary explanations which apply under different circumstances. Fazio, Zanna, and Cooper (1977) demonstrated that the degree of discrepancy between the original attitude and the behavior is critical. For situations involving a clear and relatively extreme discrepancy, people experience the aversive arousal required by dissonance theory. Such arousal does not occur when the discrepancy is smaller, and therefore the self-perception theory appears to supply the correct interpretation of the consequent attitude change.

Summary

1. Attitudes are evaluations of various attitude objects. They are based on beliefs and often have import for guiding behavior. Any given attitude is a summary of the evaluations made of different characteristics of the attitude object.

2. The scientific study of attitudes requires that they be measured. In order to measure attitudes, evaluations must be translated into some number system. For some purposes, it is adequate to measure attitudes with only two categories ("favorable" and "unfavorable"). For most purposes, it is desirable to have additional categories to reflect degrees of evaluation.

3. Most attitude measures use self-report techniques. Public

opinion (attitude) polling is a familiar example of a self-report method. Polls are taken in order to predict behavior or to gather information.

4. Polls use random samples of respondents chosen from the population of interest. The accuracy of polling results depends upon the wording of the questions asked, the way the questions are administered to the respondents, and the categories used to measure responses.

5. Attitude scales try to get a precise measurement of the extremity of people's attitudes. Scales use many items that are all related to the same attitude topic.

6. The accuracy of self-report methods is limited to the respondents' willingness and ability to express their attitudes. The search for attitude measures that are not under the voluntary control of respondents has led to investigating the body's physiological responses. These efforts have not so far been successful, but the most recent research on electromyographic (EMG) recordings of the major facial muscles appears to hold promise.

7. The formation and change of attitudes has been explained in terms of principles of learning, cognitive consistency, and cognitive responses to persuasive messages. The learning theories use concepts from classical conditioning and instrumental conditioning. The consistency theories emphasize the tendency to keep ideas about two or more attitude objects in harmony, or balance. Cognitive-response approaches focus on the respondent's cognitive organization in determining how information is interpreted, remembered, and used.

8. The basic sequence in attitude change involves a source sending a message to a receiver. In order to determine an attitude change, the receiver's attitude position after receiving a message is compared to the position held before the message was sent. The form of an attitude-change sequence is virtually identical to the form of an influence sequence.

9. Characteristics of the source which have been shown to increase attitude change are high credibility and high attractiveness. Attractiveness can be physical attractiveness, likability, or perceived similarity to the receiver.

10. Characteristics of the message which affect attitude change include suggestion, appeals to fear, and one-sided versus two-sided messages.

11. All receivers do not necessarily respond in the same way to any given message. Some receivers are more easily influenced than others. Furthermore, the impact of a message on a particular receiver depends upon how different the message is from the receiver's position and whether the receiver has had practice in defending his or her position.

12. The basic justification for the study of attitudes is that they are

reputed to guide behavior, but some studies have found no link between people's attitudes and their behavior. However, attitudes do predict behavior under certain conditions. Chief among such conditions are a correspondence between the measure of attitude and the behavior that is observed, the presence of direct experience in the formation of the attitude, and the relevance of the attitude issue to the person whose behavior is being observed.

13. In the absence of clear situational incentives and strongly held attitudes, people may deduce their attitude positions by observing their own behavior.

Terms to Know

One way to test your mastery of the material in this chapter is to see whether you know what is meant by the following terms.

Attitudes (383)
Beliefs (383)
Self-report method (384)
Public opinion (attitude) polling (384)
Target population (385)
Random sampling (385)
Attitude item (385)
Attitude scales (387)
Consistency theories (390)
Balance theory (390)
Cognitive dissonance (391)
Counterattitudinal role-playing (392)
Cognitive-response approaches (392)
Source credibility (395)
Suggestion (397)
Prestige suggestion (397)
Defensive avoidance (397)
Immunization (401)

Suggestions for Further Reading

The topic of attitudes is treated in greater depth in any social psychology text. In addition, there are a number of excellent books that are devoted exclusively to this area. A clear exposition of the theoretical issues and research findings pertaining to attitudes is contained in Richard E. Petty and John T. Cacioppo's *Attitudes and Persuasion: Classic and Contemporary Approaches* (Dubuque, IA: Wm. C. Brown, 1981). For the reader who wishes to learn about the applications of attitude research to the area of marketing, while also getting a comprehensive theoretical integration of research, Mary John Smith's *Persuasion and Human Action* (Belmont, CA: Wadsworth, 1982) is suggested.

It is also possible to become familiar with the current status of research and theory in this area, especially with respect to advances in the cognitive-response approach, through a book edited by Richard E. Petty, Thomas M. Ostrom, and Timothy C. Brock, *Cognitive Responses in Persuasion* (Hillsdale, NJ: Lawrence Erlbaum, 1981). Several authors have contributed chapters written for a professional audience; it is not for the fainthearted.

Development During Infancy and Childhood

HEADING OUTLINE

Chapter

[Charles Darwin, from an 1877 "baby biography" of his first-born, William Erasmus (Doddy)]

When two years and three months old, [Doddy] became . . . adept at throwing books or sticks, etc., at anyone who offended him; and so it was with some of my other sons. On the other hand, I could never see a trace of such aptitude in my infant daughters; and this makes me think that a tendency to throw objects is inherited by boys. [Darwin (1877), p. 288.] [Photo: The Library of Congress.]

(Sigmund Freud, from his 1905 work, *Three Contributions to the Sexual Theory*)

The pleasure-sucking is connected with an entire exhaustion of attention and leads to sleep or even to a motor reaction in the form of an orgasm. Pleasure-sucking is often combined with a rubbing contact with certain sensitive parts of the body, such as the breast and external genitals. It is by this road that many children go from thumb-sucking to masturbation. . . . No investigator has yet doubted the sexual nature of this action. . . . He who sees a satisfied child sink back from the mother's breast, and merge into sleep with reddened cheeks and blissful smile, will have to admit that this picture remains as a guide for the expression of sexual gratification in later life. [Freud (1905/1910), pp. 40, 41, 42.] [Photo: The National Library of Medicine.]

(John B. Watson, from his 1928 book, *Psychological Care of Infant and Child*)

. . . all we have to start with in building a human being is a lively squirming bit of flesh, capable of making a few simple responses such as movements of the hands and arms and fingers and toes, crying and smiling, making certain sounds with its throat. . . . parents take this raw material and begin to fashion it in ways to suit themselves. . . . It is especially easy to shape the emotional life at this early age. I might make this simple comparison: The fabricator of metal takes his heated mass, places it upon the anvil and begins to shape it according to patterns of his own. . . . So inevitably do we begin at birth to shape the emotional life of our children.

Once at the close of a lecture before parents, a dear old lady got up and said, "Thank God, my children are grown up and I had a chance to enjoy them before I met you."

Doesn't she express here the weakness in our modern way of bringing up children? We have children to enjoy them. We need to express our love in some way. The honeymoon period doesn't last forever with all husbands and wives, and we eke it out in a way we think is harmless by loving our children to death.

There is a sensible way of treating children. Treat them as though they were young adults. Dress them, bathe them with care and circumspection. Let your behavior always be objective and kindly firm. Never hug and kiss them, never let them sit in your lap. If you must, kiss them once on the forehead when they say good night. Shake hands with them in the morning. Give them a pat on the head if they have made an extraordinarily good job of a difficult task. Try it out. In a week's time you will find how easy it is to be perfectly objective with your child and at the same time kindly. [Watson (1928), pp. 45–46, p. 69, pp. 81–82.] [Photo: Archives of the History of American Psychology.]

(Jean Piaget, from a 1970 interview in *Psychology Today* and a 1970 article, "Piaget's Theory")

Many theories of some schools that I will not name are based on the rat. It is not enough for me. . . . Biologists have shown that the organism constantly interacts with its environment; the view that it submits passively to the environment has become untenable. . . .

As for teaching children concepts that they have not attained in their spontaneous development, it is completely useless. . . .

Remember . . . that each time one prematurely teaches a child something he could have discovered for himself, that child is kept from inventing it and consequently from understanding it completely. [Hall, 1970 (May), p. 30, p. 32.] [Photo: Anderson/Monkmeyer Press Photo Service.]

In the preceding quotations, we see a cascade of opinion on childhood, some from very famous figures in the history of psychology. These people had strong views on the nature of the developing child and on child rearing. Some of the views seem farfetched today, yet they were taken quite seriously by many scholars—and by many parents—during parts of the past century. The mildly outrageous quotations from Watson, for example, are drawn from his handbook of advice to parents. Somewhere within the advice and opinions of Watson and the many others who have studied children are pieces of the puzzle that is childhood. For several decades now, psychologists have been working to put these pieces—the facts of childhood—together. Their efforts have begun to yield a richly detailed, though certainly unfinished, picture of the developing person. This emerging picture, and the methods used to generate it, form the field of study called developmental psychology. We explore this field in the next two chapters.

WHAT are children for? Today the very question seems crass, but this has not always been the case. In times and places not so far away, children were valued partly for their economic usefulness. Even in as civilized a country as England, and as recently as the mid-1800s, 5- and 6-year-old boys and girls worked up to 14 hours a day in coal mines. In addition to being cheap labor, the children were small enough to fit into the narrow seams of the underground mines. At other times and in other places, children have fared much better, sometimes being carefully nurtured and sometimes even considered divine. But one of the most intriguing ideas ever advanced about children was that they give us a prototype of human evolution—that by studying children we can unlock the mysteries of our species in general. Charles Darwin helped introduce this idea in his 1871 book, *The Descent of Man*. In the decades that followed, "the child became the best natural laboratory for the study of evolution, and the idea of *development* dominated the science of man" (Kessen, 1965, p. 6).

Most scientists eventually decided that child development is not a mini-replay of evolution, but most continue to believe that studying children and how they develop can tell us a lot about human beings in general. This belief helped to spark a scientific field now known as *developmental psychology*—a field devoted to the study of development, from conception through childhood and beyond. In this chapter and the next, we will examine this field, and we will explore the work of developmental psychologists, the people who study development. We will also delve into some of the issues that have for years sparked conflict among developmental psychologists. We begin our discussion with three of these issues.

Recurring Conflicts

As the quotations at the beginning of this chapter suggest, the experts have often differed sharply about the how and why of children's development. Many of the disagreements have involved one of the three major issues outlined below. Perhaps the most enduring of these is the heredity-versus-environment issue, already familiar to most of us. It is sometimes called the *nature-nurture controversy*.

NATURE VERSUS NURTURE

We are what we are partly because of our inherited biological characteristics, our *nature*. Early evidence of this fact can be seen in babies; they show several reflexes that have clearly been built into their biological systems. Later on, children's learning of their first words may be primed by another part of their nature—a sort of inner clock for language development. This is suggested by the fact that certain steps in language development seem to take place at similar ages around the world. The broad impact that many have attributed to our inherited nature is nicely illustrated by the quotation from Darwin at the beginning of this chapter: ". . . a tendency to throw objects is inherited by boys."

Of course, the *nurture* side of the nature-nurture debate had its strong advocates too—people who believed that environmental forces have a more powerful influence on our development than does heredity (see Figure 11.1). The quotation from John Watson in our chapter opening, for example, compares the child to hot metal on an anvil, waiting to be shaped by parents and others.

The nature-nurture debate really concerns the *relative* impact of heredity and environment. Virtually no one believes that nature alone, or nurture alone, completely determines the course of our development. Psychologists today agree that development is shaped by the *interaction* of heredity and environment. Within this interaction, our genetic endowment for many characteristics provides us with a *reaction range*—that is, a range of possible levels that we may ultimately reach, depending on the quality of our experience in the environment. One picturesque way of describing this reaction range is geneticist Curt Stern's "rubber band" analogy.

> Stern compared the genetic endowment for any particular trait to the amount of "stretch" that is possible in a rubber band. Some rubber bands have a great deal of potential stretch built into them; others have relatively little. The amount of stretch a particular rubber band actually shows will depend upon *both* its native endowment—its basic "stretchability"—*and* the amount of pressure, or "pull," exerted by the people in its environment. Similarly, the amount of any particular trait that a person will show depends upon that person's inborn potential, or "stretch," for the trait and the degree to which the person's environment "pulls for" the development of that trait.

The nature-nurture debate is still alive in developmental psychology. But it now centers on questions such as: How much stretch is possible if we improve the quality of people's environment? For example: How much can we enhance school achievement in disadvantaged children if we give them extra preschool training? Questions like this are important to people who decide the fate of social programs such as Head Start.

PASSIVITY VERSUS ACTIVITY

As we develop, we interact with our environment; but how much of that interaction is spurred by *us*, and how much is spurred by the environment? Some psychologists picture us as fairly passive, doing what we do largely because of the environmental forces around us. John Watson, as you saw earlier, described the child as raw material waiting to be shaped by parents and others. B. F. Skinner, one of the most influential of modern psychologists, describes human behavior and development as a process of responding to rewards and punishments in the environment. (See, for example, his analysis of poetry writing on page 595.)

A different point of view has been taken by psychologists such as Jean Piaget. In his statement at the beginning of this chapter, Piaget attacked the view that the developing person merely "submits passively to the environment." Instead, Piaget argued, people actively manipulate the objects and events around them. We do not merely copy or learn

(a)

(b)
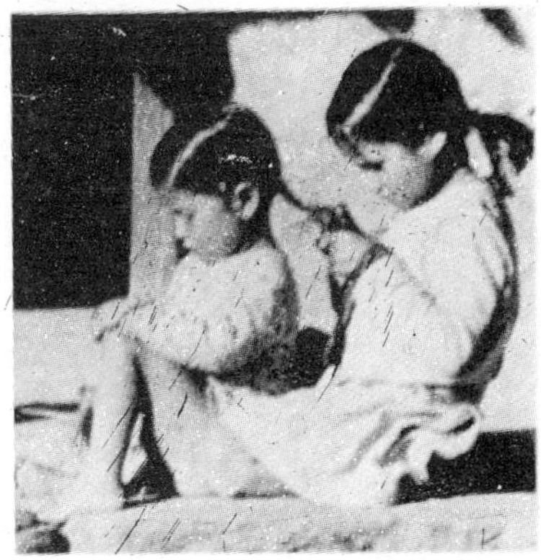
(c)

(d)

(e)

(f)

(g)

Figure 11.1

Many environmental forces help to nurture the development of the child; but all these forces operate within limits set by the child's biological makeup, or *nature*. Nature and nurture interact to shape development. (*a*, Erika Stone/Peter Arnold. Inc.; *b*, James Karales/Peter Arnold, Inc.; *c*, courtesy United Nations/Ray Witlin; *d*, Erika Stone/Peter Arnold, Inc.; *e*, Erika Stone/Peter Arnold, Inc.; *f*, Sybil Shelton/Peter Arnold, Inc.; *g*, Erika Stone/Peter Arnold, Inc.)

about reality as we develop. Instead, we construct our own ways of understanding the world; psychologically speaking, we each invent our own reality.

The tension between passive and active views of the developing individual becomes important in a practical way when educational decisions have to be made. Should parents seek out schools where their children will be taught facts and skills in a highly structured way? Or should parents look for schools where their children will be encouraged to

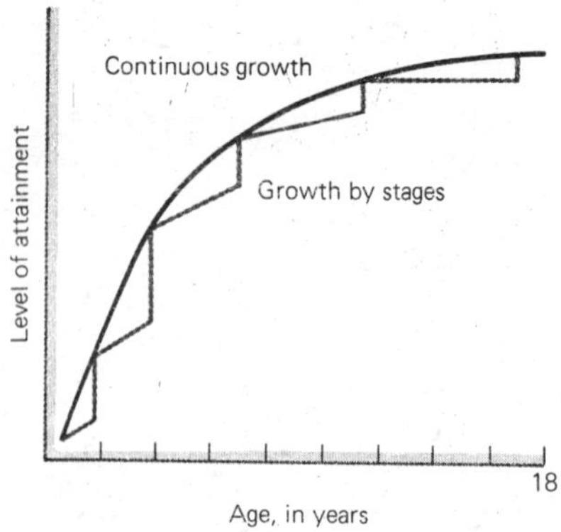

Figure 11.2

Some theories of development assume continuous progress; others postulate significant discontinuities (stages).

experiment to discover facts and develop skills for themselves? What would you advise a parent who faces such a decision? Your advice may reflect your views on how active or passive developing children are in their interactions with the world.

CONTINUOUS VERSUS DISCONTINUOUS DEVELOPMENT

Developmental psychologists also disagree as to the best way of describing development. Some see development as a sort of continuous progression—that is, a steady accumulation of skills, knowledge, and maturity. According to this view, development is best viewed as a smooth curve, like the one in Figure 11.2; and development is best measured in quantitative ways—that is, ways that tell us *how much* of a particular ability the child has. The major intelligence tests for children (see Chapter 13, page 529) reflect this view of development as a smooth progression. The tests measure intellectual maturity by assessing, for example, how many numbers a child can remember in a series and how many increasingly complex math problems a child can solve. Other psychologists see development as a discontinuous progression—that is, as a sequence of leaps from one stage to another. The resulting picture of development looks roughly like the stair-step curve shown in Figure 11.2—a series of upward steps in maturity, each followed by a leveling off, or plateau. Psychologists of this persuasion prefer to measure developmental changes in qualitative ways—that is, in terms of the characteristics of people's behavior. For example, Piaget's theory of intellectual development (described later in this chapter) involves a series of stages, each one involving not just more intelligence but also a *different way of thinking* than the previous stage. This theory defines a child's level of development in terms of which stage the child has attained. To assess a child's stage level, we need information about *how* a child goes about solving math problems, not *how many* math problems the child can solve.

Methods of Studying Development

Developmental psychologists focus on time and transformation. They study the changes that occur as the developing individual unfolds—changes in processes as basic as perception and as complex as forming a self-concept. Indeed, they may focus on any of the specialty areas discussed in this text. What developmental psychologists share with one another is an interest in exploring maturation and change. In this exploration, they rely on research methods geared specifically to the study of development. Two of the most important are the longitudinal method and the cross-sectional method.

THE LONGITUDINAL METHOD

A psychologist using the *longitudinal method* observes the same individuals at different points in time. The individuals may be the children of oil barons and migrant workers studied at yearly intervals from birth, West Point cadets tested for judgment under stressful conditions every 5 years after graduation, or even mice given monthly learning tests. If the "individuals" are animals such as mice, the investigator may control the environment in a laboratory and study each individual throughout its brief life span. With human beings, longitudinal research can be much more difficult. People who enlist in a study may move away, lose interest, or for other reasons be unavailable for later observation or testing. This is a logistical problem for the investigator, and it is a source of bias; it might mean that the findings of the completed study would apply only to people who rarely move and who are interested in research. Another risk of longitudinal research is that a study will seem less important or sophisticated at its end than it did at its beginning; this is because the central issues and the preferred research methods of psychology are continually shifting. Decades ago, a longitudinal researcher might have set out to study the effects of strict toilet training—a hot issue at the time—by asking parents to recall how they trained their children. Today, toilet training is not a central issue; and many believe that parents' memories of their child-rearing practices may not be very reliable anyway.

Carefully conducted longitudinal research, despite its problems, is highly regarded by most developmental psychologists, who recognize the value of repeatedly observing the same individuals as they mature. A particularly well-known example of long-term longitudinal research is the Fels Longitudinal Study. In Ohio, beginning in 1929, the Fels investigators repeatedly observed and tested 44 boys and 45 girls from birth through age 17. Most of the children were from Protestant, middle-class families. One set of measures was designed to assess how dependent the youngsters were at home, at school, and at summer camp. About 3 decades after the Fels study began. Kagan and Moss (1962) assessed dependency in 71 of the original Fels subjects, who, by then, were all young adults. Their primary interest was in finding out how stable the quality of dependency is. Does it change unpredictably from time to time? Or is there a particular age at which dependent behavior in childhood is a good predictor of dependency in adulthood? Kagan and Moss found that by the time girls were in the 6- to 10-year age range, their behavior reliably predicted their adult patterns. This was not so true for boys. In other words, dependent girls generally grew up to be passive, dependent young women; but dependency was not so predictable for males. This kind of conclusion would be very difficult to draw without longitudinal research. Of course, longitudinal studies need not span many years; some involve periods as brief as a few months.

THE CROSS-SECTIONAL METHOD

Most developmental research involves the *cross-sectional method*. In studying dependency, for example, many investigators simply compare representative samples of youngsters at two or more age levels on the same measures. Ruble and Nakamura (1973), for instance, compared kindergarten, first-, second-, and third-grade children on a measure of dependency. They found large group differences, with dependency most pronounced in the youngest children and least pronounced in the oldest. This, of course, suggests that dependency, as measured by these researchers, *probably* declines from the early- to the mid-elementary years. Such cross-sectional research is an efficient way of spotting age-group differences as such. It has its disadvantages, though. Because it does not involve repeated measurements of the same individuals, it cannot tell us how stable people's characteristics are as they mature.

OTHER METHODS AND METHODOLOGICAL ISSUES

For developmental psychologists, choosing a research method can be a complicated process. One reason is that more methods are available than the two outlined above. For example, some researchers now use research designs that combine aspects of the longitudinal and cross-sectional approaches (for example, Baltes et al., 1979). In addition to choosing from among longitudinal, cross-sectional, and more complex designs, developmental researchers must also decide whether their research will involve a controlled laboratory situation or naturalistic observations in "real life." Another particularly important consideration is whether the research subjects, often young children, will be exposed to discomfort or risks. Figure 11.3 shows a set of ethical principles proposed to protect children from such risks.

Applying guidelines like those in Figure 11.3 is not always easy. Often, risks must be weighed against the potential benefits. Suppose, for example, that researchers have developed a new method of training children to keep trying when they fail at math problems. The method involves cognitive retraining—that is, teaching children to analyze their failures, learn from them, and try a new approach. The researchers think this method may only work well for older children. To find out, they need to carry out a study in which both older and younger children are exposed to failure and then given cognitive retraining. But failure can be psychologically painful to children; thus there is a risk that guideline 5 (Figure 11.3) might be violated by the research. On the other hand, learning to profit from failure and to keep trying can be extremely valuable to children; so the proposed research might help children a lot in the long run. Should this experiment be carried out? The answer depends upon whether the potential benefits outweigh the risks when those risks have

Figure 11.3

A children's bill of rights for the conduct of research. (From Hetherington & Parke, 1979.)

A Children's Bill of Research Rights

1 *The right to be fully informed*: Each child participant has the right to full and truthful information about the purposes of the study and the procedures to be employed.

2 *The right to informed and voluntary consent of participation*: Each child participant has the right to either verbally or in written form agree to participate in a research project. In the case of children who are too young to understand the aims and procedures and to make an informed decision about participation, parental consent should always be secured.

3 *The right to voluntary withdrawal*: Each child participant has the right to withdraw at any time from continued participation in any research project.

4 *The right to full compensation*: Each child participant has the right to be fully compensated for his or her time and effort as a research subject, even if he or she withdraws and does not complete participation in the project.

5 *The right to nonharmful treatment*: Each child participant has the right to expect that he or she will not experience any harm or damage-producing events during the course of the research procedure.

6 *The right to knowledge of results*: Each child participant has the right to new information concerning the results of the research project. In the case of young children, their parents have the right to be provided this information. Often, this information will take the form of the group scores on a task, rather than the individual participant's own score.

7 *The right to confidentiality of their research data*: Each child participant has the right to expect that personal information gathered as part of the research project will remain private and confidential. Nor will any information about individual research participants be available to any other individuals or agencies.

been reduced to the lowest level possible. This, like other ethical problems in developmental research, requires a judgment call—one that is likely to be made by both the investigators and a "human subjects" committee in the institution where they work.

Some developmental researchers face yet another problem. Their youngest subjects, babies, cannot talk. We will explore how researchers cope with this problem and others as we turn now to the study of infancy.

Infancy: Early Steps in the March to Maturity

For centuries, the deeply private world of the infant was cloaked in mystery. Because babies could not talk, the adults in their world were reduced to guesswork and speculation about them. In recent decades, however, ingenious investigators have figured out ways of peering into the infant's world. Here we will examine what they have seen. We focus first on the *neonatal* (newborn) period, the first 4 weeks after birth. This is a time of transition from the total dependency of prenatal life to a more independent, creative existence. It is a time when rhythms of breathing, feeding, sleeping, and elimination are established and when babies and parents make some critical adaptations to one another. Next, we will consider the full span of *infancy,* a period that lasts from about 4 weeks to about 18 months, when language appears. Infancy is a time of fast-paced change in the youngster's ways of perceiving and interacting with the world.

THE NEONATE

The psychologist and philosopher William James once described the newborn's world as "nothing more than a blooming, buzzing confusion." Was he right? Most experts on infancy would now say no. Most now agree that neonates are born with abilities to perceive and respond to some parts of their world in an organized and effective way. For example, reflexes that are in place at birth permit the neonate to grope, or "root," for the breast (Figure 11.4), to suck when an object is placed in its mouth, and to swallow milk and other liquids. These three reflexes are obviously essential to feeding. Other inborn reflexes with obvious adaptive value are breathing, blinking, coughing, sneezing, vomiting, and withdrawing from painful stimuli. Another class of reflexes is attributed to the immaturity of certain parts of the brain. One example is the *Moro reflex:* When support is suddenly removed from the back of its head, the neonate will fling its arms to the side, extend its fingers, and bring its arms inward in a sort of embracing movement. The Moro reflex and other reflexes in this category normally disappear in early infancy, as the brain matures. If these reflexes persist for too long, it may mean that there is a problem with the infant's developing central nervous system.

Neonates show perceptual abilities that would surprise most people. They show positive reactions to certain sweet tastes and negative reactions to certain sour, bitter, or salty tastes. They turn their heads to avoid some strong odors; and they turn in the direction of certain sounds, including human speech. Some of the most exciting findings about neonates involve their visual abilities. They not only orient toward light but they can, under the right conditions, actually follow a light or an object

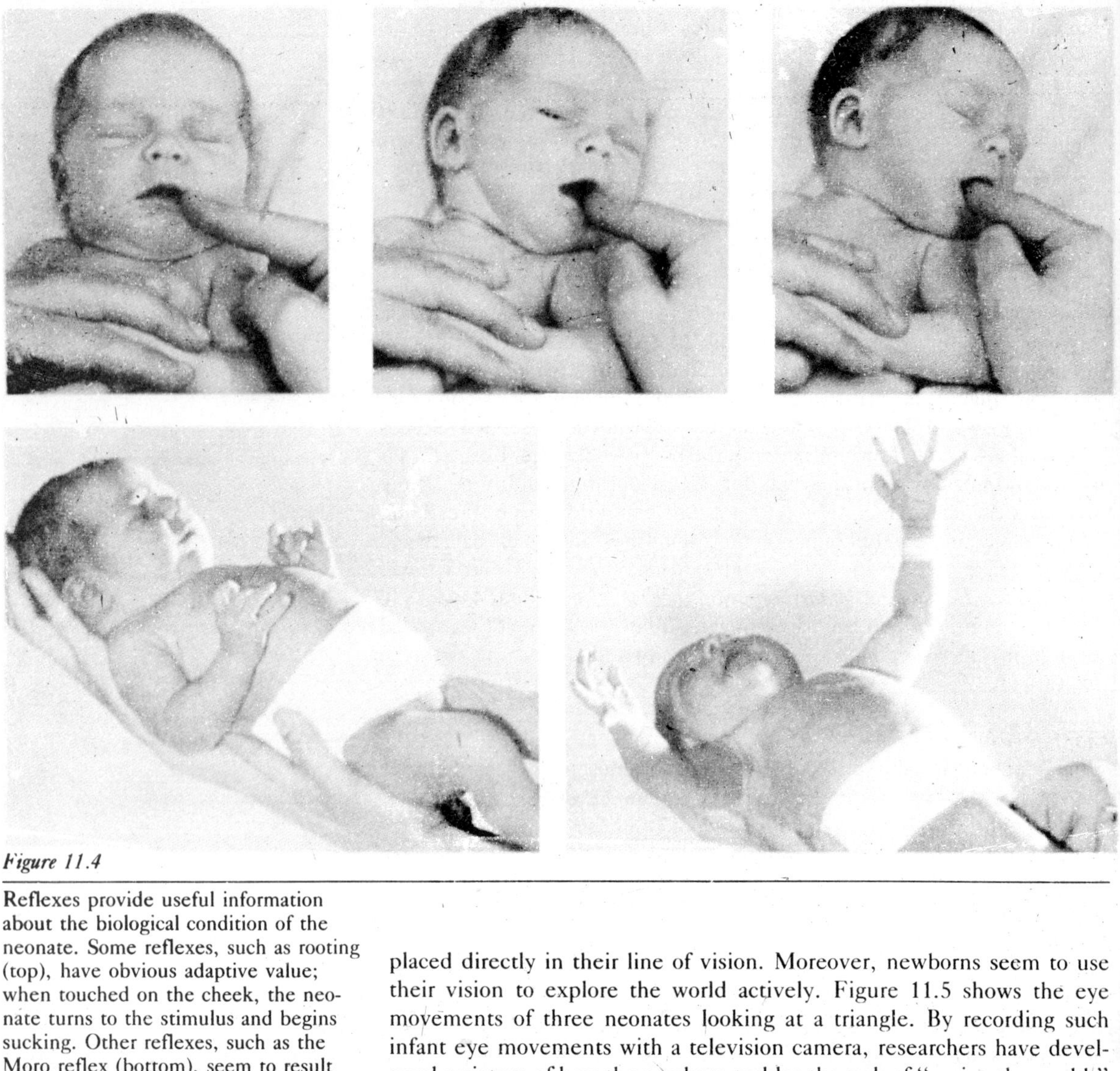

Figure 11.4

Reflexes provide useful information about the biological condition of the neonate. Some reflexes, such as rooting (top), have obvious adaptive value; when touched on the cheek, the neonate turns to the stimulus and begins sucking. Other reflexes, such as the Moro reflex (bottom), seem to result from the relative immaturity of parts of the brain. (H. F. R. Prechtl, *The Neurological Examination of the Full-Term Newborn Infant.* London: SIMP with Heinemann Medical; Philadelphia: Lippincott, 1977.)

placed directly in their line of vision. Moreover, newborns seem to use their vision to explore the world actively. Figure 11.5 shows the eye movements of three neonates looking at a triangle. By recording such infant eye movements with a television camera, researchers have developed a picture of how the newborn tackles the task of "seeing the world."

Haith (1976) described this perceptual process in terms of a set of "rules."

1. If I am awake and the light is not too bright, I will open my eyes.

2. If the area I am looking at is dark, I will begin a search for shadows or objects.

3. If I find an area that has light but no edges, I will start a broad and sometimes uncontrolled search.

4. If I find an edge, I will keep looking at it, trying to cross back and forth.

Infants are certainly not aware of these rules; their visual activity, though, seems to follow them in a fairly predictable way, suggesting that the rules may be programmed, or "wired in," to their nervous systems.

Under carefully arranged conditions, newborns can also learn via classical and instrumental conditioning (Chapter 4). For example, if they are allowed to suck a sweet liquid when they turn their heads to the right, they will increase the frequency of their right turns; they will reverse their turns if the sweet liquid is given for left turns. Neonates only a few days old have also been taught to turn their heads in response to one sound and not to another (Lipsitt, 1982).

Some surprising findings of a study conducted by Meltzoff and Moore (1977) suggested that neonates are even capable of imitation (Chapter 4, page 172). The research, illustrated in Figure 11.6, appeared to show that babies as young as 2 to 3 weeks can mimic certain adult behaviors, such as facial expressions. To many, this seemed remarkable because most experts believed at the time that imitation was not possible until about the end of the first year of life. Consequently, the study triggered a controversy. Some researchers tried and failed to replicate the Meltzoff-Moore findings (Hayes & Watson, 1981), but others succeeded—in one case, with infants averaging only 36 hours old (Field et al., 1982)! In fact, some researchers now suspect that the apparent imitation may be *most* pronounced among very young infants and that it may be reflexive—something like the early rooting and Moro reflexes discussed earlier in this chapter (Abravanel & Sigafoos, 1984). Perhaps some kinds of facial imitation are triggered automatically in early infancy only to fade out later; they may then reappear in the latter part of the first year, this time under the infant's *voluntary* control. Future research should help us test this possibility.

In the first few weeks of life outside the womb, glimmers of "personality" can be seen in the temperament babies display. For example, some babies are "difficult" even in the first weeks after their birth. They may show irregularities in their sleeping, feeding, or elimination patterns. They may be easily distressed and irritable, and prone to cry. Evidence suggests that youngsters who show this "difficult-child syndrome" are more likely than are other infants to develop behavior disorders in their later life. Genetic and other biological factors seem to influence such characteristics of temperament. However, there is also some evidence that by responding to their difficult child calmly and in good humor, parents can lower the risk of later behavior problems (Thomas & Chess, 1977, 1980). An important idea is suggested by much of this research. It is that babies are not simply the products of what their parents do to them but, instead, babies begin life with certain psychological characteristics of their own. Parents respond to those characteristics, adding input of their own. The baby's development is thus shaped by the interplay of its inborn characteristics and its parents' behavior. This is one example of the interaction between heredity and environment—nature and nurture—that we discussed earlier.

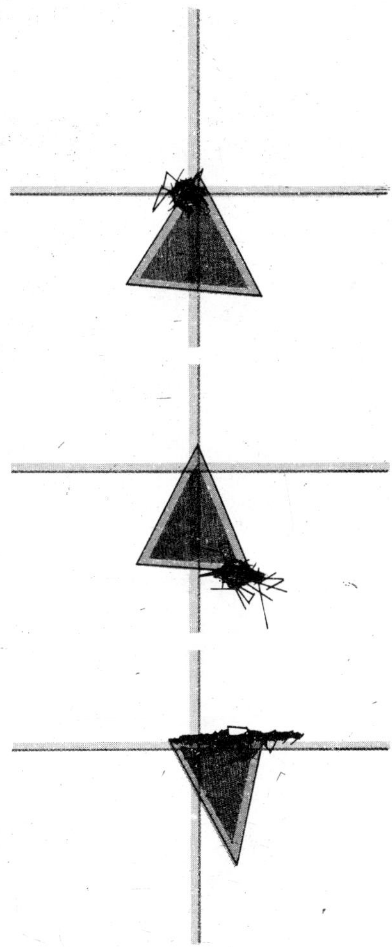

Figure 11.5

The black lines at the edges of these triangles show the eye movements of three newborn babies. They illustrate the tendency of neonates to explore edges, contours, and other areas of high color contrast. (Adapted from Kessen, 1967.)

MOTOR DEVELOPMENT

The development of motor activity (that is, movements involving muscle action) in the period of infancy has been studied extensively. Investiga-

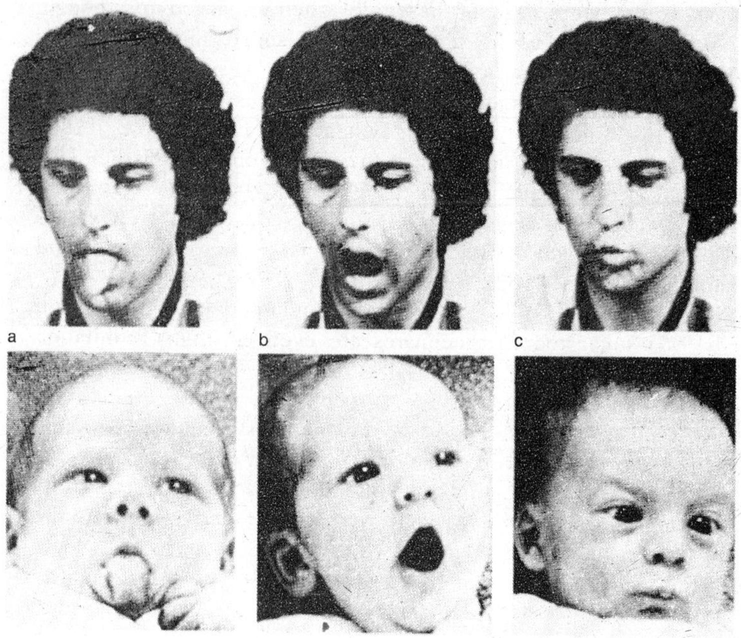

Figure 11.6

Sample photographs from videotape recordings of 2- to 3- week-old infants imitating (*a*) tongue protrusion, (*b*) mouth opening, and (*c*) lip protrusion demonstrated by an adult experimenter. (Copyright © 1977. Meltzoff, A. N., and Moore, M. K., "Imitation of Facial and Manual Gestures by Human Neonates," *Science*, vol. 198, pp. 75–78, fig. 1, 7 Oct. 1977.)

tors have built up a rich fund of normative data on the ages at which certain motor milestones are attained. Figure 11.7 shows the norms for several such milestones. The figure shows that although there is a fairly broad age range within which individual infants may reach each milestone, the *order* in which the milestones are reached rarely differs.

The same can be said of the steps involved in learning to move one's body around. The development of walking, in particular, involves a predictable series of milestones, shown in Figure 11.8. The order of events is quite consistent, but the age at which each milestone will be reached is hard to prediet for a given child. For example, Figure 11.7 shows that 5 percent of infants walk alone by the age of 9 months but that

Figure 11.7

Motor development from birth to 18 months. The heavy line gives the average age at which the behavior appears; the zones on either side show the range from the age at which 5 percent of infants show the behavior to the age at which 95 percent do. Remember that although the sequence is relatively fixed, the age at which babies show these behaviors varies considerably. (Based on data from Bayley, 1969.)

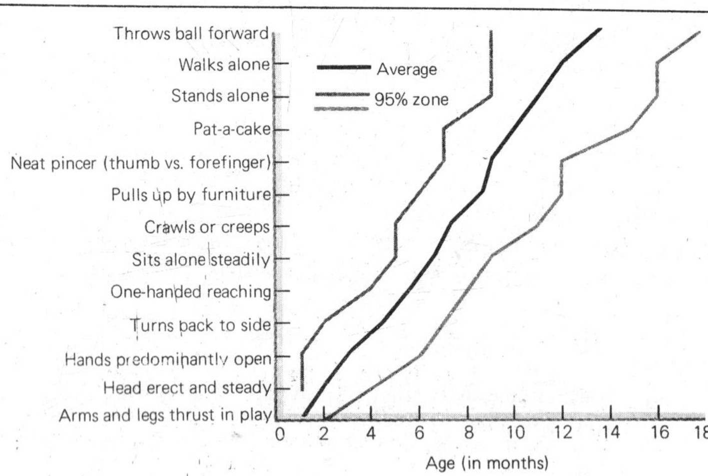

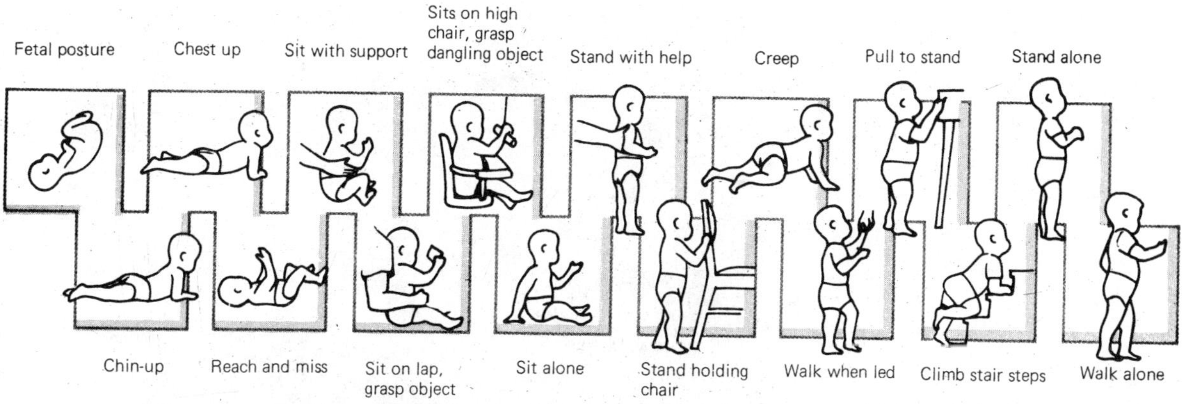

Figure 11.8

The sequence of development that leads to walking. (After Shirley, 1933.)

another 5 percent do not walk alone until after their sixteenth month. Walking is another good example of the interaction of nature and nurture; although it seems to be a wired-in developmental sequence, it can be speeded up or slowed down by variations in the infant's experience. For example, infants in some institutional settings who have had few opportunities to practice their motor skills show retarded motor development and delayed walking. By contrast, infants given a few minutes a day of "practice walking" during the first 2 months of their life walk earlier than infants given no such practice (Zelazo et al., 1972).

Prehension, the use of the hands as tools, shows another predictable developmental sequence. It begins with infants thrusting their hands in the direction of a target object, essentially "taking a swipe" at the object. This is followed by crude grasping involving only the palm of the hand. Then there is a sequence of increasingly well-coordinated finger and thumb movements. Late in the first year of life, most infants can combine thumb and finger action into a pincer motion that allows them to pick up a single chocolate chip from a tabletop.

What they will then do with that chocolate chip depends upon the state of yet another motor system, *mouthing.* The most common form of mouthing in infancy is sucking. Sucking is first linked to rooting, and both occur as relatively unrefined reflexes. Once an object that can be sucked is in its mouth, an infant in the first few weeks of life sucks with a consistent pressure and at the predictable rate of about 2 sucks per second. By the age of about a month, though, the baby's sucking will change in response to environmental events. Infants can vary their rate of sucking and the length of pauses between sucks. They can even shift from sucking to lapping if that does a better job of getting the food into their mouths. By 2 to 3 months of age, they can coordinate sucking with perceptual systems such as vision. For example, they can learn to suck faster or slower if a change in rate will sharpen the focus of a blurred film they are viewing (Kalnins & Bruner, 1973). Sucking figures prominently in several theories of development; later in this chapter, we will consider several ideas about what this activity implies about infants.

DEVELOPMENT OF PERCEPTION

The past 2 decades have seen an explosion of research on infant perception, particularly visual perception. (See also Chapter 3.) We have learned

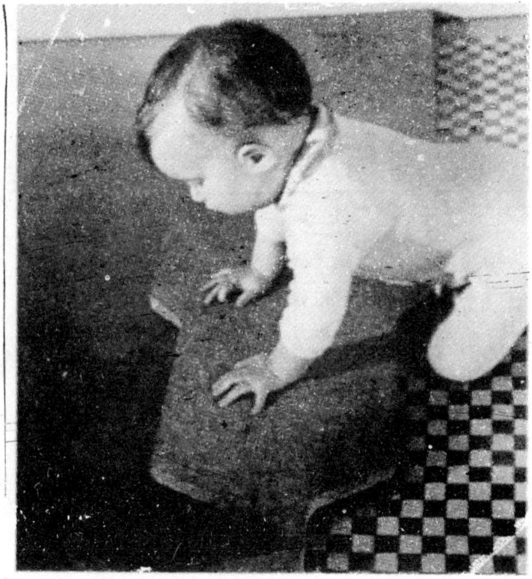

Figure 11.9

The *visual cliff*, a test of depth perception that can be used with infants and animal babies as soon as they can crawl or walk. Infants and baby animals with depth perception avoid the side that looks deep to them. (Richard D. Walk.)

much about the ways in which infants organize and interpret what they see. One example of such research is a study of depth perception conducted by Gibson and Walk (1960). To judge whether infants can read the perceptual cues that adults use to judge depth, these researchers used the *visual cliff* shown in Figure 11.9. It involved an apparent drop-off made safe by a clear glass cover. Despite the cover, Gibson and Walk found that none of the 6- to 14-month-old infants they tested would cross the "deep" area to get to their mothers. Yet all 36 of them eagerly crawled to their mothers when the moms were stationed on the "shallow" side. This strongly suggests that even 6-month-old infants have depth perception. But what about babies who are too young to crawl and thus unfit for the Gibson-Walk test? In a clever extension of the visual-cliff experiment, Campos, Langer, and Krowitz (1970) simply placed infants too young to crawl on either the shallow or the deep side and then measured changes in their heart rates. Even 1½-month-old infants showed heart-rate increases when they were placed over the deep side. They were evidently responding to depth cues.

From the parents' point of view, one of the most important infant perceptual activities is looking at adult faces, particularly when the baby and parent make eye contact. By following eye movements, as in the study of triangle perception (Figure 11.5), investigators have traced significant developmental changes in face watching. One-month-olds show only a modest interest in real human faces; when they do focus on a face (which, in one study, was less than a quarter of the available time), they focus mostly on edges and points of light-dark contrast (a bit like the way neonates look at triangles). Two-month-olds, by contrast, spend more time looking at the interior of the face, especially the eyes, than at the outer edges. Most researchers agree that by the fourth or fifth month, infants can "assemble" parts of a face into a meaningful whole. By 5 months, for instance, babies can distinguish between two dissimilar faces (Cohen et al. 1978).

COGNITIVE DEVELOPMENT—PIAGET'S THEORY

In organizing this chapter, we have separated motor, perceptual, and cognitive development from one another. This separation is quite artificial, though, particularly in a discussion of babies. For the infant, cognitive development is expressed *through* perceptual and motor activity. When a baby looks intently at the points and contrasts of a triangle (Figure 11.5) or inspects her father's face, she is manifesting one of her few means of "thinking about" or "knowing" the triangle or the face. When another infant sucks on the handle of his rattle, this motor activity is his way of knowing, or "understanding," that rattle.

This point has been emphasized by Jean Piaget (1970), a Swiss biologist, philosopher, and psychologist who has developed the most detailed and comprehensive theory of cognitive development. Piaget called his approach *genetic epistemology*. *Epistemology* is the study of the nature and acquisition of knowledge; Piaget's approach was "genetic" in the sense that it focused on origins *(genesis)* and development. In this case, the word does not refer to genes and heredity. In Piaget's view, the development of knowledge is a form of adaptation and, as such, involves the interplay of two processes, assimilation and accommodation. *Assimilation* means modifying one's environment so that it fits into one's already developed ways of thinking and acting. For example, when a child hoists a banana and runs around in a circle shouting "Look—it's a jet," the child is assimilating the banana into ways of thinking and behaving that are already in place. *Accommodation* means modifying oneself so as to fit in with existing characteristics of the environment. The child who, for the first time, manages to peel a banana and adjust his mouth so that the banana will fit into it has accommodated his ways of thinking and behaving to the banana as it really is. Of course, most steps in development involve some blend of assimilation and accommodation. The child who accommodates to the banana (adjusting his ways of holding his hand and his mouth) also assimilates the banana (by chewing and swallowing it). Piaget also spoke of *equilibration*—the tendency of the developing individual to stay "in balance" intellectually by filling in gaps in knowledge and by restructuring beliefs when they fail to test out against reality.

According to Piaget, the processes of assimilation, accommodation, and equilibration operate in different ways at different age levels. One result is that our ways of thinking about, or knowing, the world pass through certain predictable stages. Table 11.1 shows the major stages described by Piaget. The ages specified for each of the stages are approximate and can vary from person to person, but this is not so true of the *order* in which the stages occur. Research indicates that children in most cultures tend to pass through the stages in a similar order, although exceptions can be found. We will discuss the characteristics of each of the stage levels and the corresponding age levels in this chapter and in Chapter 12.

Piaget called the period of infancy the *sensorimotor stage*. This label reflects something we mentioned above: the infant's ways of knowing the

Table 11.1

The stages of cognitive growth according to Piaget

SENSORIMOTOR STAGE (FIRST 2 YEARS)

Characterized by incorporation of reflex patterns into intentional movements designed first only to repeat, later to maintain, and then to produce new changes in the environment; increasing understanding of means-end relationships. Object constancy is achieved, and the beginning of true "thought" and internalized problem solving are seen; but the child still operates very much in the here and now.

PREOPERATIONAL STAGE (2 TO 7 YEARS)

Characterized by unsystematic reasoning. Impressive development of internal representations and language. Thought characterized by egocentrism, animism, and faulty reasoning about cause-effect relationships.

Preconceptual Substage (2 to 4 Years)
Rapid development of language. Begins to engage in symbolic play. Tends to use classes inaccurately (for example, calls all men "Daddy").

Perceptual, or Intuitive, Substage (4 to 7 Years)
"Reasoning" appears but remains centered on appearances rather than implications. Tends to center on the most noticeable aspects of things observed and therefore fails to "conserve" identities in volume, number, and mass. May discover true relationships through trial and error but is unable to think in flexible ways that involve reversibility. Confuses reality and fantasy but tries to test which is which.

CONCRETE OPERATIONAL STAGE (7 TO 12 YEARS)

Systematic reasoning appears; thought processes are logical and reversible but limited to a child's area of concrete experience. Alternative strategies are invented (for example, two ways of getting to the store). Can coordinate part-whole, hierarchical classifications. Comprehends conservation of number, mass, and other properties.

FORMAL OPERATIONAL STAGE (12 YEARS ONWARD)

Characterized by logic, reasoning from hypothetical propositions, evaluating hypotheses through testing all possible conclusions. Present reality seen as only one alternative in an array of possibilities. Can think about thinking and uses theories to guide thought.

world are sensory, perceptual, and motoric. Piaget called each specific "way of knowing" a scheme. A *scheme* is an action sequence guided by thought. For example, when infants suck, they are exercising a sucking scheme. Their first sucking is primitive and not very flexible in style; they need to adjust the way they hold their mouths so as to fit the object being sucked (for example, a nipple). In making the necessary adjustments, they accommodate their sucking scheme to the shape of the nipple. This allows them to assimilate the nipple into their sucking scheme. This combination of assimilation and accommodation results in adaptive behavior (that is, sucking effectively) that helps the infant survive. It is also a simple prototype of the way cognitive development takes place throughout infancy.

Piaget described many specific cognitive changes that take place during the sensorimotor stage. One of the best-documented changes is illustrated in Figure 11.10. When young infants see an object and the object is then hidden, they seem unaware that the object continues to exist. You can demonstrate this for yourself if you are on good terms with an infant less than about 8 months old. Hold an object within view of the baby until he or she is clearly interested and is reaching for it; then quickly cover the object with a cloth. Chances are that the baby will stop

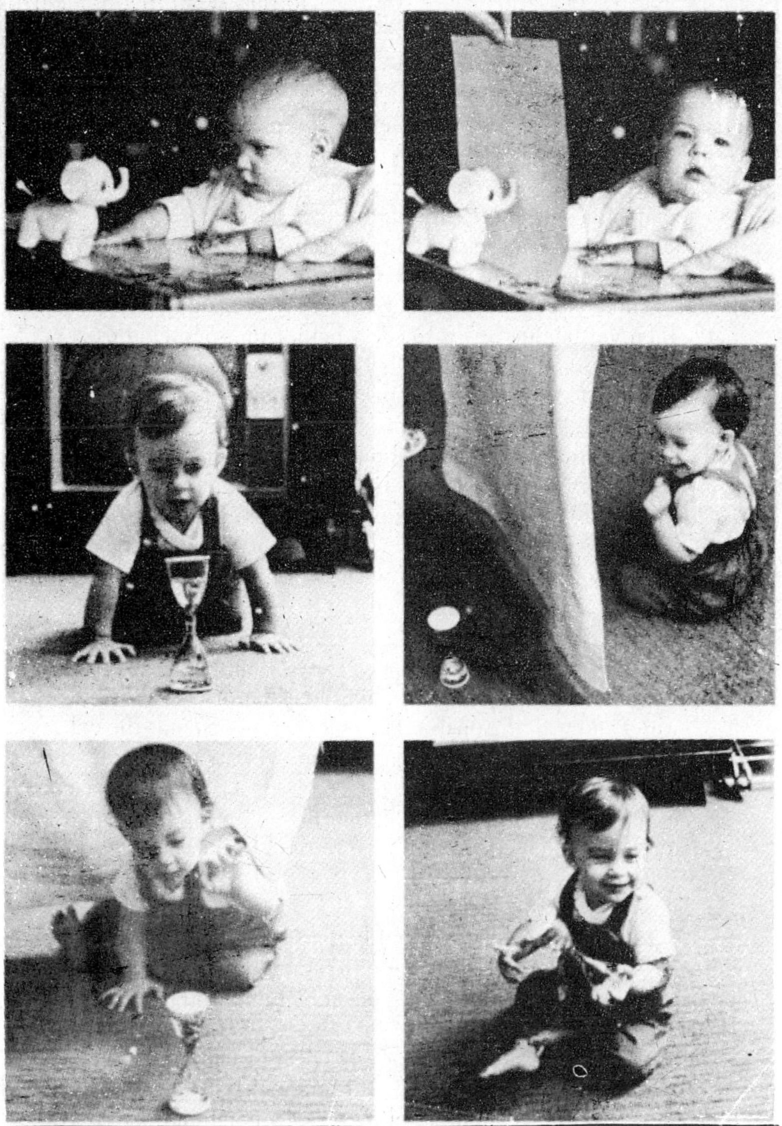

Figure 11.10

Two children demonstrate the development of *object permanence*. Child *A* (at the top) fails to search for the toy once his view of it has been blocked. This suggests that he does not know that such objects continue to exist when they cannot be seen. Child *B* (below) is undaunted when the hourglass is hidden from view; he charges through the barrier and seizes the object, showing that he has attained object permanence. (Zimbel/Monkmeyer Press Photo Service.)

in mid-reach and will not search for the object at all. Repeat the same experiment with the same youngster 3 or 4 months later, and you are apt to see the baby search for the hidden object. The search suggests that the baby has attained what Piaget called *object permanence*—the idea that objects continue to exist even when we can no longer see them. To understand the importance of object permanence, remember that the mother and father are also objects in a baby's world. Evidence suggests that very young infants are not aware of the permanence of people—even their own parents—when these people are hidden from view. This same research, though, has shown that infants tend to attain object permanence with respect to people before they do with respect to inanimate objects and that they attain it with respect to their mothers before they do with respect to other people (Gouin-Décarie, 1965).

During the sensorimotor period, babies broaden their repertoire of sensory and motor activities. The infant's first month is spent largely in practicing reflexes (for example, sucking) that were present at birth. Several subsequent months are devoted to producing new effects on the world and then trying to repeat or prolong these effects (for example, scratching a sheet with one finger, thus producing an interesting sound; then scratching repeatedly in order to recreate the sound). At the age of between 12 and 18 months, infants begin to vary these efforts to produce effects—varying them in ways that make it seem as if they were conducting experiments to see what will happen (for example, scratching very hard, then very lightly, as if to see whether the sound varies).

From the age of about 18 months and on, a keen observer can detect a subtle but crucial shift in the nature of the child's thought. Thought that has been carried on through sensory and motor activity is increasingly carried on internally. This transition was nicely illustrated by an interaction between Piaget and his 16-month-old daughter, Lucienne. Piaget placed a watch chain inside an empty match box and left the box partly open. Lucienne, who had not seen Piaget open or close the box, reached into the opening and groped for the chain. But the opening was too small. She failed. Here is what happened next:

> She looks at the slit with great attention; then, several times in succession, she opens and shuts her mouth, at first slightly, then wider and wider! Apparently Lucienne understands the existence of a cavity subjacent to the slit and wishes to enlarge that cavity. . . . Soon after this phase of plastic reflection, Lucienne unhesitatingly puts her finger in the slit and, instead of trying as before to reach the chain, she pulls so as to enlarge the opening. (Piaget, 1952, p. 338)

In this example, Lucienne's thinking is sensorimotor in that she apparently uses movements of her mouth to represent movements of the match box. Yet other aspects of her thinking are internalized; several parts of her final solution to the problems (for example, reaching, grasping) are not represented by physical action as she plans what to do. The fact that sensorimotor thought is mingled with internalized thought suggests that Lucienne is on the verge of entering the preoperational stage of development, a stage which we will discuss later in this chapter.

SOCIAL DEVELOPMENT

The first "social" relationship most infants form is with a parent, and in most cultures that parent is the mother. Various theorists have offered various ideas about the psychological significance of that relationship.

As we have just seen, Piaget emphasized the cognitive aspects of infancy. In the infant's ways of "relating" to parents and others, Piaget saw signs of sensorimotor intelligence. Sigmund Freud's view (detailed in Chapter 14) was quite different. He saw infancy, the oral stage, as a time when issues of dependency were being dealt with and when physical satisfaction was derived from stimulation in the oral region of the body. As the quotation from Freud at the beginning of this chapter suggests, he saw at least some parts of the infant-mother relationship as sensual in nature. Another theorist, Erik Erikson (Chapter 12, page 473) argued that mother-infant interaction is a context for the baby's basic conflict between trust and distrust of the world.

Despite their differences, all three theorists would have agreed that infants typically form intimate attachments to their mothers. If we are to understand social development in infancy, we must understand how these attachments develop.

Attachment *Attachment* is an early, stable, affectional relationship between a child and another person, usually a parent. Early efforts to study this relationship were clinical and somewhat informal. For example, John Bowlby and his colleagues (1966) set out in the 1940s and 1950s to study the consequences of early mother-child separation by observing children in institutions. The children they saw had been separated from their mothers quite early and lived in nurseries or hospitals where no stable mother substitute was available. They described these youngsters as unable to relate to other people, afraid to explore or play, and, generally, morose. From these and other clinical observations, Bowlby concluded that it is "essential for mental health" that "the infant and young child should experience a warm, intimate and continuous relationship with his mother (or permanent mother-substitute) in which both find satisfaction and enjoyment" (1966, p. 11). Bowlby went on to make clinical observations of the normal course of attachment in early life.

Other researchers studied attachment in a more structured way. Their work yielded a surprisingly consistent picture.

1. Initially, the infant develops an attraction to social objects in general and to humans in particular; the baby shows proximity-maintaining behaviors (crying, clinging, and other behaviors that serve to keep humans nearby).

2. Next, the baby distinguishes familiar from unfamiliar people and the primary caretaker (usually the mother) from other familiar people; then proximity-maintaining behaviors begin to be aimed more directly at familiar persons, particularly at the primary caretaker.

3. By the second half of their first year, most infants develop a true attachment to the primary caretaker; they recognize that person and direct proximity-maintaining behaviors toward that person and not toward others.

4. By the first birthday, the attachment is so strong that children react negatively to separation from the primary caretaker; they grow fearful and tearful, for example, when their parent leaves them with a sitter.

The picture of attachment presented by Bowlby and others raises some difficult questions for the times we live in. First, consider Bowlby's emphasis on the pivotal role of the mother. Men who would like to join their wives as equal partners in child care may wonder whether infants will grow as attached to their fathers as to their mothers. A study by Lamb (1976) indicated that in nonthreatening situations, such as play involving the mother, father, and infant alone, the infant shows equal attachment to both parents; but when a stranger enters the room, the infant shows stronger attachment to the mother. This may reflect differences in the ways fathers and mothers interact with their infants. Throughout infancy, mothers are more likely to interact verbally with their babies and to play the role of caretaker; fathers are more likely to interact physically with their babies and to fill the role of playmate (Yogman et al., 1977). Perhaps in response to these differences, infants orient toward their mothers in times of stress and toward their fathers at playtime.

A second question may arise in families where both parents pursue a career: Does regular day-care for infants and young children interfere with attachment to parents? Many followers of Bowlby have predicted dire consequences of early day-care. Although most of the research thus far has not found major differences in attachment between day-care and non-day-care youngsters (Belsky & Steinberg, 1978; Clarke-Stewart & Fein, 1983), several studies do suggest that children in day-care may keep more physically distant from their mothers than do youngsters whose mothers stay at home.

We have not yet posed one of the most basic questions about attachment: Why does it happen at all? The view held by some theorists is that infants become attached to the adults who care for them because these adults are associated with feeding. According to this view, feeding has this effect because it reduces the primary drive of hunger—that is, it relieves the painful tension of being hungry.

Ethologists (Chapter 2, page 39), scientists who study living things in their natural habitat, offer a different theory. They point out that in many species of animals, infants become imprinted on a mother figure—clinging, following closely, or showing other behavior that resembles attachment in human infants. Ethologists note that during certain sensitive periods (for example, the first 3 days after hatching in some species of ducklings), infants of many species are biologically programmed to imprint on some object in the environment. Imprinting is triggered by specific releasers in the environment. For example, birdlike strutting is a releaser which can produce imprinting in certain baby birds. If the object doing the strutting during the baby birds' sensitive period is a human being, they can become imprinted on that person rather than on an adult bird (Lorenz, 1957). Some ethologists argue that attachment in humans operates something like imprinting in other animals. We should be cautious about such a conclusion, yet some of the apparent similarities between imprinting and attachment are certainly worth exploring. One possibility that some ethologists have explored is that attachment, like imprinting, is triggered by releasers. Researcher Harry Harlow (1958) has done work that sheds some light on this issue.

Contact Comfort Harlow studied monkeys. Monkeys' infancy resembles human infancy in some important ways, but it is possible to manipulate their rearing conditions in ways that would be ethically unacceptable for human beings. The rearing conditions designed by Harlow for his baby monkeys involved surrogate mothers—artificial stand-ins for the monkeys' real mothers.

Each monkey was equipped with a surrogate mother. One was a cylinder made of wire mesh with a block of wood as its head; the

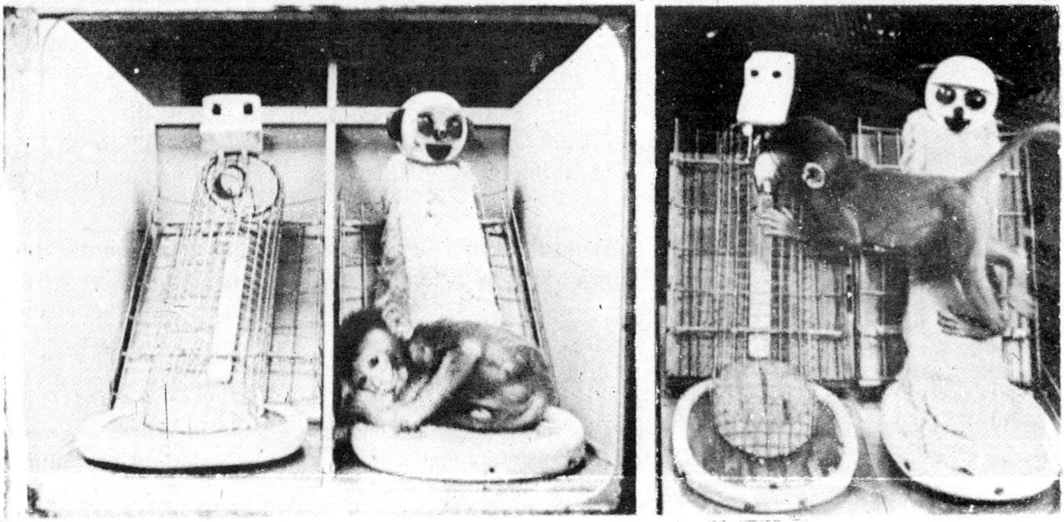

Figure 11.11

Left, mother "stand-ins," or surrogates, made of wire or cloth, used in experiments on the need for contact comfort in monkeys. Right, a baby monkey maintains contact with a cloth surrogate mother while feeding from the wire-mother's bottle. (From Wisconsin Primate Research Center.)

other was a block of wood covered with sponge rubber and then terry cloth. Behind each "mother" was a light bulb that provided radiant heat for the infants. For one group of babies, the wire "mother" had a nursing bottle, placed at the center of her "breast." For a second group of babies, the cloth "mother" had the nursing bottle. When observed with both "mothers" present, monkeys spent almost all the time with the cloth "mother," regardless of which mother fed them. (See Figure 11.11.)

Harlow's research has also shown that monkeys reared with cloth "mothers" react very differently when frightened than do monkeys reared with wire "mothers." For example, when placed in a strange test room, together with their "mothers," the cloth-reared monkeys clung tightly to their "moms." The wire-reared monkeys, in contrast, made little effort to go to their "mothers." Instead, they threw themselves on the floor, cried, grimaced, or huddled against the wall, rocking back and forth while covering their faces with their hands. Evidently the cloth-fed monkeys had developed a strong attachment to their surrogate mothers, but the wire-fed monkeys had not. The wire-fed group also showed strange and occasionally self-destructive behavior. Some rocked back and forth or paced their cages for hours on end, and many bit themselves or pulled out their own hair until their flesh was raw.

Further research by the Harlow group also showed that many of the adverse effects of parent deprivation are reversible. In one study (Suomi et al., 1974), monkeys that had been isolated for 6 months and had begun showing the strange behavior described above were given "therapy." The "therapists" were 3-month-old normal, energetic, sociable female monkeys. Six months of living with these peer "therapists" resulted in major improvements in the behavior of the socially deprived misfits, and another six months led to virtually complete recovery.

The Harlow studies make some important points about attachment and adaptation.

1. They show a situation in which attachment was not merely due to feeding. In the surrogate-mother research, the monkeys' attachments did not depend upon which mother "fed" them. This casts doubt on the idea that attachment is caused by hunger reduction.

2. The studies seem to reveal a basic need for contact with a soft, warm surface; Harlow called this a need for *contact comfort*.

3. The studies suggest that contact comfort may operate like a releaser. With contact comfort, attachments were formed; without it, attachments were not formed.

4. The research suggests that many of the ill effects of being reared without parents may sometimes be reversed, or prevented, by close peer relationships.

One very moving case study indicates that peer attachments in human beings, too, may help make up for the absence of parents (A. Freud, 1951).

In World War II, six Jewish infants, orphaned when their parents were killed in gas chambers, formed a close attachment to one another in the concentration camp where they were kept. After the war, Anna Freud (Sigmund's daughter) and Sophie Dann brought the children, by then aged 3 to 4, to Bulldogs Bank, England, for care. At first, the children were wild and hostile toward adults—sometimes even biting and spitting at them (see Figure 11.12). But they were remarkably affectionate and attached to one another.

Figure 11.12

When the six orphans from a Nazi concentration camp arrived at Bulldogs Bank, England, in 1945, they were an unruly but close-knit gang (left). Less than a year later, they had become better socialized and more approachable, but they still remained closely attached to one another (right). This close attachment may be one of the strengths that helped them adjust, cope with their stressful circumstances, and develop into normally functioning adults. (Courtesy of Sophie Dann.)

They were extremely generous in sharing food and warm clothing with each other, and they insisted on being together at all times. For example, no child would go for a walk unless the other five went along. Gradually, the youngsters warmed up to the adults at Bulldogs Bank, forming attachments to those who cared for them and adjusting to the social demands of polite society. Now in their forties, all are reportedly effective adults (Hartup, 1983). Evidently, the close bonds these six formed early in life helped soften the impact of their severe deprivation as infants.

The Bulldogs Bank story suggests that attachment may be closely related to the aspect of human development that we will consider next: emotion.

EMOTIONAL DEVELOPMENT

When babies smile, does it mean they are happy? This seemingly simple question is actually very complicated because what looks like an emotion may not always be one. This fact has helped to make the study of emotional development in infancy challenging, to say the least. To illustrate the challenging but intriguing nature of the subject, let us focus on smiling, and in particular on the question of what makes it happen.

Evidently, smiling happens for different reasons at different ages. Some smiling is seen even in newborns, but much of this seems automatic and hardly "emotional." For example, some smiling seems to be triggered merely by the infant's bodily state, as when babies break into grins during REM sleep (Chapter 2, page 56) in the first few days after their birth. Other early smiles operate like reflexes, as when neonates smile when someone strokes their lips. In the second month, smiles can be brought on by events in the environment—particularly the sound of a human voice or the sight of a human face. A powerful smile evoker is a combination of a voice and a moving face, particularly if the voice is high-pitched. By the third or fourth month, babies smile more for their mothers than for an equally encouraging female stranger. By the beginning of their fifth month, most babies have begun to combine smiling with laughing. From that point on, there is a series of marked changes in the kind of stimulation that sparks a laugh. For example, in the middle of their first year, babies will laugh at stimulation that involves touching—such as when Dad tickles them or buzzes his lips on their stomachs. By their first birthdays, such tactile fun evokes fewer laughs; but interesting visual displays, like a human mask, get more laughs.

One interpretation of this developmental change is that emotions like happiness and delight develop hand-in-hand with a child's intellect. Simple pleasures that are not cognitively demanding may please an immature infant; but as cognitive development proceeds, a baby needs to be increasingly challenged intellectually in order to experience pleasure and humor. Some investigators champion the perceptual–recognition hypothesis to account for smiling and laughing in infancy. They argue that babies make sense of the world around them by forming mental representations, or *schemas,* of certain kinds of objects. When they see an object and are able to match it to their schema (and thus "recognize" it), the result is pleasure, which is signaled by a smile or a laugh. They note, for example, that 4-month-olds are near the point of having an established schema for the human face. This, they argue, explains why babies of this age tend to look intently at a picture or sculpture of a face for 3 to 5 seconds and then break into a smile. It is as if they are carefully fitting this new face into their emerging face schema and then signaling the satisfaction of a good fit with a smile that says "Voilà!"

ADJUSTMENT PROBLEMS IN INFANCY

In an ideal world, infancy would be a time when baby and parent would quickly adjust to one another and develop a smooth harmony of styles that Stern (1974) called a "waltz." Unfortunately, the waltz is harder to learn for some parent-infant pairs than for others. Quite common in the first year of life are infant feeding problems—especially a digestive discomfort known as colic—and vomiting. Constipation and diarrhea, irregular sleep patterns, and mystifying bursts of crying also occur very often in the first year. Near the end of the first year and well into the second, the problems most often involve a conflict between the baby's growing physical and mental processes and the parents' efforts to regulate behavior that seems to them to be aggressive or dangerous. Two of the most common child-behavior problems parents report at this age are "stubbornness" and "temper" (Achenbach & Edelbrock, 1981).

Sometimes, the stresses involved in coping with an infant's problems can make a parent feel inadequate or even angry at the baby. Such feelings, particularly when combined with money problems and worries about unemployment, can set the stage for child abuse. (See Figure 11.13.) Many communities now have services such as telephone hot lines, emergency counseling, and Parents Anonymous groups to assist potentially abusive parents. Most experts agree that these services work best when they are utilized early in the development of family distress.

A number of clinical disorders make their first appearance during infancy. Among these are several that are known to be caused by genetic or other biological factors. *Down syndrome* (Chapter 13, page 536), for example, involves mental retardation and a characteristic physical appearance noticeable even in the newborn. Although the syndrome is known to be caused by a chromosomal abnormality, it usually does not run in families. Mental retardation, often accompanied by physical abnormalities, can also be caused by infectious diseases in the mother, by various metabolic disorders, or by brain damage.

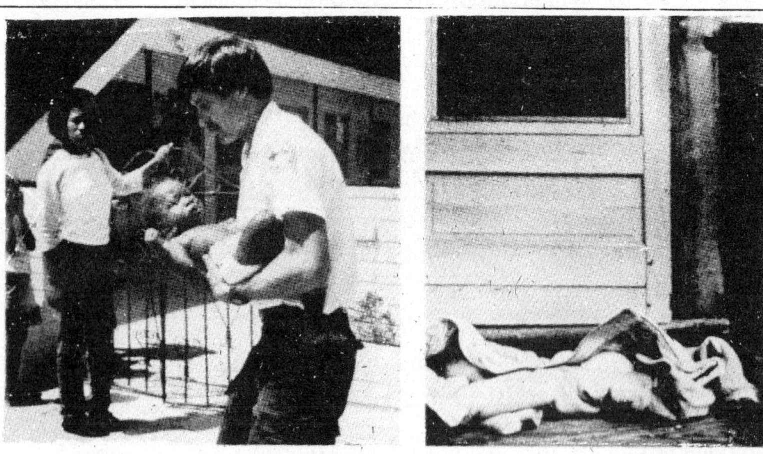

Figure 11.13

Abuse and neglect pose a serious risk for many infants and children. Here are two examples. The infant at the left is being carried by a paramedic after being hit by a baseball bat during an argument between adults over how best to rear children. The year-old infant at the right was left on a back door stoop, wrapped in a jacket—the result of a parental argument. The child was later hospitalized and treated for exposure. (AP/Wide World Photos.)

Early signs of the disorder known as *infantile autism* (Chapter 15 page 631) make their appearance during the first year and a half of life. Autistic youngsters fail to show several of the landmark features of infancy that we discussed above. They fail to focus on other peoples' eyes, they do not smile regularly in response to peoples' faces or voices, and they do not show such key signs of attachment as protest when a parent leaves them. Later, difficulties with social interaction and especially with language will plague these children into their adulthood. Infants suffering from a *failure to thrive* show apathy, a lack of normal social interest, and stunted growth despite seemingly adequate nutrition. What little evidence we now have suggests that this disorder often results when rearing situations leave infants neglected, abused, or poorly stimulated.

Early Childhood: Play, Preschool, and Preoperations

From the age of about 18 months through the age of 6, the comfortable confines of the child's family give way to the world of peers. The play that goes on in that world may seem frivolous to many adults, but we are now coming to recognize it as, to use Piaget's expression, "the work of the child." In the context of play, children make the transition from sensorimotor thinking to thinking that involves internal manipulation of symbols. The elegant symbol system known as language takes shape at a pace that leaves even experienced parents dazzled. The frequency and intensity of peer interaction force the child to deal with interpersonal issues, such as coping with aggressive impulses and learning how to help. The world grows more structured as the child moves into more and more formal educational settings. By the age of 3, about one-fifth of all U.S. children are in some structured nursery school or preschool program with educational objectives; by the age of 5, the ratio is closer to four-fifths. As children prepare to enter the first grade, they bear the marks of seasoned young veterans who have been through a lot and have grown a lot in the process.

COGNITIVE DEVELOPMENT

The period between about the ages of 2 and 7 was labeled the *preoperational stage* by Piaget (1970). By this label, he meant that these years are preliminary to the development of truly logical operations.

What are *operations?* They are flexible mental actions that can be combined with one another to solve problems. To understand what a difference having logical operations can make, consider the following simple problem as handled by children who do and children who do not have such operations.

> We start with two groups, one of 4-year-olds and one of 8-year-olds. We first show each child two sticks, A and B, pictured in Figure 11.14, step 1. Each child is asked to compare the two sticks, then to tell whether one is longer or both are the same. Assuming the children all recognize that A and B are the same length, we then move stick B into the position shown at the bottom of step 2. This presents the children with the Müller-Lyer illusion (Chapter 3, page

Figure 11.14

A simple situation for testing whether children's thinking shows reversibility. In step 1, a child agrees that sticks A and B are equal in length. In step 2, stick B is made to *appear* longer than A by use of the Müller-Lyer illusion. Children who are capable of mentally reversing the path of stick B recognize that a return to its original configuration would show it to be the same as A. Children whose thinking is not reversible maintain that B is longer than A.

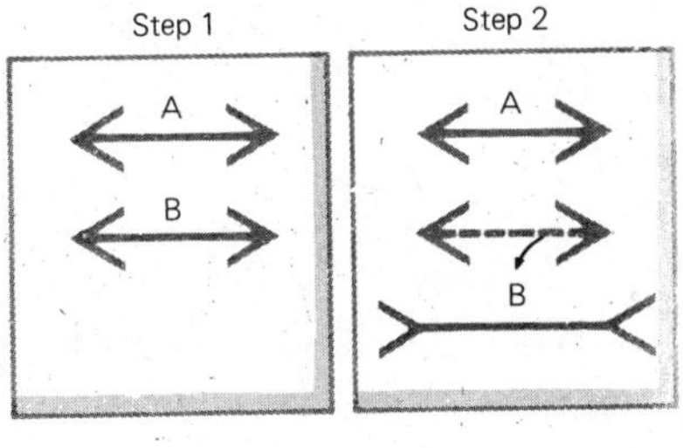

> 120); B now *looks* longer than A. We then ask the children whether one of the sticks is now longer than the other or whether they are both the same. Most of the 4-year-olds will say that B is longer; most of the 8-year-olds will say that the two sticks are the same length.

Why do the two age groups differ so in their answers? One reason is that the older children have flexible logical operations that allow them to reverse the movement of the sticks mentally. They can imagine returning the sticks to their original configuration (step 1); and they know that if they did so, A and B would be identical. The younger children, by contrast, do not have flexible logical operations. Their thinking suffers from irreversibility, and their judgments are thus dominated by perceptual appearances; in step 2, B does *look* longer than A, and that is enough to convince the preoperational child.

Even though children as old as 5 or 6 may be fooled by the Müller-Lyer illusion in Figure 11.14, they are apt to show surprise that the stick is suddenly longer than it seemed to be. Children younger than about age 4 typically do not show such reactions of surprise (Achenbach, 1973). Think for a minute about what surprise means in an experiment like this. It seems to reflect a primitive *identity concept*—that is, an idea that characteristics like the length of a rigid object should not change just because the object has been moved to a new location. We call the identity concept "primitive" because it gives way so easily—that is, though the children show surprise, they still conclude that line B is longer.

The primitive identity concept is an important milestone. One reason is that it enters into the way children think about their gender identity, as we will discuss in a later section on sex roles. Another reason is that identity concepts seem to be necessary steps on the way to conservation, a defining feature of the next major Piagetian stage, concrete operations (page 450). Finally, these early object-identity concepts may be linked to a more personal sense of identity—that is, the self-concept.

> To study the self concept, Lewis and his colleagues secretly placed a spot of rouge on children's noses. The children, aged 9 to 24 months, were then placed in front of a mirror. (See Lewis & Brooks-Gunn, 1979.) The 9- to 12-month-olds did not seem surprised at their mirror images, but most of the children over the age of 16 months reached up to touch the rouge spot on their noses. This suggests that the older children had some awareness that they had a "self" with a certain identity (in this case, with no red spot on the nose).

Another important development in the preoperational period is *representational thought*—the ability to form mental symbols to represent objects or events that are not present. As early evidence of representational thought, Piaget cites delayed imitation. Consider this example from Piaget's daughter:

> Jacqueline had a visit from a little boy . . . who, in the course of the afternoon got into a terrible temper. He screamed as he tried to get out of a play pen and pushed it backwards, stamping his feet. Jacqueline stood watching him in amazement, never having wit-

> nessed such a scene before. The next day, she herself screamed in her play pen and tried to move it, stamping her foot lightly several times in succession. (Piaget, 1962, p. 63)

Piaget pointed out that if this imitation had been immediate, he would not have judged that mental symbols were involved; but because the imitation came one day after Jacqueline had witnessed the scene, Piaget reasoned, Jacqueline must necessarily have stored away some internal representation of what she had seen.

Early in the preoperational stage, reasoning is not truly deductive (it does not proceed from the general to the particular) nor is it truly inductive (it does not proceed from the particular to the general). Instead, very young children show *transductive reasoning;* that is, they reason from the particular to the particular, often in ways that are influenced by their desires. Consider this reasoning by Piaget's daughter, Jacqueline, at the age of 2 years and 10 months:

> Jacqueline had a temperature and wanted oranges. It was too early in the season for oranges to be in the shops and we tried to explain . . . 'they're still green. We can't eat them. They haven't yet got their lovely yellow colour.' Jacqueline seemed to accept this, but a moment later as she was drinking her camomile tea, she said: 'Camomile isn't green, it's yellow already . . . give me some oranges!' (Piaget, 1962, p. 231)

In this example, we can see a glimmer of logical reasoning in the *form* of what Jacqueline says—that is, if A, then B. The problem is that Jacqueline's A (camomile is yellow) does not logically imply her B (oranges are yellow). Some transductive reasoning persists throughout the preoperational period, but it gets mingled, more and more, with instances of truly valid "If A, then B" statements. It is also sprinkled with episodes where children suddenly seem to spot the logical gaps in their transductions, as in the following example from Piaget's conversation with Jacqueline, then age 4.

> *Jacqueline:* It's a horse because it has a mane.
> *Piaget:* Haven't mules got manes?
> *Jacqueline:* Yes.
> *Piaget:* Well then?
> *Jacqueline:* (silence). (Piaget, 1962, p. 233)

Such experiences of being "stumped," of seeing the inadequacies in one's reasoning, may act as spurs, stimulating further development. Once the inadequacies have been seen, the child's tendency toward equilibration (page 425) makes it very hard for her to remain satisfied with her current ways of thinking.

Some other characteristics of preoperational thought can be surveyed briefly. *Egocentrism,* means an inability to take the point of view of another person. Preoperational children tend to assume that others see the world just as they themselves see it. Note that egocentrism, as thus defined, does not mean selfishness; instead, it refers to an *intellectual* limitation. Preoperational children also display *animism,* the belief that inanimate objects which have certain characteristics of living things (such as motion) are, in fact, alive. Finally, preoperational children do not understand cause-effect relationships very well. They tend to see unrelated events and objects as causally related to one another. In fact, they tend to believe that each event has a clearly identifiable cause, and thus they often fail to recognize the operation of chance and luck (Weisz, 1983). Young children often come up with simple causal explanations for random events in nature. For example, if asked why clouds move across the sky, they are apt to answer, "They are following us." Another manifestation may be the persistent "Why?" questions that adults cannot answer because they involve events that are essentially random—for example, "Why do you have such a long nose when you are so short?" In many ways, then, the reasoning of the preoperational child falls short of what we consider mature thought.

LANGUAGE DEVELOPMENT

Achieving mature thought requires achieving a mature use of language. The steps involved in acquiring language look quite similar in children from a variety of cultures. In fact, as researchers began to notice this similarity across cultures, the nature-nurture issue resurfaced. Language expert Noam Chomsky (1968) proposed that there are universal grammatical rules used by all children everywhere and that this universal grammar is stimulated by an inborn language-acquisition device. More recent evidence, though (Braine, 1976), indicates that children may not really use the same underlying grammatical rules the world over. However, children do seem to follow some remarkably similar steps as they begin to combine words in order to express ideas.

One can view the course of language development as either continuous or discontinuous (recall the earlier discussion of this recurring conflict). Vocabulary development appears to be a fairly smooth, continuous process. In English-speaking cultures, the infant's first legitimate English word usually appears around the time of the first birthday. By the age of 2, the vocabulary has usually expanded to about 50 words; and by age 3, it consists of about 1,000 words. By their sixth birthdays, most children can use between 8,000 and 14,000 words (Carey, 1978).

Language development looks more discontinuous, or stagelike, when we focus on *syntax*—the formation of grammatical rules for assembling words into sentences. Table 11.2 shows the sequence of stages through which children pass as they master the syntax of their language. The ages shown in the table are approximate; they even overlap in places. This is because there are large differences among children in their rate of development and because children (like adults) do not always use their most advanced forms of language. In many children, syntactic development actually begins before stage 1 in the table. Halliday (1975) showed that even before their first real English word, children can often communicate with their parents by using private expressions such as *bi* for bird, and *moosh* for milk.

When the first legitimate words are learned, stage 1 emerges quickly. Children get maximum mileage out of these single words by altering inflection and gesture. When a child reaches for her cup and shouts "Milk!" in an urgent tone of voice, she is actually communicating the sentence, "Give me my milk!" A single word thus used as a sentence

Table 11.2

Six stages in children's syntactic development

STAGE OF DEVELOPMENT	NATURE OF DEVELOPMENT	SAMPLE UTTERANCES
1. Sentencelike word (12 to 18 months)	The word is combined with nonverbal cues (gestures and inflections).	"Mommy." "Mommy!" "Mommy?"
2. Modification (18 months to 2 years)	Modifiers are joined to topic words to form declarative, question, negative, and imperative structures.	"Pretty baby." (declarative) "Where Daddy?" (question) "No play." (negative) "More milk!" (imperative)
3. Structure (2 to 3 years)	Both a subject and a predicate are included in the sentence types.	"She's a pretty baby." (declarative) "Where Daddy is?" (question) "I no can play." (negative) "I want more milk!" (imperative)
4. Operational changes (2½ to 4 years)	Elements are added, embedded, and permuted within sentences.	"Read it, my book." (conjunction) "Where is Daddy?" (embedding) "I can't play." (permutation)
5. Categorization (3½ to 7 years)	Word classes (nouns, verbs, and prepositions) are subdivided.	"I would like some milk." (use of "some" with mass noun) "Take me to the store." (use of preposition of place)
6. Complex structures (5 to 10 years)	Complex structural distinctions made, as with "ask-tell" and "promise."	"Ask what time it is." "He promised to help her."

(*Source:* Modified from B. S. Wood's *Children and Communication: Verbal and Nonverbal Language Development,* © 1976, pp. 129 and 148. Reprinted by permission of Prentice-Hall, Englewood Cliffs, NJ.)

is called a ***holophrase.*** Children usually move so rapidly through the next four stages of syntax that by age 4 they are using grammar much like that of adults in informal conversation.

The recurring conflict between active and passive views of the developing person can be seen in the study of language development. Some theorists, notably B. F. Skinner (1957), have argued that children *learn* language by trying various combinations of sounds and being rewarded (for example, with praise and attention) by their parents and others for those sounds that represent true language. Others, such as Piaget, have argued that children *create* their language by *constructing* their own rules and revising them as needed. There can be little doubt that some of children's language acquisition comes from being rewarded or encouraged by others; all of us have seen this process in action. Yet it also is hard to deny that children are active builders of their own language. One line of evidence often used to support this view is the erroneous language that children use—language that reveals rules the children have constructed but that is not likely to have been rewarded. Children who have found out that the plural of *dog* is *dogs* and that the plural of *box* is *boxes* are apt to decide on their own that more than one mouse should be called *mouses*. Children who do this are showing they have understood a rule governing grammar; they are obviously not just repeating what they have been rewarded for saying, nor are they imitating an adult model. Far from being discouraged by such errors, parents should take heart from the knowledge that their child has mastered a rule, even if the child clings longer than the parents might like to a mistaken use of the rule. Consider, for example, this transcript of a conversation between a researcher, Dr. Gleason, and a 4-year-old.

> *Child:* My teacher holded the baby rabbits and we patted them.
> *Gleason:* Did you say your teacher held the baby rabbits?
> *Child:* Yes.
> *Gleason:* What did you say she did?
> *Child:* She holded the baby rabbits and we patted them.
> *Gleason:* Did you say she held them tightly?
> *Child:* No, she holded them loosely.
> (Gleason, 1967, p. 1)

In this case, the child had in mind a rule about how to form past tenses (add *ed*). However, she applied this rule to a verb which has an irregular past tense. Eventually, she will accommodate (page 425) to this arbitrary feature of her language. Most children do, but often they accommodate in ways that show just how arbitrary they perceive these matters to be. An example of such accommodation to the arbitrary (an incorrect accommodation in this case) has been reported by Warner, as cited in Biehler (1981).

> A girl, aged three, was looking at the illustrated book, *Three Billy Goats Gruff*. She pointed to each goat and said "Look! Three billy goats eated." Her mother replied, "Ate." The child re-checked the goats in the picture, then said, "Three billy goats eated." Again her mother replied, "Ate." Exasperated, the little girl shrugged her shoulders and said, "OK! *Eight* billy goats eated."

In this example, we can see the truth of both the active and passive views of the child's language development. Children actively construct rules of language usage. Yet there are many cases where seemingly good rules simply do not apply, and the child learns that arbitrary exceptions must be passively accepted. Through this interplay of activity and passivity, a mature language system takes shape in a few short years.

This, at least, is what happens in a normal, language-rich environment. But what happens to people who are not exposed to language during childhood? Do they develop good language skills, only later? Or is childhood a "critical period" for language? In Inquiry 11.1, we discuss this question by exploring the case of a tragically deprived child.

SOCIAL DEVELOPMENT

Along with the increasing mobility and accelerating language skills of the preschool child comes an expanding social world. The process by which the child's behavior and attitudes are brought into harmony with that world is called *socialization*.

Freud's theory (Chapter 14, page 576) focused mainly on the child's socialization with respect to parents during this period. Freud believed that during the anal stage, roughly the second year of life, key interactions center around toilet training. The child takes physical satisfaction from stimulation in the anal region of the body, and social issues including self-control and orderliness are confronted. During the phallic stage, roughly ages 3–6, children find physical satisfaction in stimulation of their sexual organs and are attracted to the parent of the opposite sex. Out of the experiences of this stage, Freud believed that children forge a lasting identity with their same-sex parent.

Erikson (Chapter 12, page 473) saw the second year of life and the toilet-training experience as a time of conflict between autonomy, on the one hand, and shame and doubt, on the other. During the next 3 or 4 years, Erikson argued, the child's core conflict is between the urge to be industrious in school and elsewhere and the risk of feeling inferior. We will begin our discussion, as Freud and Erikson did theirs, with a focus on the parent-child relationship.

The parent-child relationship The first part of early childhood has been dubbed "the terrible 2s." One reason for this label is that the child's increasing physical prowess, intellectual power, and language skill transform the nature of the parent-child relationship; the child becomes less compliant and manageable than before.

In addition, a major task of socialization must be confronted: toilet training. Parents who have been largely nurturers and caretakers of their children become teachers and enforcers, active agents of socialization. In teaching specific skills to their children, parents may profit from the work of behavioral psychologists.

> For example, Azrin and Foxx (1974) developed a set of behavior-modification techniques that they say can produce appropriate toilet behavior very quickly. The procedures included parents' use of a doll to show their child the appropriate behavior. Children were also given large quantities of their favorite liquids and encouraged to follow the doll's example on the potty. As soon as the child urinated, the parents responded with rewards in the form of favorite treats, lavish praise, and references to how pleased special people in the child's life (for example, Granddad) would be. When applied carefully with children who are physiologically ready for bladder control, the technique can reportedly produce *Toilet Training in Less Than a Day* (the title of Azrin and Foxx's book).

In addition to teaching specific skills, the parent during this period is called upon to be a disciplinarian. But how should parents go about telling their child no? Research findings offer some guidelines.

First, a combination of general parental warmth and specific explanations for specific prohibitions seems to promote effective discipline. Parental warmth seems to make the child eager to maintain the parent's approval and to understand the parent's reasons for the prohibition.

Inquiry II.1

IS CHILDHOOD A CRITICAL PERIOD FOR LANGUAGE? THE CASE OF GENIE

In 1970, officials in California discovered a 13-year-old girl who had been reared, for all intents and purposes, as an animal. The youngster, called *Genie* to protect her privacy, had been kept in almost total isolation from her twentieth month until the age of 13. Evidently her parents and brother had kept her in a small room throughout this entire period, never interacting with her except to bring her food (mostly baby food) and drink (mostly milk). Genie was kept naked. She was not toilet-trained. She was restrained by a harness that forced her to sit throughout each day on a potty-chair, able to move only her hands and feet. At night she was often wrapped in a sort of straitjacket and secured in a "crib" that had wire-mesh sides and a covered top.

Genie's father never spoke to her; his only vocal communications with his daughter were growls and barking sounds. He forbade his wife and son to talk to Genie; both, terrified of the man, complied with his orders. The two even spoke to each other in very quiet tones lest they disturb Genie's father. As a result of all this, Genie passed through most of her childhood without any real opportunity for language development, not to mention normal development in other areas.

Among the many specialists who worked with Genie after she was discovered was Susan Curtiss, an expert in language development. Curtiss described her work in a 1977 book, *Genie: A Psycholinguistic Study of a Modern-Day "Wild Child."* One of the questions Curtiss hoped to answer in her work with Genie was whether there might be a *critical period* for language acquisition—that is, a period of time during which language development must take place if it is to happen at all. Critical-period hypotheses are rarities in modern psychology, and as such they deserve special attention. Earlier in this chapter, we used the more popular, and more cautious, term *sensitive periods;* this term refers to segments of the life cycle during which the organism is especially susceptible to, or *ready for,* certain key cognitive or social developments.

In the area of language development, though, psychologist Eric Lenneberg (1967) proposed a true critical-period hypothesis: Before the age of 2, he argued, a child's brain has not matured enough for language acquisition to take place, and after puberty the brain has lost the flexibility necessary for the acquisition of a first language. Genie's tragic circumstances provided a test of Lenneberg's hypothesis because her language deprivation began before the age of 2 and continued until after puberty. In other words, according to Lenneberg's hypothesis, Genie failed to develop language during the only period when language development is possible. According to this reasoning, Genie's advanced age at the time she was discovered should have meant that she would be unable to acquire language.

Did Genie develop language? Yes and no. With training, she progressed through the normal developmental sequence from one-word utterances to two-word sets (for example, "big teeth") and then to three-word chains (for example, "small two cup"). In this respect she did develop *some* language, and this fact seems inconsistent with a strict interpretation of Lenneberg's hypothesis. On the other hand, Genie did not become really fluent. She did not learn to distinguish properly between various pronouns, to differentiate between active and passive verbs, to ask questions, or to do a number of other things that many linguists see as evidence of a truly human language. Consequently, some would say that Genie's language deficits support Lenneberg's view that there *is* a critical period for true language acquisition. In summary, the case of Genie could be interpreted as evidence either for *or* against Lenneberg's critical-period hypothesis.

Curtiss dealt with this standoff by proposing a modified critical-period hypothesis. She noted that several tests had revealed something very unusual about Genie's brain activity: Genie, unlike nearly all other right-handed people, used the right side of her brain for language (see Chapter 2, page 69). Moreover, normal right-hemisphere activities (such as face recognition) were well developed in this very deprived adolescent. In general, Genie's right hemisphere seemed to be functioning fairly well when it came to mental activities normally associated with that hemisphere. On the other hand, Genie did poorly at activities normally handled by the left hemisphere; some of these activities, including language, appeared to have been "adopted" by her right hemisphere.

Such information led Curtiss to suggest that there might be critical periods for the development of the left hemisphere. It might be, she suggested, that left-hemisphere brain tissue normally reserved for language has to be used during a particular period of time (perhaps the age span specified by Lenneberg). If that critical period is missed, brain tissue in the language-designated area may lose its capacity for learning. After this has happened, any subsequent language activity may have to be relegated to the right hemisphere and may well be less effective than language that is governed by the left hemisphere. Genie, Curtiss suggested, may have missed the critical period during which her language functioning would have been located in the left hemisphere and thus have developed normally.

The notion of critical periods in development is an extremely important one. It has generated a great deal of controversy, especially when applied to such extreme cases as Genie's. In this instance, the controversy has even extended to the courtroom. Genie's mother has filed suit against Curtiss and a hospital that was involved in Genie's treatment. She charges that confidential information about her daughter and herself have been disclosed for "prestige and profit." Her lawyer estimates that damages may total $500,000.

REFERENCES

Curtiss, S. (1977). *Genie: A psycholinguistic study of a modern-day "wild child."* New York: Academic Press.

Pines, M. (1981, September). The civilizing of Genie. *Psychology Today, 15*, pp. 28–34.

Giving reasons, in turn, helps the child form concepts of what is appropriate—concepts that can then generalize to new situations. Second, consistency—over time and between parents—is very important. Many studies have shown that parental inconsistency and conflict are linked to maladjustment in children, especially to aggression and delinquency.

A third guideline concerns the question of how controlling parents should be with their children. This is one of the questions addressed in research by Baumrind (1967, 1980).

> In one part of this research, observers spent 14 weeks recording the behavior of preschool children. In addition, they interviewed the children's parents and observed them while interacting with their children. They found that parental style seems to be related to children's behavior in several ways. Parents with a style Baumrind labeled "authoritative" were firm about rules and expected high levels of maturity and achievement, but were also nurturant and responsive to their children. These parents had children who were especially energetic, friendly, curious, and self-reliant. The children of very "permissive" parents, by contrast, showed little self-reliance and poor self-control; many of these children were judged to be impulsive and aggressive. Finally, a highly controlling "authoritarian" parental style was associated with children who were apprehensive, withdrawn, unhappy, and distrustful.

Baumrind's study, like several other investigations in this area, suggests that parenting which combines warmth with moderate restrictiveness and an authoritative style will foster independence and social maturity in children even as early as the preschool years.

Baumrind's (1980) research also suggests that identical parental styles may foster different behavior patterns in boys and in girls. For instance, boys with punishing fathers seem to have difficulty in forming good peer relationships, but girls with such fathers seem especially likely to be independent and self-reliant. It has been suggested that girls with very warm and tolerant fathers may be missing out on the parent-child tension that can stimulate self-assertion and autonomy.

Before leaving this discussion of parental styles and child behavior, let us reemphasize a key point: Children are not formless lumps of hot metal waiting to be shaped by parents—regardless of what John Watson (quoted at the beginning of this chapter) may have once said. The longitudinal research we discussed earlier (Thomas & Chess, 1977, 1980) indicates that children may be born with "styles" of their own in the form of basic temperament patterns. Parental style may influence the way these patterns are expressed, but parental style is also partly a *response* to the child's style. Parent behavior and child behavior influence one another in an ongoing cycle. To view the parent as "the cause" and the child as "the effect" would certainly be a serious mistake, just as it would be a mistake to think that every troubled child is a product of poor parenting.

Sex roles Children's identification with their parents influences their ideas about sex roles. Children of *both* sexes may initially adopt many traditionally feminine and maternal behavior patterns (Sears et al., 1965); but by the age of 4 or 5, boys have already begun to show traditional male types of behavior. The toys they choose and the roles they play in games become increasingly masculine. At about the same age, girls intensify the feminine sex-typing of their play. One reason for the divergence of boys and girls is that children pick up sex-typed behavior through *observational learning*—that is, boys observe and imitate males, particularly their fathers, while girls observe and imitate females, particularly their mothers.

Why does such differential imitation of males and females not show up strongly until children are 4 or 5 years old? One reason seems to be that children's awareness of sex differences is influenced by their cognitive development. Before the age of 4 or 5, as we saw in the preceding section on cognitive development, most children do not understand the *principle of identity*—that is, they do not recognize that certain characteristics of objects remain fixed even when the appearance of the object changes. In the same way, young children do not recognize that one of their own key characteristics—their gender identity—remains fixed even though their appearance changes. This intellectual limitation can be seen in the following conversation between two boys, one of whom has not yet understood the fixedness of this key aspect of himself.

> *Johnny* (age 4½): I'm going to be an airplane builder when I grow up.
> *Jimmy* (age 4): When I grow up, I'll be a Mommy.
> *Johnny:* No, you can't be a Mommy. You have to be a Daddy.
> *Jimmy:* No, I'm going to be a Mommy.
> *Johnny:* No, you're not a girl, you can't be a Mommy.
> *Jimmy:* Yes I can.
> (Kohlberg, 1966, p. 95)

Children like Johnny, who understand that sex is a fixed attribute, are more likely than are children like Jimmy to focus their observational learning on models of their own sex. Evidently, cognitive development *and* observational learning both contribute to sex-typed behavior. In addition, a broad range of environmental influences can be identified. For generations, parents have tended to treat boys and girls differently, differentially rewarding them for "sex-appropriate" behavior and even decorating boys' rooms with "educational" materials while decorating girls' rooms with dolls and ruffles (Rheingold & Cook, 1975). Sex typing has also been common in children's books and in television portrayals of male and female characters. These trends are changing now, but not as quickly as many would like.

In addition to these cognitive and environmental factors, there seem also to be biological causes for sex-role development. This is illustrated by Money and Ehrhardt's (1972) study of "androgenized" girls.

> Money and Ehrhardt found 25 girls, aged 4 through 16, who had been exposed to high levels of an androgenlike hormone (progestin) as fetuses. Their mothers had used the hormone to minimize the risk of a miscarriage, but it also turned out to have some powerful "masculinizing" effects on the fetuses. For example, some of the girls were born with male-like genitals. These physical effects were corrected surgically, and the girls grew up looking female. But apparently they were psychologically different from other girls. When the 25 "androgenized" girls were compared to 25 untreated girls of the same age, race, IQ, and socioeconomic background, some key differences were found in sex-role attitudes and behavior. The "androgenized" girls were more often considered "tomboys"; they were much more likely than the comparison group to be involved in vigorous, competitive activities, to prefer functional clothing over fashionable dresses, and to prefer male-type toys like miniature trucks and guns. Unlike the other girls, the "androgenized" group tended to be indifferent to dolls and, later, to human infants, and when asked whether they would prefer to have careers or to be married housewives, most chose a career; most of the "nonandrogenized" girls said that marriage was their most important long-range goal.

Money and Ehrhardt argued that much of the tomboyishness seen in the "androgenized" girls was most likely a result of the effects of the hormone on the brains of the girls when they were fetuses (Chapter 7, page 279). The Money-Ehrhardt findings suggest that biological factors, like environmental factors and cognitive development, can play a significant role in producing sex-role differences in early childhood.

Peers and play As children mature, their relationships with their parents are increasingly rivaled by their relationships with their peers. The nature of child-to-child interaction in the context of play changes in predictable ways over the early childhood years.

Initially, children engage in *solitary play*. They may show a preference for being near other children and show some interest in what those others are doing, but their own individual play runs an independent course. Solitary play is eventually replaced by *parallel play*, in which children use similar materials (such as, a pail and toy shovel) and engage in similar activity (such as, digging sand), typically near one another; but they hardly interact at all. By age 3, most children show at least some *cooperative play*, a form that involves direct child-to-child interaction and requires some complementary role taking. Examples of such role taking can be found in the "pretend" games that children use to explore such mysteries as adult relationships (for example, games of "Mommy and Daddy") and other children's anatomy (for example, games of "Doctor"). Additional signs of the youngster's growing awareness of peers can be seen at about age 3 or 4. At this age, at least some children begin showing a special faithfulness to one other child when they choose playmates or friends. This faithfulness may last only a few days, but it is still a significant advance; prior to age 3, children's choices of playmates and friends may change almost randomly from day to day. It is also at about age 3 or 4 that children begin to prefer playmates of the same gender; this may be a preliminary step on the way to the stable sense of gender identity which emerges at age 4 or 5.

Aggression In early childhood, boys and girls face an important new task: learning to express unpleasant feelings in socially acceptable ways. Often the feelings are vented in the form of aggressive behavior. Studies show that aggressive behavior, across many cultures, is more common among boys than among girls; it is also more common in early childhood than later, at least in many of its physical forms. Why do children show aggression, and why does it take the forms that it does? Many have argued that frustration provokes aggression (Chapter 7, page 290), and that the forms it takes will depend on the child's previous learning.

Bandura et al. (1963) tested this idea, focusing specifically on observational learning.

They arranged for 5-year-olds to observe models attacking an inflated "Bobo doll," using the techniques shown in Figure 11.15.

Figure 11.15

Imitation of aggressive behavior. Top: An adult model acts aggressively toward a large plastic doll. Middle: A boy who has watched the adult model shows similar aggressive behavior. Bottom: A girl also imitates the model's behavior. (From Bandura et al., 1963. *Journal of Abnormal and Social Psychology*.)

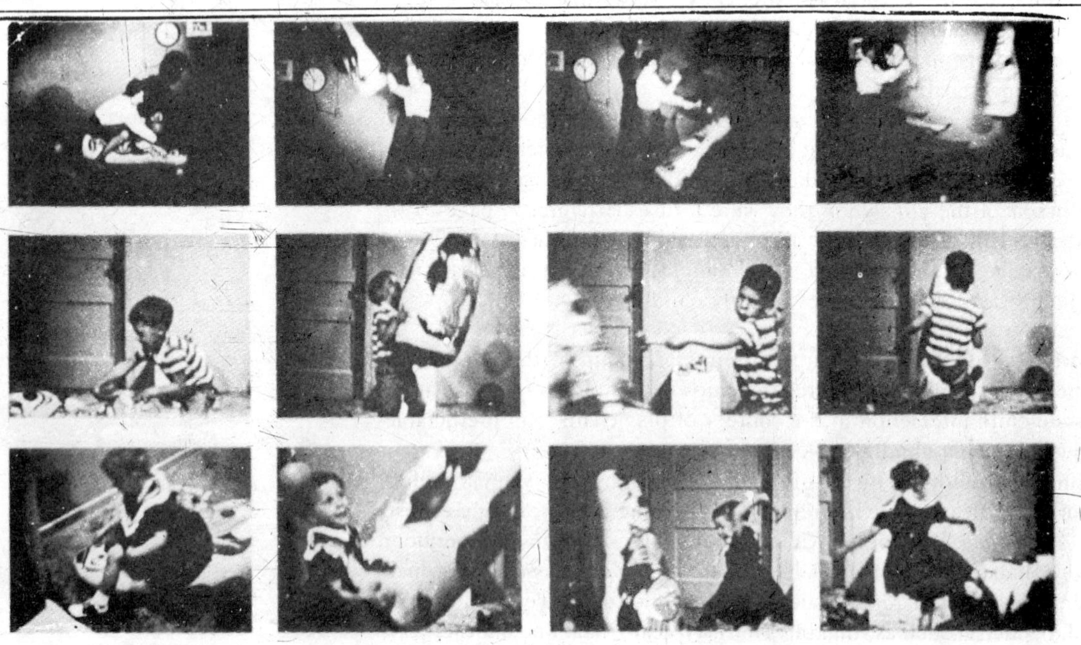

Another group of 5-year-olds did not see the models attack the dolls. Children in both groups were then frustrated. Each child was allowed to start playing with some attractive toys, only to be interrupted and taken to a room that contained a Bobo doll and other toys. Children in both groups reacted with aggressive behavior, but the children in the observational-learning group focused their aggression on the Bobo doll much more than did the other children. In fact, some of the children who had seen the models attack the doll attacked it in exactly the same ways and in the same sequence as the model had (see Figure 11.15).

Note that children in both groups reacted to frustration with aggressive behavior, but the two groups showed different patterns, or styles, of aggression. Evidently, the specific forms of aggression that children choose may have much to do with what they have learned, often through observation. This has led to some concern about the possible effects on children of the aggressive and violent behavior they see on television programs. We discuss these concerns and whether they are valid in Chapter 7 (pages 291–292).

Aggressive behavior may be fostered not only by observational learning but also by direct reinforcement, or reward. In many settings where children play, the aggressive children often triumph over others, have easier access to preferred toys, and even get extra attention from adults who are encouraging them to be less combative. Many youngsters are also rewarded for their aggressive behavior at home. Parents often respond to such behavior by paying special attention to the child and even by giving in to the child's demands "just to get a little peace and quiet" (Patterson, 1976).

Prosocial behavior Preschoolers can be aggressive, but they can also be touchingly helpful, generous, and comforting. Such behavior is called *prosocial.* It is often seen in the same children who tend to be aggressive. Some have argued that these children are motivated to be involved with other children, and whether the involvement is aggressive or prosocial will depend upon the situation. Others argue that aggressive children, who themselves are easily upset, "find it easier to empathize with others who are upset."

According to Hoffman (1976), children pass through four predictable stages in the development of the empathy that makes prosocial behavior possible. In the first stage, infants have trouble differentiating self from others. Their behavior is triggered by, and often looks like, the strong emotional displays of others. They often cry when others cry and laugh when others laugh. It is almost as if they were directly wired to the other person's emotional system. People who work in nurseries often see this crude form of empathy in the form of a chain reaction; one baby's crying triggers another baby, whose crying triggers an outburst on the other side of the room. After the first year, children gradually develop a sense of self as different from others, and at that point they enter a second stage. Although they have come to recognize that another person is, in fact, another person, their egocentric thinking (page 437) leads them to "help" the other person in ways that they themselves would want to be helped. So a boy whose mother is upset may bring her his favorite blanket or his teddy bear. In the third stage, children recognize that a distressed person may have feelings and needs that are different from their own. Their efforts to help become aimed at figuring out what the distressed person really needs, even if the need is different from the child's own personal preference. During this stage, which lasts through early childhood, children are limited to empathy for others who show specific expressions of emotion. Their empathy is situation-specific. It is only in later childhood, when the fourth stage is reached, that children come to relate one expression of distress to another and to be concerned for the general condition of others. It is only in this fourth stage that children are likely to empathize with and seek to help, say, an unpopular child who seems generally morose or withdrawn.

Children's level of development seems to influence their competence at empathy and prosocial behavior; but whether they actually *try* to show such behavior may depend heavily on their learning. There is good evidence that children can learn helping behavior by imitating models; this is particularly effective if the models express happiness about their prosocial behavior (saying, for example, "I really felt great when I helped that boy pick up his bike") and if the models are able to dispense rewards that the observing children value.

It may have occurred to you that television viewing can have good as well as bad effects—that it can enhance prosocial behavior through positive observational learning (see Figure 11.16). There is some evidence that this happens when nursery-school children watch "Mr. Rogers' Neighborhood" (Stein & Friedrich, 1975). However, "Mr. Rogers" is a gentle, low-keyed program. It is less riveting than are the more exciting, aggression-laden programs. Its prosocial messages have been shown to have an impact only when children were asked to comment on them or act them out. There may be several positive effects of television watching in early childhood, but the effects found in research thus far have not been dramatic or clear-cut.

ADJUSTMENT PROBLEMS IN EARLY CHILDHOOD

In the preschool years, children acquire a risky combination: mobility, language, and immature judgment. Their limited powers of reasoning

Figure 11.16

The power of television to capture the child's attention (left) is undeniable. Some have argued that this power, via such programs as *Mr. Rogers' Neighborhood* (right), can be harnessed to promote prosocial behavior. Research to date has shown that such prosocial effects can be produced only when the children are given extra help in thinking about the meaning of what they have seen. (*Left:* Mimi Forsyth/Monkmeyer Press Photo Service; *right: Mister Rogers' Neighborhood.*)

make it hard for them to foresee the consequences of their physical activity. They are physically able to cross the street but unable to envision all the dangers that crossing the street poses. Parents' mysterious sanctions are often merely the stuff of games, as when 30-month-old Julia was found dropping eggs on the floor while scolding herself, "No No No. Mustn't dood it. No No No *Mustn't* dood it" (Fraiberg, 1959, p. 135). Preschoolers also use their newfound language skills with a distinct lack of restraint. Their cognitive egocentrism (page 437) prevents them from taking the perspective of their listener; the result can be painfully honest comments such as, "Hello, fat lady" or "You have ugly teeth."

Preschoolers pay a price for their powers of representational thought. That price is a lively imagination that can careen out of control at times. Shadows on the wall at bedtime can become burglars, kidnappers, or ghosts. There is a perpetual tension between the rational and irrational uses of imagination. This tension was nicely illustrated by the comment of a 6-year-old boy whose fear of ghosts had him wide-eyed in the middle of the night: "I know they aren't real, but what if they are!" A common fear among preschoolers is that something under the bed will grab a hand if it hangs free. Perhaps you remember thinking to yourself, "There's probably nothing under the bed, but why take a chance, right?" Surveys of parents show that fears are among the most common behavior problems of early childhood, but what children fear changes markedly during this period. As Figure 11.17 shows, fears involving thought and imagination increase sharply during the preschool years; in contrast, the percentage of fears related to concrete, physical sensations generally declines. This illustrates the shift away from the sensorimotor "thought" of infancy and toward the internalized, representational thought of early childhood.

The other problems mentioned most often in parent surveys reflect their children's unsocialized use of physical and verbal powers. "Overac-

Figure 11.17

Age changes in what children fear. Fears involving concrete sensory and physical experiences generally decline with age during the preschool years (left). During those same years, though, children's growing powers of representational thought lead to an increase in fears involving thought and imagination (right). (From *Children's Fears* by A. T. Jersild & F. B. Holmes, New York: Teachers College Press, 1935.)

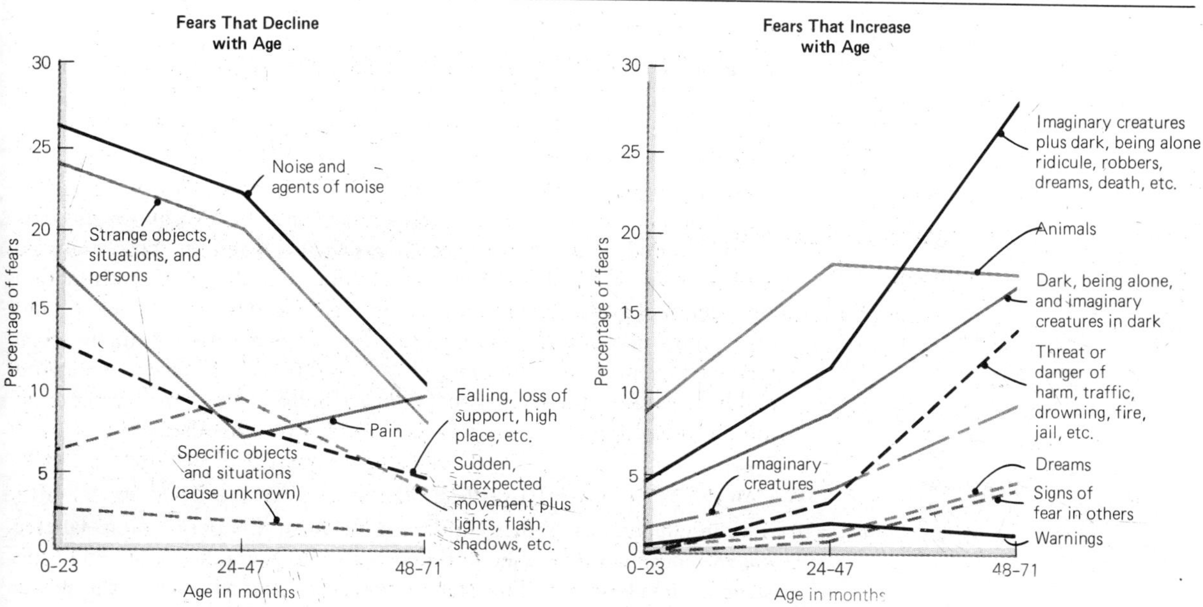

tive," "shows off," "talks too much," "temper," "whines," "argues," "demands attention," "disobedient," and "resists bedtime," are terms that appear often in these surveys. Some of these problems are common throughout the preschool years; contrary and stubborn behavior, for example, are reported for about 60 percent of children at ages 2, 3, 4, and 5 (Heinstein, 1969). Other problems, such as temper tantrums, decline over the preschool years.

Some of these problems become exaggerated enough in some children to require clinical intervention. Two of the most common causes of clinic referral are unsocialized behavior and *phobias*—fears so irrational, persistent, or intense that they interfere significantly with normal functioning. Speech problems are a third major reason for clinic referral, especially in the 2- to 5-year age group. Poor hearing, faulty brain functioning, or mere lags in development can interfere with children's ability to understand speech or to produce it. Stuttering can also begin in early childhood. Sometimes it can be traced to common early problems in pronunciation that lead children to worry about their ability to talk. At other times, the cause is harder to find. Stuttering usually fades out with development, but in some cases it lingers throughout life. In some cases, difficulties in producing language sounds or other anxiety-laden circumstances in a child's life can lead a youngster to stop speaking altogether, a problem called *elective mutism.* Finally, infantile autism, which we discussed in the section on infancy, may not be actually identified until it shows up in certain speech peculiarities in early childhood. One example is *echolalia,* a tendency to repeat, or "echo," what another speaker has just said rather than to use speech for true communication. Many researchers who study autism now believe it is especially ominous if a child reaches the age of 5 and still cannot use speech to communicate. This fact reminds us that an ability to blend thought and language effectively is virtually a litmus test of successful development in early childhood.

Later Childhood: Cognitive Tools, Social Rules, Schools

The elementary school years, the years 6 through 12 in a child's life, are sometimes referred to as the latency period. The term comes from Freud's psychoanalytic theory (see Chapter 14, page 576), which holds that important conflicts, particularly sexual conflicts, are submerged during this period. However, in many areas of development, these years are actually action-packed, not latent at all. They are filled with both motion and emotion as the child confronts the diverse demands of school and entry into a rule-bound society. A capacity for increasingly intimate social relationships promises important rewards but poses real risks as well. And children acquire intellectual tools during these years that give them an unprecedented grasp of the way the world is put together.

COGNITIVE DEVELOPMENT

The intellectual tools that children develop in this period were labeled *concrete operations* by Piaget, and that is also the name he has given to this stage of development. This stage involves a major advance in the power of the child's reasoning. Recall that in our imaginary experiment with the Müller-Lyer illusion (Figure 11.14, page 435), preoperational children were fooled by perceptual appearances. In the period of concrete operations, children are no longer fooled. They recognize that the length of a particular stick will remain fixed so long as nothing is added or taken away from it. Thus, they have learned one of nature's constants: the principle of conservation of length. They believe in this principle because their mental actions now show *reversibility*. This means that they can now imagine picking up the stick that *looks* longer and returning it to its original position, shown in step 1 of Figure 11.14. They know that if they were to do so, stick B would turn out to be the same length as stick A. Their reversible operations also help them understand, for the first time in their lives, that subtraction reverses and compensates for addition and that multiplication inverts and compensates for division. In this way, a variety of mental activities are seen *in relation to* one another. When these mental activities can be used in flexible ways (for example, with reversibility) and understood in relation to one another, they qualify as true operations (page 435).

With the advent of these operations, children's awareness of the ways the world is organized begins to mushroom. They understand not only conservation of length but conservation of other physical entities—like mass, number, and area. You can conduct your own experiments with children to test their conservation concepts; Figure 11.18 shows some of the common experimental procedures used for these purposes.

The concrete-operational child organizes the world by using hierarchies. In these hierarchies, a given "thing" can fall somewhere on more than one dimension at the same time. To understand how hierarchical thinking works, imagine that we have a picture of seven people in a group, two adults and five children. We show the picture to some 4-year-olds and some 9-year-olds, asking "Are there more children or more people?" Most 4-year-olds will say "More children"; most 9-year-olds will say "More people." Why the difference? It happens because we are asking the children to think about two dimensions in a hierarchy. One dimension is people versus nonpeople; another is children versus adults. The 4-year-olds, being preoperational, can focus on only one dimension at a time. Because our question mentioned children first, our young subjects used the child-versus-adult dimension and concluded that there were more children. The 9-year-olds recognized that both children and adults fall on the "people" end of the people-versus-nonpeople dimension regardless of their position on the children-versus-adults dimension. These older children answered correctly because their flexible operations allow them to think in terms of a hierarchy involving two dimensions, one broader than the other.

In many ways, then, the concrete-operational child's thinking shows a power and versatility that would have been literally unthinkable in the preoperational period. But even this more advanced level of thought has its limitations. The operations are concrete in the sense that they are tied to the real world of objects and events. The children can think clearly about things that are real, but not very clearly about more hypothetical propositions such as, "Suppose that people hatched from eggs" or "Suppose that water boiled at 32°." It is also hard for the concrete-operational child to grasp the broad meaning of abstract concepts such as freedom,

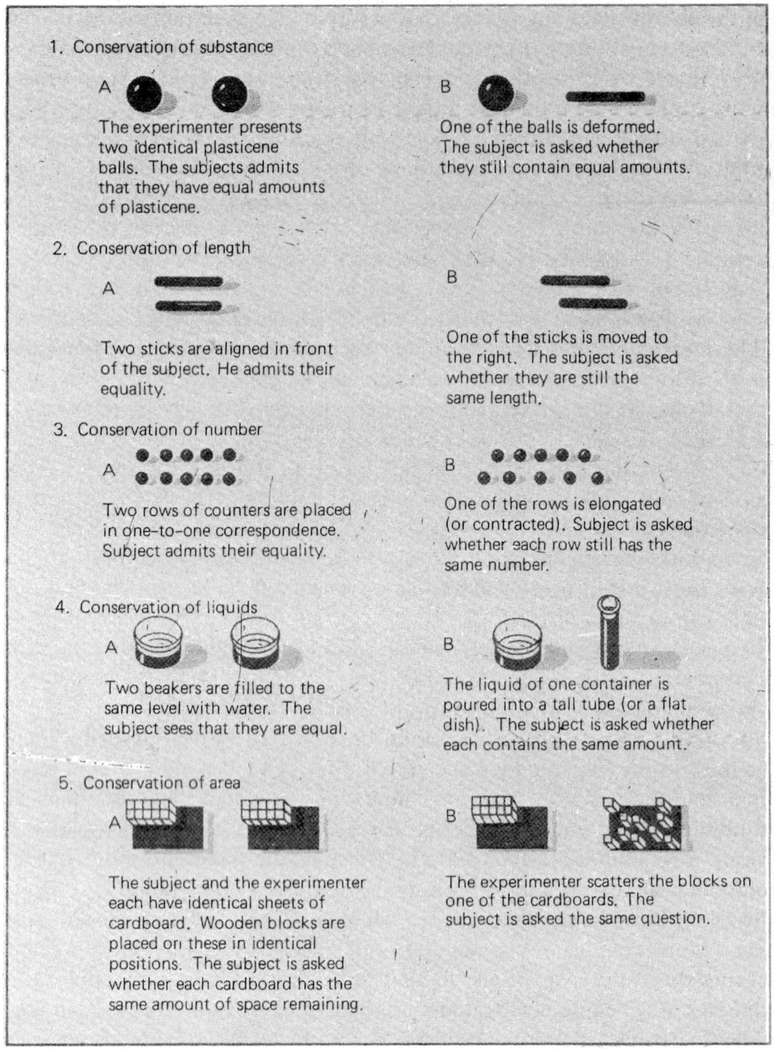

Figure 11.18

Some tests use to assess conservation. (From *Of Children* by Guy R. Lefrancois; © 1980 by Wadsworth, Inc. Reprinted by permission of publisher.)

integrity, or truth. These limitations mean that intellectual growth in later childhood, though impressive, is still incomplete.

SOCIAL DEVELOPMENT

As their social world expands to include classmates and teachers, children's ways of thinking about people show a corresponding change. Studies of "person perception" (Chapter 9, page 343) show that a child even as old as 6 or 7 will describe others in egocentric ways, referring to what the other people do to or for the child. Descriptions at this age also focus on concrete, observable characteristics of others, such as their physical appearance or their outward behavior. Here are examples from two 7-year-olds:

> She is very nice because she gives my friends and me toffee. She lives by the main road. She has fair hair and she wears glasses. (Livesley & Bromley, 1973, p. 214)

> Max sits next to me, his eyes are hazel and he is tall. He hasn't got a very big head, he's got a big pointed nose.
> (Livesley & Bromley, 1973, p. 213)

During the next few years, children begin to use more and more descriptive statements involving psychological characteristics—statements that require some inference about the other person. Note the psychological inferences (italicized) made by this 9-year-old:

> He smells very much and is very nasty. He has *no sense of humor* and is very *dull*. He is always fighting and is *cruel*. He does silly things and is very *stupid*. He has brown hair and cruel eyes. He is *sulky* and 11 years old and has lots of sisters. I think he is the most horrible boy in the class. He has a croaky voice and always chews his pencil and picks his teeth and I think he is disgusting.
> (Livesley & Bromley, 1973, p. 217, italics added)

Did you keep a diary during your elementary school years? If so, you may be able to spot in it signs of similar developmental changes in "person perception" in your descriptions of other people, or even in your comments about yourself.

Friendship The development of "person perception" goes hand in hand with changes in the nature of friendship. In the early preschool years, children have momentary playmates but not ongoing, reciprocal friendships. Some time between the ages of 4 and 9, most children develop an ongoing friendship, or perhaps several. Their first friendships tend to be self-serving; a friend is someone who "does what I want." Later, during the elementary school years, friendships become not only outgoing but reciprocal as well; friends are seen as people who "do things for each other" (Selman, 1980). Try to remember the first "friend" that you thought about in this way. Chances are that the first friend you would go out of your way to help, and who would do the same for you, was someone you knew in elementary school, probably in the middle grades. Chances are also that this person was the same sex as you; friendships are almost exclusively boy-boy and girl-girl in elementary school. Finally, the chances are good that most of your elementary school friendships were somewhat exclusive—you and your friend saw your relationship as ruling out equally close friendships with other children. This quality of exclusion, or possessiveness, goes along with many friendships in the middle and late elementary years, and also in adolescence.

Groups At the same time that children are learning to form one-to-one relationships with friends, they are learning to organize themselves into groups. Groups have certain defining characteristics: goals shared by its members, rules of conduct (often merely implied or understood), and a hierarchical structure. The structure resembles the organizational chart of a corporation. There are leaders at the apex and followers at the lower levels; each individual member has some identifiable relationship to other members. Psychologists have tried to learn what conditions cause group structures to take shape in later childhood. Some of the most interesting answers have come from studies of summer camps (Figure 11.19).

Figure 11.19

Boys and girls at summer camp, where groups can form quickly and peer relationships can be intense. [(*a*) David Strickler/Monkmeyer Press Photo Service; (*b*) Wide World Photos.]

A classic camp study by Sherif and others (1961) showed that group formation in preadolescents is stimulated by the experience of living together, sharing pleasant experiences, cooperating in ventures that involve shared goals, and, especially, competing with other groups.

When antagonism between competing groups of campers escalated to name calling and insults, Sherif engineered "experimental crises," such as a mysterious shutoff of the water supply. Forced by such adverse circumstances to work together, the competing groups combined forces and engineered a solution. Afterward, ill will between the groups faded. The Sherif study, like others focused on this age range, suggests that shared adversity and joint problem solving can stimulate group formation and reduce antagonism between groups.

Peer versus Adult Influence During the elementary school years, as we have just seen, friends and groups of peers take on central importance in a child's social life. Does this mean that the influence of parents wanes? Yes, at least in the U.S., which is perhaps one of the most peer-oriented of cultures. By the late elementary school period, there are many situations in which American youngsters prefer relying on peers to relying on parents. Perhaps more importantly, there are many situations in which children, if forced to choose, will opt for behavior approved by their peers rather than behavior approved by their parents and other adults.

Studies of other cultures, though, show that these tendencies are not inevitable in late childhood. Instead, they seem to depend upon the socializing experiences children are exposed to as they grow up. In the Soviet Union, for example, peers are used throughout the school years as enforcers of adult norms (Figure 11.20). Peer monitors keep track of how well each classmate adheres to these adult norms; peer groups meet to decide how to punish children for deviations from the adult-established standards or reward them for meeting the standards. Members of the groups, or "links," work together in ways that Sherif would tell us are likely to promote group solidarity; the solidarity, though, is in the service of goals that adults endorse. What impact do these socialization practices have on Soviet youngsters?

> To find out, Bronfenbrenner (1967) asked a group of Soviet children and a group of American children to tell what they would do in response to several dilemmas. The dilemmas pitted adult standards against peer standards. One, for example, asked the youngsters whether they would go to a movie that their friends recommended but that their parents disapproved of. Some of the children in each group were told that their answers would later be shown to peers in their class. This information had precisely opposite effects on children from the two cultures. American children who thought their peers would see their answers were especially likely to violate the adult norms; Soviet children given the same instructions were especially likely to *follow* the adult norms.

Apparently, peers have a powerful impact in later childhood; but whether this impact works for or *against* adult values and teachings will depend upon the previous socialization of the children.

SCHOOL ACHIEVEMENT

The cognitive and social themes discussed above are often played out within one of civilization's most remarkable creations: the elementary school. Formal schooling in almost all countries begins around the age of 6 or 7. By that age, the major limitations of preoperational thought (page 437) have been left behind, language has matured to a point where a teacher can communicate with children in groups, and the perceptual and motor skills needed for such activities as writing with a pencil are in place. During the early school years, most children develop a broad array of basic skills; in fact, a person who has reached the average fourth-grade achievement level in reading, writing, and arithmetic is considered "literate" in the United States.

Figure 11.20

Peer groups in the Soviet Union, unlike those in the United States, support and enforce the values of the adult culture. Soviet youngsters are urged to support adult standards even if it means reporting peers, as shown in the wall poster (left). Right, a council of youths meets to decide what punishment should be given to some peers who went swimming without supervision. (From *Two Worlds of Childhood: U.S. and U.S.S.R.* by Urie Bronfenbrenner. Copyright © 1970 Russell Sage Foundation. Reprinted by permission of Basic Books, Inc., Publishers.)

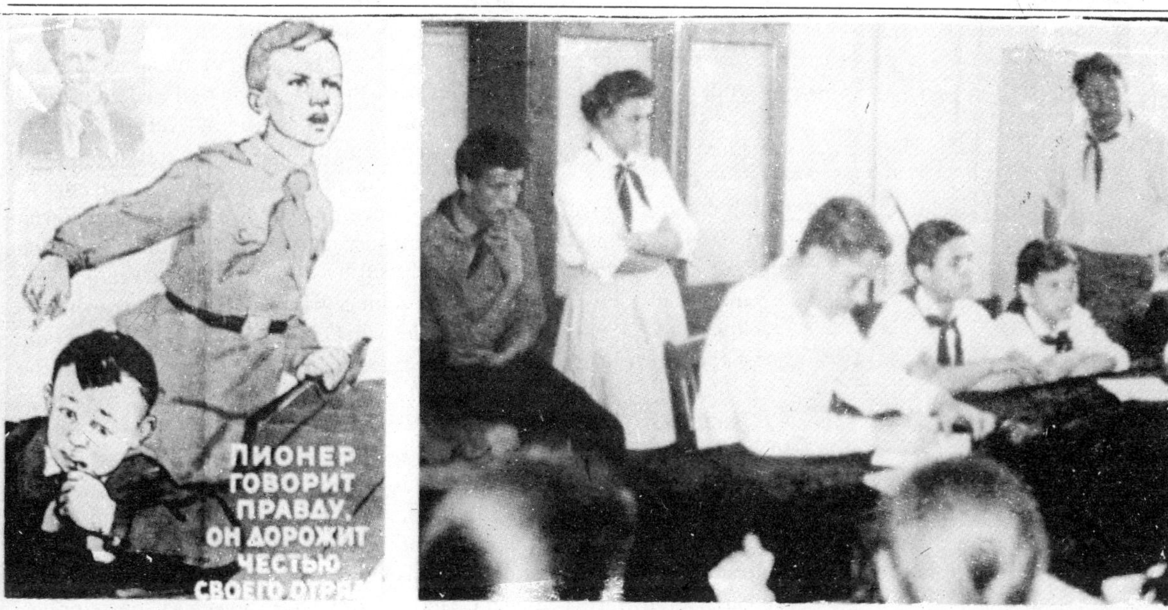

How do schools teach, and how well do they teach? The major mission of the school is to promote systematic learning—learning of specific material in a more or less prescribed order. Learning in school can be very different from informal, everyday learning. In the latter, children often learn by immersion: they jump in feet first, try their best, learn from their mistakes, and try to "get the hang of it"—often without learning any general rules until much later. Most of us learned to ride a bicycle well before we could state a rule for how to do it. Most of us followed some basic grammatical rules before we could articulate what these rules were.

In school learning, the process is often reversed: the teacher presents a general rule first, and the children later figure out how the rule applies to concrete examples in real life (Scribner & Cole, 1981). Remember the old rule, "A noun is the name of a person, place, thing, or idea"? Most children who learn this rule in school cannot actually point out the nouns in sentences until some time later (Kenny, 1983).

Research findings like this raise a question as to the effectiveness of schools. Does their more formalized, "rule-first" approach actually promote effective learning, or would children be about as well off if they did all their learning informally, outside of schools? The question is difficult to answer in countries where virtually every child goes to school; but in many countries it is still possible to compare large numbers of unschooled children to their schooled counterparts. These comparisons, by and large, show that schooling has some strong positive effects. In one study (Stevenson, 1982), Peruvian first-graders were compared to unschooled children of the same age and social class on 15 different cognitive skills (for example, memory and classification). The first-graders proved superior on every skill, suggesting that even a year or so of schooling may make quite a difference in cognitive performance. Other studies suggest that in addition to teaching specific cognitive skills, schools are particularly good at teaching general problem-solving techniques (Fischer & Lazerson, 1984); such techniques as labeling the parts of a problem and constructing a general solution rule can give the schooled youngster a real edge in tackling new and unfamiliar problems.

Given the importance of peers in the lives of elementary-schoolers, one might reasonably ask whether peer-to-peer instruction makes sense. Several studies show that it does. Tutoring by peers a few years older than the tutee has been found to produce significant learning in the tutee, and often in the tutor as well. Finally, recent research has shown that many children can improve their school achievement by monitoring their own work, giving themselves periodic self-instructions (for example, silently repeating "Go slowly, and think carefully about each answer"), and rewarding themselves for reaching certain goals. In addition to the work of teachers, there seems to be a good deal that children of this age can do to boost their own learning in school.

ADJUSTMENT PROBLEMS IN LATER CHILDHOOD

In later childhood, the classroom is the stage upon which some of life's most important dramas are played out. School experience can be exhilarating for a child who "fits in" academically and socially; but it can be sheer misery for children who do not. The children, teachers, and literature in most North American schools are predominantly white and middle-class. Nonwhite youngsters and children from lower socioeconomic levels may feel out of place and uncomfortable in these settings. Half of all children from the lowest socioeconomic level drop out before they finish high school. Person-to-person comparison becomes intense during the elementary school years; children who do badly in their schoolwork may have a pretty good idea, by the third or fourth grade, where they rank in relation to their peers. With this knowledge comes a newfound capacity for feelings of inferiority. This is particularly difficult for children who suffer from cultural-familial mental retardation (Chapter 13, page 536). Because their retarded functioning is not caused by any known physiological disorder, their intellectual problems may not be noticed until they enter school.

For some children, the social demands of school are harder to satisfy than the academic demands. Often the setting seems to call for impressing one's peers and teacher; this may help to explain some of the most commonly reported problem behaviors of this age—"argues," "brags," "shows off," "talks too much," "self-conscious". These problems are continuations of behaviors seen in preschool. Clusters of these problems, taken to extremes, can lead parents to refer their children to clinics for special professional attention. The rate of such referrals goes up sharply during the elementary school years, and mostly because of problems demonstrated in school. Self-assertive behavior carried to the point of aggression is one of the most common causes of clinic referral during these years. At the other extreme are children who respond to the social demands of school with self-consciousness carried to the point of social withdrawal; this is another comon cause of referral.

A problem known as school phobia is seen early in the school years. It is an extreme form of normal school anxiety—school-phobic boys and girls often panic and even show physical reactions like vomiting when it is time to go to school. Many psychologists now agree that the phobia is often a fear of separation from the parents rather than a fear of school itself and that a quick reentry into school is usually the best treatment.

School phobias combined with generalized anxiety, sadness, or shyness can form an ongoing pattern called internalizing problems. Aggressive behavior combined with other kinds of antisocial behavior and attitudes can form an ongoing pattern called externalizing problems. Persistent internalizing or externalizing patterns can lead to many long-term difficulties, but the externalizing patterns pose the most serious threat to long-term adult adjustment (Robins, 1979).

Two other common causes of clinic referral during later childhood are learning problems and hyperactivity. The most frequent learning problem is *dyslexia,* difficulty with reading. Other children are said to have *learning disabilities* or *specific development disorders.* These are umbrella terms that cover a variety of specific learning difficulties; their common denominator is that all involve islands of poor functioning in children whose performance in other areas of schoolwork is average or above average. Such problems often accompany hyperactivity. Youngsters with this problem are impulsive and overactive. In some settings, these children are diagnosed as having an attention deficit disorder. Many are taken to physicians and given drugs to calm them down and help them pay attention in class. Recently, this use of drugs has come under attack. Critics note that the long-term effects of the drugs are not yet known. Critics also note, correctly, that many overactive children are able to manage their problems without being medicated if they are given careful training in self-control.

Summary

1. Developmental psychologists study the changes that occur, during all or part of the life span, in the processes of perception, learning, thinking, social activity, and other aspects of human behavior.

2. Three major issues in developmental psychology have stimulated recurring conflicts. One is the question of how much nature and nurture influence the developing individual. A second is the question of whether individuals play passive or active roles in development. A third is the question of whether the accumulation of skills and knowledge is continuous or discontinuous and stagelike.

3. Studies of the changes in human behavior sometimes use the longitudinal method, in which the same individuals are tested or observed at different points in time. More often, though, developmental differences are studied via the cross-sectional method, in which groups from two or more age levels are compared.

4. Infants in the first month after their birth are called neonates. Neonates use several reflexes automatically, but they also show perceptual and learning activity that involves active exploration of, and interaction with, their world.

5. During infancy (from birth to about 18 months), locomotion, prehension, and other dimensions of motor development show highly predictable stage sequences; visual-perception skills develop rapidly.

6. A major contributor to the study of infancy and later development was Jean Piaget. His approach, labeled genetic epistemology, focused on the development of knowledge through the interplay of assimilation (modifying the environment to fit one's established patterns of thinking or doing) and accommodation (modifying oneself to fit the environment).

7. In Piaget's system, infancy corresponds to the sensorimotor stage; thought in this stage typically involves sensory and motor activity (for example, sucking), but some internalized thinking does eventually occur. An important sensorimotor milestone is the development of object permanance, the awareness that objects continue to exist even when they cannot be seen.

8. Most reseach on social development in infancy has focused on the baby's attachments to "significant others." Attachment formation appears to follow a predictable sequence and may be facilitated by contact comfort from a soft, warm surface.

9. Early childhood, from the age of about 18 months through the sixth year, was labeled the preoperational stage by Piaget. Preoperational thought has major limitations; but children in this stage do develop significant strengths, such as representational thought—the capacity to form mental symbols for absent events or objects.

10. A centerpiece of early childhood is the development of lan-

guage. The process seems to combine learning via rewards with active construction of rules by children themselves. By the age of 4, most children are using adultlike grammar.

11. Child-parent relationships take on a new and different tone as the parent begins to administer discipline. A parental style that combines warmth with moderate restrictiveness and an authoritative manner appears to enhance independence and social maturity in preschoolers.

12. In early childhood, play begins as a solitary activity. This is followed by parallel play, cooperative play, and, finally, true peer interaction.

13. Aggression can become a problem in early childhood. Social influences such as television may, through modeling, encourage aggression. Some have argued that television can also encourage empathy and prosocial behavior, but research thus far has not shown strong prosocial effects of television.

14. Later childhood, from the age of about 6 or 7 through 12, has been called the concrete-operations stage by Piaget. During this stage, children use logic and begin to grasp such important principles of nature as the conservation of mass and length.

15. As the child's social world expands to include school peers and teachers, there is a corresponding expansion in ways of thinking about people. Children at the age of 6 or 7 describe people in terms of their tangible characteristics (for example, hair color) and their relation to themselves (for example, "She gives me things"); children a few years older include psychological inferences in their descriptions (for example, "He has no sense of humor").

16. Corresponding changes take place in friendships. At the beginning of later childhood, friendships tend to be self-serving; but by the late elementary school years, friendships are seen as reciprocal—friends are people who "do things for each other." As children learn to form relationships with individuals, they also learn to organize themselves into groups that have shared goals and hierarchical structures.

17. Schools are central to the lives of children during later childhood. Evidence suggests that schools improve children's thinking in a number of general ways.

Terms to Know

One way to test your mastery of the material in this chapter is to see whether you know what is meant by the following terms.

Developmental psychology (413)
Nature-nurture controversy (413)
Nature (413)
Nurture (414)
Reaction range (414)
Passivity versus activity (414)
Continuous versus discontinuous development (416)
Longitudinal method (416)
Cross-sectional method (417)

Infancy (419)
Neonate (419)
Moro reflex (419)
Prehension (423)
Mouthing (423)
"Visual cliff" (424)
Cognitive development (425)
Genetic epistemology (425)
Epistemology (425)
Assimilation (425)
Accommodation (425)
Equilibration (425)
Sensorimotor stage (425)
Scheme (426)
Object permanence (427)
Attachment (429)
Ethologists (430)
Contact comfort (430)
Schemas (433)
Down syndrome (434)
Infantile autism (435)
Failure to thrive (435)
Preoperational stage (435)
Operations (435)
Identity concept (436)
Representational thought (436)
Transductive reasoning (437)
Egocentrism (437)
Animism (437)
Syntax (438)
Holophrase (439)
Socialization (441)
Critical period (442)
Sensitive period (442)
Observational learning (444)
Principle of identity (444)
Solitary play (445)
Parallel play (445)
Cooperative play (446)
Prosocial behavior (447)
Phobias (450)
Elective mutism (450)
Echolalia (450)
Concrete operations (450)
Reversibility (451)
Dyslexia (457)
Learning disabilities, specific development disorders (457)

Suggestions for Further Reading

Two excellent textbooks provide a detailed look at the current state of child-development research and theory. Howard Gardner's *Developmental Psychology: An Introduction*, 2nd ed. (Boston: Little, Brown, 1982) takes a chronological approach, tracing characteristics of the developing person from birth through adolescence. E. Mavis Hetherington and Ross D. Parke have adopted a topical approach in *Child Psychology: A Contemporary Viewpoint*, 2nd ed. (New York: McGraw-Hill, 1979). In each of their chapters, they trace developmental changes within one area of human functioning (for example, learning, emotional development, language, and communication). Readers interested in an in-depth look at atypical development and psychopathology from infancy through adolescence will profit from reading Thomas M. Achenbach's *Developmental Psychopathology*, 2nd ed. (New York: Wiley, 1982). Those interested in learning more about Piaget's theory should read Herbert Ginsburg and Sylvia Opper's *Piaget's Theory of Intellectual Development*, 2nd ed. (Englewood Cliffs, N.J.: Prentice-Hall, 1978). For an engrossing historical perspective on childhood, see *The Child* (New York: Wiley, 1965), edited by William Kessen. Finally, for an informative cross-cultural perspective on development, we recommend Urie Bronfenbrenner's *Two Worlds of Childhood: U.S. and U.S.S.R.* (New York: Russell Sage Foundation, 1970).

Development During Adolescence, Adulthood, and Old Age

HEADING OUTLINE

Chapter

One of the benefits of adolescence and adulthood is a revamped intellect that takes us beyond the limitations of childhood thought. Several of the broad intellectual capacities of adolescent and adult thinking have been grouped together by Piaget and labeled formal operations, the final stage of intellectual development. To capture the flavor of your own formal-operational thinking skills, try them out by tackling these problems.

INTERNATIONAL KALEIDOSCOPE

You live in a very cosmopolitan college town. Across the street from you, there are five houses, each painted a different color. The houses are inhabited by men from five different countries, who use five different modes of transportation to get them to the campus every day. The men also play different sports and prefer different ice-cream flavors.

1. The Englishman lives in the red house.
2. The Spaniard plays basketball.
3. Chocolate chip is the preferred flavor in the green house.
4. The Kenyan prefers fudge ripple.
5. The green house is to the right (your right) of the ivory house.

6. The bicyclist plays racquetball.
7. The car is owned by the inhabitant of the yellow house.
8. Peach ice cream is preferred in the middle house.
9. The Iraqi lives in the first house on the left.
10. The van driver lives in the house next to that of the man who plays polo.
11. The driver of the car lives in the house next to that of the hockey player.
12. The man who roller-skates prefers pralines and cream.
13. The American rides a motorcycle.
14. The Iraqi lives next to the blue house.
    A. WHO PREFERS BUTTER PECAN?
    B. WHO PLAYS SQUASH?

LIGHTS AND SWITCHES

Imagine that you have walked into a large room containing eight light switches (1–8) and eight light bulbs (A–H), arranged as shown in the diagram below. Your task is to figure out, in the most efficient way possible, which switch will turn on light bulb *E.* You may use as many "moves" as you need; each move will involve turning on as many or as few light switches as you wish, all at the same time. The objective is to use the smallest number of moves that will be sure to yield the solution. What will your moves be?

Answers to these two problems are given in the section on formal operations on page 470. As you think about the problems, bear in mind that they are not designed to be "tests" of how smart you are or to classify you into one of "Piaget's stages." After all, Piaget's stages have much more to do with *how* you think than whether you get a correct answer. The problems are merely illustrations; they are ways of getting you to use some key mental processes. To understand these processes and how they differ from childlike thinking, read on.

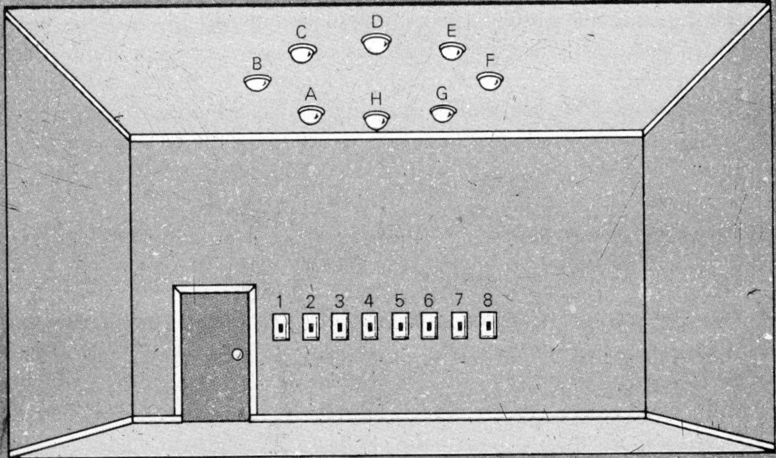

IN this chapter, we will focus on three of the most dramatic life transitions that people in our culture make. One of these is the transition from childhood into adolescence. In a stunningly short time, the child takes on an adultlike physique and intellect and is dubbed a "teenager." Some theorists portray the adolescent's world as especially peaceful, a time to make decisions about who one is and how one is to live one's life. Others portray adolescence as a time of great stress—as a struggle to keep one's psychological balance in the face of the crisis of puberty. Whatever your memories of those teenage years, the chances are you will enjoy comparing your memories with the theories and evidence discussed in this chapter. A second transition we will consider is the movement into the responsible world of the true adult—the world of careers, marriage, and family. Finally, we will consider the aging process and the period we call old age. This period, as we shall see, involves some kinds of decline; but it also offers the prospect of intellectual mellowing and a special quality called "wisdom." In this chapter, we cut a broad swath; broad enough, in fact, to give most of us room to reflect on our past and contemplate our future.

Adolescence: Storm and Stress or Smooth Sailing?

"It's when you have to pay adult prices for movies, but you can't see adult movies."

With these words, a 13-year-old named Dawn summed up her views on early adolescence—a time when society sends mixed signals to its youngsters. Technically, *adolescence* is the period from the beginning of sexual maturity (*puberty*) to the completion of physical growth. As we shall see below, however, the exact ages spanned by adolescence vary from one person to the next. Moreover, the psychological impact of the transition to adolescence may differ across individuals and perhaps even across cultures. Some who have studied adolescence view it as a period of "storm and stress." This was the view of G. Stanley Hall, an American psychologist whose 1904 book, *Adolescence,* helped make this age period a focus of scientific study. Hall saw adolescence partly as an upheaval, a disruption of peaceful growth. So did Anna Freud, a prominent theorist and daughter of Sigmund Freud; she even argued that those adolescents who maintain their psychological balance during adolescence may be abnormal.

Do "storm and stress" accompany puberty in every culture? Not according to the late anthropologist Margaret Mead. (See Figure 12.1.) Her early observations of adolescent girls in Samoa, for example, led her to conclude that for these girls:

> Adolescence represented no period of crisis or stress, but was instead an orderly developing of a set of slowly maturing interests and activities. The girls' minds were perplexed by no conflicts, troubled

by no psychological queries, beset by no remote ambitions. (Mead, 1928/1939, p. 157)

Mead described Samoan adolescence as a time free of sexual anxiety and repression; in Samoan society, as she saw it,

> . . . there are no neurotic pictures, no frigidity, no impotence, except as the temporary result of severe illness, and the capacity for intercourse only once in a night is counted as senility. (Mead, 1928/1939, p. 157)

Mead's description of Samoan adolescence as peaceful, untroubled, and uninhibited was challenged recently by the Australian anthropologist Derek Freeman (1983, see Figure 12.1). He argued that Mead was misled by her Samoan informants. Samoan adolescents, Freeman argued, are actually plagued by guilt and quite conflicted about sexuality. The Samoan adolescence described by Freeman sounds very much like the traditional "storm and stress." So the issue of whether adolescence involves psychological stress and upheaval in *all* cultures remains controversial.

In many western cultures, though, it seems clear that stressful conflicts are often a part of adolescence. Some of these conflicts pit the teenager's movement toward adulthood against the limits imposed by society. Our society forces adolescents to go to school, limits the labor they can do, and governs them with laws that apply to minors but not to adults (see page 477). Society controls the age at which adolescents may vote, drink, drive, enlist in the military, and even enter into contracts. Often these age limits bear little relation to the biological or psychological development of the adolescents involved. In the following sections, we will examine both kinds of development and the social context in which they take place.

BIOLOGICAL DEVELOPMENT AND PUBERTY

In a physical sense, the events of puberty mark the transition from child

Figure 12.1

Margaret Mead as a young anthropologist is pictured at the left with two adolescents she studied for *Coming of Age in Samoa*. In that book, she portrayed Samoan adolescence as a period of orderly development, devoid of serious stress or crisis. Recently, Derek Freeman (right) criticized Mead's work and attacked her conclusions. Freeman has argued that Samoan adolescence is riddled with guilt, neuroses, and conflicts over sexuality. (*Left*, South Pacific Ethnographic Archives, Margaret Mead Papers Manuscript Division, Library of Congress, and The Institute for Intercultural Studies; *right*, Rob Little.)

Figure 12.2

Anyone who doubts that girls have an earlier growth spurt than boys need only visit a junior high school dance. (Donald Dietz, Stock, Boston.)

to adult. These events are triggered by a signal from the region of the brain known as the *hypothalamus* (see Chapter 2, page 57). The signal stimulates the *pituitary gland*, which sends extra growth hormone throughout the body. One result is the familiar growth spurt discussed below. Another result is a shift in the balance of sex-linked hormones in boys and girls. During childhood, boys and girls produce small, roughly equal amounts of *androgens* (male hormones) and *estrogens* (female hormones). At puberty, the pituitary, through its hormones, stirs the adrenal glands and testes of boys and the ovaries and adrenal glands of girls into action; sex-linked hormones are secreted into the bloodstream. Suddenly, boys have high levels of androgens, girls have high levels of estrogens, and dramatic sex differences in bodily development begin. (See Chapter 7, page 279, for a discussion of the organizational effects of sex-linked hormones on the body.)

The Growth Spurt Perhaps the earliest outward evidence that adolescence has started is the *growth spurt*, when girls' and boys' growth rates double. Boys in their peak year, most often their fourteenth year, grow from 3 to 5 inches in height. Girls in their peak year, usually their eleventh or twelfth year, grow an average of from 2 to 4 inches. Boy-girl differences in the timing of the growth spurt produce some interesting shifts in size and maturity of the two sexes. (See Figures 12.2 and 12.3.) By about the age of 14, though, boys have caught up and begun to move beyond girls in both body size and muscular strength.

Maturing Sexually A hallmark of adolescence is a cluster of key changes in the *primary sexual characteristics*. Most notably, reproductive organs gradually increase in size. For girls, these organs include the ovaries and uterus. For boys, these organs include the penis, prostate gland, and seminal vesicles. (See Figure 12.4.)

As boys begin their growth spurt, the penis begins a 2-year growth program of its own. Most boys can ejaculate after about 1 year of penis

;rowth; but another year will pass before sperm cells in the seminal fluid .re concentrated enough for real fertility. About 1½ years after the growth purt has peaked, girls experience *menarche,* the onset of menstruation. nitial menstrual periods are often irregular, and ovulation (the release of a nature egg) is unlikely for another year or so.

Along with changes in the primary sexual characteristics, boys and irls show changes in the *secondary sexual characteristics*—for example, ody proportion, hair quality and distribution, voice, and other physical eatures not directly related to reproduction.

ge at Puberty: Generational and Individual Differences In many ountries, the age at which boys and girls reach puberty has been dropping over the past several generations. In the mid-1800s, girls in the U.S. ached menarche at about age 17; since then, this average age has ropped by about 4 months per decade, and American girls now reach enarche at about 12½. Parallel trends have been found for puberty in oys, although in each generation boys mature about 2 years later than do rls. Similar trends have been found for the growth spurt in boys and rls; growth now spurts a few years earlier than it did 100 years ago. Why ese changes? Perhaps partly because nutrition and health care have nproved. Malnutrition, disease, and low body weight can delay physical evelopment, including sexual maturation. Key developments such as enarche seem to correlate with body weight and height. In fact, some searchers have proposed that when a girl reaches a critical weight of proximately 100 pounds, her body initiates the preprogrammed anges of adolescence (Frisch, 1974). Perhaps puberty comes earlier and rlier because children are growing faster, and healthier, with each neration.

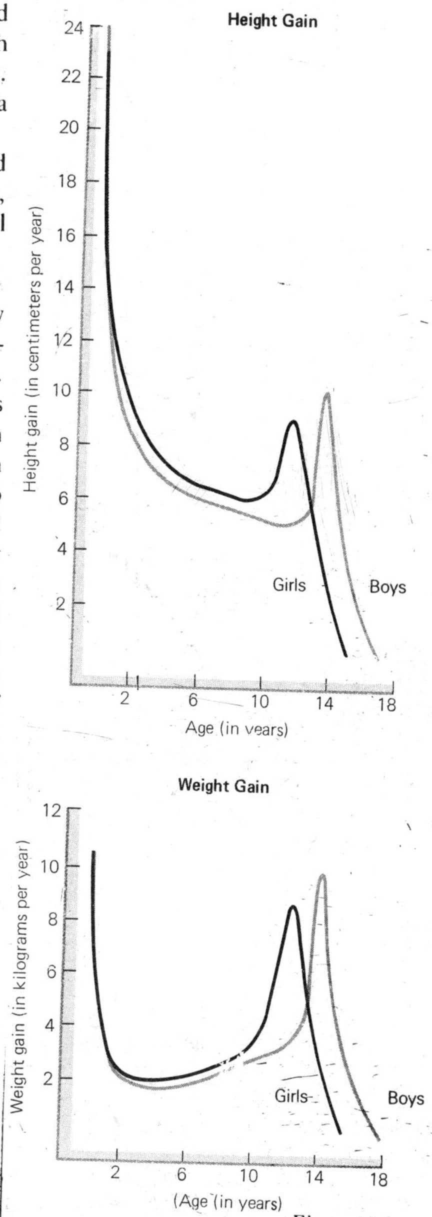

Figure 12.3

Annual gains in height and weight. Note that girls begin and end their growth spurt before boys do. (Modified slightly from Tanner et al., 1966.)

)GNITIVE DEVELOPMENT

ong with the bodily changes of adolescence come major intellectual anges. Remembering your own leap from child to teenager, you may all a new preoccupation with such cosmic intellectual issues as "the aning of life," "injustice in the world," and "What lies beyond our iverse?" You may have tried to solve ethical problems by relying, more d more, on abstract moral principles. You may also have grown much tter at solving complicated riddles or "posers," playing games like dge or chess by planning several moves in advance, and figuring out wers to purely hypothetical questions. Many of these features of olescent thought can be understood from the perspective of Jean get's theory of intellectual development (Chapter 11, page 425).

rmal Operations: Thinking Abstractly At around the age of 12, most ungsters begin the final major stage of cognitive development discussed Piaget: *formal operations.* In this stage, thinking becomes quite ıltlike; in fact, most adult capabilities are thought to be in place by ut the age of 16. This does not mean that we learn no new *facts* or *skills* r 16; it means that the basic *processes* we use to think do not change ch beyond this age.

A general feature of formal-operational thought is the ability to think terms of the abstract concepts that link concrete objects or actions

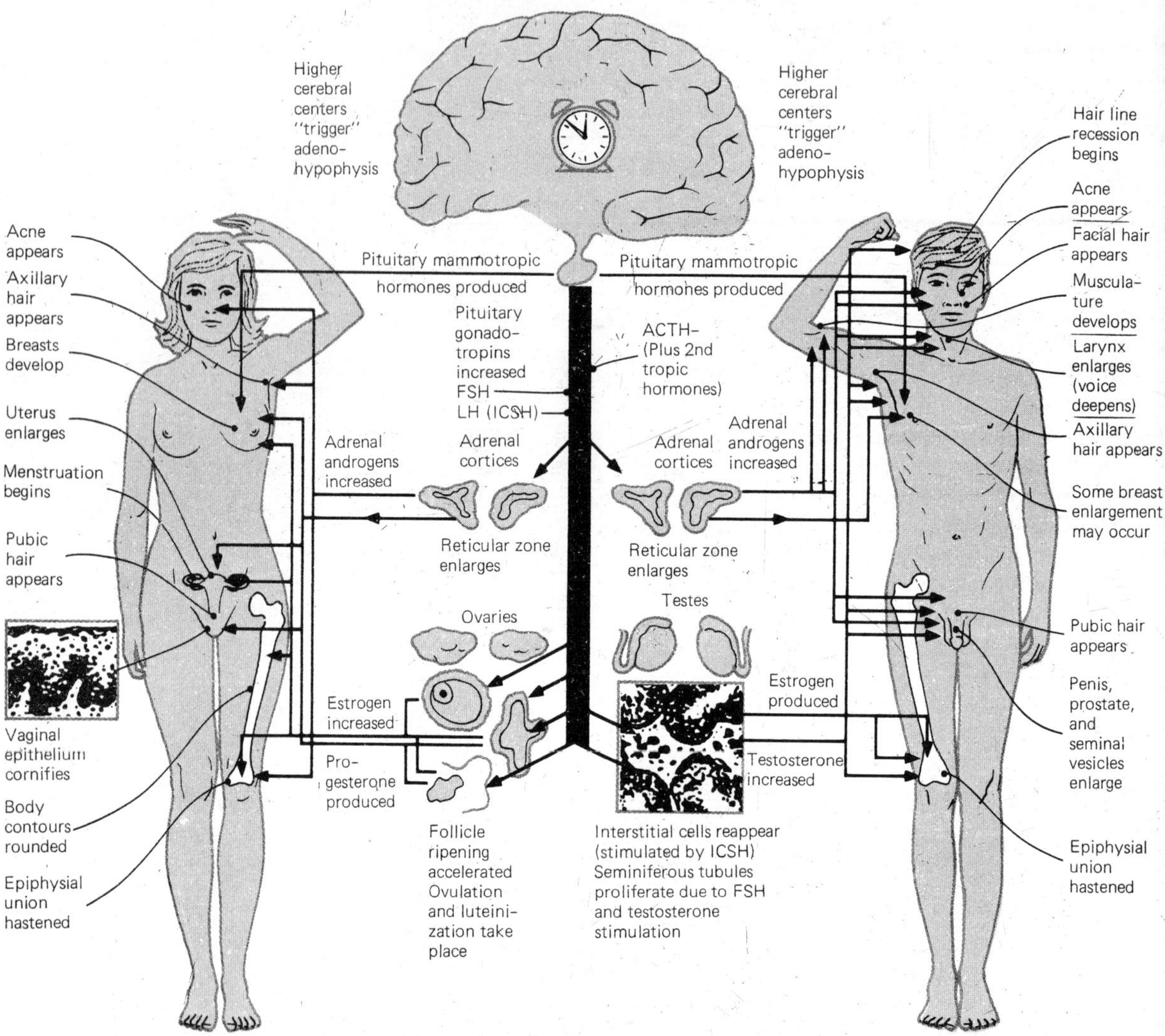

Figure 12.4

Pubertal changes in boys and girls. (Adapted from an original painting by Frank H. Netter, M.D., from the *CIBA Collection of Medical Illustrations*, copyright by CIBA Pharmaceutical Company, Division of CIBA-Geigy Corporation.)

together. For example, when asked about the purpose of laws, children tend to mention concrete examples such as "keeping people from speeding or stealing"; but adolescents can see broad, abstract purposes such as "keeping us safe and free," or "helping people live in harmony" (Adelson et al., 1969; Gallatin, 1980). Asked what they like about their mothers, children tended to mention concrete acts, such as "she fixes me chili"; adolescents, who see the abstract principles that underlie their mothers' behavior, gave answers such as "She sets a good example" or "She always cares" (Weisz, 1980). In addition to abstract thinking, several other interlocking capacities contribute to formal operations. We will discuss a few of these.

Hypothetical Thinking With formal operations, boys and girls move from the world of the actual to the world of the hypothetical. They can still think about the way things *are*, but they become much more skilled at thinking about how things *might be* if certain changes took place. Such thinking allows adolescents to judge the "reasonableness" of a purely hypothetical line of reasoning. To illustrate, suppose we ask a group of 9-year-olds and a group of 17-year-olds what they think about the following statement:

> I am glad I don't like onions, because if I liked them I would always be eating them, and I hate eating unpleasant things.

The 9-year-olds would probably focus on the concrete aspects of what was said. They would give answers like, "True, onions taste awful," or "Onions aren't so bad." The 17-year-olds would be much more likely to focus on the logic of the hypothetical situation, noting that "*If* I liked onions, *then* they wouldn't be unpleasant, would they?" (Hunt, 1961). The difference between the two groups is that adolescents are better able to think through a chain of purely hypothetical events. This ability is another part of formal-operational thinking.

Deduction and Induction Hypothetical and abstract thinking make sophisticated deduction and induction possible. *Deduction* is reasoning from abstract, general principles to specific hypotheses that follow from these principles. *Inductive thinking* is the complementary process of observing a number of specific events or instances and inferring an abstact, general principle to explain those instances. The two processes can be seen in the adolescent's reasoning about nature, science, and even social problems. For example, you may recall trying to figure out, from a number of specific observations, some general principles by which people can become attractive to the opposite sex (*inductive reasoning*); and you may have tried to use those general principles to generate specific hypotheses about how *you* should behave in order to attract *one particular* member of the opposite sex (*deductive reasoning*).

Interpropositional Logic As the onion example (above) illustrates, formal operations involve the ability to judge whether propositions are logically connected to one another, regardless of whether the propositions are true. This is called *interpropositional logic*. The concrete-operational child (Chapter 11, page 450) is able to test the factual truth of a single proposition, such as, "All college students are green." But the concrete-operational youngster has trouble recognizing that the following line of reasoning is *logically* correct:

> All college students are green.
>
> Sylvia is a college student.
>
> Therefore, Sylvia is green.

Prior to formal operations, children may dismiss this reasoning as false because they judge the factual accuracy of each individual proposition. The formal-operational person can recognize that *if* all college students are green, and *if* Sylvia is a college student, then, logically, Sylvia has to be green. This ability to use interpropositional logic really involves judging the formal relationships among propositions. This is one reason this stage of intellectual development is called *formal operations.* You needed interpropositional logic to solve the international kaleidoscope problem at the beginning of this chapter (pages 462–463). Did you figure out that the Iraqi student likes butter-pecan ice cream and that the American student plays squash? If so, chances are you used interpropositional logic—you focused on the logical relationships among all the propositions presented. In other words, you used an important skill involved in formal-operational thinking.

Reflective Thinking Another formal-operational skill that you probably used in the kaleidoscope problem is *reflective thinking,* the process of evaluating or testing your own reasoning. Reflective thinking allows the formal-operational person to be his or her own critic, to evaluate a process, idea, or solution from the perspective of an outsider and to find errors or weak spots in it. The reflective thinker can then sharpen plans, arguments, or points of view—making them more effective, more powerful. With the kaleidoscope problem, you may have developed tentative notions about who eats what, who plays what sport, and so on; tested these notions for errors; and corrected them when you found weaknesses in them. Such reflective thinking can make the formal-operational thinker a tough opponent in games of strategy, such as chess, or in debates on such social issues as the morality of abortion or the wisdom of a nuclear freeze.

Reflective thinking can also make the adolescent a powerful experimenter and problem solver. Why? Because it involves the ability to think through a number of possible strategies or "experiments" and to decide which one will yield the most information.

To illustrate, consider the light-bulb problem given at the beginning of this chapter. If you tackled this problem using reflective thinking, you would have generated several strategies—several sequences of moves—and then contemplated which strategy would yield the most information with the fewest number of moves. With a little reflective thinking, you would have settled on a strategy that involved testing half the potentially correct switches, say 1–4, all at once. If you tried these four switches and bulb E did not light up, then you would know that the "correct" switch has to be 5, 6, 7, or 8; thus your next move might be to turn on 5 and 6 at the same time. If bulb E lit up, then you would need only try 5 or 6 to know which was the correct switch. By testing half the potentially correct switches in each move, you would be assured of reaching a solution in three moves. A formal-operational thinker might well select such a strategy before tackling the problem.

MORAL JUDGMENT: DECIDING WHAT IS RIGHT AND WRONG

Practically every day we have to make judgments about "right" and "wrong." When we do, we are reasoning about moral issues. The kinds of moral reasoning done by most adolescents and adults is often quite different from the child's moral reasoning. In fact, some of Piaget's (1932) early work suggested that people pass through steps in the development of their moral reasoning much as they pass through stages of cognitive development. Building on Piaget's work, Lawrence Kohlberg (1976) and his associates (Colby et al., 1983) studied the development of moral reasoning by asking people of various ages to resolve moral dilemmas.

In one dilemma, a woman is dying of cancer, and a newly developed drug can save her life. The druggist who invented this drug is asking a very high price for it. Heinz, the husband of the cancer patient, cannot raise the money, and the druggist refuses to let him pay later. Should Heinz steal the drug to save his wife? In this dilemma, as in the others Kohlberg used, there are no clearly "right" or "wrong" answers. Kohlberg's main interest was not in what people said the characters in each dilemma should do; instead, he focused on the *reasons* people gave for their decisions.

Levels of Moral Development Moral reasoning, Kohlberg argues, passes through three different levels as people mature. At first, children reason at the *preconventional level.* They think in ways that fall short of the customary moral concerns of society. Their reasoning is somewhat egocentric; it focuses on the personal consequences of the individual's behavior. For example, in the cancer-drug dilemma, they may focus on avoiding punishment and say that the husband should not steal the drug because he might get caught and put in jail. Later, children enter the *conventional level.* Their reasoning fits what many societies consider to be acceptable moral rules. For example, they might say that Heinz should not steal because it is against the law. Still later, perhaps in adolescence, people may enter the *postconventional level,* in which they rely on abstract principles that go beyond commonplace views of ethics and morality. For example, they might say that Heinz should not steal the drug because if everyone took such actions, social order could break down ("You can't have everyone stealing just because they get desperate").

Kohlberg believes that everyone passes through the levels of moral reasoning in the same order, in part because each level is more logically advanced than is its predecessor. On the other hand, Kohlberg now thinks that most people fail to reach the postconventional level. The specifics of Kohlberg's theory have changed quite a lot in recent years (Colby et al., 1983; Kohlberg, 1976), but the general framework outlined above remains intact. Moral reasoning is still thought to pass through a predictable sequence of levels, beginning with egocentric moral judgments and moving toward abstract moral principles.

Like many theories in psychology, Kohlberg's theory has its fans and its critics. Some, for example, have noted that the theory rests almost entirely on interviews with males and may not apply so well to females (for example, Gilligan, 1982). However, most agree that as both males and females move from childhood into adolescence, they use less and less egocentric reasoning; and some come to rely almost entirely on abstract moral principles. We should recognize, however, that individuals using even very advanced moral reasoning may come to entirely different decisions about what course of action is "right." It is partly for this reason that moral issues such as abortion, sexual ethics, and the nuclear freeze can be so involving and so troubling to adolescents and adults. They may frame their judgments about these issues in terms of broad, abstract moral principles; but they also recognize that others who share these principles may reach very different conclusions about what is "right."

Are Moral Thought and Moral Action Linked? Because similar mora principles may lead to different judgments and behavior in differer people, the level of a person's moral thinking may not tell us much about what actions that person will choose. Does this mean that there is no real connection between how we think about a moral issue and what we will actually do? Some researchers have tried to answer this question. They have studied whether people's levels of moral reasoning can predict such behavior as cheating, campus activism, fairness in distributing valued goods, and even abortion (for example, Gilligan, 1977; Haan et al., 1968). Most of the research has found some relationships between moral reasoning and behavior, but the relationships have often been complex. For example, campus activism was found to be especially likely among college students who had reached Kohlberg's postconventional level, as was expected, but it was also likely among students who still reasoned at the preconventional level. The researchers suggested that the postconventional students may have been activists defending abstract principles and the rights of all whereas the preconventional students may have become activists for egocentric reasons—defending their own personal rights. Evidently, there are some connections between moral reasoning and moral behavior, but the precise connection may vary from one individual to the next and from one situation to the next.

In this discussion of moral development, we have focused on the ways people *reason* about ethical issues and on the connections between this reasoning and their actual behavior. Does something seem to be missing? What has been omitted is the concept of conscience, or what some would call superego. Because this concept has been so closely tied to major personality theories, we will discuss it in Chapter 14 (page 578), when we discuss personality development.

ACHIEVING IDENTITY: A KEY TASK OF ADOLESCENCE

Some theorists believe that the key developmental task for the adolescent is answering the question "Who am I?" In Erik Erikson's developmental theory, summarized in Table 12.1, the core conflict of adolescence is the tension between role confusion and identity. *Seeking identity* involves searching for continuity and sameness in oneself—trying to get a clear sense of what one's skills and personal attributes are (Figure 12.5), to discover where one is headed in life, and to believe that one can count on recognition from "significant others." The adolescent who forms a sense of identity gains two key benefits, according to Erikson: (1) "a feeling of being at home in one's body" and (2) "a sense of psychological well-being" (1968, p. 165).

Adolescents who fail to achieve a sense of identity may face confusion over what roles they can or should be playing in life. They may delay any commitment to adult roles, a delay which Erikson calls a *psychosocial moratorium*. Erikson himself went through such a moratorium. After finishing secondary school, he spent several years wandering around Europe, avoiding any firm decisions about what sort of career he might pursue. His experience led him to see the psychological moratorium as both promising and risky. It can be a valuable period of information gathering or it can involve rebellion—an attempt to do precisely the opposite of what parents and others think is proper and desirable. Erikson calls this rebellious pattern the pursuit of *negative identity*.

In searching for an identity, the adolescent also tackles the question, "What do I really believe in?" With the development of formal operations, the adolescent can appreciate and cultivate abstract values and

Table 12.1

Erikson's psychosocial development stages

BASIC CONFLICT	OPTIMUM OUTCOME
1. Basic trust vs. basic mistrust (infant)	Trust is the faith that things will be "all right." It develops from good care provided by reliable others. A favorable ratio of trust to mistrust results in hope.
2. Autonomy vs. shame and doubt (toddler)	Without a sense of self-control (autonomy), children feel shame and doubt. A favorable ratio of autonomy to shame-and-doubt results in self-direction with self-esteem.
3. Initiative vs. guilt (preschooler)	Initiative adds to autonomy the quality of doing things just to be doing them. A sense of guilt is often experienced over things contemplated or actually done. A favorable ratio of initiative to guilt results in a sense of purpose.
4. Industry vs. inferiority (schoolchild)	Grade-school children learn to win approval by making things and doing things approved of in the culture. In literate societies, they learn to read; in preliterate societies, they learn the skills necessary for survival. Failure to produce or do valued things leads to a sense of inferiority. A favorable ratio of industry to inferiority leads to a sense of competence and pleasure in work.
5. Identity vs. role confusion (adolescent)	Identity refers to the "Who am I?" and "What am I going to do with my life?" questions of adolescence. Difficulty in answering such questions leads to role confusion. A favorable ratio of identity to role confusion leads to a sense of consistency.
6. Intimacy vs. isolation (young adult)	Here the task is to establish lasting and loving relationships with other people. Love is the outcome of a favorable ratio of intimacy to isolation.
7. Generativity vs. stagnation (middle adult)	Generativity includes productivity and creativity, but here it refers primarily to preparing the next generation for life in the culture. Care is the outcome of a favorable generativity-to-stagnation ratio.
8. Ego integrity vs. despair (older person)	Ego integrity has many facets. In part, it refers to one's acceptance of one's life as what it had to be. Despair, on the other hand, includes the feelings that life is too short to do much and that integrity cannot be achieved. A favorable ratio of ego integrity to despair brings wisdom and the ability to face death calmly.

Figure 12.5

Left, adolescence, according to Erikson, is a time of searching for one's identity. (George W. Gardner.) *Right*, sometimes this search can lead to a rebellious pattern that Erikson calls "negative identity." (Peter Marlow/Magnum Photos, Inc.)

principles. Abstract thinking makes it possible to love freedom and hate greed. By blending abstract ideals with the information drawn from admired models in real life, the adolescent can generate a broad array of possible roles. In the best of cases, this array gets thinned out so that by the end of adolescence, a satisfying self-definition has begun to take shape.

SOCIAL DEVELOPMENT

Because it marks the transition from childhood to adulthood, adolescence requires the redefining of some basic social relationships. Relationships with family members at this time involve increasing independence for the adolescent and usually involve increased conflict, too. Relationships with peers may become much more intimate and vital than they were in childhood. Finally, relationships with the opposite sex have new overtones of sensuality. All these shifts combine to make the adolescent's social world complex and electric.

Family Relationships In relationships between adolescents and their parents, a central theme is often that of testing limits. Most teen-parent arguments concern the timing of rights and responsibilities (Hartup, 1983). Teenagers see themselves becoming adults, and they press for the freedom and privileges of true adulthood. Also, the emergence of formal operations and more advanced moral reasoning means that the adolescent can think of reasonable alternatives to parental rules. As these alternatives occur to the teenager, parents can expect less automatic obedience and more resistance. These are signs of the adolescent's cognitive growth and should be respected as such. Parents who respond to disagreements with open discussion are encouraging their children to do the same and thus to practice a valuable social skill. Some studies suggest that self-reliance, independence, and social responsibility are most effectively promoted by parents who are flexible and encourage discussion—*if* the parents also hold their youngsters to clear, consistent standards (Baumrind, 1980). This parental style, unlike extreme parental dominance or extreme permissiveness, gives the adolescent some decision-making experience while insuring a degree of guidance and control.

In most families, adolescent girls have to struggle much harder for their independence than do adolescent boys. Teenage girls report more conflicts with their parents than do teenage boys, and the conflicts they report more often involve emotional flareups. Parents seem to place more restrictions on their teenage daughters than on their teenage sons; they worry more about their daughters' safety and especially about their sexual activity and the risk of pregnancy. Their daughters, unfortunately, do not see these restrictions as "protection." One of the most common explanations teenage girls give for their conflicts with their parents is that their parents do not respect their maturity (Konopka, 1976).

Boys' conflicts with their parents tend to involve more objective issues of authority and privilege, such as access to the family car. Boys are more likely than are girls to report that they are disciplined primarily by their fathers and that they receive affection primarily from their mothers. The boy's relationship with his mother and father appears to shift significantly around the time of puberty. Research based on videotaped observations (Steinberg, 1979) documented some of these shifts. Just prior to puberty, mothers and fathers seem to have about the same influence over their sons' behavior. With the advent of puberty, boys act more assertively toward their mothers; some tension and coolness persist between mother and son into mid-adolescence, with mother and son reluctant to defer to each other. As the mother-son tension persists, the father's role seems to shift to that of mother's ally. Fathers show increased efforts to restrain their sons' assertiveness, but with only partial success. With puberty completed, boys seem clearly more influential in family decision making than they were prior to puberty.

Although parent-child relationships change during adolescence, sometimes dramatically, we must not assume that they are uniformly poor or unsatisfying. In a 1977 Gallup poll, adolescents were asked, "How well would you say you get on with your parents. . . ?" Some 56 percent said "very well" and another 41 percent said "fairly well"; only 2 percent said "not at all well." Recent surveys of personal values, political views, moral development, and occupational choice indicate that high school students are likely to be influenced more in each of these areas by their parents than by their peers (for example, Feather, 1980). For most adolescents, though, the influence of family clearly interacts and competes with the strong influence of peers.

Peer Relationships Peer-group membership assumes more importance during adolescence than at any other period of life. A peer group can provide a refuge and a source of support for youngsters in conflict with their families. Moreover, being part of a clearly identified group can help answer the burning question, "Who am I?" For these and other reasons, adolescents spend a great deal of time with other adolescents (Figure 12.6). One study, which used beepers to contact adolescents at random times (Larson et al., 1977), found that they spend more time talking with peers than doing academic work, being with their families, or being alone.

The structure of peer groups seems to change over the course of adolescence (Coleman, 1980; Dunphy, 1963). For teens around the age of 13 or so, the peer group is usually a clique consisting of half a dozen or fewer youngsters of the same sex. These unisex cliques then begin to

Figure 12.6

Peer relationships take on a special importance in adolescence. (Paul Conklin/Monkmeyer Press Photo Service.)

interact with cliques that include members of the opposite sex. There is cross-clique kidding and conversation, but only from the safety of the unisex home base. During the high school years, a number of adolescents belong to both unisex and mixed-sex groups; for example, three or four boys who are members of the same unisex clique may regularly interact with three or four girls from an all-girl clique, thus forming a mixed-sex clique. Gradually, two, three, or four cliques merge into larger groups. By late adolescence, group unity has begun to weaken, and loosely linked groups of couples have begun to form (see Figure 12.7). One advantage of this overall sequence is obvious: Boys and girls use the initial security of their unisex cliques to move—gingerly, nervously, step-by-step—into contact with members of the opposite sex.

Another structural feature of the adolescent peer group that changes during adolescence is the dominance hierarchy (Hartup, 1983; Savin-Williams, 1980). In middle to later childhood, youngsters who are skillful in directing play and games emerge as leaders. With the transition to early adolescence, the dominant youngsters are those with athletic and social skills *and* those who show the physical signs of puberty. By later adolescence, though, such physical factors are no longer so important; the leaders tend to be those who are bright, creative, and well-liked.

Heterosexual Relationships and Adolescent Sexuality On the average,

Figure 12.7

The structure of peer groups is said to pass through five different stages during the adolescent years. (Dunphy, 1963.)

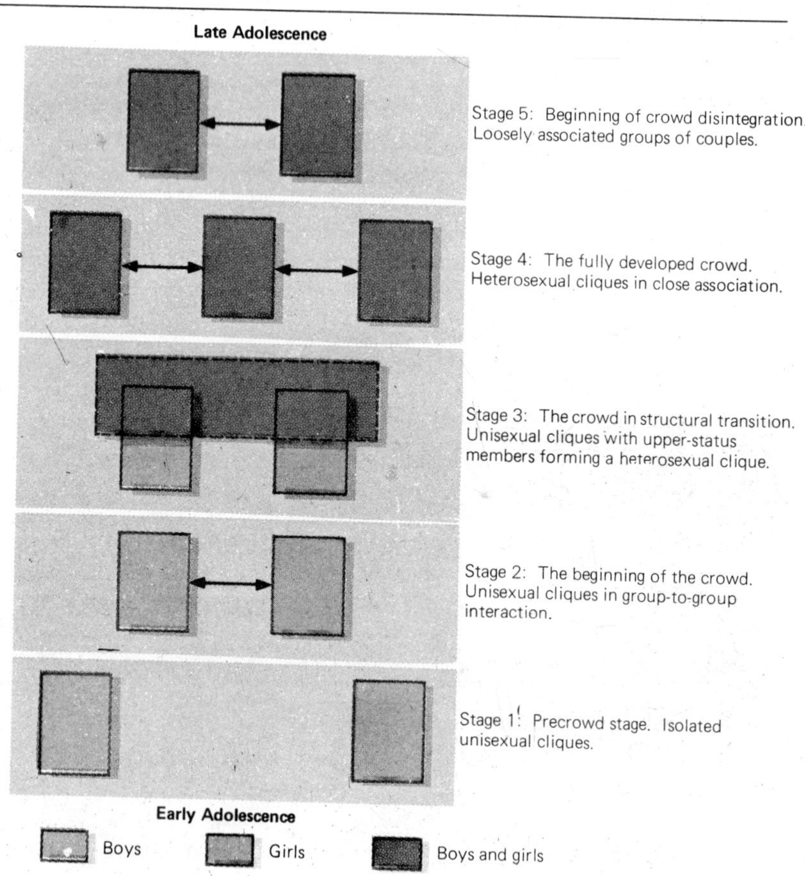

American girls begin dating at about age 14, and American boys begin at about 15. Of course, many begin earlier; and others are not dating even as high school seniors. Evidence suggests that these nondaters, at least the girls, may be more insecure and dependent on their families than those who do date in high school (Douvan & Adelson, 1966). Like peer-group relationships, dating helps the young person find a sense of identity—knowing what kind of person you can attract helps you know what kind of person you are. Dating is also a proving ground for sexual values and behavior, as shown in Figure 12.8. The figure is from a 1981 report on the percentage of adolescent boys and girls who have had sexual intercourse. The percentages are low at age 13, but they rise sharply in each subsequent year. By age 19, more than two-thirds of girls and more than three-fourths of boys have had sex.

This represents a major change. Even as recently as 1973, only 45 percent of girls and 59 percent of boys had had intercourse before the age of 20 (Sorensen,1973); in 1953, the figures were 20 percent for girls and 45 percent for boys (Kinsey et al., 1953). Clearly, sexual attitudes and behavior have changed. For many of today's adolescents movement toward sexual maturity is fast-paced and risky, as we shall see in the following section.

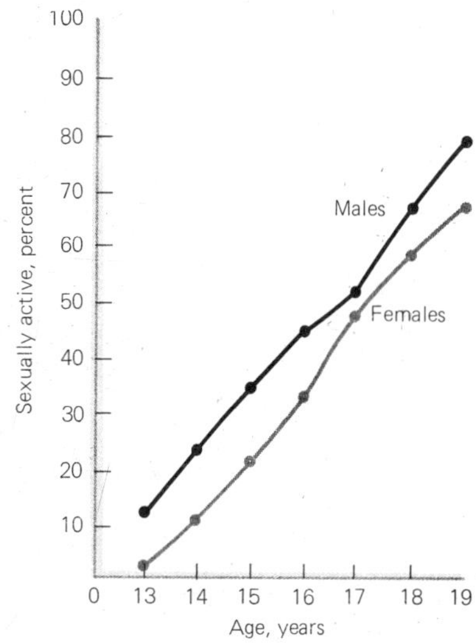

Figure 12.8

By age 17, about half of all boys and girls in the United States are sexually active. (Alan Guttmacher Institute, 1981.)

ADJUSTMENT PROBLEMS IN ADOLESCENCE

When we look back on our teenage years and recall the uncertainties, the conflicts, and sometimes the loneliness, it is easy to recognize that adolescence is a time of real vulnerability. Nowhere is this truer than in the area of adolescent sexuality. Although sexual activity has increased dramatically over the past decade or two, sex education has not. As of 1980, most United States high schools did not offer sex-education courses. Adolescent girls (not to mention boys) are sadly unaware of such crucial basics as how to figure out the high-risk period for conception. A majority of teenage girls now use contraceptive devices when they have intercourse, but a majority of them did *not* at the time they *first* had intercourse. Girls who delay the use of contraceptives beyond their early sexual experiences are about three times as likely to get pregnant as girls who use protection from the beginning. The bottom line is a million-plus teenage pregnancies a year (Figure 12.9). About 400,000 of these end in abortion. In most of the remaining cases, the result is a new teenage mother, and over 90 percent of these mothers choose to keep their babies. This choice often sets in motion a cycle of educational and economic disadvantages, an increased likelihood of child abuse, and an increased risk of psychological problems in both the mother and the child (Clarke-Stewart & Koch, 1983).

Figure 12.9

The newfound sexuality of adolescence poses significant risks, some as old as humanity itself. (Terry Evans/Magnum Photos, Inc.)

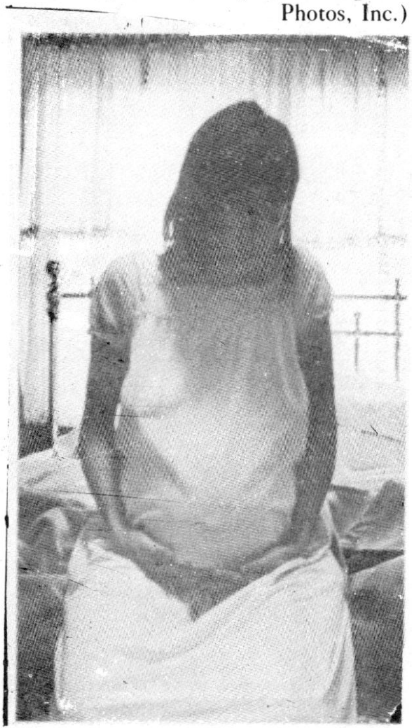

Sexually active adolescents who manage to avoid pregnancy still face risks of venereal disease, risks that have increased sharply over recent years. In addition to the diseases that are treatable with antibiotics, there is a new strain of gonorrhea that is not. There is also herpes simplex virus, type 2, otherwise known as genital herpes; it is thus far incurable and is thought to infect as many as 20 million Americans. Adolescents who have multiple sexual partners are especially at risk; about three-fourths of those who contract venereal diseases are between 15 and 24 years of age.

Illegal behavior is defined broadly for adolescents. It includes both *status offenses*, acts that are illegal only for minors (for example, truancy, running away from home, and drinking), and more serious offenses that would be illegal at any age (for example, assault, vandalism, and illicit drug use). About 80 percent of American adolescents admit to having committed at least one illegal act. Especially common is drug use, as Figure 12.10 shows. In addition, about 20 percent of violent crimes are committed by people under 18. Contrary to popular opinion, seriously delinquent adolescents do not come disproportionately from lower-income families or from minority-group families (Gold & Petronio, 1980). However, the delinquent youths who do come from such families are more likely to go on to careers of crime. Many argue that this is because middle-class delinquents are more likely to be saved by their family's resources—money for an attorney, influence with the police, and so forth.

The beginning of adolescence can mean facing up to some very adult psychological problems. One of these is *depression*, which is characterized by feelings of guilt, a loss of interest in activities, sleep problems, and even suicidal thoughts (Chapter 15, page 647). Successful suicides in the U.S. increase sharply over the adolescent years; there are 170 per year among 10- to 14-year-olds but about 1,600 per year among 15- to 19-year-olds (Holinger, 1979). Another life-threatening disorder that surfaces in adolescence is *anorexia nervosa*, a form of self-starvation discussed in Inquiry 12.1. (Also see Inquiry 7.1, page 275, for a discussion of the other side of eating difficulties—overeating and obesity.) Still other adolescents have an escalating sense of confusion about things around them; they feel that "things are not real" or that they are actually outside of themselves. Distortions in thinking may develop into irrational belief systems *(delu-*

Figure 12.10

Percentage of adolescents graduating from high school in 1975–1980 who had used various drugs without a doctor's order. (Johnston et al., 1980.)

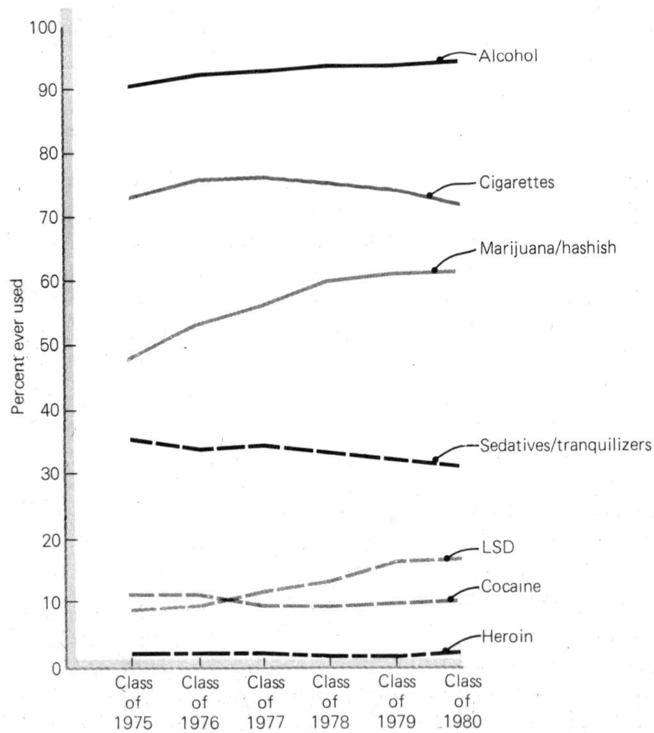

sions) or into perceptual experiences that seem to be, but are not, real *(hallucinations)*. These and other problems can combine to form schizophrenia, a disorder discussed in detail in Chapter 15 (page 637).

Youth

The part of the life cycle known as adolescence did not become recognized as a "life phase" until the nineteenth century. During that century, a noticeable gap developed between the onset of puberty and the beginning of life as a working adult; the years that filled that gap came to be called adolescence.

The number of years filling that gap increased even more during the twentieth century—so much so, in fact, that Keniston (1970) proposed the recognition of yet another "life phase": youth. *Youth* is essentially a period of "studenthood"; it exists only for those who move on to post-secondary education before settling into full-time work (Figure 12.11). Whatever the length of the period, the time it affords can be valuable. It can serve as a kind of lull, a time for serious experimentation without the need for a long-term commitment to a single course of adult life. Because the peer group is no longer such a dominant influence, the individual has a new freedom to develop individually—to shape a personal perspective on life and a sense of direction before tackling the challenges of true adulthood.

Early and Middle Adulthood

The longest part of the life cycle—the period from about the early twenties to the early sixties—is actually the part that developmental psychologists have studied least. Why? Perhaps because the dramatic transformations of infancy, childhood, and adolescence distracted them from the often subtle changes of adulthood. Perhaps also because many of

Figure 12.11

Whatever the length of the period known as youth, the time it affords can be a valuable asset—a time for thought and experimentation with no need for long-term commitment to a single life course. (David S. Strickler/Monkmeyer Press Photo Service.)

Inquiry 12.1

WHY DO THEY STARVE THEMSELVES?

For Alma, puberty came at age 12. After menarche, she continued to be healthy and well developed. By age 15, she was 5 feet tall and weighed 120 pounds. At this time, her mother urged her to switch to a school that was more demanding academically. Alma resisted. At about the same time, Alma's father suggested that she watch her weight; she complied immediately, and with a vengeance. Under a rigid diet combined with a strenuous program of swimming, tennis, and calisthenics, Alma lost weight rapidly, dropping eventually to less than 70 pounds. She looked "like a walking skeleton," her mother complained. "When I put my arms around her, I feel nothing but bones, like a frightened little bird." With the extreme weight loss, Alma had stopped menstruating, and her face took on the hollow look "of a shriveled-up old woman with a wasting disease." Nonetheless, Alma insisted that she looked just fine, and she stressed, "I enjoy having this disease, and I want it." (Adapted from Bruch, 1978, pp. 1–2)

Alma has *anorexia nervosa*, an eating disorder most likely to be found among adolescent girls. Anorexia means loss of appetite, but girls suffering from this disorder do not so much lose their appetite as fight it off. In fact, many are preoccupied with food—hoarding it, reading about it, learning to cook, and talking incessantly with friends about it—but they do not eat it, at least not very much of it. In addition, they may exercise to excess and resort to self-induced vomiting and laxative abuse. As a result, they lose weight dramatically—a loss of 25 percent of original body weight is the usual criterion for a diagnosis of anorexia. In addition to severe weight loss, a near-universal symptom of anorexia is amenorrhea, cessation of menstruation. Estimated death rates from anorexia range from 3 percent to as much as 25 percent.

Why does anorexia occur? To answer this question, many have taken as a clue the fact that the disorder usually begins in adolescence and occurs primarily among girls (less than 15 percent of the cases are male). One widely held view is that anorexia begins in response to anxiety over one's emerging sexuality. Self-starvation suppresses most of the physical changes of puberty and may seem to provide a measure of "protection" against it. This suppression also keeps girls looking small, fragile, and childlike; and some have argued that anorexia often results from a fear of maturity in general, or from a desire to remain "Mom and Dad's little girl." Yet another view is that anorexia can begin with the normal dieting and the preoccupation with slimness that is almost universal among teenage girls, particularly in western cultures. Some argue that the dieting gets out of hand in some youngsters because of their desire for perfection and high achievement; many of these girls have been very orderly and eager to do things "exactly right" since childhood, and they may view dieting as another achievement task demanding their best effort. Another view is that with adolescence, a conflict over power and control emerges between some adolescents and their parents and that this conflict sometimes becomes focused on eating habits; self-starvation may give the adolescent a form of control that parents cannot take away. It seems likely that there are many different paths to anorexia and that each of the explanations given above applies to some cases.

How can anorexia be treated? One of the first steps is often medical attention, together with firm rules about weight maintenance. Obviously, there must be consequences for rule violation; these consequences can take the form of hospital admission or even forced or intravenous feeding. Psychological approaches to treatment have included individual psychotherapy, family therapy, and instrumental conditioning methods (see Chapter 16.) Of the family therapy techniques, the most successful seem to be those that treat the eating disorder as just one part of a larger family problem that needs attention. Reasonably good success, at least in the short run, has been reported for treatment with instrumental conditioning techniques—for example, making special privileges depend upon small, regular increases in weight. Despite some modest success, most who have worked with these youngsters view the disorder as very difficult to treat and very likely to recur. Anorexia nervosa is clearly one of the most perplexing and frightening problems that parents of an adolescent can face.

REFERENCES

Bemis, K. (1978). Current approaches to the etiology

and treatment of anorexia nervosa. *Psychological Bulletin, 85,* 593–617.
Bruch, H. (1978). *The golden cage: The enigma of anorexia nervosa.* Cambridge, MA: Harvard University Press.
Kellerman, J. (1977). Anorexia nervosa: The efficacy of behavior therapy. *Journal of Behavior Therapy and Experimental Psychiatry, 8,* 387–390.
MacLeod, S. (1981). *The art of starvation.* London. Virago.

them saw adulthood as a time of stability, not a period of development. Recently, this viewpoint has been challenged, particularly by the *life-span developmental psychologists*. These researchers are committed to the view that development occurs at *every* point along the life span, including adulthood. Consequently, they have made the full life span—from conception to death—a focus of study. In this section, we will examine what these psychologists and other researchers have found out about adulthood. Our coverage will be brief, in part because so many aspects of adult behavior are discussed in our other chapters. But we emphasize that research on adult development is now a major force in psychology and that it is already reshaping our view of adulthood.

BIOLOGICAL CHANGE IN ADULTHOOD

During their twenties, most people reach their peak of strength, agility, reaction time, and manual dexterity. All four attributes decline gradually over the next decades, but most people are reasonably healthy and physically sound into their fifties and sixties. Muscular strength, for instance, peaks between the ages of 25 and 30, but there is only about a 10 to 15 percent loss of strength by age 60. Aging also revamps our physical appearance, as visitors to any 10- or 20-year class reunion can testify. Weight is redistributed, hairlines may recede, hair grays, skin texture changes as drying and wrinkling begin, and often the structure of the face becomes modified. In women, one of the most dramatic physical changes is *menopause,* the cessation of menstruation. This usually occurs between the ages of 45 and 55; it signals the *climacteric*—the end of ovulation and the termination of reproductive capacity. There is no parallel event for men; men can produce viable sperm at all ages, but their reproductive capacity does decline gradually over their adult years.

In thinking about biological changes in adulthood, it is important to keep an important fact in mind: Not all of the changes have a direct impact on behavior. For example, some research has shown that it is only among very unhealthy adults that heart and lung functioning are highly correlated with performance at work or activities at home. Generally, the physiological changes that accompany early and middle adulthood seem not to have major effects on work or other behavior, except where physical performance expectations are very high, as in professional athletics.

COGNITIVE DEVELOPMENT IN ADULTHOOD: STILL GROWING AFTER ALL THESE YEARS?

For years, a traditional viewpoint was that cognitive development streaked upward from childhood into the college years, where it peaked; in the adult years, according to this view, the intellect held its own initially, then began slipping downhill as adults did their best to hold on to what intellectual powers they could. Recent findings raise questions about this view. These findings come mostly from two groups of researchers. One group, working within the Piagetian tradition (Chapter 11, page 425), has focused on the structure of thinking, including cognitive stages; the second group, the life-span developmental psychologists we mentioned earlier, has focused on the development of specific cognitive processes, such as memory and learning. In this section, we will consider the work of both groups.

Piagetian Perspectives on Cognitive Development: Formal Operations and Beyond The Piagetian group has tackled two intriguing questions.

1. How effective are adults, as they mature, in using formal operations?
2. Could there be a cognitive stage that goes beyond formal operations?

Evidence on the first question has been somewhat mixed. For example, one study found that most adult women in a supermarket are unable to solve formal-operational reasoning problems. This is true even when the content of the problems is relevant to shopping—for example, figuring out which of two deodorant offers is the better buy. On the other hand, some research suggests that when older and younger people are matched carefully for educational level, there is little difference in their use of formal operations (Blackburn, 1984). Research on this issue is continuing.

The question of whether cognitive structure changes after formal operations has stirred real excitement among developmental psychologists. Piaget (1970) did not identify any stage levels beyond formal operations, but he did note that reasoning may operate differently in adults than in adolescents. Adolescents often have the luxury of using hypothetical reasoning "playfully," to solve hypothetical "What if . . ." problems—the essence of pure formal operations. Adults, on the other hand, often have to fit their reasoning into the dimensions of real life. As a result, they may often think in ways that are less abstract and less purely logical. That is, they may combine abstract thinking with realistic thinking about the way life actually works. In choosing between two political candidates, for example, an adult may evaluate not only their ideals and the logic of their positions but also their abilities to wheel and deal, push legislation through committees, and so forth. Some researchers (for example, Labouvie-Vief, 1982) have suggested that the adult's integration of abstract logical principles with realism represents a real advance that goes beyond formal operations. Others have begun to study this kind of integration, focusing, for example, on how adults think when they serve on juries (Kuhn et al., 1983).

Life-Span Perspectives on Cognitive Development: Aging versus Cohort Effects A second general approach to adult cognitive development has been taken by the life-span developmental psychologists. One aim of this group has been to learn what kinds of age changes adults show in intelligence test scores (see Chapter 13) and in specific cognitive processes such as memory and learning. In this task, they have used research methods that combine the cross-sectional and longitudinal approaches we discussed in Chapter 11 (pages 416–418). One example is shown in Figure

12.12. The top half of this figure gives a *cross-sectional* picture of the changes in intelligence test scores with age. People of different ages were tested in 1956, 1963, and 1970. On each testing, the results obtained using this cross-sectional approach *seem* to show a sharp decline in intelligence test scores with advancing age. But take a look at the bottom half of this figure, which shows a *longitudinal* picture of changes in intelligence test scores from the first testing in 1956 to the last one in 1970. In this longitudinal approach, cohorts of subjects were studied. A *cohort* is a group of people who were born about the same time. The lower half of Figure 12.12 shows the results for seven cohorts—for example, those people who were 25 at the first testing in 1956, those who were 32 at the first testing, and so on up to those who were 67 in 1956. Each cohort was then tested again in 1963 and 1970. Results with this longitudinal approach show that *within cohorts* tested intelligence actually changes very little until about age 53. In fact, some of the scores for the younger cohorts actually increased from one testing to the next. On the other hand, the three cohorts tested at age 53 and older—the cohorts shown in the lower-right portion of the figure—declined markedly at their second and third testings.

Research of this sort by the life-span development group points to an important general truth about intellectual development in adulthood: Adults in different generations may differ in their levels of intellectual performance, but a given individual will probably change very little throughout early and middle adulthood (Kuhn et al., 1983; Schaie, 1983). We may outscore our parents, and our children may outscore us; but if so, the differences would only tell us how the generations, or cohorts, differ from one another. If we follow the same individual year after year, we are not likely to see real, significant declines in intelligence during early or middle adulthood.

Figure 12.12

Cross-sectional and within-cohort age differences in a composite measure of intellectual ability. (Modified from Schaie et al., 1973.)

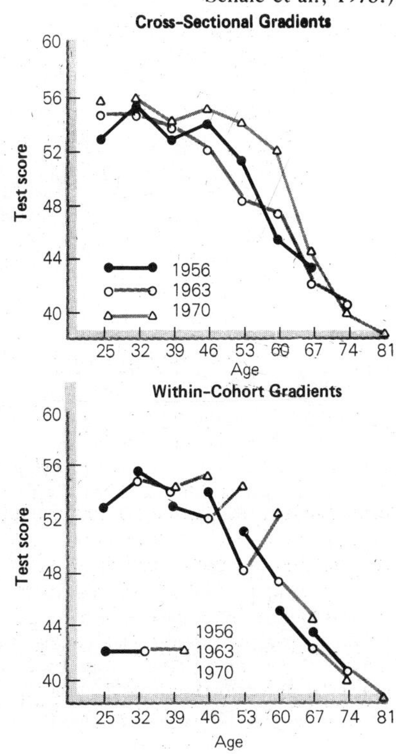

Multiple Types of Intelligence Investigators of adult development have emphasized that when age changes *do* occur, they may point in different directions for different types of intelligence. One example of this viewpoint is Horn and Donaldson's (1980) developmental model of intelligence; a simplified illustration of their model is shown in Figure 12.13 (top). Horn and Donaldson propose that fluid intelligence may well decline over the period from young adulthood to old age but crystallized intelligence actually increases well into old age. By *fluid intelligence,* they mean flexible thought that is independent of culturally based content—the kind of thought required to solve item A in Figure 12.13. By *crystallized intelligence,* they mean the degree to which an individual has mastered the knowledge and skills of the culture; this kind of mastery is assessed by problems like item B in Figure 12.13. Thus far, this fluid-crystallized model of age change is not much more than a hypothesis, supported only by a few cross-sectional studies. Nonetheless, it is an intriguing notion, one that should be tested in longitudinal research.

Some investigators (for example, Schaie, 1983) distinguish between types of intellectual activity that require speed of performance and those that do not. Real age declines, when they are found, seem most likely to be found on timed tasks—for example, naming all the words you can think of in 1 minute that begin with the letter "G." Even on timed tasks, age declines are usually confined to people over 50 and those with health problems (Schaie, 1983).

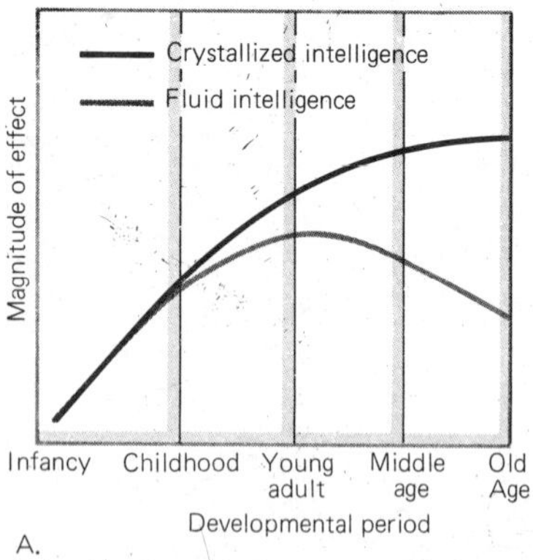

A.

Below is a geometric figure. Beneath the figure are several problems. Each problem consists of a row of five shaded pieces. Your task is to decide which of the five shaded pieces will make the complete figure when put together. Any number of shaded pieces from two to five may be used to make the complete figure. Each piece may be turned around to any position but it cannot be turned over.

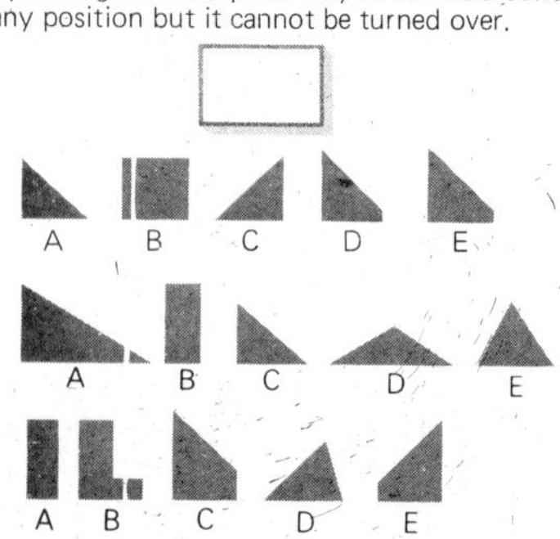

B.

Choose one of the four words in the right-hand box which has the same meaning as the word in the left-hand box.

1.	bizarre	market odd	conventional imaginative
2.	pecuniary	involving money trifling	esthetic unusual
3.	germane	microbe relevant	contagious different

Figure 12.13

Top, Horn and Donaldson's life-span developmental model of fluid and crystallized intelligence. (Modified from Horn & Donaldson, 1980.) *Below*, item A shows an example of how fluid intelligence might be assessed. (From *Form Board Test, VZ-1*; Educational Testing Service, 1963, 1976.) Item B shows a sample problem designed to test crystallized intelligence. (From *Vocabulary Test, V-5*; Educational Testing Service, 1963, 1976.) (Items A and B reprinted by permission of Educational Testing Service, the copyright owner.)

Creativity and Productivity What about changes during adulthood in *creativity*—the use of one's intellect to produce novel, meaningful ideas? As you might expect, developmental changes in this area differ with the type of creativity involved. College students specializing in such fields as mathematics or physics are on the verge of their most creative period; major innovations in these fields most often come from people in their twenties. Einstein published his theory of relativity when he was 26; Newton developed the calculus at age 28; and Galileo was only 26 when he proposed the theory of falling bodies. Some researchers have theorized that the common denominator in these achievements is the appearance of a burst of fresh insight—a bright idea by people whose thinking is not weighted down with the givens of established knowledge. (See Chapter 6, page 244.) Creative contributions in other areas—such as philosophy, history, and literature—seem to require a thoughtful synthesis of accumulated knowledge. It is perhaps for this reason that major achievements in these fields are usually made by people 40 and older; in fact, creative contributions of this type seem to grow increasingly likely from age 40 through the sixties and often well into the seventies (see page 494).

DEVELOPMENTAL TASKS OF ADULTHOOD

The developmental tasks of adulthood are many and varied. Here we will consider a few of the most important. First, though, we will set the stage by reviewing two theoretical perspectives on the entry into the young-adult and middle-adult years.

Entering Young Adulthood Erikson (1963; also see Table 12.1, page 473) depicted young adulthood as a time of tension between isolation and intimacy. If the tension is handled well, mature love can result. Erikson stressed that commitments to others can ultimately help society flourish. Traditionally, this is accomplished through intimate heterosexual relationships that are socially sanctioned through marriage and expanded with the birth of children. Alternative patterns are much in evidence now; but even today there can be little doubt that forming close, stable, and unselfish relationships is an important task for the young adult. To the extent that partners in such relationships find satisfaction in caring and mutual sharing, the intimate relationship can mature and endure. When one partner, or both, cannot sustain an unselfish relationship of mutual sharing—or when an individual cannot experience such a relationship—the result can be a sense of isolation, "aloneness."

Daniel Levinson (1978) interviewed 40 men (no women) over a 2-year period and developed a model of the major tasks at each phase, or "season," of adulthood—a model that is something of a complement to Erikson's perspective. Levinson focused on four tasks for young adulthood. Perhaps the most overarching is forming a "dream"—that is, a sense of how one will fit into the adult world. In addition, Levinson argued, the young adult needs to establish an intimate relationship and enter the realm of parenthood. A third task involves selecting from among one's interests, picking out an occupation, and acquiring the skills and credentials that will foster success in that occupation. The fourth task of young adulthood is related to the others: The developing adult needs to find a mentor. The mentor is usually older and more experienced than the young adult and often (but not always) a colleague in the work place. The mentor may serve as teacher, sponsor, or counselor, facilitating the young adult's pursuit of the "dream" mentioned above.

The Transition to Middle Adulthood According to Erikson (1963; also see Table 12.1, page 473), adults in their middle years confront a core conflict between generativity and stagnation. The generative adult is productive, creative, and involved in preparing the next generation for life within the culture. The adult who fails to generate in these ways may show a stunting of personal growth, a sort of psychological shriveling that Erikson called "stagnation."

Levinson (1978) described the *midlife transition* as a period stretching from the late thirties to the early forties and involving two fundamental tasks: (1) a reappraisal of one's life as it is being lived and (2) a decision about whether to "stay the course" or shift to new patterns. At this point, some may shift to new career tracks, new marital relationships, or new lifestyles. For about 80 percent of the men Levinson interviewed, the midlife transition was judged to be a time of moderate or severe crisis. Levinson's report fueled the popular notion that most adults have a midlife crisis in their middle years—a period when "Every aspect of their lives comes into question, and they are horrified by much that is revealed" (Levinson, 1978, p. 199). As we will see later in this chapter, the midlife crisis is a controversial notion that many do not accept.

Adult Developmental Tasks in Perspective The picture of adult development presented by Erikson (1963) was based largely on his clinical observations of adults; and Levinson's (1978) views, as we noted earlier, were derived from interviews with 40 men. Thus, neither of these developmental perspectives is well established scientifically, and one could question the relevance of Levinson's scheme to women (Gilligan, 1982). On the other hand, there can be little doubt that the themes of achievement in the work place and intimate relationships, both dealt with by Erikson and Levinson, are central for many adults. In the following sections, we will focus on both of these themes; then we will examine the evidence both for and against the controversial midlife crisis.

Vocational Development How important is one's vocation during adulthood? So important that people are often defined largely by their answer to the question, "What do you do?" Most working adults spend more time at their occupations than at any other single activity. The jobs held by parents largely determine the family's social status, place of residence, how often they move, who cares for the children, when family members sleep and eat, and when (and how much) they are together. All these facts made vocations important when most workers were male; they make vocations even more important now, when over one-half of all American women of employable age are also working.

Given the importance of people's vocational choices, one might reasonably ask how people make that choice. How do people select *one* from among the 25,000 or so different occupations that exist today? Personality factors seem to play a role. For example, some researchers have found a modest but fairly consistent linkage between achievement motivation (Chapter 7, page 283) and a career in science, and between power motivation (Chapter 7, page 287) and an executive or managerial career (Veroff & Feld, 1970). However, our evidence on personality and career choice is sketchy at best. In an effort to fill out the picture, Holland (1973) studied the vocational preferences of high school and college students and developed a personality-environment model for career selection: The model includes the six personality types shown in Figure 12.14, with each type suited to a particular kind of work environment. For example, people who are energetic, talkative, ambitious, and otherwise enterprising are said to be inclined toward careers that reward people for achieving power, status, and money—careers such as sales. Holland's model is interesting, but it has not yet been rigorously tested.

Work Choice, Work Change, and Life Satisfaction One reason that early career choices are considered important is that each time we open one career door, we, in effect, close others. When we choose to sharpen one set of skills, we leave other skills undeveloped. Each choice moves us closer to being locked into, and out of, certain career options (Abeles et al., 1980). On the other hand, no career choice has to be final. By one estimate, the average adult changes careers three to five times (Hultsch & Deutsch, 1981), and, by the mid-1970s, as many middle-aged workers as younger workers were completing job-retraining programs. (Siegler & Edelman, 1977).

Adults who shift careers voluntarily reflect a combination of personality and situational factors that we are just now beginning to understand. Some research has suggested that personality factors such as risk-taking tendencies and a sense of control over one's own destiny may contribute. Other evidence has pointed to situational factors such as increasing disenchantment with one's present career (or "burnout"), discovery of an alternative occupation that promises greater satisfaction, and pivotal events (such as, divorce or death of a loved one) that lead one to shift life goals and priorities. Sometimes the career shift makes very good sense, but often it reflects unrealistic expectations for job gratification and job success (Freudenberger & Richelson, 1980).

Even adults who stick with their initial vocational choices for a lifetime may show marked changes in their career orientation over time.

Figure 12.14

Holland's hexagonal model, showing personality types (bold type), associated traits, and a sample career choice which might be appropriate for each type (in parentheses). (Adapted from Holland, 1973, by Whitbourne & Weinstock, 1979.) Reprinted by permission.

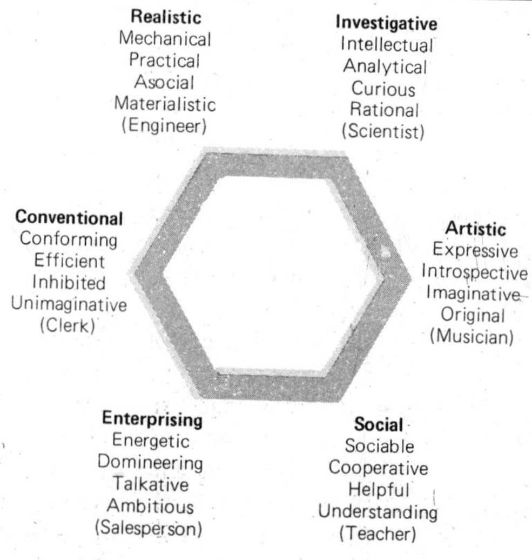

Some grow increasingly wedded to their work; others grow increasingly divorced from theirs. A 20-year longitudinal study of Bell Telephone System employees illustrated these two patterns (Bray & Howard, 1980). Among middle-level managers, the most successful were those for whom work increased in importance over the 20-year span. Over time, these people put longer and longer hours into their jobs and showed increased identification with the values of the Bell system. By contrast, less successful managers showed decreasing involvement with Bell and increasing commitment to family and recreation. Not surprisingly, the more successful managers expressed greater satisfaction with their careers and with their specific jobs than did the less successful managers. However, the two groups did not differ on measures of marital satisfaction, marital stability, family worries, cynicism, feelings of crisis, or adjustment. The study suggested an important idea: People may differ in their patterns of job committment and success over time, but these individual differences are not necessarily related to their overall adjustment and satisfaction. Work, while it is an important part of life, is only *one* part. When careers fail to provide a full measure of gratification, some change jobs, but others stay put and find satisfactions outside the world of work.

Intimate Partnerships For many, the greatest satisfactions in life come from their intimate relationships. Today, some of these relationships involve cohabitation—unmarried couples living together. However, over 95 percent of all Americans do eventually marry. Despite the prolonged dating and sexual intimacy of adolescence, youth, and young adulthood, the decision to marry is often made quickly. Yet couples still tend toward what geneticists call *assortative mating*—that is, partners resemble each other at greater-than-chance levels. Partners tend to be similar in their physical characteristics, their intelligence and education, their social and ethnic backgrounds, their religion, their temperaments, and their life outlook. Moreover, the more similar partners are in these ways, the more stable their marriages tend to be.

Marriages, once formed, often evolve toward one of three categories identified by experts as particularly common in the United States. (Turner & Helms, 1982):

1. *Traditional marriage*—in which the husband is undisputed head of the family, traditional sex roles are maintained, and the wife shows deference to her husband (see Figure 12.15).

2. *Companionship marriage*—in which the male and female roles are not regarded as fixed, and husbands and wives freely assume the rights and obligations of their partners, depending upon the demands of the situation.

3. *Collegial marriage*—in which comradeship and sharing are emphasized, but husband and wife assume responsibility for different roles in the marriage, with each respecting the individual abilities and interests of the other.

A given individual may prefer one of these types or some other arrangement, but satisfaction with marriage is apt to depend in part upon on how well the expectations and preferences of the two partners match up.

"It's taken her thirty-one years, but, by God, she's got our marriage working again."

Figure 12.15

Extreme forms of traditional marriage may work better for some family members than for others. (Drawing by Geo. Price; copyright © 1983 The New Yorker Magazine, Inc.)

Parenthood and Transition to the "Empty Nest" As marital partners become parents, a new process of growth begins—and not only in the children. As author Peter DeVries put it, "The value of marriage is not that adults produce children, but that children produce adults." Learning to care for infants and young children is a developmental task of enormous magnitude. In fact, couples tend to rate the arrival of their first child as at least "moderately stressful" (Hobbs & Cole, 1976). Parenthood usually means a decline in the couple's sexual activity and in most other one-to-one adult interactions. It also often means movement toward a more traditional household, with the wife setting her career aside for a time and entering the world of children and housework. Despite such potentially stressful changes, couples often put children at the top of their list of marital satisfactions. In fact, one study found that 63 percent of couples who were dissatisfied with their own relationship listed children as their *only* satisfaction in the marriage (Luckey & Bain, 1970).

As children mature into adolescents and eventually move away from home, their parents face the prospect of an "empty nest." The experience can be painful and depressing for some parents. For most, however, the departure of children is a time of mixed emotions, including a tinge of sadness that soon fades away. As one parent put it:

> I guess it's rather a bittersweet thing. It's not that it's either good or bad, it's just that it's an era that's coming to an end and, in many ways, it was a nice era. So there's some sadness in it, and I guess I feel a little lost sometimes. But it's no big thing; it comes and goes. [Suddenly straightening in her chair and laughing] Mostly, it goes. (Rubin, 1979, p. 16)

As "it goes," most parents take some satisfaction in their reduced ties and their newfound freedom to pursue personal goals. The empty nest allows nurturance to be redirected; husband and wife can offer each other much of the support and companionship that had been focused on their children when all the family members lived together. For most parents, personal and marital satisfaction increase after the parent-child ties have been loosened (Rubin, 1979).

The Midlife Crisis: How Real and How Common? As we noted earlier, Levinson (1978) described the middle adult years as a period when crisis is very likely. He described the midlife crisis as a period of self-examination, self-recrimination, and questions like: How did I fail? How can I break out of this unbearable lifestyle? Where do I go from here? The crisis is depicted as a time of lost illusions. For example, the adult who believed for years that "if I become vice president of Consolidated and get a house in Forest Pointe, I will be truly happy" may now have the position and the house but not the happiness. In the wake of the midlife crisis, adults may shift their careers or relationships, resign themselves to failure within the old ones, or renew their commitment to the careers and relationships they have had all along. Several other theorists seem to share Levinson's view that the middle years are a time of trauma and crisis (for example, Erikson, 1963; Gould, 1978).

An alternative viewpoint held by many psychologists (for example, Brim, 1976; Costa & McCrae, 1980) is that midlife crisis is the exception, not the rule. Most carefully done longitudinal studies have not found midlife crisis to be a common occurrence. Similar findings have emerged from cross-sectional research with large samples. For example, Costa and McCrae (1980) developed a scale to measure the distress and turmoil that many have attributed to the midlife crisis and administered the scale to a large sample of men aged 33 to 70. They found *no* age differences—no peak in distress or turmoil at midlife. These investigators concluded that while some people do experience a crisis at midlife, most do not. Moreover, many who do have a midlife crisis may be prone to crises at other points in their lives as well. The middle adult years are certainly a time of change; but for most adults, the changes that occur in the middle years are apparently no more traumatic than the changes that occur at other periods in the life span.

ADJUSTMENT PROBLEMS IN ADULTHOOD

While full-fledged midlife crises may not be common, early and middle adulthood are still periods of significant risk. One risk is apparent when one considers that about 40 percent of all U.S. marriages now end in divorce. This does not mean that Americans are giving up on marriage; 75 percent of divorced people remarry within 3 years. However, the high divorce rate does mean that a high proportion of U.S. families undergo the multiple strains of a family breakup. A major study of 144 divorced and married couples and their children recently reached an important conclusion: Divorce is not victimless (Hetherington et al., 1979). What this means is that among the divorced families, at least one member showed significant distress and behavior disruption. While divorce may often be better than persistence in a destructive relationship, divorce and its aftermath seem almost certain to leave scars.

When families remain intact despite multiple problems, individual members may suffer anyway. Children may simply be underattended, as in the case of America's "latchkey children," the many thousands of children who use their own keys to enter their empty homes alone every day after school, while their parents are still working. In more serious cases, parents struggling to make ends meet and trying to cope with personal problems may neglect their children emotionally or physically, abuse their children physically, or do violence to one another. It is estimated that 2 million children are physically abused by their parents each year; about 28 million wives and 12 million husbands have reportedly been abused by their spouses at some point during their marriage. There are increasing reports of parent abuse by teenagers and of grandparent abuse by younger family members. Family life, a source of strength and security for many, can be a source of anxiety and danger for others.

Besides the risk of family problems, the adult years include risks outside the family. Many people simply do not make a success of adult life. Those who were deviant or underachieving children may crash miserably in the adult social world and the world of work. Sometimes these failures are linked to serious thought disorders such as schizophrenia, to substance-abuse problems such as alcoholism, or to long-standing personality disorders—all of which become more common in the adult years (see Chapter 15 for a detailed discussion). These and other vulnerabilities of the middle years are particularly ominous because their impact is likely to be felt not only by the troubled adults themselves but also by their children and others who have grown to depend on them.

Old Age

"How old would you be if you didn't know how old you was?" This question, posed by folk philosopher Satchel Paige, raises an important point for psychologists to consider: Age, and especially "old age," is partly a matter of subjective perception (Figure 12.16). The boundary between middle and old age is not clearly marked by any physical or intellectual transformation. In North America, retirement, which occurs at about the age of 65, is generally viewed as the beginning of old age. In eastern Europe, by contrast, women usually retire at 55 and men at age 60. In the agrarian societies of developing African countries, retirement is virtually unheard of; there, the attainment of "old age" is a gradual process marked by subtle changes in physical appearance, and the timing varies from one individual to the next.

The U.S. population is getting older. The proportion of elderly people in our population has increased from 4 percent in 1900 to about 10 percent today. In 1983, the number of Americans over 65 (27.4 million)

Figure 12.16

Age is partly a matter of subjective perception. (Peter Simon/Peter Arnold, Inc.)

edged ahead of the number of teenagers (26.5 million). The change results partly from the fact that people are living longer than they used to; U.S. life expectancy is now up to 74. However, the major cause of the change is a dramatic decline in the birthrate over the past 2 decades. As a result, the population has been redistributed, with older people making up a much bigger portion of the population than before. The elderly are noticeable not only because of their numbers but also because of some characteristic physical changes.

PHYSICAL CHANGES

Many of the physical changes that come with age are familiar. Hair whitens and becomes sparse; skin dries and wrinkles; gums recede, and teeth are lost—more than half of the elderly have none of their own teeth; the facial configuration shifts; the spine bows; and strength and agility fade somewhat. Sensory capacity declines as well. After age 60, most people need glasses. Impaired hearing is five times as common among 65- to 79-year-olds as among 45- to 64-year-olds. Aging even has subtle effects on autonomic nervous system arousal. (See Chapter 8, page 315.) In younger adults, such arousal peaks quickly when alertness or a quick response is needed and fades quickly as the need disappears. In older adults, arousal may still peak quickly, but it fades slowly. Thus the older adult remains "wired" longer and may seem overly nervous, vigilant, or cautious when events are moving quickly.

Although the story of aging is partly about bodily decline, there is also some good news about aging bodies. Today's college students can probably look forward to sounder bodies and a healthier life in old age than any previous generation. Researcher James Fries (1980) has found some very encouraging trends in U.S. data on aging and health. Each year during the 1970s, life expectancy from birth increased by about 4 months, while life expectancy from age 65 showed a 1-month increase. The two trends intersect in the year 2018 at a mean age at death of about 85. Fries noted that most of the quick killers, such as smallpox and polio, have been virtually wiped out and that rapid progress is being made in the areas of cancer, lung disease, and diabetes. He concluded that average life expectancy may never get beyond about 85 but that the years people do live are likely to be increasingly disease-free. More and more, he predicted, people will live "physically, emotionally, and intellectually vigorous" lives until shortly before the end, when their organs will simply wear out and be incapable of renewal; at this point, people will die quickly and mercifully, without lingering illness.

COGNITIVE DEVELOPMENT IN OLD AGE: STABILITIES, CHANGES, AND MELLOWINGS

The cognitive developments in old age can best be described as a "mixed bag." Some intellectual skills stay on a more or less smooth plateau year after year; other skills decline somewhat. Beyond the level of specific skills, elderly people often show subtle changes in the ways they use their intellect—changes that might be described as "mellowing." To sample the array of intellectual developments during this period, we will consider several different research perspectives, beginning with the Piagetian viewpoint (Chapter 11, page 425).

Piagetian Perspectives Considerable research and theorizing within the Piagetian tradition has been focused on the performance of older people on traditional Piagetian tests of concrete and formal operations. Much of the research evidence (reviewed by Mussen et al., 1979) has shown that the elderly do not do as well on many tests as do adolescents and younger adults. In fact, some studies even showed poorer performance by the elderly than by older children on conservation tasks like those shown in Figure 11.18 on page 452. The easy conclusion to draw from such findings is that as people grow old, they regress cognitively—that is, they slip back to earlier, less adequate levels of thinking. Some have drawn this conclusion, and they may be right.

On the other hand, some have raised questions about what Piagetian tests really tell us about the elderly and whether these tests are really appropriate ways of measuring ability in older people. A quick look at page 452 will show that most Piagetian tests were designed for children. They use child-oriented materials (for example, wooden blocks, balls of clay) and child-oriented language. The tests may seem silly or uninteresting to some adults and unfamiliar or confusing to others. As a result, older adults who take the tests may not do so well at first, despite adequate ability. One way to explore this possibility is to give older adults practice with some Piagetian measures and then test them with other, similar Piagetian measures. When such training was provided, older people generally showed marked improvement in their performance (for example, Hornblum & Overton, 1976). The performance of the elderly seemed to be quite improvable. This, in turn, suggests that the true underlying ability of older people may often be more advanced than their initial performance on unfamiliar tasks indicates.

Another question raised by some investigators is whether the health problems and educational limitations of older adults may have hampered their performance on Piagetian tests. One investigator (Blackburn, 1984) tried to control for this problem. He selected adults aged 63–75 who were all healthy, all college-educated, and all currently taking college-course work. When these elderly adults were compared to college students aged 19–24, there were *no* significant differences between the groups on nine different tests of formal-operational thinking. Could it be that the poor performance of the elderly in earlier research resulted partly from generational differences in health and education? This is a very real possibility. To check it out, we will need some carefully controlled research.

Learning and Memory Because learning and memory are closely intertwined and both are crucial to intellectual functioning, it is important to find out how the two operate in old age. Many investigators have worked at this task, and some general trends are emerging from their work (Perlmutter & List, 1982).

First, although older adults can certainly learn via classical and instrumental conditioning as described in Chapter 4, some kinds of classical conditioning take longer in older people than in younger adults. This may be due partly to the fact that older people tend to have more problems with their sight, hearing, and other sensory functions.

Second, in verbal learning—learning lists of words, for example—adults over 60 generally do not perform as well as do young adults, but experts are now debating whether these *performance* differences reflect real differences in learning *ability*. Some attribute the inferior performance of older adults to pacing problems. It is on timed tasks that older people perform most poorly. This, in turn, may be related to their cautious style of responding, which we discussed earlier. Older adults tend to make errors of omission rather than to give incorrect responses; perhaps they are just being careful—taking a lot of time to make sure their answers are right—and this lowers their score on tests where speed is crucial.

Third, overcaution and pacing problems may also contribute to some of the memory-performance deficits older people demonstrate (Salthouse & Kail, 1983). These deficits are not found in all aspects of their memory but are largely confined to what is called *secondary memory*, a term applied to our system for processing information. Secondary memory includes three processes: (1) encoding, putting the to-be-remembered material into our memory systems; (2) storage, retaining the material until it is needed; and (3) retrieval, pulling the information out of storage (Chapter 5, page 185). Older people show deficits in the first and the third processes: encoding and retrieval. They are less effective than are younger adults in coming up with good strategies for organizing and rehearsing the memories they are encoding, and they often take longer than do younger adults to draw memories out of storage (Erber, 1982). Memories—be they words, pictures, or concepts—seem to be recognized about as effectively by older as by younger adults; it is recall, remembering from scratch or from partial cues, that tends to suffer with old age. Note, however, that the memories are rarely lost; it just takes longer "to find them in the file."

Now, a word of caution. Almost all our research on learning and memory is cross-sectional. One of the few relevant longitudinal studies suggests that there may be some real age declines in certain forms of learning, at least after age 60 (Arenberg & Robertson-Tchabo, 1977); but most evidence thus far tells us only that there are age-group differences. As we have stressed repeatedly, such differences may reflect the differences between generations, or cohorts, rather than true effects of aging. Thus far, we just cannot be sure.

Life-Span Research on Intelligence-Test Performance The distinction between cohort differences and true effects of aging has been emphasized by the life-span developmental researchers. By repeatedly assessing intellectual functioning in the same adults as they grow older, these investigators have reached some important, though tentative, conclusions about true age effects (Schaie, 1983). First, true intellectual declines before the late fifties are unusual; when declines do occur before 50, they usually reflect some pathological process, such as an illness or a degenerative disorder. Second, from the late fifties on, there is often a decline in abilities that involve speed of response. Third, beyond age 80, performance declines of some sort become the rule rather than the exception.

The fourth and perhaps most intriguing conclusion is that many of the intellectual limitations found in older adults reflect obsolescence, not decline. In other words, big generational differences exist in the approaches people take to the tasks of processing information and solving problems. Many older people seem to use the same reasoning strategies they used in their younger days, but these strategies are often not so effective as those used by more recent generations. Some life-span researchers (for example, Schaie, 1983) have stressed that this may be very good news; it may mean that what were once thought to be age declines are simply outdated skills, skills that can be upgraded with new training and experience.

Creativity in the Elderly Whatever their performance on various tests and tasks, it is clear that many people make major creative contributions well into their so-called retirement years. Consider these examples of intellectual creation and fulfillment:

1. Goethe completed *Faust* at age 82.
2. Grandma Moses was painting at 100.
3. Sophocles completed *Oedipus Rex* at 70 and *Electra* at 90.
4. Pablo Casals was giving cello concerts at 88.
5. Verdi composed the operas *Otello* at 72 and *Falstaff* at 77.
6. Mary Baker Eddy was director of the Christian Science Church at 89.
7. Cecil B. De Mille's movie, *The Ten Commandments*, was made when he was 70.
8. George Burns won an Academy Award for his performance in *The Sunshine Boys* at age 80.
9. Gen. Douglas MacArthur became commander of the United Nations armed forces in Korea when he was 70.

10. Mohandas Gandhi led India's opposition to British occupation in his late seventies.

11. Ronald Reagan celebrated his seventieth birthday as a resident of the White House, having initiated a set of new policies that would come to be called "Reaganomics." [Adapted from Turner & Helms (1982), p. 252]

As these examples suggest, the years past 65 can be a time of real consolidation—a time when the ideas and skills developed over most of a lifetime can be brought together to produce major new achievements. Most people will not achieve at the level of these examples at any age. Yet most will have many opportunities for personally gratifying consolidation and creation well into their later years.

DEVELOPMENTAL TASKS OF OLD AGE

Erikson's View In Erikson's (1963) account of the life course (page 473), old age brings on a core developmental conflict: integrity versus despair. To achieve *integrity,* in Erikson's sense of this term, means to integrate one's attitudes, beliefs, motives, and experiences in such a way that they fit together comfortably and form a coherent whole. One result is a feeling of satisfaction with a life well lived (Figure 12.17). Such integrity, Erikson believed, is most likely among those who have "taken care of things and people" and have adapted themselves "to the triumphs and disappointments adherent to being" (1963, p. 268). Without this integrity, the older person feels a growing sense of *despair,* a fear that time is running out

Figure 12.17

One who has achieved *integrity*, in Erikson's sense of the term, can reflect with satisfaction on a life well-lived. (Roger Malloch/Magnum Photos, Inc.)

Figure 12.18

In many cultures elders are called upon to bring balanced, realistic judgment to bear on complex problems. The attribute that makes such judgment possible is often called wisdom. (*a*, AP/Wide World Photos; *b* and *c*, courtesy of the United Nations.)

before the pieces of life's puzzle can be assembled in a satisfying way. This despair can show up in various ways—as perpetual irritability and disgust or as a nagging fear of death—but at its core is a sense of incompleteness, of a life that is not yet whole.

Wisdom Those who do achieve a sense of wholeness and integrity may develop one of the hallmarks of successful aging: wisdom. Many cultures traditionally rely on selected elderly people for advice about complex life problems (Gutmann, 1977; also see Figure 12.18). One reason may be that older people who have been attentive to their life experience often have a perspective on reality that is richer and more informed than the view most younger people take. The philosopher Schopenhauer depicted life as an embroidery viewed differently at different ages:

> Life could be compared to an embroidery of which we see the right side during the first half of life, but the back in the last half. This back side is less scintillating but more instructive: it reveals the interpatterning of the threads. (Translated from a quotation in Seitelberger, 1978, p. 215)

A clear, illusion-free view of life's "interpatterned threads" is one part of what many mean by the concept of wisdom. Some have also suggested that the wise person is one who has a "balanced investment in self as well as in others" and who combines "experience, reflectiveness, and emotional balance" (Birren & Renner, 1980, p. 28).

Life Review Some elderly adults inject their wisdom into memoirs or autobiographies that summarize hard-earned lessons from the school of life. Most, however, simply conduct informal life reviews through long conversations with their families and friends, sometimes sparked by images in the family album or a chance recollection. Through these life reviews, the older person can detect common threads of meaning running through the diverse experiences of their 60-plus years. These threads of meaning may, in turn, provide a legacy of insight for those who take the time to listen.

Paradoxically, the process of life review goes on at the same time that much of life is still unfolding. Many issues in the areas of occupation and personal relationships must still be confronted after the age of 60. We will consider some of these issues next.

Retirement Retirement, a traditional marker of old age in many western cultures, is not nearly the painful event that some popular stereotypes suggest. Certainly there are some people for whom retirement signifies loss—loss of a familiar daily routine, loss of valued social interactions, loss of a well-established role, and even loss of income. For many others, though, retirement is a welcome transition, one that offers new opportunities and new freedom (Figure 12.19). The individual response to retirement depends on many factors, but two of the most important seem to be job satisfaction and income. People who find their jobs unrewarding are likely to welcome retirement (Barfield & Morgan, 1978). So are people whose postretirement incomes will sustain their preretirement standards of living (Glamser, 1976). People who are wedded to their work and people who fear a drop in their standard of living, on the other hand, may well dread separation from their jobs; in fact, some of these people will cope by seeking new postretirement jobs. Finally, we need to distinguish between scheduled, long-anticipated retirement and unscheduled retirement caused by a sudden illness or a demand from supervisors that one quit the job. Such unscheduled retirements tend to produce anxiety and depression; scheduled retirements do not (Pearlin, 1980).

Although different people react differently to retirement, recent evidence suggests that it is welcomed by more than two-thirds of elderly working people and that it has few really negative effects. A recent survey of six longitudinal studies (Palmore et al., 1984) found that retirement has few, if any, adverse effects on social activity, health, happiness, or life

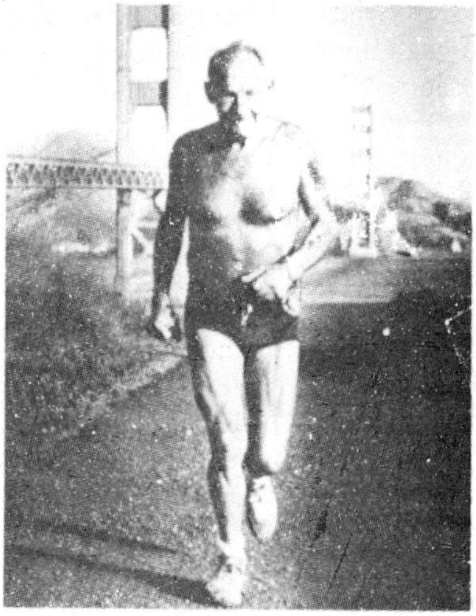

Figure 12.19

Retirement from one's job does not necessarily mean retirement from activity. In fact, activity level may *increase* when job demands no longer interfere. (Bottom photo, Mariette Pathy Allen/Peter Arnold, Inc. All other photos AP/Wide World Photos.)

satisfaction. And "real-life" evidence that retirement is attractive continues to mount. For example, a major court decision in 1978 gave many Americans the option of working beyond age 65, but today only about 15 percent take that option. In fact, most U.S. workers now retire voluntarily *before* 65, and the average age of voluntary retirement keeps dropping. Once retired, most seem to adjust well; and those with good health and an adequate income seem to thrive.

Retirement often brings dramatic changes in the nature of interpersonal relationships; it can alter one's collection of daily acquaintances, and it can greatly increase the time husband and wife spend together. Both changes may bring on major adjustments and new personal growth. For example, as elderly people share more activities and duties with their spouses, the couple may become increasingly androgenous—that is, less bound by traditional sex-role stereotypes in their beliefs and behavior, and more capable of showing both "masculine" and "feminine" attributes (Sinnott, 1982). Once spouses have become adjusted to the increased togetherness of retirement, their relationship can flower in ways that were not possible when the demands of employment interfered.

As the marital relationship changes during old age, are there corresponding changes in the role of sexuality? Do sexual interest and activity decline or hold steady in old age? We consider such questions in Inquiry 12.2.

Grandparenthood For many, one of the special delights of old age is having grandchildren (see Figure 12.20). In many western societies, though, grandparenthood is a "roleless role" (Clavan, 1978); that is, it carries few or no clearly defined rights or responsibilities. As a consequence, the role of grandparent may differ from one family to the next.

One study (Neugarten & Weinstein, 1964) identified several different "styles" of filling the grandparent role. Most common of all is the formal style, in which grandparents take an ongoing interest in their grandchild and occasionally give the child special treats but carefully limit their role so as not to interfere with the parents. The fun-seeker style—the second most common—is an informal, playful approach in which grandchildren are seen as a source of leisure activity and mutual fun. The third most common pattern is the distant-figure style, in which grandparents are benevolent but have only brief, infrequent contact with their grandchildren.

In another look at grandparenthood, Robertson (1977) focused only on grandmothers, distinguishing between a social and a personal orientation. Grandmothers who adopt the social orientation focus on their perceived duties—to set a good example, encourage the grandchild to be honest, and so on. Those who adopt the personal orientation focus more on the joys and rewards of grandparenthood—("grandchildren will keep me happy and youthful," for example). The orientation taken by any particular grandmother depends partly on her life circumstances. For example, younger grandmothers, who have jobs and whose husbands are living, emphasize the social orientation and are not deeply involved in the grandmother role. Older grandmothers, most of whom are widowed and who do not have jobs, emphasize the personal orientation and are more involved with their grandchildren.

Inquiry 12.2

IS THERE SEX AFTER 60?

With advancing age, a number of biological changes may make some men and women feel less "sexual." Between about 45 and 60, women pass through the climacteric, which leads to decreased vaginal lubrication and shrinkage of the vagina itself. In men, sperm production gradually declines, erections occur more slowly, the volume of seminal fluid is reduced, the force of ejaculation decreases, erections are lost sooner after ejaculation, and it takes longer to achieve another erection afterward. Changes like these, together with the popular conception of "the celibate senior citizen," may make some older people think of their sex life as a flame that is flickering and ready to die. Some of the elderly do have trouble engaging in or enjoying sexual intercourse, and some of their problems are directly caused by physiological factors; such problems are termed primary sexual dysfunction. Much more numerous, though, are sexual difficulties caused by psychological factors—for example, a loss of confidence or a perception that one is no longer sexual; problems of this type, caused by psychological distress, are called secondary sexual dysfunction.

For many of the elderly, though, these changes are more than counterbalanced by positive changes. For example, menopause can be sexually liberating for some women because it removes the risk of pregnancy. In men, the declining potency of ejaculation permits better self-control; ejaculation can be delayed and coitus prolonged, making older men potentially better sex partners.

Such factors may help to explain why, for many adults, the sexual flame continues to burn well into the later years. To find out just how widespread sexual feelings and activity are in the mid- to late-adult years, Pfeiffer, Verwoerdt, and Davis (1972) surveyed 502 middle-aged and elderly men and women. They found that self-reported sexual interest and activity were lower among older adults than among the middle-aged, as shown in the two accompanying tables. However, only among women was this age difference very pronounced, and there were some mitigating factors in the findings for women. For example, many women over 60 said they had stopped having intercourse because their husbands had lost interest in sex

AGE GROUP	NUMBER	NONE	MILD	MODERATE	STRONG
MEN					
46–50	43	0	9	63	28
51–55	41	0	19	71	10
56–60	61	5	26	57	12
61–65	54	11	37	48	4
66–71	62	10	32	48	10
Total	261	6	26	56	12
WOMEN					
46–50	43	7	23	61	9
51–55	41	20	24	51	5
56–60	48	31	25	44	0
61–65	43	51	37	12	0
66–71	54	50	26	22	2
Total	229	33	27	37	3

Current level of sexual interest (percentage).

(*Source:* From Pfeiffer, Verwoerdt, & Davis [1972]. Copyright 1972, the American Psychiatric Association. Reprinted by permission.)

AGE GROUP	NUMBER	NONE	ONCE A MO.	ONCE A WK.	2-3 TIMES A WK.	MORE THAN 3 TIMES A WK.
MEN						
46–50	43	0	5	62	26	7
51–55	41	5	29	49	17	0
56–60	61	7	38	44	11	0
61–65	54	20	43	30	7	0
66–71	62	24	48	26	2	0
Total	261	12	34	41	12	1
WOMEN						
46–50	43	14	26	39	21	0
51–55	41	20	41	32	5	2
56–60	48	42	27	25	4	2
61–65	44	61	29	5	5	0
66–71	55	73	16	11	0	0
Total	231	44	27	22	6	1

Current frequency of sexual intercourse (percentage).

(*Source:* From Pfeiffer, Verwoerdt, & Davis [1972]. Copyright 1972, the American Psychiatric Association. Reprinted by permission.)

or the ability to have erections; in some cases, the husbands had grown ill or died. Evidently the majority of women and men who have able partners continue to be both interested and involved in sex beyond 60.

Finally, we should note that the two tables are from cross-sectional research; thus, they reflect age group differences, not true age changes. True longitudinal research with 60- to 94-year-olds, spanning a 10-year period, showed many individual patterns of age change (Verwoerdt et al., 1969). Many of the adults sampled reported declining sexual interest and activity, but many reported very stable patterns across time. In addition, about 20 percent of the men actually reported increasing sexual interest and activity as they grew older. One of the best predictors of sexual interest after 60 turns out to be sexual interest *before* 60.

REFERENCES

Hyde, J. S. (1979). *Understanding human sexuality*. New York: McGraw-Hill.

Masters, W. H., & Johnson, V. E. (1970). *Human sexual inadequacy*. Boston: Little, Brown.

Pfeiffer, E., Verwoerdt, A., & Davis, G. C. (1972). Sexual behavior in middle life. *The American Journal of Psychiatry, 128,* 1262–1267.

Verwoerdt, A., Pfeiffer, E., & Wang, H. S. (1969). Sexual behavior in senescence—changes in sexual activity and interest of aging men and women. *Journal of Geriatric Psychiatry, 2,* 163–180.

Studies like the two we have just described show that there are various ways of filling the grandparent role. Yet across the many variations in its form, the grandparent-grandchild relationship remains one of the most important institutions in human society. It is a bridge across generations, one that is valued by grandchildren even after they mature into adulthood (Robertson, 1976). For grandparents, the relationship can provide one path to the kind of generativity discussed by Erikson (1963)—that is, an opportunity to "pass the torch" to younger members of the family team.

Figure 12.20

Grandparents play a valuable role in bridging the gap across generations. Styles of playing the role can differ greatly from one grandparent to the next. (*a*, courtesy of John Weisz; *b*, Erika Stone/Peter Arnold, Inc.; *c*, Erika Stone/Peter Arnold, Inc.; *d*, Mariette Pathy Allen/Peter Arnold, Inc.; *e*, Sybil Shelton/Peter Arnold, Inc.; *f*, Sybil Shelton/Peter Arnold, Inc.)

Widows and Widowers: Coping with Loss One of the painful inevitabilities of intimate relationships in old age is that one partner will lose the other and face the pain of bereavement. Because women live longer than men (and often marry men older than they are in the first place), the surviving partner is usually female. In the U.S., widows outnumber widowers by more than 4 to 1, and more than 10 percent of all women are widows. As difficult as it is to develop and sustain a warm, intimate relationship with a single partner, it can be even more difficult to face life without that partner.

The difficulties can be serious for both widows and widowers, but women in general seem to cope better than men.

> One study (Barrett, 1978), for example, compared widows and widowers in their sixties or older who were living in the community (that is, not in institutions). Special attention was given to six major areas of their life functioning: health care, household roles, education, nutrition, psychosocial needs, and transportation. Compared to widows, widowers reported feeling lonelier and more dissatisfied with life, were more likely to think community services inadequate, needed more help with household duties, did not eat as well, had more negative attitudes about continued learning, and showed poorer morale overall. Widowers were also less willing to talk about the loss of their spouses or about death.

Although women tend to cope better than men in general, some groups of women are especially at risk for problems. In particular, middle-class women with a strong investment in their roles as wives report "strong disruption after the death of the husband" (Lopata, 1975, p. 229). However, women who, in addition to their marriages, have had active lives in the community or workplace report loneliness but relatively less disruption in their lifestyles. There is also evidence that older widows, particularly those who had advance warning of their husband's impending death, adjust better than do younger widows and those for whom the death was unexpected (Balkwell, 1981).

Some widows and widowers eventually cope with their loss by remarrying (see Figure 12.21). How successful are these new relationships?

> To find out, Vinick (1978) interviewed people aged 60–84 who had remarried. She asked them about how and when they had decided to marry again and about their level of satisfaction with the new marriage. Most said that they had originally intended to live alone after their spouse's death. However, over half the men remarried within a year of becoming widowers. The women waited longer, nearly all delaying more than 2 years. Most of the new couples had either known each other during the earlier marriage or had been introduced by a mutual friend or relative. In most cases, the man took the initiative in promoting the new relationship.
>
> The most common reason given for the remarriage was a desire for companionship. In most cases, the companionship seemed to work well. Some 80 percent of the women and 87 percent of the men said they were "satisfied" or "very satisfied" with the

Figure 12.21

Some elderly people choose to remarry after the death of their spouse. The evidence thus far suggests that these marriages are usually successful. (AP/Wide World Photos.)

new marriage. In fact, Vinick described the marriages as more tranquil and harmonious than those of most young couples. Vinick's (1978) positive findings led her to conclude that remarriage is a workable option for many older adults.

Vinick's conclusion may be more comforting to older men than to their female counterparts because the pool of available spouses is much larger for elderly males. Perhaps this is for the best in at least one respect. Older men have more difficulty than do older women in coping after the death of a spouse; thus older men, in general, may have a greater need for the remarriage option.

Whatever the coping strategies an older man or woman adopts for dealing with the death of a spouse, one implication of that death is difficult to deny: The surviving spouse, too, is undergoing an aging process that will culminate in death. Facing up to this process and the inevitability of one's own death is the final aspect of old age that we will consider.

Facing Mortality and Death Mark Twain once admitted that although he knew everyone had to die, he had always felt that an exception would be made in his case. Deep down, many share Mark Twain's feeling that "It can't really happen to me." Facing up to the inevitability of death is a major developmental task of old age. For some elderly people, the awareness grows gradually and eventually fits as warmly and comfortably as a familiar sweater. For others, though, the knowledge dawns starkly, sometimes with the diagnosis of a terminal illness.

Psychiatrist Elisabeth Kübler-Ross (1969, 1975) worked with people who were facing up to their own impending deaths. She proposed that the psychology of the dying process involves five stages.

1. The first is denial; informed of a terminal illness, the individual reacts with shock and disbelief ("No, it can't be me").

2. The second is rage and anger, particularly over the idea that others will live while he or she will not.

3. Stage three involves bargaining; the person accepts the inevitability of death but pleads for a bit more life, sometimes trying to "negotiate" with God for a few extra months.

4. The fourth stage is depression, a kind of anticipatory self-mourning.

5. The fifth and final stage is called acceptance; anger and depression subside, and the person becomes quietly expectant—not happy about death, but ready for it nonetheless.

Kübler-Ross originally proposed that these five stages always occur, and in the same order, for every person who knows that he or she is dying. She has softened this stance in recent years, and others who work with dying people have doubts about whether the stages are really universal. However, most agree that many who are facing death experience some of the feelings described by Kübler-Ross.

Regardless of whether a particular elderly person experiences these stages or not, acceptance of death is likely to come easier if he or she can reflect positively on a life well lived. This capacity is one part of Erikson's (1963) notion of ego integrity, which we discussed on page 495. A similar perspective is offered by Levinson (1978). In his account of the life course, he suggested that reaching and passing the age period from 60 to 65 occasions a "view from the bridge" at the end of the life cycle. If this view offers a satisfying sense of what one's life has been about and what one has made of it, the result may be a sense of satisfaction and a willingness to let go when one's time arrives.

ADJUSTMENT PROBLEMS IN OLD AGE

"Every man desires to live long, but no man wants to be old." This maxim from Jonathan Swift may help to explain why, even today, old age can take us by surprise. People often avoid thinking about their upcoming transition to "senior citizen" and "retiree" for so long that they fail to prepare for it. Inadequate financial planning can leave older people "newly poor" shortly after retirement and thus poor for the rest of their lives. In fact, while the elderly constitute about 10 percent of the United States population, they constitute 20 percent of the poor in the United States. Like other age groups, the elderly like to have friends their own age and recreation to enjoy with those friends. However, retirement communities that offer these attractions are mostly for the financially secure. Other elderly people may live lives of dreary loneliness, eking out an existence in substandard housing and on substandard diets.

One of the most ominous threats to the elderly is prolonged illness. The aging body is highly vulnerable to the ravages of disease and injury. Among the most frightening threats are diseases that attack the brain—*Alzheimer's disease* (Chapter 5, page 212) and Pick's disease, for example, both of which cause disorientation, poor judgment, and death within 5 years; there is no known cure for either. *Senile brain disorders* cause tissue loss and shrinkage in the brain, which in turn triggers mental and physical deterioration. *Cerebral arteriosclerosis,* or hardening of the arteries in the brain, blocks the supply of oxygen and nutrients to the brain; brain tissue thus degenerates, and severe intellectual slippage follows. In addition to these physiologically based disorders, the psychological stressors of old age can provoke various psychological disorders. Many of these are discussed in Chapter 15, but one that deserves special attention here is *depression.* In the elderly, bleak circumstances and the internal state that Erikson referred to as despair (page 473) can combine to make life seem not worth living. This happens especially to elderly people who are enfeebled and dependent at home or in nursing facilities that give little reinforcement for responsible, adaptive behavior. Rates of depression are high among the elderly, and suicide rates are considerably higher than among younger adults. Old age is inherently a period of high risk, but some who have studied the elderly (for example, Langer, 1981) report that the risks are especially high in extremely youth-oriented societies—societies that do not so much abuse their elderly as underestimate and overlook them.

Summary

1. With the transition from childhood to adolescence comes a period of fast-paced physical and intellectual change; experts cannot agree on whether this change necessarily involves psychological "storm and stress."

2. At a signal from the hypothalamus, the pituitary gland triggers the growth spurt, and puberty follows. For girls, the peak growth year is usually the eleventh or twelfth; boys peak at about 14.

3. Adolescence ushers the youngster into the intellectual realm of formal operations (Piaget's term). Thinking expands to encompass the world of the possible as well as the actual; formal thought involves hypothetical reasoning, induction, deduction, and a capacity to reason about abstractions and to judge the logical correctness of a chain of reasoning.

4. For some adolescents, the stage of formal operations coincides with new forms of moral judgment—forms which Kohlberg calls postconventional because they involve a reliance not on conventional rules but on abstract ethical principles.

5. According to Erik Erikson, adolescence is a time of searching; the objective is a sense of identity, an answer to the question, "Who am I?" In seeking a satisfying answer, adolescents may delay their

commitment to adult roles, an action that Erikson has dubbed the "psychosocial moratorium."

6. Adolescents' relationships with their parents become a context for learning independence and a sense of control over one's own destiny. Adolescents' relationships with their peers change in several ways over the teen years, reflecting increasing dominance by bright, creative youngsters and increasingly close interactions between boys and girls.

7. Over the past few decades, adolescent sexuality has virtually exploded; now, more than two-thirds of all girls and three-fourths of all boys have had sexual intercourse by the age of 19.

8. Adolescent sexuality and the inclination to experiment carry significant risks—both physical and psychological. The former include the risks of unplanned pregnancy and the contraction of sexually transmitted diseases. The latter include problems of sexual preference and delinquency as well as psychological disorders such as anorexia nervosa, schizophrenia, and depression (with a sharply increased risk of suicide).

9. During early and middle adulthood, bodies age largely because of characteristics programmed into the genes. In healthy adults, though, the bodily changes have relatively little impact on job performance or other behavior.

10. Most of the research on intellectual change in adulthood has involved cross-sectional methods, and most has shown that adolescents and younger adults outperform older adults on intellectual tasks. However, some psychologists of the Piagetian school have argued that adults show more mature, more realistic reasoning as they mature. Moreover, recent longitudinal studies suggest that when the same people are tested repeatedly, general intelligence may show little overall change until age 55 or 60.

11. The major developmental tasks of adulthood have been construed as "seasons" by Levinson, and tensions, or "psychosocial conflicts," by Erikson. Both theorists focus on the importance of relations with the world through one's vocation and relations with the family as a spouse and a parent.

12. Attempting to find out how people choose one vocation from among the 25,000 or so possibilities, Holland and other psychologists studied how interests and personality types match up to the characteristics of various occupations. The study of vocational development has broadened in recent years to include both career change and changing levels of commitment to one's career over the adult years. The so-called midlife crisis described in the popular press has not been well documented in careful research.

13. Marital choices follow assortative mating patterns, with partners resembling each other at greater-than-chance levels. In addition, most marriages fall into one of three patterns: Traditional, with the husband dominant; companionship, with the "male" and "female" roles taken on interchangeably by husband and wife; and

collegial, with comradeship and sharing emphasized but with husbands and wives taking responsibility for different roles.

14. The vulnerabilities of adulthood are related primarily to marriage and career. Some 40 percent of all U.S. marriages now end in divorce, an event that is rarely victimless. In addition, many intact families face the risk of in-house violence: Child abuse, spouse abuse, and even parent and grandparent abuse are now widely reported. Some adults must face up to career failure or a failure to meet the social demands of adulthood.

15. With old age, strength and agility fade, resistance to disease declines, and autonomic nervous system arousal grows less responsive to situational demands. However, current trends point to increasingly vigorous and healthy lives for each successive generation of elderly people for at least the next three or four decades.

16. Except in cases of serious disease, older people apparently hold their own well into their seventies in many areas of intellectual functioning, especially in recognition memory and on a number of standardized cognitive tasks. By contrast, verbal learning, recall memory, and tasks requiring speed of response often show declines from the late fifties on. Some of these declines, though, may be due to increased caution or physical problems, including poor vision or hearing.

17. A key developmental task of old age, according to Erikson, is the achievement of ego integrity, the capacity to reflect positively on a life well lived. Those who do achieve a sense of wholeness and integrity may also develop the kind of rich, informed, and emotionally balanced perspective on reality that some have called wisdom.

18. Other developmental tasks of old age include retiring (a more positive experience than popular stereotypes suggest); modifying interpersonal relationships (marriages often grow richer); grandparenting (styles differ greatly from one grandparent to the next); coping with the loss of loved ones; and facing up to one's own mortality.

19. Old age carries special risks, not the least of which is poverty at a fixed income. In addition, the aging body is highly vulnerable to disease, including organic brain syndromes and cerebral arteriosclerosis. Finally, the bleak circumstances and loneliness of some can produce psychological disturbances, including depression severe enough to provoke suicide.

Terms to Know

One way to test your mastery of the material in this chapter is to see whether you know what is meant by the following terms.

Adolescence (464)
Puberty (464)
Hypothalamus (466)
Pituitary gland (466)

Androgens (466)
Estrogens (466)
Growth spurt (466)
Primary sexual characteristics (466)
Menarche (467)
Secondary sexual characteristics (467)
Formal operations (467)
Deduction (469)
Inductive thinking (469)
Interpropositional logic (469)
Reflective thinking (470)
Preconventional level (471)
Conventional level (471)
Postconventional level (471)
Seeking identity (472)
Psychosocial moratorium (472)
Negative identity (472)
Status offenses (477)
Depression (478, 506)
Anorexia nervosa (478)
Delusions (478)
Hallucinations (479)
Youth (479)
Life-span developmental psychologists (481)
Menopause (481)
Climacteric (481)
Cohort (483)
Fluid intelligence (483)
Crystallized intelligence (483)
Creativity (484)
Midlife transition (485)
Assortative mating (487)
Traditional marriage (487)
Companionship marriage (487)
Collegial marriage (487)
Secondary memory (493)
Integrity (495)
Despair (495)
Alzheimer's disease (506)
Senile brain disorders (506)
Cerebral arteriosclerosis (506)

Suggestions for Further Reading

Alison Clarke-Stewart and Joanne Koch have written a fine new text entitled *Children: Development through Adolescence* (New York: Wiley, 1983). Another excellent text is Kurt Fischer and Arlyne Lazerson's *Human Development: From Conception Through Adolescence.* (New York: Freeman, 1984). Students interested in learning more about adolescence should see the thoughtful treatments of this age period in these two texts. Those interested in exploring adult development further might see two of the several fine texts in this area: David Hultsch and Francine Deutsch's *Adult Development and Aging: A Life-Span Perspective* (New York: McGraw-Hill, 1981) and Jeffrey Turner and Donald Helms's *Contemporary Adulthood*, 2d ed., (New York: Holt, Rinehart, and Winston, 1982). Erik Erikson's *Identity: Youth and Crisis*, 2d ed., (New York: Norton, 1968) is regarded by many as a classic analysis of adolescence and the search for a sense of identity. Finally, students interested in adolescence, cultural comparisons, and controversy might find Derek Freeman's *Margaret Mead and Samoa: The Making and Unmaking of an Anthropological Myth* (Cambridge, MA: Harvard University Press, 1983) exciting reading. This book has stirred a controversy over Margaret Mead's classic analysis of Samoan adolescence in *Coming of Age in Samoa* (New York: Morrow, 1928).

Psychological Assessment and Testing

HEADING OUTLINE

Chapter 13

A farmer is tending 17 sheep. All but nine escape through a break in the fence. How many sheep are left?

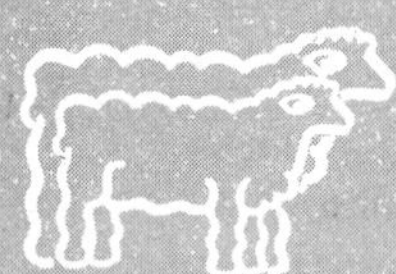

The pride of State University is its lily pond. Every summer the lilies spread fast, doubling in area with each passing day. On the first day of summer, there is one lily on the pond. It takes 60 days for the lake to be covered completely. On what day is the lake half covered?

One particular bottle of wine costs \$10. The wine itself is worth \$9 more than the bottle it comes in. What is the value of the bottle alone?

IN YOUR OPINION . . .

1. Which of these problems would do the best job, and which would do the poorest job, of distinguishing between people at various IQ levels?

2. Which of the problems would be the hardest for most people to solve?

This chapter is about individual differences and how they are measured. Of special interest here are differences in intelligence and personality. Which of the two do you think these three problems tap? Problems like these are commonly called brain teasers, and they require a burst of insight to be solved correctly. Insight seems to be a combination of intellectual *and* personality processes. It involves reasoning and judgment, of course; but it also seems to involve patience and diligence, a willingness to reflect on the information given rather than jump to hasty conclusions.

Some people are too hasty in tackling the sheep problem. They subtract 9 from 17 and come up with the answer 8; they fail to reflect carefully on the information given and thus fail to realize that the answer has already been provided: "All but nine escape." People's ability to solve this problem is not very closely related to their IQs, perhaps because such personality factors as patience and carefulness have so much to do with success.

Individual differences in success at the water-lily problem *do* correspond well to individual differences in IQ. People who do well on this problem tend to have high IQs. To solve the problem, they have to focus selectively on the information provided and recognize that the most important bit of information is that the lilies double every day. Given this fact, the pond has to be half covered on day 59.

The toughest problem of the three is the wine-bottle task. It was solved correctly by only 7 percent of a sample of adults from New Haven, Connecticut. Most people who failed to arrive at the correct answer (50 cents) apparently misinterpreted the word "more."

Most of us can devise problems like these that will reveal individual differences among people. The trick is to reveal differences that tell us something more about the people involved than how they perform on a problem or two—something about their levels of intelligence, their personality characteristics, or their typical behavior. This kind of information is the goal of psychological assessment. In this chapter we learn what psychological assessment is and how it is done.

[Source for all brain-teaser information: Sternberg & Davidson, 1982. Reprinted with permission from *Psychology Today* magazine, copyright © 1982 (APA).]

IN 1884, visitors to England's International Health Exhibition were treated to an unusual sight: an "anthropometric laboratory," where people were measured in some very novel ways. Francis Galton, an eminent Englishman who was also the cousin of Charles Darwin, had devised procedures by which "a man may, when he pleases, get himself and his children weighed, measured, and rightly photographed, and have their bodily faculties tested by the best methods known to modern science" (Galton, 1908, p. 244). Galton described his measurement procedures and his experience with them in the following way:

> The measurements dealt with Keenness of Sight and of Hearing; Colour Sense; Judgment of Eye; Breathing Power; Reaction Time; Strength of Pull and Squeeze; Force of Blow; Span of Arms; Height, both standing and sitting; and Weight. The ease of working the instruments that were used was so great that an applicant could be measured in all these respects, a card containing the results furnished him, and a duplicate made and kept for statistical purposes, at the total cost of the threepenny fee. . . .
>
> It is by no means easy to select suitable instruments for such a purpose. They must be strong, easily legible, and very simple, the stupidity and wrong-headedness of many men and women being so great as to be scarcely credible.
>
> (Galton, 1908, p. 249)

Figure 13.1

Poster used by Francis Galton to draw people to his Anthropometric Laboratory. Galton believed that the sensory and motor measurements he took in this laboratory provided a means of assessing intelligence. Later evidence proved him wrong. (From *The Life, Letters, and Labours of Francis Galton*, Vol. 2, by K. Pearson, 1924, Cambridge, England: Cambridge University Press. The Granger Collection, N.Y.)

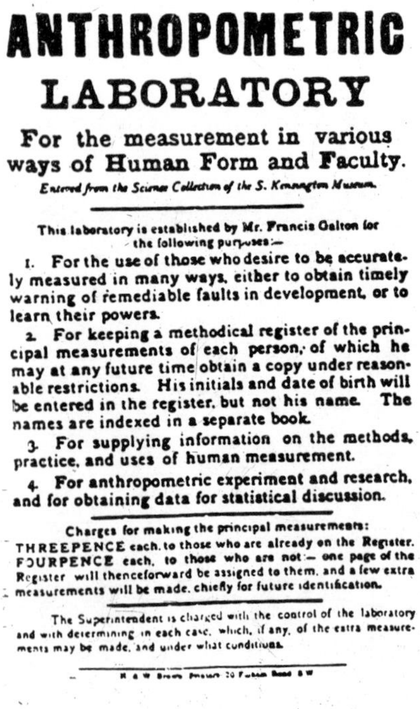

ANTHROPOMETRIC

LABORATORY

For the measurement in various ways of Human Form and Faculty.

Entered from the Science Collection of the S. Kensington Museum.

This laboratory is established by Mr. Francis Galton for the following purposes:—

1. For the use of those who desire to be accurately measured in many ways, either to obtain timely warning of remediable faults in development, or to learn their powers.

2. For keeping a methodical register of the principal measurements of each person, of which he may at any future time obtain a copy under reasonable restrictions. His initials and date of birth will be entered in the register, but not his name. The names are indexed in a separate book.

3. For supplying information on the methods, practice, and uses of human measurement.

4. For anthropometric experiment and research, and for obtaining data for statistical discussion.

Charges for making the principal measurements:
THREEPENCE each, to those who are already on the Register.
FOURPENCE each, to those who are not — one page of the Register will thenceforward be assigned to them, and a few extra measurements will be made, chiefly for future identification.

The Superintendent is charged with the control of the laboratory and with determining in each case, which, if any, of the extra measurements may be made, and under what conditions.

When the exhibition closed in 1885, Galton relocated. He moved his laboratory to a room in the Science Galleries of the South Kensington Museum, keeping it in operation there for 6 more years. He used posters like the one shown in Figure 13.1 to draw people to his laboratory for testing. Galton believed that he was measuring the individual building blocks of an extremely important human trait: intelligence. This belief had grown partly out of Galton's observations of mentally retarded people; he became convinced that their sense of pain was very poor and that this reflected poor sensory-discrimination ability. He reasoned that poor sensory discrimination hampers people's ability to profit from experience—in other words, their intelligence. Thus, he concluded that sensory and physiological tests might provide a means of measuring intelligence.

Unfortunately, research findings soon revealed that Galton was mistaken. Evidence showed that scores on the various sensory and motor tests did not correlate well with teachers' ratings of intelligence, with the subjects' school grades, or even with one another. Evidently, Galton's tests were not valid measures of intelligence.

While Galton was trying to measure intelligence by means of sensory and motor tasks, a Frenchman named Alfred Binet began trying a different approach that was to prove much more successful. Binet devised measures of 11 different faculties—higher mental processes such as comprehension, imagery, suggestibility, and judgment of visual space. This work set the stage for a very important assignment Binet was to receive

In 1904, Binet and his colleague, Theodore Simon, were asked by a French education commission to devise procedures for identifying children who needed special classes—children who were intellectually unable to profit from ordinary instruction.

Binet and Simon, in accepting this assignment, were eager to insure that children would not be placed in special classes simply on the basis of teachers' opinions or the views of other adults. Such judgments, they noted, might be biased or simply inaccurate. Fairness and accuracy were especially important here, they argued, because

> To be a member of a special class can never be a mark of distinction, and such as do not merit it, must be spared the record. . . . We are convinced . . . that the precision and exactness of science should be introduced into our practice whenever possible, and in the great majority of the cases it is possible.
> (Binet & Simon, 1905/1916, pp. 9–10)

Thus, a concern for fairness and accuracy in the education of children was a major factor in the development of the first viable intelligence test. This is worth bearing in mind. Later we will discuss concerns about the possible unfairness of some tests. The concerns may be well founded in some respects, but they should be tempered by historical perspective. That is, we should remember that testing began in part as an effort to *promote* fairness. In this chapter we will focus on some of the tests that followed from the early efforts of Binet and others. We will also address issues and controversies bearing on these tests. Finally, we will consider some alternative approaches to psychological assessment—approaches that do not involve tests.

Psychological Tests

Some of the most interesting things about people cannot be seen easily by a casual observer. Attitudes, personality characteristics, and abilities, for example, cannot be viewed directly. What we can do, however, is observe people's behavior in a systematic way and make inferences about the underlying attributes that stimulate that behavior. Psychological tests help us do these things.

WHAT IS A PSYCHOLOGICAL TEST?

A *psychological test* is a structured technique used to generate a carefully selected sample of behavior. This behavior sample is used, in turn, to make inferences about the psychological attributes of the people who have been tested—attributes such as intelligence, self-esteem, and so forth. Tests come in many different forms. Some involve open-ended situations with standard stimuli (such as a set of pictures); these are often used to bring out highly individualistic responses (such as stories composed in response to the pictures). Other tests involve very structured situations in which the range of possible responses is narrow and the answers are either right or wrong.

Tests are not magical or even very mysterious. They are merely standard ways of generating samples of people's behavior. However, assessments of these behavior samples are considerably more informative than are assessments based on random observations of someone's behavior. Their special value lies in the fact that they are

1. *Uniform*—The procedures are specified precisely so that different testers will follow the same steps every time they administer the tests (see Figure 13.2). This means that the test performances of different people (or of the same person tested at different times) can be compared directly.

2. *Objective*—The rules for scoring are spelled out, like the rules for test administration. Thus, the subjective input of the individual tester is minimized and the potential impact of personal biases on test scores is kept under control.

3. *Interpretable*—The better tests have been subjected to research that makes test scores meaningful to the psychologist. The research may show, for example, what characteristics of people tend to be associated with low or high scores.

TYPES OF TESTS

Tests have been developed to measure many different human characteristics. Among the most widely used tests are those designed to measure what people have learned—skills such as reading and arithmetic and the general information that people have acquired. These *achievement tests* have been developed and standardized for educational levels ranging from preschool through college. Because achievement tests focus on specific educational attainments rather than on psychological attributes per se, they are used more often by educators than they are by psychologists and therefore will not be discussed in detail here. However, it is worth noting that psychologists do sometimes use achievement tests in their efforts to develop a well-rounded picture of the people they work with in research or psychotherapy. Psychologists are much more likely to use two other kinds of tests, though: tests of ability and tests of personality.

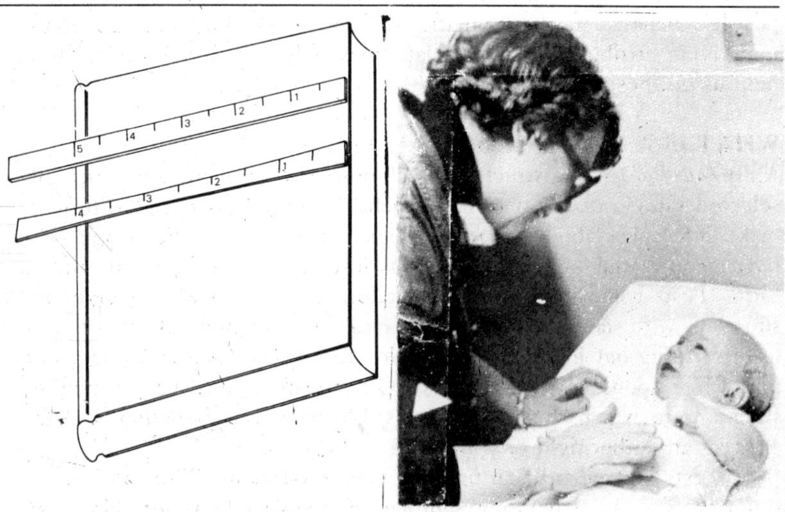

Figure 13.2

Left: A wooden ruler can be used to measure a line reliably; an elastic ruler cannot. Rulers are calibrated to a standard so that people all over the world will obtain comparable measurements. *Right*: Psychological testers similarly strive for a high level of uniformity—that is, for measurement procedures that are the same for all testers and test takers. Tests of infant development, for example, specify the exact steps to be taken by the tester with all infants tested. (Right figure from the University of Washington Child Development and Mental Retardation Center.)

Ability Tests Ability testing focuses on the question of what people *can do* when they are at their very best. In other words, *ability tests* are designed to measure capacity or potential rather than actual achievement. On the other hand, even the best ability test can measure no more than what a person does on the test itself. Thus, in one sense, every test is an achievement test. To get beyond this problem, ability-test constructors often try to measure skills or knowledge that most of the test takers have had a roughly equal opportunity to learn. For example, questions calling for skill in solving familiar problems or knowledge of one's native language can help to distinguish people of high ability in those areas from people of lower ability who have had a similar opportunity to learn the relevant skills. Another approach is to include tasks that are equally unfamiliar to most test takers. For example, people might be asked to memorize and use lists of nonsense symbols that have been assigned some arbitrary meaning. The symbol δ might stand for dog, the symbol α might stand for cat, and so forth. Since everyone would be learning these symbols for the first time, any individual differences in performance should reflect individual differences in ability at this kind of task. With methods like these, ability-test constructors try to minimize the effects of people's past experience so as to better measure their true capacity or potential.

Few tests of ability are actually called that. Most are called *tests of intelligence* or *tests of aptitude.* These two terms need to be distinguished from one another. *Intelligence* is probably made up of many abilities, but the term is most often used to refer to overall capacity for learning and problem solving. A good intelligence test measures capacities such as a child's potential for school learning or an adult's ability to cope with general intellectual problems. *Aptitude* usually refers to the ability to learn a particular kind of skill required in a specific situation. For example, we might test a person's aptitude for learning how to do mechanical drawing, learning how to pilot an airplane, or learning how to sell insurance.

Personality Tests As Chapter 14 will explain, an individual's *personality* includes his or her characteristic ways of thinking, feeling, or behaving. *Personality tests* are designed to reveal some of these characteristic patterns. Some personality tests measure attitudes—that is, the way a person responds emotionally and cognitively to another person, thing, or situation. Some personality tests measure interests—vocational interests, for example. Still other personality tests are designed to assess underlying thought processes, emotional states, or behavior patterns that are abnormal and may reflect psychological disorders.

CHARACTERISTICS OF A GOOD TEST

Many facets of ability and personality interest psychologists, and tests have been devised to measure many of these facets (see Buros, 1978, and Sundberg et al., 1983, for multiple examples). Thus, an important first step in most psychological testing is to select a test that is clearly focused on the ability or personality attributes of interest. Another important step is to ensure that the test selected is a sound one that will yield information of real value. This brings us to a key question: How do we decide whether a test is sound or not?

Is there any way to be absolutely certain that a given test is sound and has real value? No, but there are some important characteristics to look for—characteristics that a test should have if it is to be really trustworthy and informative. Three of the most important of these characteristics are reliability, validity, and norms (for a detailed discussion, see Anastasi, 1982).

Reliability A good test should be highly *reliable*. This means that the test should give similar results even though different testers administer it, different people score it, different forms of the test are given, and the same person takes the test at two or more different times. Reliability is usually checked by comparing different sets of scores, as shown in Figure 13.3*a*, for example. The figure shows some reliability information on two intelligence tests devised by "Smith and Jones"—two fictitious test constructors. Smith's test uses a series of analogies to test intelligence. Jones's test involves anagrams. To assess the reliability of both tests, we give each one to a group of people in September, then again to the same people in November. This is an assessment of test-retest reliability. As the figure shows, people who take Smith's test twice tend to score in about the same position relative to one another at both testings; alas, the same cannot be said of Jones's test. Another way of putting this is that the test-retest correlation is high for Smith's test but not for Jones's, which means that only Smith's test has high test-retest reliability.

In actual practice, psychological tests are never perfectly reliable. One reason is that real, meaningful changes do occur in individuals over time; for example, a person who scores low in her group at an initial testing may develop new skills that raise her to a higher position in the group by the time of the second testing. Despite such real changes, the best intelligence tests usually yield reliability correlation coefficients of .90 or higher (where 1.00 indicates perfect correspondence and 0.00 indicates no correspondence whatever). (See Appendix 1 for a discussion of correlation and correlation coefficients.) Almost all personality tests

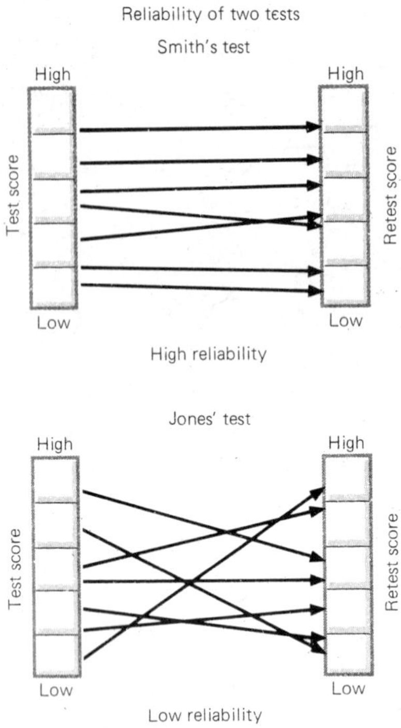

(A)

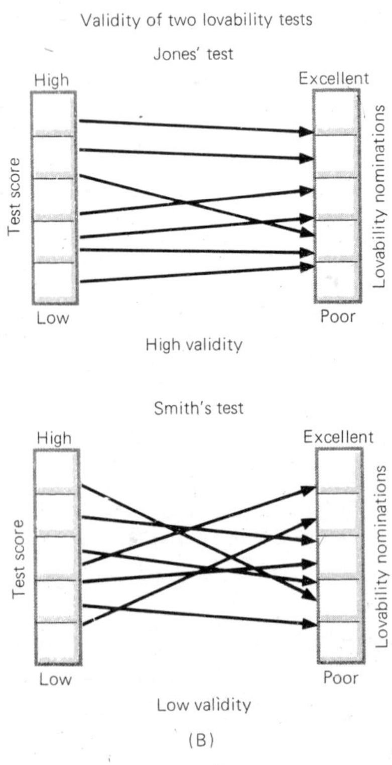

(B)

Figure 13.3a

A schematic view of test reliability for Smith's (top) and Jones's (bottom) intelligence tests. The left side of each diagram shows the test scores of seven different people, with scores rank-ordered from lowest to highest. The right side of each diagram shows the scores obtained by those same individuals when they took the same test a second time. The top diagram shows the results for Smith's test: *high reliability*—that is, the two sets of scores on Smith's test correspond closely to one another. The lower diagram shows the results for Jones's test: *low reliability*—that is, the two sets of scores do not correspond well at all. (Diagrams adapted from Braun & Linder, 1979.)

Figure 13.3b

A schematic view of test validity for Jones's (top) and Smith's (bottom) tests of "lovability." The left side of each diagram shows the test scores of seven different people rank-ordered from "most lovable" to "least lovable." The right side of each diagram shows the scores of those same people on a "lovability" criterion (in this case, lovability nominations by classmates). The upper diagram shows the results for Jones's test: *high validity*—that is, test scores correspond closely to scores on the criterion measure. The lower diagram shows the results for Smith's test: *low validity*—that is, test scores do not correspond to the scores on the criterion measure. (Diagrams adapted from Braun & Linder, 1979.)

have lower reliability than this; this may be due partly to the instability of things like attitudes and feelings, which personality tests are designed to measure. Low reliability usually means that the test in question has a high probability of error; this means that any predictions we make about people based on such tests will be risky. If tests with low reliability are used at all, their scores should be interpreted with caution.

What can be done to improve reliability? What might Jones, our frustrated test constructor in Figure 13.3*a*, do to improve the figures on his test? One possibility would be to ensure that the test was administered and scored by a truly standard procedure. Testers who do not know the test procedures well or who put the test instructions in their own words may be preventing the test experience from being uniform across all testings. In effect, they may be giving a somewhat different version of the test each time they administer it; and highly variable, unreliable scores may have resulted. Making the test procedures uniform might make the test more reliable.

Validity A second key characteristic of a good test is *validity*—the test must really measure what it has been designed to measure. Validity is most often assessed by exploring how the test scores correspond to some *criterion*—that is, some behavior, personal accomplishment, or characteristic that reflects the attribute the test is designed to gauge.

To illustrate, let us suppose that our intrepid test constructors, Smith and Jones, have been at work once more, this time trying to devise a test of lovability. Smith's test includes a number of questions asking respondents to state how they would solve complex interpersonal problems; Jones's test asks people to report on their attitudes toward and beliefs about people in general, friends, and dating partners. To assess validity, Smith and Jones agree to compare the scores on each test to a validity criterion that both agree makes sense: The number of nominations each test taker receives from classmates on a lovability form that reads, "Please name the three people in your class whom you find most lovable." As Figure 13.3*b* shows, scores on Smith's test do not correspond well to those lovability nominations; people who rank high on Smith's test do not necessarily rank high on nominations. Jones, though, has done well; people who scored as most lovable on his test tend to be considered most lovable by their classmates as well. Thus, the validity of Jones's test is supported, but the validity of Smith's test is not.

Of course, several different validity criteria might be used. Some might argue that true lovability is reflected best by the feelings of one's family, close friends, and dating partners—not by the feelings of one's classmates. Others might argue that the peer-nomination measure can assess only people's *reported* attitudes toward the test takers, and that people's actual behavior toward the test takers would be a better validity criterion. In other words, the validity of the lovability tests might be better assessed by observing the test takers in their everyday interactions; this might reveal whether people who score high on the measures also have lots of loving behavior directed toward them. What we are emphasizing here is that assessing the validity of any test requires careful selection of appropriate criterion measures and that reasonable people may disagree as to which criterion measure is best. This is equally true of intelligence tests. Reasonable people may disagree as to whether the best criterion measure of intelligence is school grades, teacher ratings, or some other measure. Nonetheless, if we are to check on the validity of a test, we must settle on one or more criterion measures of the attribute the test is designed to tap, whether it is lovability, intelligence, mechanical aptitude, or some other attribute. Once the criterion measures have been identified, people's scores on the measures can be compared to their scores on the test and the degree of correspondence can be examined for what it tells us about the validity of the test.

Only valid tests can give useful information about people, but the correlation coefficients for validity are never as high as are those for reliability. Though we try for reliabilities of .90 or more, we are usually satisfied with validities of .50 or .60; *validities*—correlations between test scores and criterion measures—seldom run higher than that. Validities of .30 and .40 are common. Tests with validities this low actually relate very poorly to the criterion. However, several tests with low but significant validity can sometimes be useful if they are given together as a battery and their results are considered together.

One reason that validity coefficients are lower than reliability coefficients is that the reliability of a test sets limits on how valid the test can be. A test that cannot give us reliable scores from one testing to the next is not likely to show dependable correlations with any validity-criterion measure either. On the other hand, high reliability is no guarantee that a test is valid. The relationship between reliability and validity can be understood by harking back to Sir Francis Galton's efforts, discussed earlier. Galton's sensory and motor measures could never have been valid if they had not been reliable; that is, if they had not yielded similar scores for different testers, different times of testing, and so on. Yet even though some of Galton's measures turned out to be very reliable, later evidence showed that they were not valid as measures of intelligence. The measures yielded similar scores time after time, but those scores were poorly correlated with validity criteria such as school grades (Wissler, 1901) and teacher ratings of intelligence (Bolton, 1892).

Norms In addition to reliability and validity, good tests need norms. *Norms* are sets of scores obtained by representative groups of people for whom the test is intended. The scores obtained by these groups provide a basis for interpreting any individual's score.

To understand why norms are important, we need only imagine a test that doesn't have any. Suppose a person takes a newly developed intellectual aptitude test and receives a score of 1437. Is this a "good score"? Should the person be elated or deflated? Obviously, a score without any basis for comparison is not very useful. In fact, one of the first things a person in this fictitious situation might do is seek out others who have taken the test to find out how his or her score compares to theirs. Psychologists do something similar to this when they develop norms. They seek out comparison groups whose performance on the test can serve as a standard of comparison for each individual who takes the test later.

To illustrate how this works, let us focus on a familiar standardized test, the Scholastic Aptitude Test (SAT). You probably took the SAT or a test like it before you entered college. When the results came back to you, you were told not only your scores, which by themselves may not have meant much to you, but also your percentile compared with certain groups. (A *percentile score* gives the percentage of scores that yours equals or exceeds.) One norm is the scores of all high school students taking the test. Another norm is the scores of all students who later went to college. Sometimes you can also get your percentile relative to entering freshmen in specific colleges you are considering. Each of these scores will help you understand the implications of your test results. Of course, the scores will not tell you everything you need to know. For example, you may over-achieve because you work very hard; thus, you may do better in college than your scores, and percentile, suggest.

Psychologists have developed only a few well-standardized tests like the SAT. One reason is that enormous expense and effort are involved in obtaining and updating national norms. In most cases, the normative sample needs to be representative of the population with which the test will be used. This is not a major problem for populations such as school-aged children, who can readily be found in schools, but it is a very difficult problem for adults. Furthermore, time does not stand still; norms go out of date. In the past 2 decades, intelligence-test scores for preschool children (Garfinkel & Thorndike, 1976) and adults (Schaie et al., 1973) have gone up; while at the high school and college levels, the SAT, GRE (Graduate Record Exam), and others have gone down. Although the reasons for these trends are a matter of debate, such changes point up the necessity of restandardizing tests from time to time.

WHO MAKES DECISIONS FROM TESTS, AND HOW?

While tests are often used in research, they are used much more often in practical situations where decisions need to be made about people—for example, decisions about employment or about admission to certain educational programs. Because these decisions are often quite important to the people involved, tests must be selected and used with great care. This means that the people who make the decisions must pay close attention to test content, selecting tests focused precisely on the relevant person attributes. It also means that only tests with demonstrated reliability and validity should be used. Finally, it means that even tests which meet all these criteria should be used as *only one part* of the decision-making process. The most informed decisions about people tend to be those involving a healthy mix of information. Test data should be combined with information about a person's background, interests, and past performance in order to flesh out the picture of the individual in ways that test scores alone could never do. The result can be informed decision making that is aided but not dominated by testing.

Tests are used to select people for some very significant life options. Colleges almost all require scores on some ability test like the SAT. Professional schools—law and medical schools, for example—use tests devised especially for their purposes. Most graduate school programs require the GRE.

Employers, too, use ability tests and personality tests; but they often use *achievement tests* as well—tests of typing speed, mechanical knowledge, or other job-relevant skills. In practice, tests are most often used in situations where there are more applicants than positions available. Often, people are initially considered on the basis of their previous education or employment, the recommendations of others, personal interviews, and other such "real-life" input; if, after all these factors have been considered, there still remains an oversupply of qualified applicants, test results may be added to the blend of information used to make the final decision. Although tests are most often used in situations where there is an oversupply of applicants, an organization might also require that all of its employees exceed a certain minimum test score and might actually leave positions unfilled if no applicants meet this minimum standard. This sometimes happens because previous experience has shown that applicants scoring below the cutoff have a higher risk of failure than the organization is prepared to accept.

At the other extreme, test scores at the upper end of the scale may call attention to particularly competent individuals who might otherwise have been overlooked. Often these are people who do not score well on achievement tests but who have aptitude that is waiting to be tapped. Testing in the military, for example, sometimes identifies "diamonds in the rough"—people who score unexpectedly well on specific aptitude tests and can be steered into training programs that build on their newly discovered potential. One example is the case of Ben.

> Ben joined the Marines at the outbreak of World War II, before completing high school. Over the previous 10 years, he had been in a series of foster homes and a juvenile correction facility. Ben had a very inconsistent high school record—a few high grades mixed with low ones, many absences, and a reputation for being something of a "wise guy." Military service seemed to offer him an attractive way out of a life without much promise. On entry, Ben was given the Army General Classification Test (AGCT). He scored exceptionally high. He thus came to the attention of a personnel officer who saw to it that he was sent to a school offering intensive and challenging training in ballistics and meteorology. Eventually, Ben became an officer. Following discharge in 1945, he was much more confident of his scholastic abilities. He used his veterans' benefits to attend a top-level college, then entered graduate school, and eventually became a college professor.

Identification and special treatment of extremely high and extremely low scorers actually begins well before the years of military service. In school systems, selection for special class placements usually involves some use of tests—a practice which began with the work of Binet and Simon (1905/1916). Often standardized intelligence tests are required before children can be placed in classes for the mentally retarded, the intellectually gifted, the learning disabled, or the emotionally handicapped. However, such special class placements are rarely made on the basis of test scores alone; in fact, it would be illegal to do so in most cases. Often standardized tests are administered only after the children's performance in their regular classes has shown that they do not fit well—that is, that they are moving much more slowly or much more quickly than most in the class or they have extreme difficulty with certain kinds of learning. Tests can also be used in quite a different way; sometimes a child is thought to be mentally retarded or learning disabled, but tests suggest otherwise. In some cases, what seemed to be a problem of intellectual functioning turns out to be a visual or auditory handicap or a motivational problem.

Tests can also be used to help make decisions involving commitments or to make judicial decisions about people who are in trouble. Sometimes a person behaves so strangely or seems so disturbed that family members or others believe the person should be committed to a mental hospital. Practically every state requires that some combination of psychological testing and interviewing by professionals be done before any decision is made about involuntary commitment. Most require that the interviewing and testing show the person to be in danger of harming self or others before a forcible commitment can be made. Another kind of legal issue that may stimulate testing is the insanity defense (see Chapter 15). If a person accused of a crime pleads "not guilty by reason of insanity," a psychologist may be called on to testify and may base the testimony partly on the results of psychological testing. Decisions about the treatment of juvenile offenders may also be based in part on input from psychologists and their tests. Finally, decisions as diverse as those involving child custody and whether a person qualifies for social security disability payments may hinge, in part, on the results of psychological testing.

Is it reasonable and fair to let so many crucial decisions be influenced by tests? The answer to this important question depends partly on the tests themselves—their strengths and their limitations. In the following section we will examine some of the major tests, beginning with those designed to measure intelligence. Before we look at intelligence tests, though, we must return to the basic question tackled by Galton and Binet decades ago: What *is* intelligence?

The Nature of Intelligence

> What is intelligence, anyway? When I was in the Army I received a kind of aptitude test that all soldiers took and, against a normal of 100, scored 160. No one at the base had ever seen a figure like that and for two hours they made a big fuss over me. (It didn't mean anything. The next day I was still a buck private with KP as my highest duty.)
>
> All my life I've been registering scores like that, so that I have the complacent feeling that I'm highly intelligent, and I expect other people to think so, too. Actually, though, don't such scores simply mean that I am very good at answering the type of academic questions that are considered worthy of answers by the people who make up the intelligence tests—people with intellectual bents similar to mine? . . .
>
> I had an auto repairman once, who, on these intelligence tests, could not possibly have scored more than 80, by my estimate. I always took it for granted that I was far more intelligent than he was. Yet, when anything went wrong with my car I hastened to him with it, watched him anxiously as he explored its vitals, and listened to his pronouncements as though they were divine oracles—and he always fixed my car. . . .
>
> He had a habit of telling me jokes whenever he saw me. One

time he raised his head from the automobile hood to say: "Doc, a deaf-and-dumb guy went into a hardware store to ask for some nails. He put two fingers together on the counter and made hammering motions with the other hand. The clerk brought him a hammer. He shook his head and pointed to the two fingers he was hammering. The clerk brought him nails. He picked out the sizes he wanted, and left. Well, doc, the next guy who came in was a blind man. He wanted scissors. How do you suppose he asked for them?"

Indulgently, I lifted my right hand and made scissoring motions with my first two fingers. Whereupon the auto repairman laughed raucously and said, "Why, you dumb jerk, he used his voice and asked for them." Then he said, smugly, "I've been trying that on all my customers today." "Did you catch many?" I asked. "Quite a few," he said. "But I knew for sure I'd catch you." "Why is that?" I asked. "Because you're so . . . educated, doc, I knew you couldn't be very smart."

And I have an uneasy feeling he had something there.

[Isaac Asimov, from Braun, J. J. & Linder, D. E. (1979). *Psychology Today: An Introduction* (4th ed.). New York: Random House. Copyright © 1979 by Random House, Inc. Reprinted with permission.]

Figure 13.4

Developing a theory of intelligence is a little like coping with a Loch Ness monster. Sailors who spot the "monster" shown in *a* (top) may form various theories about what lies beneath the surface. They may conclude that all the visible parts belong to a single serpent (*b*); this is analogous to the view that intelligence is a single, general ability. They may conclude, instead, that a few different serpents lie beneath the surface (*c*); this is analogous to the view that intelligence is composed of several separate abilities. Or they might conclude that every visible part belongs to a separate creature (*d*); this is analogous to the view that every intellectual task involves a different ability. (Modified from Gleitman, 1981.)

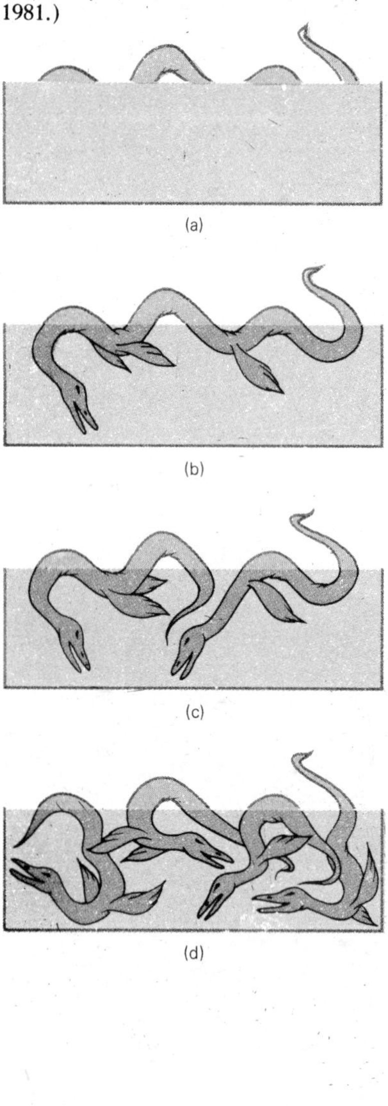

(a) (b) (c) (d)

As Isaac Asimov's story suggests, intelligence is hard to define. When several people in a group speak of intelligence, they are likely to nod knowingly as if they all share a common definition. Different people are likely to agree fairly well on who the bright people in their work group or social circle are. However, there are wide variations in lay people's definitions of intelligence. Psychologists, too, differ from one another in their definitions of the concept. To illustrate these differences, we will sample a few theories of intelligence from two different groups of theorists. One group consists of theorists who have studied the organization of mental ability; their primary interest is in identifying the factor or factors which constitute intelligence. The theories that have emerged from these efforts are called *factor theories*. The second group of theorists has focused not on the component parts of the intellect but on the *processes involved in intellectual activity*—that is, the processes involved in solving problems or planning how to remember something. The result of their efforts has been a group of *process-oriented theories of intelligence*. We turn first to the factor theories.

FACTOR THEORIES OF INTELLIGENCE

Is intelligence a single characteristic, or is it a collection of specific, distinguishable abilities? This question has been a centerpiece of discussion and debate among the factor theorists for many years. The question is hard to answer, partly because it involves underlying intellectual capacities that we cannot see directly. We can only infer these underlying capacities from people's observable surface behavior—on intellectual tests, for example. The situation has been compared to that of a man who sees what *may* be a serpent in a lake (Gleitman, 1981). Figure 13.4*a* shows what the man actually sees. Figure 13.4*b* shows one possible interpretation the man might make—that is, that all the visible parts belong to a single creature. This is analogous to the view that there is a single ability—intelligence—underlying the various intelligent behaviors that people show on the surface. Parts *c* and *d* of the figure show two alternative interpretations, or theories, that the man might form about what he sees. The view that there are two or three serpents *(c)* is analogous to the view that intelligence consists of a few major factors. The view that there are as many serpents as there are visible parts *(d)* is analogous to the view that every intellectual task involves a totally different ability. Like the man observing the creature(s) in the lake, investigators who study intelligence must decide how those things they can observe fit together, or if they fit together at all.

In making such decisions about intelligence, many have used a statistical technique known as *factor analysis*. The technique is a way of identifying groups of abilities or behaviors or traits that are related to one another. In the area of intelligence testing, the technique is usually applied to several specific subtests, each designed to measure one specific cognitive ability. For example, one subtest might require the test taker to remember a series of numbers called out by the tester; another subtest might require the test taker to arrange pictures in an order that tells a coherent story, and so forth. Correlation coefficients are then computed among the various subtests to determine which ones are most closely related to each other. Subtests that correlate well with each other but not with other subtests are said to represent a common factor. By inspecting the subtests that form a common factor, psychologists can make a judgment about the nature of that factor. Suppose, for instance, that one factor has been found that includes the following subtests: remembering a series of numbers, recalling details of a story, reconstructing a design from memory, and recalling a series of steps in a production manual. The factor that includes these four tests might reasonably be considered a memory factor.

Factor analysis poses several problems for the investigator. For example, different methods of factor analysis can yield different factors, and it is often hard to judge which factors are best. An even more basic problem is that what comes out of factor analysis depends on what the psychologist puts into it. A psychologist who read Isaac Asimov's story (above) might decide to put several tests of mechanical ability into a collection of intelligence subtests. If so, a mechanical-ability factor might well be found, meaning that the mechanical-ability subtests correlate well with one another. Does this mean that mechanical ability is an integral part of intelligence? Not necessarily. Factor analysis can help us form coherent groups of subtests, but it cannot tell us what subtests to include in the first place.

As the preceding discussion shows, factor analysis involves a number of subjective judgments. Different theorists make these judgments differently. Thus, even though several theorists may all be using factor analysis to study, say, the nature of intelligence, the theories they come up with may be quite different from one another. This is illustrated by the theories we consider next; each one is noticeably different from the others, despite the fact that all were developed partly through factor analysis.

G-Factor Theory The factor theories of intelligence, like the "serpent theories" in Figure 13.4, differ in their perspective on what lies "beneath the surface." In one of the earliest and most influential factor theories, British psychologist Charles Spearman (1927) proposed that a broad general intelligence (G) factor lay beneath the surface. Spearman noted that a number of different cognitive tasks and intellectual measures tend to be correlated with one another—that is, people who score high on one tend to score high on the others as well. Using an early version of factor analysis, Spearman found what he believed to be a single common factor, G, shared by the various tests. Spearman argued that each individual intellectual task taps both general intelligence, or G, and some other ability, S, specific to that particular task. For example, an arithmetic test might tap both G and a specific mathematical ability. Spearman's views, which have come to be called *G-factor theory*, are reflected in intelligence tests that yield a single score, such as an IQ (see pages 529–534). The notion that intelligence can be largely summed up with one score is closely related to Spearman's idea that intelligence consists of one general ability factor.

Multifactor Theories In contrast to Spearman, several theorists have concluded that intelligence has multiple components. These theorists generally agree with Spearman that diverse intellectual tasks are usually correlated with one another—the basic fact that led to G-factor theory in the first place. However, they go beyond this fact, noting that certain clusters of tests show higher correlations with one another than with other tests. For example, memory tests tend to show higher correlations with each other than with other tests; and tests that involve calculating numbers are better correlated with each other than with tests that are not numerical. Does this means that intelligence includes a memory factor, a numerical-ability factor, and perhaps other factors as well? Yes, say a number of the theorists who disagree with Spearman. The theories they have proposed are known as *multifactor theories*.

One of the most influential of the multifactor theories grew out of the work of factor analyst L. L. Thurstone (1938). Thurstone began with a set of 56 tests; from the patterns of correlation among these tests, he identified factors which he called *primary mental abilities*. They included verbal comprehension, word fluency, perceptual speed, memory, numerical ability, spatial ability, and reasoning. Thurstone assembled a battery of tests to measure these abilities. This Primary Mental Abilities test (PMA) is still widely used.

A multifactor theory considerably more complex than Thurstone's was proposed by J. P. Guilford (1967). This three-dimensional theory grew out of a massive analysis of a great many existing tests. It resulted in the cubical model seen in Figure 13.5. This model provides for 120 factors of intelligence. Each factor is represented by a cell in the cube and is some combination of these three dimensions: (1) five kinds of operations, (2) six kinds of products, and (3) four kinds of contents ($5 \times 6 \times 4 = 120$). That seems like a lot, but one or more tests were devised to measure most of these factors. Guilford's concept of intelligence also includes what he calls *divergent thinking*, which is closely related to creative, or original, problem solving (Chapter 6, page 246), as opposed to *convergent thinking*, which is involved in solving problems with a single correct answer.

Hierarchical Theory We still do not know which of the various factor theories provides the best account of intelligence. However, there does

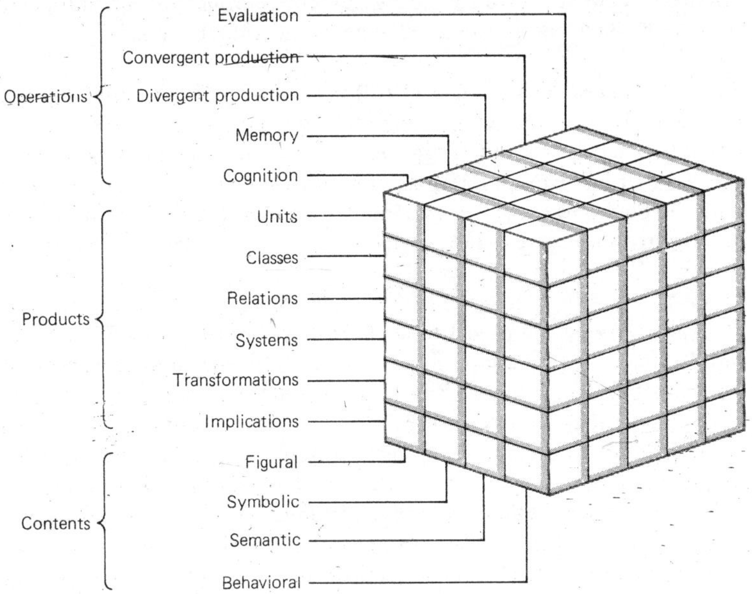

Figure 13.5

The cubical model of intelligence. Each of the 120 small cubes represents a primary ability that is some combination of operations, products, and contents. (Guilford, 1961.)

seem to be some truth in *both* G-factor theory and those theories that propose multiple factors. We can identify some ability factors that are *relatively* independent of one another; but when we do, we usually find some significant correlations among the factors, indicating that they share some sort of general intelligence factor. Consequently, some (for example, Vernon, 1950) have proposed that elements of G-factor theory and the multifactor theories be combined to form a *hierarchical theory*. In such a theory, intelligence is pictured as a sort of pyramid. At the top of the pyramid is G, general intelligence, which shows up in virtually all kinds of intellectual activity. Underneath it are several moderately specific ability factors like Thurstone's primary mental abilities. At the bottom of the pyramid are a larger number of highly specific abilities, similar to Spearman's (1927) S factors—abilities that may come into play on one particular task. This hierarchical theory borrows from several factor theories to form a multilayered view of intelligence—a view that may turn out to be the most reasonable of all.

PROCESS-ORIENTED THEORIES OF INTELLIGENCE

Each of the preceding theories is an attempt to unravel intelligence—to find its component parts and describe how those parts fit together. This is not the only path to an understanding of intelligence. An alternate approach taken by several influential theorists is to focus on *intellectual processes*—the patterns of thinking that people use when they reason and solve problems. These theorists use a somewhat different vocabulary than the theorists described above—for example, they often speak of *cognition* and *cognitive processes* rather than *intelligence*. Also, they are often more interested in *how* people go about solving problems and figuring out answers than in *how many* right answers people get. Finally, the process-oriented theorists tend to focus on the *development* of intellectual processes—how the processes change as individuals mature.

Piaget's Theory Jean Piaget (1970) is a particularly prominent process theorist. His stage theory of cognitive development was discussed in detail in Chapters 11 and 12. In Piaget's view, intelligence is an adaptive process that involves an interplay of biological maturation and interaction with the environment. He views intellectual development as an evolution of cognitive processes such as understanding the laws of nature, the principles of grammar, and mathematical rules.

Bruner's Theory Jerome Bruner (1973) is a process theorist who sees intellectual development partly as a growing reliance on internal representation. Babies, according to Bruner, have a highly action-oriented form of intelligence; they "know" an object only to the extent that they can act on it. Young children know things by perceiving them and are consequently strongly influenced by the vivid perceptual characteristics of objects and events. Older children and adolescents know things *internally* and *symbolically;* this means that they are able to devise internal symbols, or representations, of objects and actions and to hold these mental images in mind. Bruner is interested in how these growing abilities are influenced by the environment—especially by the rewards and punishments people receive for using particular intellectual skills in particular ways.

Information-Processing Theories Among the most influential process-oriented approaches to intelligence are those known as *information-processing theories*. These theories break intelligence down into various basic skills that people employ to take in information, process it, and then use it to reason and solve problems. These basic skills may be as simple as the ability to distinguish between two tones or as complex as the ability to plan how to remember a long list of names. In one approach, componential analysis, Robert Sternberg (1984) distinguishes between information-processing "components" and "metacomponents." Components are the steps one goes through to solve a problem; metacomponents are the kinds of knowledge one has about how to solve the problem. Sternberg's idea is that we use metacomponents to plan and regulate our behavior and that metacomponents are closely related to the kind of general intelligence proposed by Spearman. Sternberg is researching how the components and metacomponents relate to each other during various intellectual activities and how both grow more complex during development.

There are many more information-processing theories than we can review here, but a few additional examples will help fill out the picture. A group sometimes called the neo-Piagetians has virtually rewritten Piaget's theory in information-processing terms—terms often resembling those used to describe computers. One of these theorists, Juan Pascual-Leone (1983), has expanded Piaget's notion of the scheme (Chapter 11, page 426), distinguishing, for example, between action schemes, which are similar to specific, repeatable intellectual sequences, and executive schemes, similar to plans and strategies. Such schemes are referred to as intellectual software, as distinguished from intellectual hardware, resources such as attention and memory. The neo-Piagetians hypothesize that people's software grows more sophisticated as they mature, with their

schemes expanding in complexity and their amount of a
energy increasing. Such changes, they claim, promote
intelligence. Other information-processing approaches focus
involved in intelligent behavior (for example, Siegler, 1983)
required for various tasks (for example, Fischer, 1980). De
differences, all of the various approaches break intelligence into
nent processes and explore how the processes change over the c
development.

Assessing Intelligence

Each theory about the organization and nature of intelligence of course implies a somewhat different way of sampling people's behavior to estimate their mental ability. The G-factor theory, for example, suggests that a single score will represent intelligence adequately. Multifactor theories point to a need for separate subtests to tap the various ability factors. For instance, Guilford and his associates have been working diligently for years to develop a subtest for each of the 120 cells in his three-dimensional model (Figure 13.5). By the same token, the process-oriented theories point to specific patterns, component processes, capacities, or skills that would need to be measured in a test of intelligence. The best-known and most widely used intelligence tests are not *deeply* rooted in any particular theory, but they do use subtests, and they yield an overall summary score (the IQ, to be discussed below)—two features that are consistent with the factor theories of intelligence. Among the most important of the intelligence tests are the Stanford-Binet Intelligence Scale and three tests developed by David Wechsler for three different age groups. We will begin with the Stanford-Binet, then turn to the Wechsler scales.

STANFORD-BINET INTELLIGENCE SCALE

The test developed by Binet and Simon to identify mentally retarded children in French schools served its purpose well. Subsequently, several English-language versions of the test were produced. The most successful was brought out in 1916 by Lewis Terman of Stanford University. Terman's scale, known as the Stanford-Binet, became the model for many intelligence tests and has itself been revised several times (Terman & Merrill, 1973).

Binet devised his test by age levels. This was because he observed that mentally retarded students seemed to think like nonretarded children at younger ages. Following Binet's lead, other test constructors have also produced age scales. Within these scales, the tasks at each level are those which average children of that age should find moderately difficult. Children are given only the levels in their range. For testing purposes, the highest level at which all items are passed by a given child is that child's basal age. Starting with that basal age, the tester adds additional credit for each item the child passes until the child reaches a ceiling age—that is, the lowest level at which all items within the level are failed. The Stanford-Binet was so constructed that a random population of children of a given *chronological age* (*CA*) obtains an average score, or *mental age* (*MA*), equal to their CA. An individual's performance on the test can therefore be expressed as a mental-age score.

Table 13.1 presents some of the items from the present version of

Table 13:1
Some items from the Stanford-Binet Intelligence Scale. On the average,
items should be passed at about
indicated.

…hese … the ages

AGE	TYPE OF ITEM	DESCRIPTION
2	3-hole form board	Places forms (circle, triangle, square) in correct holes after demonstration.
	Block building: tower	Builds a 4-block tower from model after demonstration.
3	Block building: bridge	Builds a bridge consisting of 2 side blocks and 1 top block from model after demonstration.
	Pure vocabulary	Names 10 of 18 line drawings.
4	Naming objects from memory	One of 3 objects (for example, car, dog, or shoe) is covered after child has seen them; child then names object from memory.
	Picture identification	Points to correct pictures of objects on a card in response to "Show me what we cook on" or "What do we carry when it is raining?"
7	Similarities	Answers such questions as "In what way are coal and wood alike? Ship and automobile?"
	Copying a diamond	Draws 3 diamonds following a printed sample.
8	Vocabulary	Defines 8 words from a list.
	Memory for stories	Listens to a story, then answers questions about it.
9	Verbal absurdities	Says what is foolish about stories similar to "I saw a well-dressed young man who was walking down the street with his hands in his pockets and twirling a brand new cane."
	Digit reversal	Repeats 4 digits backward.
Average adult	Vocabulary	Defines 20 words from a list (same list as at age 8, above).
	Proverbs	Explains in own words the meaning of two or more common proverbs.
	Orientation	Answers questions similar to "Which direction would you have to face so your left hand would be toward the south?"

(*Source:* Terman and Merrill, 1973.)

the Stanford-Binet Scale. The first thing to notice is the wide variety of abilities that are tapped. The items chosen from among a large pool of "good bets" were those which correlated best with scores on the scale as a whole. Thus, the test makers tried to select items which revealed a general dimension. Binet and Terman worked from a notion of intelligence as an overall ability related to abstract reasoning and problem solving; judgment was the term Binet used. The items become less concrete and more verbal as one goes up the age scale.

We have noted that an individual's score is expressed as a *mental age* (*MA*). It did not take psychologists long to note that the MA could be expressed in relation to the CA in order to estimate the *rate* of development. The MA/CA ratio yields the *Intelligence Quotient* (*IQ*), a concept proposed by psychologist William Stern in 1912. If two children both obtain an MA of 5 years on an intelligence test, but one child is 4 years old and the other is 6, obviously the younger child is developing intellectually at a much faster rate. To express this fact in the form of IQs, we take the ratio of MA to CA and multiply by 100 to eliminate decimals.

$$IQ = \frac{MA}{CA} \times 100$$

Child 1	Child 2
$IQ = \frac{5}{4} \times 100$	$IQ = \frac{5}{6} \times 100$

Thus, the bright child mentioned above earns an IQ of 125, and the slower child earns an IQ of 83.

The *ratio IQ*, as this is called, makes good logical sense, but it has a number of problems. One problem is that mental age does not increase in a rapid, orderly fashion after the middle teens. The concept of mental age in adults—say, a mental age of 21 or 37—is meaningless. A ratio IQ is therefore not very useful with adults. The Stanford-Binet test now uses another kind of IQ, the deviation IQ, which is described in an upcoming section.

WECHSLER TESTS

David Wechsler developed a family of tests for people at various age levels. The tests include the Wechsler Adult Intelligence Scale, Revised (WAIS-R, 1981), the Wechsler Preschool and Primary Scale of Intelligence (WPPSI, 1967), and the Wechsler Intelligence Scale for Children, Revised (WISC-R, 1974). (See Figure 13.6.) These are all individual tests made up of a variety of tasks. They do not, however, have separate levels for different ages. Instead, all the tasks of a single kind are grouped together in a subtest. The subtests are short and therefore not very reliable so that differences must be large to be taken seriously.

The subtests can be grouped into two categories, verbal and performance, as shown in Table 13.2. This feature is often helpful in testing

Figure 13.6

An item from the Wechsler Intelligence Scale for Children, Revised. (University of Washington Child Development and Mental Retardation Center.)

Table 13.2

Subtests of the Wechsler Adult Intelligence Scale, Revised.*

	VERBAL SUBTESTS
Information	Factual knowledge about nature, geography, and historical events (for example, On what continent is the Rhine River located?).
Comprehension	Understanding of social conventions, rules, familiar sayings, and so on (for example, What does this saying mean? "People who live in glass houses should not throw stones").
Memory span	Simple rote memory for a series (for example, When I finish reading this list, repeat it after me. "Dog, Susan, go, box, dice, cannot, chair, open, beneath").
Arithmetic	Mathematical reasoning and computation (for example, Suppose you buy a coat that was $50 but is now marked "20% off." How much should you pay?).
Similarities	Detecting relationships among objects and concepts (for example, How are a tree and a blade of grass alike?).
Vocabulary	Defining words of varying difficulty (for example, What does *perverse* mean?).
	PERFORMANCE SUBTESTS
Picture arrangement	Putting a set of pictures in order so that they tell a coherent story.
Picture completion	Finding incomplete or missing parts of pictures that are otherwise complete.
Block design	Arranging colored blocks into a design that matches one that is pictured on a card.
Object assembly	Putting pieces of a jigsawlike puzzle together correctly.
Digit symbol	Learning to use a coding system in which nonsense symbols (for example, ♂ and ⇧) represent numbers.

* The illustrations given in parentheses are not actual items in the Wechsler scale.

people with limited verbal skills, foreign backgrounds, or poor education, provided they can understand the instructions. Such individuals frequently do better on performance tests than on verbal tests. The verbal-performance distinction is also helpful in testing people who are experiencing mental impairment because of brain damage or emotional disturbance. Such people sometimes perform quite differently on verbal than on performance tests, and the nature of the discrepancy can provide clues about the nature of their problems.

Wechsler devised the deviation IQ, which is now in general use with well-standardized intelligence tests. To explain the deviation IQ, we must first discuss the statistic known as the *standard deviation;* this is a measure of the spread, or variability, of scores for a group of people. (See Appendix 1.) A large standard deviation indicates that a number of the scores deviate quite a bit from the average, or *mean;* a small standard deviation tells us that there is less variability. For example, on the first quiz of the semester, the mean of a psychology class might be 70 and the standard deviation 10; on the second quiz, the mean might also be 70 but the standard deviation 15. The scores were more variable—that is, more spread out—with more extremes—that is, more high and low scores—on the second quiz (Figure 13.7).

Any individual's score can be expressed in terms of how it compares to the standard deviation—that is, the number of standard deviations by which it differs, or *deviates,* from the mean. Suppose your score on the first quiz was 80; since the standard deviation is 10 and your score is 10 points above the mean, your score is exactly one standard deviation above the mean. If you also made a score of 80 on the second quiz, where the mean was 70 and the standard deviation was 15, your score would be 0.67 standard deviations above the mean (Figure 13.7). Compared with the rest of the class, you did not do quite so well on the second quiz as on the first one, even though your score was the same on both quizes. Scores expressed in terms of standard deviation units are called *standard scores* because they make it possible to compare scores on different tests. Your standard score on the first test would be 1.00. Your standard score on the second test would be 0.67.

Analysis from the cupid pod?
bell shaped curve
normal probability curve.

The deviation IQ is a type of *standard score*—that is, an IQ expressed in standard deviation units. Wechsler (and, since him, other test makers) converted to standard scores the actual raw scores obtained by various age groups of subjects on whom the tests were developed. No matter what the average score obtained by persons in an age group, Wechsler changed it to 100; no matter what the actual standard deviation of that group, he changed it to 15. Each score (number of points) a person received could then be expressed as a standard score, or *deviation IQ*. Wechsler's tests yield three different deviation IQs, one for the verbal subtests, another for the performance subtests, and a third full-scale IQ.

To illustrate how deviation IQs are calculated, let's go back to the hypothetical set of test scores shown in Figure 13.7. The first step is to calculate, in actual raw scores, how far an individual's score deviates from the average. Then that score is compared with the standard deviation of the scores. Remember on the first test, a score of 80 was equal to +1.00 standard deviation units, and on the second test, a score of 80 was equal to +0.67 standard deviation units. We can express this procedure as a formula.

$$\text{Standard score} = \frac{X - M}{SD}$$

where X is the individual's score, M is the mean, and SD is the standard deviation. Thus, as you will recall, your standard score on the first quiz was +1.00; on the second quiz, it was +0.67.

Once we have calculated the standard score, we can convert it to any scale we choose. We could, like the makers of the SAT, make the mean

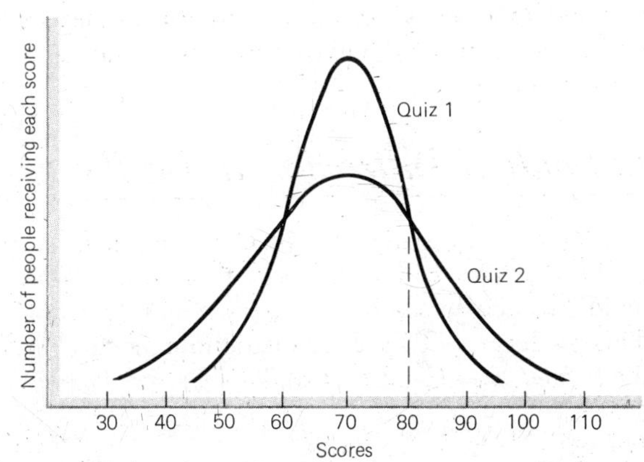

Figure 13.7

Two distributions of quiz scores, one with a broader spread than the other. If a student obtained a score of 80 on each quiz, we can say that, relative to the rest of the class, the student did better on the first quiz than on the second.

equal to 500 and the standard deviation equal to 100. In that case, the first quiz score would be converted to 600 and the second to 567. Or like Wechsler, we could make the mean equal to 100 and the standard deviation equal to 15. Then, the first psychology test score would be expressed as 115 and the second as 110.

PROCESS-ORIENTED ASSESSMENT OF INTELLECTUAL DEVELOPMENT

Process-oriented theorists (page 527) have now begun to develop assessment methods of their own, but these methods are still in their infancy. Thus, perhaps it is only fitting that the example we will discuss here involves infant assessment. Ina Uzgiris and J. McV. Hunt (1975) developed a set of six developmental scales intended to measure "progressive levels of cognitive organization" in the first 2 years of life. The scales are rooted in cognitive developmental theory, particularly Piaget's theory. Because this theory deals with stages of intellectual development, the scales developed by Uzgiris and Hunt are stagelike. They are designed to indicate where a particular baby is at a given point in time within a particular sequence of developmental stages. Uzgiris and Hunt did not standardize their scales. Although representative samples of infants could be sampled and age norms for each scale could thus be established, Uzgiris and Hunt chose not to do so. One reason is that their strong process orientation leads them to focus on the question of where a given infant is in relation to a sequential process of development, *not* how the infant compares to other babies of the same age.

The six Uzgiris-Hunt scales are designed to capture six different processes of cognitive development, all occurring within what Piaget labeled the *sensorimotor stage* (Chapter 11, page 425). For example, one scale is focused on development of the concept of the permanent object, as discussed in Chapter 11. Another scale is designed to assess ways of relating to objects (ranging from mouthing them, at about 2 months, to calling them by name, at about 18 months).

The Uzgiris-Hunt scales are useful as an illustration of process-oriented assessment. This form of assessment is a means of understanding individuals and where they stand with respect to specific cognitive processes, particularly those that are developmental in nature. Such assessment has been especially useful in revealing what intellectual skills individual handicapped children possess and what skills they are prepared to develop next (Dawson & Adams, 1984); such information can be valuable to both teachers and parents. On the other hand, process-oriented approaches like the Uzgiris-Hunt scales are not as useful as are traditional IQ tests when it comes to the study of individual differences, our next topic.

Individual Differences in Intelligence

Intelligence (like many other psychological traits) seems to be distributed in the population in such a way that most people make scores in the middle range while only a few people make very high or very low scores. This produces a bell-shaped distribution, a curve which statisticians call the *normal curve* (Figure 13.8).

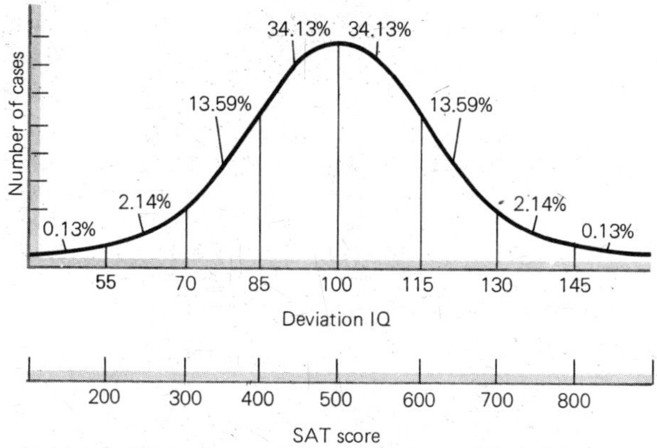

Figure 13.8

Norms for IQ and SAT (Scholastic Aptitude Test) scores. Norms for large groups of measurements often approximate the normal curve (top). The numbers above the curve give the percentages of people in each of the indicated segments. For instance, 13.59 percent of the population obtain IQ scores between 115 and 130.

Differences in intelligence greatly affect people's ability to cope with the demands of society. This is particularly true in a technologically sophisticated, mobile, and competitive society like our own. Nowhere are the demands as strict as they are during school, where children are expected to master complex tasks at a pace determined by the development of the average child. Some children learn quickly, others slowly. While high intelligence is no guarantee of the "good life," low intelligence creates enormously difficult barriers to full participation in society and the attainment of a high standard of living.

MENTAL SUBNORMALITY

People are appropriately regarded as *mentally retarded* if (1) they attain IQs below 70 on an appropriate intelligence test *and* (2) their adaptive skills are inadequate to cope with ordinary daily tasks. During early childhood, a person's adaptation is judged by attainment of developmental skills such as walking and talking; during school, by academic skills and coping skills such as telling time and using money; and during adulthood, by vocational performance and social responsibilities.

While low intelligence is often accompanied by an inability to cope with life's demands, there are many exceptions, particularly among those whose IQs are in the mildly retarded range (IQ 55 to 69). On the basis of low IQ alone, over 2 percent of our population would be regarded as retarded, but in fact the percentage is somewhat lower because many mildly retarded people are able to blend into the society and to function with at least some independence.

Levels of Mental Retardation The categories listed in Table 13.3 are those with the widest currency in the United States and Canada. They have been recommended by the American Association on Mental Deficiency (Grossman, 1977) on the basis of the statistical distribution of IQs. We can predict the approximate maturity levels children at each IQ level will reach as adults. A mildly retarded adult generally falls in the MA range of 8½ to 11 years; a moderately retarded adult, 6 to 8½ years; a severely retarded adult, 3¾ to 6 years; and a profoundly retarded adult, below that level.

Table 13.3

Normal distribution of IQs for a test with a standard deviation of 15.

IQ	DESCRIPTION	% OF POPULATION
130 and above	Very superior	2.2
120–129	Superior	6.7
110–119	Bright normal	16.1
85–109	Average	59.1
70–84	Borderline	13.6
55–69	Mildly mentally retarded	2.1
40–54	Moderately mentally retarded	0.1
25–39	Severely mentally retarded	0.003
Below 25	Profoundly mentally retarded	0.0000005

(*Sources:* Wechsler, 1955; Grossman, 1977.)

Causes of Mental Retardation There are two general causes of mental retardation. The majority of retarded persons, about 75 percent, are those whose IQs fall by chance (or at least without some identifiable physiological cause) into the lower ranges of the bell-shaped normal curve (Figure 13.8). Mental retardation of this type, usually involving IQs above 50, is called *cultural-familial,* or *sociocultural, retardation* when it occurs in people who have at least one retarded parent. A bell-shaped distribution results when scores are determined by the action of many somewhat independent factors, each of which can exert only a little influence. In the case of intelligence, the factors probably include numerous gene pairs and a broad variety of environmental events. Most people are in the middle of the curve, some are lucky and fall heir to favorable combinations, but others are unlucky and receive unfavorable combinations.

As we might expect, most retarded people are mildly retarded. Table 13.3 shows how very few persons would fall within the more seriously retarded ranges if it were simply a matter of chance. In addition, however, retardation can stem from any of a number of catastrophes which, by themselves, prevent normal development. The catastrophe might be a genetic or a chromosomal defect. *Down syndrome,* for example, is usually caused by a failure of the mother's twenty-first chromosome pair to separate. The two chromosomes join with the single twenty-first chromosome from the father, forming an abnormal condition known as trisomy 21. The result is a distinctive physical appearance (Figure 13.9) and mental retardation.

Other physiological causes of retardation are environmental, not genetic. For example, pregnant women who contract rubella (German measles), scarlet fever, syphilis, or even mumps may give birth to infants who have suffered brain damage as a result. Also, insult or injury to the brain or nervous system before or after birth may result in retardation. Such damage may be done by x-rays, by inappropriate drugs, by severe pressure on the infant's head during birth, by oxygen shortages during or after birth, and even by severe maternal malnutrition.

There are many more persons with IQs at the moderately to profoundly retarded level than would be expected by chance. Most, but not all, show physical evidence of their handicap. Because such catastrophes

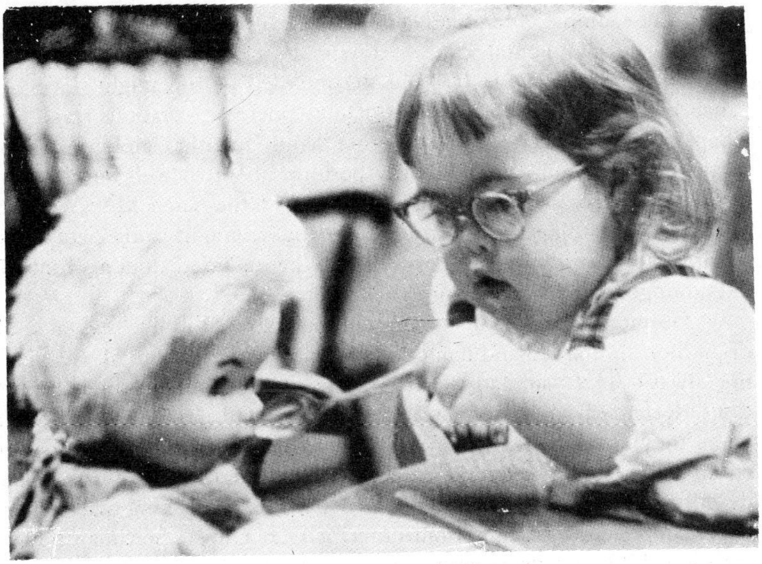

Figure 13.9

This little girl has Down syndrome, a condition due to a chromosomal error that generally results in significant mental retardation. She has been enrolled in a special program designed to maximize her capabilities, including her ability to enjoy imaginative games with her dolls. (University of Washington Child Development and Mental Retardation Center.)

can happen to any family, the backgrounds of such children are only slightly weighted toward the lower end of the socioeconomic scale.

By and large, the progress made in preventing mental retardation has occurred with this "catastrophic" group. It has involved such steps as reducing the incidence of rubella and diagnosing some genetic disorders early in pregnancy. Many such diagnoses can now be made as early as the second month of pregnancy through chorionic villi sampling (CVS); a tissue sample is taken from near the embryo and checked for genetic abnormalities. In the case of cretinism, a once-mysterious disorder that leads to retardation and characteristic physical symptoms, thyroid deficiency was found to be the cause. The discovery made this particular form of mental retardation treatable with thyroid extract. In contrast to such success stories with catastrophic forms of retardation, efforts to prevent retardation in the cultural-familial group have been only moderately effective. Thus far, the most effective strategies for this latter group appear to involve social interventions that improve the standard of living, educational climate, and coping ability of at-risk families (Ramey & Campbell, 1979).

Intellectual Development and Behavioral Characteristics of the Mentally Retarded A number of studies have been carried out in an effort to identify ways in which retarded people are, and are not, different from nonretarded people. Some of these studies have, in effect, brought the intelligence testing and process-oriented approaches together. Results of studies focused primarily on cultural-familial retarded persons have suggested that while these persons are developing intellectually, they pass through the major stages of cognitive development as described by Piaget (1970)—see Chapter 11, page 426—in much the same way as the nonretarded population; the only major differences seem to be that retarded people pass through the stages at a slower pace and cease developing cognitively at a lower stage level than do most of us (Weisz & Zigler,

1979). For example, a retarded child with an IQ of 67 might pass through cognitive developmental stages only about two-thirds as fast as a nonretarded child of average IQ; and the retarded person might never advance beyond about the midpoint of the concrete-operations stage (see Chapter 11, page 450). On the other hand, retarded children at any particular MA level can apparently perform about as well on cognitive tasks like Piaget's as nonretarded children of the same MA (Weisz & Yeates, 1981). Binet, if he were alive today, might appreciate these findings. They fit his original observation that retarded students function like younger intellectually normal children.

Despite the apparent similarity in their ability to solve several kinds of problems, retarded and nonretarded children at the same MA levels do not always *behave* similarly in achievement and problem-solving situations. Retarded youngsters are often passive and dependent on others when they are asked to solve problems—a tendency that seems to grow stronger when they experience failure. Retarded children have been described as outer-directed (Zigler & Balla, 1982), meaning that they are overly dependent on cues from other people. One reason may be that retarded children have experienced so much failure in their lives that they begin each new activity with a relatively low expectancy of success (Zigler & Balla, 1982). Another reason may be that adults tend to accept passive and helpless behavior in retarded children who, after all, are considered low in ability. Recent studies (for example, Weisz, 1981) compared adults' responses to problem-solving failure in a child described as mentally retarded with problem-solving failure in another child who was not; the children in these studies were described as having identical MAs on an IQ test, and many adults might read this as a sign of similar ability. Nonetheless, the adults in both studies made different judgments about the two children. They considered "low ability" to be a more significant cause of the retarded child's failure, rated the retarded child as less likely to succeed in future attempts to solve the problem, and rated themselves as less likely to urge the retarded child to persist at the problem. Findings like these suggest that adults may sometimes underestimate what retarded people can do and may consequently "go easy" on them—often in ways that are not actually in the best interests of the retarded people.

Education and "Treatment" for Mental Retardation Many parents who are told that their child is mentally retarded hope for a cure—for some means of making their child "normal." Dramatic stories in the press sometimes give the impression that such cures are possible. Unfortunately, though, once serious retardation has been identified, there is usually no way to undo it. However, special training can sometimes produce modest changes in IQ and adaptive behavior (see Figure 13.9). Training can also enhance the retarded person's all-important social skills. A friendly style and an endearing smile can be major assets for a Down-syndrome child like the one shown in Figure 13.9; such simple social skills may go a long way toward ensuring that people outside the family meet the child's legitimate needs for help and affection and that peers will accept the child. Later in life, the retarded person's ability to live in community settings such as neighborhood group homes will depend partly on the social and self-help skills the individual has developed (Schalock et al., 1981). In short, education and training are not likely to cure mental retardation, but they can make a big difference in the personal, social, and occupational adjustment of the retarded person.

THE MENTALLY GIFTED

At the upper end of the IQ distribution are the intellectually gifted, a group not so clearly defined as the mentally retarded. There is simply no agreed-upon definition of giftedness. Some researchers and educators regard IQs of 120 and up to be sufficient evidence of exceptionality; others require IQs of 140 or higher. Such differences are far from trivial because there do seem to be real differences in capacity as we ascend the IQ scale. An IQ of 115 or 120, for example, usually indicates a level of mental ability quite adequate for average or above-average work in a good state university. As Figure 13.8 shows, about one person in six has an IQ of 115 or higher. By contrast, only about 2 percent of the population has an IQ of 130 to 145; IQs in this range are commonly found among M.D. and Ph.D. candidates. IQs in the upper strata of the distribution, those above 145, are rare indeed; the normal curve suggests that only one person in a thousand should score this high, but the actual incidence of 145-plus IQs is slightly higher than that.

High IQ scores begin to be predictive of adjustment as early as the elementary school years. A child who is moderately bright is likely to be one of the class "stars," but the youngster with an extraordinarily high IQ may often be a misfit, misunderstood by peers and teachers and often regarded by both as impudent. A classic study by Lewis Terman (1925) found that children with IQs above 140 were generally quite well adjusted, but another classic study by Leta Hollingsworth (1942) found that many children with IQs above 180 were very poorly adjusted. Many of these children were gross underachievers; many were extremely unhappy and some were even suicidal. One problem seems to be that such extremely bright children are trapped in a world with few real peers; they are "out of synch" intellectually with children their own age, and "out of synch" physically and socially with the older people who are their intellectual equals. The picture grows brighter, though, as these youngsters mature. They are increasingly able to find settings, social groups, and work in which their abilities prove a real asset. In fact, gifted adults appear to be happier and better adjusted than most other people (Sears & Barbee, 1977; Terman & Oden, 1959).

GENETIC AND ENVIRONMENTAL INFLUENCES ON INTELLIGENCE

Is intelligence determined by heredity or by environment? As we suggested in Chapter 11, the most appropriate answer is: both—the two factors interact (see page 414, esp. Stern's rubber-band analogy). There can be little doubt that genetic factors play some role in determining intelligence, but the precise nature of that role is hard to establish. One of the difficulties is that many different genes probably play a role; this means that the genetic mystery cannot be unraveled merely by carefully studying the action of one or two genes. Another limitation is that we obviously cannot conduct with humans the kinds of controlled-breeding studies we might use in genetic research with plants or animals. Instead, we must rely on analyses of intelligence as it shows up naturally over the course of people's development.

For these analyses to be of value, they must involve comparisons of the IQs of people who differ in their degrees of genetic relatedness. For example, a researcher might compare the IQ similarity of fraternal twins to the IQ similarity of identical twins. Fraternal-twin pairs, on the average, are identical in 50 percent of their genes; identical twin pairs are identical in 100 percent of their genes. If IQ is influenced by genes, we should expect pairs of identical twins to be more similar in IQ than pairs of fraternal twins. Substantial differences in precisely this direction have been found in about 20 studies comparing fraternal- to identical-twin pairs (Scarr, 1981). These studies, conducted in Britain, Finland, France, Germany, Sweden, and the U.S., all support the notion that there is a considerable genetic influence on IQ.

Other kinds of research point to a similar conclusion. Some investigators have been able to locate identical twins who had been separated in childhood and reared in different environments; the aim has been to assess how similar the twins' IQs were despite their environmental differences. Studies of this type in three different countries have found correlations between separated twins to range from 0.67 to 0.78 on a scale that ranges from 0.00 to 1.00. These very high correlations suggest, again, that genetic similarity leads to similarity in intelligence.

Still another approach has been to study IQ similarity between parents and their adopted children. The question in this research has been, Do children's IQ scores correlate more strongly with the IQs of their biological parents or the IQs of their adoptive parents? A strong correlation with the IQs of the biological parents would point to genetic factors, because the adopted children share genes, but not environment, with these parents. A strong correlation with the adoptive parents would, of course, point to the importance of environmental factors. Most of this research has shown that children's IQs correlate more strongly with those of their biological than those of their adoptive parents (Scarr, 1981).

All of the above research, particularly the twin studies, can be criticized on some methodological grounds (Farber, 1981; Eysenck & Kamin, 1981); and some of the earliest genetic evidence may even have been fraudulent (see, for example, Kamin, 1974). All of the above findings certainly need to be interpreted with real caution. The adoption studies, for example, may yield a high correspondence between biological parents and their children in part because they shared a common environment during pregnancy and at the time of birth—thus, the findings may not be due exclusively to genetic factors. Despite all the qualifications that might be raised about the individual studies, however, the fact remains that most of the studies point in roughly the same direction. They generally support the view that genetic factors play a significant role in the transmission of intelligence.

Having said this, we must emphasize that environmental conditions also play a substantial role. In studies of identical twins, sibling pairs who are not twins, and pairs of unrelated persons, it is generally found that the pairs reared together correspond more closely in IQ than do the pairs reared apart. It has also been found that extremely poor rearing conditions are associated with low IQs and that enriched rearing and educational conditions are associated with higher IQs (Ramey & Campbell, 1979; Skeels, 1966). Many who study the nature-nurture issue with respect to intelligence now believe that our genes endow us with a *reaction range*—that is, a range of possible intellectual levels that we may attain, depending in part on the nature and quality of the environment into which we are born and within which we mature (see Chapter 11, page 414). Our genetic endowment may set limits on what we can attain; our environment may have a major influence on how much of our potential we actually realize. (See Application 13.1.)

GROUP DIFFERENCES IN INTELLIGENCE

Everyday thinking is influenced by notions about the abilities of different groups of people. Some people think that blacks and foreigners are not so intelligent as white Americans. Some regard older people as wiser than, if not so quick as, younger people. Some employers may think women and young people are more suitable for certain positions, men and older people for others. Psychological research indicates that differences among groups do exist, but often these differences are not the ones people imagine. Seldom are the differences between groups as large as the differences within the groups.

Sex Differences The overall IQs of males and females at any age are virtually the same. In part, this is because makers of intelligence tests have deliberately omitted items on which there are sex differences. In part, however, it is also due to the averaging out of differences on subtests of the intelligence scales. During childhood, there are few impressive sex differences on intellectual tasks, although girls do show an early and increasing superiority in verbal behavior.

Differences become more noticeable about the time of adolescence. Girls and women generally do better on tasks that call for verbal expression and fluency, the perception of details quickly and accurately, and rapid, accurate manual movements. Boys and men surpass females on spatial, numerical, and many mechanical tasks. Some, but not all, of these differences correspond to our common impressions of what each sex does best. (See Figure 13.10.)

Dog is to wolf as pig is to________

(a) pork

(b) bacon

(c) sheep

(d) boar

(e) panther

Can you locate each figure shown at the top within the corresponding pattern below?

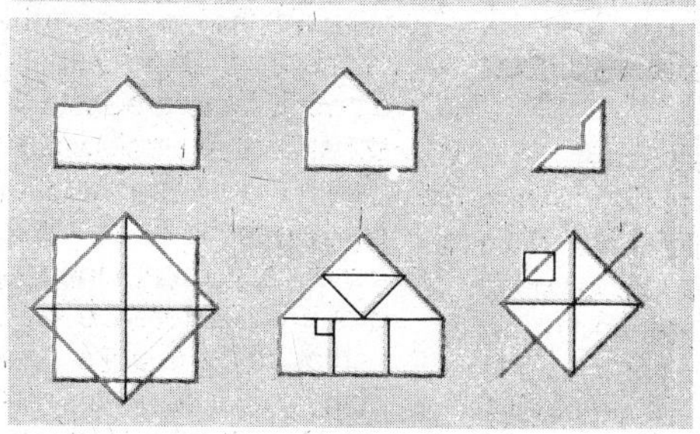

Figure 13.10

Sex differences in intellectual functioning show up on problems like these. On verbal tasks, like the analogy problem shown at the top, females generally outperform males, especially in adolescence and beyond. On spatial tasks, like the embedded-figures problem, males generally outperform females, especially from adolescence on. (Embedded figures diagrams from Consulting Psychologists Press, Inc., *The Embedded Figures Test*, by Herman A. Witkin. Copyright 1969 and reproduced by permission.)

Differences Related to Home Environment Obviously the intellectual environment of children of professional parents is quite different from that of unskilled workers' children. What is the effect of home environment on intelligence? One way to investigate this question is to look for specific factors in the home which seem to relate to differences in mental ability (Bradley & Caldwell, 1976).

> A group of 77 normal children was given an infant-development test and a home-assessment inventory at age 6 months and the Stanford-Binet at age 3 years. It was found that the home inventory predicted IQ at age 3 better than did the infants' own mental development at 6 months! Children with increasing scores had mothers who were involved with them and provided appropriate play materials; those with decreasing scores tended to live in homes where material things and daily events were disorganized.

Racial Differences There is no more sensitive area in all of psychology than the controversy over variations in IQ and intellectual achievement among racial groups. When relatively comparable groups have been tested, Caucasian and Asian Americans have tended to do well, blacks and native Americans have done less well, and groups such as Mexican Americans and Puerto Ricans have tended to fall somewhere in the middle. The size of the differences has tended to vary depending on the part of the country, the social class, and the educational levels of the population samples, but the overall findings have remained relatively consistent. Most people do not like these findings. Persistent group differences stand as a barrier to a true equal-opportunity society, even though, as is clearly the case, *the differences within each racial group are far greater than are the differences between groups*. In Application 13.1, we discuss this problem from a genetic point of view and conclude that the existence of genetic differences in intelligence among the races is questionable.

Application 13.1

BLACKS, WHITES, AND GENES: THE IQ DEBATE

It is a stark fact that blacks in the United States, as a group, average about 15 IQ points lower than whites as a group. A difference of 15 IQ points is not trivial because an IQ score predicts how far people will go in school and how much they will learn there, and many high-prestige and high-paying occupations in the modern world require a certain amount of schooling. Thus people with lower IQs (and poorer educational achievement) are at a disadvantage in the competition for high-level jobs and the cultural and economic advantages that go with them. These facts are not in dispute; facts are facts. The debate is about the *reasons* for the black-white group difference in IQ. Because the evidence indicates that there is probably a partly genetic basis for IQ and because race itself is genetically determined, some have argued that the group IQ difference is due to the different genetic heritages of blacks and whites. Others, including many social scientists, are impressed by the evidence showing the contribution of environmental factors to measured IQ. In other words, the debate revolves around the roles of nature and nurture and their interaction in the determination of measured IQ.

There is little direct evidence of a genetic basis for the black-white group IQ difference. The argument is indirect and usually runs something like this: Racial differences are genetic, and genetic influences play a role in the determination of IQ; therefore perhaps it is reasonable to assume that racial differences in IQ are genetic.

This is the type of argument used in part by Arthur Jensen in his controversial article, "How Much Can We Boost IQ and Scholastic Achievement?" Since Jensen was arguing the case for a strong genetic basis of intelligence, his answer to his own question was "not much." But Jensen's argument did not go unchallenged for long. A number of counterarguments have been arrayed against it by those who believe that the black-white IQ difference is caused mainly by environmental factors.

It has been observed, for instance, that the black-white IQ difference almost vanishes when cultural factors and learning opportunities are equated for blacks and whites. Furthermore, the contribution of genes to IQ is lower for people from culturally disadvantaged backgrounds. Since, on the whole, blacks are less well-off than whites, environmental influences are more important and genetic influences are less important for blacks than for whites. This means that for blacks as a group, environmental conditions are apt to be especially potent in determining measured IQ. A relatively weak argument that is often heard is that IQ tests are biased against blacks because the items on them are drawn from the white middle-class culture; while this may be true for a few items, it does not apply to most. Moreover, since the tests do predict success in school work and vocations, the forms of intelligence they measure are evidently important within our society for both blacks and whites.

Perhaps the strongest arguments against an innate basis for the black-white IQ difference are those based on the interaction of genes and environment. *Nature*, or the *genotype*, interacts with *nurture*, or the environment, to produce the characteristic actually observed—the *phenotype*. Within limits, the genotype can be influenced by environmental factors to give rise to a range of results in the phenotype; this is known as the *reaction range*, a concept we discussed in Chapter 11 (page 414). Observations on the reaction range give us some idea of the impact environmental differences can have when they interact with a genotype. In the case of the phenotype of measured IQ, it turns out that the reaction range is close to the 15-point difference separating the black and white groups. For instance, black children who are adopted at an early age into white middle-class families score, on the average, about 15 points higher than does a comparable group reared in their own families. Such results strongly suggest that the15-point IQ difference between the black and white groups may be largely environmental in origin. With these observations in mind, the answer to Jensen's question, "How much can we boost IQ?" may be "Quite a bit, perhaps even by 15 points."

REFERENCES

Jensen, A. R. (1969). How much can we boost IQ and scholastic achievement? *Harvard Educational Review, 39,* 1–123.

Loehlin, J. C., Lindzey, G., & Spuhler, J. N. (1975). *Race differences in intelligence.* San Francisco: W. H. Freeman.

Scarr, S. (1981). *Race, social class, and individual differences in IQ: New studies of old issues.* Hillsdale, NJ: Erlbaum.

Scarr, S., & Weinberg, R. A. (1976). IQ test performance of black children adopted by white families. *American Psychologist, 31,* 726–739.

Environmental differences among the races, however, are not an open question—large differences exist. It is surely not happenstance that socio-economically favored groups score higher on intelligence tests. A cycle is perpetuated in which some groups live, work, and raise children in circumstances much more conducive to intellectual achievement than do others. Despite efforts to integrate schools, jobs, and neighborhoods, we have, in some respects, a persistent caste system. Even when roughly equated for socioeconomic class, the cultures of some groups, particularly blacks and whites, remain markedly different.

Whatever group differences may exist, two conclusions are clear.

1. Social inequities are deep and pervasive; we should judge tests by how well they do their job, which is predicting academic or

vocational success within a particular society, not by whether they manage to "equalize" social groups which have never been treated equally.

2. Whatever the group differences may be, our major focus should be on the individual; between-group differences are small compared with individual differences among group members. In a fair society, individual differences are recognized and opportunities extended to those able to profit from them—whatever their race or background.

Testing for Special Aptitudes

There is no clear dividing line between intelligence tests and aptitude tests. In general, though, we use *intelligence tests* to give us a broad assessment of intellectual capacity and *aptitude tests* to measure the more specialized abilities required in specific occupations and activities.

SCHOLASTIC APTITUDES

If we are trying to predict success in academic training, we speak of *scholastic aptitude*. Perhaps the best-known test of this type is the *Scholastic Aptitude Test* (*SAT*), given to students who want to enter liberal arts colleges. Similar tests are used for schools of medicine, dentistry, nursing, law, and several other professions; each is focused partly on specific abilities thought to be important to the profession. The *Graduate Record Examination* (*GRE*) has been designed for students who plan to do graduate work in a number of fields in the arts and sciences. The *Miller Analogies Test* (*MAT*) is also used to predict success in graduate school. Many graduate and professional schools require applicants to take an appropriate aptitude test.

VOCATIONAL APTITUDES AND INTERESTS

Psychologists often refer to the abilities tested by intelligence and scholastic aptitude tests as cognitive abilities. Such abilities are necessary for getting along in school, and a certain level of schooling is a requirement for entering certain occupations. Once a person is in an occupation however, these abilities become less important. In many occupations they may not count at all, and even professions such as medicine require skills—such as reading x-rays, judging the color of throats, feeling for lumps, using surgical instruments, and tying sutures—that have large perceptual-motor components. Such physical and perceptual skills are known as noncognitive abilities.

Many tests are intended for specific jobs. For example, tests for mechanics, machine operators, and assembly-line workers, measure mechanical knowledge or ability to manipulate objects. These tests make up the general class of mechanical-ability tests. People who score high on one mechanical-ability test tend to do so on another. But since different jobs require different combinations of mechanical abilities, the number of different tests has mushroomed. Some examples are given in Figure 13.11 and 13.12.

Psychomotor tests are a second general class of *vocational- ptitud test*. They involve such psychomotor tasks as manual dexterity, stead

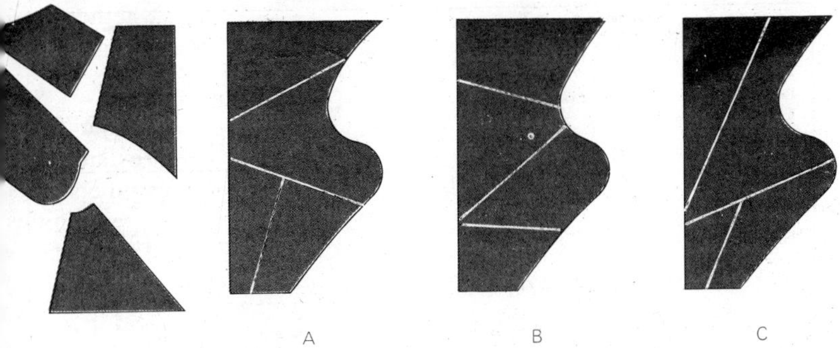

Figure 13.11

A sample from the Minnesota Paper Form Board Test, a mechanical-aptitude test. The examinee looks at the pieces on the left and indicates whether they fit together to make *A*, *B*, or *C*.

iness, muscular strength, speed of response to a signal, and coordination of many movements into a unified whole. So far there is little evidence of a general motor ability comparable to mechanical ability. A person who has good manual dexterity, for example, is not necessarily good at the kind of coordination involved in operating a tractor or an airplane. So psychomotor tests must be developed and proven valid for particular jobs and occupations.

Vocational-aptitude tests are used both by employers to select employees and by vocational counselors to help people assess their aptitudes for different types of work. The same tests are usually not suitable for both purposes. Employers, who know exactly what jobs are open, want tests that will forecast success in them as accurately as possible. They need tests designed specifically for that purpose—for instance, to select electronics technicians, electrical welders, or lathe operators. Counselors, on the other hand, are trying to help people make a choice—usually a fairly general choice—among broad lines of work. For this purpose, counselors want more generalized tests that sample many different aspects of specific aptitudes.

Counselors often administer *test batteries*, combinations of tests covering a wide spectrum of abilities. One such battery, designed especially for counseling high school students and noncollege adults, is called the *Differential Aptitude Tests (DAT)*. Sample items are shown in Figure 13.13. The scores on each subtest are plotted as a profile of scores on a special chart. The scores, either singly or in various combinations, can be used to predict scholastic success as well as success in tasks requiring specific aptitudes.

Since a person's interest in various occupations is important in career choice, vocational counselors often use information from tests designed especially to measure vocational interests. The Strong-Campbell Vocational Interest Inventory (Campbell, 1981) and the Kuder Occupational Interest Survey (Kuder, 1979) are examples.

Vocational aptitude and interest tests are frequently available in schools and communities. If a college has a psychological clinic or student counseling service, this office is usually prepared to administer such tests. The U.S. Vocational Rehabilitation Service and the Veterans Administration provide testing services for those who qualify for assistance. In larger cities there are usually several independent agencies and individuals that offer competent testing for a reasonable fee.

Figure 13.12

Which shears would be better for cutting metal? This is an example of the type of question included in the Bennett-Fry Test of Mechanical Comprehension. Actual test items are generally more difficult than this sample item. (The Psychological Corporation.)

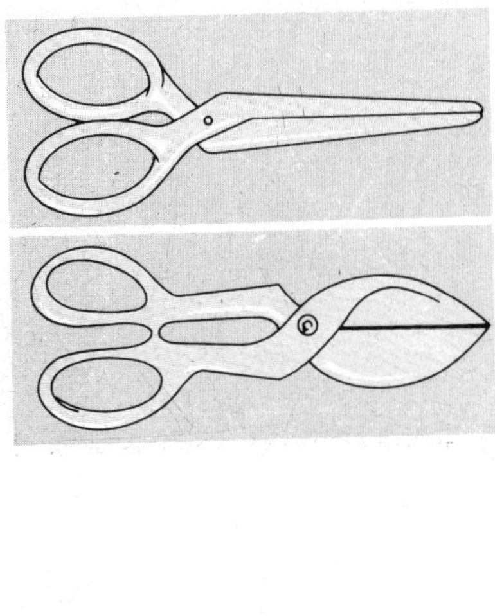

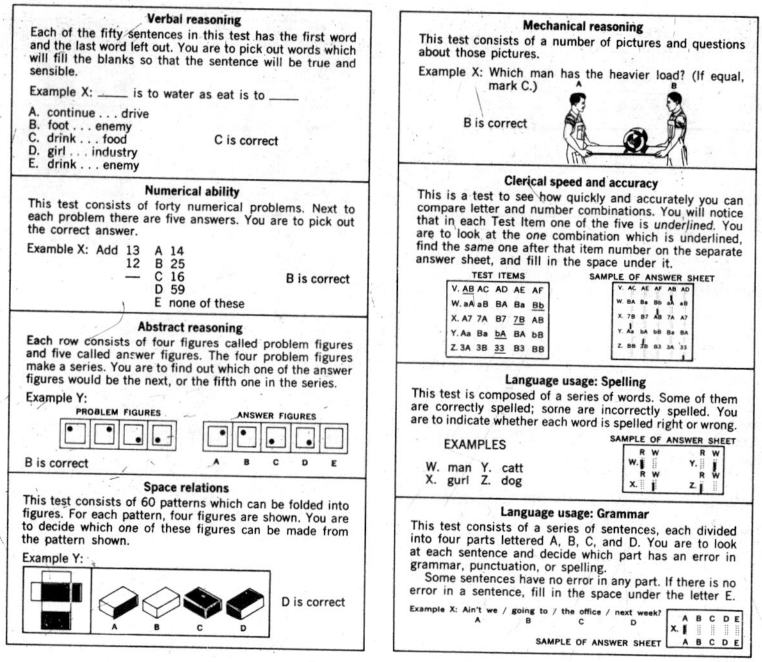

Figure 13.13

Sample items illustrating the eight tests of the Differential Aptitude Tests (DAT) battery. These sample items are generally easier than those on the tests themselves (The Psychological Corporation.)

Personality Assessment

True or False

1. When I was younger, I used to tease vegetables.
2. Sometimes I am unable to prevent clean thoughts from entering my mind.
3. I am not unwilling to work for a jackass.
4. It is important to wash your hands before washing your hands.
5. The sight of blood no longer excites me.
6. I believe I smell as good as most people.

(Art Buchwald and others, *American Psychologist*, 1965, *20*, page 990)

These questions are part of a "personality test" devised in part by humorist Art Buchwald. Neither we nor Mr. Buchwald can be certain just what aspect of "personality" the test was meant to measure; the purpose of the "test" was to illustrate, not measure. It illustrates, first, that questions on personality tests can seem quirky, silly, or even pointless to the test taker. It also illustrates a key difference between personality tests, on the one hand, and tests of intelligence, special aptitude, or achievement, on the other. Tests in this latter group involve "right" and "wrong" answers and objective standards of performance quality; they are designed to reveal a picture of people at their best, performing as close to the upper limits of their abilities as possible. Personality tests have very different objectives. Personality testing does not involve levels of success or even "right" or "wrong" answers; its objective is not to gauge how successful a person will be but, rather, what the person is ***usually*** like (in thoughts, feelings, and behavior patterns). In short, personality testing aims for ***typicality***, for a representative picture of individuals as they usually are.

Personality testing is done for many reasons. A personnel psychologist may want to identify people whose personality characteristics should make them good salespeople. A military psychologist may want to measure tendencies that would make a person fit, or unfit, for a sensitive assignment. An experimental psychologist may want to measure anxiety in order to control its influence in experiments on perception or learning. A clinical psychologist often uses personality tests to evaluate psychological disorders. (See Application 13.2.) A variety of methods exist to suit these diverse purposes.

PENCIL-AND-PAPER TESTS

The most convenient kind of measure to use for almost any psychological purpose is a pencil-and-paper test in the form of a questionnaire or inventory. Such tests can be given cheaply and quickly to large groups of people, and consequently psychologists have constructed a wide variety of them.

Questionnaires Pencil-and-paper tests of personality characteristics are usually questionnaires which ask questions or give simple statements to be marked yes or no, true or false. The questions are usually more reasonable than were Art Buchwald's, but their precise purpose may not be self-evident. Some examples:

> I generally prefer to attend movies alone.
>
> I occasionally cross the street to avoid meeting someone I know.
>
> I seldom or never go out on double dates.

Some questionnaires offer people the option of answering "doubtful" or "uncertain" to questions like these.

This type of pencil-and-paper personality test was first used widely during World War I to weed out emotionally unstable draftees. The statements, or items, were chosen to reflect psychological states or processes that might predict future emotional breakdowns. Items such as the following were included:

> I consider myself a very nervous person.
>
> I frequently feel moody and depressed.

Do items like these really test what the examiner thinks they test? This is the validity question that we discussed earlier (page 519). Questions of validity and reliability are just as important in personality testing as they are in other kinds of psychological assessment. In fact, it might be argued that validity issues are particularly important in personality testing because there are so many personality tests of uncertain validity.

For example, count the number of "personality tests" that appear in popular magazines—tests that purport to tell us whether we are good lovers, optimists, introverts, self-actualized people, and so forth. The fact is that most of these tests have not been validated. Nor has validity been established for some of the tests that are used in selecting employees, deselecting trainees, and identifying "executive material" within many companies. Such tests are sometimes selected for use because their items *look* valid—that is, they have what psychologists call *face validity*. But looks can deceive. As many would-be test developers can testify, proven validity is hard to come by, particularly in the area of personality testing. In many employment situations, tests for which validity cannot be demonstrated can now be challenged in court and legally barred. For the would-be consumer, in employment situations and elsewhere, caution is in order. The safest assumption to make is that a personality test is invalid unless it has been proved otherwise.

Minnesota Multiphasic Personality Inventory (MMPI) One personality test that has proven to be valid for many purposes is the Minnesota Multiphasic Personality Inventory, or MMPI (Dahlstrom & Dahlstrom, 1980). The MMPI contains 566 statements, or items, for people to answer about themselves. The items, similar to those shown in Figure 13.14, can be answered "true," "false," or "cannot say."

One obvious way to design a personality test would be to focus on

Figure 13.14

Items Like Those from the MMPI

1. I enjoy books and magazines about electronics.
2. I urinate regularly.
3. I almost never have bad dreams when I sleep.
4. My body is sometimes inhabited by a sinister force.
5. In public places I sometimes see people whispering about me.

The Ten Basic Clinical Scales of the MMPI

1. Hs: Hypochondriasis
2. D: Depression
3. Hy: Hysteria
4. Pd: Psychopathic deviate
5. Mf: Masculinity/femininity
6. Pa: Paranoia
7. Pt: Psychasthenia
8. Sc: Schizophrenia
9. Ma: Hypomania
10. Si: Social introversion

A Sample MMPI Profile

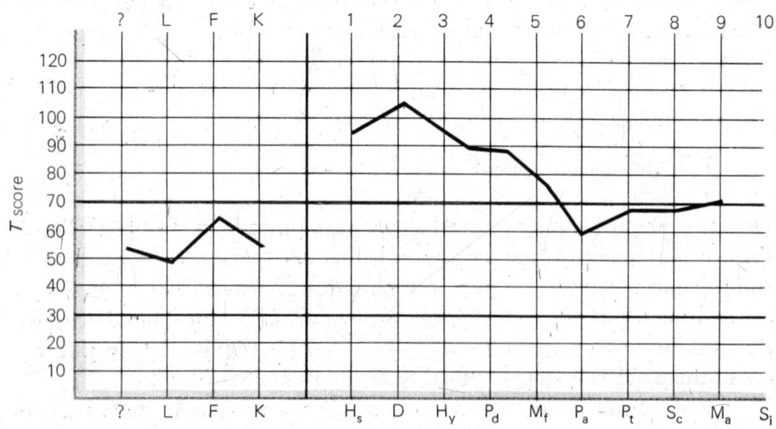

Items similar to those on the Minnesota Multiphasic Personality Inventory (MMPI), the 10 MMPI clinical scales, and a sample MMPI profile. The profile shows one individual's scores. The scores on each scale are expressed as T scores; these are standard scores (see page 533) with a mean of 50 and a standard deviation of 10. The heavy horizontal lines indicate scores that are 2 standard deviations above and below the mean; these are roughly equivalent to percentile scores of 98 percent and 2 percent, respectively. This profile was obtained from a person who suffered from a physical complaint that actually had a psychological source. The scores on the four validity scales at the left indicate that the person gave careful, frank answers to the items on the test. The profile of scores for the 10 clinical scales shows a particularly high elevation on the hypochondriasis, depression, and hysteria scales—the so-called neurotic triad; this pattern tends to confirm that the individual's physical problems do, indeed, have a psychological basis. The relatively high elevation on the fourth clinical scale (psychopathic deviate) suggests that this person may tend to manipulate others for selfish reasons. (Modified from Dahlstrom & Welsh, 1960.)

One obvious way to design a personality test would be to focus on the content of the items people mark as "true" or "false," but the makers of the MMPI did not follow this strategy. Instead, they used what is known as the *empirical approach* to test construction. They ignored the content of people's answers on the test, choosing instead to match the patterns of people's answers to the patterns shown by "criterion" groups of people with known characteristics of special interest. Since the MMPI was designed to identify people with tendencies toward certain psychological disorders, most of the criterion groups consisted of people with one particular psychological diagnosis or another, such as depression, paranoia, or schizophrenia. (See Chapter 15.) The test makers analyzed the way each criterion group answered the items, then used this information to form the clinical scales shown in Figure 13.14. For instance, if depressed people, more often than other groups, answered yes to the statement, "My favorite teddy bear was green" (a fictitious item), this statement would be placed on the Depression (D) scale. The test constructors would not stop to wonder what strange parents would buy green teddy bears, whether they actually had, or whether the individuals taking the test ever had a teddy bear at all. The only thing that counts in this empirical approach to test making is the fact that depressed people say yes to this item more often than do other groups. The test makers used criterion groups to develop scales for feminine and masculine interests and for social introversion as well as for the other clinical scales shown at the right in Figure 13.14.

When a person takes the MMPI, one or a few items answered in an unusual fashion will make little difference. But a person who answers many of the statements in the same way as the people in some criterion group—say, depressed people—will obtain a high score on the relevant subscale; this score might indicate a tendency for the person to show other characteristics of people in the criterion group. For example, a high score on the depression scale might indicate a tendency toward depression; however, as we shall see below, the interpretation of MMPI results is more complex than this.

In addition to its 10 clinical scales, the MMPI also has several validity scales—scales designed to assess a test taker's frankness and thoroughness in answering the items and to check for defensiveness or other attitudes that might influence his or her answers. With these validity scales, deceptiveness and attempts to create especially good or bad impressions can often be detected.

Sometimes the MMPI is used as an aid in determining what diagnosis will be given to people who have psychological problems. However, the test is also frequently used to assess the personality characteristics of people whose behavior is not blatantly disordered. To interpret the test, the psychologist looks at the total profile, not just the separate scales. (See Figure 13.14.) Then the information from the validity scales is considered, along with such features as the overall elevation of the profile above the average level, the highest and lowest scores and their relationships to one another, and so on. Because of all the factors that have to be considered, MMPI interpretation requires considerable training and skill. In untrained hands, the test can lead to serious misjudgments about an individual's personality.

In the right hands, though, the MMPI is a powerful tool. It is the most widely used of all pencil-and-paper personality tests, both for clinical and research purposes. It has already been used in over 6,000 studies and is likely to be used in several thousand more. It has also spawned new personality tests—tests built in whole or in part from MMPI items. One of these is the Manifest Anxiety Scale (Taylor, 1953), whose purpose is evident from its title. Another is the California Psychological Inventory (CPI) (Gough, 1975), designed to assess normal personality processes like dominance, sociability, and self-control.

The 16 Personality Factor Questionnaire (16PF) Another example of a pencil-and-paper test is the 16 Personality Factor Questionnaire, (16PF). In developing the 16PF, Raymond Cattell and his associates used an approach quite different from the MMPI procedure (Cattell et al., 1970). They began with a list of over 4,500 adjectives applicable to human behavior and then reduced the list to 170 that they believed captured most of the major meanings in the original list. They then had college students describe their acquaintances using the 170 adjectives and used a statistical procedure called *factor analysis* (see page 525) to identify groupings, or factors, among the items. The 16 factors thus identified were said by Cattell to reflect key characteristics, or source traits, of the human personality. Some consider the 16PF the most comprehensive approach to trait assessment yet developed. The 16 bipolar traits (for instance, cool-warm, trusting-suspicious) are shown on the 16PF test profile in Figure 13.15.

PROJECTIVE METHODS

Unlike paper-and-pencil tests, in which test takers are asked to select from among a set of alternatives (such as yes, no, or "cannot say"), some personality tests are deliberately designed to evoke highly individual responses. Most of the tests in this latter group are called *projective methods*. They call for the test taker to respond to stimuli such as inkblots or pictures but provide few guidelines as to what the response should be. The scoring procedures for projective-method tests are also generally less structured than are those for paper-and-pencil measures; the interpreter must often rely heavily on a subjective evaluation of the responses.

Projective methods are based on the *projective hypothesis*, derived from Freud's personality theory (see Chapter 14). The basic idea is that the way people respond to a vague or ambiguous situation is often a projection of their underlying feelings and motives. For example, a man who interprets a woman's smile as a sexual come-on may be projecting his own sexual feelings onto the woman and thus revealing a good deal about himself (see Chapter 14, page 589). A related assumption about projective tests is that the test taker responds to the relatively unstructured test stimuli in ways that give meaning to the stimuli, and that much of that meaning comes from within the person responding. Often the people responding are not aware of any underlying meaning in their responses, but people trained to interpret projective tests are taught ways of inferring such meaning and judging personality characteristics accordingly. Thus, projective methods are intended to provide access to unconscious impulses and other aspects of personality of which the test takers themselves may not be aware.

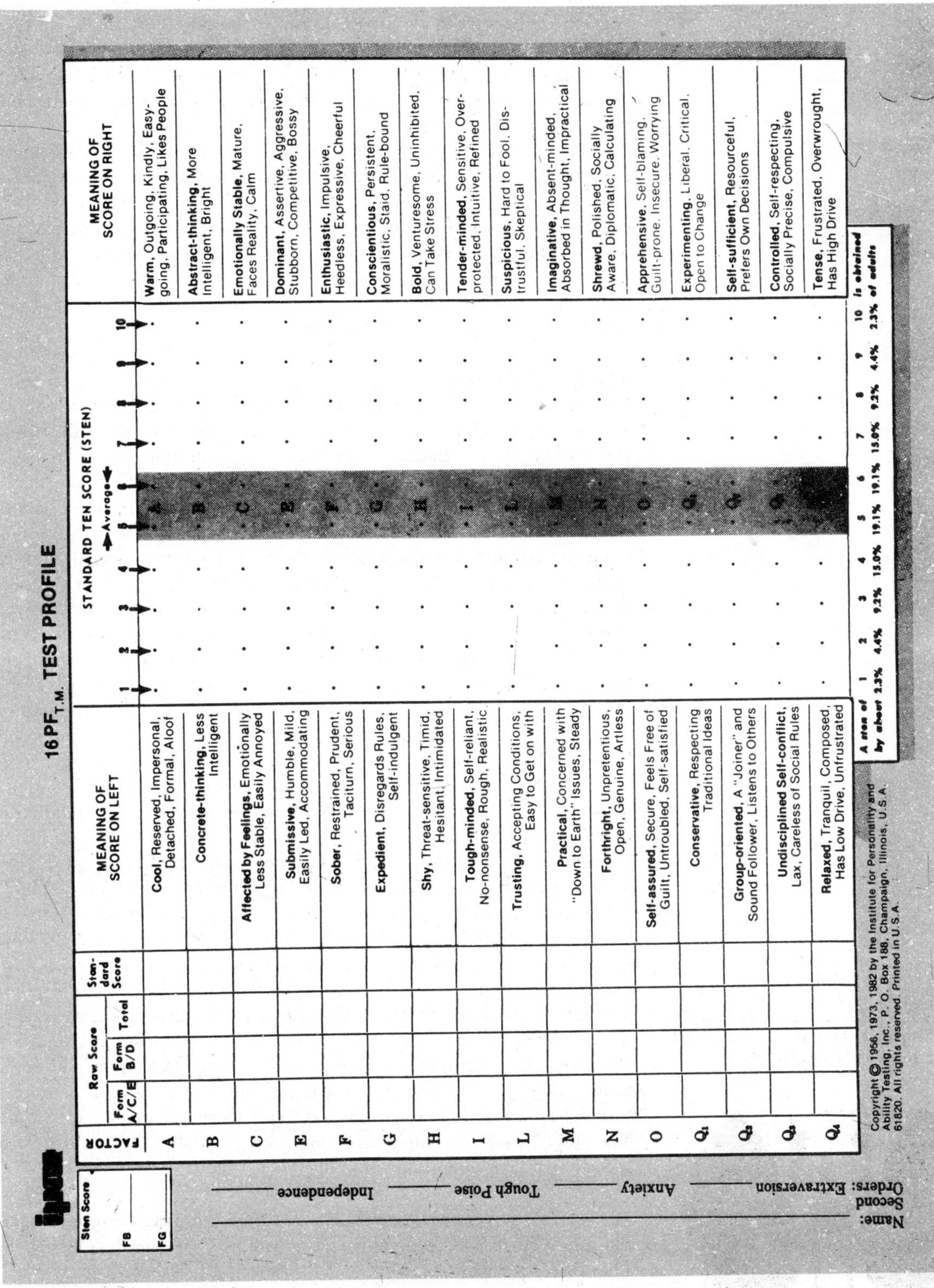

Sten Score FB ____ FG ____

Name: ____

Second Orders: Extraversion ____ Anxiety ____ Tough Poise ____ Independence ____

$16PF_{T.M.}$ TEST PROFILE

FACTOR	Raw Score Form A/C/E	Raw Score Form B/D	Raw Score Total	Standard Score	MEANING OF SCORE ON LEFT	STANDARD TEN SCORE (STEN) 1 2 3 4 ← Average → 5 6 7 8 9 10	MEANING OF SCORE ON RIGHT
A					**Cool**, Reserved, Impersonal, Detached, Formal, Aloof	A	**Warm**, Outgoing, Kindly, Easy-going, Participating, Likes People
B					**Concrete-thinking**, Less Intelligent	B	**Abstract-thinking**, More Intelligent, Bright
C					**Affected by Feelings**, Emotionally Less Stable, Easily Annoyed	C	**Emotionally Stable**, Mature, Faces Reality, Calm
E					**Submissive**, Humble, Mild, Easily Led, Accommodating	E	**Dominant**, Assertive, Aggressive, Stubborn, Competitive, Bossy
F					**Sober**, Restrained, Prudent, Taciturn, Serious	F	**Enthusiastic**, Impulsive, Heedless, Expressive, Cheerful
G					**Expedient**, Disregards Rules, Self-indulgent	G	**Conscientious**, Persistent, Moralistic, Staid, Rule-bound
H					**Shy**, Threat-sensitive, Timid, Hesitant, Intimidated	H	**Bold**, Venturesome, Uninhibited, Can Take Stress
I					**Tough-minded**, Self-reliant, No-nonsense, Rough, Realistic	I	**Tender-minded**, Sensitive, Over-protected, Intuitive, Refined
L					**Trusting**, Accepting Conditions, Easy to Get on with	L	**Suspicious**, Hard to Fool, Distrustful, Skeptical
M					**Practical**, Concerned with "Down to Earth" Issues, Steady	M	**Imaginative**, Absent-minded, Absorbed in Thought, Impractical
N					**Forthright**, Unpretentious, Open, Genuine, Artless	N	**Shrewd**, Polished, Socially Aware, Diplomatic, Calculating
O					**Self-assured**, Secure, Feels Free of Guilt, Untroubled, Self-satisfied	O	**Apprehensive**, Self-blaming, Guilt-prone, Insecure, Worrying
Q_1					**Conservative**, Respecting Traditional Ideas	Q_1	**Experimenting**, Liberal, Critical, Open to Change
Q_2					**Group-oriented**, A "Joiner" and Sound Follower, Listens to Others	Q_2	**Self-sufficient**, Resourceful, Prefers Own Decisions
Q_3					**Undisciplined Self-conflict**, Lax, Careless of Social Rules	Q_3	**Controlled**, Self-respecting, Socially Precise, Compulsive
Q_4					**Relaxed**, Tranquil, Composed, Has Low Drive, Unfrustrated	Q_4	**Tense**, Frustrated, Overwrought, Has High Drive

A sten of	1	2	3	4	5	6	7	8	9	10	is obtained
by about	2.3%	4.4%	9.2%	15.0%	19.1%	19.1%	15.0%	9.2%	4.4%	2.3%	of adults

Score sheet used for the 16PF test profile. (Reproduced by permission of the Institute for Personality and Ability Testing, Inc.)

Figure 13.15

Projective methods are many and varied. Some are completion techniques, tests that call for subjects to finish some product—perhaps a sentence or a story—presented to them in fragmentary form. Others are expressive techniques, tests that call on subjects to express themselves via play, drawing, or drama.

Many psychologists question the validity of the projective hypothesis, the projective methods themselves, and the inferences about personality that are made on the basis of these methods. In most cases, the evidence for validity is quite modest. On the other hand, many psychologists in clinical practice rely heavily on projective methods to enrich their understanding of the clients they are trying to help. In this section, we focus on two of the most frequently used projective tests, one involving inkblots and the other involving stories.

The Rorschach Inkblot Technique As early as the fifteenth century, Leonardo da Vinci observed that many different feelings could be brought out by simply looking at a blot that a sponge had made on a wall. It took about 500 years for students of personality to absorb this idea and translate it into an official technique. One of the most acclaimed efforts was that of Hermann Rorschach (1921/1942). He produced a set of 10 inkblots similar to the one shown in Figure 13.16. The blots, some black and white, some multicolored, appear on separate cards. Subjects are presented with the cards, one at a time, and asked questions such as, "What might this be?" or "What does this remind you of?" After writing down as many answers as the subject cares to give for each blot, the tester goes back through the set, asking the subject for more details, including what it was about the blot that determined the subject's response. The first phase of the test is called the free-association phase; the second phase is called the inquiry.

Several methods for scoring the Rorschach have been proposed. Most combine objective and subjective procedures. One example of an objective approach might be counting the number of times the subject responded to only part of the blot and comparing this to the number of times the subject integrated all parts of the blot into one response. For example, some test takers might focus only on the central part of the blot in Figure 13.16 and see "a cheerleader, with two pom-poms raised high"; others might focus on the entire blot, turn it upside down (this is permissible under some procedures), and see "two one-armed, intergalactic creatures with arms raised, swearing allegiance to their emperor, who has a pointed hat, a large penis, and even larger feet." The second response uses all of the blot; the first uses only part. The content and style of responses become grist for the interpreter's mill in the subjective aspect of scoring. For example, the second response involves themes of dominance, allegiance, quasihuman creatures, and distorted body parts, including genitalia. If one or more of these themes emerged on several other blots, the tester might take note and judge the theme(s) to be of special significance and meaning for this particular subject.

Figure 13.16

An example of the type of figure used on the Rorschach inkblot test.

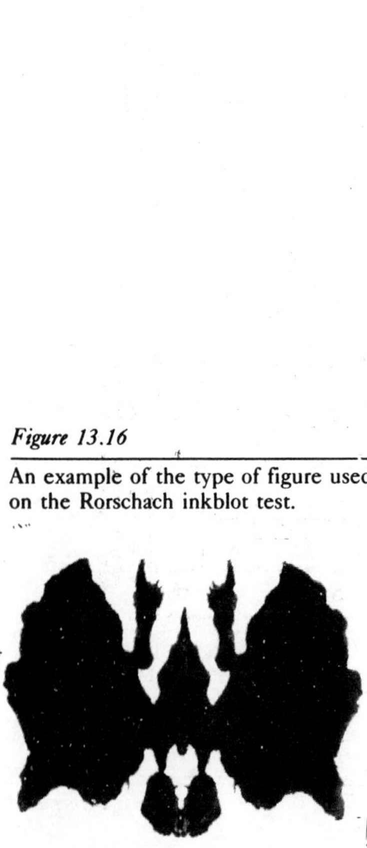

Thematic Apperception Test (TAT) A different projective approach was taken by Christina Morgan and Henry Murray (1938) in developing the Thematic Apperception Test (TAT). The TAT is based on Murray's theory of needs (Chapter 7, page 282). It is designed to ferret out people's basic needs by having them tell stories. To guide story production, the tester presents a series of pictures, some like the one shown in Figure 13.17, and asks the subject to make up a story about what is happening, what went before, what is going to happen, and what the people involved are thinking and feeling.

The TAT includes a standard set of 30 pictures, but it is rare that all 30 are included in any one testing. Instead, the tester selects pictures involving themes of special significance to the test taker. For example, a person who is worried about the future might be given a picture like the one shown in Figure 13.17, whereas a person whose difficulties are family-oriented might be given pictures that involve parent-child relationships or larger family-like scenes. The test is built on the assumption that people's stories reveal important aspects of their needs and self-perceptions as well as their views about "significant others" in their lives. Indeed, the stories given do often seem to be quite revealing. Consider, for example, these two segments of stories, told in response to a picture similar to Figure 13.17:

> STORY A: This is sad. This man is remembering the past 20 years of his life, his failed career, his broken marriage, and the rotten way his children have turned out. He knows that he tried his best, and he tries not to blame himself, but he can't help it. One thing he has learned, though. You can't count on people; as soon as you really trust them, they let you down.

> STORY B: Here is this guy—he's tired, rough day at the office and all that. But it's a good kind of tired, if you know what I mean. I mean, he knows he's working hard, and it's really paying off. He's rising fast in his company. People have treated him well. They appreciate his work, and they show it with promotions, raises, and

Figure 13.17

A picture similar to those used on the Thematic Apperception Test (TAT). A person is shown a card and asked to tell a story about what is happening.

respect. He's daydreaming now. If he keeps doing good work, he figures it's only a matter of time before he's in the "inner circle"; and someday, maybe not too far away, he's going to be running the whole show.

With stories as dramatically different as these two, it is tempting to infer some basic personality differences between the person who told story A and the person who told story B. Story A seems to reveal a pessimistic outlook on life, a feeling that life is unfair, and also a need to trust others, but a real reluctance to do so. Story B seems to show an upbeat outlook, a belief that life rewards those who do good work, and also a need for achievement. Before we make any other personality inferences, though, we should point out an important fact: story A and story B were told by the same person, and both stories were told within the same hour. Sometimes stories *are* closely related to the personality of the storyteller, but sometimes they are only fantasy productions with little assessment value. Even when the stories relate to the inner state of the storyteller, that inner state may change over time—from depression to elation, from warmth to hostility, and so on. For these reasons, trained psychologists who use the TAT usually look for themes that surface several times in response to several different pictures (or on other tests); and even then, they make personality inferences sparingly and cautiously.

Validity and Usefulness of Projective Techniques Are projective techniques useful and valid? There is no easy answer to this question. The techniques are so numerous and the ways they are used are so varied that any judgment made on projective methods in general would be an oversimplification. However, a few points can and should be made. Clinical psychologists use projective tests often, but interpretation of most such tests remains more an art than a science. Most of the available evidence casts doubt on the validity of most projective tests. Consequently, some psychologists question whether the tests are worth the time and effort needed to administer and score them. Nonetheless, psychologists who use projective tests regularly feel that the tests are useful *in the right hands*—and that theirs *are* the right hands.

To illustrate how a skilled clinician might use projective techniques together with paper-and-pencil tests, we have included a case report in Application 13.2. The report is condensed, but it portrays a typical sequence of events: An adolescent gets into trouble; a psychologist is called in; specific questions are asked about the youth, the causes of his behavior, and his likely future behavior; personality tests are administered to address these questions; personality inferences are made, partly on the basis of the test results; and specific recommendations are made for helping the youngster.

As Application 13.2 illustrates, projective techniques and pencil-and-paper tests are often used in combination with interviews to make judgments about people; this is especially true when personality assessment takes place in a clinical setting. Interviews, in fact, are probably the oldest and still the most widely used approach to personality assessment. Interviewing is also frequently combined with yet another approach to assessment, a relatively new but important approach that we examine in the final section of this chapter.

Application 13.2

PERSONALITY TESTS HELP ANSWER CLINICAL QUESTIONS

Many clinical psychologists work with clients whose subjective feelings are disturbing to them or whose behavior is getting them into trouble. Even after the psychologist has talked with the individual and has gotten as much other information as possible, a number of questions may still need to be answered. Here is an example of one instance in which tests were useful shortcuts in getting to know a troubled young man.

Robert, age 16, was referred to a clinic for adolescents by the juvenile court because he had made obscene telephone calls to his high school counselor and had broken into and vandalized a grocery store. Robert lived with his mother, his stepfather, and a younger brother. His teachers reported that he was not generally a disciplinary problem, except for some rude language and occasional outbursts of fighting with fellow students. School records indicated average intelligence but low grades and scores on achievement tests, which tended to average about 1 year below his grade level.

When Robert was interviewed, he was sullen and reluctant to talk. He denied making any obscene telephone calls and said that he had vandalized the store in order to get back at the owner's son, who had laughed at him in school. He described himself as shy and said he had only a few friends, all younger than himself.

The psychologist who administered the tests did so with the following questions in mind:

1. Is Robert's socially unacceptable behavior symptomatic of a severe behavior disorder or psychosis? Is it a plea for help? Is he an unsocialized personality?

2. What form of intervention is appropriate? If psychotherapy is indicated, should it be on a residential or on an outpatient basis?

3. Is Robert likely to commit further crimes? If so, are they likely to be more violent than those he has committed so far?

Robert's MMPI profile showed a general elevation on scales indicating impulsive and restless behavior, suspiciousness, and asocial orientation. This profile is characteristic of individuals with a severe behavior disorder known as paranoia. (See Chapter 15, page 644.) The test suggested that he would show poor judgment in difficult social situations and would be unpredictable and maladaptive in expressing his emotions.

Robert's Rorschach record included several peculiar, disorganized responses that did not fit the inkblots very well. Although he gave a reasonable number of detailed responses, he became somewhat excited when he gave personal responses such as "my dog" or violent responses such as "blood" or "a gash" on the colored blots.

On the TAT, his stories seethed with conflict and anger. He described a succession of violent deaths—a shooting, a heart attack, a strangling, and four suicides—in morbid and gruesome detail in one story. Important themes included parents disappointing their sons. One story in particular seemed rather ominous: A woman getting out of prison after 15 years kills the two doctors who had turned her in to the police for a shooting.

The overall impression was of a disturbed, angry teenager with an unhealthy preoccupation with violence and rather disorganized thought processes. The staff recommended that Robert be placed in a residential treatment facility where he could receive intensive psychotherapy.

Behavioral Assessment

The last general approach that we will discuss grows out of the learning theories discussed in Chapter 4 and the behavioral school of psychology discussed in Chapters 14, 15, and 16. *Behavioral assessment* includes several methods, each one aimed at some aspect of the individual's observable behavior. In general, the methods are not concerned with underlying traits or mental processes, largely because such traits and processes cannot be seen directly. Behavioral assessment methods have two other characteristics in common:

1. They are designed to reveal the stimulus conditions associated with the specific target behaviors—for example: What circumstances lead to outbursts of temper?

2. They involve direct scrutiny—observing the person's behavior as it unfolds, or at least seeking a specific description of the behavior and the situations in which it happens.

PROBLEM CHECKLISTS

One behavioral method that is growing in popularity is the problem checklist. Some checklists ask for specific details of a person's difficulties in one particular problem area. For example, a fear checklist might list objects and situations that many people avoid (for example, public speaking, snakes, going to the dentist) and ask the person to indicate which ones he or she avoids. Behavior therapists (Chapter 16, page 693) often ask patients to fill out checklists of this sort just prior to therapy in order to help the therapist decide which specific problem behaviors need to be treated and in which order.

Other checklists, like the Child Behavior Checklist shown in Figure 13.18, cast a wide net, surveying problem behaviors of many different types. One advantage of such broadly-cast measures is that they are likely to include more clinically significant problems than psychologists, clients, or the relatives of clients might ordinarily think to mention during an interview. Another advantage, from the behavioral perspective, is that the problems can be listed in explicit, observable terms that make them well suited to behavioral therapies. Checklists may be structured to provide for reports from clients themselves, from family members of clients (as in Figure 13.18), or from "significant others," such as teachers or therapists.

BEHAVIOR SAMPLING TECHNIQUES

Most behavioral assessment experts believe it is important to observe a person's naturally occurring behavior in "real-life" settings. Because such direct observation is not always possible, psychologists sometimes arrange to observe behavior in contrived laboratory situations designed to be experimental analogues of real life. Neither "real-life" nor laboratory observations are entirely problem-free. For example, the mere presence of observers can influence the behavior of the individual being observed; thus the behavior seen may not actually be a representative sample of the person's usual behavior. Despite this and other problems, direct observation of actual behavior adds so much to the assessment process that it is considered almost indispensible by many behaviorally oriented psychologists. Here is an example of behavior sampling (Allen et al., 1964):

> The subject was a 4-year-old girl named Ann. In nursery school, Ann avoided other children and used inappropriate behavior to get adult attention. The treatment designed for Ann used adult attention as a reward, or reinforcer. Adults showed her maximum atten-

tion when she played with another youngster and minimum attention when she sought out adults or avoided other children. To see how Ann behaved before, during, and after this intervention, observers recorded her behavior at 10-second intervals on selected days. Figure 13.19 shows what this behavior sampling procedure revealed. During the baseline period, the days before the intervention began, a very low percentage of Ann's behavior involved interactions with children and a very high percentage involved adults. On days 6–11, with the reinforcement procedure in place, Ann's behavior changed markedly; more than half the observations showed her to be interacting with children, while only about 20 percent showed her interacting with adults. When the reinforcement procedure was temporarily withdrawn during days 12–16, Ann's behavior reverted to its baseline state. On days 17–25, the intervention procedure was reinstated, but with only intermittent, or scheduled, reinforcement—a procedure that often has more lasting effects (see Chapter 4, page 159). The intermittent reinforcement apparently did the job in Ann's case. Post checks on days 31, 38, 40, and 51 showed that the desired behavior was holding up well.

Behavioral assessment methods like those discussed above have shown high reliability and good validity. Critics of behavioral assessment might argue that such methods do little more than scratch the surface and that they tell us little about the inner state or personality of the individual. Advocates of behavioral assessment might reply that we should learn to assess observable behavior before we try to infer inner characteristics that we cannot see. The ongoing debate about assessment, what it can and should be, will certainly continue for years to come.

Figure 13.18

Thirty of the more than 100 problems listed on the Child Behavior Checklist. Parents or guardians indicate the degree to which each problem is characteristic of their child by circling 0, 1, or 2 for each item. Such problem checklists are an increasingly popular approach to behavioral assessment. (Courtesy of Dr. Thomas M. Achenbach.)

Below is a list of items that describe children. For each item that describes your child *now* or *within the past 6 months*, please circle the *2* if the item is *very true* or *often true* of your child. Circle the *1* if the item is *somewhat* or *sometimes true* of your child. If the item is *not true* of your child, circle the *0*. Please answer all items as well as you can, even if some do not seem to apply to your child.

0 = Not True (as far as you know) **1 = Somewhat or Sometimes True** **2 = Very True or Often True**

0	1	2	1. Acts too young for his/her age
0	1	2	2. Allergy (describe): ____________
0	1	2	3. Argues a lot
0	1	2	4. Asthma
0	1	2	5. Behaves like opposite sex
0	1	2	6. Bowel movements outside toilet
0	1	2	7. Bragging, boasting
0	1	2	8. Can't concentrate, can't pay attention for long
0	1	2	9. Can't get his/her mind off certain thoughts; obsessions (describe): ____________
0	1	2	10. Can't sit still, restless, or hyperactive
0	1	2	11. Clings to adults or too dependent
0	1	2	12. Complains of loneliness
0	1	2	13. Confused or seems to be in a fog
0	1	2	14. Cries a lot
0	1	2	15. Cruel to animals
0	1	2	16. Cruelty, bullying, or meanness to others
0	1	2	17. Daydreams or gets lost in his/her thoughts
0	1	2	18. Deliberately harms self or attempts suicide
0	1	2	19. Demands a lot of attention
0	1	2	20. Destroys his/her own things
0	1	2	21. Destroys things belonging to his/her family or other children
0	1	2	22. Disobedient at home
0	1	2	23. Disobedient at school
0	1	2	24. Doesn't eat well
0	1	2	25. Doesn't get along with other children
0	1	2	26. Doesn't seem to feel guilty after misbehaving
0	1	2	27. Easily jealous
0	1	2	28. Eats or drinks things that are not food (describe): ____________
0	1	2	29. Fears certain animals, situations, or places other than school (describe): ____________
0	1	2	30. Fears going to school

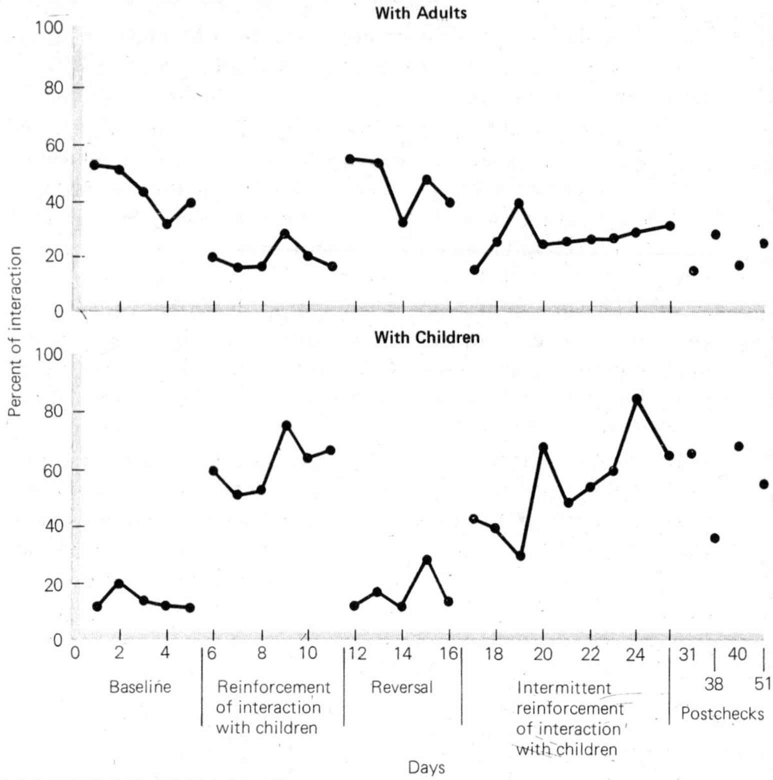

Figure 13.19

Behavioral assessment through direct observation, or behavior sampling. The percentage of time during nursery-school observation periods that Ann spent interacting with adults and with children was plotted over a 51-day period. Reinforcement (see text) resulted in increased time spent with children and decreased time with adults. (Modified slightly from Allen et al., 1964.)

Summary

1. An early pioneer in psychological assessment was Francis Galton, who attempted unsuccessfully to measure mental ability using physiological measures. The first real success in intelligence testing was the Binet-Simon test, developed in France to identify mentally retarded children.

2. Psychological tests are behavior samples that are uniform, objective, and interpretable. They are designed to gauge individual differences, and norms are often developed to help interpret these differences.

3. Achievement tests tap skills and information already acquired. Ability tests (intelligence and aptitude tests) are designed to minimize differences in achievement and to reveal what people can do at their very best; in other words, ability tests are aimed at assessing capacity, or potential.

4. A good test is reliable; that is, it yields similar results with different testers, different scorers, different forms of the test, and on repeated testings of the same person. A good test is valid; that is, it measures well what it was designed to measure. A good standardized test has norms based on a representative sample of the population for which it is intended.

5. A number of theories define intelligence in terms of its organization (for example, the G-factor theory and multifactor theories). The Stanford-Binet Intelligence Scale, which is arranged by age levels, and the Wechsler tests, which are arranged by type of item, are among the best-known intelligence tests. These tests now use deviation IQs. Both do a good job of measuring the intellectual abilities needed for school achievement.

6. Some theorists define intelligence in terms of cognitive processes (for example, Piaget, Bruner, and the information-processing theorists). Their theories have not yet led to comprehensive tests like the Stanford-Binet, but the Uzgiris-Hunt infant-development scales illustrate the process-oriented approach to assessment.

7. The intellectual development of mentally retarded people proceeds more slowly and ceases at a less-advanced level than does that of nonretarded people. Some become retarded because of specific biological catastrophes, but most retardation results from unfavorable combinations of environmental and genetic factors. Many retarded people can be helped to lead productive lives, but "cures" are extremely rare.

8. Some studies of people with IQs higher than 140 show that most are well adjusted from childhood through adulthood. Other studies of very bright people with IQs higher than 180 show that they may be misfits as children but that their life adjustment as adults is generally superior.

9. Reliable IQ differences can be identified among groups differing in home environment and race, but differences within the groups tend to be far larger than these between-group differences.

10. Aptitude tests tap specific abilities, usually for specific types of training, education, or vocation.

11. Personality tests do not have "right" or "wrong" answers. Instead, they seek answers that will reveal people's characteristic tendencies or behaviors. In this respect, they differ from achievement and ability tests, which measure optimum performance and judge it against established performance standards. Personality tests can be divided roughly into pencil-and-paper questionnaires and projective tests.

12. The validity of many pencil-and-paper tests is doubtful. However some, like the Minnesota Multiphasic Personality Inventory, have been validated in research; criterion groups were used to establish the validity of the subscales of this test.

13. Projective methods of assessing personality employ standard sets of somewhat ambiguous stimuli (for example, inkblots, vague pictures) and give test takers little guidance as to the specific form their responses should take. The expectation is that people will "project" their needs, feelings, or beliefs onto the stimuli and respond in ways that reveal important aspects of their personalities.

14. Behavioral assessment includes such approaches as problem checklists and direct observation of people's behavior. Observa-

tional methods permit comparison of behavior patterns prior to and after therapeutic interventions.

Terms to Know

One way to test your mastery of the material in this chapter is to see whether you know what is meant by the following terms.

Psychological tests (515)
Achievement tests (516)
Ability tests (517)
Intelligence (517, 524, 527)
Aptitude (517, 544)
Personality (517)
Personality tests (517, 547)
Reliability (518)
Validity (519)
Criterion (519)
Norms (520)
Percentile score (521)
Factor theories (524)
Process-oriented theories of intelligence (524, 527)
Factor analysis (525)
G-factor theory (525)
Multifactor theories (526)
Primary mental abilities (526)
Divergent thinking (526)
Convergent thinking (526)
Hierarchical theory (527)
Information-processing theories of intelligence (528)
Stanford-Binet Intelligence Scale (529)
Wechsler tests (531)
Chronological age (CA) (529)
Mental Age (MA) (529)
Intelligence Quotient (IQ) (530)
Ratio IQ (531)
Standard deviation (532)
Mean (532)
Standard scores (533)
Deviation IQ (533)
Normal curve (534)
Mental retardation (535)
Cultural-familial (or sociocultural) retardation (536)
Down syndrome (536)
Reaction range (541, 543)
Nature (543)
Genotype (543)
Nurture (543)
Phenotype (543)
Aptitude tests (544)
Scholastic aptitudes (544)
Vocational-aptitude test (544)
Test batteries (545)
Differential Aptitude Tests (DAT) (545)
Pencil-and-paper tests (547)
Face validity (548)
Minnesota Multiphasic Personality Inventory (MMPI) (548)
Empirical approach (549)
The 16 Personality Factor Questionnaire (16PF) (550)
Projective methods (550)
Projective hypothesis (550)
Rorschach inkblot technique (552)
Thematic Apperception Test (TAT) (552)
Behavioral assessment (555)

Suggestions for Further Reading

The early history of psychological testing makes interesting reading. In *Memories of My Life* (London: Methuen, 1908), Sir Francis Galton described some of his earliest efforts to test intelligence with measures like "strength of pull and squeeze." Alfred Binet and Theodore Simon described their more successful efforts in *The Development of Intelligence in Children*, reprint ed. (New York: Arno Press, 1973). Leon Kamin recounts early uses and abuses of the American translation of the Binet-Simon scale, including testing at Ellis Island to establish immigration quotas and the use of test data to promote the eugenics movement (compulsory sterilization of the retarded). Kamin's account is found in *The Science and Politics of IQ* (Potomac, MD.: Lawrence Erlbaum, 1974). By reading this book and another by H. J. Eysenck and Kamin, *The Intelligence Controversy*, (New York: Wiley, 1981), one can capture the flavor of the ongoing debate over IQ tests and the genetics of intelligence.

For a richer understanding of life at the two ends of the IQ distribution, see James J. Gallagher's, *Teaching the Gifted Child*, 3d ed. (Boston: Allyn and Bacon, 1985) and Nancy M. and H. B. Robinson's *The Mentally Retarded Child: A Psychological Approach*, 2d ed. (New York: McGraw-Hill, 1976). For a fascinating account of how mentally retarded adults discharged from a California institution coped with life on the outside, see Robert B. Edgerton's *The Cloak of Competence* (Berkeley, CA: University of California Press, 1967).

Personality and behavioral assessment and their relation to personality theories are nicely described in two texts: Lawrence A. Fehr's *Introduction to Personality* (New York: Macmillan, 1983) and Seymour Feshbach and Bernard Weiner's *Personality* (Lexington, MA: Heath, 1982). Howard B. Lyman has described some of the fundamentals involved in using and interpreting psychological tests in *Test Scores and What They Mean*, 3d ed. (Englewood Cliffs, NJ: Prentice-Hall, 1978).

Personality

HEADING OUTLINE

Chapter

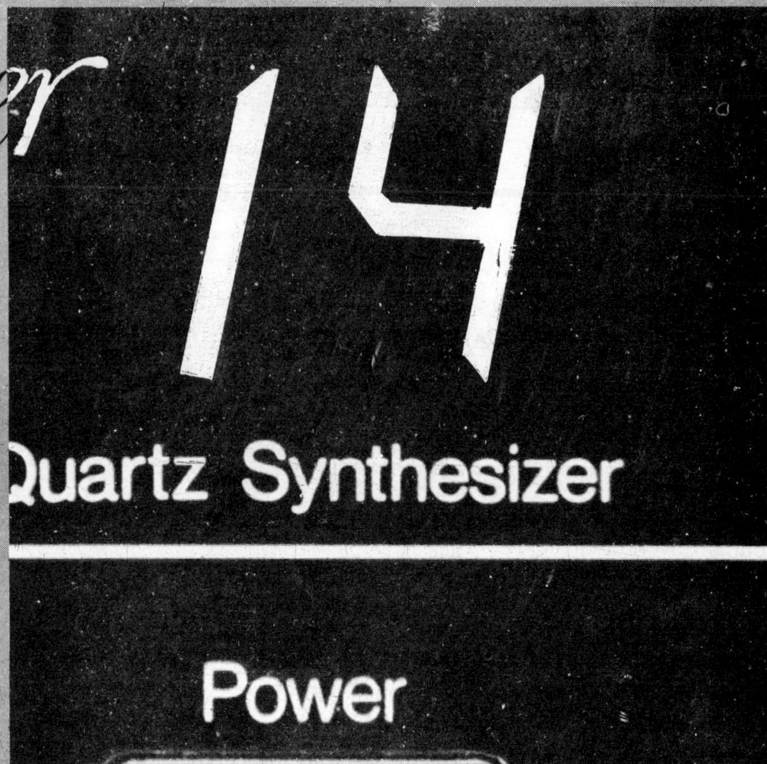

GEORGE

He was born in 1930 and grew up near the Hudson River docks in Hoboken, New Jersey. He was the older of two children, the son of a successful Manhattan attorney. As a child, George was shy, sickly, and often fearful. He feared authority figures such as his mother and his paternal grandmother. He feared giant noisemaking machines, such as the truck-mounted vacuum that sometimes appeared on his street to clean furnaces. Disgusted with himself for letting fear dominate his life, the boy took inspiration from Adolf Hitler, leader of the Nazi party in far-off Germany—a man whom the family's German maid had praised for freeing her homeland from fear. Young George set out to free *himself* from his fears, one at a time.

In one of his first battles, George tackled his fear of rats. One day in 1941, he discovered a dead rat that the family cat had left on the kitchen floor. To demonstrate to himself that he need not fear the rat, George picked it up with his hands instead of using a stick. Then, to stamp out "forever" any remaining fear or dread, George forced himself to roast the rat over a fire, cut off the haunches with his scout knife, and eat them. He drew courage from this experience; after all, he reasoned, the rats of the world should now fear *him*, an *eater* of rats.

Another episode in George's battle against fear involved thunderstorms. Late one afternoon, as a violent storm approached, young George ascended a 75-foot pin oak and lashed himself to the trunk, near the top. With the gathering darkness, the storm arrived; the wind bent the tree trunk, and lightning flashed all around him, turning everything "strobe blue." Throughout the storm, George commanded himself, "Open your eyes." He wanted there to be no doubt that he had faced death openly.

In 1957, after finishing Fordham Law School, George passed the New York state bar exam. In the same year, he became an FBI agent. He also intended to marry within the year. Frances, "the woman I wanted to bear my children," was, as George described her, "a Teuton/Celt of high intelligence." Before marrying her, though, George followed what he thought to be appropriate FBI procedures; he submitted a memo on Frances to the FBI and "had her checked." The result of the check, according to George? "She was clean." At the wedding, George followed FBI principles to the letter: He packed a gun inside his morning coat.

[Before you learn more about George (in Chapter 14), would you care to venture a prediction? Based on what you now know of his personality, what sorts of "missions" do you think would appeal to him? How vigorous is he likely to be in carrying out those missions? Is he likely to be successful, or is he likely to get into trouble?]

CRYSTAL

She was born in 1940 and grew up in the textile-mill country of rural North Carolina. Although her parents were low-status mill workers, Crystal established herself early in life as a youngster with self-confidence, even spunk. She was also an organizer. In the second grade, for example, she turned an unused chicken coop into a "school," where she enrolled her neighborhood friends. When peers didn't go along with Crystal's ideas, she was not averse to fighting. She could handle pain, too. Once, in the sixth grade, while playing baseball, she was hit in the eye by a fly ball; it hurt, but she steadfastly refused to cry.

It was also in the sixth grade that Crystal began to understand the concept of "social class." She noticed for the first time that the children of lawyers and doctors tended to stick together in her school, while "the poor children" were put into a classroom "all by themselves." She also noticed that "I was one of the poor children." Inequities, and the distinction between the powerful and the powerless, would continue to trouble her for many years.

Crystal's father had a strong influence on her personality. He was strong, decisive, and spontaneous. When he decided on some course of action, there was no stopping him. Sometimes, when he would take a notion to go for a ride, he'd pile Crystal into his old Dodge, (or, later, his Cadillac) and just take off. Once, when the steam boiler blew at the mill and workers were laid off for a couple of days, he zoomed home, told Crystal to "Get your rags together," and left a note for the rest of the family: "Gone to the beach. Be back tomorrow." He and Crystal then drove 250 miles to Virginia Beach, where he bought her dinner and a cheap ring, told her jokes, and took her for walks on the sandy beach. Crystal, like her dad, grew to love spontaneous decisions and quick action.

The worlds of work and marriage held many disappointments for Crystal. She had wanted to go to beauticians' school in Raleigh, but family funds could not cover her expenses there; so Crystal, at age 17, went to work at the mill, like her parents, uncles, and aunts before her. Her first marriage, to a man with some of her father's qualities, ended with a car crash that left her a widow and single parent. Her second marriage, to a quiet, passive man, left her bored and weary. To make ends meet, Crystal continued her work at the mill. There, when she was in her early thirties, she felt increasingly exploited and often fell into a state of mind that could only be called despair.

[Care to venture another prediction? From what you know about Crystal's personality, how do you think things will turn out for her? Will she eventually come to accept her fate and learn to live with it? Or will she take some action to shake things up and change her life? Is she headed for a major success, a major tragedy, or neither?]

PERSONALITY may be hard to define, but we know it when we see it. We all make personality judgments about the people we know. A major part of coming to understand ourselves is developing a sense of what our personality characteristics are. We even form impressions about the personalities of people we do not know but have only read about.

Consider George and Crystal, for example. You were given only a little information about them—a few life episodes and a couple of generalizations—but chances are you made some personality inferences about them nonetheless. You may also have ventured a few predictions based on those inferences. If so, let's see how accurate your predictions are.

If you predicted that George would like high-risk missions, that he would pursue them with vigor, and that he would eventually get himself into trouble, you were right. George's full name is actually George Gordon Liddy, but he prefers to be called G. Gordon Liddy. In the fall of 1971, Liddy was asked by John Dean, counsel to then-President Nixon, to set up a "first-class intelligence operation" that would help insure the reelection of the President. Liddy went about his task with a vengeance, developing plans for infiltrating the Democratic camp with spies, for conducting electronic surveillance of key Democrats, for using prostitutes to compromise Democratic leaders, and even for assassinating columnist Jack Anderson. A number of Liddy's most flamboyant plans were nipped in the bud by higher-ups, but one that was approved was breaking into Democratic National Headquarters in the now-famous Watergate complex. Later, when brought before the Senate Watergate Committee to testify and asked by Sam Ervin, "Do you solemnly swear to tell the truth, the whole truth, and nothing but the truth, so help you, God?" Liddy answered: "No." For his role in the Watergate break-in, Liddy spent 4½ years behind bars. (See Liddy's 1980 autobiography, *Will,* for further details.)

If you predicted that Crystal (actually Crystal Lee Jordan) would take some action to change her life and that she was headed for a major success, you were right again. Shortly after our narrative left off, Crystal joined forces with a union organizer from "up north." She decided that she had had enough of the unfairness of mill work and its caste system. She went on the attack by helping to organize the J. P. Stevens mill where she worked. Her efforts exposed her to real personal risk and left her jobless, but a hard-fought year of organizing paid off: The Stevens workers, for the first time ever, voted to unionize. Crystal's role in this process, and her life leading up to it, became the subject of a major motion picture, *Norma Rae.* Actress Sally Field won an Academy Award for her portrayal of this humble, earthy, and tenacious woman. (See Leifermann, 1975, for further details.)

Some might refer to G. Gordon Liddy and Crystal Lee Jordan as "real personalities," meaning that they are unusual people. Others might use the term personality to refer to public figures, as in "a major television personality," or to compliment or insult someone, as in "she has lots of [or absolutely no] personality." As we shall see, these everyday uses of the term are quite different from the meaning psychologists give to the term personality.

Ways of Defining and Thinking About Personality

When psychologists define *personality*, they tend to refer to qualities within a person, characteristics of a person's behavior, or both. In a now-famous definition, psychologist Gordon Allport (1937) mentioned both inner qualities and behavior, but he emphasized the inner qualities: "Personality is the dynamic organization within the individual of those psychophysical systems that determine his unique adjustments to his environment" (p. 48). In a more recent definition, psychologist Walter Mischel (1976) mentioned both inner processes and behavior but emphasized behavior. Personality, he wrote, consists of "the distinctive patterns of behavior (including thoughts and emotions) that characterize each individual's adaptation to the situations of his or her life" (p. 2). No single definition of *personality* is acceptable to all psychologists. However, most agree that personality includes the behavior patterns a person shows across situations or the psychological characteristics of the person that lead to those behavior patterns.

Personality has been studied in a number of different ways. Some have developed broad theories to explain the origins and makeup of personality. Others have focused only on one or two issues, such as the influence of heredity on personality.

The first approach, theory construction, was popular for many years. As a result, we have many broad personality theories; much of this chapter will focus on them. Most of these broad theories can be grouped into four categories: (1) *type and trait approaches*, which focus on people's characteristics—stubbornness, shyness, and so forth—and how these characteristics are organized into systems; (2) *dynamic approaches*, which emphasize on-going interactions among motives, impulses, and psychological processes; (3) *learning and behavioral approaches*, which emphasize the ways habits are acquired through basic conditioning or learning processes; and (4) *humanistic approaches*, which emphasize the self and the importance of the individual's subjective view of the world.

Which of these theories is best? Which gives the most accurate picture of "true" personality? The honest answer is that we do not really know; one reason is that we have not yet found acceptable ways to test entire theories. We can test specific implications of particular theories, one or two at a time, but this is a slow process, and it has not yet brought us anywhere near a final answer. What does seem likely, though, is that there is some truth and some useful insight in many of the major personality theories. In this chapter we will delve into several of these theories, focusing first on type and trait approaches and then moving to the other three categories listed above.

Type and Trait Theories of Personality

Type and trait theories of personality both focus on people's personal characteristics. However, various type theorists and trait theorists differ in the ways they use those characteristics to describe people. We will begin to examine these differences by returning briefly to George and Crystal and considering what "types" of people they are.

TYPE THEORIES

Reading about George and Crystal, you may have found yourself classifying them into "types." You may have seen Liddy as the "gung-ho type"; Crystal may have seemed like the "spontaneous type," or perhaps the "energetic type." Classifying people into types is one device many of us use to try to make sense out of others' behavior and to anticipate how they will act in the future. The notion that people can be classified into types is one of the oldest ideas about personality—over 2,000 years old, in fact.

One of the first type theories that we know of was proposed about 400 B.C. by Hippocrates, a Greek physician now known as the father of medicine. He grouped people into four temperament types: sanguine—cheerful, vigorous, confidently optimistic; melancholic—depressed, morose; choleric—hot-tempered; and phlegmatic—slow-moving, calm, unexcitable. Since the time of Hippocrates, countless other ways of grouping people into types have been tried. The groupings or sets of types are called typologies; they can be found in the folk wisdom of most cultures. In American culture, for example, people are often classified as leaders or followers, liberals or conservatives, losers or winners, and so forth; on some college campuses, the typologies include jocks, preps, Greeks, necks, heads, and punks. Typologies have been with us for centuries, and they are very much with us today.

What are these types that we put together to form typologies? A *type* is simply a class of individuals said to share a common collection of characteristics. For example, *introverts* could be described as people who share characteristics such as shyness, social withdrawal, and a tendency not to talk much; while *extraverts* share a tendency to be outgoing, friendly, and talkative. Usually, sharp typologies do not work well scientifically. A major reason is that almost every dimension of personality—friendliness, sociability, and so on—is distributed according to a normal bell-shaped curve, as discussed in Chapter 13, page 534. If people fell naturally into distinctly different types, we would see not bell-shaped curves but humps corresponding to the types, as shown in the left side of Figure 14.1. On most human characteristics and especially psychological characteristics—such as sociability—people tend to be bunched near the middle of the range, and only a few stand out at the extremes, as shown in the right side of Figure 14.1.

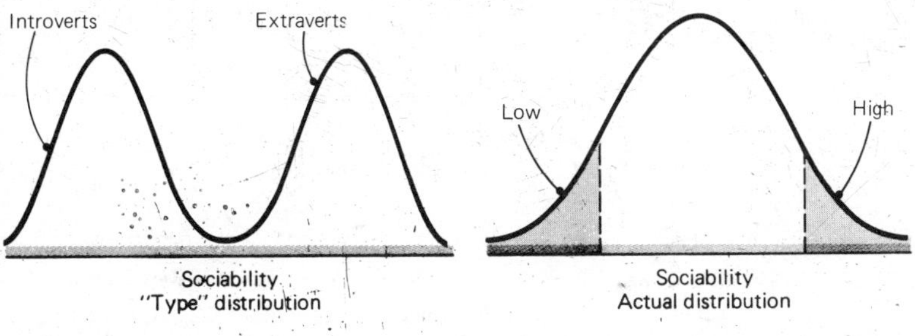

If people clustered into types, measures of their characteristics should also tend to cluster. The distriubtion on the left shows, in exaggerated form, how one kind of measure would be distributed if everyone could be categorized as either *introvert* or *extravert*. On the right, the bell-shaped, normal distribution shows that most people are reasonably close to average levels of sociability, with only the extremes of sociability seeming like "types" to us. Although we can all call to mind people who are extremely introverted or extremely extraverted, in reality most people are neither.

Figure 14.1

Since people do not fall naturally into distinct personality types, psychologists who want to develop type theories have to take special steps. One approach has been to develop theories about specific types by treating each type as a personality dimension. Individuals can then be scored or rated to determine their position on each dimension—that is, how close they come to fitting one of the types. One theorist who used this approach is H. J. Eysenck. We will examine his theory first.

Eysenck's hierarchical theory Eysenck (1967) identified the major components of personality as a small number of personality types. Each type is made up of a set of personality characteristics. For example, people who fit Eysenck's extraverted type are said to have such characteristics as sociability, liveliness, and excitability. Each one of these characteristics, according to Eysenck, can be broken down into certain habitual-response patterns that apply to several situations; each of these habitual-response patterns can be broken down further into specific responses within specific situations. This progression from broad, global types down to specific, situation-bound responses is what makes Eysenck's approach a hierarchical theory.

To develop his theory, Eysenck used a kind of steam-shovel approach to personality assessment. He scooped up a massive amount of data from many tests and measures administered to some 10,000 people. Applying factor analysis (Chapter 13, page 525) to the data, he identified (among other things) the typology shown in Figure 14.2. The vertical dimension shows people high in neuroticism at the upper end and people

Figure 14.2

Scores obtained by various neurotic, criminal, and normal groups on Eysenck's neuroticism-stability and introversion-extraversion factors. Dysthymics are moderately depressed people with impaired energy or effectiveness. Hysterics are people with physical symptoms that appear to result from psychological conflict or need. (Modified slightly from Eysenck, H. J. (1967). *The Biological Basis of Personality*. Springfield, IL: Thomas. Courtesy of Charles C Thomas, Publisher, Springfield, Illinois and H. J. Eysenck.)

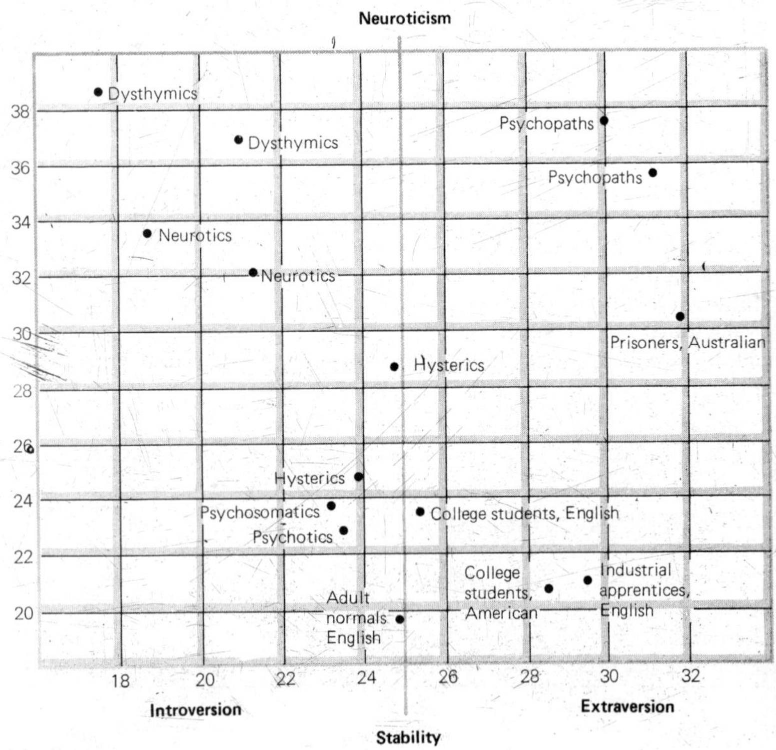

high in stability at the lower end. The horizonal dimension shows people high in introversion at the extreme left and people high in extraversion at the extreme right. The filled circles inside the figure show the average locations of various groups, based on their test results. For example, American college students scored as very stable and relatively extraverted, whereas college students from England tested out as slightly less stable and less extraverted.

Figure 14.3

In baseball, the strike zone is over the plate and at a specific height—between the batter's knees and just below the shoulders. Anywhere else is the ball zone. Similarly, if people's trait characteristics all fall within a given "type zone," fitting a particular pattern or cluster, we can speak of a psychological *type*.

Focusing on extreme cases and "good fits": The "strike-zone" approach Eysenck developed a type theory in which the types are actually personality dimensions, and every individual is scored or rated for his or her position on each dimension. A second general approach to personality types involves specifying certain key characteristics or extreme scores that must be manifest before any individual is said to fit the type. In this approach, people who do not fit the type are simply ignored, and attention is focused on the relatively "pure" cases who fit the "strike zone" for the type in question (see Figure 14.3). This approach is commonly used in diagnosing psychological disorders (see Chapter 15); people must show certain specific personality characteristics to a certain degree before they are typed as having, say, a schizophrenic disorder. The strike-zone approach is also used to identify *Type A* and *Type B* people—two groups who differ in their susceptibility to the major cause of death in the United States, heart disease. For details, see Application 14.1.

TRAIT THEORIES

Perhaps when you read about George and Crystal at the beginning of this chapter you did not think of them in terms of types at all. Instead, you may have seen Liddy as willful, determined, flamboyant, and perhaps "inclined to go to extremes to prove himself or make a point." Crystal may have seemed impulsive, strong, and "inclined to make quick decisions and stick by them." Descriptive terms like these represent *traits*—characteristics that lead people to behave in more or less distinctive and consistent ways across situations.

Allport's Theory Years ago, psychologist Gordon Allport counted about 18,000 traitlike terms in the English language—terms that designated "distinctive and personal forms of behavior" (1937, pp. 303–304). These terms, mostly adjectives, describe how people act, think, perceive, and feel. Not all of these terms reflect personality traits, but several thousand of them do. Allport (1961) believed that this rich collection of traitlike terms provided a way of capturing the uniqueness of each individual. He believed that this uniqueness could be described well in terms of the individual's traits, or "personal dispositions," at three levels of generality.

Cardinal traits he defined as those which are so dominant that nearly all of the individual's actions can be traced back to them. These broad, highly influential traits are often called by names drawn from key historical figures. For instance, one person might be described as Christlike, another as Machiavellian, and still another, in recent years, as Kennedyesque or Nixonian. Each term describes a trait so broad and so deep in its impact that it overshadows the influence of other traits in the same individual. Allport believed that most people have *no* true cardinal traits but that when someone *does* have a cardinal trait it shows itself in virtually all of that person's behavior.

For most people, who are without a cardinal trait, central traits

Application 14.1

PERSONALITY TYPES AND HEART DISEASE

We are inclined to think of heart attacks as a purely medical problem, although we all recognize their connection with lifestyle factors such as smoking, obesity, and inactivity. During recent years, however, a connection with personality type has also emerged. Two specific behavior-pattern types are now known to be associated with increased and decreased likelihood of coronary-artery disease. Indeed, these behavior-pattern types are apparently better at predicting heart disease than are elaborate medical diagnostic methods such as the examination of a person's blood cholesterol level.

Type A persons are hard-driving and competitive. They live under constant pressure, largely of their own making. They seek recognition and advancement and take on multiple activities with deadlines to meet. Much of the time they may function well as alert, competent, efficient people who get things done. When put under stressful conditions they cannot control, however, they are likely to become hostile, impatient, anxious, and disorganized. They may fume at a slow elevator or a poorly informed salesperson who interferes with their tight schedule, for example.

Type B persons are quite the opposite. They are easy going, noncompetitive, placid, unflappable. They weather stress easily. In some ways, they are like the tortoise in the tale of the tortoise and the hare. Type B's may be a little dull, but they are likely to live longer than the harelike Type A's.

Type A's can be distinguished from Type B's by their answers to personality questionnaires and also by their behavior. Given a task to do, Type A's tend to perform near their maximum capacity no matter what the situation calls for. They work hard at arithmetic problems whether or not a deadline is imposed; Type B's work harder when given a deadline. On a treadmill test (continuous walking on a motorized treadmill), Type A's expend more energy and use a greater proportion of their oxygen capacity than do Type B's, yet they rate their fatigue as less severe. If asked to judge when a minute has elapsed, Type A's judge the periods as significantly shorter than do Type B's. In other words, Type A's show a push toward achievement, a suppression of the cost (fatigue) to themselves, and impatience with delay.

It is particularly interesting, however, that when placed in long-lasting stressful situations over which they have little control, Type A's tend to give up. They show a kind of helplessness (Chapter 15, page 650) and become less responsive and less effective than Type B's. At first, they struggle to control the situation; but when they fail to do so, they stop coping.

There is, of course, a rational explanation for the link between personality type and heart disease. It probably lies in the chemical substances released by the autonomic nervous system (Chapter 8, page 315) in response to stressors. There is some evidence that victims of heart attacks have different reactions to stressors. Either their blood-chemistry response is different or the *timing* of their response is different. There is evidence that their responses are delayed and protracted rather than immediate. All this suggests that programs to help Type A patients cope more constructively with stressors might be helpful. Indeed, a number of psychologists are pursuing that very idea, apparently with some success.

REFERENCES

Blass, D. C. (1977). Stress, behavior patterns, and coronary disease. *American Scientist, 65,* 177–187.

Cromwell, R. L., Butterfield, E. C, Brayfield, F. M., & Curry, J. J. (1977). *Acute myocardial infarction: Reaction and recovery.* St. Louis, MO: Mosby.

Friedman, M., & Rosenman, R. H. (1974). *Type A behavior and your heart.* New York: Knopf.

Matthews, K. A. (1982). Psychological perspectives on the Type A behavior pattern. *Psychological Bulletin, 91,* 293–323.

become crucial. *Central traits* he described as characterizing an individual's behavior to some extent but not in such a complete way as cardinal traits. Allport described central traits as those that might be mentioned in a careful letter of recommendation or checked off on a rating scale where the rater is asked to select the outstanding characteristics of an individual. It would be rare, he wrote, for an individual to have more than 10 or 12 such central traits.

Finally, the least generalized characteristics of the person he labeled *secondary traits*. These are traits such as "likes chocolate" or "prefers foreign cars"—traits that are influential but only within a narrow range of situations. Allport recommended that cardinal, central, and secondary traits be used to assemble what he called psychological life histories and that information about these traits come from materials produced by the individuals themselves—materials such as letters, diaries, or personal journals.

In recommending that people be described in terms of the traits that capture their uniqueness, Allport was subscribing to the *idiographic approach*. This approach involves the psychological study of the individual case; it entails efforts to understand, explain, and sometimes predict an individual's behavior in various situations. A contrasting approach, which Allport also valued (though somewhat less), is called *nomothetic*, or *dimensional*; this approach is aimed at the discovery of personality principles that apply to people in general. The idiographic approach involves a search for consistencies within particular individuals; the dimensional approach involves a search for consistencies and general principles that apply *across* individuals.

Single-Trait Research Allport's approach involved multiple traits, but not all trait theorists and researchers cast such a broad net. Some try to focus carefully on one single trait. For example, many have studied *locus of control*—the degree to which we believe that we cause, or control, the events in our lives. If we believe that we are the cause of most events, we have a highly internal locus of control; if we believe that most events are caused by luck, fate, or powerful others, we have a highly external locus of control. Julian Rotter (1966) developed a questionnaire to measure internal versus external locus of control. It required people to choose between pairs of alternatives such as the following:

1*a*. Students who try their best should have little difficulty making good grades in most classes.
b. Grades in most classes are strongly influenced by teacher bias and other factors beyond the student's control.
2*a*. I am the master of my fate.
b. Most of the things that happen to me in life are caused by forces more powerful than I.

People who are in life circumstances that reduce their "control over reinforcement"—for example, minority groups or new army recruits—tend to choose external responses such as 1*b* and 2*b* more often than do people in more powerful circumstances (see Figure 14.4). Moreover, people who score as highly internal on this questionnaire often seek out learning experiences relevent to their life circumstances. For example,

Figure 14.4

Which of these two people is likely to have a more internally perceived locus of control? (Burke Uzzle/Woodfin Camp & Associates.)

internal patients with life-threatening illnesses have been found more likely than external patients to seek out information about their illnesses and to look for ways of treating it. As research on locus of control illustrates, we can learn useful things about people by focusing on a single trait.

QUESTIONS ABOUT TYPE AND TRAIT THEORIES

Type and trait approaches to personality are close enough to our everyday way of thinking about people as to have a certain ring of truth. Yet truth can be elusive, especially when our quarry is as complex as the human personality. Although type and trait theories have enhanced our understanding of human behavior, critics have questioned whether these approaches give us a complete and accurate picture of personality. Their concerns have fallen into two broad areas: questions about methodology and a philosophical question.

Methodological Questions One of the most basic questions about methodology concerns reliability (see Chapter 13, page 518). Generally, reliability in the form of agreement among observers is not a major problem, provided the observers are given careful training and clear rules for judging trait (or type) dimensions. This is true whether we are speaking of well-developed schemes for observing behavior directly or well-developed tests that yield scores on various traits. Problems of interjudge reliability do arise, however, when the procedures for inferring traits are not clear or when very subjective judgments are involved.

A second question concerns the validity of trait assessments—that is, whether the assessments mean what they are supposed to mean. We know from experience that people may give answers on tests and questionnaires to present themselves in a certain light. Some may try to make themselves "look good." Others may try to make themselves look troubled in order to get help from others. To the degree that such response sets (for example, social desirability or help seeking) influence the scores people receive on trait or type measures, the measures are not doing a valid job of assessing traits and types. Fortunately, some of our measures (for example, the MMPI discussed in Chapter 13) have built-in ways of assessing such response sets.

A third methodological question concerns type and trait consistency. Types and traits are generally thought of as sources of stability in our behavior that lead us to behave in consistent ways across different situations. Yet many research findings over the past 60 years suggest otherwise: People who are honest in one situation are often not so honest in another; people who are calm in one situation may be very anxious in another, and so forth. Findings like these have led some to conclude that type and trait theories may be a bit misleading—that they may depict more consistency in human behavior than really exists. This line of thinking has helped to promote *situationism*—the view that behavior is more a product of the particular situation we find ourselves in than a product of enduring "person" characteristics like traits or types. We will consider the person-versus-situation debate later in this chapter (pages 606–608).

A Philosophical Question One question about trait and type approaches is purely philosophical in nature: Is it really adequate to think of our personality as the sum of our traits or as the particular type we fit? Some argue that personality is much more—that it is an active, dynamic interplay of motives, thoughts, and feelings and that it is best described not as a set of traits or types but as a set of processes by which people cope with life. We turn now to some theories that present this dynamic, process-oriented view of personality.

Dynamic Personality Theories

Before Gordon Allport became a famous trait theorist, he visited an already-famous theorist of a different kind: Sigmund Freud. At the ripe old age of 22, and fresh out of college, Allport wrote a note to Freud asking if he might see the famous psychoanalyst when he, Allport, visited Vienna. Freud very kindly replied with an invitation to visit. Young Gordon was ushered into "the famous red burlap room" which had "pictures of dreams on the walls"; then he was summoned to Freud's inner office. Here, in Allport's words, is what happened next:

> He did not speak to me but sat in expectant silence for me to state my mission. I was not prepared for silence and had to think fast to find a suitable conversational gambit. I told him of an episode on the tram car on my way to his office. A small boy about four years of age had displayed a conspicuous dirt phobia. He kept saying to his mother, "I don't want to sit there . . . don't let that dirty man sit beside me." To him everything was *schmutzig* (filthy). His mother was a well-starched Hausfrau, so dominant and purposive looking that I thought the cause and effect apparent.
>
> When I finished my story Freud fixed his kindly therapeutic eyes upon me and said, "And was that little boy you?" (Allport, 1968, p. 383)

The little boy, in fact, was *not* Gordon Allport. Allport's story was, in fact, precisely what he presented it to be: a description of an actual episode on an actual tramcar, involving an actual other person. But Freud's attempt to probe for a deeper meaning reveals an important difference between the type and trait theories of personality that we have just discussed and the dynamic theories to which we now turn our attention. *Type and trait approaches* involve a search for separable components of personality and for the ways those components fit together to form a personality structure. *Dynamic approaches* involve a search for the processes by which needs, motives, and impulses—often hidden from view—interact to produce the individual's behavior. Freud's question to Allport revealed just such a search—that is, a search for any less-than-apparent needs, motives, or impulses which might have led Allport to tell that particular story at that particular time.

Freud's question also reveals a tendency for which dynamic theorists are often criticized: a tendency to overinterpret—that is, to attribute deeper meaning to behavior than the behavior really warrants. A related criticism is that dynamic theorists often make claims about personality processes without testing them scientifically. Both criticisms are often applied to Freud's theory, which remains in some respects one of the least substantiated of the major personality theories. However, Freud's is also the most influential and the most comprehensive theory of personality. It has had a major impact on psychology, sociology, anthropology, and literature—indeed, on western thought in general. Currently its influence and popularity have faded somewhat among some psychologists, partly because there is so little hard, scientific evidence to support the theory. However, recent surveys of psychotherapists (Chapter 16) reveal that more of them subscribe to dynamic theories, including Freud's, than to any other approach. Certainly there is much about Freud's theory that is still widely accepted, and no examination of personality theories would be complete without a careful look at his monumental contribution.

FREUD'S PSYCHOANALYTIC THEORY

We begin our study of Freud's approach with a look at the raw material he used to shape his theory: the thoughts and recollections of his patients. Here is a sample from the analysis of "Frau Emmy von N.," a patient Freud began treating in 1889.

> "When I was a child, it often happened that out of naughtiness I refused to eat my meat at dinner. My mother was very severe about this and . . . I was obliged two hours later to eat the meat, which had been left standing on the same plate. The meat was quite cold by then and the fat was set so hard" (she showed her disgust). . . . "Whenever I sit down to a meal I see the plates before me with the cold meat and fat on them. . . ."
>
> "Many years later, I lived with my brother who was an officer and had that horrible disease [venereal disease]. I knew that it was contagious and was terribly afraid of making a mistake and picking up his knife and fork" (she shuddered) ". . . soon after that, I nursed my other brother when he had consumption [tuberculosis] so badly. We sat by the side of his bed and the spittoon always stood on the table, open" (she shuddered again) ". . . and he had a

habit of spitting across the plates into the spittoon. This always made me feel so sick, but I couldn't show it for fear of hurting his feelings."
(Breuer & Freud, 1893–1895/1955, p. 82)

As "Frau Emmy's" recollections spilled out, Freud's mind was churning, piecing together a complex psychological puzzle. Frau Emmy suffered from anorexia (see Chapter 12, page 480)—she avoided food and drink. Emmy also stuttered and had periodic tics—seemingly uncontrollable facial contortions—and bursts of inappropriate speech. For example, in the middle of a pleasant conversation, Frau Emmy's face would sometimes twist into a grimace of disgust, and she would say, "Keep still! Don't say anything! Don't touch me!" To understand these "symptoms" and their origin, Freud tried in whatever ways he could to unravel the complex fabric of Frau Emmy's personality. Gradually, it occurred to Freud that Emmy's failure to eat and her symptoms of disgust and revulsion might all relate to deeply buried memories—memories like those we quoted above—in which food became disgusting. Unraveling current problems meant digging such memories out of the unconscious.

For this troubled woman, and for most of Freud's other patients, the unraveling and digging process involved reconstructing and even reliving the past—often going as far back as early childhood. Freud's methods evolved over time. He favored hypnosis early in his career (Frau Emmy's recollections, above, were brought out under hypnosis); but he moved gradually toward free association, in which the patient said *everything* that came to mind, no matter how silly, illogical, or even forbidden it might seem. Applying such techniques to a steady stream of patients, Freud pieced together a picture of personality that shook the western world and shaped much of the field of psychology for decades.

Freud's psychoanalysis (Freud, 1920, 1938, 1940/1949) became both a theory of personality and a method of psychotherapy (see Chapter 16, page 681). *Psychoanalytic theory* has three major parts: (1) a theory of the structure of personality, in which the ego, id, and superego are the principal parts; (2) a theory of personality dynamics, in which conscious and unconscious motivation and ego-defense mechanisms play a major role; and (3) a theory of psychosexual development, in which different motives and body regions influence the child at different stages of growth, with effects persisting in the form of adult personality traits.

Personality Structure: Id, Ego, and Superego Freud constructed a model of personality with three interlocking parts: the id, the ego, and the superego.

The *id,* the most primitive part, can be thought of as a sort of storehouse of biologically based urges: the urge to eat, drink, eliminate, and, especially, to be sexually stimulated. The sexual energy that underlies these urges is called the *libido.* The id operates according to what Freud called the pleasure principle. That is, left to itself, the id would satisfy its fundamental urges immediately and reflexively as they arose, without regard to rules, the realities of life, or morals of any kind.

The id, however, is usually bridled and managed by the *ego.* The ego consists of elaborate ways of behaving and thinking which constitute the "executive function" of the person. The ego delays satisfying id motives and channels behavior into more socially acceptable outlets. It keeps a person working for a living, getting along with people, and generally adjusting to the realities of life. Indeed, Freud characterized the ego as working "in the service of the reality principle." That is, the ego tries to satisfy the id's urge for pleasure but only in realistic ways that take account of what is possible in the real world. The ongoing tension between the insistent urges of the id and the constraints of reality helps the ego develop more and more sophisticated thinking skills.

The *superego* corresponds closely to what we commonly call the conscience. It consists mainly of prohibitions learned from parents and other authorities. The superego may condemn as "wrong" certain things which the ego would otherwise do to satisfy the id. But the superego is not all fire and brimstone. Its conscience-like proddings are also guided by what Freud called the ego ideal, a set of positive values and moral ideals that are pursued because they are believed to be worthy.

Personality Dynamics and Levels of Consciousness Freud did not intend to divide personality into three separate compartments but rather to convey a lively, ongoing interplay among the id, the ego, and the superego. In this interplay, he saw the ego acting as a sort of mediator between the id—with its blind demands for instant gratification—and the superego—with its rigid, often irrational rules, prohibitions, and ideals. The ego's task of satisfying both id and superego requires a somewhat risky balancing act. For example, if the ego yields to the id's desire for something that is morally forbidden, the superego may "punish" the ego with feelings of guilt. Thus the ego's task often involves finding a compromise between the instinctual gratification sought by the id and the strict rule-following sought by the superego. Freud's general notion that our behavior is influenced by biological drives (*id*), social rules (*superego*), and mediating thought processes (*ego*) may not seem farfetched. However, his heavy emphasis on the primitive, sexual nature of human drives and energy (*libido*) helped make his theory very controversial.

Figure 14.5

One can imagine the components of personality, in psychoanalytic theory, something like this. Note how important Freud thought the id to be (its size relative to the other components) and how large a part of the personality remains unconscious.

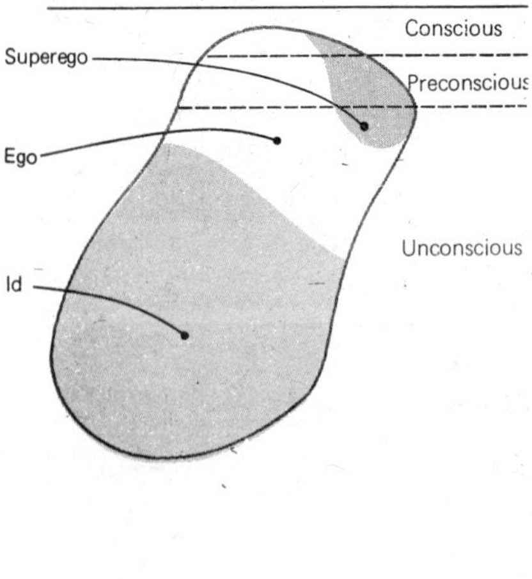

Less controversial but equally novel was Freud's notion of unconscious processes. He used this concept to explain why people often act in ways that seem irrational. Freud proposed three levels of consciousness, or awareness: the *conscious*, the *preconscious*, and the *unconscious*. At the conscious level, we are aware of certain things around us and of certain thoughts. At the preconscious level are memories or thoughts that are easily available with a moment's reflection—for example, what we had for breakfast, or our parents' first names. In contrast, the unconscious contains memories, thoughts, and motives (Chapter 7) which we cannot easily call up. All of the id is unconscious; the ego and superego include material at all three levels of consciousness (see Figure 14.5).

Why do some ideas and feelings become unconscious? There are several answers, but one that became a centerpiece of Freud's theory is summed up in the word *repression* (discussed further on page 588). We *repress*, or banish from consciousness, ideas, memories, feelings, or motives that are especially disturbing, forbidden, or otherwise unacceptable to us. The process of repression is itself unconscious and automatic. We do not *choose* to repress an idea or impulse—it just happens, whenever the idea or impulse is so painful and anxiety-arousing that we must escape from it. In such cases, our anxiety triggers repression, and the unacceptable material is buried in the unconscious.

The material we repress is usually something that clashes painfully with our ethical standards or self-image. For example, one of Freud's patients, Fräulein Elisabeth von R., had experienced what she knew was an unacceptable desire for her sister's husband. Her sister had grown very ill, and one day Fräulein Elisabeth arrived at the sickbed to find that her sister had died. Freud described Elisabeth's reaction.

> At that moment of dreadful certainty that her beloved sister was dead . . . another thought had shot through Elisabeth's mind, and now forced itself irresistibly upon her once more, like a flash of lightning in the dark: "Now he is free again and I can be his wife." (Breuer & Freud, 1893–1895/1955, p. 156)

Here was a shocking idea, totally foreign to Elisabeth's moral standards and to her view of herself as an ethical person. Not only had she desired her sister's husband while her sister lay ill and defenseless, but even at the moment of her sister's death her first thought had been one of selfish joy—that she might profit from the death by gaining a husband. This whole set of ideas and motives conflicted so painfully with Elisabeth's moral standards and her self-image that it was repressed and exiled to her unconscious. She never thought about it again, at least not until Freud brought the memories back through psychoanalysis.

Why bring the memories back, you might ask. Why not leave them safely tucked away in the unconscious? The answer, according to Freud's theory, is that repressed material does not remain *safely* tucked away. Instead, it continues to operate underground, often converting the repressed conflicts into *neurosis*—disturbed behavior involving anxiety or defenses against anxiety. Neurotic symptoms often bear a *symbolic* relationship to the repressed material that is causing them. For example, Fräulein Elisabeth was first attracted to her brother-in-law when the two went for a long walk at a summer resort. After her sister's death, she developed severe pains in her legs and was eventually unable to walk or even to stand up. It was as if she was being punished for the feelings she developed during a walk by developing symptoms that would prevent her from walking. Freud believed that for Elisabeth and other people as well, repressed material is not dormant but active—in unconscious, often painful ways.

Unconscious processes also figured prominently in Freud's ideas about dreams and "accidents." (See Chapter 5, page 207.) He saw dreams as disguised manifestations of id motives, and he described dreams as "the royal road to the unconscious." In everyday life, he said, unconscious thoughts and forbidden impulses are revealed by accidents and by slips of the tongue or the pen. For example, Freud told of a certain Social-Democratic newspaper that reported on activities of "the Clown Prince" instead of "the Crown Prince." The paper ran an apology the next day, correcting the misprint so that it read "the Crow Prince." In another example, Freud described a letter in which a patient attributed his nervousness to a cold wave that had killed off the crops. But instead of writing "cold wave," the patient wrote "cold wife." This "slip," Freud said, grew out of the patient's unconscious anger toward his wife for her "marital frigidity." Even today, slips in letters may point to unconscious motives; take a look at Figure 14.6 and see how many revealing slips you can find.

Psychosexual Stages of Development Freud put heavy emphasis on biological development in general and on sexual development in particular. In his theory of child development, Freud described a succession of stages revolving around body zones. Freud's idea was that from birth on, we have an innate tendency to seek pleasure, especially through physical stimulation and particularly through stimulation of parts of the body that are sensitive to touch: the mouth, the anus, and the genitals. It was no accident that Freud referred to these parts of the body as *erogenous zones*. He saw the pleasure derived from touching and rubbing these sensitive zones as *erotic*, or sexual, for infants as well as adults. Flip back to page 410, and review the quotation from Freud. There he described the baby's sucking as a sensual activity that leads to something very much like an orgasm. For babies, the most sensitive erogenous zone is the mouth. With advancing age, however, other body zones become especially sensitive to stimulation. With each of these shifts in the focus of sexual stimulation come parallel shifts in the dominant psychological issues faced by the person. Thus Freud called each step in this process a *psychosexual stage*.

Freud believed that if a child's needs at one of the psychosexual stages were either unsatisfied or oversatisfied, *fixation* would take place. That is, the child would show continued attachment to an old stage even after moving on to a new one. As a result, behavior patterns—and problems—from the fixated stage would persist, often into adulthood. By observing an adult's behavior carefully, one could recognize the psychosexual stage in childhood at which the adult had fixated. We turn now to a description of the psychosexual stages and some of their implications for adult personality.

Figure 14.6

An exercise in the detection of Freudian slips—in this case, slips of the pen. A woman writes to her son after receiving the news of his engagement to a woman from Richmond, Virginia. In commenting on this "delightful surprise," she may be telling her son more of her true feelings than she realizes. How many slips of the pen can you spot? (From *Psychology and Life* (10th ed.) By P. G. Zimbardo. Copyright © 1979 by Scott, Foresman and Company. Reprinted by permission.)

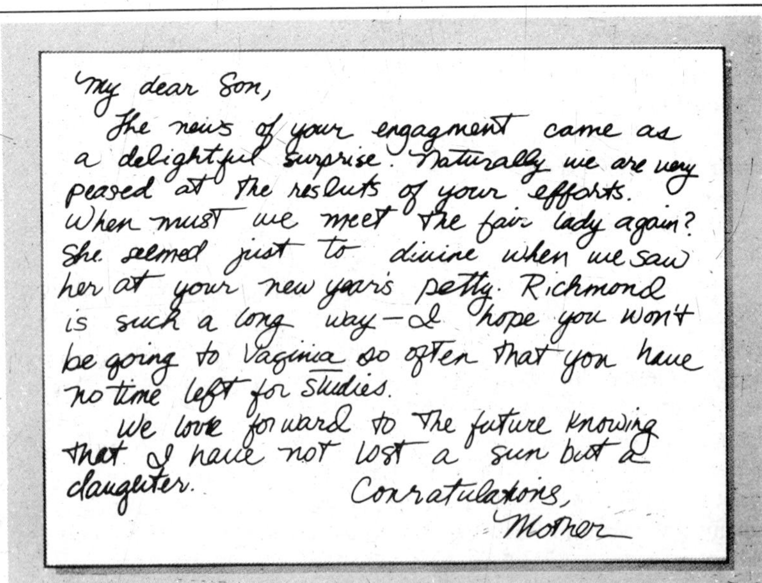
My dear Son,
The news of your engagment came as
a delightful surprise. Naturally we are very
peased at the resluts of your efforts.
When must we meet the fair lady again?
She seemed just to divine when we saw
her at your new year's petty. Richmond
is such a long way—I hope you won't
be going to Vaginia so often that you have
no time left for studies.
We love forward to the future knowing
that I have not lost a sun but a
daughter.
Conratulations,
Mother

In the *oral stage* (birth to about age 1), according to Freud's theory, the infant obtains sensual pleasure first by sucking and later by biting. Feeding and contact with the mother, mouthing new objects, and even relief of teething pain by biting—all help to make the mouth the focus of pleasure during the first year. A baby given too little opportunity to suck (or too much), or made anxious about it, may acquire an oral fixation which, in adulthood, may foster excessive oral behavior—for example, "taking in" in concrete forms such as smoking or in psychological forms such as dependency. (See Figure 14.7.) Fixation during the oral biting stage may produce a critical, "biting" personality.

The *anal stage* (ages 1–3) occurs when parents are toilet training their children and teaching them to avoid prohibited behavior connected with excretion. At this time the region around the anus becomes highly sensitive to the stimulation of "holding on" and "letting go." In our society, toilet training is ordinarily the child's first encounter with authority and the first time the id must be brought under control of the emerging ego. Psychoanalytic theory holds that the first part of the anal stage involves pleasure from expulsion of feces; the latter part involves pleasure from retention. Freud maintained that fixation at the first substage results in adult characteristics of messiness and disorder; fixation at the latter substage results in excessive compulsiveness, overconformity, and exaggerated self-control.

After they master toileting, children grow more interested in their genitals. In fact, Freud believed that genitals become the major focus of sexual excitement in the *phallic stage* (ages 3–5). It is also at this time that he felt children developed sensual feelings toward the parent of the opposite sex. Freud called these thoughts and feelings in boys the *Oedipus complex* after the mythical story of Oedipus, who unknowingly killed his father and married his mother; in girls, he called the thoughts and feelings the *Electra complex* after Agamemnon's daughter, who arranged for her mother to be murdered.

In the phallic stage, the boy's desire for stimulation of his penis is associated with his attachment to his mother, and he comes to desire her in a sexual, or at least a sensual, way. Eventually, the boy becomes fearful that his father knows of these feelings and will be furious. He fears retaliation by his father, retaliation aimed at the "source of the problem"—the boy's penis. In other words, the boy fears castration. To cope with the severe anxiety that this fear brings on, the boy begins to construct psychic defenses, especially the defense of *identification*—the boy tries to become like his father. Anxiety is reduced because a threatening father would not be likely to harm someone just like him—a "chip off the old block." In the process of identifying with his father, the boy not only takes on his father's behavior patterns but also his father's ideas of right and wrong. Thus, it is through identification in the phallic stage that the boy's superego begins to form.

For the girl, the sequence begins with an erotic focus on the father. But, in addition, the girl notices that she does not have the sexual organs of her father or brothers and she experiences "penis envy." She suspects that she may actually have been castrated by her mother; this makes her angry, and she comes to resent and devalue her mother. Nonetheless, she eventually identifies with her mother—partly because she suspects that if she takes on her mother's characteristics she will stand a better chance in

Figure 14.7

Changes occur in the normal expression of oral behavior as a person grows older. Fixation at the oral stage may produce an oral syndrome characterized by excessive oral behavior and such traits as dependency and passivity. (After Wattenberg, 1955.)

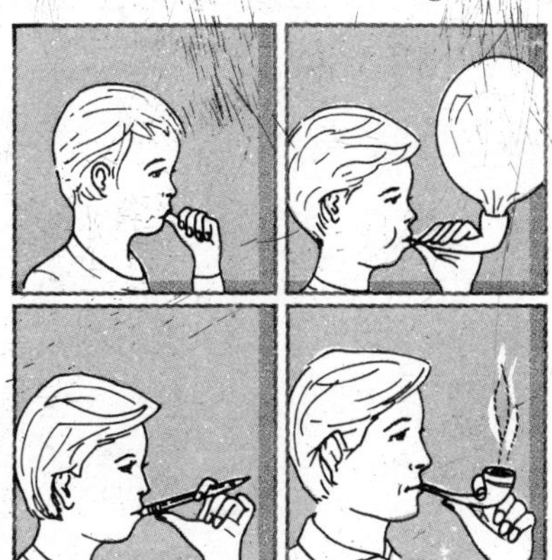

her own "romantic relationship" with her father. Thus, in spite of her affection for her father and her resentment of her mother, the little girl identifies with her mother, behaving like her and incorporating her values. In this way, the girl develops her superego.

Of course, the specifics of the phallic stage vary from child to child. To illustrate, we have included one of Freud's most famous case histories. "Little Hans," discussed in Application 14.2, was, in Freud's opinion, caught up in the throes of the Oedipus complex.

Freud believed that superego strength—that is, how harsh one's conscience is in punishing misbehavior—depended largely on events of the phallic stage. Very punishing, anxiety-ridden experiences were said to produce a strong, punishing superego.

The *latency period* (6 through puberty) was not considered by Freud to be very important to the development of personality. As the child learns more about the world, sexuality is largely repressed and the ego expands.

Puberty marks the beginning of the *genital stage* (adolescence and beyond), when mature heterosexual interests appear. There are three major sources of sexual arousal during this period: memories and sensations from earlier childhood periods, physical manipulation of genitals and other erogenous zones, and hormonal secretions. Many of the themes and anxieties of earlier stages resurface, but in new and more mature forms. In particular, the targets of sexual arousal now lie outside the tiny circle of self and family. Mature heterosexual relations emerge, with the species-preserving possibility of procreation now very real. The stage is set for responsible enjoyment of adult sexuality—which was, for Freud, the epitome of healthy development.

JUNG'S ANALYTICAL PSYCHOLOGY

Carl Gustav Jung (pronounced "Yoong") was, at one time, considered by Freud to be his heir apparent. In 1907, Freud wrote to Jung, "I could hope for no one better than yourself . . . to continue and complete my work" (McGuire, 1974, p. 27). Yet the direction that Jung took with his *analytical psychology* (Jung, 1928) differed from psychoanalysis in some very significant ways, and Freud's affection for Jung faded. One of the main differences between the theorists was that Jung thought childhood psychosexual development to be not nearly so important to adult adjustment as Freud did. Jung placed much less emphasis than Freud on sexual and aggressive impulses arising from past conflicts and much more emphasis on people's future-oriented goals, hopes, and plans. The two men also had quite different ideas about the nature of the unconscious. To appreciate Jung's ideas on this subject and others, we need to first consider his background, beginning with his childhood.

From childhood on, Jung was extremely interested in dreams and fantasy. Some of his boyhood dreams even terrified him. Young Carl's minister father was often depressed and prone to doubt his faith, and some of Carl's most powerful dreams involved religious images and symbols. These dreams were often laced with forbidden and sacrilegious overtones—for example, a godlike figure that looked like a giant penis. Such boyhood dreams helped give Jung, the theorist, his lifelong fascination with the content and meaning of dreams and visions in people's lives.

In developing his theory, Jung drew on both his own dream and

Application 14.2

A PSYCHOANALYTIC CASE HISTORY

One of the most famous cases in the history of psychoanalysis is that of "Little Hans," a 5-year-old who revealed many of his perceptions, fantasies, and fears to his physician father, who, in turn, reported them to Sigmund Freud. The case of Little Hans illustrates many features of the phallic stage of development as well as some ideas of psychoanalytic theory about phobias, or irrational fears.

Little Hans, from the age of 3 and on, showed a lively interest in that part of his body which he called his widdler. He had an active sexual curiosity that was particularly directed toward members of his family. For a time, he maintained that "widdlers" were possessed by all animate objects and were the feature which differentiated animate from inanimate objects. Naturally, he surmised that girls and women had widdlers, only small ones. Even when he viewed his newborn sister, the evidence did not dissuade him. His frequent conversations about widdlers revealed his misinterpretations of things he saw. For example, he thought a monkey's tail was a very long widdler. At a railroad station, when he saw an engine discharging steam, he assumed that it was using its widdler. He sometimes fondled his widdler, and when he was 3½ his mother threatened him with castration if he continued this behavior. Some guilt about masturbation dated from that time.

As one would predict from psychoanalytic theory, Han's sexual interest and focus on his widdler began to intensify at about age 4. When he was 4½, for instance, he made it clear that having his knickers unbuttoned and his widdler taken out was quite pleasurable, and he finally acknowledged that there was, indeed, a distinction between male and female genitals. At age 5, Hans awoke one morning in tears after an anxiety dream in which his mother was gone. His anxiety—together with a fantasy about a big giraffe and a crumpled giraffe (interpreted as his father's penis and his mother's genitals)—was thought to express an erotic longing for his mother.

Hans began to fear horses; this fear grew to a point where he would not leave the house (horse-drawn carriages were the major means of transportation in Freud's Vienna). The immediate event which precipitated this phobia was seeing a big, heavy horse fall down. Freud interpreted this to mean that Hans, at that moment, perceived his own wish that his father would fall down (die). Then Hans, a little Oedipus, could take his father's place with his beautiful mother. Another part of the fear derived from the large size of horses (and their widdlers), which Hans unconsciously identified with the great power of his father. He expressed the fear that a horse would come into his room. He also became afraid not only of horses but of carts, furniture vans, and buses This revealed, to the psychoanalyst, still another aspect of Hans's unconscious fantasies: that the falling-down horse was not only his father but his mother in childbirth (with the boxlike carts and vehicles representing the womb). All these complicated, repressed (and, in psychoanalytic theory, universal) feelings and perceptions were thus incorporated into a single phobia.

It is important to note that Little Hans was basically a straightforward, cheerful child; he experienced normal psychosexual development, marred only by the episode of the phobia, from which he recovered rather promptly. Fourteen years later, a 19-year-old Hans came to see Freud. He had continued to develop well and had survived, without unusual difficulty, the divorce and remarriage of both parents. The problems of his childhood (which by then he could not remember) were used by Freud to illustrate the normal process of psychosexual development—the complex, intense, erotic drama of early childhood.

REFERENCE

Freud, S. (1925). Analysis of a phobia in a five-year-old boy. In *Sigmund Freud: Collected Papers* (Vol. 3). (A. Strachey & J. Strachey, Trans.). London: Hogarth Press. (Original work published 1909)

ıntasy material and that of his patients; but he also drew on findings from ıch experimental techniques as the word-association test, in which peo-le were read a standard array of 100 terms (for example, "head," "to sin," "to pray," "bride," "to abuse") and instructed to respond to each term "as quickly as possible with the first word that occurs to you." In addition to the content of the test taker's associations, Jung recorded how long it took them to respond and whether certain words led to particular patterns of breathing or even perspiration. Stimulus terms that resulted in long delays, an inability to respond, or certain other key signs were thought by Jung to be parts of what he called *complexes*. He defined a *complex* as a network of ideas bound together by a common emotion or set of feelings. For example, Freud's patient, Frau Emmy (page 576), might be said to have had an eating complex.

In using his word-association test to diagnose complexes, Jung believed he was exploring the unconscious, much as Freud had done with hypnosis and free association. Yet Jung's findings with his own techniques led him away from Freud's ideas in several respects. As we noted above, Freud emphasized sexual impulses and motives as the motor of human behavior—a prime source of the libido's energy. But Jung saw the driving force that fueled the libido as "only a continuous life urge," a striving to live and insure the survival of one's species. In addition, Jung's work led him to emphasize some personality structures whose existence Freud either doubted or considered unimportant.

One such structure was the collective unconscious, which Jung believed to be a part of the unconscious mind that went beyond the personal experiences of the individual. The *collective unconscious*, he said, grows out of the past experiences of the human race. Stored within this collective unconscious are primitive fundamental images, impressions, or predispositions that were common to earlier members of the human race. He called these images, impressions, or predispositions *archetypes*. Archetypes are hard to define. They do not have real concrete content, and they are not quite like memories. They are more like subjective reactions that originated in our ancestors in response to certain universal experiences—such as the recurrent rising and setting of the sun, the presence of males and females, and so forth. We have inherited tendencies toward similar subjective reactions, and these tendencies are our archetypes. Put another way, archetypes are inherited ways of organizing, or reacting to, our experiences with the world. Some of the most common archetypes involve God, rebirth (or resurrection), devil (or the embodiment of evil), wise old person, mother, trickster or magician, hero, *animus* (a male archetype present in females), and *anima* (a female archetype present in males). Because each archetype involves strong emotion, Jung believed that people's emotion-generated behavior could often be explained by identifying the key archetype that was influencing them. He applied this principle to groups as well; for example, the ascendence of the God archetype might have played a role in the Crusades; the ascendence of the hero archetype (or perhaps the devil archetype) in post-World War I Germany might have played a role in Adolf Hitler's rise to power.

In addition to the collective unconscious, Jung credited each individual with a *personal unconscious*, which developed out of any of the individual's conscious experiences that had been repressed (page 578). Psychologically healthy people were said to gradually come into contact with the unconscious parts of their personalities, integrating the unconscious or "shadow" side with their "conscious ego." In this way, all major components of the personality could eventually work in concert to form a fully realized, purposeful self. This process of harmonizing one's conscious and unconscious components happened in a unique way for each person, he said, and led to a unique pattern of behavior. Jung called this process *individuation*; he saw it as the means by which each of us becomes an individual distinct from others.

Jung's ideas, often mystical and complex, have been difficult to test experimentally. However, the ideas have intrigued and attracted many—as evidenced by the numerous "Jung societies" that have sprung up in North America and Europe. In addition, some of Jung's concepts, such as "the complex," have found their way into everyday use. Other Jungian concepts have spurred important work by later theorists. For example, Jung proposed the concepts of *introversion* and *extraversion*—turning inward toward contemplation or outward toward others. These two concepts were later picked up and revised by H. J. Eysenck, eventually becoming a key part of his personality theory (page 570). Jung's ideas have certainly outlived him, and they continue to provoke interest.

ADLER'S INDIVIDUAL PSYCHOLOGY

The third dynamic theorist we will consider is Alfred Adler. He was a colleague of Freud and Jung, but his theory took a different direction from theirs (Adler, 1929, 1931). To explain Adler's focus, we will begin with a story that he liked to tell.

> . . . three children . . . were taken to the zoo for the first time. As they stood before the lion's cage, one of them shrank behind his mother's skirts and said, "I want to go home." The second child stood where he was, very pale and trembling, and said, "I'm not a bit frightened." The third glared at the lion fiercely and asked his mother, "Shall I spit at it?" The three children really felt inferior, but each expressed his feelings in his own way, consonant with his style of life.
> (Adler, 1931, p. 50)

Adler's story illustrates some key differences between his dynamic theory and the theories of Freud and Jung. First, Adler believed that like the hypothetical children in this story, people are forever struggling to overcome their feelings of inferiority. In fact, in a sharp departure from Freud and Jung, Adler identified this struggle as *the* most basic life urge. Acting on this urge, people strive continually, he said, for "superiority" and "mastery of the external world." When someone fails repeatedly to overcome weaknesses and achieve some mastery or simply places excessive emphasis on some particular inferiority, the result can be what Adler labeled an *inferiority complex*, a term that is now part of our everyday language. Adler also stressed the concepts of *compensation* and *overcompensation*—the pursuit of activities designed to make up for or to overcome inferiority (for example, spitting at the lion in the story above). These concepts had great personal significance for Adler; as a child, he was chronically ill, weak, and intensely afraid of death. He compensated, in part, by his career choice: "I came to choose the occupation of physician in order to overcome death and the fear of death" (Adler, in Ansbacher & Ansbacher, 1956, p. 199).

A second distinctive concept of Adler's that is reflected in the zoo story is the *style of life.* Each individual creates his or her own personal approach to living; this highly individualized style grows out of the individual's unique sense of his or her own inferiorities and the strategies he or she develops to overcome these inferiorities. Just as each child in the zoo story coped differently with fear of the lion, each of us develops a unique and relatively lasting method of coping with life. This concept, together with the concept of the subjective nature of individual goals, led Adler to label his theory *individual psychology*.

A third contribution of Adler's theory is the close attention he gave to birth order as an influence on personality development. First-borns, said Adler, begin life as the exclusive focus of their parents' attention and then are often abruptly dethroned with the birth of their first siblings. The result may be that the child feels cheated and later becomes an unruly misfit. A more positive possibility is that first-borns may develop a style of life in which they behave in a parentlike way toward their siblings and become particularly responsible adults. Adler described other possibilities for second-born, last-born, and only children. Many of his specific ideas have not been borne out, but his ideas did help stimulate some important research. Thanks to this research (for example, Dunn & Kendrick, 1982; Sutton-Smith & Rosenberg, 1970), we now know that first-borns do have especially intense relationships with their parents and that they tend to be more achievement-oriented and self-controlled than do later-born children. Adult firstborns are also more likely to be listed in *Who's Who*, while later-born children tend to be less conforming, more sociable, and more popular. So Adler's general idea that birth order is related to personality does seem to be correct.

ADLER VERSUS FREUD VERSUS JUNG: FIREWORKS AMONG THE FOUNDERS

For his ideas, at least those that differed from Freud's, Adler was rewarded with rebuke. He joined Freud's inner circle in 1902 and rose to the presidency of the Vienna Psychoanalytic Society and the coeditorship of the psychoanalytic journal *Zentralblatt.* By 1911, though, the conflict between Freud and Adler had become so intense that Adler resigned both the presidency and the coeditorship and severed all contact with the Society. After Adler's resignation, Freud wrote to Jung,

> I have finally got rid of Adler. . . . The damage [from his loss] is not very great. Paranoid intelligences are not rare and are more dangerous than useful. As a paranoiac of course he is right about many things, though wrong about everything.
> (McGuire, 1974, p. 428)

Jung had his own opinions about both Adler's and Freud's theories. He believed that both had missed some of the key issues in personality and that both had misunderstood the true nature of the unconscious. Moreover, he incorporated Freud and Adler into his theoretical concept of "introversion-extraversion." Jung believed that Freud's basic extraversion led him into a theory that emphasized the individual's relationship to people and things in the external world; Adler's basic introversion, Jung thought, led him into a theory focused on the inner self, strivings to overcome perceived personal inferiority, and so forth. Freud, for his part, eventually came to see Jung as both disloyal and wrong, and they too parted company.

In an angry 1913 letter to "Dear Doctor" (Jung), Freud wrote

> I propose that we abandon our personal relations entirely. I shall lose nothing by it, for my only emotional tie with you has been a thin thread—the lingering effect of past disappointments.
> (McGuire, 1974, p. 539)

Jung replied

> I accede to your wish that we abandon our personal relations, for I never thrust my friendships on anyone. You yourself are the best judge of what this moment means to you.
> (McGuire, 1974, p. 540)

So after years of collaboration, Freud, Jung, and Adler went off in separate directions, each cultivating his own band of followers.

HORNEY'S PSYCHOANALYTIC INTERPERSONAL THEORY

One psychoanalytic theorist who was not a part of this trio but who also dissented from Freud on key issues was Karen Horney (pronounced "horn- eye"). Trained in Berlin, she emigrated to the United States in the 1930s. An early feminist, she violated the prevailing rule that medicine and psychiatry were "male-only" professions. She took sharp issue with Freud on some of his views about "feminine psychology." For example, she argued that "penis envy" was not a normal development in females but rather an unusual and pathological occurrence. She also countered that some of her male patients envied women *their* capacity for pregnancy, childbirth, motherhood, breast development, and suckling. Her views on females, sex differences, and Freud's theory were compiled in the book, *Feminine Psychology* (1967). Horney did not deny the existence of sex differences altogether, but she stressed the striking similarities of the two sexes as members of the human race facing similar challenges; she proposed that what psychiatry and personality theory needed was a "psychology of persons."

Two major components of her "person" psychology were the twin notions of basic anxiety and basic hostility (Horney, 1937).

Basic anxiety, she said, is what arises in childhood when the child feels helpless in a threatening world. Children learn that they are relatively weak and powerless, dependent on their parents for safety and satisfaction. Loving and reliable parents can create a feeling of security (even serenity), but erratic, indifferent, or rejecting parents may sharpen the child's sense of helplessness and vulnerability. This sets the stage for basic anxiety.

Basic hostility, in Horney's view, is what usually accompanies basic anxiety and grows out of resentment over the parental behavior that led to anxiety in the first place. Because the hostility cannot be expressed directly to parents, it is typically repressed, which only increases the child's anxiety. Children caught in this bind—dependent on their parents, anxious because of their parents, hostile toward their parents, and unable to express their true feelings directly—tend to rely heavily on one of three modes of social behavior, each of which might well carry into adulthood.

One such mode—*moving toward others*—involves excessive compliance. Security is sought by making oneself indiscriminately compliant, subject to the will of others, and inclined to do whatever they wish in order to gain their approval and affection. The result may be a kind of security, but a costly kind that involves total repression of basic hostility and leaves the individual feeling depleted, exploited, and unhappy.

A second mode of social behavior—*moving against others*—involves pursuit of satisfaction through ascendance and domination of others. Self-protection is provided via one's power over others. Basic hostility may be expressed, but basic anxiety is usually denied. As a result, feelings of weakness and vulnerability are neither explored nor resolved.

A third approach—*moving away from others*—is self-protection by withdrawal. Some people avoid the risk and pain of social relationships by avoiding relationships in the first place. This strategy does provide some protection, but it also cuts short any real prospect for growth in the social realm.

Horney believed that normal people use all three modes of social interaction at times but in a relatively balanced and flexible manner, adjusting their approach to situational demands. Neurotic people, she argued, allow one approach to dominate their social interactions, and this rigidity gets them into trouble.

DEFENSE MECHANISMS IN THE DYNAMIC THEORIES

Thus far, we have emphasized disagreements among the dynamic theorists, but there are areas of agreement. One of these is the notion that people use *defense mechanisms* to reduce their anxiety and guilt (A. Freud, 1946). Psychoanalytic theory holds that because the id's unconscious demands are instinctual, infantile, and amoral, they must often be blocked by the ego and superego. Because of this conflict and the persistence of unsatisfied demands, *anxiety* (vague fearfulness) and guilt are aroused. The person then seeks ways to protect the ego from this anxiety and guilt by setting up defenses. Freud described several defense mechanisms by which the ego disguises, redirects, hides, and otherwise copes with the id's urges. Others have been added by the dynamic theorists who followed Freud.

Many psychologists do not agree with Freud's view that defense mechanisms originate in conflicts among the id, ego, and superego. However many do agree that these mechanisms account for some of the ways people cope with their problems. Thus, defense mechanisms—an intellectual bequest from the dynamic theories—are generally accepted as a useful way of looking at how people handle stressful situations and conflicts. (See Chapter 7, page 300.)

Repression In Freud's theory, *repression* is the fundamental technique people use to allay anxiety caused by conflicts. As we noted on page 578, repression is an active mental process by which a person "forgets" by "pushing down" into the unconscious any thoughts that arouse anxiety. In other words, as a therapist and theorist named R. D. Laing put it, we forget and then forget that we forgot. In terms of psychic energy, repression is an expensive defense mechanism. The unconscious memories or urges continue to seek expression and may emerge in the form of "accidents," "slips," or neurotic symptoms (page 579).

Reaction Formation Reversal of motives is another method by which people attempt to cope with conflict. A motive that would arouse unbearable anxiety if it were recognized is converted into its opposite. For example, a person who was reared to believe that sex is evil and "dirty" may be painfully anxious every time sexual feelings surge to the surface. The person may defend against that anxiety by dressing and behaving in very puritanical ways and perhaps even by joining organized crusades against sex in the media. The implicit principle seems to be that "the best defense is a good offense." Similarly, if people are *too* modest, *too* solicitous, *too* affectionate, or *too* strident in their crusades against an "evil" such as alcoholism, homosexuality, or child abuse, it is possible that they are unconsciously harboring the opposite feelings. Thus disguised, the unwanted motives can be controlled. A quotation from Shakespeare captures the idea of reaction formation: "The lady doth protest too much methinks."

Projection Blaming others, or *projection,* is a way of coping with one's unwanted motives by shifting them on to someone else. The anxiety arising from the internal conflict can then be lessened and the problem dealt with as though it were in the external world. For example, an insecure student may have a strong desire to cheat on an examination, but his conscience will not allow him even to consider such a thing. He may then suspect that other students are cheating when they in fact are not. Carried to the extreme, projection is the mark of a behavior disorder known as paranoia (Chapter 15, page 644). People with this disorder may project their own unacceptable hostile feelings about others into a whole system of thinking in which they feel that others are out to get them.

Rationalization This defense mechanism substitutes an acceptable conscious motive for an unacceptable unconscious one. Put another way, we "make excuses," giving a reason different from the real one for what we are doing. Rationalization is not lying; we believe our explanations. Examples range from the innocent to the serious. (See Figure 14.8.) The long lines at the movies during finals weeks are populated by students

Figure 14.8

Rationalization is an everyday occurrence. (Copyright 1977 by King Features Syndicate, Inc.)

who "need to relax" to do a good job on their tests. A tense father who strikes a rambunctious child may rationalize that he is acting for the child's good. Aesop's fable of the fox and the sour grapes is another example of rationalization: Something we cannot get becomes something we did not want anyway. Rationalization is a common mechanism we all use to bolster our self-esteem when we have done something foolish. If overused, however, it can prevent us from coping with a situation head-on. For example, a person with an unconscious fear of intimate relationships may find a succession of potential mates unacceptable for different reasons and, as a result, spend the rest of life alone.

Intellectualization Related to rationalization is *intellectualization*, another defense mechanism which involves reasoning. In intellectualization, however, the intensity of the anxiety is reduced by a retreat into detached, unemotional, abstract language. Professionals who deal with troubled people may intellectualize in order to remain helpful without being overwhelmed by sympathetic involvement. For example, a nurse may describe in an intellectual fashion an encounter with a dying or angry patient. Some adolescents discuss their new experiences with sex and independence on an abstract and impersonal plane. Temporarily separating emotional and cognitive components sometimes helps the individual to deal with parts of an experience when the whole is too much to handle.

Displacement In *displacement,* the motive remains unaltered but the person substitutes a different goal object for the original one. Often the motive is aggression that for some reason, the person cannot vent on the source of the anger. A person who is angry with his or her boss but cannot show it for fear of being fired may come home, bawl out the children, and kick the dog. When a new baby is the center of attention, an older child may become jealous; prevented from harming the baby, the child demolishes a doll. Thus by displacing aggression, the child finds a substitute outlet.

Regression In the face of a threat, one may retreat to an earlier pattern of adaptation, possibly a childish or primitive one. This is called *regression.* Faced with the upsetting arrival of a new baby or going to school for the first time, a 5-year-old may have toilet accidents, revert to "baby talk," demand cuddling, or suck her thumb. (See Figure 14.9.) Adults, too, sometimes revert, when in stress-producing situations, to childish episodes of exaggerated dependency. Such behavior may ward off anxiety by focusing attention on earlier ways of achieving tranquility.

Sublimation For Freud, sublimation was the highest level of ego defense. It consists of a redirection of sexual impulses to socially valued activities and goals. For example, a writer may divert some of his or her libido from sexual activity to the creation of a poem or novel, thus indirectly satisfying the same drives. Freud believed that much of our cultural heritage—literature, music, art—is the product of sublimation. He also believed that satisfactory sublimation could only be achieved by an individual whose sexual impulses were being at least partly gratified and whose ego was healthy and mature.

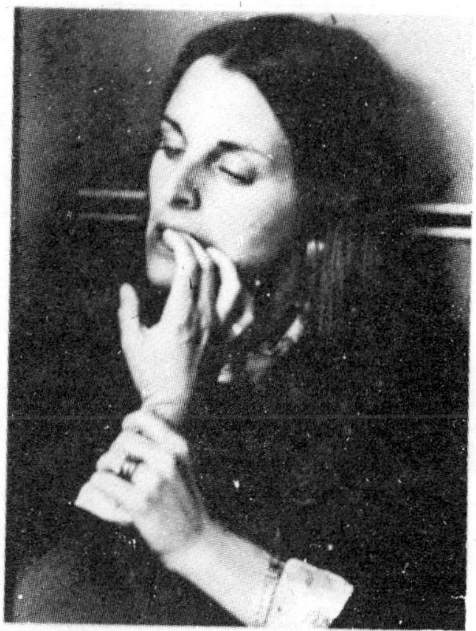

Figure 14.9

When life gets hectic or stressful, children and adults may regress to an earlier (for example, oral) stage of development. (Left photo from Harvey Stein, copyright © 1981; right photo from Dan Budnik/Woodfin Camp & Associates.)

The Use of Defense Mechanisms Most readers will recognize in themselves a number of the coping patterns described above. Everyone resorts to them from time to time, and, when they are used sparingly and without cost to others, they are nothing to worry about. If they allow us to feel more comfortable, as they often do, their value in reducing tension and letting us get on with important problems more than offsets the trivial self-deceptions they entail. However if a person comes to depend on them too much, then these defensive patterns may be harmful. They do not solve the real problem; they only relieve anxiety about it. The more aware we are of our use of these mechanisms, the more rationally we can assess our behavior and come to terms with the unknown sides of ourselves.

QUESTIONS ABOUT DYNAMIC THEORIES

Without a doubt, dynamic theories have had a sweeping influence both within and without psychology. Some of the concepts spawned by these theories—defense mechanisms, for example—have stimulated research and gained broad acceptance. The idea that early childhood experiences influence later personality, the notion that the true motives for our behavior may be unconscious—these are direct outgrowths of dynamic theorizing, particularly Freud's. All of the dynamic theories we have discussed here have also stimulated influential schools of psychotherapy. (See Chapter 16 for a discussion of some of these.)

On the other hand, each of the theories includes ideas that are difficult to prove or disprove. How can we fairly test whether boys go through the Oedipus complex or girls experience penis envy? How can we test the validity of Freud's judgments about Little Hans's phobias and what caused them (see Application 14.2)? When we find supporting evidence in a case report from one of the dynamic theorists, it is often hard to judge whether "truth" has been found or the case has been forced into the theory's pigeonholes. Most dynamic theories involve disguised motives, hidden goals, and indirect ways of coping—all of which means that the ordinary "rules of evidence" cannot really be applied. Yet without these rules, we cannot really test the theories in a scientific way.

Dynamic theories, like all theories, are, to some extent, products of the cultural environment in which they were formed. Freud's theory, for example, emerged in straightlaced, sexually repressive nineteenth-century Vienna; his emphasis on the significance and pervasiveness of repressed sexuality may have been more apt for that society and era than it would for other places and times. We should also remember that most of the major dynamic theories were based largely on evidence obtained from disturbed adults—people who may not have represented the mainstream of their culture. Given all these considerations, it is probably fortunate that Freud's original ideas were modified and revised by new theorists who came on the scene. Jung, Adler, and Horney each broadened the dynamic perspective in significant ways; so did Erik Erikson (see Chapters 11 and 12) and others in a group known as ego psychologists. While these dynamic theorists went beyond Freud in some respects, their theories continue, by and large, to share the limitation of Freud's approach: They are difficult to confirm or disprove in experimental research. The theories we will consider next do not suffer from this limitation.

Learning and Behavioral Theories of Personality

The learning and behavioral theories of personality, which we will examine now, are specifically structured to be tested. In fact, as we shall see, the theories themselves emerged from experiments in classical conditioning, instrumental conditioning, and cognitive learning. At this point, you may wish to review these forms of learning as they are described in Chapter 4.

Psychologists who build their theories on learning and behavioral principles share some important assumptions and practices. One shared assumption is that many of the behaviors that make up personality are conditioned, or learned. This means, first, that many such behaviors originate somewhere in the learning history of the individual, often as early as childhood. A second assumption is that current conditions in the individual's environment help maintain these behaviors. Thus learning theorists seek to understand people's behavior by studying their learning history, their current environment, or both.

Learning and behavioral theorists also believe in testing their theories. Their major propositions are generally stated in ways designed to be clear—so clear that others can know how to test them. Striving for clarity, in turn, leads these theorists to focus on *observable* events and behavior. In cases where they deal with unseen processes, such as anxiety, the theorists are violating strict behaviorism, which avoids all mention of unobservable events "inside the person." (See page 27.) However when the theorists do discuss such "unseen" processes, they typically translate the processes into language designed to make them as observable as possible.

DOLLARD AND MILLER: EARLY SOCIAL LEARNING THEORY

Two of the best examples of learning theorists are Neal Miller (lower picture), an experimental psychologist, and sociologist John Dollard (upper picture). They joined forces in the 1930s to test the basic idea that individual and social behavior can be explained by means of basic learning principles. Dollard and Miller took this basic premise to extremes that few experimentalists would have dreamed of. They even tried to translate Freud's psychoanalytic concepts into the language of learning theory and to test his concepts in the laboratory—with rats as their subjects! As we noted above, Freud believed that neurotic, disturbed behavior involves conflict between id demands and ego/superego restraints. Dollard and Miller drew an analogy between this conflict and the conflict between approach and avoidance tendencies, as discussed in Chapter 7 (pages 301–303). They argued that we may act indecisive and "neurotic" when we are torn between approaching and avoiding a certain course of action—as, for example, when we want to get a tooth filled but we fear the pain. In such cases, the tendency to approach (get the filling, for example) is often stronger than the avoidance tendency at first; but the closer we get to the "moment of truth," the more likely it is that the avoidance tendency will win out and we will retreat from the planned action (we will cancel our dental appointment, for example).

Dollard and Miller (1950) applied ideas like this to a variety of human problems. One example is their analysis of "Mrs. A.," a woman treated by a psychotherapist in New York.

> One of Mrs. A.'s most serious problems was a severe obsession: She was tormented, especially when she left her house alone, by the idea that her heart would stop unless she counted its beats. She also feared that "something" might happen to her while she shopped and that no one would know where she was. She was so obsessed with these worries that her friends and relatives began to avoid her, and her husband threatened to leave her.
>
> Careful probing by a therapist revealed that these "neurotic symptoms" might have grown out of Mrs. A.'s conflicting feelings about sex. On the one hand, she had a strong tendency to avoid sex. Her foster mother had repeatedly stressed that sex was evil and "unclean." Moreover, Mrs. A. had married a man more educated than she and higher in social status, and she was afraid that openly enjoying sex with him might make her seem "low-class." So she was extremely restrained, almost prudish, in her relations with her husband. On the other hand, Mrs. A. also had a strong sex drive—a strong "approach tendency." Before she married, a dozen men had tried to seduce her, and, as Dollard and Miller put it, "most of them had succeeded" (Dollard & Miller, 1950, p. 18). After marriage, her sexual appetites often led her to literally "flirt with danger." For example, she would sometimes hitch rides with truck drivers.
>
> Dollard and Miller saw Mrs. A.'s difficulties this way: She felt a strong conflict between her sexual urges *(approach)* and her fear of being openly sexual *(avoidance)*. Each time she got very close to a real sexual experience, her fear of sex overpowered her desire for it

and she shied away from actual intercourse. As the situations Mrs. A. got herself into grew more and more dangerous, she got very anxious and she eventually found another way of avoiding sex: She developed intense fears that kept her mind totally occupied—that is, she feared that "something" would happen to her and that her heart might stop, and she began counting every heartbeat. This prevented sexual thoughts and feelings from arising. This approach, though it was upsetting in many ways, was rewarding in one important way: As soon as Mrs. A. started counting, she "felt better." Was Mrs. A. repressing her sexuality? Yes, Dollard and Miller said. They defined Freud's notion of *repression* in learning-theory terms: *Repression* is a learned response of "not thinking about" something that is anxiety-provoking. For Mrs. A., this response was rewarded by the reduction in anxiety that it produced.

Overall, as the example of Mrs. A. suggests, Dollard and Miller were not out to bury Freud or to praise him, but rather to redefine him. Their aim was to engineer, if not a merger, at least a bridge between the dynamic and learning perspectives on personality.

SKINNER'S RADICAL BEHAVIORISM

Bridge building is not what B. F. Skinner (1953, 1971, 1972) had in mind in advancing his radical behavioral perspective. Skinner's views, like those of Dollard and Miller, are derived from experiments. But Dollard and Miller borrowed from both the classical and instrumental conditioning models (Chapter 4), whereas Skinner's approach is exclusively *instrumental*, or *operant*—that is, it deals only with the processes by which *reinforcement* (reward) and *punishment* influence the likelihood of behaviors (Chapter 4). Ruled out of Skinner's analysis wherever possible are such "unobservables" as drives, motives, and emotions. Also ruled out are the trait and type notions discussed in our first section. What, we might ask, is left to constitute the human personality? For Skinner, the answer seems to be that what most people call personality is actually a collection of reinforced responses. We are the persons that we are because we behave in certain ways, and we behave in certain ways because of the reinforcement contingencies we experience. In Skinner's view, there is really no need for a concept of "personality" or for such "excess baggage" as traits and types.

Consider Mrs. A., for example. We might describe her in trait terms as being "seductive and flirtatious," in type terms as being "a tease," or we might describe her in terms of her approach-and-avoidance tendencies—her drives. From a Skinnerian perspective, none of these descriptions would add to the understanding we can achieve by simply describing Mrs. A.'s observable behavior. She flirted with men, was sexually suggestive, but usually stopped short of actual intercourse. Does the trait, type, and drive terminology help us to understand *why* Mrs. A. behaved as she did? No, Skinner would say. To say that Mrs. A. flirted with men because she had a strong sex drive is simply redundant; it does not really *explain* her flirting any more than "a strong dependency drive" explains why a child clings to its mother or a "strong art drive" explains why Picasso painted so much. Skinner prefers to focus on the reinforcement that accompanies behavior. In the case of Mrs. A., we might observe that

Inquiry 14.1

WHEN WE ARE "CREATIVE," ARE WE REALLY CREATING?

B. F. Skinner has been accused of many things, but never cowardice. In a 1971 lecture to an audience of poets (Skinner, 1972), he risked the wrath, not only of poets but of mothers as well! Those of us who have written a poem, composed a song, or produced a work of art (however humble) tend to think of our actions as inventive and original and ourselves as possessing a "creative urge" or a trait of "creativity." Skinner suggested that we are much too generous in these self-assessments. He also suggested that we, like many mothers, take all too much personal credit for such "achievements." As Skinner put it,

Only a person who truly initiates his behavior can claim that he is free to do so and that he deserves credit for any achievement. If the environment is the initiating force, he is not free, and the environment must get the credit.
(Skinner, 1972, p. 352)

Most of our "creative products," Skinner argued, are primarily products of our environment. To make his point, Skinner drew a parallel between an act of biological creation—having a baby—and an act of mental creation—"having" a poem.

We usually say simply that a woman "has" a baby where "has" means little more than possesses. . . . What is the nature of [the mother's] contribution? She is not responsible for the skin color, eye color, strength, size, intelligence, talents, or any other feature of her baby. She gave it half her genes, but she got those from her parents. She could, of course, have damaged the baby [for example, by abortion, catching rubella at the wrong time, etc.]. . . . *But she made no positive contribution.*

A biologist has no difficulty in describing the role of the mother. She is a place, a locus in which a very important biological process takes place.
(Skinner, 1972, p. 352)

So much for the "creative" role of mothers. Now on to poets.

The poet is also a locus, a place in which certain genetic and environmental causes come together to have a common effect. . . . A poem seldom makes its appearance in a completed form. Bits and pieces *occur* to the poet, who rejects or allows them to stand, and who puts them together to *compose* a poem. But they come from his past history, verbal and otherwise, and he has had to learn how to put them together. The act of composition is no more an act of creation than "having" the bits and pieces composed.
(Skinner, 1972, p. 352)

Summing up, Skinner offered yet another biological analogy.

A person produces a poem and a woman produces a baby, and we call the person a poet and the woman a mother. . . . Writing a poem is the sort of thing men and women do as men and women, having a baby is the sort of thing a woman does as a woman, and laying an egg is the sort of thing a hen does as a hen.
(Skinner, 1972, p. 354)

When a hen lays an egg, Skinner was asking, do we call the hen *creative?*

There is another, more general implication of Skinner's point, one that touches us all. When we "have an idea" or when "an idea occurs to us," we tend to take credit for the idea as if we chose to produce it by an act of free will. In fact, said Skinner, we "have" ideas in the same sense that women "have" babies—that is, we are merely a place in which the ideas happen. Taking this line of thought to its logical conclusion, we should not give Skinner very much credit for the ideas he presented in this speech. Skinner agrees, as he noted at the end of his lecture.

And now my labor is over. I have had my lecture. I have no sense of fatherhood. If my genetic and personal histories had been different, I should have come into possession of a different lecture. If I deserve any credit at all, it is simply for having served as a place in which certain processes could take place. I shall interpret your polite applause in that light.
(Skinner, 1972, p. 355)

REFERENCE

Skinner, B. F. (1972). A lecture on "having" a poem. In *Cumulative record: A selection of papers.* (3rd ed.). New York: Appleton-Century-Crofts.

flirting, both before and after marriage, was repeatedly rewarded with male attention. We might also observe that her later avoidance of going out alone and her counting of heartbeats protected her from real objective dangers and were further rewarded, at least at first, by increased attention from her husband and relatives. Finally, we might observe that when this increased attention stopped—when, in fact, her relatives began to avoid her—Mrs. A. reported for psychotherapy, presumably to change her behavior because its consequences were no longer rewarding.

With such accounts of people's behavior, described entirely in terms of objectively observable events, Skinner generated what some consider a "personalityless" view of personality. Many are offended by Skinner's efforts to reduce the seeming richness of human personality to nothing more than a set of responses strengthened by reinforcers (see Inquiry 14.1). Others see Skinner's work as refreshing because it is an effort to be clear, precise, and parsimonious—that is, to rely on the smallest possible number of theoretical concepts and assumptions.

BANDURA AND WALTERS: LATER SOCIAL LEARNING THEORY

The third learning-theory perspective we will review is built on the perceived limitations of the first two. Albert Bandura (see picture) and Richard Walters (1963) saw the animal-derived principles of Dollard, Miller, and Skinner as simply too limited to account for important aspects of real human behavior. For example, they thought the animal experiments involved artificially "safe" laboratory situations. In real life, they argued, people (and rats, for that matter!) often do not have the luxury of learning through instrumental or operant conditioning—trial-and-error learning in a protected laboratory where only "correct" responses get rewarded. In real life, it is often too costly to risk "incorrect" responses. Would we teach our children to swim or to cross busy intersections by means of instrumental learning procedures? Not likely because an "incorrect" response would have serious consequences, including a risk of death. By the same token, we would not rely solely on instrumental learning procedures to teach adolescents how to drive cars, medical students how to perform heart surgery, or CIA recruits how to conduct covert operations. In these situations, and in many others, instrumental conditioning may be too inefficient and too dangerous to be relied upon.

The approach that Bandura and Walters took focused on the highly efficient form of learning known as *observational learning*, or *imitation*. (See Chapter 4, page 172; also Chapter 7, page 291, and Chapter 11, page 446.) They viewed observational learning as requiring no direct reinforcement to the learner (although they acknowledged that *persistence* of the learned behavior may depend upon reinforcement, or at least *anticipated* reinforcement). Observational learning generally takes place in a social situation involving a model and an imitator (see Figure 14.10). The imitator observes the model and experiences the model's behavior and its consequences vicariously; this process is called *vicarious reinforcement*. Bandura maintains that nearly all learning that can take place directly with instrumental learning procedures can also take place vicariously through modeling. When seeking to explain, say, the behavior of Little Hans (Application 14.2, page 583), Skinner might look to the reinforcements for fearful behavior in the boy's life, but Bandura and Walters would look

Figure 14.10

Imitation in the family. This boy is learning to act like his father, who is an accomplished pianist. Perhaps more important, he is learning to value many of the same things. (Suzanne Szasz.)

for fearful models and observational learning opportunities in the boy's history. Similarly, Bandura and Walters might explain the two case examples at the beginning of this chapter by looking partly to the models who influenced G. Gordon Liddy (for example, Hitler) and Crystal Lee Jordan (for example, her father).

The powerful influence of observational learning is evident in both clinical and everyday settings (Bryan & Test, 1967).

> One study showed the effects on everyday behavior of "good Samaritanism." The investigators stationed a woman beside a car with a flat tire and checked to see how many drivers would stop to help her. In some cases, the drivers were given a modeling opportunity. A quarter of a mile ahead of the "distressed" woman, these cars passed a woman who was being helped with her flat by a young man (seemingly a "good Samaritan" but actually an experimental confederate). The drivers who had passed this "helpful" model were significantly more likely to stop for the "distressed" woman than were the drivers who had *not* seen the model.

As we shall see in Chapter 16, observational learning (especially as promoted by Bandura) has also been used to treat such troubling problems as severe phobias and to explain a variety of both normal and abnormal personality patterns.

QUESTIONS ABOUT LEARNING AND BEHAVIORAL APPROACHES

Few psychologists would deny that many of the behavior patterns we know as "personality" come about through learning and conditioning. However there is real controversy about whether the various learning and behavioral accounts tell us all we need to know about personality development. Some argue that a strict learning-theory approach leads only to an understanding of behavior in specific situations and that such "situationism" ignores the consistencies that many people show from one situation to the next (Bowers, 1973). Others argue that focusing on behaviors one can observe outwardly and emphasizing concepts such as "reinforcement" seem to diminish the *person* in *person*ality.

Learning theorists counter such arguments in several ways. First, they note that conditioned responses can be quite complex. Human learning is not just a matter of acquiring specific muscle twitches in response to certain stimuli. Instead, because of the great adaptability of human beings, a rich, interwoven tapestry of behaviors is learned. This tapestry is intricate enough to account for many of the subtleties of human personality.

Second, such social learning theorists as Bandura and Walters do use concepts about internal cognitive processes to explain imitation and delayed performance. In fact, with the exception of Skinner and other radical behaviorists, learning theorists today acknowledge the influences of at least some powerful variables within the learner and portray learners as far more than passive recipients of environmental influence. One way that human learners demonstrate their nonpassivity is by making judgments about the causes of their reinforcement, as our earlier discussion of locus of control (page 573) showed.

Humanistic Theories: Personality as the Self

If we stop studying other people's behavior for a moment and pause to think about our own, we become aware of a set of feelings and attitudes and a certain sense of who we are. A number of theorists have focused their work upon this entity known as the *self*. Generally speaking, the term *self* has two distinct sets of meanings. One set has to do with

Figure 14.11

Our self-image develops on the basis of information about the way we are and the way others see us. (Ken Heyman.)

people's attitudes about themselves; their picture of the way they look and act; the impact they believe they have on others; and their perceived traits, abilities, foibles, and weaknesses. This collection constitutes what is known as the *self-concept*, or *self-image*, "attitudes, feelings, perceptions, and evaluations of . . . self as an object" (Hall & Lindzey, 1970, p. 516). (See Figure 14.11.) The second set of meanings relates to the *executive functions*—processes by which the individual manages, copes, thinks, remembers, perceives, and plans. These two meanings, *self* as "object" and *self* as "process," are seen in most theories involving the notion of self.

We have already considered one concept similar to the "self as process." In psychoanalytic theory, the ego comprises the personality's executive functions. In fact, for many of the "ego psychologists" who came after Freud, the ego plays *the* paramount role; Freud's own daughter, Anna Freud (1946), was a leader in this movement. Self-concepts are central to *humanistic theories*—theories which focus upon the individual's subjective perceptions of self, the world, and the self within the world. Two of the most influential humanistic theorists are Carl Rogers and Abraham Maslow.

ROGERS' SELF THEORY

Carl Rogers' theory (Rogers, 1959, 1961) grew out of his client-centered approach to psychotherapy and behavior change (see Chapter 16, page 688). Like psychoanalysis, the theory grew from efforts to help troubled people, but Rogers' theory does not involve complex personality structures or stages of development. Instead, Rogers emphasized the whole of experience, the *phenomenal field.* This is the individual's subjective frame of reference; it may or may not correspond to external reality.

The Self Out of the phenomenal field, a self or self-concept gradually develops. Rogers did not start out to make the self a central idea in his theory, but he found that clients spontaneously thought in such terms. "It seemed clear . . . that the self was an important element in the experience of the client, and that in some odd sense his goal was to become his 'real self'" (Rogers, 1959, p. 201). Thus, in addition to the present self, there is also an *ideal self,* the self the person would like to be. Trouble occurs when there are mismatches, or *incongruences.* For example, the perceived self may not match the ideal self, and this can be very disturbing. Because we need self-esteem, we may distort our perception of our experiences in self-serving ways. In doing this, we may use conditions of worth—characteristics of our behavior that we learned in childhood to associate with acceptance (versus disapproval). This brings us to one of Rogers' key notions about personality development.

Personality Development As children grow, parents and others react to their behavior, sometimes in a positive way and sometimes with disapproval. Children thus learn to regard some of their actions, thoughts, and feelings as unworthy, and they often react by distorting or denying these unworthy aspects of self.

Consider Dollard and Miller's "Mrs. A." for example (page 593). She had grown up in an atmosphere of severe strictures against sexual expression of any kind, even verbal expression. Her foster mother repeatedly emphasized that sex was wrong and "dirty." She did not even permit terms like *pregnant* to be used in the house. Instead, she referred to pregnant women as "big." Thus, Mrs. A. grew up regarding her own sexual thoughts and feelings as wrong, "dirty," and unworthy. She learned to cope by oscillating between distortion on the one hand and denial on the other. She distorted by sometimes adopting the view that she was a "dirty," evil person and proving her point by going on binges of flirtation and sexual seduction (the truck-driver episodes, for example). She denied by sometimes behaving in an overly straightlaced, conservative, asexual manner (for instance, in relations with her husband). Neither pattern—distortion or denial—was an accurate reflection of Mrs. A.'s true self. Consequently, both patterns led to maladaptive behavior and eventually unhappiness. Rogers believes that in mature, adjusted people, there is congruence between the total person and the self and that well-adjusted people can accept the full range of their experiences without distorting or avoiding them.

Rogers encouraged research to test his hypotheses. He was, in fact, among the first to subject the intimate experience of psychotherapy to the cool eye of evaluation when he began tape-recording sessions so that they could be coded and evaluated. He also helped to popularize the *Q-sort technique*, a method of self-description, in personality research. In this technique, the individual is given a large number of descriptions and is asked to sort them into categories from "least characteristic" to "most characteristic." A person could sort the statements, for example, to indicate "how I am now" and "how I would like to be," yielding a measure of the degree of congruence between the perceived self and the ideal self.

Research findings have supported some, but not all, of Rogers' theoretical notions. In support of his theory it has been found that people who are seeking psychotherapy indicate more of a disparity between their real and ideal selves than do people who are not seeking help. Moreover, this disparity tends to shrink in the course of successful therapy. On the other hand, some findings have suggested that acknowledging a disparity between one's real and ideal self may be a sign of maturity. Among children and adolescents, for example, it has been found that disparity

MASLOW'S SELF-ACTUALIZATION THEORY

Another phenomenological theorist who emphasized development of the self was Abraham Maslow. Maslow believed that each person has an essential nature that "presses" to emerge, like the "press" within an acorn to become an oak tree. In his view, we all have higher-level growth needs—such as the need for self-actualization and understanding of ourselves—but that these higher needs only assume a dominant role in our lives after our more primitive needs (physiological needs, safety needs, needs for love and "belongingness," and self-esteem needs) are satisfied (see Figure 14.12). (See also Chapter 7, page 298.) The growth needs, Maslow believed, help make us distinctly human. Maslow stressed that "the human being is not a white rat" and emphasized that "man has a higher and transcendent nature" (Maslow, 1971, p. 349).

To understand this transcendent nature, Maslow believed we should study transcendent people. He worried that the exclusive study of emotionally disturbed people—as in Freud's theory and others—was

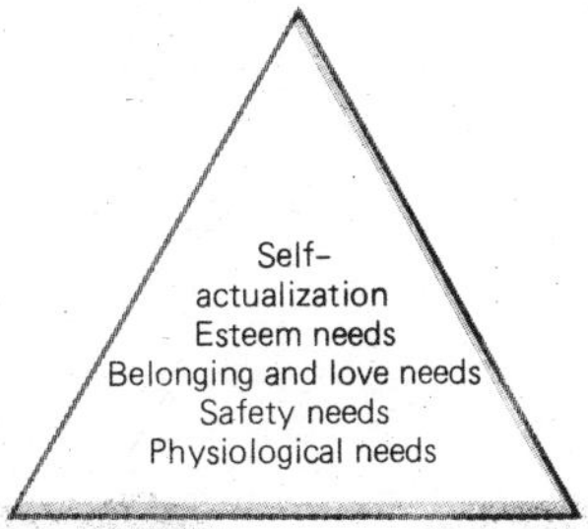

Figure 14.12

(*Left:*) Maslow's hierarchy of needs. Shown at the bottom are basic biological needs; at the top are distinctly human psychological needs. Needs lower in the hierarchy must be met before "higher-level" needs can assume a dominant role. (*Right:*) In seeking examples of *self-actualized people*—people who functioned at the upper level of the hierarchy—Maslow looked to such "optimal" people as Abraham Lincoln, shown here with his son, Todd. (Photograph by Mathew Brady from The Library of Congress.)

bound to produce a distorted psychology. Instead, he studied models of *self-actualized people*—people who appeared to have fulfilled their basic potentialities. He found some of his subjects in history (Lincoln [Figure 14.12], Jefferson, Thoreau, Beethoven) and others from among his contemporaries (Eleanor Roosevelt, Einstein, a friend who was an unusually creative housewife, another who was a clinical psychologist, and others who were in business, sports, and the arts). Maslow (1967) found that this group of "optimal" people shared some distinguishing characteristics.

1. They were open to experience "vividly, selflessly, with full concentration and total absorption."
2. They were in tune with themselves, their inner beings.
3. They were spontaneous, autonomous, independent, with a fresh, unstereotyped appreciation of people and events.
4. They devoted total effort to their goals, wanting to be first-rate, or at least as good as they could be.
5. They were dedicated, fully and creatively, to some cause outside themselves.
6. They related to a few specially loved others on a deep emotional plane.
7. They resisted conformity to the culture; they could be detached and private.

Few people can be labeled *self-actualizing* in this complete sense. Yet most of us have had moments of true self-actualization, or what Maslow referred to as *peak experiences*—a burst of insight, a betrothal, the birth of a baby (Figure 14.13), a mountaintop sunrise. During these highly focused, vivid moments, there is often a disorientation in time and space, a feeling of richness and unity. The accompanying emotional reaction "has a special flavor of wonder, of awe, of reverence, of humility and surrender before the experience as before something great" (Maslow,

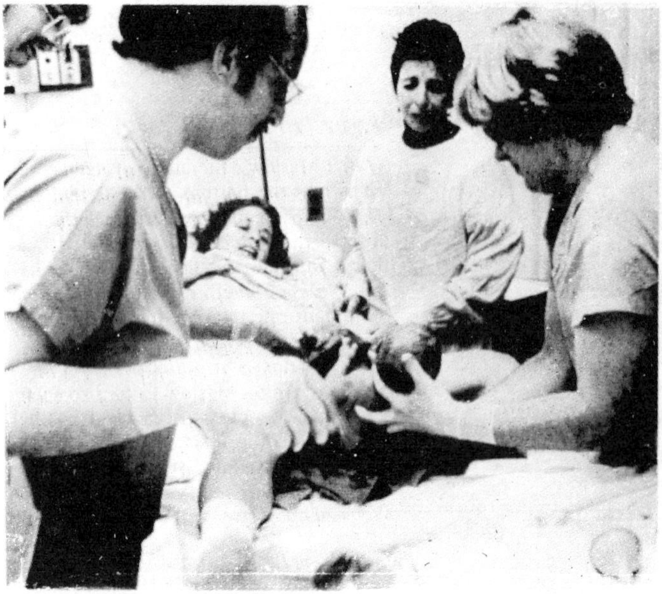

Figure 14.13

Each of us has had peak experiences that transcend our usual lives and approach what Maslow called *self-actualization*. (Birth photo from Hella Hammid/Photo Researchers; hang-gliding photo from Perer Menzel/Stock, Boston.)

1968, p. 82). "The person at the peak is godlike not only in senses . . . but . . . in the complete, loving, uncondemning, compassionate and perhaps amused acceptance of the world and of the person" (pp. 87–88).

QUESTIONS ABOUT HUMANISTIC THEORIES

There is no single humanistic theory; there are many. Yet all of them share a view of personality that is focused on people's internal perceptions, or introspections. Psychologists have waxed and waned over the years in their acceptance of personal reports based on such "inner experience." Nineteenth-century psychologists relied heavily on self-reports. The advent of behaviorism, with its emphasis on observable events, put an end to that, but lately there has been a resurgence of respect for this approach because self-reports sometimes prove as valid in research as do more complex measures. While some psychologists still scoff at "naïve" or "gullible" views built on self theories, others are now more respectful of the insights people report.

At the same time, however, both Rogers and Maslow can be faulted for the developmental aspects of their theories, which have only the flimsiest research foundations. Their evidence on the dynamics of the self concept is made up mainly of case reports and anecdotes. It is probably best to regard these self theories of development as working sets of hypotheses—thought-provoking, but not to be taken too literally.

Issues and Controversies in Personality Theory and Research

The study of personality involves much more than the "grand theories" we have studied thus far. In fact, many psychologists have become disenchanted with sweeping efforts to capture "the whole person" in a single theory and have turned, instead, to focused research on specific issues in personality. To fill out our picture of personality psychology, we will consider two of these issues.

GENES AND PERSONALITY

One is perhaps the oldest issue in psychology: nature versus nurture. (See Chapter 11, page 413.) Most of the theories we have considered so far emphasize the acquisition of personality *after* birth; consistent with this emphasis, a recent view of genetic research noted that "investigators who have studied . . . twins are unanimous in concluding that personality is more affected by environment than any other area of human functioning" (Farber, 1981, p. 269). Nonetheless, some investigators have asked whether genetic endowment might not *also* play an influential role. The answer seems to be a cautious yes, but we are still trying to learn how much genes influence personality and the specific ways in which they make their mark.

Studies of Temperament One line of research concerns *temperament*—the aspect of personality that includes mood, activity level, and emotion, and the variability of each. Studies with animals have shown that selective breeding can heighten or diminish characteristics like emotionality over successive generations. Twin studies with humans also seem to show a genetic influence.

For example, one investigative team (Buss et al., 1973) had mothers of identical, or *monozygotic* (*MZ*), and fraternal, or *dizygotic* (*DZ*), twins fill out temperament questionnaires about their children. Each mother rated each child on four dimensions: (1) emotionality (e.g., "Child has a quick temper"), (2) activity (e.g., "Child is always on the go"), (3) sociability (e.g., "Child tends to be shy"), and (4) impulsivity (e.g., "Child goes from toy to toy quickly"). On all four dimensions—indeed on virtually all of the 20 individual items—correlations were much higher for MZ than for DZ twins. Since MZ twins have 100 percent identical genes while DZ twins average only 50 percent identical genes, the higher correlations for MZ-twin pairs seem to show that shared genes make for similar temperament. Subsequent twin research (Plomin & Rowe, 1977, for example) has continued to show MZ-DZ differences and thus seemingly to support the notion of genetic influence. On the other hand, these studies have involved ratings by mothers who know whether their children are MZ or DZ twins and whose ratings therefore may be influenced by their preconceived ideas about how similar the two types of twins *should* be. Until this problem is controlled, we had better regard any conclusions as tentative.

Traits, Types, and Chromosomes Moving beyond child temperament, we can find some evidence of a genetic contribution to a large number of adult trait dimensions and to at least one typology. Some of this evidence is summarized in Table 14.1. The table lists the 18 scales of the California Psychological Inventory; it shows, for each scale, the average correlations obtained across several studies of MZ and DZ pairs. On every scale, MZ correlations average much higher than DZ correlations. By themselves, these correlations do not show, conclusively, that personality traits have a genetic basis; one reason is that MZ twins are often more similar *in* *environment* than are DZ twins because parents tend to treat their identical twins alike, to dress them alike, and so forth. However, at least one study (Loehlin & Nichols, 1976) showed that similarity of upbringing was not a major cause of higher correlations among MZ twins; this, of course, strengthens the case for genetic influence.

If the pattern shown in Table 14.1 reflects a real genetic influence, how does this influence operate? Obviously we do not inherit genes for "dominance," "sociability," and so forth. What *do* we inherit that might influence such traits? There is no widely accepted answer to this question, but some interesting hypotheses have been proposed. For example, H. J. Eysenck, whose theory we considered earlier (page 570), proposed that traits related to his introversion-extraversion type dimension are linked to inherited characteristics of the *reticular formation* (Chapter 8, page 317), the part of the brain that influences an individual's level of arousal. Eysenck believes that introverts inherit more of a tendency to be aroused, "revved up," than do extraverts and that they have a basic need to inhibit, or "damp down," their arousal. As a result, introverts tend to avoid extreme excitement; seek out calm, quiet conditions; and shy away from the activation caused by social interaction. Extraverts, on the other hand, genetically predisposed to be *under*aroused, are attracted to excitement and social interaction. Tests of introversion-extraversion do show fairly strong evidence of heritability. Moreover, four of the CPI scales in Table 14.1 that show very pronounced MZ-DZ differences—dominance, sociability, social presence, and self-acceptance—are considered to be components of the introversion-extraversion dimension.

Table 14.1

Mean MZ and DZ twin-pair correlations for the 18 CPI scales.

CPI SCALE	MZ	DZ
1. Dominance	.59	.31
2. Capacity for status	.61	.46
3. Sociability	.54	.21
4. Social presence	.54	.28
5. Self-acceptance	.51	.23
6. Well-being	.46	.31
7. Responsibility	.49	.33
8. Socialization	.47	.25
9. Self-control	.52	.30
10. Tolerance	.52	.32
11. Good impression	.48	.23
12. Communality	.30	.12
13. Achievement (conformance)	.40	.25
14. Achievement (independence)	.53	.37
15. Intellectual efficiency	.54	.38
16. Psychological mindedness	.43	.21
17. Flexibility	.50	.30
18. Femininity	.42	.25

(*Source:* Carey et al., 1978, p. 307)

In addition to similarities in their test scores, identical twins show a remarkable number of subtle behavioral similarities (see Figure 14.14). Even twins who have been reared apart tend to laugh alike, smoke similar numbers of cigarettes per day, and show similar nervous mannerisms (Farber, 1981). For the past several years, Thomas Bouchard and a research team from the University of Minnesota have been recruiting MZ-twin pairs who have been reared apart (Bouchard, 1984). Several of the pairs studied had never met until Bouchard brought them together (Bartlett, 1981).

> "Bridget" and "Dorothy," for instance, met for the first time on the plane bringing them from England to Minneapolis. Both wore seven rings, and each had several bracelets on each wrist; one had named her children Richard Andrew and Katherine Louise, and the other had named hers Andrew Richard and Karen Louise.
>
> A pair of male twins both turned out to love carpentry, to have workshops in their homes, and to have built very similar circular benches around trees in their yards—both, for some reason, found round benches very appealing.
>
> One of the most interesting pairs had strikingly different backgrounds. "Oscar" had been raised as a German Catholic and had once joined the Hitler Youth. Oscar's twin, "Jack," had been reared as a Jew in Trinidad. Now in their fifties, the two showed up for Bouchard's interviews in blue shirts with epaulets and nearly identical aviator glasses. Both kept rubber bands on their wrists, both had a habit of flushing the toilet before and after use, and both enjoyed startling people in elevators by sneezing.

The above anecdotes may represent rare extremes, or they may reflect factors other than genetic influence. But until we can find good

Figure 14.14

Identical twins are, of course, strikingly alike in physical characteristics, but evidence suggests that they may also be surprisingly similar in personality. Take a close look at this picture. Notice that without special instructions, these pairs of twins have assumed remarkably similar facial expressions. Notice also the positioning of their hands. (From *Human Genetics* by E. Novitski, 1977. New York: Macmillan.)

counter-explanations, we must seriously consider an important possibility: Genes may influence personality in more diverse and dramatic ways than most have thought.

THE PERSON-SITUATION CONTROVERSY

Another major issue in personality research is called the person-situation controversy. It centers on one key question: To what extent is our behavior caused by characteristics inside of us and to what extent by characteristics of the situation in which we find ourselves? To examine this question, let us focus on a key principle: When we think about people's behavior, we often tend to reduce complex reality to a few simple generalities. The processes by which we simplify complexity were nicknamed our "reducing valve" by Walter Mischel (1969). He gives the following illustration:

> When we observe a woman who seems hostile and fiercely independent some of the time but passive, dependent, and feminine on other occasions, our reducing valve usually makes us choose between the two syndromes. We decide that one pattern is in service of the other, or that both are in the service of a third motive. She must be a really castrating lady with a facade of passivity—or perhaps she is a warm, passive-dependent woman with a surface defense of aggressiveness.
> (Mischel, 1969, p. 1015)

From this example, it is easy to see that *traits* ("hostile," "fiercely independent," "passive") and *types* ("castrating lady") are often used as simplifying devices. Advocates of trait and type approaches to personality tend to believe that people show certain consistencies across situations and across time and that trait and type descriptions are useful ways of identifying the "person" characteristics that produce such consistencies.

On the other hand, some have argued that such trait and type descriptors may generate only an *illusion* of consistency and that people's behavior is actually quite diverse and situationally determined. Mischel went on to make the following point with regard to the "complex woman" we introduced above:

> But perhaps nature is bigger than our concepts and it is possible for this lady to be a hostile, fiercely independent, passive, dependent, feminine, aggressive, warm, castrating person all-in-one. Of course which of these she is at any particular moment would not be random and capricious—it would depend on who she is with, when, how, and much, much more.
> (Mischel, 1969, p. 1015)

Illustrations such as this have become part of a position called *situationism* because it emphasizes situational, environmental determinants of people's behavior. The position has often taken the form of outright attacks on trait and type notions—for example:

> What happens when the mother-dependent child finds that his preschool peers now consistently have little patience for his whining, attention-getting bids, and instead respect independence and

> self-confidence? Generally the child's behavior changes in accord with the new contingencies, and if the contingencies shift so does the behavior—if the contingencies remain stable so does the new syndrome that the child now displays. Then what has happened to the child's dependency trait?
> (Mischel, 1969, p. 1017)

Extreme Positions As is true of most debates, the person-situation controversy has involved a small number of theorists at both extremes and many occupying the middle ground. At the extreme "person" end are a few theorists, often of the dynamic group, who believe that people possess certain core drives, motives, conflicts, or tendencies that produce consistency in their behavior across various situations. At the extreme situational end of the controversy are theorists like B. F. Skinner, who believe that notions like "traits of character" and "free will" should be dropped and who favor a detailed account of the environment and how it determines our behavior.

The Interaction Position: Predicting Some of the People Some of the Time There are several less extreme positions, however. They have developed, in part, because much of the relevant research has been inconclusive. Some studies seem to show that people's behavior depends more on who they are and what their characteristics are while other studies show that people's behavior depends more on the situations they find themselves in. The findings seem to depend upon the particular situations and the particular people the investigator studies. For instance, a researcher who studies people's behavior at a football game, a fraternity party, a church service, and an "all-you-can-eat" fruit-and-salad bar will find that behavior is strongly influenced by which of these situations people find themselves in. However a researcher who compares the behavior of a cheerleader, a professional wrestler, a television talk-show host, and a Buddhist monk may well find that behavior is very much affected by "person" characteristics. If we look at a large number of studies that are not so extreme as these examples, we find that "person" characteristics and situation characteristics both influence behavior; but neither factor alone is so important as is the *interaction* of the person and the situation (Bowers, 1973). What this means, in part, is that the influence of "person" factors and situation factors depends on which persons and which situations we observe.

A study by Bem and Allen (1974) illustrates this sort of interaction. It suggests that we can only use "person" characteristics or traits to predict "some of the people some of the time."

> Bem and Allen asked college students to tell them how consistent they were on two traits: friendliness and conscientiousness. Then they gathered information on how friendly and how conscientious these students actually were in a variety of situations. Their findings regarding friendliness are illustrative. The students who rated themselves "not so consistent" in friendliness turned out to be friendly in some situations and not so friendly in others; for these students, situations mattered a lot, but the "person" characteristic of friendliness did not. On the other hand, students who rated themselves as consistent in friendliness came closer to being equally friendly in all

Figure 14.15

Some theorists now think of situations as having "personalities" of their own. Some situations (left) are considered "powerful" because they offer a very small range of "acceptable" behaviors and therefore exert great control over people's actions. Other situations (right) are considered "weak" because they allow a broad range of behaviors. Personality traits seem to be better predictors of behavior in relatively weak situations than in powerful situations. (Both photographs copyright © George W. Gardner.)

situations; for these students, the trait of friendliness mattered a lot, but situation characteristics did not.

Some researchers have studied person-situation interactions by focusing on "the personality of situations" (Funder & Bem, 1977). Some situations, it seems, have a stronger impact on people's behavior than do others. For example, at a funeral or in a job interview, there is a very small range of "appropriate behaviors"; such "powerful situations" induce fairly similar expectations and fairly similar behavior in most people (Mischel, 1977). By contrast, situations such as a family picnic, a visit to a new city, or an uncommitted Saturday night are "weak"—they permit a broad range of expectations and behavior in different people. In "powerful situations," the impact of the situations will be especially strong; in "weak situations," person characteristics like traits are apt to have especially strong effects. (See Figure 14.15.)

So the person-situation controversy has moved toward an interactionist viewpoint. Most personality researchers now agree that the relative impact of "person" characteristics and situation characteristics depends on *which* persons and *which* situations we choose to study.

Summary

1. The study of personality emphasizes normal individual variation. Definitions of personality vary from one theorist to the next, but most agree that it consists of distinctive patterns of behavior (including thoughts and emotions) that characterize a person's adaptation to the situations of his or her life.

2. Theories about personality can be grouped into those which emphasize types and traits, dynamic processes, learning and conditioning, or the humanistic perspective.

3. Types are collections of individuals said to share some common characteristics. Most type theories have not worked well because such collections of individuals typically leave out much of the population. Type theorists have dealt with this problem in two ways. Some, such as Eysenck, have defined their "types" in terms of personality dimensions, with every individual falling somewhere

on each dimension. Other typologies, such as the Type A-Type B distinction, focus on extreme cases that fit a specific definition, or "strike zone," and simply ignore individuals who do not fit.

4. Traits are tendencies to behave in relatively consistent and distinctive ways across situations. Allport judged people's cardinal, central, and secondary traits by examining their personal documents, such as letters and diaries. Others have focused on a single trait, such as locus of control.

5. Questions arise with regard to the adequacy of trait and type assessment, because (*a*) people's behavior often varies with the situation they are in and (*b*) people's performance on various measures can be influenced by response sets, including the desire to make a certain impression.

6. The most influential dynamic theory of personality is Sigmund Freud's psychoanalytic theory. It includes a theory of personality structure, with the id as storehouse of unconscious drives and impulses; the superego as conscience; and the ego as executive force, or mediator, balancing the pressures of id and superego with the constraints of reality. Freud also described stages of psychosexual development (oral, anal, phallic, latency, and genital) and proposed that puzzling events such as dreams and "slips of the tongue" reveal unconscious impulses and conflicts.

7. Among the theorists who built on and modified Freud's ideas were Jung ("collective unconscious," "archetypes"), Adler ("individual psychology," "inferiority complex," "style of life"), and Horney ("basic anxiety and basic hostility," "moving toward, against, and away from others"). Many others have modified the original notions of psychoanalysis to fit contemporary society.

8. A major contribution of the dynamic approach has been the concept of defense mechanisms, which are used to cope with anxiety. Among these defense mechanisms are repression, reaction formation, projection, rationalization, intellectualization, displacement, regression, and sublimation.

9. Though influential, dynamic theories are limited in that many of their concepts and interpretations are difficult to prove or disprove. The dynamic approach, in general, has been rich in ideas but poor in experimental tests of those ideas.

10. Learning and conditioning—in classical, instrumental, and cognitive forms—are highly relevant to personality and its development. Dollard and Miller used animal experiments to develop and test selected Freudian notions (for example, conflict and repression), thus advancing early social learning theory. Bandura and Walters extended social learning theory into the domain of observational learning. Skinner's radical behaviorism uses instrumental conditioning principles to explain the ways in which environmental conditions influence people's behavior.

11. Learning and behavioral theorists have been accused of diminishing the *person* in *person*ality, but many value their approaches for their clarity and their experimental "testability."

12. Humanistic theories emphasize the importance of people's subjective attitudes, feelings, and beliefs, especially with regard to the self. Carl Rogers' theory focuses on the impact of disparity between a person's ideal self and perceived real self. Maslow focuses on the significance of self-actualization. Theories of this type are often criticized for their heavy reliance on subjective self-report data.

13. Research, especially that involving twins, has uncovered evidence for a genetic contribution to personality. Two general personality manifestations have been studied: child temperament and adult traits (including a collection of traits linked to introversion-extraversion).

14. The person-situation controversy revolves around a key question: Are people's behaviors determined primarily by such "person" factors as traits or primarily by the situations people find themselves in? Recent research suggests an answer: "It depends." Some people are strongly influenced by some traits that do not greatly influence other people. Moreover, some situations are "powerful," involving heavy demands for specific kinds of behavior. Other situations are less structured and thus "weak"; they permit personality traits to have a stronger influence on people's behavior.

Terms to Know

One way to test your mastery of the material in this chapter is to see whether you know what is meant by the following terms.

Personality (568)
Type and trait approaches (568)
Dynamic approaches (568, 576)
Learning and behavioral approaches (568, 592)
Humanistic approaches (568, 598)
Type (569)
Introvert (569)
Extravert (569)
Traits (571)
Cardinal traits (571)
Type A and Type B (572)
Central traits (573)
Secondary traits (573)
Idiographic approach (573)
Nomothetic, or dimensional, approach (573)
Locus of control (573)
Situationism (575, 606)
Psychoanalytic theory (576)
Id (577)
Libido (577)
Ego (577)
Superego (578)
Personality dynamics (578)
Conscious (578)
Preconscious (578)
Unconscious (578)
Repression (578, 579, 588)
Neurosis (579)
Erogenous zones (580)
Psychosexual stage (580)
Fixation (580)
Oral stage (581)
Anal stage (581)
Phallic stage (581)
Oedipus complex (581)
Electra complex (581)
Identification (581)
Latency period (582)
Genital stage (582)

Analytical psychology (582)
Complex (584)
Collective unconscious (584)
Archetypes (584)
Animus (584)
Anima (584)
Personal unconscious (584)
Individuation (585)
Introversion (585)
Extraversion (585)
Inferiority complex (585)
Compensation and overcompensation (585)
Style of life (586)
Individual psychology (586)
Basic anxiety and basic hostility (587)
Moving toward, against, and away from others (588)
Defense mechanisms (588)
Anxiety (588)
Reaction formation (589)
Projection (589)
Rationalization (589)
Intellectualization (590)
Displacement (590)
Regression (590)
Sublimation (590)
Instrumental conditioning/operant conditioning (594)
Reinforcement (594)
Punishment (594)
Observational learning (596)
Imitation (596)
Vicarious reinforcement (596)
Self (598)
Self-concept, or self-image (599)
Executive functions of self (599)
Humanistic theories (599)
Phenomenal field (599)
Ideal self (599)
Incongruences (599)
Q-sort technique (600)
Self-actualizing (601)
Peak experiences (601)
Temperament (603)
Monozygotic (MZ) twins (603)
Dizygotic (DZ) twins (603)
Interaction position (607)

Suggestions for Further Reading

Some of the personality theories we have discussed here are dealt with briefly but informatively in the following articles and short books: S. Freud's *A General Introduction to Psychoanalysis* (New York: Washington Square Press, 1960, Paperback); C. S. Hall's *A Primer of Freudian Psychology* (New York: Mentor, 1954, Paperback); C. S. Hall and V. J. Nordby's *A Primer of Jungian Psychology* (New York: Taplinger, 1973); K. Horney's *Feminine Psychology* (New York: Norton, 1967); A. H. Maslow's *Toward a Psychology of Being*, 2d. ed. (Princeton, NJ: Van Nostrand Reinhold, 1968, Paperback); and B. F. Skinner's *Beyond Freedom and Dignity* (New York: Bantam Books, 1971, Paperback).

In addition to these offerings, you may want to read Christopher Monte's *Beneath the Mask: An Introduction to Theories of Personality* (New York: Praeger, 1977), a book which weaves together interesting accounts of major personality theories and the life circumstances of the theorists which may have given rise to their theories. For a well-written, carefully researched overview of the broad field of personality, we recommend Seymour Feshbach and Bernard Weiner's *Personality* (Lexington, MA: Heath, 1982); these authors cover not only the major personality theories but also such wide-ranging applications of personality theory and research as aggression, achievement strivings, emotion, and altered states of consciousness.

Abnormal Psychology

HEADING OUTLINE

Chapter 15

The letter and drawing on the next page were produced by David Berkowitz, also known as Son of Sam. Berkowitz (pictured in Figure 15.1 on the night of his arrest) terrorized New York City from July 1976 to August 1977—a period during which he is thought to have killed six young people—five of them women between the ages of 18 and 26—and wounded seven others. In most cases, he used the .44 that led reporters to dub him "the .44-caliber killer."

Berkowitz apparently had problems throughout his life in establishing a secure identity. At birth, he was given the name Richard David Falco and then put up for adoption. A Bronx couple, Nat and Pearl Berkowitz, took him in and renamed him David Richard Berkowitz. While he was growing up, David reportedly suffered from severe fears—especially of monsters—and a painful sense of rejection. He learned to cope by bragging to others about imagined feats of bravery, strength, and sexual prowess. These imaginings blossomed into a grandiose fantasy world, where Berkowitz found the popularity and success that he lacked in real life.

As Berkowitz neared young adulthood, his fantasies apparently turned sinister; demons, he says, told him to kill young women. He responded, in 1975, by prowling Co-op City, a huge apartment complex in the Bronx; he carried a knife. The demons helped him spot "the right person" and then commanded him to "get her." He first attacked a woman who was leaving a supermarket carrying groceries. As he stabbed her, she screamed. This confused him. After all, he said later, "I wasn't going to rob her, or touch her, or rape her. I just wanted to kill her" (Klausner, 1981). The prolonged screams, and close contact with the blood of his victims, convinced Berkowitz that killing with a knife was not for him. He switched to a Charter Arms .44-caliber Bulldog and began a 1-year rampage of shootings that ended with his arrest in August of 1977.

The shootings were evidently triggered by increasingly bizarre ideas. Berkowitz reported that his landlords and neighbors were actually nonhumans in league with forces of evil such as "the Blood Monster," "the Duke of Death," and "General Jack Cosmo." A central figure in Berkowitz's fantasies was Sam Carr, a neighbor whom Berkowitz believed to be two people: (1) Sam Carr, businessman; and (2) a creature named Sam who was "a high official of the Devil's legion" and sent demons to torment David at night and prevent him from sleeping. David tried to do away with the demons by throwing a Molotov cocktail into Sam Carr's backyard at 5 A.M. The demons only laughed. After this experience, David yielded, becoming a self-described "Son of Sam" and following the demons' instructions to kill.

After his apprehension by police, Berkowitz was examined by psychiatrists. One report diagnosed him as "paranoid" (see page 644) and unable to understand the court proceedings or to assist in his own defense. Another report acknowledged "paranoid traits" but maintained that Berkowitz was competent to stand trial. The author of the second psychiatric report also testified, "Your Honor, the defendant is as normal as anyone else. Maybe a little neurotic" (see page 625) (Klausner, 1981, p. 375). On the trial date, Berkowitz entered a plea of guilty. He was given consecutive sentences totaling 365 years. The "Son of Sam" is now a prisoner in New York's Attica Correctional Facility.

This is not the end of the story, however. After Berkowitz's trial, David Abrahamsen, a psychiatrist who argued that Berkowitz was not insane, wrote an article defending his views. In his article, entitled, "Unmasking 'Son of Sam's' demons," Abrahamsen argued that Berkowitz had been lying, that the "demons" were created by Berkowitz as a deliberate fiction and that the "Son of Sam" had admitted as much in a letter to Abrahamsen after the trial. Yet two intriguing questions remain. How could Berkowitz's story have so thoroughly convinced other psychiatrists that he was insane? And if Berkowitz was trying to portray himself as insane, why did he plead guilty rather than "not guilty by reason of insanity"? The controversy continues. For further details, see Abrahamsen (1979) and Klausner (1981).

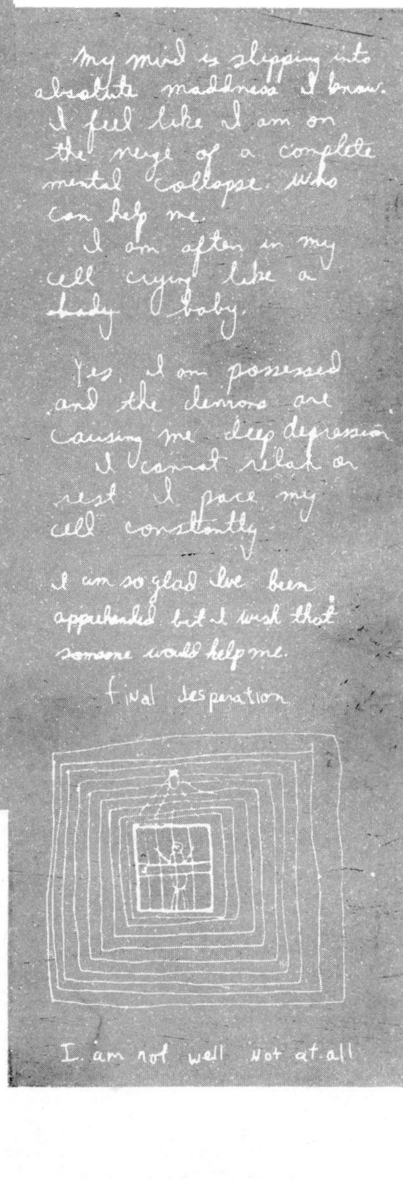
my mind is slipping into absolute madness I know. I feel like I am on the verge of a complete mental collapse. Who can help me.
I am often in my cell crying like a baby baby.

Yes, I am possessed and the demons are causing me deep depression. I cannot relax or rest. I pace my cell constantly.

I am so glad I've been apprehended but I wish that someone would help me.

final desperation

I am not well not at all

HOW are we to decide who is "insane"? This is one of the most difficult questions faced by psychologists who testify in court. David Berkowitz (pictured in Figure 15.1), despite what appeared to be a bizarre set of beliefs, was judged sane enough to be sentenced to prison for his actions, but John Hinckley, Jr. (also pictured in Figure 15.1) was judged "not guilty by reason of insanity" when tried for his 1981 attempt to assassinate President Reagan.

Hinckley had some unusual ideas of his own. His obsession with the movie *Taxi Driver*, for example, led him to believe that he could win star Jodie Foster's affection by killing the President. After the shooting, Hinckley seemed not fully aware of its implications. For example, responding to a written question from *Newsweek* concerning his views on President Reagan, Hinckley wrote

> I believe all Americans will eventually be helped by Mr. Reagan's changes. He is the best president we've had this century. Let's give the man a chance. (*Newsweek*, Oct. 12, 1981, p. 51)

So Hinckley, who had done his part *not* to give the President "a chance," had unusual thoughts and islands of "unawareness"; but was he more "insane" than Berkowitz? The question is difficult to answer, partly because there are no absolute, clear-cut rules for judging insanity.

Abnormality in Everyday Life

The question of what is and is not "sane" extends to everyday life, among everyday people. Consider these examples:

> Susan has broken up with her boyfriend, the young man she had been sure she would marry. Now she constantly feels tired, weak, and very depressed. She can't seem to sleep at night, and she can hardly taste what little food she eats. Her friends have begun to worry about her mood and her weight loss. Susan drags herself to classes, but she can't concentrate when she tries to study. As a result, her grades have begun to drop.

Is Susan showing a normal reaction or a serious psychological disturbance? And what about Mark?

> Three weeks ago, Mark returned to college after attending his sister's funeral. She had been killed in a highway collision. Mark and his sister had always been very close. Last night, something very strange happened. Alone in his dorm room, Mark suddenly saw his dead sister and heard her speak to him.

Is Mark's behavior a sign of insanity? It's hard to say. People who are otherwise quite normal sometimes think they see or hear a loved one who

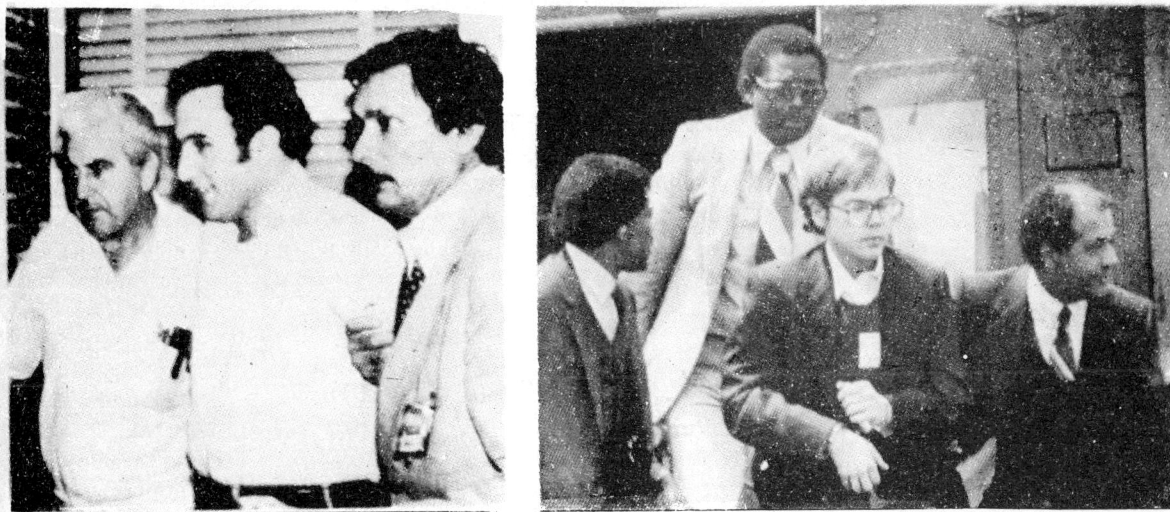

Figure 15.1

David Berkowitz (left) murdered a number of innocent young people on instructions from "demons"; despite what seemed to be delusions and hallucinations, Berkowitz entered a guilty plea in his murder trial, and he was sentenced to prison. John W. Hinckley, Jr., (right) tried to assassinate President Reagan, apparently to impress a young actress; Hinckley was found "not guilty by reason of insanity" and thus received treatment, not prison. The contrast between these two young men highlights the difficulty our society faces in making judgments about insanity. (Left, United Press International; right, AP/Wide World Photos.)

has died weeks earlier (Schulz, 1978). On the other hand, if Mark has a history of seeing or hearing things that aren't really there, or if he begins to show highly irrational behavior, psychologists might become concerned. Similarly, psychologists might judge Susan's behavior on the basis of how extreme and how persistent it is.

Questions about sanity may begin to feel quite personal as you get further into this chapter. As you read descriptions of various kinds of abnormal behavior, you may occasionally get an uneasy feeling that "this sounds a lot like me!" That shock of recognition, those nagging questions—Could I be paranoid? (page 644) or Do I have an antisocial personality disorder? (page 660)—are part of the "medical-student syndrome" and common among people studying abnormality of many different types. The syndrome typically fades after a few weeks, but it does illustrate an important point: There is at least a tinge of "abnormality" in most of us, and this makes questions about abnormality and sanity personally relevant to everyday people in everyday life.

The Language of Abnormality

In everyday life, the terms insanity, mental illness, psychopathology, emotional disturbance, behavior disorder, mental disorder, abnormality, and psychological disorder have roughly similar meanings. All are often applied to behavior that is highly unusual. In this chapter, though, we will be more selective in the terminology we use, and we want to explain why.

The words sanity and insanity are primarily legal terms that require an either/or judgment. As we suggested above, clear-cut "sane versus insane" distinctions are difficult for even experts to make reliably. Furthermore, this chapter is not intended to encourage *you* to make such distinctions. Thus we will not emphasize the terms sanity and insanity. We will not use terms such as mental illness and psychopathology because such terms seem to imply that people with serious psychological problems are "ill." This idea, once widely accepted, is now hotly debated; we discuss this debate over the "medical model" on page 627. The terms emotional disturbance, behavior disorder, and mental disorder are too narrow for use in this chapter; the problems we discuss here may involve any or all of these three features. Thus we will emphasize two somewhat-broader terms, abnormality and psychological disorder. We define *abnormality* as significant deviation from commonly accepted patterns of behavior, emotion, or thought. We define *psychological disorders* as patterns of abnormal behavior, emotions, or thought that significantly interfere with an individual's adaptation to important life demands and often cause distress in the individual or in others.

Judging Abnormality and Psychological Disorder: Some Key Factors

As the preceding discussion suggests, judging whether a person's functioning is abnormal and whether the person has a psychological disorder can be a complex process that involves weighing several factors. In the following section, we will discuss some of these factors, the subjective ways in which they are used, and the ways they are sometimes misused.

SOCIAL CONTEXT

Very few behaviors are inherently abnormal. Abnormality must be judged in the context of a particular social situation. Alone at home, you may safely strip to your underwear, sing an aria into the mirror, and deliver your acceptance speech for the Nobel Peace Prize. But try this in your introductory psychology class, and your behavior will be considered abnormal. Try it repeatedly, and people may suggest that you seek "professional help"—not because of your behavior alone, but because you have chosen inappropriate times and places to display it.

The same general principle applies to behavior across different cultures. Actions that are normal, expected, and acceptable in one culture may be considered deviant and be quite unacceptable in another (see Figure 15.2). An American visiting an Asian country might delight in eating a roast, only to vomit when told he has eaten dog. In some Asian countries, the American's behavior might be considered deranged. However in many western countries, a host who served his dinner guests roast dog might be considered deranged. By the same token, a belief that one is being tormented by demons—a belief apparently held by David Berkowitz—would not be considered abnormal, or even incorrect, in some cultures. To be considered abnormal, behavior must deviate markedly from the accepted rules and practices of the society in which it occurs.

Theorists Thomas Szasz (1960, 1970) and Thomas Scheff (1966) have argued that social context should be the centerpiece of our thinking about abnormal behavior. According to both theorists, people labeled mentally ill are not so much "sick" as they are different—they deviate from the accepted social norms. Though these norms are unwritten, they are understood intuitively and taken for granted by most of the people in the society. For example, there are unwritten rules about how loudly one should speak and how close one's face should be to another person's face

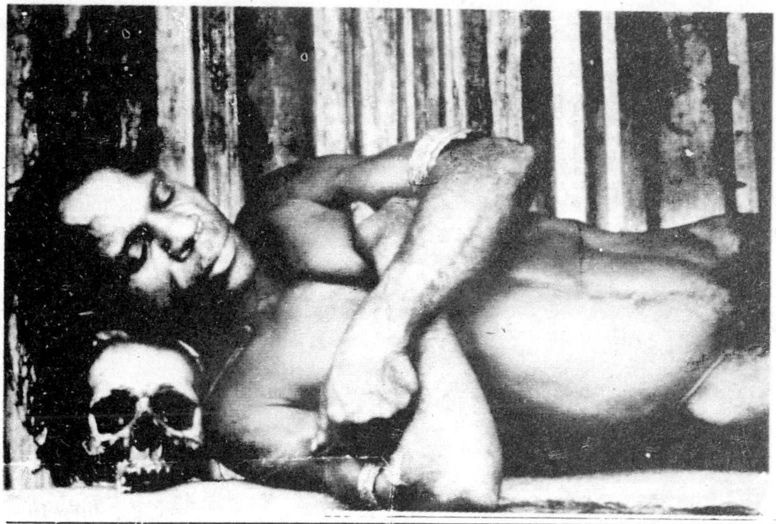

Figure 15.2

This New Guinea tribesman is a headhunter. He sleeps on the skull of his victim to ward off the victim's spirits. Is this behavior normal or insane? In this man's social context, it is normal; in our own, it would be considered insane. (Malcolm S. Kirk.)

when carrying on a conversation. When people violate these rules by speaking too loudly or too softly or by getting too close, they risk being seen as abnormal and, possibly, "psychologically disordered." Scheff believes that people with little social power—people from low socioeconomic levels, for instance—are especially likely to be labeled abnormal or mentally ill if they violate norms. People with greater social power are more likely to be considered "eccentric."

DISTRESS IN OTHERS

Scheff's (1966) ideas suggest that deviance from social norms may be risky, in part, because it causes discomfort and distress in others. People identified as having psychological disorders are often those whose behavior disturbs others. People who walk the streets looking dirty, making bizarre body movements, shouting at strangers, or telling loud stories are especially likely to be "spotted" and ushered into some form of mental-health care. Likewise, family members whose poor judgment puts them on the brink of squandering the family wealth tend to receive quick attention.

In many countries, people who cause distress in government leaders may also be considered abnormal or disordered. This is true to some extent in western countries like the United States (see Torrey, 1983), but more extreme examples can be found in certain totalitarian states. The problem exists in the Soviet Union, for instance. There, people whose behavior disturbs government officials may be diagnosed as having a psychological disorder and sent to a Soviet prison hospital or "institute of forensic psychiatry" (Nekipelov, 1980). One study (Bloch & Reddaway, 1977) reported 210 such cases. An example (reported by Fireside, 1979) is the case of P. G. Grigorenko (Figure 15.3).

> General Grigorenko had served in the Red Army for 35 years and was well respected. When he was in his mid-fifties, he began to express some concerns about Communist party policies. Instead of exploring his concerns, officials began exploring Grigorenko's sanity. He

Figure 15.3

Behavior that is distressing to others puts a person at risk of being considered abnormal or psychologically disordered. Soviet General P. G. Grigorenko (left) of the Red Army distressed government officials when he questioned Communist party policies. He was examined by Soviet psychiatrists, diagnosed, and sent to a Soviet "psychoprison." One such institution, the prison hospital at Oryal, is shown at the right. (Left, reproduced on behalf of Petro G. Grigorenko, *Memoirs*, 1982, with the permission of the publisher, W. W. Norton & Company, Inc.; right, courtesy of Aid to Russian Christians.)

> was examined by Soviet psychiatrists and diagnosed as having "paranoid development of the personality, with reformist ideas arising in the personality, with psychopathic features of the character and the presence of symptoms of arteriosclerosis of the brain." However a second team of psychiatrists found Grigorenko sane. The second opinion was reviewed by an official commission and overruled. Grigorenko spent a total of 6 years in three Soviet "prison-hospitals" but was eventually allowed to emigrate to the U.S.

The psychiatrists who diagnosed Grigorenko as "paranoid" may well have been quite sincere. In fact, such sincerity would merely underscore the point that people who violate accepted norms in disturbing ways risk being judged abnormal or disordered.

SUBJECTIVE DISTRESS OF THE INDIVIDUAL

Judgments about psychological disorders are also influenced by our perceptions of the subjective distress of the individual in question. Such distress may include (1) feelings of dissatisfaction, sadness, anxiety, or lethargy; (2) physical complaints, such as nausea or headaches; or (3) unwanted thoughts or impulses. In fact, for some of the "milder" disorders, the discomfort that people report is often the most obvious sign that something is wrong. Problems that do not seem major to an outside observer can overwhelm certain people.

In recent years, subjective distress has become an increasingly important factor in judging whether *unusual* behavior represents a *disorder*. For example, homosexuality—a statistically unusual behavior pattern—was only recently removed from most standard lists of psychological disorders. Most mental-health professionals now regard homosexuality as a disorder only if it is subjectively distressing to the individual—that is, only if the homosexual feelings are unwanted and upsetting to the person who has them. (American Psychiatric Association, 1980.)

Of course, the *absence* of subjective distress can also be a telling sign. People who blithely maim or murder others or who are unconcerned when they hear voices that are not really there are showing deviant behavior that needs attention. So while the presence of distress may signal some disorders, the absence of *appropriate* distress may signal other kinds of disorders.

A MATTER OF DEGREE

The frequency, degree, or intensity of certain behaviors also helps determine how abnormal they are judged to be. For example, all of us are nervous or anxious at times, but some people are judged to be disturbed because their anxiety is intense, persistent, and handicapping. Similarly, most of us have daydreams; some of these daydreams even involve unrealistic fantasies in which we identify with movie characters. But some people carry such fantasies to extremes. John Hinckley, Jr., for example, became obsessed with the film *Taxi Driver*. He identified with its chief character, Travis Bickle, to such an extent that he set out to mimic Bickle's assassination attempts. When fantasies lead to behavior this extreme, we regard the fantasies as abnormal; and we may decide that the person with such fantasies has a psychological disorder.

Behavior that is extremely *overdone* or *underdone* may be considered abnormal. For example, some people are so overwhelmed with guilt over minor infractions that they cannot cope with the ordinary demands of living. Others feel no guilt even when they cause serious injury to others (Figure 15.1, for example, shows David Berkowitz smiling after his arrest for murder). Socially, some people are so dependent that they cannot function alone; others show an abnormal *lack* of interest in social relationships. The ancients advised "moderation in all things." Behavior that follows this ancient precept tends to be considered "normal." Behavior that violates the precept—behavior that goes too far or not far enough—may deviate significantly from social norms and thus be judged abnormal.

IMPAIRMENT OF ADAPTIVE FUNCTIONING

We consider some behaviors abnormal because they pose a threat to adaptation. The term *adaptation*, as used here, means meeting the performance requirements or role demands of one's situation. Adults who withdraw into a private fantasy world may find it difficult to meet the role demands of their job. Children and adolescents who express their anger by physically attacking others will have trouble meeting the social expectations of peer relationships and the behavioral requirements of school.

Adaptive risks are also posed by various forms of self-injury. For example, some children, for reasons we don't yet know, repeatedly strike themselves in the head. These children must wear helmets virtually all day to prevent the skull and brain damage that would otherwise result. Some adolescents and adults burn holes in their flesh with cigarettes unless they are watched closely. In these and many other ways, abnormal behavior may pose adaptive risks. When the risks are serious, the behavior may well be taken as evidence of a psychological disorder.

Judgments about what is adaptive are sometimes very difficult to make. For example, many people would list suicide as an example of maladaptive behavior. However, many also believe that suicide may be the most adaptive course one can follow in cases of extremely painful, protracted illness. Judgments about what is adaptive will certainly influence people's judgments about abnormality and psychological disorder, but judgments about which behavior is adaptive and which is not will vary considerably from situation to situation and from person to person.

This same word of caution applies to all the factors we have discussed in this section. Each factor—violation of social norms, distress in self or others, degree or intensity of behavior, and adaptive impairment—can influence judgments about abnormality and disorder, but each factor can influence different people in different ways. Any two people might well disagree about which behavior represents a serious violation of social norms, which behavior is too excessive to be normal, and so forth. Furthermore, no single factor would be an adequate basis for judgment in *all* cases; and in any single case, two people might well disagree about which factors mattered. So for many reasons, different people often disagree as to which behavior is abnormal and which constitutes a psychological disorder. Similarly, experts often disagree about which list of disorders makes the most sense. We will learn more about this in the following section.

Classifying Psychological Disorders

For years, psychologists and psychiatrists have tried to come up with a reasonable list of psychological disorders. Several different classification systems have been tried, each with various disorders named, defined, and grouped into categories. One purpose was (and is) to give precise meanings to terms that are tossed around in our everyday language without much precision—terms like "paranoid" and "schizophrenic." Giving clear definitions of such terms is important because they are often used in making crucial decisions. For example, when expert witnesses classified John Hinckley, Jr., (Figure 15.1) as schizophrenic, the jury ruled that he was "not guilty by reason of insanity." Over the years, efforts to define and classify psychological disorders have taken two general forms: empirical and clinical-consensual.

THE EMPIRICAL APPROACH TO CLASSIFICATION

The *empirical approach to classification* is often linked to behavioral assessment, as discussed in Chapter 13 (pages 555–557), and often relies on factor analysis, as discussed on page 525. Disorders are identified by finding groups of problems that often occur together. How can we tell which problems occur together? One approach is to use behavior checklists, like the one shown on page 557.

The investigators working with that checklist (Achenbach & Edelbrock, 1983) located hundreds of children who were being treated for psychological problems. The parents of these children filled in the checklist, indicating which of the more than 100 problems on the list their child demonstrated. Then the investigators computed the correlations between each problem on the list and each of the other problems. When several problems showed high correlations with each other, they were said to form a *syndrome*—a cluster of problems that often occur together. Each syndrome can be seen as a sort of "psychological disorder." To illustrate, let us examine the syndromes, or disorders, these investigators found for one particular group, adolescent girls.

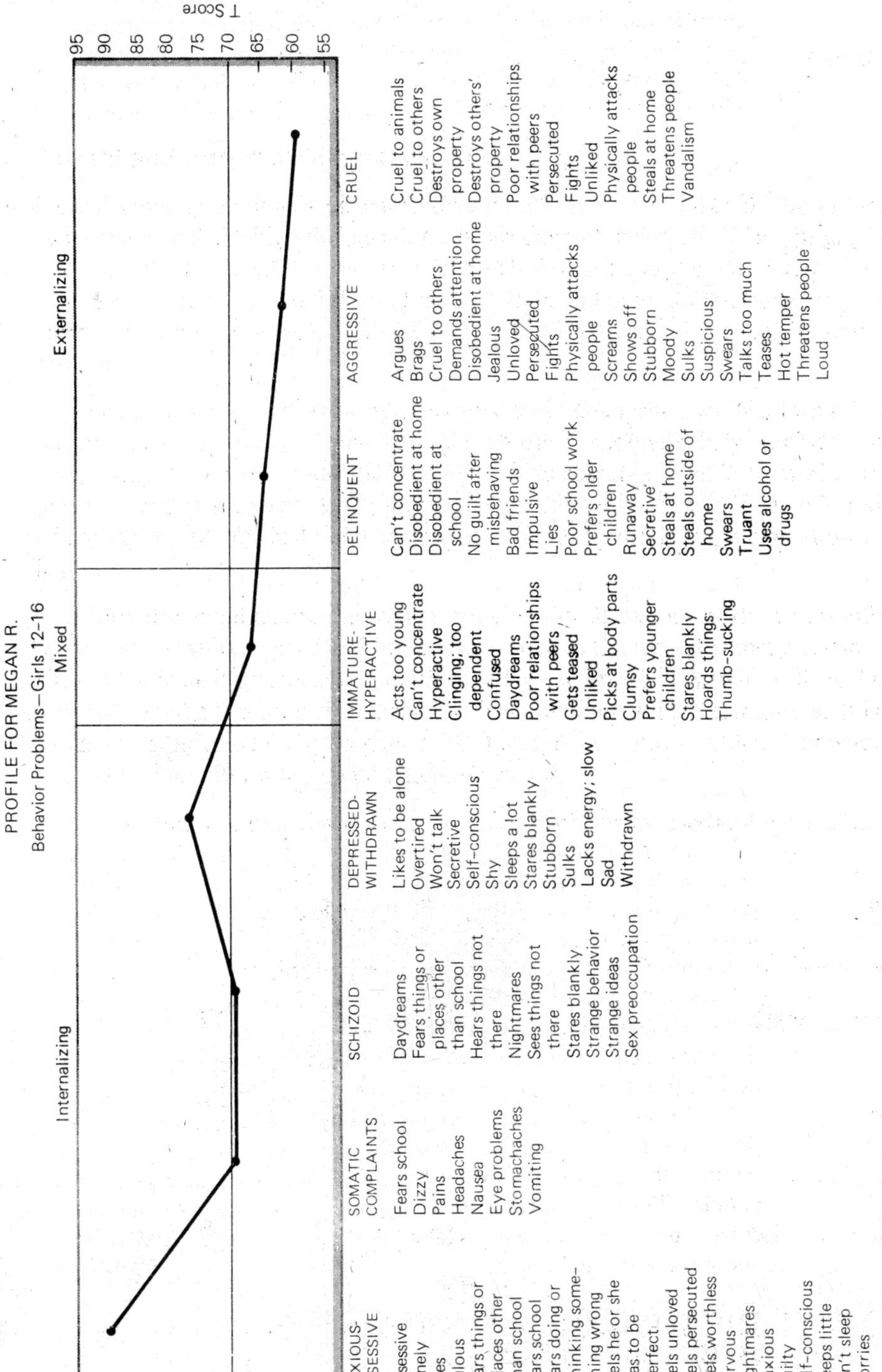

The empirical approach to classification is illustrated by this profile for Megan R., an adolescent girl. Factor analysis yielded the factors, or "syndromes," listed across the center of the figure; the dots indicate Megan's standing on each of these factors. (Modified from Achenbach & Edelbrock, 1983.)

Figure 15.4

Figure 15.4 shows eight syndromes for adolescent girls. The names given to the syndromes are listed across the center of the figure—"anxious-obsessive", "somatic complaints," and so on. Underneath each of these names are the individual problems from the checklist that correlate with each other to form that particular syndrome. For example, the "depressed-withdrawn" syndrome includes such problems as "likes to be alone," "overtired," and "sad." This means that these problems (and the others listed with them) tend to occur together fairly often and can therefore be thought of as forming a *syndrome.* The "depressed-withdrawn" syndrome, or disorder, is thus defined by the problems listed beneath it.

The dots shown in the top half of the figure indicate how one particular adolescent, "Megan R.," scored on each of the syndromes. As you can see, she scored especially high on the anxious-obsessive syndrome, but she was also above the 98th percentile for adolescent girls on the depressed-withdrawn syndrome. One advantage of an empirical approach of this sort is that it gives Megan's parents, or the clinician who works with her, a pretty good idea of what each of her high scores means, because the problems that define each disorder are listed directly under its name. Another advantage is that after the checklist has been filled out, Megan's problems can be grouped mathematically by computing her score on each factor. Thus "diagnoses" made with this approach tend to be more reliable (see Chapter 13, page 518) than diagnoses based entirely on the judgment of mental-health workers. For example, when different adults independently fill in the Child Behavior Checklist for a particular child, the syndromes that show up for the child tend to show good agreement from one adult to the next (Achenbach & Edelbrock, 1983). Diagnoses made with other classification systems have not proven so reliable (see page 629).

Empirical approaches also have significant limitations, however. To begin with, the information they produce is only as good as the information put into them. If Megan, for example, has important problems not listed anywhere on this particular checklist, then those problems cannot enter into her scores on any of the disorders. Also, although the empirical approaches are good at identifying *common* clusters of problems, they are not so good at finding *rare* clusters. This means that rare but important disorders such as infantile autism (see pages 631–634) would probably not be identified by research using the empirical approach.

Despite these and other limitations, however, the empirical approach has much to offer, and, with the increasing availability of computer scoring, this approach to classification is almost certain to grow more popular. Thus far, however, it is used primarily for research purposes. For clinical purposes, the most widely used approach to classification is the one we will examine next.

THE CLINICAL-CONSENSUAL APPROACH TO CLASSIFICATION

By far the oldest and most common method of classification is what some have called *the clinical-consensual approach*. Experts, usually clinicians who work with disturbed people, try to reach a consensus about which psychological disorders exist and how those disorders should be defined and diagnosed. The consensus is based on the experience and judgment of the experts rather than on statistical procedures. As our knowledge about abnormal behavior and psychological disorders has expanded, clinical-consensual classification systems have grown much more complex and detailed. In 1840, the basic classification scheme used in the United States census contained only one category for all mental disorders. Since then, classification systems have shown increasing attention to the multiple forms that abnormal behavior can take.

Traditional Classification Systems: Four Common Categories For many years, systems of classification have grouped disorders into four general categories.

One such category, *brain syndromes*, includes disorders thought to result from damage to brain tissue. The disorders may be temporary and reversible (for example, phenobarbital intoxication) or long-term and irreversible (for example, tissue damage sometimes associated with old age, as discussed in Chapter 12, page 506).

A second category, *psychosis*, includes severe disorders in which the individual's perception of reality is seriously distorted and in which psychological functioning becomes severely disorganized. Some psychoses result partly from physiological factors; the evidence for others is not so clear. Psychoses are frequently so handicapping that they prevent the individual from holding a job, attending school, or carrying out complex responsibilities. People suffering from psychoses may experience *delusions*—false beliefs that persist despite evidence showing that they are false. (John Hinckley, Jr., for example, apparently believed, despite evidence to the contrary, that he could win the heart of a famous actress by assassinating the President.) People with psychoses may also experience *hallucinations*—sensory or perceptual experiences that have no real external source, such as the demons' voices that David Berkowitz said he heard commanding him to kill. Among the disorders traditionally considered to be psychoses are the schizophrenic and paranoid disorders, both discussed later in this chapter.

A third category, *neurosis*, includes relatively mild disorders that are moderately incapacitating—disorders in which the individual's perception of reality is not grossly impaired. The concept of "neurosis" has historic ties to Sigmund Freud's psychoanalytic theory (Chapter 14, page 576). Freud thought neurotic disorders resulted from internal conflict and the anxiety it engendered. In his view, neurotic symptoms were often exaggerated efforts to defend against, or avoid, that conflict and anxiety. For example, Freud's patient, Fräulein Elizabeth (Chapter 14, page 579), felt severe conflict and anxiety when she realized, after walking alone with her sister's husband, that she wanted to have him for herself. Later she developed "neurotic" pains in her legs and was unable to walk for long periods of time. In Freud's view, she had defended herself against the conflict and anxiety that arose partly as a result of that walk by unconsciously developing neurotic symptoms that prevented her from walking at all. The concept of "neurosis" also includes cases in which people *fail* to defend themselves against anxiety and, instead, experience it directly and intensely—in the form of fear, dread, or even panic. Some problems of this sort are described in our section on anxiety disorders, page 655.

The fourth category, *personality disorders*, includes long-standing, maladaptive personality patterns. People exhibiting these disorders have habitual, relatively inflexible ways of behaving across a wide range of situations. Because these behavior patterns impair the individual's relationship with the outside world, they often interfere with social, school, or job effectiveness. Personality disorders may involve persistent mistrust of others, persistent antisocial behavior, or persistent avoidance of contact or relationships with others. Character disorders, often included within the personality-disorder category, are deficits in impulse control or socialization. We will have more to say about personality disorders on page 659.

A Current System: DSM-III Currently, the "official" United States system for classifying psychological disorders is the third edition of the *Diagnostic and Statistical Manual of Mental Disorders*, referred to as *DSM-III* (American Psychiatric Association, 1980). State governments, mental hospitals, and clinics all around the U.S. typically require that individuals receiving mental-health services be assigned a diagnosis from the DSM-III list of disorders. DSM-III may or may not be the best classification system around, but it is certainly the one most widely used in the United States.

DSM-III includes a way of reporting on more than just the person's psychological disorders. (Stressful life circumstances, such as a recent divorce, can be included in the diagnosis, for example.) But we will focus on the disorders listed in DSM-III. They are divided into two categories: (1) *clinical syndromes*, which include most of the disorders in the traditional brain syndrome, psychosis, and neurosis categories discussed above; and (2) *personality disorders*, which closely resemble the traditional category of the same name described above. Table 15.1 lists some of the clinical syndromes and personality disorders included in DSM-III; we will discuss several of them in this chapter.

An overall objective in preparing DSM-III was to describe disorders in such a way that different clinicians using the DSM-III system would usually agree as to which disorder a particular person has. Such agreement is called interjudge reliability. In two studies testing for interjudge reliability (Spitzer et al., 1979; Spitzer & Forman, 1979), different clinicians using DSM-III to diagnose the same individuals *did* arrive at the same diagnosis more often than did clinicians using earlier classification systems, such as DSM-II. However these studies have been criticized for failing to make sure that the clinicians made their diagnoses *independently* of one another (Rosenhan & Seligman, 1984). If the clinicians discussed their diagnoses among themselves while making them, this would obviously have inflated the level of agreement found in the studies. Even with this possible inflation, the actual reliabilities reported for many of the disorders were too low to be acceptable, and this was especially true of the diagnoses for children and adolescents, *most* of which have shown unacceptably low reliability (Mattison et al., 1979; Mezzich & Mezzich, 1979). If a classification system is to be maximally useful, it must have good interjudge reliability. We need more evidence on DSM-III before we can be confident that it has this property.

Reliability is not the only issue raised by classification systems. Some of the most important issues concern the classification process itself, as we will see in the following section.

RECURRING ISSUES IN CLASSIFICATION

Before we examine some of the disorders listed in DSM-III, we need to

Table 15.1

Some DSM-III categories

CLINICAL SYNDROMES	PERSONALITY DISORDERS
*Disorders usually first evident in infancy, childhood, or adolescence	*Paranoid
Organic mental disorders	Schizoid
*Substance-use disorders	*Schizotypal
*Schizophrenic disorders	*Histrionic
*Paranoid disorders	Narcissistic
Psychotic disorders not elsewhere classified	*Antisocial
*Affective disorders	Borderline
*Anxiety disorders	Avoidant
Somatoform disorders	Dependent
Dissociative disorders (or hysterical neuroses, dissociative type)	*Compulsive
Psychosexual disorders	Passive-aggressive
Factitious disorders	Atypical, mixed, or other personality disorder
Disorders of impulse control not elsewhere listed	
Adjustment disorder	

* Described in this chapter.
Source: This table is derived from, *Diagnostic and Statistical Manual of Mental Disorders,* 3d ed., Washington, DC: American Psychiatric Association.

consider three important issues bearing on these classifications. One is the degree to which psychological disorders resemble medical disorders.

The Medical Model: Is a Psychological Disorder an "Illness"? The DSM classification systems have all been based on a *medical model*—that is, they have been based on the view that psychological disorders are similar to medical disorders in significant ways. When this model is used, psychological disorders are assumed to represent "mental illnesses" revealed by the "symptoms" shown by the "patients." DSM-III, though it is less extreme than some earlier systems, frequently refers to psychological disorders as "illnesses" revealed by "symptoms." The medical model was fueled in the late 1800s and early 1900s by some dramatic discoveries about general paralysis, later called paresis, a psychological disorder involving memory loss, irrational behavior, and declining motor skills. Many felt that the disorder was caused by a disease that attacked the

Many people who hold to the medical model today believe that other psychological disorders, too, are caused by disease-related processes. As research has progressed, we have discovered that *some* psychological disorders are accompanied, if not by diseases, then at least by physical and biochemical abnormalities in the brain (see pages 642 and 652, for example); but this has *not* been found for *most* of the psychological disorders. This is one reason we defined *psychological disorders* as we did on page 618, specifically omitting any mention of underlying illness or disease.

Many people who oppose the medical model think that psychological disorders are best viewed not as illnesses but as specific combinations of problems that often occur together. According to this view, disorders should be grouped within a classification system that will help us organize our knowledge about which problems occur together often and what might cause these problems. Such causes might include severe life stress (Chapter 8); unusual processes of learning and conditioning (Chapter 4); or atypical child rearing, peer influence or other developmental influences (Chapters 11 and 12). Opponents of the medical model feel that although biological factors may well play a role in some disorders, particularly those involving severely disturbed thoughts and mood, they rarely tell the whole story and they probably play a modest role, if any, in most psychological disorders.

In the final analysis, the question of what really causes psychological disorders—and whether some reflect "illnesses"—must be settled in the scientific arena. Systematic research will certainly reveal different causal factors for different disorders. We will discuss the current status of some of this research later in this chapter.

Risks and Benefits of Labeling People In practice, classifying psychological disorders means labeling people. Some are labeled schizophrenic, others paranoid, and so forth. Because the medical model is so influential, people who have been given these labels are likely to be considered "mentally ill" by many in society. Thomas Scheff (1966) and Thomas Szasz (1960, 1970), theorists we discussed earlier, warned that this labeling can be dangerous. Once labeled, a person may be forced into the social role of "mentally ill person." Friends and family treat the person differently, potential employers shy away from hiring such a "high-risk" applicant, and the person may not even be allowed to vote or make legal contracts. On the other hand, if the person accepts the "mentally ill" role, a number of rewards may also follow: attention, warmth and sympathy, and even a certain freedom from responsibility. After all, the "mentally ill" can't be held completely responsible for their actions, some people would say.

This process can therefore operate like a self-fulfilling prophecy, with people eventually accepting their labels and behaving as the "mentally ill" are "supposed to." In cases where labeled people do not follow this path but persist in behaving "normally," they may be reminded in a number of ways that their labels are long-lived and powerful in their impact on others. One of the most vivid examples of this phenomenon occurred in 1972, when public pressure forced Senator Thomas Eagleton off the Democratic presidential ticket with Senator George McGovern. Eagleton had been an articulate and effective senator and vice-presidential candidate, but the news that he had been treated for depression years earlier created a public uproar that led to his withdrawal.

Labeling may also influence the judgments of mental-health professionals, as illustrated in a study by Langer and Abelson (1974).

> Professional clinicians were asked to view a videotaped interview of a young job applicant. Half the clinicians were told, truthfully, that the man in the tape was a job applicant; half were told that he was a "patient." After viewing the tape, all answered questions about the

interviewee (his attitudes, his outlook on life, and so forth), and their answers were used to form overall adjustment ratings. The clinicians were influenced by the label. Those who thought they had seen a "patient," judged the man to be more disturbed and less well adjusted than did those who thought they had seen a job applicant—this despite the fact that all of the clinicians had seen the same videotape. [See Figure 15.5.] An interesting aspect of these findings, though, was that not everyone was influenced by the label. Clinicians of the behavioral school (Chapter 14, page 592, and Chapter 16, page 693), who do not subscribe to the medical model, were not swayed by the label. Unlike the other clinicians, the behavioral group described the "patient" just as they described the "job applicant."

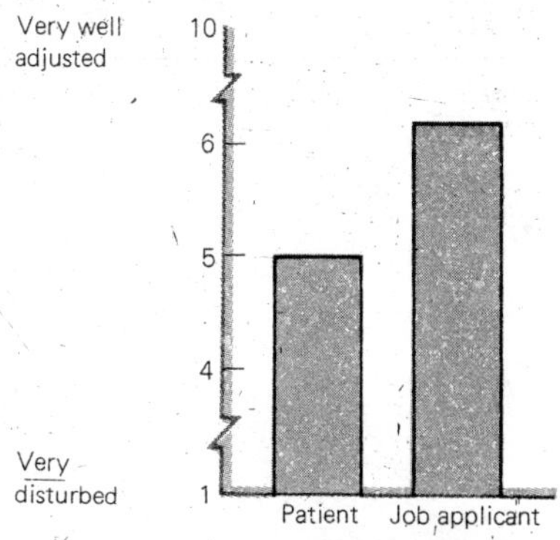

Figure 15.5

Clinicians' overall adjustment ratings for a "job applicant" and a "patient." Apparently labels can influence the judgments of professionals. The "job applicant" and the "patient" were actually the same person. (Study by Langer & Abelson, 1974; figure from Rosenhan & Seligman, 1984.)

Not all of the effects of labeling are bad. In fact, some very important benefits can accrue from careful use of diagnostic labels. An "official" diagnosis can clear the path to appropriate mental-health services for very disturbed people. A diagnostic label can open the door to a specialized treatment center and can guarantee financial help from the individual's health insurance company. Furthermore, while labels may sometimes lead to discrimination, they may at other times evoke sympathy and understanding. A person who is moody and occasionally "difficult" may be better accepted at school, on the job, or by neighbors if it is known that the person is trying to cope with some specific psychological problem.

Perhaps the most important benefit of labeling will be its long-term contribution to our understanding of specific disorders. Labels and categories help us to organize our research findings on groups of people who share certain characteristics. As we shall see later, the use of admittedly imperfect labels like schizophrenia and depression has already helped scientists accumulate a good deal of information on the people given such labels. We now know, for example, the forms of thinking that characterize people in these two groups, and we know that genetic factors may play a role in both kinds of disturbance. Over time, the use of diagnostic labels may help us understand many forms of psychological distress and find ways of alleviating that distress. Certainly our use of labels and categories should not blind us to people's individuality. But neither should our desire to treat people as individuals blind us to the value of categories.

The Fallibility of Diagnosis Some people feel that because we are so imprecise in our use of labels, the risks outweigh the benefits. According to this view, the diagnosis of psychological disorders is a primitive process, susceptible to glaring errors. One study often cited in support of this viewpoint was conducted by David Rosenhan (1973).

Rosenhan enlisted the help of pseudopatients, adults who had never shown seriously abnormal behavior but who nonetheless agreed to apply for admission to mental hospitals. The pseudopatients included a graduate student, three psychologists, a pediatrician, a housewife, a psychiatrist, and a painter. All called for appointments at various east- and west-coast mental hospitals. When interviewed, each complained of hearing voices that, though unclear, seemed to be saying "empty," "hollow," and "thud." Beyond this decep-

> tion—and falsifying their names and vocations—the pseudopatients reported openly and honestly about their past and present lives. Thus it might be expected that all would have been considered "sane," or at least considered as candidates only for outpatient counseling. In fact, all were admitted as inpatients; and all but one were diagnosed as suffering from schizophrenia (see page 637). Although the pseudopatients behaved normally after admission to the hospital, they were kept in their respective hospitals for periods ranging from 7 to 52 days. When discharged from the hospital, all but one were labeled as having schizophrenia in remission, meaning that the diagnosis was accurate but the "symptoms" had subsided. As an interesting sidelight, a number of the "real" patients in the institutions suspected that the pseudopatients were, in fact, sane and were writers or professors doing a study. In other words, the "real" patients, unlike the staff, recognized that the imposters were "normal" people. The mistaken diagnosis of schizophrenia may have blinded the staff to the normality of the pseudopatients.

Rosenhan's study has many critics. One of the most prominent is Robert Spitzer, head of the task force that produced DSM-III. Spitzer (1975) argued that the hospital staff acted responsibly when they diagnosed and admitted the pseudopatients because these people were seeking admission and reported that they heard voices. Hearing voices, after all, is generally considered a danger signal pointing to schizophrenia (page 638). A cautious hospital-staff member might well decide to admit anyone who reports hearing voices—just to be on the safe side and to permit an opportunity to observe the person carefully. In Spitzer's view, the staff members were also reasonably quick to note the absence of abnormal behavior after admission. The average hospital stay for the pseudopatients was 19 days. This might be considered a brief period if we take into account one important point: People with schizophrenia often have extended periods of normal behavior. So perhaps a person who has been hearing voices should be observed for more than just a few days before being discharged.

Rosenhan (1975) replied to Spitzer and other critics by emphasizing the important general points made by his study. One of the most important was that context strongly influences judgments about whether people have psychological disorders. The fact that his pseudopatients were in a mental-hospital context helped make even their normal behavior seem abnormal and indicative of a disorder. For example, staff notes on one of the pseudopatients included the statement, "Patient engages in writing behavior." Thus writing, which is considered normal in most places, became "writing behavior," presumably a sign of the "patient's" disorder when it occurred in a mental-hospital context. Anecdotes of this sort do not tell us how widespread such "context effects" actually were in the Rosenhan study. However the study does make an important overall point that is hard to deny: Judgments about abnormality and diagnoses of psychological disorder are imperfect and sometimes way off base.

Rosenhan went on to make a controversial recommendation about diagnosis:

> My own preference runs to omitting diagnoses entirely, for it is far

better from a scientific and treatment point of view to acknowledge ignorance than to mystify it with diagnoses that are unreliable [and] overly broad . . . (Rosenhan, 1975, p. 467)

Rosenhan is certainly right when he says that diagnoses are often unreliable, and mistaken diagnoses can certainly have harmful effects, but the argument that we should dispense with diagnoses remains a minority view. A more widely accepted conclusion from Rosenhan's study is this: The diagnosis of psychological disorders is an imperfect process at best, and any single diagnosis should be interpreted with caution.

Abnormality and Psychological Disorders: A Perspective

So far in this chapter, we have tried to encourage some critical thinking about abnormality. We have noted that judgments of abnormality and disorder depend on several factors, each one at least partially subjective. We have discussed the debate over how (and whether) to classify psychological disorders at all. And we have stressed that diagnosis is an imperfect process. Certainly the DSM-III list of disorders, to which we will turn shortly, is best seen as an educated guess—a way of identifying and classifying disorders in the 1980s but a way that will surely change as our knowledge expands. There is little doubt that understanding abnormality and psychological disorders is a long-term and a formidable task, but it is also a worthwhile task, one that must be tackled if we are to learn enough about psychological problems and their origins to help alleviate them. Despite the difficulties, we are making progress. This can be seen in the remaining sections of this chapter, where we describe some of the psychological disorders listed in DSM-III and discuss important advances in our understanding of them. In Chapter 16, we will return to some of the disorders and focus on what has been learned about treating them.

Psychological Disorders Arising in Childhood and Adolescence

Our discussion of psychological disorders will begin with those that arise early in the life span. DSM-III includes one large category of disorders that usually make their first appearance during infancy, childhood, or adolescence. We will focus on two disorders in this category: infantile autism and bulimia. Then we will turn to some disorders that are more likely to arise at the end of adolescence or after.

INFANTILE AUTISM

One of the most frightening disorders a parent can confront is the rare but devastating combination of retarded development and peculiar behavior known as *infantile autism* (Figure 15.6). The combination is illustrated by the case of Dana.

Dana, aged 11, is flapping and rocking again. She sits alone in a

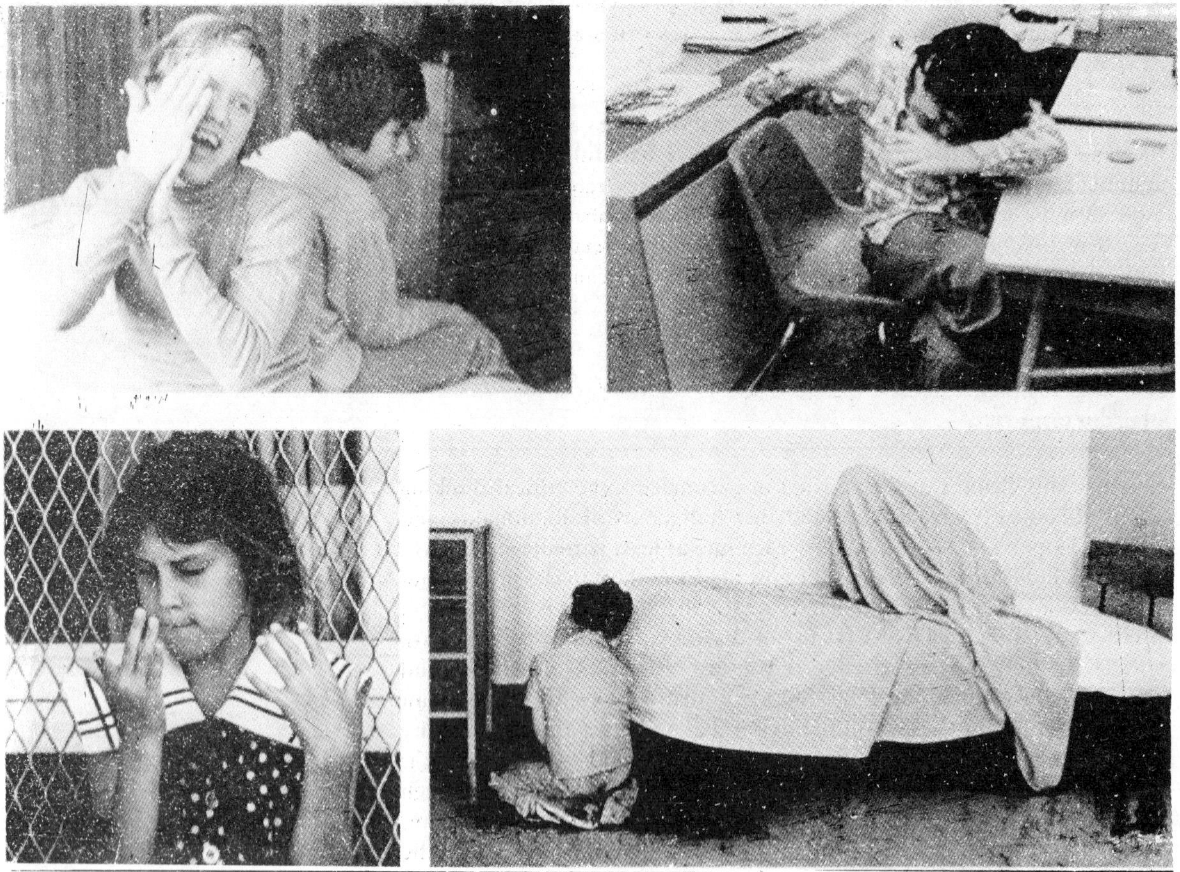

Figure 15.6

Autistic children frequently show unusual, repetitive body movements and bizarre responses to their environment. (Top photos, E. F. Bernstein Photos; bottom photos, © Allan Grant Productions.)

corner, rocking back and forth while flapping her hand in front of her face as if waving goodbye to someone beside her. She smiles, but not in response to what others are doing; the reason for the smile is locked somewhere inside Dana. Her parents now recall that even as an infant, Dana was not responsive to people; she did not keep eye contact, smile at faces, or cuddle when picked up. As she matured, her parents wondered whether she might be deaf, especially when she failed to develop normal language. Even now, Dana uses very immature grammar. She reverses pronouns, for example, saying, "You go store?" when she means, "Can I go to the store?" Her speech also shows abnormal "melody," with questionlike rises at the end of declarative statements. At school, Dana is in a special class for the "emotionally handicapped," and even in that class she is clearly the most unusual child.

Characteristics of Autism Dana shows the essential features of infantile autism: (1) a general lack of responsiveness to other people; (2) serious deficiencies in communication skills, especially language; and (3) unusual repetitive responses to the environment. Moreover, all of these problems appeared within the first 30 months of life. Autistic children seem to live in a world of their own. Even as infants, they may not raise their arms when their parents reach down to pick them up and they may not mold their bodies to their parents when they are picked up. They often fail to show normal signs of attachment to their parents (Chapter 11, page 429); for example, they tend not to seek comfort from their parents when they are upset. As they mature, autistic children have trouble perceiving other children's feelings and playing cooperative games. They generally do not form close friendships.

They also have trouble with language—learning to speak it and understand it—and the language they do develop is often peculiar. For example, they may routinely echo the last phrase they hear another person speak; or they may reverse certain pronouns—saying "I" when they mean "you" and vice versa, as in the case of Dana.

Especially noticeable are the unusual ways in which autistic youngsters interact with the objects in their environment. For example, instead of playing with toy cars by pretending to drive them, as most children do, autistic children may endlessly line the cars up in rows or twirl them round and round. Autistic youngsters may also twirl their hair, flap their hands (as did Dana), or show other repetitive body movements time and again. Even when they mature into adolescents and adults, people with autism tend to have poorly developed and peculiar language and social behavior. A few succeed at conventional jobs; but many must work in sheltered workshops and live in sheltered settings such as group homes for most of their adult lives.

Family, Physiological, and Genetic Factors in Autism What causes autism? Up until the late 1960s, the most popular answers to this question emphasized psychological causes—especially cold and rejecting behavior by parents (Despert, 1951). Such parent behavior was thought to be especially common in the middle class, and some early research did suggest that autism was unusually common in middle-class families (Lotter, 1967). More recent research (Gillberg & Schaumann, 1982; Wing, 1980) has not supported this idea, and most researchers have decided that there is not enough evidence to link autism to parental behavior (Rutter & Garmezy, 1983).

To understand what does cause autism, many investigators are now focusing on physiological factors. Several studies suggest that autistic children may have abnormally high, or low, levels of physiological arousal (DesLauriers & Carlson, 1969). Overarousal might explain why autistic children avoid looking at faces, but stare intently at less-stimulating objects, like room fixtures (see Figure 15.7): This behavior may be a way of preventing increased arousal. Underarousal, on the other hand, might explain behavior like the flapping and rocking shown by Dana; these might be ways of increasing arousal.

Some researchers have studied cerebral lateralization in autistic youngsters. As we noted in Chapter 2, (page 69), the left cerebral hemisphere is the area that usually specializes in language. There is growing evidence that this specialization may not happen for many autistic children; instead, their language functions may be centered in the right hemisphere, an area not as well suited to language (Dawson et al., 1982). This abnormal pattern might help explain why autistic children show severe language deficits. A study by Blackstock (1978) illustrates this line of research.

Figure 15.7

The percentage of time spent by autistic and nonautistic disturbed youngsters looking at environmental stimuli such as room fixtures (top) and various face models. (From Hutt & Ounsted, 1970.)

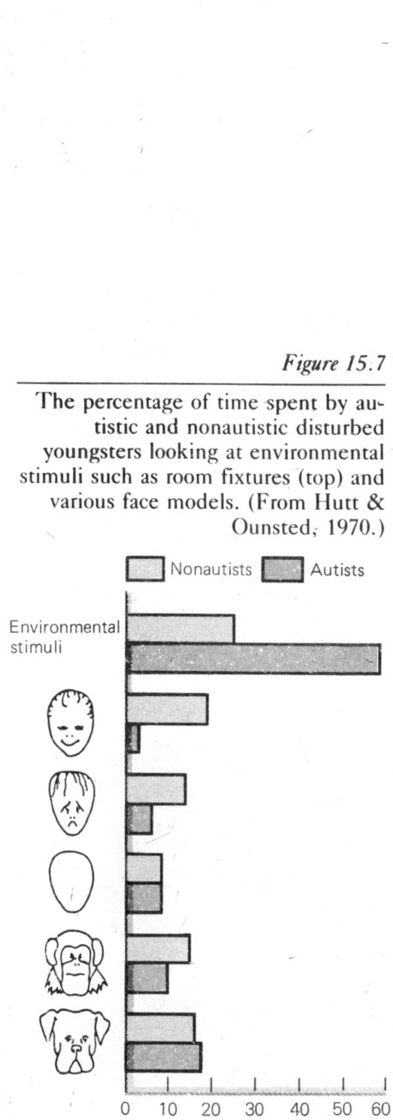

> In this study, autistic and normal children listened to stories, using special devices that allowed them to control (1) whether the stories were spoken or sung and (2) whether they heard the stories through their left or their right ears. Music listening and left-ear listening are both primarily right-hemisphere functions. The autistic children showed a much stronger preference for the musical version of the stories and for left-ear listening than did the others. This suggests that, for reasons as yet unknown, autistic children may not develop the normal left-hemisphere specialization for language; instead, some of their key language functions may be handled—and not very successfully—by the right cerebral hemisphere.

Although several studies point to a left-hemisphere deficit in autistic children, others do not (Fein et al., 1984). We may eventually find that the left-hemisphere deficit hypothesis is correct, but only for one subgroup of autistic people. For others, different causes may be involved.

Whatever physiological factors may be involved in autism—abnormal arousal, unusual cerebral-dominance patterns, or others as yet unidentified—we need to understand why these factors operate in some children and not in others. To tackle this question, some researchers have studied the role of genetic factors. There is growing evidence that hereditary transmission plays a role (Folstein & Rutter, 1977; Minton et al., 1982; Ritvo et al., 1985). However what gets inherited is probably not autism, as such, but rather a susceptibility to various cognitive and language problems. In some especially unlucky children, these problems may combine in a way that leads to autism.

BULIMIA: AN EATING DISORDER

In Chapter 12 (page 480), we discussed *anorexia nervosa*, a behavior pattern involving self-starvation that often arises in adolescence and is especially prevalent among girls. Anorexia is one of several eating disorders listed in DSM-III. Another, closely related to anorexia, is *bulimia*.

> Ginny, aged 18, has been trying to keep a secret from her family and friends: she is a binge eater. When others are not around, or after they are asleep, Ginny often gorges on cake, ice cream, candy, and other food—so fast that there is hardly time to chew. It seems that once she has begun, there is no stopping until she is so stuffed that her stomach aches or she is so exhausted that she falls asleep. After a binge, Ginny is disgusted with herself for her lack of self-control; she feels depressed and guilty. Often she makes herself vomit, then takes a double or triple dose of a laxative. To compensate for the gorging, she sometimes goes on severe fasts. Her life, she feels, is almost totally dominated by her obsession with food.

Characteristics of Bulimia Ginny's behavior, thoughts, and feelings fit the DSM-III diagnostic criteria for bulimia. The disorder is sometimes called the binge-purge syndrome, and Ginny's pattern certainly fits the label. Her struggle with her urge to eat leads, over and over again, to cycles in which she stuffs food in, purges it out with vomiting and laxatives, and sometimes goes on extreme diets. Bulimia is hard to detect because the binging and purging are hidden; the victim's public eating habits are appropriate, and body shape and weight remain within normal limits (Fairburn, 1980). Thus it is hard to know just how widespread bulimia is. The evidence we have indicates that it is most common among white females in their late teens to late twenties. One study found that about 4 percent of students treated in a university, psychiatric clinic had been diagnosed as having bulimia (Stangler & Printz, 1980). But another study reported that 13 percent of 355 students in a suburban college experienced all the major DSM-III diagnostic symptoms of bulimia (Halmi et al., 1981). Discrepancies like this suggest that there may be many people on college campuses (and perhaps elsewhere) who have bulimia but do not seek treatment.

Bulimia most often begins as a problem of overeating, usually in late adolescence (Figure 15.8). Fears of losing control over eating and body weight lead to radical efforts to regain control by extreme dieting or purging. Within a year or so, the binge-purge-diet cycle has begun. At this point, the bulimic person's life can become almost completely dominated by thoughts of food; worries about weight; and feelings of depression, shame, and self-contempt. All this can make it very hard to concentrate on schoolwork or to meet the daily demands of a job. Social relationships may also take a beating, partly because of the extreme secrecy that shrouds the bulimic person's life.

Causes of Bulimia We do not yet know the causes of bulimia, but researchers have been busy studying the psychological processes in bulimic people and their families. One common finding is that the family history usually includes some impulse-control problems (there is often an obese or alcoholic family member; Schlesier-Stropp, 1984). It may be that bulimic people inherit some susceptibility to impulse-control problems; or perhaps such problems in their families make them especially sensitive to concerns about self-control. These concerns come to center on food and body size, much as they do in anorexia (page 480). Like anorectic people, bulimic people typically have a morbid fear of becoming fat, feel overweight even when they are not, and have exaggerated ideas about

Figure 15.8

Bulimia often begins as a problem of overeating; eventually, eating binges are followed by purging (with laxatives or vomiting), and the binge-purge cycle has begun. (© 1982 Teri Leigh Stratford/Photo Researchers, Inc.)

how thin and light they "should" be. In fact, many people diagnosed as having anorexia also have bulimic episodes of gorging and purging (Casper et al., 1980). A major difference between the two disorders is that people with bulimia, unlike those with anorexia, manage to keep their body weight within normal limits—though at great psychological cost. Research is now being focused on the question of why some people's concerns about self-control, food, and body size lead to anorexia while others with similar concerns develop bulimia.

Substance-Use Disorders

As we discussed earlier, social context influences people's judgments about what behavior is abnormal or disordered. This seems especially true when it comes to the problem of substance abuse. Most societies frown on *some* forms of substance dependence, but different societies do not agree on which dependencies are normal and which are problems. For instance, few people in Western cultures consider caffeine or tobacco dependence a serious psychological disturbance, despite the fact that DSM-III lists caffeine intoxication, tobacco dependence, and tobacco withdrawal as disorders. On the other hand, alcohol and heroin dependence are widely considered to be serious disorders in our culture. They are also two of the most prevalent serious dependences—in the United States alone, there are half a million heroin addicts and over 5 million alcoholics. We will therefore focus on these two disorders.

ABUSE VERSUS DEPENDENCE

In describing disorders involving alcohol, heroin, and a variety of other substances, DSM-III distinguishes between two levels of substance misuse: abuse and dependence. *Abuse*, the milder level, involves some maladaptive use pattern—remaining intoxicated throughout the day, for example—and some impairment in social or job functioning. *Dependence* is a state of physical addiction, which involves *tolerance* (increasing amounts of the substance are needed to achieve the desired effect) and *withdrawal symptoms* (when the substance use is stopped or reduced, unpleasant physical symptoms result). Withdrawal symptoms and increased tolerance are signs that a real physical need has developed and that there are actual alterations in the body's physiology. In some cases, brain impairment occurs (see Chapter 5, page 211). Addicts who are withdrawn from a drug become so agitated, depressed, or otherwise miserable that they can think of nothing but getting their next dose. In the case of alcohol dependence, withdrawal symptoms may include tremor (the shakes), delirium, convulsions, and hallucinations; with or without convulsions, these symptoms are collectively called *delirium tremens*. Note, however, that a number of the drug or substance habits that people form do not involve any real physical dependence (see Table 2.2, page 51). In these cases—for example, marijuana and LSD habits—what develops is a strong *psychological* need for the substance (it helps the person relieve tension or feel better in some other way). This psychological need can produce a craving that, for some people, is as strong as physical dependence.

ORIGINS OF HEROIN AND ALCOHOL DEPENDENCE

Even for substances like heroin and alcohol, which produce real physiological dependence (Figure 15.9), psychological factors play a major role. The dreamy, euphoric state produced by heroin is used by many people as a way to reduce their anxiety and tension. For instance, when supplies of heroin were readily available to soldiers in the Vietnam war, men who would not otherwise have used it were drawn to it, both because it was readily available and because it reduced, for a time, their distress and anxiety. Cultural norms also play a large role in heroin addiction; heroin use may be accepted, and even expected, in certain subcultures as the norm —"the thing to do." In fact, heroin addicts who leave their subcultures and kick the habit are very likely to start using heroin again if they return to their cultural groups (Ray, 1976).

In the case of alcoholism, a number of explanations have been proposed for the psychological craving that develops. One possibility is that alcohol, like heroin, calms anxiety and helps people relax. Of course, most people who use alcohol to relax do not become alcoholics. Perhaps, as some research indicates, people's inborn responses differ, with some people inheriting a capacity to drink more before vomiting or passing out or a tendency to experience higher levels of euphoria from alcohol (Cadoret et al., 1980; Goodwin, 1979). Such people would be especially susceptible to alcohol dependence. Another idea is that since it reduces social inhibitions, alcohol permits people to act out their hostilities and sexual wishes. Still another idea is that some people who drink to excess have a need for personal power and that drinking makes them feel powerful (McClelland et al., 1972). While these explanations may cover some cases of alcohol dependence, much remains to be discovered about the reasons why some people become alcoholics while others who drink do not.

Schizophrenic Disorders

We turn now to a group of disorders that many regard as the most serious of all: the various forms of schizophrenia. In the United States alone, as

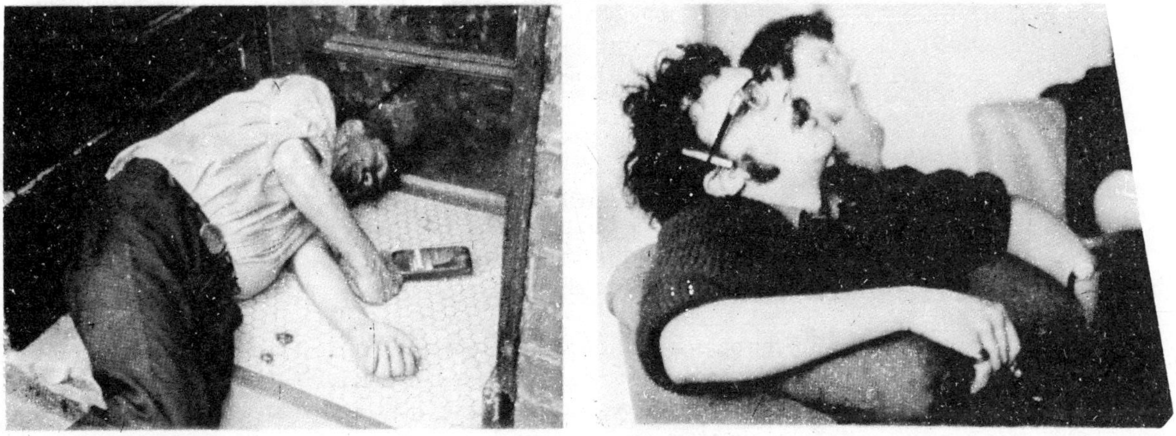

Figure 15.9

Because alcoholism and heroin addiction involve real physiological dependence, they can be extremely disabling. (Left, © 1984 Eric Kroll/Taurus Photos; right, Ian Berry/Magnum Photos, Inc.)

many as 2 million people may have been affected by schizophrenia at some time in their lives. Schizophrenia is one of the most widely misused of all psychological terms. It is often erroneously applied to behavior that is merely inconsistent, as when newspapers brand a politician "schizophrenic" for taking contradictory positions on an issue. Sometimes the term is applied to those who exhibit split personalities or multiple personalities, but this application is incorrect. (Multiple personality is a rare disorder that is not a form of schizophrenia.) Before we describe what schizophrenia is, let us illustrate it with the following interview excerpt:

> *INTERVIEWER (I)* How old are you?
> *PATIENT (P)* Why, I am centuries old, sir.
> I How long have you been here?
> P I have been now on this property on and off for a long time. I cannot say the exact time because we are absorbed by the air at night, and they bring back people. They kill up everything; they can make you lie; they can talk through your throat.
> I Who is this?
> P Why, the air.
> I What is the name of this place?
> P This place is called a star.
> I Who is the doctor in charge of your ward?
> P A body just like yours, sir. They can make you black and white. I say good morning, but he just comes through there. At first it was a colony. They said it was heaven. These buildings were not solid at the time, and I am positive this is the same place . . .
> (White, 1964, p. 514)

CHARACTERISTICS OF SCHIZOPHRENIA

This style of speech, and the thinking it reflects, illustrates some of the features of *schizophrenia*, a broad category that includes several specific psychotic disorders. All these disorders involve significant disturbances in thought. One of the most common features is incoherence, or *loosening of associations*—disjointed expressions that seem to take on a life of their own, to ramble in a more or less aimless way without ever quite reaching a logical conclusion. The interview quoted above illustrates this quality of thought, which is also called *cognitive slippage and derailment*. The personal journals of schizophrenics provide particularly clear illustrations of their tendency to skitter away from logical frameworks. Consider this excerpt from the diary of a man often hospitalized for schizophrenia:

> I confess my moral venial sins of all of the State and City Police and Dr. S. [ward psychiatrist] of orange elastic blue color any color of bloomers and panties boths and all of the reals bloomers and panties of boths and all of the real times and Father Lawlors confession and Dr. S. said go in peace.
> (Grinspoon et al., 1972, pp. 69–70)

Schizophrenics often have *delusions* (page 625). They may believe that the FBI is transmitting thoughts into their brains through the television set or that unseen forces are controlling their behavior. They may also have *hallucinations*, most commonly auditory hallucinations. They may hear a

Figure 15.10

An example of the autistic thought seen in schizophrenia. What seems to make little sense probably had deep private and personal significance for the patient who drew this picture. (Courtesy of CIBA, *State of Mind.*)

single voice continually commenting on their behavior or multiple voices arguing and shouting profanities at each other. Schizophrenic thought processes are often called *autistic,* meaning they are determined more by private, inner processes than by external reality. The drawing and text in Figure 15.10 illustrate this dominance of internal processes, expressed in the form of symbols and expressions that have private meaning and personal significance but convey little information to others.

Because schizophrenics' thought disturbances produce distortions in the way they perceive the world, the behavior growing out of their perceptions may often seem bizarre. The schizophrenic may withdraw almost completely from others, slipping into a world of private thoughts and experiences. Often, too, there are serious mood disturbances. The schizophrenic may cry, seethe with anger, laugh or giggle inappropriately, or display *flat affect*—a bland, lackluster mood, with little responsiveness to others or to ongoing events (Figure 15.11).

Schizophrenia used to be called *dementia praecox,* which means "youthful insanity," because it tends to develop in adolescence or early adulthood. Although schizophrenia may occur at any age, the highest rate of first-time admissions to mental hospitals for this disorder occurs among people in their late teens and early twenties. Schizophrenia, one of the most common psychotic disorders, is also one of the most crippling.

Different types of schizophrenia are identified in DSM-III. One of them, the *disorganized type,* is characterized by incoherent thinking (like that illustrated in the quotations above) but no systematic delusions. The *catatonic type* of schizophrenia involves various kinds of severe psychomotor disturbances. For example, the person may assume a fixed body posture—sometimes a painful-looking statuelike position—and maintain it for periods lasting several minutes, even hours (Figure 15.12). **During**

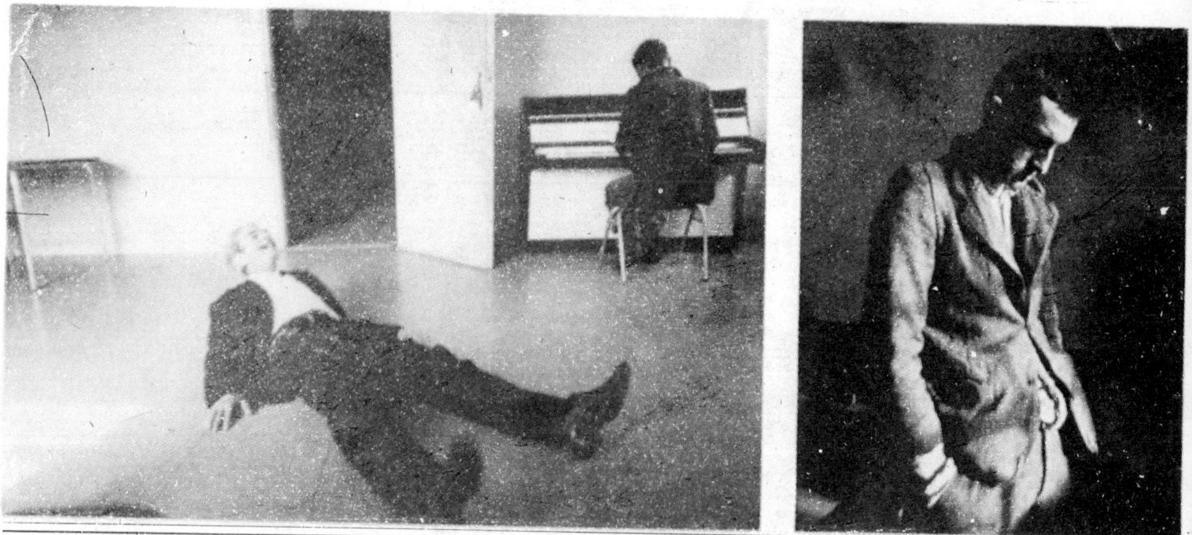

Figure 15.11

People diagnosed as schizophrenic often show mood disturbance. They may laugh or giggle even though nothing around them seems funny to others (left); or they may seem numb and emotionless—affectively "flat" (right). (Left, J. K./Magnum Photos; right, Raymond Depardon/Magnum Photos.)

such periods, the person is usually mute and the body assumes a waxy flexibility—meaning that if someone else repositions the person's arm or leg, the limb will remain in the new position. At the other extreme, the catatonic may have periods of agitation—with arms flapping or swinging and excessive talking, shouting, or even animal sounds. The hallmark of the *paranoid type* of schizophrenia is the presence of hallucinations or delusions. Both tend to involve themes of persecution or wildly grandiose ideas of one's own importance or influence. For example, a paranoid schizophrenic may believe that the devil is torturing him by shouting profanities at him through the television set.

THEORIES AND EVIDENCE ON SCHIZOPHRENIA AND ITS CAUSES

Because schizophrenia is so widespread and so handicapping, theorists and researchers have tried, for years, to understand its characteristics and causes. In the following sections, we will examine some of the resulting theories and findings.

Life Experience and Family Interaction in Schizophrenia Schizophrenia tends to run in families; people diagnosed as schizophrenic are likely to have had at least one parent with the same diagnosis. This means that many schizophrenic people have grown up with an in-house, parental model for observational learning (Chapter 14, page 596); indeed, some influential theories attribute schizophrenia to processes of modeling (Bandura, 1968) or instrumental conditioning (Ullmann & Krasner, 1975). Others (for example, Lidz, 1973) blame family conflict, especially marital schism, a conflict in which each parent tries to form an alliance with the child against the other parent; and marital skew, a conflict in which one parent passively submits to the bizarre behavior of the other.

Still other theories have focused on communication patterns in the family. One theory, for example, maintains that schizophrenic behavior results from constant *double-bind communication* from one family member to another—typically from parent to child (Bateson et al., 1956). In the double bind, one parent conveys two conflicting messages to the child at the same time, but the child's dependency on the parent prohibits confronting the parent with the contradiction or escaping it.

> For example, a young man diagnosed as schizophrenic was meeting with his mother and his therapist to discuss moving away from his mother's home into an apartment of his own. The mother remarked, "I think it is a wonderful thing for a young man to grow up, move into his own place, and leave his mother all alone." In this single sentence, the mother managed to tell her son that it was both wonderful *and shameful* for him to move away from her. When her son replied that she was making him feel guilty about moving away, she looked deeply hurt and said, "How could you say such a thing about your own mother?" Thus there were conflicting messages plus a prohibition against examining the messages closely—a double bind.

In other cases, the verbal content of a communication may conflict with the physical or emotional content (Bateson et al., 1956).

> For example, a young man whose mother visited him in his mental hospital was delighted to see her. He put his arm around her, at which point she stiffened. In response, he withdrew his arm, but then she asked, "Don't you love me anymore?" Confused and embarrassed, he blushed; and his mother said, "Dear, you must not be so easily embarrassed and afraid of your feelings."

Figure 15.12

Catatonic schizophrenia is characterized by unusual motor behavior, often including awkward postures that are held for long periods of time. (Grunnitus/ Monkmeyer Press Photo Service.)

Recent evidence (Liem, 1980) does indicate that people diagnosed as schizophrenic often grow up in families where communication is disordered and where at least one parent is critical and overcontrolling. Does this mean that these family factors cause schizophrenia? Not necessarily. An alternative possibility is that some kinds of poor communication and strained parent-child relationships *result from* having a schizophrenic person in the family. In other words, theories that focus on life experience and communication patterns face a sort of chicken-and-egg problem: Which came first—the unusual life and family patterns or the schizophrenia? In most of the research, thus far, this question has been hard to answer.

Biological Factors and Schizophrenia While the study of life experience and family patterns continues, other investigators have turned their attention to the role of biological factors. Some of these factors may be hereditary, as we will discuss below; but others may result from injury or disease, particularly during pregnancy, birth, or early infancy. Some research on schizophrenic populations has shown a high incidence of pregnancy complications, low birth weight, and oxygen deprivation at birth, all of which put children at risk. The risks posed by such problems are varied, of course. It may well be that schizophrenia is just one of several possible outcomes of biological damage before and at birth; other possible outcomes include such disorders as mental retardation and cerebral palsy.

Much of the research on biological factors in schizophrenia has focused on the biochemistry of the disorder. One prominent idea to emerge from this research is the ***dopamine hypothesis***, the notion that schizophrenia may be the result of excess dopamine activity in certain areas of the brain. You may recall from Chapter 2 (page 50) that dopamine is one of the substances that help transmit information across synapses (gaps between nerve cells) in the central nervous system. In other words, dopamine is one of the neurotransmitters.

Several lines of evidence now suggest that excess dopamine activity may, indeed, play a role in schizophrenia. One clue is that the behavior seen in paranoid schizophrenia—confusion, delusions, and often hallucinations—is strikingly similar to the behavior demonstrated by people who have taken large doses of amphetamines, or "speed." These "amphetamine psychoses" are, in fact, sometimes misdiagnosed as paranoid schizophrenia. This may be important because the drug amphetamine activates cells that use dopamine as a transmitter. Furthermore, the drugs most effective in treating schizophrenia are also the best treatment for "amphetamine psychosis." One group of these drugs—the phenothiazines (see Chapter 16, page 679)—appears to work by blocking dopamine activation of cells in the brain. This suggests that excess dopamine activity may be a big part of the problem in schizophrenia.

The dopamine hypothesis (Snyder, 1981; Snyder et al., 1974) is still a loose fabric. Researchers are gradually weaving threads of evidence together, but no firm and final proof has yet been offered. Still, many see this as one of the most promising biochemical accounts of schizophrenia.

Another major strand of biological research on schizophrenia has dealt with structural impairment of the brain itself. Some of this research has relied on postmortem examinations of the brains of schizophrenics and nonschizophrenics. Brain banks in several research centers have helped make this possible. Recent technological advances have also made it possible to study living people, using successive x-rays of very thin sections of the brain. The studies of brain structure have shown some impairment in about 20 to 35 percent of people diagnosed as schizophrenics (Seidman, 1983).

There are many different kinds of impairment, but one of the most common is enlargement of the cerebral ventricles—openings that contain cerebrospinal fluid. Some investigators have also found groups of schizophrenic people with unusually *small* ventricles. One team of researchers reports that the cases with small ventricles are especially likely to experience hallucinations and delusions and to demonstrate bizarre behavior, whereas those with abnormally *large* ventricles were more likely to be emotionally flat and unambitious (Andreasen et al., 1982). As this illustrates, the research on brain structure is pointing us toward a very plausible idea: Different forms of schizophrenia may involve different biological abnormalities.

Genetic Influence and Schizophrenia While research and debate continue over the role of biological factors and life experience in schizophrenia, a consensus is forming on one major point: Genetic factors play a role. One group of studies that has convinced many people involves identical and fraternal twins. Identical-twin pairs have identical heredity; fraternal-twin pairs do not. So if schizophrenia is strongly influenced by genes, then identical-twin pairs should show greater similarity, or more *concordance,* than fraternal-twin pairs for the presence or absence of schizophrenia. Table 15.2 shows the results of concordance studies conducted in five different countries. In each study, the investigators started with twin pairs in which at least one member had been diagnosed as schizophrenic. They then computed the percentage of cases in which the other member of the pair also suffered from schizophrenia or a similar abnormality. As the table shows, the concordance rates were much higher for identical than for fraternal twins. Such data need to be interpreted cautiously (see Chapter 14, page 603), but they certainly suggest that heredity plays some role in schizophrenia. On the other hand, note that concordance rates never approach 100 percent, even for identical twins. This suggests that the disorder cannot be accounted for by genetic factors alone.

Another way to examine the contribution of heredity versus environment is to look at schizophrenics who have been brought up in adoptive homes from an early age and therefore not reared by their biological families. One investigator (Kety, 1975) interviewed the biological (true) relatives of schizophrenic patients who had been adopted as children and of nonschizophrenic persons who had also been adopted. The rates of both schizophrenia and suicide in the relatives of the patients were three to four times higher than in the relatives of nonpatients. There was no increase, however, in the incidence of other behavior disorders. Thus, except for a higher rate of suicide, the inheritance pattern seemed to be rather specific to schizophrenia.

Multifactor Theories Probably the best accepted position today acknowledges both a genetic-biological predisposition to schizophrenia *and* the influence of life experiences. Many people believe that inherited biological deficits make certain people susceptible to schizophrenia but that, in a normal life situation, the deficits produce only modest problems that may weakly resemble schizophrenia. On the other hand, where life stress is high or family problems are severe, individuals with the inborn biological susceptibility may be especially vulnerable and may therefore become schizophrenic. People who suffer severe life stress but do not have the genetic biological susceptibility may develop milder disorders, but not schizophrenia. To summarize, these theories hold that schizophrenia develops when people who have an inherited or other biological

Table 15.2

Concordance rates of schizophrenia in twin studies

STUDY	CONCORDANCE (% OF TWIN PAIRS DIAGNOSED AS SCHIZOPHRENIC) IDENTICAL	FRATERNAL
Inouye (1963), Japan	60	18
Kringlen (1967), Norway	25	4
Tienari (1971), Finland	16	5
M. Fischer (1973), Denmark	24	10
Gottesman and Shields (1972), England	42	9

(*Source:* Mehr, 1983.)

susceptibility to the disorder experience particular kinds of severe life stress. Among experts, this is one of the most widely accepted general views on schizophrenia.

PROGNOSIS IN SCHIZOPHRENIA

The prospect for improvement in a schizophrenic person—the individual's *prognosis*—depends on a number of factors. One is the person's adjustment before he or she became disturbed enough to be diagnosed or hospitalized. The prognosis is rather good if the onset of the schizophrenic behavior was rapid; the person suffered some pronounced shock, or trauma, just before the schizophrenic "break"; and the person had been moderately well adjusted until then. But if the disorder came on slowly, the symptoms gradually increased in severity, there was no precipitating trauma, and the patient's adjustment before being diagnosed as schizophrenic was marginal, the prognosis is poor. The prognosis may be better for the first group of schizophrenics because they have learned more of the social skills necessary for normal life, they are more tuned in to their social environment, and their level of thinking is more advanced (Sappington, 1975; Watson, 1973).

Paranoid Disorders

The following account illustrates the kind of plausible delusions of persecution often seen in our next category, the relatively rare disorder known as *paranoia* (Abstracted from Swanson et al., 1970, pp. 226–227).

> Arthur, a 37-year-old officer in the Navy, had worked effectively for 13 years in radar electronics before he was apprehended by the police. He had been stopped because someone noticed a gun—in his purse. In addition to carrying a purse, he was wearing a dress and was convincingly disguised as a woman. When the police discovered that he was actually a man, they identified him and contacted his commanding officer, who had Arthur hospitalized for evaluation. Interviewed by mental-health professionals, Arthur gave a calm, coherent, and logical explanation for his behavior. He reported that his troubles began years before when he got involved in some criminal activity and double-crossed a partner. The partner then called on some of his underworld friends, who began to harass Arthur and his family. Arthur's wife confirmed some of the details of his story. Even after Arthur arranged distant duty assignments, the harassment continued. Finally, in desperation, Arthur and his wife arranged for him to "disappear" by masquerading as a woman.
>
> The story was unusual but believable. Arthur seemed quite normal on the ward, and two prominent psychiatrists called in as consultants concluded that Arthur needed a lawyer, not psychiatric care. But Arthur's lawyer was troubled by nagging doubts about his client's story, and he arranged to have the details checked. A 4-month investigation failed to support the details. Arthur's wife, when interviewed at length, revealed that her support for his story was actually based on what he had told her, not on her own indepen-

dent experience—except for two incidents that seemed like possible harassment but had perhaps been coincidences. She also reported that Arthur had grown increasingly preoccupied and isolated in recent years, an observation also made by Arthur's commanding officers. Confronted with the findings of the investigation, Arthur remained outwardly calm but then, gradually, began to describe new evidence of harassment by the underworld figures during his stay in the hospital.

The above account illustrates that the delusions involved in paranoia are often so logical and believable that intelligent people, sometimes even mental-health experts, are tempted to "buy" them. Instead of, or in addition to, persecutory delusions, people with paranoia may also show delusional jealousy. (They may believe, unjustly, that their lovers or spouses have been unfaithful and collect "evidence" in the form of rumpled clothing, spots on the sheets, or cigarette butts on the porch.) Or they may show delusions of grandeur (for example, beliefs that they are actually Jesus or Einstein reincarnated and that others are jealous of their power and intelligence).

Except for their elaborate delusional systems, paranoid people typically show no thinking disorder. They appear normal (although perhaps a bit sensitive or suspicious) until something happens to activate their delusions. In general, their delusional systems are well worked out and they do not have hallucinations. Such intact thinking and "logical" delusions contrast sharply with the disordered thinking and hallucinations seen in cases of paranoid schizophrenia.

PARANOID THINKING VERSUS PARANOIA

The logical nature of paranoid ideas means that they are often uncomfortably close to our own patterns of thinking. This, in turn, reminds us of how difficult it is to distinguish between normalcy and psychological disorder. Perhaps it is best to think of paranoid thinking as a dimension involving different levels, or *degrees*. Most people occasionally have suspicious thoughts, but not to a degree that clashes sharply with social norms, undermines adaptation, or causes severe distress. Further along the paranoid-thinking dimension are people who are habitually, and wrongly, suspicious of others and overly sensitive to what they see as "hidden motives" or "secret meanings" in the behavior of other people. If such patterns pervade a person's daily living but do not include full-fledged delusions, DSM-III applies the term *paranoid personality disorder*. *Personality disorders* (page 659) are lifestyle patterns involving inflexible, persistent, and maladaptive personality traits—suspiciousness and jealousy, in this case.

It is only when paranoid thinking involves an extreme, well-organized delusional system, with emotions and behavior that are logically appropriate to the delusion, that a diagnosis of paranoia is given. The example of Arthur (above) illustrates this extreme level. Arthur's well-organized delusional system dominated his life, provoked him into changing his identity and posing as a woman, and caused both Arthur and his wife to live in fear of "the underworld."

ORIGINS OF PARANOIA

Causal explanations of paranoia have been debated for years, but research on causal factors has been hampered by the rarity of identified cases. One theory (Cameron, 1967) focuses on the fact that people who are diagnosed as having paranoia seem to show lifelong patterns of finding fault with everyone but themselves. Trusting no one, they withdraw socially and emotionally. When a real threat or stressful situation arises, they are isolated just when they need most to confide in friends who might help them see their problems more objectively. With no one to change their minor false beliefs, they gradually reconstruct their perceptions of the world to fit their own views. Ordinary frustrations (such as a restaurant refusing service because it is about to close) are taken personally. People showing this disorder perceive the rest of the world as paying special attention to them and being out to "get" them. This false belief explains many situations for them; the explanation may be comforting and reinforcing because it makes sense out of an otherwise confusing world. This theory might well explain even an innocuous example of paranoid thinking such as the following (Abstracted from Busse & Pfeiffer, 1977, p. 187).

> An 80-year-old widow who seemed alert and aware became upset because, for unknown reasons, her neighbors were secretly putting lint in her washer and dryer. She took her complaint to the police. No one, her daughter included, could convince her that her neighbors were innocent. In most respects, she continued to get along quite well and behave normally. On wash days, however, she grew suspicious and her accusations began again.

Another example of paranoid thinking is given by Carson (1977; cited in Altrocchi, 1980).

> An elderly woman in a midwestern town complained about the boys in a nearby college dormitory. They were, she said, using electric waves to give her multiple orgasms; and the orgasms were disrupting her sleep. Her psychiatrist prescribed some "antiorgasm" tablets (in reality an antipsychotic drug), and the problem disappeared.

Examples like these can also be used to support other theoretical explanations of paranoid thinking. The fact that the delusional behavior characteristic of paranoia is most common in the elderly, and particularly in conjunction with physical deterioration, has led some to believe that biological factors can play a role. People who hypothesize biological causes also note that delusions occur frequently in people taking drugs such as amphetamines, as we pointed out earlier.

IS PARANOIA REALLY A PSYCHOLOGICAL DISORDER?

Some researchers (for example, Colby, 1981) view delusional thinking, not as a sign of psychological disorder but as a style of information processing—a style developed by people especially sensitive to blame and shame. The two brief examples given above might well be seen as efforts to project blame—for lint and orgasms—onto others. A similar argument could be made in more extreme cases, like that of Arthur (above). Does this mean that paranoia should be viewed only as an extreme self-protective style of thinking, not as a "true" psychological disorder? This idea has growing support, but not, of course, among the authors of DSM-III.

Disorders of Affect

Just as many of us show occasional traces of paranoid thinking, many of us also show occasional changes in *affect,* or mood. At times, we can be unusually sad, or down; at other times, we may be especially elated, or up. In some people, at some times, these mood states spiral out of control, becoming so extreme in degree or duration that they pose major problems. When such extreme moods are combined with particular patterns of maladaptive thinking and behavior, they are diagnosed as one of the *affective disorders*. We discuss two such disorders here, beginning with major depression.

MAJOR DEPRESSION

To be diagnosed as suffering from major depression, a person must have had one or more *major depressive episodes*—periods that involved more than just "sadness."

Characteristics of a Major Depressive Episode In a major depressive episode, "sad" feelings are accompanied by persistent problems in other areas of life—problems such as an appetite change (an increase or a decrease); altered sleep patterns (more or less sleep than usual); loss of interest or pleasure in usual activities, including sex; loss of energy; diminished ability to think or concentrate; feelings of worthlessness or self-reproach; or suicidal thoughts or acts. During a depressive episode, the person's mood and thought patterns may be strikingly negative. The depressed person often appears lost, vulnerable, detached, unable to find joy in any aspect of daily life; life seems to have lost its vitality, and a sort of numbness has set in. Often the person seems constantly on the verge of tears. Thoughts about self are likely to be overwhelmingly negative; the person may feel unattractive, inferior to others, incompetent as a spouse or parent, and unsuccessful at school or work. The future may seem almost completely hopeless, and this, of course, is one reason that suicide becomes a risk.

An estimated 20 percent of American and European women have experienced a major depressive episode at some time; for men the figure is about 10 percent. Some 6 percent of women and 3 percent of men have had episodes serious enough to require hospitalization. An estimated 46 percent of college students are thought to have experienced depressions serious enough, at some point, to have warranted professional help (Beck & Young, 1978). Many depressed people are treated as outpatients, but many others suffer in silence and solitude (see Figure 15.13). One example of the many forms depressive episodes can take is the case of Mr. A. (Abstracted from Litman, 1970, pp. 294–295).

> Mr. A., a middle-aged realtor, strongly valued financial success. It helped him deal with feelings of inadequacy stemming from his lack of a college education. During one short period, he lost out on three

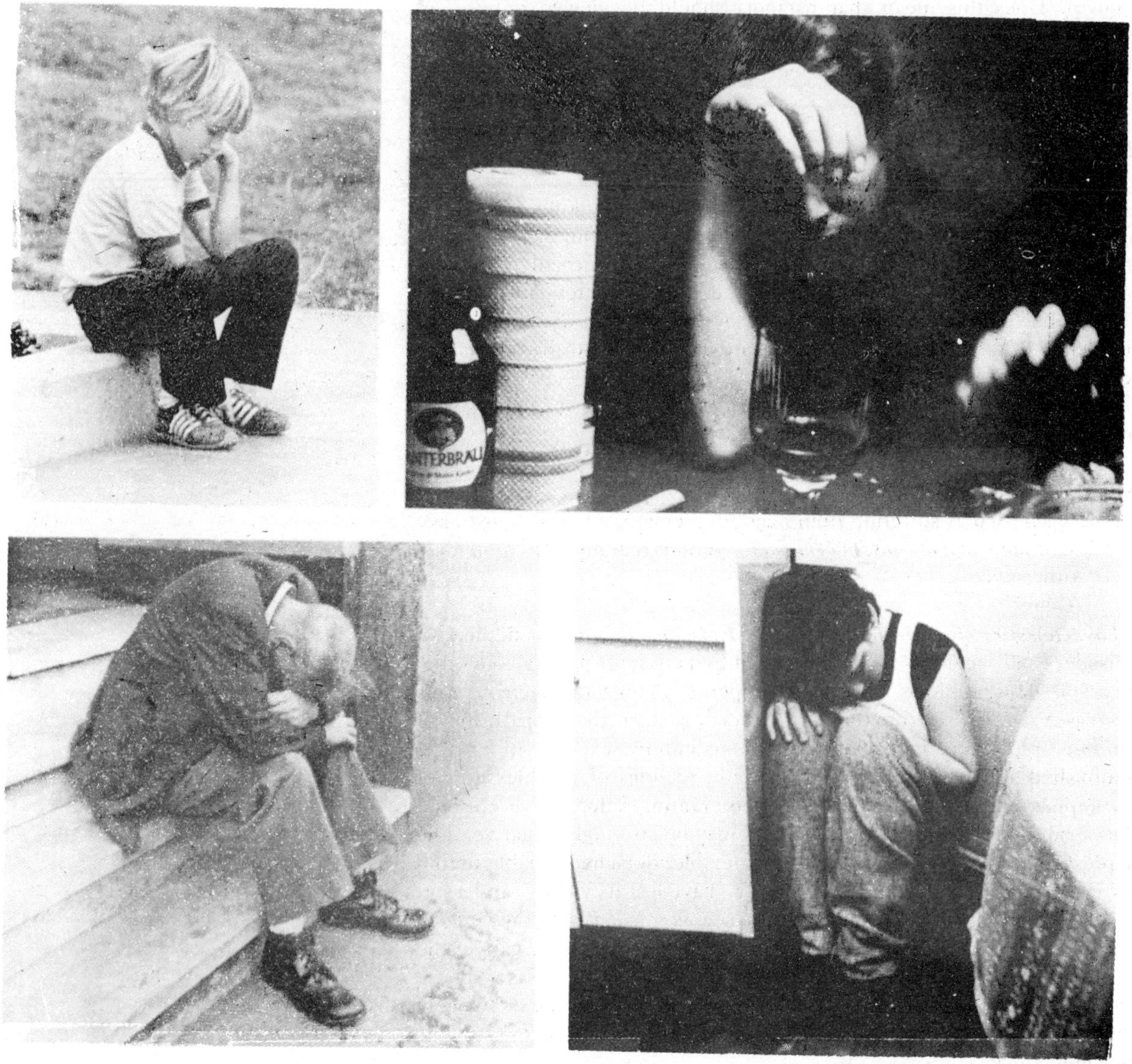

Figure 15.13

Depression is an extremely common problem, and it can be an extremely painful one, regardless of the individual's age or life circumstances. (Clockwise, from upper left, Mimi Forsyth/Monkmeyer Press Photo Service; © Gilles Peress/Magnum Photos; © J. Albertson/Stock, Boston; © Frances Cox/Stock, Boston.)

important business deals that he had counted on; he grew depressed, morose, and despondent and began to drink more heavily than before. He began talking vaguely about changing careers, and he seemed to have trouble making decisions. His children had left home, he and his wife had grown apart psychologically, and the two had had almost no sex for several years. Mrs. A. grew concerned about her husband's sleeplessness, loss of appetite, and generally hopeless attitude. One day, she consulted a physician, who said that Mr. A. should have more activity. On the same day, Mr. A. spent several hours with a clergyman, discussing his feelings of anxiety and failure. The next morning at around 5 A.M., Mr. A. awoke and went for a walk. He then unwrapped and loaded a rifle he had recently bought and left in his car. Mr. A. then returned home for breakfast. There, at Mrs. A.'s request, they both prayed. Afterward, Mr. A. went for another walk, returned to his car, and killed himself with the rifle. After the suicide, Mrs. A. found several notes, torn into small pieces and thrown in the wastebasket. When pieced together, one of the notes read, in part, "Honey, I am unable to take this any longer. . . . Sorry that life turned out this way. I hope you can find a better life without me."

Certainly not all major depressive episodes end in suicide, but some do. Even in cases where the end is not this tragic, the pain and suffering endured by the depressed person can be intense. The intensely painful nature of this disorder, and its high incidence, have spurred many to study depression and its causes.

Depressions like Mr. A.'s—occurring in middle age and after a series of setbacks—may have different causes than depressions that arise earlier in life and with no clear precipitating events. A depression that is severe enough to result in suicide may have different causes than a milder case. Bearing in mind that different kinds of depression may have different causes, let us examine some of the major causal theories and some of the evidence for them.

Life Experience, Learning, and Social Interaction in Depression In psychoanalytic theory (Chapter 14, page 576), depression is seen as resulting from an overly demanding superego (page 578)—one that sets standards too high for the person to live up to—and from early loss of attachment figures (Bowlby, 1980; Freud, 1917/1957). For example, the adult on whom a child is most dependent, usually the mother, may be lost through death or parental separation. Or there may be a serious disruption in the dependency relationship, such that the child comes to fear or fantasize such a loss. Losses of this sort can set in motion a prolonged and exaggerated grief reaction—one that includes feelings of guilt and self-blame and a loss of self-esteem. This, in turn, can set the stage for the development of depression when the child enters adulthood. There is some evidence that the death of a mother during childhood increases the risk of depression in adulthood, at least among women (Brown et al., 1977). However most research has not found strong support for this idea (Crook & Eliot, 1980).

A better-validated theory attributes depression to processes of reinforcement and social interaction. Peter Lewinsohn and his research team found that depressed people often have had especially high frequencies of unpleasant, unrewarding events in their lives; depressed people also *experience* these adverse events as more unpleasant than do nondepressed people (Lewinsohn & Talkington, 1979). Because of the low level of reinforcement they have received, depressed people tend to invest less hope and less energy in their activities, including their social interactions. This, in turn, makes them less rewarding to be around. Thus others tend to avoid depressed people, and this, in turn, deepens their depression (Lewinsohn & Arconad, 1981). Coyne (1976) illustrated the negative impact depressed people can have on others.

Coyne had 45 nondepressed people carry on phone conversations with people receiving treatment for depression, people receiving

> treatment for other disorders, or nondisturbed people not receiving treatment. After the conversations, ratings indicated that talking to the depressed people had been considerably less rewarding than talking to the other groups. After talking to depressed people, nondepressed experimental subjects were more anxious, more hostile, more depressed, and less interested in talking to the same people again than were subjects who had spoken to the other groups.

Coyne (1982) found that depressed people try to cope with stressful situations by seeking emotional support from others, a dependent style that may contribute to the discomfort others feel around them. When others *do* offer their support or aid and the depressed person does not improve, irritation may be the result. Others may then combine their reassurance and statements of support with increasing avoidance of the depressed person, thus leaving the person confused and—of course—still depressed. This body of theory and research has been useful by helping to stimulate new ideas on how to help depressed people recover. For example, depressed people might be trained to interact with others in ways that others will find rewarding rather than punishing; and others who want to help depressed people shake their depression might learn how to communicate in ways that would not be confusing or threatening (Coates & Wortman, 1980).

Cognitive Processes and Depression One prominent model holds that depression involves a kind of "giving up," or *learned helplessness* (Abramson et al., 1978). The idea is that people sometimes experience a series of setbacks or failures, conclude from the cues available that the situation is one they cannot control, and simply stop trying to make things better. According to this model, depression can result when people expect bad things to happen to them and assume that they will not be able to prevent or control them. If people attribute this lack of control to personal causes (such as incompetence), their self-esteem will be impaired. If they believe the causes are stable, then their depression will be long-lasting. If they believe their lack of control extends to many situations, then their depression will be generalized across situations. In other words, depression is colored by cognitions. Some studies support this model, showing that beliefs like those outlined above—"The causes of my failure are personal, stable, and present in many situations"—are more common in depressed than in nondepressed psychiatric clients (Raps et al., 1982); also, people who show these cognitive patterns but are *not* depressed tend to *get* depressed later, when bad things do happen to them (Peterson & Seligman, 1984).

Aaron Beck (1974, 1976) is another theorist who emphasizes the cognitive, or thought, aspects of depression. He sees depression primarily as a thought disorder and only secondarily as a mood disorder. According to Beck, depressed persons are dominated by negative views of self, the outside world, and the future. They see themselves as losers, and all their perceptions are colored by this major premise. Beck (1974) further proposes that depressed people experience major distortions of logical thought. These distortions include:

1. *Arbitrary inference,* drawing a conclusion based on too little evidence or no evidence at all. (A housewife concludes that her husband does not love her because he leaves for work every morning, for example.)

2. *Selective abstraction,* drawing a conclusion by concentrating on one detailed aspect of a situation. (After receiving compliments on the entrée, the salad, the rolls, and the dessert, the cook is sure the soup must have been terrible.)

3. *Overgeneralization,* unjustified generalizing from limited evidence. (A student who receives a low grade on a single assignment is sure that he or she is failing the course and will never graduate or get a job.)

4. *Magnification and minimization,* exaggerating or limiting the significance of information. [A dropped stitch in a sweater makes the knitter want to throw it away (magnification); an employee continues to feel incompetent even after being given a raise (minimization).]

In contrast to Beck's view, some recent research has suggested that depressed people may actually make *more* accurate assessments of themselves and certain situations than nondepressed people. For example, in one study (Lewinsohn et al., 1980), observers rated the social competence of several depressed and nondepressed people, all of whom also rated themselves. The self-ratings of the nondepressed people were considerably higher than were the ratings given them by impartial observers, but the depressed people gave themselves ratings quite close to those the observers gave them. Perhaps even more intriguing was the finding that as depressed people grew less depressed during treatment, their self-ratings became more inflated compared to those made by observers. In other words, reduced depression may go hand-in-hand with exaggerated self-perception. Perhaps one way many of us stave off depression is by being a bit unrealistic about ourselves—adopting a "warm glow" (Lewinsohn et al., 1980) that keeps us focused on our strengths and successes rather than our weaknesses and failures. Depression may be, in part, a breakdown of this process.

We should add one cautionary note about these cognitive models of depression. A careful review of the research on cognitive processes (Coyne & Gotlib, 1983) showed that the major models each have some scientific support but none has strong support. Evidently, different depressed people in different studies do somewhat different things, cognitively speaking. There may be some truth in each model; but there may also be several ways of thinking that can go along with being depressed.

Biological/Genetic Factors and Depression Whatever cognitive-learning or life-experience factors may be involved in depression, many investigators believe that biological factors also play a role. Some physiological deficit, either inherited or acquired in other ways, is thought to make some people especially vulnerable to depressive episodes. We have known for some time that hereditary factors play a role in some depression

(Liston & Jarvik, 1976). As with the schizophrenic disorders (page 642), concordance rates for depression are higher for identical twins than for other twins not genetically identical. But what specific biological mechanism, genetic or otherwise, might underlie depression?

As with schizophrenia, considerable interest has focused on the role of neurotransmitters—substances involved in the transmission of information across synapses in the nervous system. One of the neurotransmitters most often studied in regard to depression is norepinephrine (Chapter 2, page 50). For information to be transmitted from one nerve cell, A, to another, B, the norepinephrine produced in A must be discharged into the synapse between A and B, thus stimulating B to fire. Sometimes the optimum amount of norepinephrine may be unavailable for this purpose because some of it has been absorbed back into cell A or has broken down and exited from the synapse. Another idea is that there may be fewer receptors (Chapter 2, page 50) for norepinephrine on the postsynaptic (the B) cells. When one of these effects happens on a broad scale in particular sites of the brain, activity level and motivation drop, and depression can result—this, at least, is one biological theory. The theory is supported by the fact that some drugs effective in treating depression keep norepinephrine available for information transmission by preventing either its absorption or its breakdown. The essential idea is that depression is caused, not just by low levels of norepinephrine in the system but, more specifically, by low levels of norepinephrine (or other similar substances) *available* for information transmission (Sulser, 1979). The evidence favoring this idea is not conclusive, but it is encouraging enough to spark continued research.

The Multifactor Concept As is true of schizophrenia (page 643), few researchers expect to find a single "magic bullet" that will completely explain all cases of depression. Instead, investigators are piecing together an account of how multiple factors interact to produce depressive episodes. Some of the factors will likely be biological in nature, and some of the biological factors may be partly hereditary. Other factors such as life experience, social interaction, and cognitive processes seem likely to play a role when added to basic biological vulnerabilities. Finally, these pieces of the puzzle may well be put together differently for different groups of depressed people. Depression is not easy to understand, but we are making progress.

BIPOLAR DISORDERS

For reasons we do not yet understand, some people fall prey not just to depressive episodes but to sweeping mood swings that involve both depression and elation. Their elated mood states are called *manic episodes*, and the people experiencing the mood swings were once diagnosed as having manic-depressive disorders. The current term, *bipolar disorders*, reflects the tendency of these people to oscillate between two mood states that seem, in many ways, to be polar opposites.

Manic Episodes in Bipolar Disorders The manic episodes these people experience are periods of persistent elevated, expansive, or irritable moods. The periods often last a week or more and may involve a variety of extreme behaviors, each reflecting excitement and high energy. People in a manic state may sing, dance, run, or talk furiously, as if pressured to perform. They may show sharp bursts of activity—social, professional, even sexual—which may not seem pathological at first but escalate to a point that shows poor judgment and irrationality. Grandiose feelings lead them to grandiose behavior. The person may place urgent calls to the President or to a famous singer or buy a new luxury car with no thought as to how payment will be made. In fact, thoughts are often racing so fast that there is not enough time to examine or question them. Sometimes there is not even enough time to tell the thoughts to another person. This type of rushed, frenetic, manic thinking is called *flight of ideas*.

Sometimes this upbeat exuberance gives way to frustration and anger—especially when someone refuses to go along with one of the irrational plans. In such cases, the manic person may become furious or burst into tears. In fact, tearful depression often seems to hover very close to the surface in the manic person's manic mood. Some of these features of manic thought have an "all-or-nothing" quality; things are either wonderful or terrible, each person is either a close friend or a mortal enemy, and so forth. In some, but not all, manic episodes, the person's thinking and behavior get revved up to a point that is frightening and dangerous to others; language gets vile and threatening, furniture gets smashed, and the manic person may drive a car recklessly or intentionally attack others. Manic episodes can take a large variety of forms, but perhaps one specific example will be helpful (Abstracted from Pfeiffer, 1968, pp. 29–30).

> David, a middle-aged businessman, became dejected after his best friend died of a heart attack, but within a few weeks the dejection had given way to increased activity. David grew unusually talkative, made many phone calls, and talked excitedly about lucrative business deals and major trips to Hong Kong and around the world. He wrote letters to numerous travel agencies, resorts, and banks, as well as to the governor and the President, marking the letters "Important" in large writing on the envelopes. He began to spend money lavishly, on himself and even on casual acquaintances; one week, he bought a piece of lakefront property and a new sports car, even though he could not afford either. Formerly reserved and socially inhibited, he developed a loud, unstoppable conversational style, shifting quickly from one topic to another and never seeming to reach a conclusion. He now drank and smoked more than ever before, slept little, often made jokes and puns, and dismissed all concerns about his condition with the statement, "I feel like a million bucks."
>
> When friends and family finally realized that David needed help, they suggested a hospital. At first David grew angry; but then he brightened, said this was a good idea, and invited everyone in the family to join him. When the psychiatrist arrived to interview David, David greeted him warmly and offered several ideas on how to improve the hospital. Although he occasionally grew angry with the psychiatrist, David was generally upbeat and animated, often speaking with alliteration and making rhymes. Noticing the psychiatrist's fountain pen, David commented, "That's a nice pen you got, Doc. We should become pen pals. I was in the pen once for

drinking." He then burst into song with "Drink to Me Only with Thine Eyes" and asked the psychiatrist to join in. When asked about the death of his friend, David seemed on the verge of depression but then countered with, "I've got a hundred friends. I've got a thousand friends. I've got more friends than anybody in the whole world."

Manic episodes like David's are rare. When they do happen, they are almost always part of a *bipolar disorder*; that is, the people who experience manic episodes almost always have at least one depressive episode sooner or later. On the other hand, only about 5 to 10 percent of those who experience a depressive episode will also have a manic episode at some time. In the United States, between 0.4 and 1.2 percent of the population will experience both kinds of episode and thus a bipolar disorder (Weissman & Myers, 1978).

Bipolar disorders usually come on suddenly and begin with a manic episode. Both the manic and the depressive episodes may last from a few days to a few months. Manic-depressive cycles in very mild forms may actually be associated with high levels of achievement—especially during the bursts of energy and enthusiasm that come with the manic episodes. For example, Theodore Roosevelt and Winston Churchill are both thought to have had moderate manic-depressive cycles. (See Figure 15.14; see also Rosenhan & Seligman, 1984.) However, diagnosis of "true" bipolar disorders is generally reserved for severe cases, most of which are psychologically painful and severely handicapping. In fact, as many as 15 percent of those with bipolar disorders may commit suicide, and many others will attempt it.

Causes of Bipolar Disorders The rarity of bipolar disorders has hampered our efforts to find the causes. However investigators are gradually building up a fund of knowledge and following up on clues. One poten-

Figure 15.14

Theodore Roosevelt is said to have experienced mild manic and depressive episodes. (Left, The Granger Collection; right, The Bettmann Archive.)

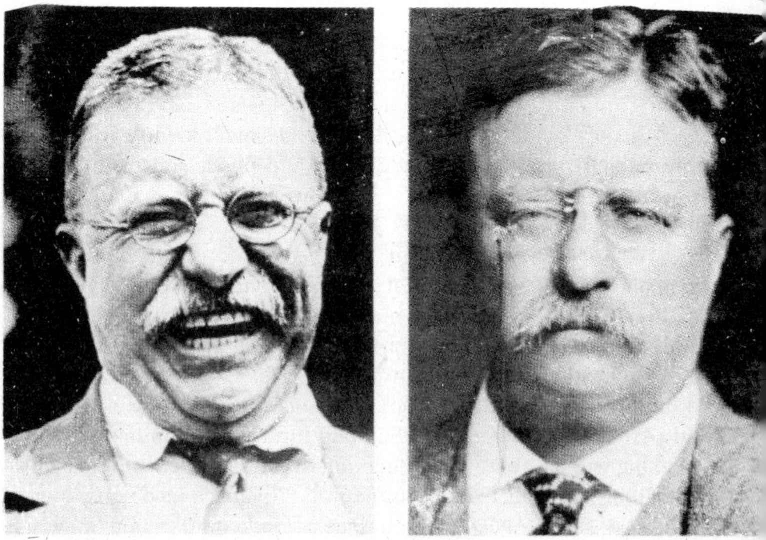

tially important clue is the fact that depression seems so close to the surface during manic episodes. A person whose manic plans have been stifled may easily disintegrate into tears and conclude that all is hopeless, only to recoup and become fully manic again in a few minutes. From this sort of pattern, many clinicians have concluded that manic episodes are a sort of extreme, though fragile, defense against the *true* problem, which is depression.

Another idea is that bipolar disorders may be rooted in the brain's biochemistry, just as some cases of depression seem to be (page 652). Unlike depression, though, which seems related to low levels of available norepinephrine (and other similar neurotransmitters) or to fewer postsynaptic receptors (page 652), manic episodes may be related to unusually *high* levels of these substances or to a proliferation of receptors. In other words, manic episodes may be related to a state of brain chemistry that could be called supercharged, or overly ready to transmit between nerve cells. Building on this idea, some have suggested that the cyclic, oscillating nature of the bipolar disorders relates to homeostasis (Chapter 7, page 272). When the brain becomes overcharged with available neurotransmitters or receptors and a manic state ensues, the body may take corrective action by drastically lowering the level of available neurotransmittersor the number of receptors. If, in this balancing process, the body overcorrects, depression may result. In other words, cycles of mania and depression might result when the basic balancing processes of homeostasis get out of hand (Bunney et al., 1972).

Whatever biological processes may be involved, genetic transmission may help cause them. Evidence thus far suggests that genetic factors are more important in bipolar disorders than they are in depression (Bertelsen et al., 1977). George Winokur (1980), who studied bipolar families, suggested that the culprit may be a gene on the X chromosome. He found that families with more than one bipolar member are also likely to have members with color blindness and members in the blood group Xg—both traits known to be carried on the X chromosome. There is evidence (for example, Hays, 1976) that this explanation may not apply to all bipolar cases but it may account for many.

Anxiety Disorders

Anxiety—an uneasy, fearful feeling—is the hallmark of many psychological disorders. It is often concealed and reduced by defensive behaviors such as avoidance or ritualistic action (hand washing, for example). In many anxiety disorders, however, intense observable anxiety or fears are the principal signs. A major United States survey by the National Institute of Mental Health (Regier et al., 1984) showed that anxiety disorders are more common in the general population than are any other disorders, including depression. To understand anxiety disorders, let us examine a few of the forms they most often take.

PHOBIC DISORDERS

An intense, persistent, irrational fear of something specific is a ***phobia***. There are many kinds of phobias: fear of enclosed places, high places, the dark, insects, animals, blood, and so on. (See Figure 15.15.) The person

Figure 15.15

Heights can provoke fear in most of us, but a person with a height phobia might react to being in these situations with sheer terror. (Left, Jeffrey Grosscup; right, Michael Weisbrot and Family.)

tries to control anxiety by avoiding the phobic object, activity, or situation.

Some otherwise normal, well-adjusted people have phobias. Their fears may be mild and infrequent; if so, they cause little difficulty. On the other hand, a phobia can be so powerful and irrational that it alters the person's life in a major way. For instance, a business executive with a strong fear of flying may have to change careers if he or she cannot overcome the fear. And people with severe phobias about being in crowds, high places, or closed-in places may become complete recluses in their efforts to avoid such situations.

GENERALIZED ANXIETY DISORDER

In *generalized anxiety disorder*, distress and uneasiness are persistent, painful, and spread across multiple situations. Such anxiety can make people thoroughly miserable (Figure 15.16) and even upset their health. These people often think they have a serious medical disorder because the "symptoms" may include trembling, fatigue, breathlessness, insomnia, sweating, nervousness, chest pain, dizziness, faintness, headache, and so on. A sense of foreboding, apprehension, and a feeling of impending doom may also be mixed with the physical symptoms.

PANIC DISORDER

Unlike generalized anxiety disorder, *panic disorder* involves specific, focused, time-bound attacks of intense fear, even terror. The panic attacks, lasting from a few minutes up to an hour or more, may include intense versions of the generalized anxiety characteristics described above but may also include such severe physical symptoms as choking or smothering sensations. The clinical picture is illustrated by the case of Mr. B. (Abstracted from Leon, 1977, pp. 113–118).

> Mr. B., aged 38, sought psychotherapy because of repeated, intense anxiety attacks in which he felt chest pains, heart palpitations, numbness, and shortness of breath—all making it seem as if he

Figure 15.16

In *The Shriek*, Edvard Munch portrayed the desperation of a person who experiences generalized anxiety. (The Bettmann Archive.)

were having a heart attack. On two separate occasions, his wife had had to rush him to the hospital because he was certain that his heart was going to stop beating. In childhood, Mr. B. had had infections that made bladder control difficult and urination very painful. Even though he had had no bladder or urination problems since he was 11, he continued to live in dread of wetting his pants. He panicked whenever he found himself in a situation with no bathroom nearby. A recent promotion had made it necessary for him to spend long hours meeting with other people, often in unfamiliar buildings. In such situations, if he was not successful in locating a men's room nearby before beginning a conference, he felt an overwhelming need to urinate and was gripped again by the fear of wetting his pants and humiliating himself. He had recently begun to feel "trapped" when involved in conversation, fearing that he might not be able to break away in time to make it to the rest room; he also developed a fear of being trapped in his car in heavy traffic or stalling in a tunnel. He had recently given up bicycling with his children on weekends because the trails took him frighteningly far away from a rest room or medical facility. After sweating through some of his more severe anxiety attacks, Mr. B. could not eat or go to work; he had recently taken a sick leave from his job because he had begun to panic in interpersonal situations.

OBSESSIVE-COMPULSIVE DISORDER

Sometimes anxiety and tension are associated with *obsessions*—persistent unwanted thoughts, impulses, or ideas—or *compulsions*—seemingly irrational behaviors repeatedly carried out in a fixed, repetitive way. People with *obsessive-compulsive disorders* find their obsessions or compulsions distressing and debilitating but feel unable to stop them.

Obsessions and compulsions tend to go together. In fact, compulsive actions may be directly related to obsessive thoughts. A person obsessed with the idea that he is dirty or guilty may wash his hands compulsively every few minutes or take six or eight showers a day. Another person obsessed with anxiety-provoking thoughts may try to blot them out by concentrating on counting all her steps. Other people are compulsive in a more general way: They strive for orderliness in thought, dress, and work. Indeed, any extreme emphasis on doing things in a particular way may be regarded as compulsive while not necessarily a sign of an obsessive-compulsive disorder (see Figure 15.17). Obsessive-compulsive people may be almost completely incapacitated by the rituals they have to perform. At the same time, they are likely to seem very inhibited and unemotional—cold and detached in their views of themselves and in their relations with others. They keep their lives under tight control.

Most of us have experienced mild and temporary forms of these problems. Just before an important test, for example, students may report hearing some popular song—or worse, some jarring radio or TV commercial—running through their minds over and over. This experience is so common that it is not ordinarily labeled "obsessive," but if it persists, it can warrant such a label. Similarly, compulsions vary greatly in their severity. The following case illustrates a compulsion that is so time-consuming and disruptive it is clearly abnormal. It also illustrates an obsessive-compulsive form of a problem superficially similar to that of Mr. B., above (Adapted from Wolpe, 1973, p. 265).

> Mr. C., an 18-year-old, had a severe washing compulsion. He feared contamination by urine, especially his own urine, and he feared he might contaminate others with it. After urinating, he sometimes

Figure 15.17

Some compulsive behavior, such as compulsive collecting, can be extreme but relatively harmless. (AP/Wide World Photos.)

spent as much as 45 minutes elaborately cleaning his genitalia and as much as 2 hours washing his hands. He showered every morning, sometimes for as long as 4 hours. In addition to these "basic requirements," he spent a great deal of time cleaning up after "incidental contaminations." Eventually, Mr. C. concluded that getting up in the morning was not worth the effort and so he began to spend most of his time in bed.

WHAT CAUSES ANXIETY DISORDERS?

As you might expect, interpretations of and explanations for anxiety disorders differ according to one's theoretical standpoint. In most psychodynamic theories (Chapter 14, page 575), anxiety disorders are attributed to inner conflict and unconscious impulses. For example, in Chapter 14 we discussed the case of Mrs. A. (page 593), who was extremely anxious about leaving her house and felt compelled to count her heartbeats because of sexual impulses she found unacceptable. Anxiety like that felt by Mrs. A. is viewed as a kind of psychic overflow—too much to live with and too much to deflect with the use of defense mechanisms (page 588). Such an overflow may be expressed in the form of one of the anxiety disorders. This view is difficult to document in research, but it continues to guide the work of many therapists who treat victims of anxiety disorders.

Other therapists and theorists—those with a behavioral perspective (Chapter 14, page 592)—view anxiety disorders as resulting from unfortunate learning and conditioning experiences. Phobic disorders, for example, they feel may be acquired through the association of painful or unpleasant events with particular situations. The life histories of people with severe phobias very often include especially frightening, or traumatic, events—events that may have triggered specific phobias by a process of classical conditioning (Chapter 4, page 146). Sometimes these events, or similar ones, recur periodically, as with the woman in Application 4.1 (page 148), who feared hypodermic syringes. Once learned, phobias may be reinforced by the relief the phobic person feels when the feared object is avoided; thus instrumental conditioning (Chapter 4, page 147) may also be involved. Finally, phobias may be acquired through observational learning (Chapter 14, page 596); a child who sees his father grow deathly afraid at the sight of a spider may come to share his or her father's fear. No matter how a particular anxiety problem develops, its origin may well be forgotten and therefore the anxiety seems completely irrational.

No specific biological cause has been identified for the anxiety disorders, but many suspect that biology plays a role. We know that tranquilizing drugs (Chapter 16, page 679) help to alleviate anxiety. Investigators are now exploring *how* these drugs help and at which sites in the brain they have their effects. This may help us to understand the body's "anxiety system." Another clue pointing to biological factors is the growing evidence that heredity plays at least some role in the anxiety disorders (Carey & Gottesman, 1981).

Personality Disorders

Many people whose behavior is abnormal are classified as having *personality disorders*. This is a diverse group of disorders in which the common thread is a lifelong pattern of unusual behavior ("just the way that person is"). These people do not show the bizarre symptoms of the schizophrenic, manic, or deeply depressed individual. The way they cope with life's problems and relate to others may be considered troublesome, strange, or perhaps tiresome to others; but the people themselves may not feel a great deal of anxiety or distress except when they encounter special crises. Many kinds of deviant behavior are included in the broad, general category of personality disorder. Note that these disorders are not considered clinical syndromes, like the disorders we have discussed thus far; personality disorders make up a separate category (see Table 15.1, page 627).

Many people with personality disorders show lifelong adjustments that resemble, in mild form, the psychological disorders we have already described. For example, as we noted earlier, people who have a *paranoid personality disorder* exhibit long-standing suspiciousness and irritability but without evidence of major delusions. People with a *schizotypal personality disorder* are generally very shy, socially awkward, and unable to form close relationships. They also tend to have rather odd thoughts and behavior. The individual with a *compulsive personality disorder* tends to be rigid, conventional, perfectionist, stingy, stubborn, emotionally inhibited, and inclined to become preoccupied or obsessed with trivial pursuits. We will take a closer look at two other personality disorders—histrionic and antisocial.

HISTRIONIC PERSONALITY DISORDER

An individual with a *histrionic personality disorder* (more often a woman than a man) exhibits immature, self-centered, seductive, attention-getting behavior (Figure 15.18). People with this disorder are likely to be manipulative; they get others to do their bidding by indirect tactics. A "southern-belle" style and a knack for making others feel guilty for one's awful sufferings are often characteristic of the histrionic personality. Such people are likely to be flamboyant, to exaggerate how "truly wonderful" good things are and how "deeply tragic" bad things are. Yet they tend to deny that anything in their lives is especially troubling, insisting, like the fictional Pollyanna, that "everything will turn out just fine." They also deny the sexual connotations of their seductive behavior. Indeed, such behavior does often seem designed to attract attention more than to initiate any real romantic or sexual relationship. The histrionic pattern may also include imagined ailments or illnesses that actually have psychological origins. Such ailments were often treated by Freud (see Chapter 14, page 576); he saw them as a part of "hysterical neurosis." In a historical carry over from Freud's time, histrionic people are sometimes referred to as hysterical personalities.

Figure 15.18

Histrionic personality disorder involves excessive flamboyance, seductiveness, and emotional display. Scarlett O'Hara in *Gone with the Wind* illustrates these traits to some degree. (Museum of Modern Art.)

ANTISOCIAL PERSONALITY DISORDER

The normal-abnormal distinction is particularly fuzzy when applied to the category of *antisocial personality disorder*, sometimes also labeled *psychopathic* or *sociopathic personality*. Such persons do not play by the usual rules of society. They behave as if a special set of rules, or no rules at all, should apply to them. Some are swindlers, embezzlers, and bad-check passers. Others are drifters, never able to hold a job for very long. They show great skill in short-term interactions, with an uncanny knack for saying just the things other people want to hear. They appear charming, confident, mature, and sincere. However, their behavior is inconsistent with their words. They do not follow through on promises or obligations; they are perfectly willing to deceive and defraud other people. People with antisocial personality disorders feel no close bonds with others but are often remarkably adept at convincing other people to help and trust them (Figure 15.19). Consider the following example:

> Staff members in a prison routinely assigned new psychology interns to interview Stan. Within a week, the interns would fervently report back that Stan had been unfairly convicted and imprisoned. They would maintain that Stan was, in fact, a retired colonel and grieving widower whose revolver, a military souvenir acquired during World War II, had been wrongly implicated in a murder. They were also sure that no one before them had ever really understood this troubled prisoner, who had finally unburdened himself only because the intern had listened with unusual sympathy. The facts were of course, that Stan was no colonel but someone with a long history of delinquencies and forgeries. He was serving a long sentence for having murdered his wife in order to collect her insurance.

People like Stan can be caught and punished time and again but will continue to commit the same crimes. They never seem to learn, probably in large part because they fail to experience anxiety or guilt about their behavior as others would.

The reasons for the development of antisocial personality patterns are elusive. Some psychologists suspect a biological defect that makes ordinary rewards and punishments ineffective in these children's upbringing. Others tend to blame two kinds of parents, or parental models. The first is the parent who is cold and distant with the child. The child learns to imitate this pattern, becoming, in turn, cold and distant in relationships

Figure 15.19

Oil baron J. R. Ewing of the television series *Dallas* and Alexis Carrington of the *Dynasty* series show several of the traits associated with antisocial personality disorder. They can appear charming and sincere, but they have no qualms about manipulating or deceiving others. (Left, Frank Edwards/ Pictorial Parade, Inc.; right, Movie Still Archives.)

with others. Because the child is treated as an object to be manipulated, he or she may learn to deal with other people in the same way. The second kind of parent applies rewards and punishments inconsistently. Because the child is initially punished frequently, he or she learns the tricks of escaping and avoiding punishment. But because rewards are sparse and inconsistent, the child never learns an appropriate social role. Some families seem to foster antisocial personalities in all their children; in other families, the "black sheep" seems to be qualitatively different from the other children. As with several of the other problems we have discussed in this chapter, it seems likely that several different causes play a role and that the relative importance of each may differ from one case to the next.

Summary

1. Deciding on the behaviors which are abnormal and constitute psychological disorders is no simple matter. These judgments are influenced by the social context in which the behavior occurs, the extent to which the behavior is subjectively distressing to the individual or to others, the degree or intensity of the behavior, and the degree to which the behavior interferes with the individual's adaptation.

2. In developing systems for classifying psychological disorders, some have used empirical, statistically based approaches. They seek to find clusters of problems that often occur together, giving each cluster a name representative of the disorder, or syndrome.

3. In a second general approach, the clinical-consensual method, experts try to reach a consensus about which psychological disorders exist and how they should be defined and diagnosed.

4. The clinical-consensual method was followed in developing the most widely used classification system, the one contained in the *Diagnostic and Statistical Manual of Mental Disorders III* (DSM-III). This manual gives specific criteria for defining and diagnosing various clinical syndromes and personality disorders.

5. The DSM-III system, like its predecessors, is based in part on the medical model—the view that clusters of symptoms form syndromes that are caused by specific underlying illnesses. Critics of the medical model have doubts concerning its application to many psychological disorders.

6. The use of diagnostic labels for psychological disorders can have adverse effects; for example, they may bias the judgments of people who interact with the labeled person, and they may serve as self-fulfilling prophesies. However, labels can also be beneficial; for example, they provide categories for scientists to use in studying the causes of psychological problems.

7. The diagnosis of psychological disorders is an imperfect process. The possibility of significant errors was illustrated by

Rosenhan's study in which pseudopatients, who were actually quite normal, were diagnosed and admitted to mental hospitals.

8. One broad group of disorders in DSM-III includes those that usually emerge in infancy, childhood, or adolescence. Two examples are infantile autism—usually identified in early childhood—and bulimia—an eating disorder often occurring in adolescent girls.

9. The substance-use disorders include substance abuse (maladaptive patterns of substance use, such as staying drunk most of the day) and substance dependence (physical addiction). Alcohol and heroin dependence are two of the most prevalent disorders in our society.

10. In schizophrenia, the most fundamental feature is a disturbance in the basic thought processes, often accompanied by withdrawal, inappropriate or overly flat emotional expression, and delusions and hallucinations. Theories about the origins of schizophrenia have focused on (*a*) unusual life experiences, particularly in the schizophrenic person's relationship with parents; (*b*) genetic and other biological processes; and (*c*) the interplay of a biological predisposition with stressful life experience, both of which may be necessary for schizophrenia to develop.

11. The paranoid disorders are characterized by delusions of grandiosity or persecution. People with these disorders often have a long history of finding fault with others. Except for their elaborate delusions, these people seem quite reasonable and sane.

12. The chief characteristic of the affective disorders is a severe disturbance of mood. In major depression, depressed states alternate with normal mood. In the bipolar disorders, episodes of elation or excitement (mania) and extreme depression appear periodically.

13. Many factors have been identified as causes of depression. Among them are (*a*) learning and social factors, especially a loss of significant reinforcers; (*b*) cognitive factors, such as distorted interpretations of one's experiences; and (*c*) biochemical and genetic factors.

14. Anxiety may play a role in many disorders, but the anxiety disorders are those in which intense, observable anxiety or fear is central to the problem. Among the major anxiety disorders are phobias (intense, irrational fears of certain objects or situations), generalized anxiety disorder, panic disorder, and obsessive-compulsive disorder.

15. Personality disorders are long-standing maladaptive behavior patterns. People with personality disorders cope with life's problems and relate to others in ways that may be considered troublesome, unusual, strange, or tiresome, but their behavior is not so anxiety-ridden or bizarre as that described in the DSM-III clinical syndromes. Some personality disorders, though, do resemble mild forms of these syndromes (for example, a schizotypal personality resembles a mild form of schizophrenia). Others (for example, antisocial personality disorder) are characterized by defects of impulse control and conscience.

Terms to Know

One way to test your mastery of the material in this chapter is to see whether you know what is meant by the following terms.

Abnormality (618)
Psychological disorder (618)
Adaptation (621)
Empirical approach to classification (622)
Syndrome (622)
Clinical-consensual approach to classification (624)
Brain syndromes (625)
Psychosis (625)
Delusions (625, 638)
Hallucinations (625, 638)
Neurosis (625)
Personality disorders (625, 626, 645, 659)
DSM-III (626)
Clinical syndromes (626)
Medical model (627)
Infantile autism (631)
Anorexia nervosa (634)
Bulimia (634)
Substance-use disorders (636)
Abuse (636)
Dependence (636)
Tolerance (636)
Withdrawal symptoms (636)
Delirium tremens (636)
Schizophrenia (638)
Loosening of associations (638)
Cognitive slippage and derailment (638)
Autistic thinking (639)
Flat affect (639)
Dementia praecox (639)
Disorganized type (639)
Catatonic type (639)
Paranoid type (640)
Double-bind communication (640)
Dopamine hypothesis (642)
Concordance (643)
Prognosis (644)
Paranoid disorders (644)
Paranoia (645)
Paranoid personality disorder (645, 660)
Affect (647)
Affective disorders (647)
Major depressive episodes (647)
Learned helplessness (650)
Manic episodes (652)
Bipolar disorders (652)
Flight of ideas (653)
Anxiety (655)
Anxiety disorders (655)
Phobias (655)
Phobic disorders (655)
Generalized anxiety disorder (656)
Panic disorder (656)
Obsessions (657)
Compulsions (657)
Obsessive-compulsive disorder (658)
Schizotypal personality disorder (660)
Compulsive personality disorder (660)
Histrionic personality disorder (660)
Antisocial personality disorder (660)
Psychopatic personality (660)
Sociopathic personality (660)

Several fine texts in abnormal psychology give detailed accounts of the various psychological disorders. Barclay Martin's *Abnormal Psychology: Clinical and Scientific Perspectives*, 2d ed. (New York: Holt, Rinehart and Winston, 1981) provides a balanced view of biological and learned causes, as does Irwin and Barbara Sarason's *Abnormal Psychology: The Problem of Maladaptive Behavior*, 4th ed. (Englewood Cliffs, NJ: Prentice-Hall, 1984). David Rosenhan, author of the pseudopatients study discussed in this chapter, teamed up with Martin Seligman, architect of the learned-helplessness model of depression, to produce a thoughtful, well-researched text, *Abnormal Psychology* (New York: Norton, 1984). Thomas Achenbach's *Developmental Psychopathology*, 2d ed. (New York: Wiley, 1982) is a comprehensive examination of child and adolescent disorders which gives careful attention to research findings.

A rich variety of fictional and biographical accounts are available for those who would like an in-depth look at a particular disorder or at society's treatment of the insane. Among these accounts are Judith Guest's *Ordinary People* (New York: Viking Press, 1976), which deals with depression; Hannah Green's *I Never Promised You a Rose Garden* (New York: New American Library, 1964), which focuses on schizophrenia; and Ken Kesey's *One Flew Over the Cuckoo's Nest* (New York: Viking Press, 1962), which deals with the labeling and treatment of patients in mental hospitals. For a cross-cultural perspective, see Viktor Nekipelov's *Institute of Fools* (New York: Farrar, Straus & Giroux, 1980), which gives an account of the author's own psychiatric examination (and the Serbsky Psychiatric Institute in which it took place) following his human-rights activism in the Soviet Union.

Therapy for Psychological Distress

HEADING OUTLINE

Chapter

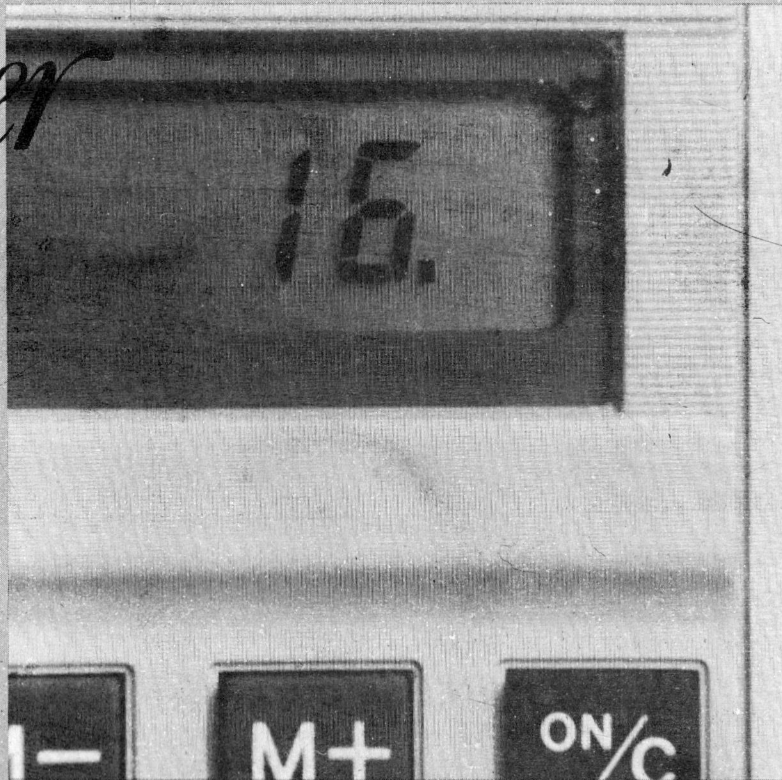

"Madness," in one form or another, has probably been recognized and "treated" for centuries. However, some of the early treatments may have done more damage than the behavior they were designed to cure. Consider trephining, for example, a procedure illustrated in the sixteenth-century painting on the facing page (top right). A hole was chipped in the skull, evidently to allow the "evil spirit" that caused the madness to escape from the victim's body. The procedure may have been used as early as the stone age. Skulls such as the one shown to the left of the painting reveal a fact that may surprise you: Some people survived trephining long enough for some healing of the skull to take place.

Trephining was humane compared to the treatment given to witches, people whose deviant behavior was thought to result from satanic forces. To protect the community, those thought to be witches were put to the torch (see facing page). This practice continued well into the seventeenth century.

As recently as the nineteenth century, the treatment methods shown at the bottom of the facing page were in use. The rotating chair was used to treat depressed people, who were spun around in a circle in an effort to make them feel better. The "crib" (bottom left) was used to restrain people whose behavior was violent or assaultive. It was a precursor of the straitjacket.

Treatment methods have come a long way since these early efforts, as we will see in this chapter.

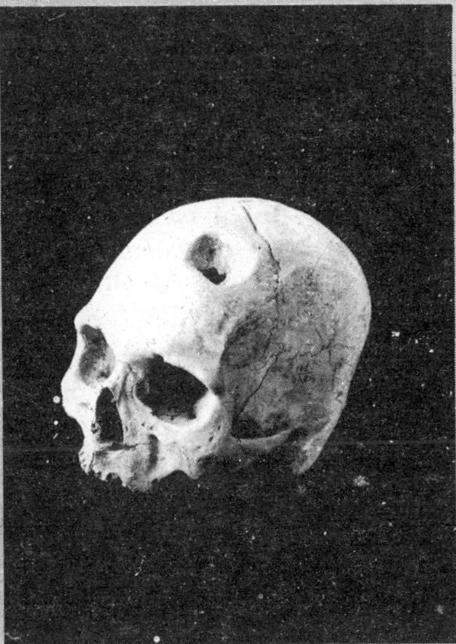
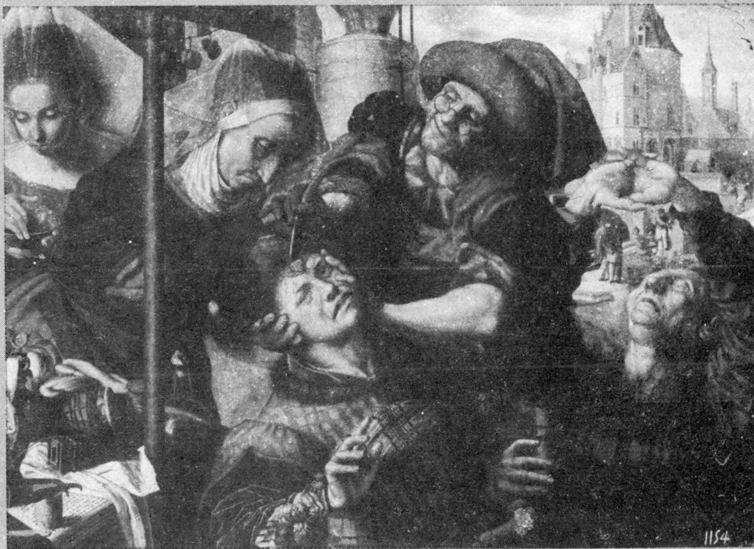
1154

OVER the centuries, as the chapter-opening illustrations suggest, approaches to treating people with psychological disorders have ranged from the benign to the barbarous. From the time of the Roman empire up until the early Middle Ages, these people were often treated with sympathy and concern. Roman law freed them from responsibility for certain crimes and required family members to care for their disturbed relatives. In the Middle Ages, a Catholic friar named Bartholomew wrote an influential book, *The Properties of Things*, in which he explained psychological disorders. The causes, he wrote, include passion, sorrow, too much work, and too-deep thought. Bartholomew recommended that disturbed people be removed from their stressful environments and given recreational and occupational therapy.

By the latter half of the Middle Ages, the influence of Roman law and Bartholomew's writings had faded. Times became much harder for psychologically disturbed people. Very disturbed behavior came to be seen as evidence that the person was "possessed" by Satan. In 1487, the printing of *Malleus Maleficarum* (*Hammer of Witches*) provided a religious rationale for executing people "possessed" by Satan. For the next 200 years, many disturbed people were kept in institutions that were little more than warehouses. In the 1400s, St. Mary of Bethlehem, a London hospital for the poor, began to admit patients with psychological disorders, who were chained, sometimes tortured, and often put on display. Wealthy Londoners paid admission fees to see patients "acting insane" (Figure 16.1). The hospital came to be called Bedlam, a corruption of Bethlehem; this is the origin of the modern term *bedlam*, which signifies the kind of uproar, chaos, and confusion that characterized that hospital and others well into the 1800s.

At the end of the eighteenth century, the picture began to brighten, at least in some parts of the western world. In 1792, Phillipe Pinel unchained patients in La Bicêtre, a Paris mental hospital. In Pinel's writings, he attacked the notion that disturbed people were guilty or that they deserved punishment. Instead, he argued that such people were "mentally ill" and that their "miserable state deserves all the consideration that is due to suffering humanity" (quoted in Zilboorg & Henry, 1941, pp. 323–324). As a result of Pinel's efforts, and those of crusaders like William Tuke in England and Dorothea Dix in the United States, hospital attendants began to be trained to treat mental patients in humane ways and physicians became more actively involved in their treatment.

These trends in the nineteenth century were all part of what has come to be called the *first mental-health revolution*. One effect of this revolution was to promote the medical model (Chapter 15, page 627) and a medically oriented approach to treatment. The legacy of this approach can be seen in today's *biomedical therapies*, efforts to modify brain and nervous system functioning by means such as surgery, electric shock, and drugs. We devote one section of this chapter to the biomedical therapies.

The *second mental-health revolution* was ushered in around the turn of the century, primarily by Sigmund Freud and his colleagues. Freud's work (Chapters 14 and 15) convinced many that unconscious motives and

Figure 16.1

Bedlam Hospital, London. A line engraving, 1735, by William Hogarth, as reworked by the artist in 1763. (From The Granger Collection.)

feelings were major causes of psychological disorders. Freud's technique, psychoanalysis, was used to help patients achieve insight into their unconscious processes. Today, techniques for unraveling such processes may or may not resemble Freud's original approach, but they do share one emphasis: understanding the dynamic processes that produce problem behavior. For this reason, they are often labeled *psychodynamic therapies.* We devote a second section of this chapter to these techniques.

Developments since Freud's time may not qualify as "mental health revolutions," but they have certainly brought about new ways of thinking about and treating abnormal behavior. In Chapter 14, we discussed the humanistic theories, those that emphasize people's subjective sense of who they are and how they fit into the world. These theories helped stimulate the *humanistic and existential therapies,* therapies designed to help people by encouraging their self-awareness, self-acceptance, and feelings of personal responsibility for how they live their lives. A third section of the chapter will focus on these therapies.

Earlier in this book we discussed the rise of behaviorism and research on instrumental conditioning, classical conditioning, and observational learning (Chapters 4 and 14). Behaviorism has spawned a distinctive set of therapies aimed at changing people's unwanted behavior without trying to intervene medically, resolve unconscious conflicts, or even understand the self. This family of *behavioral therapies* will be examined in a fourth section of our chapter.

In addition to our focus on these four broad categories of therapy, we will describe some recent trends in treatment. One of these is *behavioral medicine,* a set of psychological treatments to help people cope with their health problems. Another is *community psychology,* a family of approaches aimed at treating and preventing psychological problems in people's natural environments.

Who Needs Therapy?

Although therapies can be grouped roughly into the broad categories described above, there are literally hundreds of specific techniques, orientations, and schools of therapy available to those who need them. But who *does* need them? Who should receive therapy? This question is hard to answer, but it is important for each society to try.

Each nation needs an estimate of the need for mental-health care in its population so that it can make plans for adequate services. One approach to making these estimates has been to survey people door-to-door, interviewing them and looking for signs of psychological disorder. Researchers in the United States have made several such surveys in the past few decades. In the most ambitious of them, teams of interviewers talked to over 9,000 adults from three large cities (Regier et al., 1984). People were asked about any psychological problems they had had during the previous 6 months, and a judgment was made about whether the problems fit any of the DSM-III disorders (Chapter 15, pages 626–627). The survey showed that over a 6-month period, about 19 percent of adult Americans had experienced at least one DSM-III disorder (Myers et al., 1984). The three most common categories were anxiety disorders (Chapter 15, page 655), substance abuse or dependence (page 636), and affective disorders (page 647). An especially important finding was that only about one-fifth of those who had a disorder had tried to get therapy of any kind during the 6-month period; and most of those who had sought therapy had consulted a general physician—not a psychologist, psychiatrist, or other mental-health professional (Shapiro et al., 1984).

Surveys of school-aged children and adolescents have yielded roughly the same findings. About 10 to 15 percent of school-aged youngsters have psychological difficulties that warrant professional attention (for example, Dohrenwend et al., 1980; Rubin & Balow, 1978). Evidently only a minority of these youngsters actually receive professional help (Dohrenwend et al., 1980).

Overall, the surveys of adults and children suggest that well over 30 million Americans need mental-health services and that the majority of them are not getting the services they need. In countries where mental-health programs are less well developed, the situation is apt to be worse. Even in the U.S., the need for therapy of some kind is probably greater than the surveys have revealed. For example, many people who do not have a DSM-III disorder may have serious adjustment problems at home or work.

Who Provides Therapy?

People who seek therapy face a formidable consumer challenge. A glance at the Yellow Pages in the telephone directory reveals a broad array of mental-health specialists—psychologists, psychiatrists, social workers, and so forth. How are people to decide which specialist is right for their needs? One way to begin is to understand the differences in training and skills among the various professional groups.

Clinical psychologists (Chapter 1, page 17) have graduate degrees

(M.A.s, Ph.D.s, or Psy.D.s) in clinical psychology and have completed internships of 1 year or longer. They are trained in psychological principles and methods and the identification and treatment of psychological disorders. They are also trained in the kinds of assessment discussed in Chapter 13. *Psychiatrists* (Chapter 1, page 18) have general medical degrees and up to 3 years of postdoctoral residency training in psychiatry—training emphasizing treatment of psychological disorders with and without drugs. *Psychiatric social workers* have masters' degrees in social work and sometimes more advanced degrees; they have training focused on mental-health problems. Among their special skills are interviewing, assembling family histories of their clients, and assessing the social factors involved in problem behavior. *Psychiatric nurses* are registered nurses trained to help people with psychological disorders; these specialists are often employed in institutional settings such as mental hospitals and the psychiatric wards of general hospitals. *Psychoanalysts* (Chapter 1, page 19) may be psychiatrists, psychologists, or neither; they specialize in psychoanalysis, as practiced by Freud or his followers (Chapter 14, page 576). As part of their in-depth training, psychoanalysts have undergone psychoanalysis themselves.

In addition to these major categories, there are other specialists offering counseling for problems ranging from specific learning difficulties to career problems. Many areas have mental-health associations or similar organizations that can help people find specialists to fit their mental-health or counseling needs.

Where Does Therapy Happen? Trends over Recent Decades

The medical model (Chapter 15, page 627) was influential in the United States during the first half of the twentieth century. People with psychological problems were considered mentally ill and kept in mental hospitals for long periods. But in the 1950s and 1960s, dramatic changes took place as a result of surveys suggesting that psychological problems were widespread, even among so-called normal people living in the community. Many of these problems seemed mild enough to be dealt with by occasional visits to a therapist in the community. And it seemed clear that many people with psychological problems would not receive treatment of any kind if hospitalization were their only option. Thus, publicly funded community mental-health centers emphasizing outpatient treatment (that is, occasional visits by clients who live at home) sprang up around the country. Outpatient care was further encouraged by the advent of drug treatment (page 679). Newly developed medications made it possible for many people—even some with very serious disorders—to manage their problems while continuing to live at home. Change was also spurred by a nationwide movement for deinstitutionalization of mental patients. Some argued that public mental hospitals provided such inadequate care that they should, in the words of one writer, be taken apart "stone by stone and then, like the city of Carthage, plowed three feet under and sowed with salt" (Albee, 1968). With the convergence of all these forces, the patient population in public mental hospitals shrank sharply, as shown in Figure 16.2. Mental hospitals were not "plowed under," but in the 1980s,

Figure 16.2

Twenty-five years of change in the population of patients in county, state, and federal mental hospitals. (From Mehr, 1983.)

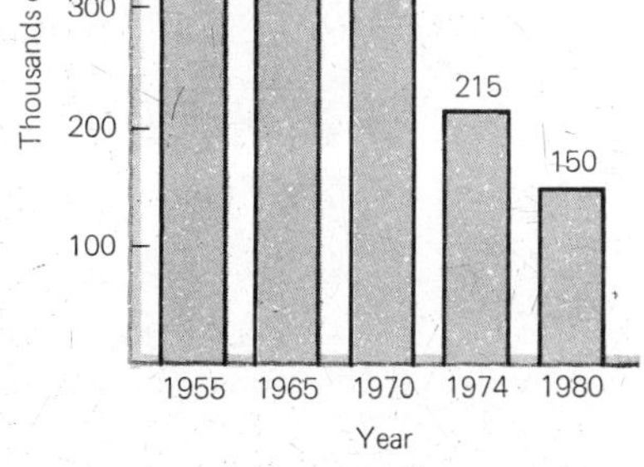

many are mere shadows of their former selves, at least in terms of the number of inpatients they serve.

To understand fully where therapy takes place today, we must ask an important question about Figure 16.2: Where did all the mental patients go? They certainly did not disappear, nor did their problems. In fact, some of these people are now struggling for survival, living in seedy hotels or on the streets of deteriorating neighborhoods (Figure 16.3; also see Edelson, 1976). A minority have managed to arrange for outpatient care in public facilities such as community mental-health centers. A larger number have found their way into privately owned nursing homes or sheltered-care facilities. Some years ago, these settings provided only nursing care to the elderly and to others with chronic physical illnesses. Today, with almost one-third of their residents under 65 and suffering from mental retardation or some psychological disorder (Bassuk & Gerson, 1978), some of these facilities are providing counseling for their psychologically disordered residents. Most, however, provide little more than room, board, and drug prescriptions. Some former mental patients who do not live in these facilities check into private mental hospitals or psychiatric wards of general hospitals for occasional short-term treatment. Finally, many who are discharged from mental hospitals eventually return. Some 40 to 50 percent of patients discharged from mental hospitals are readmitted within a year of discharge. Many face failure and rejection in the "real world." Follow-up studies have shown that only about 20 to 30 percent of former patients are able to keep full-time jobs in a competitive market (Anthony et al., 1972); finding and keeping friends and establishing other social-support systems may also be difficult. Former patients may grow despondent and hungry for the security of the hospital ward. This may help explain why, in recent years, admissions to public mental hospitals have actually begun to increase again.

While inpatient care has declined, outpatient care has shown an enormous surge. Between 1955 and 1980, a period during which inpatient

Figure 16.3

The deinstitutionalization of many mental patients has left some with survival problems they are not equipped to solve. One result: an influx of homeless, jobless street people struggling to maintain a marginal existence (© Joel Gordon, 1978.)

enrollment dropped sharply, outpatient enrollment showed the opposite trend. In 1955, outpatient care represented only 23 percent of all therapy; in 1980, the figure was almost 70 percent (Mehr, 1983). Today, outpatient care takes place in community mental-health centers, private clinics, and even in hospitals, most of which now have busy outpatient departments. The public has come to recognize that people need not be "deranged" in order to profit from professional help; completely well-adjusted people may need help in coping with a crisis or tragedy or simply in thinking a problem through.

Who Gets Therapy? Money Matters and YAVISes versus QUOIDs

Finally, a word about the people on the receiving end of therapy. In capitalistic societies, therapy is, to some extent, for sale. Professionals who make their livings in private clinics must charge fees for their services. Publicly funded mental-health centers and hospitals also charge fees, but they can usually arrange for very low income clients to pay low rates and sometimes no fee at all. It is partly because of these adjustable fees that public facilities often have more applicants for their services than they can handle; thus they often have waiting lists and can provide fewer visits with the therapist than the client might wish. Private therapists, on the other hand, may be available immediately and often; but since their services are not subsidized, they can be expensive (Figure 16.4). What all this obviously means is that in mental health, as in many other precious commodities, money matters. Those who are less-advantaged financially have more trouble finding the kinds of service they want.

Another factor that has been found to influence the therapy an individual receives is the kind of client he or she makes. Several studies have suggested that the people most likely to receive nonmedical forms of therapy are those who seem most promising as clients and those whose personalities are most appealing to the therapist. These tend to be the Young, Attractive, Verbal, Intelligent, and Successful—a group given the nickname *YAVISes*. By contrast, the group dubbed the *QUOIDs*—the Quiet, Ugly, Old, Institutionalized, and (culturally) Different—is less likely to receive intense therapy and more likely to receive brief visits and drug prescriptions. This apparent inequity troubles many professionals, but no one has yet found a solution.

Figure 16.4

Psychotherapy, like other commodities in our society, is more readily available to those who can pay than to those who cannot. (Copyright © Gil Spitzer.)

Having looked at some significant aspects of the history, personnel, politics, and finances of therapy—warts and all—we are now ready to delve into the *what* and *why* of the process. We will examine some of the many forms that therapy can take, and we will focus on the aims underlying these forms. Afterward, we will tackle what is perhaps the most important question of all: Does therapy work?

Biomedical Therapies

We will begin our discussion of therapy by focusing on the *biomedical therapies*—treatments aimed at directly altering the way the body functions. The approaches we discuss here are so "medical" that they are ordinarily conducted or supervised by psychiatrists or other physicians; psychologists and other mental-health professionals may help their clients prepare for the treatment or cope with its after-effects.

ELECTROCONVULSIVE/ELECTROSHOCK THERAPY

In the days of the Roman empire, the Greek surgeon Dioscorides used electric shocks from the torpedo fish to treat Roman soldiers' headaches; the sufferer either stood on top of the fish or the fish was clapped to his temples (Harpe, 1979). In 1781, W. Oliver, physician to King George III, gave an accidental overdose of camphor to one of his English patients, causing the patient to convulse. Afterward, Oliver judged that the convulsion had improved the patient's condition both physically and psychologically. From such humble, partly accidental beginnings came the idea that psychological distress might be treated by causing convulsions (Bini, 1938)—either with electric shock or with some swallowed or injected substance, such as insulin.

In the 1930s through the 1950s, *electroconvulsive therapy* (*ECT*), also called *electroshock therapy* (*EST*), came to be widely used in mental hospitals—especially for the treatment of psychosis and severe depression. In ECT, a full-body seizure, or convulsion, is triggered by a quick jolt of electric current. Electrodes are attached to one or both temples (Figure 16.5), and electricity (70 to 160 volts for 0.1 to 0.5 seconds) is passed directly through the brain. Typically, the patient is also given a sedative and/or a muscle relaxant before the brain shock. With the convulsion, the patient immediately loses consciousness, the body grows rigid, and the muscles twitch violently, for as long as a minute in some cases. The patient remains unconscious for several more minutes, then wakes up in a confused state, with no memory of what happened during treatment. Sometimes the memory loss extends to events occurring just before (and, more rarely, just after) the convulsion and unconsciousness. In the early days of ECT, some people were treated daily for weeks or months. Currently a full series rarely involves more than 10 treatments, and these are spread over 2 to 3 weeks.

There is considerable evidence that ECT works, at least in some cases. Because it is extreme, though, the procedure is usually reserved for severely disturbed people who have not responded to milder forms of treatment. ECT was once used to treat schizophrenia, but evidence now suggests that it has little long-term effect on this disorder (Taylor & Fleminger, 1980). ECT is now most often used to treat profound depres-

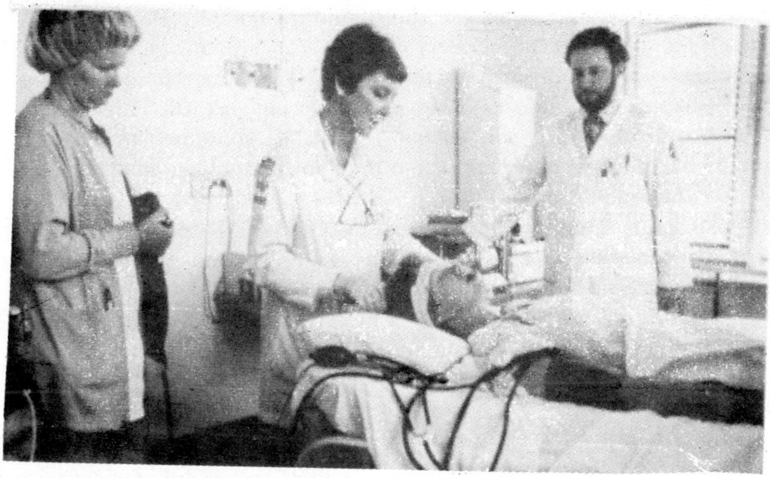

Figure 16.5

A typical arrangement for the administration of electroconvulsive therapy (*ECT*). (Will McIntyre/Science Source, Photo Researchers, Inc.)

sion. It appears to shorten depressive episodes (Chapter 15, page 647) and lead to improvement in as many as 60 to 90 percent of those treated (Weiner, 1979). When depressive episodes extend over several months, when milder treatment methods have failed, and when daily functioning—or even life itself—appears to be threatened, many feel that use of ECT is justified.

Many professionals agree that ECT can help alleviate severe depression, but there is little agreement as to *how* it works. Dozens of theories have been advanced to explain the effects. Some researchers believe that ECT increases the availability of certain neurotransmitters in a region of the brain called the limbic system (Chapter 2, page 62). Others argue that the jolt of electricity enhances feelings of well-being by producing a "cortical orgasm" (Blachly, 1977). Still others argue that ECT has *placebo effects*—that is, effects that are not physiological but psychological. The treatment helps, they feel, because the patient *expects* it to help. Friedberg (1976), for example, reported on the use of an ECT machine in a British hospital where 2 years of "successful" treatment were followed by the discovery that the machine had not been working. For 2 years, clients had been "cured" without actually receiving any electric current. A subsequent experiment (Lambourn & Gill, 1978) indicated that this experience was no fluke. The experimenters took severely depressed people through all the steps of ECT except that half of them did not receive any real electric current. The two groups did not differ in their recovery rates or in most other psychological effects. Evidently ECT *can* have placebo effects—improvement can result from purely psychological factors.

Electric shocks can also have several adverse side effects, and these have made ECT highly controversial. In addition to causing the short-term memory loss and confusion described earlier, ECT can impair the ability to acquire new memories. It can also produce dizziness and loss of appetite (Sipe, 1979). The nature and extent of ECT effects have led some to suspect that it produces brain damage (Meyer & Salmon, 1984). Neuropsychologist Karl Pribram [quoted in Friedberg (1975), p. 23] argued, "I'd rather have a small lobotomy [a kind of brain surgery, see below] than a series of electroconvulsive shock . . . I just know what the brain looks like after a series of shock, and it's not very pleasant to look at." The subjective aftereffects of ECT can also be painful, as the experience of author Ernest Hemingway illustrates. Hemingway, severely depressed, had 11 ECT sessions in December 1960. Three months later, he returned for more sessions. While he consented to the treatments, Hemingway was bitter about their apparent effect on his memory and his writing. As he put it,

> "What these shock doctors don't know is about writers . . . and what they do to them . . . What is the sense of ruining my head and erasing my memory, which is my capital, and putting me out of business? It was a brilliant cure but we lost the patient."
> (Quoted in Friedberg, 1975, p. 99)

About a month after his second ECT series, Hemingway committed suicide.

Although ECT continues to be controversial and its use has declined since 1950, it is certainly not fading away. A survey conducted in New York in the mid-seventies revealed that 30 out of 36 hospitals still used ECT. The treatment was given to 1 percent of patients in public hospitals; 5 percent in university hospitals; and 21 percent in private, for-profit hospitals. In the eighties, researchers continue to study ways of using ECT that will minimize its bad effects and maximize its benefits.

PSYCHOSURGERY

Suppose you were in severe psychological distress and you were offered a treatment that was likely to help but would also carry one significant cost: the permanent destruction of some of your brain tissue. This is the choice faced by a few thousand people each year—people offered the option of psychosurgery. In the United States, psychosurgery began to be used about 50 years ago. The techniques used in the 1930s and 1940s were crude by today's standards. They included *prefrontal lobectomy,* the removal of brain tissue from the prefrontal cortex (Chapter 2, page 63), and *prefrontal lobotomy,* the severing of the connections between the prefrontal cortex and the rest of the brain. Initially, such techniques got rave reviews, partly because they appeared to make violent and unmanageable people calm and agreeable. Unfortunately, the techniques often had serious side effects. After the surgery, people often seemed dull, apathetic, drained of emotion and motivation, and sometimes unable to care for themselves. Some even described these people after surgery as "human vegetables" or "zombies." Frontal-lobe surgery consequently grew less and less popular. Meanwhile, investigators were finding that drugs alone (see below) could produce many of the tranquilizing effects produced by psychosurgery. Consequently, in the 1950s, these drastic surgical methods lost favor and are now rarely used.

A few neurosurgeons, however, have continued to experiment with less drastic procedures. Some have tried surgery on the limbic system (Chapter 2, page 62), and some have developed techniques using laser beams or ultrasound to produce very precise, limited tissue damage—as little as 5 percent of the tissue destruction caused by the 1940s' procedures. Recent evidence (for example, Kalinowski, 1979) suggests that such techniques may be helpful with severely depressed, obsessional, and anxious individuals, but we need to be cautious about such evidence. Because the surgical procedures are so extreme and so visible, it is hard to conduct controlled experiments in which placebo effects (page 677) can be ruled out. Just as with ECT, psychosurgery may sometimes lead to improvement only because people *expect* to improve after surgery. Because the procedures are used so rarely (about 1,500 operations per year; Meyer & Salmon, 1984), it is also hard to find large groups to study. With research so sparse, we should regard psychosurgery with caution and skepticism. It is a potentially risky set of procedures that *may* be appropriate for a very small proportion of very disturbed people, but only as a last resort.

CHEMOTHERAPY

Treatment with chemical substances, or drugs, is known as *chemotherapy*. This form of treatment is now used more widely than are all the other therapeutic methods combined. Indeed, most of us self-administer substances purchased from drug stores, liquor stores, grocery stores, and bars to brighten our mood, fight pain, "calm our nerves," help us sleep, and pep us up. Even our morning cup of coffee, with its pep-inducing caffeine, is a kind of "chemotherapy." *Tranquilizers*, drugs that lower anxiety and irritability, are now a commonplace in many western countries. In fact, one of the most commonly prescribed drugs in the United States is the mild tranquilizer Valium.

Since mid-century, the variety of available drugs has expanded greatly. Commercial companies have developed diverse drugs to treat diverse problems. These developments have been spurred by research suggesting that people with certain psychological disorders may have distinct biochemical abnormalities. For example, in Chapter 15 (page 642), we noted that neurotransmitter abnormalities have been found in people suffering from schizophrenia, and we mentioned that drugs known as *phenothiazines* have been found to be effective with this group of patients. The phenothiazines (and other antipsychotic drugs) appear to reduce agitation, tension, and combativeness; decrease or eliminate delusions and hallucinations; and improve social behavior and sleep patterns. The phenothiazines certainly do not "cure" schizophrenia, but they may eliminate the major characteristics of the disorder in some victims almost completely, at least as long as the drugs are used regularly. In others, the drugs may reduce the intensity of the problem behavior enough to make the victim's life more controlled and pleasant, and this often makes possible a move from the hospital to a sheltered setting in the community. In fact, without such drugs, the sharp decline in hospital populations over the past few decades (page 673) would have been much less dramatic.

Patients with affective disorders (Chapter 15, page 647) are also considered good candidates for chemotherapy. Depression is frequently treated with a family of drugs called *tricyclics;* these drugs increase the availability of the neurotransmitter norepinephrine (page 652). Less frequently, depression is treated with a somewhat riskier class of drugs called monoamine oxidase (MAO) inhibitors; these drugs prevent the breakdown of norepinephrine and thus increase its availability (page 652). A number of studies have shown that chemotherapy, especially with tricyclics, is more effective than *placebos* (pills that contain no real drugs) in easing depression. For the manic states involved in bipolar disorders

(Chapter 15, page 652), the chemotherapy of choice is usually *lithium carbonate*, a compound that reduces extreme mood elevations in up to 80 percent of treated clients. Chemotherapy is often so effective in the treatment of affective disorders that clients come to rely heavily on it to maintain their social or job functioning.

> In one case, for example, a well-known producer of Broadway musicals was highly successful in his work when he was mildly "high" but, when extremely manic, he wasted money and alienated his friends and family members. He tried psychotherapy for years with little success; but, when he was given lithium, he improved quickly. With exactly the right dose of lithium, he was able to maintain the balance he wanted: a moderately elevated mood that maximized his creativity as a producer but prevented the wild behavior that threatened his financial and social well-being. (Abstracted from Mehr, 1983, p. 389.)

The bright side of chemotherapy is that it does help to relieve several kinds of problem behavior; the dark side is that clients may pay a price for that relief—and we are not referring to money. People can become extremely dependent on their "medications," to the extent of feeling that they cannot cope without them. The drugs for schizophrenia and depression can also damp down both emotion and motivation, thus interfering with independence and the ability to help oneself. Furthermore, virtually every form of chemotherapy has some physical side effects. Depressed people treated with tricyclics may suffer problems ranging from a perpetually dry mouth, dizziness, and constipation to agitation and tremors. Those treated with MAO inhibitors risk such side effects as dizziness, weight gain, sexual disturbance, periods of confusion, and even (rarely) death. Clients whose manic states are relieved by lithium carbonate may suffer side effects ranging from nausea and diarrhea to seizures and even (rarely) coma.

Some drugs, particularly the phenothiazines and others used in schizophrenia, carry the risk of *tardive dyskinesia*, a physical disorder that takes such varied forms as uncontrollable facial movements (lipsmacking or grimacing), limb movements like finger twitching or jerking of the arm or leg, contractions of the neck or back muscles, or an arm that is perpetually extended (Figure 16.6). Tardive dyskinesia is found most often in adults who have been treated for many years with heavy doses of antipsychotic drugs, but is also seen among adolescents and children, and it sometimes occurs after only low doses for periods as brief as a few months (Gualtieri & Guimond, 1981). In many cases, the dyskinesia is permanent and very handicapping. Sometimes the body movements are so extreme that the individual cannot hold a job and is forced into social isolation. In milder cases, people may adopt a "disguise"; for example, one man whose mouth chewed and smacked uncontrollably took to chewing gum in an effort to make the movements seem more normal to others. The appearance of tardive dyskinesia may mean that the drug treatments have physically damaged the brain, and it therefore is a very serious side effect. For this reason, some experts now recommend that antipsychotic drugs and other major tranquilizers be used only when the benefits clearly outweigh the risks. Experts also recommend periodic "drug holidays," drug-free periods of 4 months or more, to determine

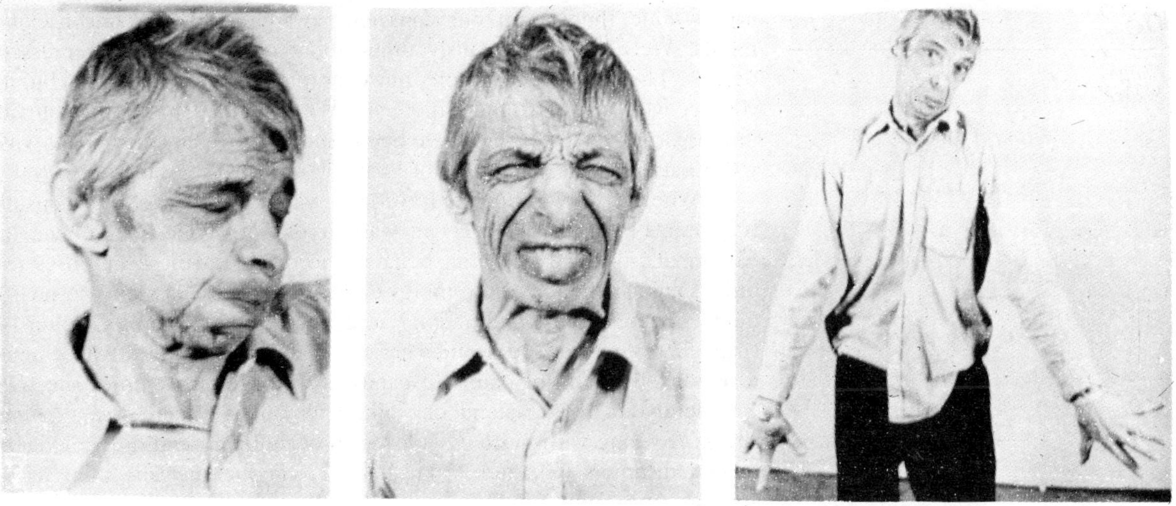

Figure 16.6

Tardive dyskinesia, a disturbing and disabling side effect of antipsychotic drugs, takes many forms. The man shown here suffers from apparently uncontrollable facial grimaces, tongue protrusion, and arm and hand movements. The movements are sometimes severe enough to prevent him from even signing his name. (Courtesy of Joseph DeVeaugh-Geiss, M.D.)

whether the patient still needs the drug and to check for early signs of tardive dyskinesia.

Psychodynamic Therapies

The risks posed by ECT, psychosurgery, and chemotherapy are worrisome—some would even say ominous. The risks certainly underscore the need for alternative treatments—treatments that do not involve any change in biological functioning, but instead emphasize psychological change. Such psychological approaches, which exist in large numbers, are known collectively as *psychotherapy*. Most psychotherapy is aimed at unraveling the psychological factors involved in a problem, sharpening people's understanding of themselves and others, or teaching people specific skills to enhance their adjustment. One family of psychotherapy methods is specifically aimed at the first objective: unraveling the dynamics of problem behavior and helping the client achieve insight into those dynamics. These methods are called *psychodynamic therapies*. Most of the psychodynamic therapies have grown directly out of psychodynamic personality theories like those we discussed in Chapter 14. Freud's psychoanalytic theory, for example, spawned a therapeutic approach called *psychoanalysis*. We will begin this section by focusing on that approach.

FREUD'S TRADITIONAL PSYCHOANALYSIS

Late in the nineteenth and early in the twentieth century, Sigmund Freud and his followers introduced psychoanalytic methods that are still in use today. A psychoanalyst, or analyst, is a therapist who uses these basic methods. The theory called psychoanalysis is based upon several specific ideas about personality and psychotherapy. As we learned in Chapter 14, the theory emphasizes the interplay of personality components known as id, ego, and superego.

According to *psychoanalytic theory*, we all experience certain sexual and aggressive urges springing from our ids. Some of these urges clash with the barriers imposed by our egos and with the taboos of society that have been incorporated into our superegos. We therefore repress the urges—hide them from our conscious awareness in our unconscious minds. We cannot consciously think about or verbalize our repressed desires. The urges are still there, however, pushing to be expressed but in conflict with our egos and superegos. We are afraid to express our id impulses and yet, after they have been repressed, we no longer ***know*** why we are afraid. This vague fear of "I know not what" is called ***anxiety***.

When anxiety is triggered, we use a variety of defense mechanisms (Chapter 14, pages 588–591) to ward it off, or at least to reduce it. Abnormal, so-called neurotic, behavior occurs when these defenses require so much of our psychic energy or distort our sense of reality to such a degree that our ability to function is impaired. For example, in Chapter 14 (page 579) we discussed Freud's patient, Fräulein Elisabeth, whose neurosis took the form of pain and paralysis in her legs. At times she was completely unable to stand or walk, even though there was nothing *physically* wrong with her legs. This kind of problem, and other anxiety-related disorders (Chapter 15, page 655), are the specialty of psychoanalysis.

The aim of analysis is to lessen anxiety and the need for neurotic behavior by developing *insight*, a deep understanding of repressed feelings and conflicts. The aim is to bring unconscious conflicts to the surface so that they can be resolved, or defused. This can take a long time, often years. The analyst must first figure out what caused the patient's neurotic behavior; this may involve tracing symbolic connections between the patient's symptoms and the unconscious processes that caused these symptoms. Then the analyst must lead the patient gradually to an understanding and acceptance of the psychoanalytic explanation for the neurotic behavior. This insight may help liberate the patient from the neurosis. The treatment of Fräulein Elisabeth illustrates the psychoanalytic process. (Abstracted from Breuer & Freud, 1893–1895/1955.)

> In the case of Fräulein Elisabeth's leg problems, Freud identified several underlying causes, all symbolically related to Elisabeth's symptoms. For example, Elisabeth had once been very attracted to a young man while walking with him late one night. She had left her ailing father—who had swollen legs—alone that night, and when she returned home he was much worse. She felt deeply guilty. Her desire for her own pleasure with the attractive man clashed painfully with her devotion to her father. Later, Elisabeth's leg pains began in precisely the spot on her thigh where her sick father had rested his leg when she bandaged it. Elisabeth had also felt attracted to her sister's husband—once while walking with him in the woods and once while standing beside him at her sister's deathbed. Freud saw Elisabeth's leg pains and paralysis as a kind of neurotic self-punishment; the symptoms were symbolically connected to her repressed impulses—legs, standing, and walking were all linked in her mind with forbidden romantic and sexual feelings. She developed symptoms that punished her legs and prevented her from standing or walking. After finding these connections, Freud's task was to help Elisabeth see and accept them. In doing so, she achieved insight, and this was the key to her recovery. As Freud put it, "Her condition was . . . improved and there had been no more talk of her pains since we had been investigating their causes" (p. 159).

Like Freud's work with Fräulein Elisabeth, most traditional psychoanalysis is complex and multifaceted. In the following section, we will examine some of these facets—the component processes that fit together to form psychoanalysis.

Probing the Unconscious: Free Association and Dream Analysis Because unconscious impulses and conflicts are assumed to be the cause of neurotic behavior, the analyst uses several techniques to discover these unconscious processes and nudge them into view. One of these techniques is *free association*, a technique in which patients are asked to let their thoughts run free, without censorship, reporting them as they occur. To facilitate their free association, patients are asked to relax on a couch and look away from the analyst. (Freud's couch and consulting room are shown in Figure 16.7). Unexpected trains of thought, sudden memory lapses, and unusual statements give the analyst clues to the patient's repressed thoughts, feelings, and conflicts.

Despite the name of this technique, Freud certainly did not think the patient's associations were "free," or random. Instead, he believed that they flowed directly from specific characteristics of that patient and issues of psychological importance to that patient. For example, one young man treated by Freud found that despite his many relationships with young women, he could not remember any of their names; instead, the seemingly meaningless name "Albine" kept popping into his head. The young man knew no one with that name, nor did the term have any conscious meaning to him. At this point Freud noted

> One might infer that the analysis had failed; but no, it was already complete, and no further association was required. The man himself was unusually fair in colouring, and whilst talking with him in analysis I had often jokingly called him an *albino;* moreover, we were just in the midst of tracing the *feminine* element in his nature. So it was he himself who was the female albino, the "woman" who interested him most at the moment.
> (Freud, 1920/1953, pp. 113–114)

Figure 16.7

Sigmund Freud's consulting room. Freud's patients reclined on the couch; Freud sat behind the couch, out of sight so as not to impede the patient's flow of free association. This, at least, was Freud's "official" explanation. He also admitted, though, that "I cannot bear to be gazed at for eight hours a day." (Photo by Edmund Engleman.)

Freud believed that another pathway to the unconscious was *dream analysis*. He felt that id urges and unconscious conflicts push for expression even when we are asleep and that our defense mechanisms even relax a bit during sleep. Yet if our conscious urges and wishes were expressed directly, even in dreams, they would be very disturbing; so we express them in a disguised form. In analyzing dreams, the psychoanalyst tries to understand the particular urges that a patient has repressed and the conflicts that are hidden from view.

Dream interpretation is a difficult art, at best. It is complicated by several problems. One is that the symbols we use in our dreams have highly personal meanings, though analysts believe that we use certain common symbols to stand for particular ideas. Another is that parts of the dream may seem illogical or confusing. Still another is that parts of dreams are forgotten. The patient may try to correct for these "faults" when reporting the dream, and the result can be a distortion of the dream as actually experienced. A skillful analyst, however, can turn these problems to advantage. By figuring out where distortions and forgetting occur, an analyst can judge what id urges are so strong, and what conflicts so disturbing, that they must be defended against even in dreams. Moreover, the confusing and seemingly illogical events of a dream can become grist for the analyst's mill. Freud's own self-analysis provides an example. (Abstracted from Freud, 1900/1953.)

> Freud recalled a boyhood dream in which his sleeping mother was carried into a room and laid on a bed by strange people with bird beaks. The beaked creatures reminded Freud of some illustrations he had seen in an edition of the Bible called *Philippson's*. With still more analysis, Freud associated the name *Philippson* with Philipp, an "ill-mannered boy" who had taught Freud a vulgar German word for sexual intercourse. This word, *vögeln*, is derived from *Vogel*, which means "bird". Thus Freud made sense of the birdlike humans in his dream; they, and their behavior, represented his own sexual feelings toward his mother.

Coping with Resistance, Transference, and Countertransference Often patients will do things that interfere with or undermine the process of analysis. For example, Freud's patient, Fräulein Elisabeth, at certain crucial points in her analysis, said she had no thoughts and nothing to tell him. Yet her tense expression told Freud that there was much on her mind. (Abstracted from Breuer & Freud, 1893–1895/1955.)

> When Freud persisted, and Elisabeth finally told him what was really on her mind, she occasionally made comments like, "I could have said it to you the first time." When Freud asked why she hadn't, Elisabeth replied, "I thought I could avoid it, but it came back each time" or "I thought it wasn't what was wanted" (p. 154).

At another point in the analysis, Freud gave Elisabeth his interpretations of her repressed feelings, and she tried to reject his ideas.

> I put the situation drily before her with the words: "So for a long time you had been in love with your brother-in-law." She com-

plained at this moment of the most frightful pains, and made one last desperate effort to reject the explanation: it was not true, I had talked her into it, it *could* not be true, she was incapable of such wickedness.
(Breuer & Freud, 1893–1895/1955, p. 157)

Such efforts to avoid, or evade—by rejecting the analyst's interpretations, holding back crucial information, or failing to remember—were labeled *resistance* by Freud. He saw resistance as the patient's unconscious struggle to prevent painful material from being brought to the surface and faced directly. The analyst copes with resistance in several ways. Often the resistance is interpreted for the patient; the analyst explains why the patient may be rejecting an interpretation or "forgetting," and the analyst may interpret the repressed feelings in ways that are easier to accept. For example, Freud told Elisabeth that "we are not responsible for our feelings" (Breuer & Freud, 1893–1895/1955, p. 157) and the very fact that she had punished herself for her romantic feelings by developing leg pains and paralysis "was sufficient evidence of her moral character" (Breuer & Freud, p. 157).

The analyst also interprets and builds on the *transference* that develops during analysis. The patient's attitudes and feelings toward people outside analysis can resurface in the patient's relationship with the analyst. Parent-child relationships are often replayed in this way. The analyst may become a sort of parent figure to the patient and may thus be the target of emotions that the patient actually feels, or felt, toward his or her own parents. Positive transference occurs when emotions such as affection and dependence are involved; when emotions like hostility are involved, it is called negative transference.

Transference is useful in several ways. If it is positive, it can help patients overcome the kinds of resistance we described above; their warm feelings toward the analyst may make them feel safe enough to explore repressed material. Transference also helps the analyst understand a patient's problem. Conflicts within the patient, or between the patient and others, that are reenacted in the analysis sessions can be exposed for the patient to see and understand.

Sometimes the analyst develops a *countertransference reaction* with respect to the patient, meaning that the analyst transfers onto his or her relationship with the patient certain attitudes and feelings that originated in the analyst's relationships outside the analysis. Countertransference can hamper effective analysis. One reason analysts go through years of analysis themselves is to understand themselves well enough to deter, or at least recognize and cope with, countertransference.

Insight, "Working Through," and Termination The last stages of analysis are reached when the patient achieves insight into the sources of his or her anxiety and neurotic behavior. Intellectual insight is not enough. The patient must not only recognize the repressed conflicts that caused the neurotic behavior but must *recapture the feelings* that went with these conflicts. Remembering without feeling is not very therapeutic, or so Freud believed.

When deeply buried conflicts and emotions bubble up to the surface, often in the form of emotional outbursts, the analyst may welcome them. These are opportunities for the patient to learn a crucial lesson: It is now safe to let conflicts and feelings out into the open. This process of repeatedly bringing repressed material to the surface, interpreting it, and dealing with it is a part of what analysts call *working through*. By working through conflicts, the patient lessens the need to repress them; as anxiety is reduced and repression weakened, the neurotic behavior that brought the patient into treatment is gradually diminished. Upon the completion of this process, the transference can be resolved, an appropriate doctor-patient relationship reestablished, and analysis terminated. But there may be flare-ups after termination. For example, even Freud's patient Fräulein Elisabeth, who went on to a normal married life, still suffered occasional slight pains. But, as with Elisabeth, the analytic patient should come to face life's demands without feeling overwhelmed by anxiety or the excess baggage of severe neurotic behavior.

TRADITIONAL VERSUS CONTEMPORARY PSYCHOANALYSIS

A number of analysts still follow Freud's original procedures very closely, but an increasing number have made adjustments. For example, analysis in its traditional form can involve five 50-minute sessions per week, and these typically run for several years. Many who continue to believe in Freud's basic principles think that this much analysis makes patients too dependent on their analysts (Figure 16.8). Some also believe that lying on a couch encourages this dependency. Thus, many contemporary clinicians work out time limits with their patients and treat them as face-to-face partners, with both sitting upright. To expedite and focus the process, a therapist may negotiate specific treatment goals with the patient (we will meet twice a week, and we agree that the process will end within 6 months, for example).

Other modern-day revisions of Freud's approach involve additions to the basic processes of analysis. For example, some now believe that processes such as gaining insight and working through a problem in the

Figure 16.8

Some contemporary therapists are concerned that traditional psychoanalysis may make patients too dependent on their analysts. (Drawing by Lorenz; © 1973, The New Yorker Magazine, Inc.)

"Mr. Frentice is not your father. Alex Binster is not your brother. The anxiety you feel is not genuine. Dr. Froelich will return from vacation September 15th. Hang on."

analyst's office may not necessarily lead to improved behavior outside the office. After analysis, a person may understand—intellectually and emotionally—why he fears authority figures; but he may still be afraid to question his teacher about an exam grade or ask his boss for a raise. To improve such "real-life" behavior, many analysts now combine traditional Freudian approaches with a focus on the *current* living problems of the patient and "real-life" coping outside the analyst's office. About 20 percent of all clinical psychologists, and a higher percentage of psychiatrists, use psychoanalytic methods, at least in part (Garfield, 1981); but most have made some adjustments in Freud's original approach—thus accommodating to life in the latter half of the twentieth century.

DISSIDENT THEORISTS AND THEIR TECHNIQUES: JUNG, ADLER, AND HORNEY

Although many psychodynamic therapists have followed or built on Freud's basic tenets and techniques, others have drawn from the work of Freud's rebellious successors—Jung, Adler, and Horney, for example (Chapter 14, pages 582–588). While their approaches are *psychodynamic*—they focus on the underlying causes of a disorder and strive for insight—their objectives and techniques bear the unmistakable imprint of the influential theorists who dissented from Freud.

Jung's approach, known as *analytical therapy*, like his personality theory (Chapter 14, page 582), placed much less emphasis on repressed sexual impulses and conflicts than did Freud's approach. Instead, Jung stressed each person's need for *individuation*, which he defined as "becoming a single, homogeneous being . . . coming to selfhood or self-realization" (Jung, 1928, p. 171). Individuation involves putting the parts of one's psychological self together into a unified whole. Jung believed that so long as the intuitive, feeling parts of oneself are hidden away in the unconscious, those parts cannot be integrated with the more rational, practical parts of the self to form a fully functioning, balanced person. People may become so rational, reasonable, and logical that they cut themselves off from their deeper, intuitive side. To help his patients explore their "deeper side," Jung used word association (Chapter 14, page 583) and dream analysis. In analyzing dreams, Jung looked for messages from the unconscious that had been condensed into symbols. Dreams about water, he said, could symbolize the unconscious or wisdom; dreams about a descent may symbolize a psychic setback; and so forth. Some symbols he interpreted as quite personal and specific to a particular patient. By delving into such symbols, Jung could begin to map out the patient's "deeper self."

Like Jung himself, Jungian analysts today often use dream analysis and encourage their patients to learn the skill of analyzing dreams. By bringing the unconscious and intuitive parts of the self into view, patients may come to discover and accept a broader, fuller view of who they are. As Jung put it, "the patient becomes what he really is" (Jung, 1954, p. 10). This expanded awareness is supposed to bring a sense of peace and harmony that will make life more understandable and thus more satisfying. The Jungian approach, overall, has a mystical and even religious flavor that sets it apart from traditional psychoanalysis and gives it a special appeal to its followers.

Alfred Adler's form of analysis, known as *individual therapy*, placed special emphasis on social and interpersonal factors. Adlerian therapists concentrate less on unconscious processes and biological drives than on the patient's need for meaning, personal freedom, and a fulfilling "style of life" (Chapter 14, page 586). Karen Horney's form of analysis, like Adler's, emphasized social factors. For example, she saw basic anxiety (Chapter 14, page 587) as a social rather than a biological experience—an experience that grows out of childhood feelings of isolation and helplessness. Basic anxiety can lead to various ineffective interpersonal coping strategies, she felt. For example, the anxious person might adopt a moving-toward-others strategy (Chapter 14, page 588)—becoming self-effacing, compliant, and dependent on others for security. Like Horney herself, the therapists who use her approach today try to help their patients identify their maladaptive interpersonal strategies and their reasons for using these strategies; then they try to lead their patients toward more constructive interpersonal styles and greater self-reliance. The approaches adopted by Adler, Horney, and their followers all emphasize the patient's self-development in a social context. This emphasis is even more pronounced in the therapeutic approaches we will examine next.

Humanistic and Existential Therapies

The humanistic perspective on personality that we discussed in Chapter 14 (page 598) has given rise to specific therapies. These therapies place special emphasis on sharpening the individual's self-awareness and self-acceptance. They often stress the need for people to feel personally responsible for how they live their lives. This emphasis on shaping one's own existence is also a part of what is called the existential approach to therapy. In practice, humanistic and existential elements are often blended. One therapist whose work certainly reflects such a blending is Carl Rogers; we discussed his personality theory in Chapter 14 (page 599).

CLIENT-CENTERED THERAPY

Carl Rogers (1951, 1959, 1961) developed an approach called *client-centered therapy*. Rogers called attention to the discrepancies that develop between the way people perceive reality and the way the world actually is. He noted that discrepancies often arise between people's "ideal" selves and their imperfect "real" selves. In Rogers' view, maladjustment occurs when these discrepancies are sizable and painful. Reducing these discrepancies and the pain they provoke is one aim of client-centered therapy.

Rogers holds that people have the resources and strength to resolve their own problems. Given a little help and support from a therapist, people can resolve many problems for themselves. The therapist should not take charge but should provide the opportunity for people to develop their own improved ways of coping. The individual seeking assistance is not called a *patient*—a more or less passive sick person who needs to be healed by a doctor—but a *client*—one who commissions and pays for a professional service and who therefore has a right to shape the objectives and the course of that service. The process is called *client-centered* because the client shapes the process.

The client-therapist relationship is designed to support the client's own pursuit of wholeness. Rogers, like Abraham Maslow, whose humanistic theory we discussed in Chapter 14 (page 600), believes that people are naturally motivated to fulfill their potential and to become self-actualized. Unfortunately, though, the path to self-actualization is often blocked; people may be unable to grow because they are out of touch with themselves or with others. The client-centered therapist tries to help by facilitating self-awareness in the client and by nurturing a humane client-therapist relationship. Specifically, the relationship must have the following qualities:

1. The therapist must have *empathy* for the client—that is, an ability to understand the client's views and feelings. The therapist must become immersed in the client's world and view things from the client's point of view. In this way, the therapist can understand how the client perceives things and can help clarify any distortions.

2. The therapist must give sensitive, unconditional positive regard, never criticizing, always accepting; therapists do not judge, probe, or disapprove. This accepting attitude will create an atmosphere in which clients find the courage to perceive and accept their denied experiences and to examine, reevaluate, and feel more positive about themselves.

3. The therapist must be genuine—that is, open, spontaneous, and caring.

The client-centered therapist tries to keep the reins in the client's hands by using various nondirective techniques. One of these techniques is called *reflection of feeling.* The therapist facilitates clients' awareness by reflecting the essence of the feelings they are expressing. When therapists succeed at seeing the world as their clients do, the interchanges can be thoughtful and productive. Here is one illustration. (Abstracted from Barton, 1974, pp. 223–224.)

Mary, a 32-year old physician, is seeing Dr. G., a client-centered therapist. She begins one of her sessions by explaining that she has been considering talking with him about sex but she isn't sure how to begin. Dr. G. says, "You feel that you want to bring this up with me, but you're also feeling uncomfortable about how to proceed."

A brief silence follows. Then Mary asks, "Can't you help me with this somehow?"

"You are really feeling a bit helpless, unable to go ahead, and you're wishing I would break the impasse for you. Is **that** how you're feeling?"

"Yes." Mary feels tense, uneasy, then angry at Dr. G. because he won't help her get started. She turns and looks him in the face, pleading without saying anything.

Dr. G.'s only reply is, "You're just feeling stuck and helpless."

"Can't you help?"

"You're even mad at me, feeling that I'm deliberately holding out."

> Now Mary is frustrated. "You're always leaving it up to me, telling me that I can do it. Why do I always have to do everything by myself?"
>
> "It's like everything is always your lonely, solitary burden, all your life long."
>
> This statement strikes home for Mary. She cries, and then says softly, "Oh, damn it, damn it, damn it!" She glances up at Dr. G. and sees him completely attentive to her, empathic, feeling what she feels. She sighs, relaxes, and says, "You don't go away just because I'm hard to get along with. You're good to me."
>
> "You're always afraid that if you aren't nice, the other person will go away. It's nice to be able to be a pain and a burden without having to be afraid of desertion."
>
> This statement from Dr. G. seems to loosen Mary's tongue. She talks for 20 minutes about her sexual experiences and anxieties. She mentions her "intense relations" with girl friends. Then she blurts out, "I'm afraid I'm queer."
>
> Dr. G. replies with understanding: "In the background now for some time, you've been living with this fear that you are sexually abnormal. . . . Just the way you say it, I can feel how very much of a haunting, spooky fear is connected with that for you."

As this interaction illustrates, the client-centered therapist tries to support the client's efforts at self-discovery and growth. The therapist is not judgmental or pushy. The client takes the lead, sets the agenda, and decides when the time is right to bring up painful topics. Throughout the process, the therapist tries to nurture the vital force of self-actualization that motivates each person to grow.

Rogers was one of the first theorists to carry out systematic research on his therapeutic methods. He kept lengthy transcripts of therapy sessions, studied the processes they revealed, and asked hard questions about whether the therapy was really helping people. From his work, he concluded that therapists cannot really be completely nondirective—that even a simple smile or tilt of the head may influence what a client does. He also decided that the specific techniques a therapist uses may be less important than the therapist-client relationship—the empathy, unconditional positive regard, and genuineness that the therapist conveys to the client. Through the understanding and acceptance of the therapist, the client can come to understand and accept himself or herself. Rogers' ideas, rich in warmth and optimism, have influenced not only his followers but many other schools of therapy as well.

GESTALT THERAPY

Another approach that has both humanistic and existential elements is *gestalt therapy*. This approach is designed to make people "whole" by encouraging them to shed their defenses, unlock their potential, let go of their pent-up feelings, accept responsibility for the way they are, and, above all, focus on the here and now. As Fritz Perls (Figure 16.9), the founder of gestalt therapy, put it, "The past is no more and the future is not yet" (1970, p. 14); thus "To me, nothing exists except the now" (p. 4). Not surprisingly, the gestalt-therapy experience involves repeated

(a) (b) (c)

Figure 16.9

Some therapeutic innovators. (*a*) Fritz Perls, founder of gestalt therapy. (*b*) Viktor Frankl and (*c*) Rollo May, existential therapists. (*a*, Photo from Deke Simon, Real People Press; (*b*) Harvard University Archives; (*c*) courtesy of Dr. May.)

emphasis on staying in touch with one's feelings *right now* and expressing those feelings to others. Ruminating about the past, contemplating the future—these are discouraged, in most cases.

Gestalt therapy includes a variety of techniques for stimulating self-awareness and feelings of personal responsibility for one's actions. People are encouraged to use direct, first-person language—"I feel," "I am," and so on—and they may even be required to add a phrase like "and I take responsibility for that" after statements about how they feel or how they are. For example, a man who complains, "I feel miserable" might be required to add ". . . and I take responsibility for that"—thus emphasizing his own role in these feelings and his own potential to change them (Phares, 1979).

An excerpt from a gestalt-therapy session will serve to illustrate the emphasis on first-person language, firsthand experience, and acceptance of responsibility. The gestalt therapist in this excerpt has discovered that his client, Mrs. R., has some "unfinished business" with her deceased mother. He therefore leads her through a release of her emotion and then a release of the blame she has hung on her mother. All the while, he keeps the discussion in the first person and the here and now by arranging for a "conversation" between Mrs. R. and her mother. (Adapted from Stevens, 1975, pp. 126–128.)

> Mrs. R: Well . . . if only she had loved me, things would be different. But she didn't and . . . and I've never had any real mother love (crying).
> Th.: Put your mother in that chair and say that to her. . . . Can you imagine her sitting there in front of you?
> Mrs. R.: Yes, I see her. . . . Mother, if you had only loved me. Why couldn't you ever tell me you loved me? Why did you always criticize me? (almost a wail, more tears).

Th.: Now switch over to the other chair and play your mother. (She moves.) . . . What do you experience as your mother?

Mrs. R.: (As mother) I don't know what to say to you. I *never* knew what to say to you. I really did love you, you know that.

Th.: Now switch back and reply as yourself.

Mrs. R.: Loved me! All you ever did was criticize me. Nothing I ever did was good enough! (voice beginning to sound more whiny). When I got married to J, you disapproved, you were always coming over and telling me what I was doing wrong with the kids. . . . You made my life miserable. . . . (breaks down and starts to cry).

Th.: Did you hear your voice?

Mrs. R.: Yes.

Th.: What did you hear in it?

Mrs. R.: Well, I guess I sounded kind of complaining, like I'm feeling sor . . . like I'm feeling mad.

Th.: You sounded more like feeling self-pity. Try this on for size: say to your mother, "Look what you've done to me. It's all your fault."

Mrs. R.: Look what you've done. Everything's your fault.

Th.: Now let yourself switch back and forth . . . changing roles.

Mrs. R.: (As mother) Come on, stop blaming me for everything. You are always complaining about something. . . . I . . . I guess I didn't show you much love. I really felt it, but I was unhappy and bitter. You know all I had to go through with your father and brother. You were the only one I could talk to. I'm sorry . . .

Mrs. R.: . . . I know you did love me, Mother, I know you were unhappy (voice much softer now, but sounding real, not whiny or mechanical). . . . I guess there's no use blaming you. You're not around any more.

As the session ends, Mrs. R. has managed to express some of her feelings openly and to begin accepting responsibility for her own life. She no longer blames all of her difficulties on her mother. From a gestalt-therapy perspective, this is movement in the right direction.

EXISTENTIAL THERAPY

Gestalt therapy is not the only therapy to put heavy emphasis on taking responsibility for one's own existence. In fact, some therapies make this existential theme such a central focus that they are called, collectively, *existential therapies* (Frankl, 1963; May, 1958). Existential therapists such as Viktor Frankl and Rollo May, both pictured in Figure 16.9, differ widely in the particular techniques they use, but most try to make their clients aware that they always have choices—and thus some control over their fate. This idea extends to the problems, or symptoms, that bring people in for therapy. For example, an overweight person who complains that he cannot control his eating is probably wrong. To underscore this fact, Viktor Frankl might require the overindulgent eater to *increase* his daily food intake—thus demonstrating vividly that he *can* control it. Similarly, the client who has a hand-washing compulsion and washes 25 times a day might be told to aim for 50, or perhaps 75. This technique of encouraging clients to exercise and even exaggerate their problem behavior is called "paradoxical intention" (Frankl, 1975). Its objective is to emphasize that the client *can* have control over the problem and is thus responsible for overcoming it.

Through techniques of this sort, existential therapists try to prod people into seizing control of their lives. Ideally, clients will come to see their problems as resulting from their own free choices of how to behave. They will thus learn that they are free—even obligated—to choose better ways of coping. For the existentialist, an important part of healthy coping is a well-developed system of values—a clear sense of what one is living for. This emphasis reveals the close link between existential therapy and the philosophical perspective known as existentialism. In fact, existential therapy is not so much a specific set of techniques as it is an intriguing blend of philosophy and psychology.

Behavioral Therapies

Unlike psychodynamic and humanistic/existential therapies, several approaches to therapy place little emphasis on insight, personality dynamics, or the self, but instead, draw heavily on learning principles (Chapter 4) and are aimed at helping people change their behavior. These approaches, collectively called *behavior therapy*, or *behavior modification*, are generally focused on changing observable behavior. One of their major strengths is their emphasis on the direct measurement of behavior and behavior change. Behavior therapists are trained to measure the behavior of their clients with and without therapy so as to assess the effectiveness of the therapy; an example of such an assessment is given in Application 16.1 below. (Another example appeared in Chapter 13, page 556.) When assessment shows that the therapist's method is not changing behavior as had been hoped, an alternative method is tried. This emphasis on assessing and shifting, as need be, has sparked the development of several effective techniques—techniques that work so well that even nonbehavioral therapists have begun to use them.

At the heart of most behavior therapies is this core concept: Psychological problems that come about through learning or conditioning can be undone via the same processes. For example, we know from the study of instrumental conditioning that behavior followed by reward tends to be repeated while behavior that goes unrewarded tends not to be repeated (Chapter 4, pages 147–167). Behaviors that seem maladaptive or abnormal are assumed to follow the same rule: Since they persist because they are being rewarded in some way, they should be reduced or eliminated if they are made less rewarding. Similarly, principles from classical conditioning (Chapter 4, page 140) and observational learning (Chapter 14, page 596) can be applied to the understanding and treatment of problem behavior.

Behavior therapists, for the most part, do not believe that concepts such as "unconscious conflict" or "mental illness" are necessary. Instead, they view their clients as suffering from acquired behavior patterns—patterns that are psychologically costly and need to be unlearned. To help their clients unlearn their costly behaviors and learn more adaptive alternatives, behavior therapists have a rich collection of techniques, a few of which we will describe here.

INSTRUMENTAL, OR OPERANT, CONDITIONING TECHNIQUES

We will focus first on techniques that involve *instrumental*, or *operant, conditioning* (Chapter 4, page 147). The essence of instrumental conditioning is the concept of "payoff." There is enormous therapeutic significance in the simple fact that rewarded responses tend to be repeated while unrewarded responses do not. Therapists who use instrumental conditioning techniques bank on this fact; they work with clients to set up reward contingencies that will stimulate and maintain desired behavior while extinguishing undesired behavior. These therapists usually focus on specific target behaviors rather than trying to eliminate entire psychological disorders, or syndromes, as discussed in Chapter 15, page 622. For example, rather than trying to cure infantile autism (page 631), operant therapists try to help autistic children change specific behaviors that interfere with their adjustment. Similarly, rather than trying to stamp out schizophrenia, operant therapists target specific behaviors for change—the behaviors that pose the biggest social or job difficulties for schizophrenic people.

Functional Analysis of Behavior Often, one of the first steps in planning instrumental, or operant, therapy is a *functional analysis of behavior.* The therapist studies the relationship between the client's behavior and the conditions and events in his or her environment.

> The analysis is often described in A-B-C terms, as follows:
> Antecedents → Behavior → Consequences

Antecedents are discriminative stimuli (Chapter 4, page 160) that act as cues to the individual, signaling when rewards are available and what particular behavior is likely to be reinforced. Think back to Mrs. A., the troubled woman described by Dollard and Miller (Chapter 14, page 593). For her, the presence of attractive men was an *antecedent,* a discriminative stimulus signaling the availability of attention and sexual titillation if she flirted. When Mrs. A. did flirt (that is, when she produced the *behavior* in the A-B-C analysis), she received rewarding *consequences* in the form of attention and sexual stimulation.

Each of us can apply functional analyses to our own maladaptive behavior. Some people smoke cigarettes when under stress. For these people, stressful situations (antecedents) serve as discriminative stimuli for smoking (behavior), which leads to a sense of relief, or reduced tension (consequences). Other people may be overweight in part because situations in which they feel sad or lonely (antecedents) are discriminative stimuli for eating (behavior), which leads to feelings of satisfaction and reduced distress (consequences). In all three examples—Mrs. A., smoking, and obesity—a functional analysis helps explain why the behavior is repeated in spite of its long-term damaging effects.

Identifying Positive and Negative Reinforcers Another critical element in designing instrumental, or operant, therapies is identifying positive and negative reinforcers for the individual client. *Positive reinforcers* (Chapter 4, page 149) are events which, when contingent on a certain response, increase the likelihood that the response will be repeated by the individual. *Negative reinforcers* (Chapter 4, page 149) are events that increase the likelihood of a response when their *termination* is contingent on the response. Using these two definitions, the operant therapist may judge whether particular events are positive or negative reinforcers for the individual client. This is not nearly so simple as it sounds; an event that is a positive reinforcer for one person—a smile or a hug, for example—may not be rewarding at all for another person. The operant therapist must learn to see reinforcement contingencies through the eyes of the individual client, often by directly observing the effects of various events on the client's behavior.

Extinction *Extinction* (Chapter 4, page 156) takes place when the reinforcement for a particular response is withdrawn. Unreinforced, the behavior usually stops (see Application 16.1). The results are sometimes dramatic, even with some problems that appear to be medical, as in the following case report (Walton, 1960).

> A 20-year-old woman had neurodermatitis on her neck, and medical treatments had failed to resolve the problem. A functional analysis of her situation showed that she received attention from her family and fiancé whenever she scratched her neck. The psychologist instructed everyone around her to ignore the scratching. When they did, the scratching ceased and the dermatitis disappeared.

Another technique related to extinction—one often used with children in classrooms—is called *time-out from reinforcement,* or, simply, *time-out.* It is also called *omission training* (Chapter 4, page 150). A place is prepared which is safe and comfortable but very uninteresting—in short, a place in which positive reinforcement is at a minimum (Figure 16.10). Sometimes the time-out area is an empty room, sometimes merely a chair turned toward the wall. As long as children behave appropriately, they have full access to the positive reinforcers available in the classroom (other children, books, ongoing activities, and so on), but this access is contingent upon their continuation of appropriate behavior. Inappropriate behavior leads to a specified period of time in the time-out area; this reduction in positive reinforcement often decreases inappropriate behavior. Many teachers have found time-out procedures to be effective *if* used consistently.

Figure 16.10

In time-out procedures, positive reinforcement is markedly reduced. This child has been removed from the other children and the interesting activities of the classroom for a few minutes. (University of Washington Child Development and Mental Retardation Center.)

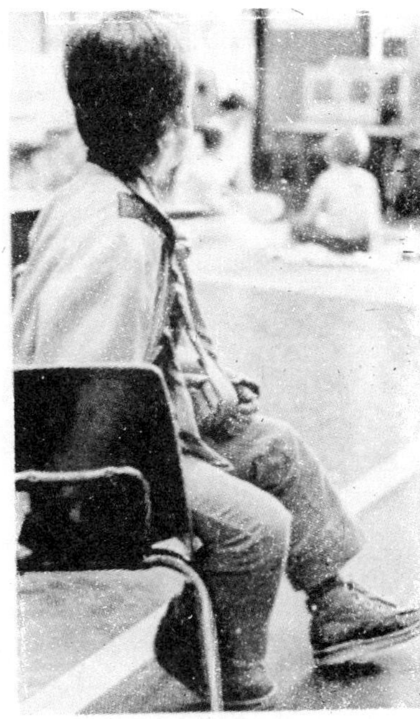

Differential Reinforcement Once a functional analysis has been carried out, positive and negative reinforcers identified, and decisions made about which behaviors to encourage and which to extinguish, it is time to use the operant-therapist's stock in trade: *differential reinforcement.* This consists of giving positive reinforcement for desired behaviors and withholding it in their absence.

Differential reinforcement is used in a very ambitious program for autistic children at University of California, Los Angeles (Lovaas, 1977, 1978). The program is aimed at extinguishing specific undesirable autistic behaviors, such as hand flapping or hair twirling. (See Chapter 15, page 632.) Sometimes these behaviors are extinguished by withholding attention or sweets. Reinforcers such as sherbet or frosted flakes are made contingent on specific nonautistic behaviors, including appropriate speech and social interaction (Figure 16.11).

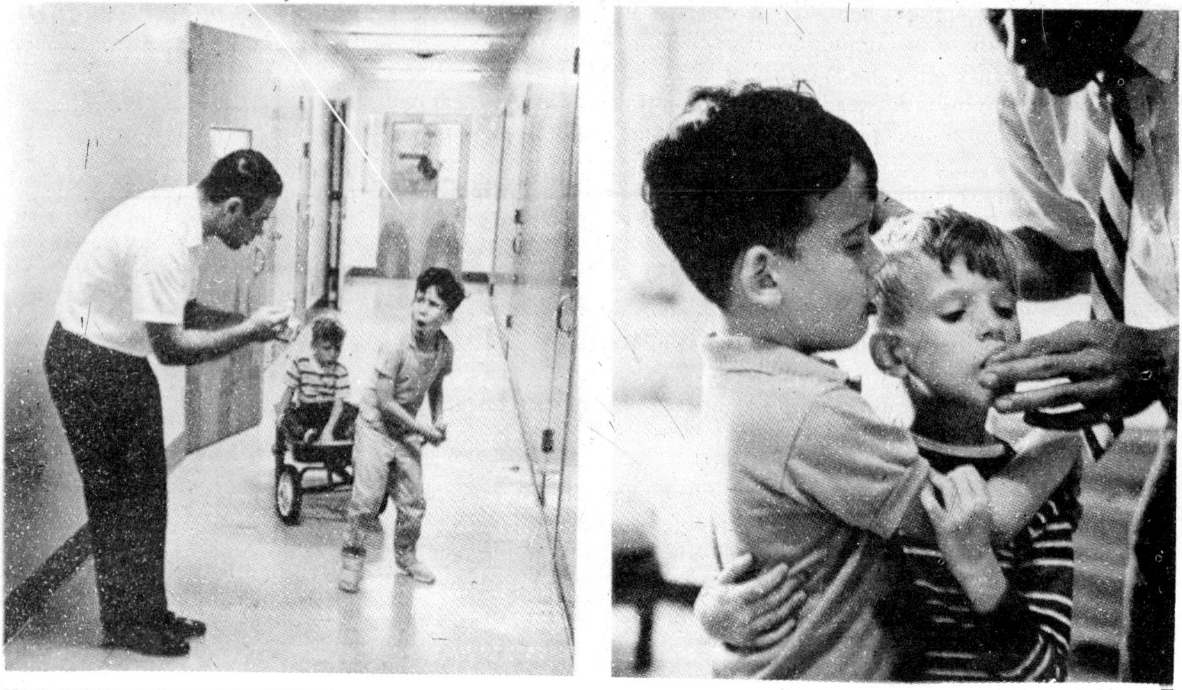

Figure 16.11

Two of the youngsters treated in Lovaas's instrumental-conditioning program for autistic children. Initially reluctant to interact socially, the boys were provided with sherbet contingent on cooperative, interactive play. Under this reinforcement contingency, the boys became less likely to scream and pull away, and more likely to accept closeness and mutual play. (Photos by Allan Grant, *Life* magazine.)

Shaping In the UCLA program, therapists may also provide the child with a prompt, a small sample of the desired behavior, in order to get appropriate learning started. The therapist may pronounce a particular word or demonstrate a specific social behavior, for example, and then encourage the child to try it. If the child's early efforts are imperfect, the therapist may reward successive approximations, closer and closer approaches to the desired response, until the youngster gets it right. This use of prompts and successive approximations is called *shaping* (Chapter 4, page 154)—a sort of gradual nurturing of correct responses. The UCLA shaping procedures have produced significant improvements in speech and social behavior among autistic children. On the other hand, these improvements do not hold up so well unless the contingencies of reinforcement are maintained—for example, by training parents to continue the operant procedures at home (Lovaas, 1978). Autistic youngsters who leave the treatment program and go into settings that do not have these operant procedures often lose many of the improvements they showed. This general problem is found with many operant-treatment programs; improvements in behavior often fade away when the carefully arranged contingencies of reinforcement are no longer in effect.

Differential reinforcement, and other tools of instrumental conditioning, have been applied to many problems other than psychological disorders. For example, instrumental conditioning procedures have been used in an effort to help cancer patients cope with their physical disorder and the painful treatment it requires. In Application 16.1, we describe two such efforts.

Token Economies In a *token economy*, people "earn" objects (tokens), which they can exchange for desirable items, services, or privileges. The tokens are made contingent on appropriate behavior in much the same way that wages are contingent on job duties for most adults. Token economies have been used in institutions, classrooms, sheltered workshops, and in numerous other settings where people are being encouraged to behave in more appropriate or adaptive ways. Here is an example (Paul & Lentz, 1977):

> All the patients in this example had been diagnosed as schizophrenic and institutionalized for at least 2 years. Among the behaviors rewarded with tokens were a number of personal-care and grooming behaviors. Three times a day, the residents were checked for appearance criteria such as the following:
>
> 1. Proper use of makeup
> 2. Clean fingernails
> 3. Hair combed
> 4. Teeth brushed
> 5. Clothing buttoned, zipped, tucked
> 6. Body clean
> 7. No odor
> 8. Shaven
>
> Tokens could also be earned for appropriate classroom and social behavior. With the tokens they earned, residents could purchase normal meals and longer, more leisurely mealtimes. During meals, more tokens could be earned by appropriate table manners; grabbing food off a neighbor's plate, on the other hand, could result in the loss of a token already earned. Residents whose behavior was consistently so inappropriate that they never had enough tokens to buy a normal meal were given food, but not in a very appetizing form. They received "medical meals," the foods in the regular meal processed through a blender, dyed a grayish purple with food coloring, and served in a single container. In addition to buying normal meals, the residents could use their tokens to buy games, snacks, and cosmetic items and to rent privacy aids, such as screens to put around their beds at night, or even private rooms.
>
> The token economy, combined with several other behavioral techniques, resulted in dramatic improvements in the patients' behavior. Appropriate behavior increased sharply during the first 6 months of the program and held up well for 4 years. At the end of the program, 27 of the 28 treated residents were accepted by sheltered-care facilities in the community, and they were still in those facilites 18 months later. By contrast, this was true for only 13 of the 28 chronic mental patients in an untreated control group. Another important group difference: At the end of the treatment program, only 11 percent of the token group were receiving chemotherapy, but 100 percent of the untreated, comparison group were on drugs.

Token-economy programs like this can be quite effective, but they

Application 16.1

USING INSTRUMENTAL CONDITIONING TO HELP CANCER PATIENTS

In recent years, behavioral psychologists have taken their methods beyond the psychological clinic and into medical settings in an effort to help people who suffer from real physical diseases (see section on behavioral medicine, page 708). Among the diseases targeted, none is more devastating psychologically than cancer. While the treatment of cancer itself is a medical enterprise, both the disease and the medical treatments for it can produce serious psychological problems that may be reduced by behavioral treatment. For example, 25 percent of the patients given anticancer drugs develop persistent nausea or vomiting. Others show persistent coughing. William Redd of the University of Illinois is one of several behaviorally oriented psychologists trying to help cancer patients with such problems. In this discussion, we will give two examples of Redd's work, applying instrumental conditioning principles to two somewhat different cancer-related problems.

Two leukemia patients, a 24-year-old man and a 63-year-old woman, were admitted to the hospital for chemotherapy and radiation treatment. Because chemotherapy often reduces the individual's resistance to disease and infection, the patients were placed in private rooms that were declared off limits to other patients; and the patients were allowed only limited contact with their families and friends. All visitors had to be masked, and even the hospital staff had to take elaborate protective measures before entering the room. This acted to discourage even the staff from making brief, casual visits. In his first days in isolation, the male patient developed a "deep, raspy cough" that he said left him feeling weak. Physical examination and laboratory tests revealed no physiological cause for the cough, and yet it seemed to be getting worse and more frequent. The woman, early in her treatment, regurgitated saliva repeatedly. This symptom often occurs early in chemotherapy and is usually accompanied by mouth ulcers. Both problems usually fade away in time. However this patient's regurgitating did not stop even after her mouth ulcers had healed. It seemed possible that the problems developed by this woman and by the man might have psychological rather than medical causes.

Redd explored this possibility by conducting a functional analysis (page 694) of the coughing and regurgitating. He placed tape recorders in the patients' rooms to detect the sounds of coughing and regurgitating; recordings were made during times when the patients were alone and also when a nurse was present. Some revealing patterns emerged. It turned out that the man coughed 25 percent of the time when he was alone but 75 percent of the time when a nurse was present. The woman regurgitated 18 percent of the time when she was alone but 82 percent of the time when a nurse was present. These findings strongly suggested that the coughing and regurgitating were contingent on the nurses' presence. Possibly the symptoms were being positively reinforced by attention from the nurses. For patients so deprived of visitors, having a nurse in the room could be very rewarding. Redd's intervention for both patients was built on this general idea. His plan was as follows:

> "Nurses were instructed to follow their usual procedures, but not to focus patient-nurse discussion on the targeted symptom. . . . If the symptom behavior continued during a procedure, the nurse was to complete the required nursing treatment and immediately leave the room. However, if the symptom ceased or did not occur, the nurse was to remain in the patient's room for at least 10 minutes after completing the required procedures and talk with the patient."
> (Redd, 1980, p. 450)

This operant treatment, then, involved both extinction (eliminating positive reinforcement for the undesired behaviors in an effort to end them) and differential reinforcement (leaving the room when the target behaviors occurred but remaining to talk when the behaviors did not occur). Seven nurses were trained to use these procedures with both patients. Schedules were arranged so that the first nurse cared for one patient daily until the patient had four successive symptom-free nurse visits (that is, no coughing or regurgitating); then the second nurse took over until the same success criterion was attained, and so on. The results are shown in the bar graph. The darkened

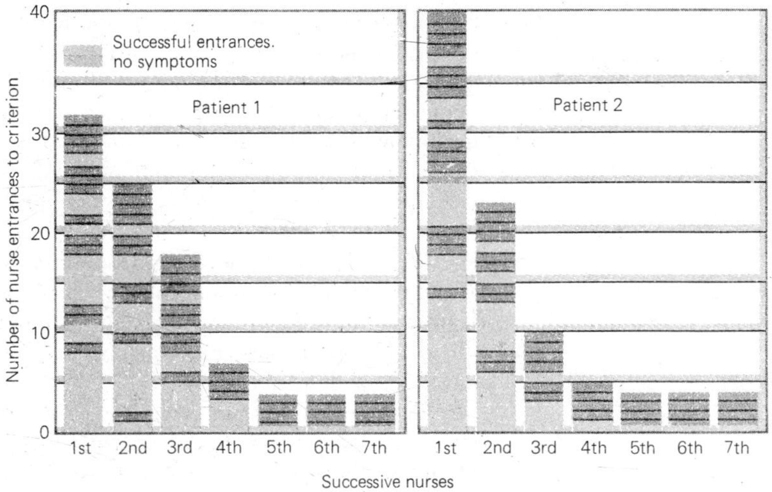

sections in each bar show symptom-free visits. As the graph shows, the man with the cough (patient I) required 32 visits from nurse 1 before he had four symptom-free visits in a row, but his coughing diminished sharply thereafter and was absent completely with the last three nurses. Similarly, the woman had four successive symptom-free visits only after 40 visits from nurse 1 but then improved quickly and did not regurgitate at all with the last three nurses.

Immediately after the instrumental conditioning procedures ended, Redd again assessed the frequency of coughing (Patient 1) and regurgitating (Patient 2) to determine whether they were still contingent on nurse presence. They were not. The man coughed 12 percent of the time when he was alone and 14 percent of the time when a nurse was present. The woman regurgitated 6 percent of the time when she was alone and 10 percent of the time when a nurse was present. At 2-week and 6-month follow-ups, the problem behaviors had not recurred in either patient, and no new problems had developed.

REFERENCES

Redd, W. H. (1980). Stimulus control and extinction of psychosomatic symptoms in cancer patients in protective isolation. *Journal of Consulting and Clinical Psychology, 48,* 448–455.

Redd, W. H., & Andrykowski, M. A. (1982). Behavioral intervention in cancer treatment: Controlling aversive reactions to chemotherapy. *Journal of Consulting and Clinical Psychology, 50,* 1018–1029.

have been questioned for both practical and ethical reasons. A practical problem is that the improvements people make when such token systems are operating often do not hold up when tokens are no longer being given (Condry, 1977). Also, a significant ethical issue arises over the question of human freedom and control. Are token economies so controlling as to be dehumanizing? Many think so; one could argue, for example, that "normal" meals and pleasant sleeping accommodations are basic rights that should not be denied people like the mental patients described above. On the other hand, some argue that strict procedures like those described above are often necessary if real progress is to be made by very seriously disturbed people. The essence of the problem is that there is some truth in both viewpoints. Token economies can produce improved behavior, but they do exact a cost in human freedom. Thus, the procedures bring our society face-to-face with hard questions about our basic values.

Punishment The ethical issues raised by instrumental, or operant, conditioning approaches are perhaps even more sharply drawn in cases where punishment (Chapter 4, page 164) is used. Most operant therapists are concerned about the ethics of punishment; most are also concerned that punishment can interfere with effective learning. Thus most prefer not to use punishment. However, they occasionally do use it as a last resort to stop behavior that is so disordered and dangerous that it must be ended quickly. One example was given by Lang and Melamed (1969).

> The life of a 9-month-old infant was seriously threatened by persistent vomiting and chronic chewing of the vomited food. Several treatments, including dietary changes, administration of antinausea drugs, and various mechanical maneuvers to improve feeding, had been tried without success.
>
> Punishment was applied as a last resort. Electromyographic recordings (electric readings of muscular activity) revealed increased activity of the baby's mouth and throat muscles just before vomiting. A painful electric shock was administered contingent on these mouth and throat movements and was continued until the vomiting stopped. Success was achieved in little more than a week; the punishment led to suppression of the vomiting response. After the vomiting and chewing stopped, the baby gained weight and showed greatly increased activity and social responsiveness.

The pictures in Figure 16.12 show this infant before and after treatment. It seems clear that the punishment worked and that without it the infant might well have died. Note that the electric shocks did not teach a new behavior so much as suppress an ongoing, dangerous one.

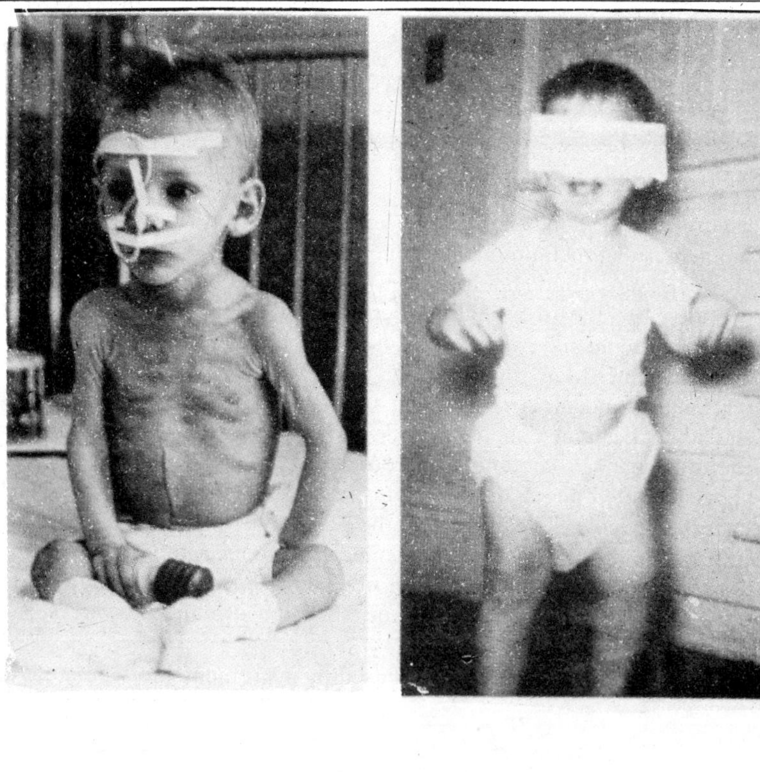

Figure 16.12

A boy who suffered from persistent vomiting and chewing of the vomited food, shown here before (left) and after treatment involving punishment with electric shock. (From Lang & Melamed, 1969. Copyright © by the American Psychological Association. Reprinted by permission of the publishers and authors.)

Covert Sensitization In some cases, a useful alternative to physical punishment is *covert sensitization*. In this technique, the unwanted behavior is *imagined* together with its *imaginary* punishing consequences. For example, an alcoholic is trained to imagine drinking and its consequences, such as vomiting or being fired. Covert-sensitization treatments are also used to treat obesity, smoking, excessive gambling, and compulsive behaviors such as excessive handwashing. The advantage of the technique is that it can be used anytime, anywhere, not just in the therapist's office; the client can imagine the behavior and its consequences privately while riding to work or sitting alone in a library. For this reason, the technique is often used as an extra—a kind of self-therapy to help the client keep things under control in the intervals between visits to the therapist.

CLASSICAL CONDITIONING TECHNIQUES

In Chapter 4 (page 143), we noted that classical conditioning occurs when two stimuli are paired. An unconditioned stimulus is one that produces a response even before the conditioning begins. A conditioned stimulus produces no response before conditioning; but after it has been paired with the unconditioned stimulus a number of times, it begins to produce a response that is often (though not necessarily) similar to the response previously given only to the unconditioned stimulus. After pairing, the response produced by the previously ineffective conditioned stimulus is called the conditioned response. It is these conditioned responses that are acquired in classical conditioning. Conditioned responses can also be extinguished if we present the conditioned stimulus by itself repeatedly. To treat certain behavior problems, classical conditioning can be used to teach new and desired conditioned responses or to extinguish unwanted ones.

Systematic Desensitization Suppose a person experiences distress in specific situations, such as public speaking, or in the presence of specific stimuli, such as snakes or spiders. The person may feel fearful, and the person's body may show responses such as muscle tension and a pounding heart. Sometimes classical conditioning can be used to undo or relieve these bodily responses and fearful feelings. One popular classical conditioning technique used for this purpose is *systematic desensitization*, a procedure in which relaxation and pleasant feelings are learned as conditioned responses to stimuli that once acted as fear producers.

Ordinarily, systematic desensitization begins with some form of relaxation training. In the form known as progressive relaxation, for instance, people are trained to alternately flex and relax their muscles so as to gain control over the relaxation response. To get an idea of how this works, try clenching your fist very tightly, holding it tightly clenched for 10 to 15 seconds; then *very gradually* let the muscles relax, concentrating on the feeling of "letting go." Imagine doing this repeatedly, focusing on various muscle groups throughout the body, and each time learning how it feels to relax specific muscle groups. Another way of teaching relaxation is called autogenic training; the client's attention is focused on specific parts of the body together with such mental images as heaviness and warmth. The therapist may make statements like, "Notice how heavy your arms are beginning to feel, and concentrate on the warm glow spreading from your shoulders down into your arm." Does this sound like your image of hypnotic induction? Hypnosis does appear to be essentially a very deep and exaggerated form of relaxation (see page 708). Regardless of the specific relaxation training techniques therapists may use, most place their clients in a comfortable position in a quiet setting (Figure 16.13), ask them to close their eyes, and employ some sort of repetitive instruction. After a few sessions, most clients can readily put themselves into a state of calm relaxation. Even the therapist's voice or office can become a stimulus for relaxation. At this point, the actual systematic desensitization conditioning can begin.

The conditioning involves systematic pairing of the feared stimuli with a relaxed bodily state. Relaxation is not compatible with the kinds of bodily activity involved in fear and anxiety—we cannot be both relaxed *and* tense at the same time. Thus the fear and anxiety weaken if the client remains relaxed. The principle involved is called *reciprocal inhibition*; it is the notion that two incompatible responses cannot occur at the same time and that the stronger of the two will replace the weaker. To maximize the chance that the relaxation response will be stronger than the competing fear response, the therapist begins by pairing relaxation with very weak fears, then increases the strength of the fears gradually. To begin this process, the client constructs a fear gradient, or anxiety hierarchy; the objects or scenes that the person fears are ranked from least-feared to most-feared. For example, a person who fears cockroaches might come up with an anxiety hierarchy like this:

Cockroach Anxiety Hierarchy

Least-feared:	Hearing a friend mention a cockroach seen in a classroom
↓	Seeing an ad on television for a cockroach spray
	Seeing a cockroach run across the sidewalk 10 yards ahead of me
	.
To	.
	.
	Opening my dormitory-room door and seeing cockroaches scurry everywhere
↓	Waking up in the middle of the night and finding a cockroach in my hair
Most-feared:	Finding a cockroach swimming in my morning coffee, after I've finished half the cup

The skillful therapist first induces relaxation, then introduces images of fear-producing stimuli, starting at the least-feared, or "weak," end of the hierarchy. The images, presented verbally by the therapist or in film or slides (as shown in Figure 16.13), are sequenced to progress toward increasingly strong forms of the feared stimulus. At any point when fear threatens to overcome the relaxation, the client signals the therapist, by raising a finger, for example; at these points, additional relaxation instructions may be given, and they may be paired with "weaker" items in the fear hierarchy. After this "backing off" has reestablished the relaxation

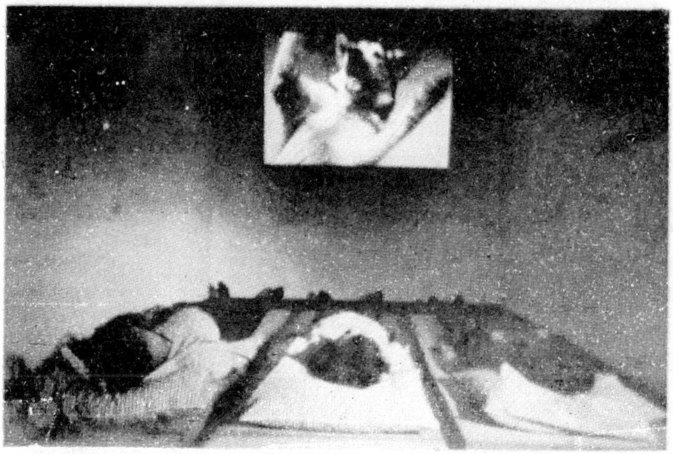

Figure 16.13

An example of systematic desensitization. The people lying on the mats are looking at color-slide projections designed to help them overcome their fear of dogs. At first, they view slides of small, friendly looking dogs. Later, they look at slides of bigger and bigger dogs. Finally, they are exposed to slides showing large, lunging dogs. In this way, they learn how to relax in the face of stimuli that had previously made them afraid. (From Sills. *The New York Times.*)

response, the therapist and client resume the progressive march through the hierarchy; eventually, relaxation should be conditioned to even extreme forms of the stimulus that would previously have produced intense fear. (See also Application 4.1, page 148.) In recent years, relaxation training has been used to treat problems other than fear. We give a few examples in our later section on behavioral medicine (page 708).

Flooding Systematic desensitization works by conditioning a new response (relaxation) to a previously feared stimulus. Another approach to lessening fears is straightforward extinction: A fear-producing stimulus is presented by itself over and over again. Such direct presentation of a high-strength conditioned stimulus, either in imagination or in reality, is called *flooding*. For example, a person who fears heights might be required to spend a certain amount of time each day viewing scenes of high places in still photos or in the movies. Or he or she might actually be required to ride elevators (even the outdoor type, if available) to the tops of tall buildings. The idea is that repeated exposure to the feared stimulus in the absence of any adverse unconditioned stimulus will weaken the classically conditioned fear response. Flooding does sometimes work quite well, but there is a risk that new conditioning may actually worsen the unwanted conditioned response. For example, repeated exposure to extreme heights can sometimes intensify the fear response.

Aversion Therapy Another classical conditioning technique involving unpleasant stimulation is *aversion therapy*. Its objective is not to *undo* fear or revulsion but to *induce* such feelings, specifically in relation to stimuli that trigger unwanted behavior. For example, alcoholics may be given a drug that causes prolonged and severe nausea and vomiting or a drug that produces feelings of suffocation and terror. They are then told to smell and sip their favorite drink. The idea is to condition the alcoholic to feel nausea or panic when exposed to the sight, smell, or taste of the most tempting forms of alcohol.

Aversion therapy should not be confused with punishment. In punishment (Chapter 4, page 150), the noxious event is contingent on the individual's performing a certain response. In aversion therapy, the aim is to condition unpleasant feelings in response to a stimulus. This distinction is made clear by two kinds of drug treatment for alcoholism. As we noted above, in aversion therapy a sickness- or fear-producing drug is paired with the sight, smell, and taste of alcohol. Just being around alcohol makes the person feel ill or frightened; thus he or she avoids drinking. A punishment (instrumental, or operant) approach involves giving people a drug that makes them sick only *after* they have actually taken a drink; sickness is thus contingent on drinking. This kind of therapy works in some cases because the punishment of being sick suppresses the drinking behavior.

A major problem with aversion therapy is that the conditioned aversion responses extinguish unless conditioning sessions are repeated frequently. Theoretically, we could achieve conditioning that would resist extinction, but the conditioning sessions would have to be so extreme and so unpleasant that they would be ethically unacceptable. Those who have seen the movie *A Clockwork Orange* will recall an example of the extremes to which aversion therapy could be taken. In real life, though, aversion therapy is actually most useful in conjunction with other therapies. It helps people avoid unwanted behaviors while other therapeutic techniques are teaching them new, more adaptive responses. To "kick the smoking habit," for example, some form of aversion therapy may help during the first brief period when the going is toughest; then other approaches must be used to help people learn to live their daily lives without cigarettes.

MODELING TECHNIQUES

The form of behavior therapy known as *modeling* is based on the type of cognitive learning (Chapter 4, page 168) known as *observational learning.* As we noted in Chapter 14 (page 596), modeling is often the most efficient means of learning complex skills. Modeling is also a valuable therapeutic tool, especially with phobias. There are at least two ways that observational learning helps people acquire new behavior. First, it provides information on the how of the behavior—the specific steps by which others are able to perform it. Second, it gives evidence of the "do-ableness" of the behavior; the fact that others can do it helps demystify the behavior, makes it less frightening, and encourages the belief that "I can do it, too."

The "I can" notion has become especially important in theory and research on modeling. Bandura (1982) has argued that people's feelings of self-efficacy—that is, their expectations about what behaviors they are able to perform—help to determine what behaviors they will try and how hard they will try. A person's sense of self-efficacy may influence the effects of therapies that use modeling. When modeling is used to overcome a fear, the modeling may be most successful if it changes the fearful person's expectations about what he or she can do (Bandura et al., 1977). Here is an experimental example:

> The people in this experiment were afraid of snakes. One group received modeling-only therapy; they watched the therapist carry out increasingly threatening activities with a boa constrictor, including putting a hand in front of the snake's head as it slithered around the room and holding the snake at eye level. Therapy for a second

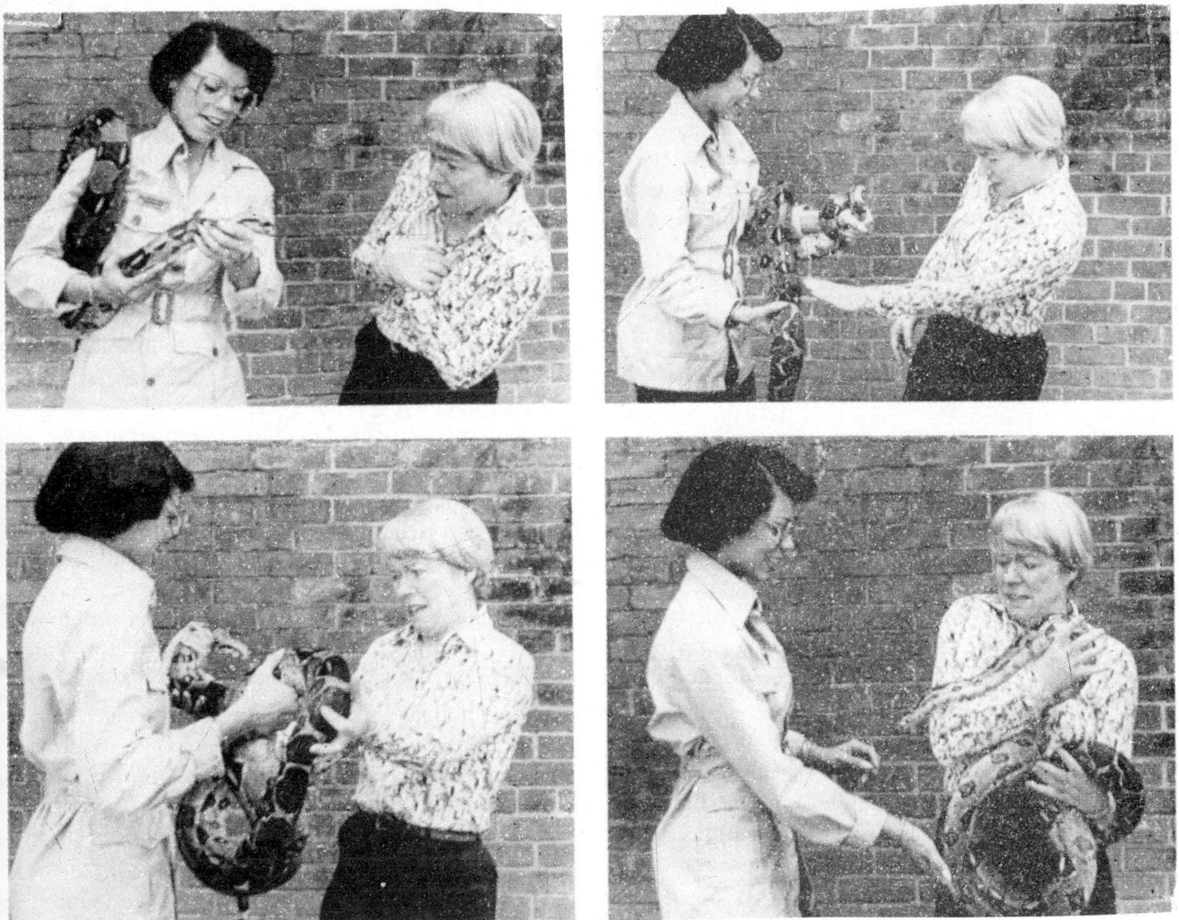

Figure 16.14

In participant modeling, the client goes through a series of steps that are first modeled by the therapist. The treatment is especially effective with specific phobias—debilitating fears of snakes, rats, or insects, for example. Recent evidence suggests that treatment effects are mediated by changes in the client's self-efficacy—that is, the expectation that "I can do it." (Susan Rosenberg/Photo Researchers, Inc.)

group was participant modeling; they not only observed but also began imitating the therapist's actions with the snake (Figure 16.14). As expected, the people in the second group experienced steeper increases in their self-efficacy expectations—that is, confidence in their ability to cope with snakes—than did the people in the modeling-only group. In subsequent tests, the participant-modeling group also showed more success in coping with boas and other snakes than did the modeling-only group. Both treatment groups showed greater increases in both self-efficacy and snake coping than did a no-treatment control group. Overall, regardless of which treatment condition the people were in, changes in their efficacy expectations corresponded closely to the degree to which they learned to approach snakes.

As this study illustrates, the "state of the art" in the modeling approach to therapy has shifted in recent years. For some time, most research in this area focused on what kinds of models and modeling experiences were most effective in changing people's behavior. Bandura's recent work (1982) has broadened the focus of modeling to include research into the thought processes within the individual and how these processes relate to behavior change.

The growing emphasis on cognitive processes, as seen in the area of modeling, is mirrored in other approaches to therapy. In fact, several approaches now emphasize cognitive processes so heavily that they are often grouped under the heading *cognitive therapy*, or, when their behavioral emphasis is strong, *cognitive-behavior therapy*. The thread that stitches these approaches together is the concept that maladaptive behavior comes partly from maladaptive ideas, or cognitions, and that therapy should focus on modifying these cognitions. This general therapeutic approach is sometimes called *cognitive restructuring*.

ELLIS'S RATIONAL-EMOTIVE THERAPY

One of the earliest cognitive-restructuring approaches was Albert Ellis's (1962) *rational-emotive therapy* (*RET*). RET is designed to reveal and break down irrational beliefs that lead to distress. The RET therapist probes the client's behavior and belief system for irrational beliefs that may have been unthinkingly accepted for years and may be causing real suffering. For example, a depressed woman may believe, without ever quite saying it, "I must be completely competent in everything I do." As a consequence, she never lives up to her impossible standards, never feels completely satisfied with her accomplishments, and goes through life feeling like a failure—even though she may have achieved many good things. Or a man who often loses his erection during sex may believe, without really thinking about it, "My failure to perform is a catastrophe; it means I'm a wimp, not a real man." By uncovering such damaging cognitions, stating them clearly, and exposing their irrationality, the rational-emotive therapist tries to help clients structure a more realistic, less punishing belief system.

BECK'S COGNITIVE THERAPY

This same objective is shared by therapists who use Aaron Beck's cognitive therapy (Beck, 1976; Beck, et al., 1979). Beck's approach is designed primarily for the treatment of depression (Chapter 15, page 647). In Beck's approach, the therapist uses "pointed, but friendly, questioning" to root out depressed people's faulty "depressogenic cognitions." For example, one 33-year-old woman, treated for depression following a divorce, discovered a previously unexamined belief that had fueled much of her distress. In therapy, she was discussing her feelings about being a "bad mother"; she felt this way because one of her children was having behavior problems.

> Therapist: Your automatic thought was, "Your children shouldn't fight and act up." And because they do, "I must be a rotten mother." Why shouldn't your children act up?
> Patient: They shouldn't act up because . . . I'm so nice to them.
> T: What do you mean?
> P: Well, if you're nice, bad things shouldn't happen to you. (At this point the patient's eyes lit up.)
> (Beck et al., 1979, p. 251)

Suddenly it had dawned on this woman that she had been basing her expectations on a false belief: "If I'm nice, bad things won't happen to me." A logical extension of this belief was "If bad things happen to me, it means I'm not nice." These were "depressogenic cognitions." They had led the woman to blame herself unfairly for the adverse events in her life—divorce, her children's problem behavior, and so forth.

In addition to examining cognitive errors and distortions, Beck's approach calls for specific kinds of behavior change. "Homework" is assigned after therapy sessions; weekly activities at home are often scheduled in advance during therapy, sometimes on an hour-by-hour basis. One aim is to counter the loss of motivation experienced by many depressed people. Another aim is to ensure that the depressed person will engage in at least *some* rewarding activities at home. Many depressed people actively avoid enjoyable activities. Thus the therapist may require the client to spend a specified number of minutes each day at a potentially enjoyable activity and to keep a record of any changes in mood and cognitions. Unlike behavioral approaches that regard behavior change as an end in itself, Beck's cognitive therapy treats these modified behaviors as a means to an important end: altering depression-generating cognitions (Beck et al., 1979).

MEICHENBAUM'S SELF-INSTRUCTIONAL TRAINING

Donald Meichenbaum's (1977) cognitive approach is even more structured and directive than Beck's. Meichenbaum uses self-instructional training to help clients replace their maladaptive cognitions with rational, positive thoughts, particularly when they are in stressful situations. For example, a college student plagued by test anxiety, when confronted with an exam, may fall into agonizing, self-doubting thought patterns like the following:

> Here we go again—I'm scared to death! I studied, but my mind's a blank. What if I forget the most important stuff, and I bomb on this exam? I could get a D or flunk. If I do, there'll be no chance of a decent grade for the course. If I keep this up, there goes med school.

Obviously this running litany of self-defeat is neither pleasant nor calming. It certainly will not enhance the anxious student's concentration or performance on the exam. Self-instructional training might be used to teach the student monologues like this:

> Okay, how to cope with this exam? First, I'll take long deep breaths and feel myself calming . . . Then I'll tackle the questions one at a time. Now let's try the first one. I'll read it carefully, ponder it for a moment, then scan my memory for the answer . . . If I get anxious, I'll rate my fear from 0 to 10 and watch it drop while I take more deep breaths.

Training in the use of such positive "self-talk" has been used to treat many problems involving achievement deficits, social anxiety, and poor self-control. There is growing evidence to support the effectiveness of several of the cognitive therapies (Miller & Berman 1983; Shapiro & Shapiro, 1982), including Beck's approach to depression and Meichenbaum's approach to problems in which high anxiety or poor self-control interfere with coping.

The Emerging Field of Behavioral Medicine

As the psychologist's collection of methods has been extended to include people with medical problems, a new field has begun to take shape. This emerging field, now often called *behavioral medicine* or *health psychology*, involves the application of behavioral-science knowledge and techniques to problems involving health and physical illness (Agras, 1982; Blanchard, 1982; Pomerleau, 1982). It is not focused on psychological disorders so much as on the behavioral aspects of medical problems—problems such as hypertension, asthma, chronic headache, heart ailments, diabetes, obesity, insomnia, cancer (page 698), and a variety of other physical maladies. In this section, we will give a few examples of the behavioral-medicine approach to helping.

RELAXATION TRAINING

Earlier, when we discussed relaxation training for systematic desensitization of fears, we noted that relaxation training has many other applications. One of them is the treatment of chronic pain. Many kinds of pain involve muscle tension which is incompatible with relaxation. Thus the principle of reciprocal inhibition (page 702) can be put to use by training people to relax those muscles contributing to the pain. Consider chronic headaches, for example, which may affect up to 40 percent of the adult population (Ziegler et al., 1977). Such headaches are often associated with tensions in the facial, neck, and shoulder muscles. In a recent illustrative study, Blanchard et al. (1982) used relaxation training focused on these muscles to treat headache sufferers.

> The therapists used progressive muscle relaxation, with training gradually zooming in on the shoulders, neck, and face. They also used cue-controlled relaxation; clients silently repeated a cue word such as "relax" while taking a deep breath and recalling their sensations during the progressive relaxation training. Clients were encouraged to practice twice daily; audiotapes of their therapists' relaxation instructions were provided for that purpose. After 10 training sessions, spread over 8 to 12 weeks, it was found that both the frequency and intensity of the headaches had dropped sharply.

Sometimes relaxation instructions are supplemented by guided imagery, with the therapist directing the client to focus on pleasant, relaxing scenes—lying on the beach under a warm sun, feeling the ocean spray, and so forth. The client may also be asked in advance to give the therapist some scenes that are especially tranquilizing for him or her.

HYPNOSIS

Sometimes relaxation instructions aim for deep-muscle relaxation and include suggestions of drowsiness or deep sleep. The result, in some but not all clients, is a deep trancelike state in which the clients are very suggestible. This state is sometimes called *hypnosis* (see Wadden & Anderton, 1982, for a discussion of controversies over just what *hypnosis* is). Some therapists use hypnosis to "uncover"—that is, to unravel—the psychological causes of a problem. Within the field of behavioral medicine, however, hypnosis is used for two other purposes.

First, the heightened suggestibility is used to get clients to comply with direct suggestions for behavior change. This is particularly true when hypnotic treatments are used to stop health-threatening habits such as smoking or overeating. To treat obesity, for example, some therapists suggest that the client will feel decreased appetite or will eat less. Others suggest the positive consequences of weight loss—the feelings of pride and confidence that will come when the trimmed-down person looks into the mirror, for example. Some even try planting the suggestion that desirable foods will nauseate the client (Miller, 1982).

Second, hypnosis is used to alleviate nausea and vomiting. In Application 16.1, we described an instrumental conditioning approach used to reduce such problems in a cancer patient undergoing chemotherapy. To round out the picture somewhat, we will illustrate how hypnosis, too, is used for this purpose. (Redd, et al., 1982)

> The patients were six women who were receiving chemotherapy for cancer; all had experienced anticipatory emesis—that is, vomiting in anticipation of chemotherapy—prior to at least three consecutive chemotherapy sessions. Hypnotic induction was used with each patient for from two to five chemotherapy sessions. The induction technique involved three steps: (1) the patient was asked to focus on a single point (for instance, a spot on the ceiling); (2) relaxation instructions were given, focusing first on muscles in the feet and then moving progressively up to the head and then back to the feet; (3) the therapist suggested relaxing imagery, including stepping down a luxurious staircase while becoming more relaxed with each step. The hypnotic sessions lasted about 30 minutes. Figure 16.15 shows the number of times anticipatory emesis (*AE*) occurred and did *not* occur (*No E*) prior to any hypnosis training (*P*), during hypnosis (*H*), and in chemotherapy sessions after the hypnosis had stopped (*N*). As the figure shows, hypnosis prevented anticipatory emesis in all 21 sessions where it was used. On the other hand, emesis occurred on all three occasions when hypnosis was withdrawn. So the hypnosis seemed to help people cope with the immediate stress of chemotherapy, but it did not seem to produce carry-over effects. There were no long-term changes in the patients that persisted beyond the hypnotic induction itself.

The fact that the hypnosis effects did not carry over to the posthypnosis chemotherapy sessions suggests that the effects do not qualify as desensitization (page 701). The patients still seemed sensitized to the unpleasant feelings of cancer chemotherapy. In fact, 6 months after chemotherapy had ended, they still reported feeling nauseated when they entered the medical facility; some reported what seemed to be classical conditioning effects (Chapter 4, page 140), such as feeling nauseated at the "medical" smell of rubbing alcohol. So the patients did not become desensitized to the associations that caused their nausea, but they did

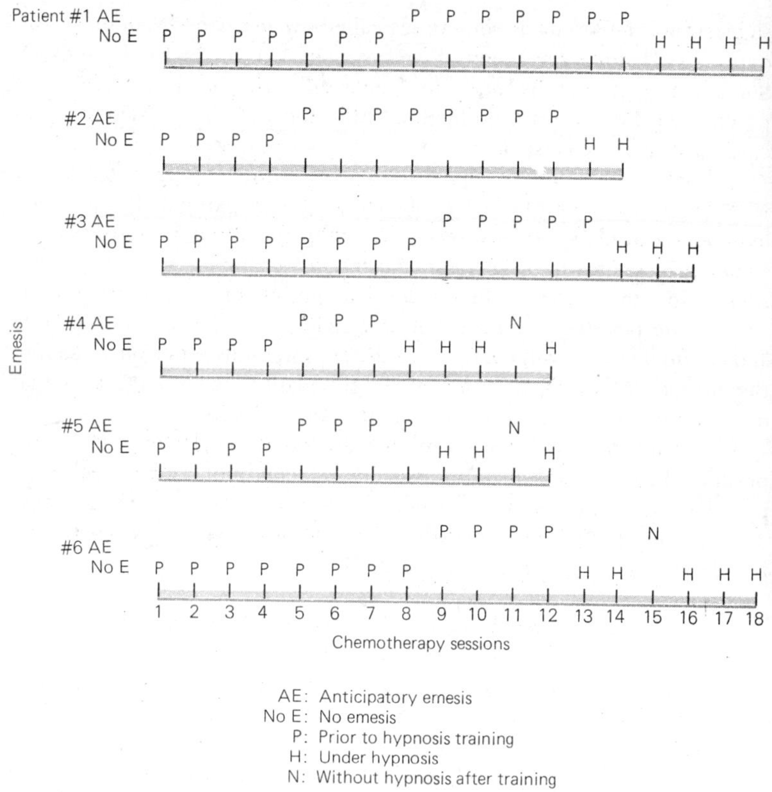

Figure 16.15

Findings of Redd et al. (1982) on the presence or absence of anticipatory emesis (vomiting) during successive cancer chemotherapy sessions with and without the aid of deep-muscle relaxation hypnosis. All six patients experienced emesis during at least three sessions prior to hypnosis; none experienced emesis during hypnosis sessions. Hypnosis was withdrawn for three sessions, and emesis recurred each time. Hypnosis clearly provided a useful means of coping with the aversive effects of the cancer treatment. (From Redd et al., 1982.)

develop a useful way of coping with the nausea. The study suggests that this method might eventually be brought under patient control, perhaps through training in self-hypnosis.

BIOFEEDBACK: IT SOMETIMES WORKS, BUT HOW?

Another technique that has become an integral part of behavioral medicine is *biofeedback*, a procedure in which people learn to modify internal responses such as heart rate and body temperature. The procedure was spurred by research (Miller, 1969) showing that involuntary physiological responses could actually be brought under voluntary control through what seems to be instrumental conditioning. In biofeedback, the conditioning takes the form of information about bodily processes (Figure 16.16). At this moment, most of us could not regulate our heart rates or body temperatures on command; however, if we were connected to a feedback device that lights up every time our heart beats or displays our skin temperature on a screen, the chances are that many of us could learn to control these two bodily processes. The discovery that some people can learn such control is one of the most important surprises of twentieth-century psychology, and it may have major significance for therapy.

One way in which biofeedback has been used is in the treatment of cardiac arrhythmia, or irregular heartbeat. A biofeedback machine flashes a green light when the heart is beating too slowly, a red light when it is beating too fast, and an amber light when the rate is within an acceptable

Figure 16.16

Learning biological control through biofeedback. These clients at the Menninger Clinic in Topeka, Kansas, are using a hand-temperature trainer to learn how to control blood flow and blood pressure, and an electromyograph to learn how to relax muscle tension. (AP/Wide World Photos.)

range. Clients go through training sessions in which they watch the light, pay close attention to subtle bodily cues, and try to keep the amber light on. Some clients learn to regulate their heart rates well enough to control it even when they are no longer getting the electronic feedback.

Biofeedback training has also been effective in relieving Raynaud's syndrome, a disorder in which blood vessels constrict and cause the hands or feet to grow cold. Some forms of headache have also been treated successfully with biofeedback, often with an apparatus giving feedback on muscle tension in the forehead and sometimes with a device giving feedback on the rate of blood flow through certain vessels. Follow-up research shows that the effects of biofeedback have been long-term in certain forms of Raynaud's syndrome and headache, but the technique has been considerably less effective when used for heart and blood pressure problems (Ford, 1982).

Biofeedback evidently works for at least some health problems, but *how* does it work? Probably the most common view is the one with which we introduced this section—the idea that biofeedback is an instrumental conditioning procedure. According to this view, feedback information rewards learners for specific physiological responses, such as speeding up their heart rate or expanding certain blood vessels.

However, several alternative explanations have been proposed (Raczynski et al., 1982). One is that biofeedback is effective because it teaches people skills in generalized relaxation. Relaxation lowers overall sympathetic nervous system activity (Chapter 8, page 314); this, in turn, moves many physiological processes (heart rate, for example) away from overactivity. Another explanation is that biofeedback effects are mediated by cognitions. For example, the cognitive behaviorist Meichenbaum (1976), whose work we discussed earlier, proposes that people learn to control internal processes by using self-statements or specific mental images; these, in turn, trigger specific physiological reactions. For example, one headache victim learned to control her temporal artery blood flow and stop her headaches by picturing in her mind "the cameo lady" from a soap advertisement she had seen.

There is some evidence to support all three views—the instrumental conditioning, relaxation, and cognitive models. We may eventually discover that *how* biofeedback works depends partly on which people and which problems it is used to treat.

Therapy for Groups

Thus far, we have focused on therapies for individuals, but there are several approaches aimed at groups. In general, these approaches use group interactions to reveal, and remedy, problems that are often interpersonal in nature. We will describe two broad group-therapy approaches, beginning with family therapy.

FAMILY THERAPY

In *family therapy*, psychological problems are interpreted and treated in a systemic fashion—that is, in ways that involve the entire family operating as a system. Disturbed behavior by one family member may be stimulated, or worsened, by disturbed relationships in the family as a whole. Furthermore, the disturbed behavior of one individual usually affects the family as a whole. Family therapy may begin with one family member being referred for treatment, but family therapists often label this individual "the identified patient," thus implying that all members of the family are "patients," or that the family as a whole is "the patient." In keeping with this view, most try to arrange sessions in which family members meet as a group to discuss problems, family background, and family relationships. These discussions, and the family interactions they reveal, become grist for the therapist's mill. They provide the therapist with an inside look at the patterns of communication, power plays, techniques of control, and reinforcement contingencies within the family.

How therapists use this inside look depends on their theoretical orientation—the school of family therapy to which they subscribe—and there are many more approaches than we can review here. (For a recent survey of representative approaches, see Gurman & Kniskern, 1981.) However, a number of the major approaches are closely related to the schools of individual psychotherapy described earlier. For example, some family therapists emphasize psychodynamics; that is, they try to uncover the underlying motives and conflicts within the family and stimulate insight into them on the part of family members. Other family therapists use more humanistic-existential approaches, and still others use a behavioral framework. Drawing on these broad traditions of individual therapy, family therapists have developed a number of specific techniques designed to reveal and sometimes reverse damaging family patterns.

One interesting example of an existential approach was developed by a team of family therapists known as the Palo Alto group. Called "prescribing the symptom," this approach resembles paradoxical intention—a technique used by some existential therapists (page 692; see also Frankl, 1975) to focus attention on problem behavior. A Palo Alto therapist might tell a son who is continually looking for pity from other family members to try everything possible to make the family feel sorry for him. Or a mother who completely dominates her family might be told to try her best to ensure that no family decisions are made without her consent. These techniques are controversial because some think they are too deceptive; however, some believe that they help to highlight the role of the "symptom" in the family system and thus ultimately to bring the symptom under the control of family members.

Therapists using behavioral approaches to family therapy aim to assess and modify harmful reinforcement contingencies in the family. These therapists may also analyze and try to restructure undesirable modeling patterns. A special strength of many behavioral approaches is their emphasis on *behavioral exchange*: The behaviors that each family member would like to see in the others are pinpointed explicitly; that information is then used to enhance the rewards of family life for each member, not just the identified patient.

GROUP THERAPY

Like family therapy, *group therapy* is intended to capture some of the essence of "real life" in a social context. Group therapy is sometimes a supplement to individual therapy, sometimes a substitute for it. It is often attractive to people who are not severely distressed—and thus would not otherwise seek treatment—but who want to gain insight or increased sensitivity through structured interaction with others. Group therapy can help people open up areas of thought and feeling that were previously sealed off, enhance their personal growth, and deepen their relationships with others.

The groups, usually consisting of about 5 to 15 people, have one or two leaders, or trainers, as the therapists are sometimes called; the leaders generally try to create an atmosphere in which emotions and feelings are stressed. Group leaders must be highly skilled in dealing with the difficult interpersonal problems that can arise when so many people are responding all at once. They must be sensitive to the anxiety that can develop in some members and capable of protecting those who are fragile. People thinking about joining such a group should investigate to make sure that the group objectives fit their own and that the group is being conducted by competent leaders. This word of caution is especially important for anyone who is considering the unusual and generally unproven group (and individual) approaches described in Application 16.2.

Community Psychology: Outreach and Prevention in Real-Life Settings

Some of life's toughest problems are hard to solve in a therapist's office—even when families or groups are there. Problems that arise in the home, school, or work place sometimes need to be tackled in those same settings. This idea is central to *community psychology*, a collection of strategies for treating and preventing psychological problems in their natural contexts. (See Chapter 1, page 22.)

SOLVING PROBLEMS WHERE THEY HAPPEN

One cluster of strategies involves getting into settings where problems have already begun—schools, homes, nursing-care centers, and so forth. Once there, the psychologist can see problems as they unfold naturally, design interventions to fit the setting, and assess the impact of those

Application 16.2

FROM AQUA ENERGETICS TO Z-PROCESS: A SAMPLER OF UNCONVENTIONAL PSYCHOTHERAPIES

Psychotherapy, as a helping enterprise, attracts the well-intentioned creativity of many who want to help others. In fact, some might argue that founding schools of psychotherapy has become a major cottage industry, particularly in North America. Each year, many new psychotherapies take root and blossom, some (but certainly not all) in the fertile, psychosensitive soil of southern California. Many of these therapies, though popular among their loyalists, remain untested by rigorous outcome research. Yet one who wants a thorough understanding of the current practice of psychotherapy cannot ignore these novel efforts. We will offer a brief introduction, consisting of quick "cuts," to these innovative psychotherapies. The few sentences we will present about each approach will barely convey the "flavor" of any of them, but perhaps your appetite for further reading will be whetted.

Proceeding alphabetically, we begin with *aqua energetics,* a kind of "marathon regression therapy." This therapy takes two forms, but both involve the pursuit of "peak experiences" by groups of people immersed in heated water. In the modest-person's version, swimsuits are worn; in the second form, they are not (see figure). The treatment takes the form of a marathon session of 20 to 24 hours. The 12 to 20 participants engage in a variety of verbal and nonverbal interactions, including passage down the human

(© Don McCoy, Rainbow.)

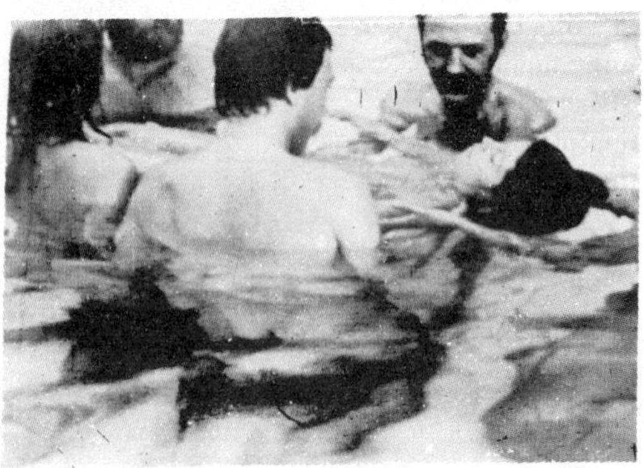

conveyor belt pictured in the figure. As they float on their backs, other group members support and rock them while underwater speakers play "intrauterine sounds" designed to "reduce the rigidity of their body armor." The overall experience is said to facilitate a "regression to infancy"—complete with "volcanic eruptions" of repressed emotion—and to stimulate "energy flow," "bliss," and "a new sense of being."

Encouragement therapy is built on the premise that regardless of the technique a therapist uses, the major reason clients change is because they are motivated to do so. The encouragement therapist tries to motivate clients by stimulating feelings of personal responsibility ("It's up to me") and self-confidence ("I have strengths, and I can do it"). Phase 1 is devoted to building the client-therapist relationship and, in part, finding the client's "claims to fame." Phase 2, cognition focusing, is aimed at stimulating perceived responsibility and confidence. Phase 3, action focusing, involves firming up plans for what the client will do, and when. In phase 4, the therapist hands responsibility for the client's life over to the client.

Natural-high therapy (*NHT*) and *provocative therapy* (*PT*) both emphasize the value of humor. NHT involves lectures, workshops, and "encouragement labs" intended to help people perceive the humorous "basic paradoxes of the human condition," develop feelings of "universal belonging," and achieve "actualization." In PT, the therapist uses pushiness, sarcasm, irony, and insult (á la Don Rickles) to prod clients into defending themselves, asserting themselves, and affirming their self-worth. In the following example, a meek woman with little self-confidence reports for therapy and asks the therapist where she should sit.

THERAPIST: (Pointing at chair next to his desk; the patient begins to sit down in the chair) Sit right there. (In a gruff tone; loudly) Hold it! (Pointing to a chair at the opposite wall) Sit over there.

PATIENT: (Shuffles over to the chair at which the therapist is pointing)

T: (In a commanding tone; looks around the office) No, wait a minute . . . (He pauses, looks uncertain) I've got it! Sit over there (pointing to a chair next to the door).

P: (Suddenly straightening up, frowning; loudly and

forcibly) Aw, go to hell! I'll sit where I want! (She plumps herself down in a chair).

T: (Throwing up his arms as though defending himself; plaintively) Okay, okay, you don't have to get violent!

P: (Bursts out laughing).
(Farrelly & Brandsma, 1974, p. 181)

Orgone therapy is built on the work of Wilhelm Reich, a prodigy who was admitted into the select Vienna Psychoanalytic Society in 1920 when he was still an undergraduate. Reich introduced the concept of character armor, the individual's chronic network of psychological defenses. Late in his career, Reich argued that character armor has a counterpart in muscular armor—that is, spasms, cramps, and various points of tension in the body are expressions of repressed ideas and feelings. Reich argued that neurosis could be attacked from the bodily side by calling the patient's attention to chronic body tensions, "loosening" them, and thus freeing up blocked emotions. Orgone therapy involves various techniques—exaggerating muscle tension so as to identify it clearly, doing breathing exercises, manipulating muscles directly, and so on—all of them designed to loosen tensions and free emotions. A key, and controversial, aspect of orgone therapy is its emphasis on freeing up the muscles and emotions involved in orgasm so as to release what Reich called "orgastic potency." The work of orgone therapy is now promoted through the American College of Orgonomy via its semiannual publication, the *Journal of Orgonomy.*

The key to success in *rebirthing,* described as a "holistic healing method," is a technique called connected breathing, a steady rhythm in which there are no pauses between filling and emptying the lungs. Such breathing is said to stimulate regression to the birth experience. Connected breathing is initially carried out in the company of a trained rebirther (therapist) and in conjunction with affirmations, statements of desired changes in one's life. Some of the affirmations are designed to rid clients, or rebirthees, of the parental-disapproval syndrome (PDS) (for example, "I now forgive my parents for their ignorant treatment of me"). Prolonged connected breathing is said to generate such altered states as "out-of-body experiences," "experiencing interactions with mythological archetypal figures," and "cosmic feelings of total bliss." Rebirthees are urged to practice connected breathing for an hour or more a day "until all major aspects of their lives work to their satisfaction."

Z-process attachment therapy, billed as "a system of human bonding," is designed to help resolve disturbances in interpersonal attachment at all points in the life cycle. The procedure is built on the premise that "psychopathology is fundamentally a disturbance of attachment." The basic technique is a holding session, in which the client reclines while various body parts are held. The client's head rests in the lap of the person who directs the session. One general goal is for the client to achieve a comfortable alignment with the "holders," then to review problems and major life issues in their presence. A more specific objective is for the client to bring emotions—particularly rage—to the foreground for expression and resolution, while maintaining face-to-face contact with "holders." The Z-process was originally known as rage reduction, and the therapy still rests on the assumption that the proper handling of rage is the key to establishing satisfying affectional bonds with others.

This sampler provides only a glimpse of a few unusual psychotherapies. Literally hundreds are available to the would-be client. Whatever your initial impression of these few and the other approaches you will encounter elsewhere, we urge you to bear in mind some venerable advice that applies to psychotherapy as well as to other "consumer products". *Caveat emptor!* ("Let the buyer beware!").

REFERENCES

For more detailed accounts of the therapies identified here and numerous others, see

Corsini, R. J. (Ed.). (1981). *Handbook of innovative psychotherapies.* New York: Wiley.

The chapters from this volume that were cited above include Baker, E. F., & Nelson, A., "Orgone therapy"; Bindrim, P., "Aqua-energetics"; Farrelly, F., & Matthews, S., "Provocative therapy"; Jones, E., "Rebirthing"; Losoncy, L., "Encouragement therapy"; O'Connell, W. E., "Natural high therapy"; Zaslow, R. W., "Z-process attachment therapy".

Farrelly, F., & Brandsma, J. (1974). *Provocative therapy.* Cupertino, CA: Meta.

interventions. Community psychologists often serve as consultants to people who work on the front line of community mental health—employers who hire people with psychological disorders, police who work with troubled individuals and families, departments of social service, and school teachers and administrators (Figure 16.17). By observing and then

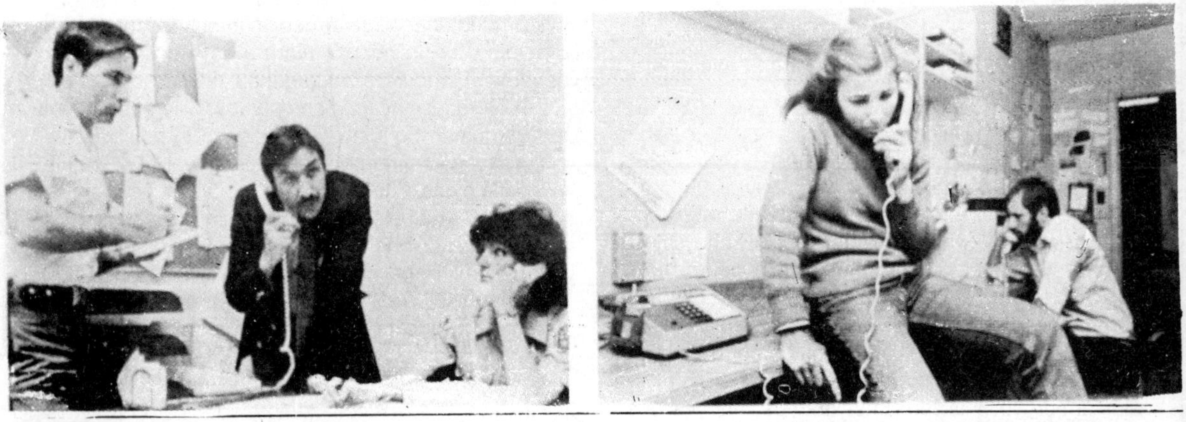

Figure 16.17

Community psychologists specialize in outreach and prevention in the "real world." *Left*, a psychologist works with police to defuse a hostage situation. *Right*, another works with a crisis hot line to counsel people suffering great personal distress or contemplating suicide. (*Left*: Courtesy of Harvey A. Goldstein, Ph.D.; *right*: Paul Fortin/ Stock, Boston.)

making suggestions, the consultant can shed new light on a problem and help generate solutions. The consultant visiting a classroom, for example, may discover that the teacher has been paying attention whenever Marissa, the most difficult child in class, is out of her seat and ignoring her when she is in it—thus rewarding precisely the wrong behavior. The consultant may suggest ways of rearranging the reinforcement contingencies to reward more appropriate behavior. This solution may generalize to the teacher's behavior with other children, thus reducing problem behavior and enhancing teacher effectiveness.

Psychologists have been increasingly active as consultants to police, helping them deal with hostage situations, support victims of violent crime, and cope constructively with family quarrels. Such quarrels are a major focus of police work and a major cause of assault and death. Police themselves are often assaulted when they try to help. In one illustrative project, psychologists trained police in a New York City precinct to calm family conflicts (Bard, 1970). The trained police formed family crisis intervention teams. The training led to big improvements and fewer assaults within families visited by those police. One statistic was particularly impressive: In the first 1,388 family crisis calls that came in after training, the police teams did not suffer one assault or see one family member killed.

PREVENTION

If an ounce of prevention is worth a pound of cure, why do we not try harder to prevent psychological problems from developing? Why spend so much time and money trying to solve problems that might have been prevented in the first place? Such questions have sparked efforts to reduce psychological distress and problems by finding and attacking their causes. This is called *primary prevention*. It includes diverse attempts to stop serious psychological problems before they happen. For example, primary prevention might include educational programs to teach junior high students the dangers of substance abuse. A closely related strategy is *secondary prevention*. Its objective is to identify problems just as they *begin* to develop and stop them before they become severe. For example, if people who have just experienced a job loss or divorce can join a support group, they may avoid the severe depression that often follows these events; or if people in deep distress have a 24-hour help-line phone number to call, they may be able to talk through their problem with a sympathetic listener and lessen the risk of suicide (Figure 16.17). Finally, *tertiary*, or third-level, *prevention* is aimed at minimizing future difficulties in people already identified as having disorders or serious problems. For example, self-help groups such as Alcoholics Anonymous and Parents Anonymous (for parents who have abused their children) can help prevent a recurrence of severe impulse-control problems.

SOCIAL SUPPORT: PERSON-TO-PERSON HELPING

Some of the examples given above involve groups that may not include a mental-health professional. The success achieved by such groups—Alcoholics Anonymous, for example—underscores a very significant point: Some of the world's best therapy is given by nonprofessionals. Sometimes, people can help others cope with their problems simply by providing warmth, empathy, acceptance, or a shoulder to lean on. This is true of college students as well as other population groups. One illustrative study compared the effects of therapy for anxious college students when it was provided by professional psychotherapists and when it was provided by warm, interested Vanderbilt University professors who were not mental-health specialists (Strupp & Hadley, 1979). Most of the measures showed that the professors were about as helpful as the professional psychotherapists. In both groups, the best treatment outcomes were obtained by "therapists" (professional or nonprofessional) who made special efforts to focus discussion on the students' problems, who kept the discussion on the "here and now" rather than on past history, and who encouraged the students to branch out into new social activities. These, of course, are all good, common sense ways of helping—things that friends, relatives, and caring acquaintances do for one another (Figure 16.18). Social support, in these and other forms, can help people cope with stressful life events and can sometimes (though not always) hold down the intensity of psychological disorders such as depression (Leavy, 1983).

Evaluating Psychotherapy: How Much Does Help Help?

As we have seen throughout this chapter, the range of therapies thought by their advocates to be effective is broad, indeed. Each approach discussed in this chapter has its proponents, and therapists naturally use the particular methods they believe will work. But if we set the therapists' opinions aside for the moment and take a cold, hard look at the research evidence, what do we find about the effects of psychotherapy? How well does it work? And do some forms work better than others?

These questions are hard to answer in a scientifically sound manner. One difficulty is that people with different problems at different levels of severity tend to seek out different therapists and different therapeutic approaches. Unless people can be randomly assigned to different forms of therapy and to no-therapy groups, it is hard to judge the relative effectiveness of different therapies or whether the clients might actually have improved on their own without any therapy at all. Another major difficulty

Figure 16.18

Relatives and friends with no special training in psychology often know a lot about what makes for "good therapy." In fact, some of the world's best therapy is probably done by nonprofessionals who care, listen, and support. (Left photo, Ann McQueen/Stock, Boston; right photo, Peter Vandermark, Boston. Cartoon from Cartoon Features Syndicate.)

is that some of the effects of psychotherapy are hard to measure. For example, it is hard to assess factors such as "well-being" and "insight."

Undaunted by these problems, psychologists continue to attempt to assess therapy outcomes, improving their methods at the same time. As these increasingly refined studies accumulate, an overall picture of the effects of therapy has begun to emerge. One pair of investigators (Smith & Glass, 1977) gathered together the findings of 375 different studies of psychotherapy and counseling. The overall analysis showed that the average client receiving psychotherapy was better off than three-quarters of the people who had similar problems but got no therapy. In other words, across various measures of psychological adjustment, the average treated client fell at the 75th percentile of the untreated people (Figure 16.19).

After these findings by Smith and Glass were published, a wave of critiques followed. Some claimed that the review included too many studies with methodological problems; others criticized the procedures used by Smith and Glass to analyze the data (Rachman & Wilson, 1980).

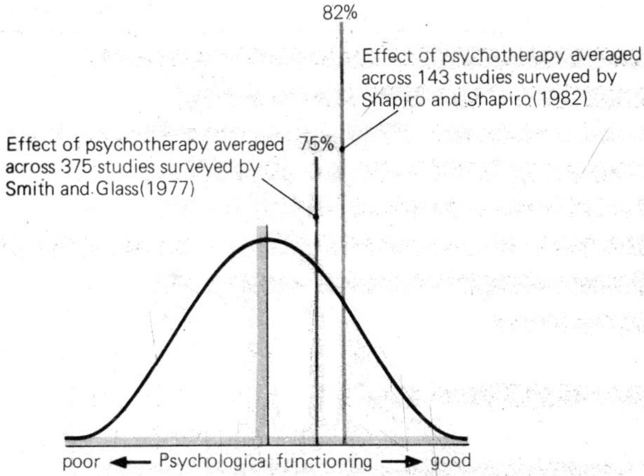

Figure 16.19

Research suggests that psychotherapy is effective. Smith and Glass (1977) surveyed 375 studies of psychotherapy effectiveness. They found that the average treated client was functioning better, after treatment, than about 75 percent of those who received no treatment. Stated another way, the average treated client lay at the 75th percentile of untreated people. In a more recent survey, Shapiro and Shapiro (1982) examined 143 carefully selected therapy studies; they found that the average treated client was at the 82nd percentile of the untreated group. Evidently, psychotherapeutic "help" really does help.

In response, other investigators conducted reviews of their own to correct the alleged problems. In one of these reviews Shapiro and Shapiro (1982) focused on only a very select group of 143 well-designed therapy studies. Yet this review, too, showed psychotherapy to be more effective than no treatment (Figure 16.19). The Smith and Glass analysis did not find significant differences among the effects of various types of therapy. (See also Parloff, 1984.) However, the Shapiro and Shapiro analysis did find significant differences. Figure 16.20 shows Shapiro and Shapiro's results for some of the forms of therapy discussed in this chapter. As you can see, therapy effects were more impressive for the behavioral and cognitive therapies than they were for the psychodynamic and humanistic therapies. Figure 16.20 also shows the size of treatment effects for four different kinds of client problems; it suggests that psychologists may be more effective, at present, in producing measurable improvements in specific phobias and physical and habit problems (for example, chronic headache and obesity) than they are in treating broad emotional problems such as anxiety and depression.

Findings such as these suggest that psychotherapy does help in the treatment of particular problems but that just how much it helps may depend on the type of problem and on the form of psychotherapy given. Eventually, we will certainly find that (1) various forms of treatment are differentially effective with different problems; (2) therapist characteristics and client expectations make a difference; and (3) factors we have not even considered as yet can influence the outcome of therapy. The picture will certainly get more complex than that shown in Figure 16.20. In other words, the evidence is not all in. There is much work to be done as we seek to learn the most effective ways of coping with psychological distress.

Summary

1. After centuries of misunderstanding and often mistreatment of psychologically disturbed people, two mental-health revolutions altered the treatment these people received in many western coun-

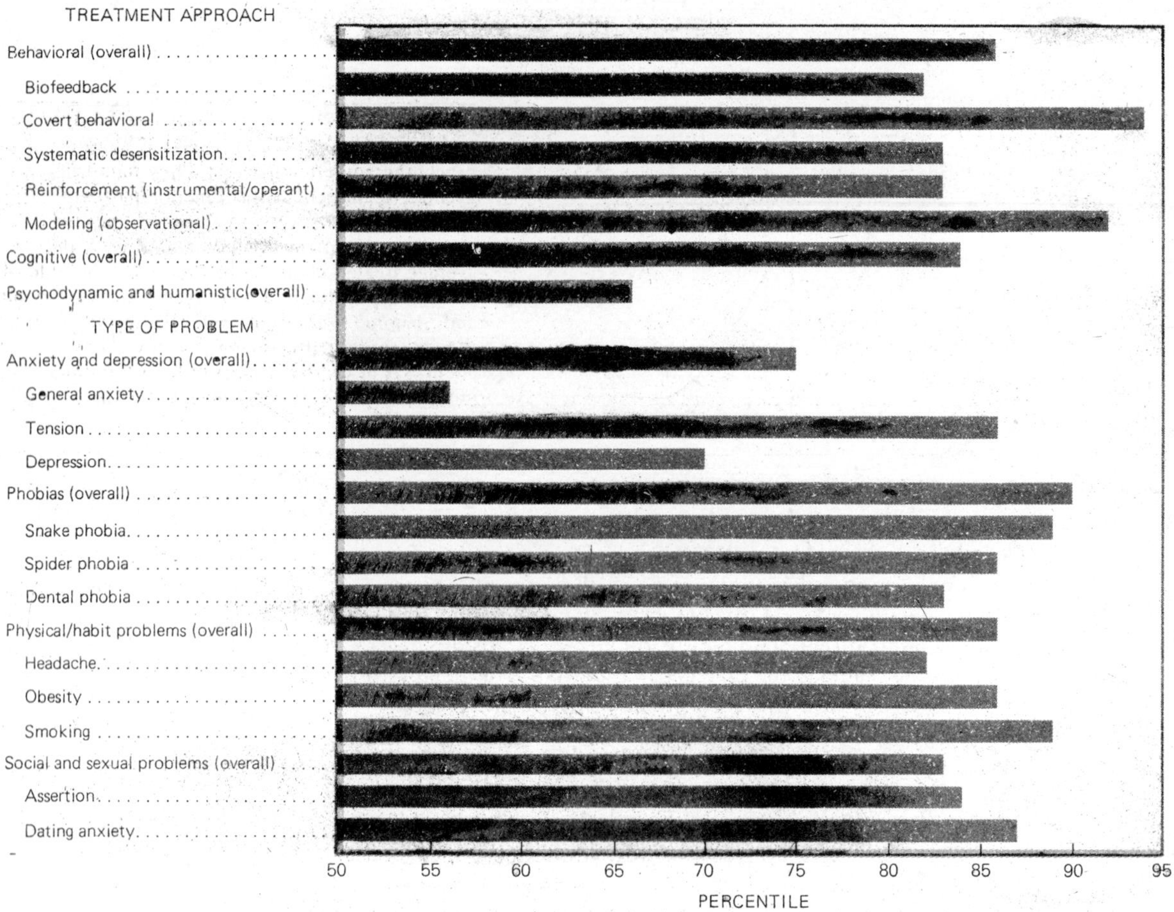

Figure 16.20

Shapiro and Shapiro's (1982) analysis of 143 well-designed psychotherapy-outcome studies indicated positive effects of therapy, but the extent of the benefits depended on the approaches the therapist used and the particular problem treated. The impact of therapy is expressed here in percentiles; for example, the top bar shows that the average client treated with behavioral methods was, after treatment, functioning better than about 86 percent of the people who received no treatment at all. The percentiles suggest that the behavioral and cognitive approaches were more effective, overall, than the psychodynamic and humanistic approaches. Also, psychotherapy appeared to be more effective in alleviating phobias, physical and habit problems, and social and sexual problems than in easing generalized anxiety or depression. (Based on Shapiro & Shapiro, 1982.)

tries. In the nineteenth century, the first mental-health revolution fostered the view that disturbed people were "ill" and promoted humane, medically-oriented treatment. At the turn of the century, Freud and his colleagues helped usher in the second revolution, emphasizing the role of inner motivations and unconscious feelings. Since then, humanistic and behavioral treatments have also risen to prominence.

2. These historical influences have stimulated the development

of biomedical, psychodynamic, humanistic-existential, and behavioral treatment approaches.

3. Who needs treatment? While the line between "normal" and "abnormal" behavior is difficult to draw, surveys indicate that well over 10 percent of children, adolescents, and adults have psychological problems serious enough to warrant professional help.

4. Most forms of treatment are provided by clinical psychologists (Ph.D. or Psy.D. degree in psychology), psychiatrists (medical degree), psychiatric social workers (master's degree in social work), psychiatric nurses (registered nurses specializing in psychological problems), or psychoanalysts (people with or without degrees who have intensive training in Freudian psychoanalysis).

5. Over the past 3 decades, inpatient hospital populations have declined sharply as outpatient treatment, often in community mental-health centers, has grown from less than one-fourth to more than two-thirds of all mental-health treatment in the U.S.

6. Among the biomedical therapies, two—electroshock therapy and psychosurgery—have narrow applicability; they are generally used only as treatments of last resort. The third, chemotherapy, is widely used today in a variety of forms for a variety of problems.

7. Psychotherapy is designed to reveal the psychological causes of behavior problems and to stimulate the development of adaptive new ways of behaving. Psychotherapies that focus on analyzing underlying motives and conflicts and achieving insight are called the psychodynamic approaches.

8. The most famous of the psychodynamic approaches is psychoanalysis. It involves techniques such as free association, dream analysis, and analysis of transference—all used to uncover unconscious urges and conflicts and reduce the need for neurotic defenses. Other psychodynamic approaches include contemporary modifications of Freud's techniques and additional approaches based on the theories of Jung, Adler, Horney, and others who dissented from Freud's views.

9. Humanistic-existential approaches emphasize development of the self and acceptance of personal responsibility for one's life. One example is client-centered therapy, a nondirective approach designed to support an individual's own positive tendency toward wholesome integration. Gestalt therapy and the various existential therapies share this goal but place special emphasis on the theme of personal responsibility.

10. Behavior therapies are built on a key principle: Problematic behavior that is acquired through learning and conditioning can be eliminated by means of techniques derived from learning and conditioning principles.

11. Instrumental, or operant, conditioning approaches emphasize the role of reinforcement in establishing and maintaining unwanted behavior. A functional analysis of behavior often suggests ways in

which behavior can be changed by modifying the antecedent-behavior-consequences pattern.

12. Among the tools of behavior modification using instrumental, or operant, principles are positive and negative reinforcement, extinction, differential reinforcement, token economies, covert sensitization, and (rarely used) punishment.

13. Classical conditioning techniques rely on the pairing of conditioned and unconditioned stimuli as the basis for therapeutic learning. Among the many such techniques are systematic desensitization, flooding, and aversion therapy.

14. Modeling therapies use observational learning as their principal means of inducing change; such learning may take several forms—including participant modeling, which seems especially effective with certain phobias. The effectiveness of modeling may depend partly on the changes it produces in the client's feelings of personal efficacy.

15. Cognitive approaches to therapy are built on the premise that maladaptive or irrational ideas can stimulate undesired behavior. A key therapeutic task is the elimination or modification of these faulty ideas through cognitive restructuring.

16. The application of psychological techniques to people with health problems has generated a new therapeutic speciality known as behavioral medicine. Techniques such as relaxation training, hypnosis, and biofeedback are used to help people cope with a variety of physical ailments.

17. Many therapists have extended their work to families and groups of unrelated people. By treating people in groups, therapists are better able to see problems in social relationships as they emerge in lifelike interactions. Solutions can also be developed and tested in the context of the group.

18. The field of community psychology involves outreach and prevention in "real-life" settings. Community psychologists may work as consultants to family groups, school systems, police forces, and others who face psychological problems that are hard to treat in a specialist's office. In addition, psychologists and others often engage in preventive efforts to head off psychological problems, in groups that are known to be at risk, before the problems actually develop.

19. Studies on the outcome of psychotherapy indicate that therapy does lead to improvement, with the average treated person falling at or above the 75th percentile of untreated people on various outcome measures. However the amount of improvement that therapy produces may depend partly on the approach used, with psychodynamic and humanistic approaches apparently less effective (in some studies) than certain behavioral and cognitive approaches. Outcomes seem also to depend on the type of problem treated; generalized anxiety and depression seem to be particularly difficult to alleviate.

One way to test your mastery of the material in this chapter is to see whether you know what is meant by the following terms.

First mental-health revolution (670)
Biomedical therapies (670, 676)
Second mental-health revolution (670)
Psychodynamic therapies (671, 681)
Humanistic and existential therapies (671, 688)
Behavioral therapies (671, 693)
Behavioral medicine (671, 708)
Community psychology (671, 713)
Clinical psychologists (672)
Psychiatrists (673)
Psychiatric social workers (673)
Psychiatric nurses (673)
Psychoanalysts (673)
Electroconvulsive/electroshock therapy (ECT/EST) (676)
Placebos (677, 679)
Psychosurgery (678)
Prefrontal lobotomy (678)
Prefrontal lobectomy (678)
Chemotherapy (679)
Tranquilizers (679)
Phenothiazines (679)
Tricyclics (679)
Lithium carbonate (680)
Tardive dyskinesia (680)
Psychotherapy (681)
Psychodynamic therapies (681)
Psychoanalysis (681)
Psychoanalytic theory (681)
Anxiety (682)
Insight (682, 685)
Free association (683)
Dream analysis (684)
Resistance (684)
Transference (685)
Countertransference reaction (685)
"Working through" (686)
Analytical therapy (687)
Individuation (687)
Individual therapy (687)
Client-centered therapy (688)
Empathy (689)
Reflection of feeling (689)
Gestalt therapy (690)
Existential therapies (692)
Behavior therapy/behavior modification (693)
Instrumental, or operant, conditioning techniques (694)
Functional analysis of behavior (694)
Positive reinforcer (694)
Negative reinforcer (694)
Extinction (695)
Time-out from reinforcement/time-out (695)
Differential reinforcement (695)
Shaping (696)
Token economy (696)
Punishment (700)
Covert sensitization (701)
Classical conditioning techniques (701)
Systematic desensitization (701)
Reciprocal inhibition (702)
Flooding (703)
Aversion therapy (703)
Modeling (704)
Observational learning (704)
Cognitive therapy/cognitive behavior therapy (706)
Cognitive restructuring (706)
Rational-emotive therapy (RET) (706)
Behavioral medicine (708)
Hypnosis (708)
Biofeedback (710)
Family therapy (712)

Behavioral exchange (713)
Group therapy (713)
Primary, secondary, and tertiary prevention (716, 717)

Suggestions for Further Reading

For those who would like to delve more deeply into the ways personality theories lead into systems of psychotherapy, Joseph F. Rychlak's text, *Introduction to Personality and Psychotherapy,* 2d ed. (Boston: Houghton Mifflin, 1981) will make for interesting reading. For a clearly written, thoughtful introduction to behavioral therapies, complete with case studies and representative research, see William H. Redd, Albert L. Porterfield, and Barbara L. Andersen's *Behavior Modification: Behavioral Approaches to Human Problems* (New York: Random House, 1979). Ralph R. Greenson, a prominent and thoughtful analyst, has written an introduction to analytic methods entitled *The Technique and Practice of Psychoanalysis,* Vol. 1 (New York: International Universities Press, 1967).

Appendix I
Statistics in Psychological Research

When psychologists make observations or do experiments, they measure events and assign numbers to them. By themselves, however, the numbers may not tell us much; to help summarize and interpret the meanings of the numbers, psychologists use various statistics. If you read articles from the psychological literature, a rudimentary knowlege of statistics will be helpful.

Descriptive Statistics: Central Tendency

Usually, many behavioral measures, or scores, are obtained in experiments or observational studies. It is very useful to have a single number that gives us the "average" of the scores. Such a number is called a *measure of central tendency*.

To illustrate measures of central tendency in more detail, Table 1 gives two sets of test scores from a beginning psychology class. The *mean* is a measure of central tendency obtained by dividing the sum of the measures, or scores, by the number of them. The mean of the first set of scores in Table 1 is 39.0; the mean of the second set of scores (the final exam scores) is 116.1. Another measure of central tendency is the *median*, which is simply the point in the group of scores above and below which half the scores fall. The median is sometimes called the *50th percentile* because 50 percent of the scores are at or above it. For the groups of measures shown in Table 1, the medians are 40.0 and 120.5. Because extremely high or low scores will bias the mean more than the median, the median is the preferred measure of central tendency when a group of scores or measures contains some extreme values. Another measure of central tendency is called the *mode*, which is defined as the score that

Table 1

Scores on the first quiz and on the final exam for a class in elementary psychology

STUDENT	FIRST QUIZ SCORE (out of a possible 50)	FINAL EXAM SCORE (out of a possible 150)	RANK* ON FIRST QUIZ	RANK* ON FINAL EXAM
A	49	130	1.0	4.0
B	47	125	2.0	9.5
C	44	127	3.5	5.5
D	44	126	3.5	7.5
E	43	132	5.5	2.0
F	43	116	5.5	13.0
G	42	126	7.5	7.5
H	42	109	7.5	16.5
I	41	121	9.5	11.0
J	41	109	9.5	16.5
K	40 —Median	127	11.5	5.5
L	40	115	11.5	14.0
M	38	125	13.5	9.5
N	38	114	13.5	15.0
O	37	108	15.0	18.0
P	35	120	17.0	12.0
Q	35	99	17.0	18.0
R	35	97	17.0	20.5
S	33	132	19.5	2.0
T	33	67	19.5	22.0
U	29	132	21.5	2.0
V	29	97	21.5	20.5

* Tied scores are given the average rank.

occurs most often. From the data in Table 1, the mode of the first quiz scores is 35. This measure is useful when many of the scores are the same, but, as you can see from Table 1, it may be misleading when the scores are spread over a wide range.

To calculate the mean with a pocket calculator, simply add the values of the measures and divide by the number of them. (Of course, all the calculations mentioned in this Appendix can also be done on a computer if you have the appropriate program or have written one yourself.) The formula for the mean is

$$M = \frac{\Sigma X}{N} \quad (1)$$

The symbols in this formula are defined in Table 2. To calculate the median, rank the measures from highest to lowest and then count down (or up) until you have counted half of them.

Table 2

Definitions of symbols

X or Y	Numerical value of a measurement
M	Mean
Σ	Greek capital sigma used to mean "sum of"
N	Number of cases in a distribution or set of measurements
SD	Standard deviation of a distribution—a descriptive statistic
p	Rho, the rank-difference coefficient of correlation—a measure of correlation when measurements are expressed as ranks
r	Pearson product-moment coefficient of correlation—a measure of correlation.

Descriptive Statistics: Variability

When we have several scores, it is also useful to have a way of describing the spread, or *variability*, of the scores. The *range*, or the interval between the highest and lowest scores, gives a rough idea of the spread of measures. However, the most commonly used measure of variability is the *standard deviation* (SD). The standard deviation is a measure of the spread of measurements around the mean value. If the measurements are grouped closely around the mean value, the standard deviation is small—perhaps one-tenth of the mean. If the measurements are widely spread out around the mean, the standard deviation is large. Figure 1 shows distributions with large and small standard deviations.

A number of formulas can be used to calculate the standard deviation with a pocket calculator. One of the easiest is

$$\text{SD} = \sqrt{\frac{\Sigma X^2}{N} - M^2} \qquad (2)$$

See Table 2 for an explanation of the symbols. Using this formula, first calculate the mean, then find the sum of the squares of the measures, and then apply the formula. Many pocket calculators will automatically add the sums of the squares and many have a square-root function. Some have an internal program that will do the whole calculation of the standard deviation in one step.

Standard scores, or *z scores* as they are known, express each score in terms of standard deviation units. (See Chapter 13, page 533.) After the mean and standard deviation of the whole group of scores have been found, the standard-score value of any particular score in the group is given by

$$z = \frac{X - M}{\text{SD}} \qquad (3)$$

Two distributions differing in variability both have the same mean, but the scores in one are grouped more closely around the mean then are the scores in the other; consequently, they have different variability and different standard deviations (SDs.).

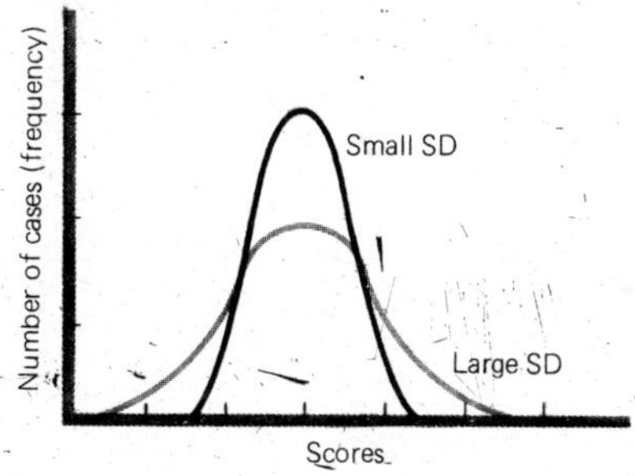

DESCRIPTIVE STATISTICS: CORRELATION

As the term implies, *correlation* refers to a co-relationship between two sets of scores. Thus, to obtain a correlation, we must have two sets of scores on the same individuals. Correlation gives us an answer to the following type of question: Do high scores on an intelligence test go along with high scores on a reading ability test? And conversely, do low scores on an intelligence test go along with low scores on a reading ability test? If so, we have some evidence that there is a correlation, or relationship, between intelligence and reading ability.

To obtain a correlation, we follow a statistical procedure in which we see how the standing of one score among its set of scores compares with its mate in the other set. If the correlation is perfect, that is, if the standing of one score is exactly the same as its mate, and if this is true of all pairs of scores in the two sets, the correlation coefficient, or number expressing this perfect relationship, is 1.00. This is the highest possible correlation. Note that a correlation of −1.00 is also perfect. The negative sign indicates the type of relationship, not the degree of relationship. With a correlation of −1.00, high scores in one set are related to low scores in the other set, and vice versa.

If, on the other hand, no correlation exists, the correlation coefficient is .00. In this case, the standing of a score in one set of scores tells nothing about the standing of its paired score in the other set of scores. The scores in the second set can be anywhere, and no prediction of placement in the second set can be made from knowledge of placement in the first set. Various degrees of correlation are expressed by numbers between .00 and 1.00, or, if the correlation is inverse, between .00 and −1.00. A correlation of .80 (−.80) or .90 (−.90) might be considered high, one of .40 (−.40) to .60 (−.60) moderate, and one of .20 (−.20) or .30 (−.30) low. But these are only general guides; the interpretation of high, moderate, or low for a particular comparison depends upon many factors.

Correlations can be presented visually on a special type of graph called a *scattergram*. Values of one measure are placed on the horizontal axis of the scattergram, while values of the other measure are on the vertical axis. Each point on the scattergram shows where an individual stands on the two measures. For instance, if a person has a score of 49 on one test and 130 on another, the scattergram dot for that person will be at the intersection of 49 and 130. The dot representing this person is circled in the scattergram of Figure 2, which shows the correlation between first quiz scores and final examination scores for the data of Table 1. The correlation between these sets of scores in this class is positive and moderate (.43). If you look at the two columns of scores in Table 1, you will see that there is a tendency for the people scoring higher on the first quiz to score higher on the final examination, while people scoring lower on the first quiz tend to score lower on the final examination. The scattergram of these data also shows this tendency; people scoring high on both tests are plotted at the upper right, while those scoring low on both tests fall at the lower left. This results in a scattergram in which the points form a roughly elliptical figure from lower left to upper right.

With a little practice, one can estimate the direction and degree of correlation from a scattergram. (See Figure 3.) If the dots tend to go from lower left to upper right, the correlation is positive; that is, high scores on measure A tend to go with high scores on measure B, while low scores go with low. But if the dots tend to go in the opposite direction, from upper left to lower right, the correlation is negative (inverse); low scores on measure A tend to go with high scores on measure B, while high scores on

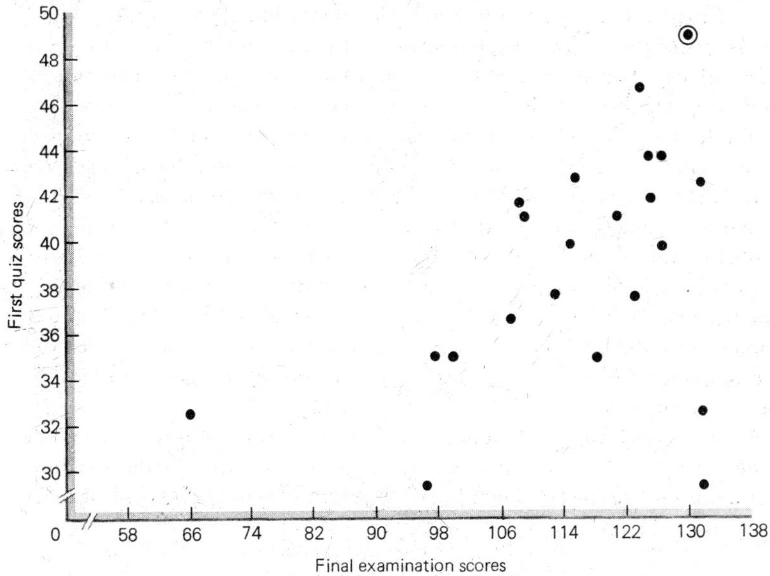

Figure 2

A scattergram of scores made by students in a beginning pscyhology class on the first quiz and the final examination. (Data from Table 1.). Each dot shows where a student stands on the two tests. For example, the circled dot in the upper right represents student A, who had a score of 49 on the first quiz and a score of 130 on the final examination.

measure A go with low scores on measure B. The degree of correlation is shown by the degree to which the dots scatter about. Perfect positive and negative correlations are shown by straight lines. For a perfect correlation, there is no scatter; a certain score on measure A always goes with a certain score on measure B. At the other extreme, a .00 correlation, the dots form a circle because any score on measure A, high or low, can go with any score on measure B. Between perfect and zero correlations, the amount of scatter indicates the degree of correlation—.20 or .80, for instance.

Scattergrams also show how correlations can be used to make predictions. If there is a perfect correlation between two measures (an almost unheard-of event in psychology), we can predict one measure from the other exactly; knowing one, we know what the other will be. Intermediate degrees of correlation lower the accuracy of prediction. For instance, the correlation between first quiz and final examination scores in the above example was .43. This means that although there is a tendency for people who score high or low on the first quiz to do the same on the final examination, there is plenty of room for change.

Figure 3

Scattergrams of several correlations. Each dot represents an individual and is placed at the point where a person's scores on the two measures intersect. (See Figure 2.) As the plot of points comes closer to a straight line, the correlation approached 1.00 or −1.00. The more the plot resembles a circle, the closer the correlation is to .00. Positive correlations trend from lower left to upper right, while negative correlations trend from upper left to lower right.

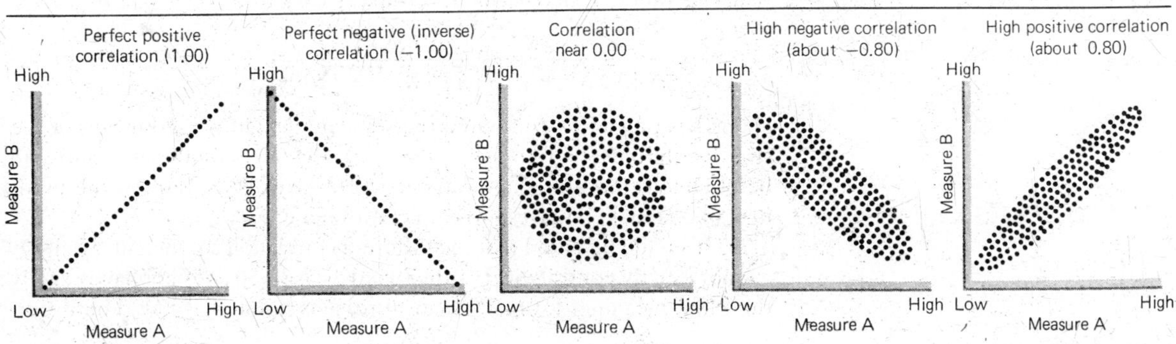

Finally, note that correlations are just what they say they are—measures of the degree of relationship between two sets of measures. By themselves, they do not indicate causation. The fact that two sets of measures are related does not prove that one causes the other. For example, there is said to be a correlation between the number of mules in a state and the average educational level in the state. It is a negative correlation: the more mules, the lower the average educational level. But it hardly seems plausible that this is a causal relationship. Both are probably the result of a common cause—economic conditions. Also, correlations are not percentages, although students sometimes think they are because they are expressed on a scale from .00 to 1.00. Again, it should be emphasized, as the scattergrams make clear, that correlations are statistics for expressing the direction and degree of a *relationship* between measures.

Two kinds of correlation coefficient are in common use. One is the *rank-difference correlation coefficient* (ρ or rho); the other is the product-moment correlation coefficient (r). The formula for the rank-difference correlation coefficient is

$$\rho = 1 - \frac{6\Sigma D^2}{N\,(N^2 - 1)} \qquad (4)$$

The D in the formula refers to a difference in ranks (see Table 3). So the first step in computing a rank-difference correlation is to rank the scores from highest to lowest. If there are tied scores, the average rank is assigned to the tied scores. Thus, in Table 3, the cartoon scores of E and F are tied at 35; since they would fall at ranks 14 and 15, they are assigned the average rank value of 14.5. The same rationale can be applied if more than two scores are tied. To compute ρ, one next obtains the rank difference (D), squares the rank differences, adds the squares, multiplies by 6, and divides by $N(N^2 - 1)$, subtracting the result from 1.

The formula for the *product-moment correlation coefficient* (r) can be written in any one of several ways, depending on what measures are also being calculated and whether a calculator or computer is available. A general formula is

$$\frac{\Sigma z_x z_y}{N} \qquad (5)$$

Writing the formula this way makes it clear that the coefficient is essentially the average of the products of z scores.

When a good calculator is available, the best formula to use is one that looks forbidding, but is nevertheless easy to solve:

$$r = \frac{N\Sigma XY - (\Sigma X)\,(\Sigma Y)}{\sqrt{[N\Sigma X^2 - (\Sigma X)^2]\,[N\Sigma Y^2 - (\Sigma Y)^2]}} \qquad (6)$$

If one has previously computed the standard deviations and means of the two distributions by formulas (1) and (2), the only additional quantity to be obtained from the measurements themselves is the sum of the products of X and Y. The rest is simple arithmetic.

It should be noted that the different correlation coefficients, ρ and r, are not exactly equivalent. In other words, a ρ of .50 does not have exactly the same meaning, for mathematical reasons, as an r of .50. The differ-

Table 3

Humor scores on a cartoon test and a limerick test for 15 people. At the right are the rank orders of these two sets of scores and the differences in ranks (D). The rank-difference correlation, computed from the differences in ranks, is about .70.

INDIVIDUAL	CARTOON SCORE	LIMERICK SCORE	CARTOON RANK	LIMERICK RANK	D
A	47	75	11	8	3
B	71	79	4	6	2
C	52	85	9	5	4
D	48	50	10	14	4
E	35	49	14.5	15	0.5
F	35	59	14.5	12	2.5
G	41	75	12.5	8	4.5
H	82	91	1	3	2
I	72	102	3	1	2
J	56	87	7	4	3
K	59	70	6	10	4
L	73	92	2	2	0
M	60	54	5	13	8
N	55	75	8	8	0
O	41	68	12.5	11	1.5

(*Source:* After Guilford, 1956.)

ences, however, are usually not very large, and methods of correction are available for determining the r that is equivalent to a particular ρ.

Inferential Statistics

When we do experiments or make systematic observations, we typically find differences between the mean values of conditions or groups. Suppose it is observed that a sample of children from an enriched environment has a mean IQ of 120, while a sample from an impoverished environment has a mean of 90. Note that these observations are made on relatively small samples drawn from the larger pools, termed the *populations*, of potential subjects. This question now arises: Is the observed difference a "real" one, or is it simply due to the "luck of the draw" in the choice of samples from the populations? If we can be confident that the difference is not due to sampling bias, or the luck of the draw, we can *infer* that the samples really represent the populations from which they were drawn. In the IQ example, we may find, using inferential statistics, that the difference between the means of 120 and 90 is "real" in the sense that the samples are representative of what we would get if we had results from the whole populations of enriched-environment and impoverished-environment children.

For a particular experiment or set of observations, researchers compute the odds that the obtained difference is due to chance sampling factors. The result is stated as a probability (p) that the difference obtained is a chance one arising from sampling bias. For instance, in the IQ example given above, the probability that the difference obtained between the groups was due to sampling bias might turn out to be less than 1 in 100 ($p < .01$). In other words, there is less than 1 chance out of 100 that the difference obtained was due to chance sampling factors. These are pretty good odds, so the psychologists who made these observations would conclude that sampling bias was not the reason for the group differences. The difference between the groups in this case would be said to be *statistically significant*, because the odds of its being due to chance sampling bias are so low. In practice, odds of 1 in 20, or p values of .05, are usually accepted as statistically significant. Differences with higher odds ($p = .10$, for example) are said to be *nonsignificant*, meaning that chance sampling factors cannot be ruled out with confidence.

Numerous methods for determining statistical significance are in common use. When you read original articles in psychology, you will run across t tests, F tests, χ^2 (chi-square) tests, Mann-Whitney U tests, and other ways of assessing statistical significance. The details for calculating these can be found in any good book on statistics for psychology. The important point is that they all give probability (p) values for the interpretation of the statistical significance of the results obtained in experiments or observations.

Appendix 2 How to Look It Up

Some of the topics mentioned in this text may have aroused interest; we hope so. Many other interesting subjects in psychology have been omitted. If you want to know more, here are a few general guides.

Perhaps the first place to look is in the *Psychological Abstracts*. Here you will find a list of subjects and authors, together with references to the journals and books in which they appear. The *Psychological Abstracts* are arranged by year, so you may have to look through several volumes before you have a fairly complete set of references. Similar index sources are available in other fields related to psychology; *Sociological Abstracts*, *Education Index*, and the *Public Affairs Information Service* provide listings in their respective fields. The *Index Medicus* and *Biological Abstracts* list articles in biology which are related to psychology, and references to articles of general scientific interest with a bearing on psychology may be found in such sources as the *Science Citation Index*. The chances are, however, that if the article has any psychological relevance at all, it will be listed in *Psychological Abstracts*. Incidentally, *Psychological Abstracts* was not started until 1927; a list of references before that date can be found in the *Psychological Index*, a supplement of the *Psychological Review*. In addition, the *Readers' Guide to Periodical Literature* may help you find popular articles on many psychological subjects. These are often useful in the first stages of becoming acquainted with a topic, but serious interest should not stop there. Information about other reference sources, indexes, and computer-based literature searches can be found in the book *How to Find Out in Psychology*, by D. H. Borchardt and R. D. Francis (Elmsford, NY: Pergamon Press, 1984). Most of these reference sources will be available in any reasonably complete college library. If they are not available, you should complain to the librarian and to your professor.

The general card catalog of the library may be of some help if books have been written about the topic in which you are interested. It often happens, however, that a shorter article is more valuable than a whole book, at least in the initial stages of study. You may also use the Suggestions for Further Reading, which are given at the end of each chapter of this book, for locating books on a subject.

When you locate a book or article, look at the list of references at the end of it. These references often prove to be invaluable in calling attention to significant work in an area, for the author has done much of the winnowing for you. For this purpose it is, of course, usually useful to have a recent book or article.

Computer-based searches of the psychological literature are readily available: PsycInfo, for psychology, and MEDLARS, for topics relevant to psychology in medicine and biology, are two of the best known. To make a computer search, start with a fairly small list of specific topics, key words, or authors. You can specify the time period for which the search should be done (1980 to 1985, for example). Several companies now provide access to various information databases. You may be able to arrange your own account number with one of these companies so that, for a fee, you can use any suitably equipped microcomputer to which you have access. Generally, though, the easiest approach may be through your college library; they will do the search for you—again for a fee.

With these hints, and the other resources and techniques you will develop for yourself, you should have no trouble finding information about almost any psychological subject. From here on, it is up to you to select and evaluate the relevance and quality of the articles and books you find. We hope this text and your beginning psychology course have enabled you to begin to do a reasonably good job of this.

References

Abeles, R. P., Steel, L., & Wise, L. L. (1980). Patterns and implications of life-course organization: Studies from Project TALENT. In P. B. Baltes & O. G. Brim, Jr. (Eds.), *Life-span development and behavior* (Vol. 3). New York: Academic Press.

Abrahamsen, D. (1979, July 1). Unmasking "Son of Sam's" demons. *New York Times Magazine*, pp. 20–22.

Abramson, L. Y., Seligman, M. E. P., & Teasdale, J. D. (1978). Learned helplessness in humans: Critique and reformulation. *Journal of Abnormal Psychology*, *87*, 49–74.

Abravanel, E., & Sigafoos, A. D. (1984). Exploring the presence of imitation during early infancy. *Child Development*, *55*, 381–392.

Achenbach, T. M. (1973). Surprise and memory as indices of concrete operational development. *Psychological Reports*, *33*, 47–57.

Achenbach, T. M. (1978). The child behavior profile I: Boys aged 6–11. *Journal of Consulting and Clinical Psychology*, *46*, 478–488.

Achenbach, T. M., & Edelbrock, C. S. (1981). Behavioral problems and competencies reported by parents of normal and disturbed children aged 4 through 16. *Monographs of the Society for Research in Child Development*, *46* (Serial No. 188).

Achenbach, T. M., & Edelbrock, C. S. *Manual for the Child Behavior Checklist and Revised Child Behavior Profile*. Burlington, VT: Queen City Printers.

Adams, J. S. (1965). Inequity in social relationships. In L. Berkowitz (Ed.), *Advances in experimental social psychology* (Vol. 2). New York: Academic Press.

Adamson, R. E. (1952). Functional fixedness as related to problem solving: A repetition of three experiments. *Journal of Experimental Psychology*, *44*, 288–291.

Adelson, J., Green, B., & O'Neil, R. P. (1969). The growth of the idea of law in adolescence. *Developmental Psychology*, *1*, 327–332.

Adler, A. (1929). *Problems of neurosis* (P. Mairet, Ed.). London: K. Paul, Trench, Trubner.

Adler, A. (1931). *What life should mean to you*. New York: Putnam.

Adorno, T. W., Frenkel-Brunswik, E., Levinson, D. J., & Sanford, R. N. (1950). *The authoritarian personality*. New York: Harper & Row.

Agras, W. S. (1982). Behavioral medicine in the 1980s: Nonrandom connections. *Journal of Consulting and Clinical Psychology*, *50*, 797–803.

Ajzen, I., & Fishbein, M. (1973). Attitudinal and normative variables as predictors of specific behaviors. *Journal of Personality and Social Psychology*, *27*, 41–57.

Alan Guttmacher Institute (1981). *Teenage pregnancy: The problem that hasn't gone away*. New York: Author.

Albee, G. W. (1968). Myths, models, and manpower. *Mental Hygiene*, *52*, 2–10.

Allen, K. E., Hart, B., Buell, J. S., Harris, F. R., & Wolf, M. M. (1964). Effects of social reinforcement on isolate behavior of a nursery school child. *Child Development*, *35*, 511–518.

Allen, R. J. (1983). *Human stress: Its nature and control*. Minneapolis, MN: Burgess Publishing Company.

Allport, G. W. (1937). *Personality: A psychological interpretation*. New York: Henry Holt & Company.

Allport, G. W. (1961). *Pattern and growth in personality*. New York: Holt, Rinehart and Winston.

Allport, G. W. (1968). *The person in psychology: Selected essays*. Boston: Beacon Press.

Allport, G. W., & Odbert, H. S. (1936). Trait names, a psycholexical study. *Psychological Monographs*, *47* (Whole No. 211).

Altrocchi, J. (1980). *Abnormal behavior*. New York: Harcourt Brace Jovanovich.

Amabile, T. M., & Glazebrook, A. H. (1982). A negativity bias in interpersonal evaluation. *Journal of Experimental Social Psychology*, *18*, 1–22.

American Psychiatric Association (1980). *Diagnostic and statistical manual of mental disorders* (3rd ed.). Washington, D. C.: Author.

Anand, B. K., & Brobeck, J. R. (1951). Hypothalamic control of food intake in rats and cats. *Yale Journal of Biology and Medicine*, *24*, 123–140.

Anastasi, A. (1982). *Psychological testing* (5th ed.). New York: Macmillan.

Andreasen, N. C., Olsen, S. A., Dennert, J. W., & Smith, M. R. (1982). Ventricular enlargement in schizophrenia: Relationship to positive and negative symptoms. *American Journal of Psychiatry*, *139*, 297–302.

Ansbacher, H. L., & Ansbacher, R. R. (1956). *The individual psychology of Alfred Adler*. New York: Harper & Row.

Anthony, W. A., Buell, G. J., Sharratt, S., & Althoff, M. E. (1972). Efficacy of psychiatric rehabilitation. *Psychological Bulletin*, *78*, 447–456.

Arenberg, D., & Robertson-Tchabo, E. A. (1977). Learning and aging. In J. E. Birren & K. W. Schaie (Eds.), *Handbook of the psychology of aging*. New York: Van Nostrand Reinhold.

Aronson, E., & Cope, V. (1968). My enemy's enemy is my friend. *Journal of Personality and Social Psychology*, *8*, 8–12.

Aronson, E., & Mills, J. (1959). The effect of severity of initiation on liking for a group. *Journal of Abnormal and Social Psychology*, *59*, 177–181.

Asch, S. E. (1951). Effects of group pressure upon the modification and distortion of judgments. In H. S. Guetzkow (Ed.), *Groups, leadership and men*. Pittsburgh: Carnegie University Press.

Asch, S. E. (1956). Studies of independence and conformity: I. A minority of one against a unanimous majority. *Psychological Monographs*, *70* (Whole No. 416).

Asch, S. E., & Zukier, H. (1984). Thinking about persons. *Journal of Personality and Social Psychology*, *46*, 1230–1240.

Atkinson, J. W. (1964). *An introduction to motivation*. New York: Van Nostrand Reinhold.

Atkinson, J. W., & Birch, D. C. (1978). *An introduction to motivation* (2nd ed.). New York: D. Van Nostrand.

Atkinson, R. C., & Shiffrin, R. M. (1968). Human memory: A proposed system and its control processes. In K. W. Spence & J. T. Spence (Eds.), *The psychology of learning and motivation* (Vol. 2). New York: Academic Press.

Ausubel, N. (1948). *A treasury of Jewish folklore.* New York: Crown Publishers.

Averbach, E., & Sperling, G. (1961). Short-term storage of information in vision. In C. Cherry (Ed.), *Information theory.* London: Butterworths.

Ax, A. F. (1953). The physiological differentiation between fear and anger in humans. *Psychosomatic Medicine, 15,* 433–442.

Azrin, N. H., & Foxx, R. M. (1974). *Toilet training in less than a day.* New York: Simon & Schuster.

Baker, E. F., & Nelson, A. (1981). Orgone therapy. In R. J. Corsini (Ed.), *Handbook of innovative psychotherapies.* New York: Wiley.

Balkwell, C. (1981). Transition to widowhood: A review of the literature. *Family Relations, 30,* 117–127.

Baltes, P. B., Cornelius, S. W., & Nesselroade, J. R. (1979). Cohort effects in developmental psychology. In J. R. Nesselroade & P. B. Baltes (Eds.), *Longitudinal research in the study of behavior and development.* New York: Academic Press.

Bammer, K., & Newberry, B. H. (1983). *Stress and cancer.* Toronto: C. J. Hogrefe.

Bancroft, J. (1978). The relationship between hormones and sexual behavior in humans. In J. B. Hutchinson (Ed.), *Biological determinants of sexual behavior.* Chichester, England: Wiley.

Bandura, A. (1968). A social learning interpretation of psychological dysfunction. In P. London & D. Rosenhan (Eds.), *Foundations of abnormal psychology.* New York: Holt, Rinehart and Winston.

Bandura, A. (1973). *Aggression: A social learning analysis.* Englewood Cliffs, NJ: Prentice-Hall.

Bandura, A. (1977). *Social learning theory.* Englewood Cliffs, NJ: Prentice-Hall.

Bandura, A. (1982). Self-efficacy mechanism in human agency. *American Psychologist, 37,* 122–147.

Bandura, A. (1983). Psychological mechanisms of aggression. In R. G. Geen & E. I. Donnerstein (Eds.), *Aggression: Theoretical and empirical reviews: Vol. 1.* New York: Academic Press.

Bandura, A., Adams, N. E., & Beyer, J. (1977). Cognitive processes mediating behavioral change. *Journal of Personality and Social Psychology, 35,* 125–139.

Bandura, A., Ross, D., & Ross, S. A. (1963). Imitation of film-mediated aggressive models. *Journal of Abnormal and Social Psychology, 66,* 3–11.

Bandura, A., & Walters, R. H. (1963). *Social learning and personality development.* New York: Holt, Rinehart and Winston.

Barash, D. P. (1979). *The whisperings within.* New York: Harper & Row.

Barash, D. P. (1982). *Sociobiology and behavior* (2nd ed.). New York: Elsevier.

Bard, M. (1970). *Training police as specialists in family crisis intervention.* Washington, DC: U. S. Government Printing Office.

Barfield, R. E., & Morgan, J. N. (1978). Trends in satisfaction with retirement. *The Gerontologist, 18,* 19–23.

Baron, R. A., (1976). The reduction of human aggression: A field study of the influence of incompatible reactions. *Journal of Applied Social Psychology, 6,* 260–274.

Baron, R. A. (1977). *Human aggression.* New York: Plenum Press.

Baron, R. A. (1983). The control of human aggression: A strategy based on incompatible responses. In R. G. Geen & E. I. Donnerstein (Eds.), *Aggression: Theoretical and empirical reviews: Vol. 2. Issues in research.* New York: Academic Press.

Baron, R. A., Byrne, D., & Kantowitz, B. H. 1977). *Psychology: Understanding behavior.* Philadelphia: Saunders.

Baron, R A., & Kepner, C. R. (1970). Model's behavior and attraction toward the model as determinants of adult aggressive behavior. *Journal of Personality and Social Psychology, 14,* 335–344.

Barrett, C. J. (1978). Effectiveness of widows' groups in facilitating change. *Journal of Consulting and Clinical Psychology, 46,* 20–31.

Barron, F. (1963). *Creativity and psychological health.* Princeton, NJ: Van Nostrand Reinhold.

Bartlett, F. C. (1932). *Remembering: An experimental and social study.* London: Cambridge University Press.

Bartlett, F. C. (1958). *Thinking: An experimental and social study.* New York: Basic Books.

Bartlett, K. (1981, October 4). Psychologist stirs up twin brouhaha. *Durham Morning Herald,* p. 20.

Barton, A. (1974). *Three worlds of therapy: An existential-phenomonological study of the therapies of Freud, Jung, and Rogers.* Palo Alto, CA: National Press Books.

Bassuk, E. L., & Gerson, S. (1978). Deinstitutionalization and mental health services. *Scientific American, 238* (2), 46–53.

Bateson, G., Jackson, D. D., Haley, J., & Weakland, J. H. (1956). Toward a theory of schizophrenia. *Behavioral Science, 1,* 251–264.

Baumrind, D. (1967). Child care practices anteceding three patterns of pre-school behavior, *Genetic Psychology Monographs, 75,* 43–88.

Baumrind, D. (1975). Early socialization and adolescent competence. In S. E. Dragastin & G. H. Elder, Jr. (Eds.), *Adolescence in the life cycle.* New York: Wiley.

Baumrind, D. (1980). New directions in socialization research. *American Psychologist, 35,* 639–652.

Bayley, N. (1969). *Manual for the Bayley Scales of Infant Development.* New York: Psychological Corporation.

Beck, A. T. (1974). The development of depression: A cognitive model. In R. J. Friedman & M. M. Katz (Eds.), *The psychology of depression: Contemporary theory and research.* Washington, DC: Winston.

Beck, A. T. (1976). *Cognitive therapy and the emotional disorders.* New York: International Universities Press.

Beck, A. T., Rush, A. J., Shaw, B. F., & Emery, G. (1979). *Cognitive therapy of depression.* New York: Guilford Press.

Beck, A. T., & Young, J. E. (1978, September). College blues. *Psychology Today, 12,* pp. 80–92.

Bell, A. P., Weinberg, M. S., & Hammersmith, S. K. (1981). *Sexual preference: Its development in men and women.* Bloomington, IN: Indiana University Press.

Belsky, J., & Steinberg, L. D. (1978). The effects of day care: A critical review. *Child Development, 49,* 929–949.

Belson, W. A. (1981). *The design and understanding of survey questions.* Aldershot, Hants, England: Gower Publishing Co.

Bem, D. J. (1965). An experimental analysis of self-persuasion. *Journal of Experimental Social Psychology, 1,* 199–218.

Bem, D. J. (1967). Self-perception: An alternative interpretation of cognitive disssonance phenomena. *Psychological Review, 74,* 183–200.

Bem, D. J. (1972). Self-perception theory. In L. Berkowitz (Ed.), *Advances in experimental social psychology* (Vol. 6). New York: Academic Press.

Bem, D. J., & Allen, A. (1974). On predicting some of the people some of the time: The search for cross-situational consistencies in behavior. *Psychological Review, 81,* 506–520.

Bemis, K. (1978). Current approaches to the etiology and treatment of anorexia nervosa. *Psychological Bulletin, 85,* 593–617.

Benson, H., Kotch, J. B., Crassweller, K. D., & Greenwood, M.

M. (1977). Historical and clinical considerations of the relaxation response. *American Scientist, 65*, 441–445.

Benton, A. L. (1959). *Right-left discrimination and finger localization.* New York: Harper & Row.

Berendt, J. (1973, November). Where were you? *Esquire*, pp. 136–137.

Berkowitz, L. (1962). *Aggression: A social psychological analysis.* New York: McGraw-Hill.

Berkowitz, L. (1983). Aversively stimulated aggression: Some parallels and differences in research with animals and humans. *American Psychologist, 38*, 1135–1144.

Berkowitz, L., & Daniels, L. R. (1963). Responsibility and dependency. *Journal of Abnormal and Social Psychology, 66*, 429–437.

Berkowitz, L., & LePage, A. (1967). Weapons as aggression-eliciting stimuli. *Journal of Personality and Social Psychology, 7*, 202–207.

Berlyne, D. E. (1971). *Aesthetics and psychobiology.* New York: Appleton-Century-Crofts.

Bermant, G., & Davidson, J. M. (1974). *Biological bases of sexual behavior.* New York: Harper & Row.

Bertelsen, A., Harvald, B., & Hauge, M. (1977). A Danish twin study of manic-depressive disorders. *British Journal of Psychiatry, 130*, 330–351.

Bever, T. G. (1973). Language and perception. In G. A. Miller (Ed.), *Communication, language, and meaning.* New York: Basic Books.

Biehler, R. F. (1981). *Child development: An introduction* (2nd ed.). Boston: Houghton Mifflin.

Bindrim, P. (1981). Aqua-energetics. In R. J. Corsini (Ed.), *Handbook of innovative psychotherapies.* New York: Wiley.

Binet, A., & Simon, T. (1916). New methods for the diagnosis of the intellectual level of subnormals. In E. S. Kite (Trans.), *The development of intelligence in children.* Baltimore: Williams & Wilkins. (Original work published 1905)

Binet, A., & Simon, T. (1973). *The development of intelligence in children.* New York: Arno Press. (Original work published, in translation, 1916)

Bini, L. (1938). Experimental researches on epileptic attacks induced by the electric current. *American Journal of Psychiatry, 94* (Supplement), 172–174.

Birren, J. E., & Renner, J. (1980). Concepts and issues of mental health and aging. In J. E. Birren & R. B. Sloane (Eds.), *Handbook of mental health and aging.* Englewood Cliffs, NJ: Prentice Hall.

Blachly, P. H. (1977). Attitudes, data, and technological promise of ECT. *Psychiatric Opinion, 14*, 9–12.

Blackburn, J.A. (1984). The influence of personality, curriculum, and memory correlates on formal reasoning in young adults and elderly persons. *Journal of Gerontology, 39*, 207–209.

Blackstock, E. G. (1978). Cerebral asymmetry and the development of early infantile autism. *Journal of Autism and Childhood Schizophrenia, 8*, 339–353.

Blakemore, C. (1974). Developmental factors in the formation of feature extracting neurons. In F. G. Worden & F. O. Schmitt (Eds.), *The neurosciences: Third study program.* Cambridge, MA: MIT Press.

Blakemore, C., & Cooper, G. F. (1970). Development of the brain depends on the visual environment. *Nature, 228*, 477–478.

Blanchard, E. B. (1982). Behavioral medicine: Past, present, and future. *Journal of Consulting and Clinical Psychology, 50*, 795–796.

Blanchard, E. B., Andrasik, F., Neff, D. F., Arena, J. G., Ahles, T. A., Jurish, S. E., Pallmeyer, T. P., Saunders, N. L., Teders, S. J., Barron, K. D., & Rodichok, L. D. (1982). Biofeedback and relaxation training with three kinds of headache: Treatment effects and their prediction. *Journal of Consulting and Clinical Psychology, 50*, 562–575.

Blasdel, G. G., Mitchell, D. E., Muir, D. W., & Pettigrew, J. D. (1977). A physiological and behavioural study in cats of the effect of early visual experience with contours of a single orientation. *Journal of Physiology, 265*, 615–636.

Blass, D. C. (1977). Stress, behavior patterns, and coronary disease. *American Scientist, 65*, 177–187.

Bloch, S., & Reddaway, P. (1977). *Psychiatric terror: How Soviet psychiatry is used to suppress dissent.* New York: Basic Books.

Block, J. H. (1976). Issues, problems, and pitfalls in assessing sex differences: A critical review of *The Psychology of Sex Differences. Merrill-Palmer Quarterly, 22*, 283–308.

Blumer, D., & Benson, D. F. (1975). Personality changes with frontal and temporal lesions. In D. F. Benson & D. Blumer (Eds.), *Psychiatric aspects of neurologic disease.* New York: Grune & Stratton.

Boden, M. (1977). *Artificial intelligence and natural man.* New York: Basic Books.

Bolles, R. C. (1970). Species-specific defense reactions and avoidance learning. *Psychological Review, 77*, 32–48.

Bolles, R. C. (1975). *Theory of motivation* (2nd ed.). New York: Harper & Row.

Bolles, R. C. (1979). *Learning theory* (2nd ed.). New York: Holt, Rinehart and Winston.

Bolton, T. L. (1892). The growth of memory in school children. *American Journal of Psychology, 4*, 362–380.

Borgida, E., & Campbell, B. (1982). Belief relevance and attitude-behavior consistency: The moderating role of personal experience. *Journal of Personality and Social Psychology, 42*, 239–247.

Boring, E. G. (1930). A new ambiguous picture. *American Journal of Psychology, 42*, 444–445.

Bouchard, T. J., Jr. (1984). Twins reared together and apart: What they tell us about human diversity. In S. W. Fox (Ed.), *Individuality and determinism.* New York: Plenum Press.

Bower, G. H., & Clark, M. C. (1969). Narrative stories as mediators for serial learning. *Psychonomic Science, 14*, 181–182.

Bower, G. H., Clark, M. C., Lesgold, A. M., & Winzenz, D. (1969). Hierarchical retrieval schemes in recall of categorized word lists. *Journal of Verbal Learning and Verbal Behavior, 8*, 323–343.

Bower, G. H., & Hilgard, E. R. (1981). *Theories of learning* (5th ed.). Englewood Cliffs, NJ: Prentice-Hall.

Bowers, K. (1973). Situationism in psychology: An analysis and a critique. *Psychological Review, 80*, 307–336.

Bowlby, J. (1966). *Maternal care and mental health.* New York: Schocken.

Bowlby, J. (1980). *Attachment and loss: Vol. 3. Loss: Sadness and depression.* New York: Basic Books.

Bradburn, N. M., & Berlew, D. E. (1961). Need for achievement and English economic growth. *Economic Development and Cultural Change, 10*, 8–20.

Bradley, R. H., & Caldwell, B. M. (1976). Early home environment and changes in mental test performance in children from 6 to 36 months. *Developmental Psychology, 12*, 93–97.

Braine, M. D. S. (1976). Children's first word combinations. *Monographs of the Society for Research in Child Development, 41* (1, Serial No. 164).

Bramel, D. (1969). Interpersonal attraction, hostility, and perception. In J. Mills (Ed.), *Experimental social psychology*. New York: Macmillan.

Bransford, J. D., Barclay, J. R., & Franks, J. J. (1972). Sentence memory: A constructive versus interpretive approach. *Cognitive Psychology*, *3*, 193–209.

Braun, J. J., & Linder, D. E. (1979). *Psychology today: An introduction* (4th ed.). New York: Random House.

Braungart, R. G. (1980). Youth movements. In J. Adelson (Ed.), *Handbook of adolescent psychology*. New York: Wiley.

Bray, D. W., & Howard, A. (1980). Career success and life satisfaction of middle-aged managers. In L. A. Bond & J. C. Rosen (Eds.), *Competence and coping during adulthood*. Hanover, NH: University Press of New England.

Breland, K., & Breland, M. (1961). The misbehavior of organisms. *American Psychologist*, *16*, 681–684.

Breuer, J., & Freud, S. (1955). Studies on hysteria. In J. Strachey (Ed. and Trans.), *The standard edition of the complete psychological works of Sigmund Freud* (Vol. 2). London: Hogarth Press. (Original work published 1893–1895)

Brim, O. G., Jr. (1976). Theories on the male mid-life crisis. *Counseling Psychologist*, *6*, 2–9.

Broadbent, D. E. (1958). *Perception and communication*. London: Pergamon Press.

Broch, O. (1927). Russenorsk. *Archiv für Slavische Philologie*, *41*, 209–262.

Bronfenbrenner, U. (1967). Response to pressure from peers versus adults among Soviet and American school children. *International Journal of Psychology*, *2*, 199–207.

Bronfenbrenner, U. (1970). *Two worlds of childhood: US and USSR*. New York: Russell Sage Foundation.

Brown, G. W., Harris, T., & Copeland, J. R. (1977). Depression and loss. *British Journal of Psychiatry*, *130*, 1–18.

Brown, P. L., & Jenkins, H. M. (1968). Auto-shaping of the pigeon's keypeck. *Journal of the Experimental Analysis of Behavior*, *11*, 1–8.

Brown, R. (1965). *Social psychology*. New York: Free Press.

Brown, R., & Kulik, J. (1977). Flashbulb memories. *Cognition*, *5*, 73–99.

Brown, R., & McNeill, D. (1966). The "tip of the tongue" phenomenon. *Journal of Verbal Learning and Verbal Behavior*, *5*, 325–337.

Bruce, C. J., Isley, M. R., & Shinkman, P. G. (1981). Visual experience and development of interocular orientation disparity in visual cortex. *Journal of Neurophysiology*, *46*, 215–228.

Bruch, H. (1978). *The golden cage: The enigma of anorexia nervosa*. Cambridge, MA: Harvard University Press.

Bruner, J. S. (1973). *Beyond the information given: Studies in the psychology of knowing*. New York: Norton.

Bruner, J. S., Goodnow, J. J., & Austin, G. A. (1956). *A study of thinking*. New York: Wiley.

Bryan, J. H., & Test, M. A. (1967). Models and helping: Naturalistic studies in aiding behavior. *Journal of Personality and Social Psychology*, *6*, 400–407.

Buckhout, R. (1974). Eyewitness testimony. *Scientific American*, *231* (6), 23–31.

Bull, P. (1983). *Body movement and interpersonal communication*. New York: Wiley.

Bunney, W. E., Murphy, D. L., Goodwin, F. K., & Borge, G. F. (1972). The "switch process" in manic depressive illness: I. A systematic study of sequential behavioral changes. *Archives of General Psychiatry*, *27*, 295–302.

Buros, O. K. (1978). *The eighth mental measurements yearbook* (Vols. 1–2). Highland Park, NJ: Gryphon Press.

Buss, A. H. (1961). *The psychology of aggression*. New York: Wiley.

Buss, A. H. (1966). Instrumentality of aggression, feedback, and frustration as determinants of physical aggression. *Journal of Personality and Social Psychology*, *3*, 153–162.

Buss, A. H., Booker, A., & Buss, E. (1972). Firing a weapon and aggression. *Journal of Personality and Social Psychology*, *22*, 296–302.

Buss, A. H., Plomin, R., & Willerman, L. (1973). The inheritance of temperaments. *Journal of Personality*, *41*, 513–524.

Busse, E. W., & Pfeiffer, E. (Eds.) (1977). *Behavior and adaptation in late life* (2nd ed.). Boston: Little, Brown.

Butler, R. A. (1953). Discrimination learning by rhesus monkeys to visual-exploration motivation. *Journal of Comparative and Physiological Psychology*, *46*, 95–98.

Butler, R. A. (1954). Incentive conditions which influence visual exploration. *Journal of Experimental Psychology*, *48*, 19–23.

Butter, C. M. (1968). *Neuropsychology: The study of brain and behavior*. Belmont, CA: Brooks/Cole.

Buzan, T. (1977). *Speed memory* (rev. ed.). Newton Abbot, United Kingdom: David & Charles.

Byrne, D., & Nelson, D. (1965). Attraction as a linear function of proportion of positive reinforcements. *Journal of Personality and Social Psychology*, *1*, 659–663.

Cacioppo, J. T., & Petty, R. E. (1979). Attitudes and cognitive response: An electrophysiological approach. *Journal of Personality and Social Psychology*, *37*, 2181–2199.

Cacioppo, J. T., & Sandman, C. A. (1981). Psychophysiological functioning, cognitive responding, and attitudes. In R. E. Petty, T. M. Ostrom, & T. C. Brock (Eds.), *Cognitive responses in persuasion*. Hillsdale, NJ: Erlbaum.

Cadoret, R. J., Cain, C. A., & Groves, W. M. (1980). Development of alcoholism in adoptees raised apart from alcoholic biologic relatives. *Archives of General Psychiatry*, *37*, 561–563.

Cameron, N. A. (1967). Paranoid reactions. In A. M. Freedman & H. I. Kaplan (Eds.), *Comprehensive textbook of psychiatry*. Baltimore: Williams & Wilkins.

Campbell, D. P. (1981). *Handbook for the Strong-Campbell Vocational Interest Inventory*. Stanford, CA: Stanford University Press.

Campbell, D. T. (1961). Conformity in psychology's theories of acquired behavioral dispositions. In I. A. Berg & B. M. Bass (Eds.), *Conformity and deviation*. New York: Harper & Row.

Campos, J. J., Langer, A., & Krowitz, A. (1970) Cardiac responses on the visual cliff in prelocomotor human infants. *Science*, *170*, 196–197.

Cantor, J. R., Zillmann, D., & Bryant, J. (1975). Enhancement of experienced sexual arousal in response to erotic stimuli through misattribution of unrelated residual excitation. *Journal of Personality and Social Psychology*, *32*, 69–75.

Capon, N., & Kuhn, D. (1979). Logical reasoning in the supermarket: Adult females' use of a proportional reasoning strategy in an everyday context. *Developmental Psychology*, *15*, 450–452.

Carey, G., Goldsmith, H. H., Tellegen, A., & Gottesman, I. I. (1978). Genetics and personality inventories: The limits of replication with twin data. *Behavior Genetics*, *8*, 299–313.

Carey, G., & Gottesman, I. I. (1981). Twin and family studies of anxiety, phobic, and obsessive disorders. In D. F. Klein & J. G Rabkin (Eds.), *Anxiety: New research and changing concepts*. New York: Raven Press.

Carey, S. (1978). The child as word learner. In M. Halle J. Bresnan, & G. A. Miller (Eds.), *Linguistic theory and psychological reality*. Cambridge, MA: MIT Press.

Carlson, N. R. (1980). *Physiology of behavior* (2nd ed.). Boston: Allyn and Bacon.

Carroll, J. B. (1964). *Language and thought.* Englewood Cliffs, NJ: Prentice-Hall.

Carson, R. C. (1977). Personal communication cited in Altrocchi, J. (1980). *Abnormal behavior.* New York: Harcourt Brace Jovanovich.

Casper, R. C., Eckert, E. D., Halmi, K. A., Goldberg, S. C., & Davis, J. M. (1980). Bulimia: Its incidence and clinical importance in patients with anorexia nervosa. *Archives of General Psychiatry, 37,* 1030–1035.

Cattell, R. B., Eber, H. W., & Tatsuoka, M. M. (1970). *Handbook for the Sixteen Personality Factor Questionnaire (16PF).* Champaign, IL: Institute for Personality and Ability Testing.

Cazden, C. B. (1968). The acquisition of noun and verb inflections. *Child Development, 39,* 433–448.

Cermak, L. S. (1975). *Improving your memory.* New York: Norton.

Chaiken, S. (1979). Communicator physical attractiveness and persuasion. *Journal of Personality and Social Psychology, 37,* 1387–1397.

Charlesworth, E. A., & Nathan, R. G. (1984). *Stress management: A comprehensive guide to wellness.* New York: Atheneum.

Chomsky, N. (1957). *Syntactic structures.* The Hague: Mouton Publishers.

Chomsky, N. (1965). *Aspects of the theory of syntax.* Cambridge, MA: MIT Press.

Chomsky, N. (1968). *Language and the mind.* New York: Harcourt Brace Jovanovich.

Christie, R., & Geis, F. L. (1970). *Studies in Machiavellianism.* New York: Academic Press.

Clark, H. H., & Clark, E. V. (1977). *Psychology and language: An introduction to psycholinguistics.* New York: Harcourt Brace Jovanovich.

Clarke-Stewart, A., & Koch, J. B. (1983). *Children: Development through adolescence.* New York: Wiley.

Clarke-Stewart, K. A., & Fein, G. G. (1983). Early childhood programs. In P. H. Mussen (Ed.), *Handbook of child psychology: Vol. 2. Infancy and developmental psychobiology* (4th ed.). New York: Wiley.

Clavan, S. (1978). The impact of social class and social trends on the role of grandparent. *The Family Coordinator, 27,* 351–358.

Coates, D., & Wortman, C. B. (1980). Depression maintenance and interpersonal control. In A. Baum & J. E. Singer (Eds.), *Advances in environmental psychology: Vol. 2. Applications of personal control.* Hillsdale, NJ: Erlbaum.

Cohen, F., & Lazarus, R. S. (1979). Coping with the stresses of illness. In G. C. Stone, F. Cohen, & N. E. Adler (Eds.), *Health psychology—A handbook.* San Francisco: Jossey-Bass.

Cohen, L. B., DeLoache, J. S., & Strauss, M. S. (1978). Infant visual perception. In J. Osofsky (Ed.), *Handbook of infancy.* New York: Wiley.

Cohen, L. H., McGowan, J., Fooskas, S., & Rose, S. (1984). Positive life events and social support and the relationship between life stress and psychological disorder. *American Journal of Community Psychology, 12,* 567–587.

Cohen, S. (1980). Aftereffects of stress on human performance and social behavior: A review of research and theory. *Psychological Bulletin, 88,* 82–108.

Colby, A., Kohlberg, L., Gibbs, J., & Lieberman, M. (1983). A longitudinal study of moral judgment. *Monographs of the Society for Research in Child Development, 48* (1–2, Serial No. 200).

Colby, K. M. (1981). Modeling a paranoid mind. *The Behavioral and Brain Sciences, 4,* 515–560.

Coleman, J. C. (1980). Friendship and the peer group in adolescence. In J. Adelson (Ed.), *Handbook of adolescent psychology.* New York: Wiley.

Collins, A. M., & Quillian, M. R. (1969). Retrieval time from semantic memory. *Journal of Verbal Learning and Verbal Behavior, 8,* 240–247.

Condry, J. (1977). Enemies of exploration: Self-initiated versus other-initiated learning. *Journal of Personality and Social Psychology, 35,* 459–477.

Conger, J. J. (1977). *Adolescence and youth: Psychological development in a changing world* (2nd ed.). New York: Harper & Row.

Conklin, H. C. (1954). *The relation of Hanunóo culture to the plant world.* Unpublished doctoral dissertation, Yale University, New Haven.

Cooper, J., & Mackie, D. (1983). Cognitive dissonance in an intergroup context. *Journal of Personality and Social Psychology, 44,* 536–544.

Cooper, J. R., Bloom, F. E., & Roth, R. H. (1974). *The biochemical basis of neuropharmacology* (2nd ed.). New York: Oxford University Press.

Coren, S. (1972). Subjective contours and apparent depth. *Psychological Review, 79,* 359–367.

Coren, S., & Girgus, J. S. (1978). *Seeing is deceiving: The psychology of visual illusions.* Hillsdale, NJ: Erlbaum.

Coren, S., Porac, C., & Ward, L. M. (1979). *Sensation and perception.* New York: Academic Press.

Cornford, F. M. (Trans.) (1935). *Plato's theory of knowledge.* London: Kegan Paul.

Corsini, R. J. (Ed.) (1981). *Handbook of innovative psychotherapies.* New York: Wiley.

Costa, P. T., Jr., & McCrae, R. R. (1980). Still stable after all these years: Personality as a key to some issues in adulthood and old age. In P. B. Baltes & O. G. Brim, Jr. (Eds.), *Life-span development and behavior* (Vol. 3). New York: Academic Press.

Costanzo, P. R., & Shaw, M. E. (1966). Conformity as a function of age level. *Child Development, 37,* 967–975.

Cottrell, N. B. (1972). Social facilitation. In C. G. McClintock (Ed.), *Experimental social psychology.* New York: Holt, Rinehart and Winston.

Coyne, J. C. (1976). Depression and the responses of others. *Journal of Abnormal Psychology, 85,* 186–193.

Coyne, J. C. (1982). A critique of cognitions as causal entities with particular reference to depression. *Cognitive Therapy and Research, 6,* 3–13.

Coyne, J. C., & Gotlib, I. H. (1983). The role of cognition in depression: A critical appraisal. *Psychological Bulletin, 94,* 472–505.

Craighead, W. E., Kazdin, A. E., & Mahoney, M. J. (Eds.) (1981). *Behavior modification: Issues and applications* (2nd ed.). Boston: Houghton Mifflin.

Craik, F. I. M., & Lockhart, R. S. (1972). Levels of processing: A framework for memory research. *Journal of Verbal Learning and Verbal Behavior, 11,* 671–684.

Craik, F. I. M., & Tulving, E. (1975). Depth of processing and the retention of words in episodic memory. *Journal of Experimental Psychology: General, 104,* 268–294.

Craik, F. I. M., & Watkins, M. J. (1973). The role of rehearsal in short-term memory. *Journal of Verbal Learning and Verbal Behavior, 12,* 599–607.

Cromwell, R. L., Butterfield, E. C., Brayfield, F. M., & Curry, J. J. (1977). *Acute myocardial infarction: Reaction and recovery*. St. Louis, MO: Mosby.

Crook, T., & Eliot, J.(1980). Parental death during childhood and adult depression: A critical review of the literature. *Psychological Bulletin, 87*, 252–259.

Curtiss, S. (1977). *Genie: A psycholinguistic study of a modern-day "wild child."* New York: Academic Press.

Dahlstrom, W. G., & Dahlstrom, L. E. (Eds.). (1980). *Basic readings on the MMPI: A new selection on personality measurement.* Minneapolis, MN: University of Minnesota Press.

Dahlstrom, W. G., & Welsh, G. S. (1960). *An MMPI handbook: A guide to use in clinical practice and research.* Minneapolis, MN: University of Minnesota Press.

Dahlstrom, W. G., Welsh, G. S., & Dahlstrom, L. E. (1975). *An MMPI handbook: Vol. 2. Research applications* (rev. ed.). Minneapolis, MN: University of Minnesota Press.

Dale, P. S. (1976). *Language development: Structure and function* (2nd ed.). New York: Holt, Rinehart and Winston.

Darwin, C. J., Turvey, M. T., & Crowder, R. G. (1972). An auditory analogue of the Sperling partial report procedure: Evidence for brief auditory storage. *Cognitive Psychology, 3*, 255–267.

Darwin, C. R. (1871). *The descent of man, and selection in relation to sex.* London: John Murray.

Darwin, C. R. (1872). *The expression of the emotions in man and animals.* London: John Murray.

Darwin, C. R. (1877). A biographical sketch of an infant. *Mind, 2*, 285–294.

Davson, H., & Eggleton, M. G. (Eds.) (1968). *Starling and Lovatt Evans principles of human physiology* (14th ed.). Philadelphia: Lea & Febiger.

Dawson, G., & Adams, A. (1984). Imitation and social responsiveness in autistic children. *Journal of Abnormal Child Psychology, 12*, 209–226.

Dawson, G., Warrenburg, S., & Fuller, P. (1982). Cerebral lateralization in individuals diagnosed as autistic in early childhood. *Brain and Language, 15*, 353–368.

Day, R. H., & McKenzie, B. E. (1977). Constancies in the perceptual world of the infant. In W. Epstein (Ed.), *Stability and constancy in visual perception: Mechanisms and processes.* New York: Wiley.

Deci, E. L. (1975). *Intrinsic motivation.* New York: Plenum Press.

DeFleur, M. L., & Westie, F. R. (1958). Verbal attitudes and overt acts: An experiment on the salience of attitudes. *American Sociological Review, 23*, 667–673.

De Longis, A., Coyne, J. C., Dakof, G., Folkman, S., & Lazarus, R. S. (1982). Relationship of daily hassles, uplifts, and major life events to health status. *Health Psychology, 1*, 119–136.

Dembroski, T. M., Lasater, T. M., & Ramirez, A. (1978). Communicator similarity, fear arousing communications, and compliance with health care recommendations. *Journal of Applied Social Psychology, 8*, 254–269.

Dement, W. C. (1974). *Some must watch while some must sleep.* San Francisco: W. H. Freeman.

Dement, W. C., & Wolpert, E. (1958). The relation of eye movements, body motility, and external stimuli to dream content. *Journal of Experimental Psychology, 55*, 543–553.

DeRicco, D. A. (1978). Effects of peer majority on drinking rate. *Addictive Behavior, 3*, 29–34.

Des Lauriers, A. M., & Carlson, C. F. (1969). *Your child is asleep.* Homewood, IL: Dorsey Press.

Despert, J. L. (1951). Some considerations relating to the genesis of autistic behavior in children. *American Journal of Orthopsychiatry, 21*, 335–350.

Deutsch, F., & Murphy, W. F. (1955). *The clinical interview: Vol. 2. Therapy.* New York: International Universities Press.

Deutsch, J. A. (1978). The stomach in food satiation and the regulation of appetite. *Progress in Neurobiology, 10*, 135–153.

Deutsch, J. A., & Hardy, W. T. (1977). Cholecystokinin produces bait shyness in rats. *Nature, 266*, 196.

DeValois, R. L., Abramov, I., & Jacobs, G. H. (1966). Analysis of response patterns of LGN cells. *Journal of the Optical Society of America, 7*, 966–977.

DeValois, R. L., & DeValois, K. K. (1975). Neural coding of color. In E. C. Carterette (Ed.), *Handbook of perception: Vol. 5. Seeing.* New York: Academic Press.

DeValois, R. L., & DeValois, K. K. (1980). Spatial vision. *Annual Review of Psychology, 31*, 309–341.

DeWied, D. (1980). Hormonal influences on motivation, learning, memory, and psychosis. In D. T. Krieger & J. C. Hughes (Eds.)., *Neuroendocrinology.* Sunderland, MA: Sinauer Associates.

Dillbeck, M. C., Aron, A. P., & Dillbeck, S. L. (1979). The transcendental meditation program as an educational technology: Research and applications. *Educational Technology, 19*, (11), 7–13.

Dodwell, P. C., & Bessant, D. E. (1960). Learning without swimming in a water maze. *Journal of Comparative and Physiological Psychology, 53*, 422–425.

Dohrenwend, B. P., Dohrenwend, B. S., Gould, M. S., Link, B., Neugebauer, R., & Wunsch-Hitzig, R. (1980). *Mental illness in the United States: Epidemiologic estimates.* New York: Praeger.

Dollard, J., Doob, L., Miller, N. E., Mowrer, O. H., & Sears, R. R. (1939). *Frustration and aggression.* New Haven, CT: Yale University Press.

Dollard, J., & Miller, N. E. (1950). *Personality and psychotherapy: An analysis in terms of learning, thinking, and culture.* New York: McGraw-Hill.

Donnerstein, E., & Wilson, D. W. (1976). Effects of noise and perceived control on ongoing and subsequent aggressive behavior. *Journal of Personality and Social Psychology, 34*, 774–781.

Doob, L. W. (1947). The behavior of attitudes. *Psychological Review, 54*, 135–156.

Douvan, E., & Adelson, J. (1966). *The adolescent experience.* New York: Wiley.

Dowling, J. E., & Boycott, B. B. (1966). Organization of the primate retina: Electron microscopy. *Proceedings of the Royal Society (London)*, Series B, *166*, 80–111.

Dunn, J., & Kendrick, C. (1982). *Siblings: Love, envy, and understanding.* Cambridge, MA: Harvard University Press.

Dunphy, D. C. (1963). The social structure of urban adolescent peer groups. *Sociometry, 26*, 230–246.

Eagly, A. H., & Chaiken, S. (1975). An attribution analysis of the effect of communicator characteristics on opinion change: The case of communicator attractiveness. *Journal of Personality and Social Psychology, 32*, 136–144.

Eccles (Parsons), J. (1983). Expectancies, values, and academic behaviors. In J. T. Spence (Ed.), *Achievement and achievement motives.* San Francisco: W. H. Freeman.

Edelson, M. (1976). Alternative living arrangements. In H. R. Lamb and Associates (Eds.), *Community survival for long-term patients.* San Francisco: Jossey-Bass.

Edwards, A. L. (1954). *The Edwards Personal Preference Schedule, manual.* New York: Psychological Corporation.

Ehrhardt, A. A. (1977). Prenatal androgenization and human psychosexual behavior. In J. Money & H. Musaph (Eds.), *Handbook of sexology.* Amsterdam, Holland: Elsevier/North-Holland.

Eimas, P. D., & Corbit, J. (1973). Selective adaptation of linguistic feature detectors. *Cognitive Psychology, 4,* 99–109.

Ekman, P. (1980). *The face of man: Expressions of universal emotions in a New Guinea village.* New York: Garland STPM Press.

Ekman, P. (Ed.) (1982). *Emotion in the human face* (2nd ed.). Cambridge, England: Cambridge University Press.

Ekman, P., & Friesen, W. V. (1969). Nonverbal leakage and clues to deception. *Psychiatry, 32,* 88–106.

Ekman, P., Levenson, R. W., & Friesen, W. V. (1983). Autonomic nervous system activity distinguishes among emotions. *Science, 221,* 1208–1210.

Ekstrom, R. B., French, J. W., Harman, H. H., & Derman, D. (1976). *Manual for kit of factor-referenced cognitive tests.* Princeton, NJ: Educational Testing Service.

Ellis, A. (1962). *Reason and emotion in psychotherapy.* New York: Lyle Stuart.

Epstein, A. N., Kissileff, H. R. & Stellar, E.(Eds.) (1973). *The neuropsychology of thirst: New findings and advances in concepts.* Washington, DC: Winston.

Erber, J. T. (1982). Memory and age. In T. M. Field, A. Huston, H. C. Quay, L. Troll, & G. E. Finley (Eds.), *Review of human development.* New York: Wiley.

Erickson, R. P., & Schiffman, S. S. (1975). The chemical senses: A systematic approach. In M. S. Gazzaniga & C. Blakemore (Eds.), *Handbook of Psychobiology.* New York: Academic Press.

Ericsson, K. A., & Chase, W. G. (1982). Exceptional memory. *American Scientist, 70,* 607–615.

Erikson, E. H. (1963). *Childhood and society* (2nd ed.). New York: Norton.

Erikson, E. H. (1968). *Identity: Youth and crisis* (2nd ed.). New York: Norton.

Estes, W. K., & Taylor, H. A. (1966). Visual detection in relation to display size and redundancy of critical elements. *Perception and Psychophysics, 1,* 9–16.

Eysenck, H. J. (1967). *The biological basis of personality.* Springfield, IL: Thomas.

Eysenck, H. J. (1970). *The structure of human personality.* London: Methuen.

Eysenck, H. J., & Kamin, L. J. (1981). *The intelligence controversy.* New York: Wiley.

Fairburn, C. G. (1980). Self-induced vomiting. *Journal of Psychosomatic Research, 24,* 193–197.

Farb, P. (1975). *Word play: What happens when people talk.* New York: Knopf.

Farber, S. L. (1981). *Identical twins reared apart: A reanalysis.* New York: Basic Books.

Farrelly, F., & Brandsma, J. (1974). *Provocative therapy.* Cupertino, CA: Meta.

Farrelly, F., & Matthews, S. (1981). Provocative therapy. InjuR. J. Corsini (Ed.), *Handbook of innovative psychotherapies.* New York: Wiley.

Fazio, R. H., Powell, M. C., & Herr, P. M. (1983). Toward a process model of the attitude-behavior relation: Accessing one's attitude upon mere observation of the attitude object. *Journal of Personality and Social Psychology, 44,* 723–735.

Fazio, R. H., & Zanna, M. P. (1981). Directed experience and attitude-behavior consistency. In L. Berkowitz (Ed.), *Advances in experimental social psychology* (Vol. 14). New York: Academic Press.

Fazio, R. H., Zanna, M. P., & Cooper, J. (1977). Dissonance and self-perception: An integrative view of each theory's proper domain of application. *Journal of Experimental Social Psychology, 13,* 464–479.

Feather, N. T. (1980). Values in adolescence. In J. Adelson (Ed.), *Handbook of adolescent psychology.* New York: Wiley.

Fehr, L. A. (1983). *Introduction to personality.* New York: Macmillan.

Fein, D., Humes, M., Kaplan, E., Lucci, D., & Waterhouse, L. (1984). The question of left hemisphere dysfunction in infantile autism. *Psychological Bulletin, 95,* 258–281.

Ferster, C. B., & Skinner, B. F. (1957). *Schedules of reinforcement.* New York: Appleton-Century-Crofts.

Feshbach, S., & Weiner, B. (1982). *Personality.* Lexington, MA: D. C. Heath.

Festinger, L. (1954). A theory of social comparison processes. *Human Relations, 7,* 117–140.

Festinger, L. (1957). *A theory of cognitive dissonance.* Evanston, IL: Row, Peterson.

Festinger, L., & Carlsmith, J. M. (1959). Cognitive consequences of forced compliance. *Journal of Abnormal and Social Psychology, 58,* 203–210.

Festinger, L., Schachter, S., & Back, K. (1950). *Social pressures in informal groups: A study of human factors in housing.* New York: Harper & Row.

Field, T. M., Woodson, R., Greenberg, R., & Cohen, D. (1982). Discrimination and imitation of facial expressions by neonates. *Science, 218,* 179–181.

Fillenbaum, S., Schiffman, H. R., & Butcher, J. (1965). Perception of off-size versions of a familiar object under conditions of rich information. *Journal of Experimental Psychology, 69,* 298–303.

Fireside, H. (1979). *Soviet psychoprisons.* New York: Norton.

Fischer, K. W. (1980). A theory of cognitive development: The control and construction of hierarchies of skills. *Psychological Review, 87,* 477–531.

Fischer, K. W., & Lazerson, A. (1984). *Human development: From conception through adolescence.* New York: W. H. Freeman.

Fischer, M. (1973). Genetic and environmental factors in schizophrenia. *Acta Psychiatrica Scandinavica,* Supplement 238.

Fishbein, M. (1967). Attitudes and the prediction of behavior. In M. Fishbein (Ed.), *Readings in attitude theory and measurement.* New York: Wiley.

Fishbein, M., & Ajzen, I. (1975). *Belief, attitude, intention, and behavior: An introduction to theory and research.* Reading, MA: Addison-Wesley.

Fiske, D. W., & Maddi, S. R. (1961). *Functions of varied experience.* Homewood, IL: Dorsey.

Fodor, J. A., & Bever, T. G. (1965). The psychological reality of linguistic segments. *Journal of Verbal Learning and Verbal Behavior, 4,* 414–420.

Fodor, J. A., & Garrett, M. F. (1967). Some syntactic determinants of sentential complexity. *Perception and Psychophysics, 2,* 289–296.

Folstein, S., & Rutter, M. (1977). Infantile autism: A genetic study of 21 twin pairs. *Journal of Child Psychology and Psychiatry, 18,* 297–321.

Ford, M. R. (1982). Biofeedback treatment for headaches, Raynaud's disease, essential hypertension, and irritable bowel syndrome: A review of the long-term follow-up literature. *Biofeedback and Self-regulation.* 7, 521–536.

Fraiberg, S. (1959). *The magic years: Understanding and handling the problems of early childhood.* New York: Scribners.

Frankl, V. E. (1963). *Man's search for meaning: An introduction to*

logotherapy (I. Lasch, Trans.). New York: Washington Square Press.
Frankl, V. E. (1975). Paradoxical intention and dereflection. *Psychotherapy: Theory, Research and Practice, 12*, 226–237.
Fredricks, A. J., and Dossett, D. L. (1983). Attitude-behavior relations: A comparison of the Fishbein-Ajzen and the Bentler-Speckart models. *Journal of Personality and Social Psychology, 45*, 501–512.
Freedman, J. L. (1975). *Crowding and behavior.* San Francisco: W. H. Freeman.
Freeman, D. (1983). *Margaret Mead and Samoa: The making and unmaking of an anthropological myth.* Cambridge, MA: Harvard University Press.
French, J. W., Ekstrom, R. B., & Price, L. A. (1963). *Kit of reference tests for cognitive factors.* Princeton, NJ: Educational Testing Service.
Freud, A. (1946). *The ego and the mechanisms of defense* (Cecil Baines, Trans.). New York: International Universities Press.
Freud, A. (with S. Dann) (1951). An experiment in group upbringing. *Psychoanalytic Study of the Child, 6*, 127–168.
Freud, S. (1910). Three contributions to the sexual theory (A. A. Brill, Trans.). *Nervous and Mental Disease Monograph Series* (Whole No. 7). (Original work published 1905)
Freud, S. (1920). *A general introduction to psycho-analysis.* New York: Boni and Liveright.
Freud, S. (1925). Analysis of a phobia in a five-year-old boy. In *Sigmund Freud: Collected papers* (Vol. 3) (A. Strachey & J. Strachey, Trans.). London: Hogarth Press. (Original work published 1909)
Freud, S. (1938). *The basic writings of Sigmund Freud* (A. A. Brill, Ed. and Trans.). New York: Random House.
Freud, S. (1949). *An outline of psychoanalysis* (J. Strachey, Trans.). New York: Norton. (Original work published 1940)
Freud, S. (1951). *Psychopathology of everyday life.* New York: Mentor Books. (Original work published 1901)
Freud, S. (1953). Fragment of an analysis of a case of hysteria. In J. Strachey (Ed. and Trans.), *The standard edition of the complete psychological works of Sigmund Freud* (Vol. 7). London: Hogarth Press. (Original work published 1905)
Freud, S. (1953). *A general introduction to psychoanalysis.* Garden City, NY: Doubleday. (Original work published 1920)
Freud, S. (1953). *The interpretation of dreams.* In J. Strachey (Ed. and Trans.), *The standard edition of the complete psychological works of Sigmund Freud* (Vols. 4–5). London: Hogarth Press. (Original work published 1900)
Freud, S. (1957). Mourning and melancholia. In J. Strachey (Ed. and Trans.), *The standard edition of the complete psychological works of Sigmund Freud* (Vol. 14). London: Hogarth Press. (Original work published 1917)
Freud, S. (1961–1963). *Introductory lectures on psychoanalysis.* In J. Strachey (Ed. and Trans.), *The standard edition of the complete psychological works of Sigmund Freud* (Vols. 15–16). London: Hogarth Press. (Original work published 1916–1917)
Freud, S. (1963). Further recommendations in the technique of psychoanalysis (J. Riviere, Trans.). In S. Freud, *Therapy and technique.* New York: Macmillan. (Original work published 1913–1916)
Freud, S., & Jung, C. G. (1974). *The Freud/Jung letters* (W. McGuire, Ed.; R. Manheim & R. F. C. Hull, Trans.). Princeton, NJ: Princeton University Press.
Freudenberger, H. J., & Richelson, G.(1980). *Burn-out: The high cost of high achievement.* Garden City, NY: Anchor Press.
Friedberg, J. (1975, August). Electroshock therapy: Let's stop blasting the brain. *Psychology Today, 9*, pp. 18–23, 98–99.
Friedberg, J. (1976). *Shock treatment is not good for your brain.* San Francisco: Glide Publications.
Friedberg, J. (1977). Shock treatment, brain damage, and memory loss: A neurological perspective. *American Journal of Psychiatry, 134*, 1010–1014.
Friedman, M., & Rosenman, R. H. (1974). *Type A behavior and your heart.* New York: Knopf.
Friedman, M. I. (1978). Hyperphagia in rats with experimental diabetes mellitus: A response to a decreased supply of utilizable fuels. *Journal of Comparative and Physiological Psychology, 92*, 109–117.
Friedman, M. I., & Stricker, E. M. (1976). The physiological psychology of hunger: A physiological perspective. *Psychological Review, 83*, 409–431.
Fries, J. F. (1980). Aging, natural death, and the compression of morbidity. *The New England Journal of Medicine, 303*, 130–135.
Frijda, N. H. (1969). Recognition of emotion. In L. Berkowitz (Ed.), *Advances in experimental social psychology* (Vol. 4). New York: Academic Press.
Frisch, R. E. (1974). Critical weight at menarche, initiation of the growth spurt, and control of puberty. In M. M. Grumbach, G. D. Grave, & F. E. Mayer (Eds.), *Control of the onset of puberty.* New York: Wiley.
Frodi, A. (1975). The effect of exposure to weapons on aggressive behavior from a cross-cultural perspective. *International Journal of Psychology, 10*, 283–292.
Funder, D. C., & Bem, D. L. (1977, August). *A proposal for assessing the personality of situations.* Paper presented at the annual convention of The American Psychological Association, San Francisco.
Funkenstein, D. H. (1955). The physiology of fear and anger. *Scientific American, 192*(5), 74–80.
Furth, H. (1971). Linguistic deficiency and thinking: Research with deaf subjects 1964–69. *Psychological Bulletin, 75*, 58–72.
Gallatin, J. (1980). Political thinking in adolescence. In J. Adelson (Ed.), *Handbook of adolescent psychology.* New York: Wiley.
Galton, F. (1871). *Hereditary genius: An inquiry into its laws and consequences.* New York: Appleton.
Galton, F. (1908). *Memories of my life.* London: Methuen.
Garcia, J., McGowan, B. K., & Green, K. F. (1972). Biological constraints on conditioning. In A. H. Black & W. F. Prokasy (Eds.), *Classical conditioning II: Current theory and research.* New York: Appleton-Century-Crofts.
Gardner, B. T., & Gardner, R. A. (1975). Evidence for sentence constituents in the early utterances of child and chimpanzee. *Journal of Experimental Psychology: General, 104*, 244–267.
Gardner, E. (1975). *Fundamentals of neurology* (6th ed.). Philadelphia: Saunders.
Gardner, H., (1975). *The shattered mind: The person after brain damage.* New York: Knopf.
Gardner, H. (1978). *Developmental psychology.* Boston: Little, Brown.
Gardner, R. A., & Gardner, B. T. (1969). Teaching sign language to a chimpanzee. *Science, 165*, 664–672.
Garfield, S. L. (1981). Psychotherapy: A 40 year appraisal. *American Psychologist, 36*, 174–183.
Garfinkel, R., & Thorndike, R. L. (1976). Binet item difficulty then and now. *Child Development, 47*, 959–965.
Garrett, M. F., Bever, T. G., & Fodor, J. A. (1966). The active use of grammar in speech perception. *Perception and Psychophysics, 1*, 30–32.
Geen, R. G. (1983). Aggression and television violence. In R. G.

Geen & E. I. Donnerstein (Eds.), *Aggression: Theoretical and empirical reviews: Vol 2. Issues in research.* New York: Academic Press.
Geis, F. L. (1978). Machiavellianism. In H. London & J. E. Exner, Jr. (Eds.), *Dimensions of personality.* New York: Wiley.
Geschwind, N. (1970). The organization of language and the brain. *Science, 170,* 940–944.
Geschwind, N. (1979). Specializations of the human brain. In the Scientific American Book, *The brain.* San Francisco: W. H. Freeman.
Gibson, E. J. (1969). *Principles of perceptual learning and development.* New York: Appleton-Century-Crofts.
Gibson, E. J., & Walk, R. D. (1960). The "visual cliff." *Scientific American, 202*(4), 64–71.
Gibson, J. J. (1950). *The perception of the visual world.* Boston: Houghton Mifflin.
Gibson, J. J. (1966). *The senses considered as perceptual systems.* Boston: Houghton Mifflin.
Giebenhain, J. E., & O'Dell, S. L. (1984). Evaluation of a parent-training manual for reducing children's fear of the dark. *Journal of Applied Behavior Analysis, 17,* 121–125.
Gilbert, S. J. (1981). Another look at the Milgram obedience studies: The role of the gradated series of shocks. *Personality and Social Psychology Bulletin,* 7, 690–695.
Gillberg, C., & Schaumann, H. (1982). Social class and infantile autism. *Journal of Autism and Developmental Disorders, 12,* 223–228.
Gilligan, C. (1977). In a different voice: Women's conceptions of self and of morality. *Harvard Educational Review, 47,* 481-517.
Gilligan, C. (1982). *In a different voice.* Cambridge, MA: Harvard University Press.
Glamser, F. D. (1976). Determinants of a positive attitude toward retirement. *Journal of Gerontology, 31,* 104–107.
Glanzer, M., & Cunitz, A. R. (1966). Two storage mechanisms in free recall. *Journal of Verbal Learning and Verbal Behavior, 5,* 351–360.
Gleason, J. B. (1967, June). Do children imitate? In *Proceedings of the International Conference on Oral Education of the Deaf* (Vol. 2).
Gleitman, H., (1981). *Psychology.* New York: Norton.
Gold, M., & Petronio, R. J. (1980). Delinquent behavior in adolescence. In J. Adelson (Ed.), *Handbook of adolescent psychology.* New York: Wiley.
Gold, R. M., Jones, A. P., Sawchenko, P. E., & Kapatos, G. (1977). Paraventricular area: Critical focus of a longitudinal neurocircuitry mediating food intake. *Physiology and Behavior, 18,* 1111–1119.
Goldblatt, M., & Munitz, H. (1976). Behavioral treatment of hysterical leg paralysis. *Journal of Behavior Therapy and Experimental Psychiatry,* 7, 259–263.
Goldin-Meadow, S., & Feldman, H. (1977). The development of language-like communication without a language model. *Science, 197,* 401–403.
Goodwin, D. W. (1979). Alcoholism and heredity: A review and hypothesis. *Archives of General Psychiatry, 36,* 57–61.
Goodwin, F. K., & Zis, A. P. (1979). Lithium in the treatment of mania. *Archives of General Psychiatry, 36,* 840–844.
Gorski, R. A., Gordon, J. H., Shryne, J. E., & Southam, A. M. (1978). Evidence for a morphological sex difference within the medial preoptic area of the rat brain. *Brain Research, 148,* 333–346.
Gottesman, I. I., & Shields, J. (1972). *Schizophrenia and genetics: A twin vantage point.* New York: Academic Press.
Gough, H. G. (1960). The Adjective Check List as a personality research technique. *Psychological Reports, 6,* 107–122.
Gough, H. G. (1969). *Manual for the California Psychological Inventory* (rev. ed.). Palo Alto, CA: Consulting Psychologists Press.
Gough, H. G. (1975). *California Psychological Inventory manual* (rev. ed.). Palo Alto, CA: Consulting Psychologists Press.
Gough, H. G., & Heilbrun, A. B. (1980). *The Adjective Check List manual: 1980 edition.* Palo Alto, CA: Consulting Psychologists Press.
Gouin-Decarie, T. (1965) *Intelligence and affectivity in early childhood* (E. P. Brandt & L. W. Brandt, Trans.). New York: International Universities Press.
Gould, J. L. (1982). *Ethology: The mechanisms and evolution of behavior.* N. Y.: Norton.
Gould, R. L. (1978). *Transformations: Growth and change in adult life.* New York: Simon & Schuster.
Gould, S. J. (1976). Biological potential vs. biological determinism. *Natural History, 85* (5), 12–22.
Gramza, A. F. (1967). Responses of brooding nighthawks to a disturbing stimulus. *The Auk, 84,* 72–86.
Greenspoon, J. (1955). The reinforcing effect of two spoken sounds on the frequency of two responses. *American Journal of Psychology, 68,* 409–416.
Greenwald, A.G. (1968). Cognitive learning, cognitive response to persuasion and attitude change. In A. G. Greenwald, T. C. Brock, & T. M. Ostrom (Eds.), *Psychological foundations of attitudes.* New York: Academic Press.
Gregory, R. L.(1978). *Eye and brain: The psychology of seeing* (3rd ed.). New York: McGraw-Hill.
Griffitt, W. (1970). Environmental effects on interpersonal affective behavior: Ambient effective temperature and attraction. *Journal of Personality and Social Psychology, 15,* 240–244.
Grinspoon, L., Ewalt, J. R., & Shader, R. I. (1972). *Schizophrenia: Pharmacotherapy and psychotherapy.* Baltimore: Williams & Wilkins.
Grossman, H. (Ed.) (1977). *Manual on terminology and classification in mental retardation, 1977 revision.* Washington, DC: American Association on Mental Deficiency.
Groth, A. N., & Birnbaum, H. J. (1979). *Men who rape: The psychology of the offender.* New York: Plenum Press.
Gualtieri, C. T., & Guimond, M. (1981). Tardive dyskinesia and the behavioral consequences of chronic neuroleptic treatment. *Developmental Medicine and Child Neurology, 23,* 255–259.
Guilford, J. P. (1956). *Fundamental statistics in psychology and education* (3rd ed.). New York: McGraw-Hill.
Guilford, J. P. (1961). Factorial angles to psychology. *Psychological Review, 68,* 1–20.
Guilford, J. P.(1967). *The nature of human intelligence.* New York: McGraw-Hill.
Guilford, J. P. (1968). Intelligence has three facets. *Science, 160,* 615–620.
Gurman, A. S., & Kniskern, D. P. (Eds.) (1981). *Handbook of family therapy.* New York: Brunner/Mazel.
Gutmann, D. (1977). The cross-cultural perspective: Notes toward a comparative psychology of aging. In J. E. Birren & K. W. Schaie (Eds.), *Handbook of the psychology of aging.* New York: Van Nostrand Reinhold.
Haan, N., Smith, M. B., & Block, J. (1968). Moral reasoning of young adults: Political-social behavior, family background, and personality correlates. *Journal of Personality and Social Psychology, 10,* 183–201.
Haith, M. M. (1976, July). *Organization of visual behavior at birth.* Paper presented at the International Congress of Psychology, Paris.
Hall, C. S. (1953). A cognitive theory of dreams. *Journal of General Psychology, 49,* 273–282.

Hall, C. S., & Lindzey, G. (1970). *Theories of personality* (2nd ed.). New York: Wiley.
Hall, E. (1970, May). A conversation with Jean Piaget and Bärbel Inhelder. *Psychology Today, 3*, pp. 25–32, 54–56.
Halliday, M. A. K. (1975). *Learning how to mean: Explorations in the development of language*. London: Edward Arnold.
Halmi, K. A., Falk, J. R., & Schwartz, E. (1981). Binge-eating and vomiting: A survey of a college population. *Psychological Medicine, 11*, 697–706.
Hamilton, D. L., & Zanna, M. P. (1972). Differential weighting of favorable and unfavorable attributes in impressions of personality. *Journal of Experimental Research in Personality, 6*, 204–212.
Hanson, H. M. (1959). Effects of discrimination training on stimulus generalization. *Journal of Experimental Psychology, 58*, 321–334.
Harlow, H. F. (1953). Motivation as a factor in the acquisition of new responses. In *Current theory and research in motivation: A symposium*. Lincoln, NE: University of Nebraska Press.
Harlow, H. F. (1958). The nature of love. *American Psychologist, 13*, 673–685.
Harpe, S. (1979). *Headaches*. Chicago: Budlong.
Harris, M. B. (1974). Mediators between frustration and aggression in a field experiment. *Journal of Experimental Social Psychology, 10*, 561–571.
Hartup, W. W. (1983). Peer relations. In P. H. Mussen (Ed.), *Handbook of child psychology: Vol. 4. Socialization, personality, and social development* (4th ed.). New York: Wiley.
Hauserman, N. M. (1975). *Combined conditioning techniques in the treatment of nocturnal enuresis*. Unpublished manuscript. Towson State University, Maryland.
Hayes, L. A., & Watson, J. S. (1981). Neonatal imitation: Fact or artifact? *Developmental Psychology, 17*, 655–660.
Hays, P. (1976). Etiological factors in manic-depressive psychoses. *Archives of General Psychiatry, 33*, 1187–1188.
Heath, R. G. (1964). Pleasure response of human subjects to direct stimulation of the brain: Physiologic and psychodynamic considerations. In R. G. Heath (Ed.), *The role of pleasure in behavior*. New York: Harper & Row.
Hebb, D. O. (1955). Drives and the C.N.S. (conceptual nervous system). *Psychological Review, 62*, 243–254.
Hécaen, H., & Albert, M. L. (1978). *Human neuropsychology*. New York: Wiley.
Heider, F. (1958). *The psychology of interpersonal relations*. New York: Wiley.
Heilman, K. M. (1979). Neglect and related disorders. In K. M. Heilman & E. Valenstein (Eds.), *Clinical neuropsychology*. New York: Oxford University Press.
Heilman, K. M., & Watson, R. T. (1977). The neglect syndrome—a unilateral deficit of the orienting response. In S. Harnad, R. W. Doty, L. Goldstein, J. Jaynes, & G. Krauthamer (Eds.), *Lateralization in the nervous system*. New York: Academic Press.
Heinstein, M. (1969). *Behavior problems of young children in California*. Berkeley, CA: California Department of Public Health.
Helmreich, R. L., Beane, W. E., Lucker, G. W., & Spence, J. T. (1978). Achievement motivation and scientific attainment. *Personality and Social Psychology Bulletin, 4*, 222–226.
Hess, E. H. (1965). Attitude and pupil size. *Scientific American, 212*(4), 46–54.
Hetherington, A. W., & Ranson, S. W. (1940). Hypothalamic lesions and adiposity in the rat. *Anatomical Record, 78*, 149–172.
Hetherington, E. M., Cox, M., & Cox, R. (1979). Stress and coping in divorce: A focus on women. In J. E. Gullahorn (Ed.), *Psychology and women: In transition*. Washington, DC: Winston.
Hetherington, E. M., & Parke, R. D. (1979). *Child psychology: A contemporary viewpoint* (2nd ed.). New York: McGraw-Hill.
Hirsch, H. V. B., & Spinelli, D. N. (1971). Modification of the distribution of receptive field orientation in cats by selective visual exposure during development. *Experimental Brain Research, 12*, 509–527.
Ho, B. T., Richards, D. W., & Chute, D. L. (Eds.) (1978). *Drug discrimination and state dependent learning*. New York: Academic Press.
Hobbs, D. F., Jr., & Cole, S. P. (1976). Transition to parenthood: A decade replication. *Journal of Marriage and the Family, 38*, 723–731.
Hoffman, M. L. (1976). Empathy, role-taking, guilt, and development of altruistic motives. In T. Likona (Ed.), *Moral development and behavior: Theory, research and social issues*. New York: Holt, Rinehart, and Winston.
Hoffman, M. L. (1980). Moral development in adolescence. In J. Adelson (Ed.), *Handbook of adolescent psychology*. New York: Wiley.
Holinger, P. C. (1979). Violent deaths among the young: Recent trends in suicide, homicide, and accidents. *American Journal of Psychiatry, 136*, 1144–1147.
Holland, J. L. (1973). *Making vocational choices*. Englewood Cliffs, NJ: Prentice-Hall.
Hollingsworth, L. (1942). *Children above 180 IQ*. Yonkers, NY: World Book.
Holmes, D. S. (1984). Meditation and somatic arousal reduction: A review of the experimental evidence. *American Psychologist, 39*, 1–10.
Holmes, T. H. (1984). *Life change events research, 1966–1978*. New York: Praeger.
Holmes, T. H., & Masuda, M. (1974). Life change and illness susceptibility. In B. S. Dohrenwend & B. P. Dohrenwend (Eds.), *Stressful life events: Their nature and effects*. New York: Wiley.
Holmes, T. H., & Rahe, R. H. (1967). The Social Readjustment Rating Scale. *Journal of Psychosomatic Research, 11*, 213–218.
Holtzman, W. H., Thorpe, J. S., Swartz, J. D., & Herron, E. W. (1961). *Inkblot perception and personality: Holtzman Inkblot Technique*. Austin, TX: University of Texas Press.
Holway, A. H., & Boring, E. G. (1941). Determinants of apparent visual size with distance variant. *American Journal of Psychology, 54*, 21–37.
Horn, J. L., & Donaldson, G. (1980). Cognitive development in adulthood. In O. G. Brim, Jr., & J. Kagan (Eds.), *Constancy and change in human development*. Cambridge, MA: Harvard University Press.
Hornblum, J. N., & Overton, W. F. (1976). Area and volume conservation among the elderly: Assessment and training. *Developmental Psychology, 12*, 68–74.
Horner, M. S. (1968). *Sex differences in achievement motivation and performance in competitive and noncompetitive situations*. Unpublished doctoral dissertation, University of Michigan, Ann Arbor.
Horney, K. (1937). *The neurotic personality of our time*. New York: Norton.
Horney, K. (1967). *Feminine psychology*. New York: Norton.
Hosobuchi, Y., Adams, J. E., & Linchitz, R. (1977). Pain relief by electrical stimulation of the central gray matter in humans and its reversal by naloxone. *Science, 197*, 183–186.
Hovland, C. I., Lumsdaine, A. H., & Sheffield, F. D. (1949).

Experiments on mass communication. Princeton, NJ: Princeton University Press.
Hovland, C. I., & Weiss, W. (1951). The influence of source credibility on communication effectiveness. *Public Opinion Quarterly, 15*, 635–650.
Hoyenga, K. B., & Hoyenga, K. T. (1979). *The question of sex differences: Psychological, cultural, and biological issues*. Boston: Little, Brown.
Hoyenga, K. B., & Hoyenga, K. T. (1984). *Motivational explanations of behavior: Evolutionary, physiological and cognitive ideas*. Monterey, CA: Brooks/Cole.
Hubel, D. H., & Wiesel, T. N. (1968). Receptive fields and functional architecture of monkey striate cortex. *Journal of Physiology, 195*, 215–243.
Hubel, D. H., & Wiesel, T. N. (1979). Brain mechanisms of vision. In Scientific American Book, *The brain*. San Francisco: W. H. Freeman.
Hull, C. L. (1943). *Principles of behavior*. New York: Appleton-Century-Crofts.
Hull, C. L. (1952). *A behavior system*. New Haven, CT: Yale University Press.
Hulse, S. H., Egeth, H., & Deese, J. (1980). *The psychology of learning* (5th ed.). New York: McGraw-Hill.
Hultsch, D. F., & Deutsch, F. (1981). *Adult development and aging: A life-span perspective*. New York: McGraw-Hill.
Humphrey, G. (1948). *Directed thinking*. New York: Dodd, Mead.
Humphrey, N. K. (1970). What the frog's eye tells the monkey's brain. *Brain, Behavior, and Evolution, 3*, 324–337.
Hunt, D. D., & Hampson, J. L. (1980). Follow-up of 17 biologic male transsexuals after sex-reassignment surgery. *American Journal of Psychiatry, 137*, 432–438.
Hunt, E., & Love, T. (1972). How good can memory be? In A. W. Melton & E. Martin (Eds.), *Coding processes in human memory*. Washington, DC: Winston.
Hunt, J. McV. (1961). *Intelligence and experience*. New York: Ronald Press.
Hutt, C., & Ounsted, C. (1970). Gaze aversion and its significance in childhood autism. In S. J. Hutt & C. Hutt (Eds.), *Behavioral studies in psychiatry*. New York: Pergamon Press.
Huttenlocher, J. (1973). Language and thought. In G. A. Miller (Ed.), *Communication, language, and meaning*. New York: Basic Books.
Hyde, J. S. (1979). *Understanding human sexuality*. New York: McGraw-Hill.
Inhelder, B., & Piaget, J. (1958). *The growth of logical thinking from childhood to adolescence*. New York: Basic Books.
Inouye, E. (1963). Similarity and dissimilarity of schizophrenia in twins. *Proceedings of the Third International Congress of Psychiatry, 1961* (Vol. 1). Montreal: University of Toronto Press.
Insko, C. A., & Schopler, J. (1972). *Experimental social psychology*. New York: Academic Press.
Iversen, L. L. (1979). The chemistry of the brain. In the Scientific American Book, *The brain*. San Francisco: W. H. Freeman.
Izard, C. E. (1971). *The face of emotion*. New York: Appleton-Century-Crofts.
Izard, C. E. (1977). *Human emotions*. New York: Plenum Press.
Jameson, D., & Hurvich, L. M. (1964). Theory of brightness and color contrast in human vision. *Vision Research, 4*, 135–154.
Janis, I. L. (1983). *Groupthink: Psychological studies of policy decisions and fiascoes* (2nd ed.). Boston: Houghton Mifflin.
Janis, I. L., & Mann, L. (1977). *Decision making: A psychological analysis of conflict, choice, and commitment*. New York: Free Press.
Jasper, H. H. (1941). Electroencephalography. In W. Penfield & T. C. Erickson (Eds.), *Epilepsy and cerebral localization*. Springfield, IL: Thomas.
Jenkins, H. M., & Moore, B. A. (1973). The form of the auto-shaped response with food or water reinforcers. *Journal of the Experimental Analysis of Behavior, 20*, 163–181.
Jenkins, J. J., Russell, W. A., & Suci, G. J. (1958). An atlas of semantic profiles of 360 words. *American Journal of Psychology, 71*, 688–699.
Jenkins, W. D., McFann, H., & Clayton, F. L. (1950). A methodological study of extinction following aperiodic and continuous reinforcement. *Journal of Comparative and Physiological Psychology, 43*, 155–167.
Jenni, D. A., & Jenni, M. A. (1976). Carrying behavior in humans: Analysis of sex differences. *Science, 194*, 859–860.
Jennings, M. K., & Niemi, R. G. (1968). The transmission of political values from parent to child. *American Political Science Review, 62*, 169–184.
Jensen, A. R. (1969). How much can we boost IQ and scholastic achievement? *Harvard Educational Review, 39*, 1–123.
Jersild, A. T., & Holmes, F. B. (1935). *Children's fears*. New York: Teachers College Press.
Johnston, L. D., Bachman, J. G., & O'Malley, P. M. (1980). *Highlights from student drug use in America, 1975–1980*. Rockville, MD: National Institute on Drug Abuse.
Jones, E. (1981). Rebirthing. In R. J. Corsini (Ed.), *Handbook of innovative psychotherapies*. New York: Wiley.
Jones, E. E., & Davis, K. E. (1965). From acts to dispositions: The attribution process in person perception. In L. Berkowitz (Ed.), *Advances in experimental social psychology* (Vol. 2). New York: Academic Press.
Jones, E. E., & McGillis. D. (1976). Correspondent inferences and the attribution cube: A comparative reappraisal. In J. H. Harvey, W. J. Ickes, & R. F. Kidd (Eds.), *New directions in attribution research*. Hillsdale, NJ: Erlbaum.
Jones, E. E., & Nisbett, R. E. (1972). The actor and observer: Divergent perceptions of the causes of behavior. In E. E. Jones, D. E. Kanouse, H. H. Kelley, R. E. Nisbett, S. Valins, & B. Weiner (Eds.), *Attribution: Perceiving the causes of behavior*. Morristown, NJ: General Learning Press.
Julien, R. M. (1981). *A primer of drug action* (3d ed.). San Francisco: W. H. Freeman.
Jung, C. G. (1928). *Contributions to analytical psychology* (H. G. Baynes & C. F. Baynes, Trans.). London: K. Paul, Trench, Trubner.
Jung, C. G. (1954). *The practice of psychotherapy*. In H. Read, M. Fordham, & G. Adler (Eds.), *The collected works of C. G. Jung* (Vol. 16) (R. F. C. Hull, Trans.). New York: Pantheon Books.
Kagan, J. (1972). A psychologist's account at mid-career. In T. S. Krawiec (Ed.), *The psychologists* (Vol. 1). New York: Oxford University Press.
Kagan, J., & Moss, H. (1962). *Birth to maturity: A study in psychological development*. New York: Wiley.
Kahneman, D. (1973). *Attention and effort*. Englewood Cliffs, NJ: Prentice-Hall.
Kahneman, D., Slovic, P., & Tversky, A. (Eds.) (1982). *Judgment under uncertainty: Heuristics and biases*. New York: Cambridge University Press.
Kalat, J. W. (1984). *Biological psychology* (2nd ed.). Belmont, CA: Wadsworth.
Kalinowsky, L. B. (1979). Psychosurgery: The past twenty years. *Psychiatric Journal of the University of Ottawa, 4*, 111–113.

Kalnins, I. V., & Bruner, J. S. (1973). The coordination of visual observation and instrumental behavior in early infancy. *Perception, 2*, 307–314.

Kamin, L. J. (1969). Predictability, surprise, attention, and conditioning. In B. A. Campbell & R. M. Church (Eds.), *Punishment and aversive behavior.* New York: Appleton-Century-Crofts.

Kamin, L. J. (1974). *The science and politics of IQ.* Potomac, MD: Erlbaum.

Kandel, E. R., & Schwartz, J. H. (1982). Molecular biology of learning: Modulation of transmitter release. *Science, 218*, 433–443.

Kanizsa, G. (1955). Margini quasi-percettivi in campi con stimolazione omogenea. *Rivista di Psicologia, 49*, 7–30.

Kanizsa, G. (1976). Subjective contours. *Scientific American, 234* (4), 48–52.

Katz, J. J. (1972). *Semantic theory.* New York: Harper & Row.

Katz, P. A., Zigler, E., & Zalk, S. R. (1975). Children's self-image disparity: The effects of age, maladjustment, and action-thought orientation. *Developmental Psychology, 11*, 546–550.

Katzman, M. A., & Wolchik, S. A. (1984). Bulimia and binge eating in college women: A comparison of personality and behavioral characteristics. *Journal of Consulting and Clinical Psychology, 52*, 423–428.

Keating, D. P.(1980). Thinking processes in adolescence. In J. Adelson (Ed.), *Handbook of adolescent psychology.* New York: Wiley.

Keesey, R. E., & Powley, T. L. (1975). Hypothalamic regulation of body weight. *American Scientist, 63*, 558–565.

Keller, F. S., & Schoenfeld, W. N. (1950). *Principles of psychology.* New York: Appleton-Century-Crofts.

Kellerman, J. (1977). Anorexia nervosa: The efficacy of behavior therapy. *Journal of Behavior Therapy and Experimental Psychiatry, 8*, 387–390.

Kelley, H. H. (1967). Attribution theory in social psychology. In D. Levine (Ed.), *Nebraska symposium on motivation.* Lincoln, NE: University of Nebraska Press.

Kelley, H. H. (1973). The process of causal attribution. *American Psychologist, 28*, 107–128.

Kelley, H. H. (1979). *Personal relationships.* Hillsdale, NJ: Erlbaum.

Kelley, H. H., & Thibaut, J. (1978). *Interpersonal relations: A theory of interdependence.* New York: Wiley.

Keniston, K. (1970). Youth as a stage of life. *American Scholarship, 39*, 631–654.

Kenny, S. L. (1983). Developmental discontinuities in childhood and adolescence. In K. W. Fischer (Ed.), *Levels and transitions in children's development.* San Francisco: Jossey-Bass.

Kessen, W. (Ed.) (1965). *The child.* New York: Wiley.

Kessen, W. (1967). Sucking and looking: Two organized congenital patterns of behavior in the human newborn. In H. W. Stevenson, E. H. Hess, & H. L. Rheingold (Eds.), *Early behavior: Comparative and developmental approaches.* New York: Wiley.

Kety, S. S. (1975). Adopted individuals who have become schizophrenic. In H. M. Van Praag (Ed.), *On the origin of schizophrenic psychoses.* Amsterdam: De Erven Bohn BV.

Key, W. B. (1976). *Media sexploitation.* New York: New American Library.

Khanna, S. M., & Leonard, D. G. B. (1982). Basilar membrane tuning in cat cochlea. *Science, 215*, 305–306.

Kimble, G. A., Garmezy, N., & Zigler, E. (1980). *Principles of general psychology* (5th ed.). New York: Wiley.

Kimble, G. A., Garmezy, N., & Zigler, E. (1984). *Principles of Psychology* (6th ed.) New York: Wiley.

King, M. S., Kimble, G. A., Gorman, J., & King, R. A. (1961). Replication report: Two failures to reproduce effects of anxiety on eyelid conditioning. *Journal of Experimental Psychology, 62*, 532–533.

Kinsey, A. C., Pomeroy, W. B., & Martin, C. E. (1948). *Sexual behavior in the human male.* Philadelphia: Saunders.

Kinsey, A. C., Pomeroy, W. B., Martin, C. E., & Gebhard, P. H. (1953). *Sexual behavior in the human female.* Philadelphia: Saunders.

Klatsky, R. L. (1980). *Human memory: Structures and processes* (2nd ed.). San Francisco: W. H. Freeman.

Klausner, L. D. (1981). *Son of Sam.* New York: McGraw-Hill.

Klein, G. S. 1970. *Perception, motives and personality.* New York: Knopf.

Kleinginna, P. R., & Kleinginna, A. M. (1981). A categorized list of emotion definitions, with suggestions for a consensual definition. *Motivation and Emotion, 5*, 345–379.

Kleinmuntz, B., & Szucko, J. J. (1984). Lie detection in ancient and modern times: A call for contemporary scientific study. *American Psychologist, 39*, 766–776.

Knapp, M. L. (1978). *Nonverbal communication in human interaction* (2nd ed.). New York: Holt, Rinehart and Winston.

Kohlberg, L. (1966). A cognitive-developmental analysis of children's sex-role concepts and attitudes. In E. Maccoby (Ed.), *The development of sex differences.* Stanford, CA: Stanford University Press.

Kohlberg, L. (1969). Stage and sequence: The cognitive-developmental approach to socialization. In D. A. Goslin (Ed.), *Handbook of socialization theory and research.* Chicago: Rand-McNally.

Kohlberg, L. (1976). Moral stages and moralization: The cognitive-developmental approach. In T. Lickona (Ed.), *Moral development and behavior: Theory, research, and social issues.* New York: Holt, Rinehart and Winston.

Köhler, W. (1925). *The mentality of apes* (E. Winter, Trans.). New York: Harcourt, Brace & World.

Kolb, B., & Whishaw, I. Q. (1980). *Fundamentals of human neuropsychology.* San Francisco: W. H. Freeman.

Kolb, B., & Whishaw, I. Q. (1985). *Fundamentals of human neuropsychology* (2nd ed.). New York: W. H. Freeman.

Konečni, V. J., & Ebbesen, E. B. (1976). Disinhibition versus the cathartic effect: Artifact and substance. *Journal of Personality and Social Psychology, 34*, 352–365.

König, J. F. R., & Klippel, R. A. (1963). *The rat brain: A stereotaxic atlas of the forebrain and lower parts of the brain stem.* Baltimore: Williams & Wilkins.

Konopka, G. (1976). *Young girls: A portrait of adolescence.* Englewood Cliffs, NJ: Prentice-Hall.

Kosslyn, S. M. (1975). Information representation in visual images. *Cognitive Psychology*, 7, 341–370.

Kosslyn, S. M. (1983). *Ghosts in the mind's machine: Creating and using images in the brain.* New York: Norton.

Kraepelin, E. (1907). *Clinical psychiatry* (A. R. Diefendorf, Trans.). New York: Macmillan.

Krause, C. A. (1978). *Guyana massacre: The eyewitness account.* New York: Berkley.

Krauskopf, K. B., & Beiser, A. (1973). *The physical universe* (3rd ed.). New York: McGraw-Hill.

Krech, D., Crutchfield, R. S., Livson, N., Wilson, W. A., Jr., & Parducci, A. (1982). *Elements of psychology* (4th ed.). New York: Knopf.

Kringlen, E. (1967). *Heredity and environment in the functional psychoses.* London: Heinemann.

Kübler-Ross, E. (1969). *On death and dying.* New York: Macmillan.

Kübler-Ross, E. (1975). Foreword. In R. A. Moody, Jr., *Life after life.* Covington, GA: Mockingbird.

Kuder, G. F. (1979). *Kuder Occupational Interest Survey: General manual, form DD.* Chicago: Science Research Associates.

Kuhn, D., Pennington, N., & Leadbeater, B. (1983). Adult thinking in developmental perspective. In P. B. Baltes & O. G. Brim (Eds.), *Life-span development and behavior.* New York: Academic Press.

Labouvie-Vief, G. (1982). Dynamic development and mature autonomy: A theoretical prologue. *Human Development, 25,* 161–191.

Lamb, M. E. (1976). Twelve-month olds and their parents: Interaction in a laboratory playroom. *Developmental Psychology, 12,* 237–244.

Lambourn, J., & Gill, D. (1978). A controlled comparison of simulated and real ECT. *British Journal of Psychiatry, 133,* 514–519.

Lang, P. J., & Melamed, B. G. (1969). Case report: Avoidance conditioning therapy in an infant with chronic ruminative vomiting. *Journal of Abnormal Psychology, 74,* 1–8.

Langer, E. J. (1981). Old Age: An Artifact? In *Biology, behavior, and aging.* New York: National Research Council.

Langer, E. J., & Abelson, R. P. (1974). A patient by any other name: Clinician group difference in labeling bias. *Journal of Consulting and Clinical Psychology, 42,* 4–9.

LaPiere, R. (1934). Attitudes versus actions. *Social Forces, 13,* 230–237.

Larson, R., Mayers, P., & Csikszentmihalyi, M. (1977, August). *Experiential sampling of adolescents' socialization: The contexts of family, friends, and being alone.* Paper presented at the Conference on Research Perspectives in the Ecology of Human Development, Cornell University, Ithaca, N.Y.

Lazarus, R. S., Averill, J. R., & Opton, E. M., Jr. (1970). Towards a cognitive theory of emotion. In M. B. Arnold (Ed.), *Feelings and emotions: The Loyola symposium.* New York: Academic Press.

Lazarus, R. S., & Folkman, S. (1984). *Stress, appraisal, and coping.* New York: Springer Publishing Co.

Leavy, R. L. (1983). Social support and psychological disorder: A review. *Journal of Community Psychology, 11,* 3–23.

Leeper, R. W. (1935). A study of a neglected portion of the field of learning. The development of sensory organization. *Journal of Genetic Psychology, 46,* 41–75.

Leeper, R. W. (1970). The motivational and perceptual properties of emotions as indicating their fundamental character and role. In M. B. Arnold (Ed.), *Feelings and emotions: The Loyola symposium.* New York: Academic Press.

Lefrancois, G. R. (1980). *Of children: An introduction to child development* (3rd ed.). Belmont, CA: Wadsworth.

Leifermann, H. P. (1975). *Crystal Lee: A woman of inheritance.* New York: Macmillan.

Leirer, V. O., Hamilton, D. L., & Carpenter, S. (1982). Common first names as cues for inferences about personality. *Personality and Social Psychology Bulletin, 8,* 712–718.

Lenneberg, E. H. (1967). *Biological foundations of language.* New York: Wiley.

Leon, G. R. (1977). Anxiety neurosis: The case of Richard Benson. In *Case histories in deviant behavior* (2nd ed.). Boston: Allyn and Bacon.

Lepper, M. R., & Greene, D. (1978). *The hidden costs of reward.* Hillsdale, NJ: Erlbaum.

Lepper, M. R., Greene, D., & Nisbett, R. E. (1973). Undermining children's intrinsic interest with extrinsic reward: A test of the "overjustification" hypothesis. *Journal of Personality and Social Psychology, 28,* 129–137.

Leventhal, G. S. (1976). Fairness in social relationships. In J. Thibaut, J. T. Spence, & R. C. Carson (Eds.), *Contemporary topics in social psychology.* Morristown, NJ: General Learning Press.

Levine, J. D., Gordon, N. C., & Fields, H. L. (1979). The role of endorphins in placebo analgesia. In J. J. Bonica, J. C. Liebeskind, & D. Albe-Fessard (Eds.), *Advances in pain research and therapy* (Vol. 3). New York: Raven Press.

Levinger, G., & Snoek, J. D. (1972). *Attraction in relationship: A new look at interpersonal attraction.* Morristown, NJ: General Learning Press.

Levinson, D. (1978). *The seasons of a man's life.* New York: Ballantine Books.

Lewinsohn, P. M., & Arconad, M. (1981). Behavioral treatment in depression: A social learning approach. In J. F. Clarkin & H. I. Glazer (Eds.), *Depression: Behavioral and directive intervention strategies.* New York: Garland STPM Press.

Lewinsohn, P. M., Mischel, W., Chaplin, W., & Barton, R. (1980). Social competence and depression: The role of illusory self-perceptions. *Journal of Abnormal Psychology, 89,* 203–212.

Lewinsohn, P. M., & Talkington, J. (1979). Studies on the measurement of unpleasant events and relations with depression. *Applied Psychological Measurement, 3,* 83–101.

Lewis, M., & Brooks-Gunn, J. (1979). *Social cognition and the acquisition of self.* New York: Plenum Press.

Liberman, A. M. (1973). The speech code. In G. A. Miller (Ed.), *Communication, language, and meaning.* New York: Basic Books.

Liberman, A. M., Mattingly, I. G., & Turvey, M. T. (1972). Language codes and memory codes. In A. W. Melton & E. Martin (Eds.), *Coding processes in human memory.* Washington, DC: Winston.

Lieberson, S., & Silverman, A. R. (1965). The precipitants and underlying conditions of race riots. *American Sociological Review, 30,* 887–898.

Liddy, G. G. (1980). *Will.* New York: St. Martin's Press.

Lidz, T. (1973). *The origin and treatment of schizophrenic disorders.* New York: Basic Books.

Liem, J. H. (1980). Family studies of schizophrenia: An update and commentary. *Schizophrenia Bulletin, 6,* 429–455.

Linden, E. (1975). *Apes, men and language.* New York: Dutton.

Lindsay, P. H., & Norman, D. A. (1977). *Human information processing: An introduction to psychology* (2nd ed.). New York: Academic Press.

Lipsitt, L. P. (1982). Infant learning. In T. M. Field, A. Huston, H. C. Quay, L. Troll, & G. E. Finley (Eds.), *Review of human development.* New York: Wiley.

Liston, E. H., & Jarvik, L. F. (1976). Genetics of schizophrenia. In M. A. Sperber & L. F. Jarvik (Eds.), *Psychiatry and genetics: Psychosocial, ethical, and legal considerations.* New York: Basic Books.

Litman, R. E. (1970). Suicide as acting out. In E. S. Shneidman, N. L. Farberow, & R. E. Litman (Eds.), *The psychology of suicide.* New York: Science House.

Livesley, W. J., & Bromley, D. B. (1973). *Person perception in childhood and adolescence.* London: Wiley.

Livingston, R. B. (1967). Brain circuitry relating to complex behavior. In G. C. Quarton, T. Melnechuck, & F. O. Schmitt (Eds.), *The neurosciences: A study program.* New York: Rockefeller University Press.

Loehlin, J. C., Lindzey, G., & Spuhler, J. N. (1975). *Race differences in intelligence.* San Francisco: W. H. Freeman.

Loehlin, J. C., & Nichols, R. C. (1976). *Heredity, environment, and personality: A study of 850 sets of twins.* Austin, TX: University of Texas Press.

Loftus, E. F. (1979). *Eyewitness testimony.* Cambridge, MA: Harvard University Press.

Loftus, E. F. (1980). *Memory: Surprising new insights into how we remember and why we forget.* Reading, MA: Addison-Wesley.

Loftus, E. F., & Loftus, G. R. (1980). On the permanence of stored information in the human brain. *American Psychologist, 35,* 409–420.

Loftus, E. F., & Palmer, J. C. (1974). Reconstruction of automobile destruction: An example of the interaction between language and memory. *Journal of Verbal Learning and Verbal Behavior, 13,* 585–589.

Loftus, G. R., & Loftus, E. F. (1976). *Human memory: The processing of information.* Hillsdale, NJ: Erlbaum.

Lopata, H. Z. (1975). Widowhood: Societal factors in life-span disruptions and alternatives. In N. Datan & L. H. Ginsberg (Eds.), *Lifespan developmental psychology: Normative life crises.* New York: Academic Press.

Lorenz, K. (1957). Companionship in bird life. In C. H. Schiller (Trans. and Ed.), *Instinctive behavior.* New York: International Universities Press.

Lorenz, K. (1966). *On aggression.* New York: Bantam Books.

Lorenz, K., & Leyhausen, P. (1973). *Motivation of human and animal behavior: An ethological view* (B. A. Tonkin, Trans.). New York: Van Nostrand Reinhold.

Losoncy, L. (1981). Encouragement therapy. In R. J. Corsini (Ed.), *Handbook of innovative psychotherapies.* New York: Wiley.

Lotter, V. (1967). Epidemiology of autistic conditions in young children. II. Some characteristics of the parents and children. *Social Psychiatry, 1,* 163–173.

Lovaas, O. I. (1977). *The autistic child: Language development through behavior modification.* New York: Irvington.

Lovaas, O. I. (1978). Parents as therapists. In M. Rutter & E. Schopler (Eds.), *Autism: A reappraisal of concepts and treatment.* New York: Plenum Press.

Luchins, A. S., & Luchins, E. H. (1959). *Rigidity of behavior.* Eugene, OR: University of Oregon Press.

Luckey, E. G., & Bain, J. K. (1970). Children: A factor in marital satisfaction. *Journal of Marriage and the Family, 32,* 43–44.

Lumsdaine, A. A., & Janis, I. L. (1953). Resistance to "counter propaganda" produced by one-sided and two-sided propaganda presentations. *Public Opinion Quarterly, 17,* 311–318.

Luria, A. R. (1968). *The mind of a mnemonist.* (L. Solotaroff, Trans.). New York: Basic Books. (Original work published 1965)

Lykken, D. T. (1981). *A tremor in the blood: Uses and abuses of the lie detector.* New York: McGraw-Hill.

Lyles, J. N., Burish, T. G., Krozely, M. G., & Oldham, R. K. (1982). Efficacy of relaxation training and guided imagery in reducing the aversiveness of cancer chemotherapy. *Journal of Consulting and Clinical Psychology, 50,* 509–524.

Maccoby, E. E., & Jacklin, C. N. (1974). *The psychology of sex differences.* Stanford, CA: Stanford University Press.

MacLean, P. D. (1949). Psychosomatic disease and the "visceral brain." Recent developments bearing on the Papez theory of emotion. *Psychosomatic Medicine, 11,* 338–353.

MacLeod, S. (1981). *The art of starvation.* London: Virago.

MacNichol, E. F., Jr. (1964). Three-pigment color vision. *Scientific American, 211*(6), 48–56.

Magoun, H. W. (1954). The ascending reticular system and wakefulness. In J. F. Delafresnaye (Ed.), *Brain mechanisms and consciousness.* Oxford, England: Blackwell.

Marks, G., & Miller, N. (1982). Target attractiveness as a mediator of assumed attitude similarity. *Personality and Social Psychology Bulletin, 8,* 728–735.

Marshall, G. D., & Zimbardo, P. G. (1979). Affective consequences of inadequately explained physiological arousal. *Journal of Personality and Social Psychology, 37,* 970–988.

Marshall, J. F., Richardson, J. S., & Teitelbaum, P. (1974). Nigrostriatal bundle damage and the lateral hypothalamic syndrome. *Journal of Comparative and Physiological Psychology, 87,* 808–830.

Martin, B. (1981). *Abnormal psychology: Clinical and scientific perspectives* (2nd ed.). New York: Holt, Rinehart and Winston.

Mash, E. J., & Terdal, L. G. (Eds.) (1981). *Behavioral assessment of childhood disorders.* New York: Guilford Press.

Maslach, C. (1979). Negative emotional biasing of unexplained arousal. *Journal of Personality and Social Psychology, 37,* 953–969.

Maslow, A. H. (1954). *Motivation and personality.* New York: Harper & Row.

Maslow, A. H. (1967). Self-actualization and beyond. In J. F. T. Bugental (Ed.), *Challenges of humanistic psychology.* New York: McGraw-Hill.

Maslow, A. H. (1968). *Toward a psychology of being* (2nd ed.). Princeton, NJ: Van Nostrand Reinhold.

Maslow, A. H. (1971). *The farther reaches of human nature.* New York: Viking.

Masters, W. H., & Johnson, V. E. (1970). *Human sexual inadequacy.* Boston: Little, Brown.

Matthews, K. A. (1982). Psychological perspectives on the Type A behavior pattern. *Psychological Bulletin, 91,* 293–323.

Mattison, R., Cantwell, D. P., Russell, A. T., & Will, L. (1979). A comparison of DSM-II and DSM-III in the diagnosis of childhood psychiatric disorders. *Archives of General Psychiatry, 36,* 1217–1222.

Maturana, H. R., Lettvin, J. Y., McCulloch, W. S., & Pitts, W. H. (1960). Anatomy and physiology of vision in the frog (*Rana pipiens*). *Journal of General Physiology, 43,* 129–176.

May, R. (1958). The origins and significance of the existential movement in psychology. In R. May, E. Angel, & H. F. Ellenberger (Eds.), *Existence: A new dimension in psychiatry and psychology.* New York: Basic Books.

Mayer, D. J., & Liebeskind, J. C. (1974). Pain reduction by focal electrical stimulation of the brain: An anatomical and behavioral analysis. *Brain Research, 68,* 73–93.

Mayer, D. J., Price, D. D., Barber, J. & Rafii, A. (1976). Acupuncture analgesia: Evidence for activation of a pain inhibitory system as a mechanism of action. In J. J. Bonica & D. Albe-Fessard (Eds.), *Advances in pain research and therapy* (Vol. 1). New York: Raven Press.

Mayer, J. (1955). Regulation of energy intake and body weight: The glucostatic theory and the lipostatic hypothesis. *Annals of the New York Academy of Science, 63,* 15–43.

McCarthy, P. R., & Knapp, S. L. (1984). Helping styles of crisis interveners, psychotherapists, and untrained individuals. *American Journal of Community Psychology, 12,* 623–627.

McClelland, D. C. (1961). *The achieving society.* Princeton, NJ: Van Nostrand Reinhold.

McClelland, D. C. (1971). *Motivational trends in society.* Morristown, NJ: General Learning Press.

McClelland, D. C. (1975). *Power: The inner experience.* New York: Irvington.

McClelland, D. C., Atkinson, J. W., Clark, R. A., & Lowell, E. L. (1953). *The achievement motive.* New York: Appleton-Century-Crofts.

McClelland, D. C., Davis, W. N., Kalin, R., & Wanner, E. (1972). *The drinking man.* New York: Free Press.

McClelland, D. C., & Winter, D. G. (1969). *Motivating economic achievement.* New York: Free Press.

McCorduck, P. (1979). *Machines who think: A personal inquiry into the history and prospects of artificial intelligence.* San Francisco: W. H. Freeman.

McGeoch, J. A., & Irion, A. L. (1952). *The psychology of human learning* (2nd ed.). New York: Longmans, Green.

McGlone, J. (1980). Sex differences in human brain asymmetry: A critical survey. *The Behavioral and Brain Sciences, 3,* 215–263.

McGuire, W. (Ed.) (1974). *The Freud/Jung letters* (R. Manheim & R. F. C. Hull, Trans.). Princeton, NJ: Princeton University Press.

McGuire, W. J. (1961). Resistance to persuasion conferred by active and passive prior refutation of the same and alternative counterarguments. *Journal of Abnormal and Social Psychology, 63,* 326–332.

Mead, M. (1939). *Coming of age in Samoa.* In *From the south seas: Studies of adolescence and sex in primitive societies.* New York: Morrow. (Original work published 1928)

Meehl, P. E. (1954). *Clinical versus statistical prediction.* Minneapolis, MN: University of Minnesota Press.

Mehr, J. (1983). *Abnormal psychology.* New York: Holt, Rinehart and Winston.

Meichenbaum, D. (1976). Cognitive factors in biofeedback therapy. *Biofeedback and Self-regulation, 29,* 373–404.

Meichenbaum, D. (1977). *Cognitive-behavior modification.* New York: Plenum Press.

Meltzoff, A. N., & Moore, M. K. (1977). Imitation of facial and manual gestures by human neonates. *Science, 198,* 75–78.

Melzack, R. (1973). *The puzzle of pain.* New York: Basic Books.

Mendelson, W. B., Gillin, J. C., & Wyatt, R. J. (1977). *Human sleep and its disorders.* New York: Plenum Press.

Menzel, E. W. (1973). Chimpanzee spatial memory organization. *Science, 182,* 943–945.

Menzel, E. W. (1978). Cognitive mapping in chimpanzees. In S. H. Hulse, H. Fowler, & W. K. Honig (Eds.), *Cognitive processes in animal behavior.* Hillsdale, NJ: Erlbaum.

Mervis, C. B., & Rosch, E. (1981). Categorization of natural objects. *Annual Review of Psychology, 32,* 89–115.

Meyer, R. G., & Salmon, P. (1984). *Abnormal psychology.* Boston: Allyn and Bacon.

Mezzich, A. C., & Mezzich, J. E. (1979, September). *Diagnostic reliability of childhood and adolescent behavior disorders.* Paper presented at the American Psychological Association meeting, New York.

Milgram, S. (1963). Behavioral study of obedience. *Journal of Abnormal and Social Psychology, 67,* 371–378.

Milgram, S. (1965). Some conditions of obedience and disobedience to authority. *Human Relations, 18,* 57–76.

Milgram, S. (1974). *Obedience to authority.* New York: Harper & Row.

Miller, G. A. (1956). The magical number seven, plus or minus two: Some limits on our capacity for processing information. *Psychological Review, 63,* 81–97.

Miller, M. M. (1982). Hypnoaversion treatment in alcoholism, nicotinism, and weight control. *Journal of the National Medical Association, 68,* 129–130.

Miller, N. E. (1944). Experimental studies of conflict. In J. M. Hunt (Ed.), *Personality and the behavior disorders* (Vol. 1). New York: Ronald Press.

Miller, N. E. (1969). Learning of visceral and glandular responses. *Science, 163,* 434–445.

Miller, P. Y., & Simon, W. (1980). The development of sexuality in adolescence. In J. Adelson (Ed.), *Handbook of adolescent psychology.* New York: Wiley.

Miller, R. C., & Berman, J. S. (1983). The efficacy of cognitive behavior therapies: A quantitative review of the research evidence. *Psychological Bulletin, 94,* 39–53.

Milner, B. (1964). Some effects of prefrontal lobectomy in man. In J. M. Warren & K. Akert (Eds.), *The frontal granular cortex and behavior.* New York: McGraw-Hill.

Milner, B. (1970). Memory and the medial temporal regions of the brain. In K. H. Pribram & D. E. Broadbent (Eds.), *Biology of memory.* New York: Academic Press.

Milner, B. (1974). Hemispheric specialization: Scope and limits. In F. O. Schmitt & F. G. Worden (Eds.), *The neurosciences: Third study program.* Cambridge, MA: MIT Press.

Milner, B., Corkin, S., & Teuber, H.-L. (1968). Further analysis of the hippocampal amnesic syndrome: 14-year follow-up study of H. M. *Neuropsychologia, 6,* 215–234.

Minton, J., Campbell, M., Green, W. H., Jennings, S., & Samit, C. (1982). Cognitive assessment of siblings of autistic children. *Journal of the American Academy of Child Psychiatry, 21,* 256–261.

Mischel, W. (1969). Continuity and change in personality. *American Psychologist, 24* 1012–1018.

Mischel, W. (1976). *Introduction to personality* (2nd ed.). New York: Holt, Rinehart and Winston.

Mischel, W. (1977). The interaction of person and situation. In D. Magnusson & N. S. Endler (Eds.), *Personality at the crossroads: Current issues in interactional psychology.* Hillsdale, NJ: Erlbaum.

Mollon, J. D. (1982). Color vision. *Annual Review of Psychology, 33,* 41–85.

Momboisse, R. M. (1967). *Riots, revolts, and insurrections.* Springfield, IL: Thomas.

Money, J., & Ehrhardt, A. A. (1972). *Man and woman, boy and girl. The differentiation and dimorphism of gender identity from conception to maturity.* Baltimore: Johns Hopkins University Press.

Montagu, A. (1976). *The nature of human aggression.* New York: Oxford University Press.

Moreland, R. L., & Zajonc, R. B. (1976). A strong test of exposure effects. *Journal of Experimental Social Psychology, 12,* 170–179.

Morgan, C. D., & Murray, H. A. (1938). Thematic Apperception Test. In H. A. Murray et al., *Explorations in personality: A clinical and experimental study of fifty men of college age.* New York: Oxford University Press.

Morgane, P. J., & Kosman, A. J. (1960). Relationship of the middle hypothalamus to amygdalar hyperphagia. *American Journal of Physiology, 198,* 1315–1318.

Moyer, K. E. (1976). *The psychobiology of aggression.* New York: Harper & Row.

Muhlenkamp, A. F., Gress, L. D., & Flood, M. A. (1975). Perception of life change events by the elderly. *Nursing Research, 24,* 109–113.

Munn, N. L. (1966). *Psychology: The fundamentals of human adjustment* (5th ed.). Boston: Houghton Mifflin.

Murray, E. J. (1964). *Motivation and emotion.* Englewood Cliffs, NJ: Prentice-Hall.

Murray, H. A., et al. (1938). *Explorations in personality: A clinical and experimental study of fifty men of college age.* New York: Oxford University Press.

Murray, J. B. (1982). What is meditation? Does it help? *Genetic Psychology Monographs, 106*, 88–115.

Mussen, P. H., Conger, J. J., Kagan, J., & Geiwitz, J. (1979). *Psychological development: A life-span approach.* New York: Harper & Row.

Myers, J. K., Weissman, M. M., Tischler, G. L., Holzer, C. E., III, Leaf, P. J., Orvaschel, H., Anthony, J. C., Boyd, J. H., Burke, J. D., Jr., Kramer, M., & Stoltzman, R. Six-month prevalence of psychiatric disorders in three communities. *Archives of General Psychiatry, 41*, 959–967.

Neisser, U. (1967). *Cognitive psychology.* New York: Appleton-Century-Crofts.

Neisser, U. (1982). *Memory observed: Remembering in natural contexts.* San Francisco: W. H. Freeman.

Neisser, U. (1982). Snapshots or benchmarks? In U. Neisser (Ed.), *Memory observed: Remembering in natural contexts.* San Francisco: W. H. Freeman.

Nekipelov, V. (1980). *Institute of fools: Notes from the Serbsky* (M. Carynnyk & M. Horgan, Eds. and Trans.). New York: Farrar, Straus & Giroux.

Netter, F. H. (1965). *The CIBA collection of medical illustrations: Vol. 4. Endocrine system.* Summit, NJ: CIBA Pharmaceutical Company.

Neugarten, B. L., & Weinstein, K. K. (1964). The changing American grandparent. *Journal of Marriage and the Family, 26*, 199–204.

Newell, A., & Simon, H. A. (1972). *Human problem solving.* Englewood Cliffs, NJ: Prentice-Hall.

Nickerson, R. S., & Adams, M. J. (1979). Long-term memory for a common object. *Cognitive Psychology, 11*, 287–307.

Nisbett, R. E., Caputo, C., Legant, P., & Marecek, J. (1973). Behavior as seen by the actor and as seen by the observer. *Journal of Personality and Social Psychology, 27*, 154–164.

Nisbett, R. E., & Wilson, T. D. (1977). Telling more than we can know: Verbal reports on mental processes. *Psychological Review, 84*, 231–259.

Norman, D. A. (1982). *Learning and memory.* San Francisco: W. H. Freeman.

Novitski, E. (1977). *Human genetics.* New York: Macmillan.

O'Connell, W. E. (1981). Natural high therapy. In R. J. Corsini (Ed.), *Handbook of innovative psychotherapies.* New York: Wiley.

Olds, M. E., & Fobes, J. L. (1981). The central basis of motivation: Intracranial self-stimulation studies. *Annual Review of Psychology, 32*, 523–574.

Olson, G., & King, R. A. (1962). Supplementary report: Stimulus generalization gradients along a luminosity continuum. *Journal of Experimental Psychology, 63*, 414–415.

Oomura, Y. (1976). Significance of glucose, insulin, and free fatty acid on the hypothalamic feeding and satiety neurons. In D. Novin, W. Wyrwicka, & G. A. Bray (Eds.), *Hunger: Basic mechanisms and clinical implications.* New York: Raven Press.

Orbison, W. D. (1939). Shape as a function of the vector field. *American Journal of Psychology, 52*, 31–45.

Osgood, C. E., Suci, G. J., & Tannenbaum, P. H. (1957). *The measurement of meaning.* Urbana, IL: University of Illinois Press.

Paivio, A. (1965). Abstractness, imagery, and meaningfulness in paired-associate learning. *Journal of Verbal Learning and Verbal Behavior, 4*, 32–38.

Paivio, A. (1971). *Imagery and verbal processes.* New York: Holt, Rinehart and Winston.

Palmore, E. B., Fillenbaum, G. G., & George, L. K. (1984). Consequences of retirement. *Journal of Gerontology, 39*, 109–116.

Parke, R. D., & O'Leary, S. E. (1976). Father-mother interaction in the newborn period: Some findings, some observations, and some unresolved issues. In K. Riegel & J. Meacham (Eds.), *The developing individual in a changing world.* The Hague: Mouton.

Parloff, M. B. (1984). Psychotherapy research and its incredible credibility crisis. *Clinical Psychology Review, 4*, 95–109.

Pascual-Leone, J. (1983). Growing into human maturity: Toward a meta-subjective theory of adulthood stages. In P. B. Baltes & O. G. Brim (Eds.), *Life-span development and behavior* (Vol. 5). New York: Academic Press.

Patterson, F., & Linden, E. (1981). *The education of Koko.* New York: Holt, Rinehart and Winston.

Patterson, G. R. (1976). The aggressive child: Victim and architect of a coercive system. In E. J. Mash, L. A. Hamerlynck, & L. C. Handy (Eds.), *Behavior modification and families.* New York: Brunner/Mazel.

Patty, R. A., & Safford, S. F. (1976). Motive to avoid success, motive to avoid failure, state-trait anxiety, and performance. In C. D. Spielberger & I. G. Sarason (Eds.), *Stress and anxiety* (Vol. 4). New York: Academic Press.

Paul, G. L., & Lentz, R. J. (1977). *Psychosocial treatment of chronic mental patients.* Cambridge, MA: Harvard University Press.

Pauls, D. L., & Kidd, K. K. (1981). Genetics of childhood behavior disorders. In B. B. Lahey & A. E. Kazdin (Eds.), *Advances in clinical child psychology* (Vol. 4). New York: Plenum Press.

Pavlov, I. P. (1960). *Conditioned reflexes* (G. V. Anrep, Trans.). New York: Dover. (Original work published 1927)

Pearlin, L. I. (1980). Life strains and psychological distress among adults. In N. J. Smelser & E. H. Erikson (Eds.), *Themes of work and love in adulthood.* Cambridge, MA: Harvard University Press.

Pearson, K. (1924). *The life, letters, and labours of Francis Galton* (Vol. 2). Cambridge, England: Cambridge University Press.

Penfield, W., & Rasmussen, T. (1950). *The cerebral cortex of man: A clinical study of localization of function.* New York: Macmillan.

Penrose, L. S., & Penrose, R. (1958). Impossible objects: A special type of visual illusion. *British Journal of Psychology, 49*, 31–33.

Perlmutter, M., & List, J. A. (1982). Learning in later adulthood. In T. M. Field, A. Huston, H. C. Quay, L. Troll, & G. E. Finley (Eds.), *Review of human development.* New York: Wiley.

Perls, F. S. (1970). Four lectures. In J. Fagan & I. L. Sheperd (Eds.), *Gestalt therapy now.* Palo Alto, CA: Science and Behavior Books.

Peterson, C., & Seligman, M. E. P. (1984). Causal explanations as a risk factor for depression: Theory and evidence. *Psychological Review, 91*, 347–374.

Petty, R. E., & Cacioppo, J. T. (1981). *Attitudes and persuasion: Classic and contemporary approaches.* Dubuque, IA: Wm. C. Brown.

Pfaffmann, C. (1964). Taste, its sensory and motivating properties. *American Scientist, 52*, 187–206.

Pfaffmann, C. (1982). Taste: A model of incentive motivation. In D. W. Pfaff (Ed.)., *The physiological mechanisms of motivation.* New York: Springer-Verlag.

Pfeiffer, E. (1968). *Disordered behavior: Basic concepts in clinical psychiatry.* New York: Oxford University Press.

Pfeiffer, E., Verwoerdt, A., & Davis, G. C. (1972). Sexual behavior in middle life. *The American Journal of Psychiatry, 128,* 1262–1267.

Phares, E. J. (1979). *Clinical psychology: Concepts, methods, and profession.* Homewood, IL: Dorsey.

Piaget, J. (1932). *The moral judgment of the child.* New York: Harcourt Brace.

Piaget, J. (1952). *The origins of intelligence in children* (M. Cook, Trans.). New York: International Universities Press.

Piaget, J. (1962). *Play, dreams, and imitation in childhood.* New York: Norton.

Piaget, J. (1970). Piaget's theory. In P. H. Mussen (Ed.), *Carmichael's manual of child psychology* (3rd ed.) (Vol. 1). New York: Wiley.

Pines, M. (1981, September). The civilizing of Genie. *Psychology Today, 15,* pp. 28–34.

Plomin, R., & Rowe, D. C. (1977). A twin study of temperament in young children. *Journal of Psychology, 97,* 107–113.

Plutchik, R. (1970). Emotions, evolution, and adaptive processes. In M. B. Arnold (Ed.), *Feelings and emotions: The Loyola symposium.* New York: Academic Press.

Plutchik, R. (1980). *Emotion: A psychoevolutionary synthesis.* New York: Harper & Row.

Polyak, S. (1957). *The vertebrate visual system.* Chicago: University of Chicago Press.

Pomerleau, O. F. (1982). A discourse on behavioral medicine: Current status and future trends. *Journal of Consulting and Clinical Psychology, 50,* 1030–1039.

Powley, T. L., & Keesey, R. E. (1970). Relationship of body weight to the lateral hypothalamic feeding syndrome. *Journal of Comparative and Physiological Psychology, 70,* 25–36.

Premack, A. J., & Premack, D. (1972). Teaching language to an ape. *Scientific American, 227*(4), 92–99.

Pribram, K. (1975). Quoted in R. G. Meyer & P. Salmon, *Abnormal psychology* (p. 89). Boston: Allyn and Bacon.

Pruitt, D. J., & Insko, C. A. (1980). Extension of the Kelley attribution model: The role of comparison-object consensus, target-object consensus, distinctiveness, and consistency. *Journal of Personality and Social Psychology, 39,* 39–58.

Quillian, M. (1966). Semantic memory. In M. L. Minsky (Ed.), *Semantic information processing.* Cambridge, MA: MIT Press.

Rachman, S. J., & Wilson, G. T. (1980). *The effects of psychological therapy* (2nd ed.). New York: Pergamon Press.

Raczynski, J. M., Thompson, J. K., & Sturgis, E. T. (1982). An evaluation of biofeedback assessment and training paradigms. *Clinical Psychology Review, 2,* 337–348.

Ramey, C. T., & Campbell, F. A. (1979). Early childhood education for psychosocially disadvantaged children: Effects on psychological processes. *American Journal of Mental Deficiency, 83,* 645–648.

Raps, C. S., Peterson, C., Reinhard, K. E., Abramson, L. Y., & Seligman, M. E. P. (1982). Attributional style among depressed patients. *Journal of Abnormal Psychology, 91,* 102–108.

Raskin, D. C. (1980, Fall). The truth about lie detectors. *Wharton Magazine, 5,* pp. 28–33.

Ray, M. B. (1976). The cycle of abstinence and relapse among heroin addicts. In R. H. Coombs, L. J. Fry, & P. G. Lewis (Eds.), *Socialization in drug abuse.* Cambridge, MA: Schenkman.

Ray, O. S. (1983). *Drugs, society, and human behavior* (3rd ed.). St. Louis, MO: Mosby.

Razran, G. (1961). The observable unconscious and the inferable conscious in current Soviet psychophysiology: Interoceptive conditioning, semantic conditioning, and the orienting reflex. *Psychological Review, 68,* 81–147.

Redd, W. H. (1980). Stimulus control and extinction of psychosomatic symptoms in cancer patients in protective isolation. *Journal of Consulting and Clinical Psychology, 48,* 448–455.

Redd, W. H., Andresen, G. V., & Minagawa, R. Y. (1982). Hypnotic control of anticipatory emesis in patients receiving cancer chemotherapy. *Journal of Consulting and Clinical Psychology, 50,* 14–19.

Redd, W. H., & Andrykowski, M. A. (1982). Behavioral intervention in cancer treatment: Controlling aversive reactions to chemotherapy. *Journal of Consulting and Clinical Psychology, 50,* 1018–1029.

Regan, D. T., & Fazio, R. (1977). On the consistency between attitudes and behavior: Look to the method of attitude formation. *Journal of Experimental Social Psychology, 13,* 28–45.

Regier, D. A., Myers, J. K., Kramer, M., Robins, L. N., Blazer, D. G., Hough, R. L., Eaton, W. W., & Locke, B. Z. (1984). The NIMH epidemiologic catchment area program. *Archives of General Psychiatry, 41,* 934–941.

Reid, J. E., & Inbau, F. E. (1977). *Truth and deception: The polygraph (lie detector) technique.* (2nd ed.). Baltimore: Williams & Wilkins.

Reis, H. T., Nezlek, J., & Wheeler, L. (1980). Physical attractiveness in social interaction. *Journal of Personality and Social Psychology, 38,* 604–617.

Reis, H. T., Wheeler, L., Spiegel, N., Kernis, M. H., Nezlek, J., & Perri, M. (1982). Physical attractiveness in social interaction: II. Why does appearance affect social experience? *Journal of Personality and Social Psychology, 43,* 979–996.

Rentzel, L. (1972). *When all the laughter died in sorrow.* New York: Saturday Review Press.

Rescorla, R. A. (1967). Inhibition of delay in Pavlovian fear conditioning. *Journal of Comparative and Physiological Psychology, 64,* 114–120.

Rescorla, R. A., & Wagner, A. R. (1972). A theory of Pavlovian conditioning: Variations in the effectiveness of reinforcement and nonreinforcement. In A. H. Black & W. F. Prokasy (Eds.), *Classical conditioning II: Current research and theory.* New York: Appleton-Century-Crofts.

Reston, J., Jr. (1981). *Our father who art in hell.* New York: Times Books.

Rheingold, H. L., & Cook, K. V. (1975). The contents of boys' and girls' rooms as an index of parents' behavior. *Child Development, 46,* 459–463.

Riesen, A. H. (1966). Sensory deprivation. In E. Stellar & J. M. Sprague (Eds.), *Progress in physiological psychology* (Vol. 1). New York: Academic Press.

Rimm, D. C., & Masters, J. C. (1979). *Behavior therapy: Techniques and empirical findings* (2nd ed.). New York: Academic Press.

Rips, L. J., Shoben, E. J., & Smith, E. E. (1973). Semantic distance and the verification of semantic relations. *Journal of Verbal Learning and Verbal Behavior, 12,* 1–20.

Ritvo, E. R., Spence, M. A., Freeman, B. J., Mason-Brothers, A., Mo, A., & Marazita, M. L. (1985). Evidence for autosomal recessive inheritance in 46 families with multiple incidences of autism. *American Journal of Psychiatry, 142,* 187–192.

Robertson, J. F. (1976). Significance of grandparents: Perceptions of young adult grandchildren. *The Gerontologist, 16,* 137–140.

Robertson, J. F. (1977). Grandmotherhood: A study of role conceptions. *Journal of Marriage and the Family, 39,* 65–174.

Robins, L. N. (1979). Follow-up studies. In H. C. Quay & J. S.

Werry (Eds.), *Psychopathological disorders of childhood* (2nd ed.). New York: Wiley.
Robins, L. N., Helzer, J. E., Weissman, M. M., Orvaschel, H., Gruenberg, E., Burke, J. D. Jr., & Regier, D. A. (1984). Lifetime prevalence of specific psychiatric disorders in three sites. *Archives of General/Psychiatry, 41*, 949–958.
Robinson, J. P., & Shaver, P. R. (1973). *Measures of social psychological attitudes* (rev. ed.). Ann Arbor, MI: Institute for Social Research.
Rodin, J. (1981). Current status of the internal-external hypothesis for obesity: What went wrong? *American Psychologist, 36*, 361–372.
Rogers, C. R. (1951). *Client-centered therapy: Its current practice, implications, and theory.* Boston: Houghton Mifflin.
Rogers, C. R. (1959). A theory of therapy, personality, and interpersonal relationships, as developed in the client-centered framework. In S. Koch (Ed.), *Psychology: A study of a science* (Vol. 3). New York: McGraw-Hill.
Rogers, C. R. (1961). *On becoming a person*. Boston: Houghton Mifflin.
Rolls, B. J., Wood, R. J., & Rolls, E. T. (1980). Thirst: The initiation, maintenance, and termination of drinking. In J. M. Sprague & A. N. Epstein (Eds.), *Progress in psychobiology and physiological psychology* (Vol. 9). New York: Academic Press.
Rorschach, H. (1942). *Psychodiagnostics* (P. Lemkau & B. Kronenberg, Trans.). Berne, Switzerland: Huber. (Original work published 1921)
Rosch, E.(1973). Natural categories. *Cognitive Psychology, 4*, 328–350.
Rosch, E. (1974). Linguistic relativity. In A. Silverstein (Ed.), *Human communication: Theoretical explorations*. New York: Halsted Press.
Rosch, E., & Mervis, C. B. (1975). Family resemblances: Studies in the internal structure of categories. *Cognitive Psychology, 7*, 573–605.
Rosenhan, D. L. (1973). On being sane in insane places. *Science, 179*, 250–258.
Rosenhan, D. L. (1975). The contextual nature of psychiatric diagnosis. *Journal of Abnormal Psychology, 84*, 462–474.
Rosenhan, D. L., & Seligman, M. E. P. (1984). *Abnormal psychology*. New York: Norton.
Rosenzweig, M. R., & Leiman, A. L. (1982). *Physiological psychology*. Lexington, MA: D. C. Heath.
Rotter, J. B. (1966). Generalized expectancies for internal versus external control of reinforcement. *Psychological Monographs, 80* (Whole No. 609).
Rubin, J. Z., Provenzano, F. J., & Luria, Z. (1974). The eye of the beholder: Parents' view on sex of newborns. *American Journal of Orthopsychiatry, 43*, 518–519, 729–731.
Rubin, L. B. (1979). *Women of a certain age: The midlife search for self.* New York: Harper & Row.
Rubin, R., & Balow, B. (1978). Prevalence of teacher-identified behavior problems. A longitudinal study. *Exceptional Children, 45*, 102–111.
Ruble, D. N., & Nakamura, C. Y. (1973). Outerdirectedness as a problem-solving approach in relation to developmental level and selected task variables. *Child Development, 44*, 519–528.
Ruch, L. O., & Holmes, T. H. (1971). Scaling of life change: Comparison of direct and indirect methods. *Journal of Psychosomatic Research, 15*, 221–227.
Rumbaugh, D. M. (Ed.) (1977). *Language learning by a chimpanzee: The Lana Project*. New York: Academic Press.
Rundus, D. (1971). Analysis of rehearsal processes in free recall. *Journal of Experimental Psychology, 89*, 63–77.
Rutter, M., & Garmezy, N. (1983). Developmental psychopathology. In P. H. Mussen (Ed.), *Handbook of child psychology: Vol 4. Socialization, personality, and social development* (4th ed.). New York: Wiley.
Saegert, S., Swap, W., & Zajonc, R. B. (1973). Exposure, context, and interpersonal attraction. *Journal of Personality and Social Psychology, 25*, 234–242.
Salthouse, T. A., & Kail, R. (1983). Memory development throughout the life span: The role of processing rate. In P. B. Baltes & O. G. Brim, Jr., *Life-span development and behavior* (Vol. 5). New York: Academic Press.
Sanders, D. (1978). *The relationship of attitude variables and explanations of perceived and actual career attainment in male and female businesspersons.* Unpublished doctoral dissertation, University of Texas at Austin.
Sanders, G. S., Baron, R. S., & Moore, D. L. (1978). Distraction and social comparison as mediators of social facilitation effects. *Journal of Experimental Social Psyhology, 14*, 291–303.
Sappington, W. (1975). Psychometric correlates of defensive style in process and reactive schizophrenics. *Journal of Consulting and Clinical Psychology, 43*, 154–156.
Savin-Williams, R. C. (1979). Dominance hierarchies in groups of early adolescents. *Child Development, 50*, 923–935.
Savin-Williams, R. C. (1980). Dominance hierarchies in groups of middle to late adolescent males. *Journal of Youth and Adolescence, 9*, 75–85.
Scarr, S. (1981). *Race, social class, and individual differences in IQ: New studies of old issues*. Hillsdale, NJ: Erlbaum.
Scarr, S., & Weinberg, R. A. (1976). IQ test performance of black children adopted by white families. *American Psychologist, 31*, 726–739.
Schachtel, E. G. (1959). *Metamorphosis*. New York: Basic Books.
Schachter, S. (1971). Some extraordinary facts about obese humans and rats. *American Psychologist, 26*, 129–144.
Schachter, S., & Singer, J. E. (1962). Cognitive, social, and physiological determinants of emotional state. *Psychological Review, 69*, 379–399.
Schaie, K. W. (1979). The primary mental abilities in adulthood: An exploration in the development of psychometric intelligence. In P. B. Baltes & O. G. Brim, Jr. (Eds.), *Life-span development and behavior* (Vol. 2). New York: Academic Press.
Schaie, K. W. (1980). Intelligence and problem solving. In J. E. Birren & R. B. Sloane (Eds.), *Handbook of mental health and aging.* Englewood Ciiffs, NJ: Prentice-Hall.
Schaie, K. W. (1983). Age changes in adult intelligence. In D. S. Woodruff & J. E. Birren (Eds.), *Aging: Scientific perspectives and social issues*. Monterey, CA: Brooks/Cole.
Schaie, K. W. (1983). The Seattle longitudinal sudy: A 21-year exploration of psychometric intelligence in adulthood. In K. W. Schaie (Ed.), *Longitudinal studies of adult psychological development.* New York: Guilford Press.
Schaie, K. W., Labouvie, G. V., & Buech, B. U. (1973). Generational and cohort-specific differences in adult cognitive functioning. *Developmental Psychology, 9*, 151–166.
Schalock, R. L., Harper, R. S., & Genung, T. (1981). Community integration of mentally retarded adults: Community placement

and program success. *American Journal of Mental Deficiency, 85,* 478–488.

Scheff, T. J. (1966). *Being mentally ill: A sociological theory.* Chicago: Aldine.

Schiffman, H. R. (1982). *Sensation and perception: An integrated approach* (2nd ed.). New York: Wiley.

Schildkraut, J. J. (1977). Biochemical research in affective disorders. In G. Usdin (Ed.), *Depression: Clinical, biological, and psychological perspectives.* New York: Brunner/Mazel.

Schlesier-Stropp, B. (1984). Bulimia: A review of the literature. *Psychological Bulletin, 95,* 247–257.

Schopler, J., & Stockdale, J. (1977). An interference analysis of crowding. *Environmental Psychology and Nonverbal Behavior, 1,* 81–88.

Schulz, R. (1978). *The psychology of death, dying, and bereavement.* Reading, MA: Addison-Wesley.

Scientific American Authors (1979). *The brain.* San Francisco: W. H. Freeman.

Scribner, S., & Cole, M. (1981). *The consequences of literacy.* Cambridge, MA: Harvard University Press.

Sears, P. S., & Barbee, A. H. (1977). Career and life satisfaction among Terman's gifted women. In J. Stanley, W. George, & C. Solano (Eds.), *The gifted and the creative: Fifty-year perspective.* Baltimore: Johns Hopkins University Press.

Sears. R. R., Rau, L., & Alpert, P. (1965). *Identification and child rearing.* Stanford, CA: Stanford University Press.

Sebeok, T. A., & Umiker-Sebeok, J. (Eds.) (1980). *Speaking of apes: A critical anthology of two-way communication with man.* New York: Plenum Press.

Seidman, L. J. (1983). Schizophrenia and brain dysfunction: An integration of recent neurodiagnostic findings. *Psychological Bulletin, 94,* 195–238.

Seitelberger, F. (1978). Lebensstadien des Gehirns—Strukturelle und funktionale Aspekte: Kontinuität und krisen. In L. Rosenmayr (Ed.), *Die menschlichen Lebensalter.* Munich: R. Piper Verlag.

Seligman, M. E. P. (1970). On the generality of the laws of learning. *Psychological Review,* 77, 406–418.

Seligman, M. E. P. (1971). Phobias and preparedness. *Behavior Therapy, 2,* 307–320.

Seligman, M. E. P. (1975). *Helplessness: On depression, development, and death.* San Francisco: W. H. Freeman.

Selman, R. L. (1980). *The growth of interpersonal understanding.* New York: Academic Press.

Selye, H. (1950). *The physiology and pathology of exposure to stress.* Montreal: Acta.

Selye, H. (1956). *The stress of life.* New York: McGraw-Hill.

Selye, H. (1976). *The stress of life* (rev. ed.). New York: McGraw-Hill.

Seward, J. P. (1949). An experimental analysis of latent learning. *Journal of Experimental Psychology, 39,* 177–186.

Shapiro, D. A., & Shapiro, D. (1982). Meta-analysis of comparative therapy outcome studies: A replication and refinement. *Psychological Bulletin, 92,* 581–604.

Shapiro, D. H., & Walsh, R. N. (Eds.) (1984). *Meditation: Classic and contemporary perspectives.* Hawthorne, NY: Aldine.

Shapiro, S., Skinner, E. A., Kessler, L. G., von Korff, M., German, P. S., Tischler, G. L., Leaf, P. J., Benham, L., Cottler, L., & Regier, D. A. (1984). Utilization of health and mental health services. *Archives of General Psychiatry, 41,* 971–978.

Sheehy, G. (1976). *Passages: Predictable crises of adult life.* New York: E. P. Dutton & Co.

Sheldon, W. H. (1942). *The varieties of temperament: A psychology of constitutional differences.* New York: Harper.

Sherif, M., Harvey, O. J., White, B. J., Hood, W. R., & Sherif, C. W. (1961). *Inter-group conflict and cooperation: The robbers cave experiment.* Norman, OK: University of Oklahoma Press.

Shiffrin, R. M., & Schneider, W. (1977). Controlled and automatic human information processing: II. Perceptual learning, automatic attending, and a general theory. *Psychological Review, 84,* 127–190.

Shirley, M. M. (1933). *The first two years: A study of twenty-five babies: Vol. 2. Intellectual development.* Minneapolis, MN: University of Minnesota Press.

Sholl, D. A. (1956). *The organization of the cerebral cortex.* London: Methuen.

Siegler, I. C., & Edelman, C. D. (1977, April). *Age discrimination in employment: The implications for psychologists.* Paper presented at the annual meeting of the Western Psychological Association, San Francisco.

Siegler, R. S. (1983). Five generalizations about cognitive development. *American Psychologist, 38,* 263–277.

Siegler, R. S. (1983). Information processing approaches to development. In P. H. Mussen (Ed.), *Handbook of child psychology: Vol 1. History, theory, and methods* (4th ed.). New York: Wiley.

Sigall, H., & Landy, D. (1973). Radiating beauty: Effects of having a physically attractive partner on person perception. *Journal of Personality and Social Psychology, 28,* 218–224.

Silva, J. M., III (1981). Normative compliance and rule violating behavior in sport. *International Journal of Sport Psychology. 12,* 10–18.

Silva, J. M., III (1983). The perceived legitimacy of rule violating behavior in sport. *Journal of Sport Psychology, 5,* 438–448.

Simmons et al. v. United States, 390 US 377 (1968).

Singer, J. L. (1976). *Daydreaming and fantasy.* London: George Allen & Unwin.

Singer, J. L. (1980). *Mind-play: The creative uses of fantasy.* Englewood Cliffs, NJ: Prentice-Hall.

Sinnott, J. D. (1975). Everyday thinking and Piagetian operativity in adults. *Human Development, 18,* 430–443.

Sinnott, J. D. (1982). Correlates of sex roles in older adults. *Journal of Gerontology, 37,* 587–594.

Sipe, N. P. (1979). Electroconvulsive therapy—No. *Illinois Issues, 12,* 12–15.

Sivacek, J., & Crano, W. D. (1982). Vested interest as a moderator of attitude-behavior consistency. *Journal of Personality and Social Psychology, 43,* 210–221.

Skeels, H. M. (1966). Adult status of children with contrasting early life experiences. *Monographs of the Society for Research in Child Development, 31* (Serial No. 105).

Skinner, B. F. (1938). *The behavior of organisms: An experimental analysis.* New York: Appleton-Century-Crofts.

Skinner, B. F. (1950). Are theories of learning necessary? *Psychological Review, 57,* 193–216.

Skinner, B. F. (1953). *Science and human behavior.* New York: Macmillan.

Skinner, B. F. (1957). *Verbal behavior.* New York: Appleton-Century-Crofts.

Skinner, B. F. (1971). *Beyond freedom and dignity.* New York: Knopf.

Skinner, B. F. (1972). A lecture on "having" a poem. In *Cumulative record: A selection of papers* (3rd ed.). New York: Appleton-Century-Crofts.

Skinner, B. F., & Vaughan, M. E. (1983). *Enjoy old age*. New York: Norton.

Slobin, D. I. (1979). *Psycholinguistics* (2nd ed.). Glenview, IL: Scott, Foresman.

Slovic, P., Fischhoff, B., & Lichtenstein, S. (1977). Behavior decision theory. *Annual Review of Psychology, 28*, 1–39.

Smith, C. P. (Ed.) (1969). *Achievement-related motives in children*. New York: Russell Sage.

Smith, G. F., & Dorfman, D. D. (1975). The effect of stimulus uncertainty on the relationship between frequency of exposure and liking. *Journal of Personality and Social Psychology, 31*, 150–155.

Smith, G. P., & Gibbs, J. (1976). Cholecystokinin and satiety: Theoretic and therapeutic implications. In D. Novin, W. Wyrwicka, & G. A. Bray (Eds.), *Hunger: Basic mechanisms and clinical implications*. New York: Raven Press.

Smith, M. J. (1982). *Persuasion and human action*. Belmont, CA: Wadsworth.

Smith, M. L., & Glass, G. V. (1977). Meta-analysis of psychotherapy outcome studies. *American Psychologist, 33*, 752–760.

Smith, S. M., Brown, H. O., Toman, J. E. P., & Goodman, L. S. (1947). The lack of central effects of d-tubocurarine. *Anesthesiology, 8*, 1–14.

Snyder, M., & Kendzierski, D. (1982). Acting on one's attitudes: Procedures for linking attitude and behavior. *Journal of Experimental Social Psychology, 18*, 165–183.

Snyder, S. H. (1981). Dopamine receptors, neuroleptics, and schizophrenia. *American Journal of Psychiatry, 138*, 460–464.

Snyder, S. H., Banerjee, S. P., Yamamura, H. I., & Greenberg, D. (1974). Drugs, neurotransmitters, and schizophrenia. *Science, 184*, 1243–1253.

Snyder, S. H., & Childers, S. R. (1979). Opiate receptors and opioid peptides. *Annual Review of Neuroscience, 2*, 35–64.

Snyder, W. U. (Ed.) (1947). *Casebook of non-directive counseling*. Boston: Houghton Mifflin.

Sociobiology Study Group of Science for the People (1976). Sociobiology—Another biological determinism. *BioScience, 26* (3), 182, 184–186.

Solomon, R. L. (1980). The opponent-process theory of acquired motivation. *American Psychologist, 35*, 691–712.

Solomon, R. L., & Corbit, J. D. (1974). An opponent-process theory of motivation: I. Temporal dynamics of affect. *Psychological Review, 81*, 119–145.

Sorensen, R. C. (1973). *Adolescent sexuality in contemporary America*. New York: World Publishing.

Spearman, C. (1927). *The abilities of man*. London: Macmillan.

Spence, J. T. (Ed.) (1983). *Achievement and achievement motives*. San Francisco: W. H. Freeman.

Spence, J. T., & Helmreich, R. L. (1983). Achievement-related motives and behaviors. In J. T. Spence (Ed.), *Achievement and achievement motives*. San Francisco: W. H. Freeman.

Sperling, G. (1960). The information available in brief visual presentations. *Psychological Monographs, 74* (Whole No. 498).

Sperling, G. (1963). A model for visual memory tasks. *Human Factors, 5*, 19–30.

Sperry, R. W. (1974). Lateral specialization in the surgically separated hemispheres. In F. O. Schmitt & F. G. Worden (Eds.), *The neurosciences: Third study program*. Cambridge, MA: MIT Press.

Spiesman, J. C., Lazarus, R. S., Mordkoff, A., & Davison, L. (1964). Experimental reduction of stress based on ego-defense theory. *Journal of Abnormal and Social Psychology, 68*, 367–380.

Spitzer, R. L. (1975). On pseudoscience in science, logic in remission, and psychiatric diagnosis: A critique of Rosenhan's "On being sane in insane places." *Journal of Abnormal Psychology, 84*, 442–452.

Spitzer, R. L., & Forman, J. B. W. (1979). DSM-III field trials: II. Initial experience with the multiaxial system. *American Journal of Psychiatry, 136*, 818–820.

Spitzer, R. L., Forman, J. B. W., & Nee, J. (1979). DSM-III field trials: I. Initial interrater diagnostic reliability. *American Journal of Psychiatry, 136*, 815-817.

Springer, S. P., & Deutsch, G. (1985). *Left brain, right brain* (rev. ed.). New York: W. H. Freeman.

Squire, L. R. (1982). The neuropsychology of human memory. *Annual Review of Neuroscience, 5*, 241–273.

Squire, L. R., & Cohen, N. J. (1982). Remote memory, retrograde amnesia, and the neuropsychology of memory. In L. Cermak (Ed.), *Human memory and amnesia*. Hillsdale, NJ: Erlbaum.

Squire, L. R., & Schlapfer, W. T. (1981). Memory and memory disorders: A biological and neurological perspective. In H. M. Van Praag (Ed.), *Handbook of biological psychiatry, Part IV*. New York: Marcel Dekker.

Srole, L., & Fischer, A. K. (1980). The Midtown Manhattan Longitudinal Study vs."the mental paradise lost" doctrine. *Archives of General Psychiatry, 37*, 209–221.

Stangler, R. S., & Printz, A. M. (1980). DSM-III: Psychiatric diagnosis in a university population. *American Journal of Psychiatry, 137*, 937–940.

Stein, A. H., & Friedrich, L. K. (1975). The effects of television content on young children. In A. D. Pick (Ed.), *Minnesota symposia on child psychology* (Vol. 9). Minneapolis, MN: University of Minnesota Press.

Steinberg, L. (1979, March). *Changes in family relations at puberty*. Paper presented at the biennial meeting of the Society for Research in Child Development, San Francisco.

Stephenson, F. D. (Comp.) (1975). *Gestalt therapy primer: Introductory readings in gestalt therapy*. Springfield, IL: Thomas.

Stern, D. N. (1974). Mother and infant at play: The dyadic interaction involving facial, vocal, and gaze behaviors. In M. Lewis & L. A. Rosenblum (Eds.), *The effect of the infant on its caregiver*. New York: Wiley.

Sternberg, R. J. (Ed.). (1984). *Mechanisms of cognitive development*. San Francisco: W. H. Freeman.

Sternberg, R. J., & Davidson, J. E. (1982, June). The mind of the puzzler. *Psychology Today, 16*, pp. 37–44.

Sternberg, R. J., & Powell, J. S. (1983). Comprehending verbal comprehension. *American Psychologist, 38*, 878–893.

Sternberg, S. (1966). High-speed scanning in human memory. *Science, 153*, 652–654.

Sternberg, S. (1975). Memory scanning: New findings and current controversies. *Quarterly Journal of Experimental Psychology, 27*, 1–32.

Stevens, J. O. (Ed.). (1975). *Gestalt is*. Moab, UT: Real People Press.

Stevenson, H. W. (1982). Influences of schooling on cognitive development. In D. A. Wagner & H. W. Stevenson (Eds.), *Cultural perspectives on child development*. San Francisco: W. H. Freeman.

Stewart, A. J., & Rubin, Z. (1976). The power motive in the dating couple. *Journal of Personality and Social Psychology, 34*, 305–309.
Stinnett, N., Farris, J. A., & Walters, J. (1974). Parent-child relationships of male and female high school students. *Journal of Genetic Psychology, 125*, 99–106.
Stokols, D., & Schopler, J. (1973). Reactions to victims under conditions of situational detachment: The effects of responsibility, severity, and expected future interaction. *Journal of Personality and Social Psychology, 25*, 199–209.
Stolz, H. R., & Stolz, L. M. (1951). *Somatic development of adolescent boys*. New York: Macmillan.
Stricker, E.M. (1977). The renin-angiotensin system and thirst. A reevaluation. II. Drinking elicited in rats by caval ligation or isoproterenol. *Journal of Comparative and Physiological Psychology, 91*, 1220–1231.
Strupp, H. H., & Hadley, S. W. (1979). Specific vs nonspecific factors in psychotherapy: A controlled study of outcome. *Archives of General Psychiatry, 36*, 1125–1136.
Stryker, M. P., & Sherk, H. (1975). Modification of cortical orientation selectivity in the cat by restricted visual experience: A reexamination. *Science, 190*, 904–906.
Stuckless, E. R., & Birch, J. W. (1966). The influence of early manual communication on the linguistic development of deaf children. *American Annals of the Deaf, 111*, 451–460.
Sulser, F. (1979). New cellular mechanisms of antidepressant drugs. In S. Fielding & R. C. Effland (Eds.), *New frontiers in psychotropic drug research*. Mount Kisco, NY: Futura.
Sundberg, N. D., Taplin, J. R., & Tyler, L. E. (1983). *Introduction to clinical psychology: Perspectives, issues, and contributions to human service*. Englewood Cliffs, NJ: Prentice-Hall.
Suomi, S. J., Harlow, H. F., & Novak, M. A. (1974). Reversal of social deficits produced by isolation rearing in monkeys. *Journal of Human Evolution, 3*, 527–534.
Sutton-Smith, B., & Rosenberg, B. G. (1970). *The sibling*. New York: Holt, Rinehart and Winston.
Svare, B. B. (Ed.) (1983). *Hormones and aggressive behavior*. New York: Plenum Press.
Swanson, D. W., Bohnert, P. J., & Smith, J. A. (1970). *The paranoid*. Boston: Little, Brown.
Szasz, T. S. (1960). The myth of mental illness. *American Psychologist, 15*, 113–118.
Szasz, T. S. (1970). *The manufacture of madness*. New York: Macmillan.
Tanner, J. M., Whitehouse, R. H., & Takaishi, M. (1966). Standards from birth to maturity for height, weight, height velocity, and weight velocity: British children, 1965. Part I. *Archives of Disease in Childhood, 41*, 454–471.
Tarpy, R. M., & Mayer, R. E. (1978). *Foundations of learning and memory*. Glenview, IL: Scott, Foresman.
Taylor, I. (1976). *Introduction to psycholinguistics*. New York: Holt, Rinehart and Winston.
Taylor, J. A. (1953). A personality scale of manifest anxiety. *Journal of Abnormal Psychology, 48*, 285–290.
Taylor, P., & Fleminger, J. J. (1980, June, 28). ECT for schizophrenia. *The Lancet I*, 1380–1383.
Teitelbaum, P. (1961). Disturbances in feeding and drinking behavior after hypothalamic lesions. In M. R. Jones (Ed.), *Nebraska symposium on motivaton*. Lincoln, NE: University of Nebraska Press.
Temerlin, M. K. (1970). Diagnostic bias in community mental health. *Community Mental Health Journal, 6*, 110–117.
Terkel, S. (1980). *American dreams: Lost and found*. New York: Pantheon Books.
Terman, L. M. (1925). *Genetic studies of genius: Vol. 1. Mental and physical traits of a thousand gifted children*. Stanford, CA: Stanford University Press.
Terman, L. M., & Merrill, M. A. (1973). *The Stanford-Binet Intelligence Scale, Third Revision*. (With 1972 tables by R. L. Thorndike.) Boston: Houghton Mifflin.
Terman, L. M., & Oden, M. H. (1959). *Genetic studies of genius: Vol. 5. The gifted group at midlife*. Stanford, CA: Stanford University Press.
Terrace, H. S. (1979). *Nim*. New York: Knopf.
Terry, W. S., & Wagner, A. R. (1975). Short-term memory for "surprising" versus "expected" unconditioned stimuli in Pavlovian conditioning. *Journal of Experimental Psychology: Animal Behavior Processes, 1*, 122–133.
Teuber, H.-L. (1964). The riddle of frontal lobe function in man. In J. M. Warren & K. Akert (Eds.), *The frontal granular cortex and behavior*. New York: McGraw-Hill.
Thibaut, J., & Kelley, H. H. (1959). *The social psychology of groups*. New York: Wiley.
Thomas, A., & Chess, S. (1977). *Temperament and development*. New York: Brunner/Mazel.
Thomas, A., & Chess, S. *The dynamics of psychological development*. New York: Brunner/Mazel.
Thomas, A., Chess, S., & Birch, H. G. (1970). The origin of personality. *Scientific American, 223*(2), 102–109.
Thorndike, E. L. (1898). Animal intelligence: An experimental study of associative processes in animals. *Psychological Monographs, 2* (Whole No. 8).
Thorndike, E.L. (1911). *Animal intelligence: Experimental studies*. New York: Macmillan.
Thurstone, L. L. (1938). *Primary mental abilities*. Chicago: University of Chicago Press.
Tiegs, E. W., & Katz, B. (1969). *Mental hygiene in education*. New York: Ronald Press.
Tienari, P. (1971). Schizophrenia and monozygotic twins. In K. A. Achte (Ed.), *Psychiatria Fennica 1971*. Helsinki, Finland: Helsinki University Central Hospital.
Tinbergen, N. (1951). *The study of instinct*. New York: Oxford University Press.
Tinbergen, N. (1952). The curious behavior of the stickleback. *Scientific American, 187*(6), 22–26.
Tobin, S. A. (1975). Saying goodbye. In J. O. Stevens (Ed.), *Gestalt is*. Moab, UT: Real People Press.
Toch, H. H., & Schulte, R. (1961). Readiness to perceive violence as a result of police training. *British Journal of Psychology, 52*, 389–393.
Tolman, E. C. (1932). *Purposive behavior in animals and men*. New York: Century.
Tomkins, S. S. (1970). Affect as a primary motivational system. In M. B. Arnold (Ed.), *Feelings and emotions: The Loyola symposium*. New York: Academic Press.
Tomkins, S. S. (1981). The quest for primary motives: Biography and autobiography of an idea. *Journal of Personality and Social Psychology, 41*, 306–329.
Torrey, E. F. (1983). *The roots of treason: Ezra Pound and the secret of St. Elizabeth's*. New York: McGraw-Hill.

Treisman, A. (1969). Strategies and models of selective attention. *Psychological Review, 76,* 282–299.

Tresemer, D. (1976). The cumulative record of research on "fear of success." *Sex Roles, 2,* 217–236.

Triplett, N. (1898). The dynamogenic factors in pacemaking and competition. *American Journal of Psychology, 9,* 507–533.

Tulving, E. (1962). Subjective organization in free recall of unrelated words. *Psychological Review, 69,* 344–354.

Tulving, E. (1972). Episodic and semantic memory. In E. Tulving & W. Donaldson (Eds.), *Organization of memory.* New York: Academic Press.

Tulving, E., & Thomson, D. M. (1973). Encoding specificity and retrieval processes in episodic memory. *Psychological Review, 80,* 352–373.

Turnage, J. R., & Logan, D. C. (1974). Treatment of a hypodermic needle phobia by *in vivo* systematic desensitization. *Journal of Behavior Therapy and Experimental Psychiatry, 5,* 67–69.

Turner, J. S., & Helms, D. B. (1982). *Contemporary adulthood* (2nd ed.). New York: Holt, Rinehart and Winston.

Tversky, A., & Kahneman, D. (1974). Judgment under uncertainty: Heuristics and biases. *Science, 185,* 1124–1131.

Ullmann, L. P., & Krasner, L. (1975). *A psychological approach to abnormal behavior* (2nd ed.). Englewood Cliffs, NJ: Prentice-Hall.

Underwood, B. J. (1957). Interference and forgetting. *Psychological Review, 64,* 49–60.

Uzgiris, I. C., & Hunt, J. McV. (1975). *Assessment in infancy: Ordinal scales of psychological development.* Urbana, IL: University of Illinois Press

Van Cantfort, T. E., & Rimpau, J. B. (1982). Sign language studies with children and chimpanzees. *Sign Language Studies, 34,* 15–72.

Vander, A. J., Sherman, J. H., & Luciano, D. S. (1975). *Human physiology: The mechanisms of body function* (2nd ed.). New York: McGraw-Hill.

Vaughan, E. D. (1977). Misconceptions about psychology among introductory psychology students. *Teaching of Psychology, 4,* 138–141.

Vernon, M. (1967). Relationship of language to the thinking process. *Archives of General Psychiatry, 16,* 325–333.

Vernon, M., & Koh, S. D. (1971). Effects of oral preschool compared to early manual communication on education and communication in deaf children. *American Annals of the Deaf, 116,* 569–574.

Vernon, P. E. (1950). *The structure of human abilities.* London: Methuen.

Veroff, J., & Feld, S. (1970). *Marriage and work in America.* New York: Van Nostrand Reinhold.

Verwoerdt, A., Pfeiffer, E., & Wang, H. S. (1969). Sexual behavior in senescence—changes in sexual activity and interest of aging men and women. *Journal of Geriatric Psychiatry, 2,* 163–180.

Vinick, B. H. (1978). Remarriage in old age. *The Family Coordinator, 27,* 359–363.

von Békésy, G. (1960.) *Experiments in hearing.* New York: McGraw-Hill.

Wadden. T. A., & Anderton, C. H. (1982). The clinical use of hypnosis. *Psychological Bulletin, 91,* 215–243.

Waid, W. M., & Orne, M. T. (1982). The physiological detection of deception. *American Scientist, 70,* 402–409.

Walen, S. R., Hauserman, N. M., & Lavin, P. J. (1977). *Clinical guide to behavior therapy.* Baltimore: Williams & Wilkins.

Walk, R. D., & Gibson, E. J. (1961). A comparative and analytical study of visual depth perception. *Psychological Monographs, 75* (Whole No. 519).

Wallace, R. K., & Benson, H. (1972). The physiology of meditation. *Scientific American, 226*(2), 84–90.

Walls, G. L. (1942). *The vertebrate eye.* Bloomfield Hills, MI: Cranbrook Institute of Science.

Walster, E., Aronson, V., Abrahams, D., & Rottman, L. (1966). Importance of physical attractiveness in dating behavior. *Journal of Personality and Social Psychology, 4,* 508–516.

Walster, E., & Walster, G. W. (1963). Effect of expecting to be liked on choice of associate. *Journal of Abnormal and Social Psychology, 67,* 402–404.

Walster, E., Walster, G. W., & Berscheid, F. (1978). *Equity: Theory and research.* Boston: Allyn and Bacon.

Walton, D. (1960). The application of learning theory to the treatment of a case of neuro-dermatitis. In H. J. Eysenck (Ed.), *Behaviour therapy and the neuroses.* New York: Pergamon.

Watson, C. G. (1973). Abstract thinking deficit and autism in process and reactive schizophrenics. *Journal of Abnormal Psychology, 82,* 399–403.

Watson, J. B. (1928). *Psychological care of infant and child.* New York: Norton.

Watson, J. D. (1968). *The double helix.* New York: Atheneum.

Wattenberg, W. W. (1955). *The adolescent years.* New York: Harcourt Brace Jovanovich.

Webb, L. J., DiClemente, C. C., Johnstone, E. E., Sanders, J. L., & Perley, R. A. (Eds.) (1981). *DSM-III training guide.* New York: Brunner/Mazel.

Webb, W. B. (1975). *Sleep: The gentle tyrant.* Englewood Cliffs, NJ: Prentice-Hall

Wechsler, D. (1955). *Wechsler Adult Intelligence Scale, manual.* New York: Psychological Corporation.

Wechsler, D. (1967). *Manual for the Wechsler Preschool and Primary Scale of Intelligence.* New York: Psychological Corporation.

Wechsler, D. (1974). *Wechsler Intelligence Scale for Children—Revised.* New York: Psychological Corporation.

Wechsler, D. (1981). *Wechsler Adult Intelligence Scale—Revised.* New York: Psychological Corporation.

Weiner, R. D. (1979). The psychiatric use of electrically induced seizures. *American Journal of Psychiatry, 136,* 1507–1517.

Weiss, J. M. (1972). Psychological factors in stress and disease. *Scientific American, 226*(6), 104–113.

Weissman, M. M., & Myers, J. K. (1978). Affective disorders in a U. S. urban community: The use of research diagnostic criteria in an epidemiological survey. *Archives of General Psychiatry, 35,* 1304–1311.

Weisz, J. R. (1980). Autonomy, control, and other reasons why "Mom is the greatest": A content analysis of children's Mother's Day letters. *Child Development, 5,* 801–807.

Weisz, J. R. (1981). Effects of the "mentally retarded" label on adult judgments about child failure. *Journal of Abnormal Psychology, 90,* 371–374.

Weisz, J. R. (1983). Can I control it? The pursuit of veridical answers across the life span. In P. B. Baltes & O. G. Brim, Jr. (Eds.), *Life-span development and behavior* (Vol. 5). New York: Academic Press.

Weisz, J. R., & Yeates, K. O. (1981). Cognitive development in retarded and nonretarded persons: Piagetian tests of the similar structure hypothesis. *Psychological Bulletin, 90,*153–178.

Weisz, J. R., & Zigler, E. (1979). Cognitive development in retarded and nonretarded persons: Piagetian tests of the similar sequence hypothesis. *Psychological Bulletin, 85,* 831–851.

Weizenbaum, J. (1976). *Computer power and human reason: From judgment to calculation.* San Francisco: W. H. Freeman.

Welsh, G. S. (1975). *Creativity and intelligence: A personality ap-*

proach. Chapel Hill, NC: Institute for Research in Social Science, University of North Carolina at Chapel Hill.

Werblin, F. S. (1973). The control of sensitivity in the retina. *Scientific American, 228*(1), 70–79.

West, M. J., King, A. P., & Eastzer, D. H. (1981). The cowbird: Reflections on development from an unlikely source. *American Scientist, 69*, 56–66.

Whitbourne, S. K., & Weinstock, C. S. (1979). *Adult Development: The differentiation of experience*. New York: Holt, Rinehart and Winston.

White, R. W. (1959). Motivation reconsidered: The concept of competence. *Psychological Review, 66*, 297–333.

White, R. W. (1964). *The abnormal personality* (3rd ed.). New York: Ronald Press.

Whitehouse, P. J., Price, D. L., Struble, R. G., Clark, A. W., Coyle, J. T., & DeLong, M. R. (1982). Alzheimer's disease and senile dementia: Loss of neurons in the basal forebrain. *Science, 215*, 1237–1239.

Whorf, B. L. (1956). *Language, thought and reality*. New York: Wiley.

Wiener, M., Devoe, S., Rubinow, S., & Geller, J. (1972). Nonverbal behavior and nonverbal communication. *Psychological Review, 79*, 185–214.

Wiesel, T. N., & Hubel, D. H. (1974). Ordered arrangement of orientation columns in monkeys lacking visual experience. *Journal of Comparative Neurology, 158*, 307–318.

Williams, R. L., & Long, J. D. (1983). *Toward a self-managed life style* (3rd ed.). Boston: Houghton Mifflin.

Wills, T. A. (1981). Downward comparison principles in social psychology. *Psychological Bulletin, 90*, 245–271.

Wilson, E. O. (1975). *Sociobiology: The new synthesis*. Cambridge, MA: Harvard University Press.

Wilson, E. O. (1976). Academic vigilantism and the political significance of sociobiology. *BioScience, 26* (3), 183, 187–190.

Wilson, E. O. (1978). *On human nature*. Cambridge, MA: Harvard University Press.

Wilson, T. D., Laser, P. S., & Stone, J. I. (1982). Judging the predictors of one's own mood: Accuracy and the use of shared theories. *Journal of Experimental Social Psychology, 18*, 537–556.

Wilson, W., & Nakajo, H. (1965). Preference for photographs as a function of frequency of presentation. *Psychonomic Science, 3*, 577–578.

Wing, L. (1980). Childhood autism and social class: A question of selection. *British Journal of Psychiatry, 137*, 410–417.

Winokur, G. (1980). Is there a common genetic factor in bipolar and unipolar disorder? *Comprehensive Psychiatry, 21*, 460–468.

Winsborough, H. H. (1979). Changes in the transition to adulthood. In M. W. Riley (Ed.), *Aging from birth to death: Interdisciplinary perspectives*. Boulder, CO: Westview Press.

Winter, D. G. (1973). *The power motive*. New York: Free Press.

Wissler, C. (1901). The correlation of mental and physical tests. *Psychological Monographs, 3* (Whole No. 16).

Witkin, H. A., & Goodenough, D. R. (1981). *Cognitive styles: Essence and origins*. New York: International Universities Press.

Wittgenstein, L. (1953). *Philosophical investigations*. Oxford: Blackwell.

Wolgin, D. L., Cytawa, J., & Teitelbaum, P. (1976). The role of activation in the regulation of food intake. In D. Novin, W. Wyrwicka, and G. A. Bray (Eds.), *Hunger: Basic mechanisms and clinical implications*. New York: Raven Press.

Wittgenstein, L. (1953). *Philosophical investigations*. Oxford: Blackwell.

Wolgin, D. L., Cytawa, J., & Teitelbaum, P. (1976). The role of activation in the regulation of food intake. In D. Novin, W. Wyrwicka, and G. A. Bray (Eds.), *Hunger: Basic mechanisms and clinical implications*. New York: Raven Press.

Wolpe, J. (1973). *The practice of behavior therapy* (2nd ed.). New York: Pergamon.

Wood, B. S. (1976). *Children and communication: Verbal and nonverbal language development*. Englewood Cliffs, NJ: Prentice-Hall.

Woodburne, R. T. (1965). *Essentials of human anatomy* (3rd ed.). New York: Oxford University Press.

Wooden, K. (1981). *The children of Jonestown*. New York: McGraw-Hill.

Woodmansee, J. J. (1970). The pupil response as a measure of social attitudes. In G. F. Summers (Ed.), *Attitude measurement*. Chicago: Rand McNally.

Worchel, S. (1974) The effect of three types of arbitrary thwarting on the instigation to aggression. *Journal of Personality, 42*, 301–318.

Wyburn, G. M., Pickford, R. W., & Hirst, R. J. (1964). *Human senses and perception*. Toronto: University of Toronto Press.

Yankelovich, D. (1974). The meaning of work. In J. M. Rosow (Ed.), *The worker and the job: Coping with change*. Englewood Cliffs, NJ: Prentice-Hall.

Yarmey, A. D. (1979). *The psychology of eyewitness testimony*. New York: Free Press.

Yogman, M. J., Dixon, S., Tronick, E., Als, H., & Brazelton, T. B. (1977, March). *The goals and structure of face to face interaction between infants and fathers*. Paper presented at the biennial meeting of the Society for Research in Child Development, New Orleans.

Zajonc, R. B. (1965). Social facilitation. *Science, 149*, 269–274.

Zajonc, R. B., & Rajecki, D. W. (1969). Exposure and affect: A field experiment. *Psychonomic Science, 17*, 216–217.

Zanna, M. P., Kiesler, C., & Pilkonis, P. A. (1970). Positive and negative attitudinal affect established by classical conditioning. *Journal of Personality and Social Psychology, 14*, 321–328.

Zaslow, R. W. (1981). Z-process attachment therapy. In R. J. Corsini (Ed.), *Handbook of innovative psychotherapies*. New York: Wiley.

Zelazo, P. R., Zelazo, N. A., & Kolb, S. (1972). "Walking" in the newborn. *Science, 176*, 314–315.

Ziegler, D. K., Hassanein, R. S., & Couch, J. R. (1977). Characteristics of life headache histories in a nonclinic population. *Neurology, 27*, 265–269.

Zigler, E., & Balla, D. (1982). Motivational and personality factors in the performance of the retarded. In E. Zigler & D. Balla (Eds.), *Mental retardation: The developmental difference controversy*. Hillsdale, NJ: Erlbaum.

Zilboorg, G., & Henry, G. W. (1941). *A history of medical psychology*. New York: Norton.

Zillmann, D. (1979). *Hostility and aggression*. Hillsdale, NJ: Erlbaum.

Zillmann, D. (1983). Arousal and aggression. In R. G. Geen & E. I. Donnerstein (Eds.), *Aggression: Theoretical and empirical reviews: Vol 1. Theoretical and methodological issues*. New York: Academic Press.

Zillmann, D., & Cantor, J. R. (1976). Effect of timing of information about mitigating circumstances on emotional responses to provocation and retaliatory behavior. *Journal of Experimental Social Psychology, 12*, 38–55.

Zimbardo, P. G. (1979). *Psychology and life* (10th ed.). Glenview, IL: Scott, Foresman.

Zuckerman, M. (1978). Sensation seeking. In H. London & J. E. Exner, Jr. (Eds.), *Dimensions of personality*. New York: Wiley.

Acknowledgments

In addition to the credits given in the text and in the References section, special acknowledgments are due to the following:

CHAPTER 1

Adapted from Vaughan, E. D. (1977). Misconceptions about psychology among introductory psychology students. *Teaching of Psychology*, *4*, 138–141. Copyright 1977 by the American Psychological Association. Adapted by permission of the author.

Figure 1.5 From Jenni, D. A., & Jenni, M. A. (1976). Carrying behavior in humans: Analysis of sex differences. *Science*, *194*, 859–860, 19 November 1976. Copyright 1976 by the American Association for the Advancement of Science.

CHAPTER 2

Page 37. Brain photograph from Omikron/Photo Researchers.

Figure 2.8 Slightly modified from Cooper, J. R., Bloom, F. E., & Roth, R. H. (1974). *The Biochemical Basis of Neuropharmacology* (2nd ed.). New York: Oxford University Press.

Figure 2.10 Based on Woodburne, R. T. (1965). *Essentials of Human Anatomy* (3rd ed.). New York: Oxford University Press. The figure used here was modified from Vander, A. J., Sherman, J. H., & Luciano, D. S. (1975). *Human Physiology: The Mechanisms of Body Function* (2nd ed.). New York: McGraw-Hill.

Lower figure of Inquiry 2.2, page 56. Modified from Dement, W. C. & Wolpert, E. (1958). The relation of eye movements, body motility, and external stimuli to dream content. *Journal of Experimental Psychology*, *55*, 543–553. Copyright 1958 by the American Psychological Association. Reprinted by permission of the publisher and the senior author.

Figure 2.16 Reprinted with permission of Macmillan Publishing Company from *The Cerebral Cortex of Man: A Clinical Study of Localization of Function* by W. Penfield & T. Rasmussen. Copyright 1950 by Macmillan Publishing Company; renewed 1978 by Theodore Rasmussen.

Figure 2.17 Modified from MacLean, P. D. (1949). Psychosomatic disease and the "visceral brain." Recent developments bearing on the Papez theory of emotion. *Psychosomatic Medicine*, *11*, 338–353. (The original figure was based on the work of W. Krieg.) Reproduced by permission of the author and Harper & Row, Hoeber Medical Division.

Figure 2.18, upper. From *Physiological Psychology* by Mark Rosenzweig & Arnold L. Leiman. Copyright © 1982 by D. C. Heath and Company. Reprinted by permission of the publisher.

Figure 2.18, lower. From *Clinical Neuropsychology* by Kenneth M. Heilman & Edward Valenstein. Copyright © 1979 by Oxford University Press, Inc. Reprinted by permission.

Figure 2.20 From Sperry, R. W. (1974). Lateral specialization in the surgically separated hemispheres. In F. O. Schmitt & F. G. Worden (Eds.), *The Neurosciences: Third Study Program*. Cambridge, MA: MIT Press. Reprinted from *The Neurosciences: Third Study Program*, edited by F. O. Schmitt & F. G. Worden. By permission of the MIT Press, Cambridge, MA. Copyright © 1974 by The Massachusetts Institute of Technology.

CHAPTER 3

Figure of cylinders on page 84. James J. Gibson: *The Perception of the Visual World*. Copyright © 1950, renewed by Houghton Mifflin Company. Used by permission.

Figure 3.3 Based on Walls, G. L. (1942). *The Vertebrate Eye*. Bloomfield Hills, MI: Cranbrook Institute of Science. Reprinted by permission.

Figure 3.4 From Dowling, J. E., & Boycott, B. B. (1966). Organization of the primate retina: Electron microscopy. *Proceedings of the Royal Society (London)*, Series B, *166*, 80–111. Repinted by permission.

Figure 3.5 From Werblin, F. S. (1973). The control of sensitivity in the retina. *Scientific American*, *228*, (1), 70–79. Copyright © 1973 by Scientific American, Inc. All rights reserved.

Figure 3.7 Modified from Polyak, S. (1957). *The Vertebrate Visual System*. Chicago: University of Chicago Press. Copyright © 1957 by the University of Chicago Press. Reprinted by permission of the University of Chicago Press. The version of the figure used here is modified from Butter, C. M. (1968). *Neuropsychology: The Study of Brain and Behavior*. Belmont, CA: Brooks/Cole. Copyright © by Wadsworth Publishing Company, Inc. Reprinted by permission of the publisher, Brooks/Cole Publishing Company.

Figure 3.9 Reproduced with permission from the *Annual Review of Psychology*, volume 33, © 1982 by Annual Reviews, Inc.

Figure 3.10 Based on DeValois, R. L., Abramov, I., & Jacobs, G. H. (1966). Analysis of response patterns of LGN cells. *Journal of the Optical Society of America*, 7, 966–977. Reproduced by permission.

Figure 3.12 Modified from Hubel, D. H., & Wiesel, T. N. (1968). Receptive fields and functional architecture of monkey striate cortex. *Journal of Physiology*, *195*, 215–243. Reprinted by permission.

Figure 3.21 From Schiffman, H. R. (1982). *Sensation and Perception: An Integrated Approach* (2nd ed.). New York: Wiley. Copyright © 1976, 1982, by John Wiley and Sons, Inc. Reprinted by permission.

Figure 3.27 Modified from Orbison, W. D. (1939). Shape as a function of the vector field. *American Journal of Psychology*, *52*, 31–45. Reprinted by permission.

Figure 3.28 From Coren, S., Porac, C., & Ward, L. M. (1979). *Sensation and Perception*. New York: Academic Press. Reprinted by permission.

Figure 3.32 From Coren, S. (1972). Subjective contours and apparent depth. *Psychological Review*, *79*, 359–367. Copyright 1972 by the American Psychological Association. Reprinted by permission of the publisher and author. Permission also obtained from G. Kanizsa and *Rivista di Psicologia*.

Figure 3.47 From Coren, S., Porac, C., & Ward, L. M. (1979). *Sensation and Perception*. New York: Academic Press. Reprinted by permission.

CHAPTER 4

Quotation on page 138. From *Principles of Psychology* by Fred S. Keller & William N. Schoenfeld. Copyright 1950 by Appleton-Century-Crofts, Inc. Reprinted by permission of Appleton-Century-Crofts and the authors.

Figure 4.11 Adapted from Ferster, C. B., & Skinner, B. F. (1957). *Schedules of Reinforcement.* New York: Appleton-Century-Crofts. Copyright © 1957. Reprinted by permission of Appleton-Century-Crofts and the authors.

Figure 4.15 Modified from Jenkins, W. D., McFann, H., & Clayton, F. L. (1950). A methodological study of extinction following aperiodic and continuous reinforcement. *Journal of Comparative and Physiological Psychology, 43,* 155–167. Copyright 1950 by the American Psychological Association. Reprinted by permission of the senior author.

Figure 4.16 Modified from Olson, G., & King, R. A. (1962). Supplementary report: Stimulus generalization gradients along a luminosity continuum. *Journal of Experimental Psychology, 63,* 414–415. Copyright 1962 by the American Psychological Association. Reprinted by permission of the authors.

Figure 4.18 After Menzel, E. W. (1973). Chimpanzee spatial memory organization. *Science, 182,* 943–945, 30 November 1973. Copyright 1973 by the American Association for the Advancement of Science.

CHAPTER 5

Quotation on page 182. From Cornford, F. M. (Trans.) (1935). *Plato's Theory of Knowledge.* London: Kegan Paul. Reprinted by permission.

Quotations on page 183. From Loftus, E. F. (1980). *Memory: Surprising New Insights into How We Remember and Why We Forget.* Reading, MA: Addison-Wesley. Reprinted by permission.

Photograph in Application 5.1 page 187. Richard Hutchings/Photo Researchers, Inc.

Photograph for Inquiry 5.1, page 188. Omikron/Photo Researchers, Inc.

Figure 5.3 From Glanzer, M., & Cunitz, A. R. (1966). Two storage mechanisms in free recall. *Journal of Verbal Learning and Verbal Behavior, 5,* 351–360. Reprinted by permission from Academic Press and the senior author.

Figure 5.6 From Collins, A. M., & Quillian, M. R. (1969). Retrieval time from semantic memory. *Journal of Verbal Learning and Verbal Behavior, 8,* 240–247. Reprinted by permission from Academic Press and M. R. Quillian.

Figure 5.7 From Nickerson, R. S., & Adams, M. J. (1979). Long-term memory for a common object. *Cognitive Psychology, 11,* 287–307. Reprinted by permission from Academic Press and the senior author.

Figure 5.8 Modified from Bower, G. H., Clark, M. C., Lesgold, A. M., & Winzenz, D. (1969). Hierarchical retrieval schemes in recall of categorized word lists. *Journal of Verbal Learning and Verbal Behavior, 8,* 323–343. Reprinted by permission from Academic Press and the senior author.

Figure 5.9 From Paivio, A. (1971). *Imagery and Verbal Processes.* New York: Holt, Rinehart and Winston. Reprinted by permission from Dr. Allan Paivio.

Figure 5.11 From Loftus, G. R., & Loftus, E. F. (1976). *Human Memory: The Processing of Information.* Hillsdale, NJ: Erlbaum. Copyright 1976 by Lawrence Erlbaum Associates, Inc. Reprinted by permission of the publisher and E. F. Loftus.

Photograph in Inquiry 5.2, page 204. Wide World Photos.

Quotation on page 204. From Brown, R., & Kulik, J. (1977). Flashbulb memories. *Cognition, 5,* 73–99. Reprinted by permission of the senior author and North-Holland Publishing Company. Amsterdam.

Quotation on page 205. From Neisser, U. (1982). Snapshots or benchmarks? In U. Neisser (Ed.), *Memory Observed: Remembering in Natural Contexts.* San Francisco: W. H. Freeman. Reprinted by permission from the publisher and author.

Diagrams on page 206. Based on McGeogh, J. A., & Irion, A. L. (1952). *The Psychology of Human Learning* (2nd ed.). New York: Longmans, Green.

Figure 5.13 Modified from Underwood, B. J. (1957). Interference and forgetting. *Psychological Review, 64,* 49–60. Copyright 1957 by the American Psychological Association. Reprinted by permission of the author.

Figure 5.15 From Squire, L. R., & Cohen, N. J. (1982). Remote memory, retrograde amnesia, and the neuropsychology of memory. In L. Cermak (Ed.), *Human Memory and Amnesia.* Hillsdale, NJ: Erlbaum. Copyright 1982 by Lawrence Erlbaum Associates, Inc. Reprinted by permission of the publisher and the senior author.

Quotation on pages 213–214. From Luria, A. R. (1968). *The Mind of a Mnemonist* (L. Solotaroff, Trans.). New York: Basic Books. (Original work published 1965.) Reprinted by permission.

Photograph in Inquiry 5.3, page 213. Bill Aron/Art Resource.

CHAPTER 6

Chapter-opening photograph: Myron Wood/Photo Researchers.

Quotation on page 234. From Sternberg, R. J., & Powell, J. S. (1983). Comprehending verbal comprehension. *American Psychologist, 38,* 878–893. Copyright 1983 by the American Psychological Association. Reprinted by permission of the senior author.

Figure 6.7 From Liberman, A.M., Mattingly, I. G., & Turvey, M. T. (1972). Language codes and memory codes. In A. W. Melton & E. Martin (Eds.), *Coding Processes in Human Memory.* Washington, DC: Winston. Reprinted by permission of the Hemisphere Publishing Corporation and the senior author.

Figure 6.9 From Carroll, J. B. (1964). *Language and Thought.* Englewood Cliffs, NJ: Prentice-Hall. Reprinted by permission of the publisher and author.

Conversation rules, pages 257–258. From Farb, P. (1975). *Word Play: What Happens When People Talk.* New York: Knopf. Copyright 1975. Reprinted by permission of the publisher.

CHAPTER 7

Quotation on pages 266–267. From Edward J. Murray, *Motivation and Emotion,* © 1964, pages 9–10. Reprinted by permission of Prentice-Hall, Inc., Englewood Cliffs, N.J., and the author.

Quotation on pages 270–271. From Solomon, R. L., & Corbit, J. D. (1974). An opponent-process theory of motivation: I. Temporal dynamics of affect. *Psychological Review, 81,* 119–145. Copyright 1974 by the American Psychological Association. Reprinted by permission of the publisher and the senior author.

Figure 7.2 From Solomon, R. L. (1980). The opponent-process theory of acquired motivation. *American Psychologist, 35,* 691–712. Copyright 1980 by the American Psychological Association. Reprinted by permission of the publisher and the author.

Figure 7.4 Modified slightly from Teitelbaum, P. (1961). Disturbances in feeding and drinking behavior after hypothalamic lesions. In M. R. Jones (Ed.), *Nebraska Symposium on Motivation.* Lincoln, NE: University of Nebraska Press. Reprinted by permission of the publisher and the author.

Figure 7.7 Modified from Bradburn, N. M., & Berlew, D. E. (1961). Need for achievement and English economic growth. *Economic Development and Cultural Change, 10,* 8–20. Published by The University of Chicago Press. Copyright 1961 by The University of Chicago.

Table 7.3 From Baron, R. A. (1977). *Human Aggression*. New York: Plenum Press. Reprinted by permission of the publisher and author.
Quotation on page 291. From Baron, R. A. (1977). *Human Aggression*. New York: Plenum Press. Reprinted by permission of the publisher and author.
Quotation on page 292. From Baron, R. A. (1977). *Human Aggression*. New York: Plenum Press. Reprinted by permission of the publisher and author.
Figure 7.10 From Harlow, H. F. (1953). Motivation as a factor in the acquisition of new responses. In *Current Theory and Research in Motivation: A Symposium*. Lincoln, NE: University of Nebraska Press. Reprinted from *The Nebraska Symposium on Motivation*, by permission of The University of Nebraska Press. Copyright 1953 by The University of Nebraska Press.

CHAPTER 8
Figure 8.1 Modified from Izard, C. E. (1971). *The Face of Emotion*. New York: Appleton-Century-Crofts. Copyright now held by Prentice-Hall, © 1971. By permission of Prentice-Hall, Inc.
Quotation, page 312. From Freud, S. (1953). Fragment of an analysis of a case of hysteria. In J. Strachey (Ed. and Trans.), *The Standard Edition of the Complete Psychological Works of Sigmund Freud* (vol. 7). London: Hogarth Press. (Original work published 1905.) Reprinted by permission.

Also from "Fragment of an analysis of a case of hysteria, part 2, the first dream." In *Sigmund Freud: Collected Papers* (volume 3, page 94), edited by Ernest Jones, M. D., authorized translation by Alix & James Strachey. Published by Basic Books, Inc., Publishers, New York, 1959, by arrangement with The Hogarth Press, Ltd., and The Institute of Psycho-Analysis, London. Reprinted by permission.
Figure 8.4 From Ekman, P., Levenson, R. W., & Friesen, W. V. (1983). Autonomic nervous system activity distinguishes among emotions. *Science, 221*, 1208–1210, 16 September 1983. Copyright 1983 by the American Association for the Advancement of Science. Reprinted by permission of the American Association for the Advancement of Science and the senior author.
Table 8.1 Reprinted by permission from the *Journal of Psychosomatic Research, 11*, Holmes, T. H., & Rahe, R. H., The Social Readjustment Rating Scale. Copyright 1967, Pergamon Press, Ltd. (The revised version of the scale is reprinted here by permission of Dr. T. H. Holmes.)
Figure 8.6 Edward A. Charlesworth, Ph.D., & Ronald G. Nathan, Ph.D., "The Distress and Wellness Cycles" from *Stress Management: A Comprehensive Guide to Wellness*. Copyright © 1982, 1984 by Edward A. Charlesworth, Ph.D., & Ronald G. Nathan, Ph.D Reprinted with permission of Atheneum Publishers, Inc., the authors, and John A. Ware.
Figure 8.9 From Weiss, J. M. (1972). Psychological factors in stress and disease. *Scientific American, 226*, (6), 104–113. Copyright © 1972 by Scientific American, Inc. All rights reserved. Reprinted by permission of W. H. Freeman & Co. and the author.
Figure 8.11 From Plutchik, R. (1970). Emotions, evolution, and adaptive processes. In M. B. Arnold (Ed.), *Feelings and Emotions: The Loyola Symposium*. New York: Academic Press. Reprinted by permission of Academic Press and the author.
Quotation, page 334. From Leeper, R. W. (1970). The motivational and perceptual properties of emotions as indicating their fundamental character and role. In M. B. Arnold (Ed.), *Feelings and Emotions: The Loyola Symposium*. New York: Academic Press. Reprinted by permission of Academic Press and the author.

CHAPTER 9
Figure 9.9 After Byrne, D., & Nelson, D. (1965). Attraction as a linear function of proportion of positive reinforcements. *Journal of Personality and Social Psychology, 1*, 659–663. Copyright 1965 by the American Psychological Association. Adapted by permission of the authors.
Figure 9.11 Modified slightly from Levinger, G., & Snoek, J. D. (1972). *Attraction in Relationship: A New Look at Interpersonal Attraction*. Morristown, NJ: General Learning Press. Reprinted by permission of the senior author.

CHAPTER 10
Chapter-opening photograph: John Trumbull, *The Declaration of Independence*, Joseph Szaszfai/Yale University Art Gallery.
Figure 10.7 Based on data reported in McGuire, W. J. (1961). Resistance to persuasion conferred by active and passive prior refutation of the same and alternative counterarguments. *Journal of Abnormal and Social Psychology, 63*, 326–332. Copyright 1961 by the American Psychological Association. Adapted by permission of the author.

CHAPTER 11
Figure 11.5 Adapted from Kessen, W. (1967). Sucking and looking: Two organized congenital patterns of behavior in the human newborn. In H. W. Stevenson, E. H. Hess, & H. L. Rheingold (Eds.), *Early Behavior: Comparative and Developmental Approaches*. New York: Wiley. Copyright 1967 by John Wiley & Sons. Reprinted by permission of the publisher and author.
Figure 11.7 Adapted from the *Bayley Scales of Infant Development*. The Psychological Corporation, 1969. Copyright © 1969 by the Psychological Corporation. All rights reserved. Used by permission.
Figure 11.8 Based on Shirley, M. M. (1933). *The First Two Years: A Study of Twenty-Five Babies: Vol. 2. Intellectual Development*. Minneapolis, MN: University of Minnesota Press. Reprinted with permission.
Quotation on page 428. From Piaget, J. (1952). *The Origins of Intelligence in Children* (M. Cook, Trans.). New York: International Universities Press. Reprinted by permission.
Quotations on pages 436 and 437. From Piaget, J. (1962). *Play, Dreams, and Imitation in Childhood*. New York: Norton. Reprinted with permission.
Figure 11.15 Modified from Bandura, A., Ross, D., & Ross, S. A. (1963). Imitation of film-mediated aggressive models. *Journal of Abnormal and Social Psychology, 66*, 3–11. Copyright 1963 by the American Psychological Association. Reprinted by permission.
Figure 11.17 Reprinted by permission of the publisher from Arthur T. Jersild & Frances B. Holmes, *Children's Fears*, p. 54. (New York: Teachers College Press, © 1935 by Teachers College, Columbia University. All rights reserved.)
Quotations on pages 452 and 453. From Livesley, W. J., & Bromley, D. B. (1973). *Person Perception in Childhood and Adolescence*. London: Wiley. Copyright 1973 by John Wiley & Sons. Reprinted by permission of John Wiley & Sons, Ltd.
Figure 11.18 From *Of Children* by Guy R. Lefrancois; © 1980 by Wadsworth, Inc. Reprinted by permission of publisher.

CHAPTER 12
Quotations on pages 464–465. From Mead, M. (1939). *Coming of Age in Samoa*. In *From the South Seas: Studies of Adolescence and Sex in Primitive Societies*. New York: Morrow. (Original work published 1928). Reprinted by permission of the publisher.
Figure 12.7 From Dunphy, D. C. (1963). The social structure of urban adolescent peer groups. *Sociometry, 26*, 230–246. Reprinted

by permission of the American Sociological Association and the author.

Figure 12.8 Reprinted with permission from *Teenage Pregnancy: The Problem That Hasn't Gone Away*, published by the Alan Guttmacher Institute, New York, 1981.

Figure 12.12 Modified slgihtly from Schaie, K. W., Labouvie, G. V., & Buech, B. U. (1973). Generational and cohort-specific differences in adult cognitive functioning. *Developmental Psychology*, *9*, 151–166. Copyright 1973 by the American Psychological Association. Adapted by permission of K. W. Schaie.

Figure 12.13, top. Modified from Horn, J. L., & Donaldson, G. (1980). Cognitive development in adulthood. In O. G. Brim, Jr., & J. Kagan (Eds.), *Constancy and Change in Human Development*. Cambridge, MA: Harvard University Press. Reprinted by permission.

Quotation on page 489. From *Women of a Certain Age: The Midlife Search for Self* by Lillian B. Rubin. Copyright © 1979 by Lillian B. Rubin. Reprinted by permission of Harper & Row, Publishers, Inc.

Quotation on pages 494–495. From Turner, J. S., & Helms, D. B. (1982). *Contemporary Adulthood* (2nd ed.). New York: Holt, Rinehart and Winston. Reprinted by permission.

Tables on pages 500 and 501. From Pfeiffer, E., Verwoerdt, A., & Davis, G. C. (1972). Sexual behavior in middle life. *The American Journal of Psychiatry*, *128*, 1262–1267. Copyright 1972, the American Psychiatric Association. Reprinted by permission.

CHAPTER 13

Quotation on page 514. From Galton, F. (1908). *Memories of My Life*. London: Methuen. The Granger Collection.

Quotation on page 515. Binet, A., & Simon, T. (1916). New methods for the diagnosis of the intellectual level of subnormals. In E. S. Kite (Trans.), *The Development of Intelligence in Children*. Baltimore: Williams & Wilkins. (Original work published 1905.) Copyright © 1916 by the Williams & Wilkins Co., Baltimore. Reprinted by permission.

Figures 13.3*a* and 13.3*b* Adapted from *Psychology Today: An Introduction*, fourth edition, by Jay Braun & Darwyn E. Linder. Copyright © 1975, 1979 by Random House, Inc. Reprinted by permission of the publisher.

Quotation on page 524. From *Psychology Today: An Introduction*, fourth edition, by Jay Braun & Darwyn E. Linder. Copyright © 1975, 1979 by Random House, Inc. Reprinted by permission of the publisher.

Figure 13.4 Reproduced from *Psychology* by Henry Gleitman by permission of W. W. Norton & Company, Inc. Copyright © 1981 by W. W. Norton & Company, Inc. Reprinted by permission.

Figure 13.5 From Guilford, J. P. (1961). Factorial angles to psychology. *Psychological Review*, *68*, 1–20. Copyright 1961 by the American Psychological Association. Reprinted by permission of the author.

Figure 13.13 From Bennett, G. K., Seashore, H. G., & Wesman, A. G. *Differential Aptitude Tests*. The Psychological Corporation. Copyright © 1947, 1961, 1962 by the Psychological Corporation. All rights reserved. Reproduced by permission.

CHAPTER 14

Figure 14.2 Modified slightly from Eysenck, H. J. (1967). *The Biological Basis of Personality*. Springfield, IL: Thomas. Courtesy of Charles C Thomas, Publishers, Springfield, Illinois, and Dr. H. J. Eysenck.

Photograph of Gordon Allport, page 571. Wide World Photos.

Quotation on page 575. From *The Person in Psychology: Selected Essays* by Gordon W. Allport. Copyright © 1968 by Gordon W. Allport. Reprinted by permission of Beacon Press.

Photograph of Sigmund Freud, page 576. Wide World Photos.

Quotation on pages 576 and 577. From Breuer, J., & Freud, S. (1955). Studies on hysteria. In J. Strachey (Ed. and Trans.), *The Standard Edition of the Complete Psychological Works of Sigmund Freud* (Vol. 2). London: Hogarth Press. (Original work published 1893–1895.)

Also from *Studies on Hysteria*, (Part 2) (1893–1895) [Case 2, Frau Emmy von N.] by Josef Breuer & Sigmund Freud. Translated and edited from the German by James S. Strachey in collaboration with Anna Freud, published in The United States of America by Basic Books, Inc. (1957), by arrangement with Hogarth Press, Ltd. Reprinted by permission of Hogarth Press, Ltd., and Basic Books, Inc.

Figure 14.6 From *Psychology and Life* (10th ed.) by P. G. Zimbardo. Copyright © 1979 by Scott, Foresman and Company. Reprinted by permission.

Photograph of Carl Gustav Jung, page 582. National Library of Medicine.

Photograph of Alfred Adler, page 585. Wide World Photos.

Photograph of Karen Horney, page 587. Lotte Jacobi.

Figure 14.7 Adapted from a drawing by Don Sibley from *The Adolescent Years* by William W. Wattenberg. Copyright © 1955 by Harcourt Brace Jovanovich, Inc.; renewed 1983 by William W. Wattenberg. Reproduced by permission of the publisher.

Quotation on page 585. from Adler, A. (1931). *What Life Should Mean to You*. New York: Putnam. London: George Allen & Unwin. Reprinted by permission.

Quotations on pages 586 and 587. From *The Freud/Jung Letters: The Correspondence Between Sigmund Freud and C. G. Jung*, edited by William McGuire, translated by Ralph Manheim & R. F. C. Hull, Bollingen Series 94. Copyright 1974 by Sigmund Freud Copyrights Ltd. and Erbengemeinschaft Prof. Dr. C. G. Jung. Reprinted by permission of Princeton University Press, Routledge & Kegan Paul, and Hogarth Press.

Photograph of John Dollard, page 593. Yale University, Office of Public Information.

Photograph of Neal Miller, page 593. Yale University, News Bureau.

Photograph of B. F. Skinner, page 594. Harvard University, News Office.

Photograph of Albert Bandura, page 596. Courtesy of Albert Bandura/Gene's Studio.

Photograph of Carl Rogers, page 599. D. Land.

Photograph of Abraham Maslow, page 600. Brandeis University.

Table 14.1 Adapted from Carey, G., Goldsmith, H. H., Tellegen, A., & Gottesman, I. I. (1978). Genetics and personality inventories: The limits of replication with twin data. *Behavior Genetics*, *8*, 299–313. Reprinted by permission.

Pages 606 and 607. Quotations from Mischel, W. (1969). Continuity and change in personality. *American Psychologist*, *24*, 1012–1018. Copyright 1969 by the American Psychological Association and reprinted by permission of the author.

CHAPTER 15

Art for chapter opening, page 615. From Klausner, L. D. (1981). *Son of Sam*. New York: McGraw-Hill. Reprinted by permission.

Quotation on page 638. From White, R. W. (1964). *The Abnormal Personality* (3rd ed.). New York: Ronald Press. Copyright 1964 by Ronald Press. Reprinted by permission.

Quotation on page 638. From Grinspoon, L., Ewalt, J. R., & Shader, R. I. (1972). *Schizophrenia: Pharmacotherapy and Psycho-*

therapy. Baltimore: Williams & Wilkins. Copyright © 1972, The Williams & Wilkins Co., Baltimore.

Table 15.2, page 643. From *Abnormal Psychology* by Joseph Mehr. Copyright 1983 by CBS College Publishing. Reprinted by permission of CBS College Publishing.

Quotation on pages 658–659. Adapted from Wolpe, J. (1973). *The Practice of Behavior Therapy* (2nd ed.). New York: Pergamon. Copyright © 1973, Pergamon Press, Ltd.

CHAPTER 16

Chapter-opening photographs, page 669: *The Stone of Madness*, painted by Sanders Hemessen (1500–1566), The Bettmann Archive; trephinated skull from Chuquitanta, Peru, and photographed by J. Otis Wheelock, courtesy of Department of Library Services, American Museum of Natural History; *Burning of Witches* from a lampoon published in 1555, The Bettman Archive; circulating swing from Hallaran, W. S., *Practical Observations on the Cause and Cure of Insanity*, Cork, 1818, courtesy of the National Library of Medicine; the crib from The Bettman Archive, Inc.

Figure 16.2 From *Abnormal Psychology* by Joseph Mehr. Copyright © 1983 by CBS College Publishing. Reprinted by permission of CBS College Publishing.

Quotation on page 683. From Freud, S. (1953). *A General Introduction to Psychoanalysis.* New York: Doubleday. (Original work published 1920.) Reprinted by permission of Liveright Publishing Corporation and Hogarth Press.

Quotation on pages 684–685. From Breuer, J., & Freud, S. (1955). Studies on hysteria. In J. Strachey (Ed. and Trans.), *The Standard Edition of the Complete Psychological Works of Sigmund Freud* (Vol. 2). London: Hogarth Press. (Original work published 1893–1895.)

Also from *Studies on Hysteria*, (Part 2) (1893–1895) [Case 5, Fräulein Elisabeth von R.], in Josef Breuer & Sigmund Freud, *The Standard Edition of the Complete Psychological Works of Sigmund Freud* (Vol. 2). Translated and edited from the German by James S. Strachey in collaboration with Anna Freud, published in The United States of America by Basic Books, Inc. (1957), by arrangement with Hogarth Press, Ltd. Reprinted by permission of Hogarth Press, Ltd., and Basic Books, Inc.

Quotations in the abstract on pages 689–690. From Barton, A. (1974). *Three Worlds of Therapy: An Existential-Phenomenological Study of the Therapies of Freud, Jung, and Rogers.* Palto Alto, CA: National Press Books. Reprinted by permission.

Quotations on pages 691–692. Adapted from Stevens, J. O. (1975). *Gestalt Is.* Moab, UT: Real People Press. Reprinted by permission of the publisher.

Quotation on page 698. From Redd, W. H. (1980). Stimulus control and extinction of psychosomatic symptoms in cancer patients in protective isolation. *Journal of Consulting and Clinical Psychology, 48*, 448–455. Copyright 1980 by the American Psychological Association. Reprinted by permission of the American Psychological Association and the author.

Figure in Application 16.1, page 699. From Redd, W. H. (1980). Stimulus control and extinction of psychosomatic symptoms in cancer patients in protective isolation. *Journal of Consulting and Clinical Psychology, 48*, 448–455. Copyright 1980 by the American Psychological Association. Reprinted by permission of the American Psychological Association and the author.

Quotation on page 706. From Beck, A. T., Rush, A. J., Shaw, B. F., & Emery, G. (1979). *Cognitive Therapy of Depression.* New York: Guilford Press. Reprinted by permission.

Figure 16.15 From Redd, W. H., Andresen, G. V., & Minagawa, R. Y. (1982). Hypnotic control of anticipatory emesis in patients receiving cancer chemotherapy. *Journal of Consulting and Clinical Psychology, 50*, 14–19. Copyright 1982 by the American Psychological Association and the first author.

Quotation in Application 16.2, pages 714–715. From Farrelly, F., & Brandsma, J. (1974). *Provocative Therapy.* Cupertino, CA: Meta. Reprinted by permission of the publisher.

Figure 16.21 Based on data in Shapiro, D. A., & Shapiro, D. (1982). Meta-analysis of comparative therapy outcome studies: A replication and refinement. *Psychological Bulletin, 92*, 581–604. Copyright 1982 by the American Psychological Association. Adapted by permission of the publisher and first author.

Glossary

An italicized word in a definition has a separate entry. If you do not know the meaning of an italicized word, you can look it up in this Glossary.

A-B-A within-subjects experimental design An experimental strategy in which a *baseline* is established, the *independent variable* is introduced, and then the independent variable is removed. The *behavior* should go back to baseline levels if the independent variable has in fact produced the observed changes.

ability A general term referring to the potential for the acquisition of a skill; the term covers *intelligence* and specific *aptitudes*.

ability tests *Tests* of potential—that is, of what an individual can learn with training. Compare *achievement tests, personality tests*. See *ability*.

abnormal behavior *Behavior* which deviates from what is considered normal; usually refers to maladaptive behavior. See *psychological disorder*.

abnormality Significant deviation from commonly accepted patterns of *behavior, emotion*, or thought. See *psychological disorder*.

abscissa The horizontal axis of a graph. Compare *ordinate*.

absolute threshold The minimum energy level permitting detection of a *stimulus*. Compare *differential threshold*. See *threshold*.

abstract word A word that evokes very little visual imagery. Compare *concrete word*. See *images*.

abuse The "milder" level of a *substance-use disorder;* involves some impairment of social or job functioning. Contrast *dependence*.

acalculia Trouble with simple arithmetic calculations; may follow *parietal-lobe* damage.

accommodation (1) The changes in eye structures, especially the *lens*, that bring light from objects at different distances to a focus on the retina. (2) In Piaget's *theory* of *cognitive development*, the modification of one's modes of *thinking* and behaving to fit in with characteristics of the environment. Contrast *assimilation*.

achievement motivation (*n* ach) See *need for achievement*.

achievement tests *Tests* used to measure present knowledge or skills, especially knowledge or skills developed through specific training. Compare *ability tests*.

acquisition curve The graphic representation of *learning* which shows that the strength of the learned response gradually increases with more and more learning trials.

activational role of sex hormones Term that refers to the role of these *hormones* in triggering sexual behavior. See *estrogens, androgens*. Compare *organizational role of sex hormones*.

acupuncture anesthesia Relief from pain accomplished by inserting and twisting needles in various regions of the body.

adaptation In the context of *psychological disorders*, a term that refers to meeting the performance requirements or *role* demands of one's situation.

adjustment A judgment *heuristic* in which *subjective probability* estimates start at a certain point and are raised or lowered depending on the circumstances. See *anchoring*.

adolescence The period of life from *puberty* to the completion of physical growth.

adrenocorticotropic (ACTH) axis A glandular system (see *endocrine gland*) of the body involved in the body's *stress* response; *stressors* cause adrenocorticotropic hormone (ACTH) to be released from the *pituitary gland*; ACTH then stimulates the adrenal gland so that *cortisol* is released to help the body deal adaptively with stressors.

adrenocorticotropic hormone (ACTH) See *adrenocrticotropic (ACTH) axis*.

affect *Mood* or *emotion*.

affective disorders One of the *DSM-III* categories of severe *psychological disorders*; the disorders are characterized by extremes of mood. See *manic episodes, major depressive episodes, bipolar disorder*.

afferent Carrying sensory information into the *central nervous system*. Compare *efferent*.

afferent code The pattern of neural activity in the *peripheral* and *central nervous systems* that corresponds to various aspects of the external stimulating environment.

age scale A *test* in which items are grouped not by type of task but by the average age at which children pass each item; scores are expressed as *mental age (MA)*. See *Stanford- Binet Intelligence Scale*.

aggression A general term applying to *behavior* aimed at hurting other people; also applies to feelings of anger or hostility. Aggression functions as a *motive*, often in response to threats, insults, or frustrations. See *frustration-aggression hypothesis, instrumental aggression, hostile aggression*.

agnosia Inability to recognize objects and their meaning; usually due to damage to the brain.

agraphia Impairment of the ability to express oneself in writing; may follow *parietal-lobe* damage.

alarm reaction The first stage of the *general adaptation syndrome;* consists of prompt responses of the body, many of them mediated by the *sympathetic system*, which prepare the *organism* to cope with *stressors*. Compare *stage of resistance, stage of exhaustion*.

alexia Difficulty in reading; may follow *perietal-lobe damage*.

algorithm In problem solving, a set of rules by which a problem may be solved more or less mechanically. Compare *heuristic*.

all-or-none law The principle which states that when a particular *neuron* is excited to fire a *nerve impulse*, the impulse is always the same size and always travels at the same rate in the *axon* of that neuron.

Alzheimer's disease A form of *primary degenerative dementia* characterized by progressive mental deterioration.

amnesia Generally, any loss of *memory*; often applied to situations in which a person forgets his or her own identity and is unable to recognize familiar people and situations.

amygdala A deep structure of the *cerebrum*; part of the *limbic system*; involved in emotion and memory formation.

anal stage The stage of *psychosexual development* in *psychoanalytic theory* in which the satisfactions of a child between 1 and 3 years center around excretion. Compare *oral stage, phallic stage, latency period, genital stage*.

analytical psychology Carl Gustav Jung's *theory* of *personality.*

analytical therapy Carl Gustav Jung's approach to *psychotherapy* which emphasizes each person's need for *individuation.*

anchoring In estimating *subjective probability,* the initial level which provides an anchor that biases the estimates. See *adjustment.*

androgens Male sex *hormones.* See *testosterone.*

angiotensin II A substance that circulates in the blood and may trigger drinking. See *renin.*

angular gyrus A region of the *cerebral cortex* at the border of the *parietal, occipital,* and *temporal lobes* important in reading.

anima A female *archetype* present in males; a concept in Jung's *analytical psychology.* Compare *animus.*

animism (1) In Piaget's *theory* of *cognitive development,* the *belief* of children in the *preoperational stage* that things are alive and move with wills of their own. (2) More generally, reasoning based on coincidences in nature.

animus A male *archetype* present in females; a concept in Jung's *analytical psychology.* Compare *anima.*

anorexia nervosa A *psychological disorder* characterized by loss of appetite, loss of weight, and in women cessation of menstruation; "self-starvation."

anterior commissure A band of white fibers that connects the *temporal lobes* of the *cerebral hemispheres.* Compare *corpus callosum.*

anterograde amnesia The inability to encode and store new *information* in memory. Compare *retrograde amnesia.* See *amnesia.*

antidiuretic hormone (ADH) A *hormone* that controls the loss of water through the kidneys.

antisocial personality disorder A *personality disorder* in which the individual displays little concern for the ordinary rules of society and little feeling for the ordinary standards of right and wrong.

anxiety A vague, objectless fear; an uneasy, fearful feeling.

anxiety disorders *Clinical syndromes* in which observable and intense *anxiety* is the principal symptom; the *anxiety* is not covered up or reduced by other processes as in some of the other *psychological disorders.*

aperiodic wave A *complex wave* made up of various amplitude and *frequency* components which occur irregularly; noise is usually aperiodic in waveform. Compare *periodic wave.*

aphasia Generally, an impairment in *language* ability; may be an impairment of comprehension (receptive aphasia) or of speech production (productive aphasia).

apparent motion Perceived motion in which no actual movement of the *stimulus* pattern over the *receptor* occurs. Compare *real motion.*

appraisal In the *cognitive-appraisal theory of emotions,* evaluation of information from the environment, from the body, and from memory which determines the *emotion* that is felt. See *reappraisal.*

approach-approach conflict A situation in which two *positive* incompatible *goals* are equally attractive at the same time.

approach-avoidance conflict A situation in which an individual is both attracted and repelled by the same *goal.*

apraxia Impairment, in the absence of a specific paralysis, of the ability to perform sensory-motor tasks; may occur after damage to the *parietal association cortex.*

aptitude The *ability* to profit by certain types of training and to do the work required in a particular situation. Compare *intelligence.* See *ability, scholastic aptitudes, vocational aptitudes.*

aptitude tests See *aptitude.*

archetypes Inherited *unconscious images* or *concepts* which represent ancestral experiences of the human race. See *animus, anima, collective unconscious.*

arcuate fasciculus The bundle of nerve fibers connecting the posterior area, or *Wernicke's* speech *area,* of *cerebral cortex* with the frontal speech area, or *Broca's area.*

arousal The amount of excitement, or the degree to which one is "stirred up"; indicated by the *electroencephalogram, galvanic skin response,* muscle tension, etc.; influenced by activity in the *ascending reticular activating system* and the *sympathetic system.*

art A skill or knack for doing something that is acquired by study, practice, and special experience. Compare *science.*

ascending reticular activating system (ARAS) The fibers and nerve cells of the *reticular formation;* an indirect sensory pathway to the *cerebral cortex;* involved in control of levels of *arousal* and the sleep-waking continuum.

assimilation In Piaget's *theory* of *cognitive development,* the modification of one's environment so that it fits into already developed ways of *thinking* and behaving. Contrast *accommodation.*

association areas Regions of the *cerebral cortex* involved in such complex psychological functions as the understanding and production of *language, thinking,* and imagery. Compare *sensory areas of cortex* and *motor areas of cortex.*

assortative mating Mating between couples who are more similar in mental or physical attributes than would be predicted by chance.

attachment The early, stable love relationship between a child and caretaker.

attention Processes that select certain inputs for inclusion in the focus of experience.

attention deficit disorder A behavioral problem of childhood characterized by a short attention span, hyperactivity, a low tolerance for frustration, and impulsivity.

attitude An evaluation; a learned predisposition to behave in a consistent evaluative manner toward a person, a group of people, an object, or a group of objects.

attitude item In measuring *attitudes,* a question or statement about the attitude object and a "format"—fixed or open-ended—for the response.

attitude scales *Self-report methods* of measuring *attitudes;* attempts to obtain a precise index of a person's attitude on a particular issue either by using calibrated statements and asking each person to indicate agreement or disagreement with each statement, or else by asking each person to specify the degree to which she or he agrees or disagrees with a group of statements. Compare *public opinion polls.*

attitude similarity Agreement of *attitudes* among people; a factor influencing the degree of attraction one person feels for another.

attributions Characteristic *traits,* intentions, and *abilities* inferred on the basis of observed *behavior;* an aspect of *social perception.*

auditory canal The canal leading from the ouside of the head to the *eardrum.*

autistic thinking Highly private *thinking* using *symbols* that have very personal meanings. Compare *directed thinking.*

autokinetic effect *Apparent motion* of a small spot of light against a dark background in a dark room.

autonomic nervous system A division of the *peripheral nervous system* serving certain glands and smooth muscles; includes the *sympathetic system* and the *parasympathetic system;* important in *emotion.* Compare *somatic nervous system.*

auto-shaping Use of *classical conditioning* procedures in the *shaping* of an animal in an *operant chamber.*

availability A judgment *heuristic* in which people estimate *subjective probability* on the basis of easily remembered events.

aversion therapy A form of *behavior therapy/behavior modification* in which the *stimuli* eliciting the *behavior* to be eliminated are paired with unpleasant states of affairs; in time, these stimuli tend to be avoided. See *classical conditioning, covert sensitization.*

avoidance-avoidance conflict A situation in which an individual is caught between two *negative goals;* as the individual tries to avoid one goal, he or she is brought closer to the other.

avoidance learning *Learning* to make a particular response to a warning signal in order to avoid a *noxious stimulus.*

axon A nerve fiber which transmits impulses from the cell body to other neurons or to muscles and glands. Compare *dendrite.*

balance theory A *theory* which predicts *attitude* formation and change on the basis of an individual's tendency to maintain consistency among the numerous attitudes he or she holds; there is a tendency toward balance, and unbalanced structures produce tension or discomfort.

basal ganglia *Nuclei* in the *cerebral hemispheres* concerned with maintaining muscle tone and other aspects of body movement.

baseline A stable and reliable level of *performance* that can be used as a basis for assessing changes in behavior caused by the introduction of an *independent variable.* See *within-subjects design.*

basic anxiety and basic hostility In Horney's *theory* of *personality,* deep-seated feelings of insecurity and anger originating in childhood which determine how a person will respond to others.

basilar membrane A membrane in the *cochlea* on which the *organ of Corti* sits.

basket nerve ending A specialized structure at the roots of hairs on the body; a sense organ for pressure or touch.

behavior Anything a person or animal does that can be observed in some way.

behavior genetics The study of the ways in which an individual's genetic constitution contributes to the determination of *behavior.*

behavior therapy/behavior modification Methods developed to alleviate *psychological disorders* which focus on changing behavioral problems by using techniques of *classical conditioning, instrumental conditioning/operant conditioning,* and *observational learning.*

behavioral assessment An approach to the study of *personality* based on the direct observation of *behavior* and the conditions under which certain behaviors occur.

behavioral medicine Psychological treatments designed to help people cope with physical health problems.

behavioral perspective A current viewpoint in *psychology* which has its roots in the older school of *behaviorism;* the emphasis is on the description, control, and understanding of what people and animals do—their *behavior.*

behavioral therapies See *behavior therapy/behavior modification.*

behaviorism The view that human and animal *behavior* can be understood, predicted, and controlled without recourse to explanations involving mental states. A school of psychology insisting that psychology be restricted to the study of *behavior.* Compare *structuralism, functionalism, Gestalt psychology.*

beliefs *Cognitions,* or thoughts, about the characteristics of objects.

belongingness and love needs *Needs* for affection, affiliation, and identification. In Maslow's theory they are fulfilled after *physiological* and *safety needs* are satisfied.

binaural Presentation of the same information to the two ears. Compare *dichotic.*

binocular cues Cues for *depth perception* that we get from both eyes working together. Compare *monocular cues.* See *retinal disparity.*

biofeedback Information provided to a person about biological events in his or her own body—heart rate, for example. The *perception* of this information may allow individuals to gain control over biological events in their bodies.

biological perspective A current viewpoint in *psychology* in which the aim is to relate *behavior* to functions of the body—the nervous and glandular systems in particular.

biomedical therapies Use of physical means in the treatment of *psychological disorders.* See *psychosurgery, electroconvulsive/electroshock therapy (ECT/EST), chemotherapy.* Compare *psychotherapy.*

bipolar disorder An *affective disorder* in which there are both *manic episodes* and *major depressive episodes.*

blind spot The region of the *retina* where the *optic nerve* fibers leave; it contains no *rods* or *cones,* and therefore what is focused on it is not seen.

body language *Communication* by gestures and movements of the body; may give messages about the *emotions* and *motives* of an individual.

bonding See *attachment.*

bouton Small bulb at the end of the *axon* of a *neuron.* Arrival of a *nerve impulse* at the bouton causes release of *neurotransmitters* into the *synaptic cleft.* See *vesicles.*

brain The part of the *central nervous system* encased in the skull.

brain comparator A hypothetical mechanism that takes account of eye and head movements in the *perception* of motion.

brain stem The division of the *brain* closest to the *spinal cord;* includes the *medulla,* the *pons,* and the *midbrain.*

brain syndromes *Psychological disorders* resulting from impairment of brain function; may be acute and reversible or chronic and irreversible.

brightness A dimension of visual experience referring to the relative degree of whiteness, grayness, or blackness. Intensity of the physical *stimulus* is the major determiner of perceived brightness. Compare *hue, saturation.*

brightness constancy A phenomenon of *perception* in which a person perceives an object as having almost the same brightness despite marked changes in the physical energy stimulating the eye.

Broca's area The portion of the *frontal lobe* of the *cerebrum* involved in programming the patterned movements necessary for speech. Compare *Wernicke's area.*

bulimia A *psychological disorder* characterized by the "binge-purge syndrome" in which people overeat and then induce vomiting or use laxatives to rid themselves of what they have eaten; extreme diets are also a feature of this disorder.

Cannon-Bard theory A *theory* about the relationship between bodily states and felt emotion stating that felt *emotion* and bodily reactions in emotion are independent of each other and triggered simultaneously by activity of lower brain areas. Compare *James-Lange theory, Schachter-Singer theory.*

cardinal traits Dominant *personality* characteristics of an individual; nearly all of a person's actions can be traced back to these traits. Compare *central traits, secondary traits.*

catatonic type A kind of *schizophrenia* characterized by negativism and the prolonged maintenance of certain bizarre postures.

categorical perception of phonemes The sounds comprising *phonemes* are perceived as belonging together as a category of sounds.

catharsis A term that refers to "venting an emotion" or "getting it out of one's system"; may sometimes be useful in lessening *aggression,* fear, or anxiety.

cell membrane A thin structure separating the fluid inside a *neuron* from that which bathes the cell on the outside; essential for the generation and conduction of *nerve impulses.*

cellular-dehydration thirst Thirst triggered by loss of water from the *osmoreceptors*. Compare *hypovolemia*. See *double-depletion hypothesis*.

central nervous system (CNS) The part of the nervous system enclosed in the bony case of the skull and backbone; the *brain* and the *spinal cord*. Compare *peripheral nervous system (PNS)*.

central sulcus A groove running obliquely from top to bottom on the side, or lateral, surface of the *cerebral cortex;* marks off the *frontal lobe* from the *perietal lobe*.

central traits *Personality* features which characterize an individual, but do so less completely than do *cardinal traits*. Compare *cardinal traits, secondary traits*.

cerebellum A structure, located toward the back of the *brain,* which is concerned with the coordination of movements.

cerebral arteriosclerosis Hardening of the arteries of the brain, especially in the *cerebrum,* which reduces the flow of blood; brain tissue degenerates and intellectual functions are impaired.

cerebral cortex The *gray matter* covering the *cerebrum*.

cerebral hemispheres The two divisions of the *cerebrum,* separated by the *longitudinal fissure*.

cerebrum The largest structure of the *forebrain;* consists of *white matter* (fiber tracts), deeper stuctures, and *cerebral cortex*.

chemotherapy The treatment of various *psychological disorders* with *drugs*.

child psychology A part of *developmental psychology* which emphasizes the changes in *behavior* which occur in the early years of life.

childhood amnesia A term that refers to the poor *memory* we have of our lives in our early years. See *amnesia*.

cholecystokinin (CCK) A *hormone* that is released into the blood when food reaches the intestine; it may be involved in the cessation of eating.

chromosomes Long chainlike structures containing *genes;* found in the nuclei of cells.

chronological age (CA) The actual age of a person in years or months.

chunking An *encoding process* in which items of *information* are grouped together in *short-term memory (STM);* increases the capacity of short-term memory.

cis-rhodopsin, trans-rhodopsin Shapes of the *rhodopsin* molecule. The nonexcited cis configuration changes to the trans configuration with absorption of light energy.

classical conditioning *Learning* that takes place when a *conditioned stimulus* is paired with an *unconditioned stimulus*. Also called respondent conditioning or Pavlovian conditioning.

classical conditioning techniques Methods of *behavior therapy/behavior modification* which use *classical conditioning*. See *classical conditioning*.

clause A verb and its associated nouns, adjectives, and other parts of speech; said to be the major unit of perceived meaning in speech.

clearness in depth perception A *monocular cue* for depth; nearer objects are generally perceived as having sharper outlines than those of distant objects.

client-centered therapy A *nondirective therapy,* developed by Carl Rogers, which typically is not so intensive or prolonged as *psychoanalysis*. It puts a person in a permissive situation where potentialities for growth and problem solving are maximized. See *nondirective therapy*.

climacteric The end of ovulation and reproductive capacity, signaled by *menopause*.

clinical-consensual approach to classification A way of grouping *psychological disorders;* an attempt to reach general agreement among experts about which psychological disorders exist and how they should be defined and diagnosed. See *DSM-III*. Compare *empirical approach to classification*.

clinical method A psychological technique which focuses on the study of an individual's behavior; it is usually used to understand behavioral problems. *Tests* and interviews are used to study the individual's behavior.

clinical psychologists See *clinical psychology*.

clinical psychology A branch of *psychology* concerned with psychological methods of recognizing and treating *psychological disorders* and research into their causes.

clinical syndromes A *DSM-III* category of *psychological disorders*. Examples are *schizophrenic disorders* and *paranoid disorders*.

closure An organizing principle in *perception* in which gaps in stimulation are filled in by perceptual processes, thus giving rise to the perception of complete and continuous forms.

cochlea A bony cavity, coiled like a snail shell, containing the *receptor* organs for hearing.

cognition Mental processes such as *thinking,* remembering, perceiving, planning, and choosing.

cognitive See *cognition*.

cognitive-appraisal theory of emotions A *theory* which states that the *emotions* we feel result from evaluations, or appraisals, of information received from the situation, from the body, and from *memories* of past encounters with similar situations. See *reappraisal*.

cognitive development The growth and changes in *thinking,* memory, perceiving, planning, and choosing which occur through the life span; Piaget's *theory* of cognitive development has been influential.

cognitive dissonance A conflict of thoughts arising when two or more ideas do not go together; people find this conflict distressing and are motivated to reduce it; *attitudes* may be changed when this conflict is reduced.

cognitive framework Categories and their perceived interrelationships used in *social perception;* included are *implicit personality theories,* relationships among *traits,* and *stereotypes*.

cognitive learning A change in the way *information* is processed as a result of experience that a person or an animal has had. See *latent learning, insight learning, imitation*.

cognitive map The learned mental representation of the environment. See *cognitive learning*.

cognitive perspective A current viewpoint in *psychology* which emphasizes information processing in the study of mind and *behavior*. See *information-processing theory*.

cognitive processes Specific mental operations occurring in *perception, learning,* or problem solving.

cognitive psychology The branch of psychology which studies *cognitive processes;* includes the study of *memory*.

cognitive-response approaches Views of *attitudes* which stress the importance of the active *information-processing* done by people in the formation of attitudes.

cognitive restructuring See *cognitive therapy/cognitive-behavior therapy*.

cognitive slippage and derailment Thought which is marked by a series of ideas which depart from a logical framework; it is characteristic of many *schizophrenic* patients.

cognitive-structural school Psychologists who argue for the importance of active interaction between the developing *organism* and the environment in determining *behavior* and *cognition;* Jean Piaget is a representative of this school.

cognitive therapy/cognitive-behavior therapy An approach to the treatment of *psychological disorders* which stresses that maladaptive *behavior* comes from maladaptive ideas; therapy should focus on changing these ideas; this approach is sometimes known as cognitive restructuring.

cohort A group of people who were born about the same time and thus may have had many experiences in common.

collective unconscious According to the *theory* of Jung, the primitive ideas and *symbols* that all people have in common; said to have grown out of the past experience of the human race. Compare *personal unconscious*. See *archetypes*.

collegial marriage A relationship in which comradeship and sharing are emphasized; husband and wife assume responsibility for different roles in the marriage, with each respecting the individual abilities and interests of the other.

commissure A band of fibers connecting the left and right sides of the *brain*. See *corpus callosum, anterior commissure*.

common fate An organizing principle in *perception* that causes a person to perceive items that move together as grouped together.

communication *Stimuli* made by one *organism* that have meaning for other organisms and thus affect their *behavior*. See *language*.

community mental health The attempt to bring public health principles to the area of mental health. Community mental health stresses crisis intervention in psychiatric emergencies; it attempts to make inexpensive specialized *psychotherapy* available to poor people; it attempts to resolve community problems that lead to *psychological disorders*.

community mental health movement See *community mental health*.

community psychology The subfield of *psychology* emphasizing application of psychological principles, ideas, and points of view to help solve social problems and to help individuals adapt to their work and living groups.

companionship marriage A relationship in which male and female roles are not regarded as fixed; husbands and wives freely assume the rights and obligations of their partners, depending on the situation.

comparison level (CL) In *social exchange theory*, a subjective standard for judging whether the *outcomes* experienced in a social relationship are satisfactory. Compare *comparison level for alternatives* *(CL_{alt})*.

comparison level for alternatives (CL_{alt}) In *social exchange theory*, an individual's standard used for judging the *outcomes* that would be received in the next best alternative relationship, or in simply being alone; when outcomes in the present relationship fall below the CL_{alt}, a person will leave the relationship in favor of the alternative. Compare *comparison level (CL)*.

compensation and overcompensation A *defense mechanism* in which an individual substitutes one activity for another in an attempt to satisfy frustrated (see *frustration) motives*. It usually implies failure or loss of self-esteem in one activity and the compensation for this loss by efforts in some other realm of endeavor.

complex In Jung's *theory* of *personality (analytical psychology)*, a network of ideas bound together by a common *emotion* or set of feelings.

complex waves *Sound waves* made up of many *frequencies;* can be *periodic* or *aperiodic*. Compare *sine waves*.

compulsions Irrational acts which are repeatedly carried out in a fixed, repetitive way. Compare *obsessions*. See *obsessive-compulsive disorder, compulsive personality disorder*.

compulsive personality A *personality disorder* in which the individual is overly rigid in *thinking* and *behavior*, often showing excessive concern for morals and social standards.

compulsive personality disorder A *psychological disorder* in which the individual tends to be perfectionistic, stingy, stubborn, emotionally inhibited, and inclined to become preoccupied with trivial pursuits.

concept A symbolic construction that represents some common and general feature or features of objects or events.

concordance The percentage of relatives of a person who show the same trait as does the person in question; concordances are often computed for *identical twins* and *fraternal twins*.

concrete operations In Piaget's *theory*, logical *operations* involving manipulation and transformation of here-and-now objects and events.

concrete word A word for which a visual *image* is easily formed. Compare *abstract word*.

conditioned emotional response (CER) Fear conditioned to *stimuli* which are associated with *noxious* events. See *classical conditioning*.

conditioned, or secondary, reinforcer A *stimulus* which has been paired with a *primary reinforcer* to become a conditioned reinforcer. The stimulus can then be used as a *reinforcer*. Compare *primary reinforcer*.

conditioned response (CR) A response produced by a *conditioned stimulus* after it has been paired with an *unconditioned stimulus;* the learned response in *classical conditioning*.

conditioned stimulus (CS) A *stimulus* that is originally ineffective, but which, after pairing with an *unconditioned stimulus*, evokes the *conditioned response*.

cones Light-sensitive *receptors* in the *retina* responsible for sharp *visual acuity*, daylight vision, and color vision; found in the *fovea*. Compare *rods*.

confabulation A *reconstructive process* which consists of filling in gaps in *memory* with plausible guesses.

conflict See *motivational conflict*.

conformity The changing of individuals' *beliefs* or *behaviors* so that they become more similar to those of other group members. Compare *deviance*.

connotative meaning The emotional and evaluative meaning of words and *concepts;* can be measured by the *semantic differential* method. Compare *denotational meaning*.

conscious Thoughts and perception of which a person is aware. Compare *preconscious, unconscious*.

consensus information The extent to which other people respond to the same *stimuli* in the same manner as the person being judged; a factor important in making *attributions*. See *consistency, distinctiveness*.

conservation In Piaget's *theory* of *cognitive development*, the *operation* of knowing that an object has not changed in fundamental properties in spite of appearance. See *identity concept*.

consistency The extent to which a particular response occurs whenever a particular *stimulus* or situation is present; a factor important in making *attributions*. See *consensus information, distinctiveness*.

consistency theories A group of *theories* about *attitudes* which focus on the individual's attempt to maintain consistency among the numerous attitudes he or she holds. See *balance theory*.

constancy (size, shape) See *perceptual constancy*.

constructive processes Modifications of the material to be remembered which take place at the time of input. Compare *reconstructive processes*.

contact comfort The gratification an infant receives from contact with a soft object; may motivate *attachment* early in life. The infant's need for contact with a warm, soft surface.

continuation An organizing principle in *perception* according to which lines that start out as straight lines are perceived as continu-

ing as straight lines, while lines that start out as curved lines are seen as continuing on a curved course.
continuous reinforcement (CRF) *Reinforcement* of all correct responses. Compare *schedule of reinforcement.*
continuous versus discontinuous development The issue of whether *behavioral* and *cognitive* development are best viewed as moving in a smooth progression (continuously) or through a series of discrete stages (discontinuously).
contours The lines of demarcation perceived by an observer whenever there is a marked difference between the brightness or color in one place and that in an adjoining region.
contralateral neglect A symptom of *parietal lobe* damage in which patients ignore the side of the body and extrapersonal space opposite the side of the brain damage. Most common after right parietal lobe damage; the patient ignores left extrapersonal space and the left side of the body.
contraprepared behaviors Responses which, for certain animal species, are almost impossible to learn. Compare *prepared behaviors, unprepared behaviors.*
contributions rule A rule employed as a standard in making judgments of fairness in social relationships based on the investments each person makes in the relationship. Compare *needs rule, equality rule.*
control groups The groups in an experiment which are equivalent to the experimental group but which do not receive the *independent variable.*
control-group design An experimental strategy which uses *control groups* to control for extraneous factors. Often experimenters will match subjects on a number of factors considered to be relevant and then assign them at random to the *experimental* and *control groups.*
control in experiments A characteristic of the *experimental method* in which extraneous factors which might affect the *dependent variable* are held constant or cancelled out in some way so that only the specified *independent variables* are allowed to change.
conventional level Type of *thinking* about moral issues in which value is placed on maintaining the conventional order and satisfying the expectancies of others. Compare *preconventional level, postconventional level.*
convergent thinking *Thinking* in which the thinker gathers information relevant to a problem and then proceeds by reasoning to arrive at the one best solution; involved in solving problems with a single correct answer. Compare *divergent thinking.*
conversion disorder A *psychological disorder* in which a *motivational conflict* has been converted into physical symptoms; the person appears to have various ailments, but these ailments have no physical basis.
conversion reaction See *conversion disorder.*
cooperative play Play requiring complementary role taking and a high degree of interaction. Compare *solitary play, parallel play.*
cornea The outermost, transparent layer of the front of the eye. See *refraction.*
corpus callosum The great *commissure* that connects one *cerebral hemisphere* with the other. See *anterior commissure.*
correlation Generally, a relationship; more specifically, a statistic summarizing the direction and degree of relationship between sets of measures. See *correlation coefficient.*
correlation coefficient A number between +1.00 and −1.00 that expresses the degree of relationship, or *correlation*, between sets of measurements. A coefficient of +1.00 (or −1.00) represents perfect correlation, a coefficient of .00 represents no correlation, and intermediate coefficients represent various degrees of correlation.
cortex See *cerebral cortex.*
corticotropin-releasing factor (CRF) A hormone-like substance secreted by the *hypothalamus* which regulates the release of *adrenocorticotropic hormone* (ACTH) from the *pituitary gland.* See *releasing factors.*
cortisol A *hormone* secreted by the outer layers of the adrenal gland under the influence of *adrenocorticotropic hormone (ACTH)*; its actions help the body deal adaptively with *stressors.* See *adrenocorticotropic (ACTH) axis.*
counseling psychology The subfield of *psychology* which stresses helping individuals with educational, vocational, family, or personal problems. See *clinical psychology.*
counterattitudinal role playing A term that refers to the state of affairs in which people express *attitudes* publicly which are opposite to their private attitudes; used in research on *cognitive dissonance.*
counterconditioning The weakening of a *conditioned response* by conditioning the *stimuli* that elicit it to other responses which are incompatible with the response to be eliminated. See *systematic desensitization.*
countertransference reaction In *psychotherapy*, the therapist may transfer to the patient *attitudes* or feelings which originated outside the therapy situation. Compare *transference.*
covert sensitization A form of *aversion therapy* in which the unpleasant, or *noxious*, events associated with unwanted *behaviors* are imagined rather than experienced.
creativity The use of one's intellect to produce novel, meaningful ideas.
criterion (1) A standard. In the evaluation of *tests*, the job or performance that a test is supposed to predict. (2) In *learning*, the level of *performance* considered to represent relatively complete learning.
critical period A time in the life cycle during which certain cognitive, perceptual, or social events must occur if they are to influence later *behavior.*
cross-sectional method The study of groups of persons or a process at a particular point in time; groups at different stages of development are contrasted. Compare *longitudinal method.*
crystallized intelligence The type of *intelligence* involved in applying what has been learned; reflects one's cultural exposure and is composed largely of knowledge and skills. Compare *fluid intelligence.*
cultural-familial (or sociocultural) retardation *Mental retardation* without an identifiable physiological cause; the majority of retarded persons fall in this category.
cumulative recorder A device for plotting responses in *instrumental conditioning/operant conditioning.*

decibel (dB) The unit of measurement used to express the *intensity* of sound.
deduction Reasoning from abstract general principles to specific hypotheses that follow from these principles. Compare *inductive reasoning.*
deep phrase structure, or deep structure The mental representation of what a person intends to say. See *transformational grammar.*
defense mechanisms Unconscious strategies used to avoid *anxiety*, resolve *conflict*, and enhance self-esteem. For examples, see *displacement, reaction formation, repression.*
defensive amnesia Forgetting which may be a way of protecting oneself from the guilt and anxiety resulting from intense, intolerable life situations and *conflicts;* a form of *motivated forgetting.* See *amnesia, defense mechanisms.*
defensive avoidance A reaction to messages appealing to fear;

avoidance of information put out by the communicator or refusal to accept the communicator's conclusions.

delirium tremens A *syndrome,* characteristically including tremors, delirium, convulsions, and *hallucinations,* produced by withdrawal from alcohol.

delusions Groundless, fixed ideas or misinterpretations of experience; characteristic of several *psychological disorders.*

dementia praecox An old term for *schizophrenia,* meaning "youthful insanity." It was so named because schizophrenic disorders tend to develop in adolescence and early adulthood.

dendrite A short, many-branched fiber of a *neuron* that receives information from other neurons. Compare *axon.*

denotational meaning The generally accepted meaning of words and *concepts;* "dictionary meaning." Compare *connotative meaning.*

dependence A "severe" form of a *substance-use disorder* in which physical addiction to the substance, *tolerance,* and *withdrawal symptoms* occur. Compare *abuse.*

dependent variable The *variable* whose value depends, or may depend, on the value of the *independent variable;* the *behavior* of a person or animal in an experiment. Compare *independent variable.*

depression An emotion characterized by "sadness," crying, withdrawal from others, and feelings of inadequacy. See *major depressive episodes.*

depth perception *Perception* of the relative distance of objects from the observer.

descriptive statistics Numbers used to describe a set of measures or the degree of relationship between sets of measures. See *mean, median, mode, standard deviation, correlation coefficient.* Compare *inferential statistics.*

despair Despondency; a fear in older people that time is running out before one can achieve a satisfying sense of wholeness and of life's meaning.

developmental perspective A current viewpoint in *psychology* concerned with characteristic changes in *behavior* that occur in people as they progress through the life span. See *developmental psychology.*

developmental psychology The branch of *psychology* which traces changes in *behavior* through the life span. See *child psychology.*

deviance *Behavior* or beliefs not in *conformity* with those of the group.

deviation IQ An *Intelligence Quotient* based on *standard scores,* so that IQs more nearly compare in meaning from one age to another. See *Intelligence Quotient (IQ).* Compare *ratio IQ.*

dichotic Presentation of different information to each ear. Compare *binaural.*

Differential Aptitude Tests (DAT) A *test battery* designed to give information about both *scholastic* and *vocational aptitudes.*

differential reinforcement (1) *Reinforcement* of a response to one stimulus but not to another: such reinforcement is used experimentally to establish a *discrimination.* See *discrimination.* (2) In *behavior therapy/behavior modification,* the *positive reinforcement* of desired responses and the *extinction* of undesired responses.

differential threshold The smallest difference that a person can perceive between two *stimuli* in the same *sensory channel.* Compare *absolute threshold.* See *threshold.*

directed thinking *Thinking* aimed at the solution of problems or the creation of something new. Compare *autistic thinking.*

discrimination (1) *Learning* in which an *organism* learns to make one response to one *stimulus* and a different response, or no response, to another stimulus. See *differential reinforcement.* (2) Treating a person or group in an unfavorable or unfair way. Compare *prejudice.*

discrimination learning *Learning* in which the subject learns to choose one *stimulus* and not another. Usually responses to one stimulus, the positive one, are *reinforced,* while responses to the other stimulus are *extinguished.* Or, stated another way, the subject learns to respond in the presence of a positive stimulus, or S^D, and not to respond in the presence of a negative stimulus, or S^Δ. See *simultaneous discrimination learning, successive discrimination learning, differential reinforcement.*

discriminative stimuli Events in the environment that signal that *reinforcement* will or will not be forthcoming when a particular response is made. See *discrimination learning,* S^D, S^Δ.

disorders In the *empirical approach to classification,* groups of specific behavioral problems that often occur together. See *syndrome.*

disorders of affect See *affective disorders.*

disorganized type A form of *schizophrenia* characterized by incoherent *thinking* but no systematic *delusions.* See *loosening of association, cognitive slippage and derailment.*

disorientation A loss of awareness of spatial, temporal, and social relationships.

displacement A *defense mechanism* in which a person copes with an anxiety-provoking *motive* by substituting another *goal* for the original one.

distinctiveness The extent to which the person being judged responds differently to different *stimuli* or situations; a factor important in making *attributions.* See *consensus information, consistency.*

divergent thinking A type of *thinking,* often used in creative thought, in which a wide variety of ideas or solutions come to mind. Compare *convergent thinking.*

dizygotic (DZ) twins See *fraternal twins.*

dopamine One of several *neurotransmitters* found in the *central nervous system; schizophrenia* may be related to disorders in nerve pathways which use dopamine as the neurotransmitter. See *serotonin, norepinephrine.*

dopamine hypothesis A biological *theory* of *schizophrenia* which says that an important causal factor in this disorder is excess activity of the *neurotransmitter dopamine* in certain areas of the brain.

dorsal root The *spinal root* toward the back which contains the sensory fibers. Compare *ventral root.*

dorsal root ganglion A cluster of nerve cell bodies that give rise to the *somatosensory* nerve fibers that carry information from the periphery of the body into the *spinal cord.*

double-bind communication The idea that a psychological cause of *schizophrenia* is to be found in the conflicting messages parents give their children.

double-depletion hypothesis The idea that *cellular dehydration* and *hypovolemia* contribute to thirst and drinking.

Down syndrome A mild to moderate form of *mental retardation* which is due to a *chromosomal* abnormality.

dream amnesia Poor *memory* of dreams. See *amnesia.*

dream analysis The analysis of dream content to obtain information about the source of a person's emotional problems; often used in *psychoanalysis.* See *wish fulfillment.*

drive theories "Push theories" of *motivation* which say that internal states within *organisms* push behavior toward *goals.* Compare *incentive theories and incentive motivation, opponent-process theory, optimal-level theories.*

drug A chemical substance which alters the structure or function of a living *organism.*

DSM-III The *Diagnostic and Statistical Manual of Mental Disorders,* third edition, of the American Psychiatric Association; the standard classification system for *psychological disorders* in the United States.

duplicity theory of vision Functional differences between *rods* and *cones.*

dyads Two-person social situations.

dynamic approaches Views of *personality* which emphasize or-going interactions among *motives*, impulses, and psychological processes; *unconscious processes* tend to be an important part of these viewpoints.

dyslexia A general term referring to difficulty in reading. See *angular gyrus.*

dysphasia A partial impairment in *language* ability; may be an impairment of comprehension (receptive dysphasia) or of speech production (productive dysphasia).

eardrum A thin membrane, also called the tympanic membrane, which separates the outer ear from the middle ear and which vibrates when sound waves reach it.

echolalia Repetition of what another speaker has just said, rather than use of speech for true *communication.*

educational psychology A field of specialization concerned with increasing the efficiency of learning in school through the application of psychological knowledge about *learning* and *motivation* to the curriculum. Compare *school psychologist, school counselor.*

EEG See *electroencephalogram.*

effectance motivation A general *motive* to act competently and effectively when interacting with the environment.

efferent Carrying information from the *central nervous system* to the organs and muscles. Compare *afferent.*

ego In *psychoanalytic theory,* a term referring to the self and to ways of behaving and thinking realistically. See *id, superego.*

egocentrism In Piaget's *theory* of *cognitive development,* the inability of children in the *preoperational stage* to adopt another person's point of view.

eidetic imagery Extremely detailed imagery; a "photographic memory." See *image.*

elaboration The degree to which incoming information is processed so that it can be tied to, or integrated with, existing memories. See *elaborative rehearsal.*

elaborative rehearsal Process of giving material organization and meaning as it is being rehearsed; an active *rehearsal* process. Compare *maintenance rehearsal.* See *rehearsal.*

elective mutism Deciding not to speak.

Electra complex In *psychoanalytic theory,* affectional responses by a girl toward her father, accompanied by jealousy of her mother. See *phallic stage.* Compare *Oedipus complex.*

electroconvulsive/electroshock therapy (ECT/EST) A form of *biomedical therapy* used principally with patients having a *major depressive episode;* consists of administering electric currents to the brain sufficient to produce convulsions and unconsciousness.

electroencephalogram (EEG) A record of the electrical activity of the brain, or "brain waves," obtained by placing electrodes on the scalp; provides indices of the depth of sleep and degree of *arousal* and an indication of some brain abnormalities.

electromagnetic spectrum The entire range of *wavelengths* of radiant energy. Compare *visible spectrum.*

electroshock therapy (EST) See *electroconvulsive/electroshock therapy (ECT/EST).*

emergency reaction, or "flight-or-fight" response The pattern of bodily changes accompanying fear and anger which help the *organism* deal with threatening situations. Compare *relaxation response.* See *sympathetic system.*

emotion A subjective feeling state which can influence *perception, thinking,* and *behavior;* usually accompanied by facial and bodily expressions; often has arousing (see *arousal*) and motivational (see *motivation*) properties.

emotional disturbance See *psychological disorder.*

empathy The ability to put one's self in another's place to understand the other person's views and feelings.

empirical Based on observation; a primary characteristic of *science.*

empirical approach See *empirical validity.*

empirical approach to classification The grouping of *psychological disorders* in terms of observable specific problems which cluster together. See *syndrome.* Contrast *clincial-consensual approach to classification.*

empirical validity *Validity* based on observations. Compare *face validity.*

empiricists Theorists who argue that *behavior* and perceptual processes (see *perception*) depend on *learning* and past experience. Compare *nativists.*

encoding processes The ways in which incoming sensory *information* is transformed into a form, or code, which can be stored in *memory.* See *storage processes.*

encounter group One name for groups which bring mentally healthy people together to enrich life. In general, these groups try to ease the lost and lonely feelings of modern life, to open up areas of thought and feeling previously blocked off, to improve mutual communication with other people, to provide an experience of trust and openness with others, and to produce conditions that will make the personality changes arising from these experiences long-lasting. See *group therapy.*

endocrine gland A ductless gland which secretes *hormones* into the bloodstream.

engram The hypothetical *memory* trace.

environmental forces Situational factors which result in *behavior.*

environmental frustration, environmental obstacles *Frustration* by something physical or by other people who prevent us from achieving our *goals.* Compare *personal frustration, unattainable goals.*

enzymes Catalysts for biological reactions. (1) Proteins necessary for the production of various substances that cells need if they are to live and grow. (2) Substances involved in the synthesis and destruction of *neurotransmitters.*

epinephrine (adrenalin) A *hormone* produced by the adrenal gland which duplicates and strengthens many of the actions of the *sympathetic system* on various bodily organs; also a *neurotransmitter.* Compare *norephinephrine.*

episodic memory A *long-term memory* store containing memories of the specific things that have happened to a person (reminiscences). Compare *semantic memory.*

epistemology The study of knowing or knowledge. See *genetic epistemology.*

equality rule A rule employed as a standard in making judgments of fairness in social relationships; requires that *outcomes* be distributed equally among participants in a relationship. Compare *contributions rule, needs rule.*

equilibration In Piaget's *theory* of *cognitive development,* the tendency of the developing individual to stay "in balance" intellectually by filling in gaps in knowledge and by restructuring *beliefs* when they fail to test out against reality.

erogenous zones Parts of the body which, when stimulated, give sexual pleasure.

escape learning *Learning* based on *negative reinforcement.* Compare *avoidance learning.*

esteem needs In Maslow's *theory, needs* for prestige, success, and self-respect; they can be fulfilled after *belongingness and love needs* are satisfied. See *safety needs.*

estradiol One of the most important *estrogens,* or female sex *hormones.*

estrogens The female *hormones* which are closely related to sexual behavior in many species of lower animals. See *estradiol.*

ethologist See *ethology.*

ethology The study of the *species-typical behavior* patterns of animals, with emphasis on the *evolution* of these patterns and thus their adaptive value.
evolution The *theory*, now generally accepted, that all species of plants and animals developed from earlier forms by hereditary transmission of slight variations through successive generations.
excitation Increases in a *neuron's* tendency to fire *nerve impulses*. Compare *inhibition*.
executive functions of self See *self*.
existential therapies *Psychotherapies* putting emphasis on taking responsibility for one's own actions and on developing a clear sense of what one is living for.
experimental controls Ways to hold constant those factors that could extraneously influence the outcome of an experiment.
experimental group The group in an experiment that receives the *independent variable*, but is otherwise equivalent to the *control group*. See *control groups*.
experimental methods Scientific methods in which the experimenter changes or varies the events which are hypothesized to have an effect, controls other variables likely to affect a result, and looks for an effect of the change or variation on the system under observation.
experimental psychology A field of *psychology* which seeks to learn more about the fundamental causes of *behavior* by investigating problems in the areas of sensation and *perception*, *learning* and *memory*, *motivation*, and the physiological bases of behavior.
extinction (1) In *classical conditioning*, the procedure of presenting the *conditioned stimulus* without the *unconditioned stimulus* to an *organism* previously conditioned; (2) in *instrumental conditioning/operant conditioning*, the procedure of omitting *reinforcement*; (3) the decreased likelihood of response resulting from these procedures.
extrasensory perception (ESP) The supposed ability of some people to gain knowledge about the world through avenues other than the *sensory channels*.
extraversion Sociable, adventurous, talkative, frank, and open *behavior* in dealing with others. Compare *introversion*.
extravert See *extraversion*.
extrinsic motivation *Motivation* directed toward *goals* external to the person. Compare *intrinsic motivation*.
face validity The appearance of *validity* in a test that "seems right"; face validity is not necessarily true validity. Compare *empirical validity*.
factor analysis A general statistical method, involving *correlation coefficients*, that isolates a few common features from a large number of *tests*, ratings, or other *measurements*.
factor theories Views of *intelligence* which emphasize the components which constitute it. See *G-factor theory*, *multifactor theories*. Compare *process-oriented theories of intelligence*.
failure to thrive A disorder of infancy characterized by apathy, a lack of normal social interest, and stunted growth despite seemingly adequate nutrition.
family resemblance structures The mental organization of word and *concept* meanings into classes, or "families," which share certain similar characteristics.
family therapy Therapeutic techniques which try to alter the disturbed behavior of a person in a family by changing the family's patterns of relationships and ways of interacting.
figure and ground *Perception* of objects or events as standing out clearly from a background.
filtering Blocking of some sensory inputs to allow processing of other sensory inputs; focusing on certain aspects and ignoring others; an important process in *attention*.
first mental-health revolution The trend toward humane treatment for mental patients, beginning in the late eighteenth century.
fissure A relatively deep crevice in the *cerebral cortex*. Compare *sulcus*, *gyrus*. See *longitudinal fissure*, *lateral fissure*.
fixation (1) A rigid habit developed by repeated *reinforcement* or as a consequence of *frustration*. (2) In *psychoanalytic theory*, failure of some personality characteristics to advance beyond a particular stage of *psychosexual development*.
fixed-action pattern (FAP) The *species-typical behavior* triggered by a *releaser*.
fixed-interval schedule (FI) A *schedule of reinforcement* in which a response made after a certain interval of time is reinforced; response rate is low after a reinforcement and increases steadily during the interval until the next reinforcement is given. See *reinforcement*.
fixed-ratio schedule (FR) A *schedule of reinforcement* in which every *n*th response is reinforced; except for pauses after each reinforcement, response rates tend to be quite high and relatively steady. See *reinforcement*.
flat affect A general impoverishment of emotional responsivity; characteristic of many *schizophrenic* patients.
flight of ideas Rushed, frenetic *thinking* characteristic of *manic episodes*.
flooding A type of *behavior therapy/behavior modification* which tries to eliminate fear by exposing the patient over and over again to the *stimuli* which arouse the fear; based on the principle of *extinction*. See *extinction*.
fluid intelligence A general relation-perceiving capacity which represents one's potential *intelligence* somewhat independent of *socialization* and education. Compare *crystalized intelligence*.
focal person (FP) The individual who is being studied in *social psychology* experiments.
forebrain The *cerebrum*, *thalamus*, and *hypothalamus*. Compare *brain stem*.
forgetting Apparent loss of *information* that has been stored in *long-term memory (LTM)*. Compare *memory*.
formal operational stage In Piaget's *theory* of *cognitive development*, the period, beginning about age 11, in which abstract logical thought is possible. See *formal operations*.
formal operations In Piaget's *theory* of *cognitive development*, mental *operations* marked by hypothetico-deductive thinking and the ability to deal with abstract ideas.
fovea The central region of the *retina* where *cones* are closely packed together and *visual acuity* is at its sharpest.
fraternal twins Twins who develop from two different fertilized eggs, and who consequently are as different in hereditary characteristics as ordinary brothers and sisters. Also called dizygotic (DZ) twins. Compare *identical twins*.
free association The technique of requiring a patient in *psychotherapy* to say whatever comes to her or his mind, regardless of how irrelevant or objectionable it may seem. More generally, *thinking* in which a person allows thoughts to drift without direction.
free nerve endings Nerve endings not associated with any special *receptor* structures; sense organs for pain, touch, and temperature.
free recall In *memory* experiments, retrieval of stored items in any order by the subjects. See *retrieval*.
frequency A dimension of vibrational stimuli; the number of cycles in a given period of time. In hearing, related to *pitch*. See *hertz (Hz)*.
frequency distribution A set of *measurements* arranged from lowest to highest (or vice versa) and accompanied by a count of the number of times each measurement or class of measurements occurs.

frontal association cortex *Cerebral cortex* of the *frontal lobes*; damage may lead to an impairment in planning ability.
frontal lobe The lobe of the *cerebrum* which lies in front of the *central sulcus*.
frustration Blocking of goal-directed *behavior*. See *environmental frustration, personal frustration, motivational conflict*.
frustration-aggression hypothesis The idea that motive *frustration* is a major cause of *aggression*.
full scale IQ An individual's combined score from all the subtests of an intelligence scale devised by Wechsler. See *verbal IQ, performance IQ, deviation IQ*.
functional analysis of behavior The breakdown of a particular sequence of *behavior* into (1) its antecedents, (2) a description of the behavior itself, and (3) the consequences of the *behavior*. The antecedents are the *discriminative stimuli* that give cues to the person that the behavior will be *reinforced* when it occurs; the consequences are the *reinforcements* for the *behavior*.
functional fixedness A *set* to use objects in the way we are accustomed to use them, even if a different use might solve a problem. See *set*.
functionalism A school of *psychology* emphasizing the study of how mind and *behavior* enable an individual to adapt to a changing environment. Compare *structuralism, Gestalt psychology, behaviorism*.
fundamental frequency The lowest *frequency* in a *periodic sound wave*. Compare *harmonic frequency*.
galvanic skin response (GSR) A decrease in the electrical resistance of the skin; an indicator of *arousal*.
gambler's fallacy Believing that odds for success are better after previous failures; the logic of probability says that if each event is independent, the odds are the same despite previous failures.
ganglion (pl. ganglia) A cluster of *neuron* cell bodies outside the central *nervous system*, such as a *dorsal root ganglion*. Compare *nuclei (sing. nucleus)*.
ganglion cells *Neurons* in the *retina* whose *axons* form the *optic* nerves; they generate *nerve impulses* which carry visual *information* to the brain.
general adaptation syndrome A term coined by Selye to refer to the stages in the body's response to *stressors;* consists of the *alarm reaction*, the *stage of resistance*, and the *stage of exhaustion*.
generalized anxiety disorder A disorder with symptoms that include persistent, vague, unfocused feelings of distress and uneasiness; physical symptoms such as trembling, fatigue, breathlessness, insomnia, sweating, nervousness, chest pains, faintness, dizziness, and headache are typical. See *anxiety disorders*. Contrast *panic disorder*.
generator potential The electrical event which triggers *nerve impulses*.
genes The essential elements in the transmission of hereditary characteristics; parts of *chromosomes*.
genetic epistemology Development of ways of knowing about the world; a description of the focus of the work of Piaget. See *espistemology*.
genital stage In *psychoanalytic theory*, the adult stage of personality. It begins in *adolescence* and is characterized by the expression of heterosexual interests. Compare *oral stage, anal stage, phallic stage, latency period*.
gentotype The genetic constitution of an *organism*. Compare *phenotype*.
Gestalt psychology A school of *psychology* emphasizing that immediate experience results from the whole pattern of sensory activity and the relationships and organizations within this pattern. Compare *structuralism, functionalism, behaviorism*.
gestalt therapy A *psychotherapy* developed by Perls that attempts to restore a person's sense of wholeness until the individual becomes strong enough for growth to take place. It strives to do this by developing an individual's awareness of *self*.
G-factor theory The *theory* that *intelligence* is composed of a single, unitary, or general (G) factor. Compare *multifactor theories*.
goal The place, condition, or object that satisfies a *motive*. *See positive goal, negative goal*.
good figure See *symmetry, good figure*.
gradient of generalization The amount of *stimulus generalization* depends on how similar the test *stimuli* are to the stimuli present during *learning*. See *stimulus generalization*.
gradients of texture One of the principal *monocular cues* for *depth perception;* consists of a gradation in the fineness of detail which can be seen at increasing distances from the viewer.
grammar, or syntax A set of rules for constructing sentences from words and *phrases*. See *transformational grammar*.
gray matter Collection of *neurons;* the *cerebral cortex*, for instance. Compare *white matter*.
group norms Standards of *behavior* or thought expected of group members; a person in a group must follow the norms set by the group or suffer the social consequences. See *conformity*.
group therapy Specialized techniques of *psychotherapy*, consisting of a group of people discussing their personal problems under the guidance of a therapist. See *encounter group*.
"groupthink" The *conformity* of opinion that arises under certain conditions in decision-making groups; often due to the reluctance of some members of the group to voice criticism.
growth spurt The period of rapid physiological growth which begins in late childhood and continues into *adolescence*.
gyrus (pl. gyri) A ridge in the *cerebral cortex* of the brain. Compare *sulcus*.
hair cells *Receptor* cells for hearing located in the *organ of Corti*.
hallucination Sensory experience in the absence of stimulation of *receptors;* sometimes present in certain *psychological disorders* such as *schizophrenia*.
harmonic frequency Components of *complex waves* that are multiples of the *fundamental frequency*. See *timbre*.
health psychology See *behavioral medicine*.
hedonistic view of motivation A view which holds that we are motivated to seek *goals* which give us pleasure and avoid those resulting in displeasure.
herz (Hz) Cycles per second; number of cyclical alternations per second.
heuristic A strategy based on past experience with problems; it is likely to lead to a solution, but does not guarantee success. Compare *algorithm*.
hierarchical theory A view of *intelligence* which incorporates ideas from both *G-factor theory* and *multifactor theories*.
hierarchical thinking A characteristic of thought which appears during development in the stage of *concrete operations;* a given "thing" can be thought of as falling somewhere on more than one dimension at a time.
hippocampus A deep structure of the *cerebrum;* part of the *limbic system;* involved in *memory* formation.
histrionic personality disorder A *personality disorder* characterized by immaturity, self-dramatization, seductiveness, and attention seeking.
holophrase A single word used as a sentence.
homeostasis The tendency of the body to maintain many of its internal physiological processes at certain optimal levels.
hormone A secretion of a specific organ, often an *endocrine gland*

into the bloodstream, where it is carried to various organs of the body to have an effect; a "chemical messenger."

hostile aggression *Behavior* aimed at harming or injuring another *organism* who is motivated to avoid such treatment. Compare *instrumental aggression*.

hue Color; largely determined by *wavelength*. Compare *brightness*, *saturation*.

humanistic and existential therapies Treatments of *psychological disorders* designed to help people by encouraging their self-awareness, self-acceptance, and personal responsibility for how they live their lives.

humanistic approaches Views of *personality* which emphasize the *self* and the importance of the individual's subjective view of the world.

humanistic perspective A current viewpoint in *psychology* which emphasizes the person's "sense of self." See *humanistic psychology*.

humanistic psychology The approach to the study of human beings that emphasizes the whole person and the internal, integrative constituents of a person's total self—*motives*, intentions, feelings, and so on.

humanistic theories See *humanistic approaches, humanistic perspective, humanistic psychology*.

hypnosis A trancelike state in which a person is very susceptible to *suggestions*.

hypnotherapy The use of *hypnosis* as an aid in therapy; especially useful in the temporary alleviation of certain symptoms and in the temporary lifting of *repression*.

hypothalamus A region of the *forebrain* which plays an important role in motivated behaviors of a biological nature; also plays a role in *emotion*. See *motivation*.

hypovolemia Decreased volume of blood plasma; one of the conditions leading to thirst. Compare *cellular-dehydration thirst*. See *double-depletion hypothesis*.

iconic image A faint copy of the visual input which persists in the visual *sensory register* for a few seconds before it gradually decays.

id In *psychoanalytic theory*, the aspect of *personality* concerned with primitive reactions. The id contains the biological instincts and seeks immediate gratification of *motives* with little regard for the consequences or the realities of life. Contrast *ego*, *superego*.

ideal self In Rogers' *self theory*, the ways in which a person would like to be regarded by others; the *self* a person would like to be.

identical twins Twins who develop from the same fertilized egg. They have exactly the same *chromosomes* and *genes* and hence the same hereditary characteristics. Also called monozygotic (MZ) twins. Compare *fraternal twins*.

identification (1) Generally, the tendency of children to model their behavior after that of appropriate adults. (2) A *defense mechanism* in which one takes in, or incorporates, aspects of someone else's behavior; in *psychoanalytic theory* it originates in the *phallic stage* and is the basis for *superego* development.

identity concept In *cognitive development*, the realization that characteristics of an object remain constant even when the appearance of the object changes or the object is hidden from view.

idiographic approach The psychological study of the single case. Compare *nomothetic, or dimensional, approach*.

illusions *Perceptions* that do not agree with other, more trustworthy, perceptions.

images Partial and altered representations of sensory experience.

imitation Copying the *behavior* of another; a response like the *stimulus* triggering the response. Also, *learning* to copy behavior, or *modeling*.

imitative learning Learning by observing the *behavior* of others. See *modeling*.

immunization Hardening of a person's *attitude* on a particular subject by giving him or her a mild exposure to an opposing attitude; exposure makes the originally held attitude resistant to change by strong further arguments.

implicit personality theory The assumed relationships among *traits;* each person has some ideas about which traits are usually related to certain other traits.

implosive therapy, flooding A form of *behavior therapy/behavior modification* which tries to eliminate unwanted emotional problems by using an *extinction* technique derived from *classical conditioning*. See *extinction, classical conditioning*.

impression formation The process of making a judgment about what a person is like.

incentive theories and incentive motivation "Pull theories" of *motivation* which say that certain qualities of goal objects pull *behavior* toward them; motivation based on the attraction of *goals*. Compare *drive theories, opponent-process theory, optimal-level theories*.

incentives The goal objects which *motivate behavior* and "pull" behavior toward them. See *incentive theories and incentive motivation*.

incongruences In Rogers' *self theory*, mismatches between the *self* as perceived and the *ideal self* the person would like to be.

independent variable A condition selected or manipulated by an experimenter to see whether it will have an effect on *behavior*. Compare *dependent variable*.

individual psychology Alfred Adler's *theory* of *personality*.

individual therapy Alfred Adler's approach to *psychotherapy* which places special emphasis on social and interpersonal factors.

individuation In Jung's *analytical psychology*, the processes by which each of us becomes an individual distinct from others. In Jung's words, "becoming a single, homogeneous being . . . coming to selfhood or self-realization."

induced movement *Apparent motion* of a stationary spot perceived when the background of the spot moves. The moon "racing" through the clouds is an example.

inductive reasoning The logical process by which general principles are inferred from particular instances.

inductive thinking See *inductive reasoning*.

industrial and organizational psychology A field of specialization concerned with the application of psychological principles to practical problems in business and industry. See *personnel psychologist*.

infancy The period of development between the *neonatal period* and the appearance of useful *language;* the upper limit is about 18 months.

infantile autism A *psychological disorder* of children characterized by failure to show landmark features of infancy, deficiencies of social responsiveness, retarded language development, and bizarre *behavior* patterns.

inferential statistics Statistical methods for finding the probability that results are due to chance sampling factors.

inferior colliculi Large relay *nuclei* of the auditory system, located at the back of the *midbrain*. Compare *superior colliculi*.

inferiority complex A concept put forth by Adler; a feeling developed out of *frustration* in striving for superiority.

information As used in *information-processing theories*, refers to sensory input from the environment.

information processing See *information-processing theory*.

information-processing theories of attention Viewpoints which stress the *filtering* of sensory inputs and the switching from one sensory channel to another in *attention*.

information-processing theories of intelligence *Theories* holding that *intelligence* should be measured in terms of such functions as sensory processing, coding strategies, *memory*, and other mental capacities. See *information-processing theory*.

information-processing theories of memory Models of *memory* which say that human beings process *information* for storage in stages. See *sensory register, short-term memory (STM), long-term memory (LTM), information-processing theory*.

information-processing theory A view of *cognitive* activity stressing the input of *information*, its transformation, its storage, and its eventual output.

inhibition Decrease in a *neuron's* tendency to fire *nerve impulses*.

insight (1) In *learning* and problem solving, the relatively sudden solution of a problem. See *insight learning*. (2) In *psychotherapy*, the understanding of one's own *motives* and their origins.

insight in creative thinking See *insight*.

insight learning *Learning* which is said to involve *perceptual reorganization;* the solution comes suddenly after a period during which little progress is made. See *cognitive learning*.

instinct *Behaviors* resulting from genetic factors; because of the role of environmental factors in the development and modification of these behaviors, the term *species-typical behavior* is preferred.

instrumental aggression *Aggression* used as a way of satisfying other, nonaggressive, motives. Compare *hostile aggression*.

instrumental conditioning *Learning* in which *reinforcement* is contingent on a particular response. Compare *classical conditioning*.

instrumental, or operant, conditioning techniques Techniques used in *behavior therapy/behavior modification* which involve *reinforcement* contingent on particular responses. See *instrumental conditioning/operant conditioning*.

insula The *cerebral cortex* in the depth of the *lateral fissure*.

integrity The individual's sustaining sense of wholeness and adequacy throughout the aging period.

intellectualization A *defense mechanism* in which a person reduces *anxiety* by thinking of the anxiety-producing situation in unemotional or abstract terms.

intelligence A general term referring to the overall capacity for learning and problem-solving; as actually administered, *tests* of intelligence measure a mixture of *abilities*.

Intelligence Quotient (IQ) The score obtained on an intelligence test. Classically, the ratio IQ is a number obtained by dividing *chronological age* into *mental age* and multiplying by 100. Now, other methods are used to compute the Intelligence Quotient. See *deviation IQ*.

intensity In audition, the amplitude of the pressure wave; related to our experience of loudness. See *decibel (dB)*.

intention movements *Species-typical behaviors*, often emotional expressions, which provide information about behavior that may ensue.

interaction position A viewpoint in the debate over whether *behavior* results from *personality* characteristics or situational characteristics; holds that behavior depends on the ways particular personality characteristics interact with particular situations. See *situationism*.

internal capsule The large fiber *tract* in the *forebrain* formed by *axons* from the *motor areas of the cortex* and sensory fibers entering the *forebrain*.

internal environment Conditions inside the body, especially the physical state and the chemical composition of the fluids which bathe body cells.

internal representation A mental *symbol* for some aspect of the world.

internalized obstacles Inner negative *valences*, usually resulting from the training in social values which a person has received; can cause *motivational conflict*.

interpersonal attraction The degree to which people are drawn toward each other; influenced by *proximity, attitude similarity, physical attractiveness*.

interposition A *monocular cue* for depth; near objects block off portions of faraway objects.

interpropositional logic The capacity to judge whether or not statements (propositions) are logically connected to one another, regardless of whether the statements are true; develops in the *formal operational stage*.

interstimulus interval In *classical conditioning*, the time between the onset of the *conditioned stimulus* and the onset of the *unconditioned stimulus*.

intrinsic motivation A person's need to feel competent and self-determining in dealing with the environment. See *effectance motivation*. Compare *extrinsic motivation*.

introspection A method of psychological experimentation in which a subject is asked to describe his or her own mental reactions to a *stimulus; perception* of one's inner feelings.

introversion *Behavior* that is withdrawn and reclusive; an introverted person is often cautious and secretive in dealings with others. Compare *extraversion*.

introvert See *introversion*.

intuitive substage In Piaget's *theory* of *cognitive development*, the second of two substages of the *preoperational stage*, from roughly 4 to 7 years of age; it is characterized by unsystematic reasoning based on perceptual appearances. Compare *preconceptual substage*.

IQ See *Intelligence Quotient*.

James-Lange theory A *theory* of the relationship between subjectively felt *emotions* and bodily changes; it states that the emotions a person feels are due to her or his *perception* of bodily reactions to *stimuli*. Compare *Cannon-Bard theory, Schachter-Singer theory*.

kinesthesis The sense informing us about the position of the limbs and the state of tension in the muscles. See *proprioceptive sense, proprioception*.

Korsakoff syndrome A set of *behavior* disturbances due to heavy alcohol use; prominent among the disturbances is *anterograde amnesia*. See *syndrome*.

language *Communication* in which word *symbols* are used in various combinations to convey meaning. See *linguistic competence*.

latency of response The time between the presentation of a *stimulus* (or the beginning of a learning trial) and a response.

latency period In *psychoanalytic theory*, the period from approximately 6 years of age to *puberty;* the middle childhood years. Characterized by the elaboration of *defense mechanisms* and *ego* growth.

latent learning *Learning* that becomes evident only when the occasion for using it arises. See *cognitive learning*.

lateral fissure A deep cleft in the *cerebral cortex* dividing the *temporal lobe* from the *frontal* and *parietal lobes*.

lateral geniculate body The portion of the *thalamus* receiving input from the eyes; here *synapses* are made, and the fibers leaving the lateral geniculate body go to the visual *cortex* in the *occipital lobe* of the brain.

lateral hypothalamus (LH) Classically considered to be the excitatory brain area for hunger. Compare *ventromedial hypothalamus*.

law of closure See *closure*.

law of complementary colors For every *hue* there is a complementary hue, and complementary hues, when mixed in appropriate proportions, produce gray or white.

law of good figure See *symmetry, good figure.*

law of proximity See *proximity, or nearness.*

learned drives Motivational states originating in the training or past experience of an *organism* which can "push" *behavior* toward certain *goals.* See *drive theories.*

learned flavor aversion The association of a food taste with illness; leads to avoidance of the food taste; can occur after one pairing of the food taste with illness; an example of a *prepared behavior.*

learned helplessness The view that the "giving up" characteristic of *depression* is a learned (see *learning*) response to unmodifiable stressful situations.

learning A general term referring to a relatively permanent change in *behavior* which occurs as a result of practice or experience. It includes *classical conditioning, instrumental conditioning/operant conditioning,* and *cognitive learning.*

learning and behavioral approaches Views of *personality* which emphasize that many of the personality characteristics of people are the result of *learning;* personality is described in terms of observable *behavior* patterns.

learning disabilities, specific development disorders Terms that cover a wide variety of specific learning difficulties in children; they refer to specific problems in certain areas of schoolwork in children whose performance in other areas is average or above average.

lens The adjustable refractive element of the eye. See *accommodation, refraction.*

level of aspiration The level at which a person sets certain *goals.*

levels-of-processing theory A view of *memory* according to which incoming *information* can be worked on at different levels of analysis; the deeper the analysis goes, the better the memory.

libido Freud's term for the instinctive drives, or energies, that *motivate* behavior; the sexual energy underlying biological urges. See *id.*

lie detector A popular name for a device measuring bodily indicators of the *arousal* presumed to accompany lying; also known as a polygraph.

life-span developmental psychologists Researchers who study the development of *behavior* at every point through the course of life.

limbic system A group of structures forming a ring around the lower portion of the *forebrain;* concerned with *emotion, motivation,* and *memory.*

linear perspective A *monocular cue* for depth; faraway objects are perceived as relatively close together, while nearby objects are perceived as relatively far apart.

linguistic competence Ability that consists of knowledge of the sounds of *language,* the *grammar* rules of language, the meaning of words, the ways to use speech to have an intended impact on others, and the rules for interpreting the speech of others.

linguistic relativity hypothesis The view that the particular *language* people use determines how they conceive of the world.

linguistics The study of *languages* as structured systems of rules; includes the study of the origin of language, the relationships among languages, how languages change over time, and the nature of language sounds.

lithium carbonate A drug used in the treatment of *bipolar disorder,* especially the *manic episodes.*

locus of control An individual's view of whether she or he is the cause of events (internal locus of control) or whether environmental forces cause events (external locus of control).

longitudinal fissure The midline crevice which divides the *cerebrum* into two symmetrical halves.

longitudinal method Study of an individual or process either continuously or at selected points in the course of development. Compare *cross-sectional method.*

long-term memory (LTM) The relatively permanent *memory* store of information which is categorized in various ways and can be drawn upon as needed. See *retrieval, episodic memory, semantic memory.*

loosening of associations A disturbance of *thinking* in which the progression of ideas is disjointed; often characteristic of the thinking of people with *schizophrenia.* See *cognitive slippage and derailment.*

Machiavellianism A term that describes people who express the *need for power* by manipulating and exploiting others in a deceptive and unscrupulous fashion.

maintenance rehearsal Going over and over what is to be remembered; does not necessarily lead to *long-term memory.* Compare *elaborative rehearsal.* See *rehearsal.*

major depressive episodes *Affective disorders* characterized by "sad" feelings and persistent problems in other areas of life. Compare *manic episodes.*

manic-depressive disorder See *bipolar disorder.*

manic episodes *Affective disorders* marked by extreme elation and activity. Compare *major depressive episodes.*

markers Cues in the flow of speech which help us make sense of what we hear.

matching of subjects A technique used in *control-group designs* to equate *experimental* and *control groups* on extraneous factors which could affect the *dependent variable.*

maturation Built-in biological developmental growth processes.

mean A *measure of central tendency* obtained by dividing the sum of the measures, or scores, by the number of them; the "average." Compare *median, mode.*

means-end analysis In problem solving, a common *heuristic* of breaking the problem down into smaller subproblems, each of which is a little closer to the end goal.

measurement The assignment of numbers to objects or events according to certain rules.

measures of central tendency *Descriptive statistics* describing the middle point of a distribution of measures. See *mean, median, mode.*

measures of variability *Descriptive statistics* describing the spread of the measures around the midpoint of a distribution. See *range, standard deviation (SD).*

mechanical ability tests *Vocational aptitude tests* for predicting success in jobs requiring mechanical skills.

medial geniculate body The auditory relay nucleus of the *thalamus.*

median A *measure of central tendency;* the point in a group of scores above and below which half the scores fall. Compare *mean, mode.*

mediation in thinking The *cognitive processes* which go between *stimuli* and responses. See *thinking.*

medical model The idea that *psychological disorders* are specific illnesses with characteristic symptoms and predictable outcomes; the view that clusters of symptoms form *syndromes* that are caused by underlying specific illnesses.

medulla The lowest part of the *brain stem,* located just above the *spinal cord;* contains *nuclei* vital for the regulation of heart rate, blood pressure, and other bodily functions; important for communication between higher parts of the *brain* and the *spinal cord.*

Meissner corpuscle A specialized structure in the skin regarded as a sense organ for pressure or touch.

memory Storage of information from past experience; closely related to *learning.*

memory consolidation Strengthening of *short-term memory (STM)* so that it becomes a part of *long-term memory (LTM).* See *rehearsal.*

menarche The first menstrual period.

meninges The three membranes that envelop the *brain* and the *spinal cord:* dura mater, arachnoid membrane, and pia mater.

menopause The cessation of menstruation, usually occurring between 45 and 55.

mental age (MA) A type of score expressing mental development in terms of the age level at which a child is performing. For example, if a 5-year-old boy does as well on an intelligence test as the average child of 7, his mental age is 7. See *Intelligence Quotient (IQ).* Compare *chronological age.*

mental deficiency Now known as *mental retardation.*

mental disorder See *psychological disorder.*

mental illness See *psychological disorder.*

mental retardation A condition marked by a deficiency in general intellectual *abilities* and inadequate coping skills; usually the IQ is below 70. In degree, may be mild, moderate, severe, or profound. See *Intelligence Quotient (IQ).*

method of loci A *mnemonic* technique; an example is imagining a place such as a building or room and then associating ideas to be remembered with parts of the building or items of furniture in the room.

method of successive approximations See *shaping.*

method of systematic observation An alternative to the *experimental method;* researchers do not willfully manipulate the *independent variable* but instead make the most exacting and systematic study they can of naturally occurring *behavior.*

midbrain The upper part of the *brain stem* above the *pons;* important in visual and auditory *reflexes.* See *inferior colliculi, superior colliculi.*

midlife transition A period of life from the late 30s to the early 40s characterized by reappraisal of one's life as it is being lived and a decision about whether to shift to a new career or lifestyle.

Minnesota Multiphasic Personality Inventory (MMPI) A widely used *pencil-and-paper test* of personality; an important feature is its *empirical validity.*

mnemonics Techniques for improving *memory.* See *method of loci, number and letter peg systems.*

mode A *measure of central tendency;* the score in a group that occurs most often. Compare *mean, median.*

modeling In general, learning to copy behavior. Specifically, a *behavior therapy/behavior modification* technique which depends on such copying. See *imitation.*

monocular cues Information for the *perception* of depth that can be obtained by one eye. Compare *binocular cues.* See *linear perspective, clearness in depth perception, interposition, shadows in depth perception, gradients of texture, movement in depth perception.*

monozygotic (MZ) twins See *identical twins.*

mood The emotional background that is relatively long- lasting and colors an individual's outlook on the world.

moro reflex A pattern of responses in the *neonate;* when support is suddenly removed from the back of a neonate's head, it will fling its arms to the side, extend its fingers, and bring its arms inward in a sort of embracing movement.

morphemes The smallest units of meaning in speech perception; can be prefixes, words, or suffixes; composed of *syllables.*

motivated forgetting Difficulty with the *retrieval* of unpleasant, anxiety-provoking information stored in *long-term memory; repression* is an example. See *forgetting.*

motivation A term referring to the driving and pulling forces which result in persistent *behavior* directed toward certain *goals.*

motivational conflict A situation that arises when two or more motives drive behavior toward incompatible *goals;* an important source of *frustration.* See *approach-approach conflict, avoidance-avoidance conflict, approach-avoidance conflict, multiple approach-avoidance conflict.*

motivational cycle A cycle including arousal of the *motive,* goal-directed *behavior,* and satisfaction. See *drive theories.*

motivational theories of emotion *Theories* emphasizing the relationship of *emotion* to *motivation;* for example, the view that emotions are best considered as motives which keep behavior going and the view that emotions amplify motives to give them their energy.

motive See *motivation.*

motoneuron A type of *neuron* involved in movement of the body. See *reflex.*

motor areas of cortex Areas of the *cerebral cortex* largely concerned with bodily movements. Compare *sensory areas of cortex, association areas.*

mouthing A complex set of reflexes in the *neonate* and infant which involve the mouth and lips.

movement in depth perception A *monocular cue* for depth; objects farther than the fixation point of vision seem to move in the same direction as a head movement; objects closer than the fixation point move opposite to the direction of head movement.

moving toward, against, and away from others In Horney's *theory* of *personality,* modes of social behavior stemming from *basic anxiety and basic hostility.*

multifactor theories Views of *intelligence* which say it is composed of many components, or factors. Compare *G-factor theory.*

multiple approach-avoidance conflict A *motivational conflict* in which several incompatible *positive* and *negative goals* are involved; characteristic of many of life's major decisions.

myelin sheath A white, fatty covering which, in many cases, surrounds the *axon,* but not the cell body or *dendrites;* increases the speed with which *nerve impulses* are conducted.

nanometer (nm) A billionth of a meter; 10^{-9} meters.

nativists Theorists arguing for the importance of *nature* in *perception.* Compare *empiricists.*

nature The genetic factors contributing to *behavior* and *perception.* Compare *nurture.*

nature-nurture controversy The argument concerning the relative roles of the contributions of *nature* and *nurture* in the development of *organisms;* an enduring question in psychology; most psychologists now favor an interaction of nature and nurture. See *nature-nurture interaction.*

nature-nurture interaction The interplay of the genetic inheritance of an individual and environmental influences to produce the characteristics actually observed. See *phenotype, reaction range.*

need (1) Any lack or deficit within an individual, either acquired or biological. (2) Sometimes used to refer to the driving state, especially when human *social motives* are under discussion.

need for achievement, (*n* ach) A *need* to succeed and to strive against standards of excellence; it serves to motivate (see *motivation*) an individual to do well.

need for power The *need* to influence the *behavior* of others.

need to affiliate The *need* to associate with other people.
needs rule A rule employed as a standard in making judgments of fairness in social relationships; based on relative needs of individuals. Compare *contributions rule, equality rule.* See *norm of social responsibility.*
negative afterimages Dark *images* which persist after a light object is viewed, and light images that persist after a dark object is viewed. Also, complementary colors perceived after a color is viewed; green after red, for example.
negative goal *Goal* which an individual tries to escape from or avoid. Compare *positive goal.*
negative halo Formation of unfavorable opinions from a few negative characteristics. Compare *positive halo effect.*
negative identity Erik Erikson's term for the rebellious *behavior* of adolescents when they do the opposite of what parents and others consider proper and desirable.
negative reinforcement See *negative reinforcer.*
negative reinforcer A *stimulus* or event which, when its termination is made contingent on a particular response, increases the likelihood of the response. Compare *positive reinforcer, punishment.*
negative transfer See *transfer.*
negative transference *Transference* marked by a hostile attitude toward the therapist. See *transference.*
neoanalyst A psychoanalytically oriented theorist who places more emphasis on social factors and less emphasis on sexuality. See *psychoanalysis.*
neoanalytic theories of personality See *neoanalyst.*
neonate The newborn from birth to 28 days of age.
nerve impulses Electrical events of short duration which move along the *axon.*
neurobiology The *science* of the nervous system; includes neuroanatomy, neurophysiology, neurochemistry, neuropharmacology, neuroembryology, physiological psychology, and other disciplines concerned with the structure, function, and development of the nervous system.
neuromuscular junction The junction between nerve and muscle fibers. Compare *synapse.*
neurons Nerve cells; a typical neuron has a cell body, *dendrites,* and an *axon.*
neurosis A *psychological disorder,* less severe than a *psychosis,* in which a person is unusually anxious, miserable, troubled, or incapacitated in his or her work and relations with other people. The person often attempts to ward off *anxiety* by using exaggerated *defense mechanisms.* This term is not used in *DSM-III.*
neurotransmitter A chemical substance stored in *vesicles* and released into *synaptic clefts* or *neuromuscular junctions* to excite or inhibit *neurons* or muscle fibers. See *dopamine, epinephrine, norepinephrine, serotonin.*
nomothetic, or dimensional, approach Attempts to discover *personality* principles that apply to people in general; concerned with finding general principles which can be applied in *psychology.* Compare *idiographic approach.*
noncommon effects Effects not common to both chosen and unchosen actions; in making *attributions,* perceivers gain more information from knowledge of the noncommon effects than from knowledge of common effects.
nondirective technique See *nondirective therapy.*
nondirective therapy *Psychotherapy* in which the client is dominant and given the greatest possible opportunity for self-expression. The method is based on the principle that the client must learn how to solve his or her own problems; the therapist cannot solve them. See *client-centered therapy.*
nonopponent cell A *neuron* in the visual system that is excited by *wavelengths* over the whole *visible spectrum.* Compare *opponent cell.*
nonverbal information See *body language.*
norepinephrine (noradrenalin) A *hormone* produced by the adrenal gland; its major effect is to constrict peripheral blood vessels and thus to raise blood pressure. Also a *neurotransmitter;* abnormalities in norepinephrine pathways may occur in *affective disorders.* Compare *epinephrine.*
norm of social responsibility A concept related to the *needs rule;* a fair *outcome* is one which meets people's legitimate needs to avoid hardship and suffering.
normal curve A bell-shaped *frequency distribution,* also called the normal probability curve, which is an ideal approximated by many distributions obtained in psychology and the biological sciences.
norms (1) Standards obtained from measurements made on selected groups of people; an individual's scores on a test are compared with these standards. (2) Standards of *behavior* agreed upon by group members which exert a powerful influence on social behavior.
noxious Perceived as painful or unpleasant.
noxious stimulus (pl. stimuli) A *stimulus* that makes an individual feel uncomfortable or fearful. See *punishment, aversion therapy.*
nuclei (sing. nucleus) Clusters of nerve cell bodies; especially such clusters in the *central nervous system; gray matter.* Compare *ganglion (pl. ganglia).*
nucleus See *nuclei.*
number and letter peg systems *Mnemonic* techniques in which to-be-remembered items are linked with a well-learned set of numbers or letters. Compare *method of loci.*
nurture Environmental factors contributing to *behavior* and *perception.* Compare *nature.*
obedience Generally, doing what another person commands us to do; the term is often used in situations where a person has the legitimate right to influence another and this person has an obligation to obey.
object permanence The child's realization that an object remains the same even though it may undergo various transformations.
observational learning See *modeling, imitation.*
obsessions Seemingly groundless ideas that constantly intrude into a person's thoughts. Compare *compulsions.* See *obsessive-compulsive disorder.*
obsessive-compulsive disorder A *psychological disorder* characterized by *obsessions* and/or *compulsions.*
occipital lobe The part of the *cerebral cortex* lying at the back of the head; contains the *primary sensory areas* for vision.
Oedipus complex In *psychoanalytic theory,* affectional responses by boys towards their mothers. See *phallic stage.* Compare *Electra complex.*
olfactory bulb The organ that receives input from smell *receptors.*
omission of reinforcement/omission training An *instrumental conditioning/operant conditioning* procedure in which *positive reinforcement* is withdrawn following a response. The effect of this procedure is to decrease the likelihood of the response which leads to removal of positive reinforcement. Compare *punishment, negative reinforcement.*
operant chamber A simple box, often called a "Skinner box," with a device which can be worked by an animal in the box to control *reinforcement.*
operant conditioning See *instrumental conditioning/operant conditioning.*

operational definition A method of defining terms and concepts by means of the observable operations performed to measure them.

operations Flexible mental actions that can be combined with one another to solve problems.

opiates Morphine and morphine-like compounds used to relieve pain; these are also *psychoactive drugs* which can produce pleasant, dreamy states (heroin is an example) and are liable to be abused. See *abuse, tolerance, substance-use disorders.*

opponent cell A *neuron* in the visual system that is excited by *wavelengths* in one part of the *visible spectrum* and inhibited by wavelengths in another part. Compare *nonopponent cell.* See *opponent-process mechanism.*

opponent-process mechanism *Neurons* in the visual system are excited by *wavelengths* in one part of the *visible spectrum* and inhibited by wavelengths in another part; provides part of the *afferent code* for color vision. Compare *nonopponent cell.*

opponent-process theory A *hedonistic view of motivation* and *emotion* which says that many emotional-motivating states are followed by an opposing, or opposite, state. Compare *drive theories, incentive theories and incentive motivation, optimal-level theories.*

optic chiasm A structure composed of the crossed fibers of the *optic nerves.* The fibers (*axons*) from the *ganglion cells* in the nasal halves of the *retinas* cross in the chiasm and run to the opposite side of the brain.

optic nerves The *axons* of the *ganglion cells*; they carry visual information into the brain.

optic radiations The *axons* of cells of the *lateral geniculate body* which project to the *cerebral cortex.*

optic tract The collection of crossed and uncrossed *ganglion-cell axons* after the *optic chiasm;* each optic tract has in it *ganglion-cell* axons from both eyes.

optimal-level theories *Hedonistic views of motivation* which say that there is a "best" level of *arousal* which is pleasurable; the *organism* will be motivated to behave in such a way as to maintain the "best" level of arousal. Compare *drive theories, incentive theories and incentive motivation, opponent-process theory.*

oral stage The stage of *psychosexual development* in *psychoanalytic theory* in which satisfactions of an infant (from birth to about 1 year) center around the mouth as an *erogenous zone.* Compare *anal stage, phallic stage, latency period, genital stage.*

ordinate The vertical axis of a graph. Compare *abscissa.*

organ of Corti The organ containing the *hair cell receptors* for hearing; located on the *basilar membrane.*

organism A person or lower animal.

organizational role of sex hormones The effect of these *hormones* on the developing structure of the body and brain. See *estrogens, androgens.* Compare *activational role of sex hormones.*

orienting reaction A reaction to a novel *stimulus* in which the muscles are tensed and there are other bodily changes to maximize the effectiveness of the stimulus.

origence A *personality* characteristic of creative thinkers; resistance to conformity and an emphasis on individualized interpretation and expression are features of this personality dimension.

osmoreceptors Nerve cells in the anterior *hypothalamus* which generate nerve impulses when they are dehydrated. This acts as a signal for thirst and drinking. See *cellular dehydration thirst.*

ossicles Three small bones in the middle ear through which vibration is conducted to the entrance of the *cochlea*—the oval window—in the inner ear.

outcomes In *social exchange theory,* the consequences of specific acts; outcomes are a joint function of the rewards experienced from a particular set of *behaviors* and the costs required to carry out those behaviors.

overlearning Extra time and effort put into *learning* material above and beyond the time and effort to just learn it.

Pacinian corpuscle A small capsule serving as a *receptor* for deep pressure.

pain "gates" Mechanisms in the *central nervous system* which may block the transmission of *nerve impulses* signaling pain.

paired-associate technique Presenting pairs of words or other items, the first element of which is the *stimulus,* the second the response; given the stimulus, the subject learns the response.

panic disorder A disturbance characterized by specific, focused, time-bound attacks of intense fear and terror. Contrast *generalized anxiety disorder.*

papillae Bumps on the tongue that are heavily populated with *taste buds.*

paradoxical sleep A stage of sleep, characterized by *electroencephalogram (EEG)* activity resembling that of waking (low-voltage, fast), in which the muscles of the body go limp, the person is very difficult to arouse, and the eyes move rapidly from side to side; associated with dreaming. Sometimes called rapid eye movement (REM) sleep.

parallel play Independent, similar activities by two or more children using matching materials and often in close proximity; develops prior to *cooperative play.* Compare *solitary play.*

paranoia See *paranoid disorders.*

paranoid disorders *Psychological disorders* marked by extreme suspiciousness of the motives of others, often taking the form of elaborate beliefs that other people are plotting against the person. In a paranoid disorder the *delusions* are usually systematized. Compare *paranoid type.*

paranoid personality disorder A *personality disorder* characterized by hypersensitivity, suspiciousness, and a tendency to blame others.

paranoid type A kind of *schizophrenia* characterized by *delusions,* often of persecution; the delusions are less systematic than those of the *paranoid disorders.*

paraprofessional In general, an individual who has not had full academic training in a particular profession, but who, with limited training, takes on some duties in the field under professional supervision; specifically, a mental health worker with a B.A. or less.

parasympathetic system The part of the *autonomic nervous system* which tends to be active when we are calm and relaxed; builds up and conserves the body's store of energy. Compare *sympathetic system.* See *relaxation response.*

parietal association cortex Area in the *parietal lobe;* damage may lead to touch *agnosia,* difficulty in spatial relations, disturbances of body perception. (See *contralateral neglect, apraxia, aphasia, alexia, acalculia,* or *agraphia.*)

parietal lobe The lobe of the *cerebrum* behind the *central sulcus;* contains the *somatosensory area of cortex* and *parietal association cortex.*

parieto-occipital fissure The cleft, visible in the medial view of the *cerebral cortex,* dividing the *parietal lobe* from the *occipital lobe.*

passive avoidance learning *Learning* to suppress responses to avoid a *noxious stimulus.* See *punishment.*

passivity versus activity Questions about the degree to which *behavior* and *cognition* are the result of the environment acting on a passive *organism* or, on the other hand, the degree to which behavior and cognition are the result of an active interchange between the *organism* and the environment.

Pavlovian conditioning See *classical conditioning.*

peak experiences For Maslow, intense experiences often accompanied by a disorientation in time and space and feelings of wholeness and oneness with the universe.

peer culture The current styles in dress, music, *language, behavior,* and ideas adopted by *adolescents;* deviation from these standards may bring ostracism.

peers People who are alike in social characteristics and age.

pencil-and-paper tests *Personality tests* which ask for written responses; given to individuals or groups.

percentile score The percentage of scores an individual's score on a *test* equals or exceeds.

perception A general term referring to the awareness of objects, qualities, or events stimulating the sense organs; refers to a person's immediate experience of the world.

perceptual-cognitive styles Generally, individual differences in the ways people typically and characteristically process *information;* more specifically, said to be involved in individual differences in *perception.*

perceptual constancy The tendency of objects to be perceived in the same way despite wide variations in the physical *stimuli.* See *size constancy, brightness constancy.*

perceptual learning An increase in the ability to extract *information* from the environment as a result of experience or practice with the stimulation coming from it; a variety of *cognitive learning.*

perceptual recognition hypothesis An attempt to explain an infant's smiling and laughing in terms of *schemas;* when an object in the environment can be matched to a schema, the result is pleasure.

perceptual reorganization Finding new relationships among objects and events; involved in *insight learning.*

performance Observed *behavior.*

performance IQ An individual's score on the performance subtests of an intelligence scale devised by Wechsler; these subtests do not require a verbal response. Compare *verbal IQ.*

periodic wave A *complex wave* having a repetitive pattern of waves. Compare *aperiodic wave.*

peripheral nervous system (PNS) The parts of the nervous system outside of the skull and spine, largely *axons* which carry information from sensory *receptors* to the *central nervous system* or from the central nervous system to the organs and muscles; includes the *somatic nervous system* and the *autonomic nervous system.* Compare *central nervous system (CNS).*

personal forces A person's characteristics or intentions which result in *behavior; attributions* about an individual are made when we perceive behavior as resulting from these forces.

personal frustration, unattainable goals *Frustration* produced by some personal characteristic of an individual; often due to a discrepancy between a person's *level of aspiration* and his or her capacity to perform. Compare *environmental frustration.*

personal modes Campbell's term for the acquisition of information about the world through direct experience; useful in understanding *conformity.* Compare *social modes.*

personal unconscious The *unconscious* ideas that depend on a person's particular life experiences. Compare *collective unconscious.* See *unconscious processes.*

personality The various enduring and distinctive patterns of *behavior* and thought that are characteristic of a particular person.

personality disorders *Psychological disorders* characterized by lifelong maladaptive behavior patterns. See *antisocial personality disorder, schizotypal personality disorder, compulsive personality disorder, histrionic personality.*

personality dynamics (1) The interactions among *personality* characteristics, especially *motives.* (2) The behavioral expression of personality characteristics in the process of adjusting to the environment. (3) In *psychoanalysis,* the management of the personality's energy system through the interactions of the *id, ego,* and *superego.*

personality structure In general, the unique organization of *traits, motives,* and ways of behaving that characterizes a particular person; in *psychoanalytic theory,* the conception of the *personality* in terms of *id, ego,* and *superego.* See *personality.*

personality tests *Tests* to measure the characteristic ways a person behaves, thinks, and feels. Compare *ability tests, achievement tests.*

personalized system of instruction (PSI) An educational application of *instrumental conditioning/operant conditioning* in which the material in a course is divided into small units, each of which must be mastered at a high level of proficiency before the next unit is attempted.

personnel psychologist An applied psychologist involved in selecting, training, and supervising people in business and industrial settings; also works at improving communications, counseling employees, and alleviating industrial strife. See *industrial and organizational psychology.*

phallic stage According to *psychoanalytic theory,* the stage of *psychosexual development* during which the child of 3 to 5 years becomes interested in the sexual organs and forms a romantic attachment to the parent of the opposite sex. See *Oedipus complex, Electra complex.* Compare *oral stage, anal stage, latency period, genital stage.*

phenomenal field The whole of an individual's experience.

phenothiazines A class of *drugs* used to reduce the symptoms of *schizophrenia.*

phenotype The observable characteristics of an *organism.* Compare *genotype.*

phobias Intense, irrational fears of specific things.

phobic disorders *Anxiety disorders* characterized by intense, irrational fear of a specific object or situation. See *phobias.*

phonemes Speech sounds which must be distinguished in the use of *language;* a basic unit of speech. Sounds which make no difference in meaning when they are substituted for each other are grouped together as phonemes.

phones The sounds of speech.

photosensitive pigments Chemical substances in the *rods* and *cones* of the *retina* that absorb light energy and initiate the visual process. See *rhodopsin, cis-rhodopsin, trans-rhodopsin.*

phrase-structure rules Principles which govern the organization of the *deep phrase structure.* See *transformational grammar, deep phrase structure.*

physical attractiveness Pleasing physical appearance; a factor influencing the degree of *interpersonal attraction.*

physiological needs *Needs* such as hunger, thirst, and sex; the lowest in Maslow's hierarchy of needs.

physiological psychology A subfield of *experimental psychology* concerned with how biological events in the body, most importantly activities in the nervous system, are related to *behavior* and experience.

pitch "High" or "low" tones; this psychological attribute of tones is related to *frequency,* but not directly proportional to it.

pituitary gland An *endocrine gland* with connections to the *hypothalamus;* its *hormones* exert control over the release of hormones from other *endocrine glands.*

place code Different portions of the *organ of Corti* on the *basilar membrane* are maximally stimulated by different *frequencies;* the experience of *pitch* depends, in part, on the place at which the organ of Corti is most stimulated.

place learning *Learning* the place where some event occurs without making a specific response and without *reinforcement.* See *cognitive learning.*

placebos Pharmacologically inert substances which nevertheless have effects on *perception* or *behavior;* a "psychologically effective" substance. Compare *drug.*

plasticity of perception Modifiability of perceptual processes (see *perception*) by *learning* or other special experiences. See *perceptual learning.*

Plutchik theory of emotions A *theory* of the relationships among *primary emotions.* See *emotion.*

polygraph See *lie detector.*

pons A region of the *brain stem* above the *medulla;* contains ascending and descending pathways, fibers projecting to and from the *cerebellum,* and many *nuclei.*

positive goal *Goal* which the individual approaches or tries to reach. Compare *negative goal.*

positive halo effect The tendency to form a generally favorable opinion of a person given only a few positive characteristics of the person. Compare *negative halo.*

positive reinforcement See *positive reinforcer.*

positive reinforcer A *stimulus* or event which, when its onset is made contingent on a particular response, increases the likelihood of that response. Compare *negative reinforcer.*

positive transfer See *transfer.*

positive transference *Transference* marked by feelings of affection and dependence toward the therapist. See *transference.*

postcentral gyrus The *gyrus* behind the *central sulcus.* See *somatosensory area of cortex.*

postconventional level Level of *thinking* about moral issues that relies on abstract principles of right and wrong. Compare *preconventional level, conventional level.*

posthypnotic suggestion A *suggestion* made by the hypnotist while a person is in a hypnotic state but carried out after the *hypnosis* has been terminated.

power motivation and social power See *need for power.*

pragmatics Term that refers to the use of *language* in order to have an impact on others.

precentral gyrus The *gyrus* directly in front of the *central sulcus;* the principal *motor area of the cortex.* Compare *postcentral gyrus.*

preconceptual substage In Piaget's *theory* of *cognitive development,* the first of two substages of the *preoperational* stage; occurs at roughly 2 to 4 years of age and is characterized by the development of *signifiers.* Compare *intuitive substage.*

preconscious Memories and thoughts of which a person is not aware at a particular time, but which may easily become *conscious.* Compare *unconscious.*

preconventional level Level of *thinking* about moral issues guided by egocentric ideas which focus on the personal consequences of *behavior.* Compare *conventional level, postconventional level.*

prefrontal cortex The portion of the *frontal lobe* farthest forward in the head. This area of the *cerebral cortex* seems to play a role in planning, sequencing of actions, flexibility of behavior, and emotional control.

prefrontal lobectomy Removal of brain tissue from the *prefrontal cortex* for the treatment of certain *psychological disorders;* seldom done nowadays; a form of *psychosurgery.*

prefrontal lobotomy The surgical interruption of pathways from the *frontal lobes* of the *cerebrum* for the purpose of alleviating various *psychological disorders;* seldom done nowadays; a form of *psychosurgery.*

prehension The grasping of objects with the hands and fingers.

prejudice An unjustified *attitude,* fairly strong, usually in an unfavorable direction, and not in line with the facts. Compare *discrimination.* See *stereotype.*

preoperational stage The second division in Piaget's *theory* of *cognitive development;* it lasts from age 2 to 7 years and is characterized by unsystematic reasoning.

preparatory set, set A person's readiness to respond to one kind of sensory input but not to other kinds; expectancy. See *set.*

prepared behaviors Responses which can be learned easily by certain animal species. Compare *contraprepared behaviors, unprepared behaviors.* See *learned flavor aversion.*

prestige suggestion A form of *suggestion* in which the message appeals to people's regard for authority or prestige.

primacy effect In *impression formation,* weighing more heavily the information obtained first.

primacy effect in memory The tendency to remember relatively well those items encountered first in a to-be-remembered list; said to contribute to the *serial position effect.* Compare *recency effect in memory.*

primary degenerative dementia A form of dementia having many of the same characteristics as *senile dementia;* an important difference is that the symptoms often begin in middle age. See *senile dementia, Alzheimer's disease.*

primary emotions *Emotions* with an evolutionary basis that are part of our species heritage. See *Plutchik theory of emotions.*

primary goal A *goal* that meets a biological *need;* an unlearned *goal.*

primary mental abilities A set of components, or factors, said to comprise *intelligence*—verbal comprehension, word fluency, perceptual speed, memory, numerical ability, spatial ability, and reasoning. See *multifactor theories.*

primary reinforcer A *stimulus* or event that acts to strengthen a response without prior association with other stimuli. Compare *conditioned, or secondary, reinforcer.* See *reinforcement.*

primary, secondary, and tertiary prevention Levels at which the prevention of *psychological disorders* can be approached. Primary prevention is directed at the causes of psychological disorders. Secondary prevention is aimed at trying to stop problems at their beginning. Tertiary prevention is aimed at minimizing future difficulties of people having behavior disorders or serious problems.

primary sensory areas Areas of the *cerebral cortex* which, as a rule, are arranged so that specific portions of them receive input from particular sensory *receptor* regions. See *topographic organization.*

primary sexual characteristics The structural or physiological characteristics of males and females which make possible sexual intercourse and reproduction. Compare *secondary sexual characteristics.*

proactive interference *Forgetting* caused by the prior *learning* of other material. Compare *retroactive interference.*

proband A member of a twin pair who has a particular characteristic; more generally, a person with a characteristic of interest for genetic studies.

problem Any conflict or difference between an existing situation and a *goal.*

process-oriented theories of intelligence Views about *intelligence* which focus on intellectual processes—the patterns of thinking that people use when they reason and solve problems. Compare *factor theories.* See *information-processing theories of intelligence.*

processing capacity The ability to deal with incoming sensory *information*. Because this ability is limited, we can only deal with a fraction of the sensory *information* available at a given time. Sensory inputs which require the most processing are said to be at the focus of *attention*.

profile of scores A chart indicating the pattern of scores on a *test battery* or a *test* with many subtests.

prognosis Prediction about the course and outcome of a disease process or a *psychological disorder*.

programmed learning Self-instruction which uses carefully designed questions or items to guide the *learning* process in small steps and which provide immediate *reinforcement*.

projection A *defense mechanism* in which conflict is dealt with by ascribing one's own anxiety-provoking motives to someone else; blaming others; prominent in *paranoid disorders*.

projective hypothesis The view that the way people respond to vague or ambiguous situations is often a projection of their underlying feelings and motives. See *projective technique*.

projective methods See *projective technique*.

projective technique Method used in the study of *personality* and *social motives* in which a subject is presented with a relatively ambiguous *stimulus* and asked to describe it in a meaningful way or to tell a story about it. See *Thematic Apperception Test (TAT)*, *Rorschach Inkblot Technique*, *projective tests*.

projective tests *Tests* in which the subject is presented with relatively ambiguous *stimuli*; from the way the individual perceives the test stimuli, a psychologist may be able to infer the *motives* and *emotions* which led to the *perceptions*. See *projective technique*.

proprioceptive sense, proprioception The *kinesthetic* and *vestibular senses* together; informs us about the orientation of the body and head.

prosocial behavior Social behavior which benefits another person; includes sharing, cooperation, and altruism.

prototypes *Concepts* which have most of the defining features of a concept class, or "family"; the best examples of a *concept* class. See *family resemblance structures*.

proximity, or nearness (1) Physical closeness; a factor influencing the degree of *interpersonal attraction* one person feels for another. (2) In *perception*, an organizing principle which says that items which are close together in space or time tend to be perceived as belonging together or forming an organized group.

psychiatric nurses Specially trained registered nurses who provide care, usually in a hospital setting, for people with *psychological disorders*.

psychiatric social workers Specially trained social workers skilled in interviewing, in assembling family histories of their clients, and in assessing social factors involved in *psychological disorders*.

psychiatrists Physicians specializing in the diagnosis and treatment of *psychological disorders*.

psychoactive drugs Drugs that affect *behavior* and *perception*; they generally work on the nervous system by influencing the actions of *neurotransmitters* at *synapses*.

psychoanalysis Primarily a method of *psychotherapy* developed by Sigmund Freud, but also a *theory* of the development and structure of *personality*. As a *psychotherapy*, it emphasizes the techniques of *free association*, the phenomenon of *transference*, and the development of *insight*.

psychoanalysts Psychotherapists (see *psychotherapy*) who practice the therapeutic techniques of Sigmund Freud and his followers. See *psychoanalysis*.

psychoanalytic perspective A current viewpoint in *psychology* which focuses on the contribution of unconscious feelings and impulses to *behavior*; sometimes known as the psychodynamic perspective.

psychoanalytic theory A dynamic approach to *personality* developed by Sigmund Freud. The theory has three major parts: (*a*) a theory of the structure of personality, in which the *id*, *ego*, and *superego* are the principal concepts; (*b*) a theory of personality dynamics, in which conscious and *unconscious motives* and ego *defense mechanisms* play a major role; and (*c*) a theory of *psychosexual development*, in which different motives and body zones influence the child at different stages of growth. See *psychoanalysis*.

psychodynamic therapies Treatments of *psychological disorders* focusing on the interplay of *motives* and urges. See *psychoanalysis*.

psycholinguistics The branch of *psychology* that studies the ways in which people generate and comprehend *language*.

psychological assessment The *measurement* of people's *behavior* and *abilities*, largely by means of *tests*.

psychological disorder Patterns of *abnormal behavior*, *emotions*, or thought that significantly interfere with an individual's adaptation to important life demands and often cause distress in the individual or in others.

psychological test See *tests*.

psychology The *science* of human and animal *behavior*, including the application of the science to human problems.

psychometric psychology The branch of *psychology* concerned with the development of *tests*, research on their usefulness, and in general, ways of measuring *behavior*.

psychopathic personality See *antisocial personality disorder*.

psychophysics The study of the relationship between variations in physical energy and reported experience or *behavior*.

psychophysiologist A scientist who studies the relationship of bodily events to *behavior*; more specifically, one who studies bodily events in *emotion*.

psychosexual development In *psychoanalysis*, the idea that the instinctual drives are expressed in different ways as children grow older. See *oral stage*, *anal stage*, *phallic stage*, *latency period*, *genital stage*.

psychosexual stage One of the steps in *psychosexual development*. See *psychosexual development*.

psychosis A severe *psychological disorder* in which the person has typically lost considerable contact with reality; *hallucinations* or *delusions* may be present; custodial care is often required. See *schizophrenia*, *affective disorders*, *paranoid disorder*.

psychosocial moratorium Erik Erikson's term for the delay of commitment to adult *roles* shown by some adolescents.

psychosomatic disorders Bodily diseases caused or aggravated by psychological factors, such as long-lasting, or chronic, emotional states and *stressors*.

psychosomatic reaction See *psychosomatic disorders*.

psychosurgery Operations on the *brain* for the purpose of alleviating various *psychological disorders*. See *prefrontal lobotomy*, *prefrontal lobectomy*.

psychotherapy The treatment of *psychological disorders* and mild adjustment problems by means of psychological techniques. Compare *biomedical therapy*. See *psychoanalysis*, *client-centered therapy*, *gestalt therapy*, *analytical therapy*.

puberty The period during which the capability for sexual reproduction is attained; it is marked by changes in both *primary* and *secondary sexual characteristics*, and is dated from *menarche* in girls and the emergence of pigmented pubic hair in boys.

public opinion (attitude) polling See *public opinion polls.*

public opinion polls *Self-report methods* of measuring *attitudes* in the population by asking many people a few questions each. Compare *attitude scales.*

punctate sensitivity Sensitivity to touch, temperature, and pain, which have separate sensitive spots on the skin.

punisher A *noxious stimulus* which, when its presentation is made contingent on a particular response, tends to decrease the likelihood of that response. Compare *negative reinforcer.* See *punishment.*

punishment The application of an unpleasant, or *noxious,* stimulus for the purpose of suppressing behavior. Compare *negative reinforcement.* See *punisher.*

pure tone The sound produced by a *sine wave.*

Q-sort technique A *personality* research technique in which a person places descriptive personality statements about herself or himself into categories ranging from "least characteristic" to "most characteristic."

questionnaire (1) A *pencil-and-paper* personality *test* that asks questions about typical performance that can be answered "yes" or "no." (2) A survey of opinions and experiences.

random sampling In *public opinion polls,* the process of choosing a sample of people from the *target population* in such a way that each person in the target population has an equal chance of being included in the sample.

range A *measure of variability;* the interval between the highest and lowest scores. Compare *standard deviation (SD).*

See *paradoxical sleep.*

ratio IQ See *Intelligence Quotient (IQ).*

rational-emotive therapy (RET) A *cognitive therapy/cognitive-behavior therapy* technique designed to break down the irrational thoughts and *beliefs* leading to psychological distress.

rationalization A *defense mechanism* in which a person "makes excuses," thus substituting an acceptable *motive* for an unacceptable or anxiety-provoking one.

reaction formation A *defense mechanism* in which a true *motive* which would provoke *anxiety* if recognized is converted into its opposite.

reaction range The limits of the observable *behaviors* possible when a given genetic endowment interacts with the environment; variability in the *phenotype* due to environmental differences interacting with a particular genetic endowment. See *nature-nurture interaction.*

real motion *Perception* of actual physical movement. Compare *apparent motion.*

realism In Piaget's *theory* of *cognitive development,* the tendency of children in the *preoperational* stage to think of *symbols* and *concepts* as real things.

reappraisal A change in the evaluation of environmental and internal information; a way of coping with prolonged stressful situations; involved in the *cognitive-appraisal theory of emotions.* See *appraisal.*

recall method A standard way of measuring *memory* in which people, after being exposed to the to-be-remembered items, are asked to call back the items from memory. Compare *recognition method.*

recency effect in memory Items encountered most recently are remembered relatively well; said to contribute to the *serial-position effect.* Compare *primacy effect in memory.*

receptor A cell, or group of cells, specialized to respond to relatively small changes in a particular kind of physical energy.

receptor potential The electrical activity generated in a *receptor* cell during *transduction.* See *transduction.*

reciprocal inhibition In *classical conditioning,* the principle that two incompatible responses cannot occur at the same time and that the stronger of the two will replace the weaker; in *systematic desensitization* the goal is to increase the strength of a desired response relative to that of an undesired one. See *systematic desensitization.*

recognition method A way of measuring *memory* in which a person is asked to recognize the to-be-remembered items when they are presented along with incorrect items. Compare *recall method.*

reconditioning The process of again pairing the *conditioned stimulus* and the *unconditioned stimulus* after *extinction.* See *classical conditioning.*

reconstructive processes Modifications of stored information at the time of *retrieval* which determine what is actually remembered. Compare *constructive processes.* See *confabulation.*

reflection of feeling A *client-centered therapy* technique in which the therapist restates what the client has said to clarify the essence of the feelings expressed.

reflective thinking The process of evaluating and testing one's own reasoning; develops in the *formal operational stage.*

reflex A simple adaptive bodily movement produced when *motoneurons* are excited by some sensory input.

refraction The bending of light rays; in vision, the bending of light rays by the *cornea* and *lens* to focus images on the *retina.*

regression A *defense mechanism* in which a person copes with *anxiety* by retreating to childish or earlier forms of *behavior;* often encountered in children and adults faced with *frustration* and *motivational conflict.*

rehearsal Focusing *attention* on an item of *information* by repetition, or processing it in some other way so as to link it up with other information which has already been stored in *memory;* thought to be an important factor in converting *short-term memory* to *long-term memory.* See *maintenance rehearsal, elaborative rehearsal.*

reinforcement (1) In *classical conditioning,* the pairing of the *conditioned stimulus* and the *unconditioned stimulus.* (2) In *instrumental conditioning/operant conditioning,* the presentation or termination of a *stimulus* or event which, when made contingent on the occurrence of a certain response, makes that response more likely to occur in the future. See *positive reinforcer, negative reinforcer.*

reinforcer See *positive reinforcer, negative reinforcer, reinforcement.*

relaxation response Bodily reactions in calm, meditative emotional states. Compare *emergency reaction. See parasympathetic system.*

releaser A *stimulus* triggering *species-typical behavior.*

releasing factors Chemicals by which the *hypothalamus* can control secretion of *hormones* from the *pituitary gland.*

reliability The consistency of a method of *measurement.* A characteristic of good psychological *tests.* Compare *validity.*

renin An *enzyme* released by the kidneys when there is a drop in blood pressure; involved in the formation of *angiotensin II,* which circulates in the blood and can trigger drinking.

replication Repetition of an observation under controlled conditions.

representational thought The capacity to form mental *symbols* which stand for objects or events not present.

representativeness In decision making, deciding whether the current situation is similar to one we have encountered before; a judgment *heuristic.*

repression A *defense mechanism* and process in which certain *memories* and *motives* are not permitted to enter awareness but are operative at an *unconscious* level; results in a failure of *retrieval* of anxiety-provoking material from *long-term memory.*

resistance In *psychoanalysis,* efforts on the part of the patient to avoid or evade the interpretation of *behavior* or ideas made by the therapist.

resistance to extinction The process of continuing to respond after *reinforcement* is stopped; tends to be greater after previous scheduled reinforcement (see *schedule of reinforcement*) than it is after previous *continuous reinforcement (CRF)*.

respondent conditioning See *classical conditioning*.

reticular formation A complex region in the center of the *brain stem* containing many small *nuclei* and fibers; present from the *medulla* up to the *midbrain*. See *ascending reticular activating system (ARAS)*.

retina The photosensitive layer of the eye; contains the visual *receptors (rods* and *cones), ganglion cells*, and other cells and connecting fibers. See *photosensitive pigments*.

retinal disparity The most important *binocular cue* for depth; a slight difference in the images of an object projected on the *retinas* of the two eyes. The images are more dissimilar when the object is close.

retrieval The process of withdrawing information from *long-term memory* or *short-term memory*. See *tip-of-the-tongue (TOT) phenomenon*.

retrieval cues Reminders which direct the search through *long-term memory*.

retrieval processes Ways in which access is gained to information stored in *memory*. See *retrieval; retrieval cues*.

retroactive interference *Forgetting* of previously learned material due to the subsequent *learning* of new material. Compare *proactive interference*.

retrograde amnesia *Forgetting* events one was exposed to in the past. Compare *anterograde amnesia*. See *amnesia*.

reversibility In Piaget's *theory* of *cognitive development*, the ability of children to understand that a changed object or state of affairs can be returned to its original state if the changes are reversed. See *identity concept*.

rhodopsin The *photosensitive pigment* found in the *rods;* also known as visual purple.

rods Photosensitive *receptors* in the *retina*, cylindrical in shape, active in dim light; contain *rhodopsin*. Compare *cones*.

role The *behavior* expected of a person who holds a certain *status* within a group. Compare *status*.

rooting reflex A pattern of responses which enables the *neonate* to find the mother's breast; when the cheek is stimulated, the head turns in that direction.

Rorschach Inkblot Technique A *projective technique* using inkblots as *stimuli*.

safety needs *Needs* for security, stability, and order. In Maslow's theory, they are fulfilled after *physiological needs* are satisfied.

safety signals *Stimuli* which are consistently paired with the absence of *noxious* events; these stimuli may act like *positive reinforcers* in *avoidance learning*.

satiety The absence of *motivation* after *goals* have been reached.

saturation The degree to which a color *(hue)* is diluted or not diluted by whiteness. Compare *brightness*.

savings method A method of measuring *forgetting* in which the subject learns again what he or she previously learned. The more that is remembered from the original learning, the fewer the trials needed to relearn the material; the amount of savings is expressed as a percentage. Compare *recognition method, recall method*.

scapegoating *Aggression* displaced toward a person or group that is the object of *prejudice*.

Schachter-Singer theory A *theory* of the relationship between felt *emotion* and bodily conditions; it states that felt emotion is based on the interpretation of the reasons for bodily arousal. Compare *James-Lange theory, Cannon-Bard theory*.

schedule of reinforcement A situation in which *reinforcement* does not follow every response. Instead, reinforcements follow certain responses according to a specified plan. See *fixed-interval schedule (FI), fixed-ratio schedule (FR), variable-interval schedule (VI), variable-ratio schedule (VR)*. Compare *continuous reinforcement (CRF)*.

schema A mental representation of objects, events, and their relationships.

schemata Plural of *schema*.

scheme In Piaget's *theory*, an action sequence guided by thought.

schizophrenia A *clinical syndrome* characterized by *cognitive slippage and derailment, hallucinations*, and *delusions*, and often by general withdrawal from contact with the environment

schizophrenic disorders See *schizophrenia*.

schizotypal personality disorder A *personality disorder* characterized by withdrawal from other people and eccentric thinking; resembles *schizophrenia*, but is not so severe. Compare *schizophrenia*.

scholastic aptitudes *Abilities* to succeed in a specified type of formal schooling. For example, college aptitude refers to aptitude for doing college work. See *aptitude*.

school counselor A person who does testing and counseling in schools. Compare *school psychologist, educational psychology*.

school psychologist A psychologist who diagnoses learning difficulties and attempts to remedy them. Compare *school counselor, educational psychology*.

science A body of systematized knowledge gathered by carefully observing and measuring events. See *empirical*. Compare *art*.

S^D The *stimulus* in the presence of which a response is reinforced in *instrumental conditioning/operant conditioning*. See *discrimination, reinforcement*. Compare S^Δ.

S^Δ The *stimulus* in the presence of which a response is not reinforced in *operant conditioning*. See *discrimination, discrimination learning, reinforcement*. Compare S^D.

second mental-health revolution The view that psychological factors (especially *unconscious motivation*) are important causes of *psychological disorders*; sparked by Freud and his followers at the turn of the twentieth century. See *psychoanalytic theory*.

secondary memory The *information-processing* aspects of memory, which include the encoding, storage, and *retrieval* of *information*.

secondary reinforcer See *conditioned, or secondary, reinforcer*.

secondary sexual characteristics Physical features such as body proportion and hair distribution, but excluding the reproductive organs, which distinguish the mature male from the mature female. Compare *primary sexual characteristics*.

secondary traits *Personality* characteristics of a person which are influential in explaining behavior only within a limited range of situations. Contrast *cardinal traits, central traits*.

seeking identity Attempting to get a clear view of what one's skills and personal characteristics are and to discover where one is heading in life.

self (1) The individual's *perception* or awareness of herself or himself—of his or her body, *abilities*, personality *traits*, and ways of doing things. (2) The executive functions by means of which an individual manages, copes, thinks, remembers, perceives, and plans.

self-actualization See *self-actualizing*.

self-actualizing According to Maslow, satisfying higher *needs;* thus a person who strives to satisfy needs for justice, beauty, order, and goodness is said to be a self-actualizing person. A person's need to develop her or his potentialities.

self-attribution Inferring certain qualities in ourselves.

self-concept, self-image A person's feelings about himself or herself; examples are self-confidence, self-esteem, and self-worthlessness. See *self*.

self-disclosure The process through which one person lets himself or herself be known by another.

self-fulfilling prophecy Behaving according to one's own expectations in such a way that this behavior influences a situation and brings about what is expected.
self-report method Measuring *attitudes* by asking people to respond to questions. Examples are *attitude scales, public opinion polls.*
self theory A theory of *personality* in which the *self* is the central idea. See *self, self-concept, self-image.*
semantic differential A method of measuring the *connotative meaning* of words and *concepts*. See *connotative meaning.*
semantic memory A *long-term memory* store containing the meanings of words and concepts and the rules for using them in language. Compare *episodic memory.*
semantics The aspect of *language* which refers to meaning.
senile brain disorders Tissue loss and shrinkage of the brain in old age which cause mental and physical deterioration. See *senile dementia.*
senile dementia A state that occurs due to brain disease in elderly persons; characterized by deficits in *memory, attention,* judgment, and abstract thought, by changes in personality, and sometimes by *delusions* and *disorientation* in time or place.
sensitive period A time in the life cycle during which the *organism* is especially susceptible to, or ready for, certain *cognitive,* perceptual (see *perception*), or social developments.
sensorimotor stage The first division in Piaget's *theory* of *cognitive development* in which the child learns to deal with objects in terms of sensory-motor *schemes;* occupies the first 2 years of life.
sensory areas of the cerebral cortex Areas of the *cerebral cortex* primarily involved in processing incoming *information*. See *primary sensory areas*. Compare *motor areas of cortex, association areas of cortex.*
sensory channel The *receptor,* nerve fibers leading from the receptor to the *central nervous system,* and the various relay stations and places of termination within the central nervous system.
sensory deprivation Experimental restriction of sensory input.
sensory register The storage of information for a brief time in a *sensory channel*. See *iconic image.*
serial-position effect The observation that in *memory* experiments using a list of items to be remembered, items at the beginning and end of the list are remembered best. See *primacy effect in memory, recency effect in memory.*
serotonin A *neurotransmitter* found in the *central nervous system.*
set Expectancy. In problem solving, a readiness to react in a certain way when confronted with a problem or *stimulus* situation; may be induced by immediately preceding experiences, by long-established practice, or by instructions which evoke old habits; also a factor in *perception*. See *functional fixedness.*
set point The level around which physiological processes are regulated to maintain *homeostasis.*
shadows in depth perception A *monocular cue* in depth perception; shadows provide cues because we are accustomed to light coming from above.
shaping In *instrumental conditioning/operant conditioning,* teaching a desired response through a series of successive steps which lead the learner to the final response. Each small step leading to the final response is reinforced (see *reinforcement*). Also called the method of successive approximations.
short-term memory (STM) The temporary store of information held in *memory* for a few seconds while it is being processed for long-term storage or use; holds information for about 30 seconds and has a very limited storage capacity. Compare *long-term memory (LTM).*
signal detection theory A way of describing judgments made in sensory situations in which a signal must be detected against a noisy background; separates sensory and nonsensory factors that enter into judgments.
signifier In Piaget's *theory* of *cognitive development,* one of the child's personal mental symbols, which develop during the *preconceptual substage*. It is concrete and imitative of the thing it symbolizes.
similarity in form perception An organizing principle causing a person to perceive similar items as grouped together.
simultaneous discrimination learning Occurs when the positive (S^D) and negative (S^Δ) stimuli are presented at the same time. See *discrimination learning*. Compare *successive discrimination learning.*
sine waves In hearing, the simplest kinds of *sound waves;* produced when a single vibrating object moves back and forth freely and changes the pressure of the air. Compare *complex waves*. See *pure tone.*
situationism The view that *behavior* is more a product of a particular situation than a product of enduring characteristics of a person such as *traits.*
Sixteen Personality Factor (16PF) Questionnaire A *test* based on *factor analysis* which measures 16 key characteristics of *personality.*
size constancy The tendency for the perceived size of objects to remain about the same despite large changes in their image size on the *retina*. See *perception, perceptual constancy.*
social comparison A proposed cause of *conformity;* we resolve ambiguity about what to do and think by observing people similar to ourselves and following their lead.
social conformity/conformity See *conformity.*
social exchange theory The idea that social relationships may be viewed as a kind of economic system in which the people engaged in a relationship become dependent on each other for the quality of the *outcomes* they experience from the relationship; involves *comparison levels (CL)* and *comparison levels for alternatives* (CL_{alt}).
social facilitation Increased *motivation* and effort arising from the *stimulus* provided by other people.
social influence The change in an individual that occurs because of contact with other people; the process of producing this change. People are affected by the presence, opinions, or *behavior* of others.
social inhibition In social groups, the retardation of action caused by the presence of other people.
social modes Campbell's term for the acquisition of information from other people; learning from other people; useful in understanding *conformity*. Compare *personal modes.*
social motivation See *social motives.*
social motives *Motives,* usually learned in a social group, that require the presence or reaction of other people for their satisfaction. See *need for achievement (n ach), need to affiliate, need for power.*
social need See *social motives.*
social perception The way we view other people, based on information obtained about others and *attributions* we make about the causes of their *behavior.*
social perspective A current viewpoint in *psychology* which emphasizes the social interactions among people and the social influences which affect *behavior*. See *social psychology.*
social psychology The scientific study of the ways in which interaction, interdependence, and influence among persons affect their *behavior* and thought.
social stereotypes Characteristics by which people are prejudged and categorized. They include race, nationality, religion, gender, or region of the country in which an individual lives.
socialization The *learning* process through which a child is trained in the *attitudes, beliefs,* and *behaviors* appropriate to her or his

culture; the process by which the child's *behavior* and *attitudes* are brought into harmony with the culture.
sociobiology The systematic study of the biological basis of social behavior.
sociometry A method of mapping social relationships of attraction and rejection among members of a group.
sociopathic personality See *antisocial personality disorder.*
solitary play Playing alone, even in the presence of *peers;* a normal stage prior to the emergence of *parallel play* and *cooperative play.*
somatic nervous system The part of the *peripheral nervous system* serving the sense organs and the skeletal muscles. Compare *autonomic nervous system.*
somatosensory Pertaining to the body senses, including touch, temperature, pain, and *kinesthesis.*
somatosensory area of cortex The *primary sensory area* involved in body sensation (see *somatosensory*); located in the posterior bank of the *central sulcus* and on the *postcentral gyrus.*
sound wave Alternating increases and decreases in pressure propagated through a medium, usually air; a vibration having a *frequency* and amplitude.
source credibility Believability of the originator of an attitude-change message; based on the perceived expertness and trustworthiness of the originator.
spatial frequency In vision, the number of alternations between light and dark within the visual field; hypothesized to be involved in form vision.
species-typical behaviors *Behavior* patterns characteristic of a particular species; behavior which all normal individuals of the species display under the appropriate circumstances.
species-typical defense reactions Behaviors "naturally" used by members of a particular species to defend themselves against threatening situations; considered to play a role in *avoidance learning.*
spike See *nerve impulses.*
spinal cord The part of the *central nervous system* encased in the backbone. It is a reflex center and a pathway for *nerve impulses* to and from the *brain.*
spinal nerves Peripheral nerves carrying sensory information into the *spinal cord* and motor commands out; one pair is associated with each of the divisions of the spinal cord.
spinal roots Groups of nerve fibers emerging from or entering the *spinal cord* which join to become the *spinal nerves.* See *dorsal root, ventral root.*
spontaneous recovery After an interval of time, an increase in the strength of a *conditioned response* which had undergone *extinction.*
stage of exhaustion The third stage of the *general adaptation syndrome* in which the body's capacity to respond to both continuous and new *stressors* is severely lessened. Compare *alarm reaction, stage of resistance.*
stage of mutuality The third stage in the growth of relationships; each individual begins to acquire feelings of responsibility for the *outcomes* the partner receives in the relationship. Compare *unilateral awareness, surface contact.*
stage of resistance The second stage of the *general adaptation syndrome* during which the body defends itself against the effects of a continuing *stressor;* however, the body's resistance to new stressors is impaired during this stage. Compare *alarm reaction, stage of exhaustion.*
stages in creative thinking A pattern of steps that is frequently involved in the solution of problems by talented and creative people; the stages are preparation, incubation, illumination, evaluation, and revision.
standard deviation (SD) A measure of the spread, or variability, of scores of a group of people.
standard scores Scores obtained by dividing the *standard deviation* into the difference between an individual's actual obtained score and the mean, or average, of the group of scores; in other words, scores expressed in terms of standard deviation units. See *deviation IQ.*
standardization The establishment of uniform conditions for administering a *test* and interpreting test results. A large number of individuals is tested in the same way to provide *norms* with which to compare any particular test score. See *norms.*
Stanford-Binet Intelligence Scale An individual *test* of *intelligence,* the descendant of the early work of Binet, mainly used with children; predicts school achievement; uses *age scales.*
state-dependent memory (state-dependent learning) *Information* stored under one drug or emotional state may not be available for *retrieval* when the individual is in another drug or emotional state.
status The position an individual holds within a group. Compare *role.*
status offences Acts which are illegal only for minors.
stereotypes Fixed sets of greatly oversimplified *beliefs* that are said to characterize members of a group.
stimulus (pl. stimuli) An event that produces responses or that results in sensory experience.
stimulus control of behavior *Discrimination* in *instrumental conditioning/operant conditioning;* the rate at which a learned response occurs depends on the *stimulus* which is present. See *discrimination.*
stimulus discrimination in instrumental conditioning See *discrimination, discrimination learning, differential reinforcement, simultaneous discimination learning, successive discrimination learning, stimulus control of behavior.*
stimulus generalization The tendency to react to *stimuli* that are different from, but somewhat similar to, a *conditioned stimulus.* See *gradient of generalization.*
stimulus generalization in instrumental conditioning The response in *instrumental conditioning/operant conditioning* is made in a particular *stimulus* situation. If the stimulus is changed, the response still occurs; the tendency to respond depends upon the degree of similarity between the original training situation and the changed one. See *gradient of generalization.*
storage processes The means by which information is actually put into *memory.* See *encoding processes.*
stress An internal state which can be caused by physical demands on the body or by environmental and social situations which are evaluated as potentially harmful, uncontrollable, or exceeding the individual's resources for coping. See *stressors, general adaptation syndrome.*
stressors The situations or events which cause *stress.*
stroboscopic motion *Apparent motion* due to successive presentations of visual *stimuli.*
structuralism An early school of psychological thought which held that conscious experience could be analyzed into mental elements. Compare *Gestalt psychology, functionalism, behaviorism.*
style of life In Adler's *individual psychology,* the idea that each person creates his or her own unique approach to living.
subjective contours *Contours* perceived in the absence of physical energy differences.
subjective organization In verbal *learning* and *memory,* organization not inherent in the material itself but imposed on the material by the learner.
subjective probability Estimate of the likelihood of various outcomes.

subjectively expected utility The combination of *utility* and *subjective probability;* according to a mathematical model of the decision process, people make decisions which will maximize subjectively expected utility. See *subjective probability, utility.*
sublimation The use of a substitute activity to gratify a frustrated *motive.* Freud believed, for example, that a frustrated (see *frustration*) sex drive could be partially gratified by channeling it into some artistic activity.
substance-use disorders *Psychological disorders* which involve the use of *drugs.*
successive discrimination learning Learning that occurs when the positive (S^D) and negative (S^Δ) stimuli are presented one after the other. See *discrimination learning.* Compare *simultaneous discrimination learning.*
suggestion Uncritical acceptance of a statement; can be important in *attitude* change.
sulcus A relatively shallow crevice in the *cerebral cortex* of the *brain.* Compare *fissure, gyrus.*
superego In *psychoanalytic theory,* that which corresponds to what is commonly called conscience; it imposes restrictions and keeps a person working toward ideals acquired in childhood. Compare *id, ego.*
superior colliculi Nuclei located just above the *inferior colliculi* in the *midbrain;* important in the coordination of eye movements and in *reflex* postural adjustments of the body to visual inputs. Compare *inferior colliculi.*
surface contact The second level of involvement in the growth of relationships; interactions are governed by general cultural norms specifying appropriate behavior and social etiquette. Compare *unilateral awareness, stage of mutuality.*
surface structure The actual expression in speech of the core ideas in the *deep phrase structure;* the deep phrase structure is converted into the surface structure by *transformation rules.* See *transformational grammar.*
syllables The smallest or shortest speech patterns we normally produce; also the smallest unit in speech *perception;* composed of several *phonemes.*
symbol (1) A *stimulus* that represents some event or item in the world; has arbitrary meaning assigned for purposes of *communication.* (2) An internal, or mental, representation of an object or event.
symmetry, good figure An organizing principle in *perception* according to which items that form a balanced or symmetrical figure are perceived as a group.
sympathetic system A subdivision of the *autonomic system* arising in the thoracic and lumbar portions of the spinal cord; most active during aroused states. Compare *parasympathetic system.* See *emergency reaction, or "flight or fight" response, arousal.*
synapse The functional connection between two *neurons.* See *synaptic cleft, neurotransmitter, vesicles.*
synaptic cleft The narrow gap separating *neurons* at a *synapse.*
syncretic reasoning The tendency of children in the *preoperational stage* to see unrelated events and objects as related.
syndrome A pattern of symptoms which cluster together so regularly that they suggest a common source.
syntax See *grammar.*
systematic desensitization A form of *behavior therapy/behavior modification* using the principles of *counterconditioning* and *reciprocal inhibition.* A situation is arranged in which a fear- or anxiety-inducing *stimulus* can be presented while a person remains relaxed; gradually the stimulus that previously produced *anxiety* is conditioned to the state of relaxation so that it produces relaxation instead of anxiety. See *counterconditioning, behavior therapy/behavior modification, reciprocal inhibition.*
tachistoscope An apparatus for presenting perceptual (see *perception*) materials for a very brief time.
tardive dyskinesia A physical disorder which may accompany prolonged use of certain antipsychotic *drugs* used to treat severe *psychological disorders;* characterized by uncontrollable facial movements, twitching or jerking of the arms or legs, and contractions of the back and neck muscles.
target population The particular group of people about which *attitude* information is sought in *public opinion polls;* the whole group, as opposed to a sample of respondents. See *random sampling.*
taste buds Clusters of specialized cells containing the *receptor* cells for taste.
temperament The aspect of *personality* that includes *mood,* activity level, and *emotion.*
temporal lobe The portion of the *cerebrum* below the *lateral fissure.* Compare *frontal lobe, parietal lobe, occipital lobe.*
temporal-occipital association cortex Areas in the *temporal* and *occipital lobes* which are strongly linked by fiber pathways; important for identification and recognition of visual *stimuli.* See *association areas.*
tests *Behavior* samples which are uniform, objective, designed to tap individual differences, and interpretable.
test batteries Groups of *tests* designed to be used together to serve a particular purpose.
testosterone The *androgen hormone* considered responsible for male sexual *behavior* in many animal species.
thalamus A region of the *forebrain* located just above the *midbrain*; concerned with relaying and integrating sensory input.
Thematic Apperception Test (TAT) A *projective technique* consisting of pictures about which a person tells stories. Compare *Rorschach Inkblot Technique.*
theory In science, a general principle summarizing many observations and predicting what we can expect to happen in new situations.
thinking The mental, or *cognitive,* rearrangement or manipulation of both *information* from the environment and of *symbols* stored in *long-term memory. Language* symbols and *images* are used, and thinking is said to *mediate,* or go between, *stimuli* and responses.
threshold (1) The critical point at which a nerve membrane changes from its resting state to generate a *nerve impulse.* (2) In *psychophysics,* the point at which a *stimulus,* or a difference between stimuli, can just be detected. See *absolute threshold, differential threshold.*
timbre The tonal quality that enables us to distinguish different musical instruments and voices that have the same *fundamental frequency;* determined by *harmonic frequencies.*
time-out from reinforcement/time-out See *omisson of reinforcement/omission training.*
tip-of-the-tongue (TOT) phenomenon A state that may occur during an attempt at *retrieval* of *information* from *long-term memory,* during which a person may retrieve incorrect information that is related in some way to the correct item; indicates that information in long-term memory is organized.
token economy The use of *conditioned, or secondary, reinforcers*—moneylike tokens—to strengthen desired behaviors in mental hospitals, prisons, and other similar institutions. By the use of these tokens, desirable behaviors which aid therapy can be shaped (see *shaping*) and maintained.
tolerance State that occurs in the *dependence* stage of a *substance-use*

disorder when increasing amounts of the substance are needed to achieve the desired effect.

topographic organization The orderly mapping of a sensory *receptor* surface on the appropriate *sensory areas of the cerebral cortex.*

tract A collection of *axons* in the *central nervous system; white matter.* Compare *nucleus.*

traditional marriage A relationship in which the husband is the undisputed head of the family, traditional sex roles are maintained, and the wife shows deference to her husband.

traits Aspects of *personality* that are reasonably characteristic of a person, are relatively consistent over time, and distinguish that person in some way from other people; sets of categories that are used for describing, remembering, and communicating characteristics of people.

tranquilizer Any one of several *drugs* used to reduce *anxiety.*

transduction The process of converting physical energy into nervous system activity; occurs at the *receptor.*

transductive reasoning A characteristic of *thinking* of the *preoperational stage* in which children reason from particular instances to other particular instances instead of from the general to the particular (deduction) or from the particular to the general (induction).

transfer, transfer of training Generally, more rapid *learning* in a new situation because of previous learning in another situation (positive transfer), or slower learning in a new situation because of previous learning (negative transfer).

transfer of training See *transfer.*

transference In *psychotherapy,* particularly *psychoanalysis,* the reenactment of previous relationships with people, especially the parent-child relationship. In *psychoanalysis,* the therapist becomes the object of transference; the transference permits the patient to express toward the therapist attitudes and feelings he or she has held toward other people. Compare *countertransference.*

transformation rules The principles guiding the conversion of the *deep phrase structure* into what is expressed—the *surface structure.* See *transformational grammar.*

transformational grammar A *linguistic* theory which says that when a person intends to communicate something the words are mentally represented in the *deep phrase structure.* By using *transformation rules,* the *deep phrase structure* is converted into what is actually expressed—the *surface structure.*

transient global amnesia *Forgetting* of past events together with the inability to form new memories; a state characterized by both *retrograde amnesia* and *anterograde amnesia.* See *amnesia.*

trans-rhodopsin See *cis-rhodopsin, trans-rhodopsin.*

trichromatic process A term that refers to the fact that color vision starts with a differential pattern of wavelength absorption in three cone types—those which absorb light best in the short, middle, or long wavelength regions of the *visible spectrum.*

tricyclics A class of antidepressant *drugs.*

type A class of individuals grouped together because they share certain personality *traits* or characteristics.

Type A and Type B Two specific behavior pattern *types* with increased and decreased likelihood of heart disease, respectively. Type A persons are hard-driving and competitive; type B persons are easy-going and handle *stressors* more calmly.

type and trait approaches Views of *personality* which focus on people's characteristics and how these characteristics are organized into systems.

unconditional positive regard The nonjudgmental, accepting attitude a counselor in *nondirective therapy* maintains toward the client.

unconditioned response (UR) The response elicited by the *unconditioned stimulus (US).* See *classical conditioning.*

unconditioned stimulus (US) A *stimulus* which consistently elicits a response, the *unconditioned response (UR).* See *classical conditioning.*

unconscious Memories and thoughts which are unavailable to *conscious* awareness. See *unconscious processes, unconscious motivation.* Compare *preconscious, conscious.*

unconscious motivation *Motivation* that can be inferred from a person's *behavior,* although the person does not realize the presence of the *motive;* an important concept in *psychoanalytic theory.*

unconscious processes Psychological processes or events of which a person is unaware.

unilateral awareness The first level of involvement in the growth of relationships; a person notices another and may make judgments evaluating the characteristics of the other. Compare *surface contact, stage of mutuality.*

unipolar affective disorder An *affective disorder* which is characterized by either recurring *manic episodes* or recurring *major depressive episodes.* Compare *bipolar disorder.*

unique colors The pure *hues* that observers judge to be uncontaminated by any other hue.

unprepared behaviors Responses which can be acquired by an animal species when *learning* procedures are applied. Compare *prepared behaviors, contraprepared behaviors.*

utility Perceived benefit or psychological value.

vacillation of behavior and thought In a *conflict* situation, *behavior* shifting to and fro between the conflicting *goals.* Common in *avoidance-avoidance conflict* and *approach-avoidance conflict.* See *motivational conflict.*

valence (1) In *attribution,* the degree to which the *noncommon effects* are "good" or "bad," "pleasant" or "unpleasant." (2) A term referring to the attraction or repulsion of a *goal;* indicated by a plus or minus sign.

validity The extent to which a method of *measurement* measures what it is supposed to measure. Validity is expressed in terms of a *correlation coefficient* representing the relationship of a set of measurements with some *criterion.* A characteristic of good psychological *tests.* See *criterion.*

variable An event or condition which can have different values; ideally, in experiments, an event or condition which can be measured and which varies quantitatively. See *independent variable, dependent variable.*

variable-interval (VI) schedule A *schedule of reinforcement* in which subjects are reinforced for a response after an interval of time which varies around a specified average; produces great *resistance to extinction* and steady rates of responding. See *reinforcement.*

variable-ratio (VR) schedule A *schedule* of *reinforcement* in which subjects are reinforced after a number of responses which varies around a specified average; produces great *resistance to extinction* and steady rates of responding. See *reinforcement.*

ventral root The *spinal root* toward the front of the body which contains the motor fibers. Compare *dorsal root.*

ventromedial hypothalamus (VMH) An area of the *hypothalamus* classically considered to be an inhibitory region for the hunger *drive.* Compare *lateral hypothalamus (LH).*

verbal IQ An individual's score on the verbal subtests of an intelligence scale devised by Wechsler. Compare *performance IQ.*

vesicles Small bodies containing *neurotransmitters;* found in the *boutons* at the ends of *axons.* See *bouton, synapse.*

vestibular sense The sense which informs us about the move-

ment and stationary position of the head; critical for balance. See *proprioceptive sense.*
vicarious reinforcement Effect that occurs when an imitator observes which *behaviors* on the part of a model are rewarded, or "pay off," and copies those behaviors; an important idea in the Bandura-Walters social learning *theory.*
visible spectrum The part of the *electromagnetic spectrum* in which the energy is visible; *wavelengths* are in the 380 to 780 *nanometer* range.
visual acuity Ability to discriminate fine differences in visual detail; visual sharpness.
visual agnosia Problems with the visual recognition and identification of complex forms, despite normal input to the visual *primary sensory area* of the *occipital lobe;* commonly follows damage to the right *temporal-occipital association cortex.*
"visual cliff" An apparatus for testing *depth perception* in young animals and babies.
visual deprivation Partial or complete restriction of visual input; used in experiments on the *plasticity of perception.*
vocational aptitude test A *test* to assess the *ability* to learn the skills of a particular job; a predictor of job success.
vocational aptitudes *Abilities* to learn the skills involved in a specific job. For example, clerical aptitude is the ability to learn clerical skills. See *aptitude.*
wavelength The distance from the peak of one wave to the peak of the next. See *electromagnetic spectrum, nanometer.* Compare *frequency.*
Wechsler Adult Intelligence Scale (WAIS) An individual *test* of *intelligence* for adults; it has 11 subtests. Compare *Stanford-Binet Intelligence Scale.* See *verbal I.Q, performance IQ.*
Wechsler Tests A family of *tests* of *intelligence* developed by the psychologist David Wechsler. See, for example, *Wechsler Adult Intelligence Scale (WAIS).*
Wernicke's area An area in the *temporal lobe* of the *cerebrum* which is necessary for the recognition of speech sounds and therefore for the comprehension of *language;* also plays a part in the formulation of meaningful speech. Compare *Broca's area.*
white matter Nerve *tracts;* the white color comes from the *myelin sheath* which covers many nerve fibers. Compare *gray matter.*
wish fulfillment In the *psychoanalytic theory* of dreams, the dream is supposed to be a disguised expression of sexual or aggressive urges or wishes; for this reason, the dream is sometimes said to be fulfilling a wish or urge.
withdrawal symptoms In *substance-use disorders,* unpleasant bodily reactions that result when substance use is stopped or reduced.
within-subjects design An experimental strategy in which subjects serve as their own controls. See *baseline, A-B-A within-subjects experimental design.*
"working through" In *psychoanalysis,* the process of repeatedly bringing repressed (see *repression*) material to the surface, interpreting it, and dealing with it; by doing this, the need to repress threatening ideas is decreased.
Young-Helmholtz theory The idea that human color vision depends on activity in three different kinds of cones—one kind which absorbs short wavelengths of the *visible spectrum* most readily, a second kind which absorbs middle wavelengths, and a third kind which is most sensitive to relatively long wavelengths.
youth The period of "studenthood" which may follow *adolescence* and which precedes full incorporation into the adult world.

Name Index

Subject Index